T0140137

Advances in Intelligent Systems and Computing

Volume 773

Series editor

Janusz Kacprzyk, Polish Academy of Sciences, Warsaw, Poland
e-mail: kacprzyk@ibspan.waw.pl

The series “Advances in Intelligent Systems and Computing” contains publications on theory, applications, and design methods of Intelligent Systems and Intelligent Computing. Virtually all disciplines such as engineering, natural sciences, computer and information science, ICT, economics, business, e-commerce, environment, healthcare, life science are covered. The list of topics spans all the areas of modern intelligent systems and computing such as: computational intelligence, soft computing including neural networks, fuzzy systems, evolutionary computing and the fusion of these paradigms, social intelligence, ambient intelligence, computational neuroscience, artificial life, virtual worlds and society, cognitive science and systems, Perception and Vision, DNA and immune based systems, self-organizing and adaptive systems, e-Learning and teaching, human-centered and human-centric computing, recommender systems, intelligent control, robotics and mechatronics including human-machine teaming, knowledge-based paradigms, learning paradigms, machine ethics, intelligent data analysis, knowledge management, intelligent agents, intelligent decision making and support, intelligent network security, trust management, interactive entertainment, Web intelligence and multimedia.

The publications within “Advances in Intelligent Systems and Computing” are primarily proceedings of important conferences, symposia and congresses. They cover significant recent developments in the field, both of a foundational and applicable character. An important characteristic feature of the series is the short publication time and world-wide distribution. This permits a rapid and broad dissemination of research results.

More information about this series at http://www.springer.com/series/11156

Leonard Barolli · Fatos Xhafa
Nadeem Javaid · Tomoya Enokido
Editors

Innovative Mobile and Internet Services in Ubiquitous Computing

Proceedings of the 12th International Conference on Innovative Mobile and Internet Services in Ubiquitous Computing (IMIS-2018)

Editors
Leonard Barolli
Faculty of Information Engineering, Department of Information and Communication Engineering
Fukuoka Institute of Technology
Fukuoka
Japan

Fatos Xhafa
Technical University of Catalonia
Barcelona
Spain

Nadeem Javaid
Department of Computer Science
COMSATS Institute of Information Technology
Islamabad
Pakistan

Tomoya Enokido
Rissho University
Tokyo
Japan

ISSN 2194-5357 ISSN 2194-5365 (electronic)
Advances in Intelligent Systems and Computing
ISBN 978-3-319-93553-9 ISBN 978-3-319-93554-6 (eBook)
https://doi.org/10.1007/978-3-319-93554-6

Library of Congress Control Number: 2018946631

Printed on acid-free paper

This Springer imprint is published by the registered company Springer International Publishing AG part of Springer Nature
The registered company address is: Gewerbestrasse 11, 6330 Cham, Switzerland

Welcome Message of IMIS-2018 International Conference Organizers

Welcome to the 12th International Conference on Innovative Mobile and Internet Services in Ubiquitous Computing (IMIS-2018), which will be held from July 4th to July 6th, 2018, at Kunibiki Messe, Matsue, Japan, in conjunction with the 12th International Conference on Complex, Intelligent and Software Intensive Systems (CISIS-2018).

This International Conference focuses on the challenges and solutions for Ubiquitous and Pervasive Computing (UPC) with an emphasis on innovative, mobile, and Internet services. With the proliferation of wireless technologies and electronic devices, there is a fast-growing interest in UPC. UPC enables to create a human-oriented computing environment where computer chips are embedded in everyday objects and interact with physical world. Through UPC, people can get online even while moving around, thus having almost permanent access to their preferred services. With a great potential to revolutionize our lives, UPC also poses new research challenges. The conference provides an opportunity for academic and industry professionals to discuss the latest issues and progress in the area of UPC.

For IMIS-2018, we received many paper submissions from all over the world. The papers included in the proceedings cover important aspects of UPC research domain. This year, we received 168 submissions and after a careful review process of 3 independent reviews per submission, 47 papers were accepted (about 28% acceptance rate).

It is impossible to organize such a successful program without the help of many individuals. We would like to express our great appreciation to the authors of the submitted papers, the program committee members, who provided timely and significant reviews, and special session chairs for their great efforts. We are grateful to Honorary Chair: Prof. Makoto Takizawa, Hosei University, Japan, for his advice and support.

This year in conjunction with IMIS-2018 we have 7 International Workshops that complemented IMIS-2018 program with contributions for specific topics. We would like to thank the Workshop Co-chairs and all workshop organizers for organizing these workshops.

We thank Donald Elmazi, Yi Liu, Miralda Cuka, and Kosuke Ozera, Fukuoka Institute of Technology, Japan, for their excellent work and support as Web Administrators.

Finally, we would like to thank: Matsue City, Shimane Prefecture, Support Center for Advanced Telecommunications Technology Research (SCAT), Foundation, Japan, for their support.

We hope that all of you enjoy IMIS-2018 and find this a productive opportunity to learn, exchange ideas, and make new contacts.

IMIS-2018 International Conference Organizers

IMIS-2018 General Chair

Leonard Barolli	Fukuoka Institute of Technology (FIT), Japan

IMIS-2018 Program Committee Co-chairs

Tomoya Enokido	Rissho University, Japan
Nadeem Javaid	COMSATS Institute of IT, Pakistan
Hsing-Chung Chen	Asia University, Taiwan

Welcome Message from IMIS-2018 Workshops Co-chairs

Welcome to the Workshops of the 12th International Conference on Innovative Mobile and Internet Services in Ubiquitous Computing (IMIS-2018), which will be held from July 4th to July 6th, 2018 at Kunibiki Messe, Matsue, Japan.

This year we have 7 workshops, which will be held together with IMIS-2018. The objective was to complement as much as possible the main themes of IMIS-2018 with the specific topics of the different workshops to cover many topics of Ubiquitous and Pervasive Computing (UPC).

The list of workshops follows below:

1. The 12th International Workshop on Advances in Information Security (**WAIS-2018**)
2. The 8th International Workshop on Mobile Commerce, Cloud Computing, Network and Communication Security (**MCNCS-2018**)
3. The 8th International Workshop on Intelligent Techniques and Algorithms for Ubiquitous Computing (**ITAUC-2018**)
4. The 8th international workshop on Future Internet and Next Generation Networks (**FINGNet-2018**)
5. The 7th International Workshop on Frontiers in Innovative Mobile and Internet Services (**FIMIS-2018**)
6. The 7th International Workshop on Sustainability Management of e-Business and Ubiquitous Commerce Engineering (**SMEUCE-2018**)
7. The 4th International Workshop on Big Data and IoT Security (**BDITS-2018**)

We would like to thank IMIS-2018 International Conference Organizers for their help and support. We are grateful to the workshop organizers for their great efforts and hard work in proposing the workshops, selecting the papers, organizing interesting programs, and for the arrangements of the workshops during the conference days.

We are grateful to Donald Elmazi, Yi Liu, Miralda Cuka, and Kosuke Ozera, Fukuoka Institute of Technology, Japan, for their excellent work and support as Web Administrators.

We hope you enjoy the workshops programs and proceedings.

Workshops Co-chairs of IMIS-2018 International Conference

Hae-Duck Joshua Jeong	Korean Bible University, Korea
Hiroaki Kikuchi	Meiji University, Japan
Fang-Yie Leu	Tunghai University, Taiwan

IMIS-2018 Organizing Committee

Honorary Chair

Makoto Takizawa	Hosei University, Japan

General Chair

Leonard Barolli	Fukuoka Institute of Technology, Japan

Program Committee Co-chairs

Tomoya Enokido	Rissho University, Japan
Nadeem Javaid	COMSATS Institute of IT, Pakistan
Hsing-Chung Chen	Asia University, Taiwan

Workshops Co-chairs

Hae-Duck Joshua Jeong	Korean Bible University, Korea
Hiroaki Kikuchi	Meiji University, Japan
Fang-Yie Leu	Tunghai University, Taiwan

Advisory Committee Members

Vincenzo Loia	University of Salerno, Italy
Arjan Durresi	IUPUI, USA
Kouichi Sakurai	Kyushu University, Japan

Award Co-chairs

Kangbin Yim	SCH University, Korea
Antonio J. Jara	HES-SO, Switzerland
Marek Ogiela	AGH University of Science and Technology, Poland

International Liaison Co-chairs

Francesco Palmieri	University of Salerno, Italy
Xiaofeng Chen	Xidian University, China
Kin Fun Li	University of Victoria, Canada

Publicity Co-chairs

Baojiang Cui	Beijing Univ. of Posts and Telecom, China
Akio Koyama	Yamagata University, Japan

Local Arrangement Co-chairs

Akimitsu Kanzaki	Shimane University, Japan
Keita Matsuo	Fukuoka Institute of Technology
Elis Kulla	Okayama University of Science, Japan

Web Administrators

Donald Elmazi	Fukuoka Institute of Technology, Japan
Yi Liu	Fukuoka Institute of Technology, Japan
Miralda Cuka	Fukuoka Institute of Technology, Japan
Kosuke Ozera	Fukuoka Institute of Technology, Japan

Track Areas and PC Members

1. Multimedia and Web Computing

Track Co-chairs

Chi-Yi Lin	Tamkang University, Taiwan
Tomoyuki Ishida	Ibaraki University, Japan

PC Members

Noriki Uchida	Fukuoka Institute of Technology, Japan
Tetsuro Ogi	Keio University, Japan
Yasuo Ebara	Kyoto University, Japan
Hideo Miyachi	Tokyo City University, Japan
Kaoru Sugita	Fukuoka Institute of Technology, Japan
Chia-Mu Yu	National Chung Hsing University, Taiwan
Chih-Lin Hu	National Central University, Taiwan
Ching-Ting Tu	Tamkang University, Taiwan
Shih-Hao Chang	Tamkang University, Taiwan

2. Context and Location-Aware Computing

Track Co-chairs

Massimo Ficco	University of Campania Luigi Vanvitelli, Italy
Jeng-Wei Lin	Tunghai University, Taiwan
Neil Yen	The University of Aizu, Japan

PC Members

Paolo Bellavista	University of Bologna, Italy
David Camacho	Universidad Autónoma de Madrid, Spain
Michal Choras	University of Science and Technology, Poland
Gianni D'Angelo	University of Benevento, Italy
Hung-Yu Kao	National Cheng Kung University, Taiwan
Ray-I Chang	National Taiwan University, Taiwan
Mu-Yen Chen	National Taichung University of Science and Technology, Taiwan
Shian-Hua Lin	National Chi Nan University, Taiwan
Chun-Hsin Wu	National University of Kaohsiung, Taiwan
Sheng-Lung Peng	National Dong Hwa University, Taiwan

3. Data Management and Big Data

Track Co-chairs

Been-Chian Chien	National University of Tainan, Taiwan
Akimitsu Kanzaki	Shimane University, Japan
Wen-Yang Lin	National University of Kaohsiung, Taiwan

PC Members

Rung-Ching Chen	Chaoyang University of Technology, Taiwan
Mong-Fong Horng	National Kaohsiung University of Applied Sciences, Taiwan
Bao-Rong Chang	National University of Kaohsiung, Taiwan
Tomoki Yoshihisa	Osaka University, Japan
Hideyuki Kawashima	The University of Tsukuba, Japan
Pruet Boonma	Chiang Mai University, Thailand
Wen-Yang Lin	National University of Kaohsiung, Taiwan
Nik Bessis	Edge Hill University, UK
James Tan	SIM University, Singapore

Kun-Ta Chuang	National Cheng Kung University, Taiwan
Jerry Chun-Wei Lin	Harbin Institute of Technology Shenzhen Graduate School, China

4. Security, Trust and Privacy

Track Co-chairs

Tianhan Gao	Northeastern University, China
Aniello Castiglione	University of Salerno, Italy
Joan Arnedo Moreno	Open University of Catalonia, Spain

PC Members

Qingshan Li	Peking University, China
Zhenhua Tan	Northeastern University, China
Zhi Guan	Peking University, China
Nan Guo	Northeastern University, China
Xibin Zhao	Tsinghua University, China
Cristina Alcaraz	Universidad de Málaga, Spain
Massimo Cafaro	University of Salento, Italy
Giuseppe Cattaneo	University of Salerno, Italy
Zhide Chen	Fujian Normal University, China
Clara Maria	Colombini, University of Milan, Italy
Richard Hill	University of Derby, UK
Dong Seong Kim	University of Canterbury, New Zealand
Victor Malyshkin	Russian Academy of Sciences, Russia
Barbara Masucci	University of Salerno, Italy
Arcangelo Castiglione	University of Salerno, Italy
Xiaofei Xing	Guangzhou University, China
Mauro Iacono	University of Campania Luigi Vanvitelli, Italy
Joan Melià	Universitat Oberta de Catalunya, Spain
Jordi Casas	Universitat Oberta de Catalunya, Spain
Jordi Herrera	Universitat Autònoma de Barcelona, Spain
Antoni Martínez	Universitat Rovira i Virgili, Spain
Francesc Sebé	Universitat de Lleida, Spain

5. Energy Aware and Pervasive Systems

Track Co-chairs

Chi Lin	Dalian University of Technology, China
Elis Kulla	Okayama University of Science, Japan

PC Members

Jiankang Ren	Dalian University of Technology, China
Qiang Lin	Dalian University of Technology, China
Peng Chen	Dalian University of Technology, China
Tomoya Enokido	Rissho University, Japan
Makoto Takizawa	Hosei University, Japan
Oda Tetsuya	Okayama University of Science, Japan
Admir Barolli	Aleksander Moisiu University of Durres, Albania
Makoto Ikeda	Fukuoka Institute of Technology, Japan
Olivier Terzo	Istituto Superiore Mario Boella, Italy

6. Modeling, Simulation and Performance Evaluation

Track Co-chairs

Tetsuya Shigeyasu	Prefectural University of Hiroshima, Japan
Bhed Bahadur Bista	Iwate Prefectural University, Japan

PC Members

Masaaki Noro	Fujitsu Laboratory, Japan
Masaaki Yamanaka	Japan Coast Gaurd Academy, Japan
Nobuyoshi Sato	Iwate Prefectural University, Japan
Tetsuya Shigeyasu	Prefectural University of Hiroshima, Japan
Jiahong Wang	Iwate Prefectural University, Japan
Shigetomo Kimura	University of Tsukuba, Japan
Chotipat Pornavalai	King Mongkut's Institute of Technology Ladkrabang, Thailand
Danda B. Rawat	Howard University, USA
Gongjun Yan	University of Southern Indiana, USA

7. Wireless and Mobile Networks

Track Co-chairs

Luigi Catuogno	University of Salerno, Italy
Hwamin Lee	Soonchunhyang University, Korea

PC Members

Aniello Del Sorbo	Orange Labs, Orange Innovation, UK
Clemente Galdi	University of Naples Federico II, Italy
Stefano Turchi	University of Florence, Italy
Ermelindo Mauriello	Deloitte Spa, Italy
Gianluca Roscigno	University of Salerno, Italy
DaeWon Lee	Seokyoung University, Korea
JongHyuk Lee	Samsung Electronics, Korea
SungHo Chin	LG Electronics, Korea
JiSu Park	Korea University, Korea
Jaehwa Chung	Korea National Open University, Korea

8. Intelligent Technologies and Applications

Track Co-chairs

Marek Ogiela	AGH University of Science and Technology, Krakow, Poland
Yong-Hwan Lee	Wonkwang University, Korea
Hsing-Chung Chen	Asia University, Taiwan

PC Members

Gangman Yi	Dongguk University, Korea
Hoon Ko	J. E. Purkinje University, Czech Republic
Lidia Ogiela	AGH University of Science and Technology, Poland
Libor Mesicek	J. E. Purkinje University, Czech Republic
Rung-Ching Chen	Chaoyang University of Technology, Taiwan
Mong-Fong Horng	National Kaohsiung University of Applied Sciences, Taiwan
Bao-Rong Chang	National University of Kaohsiung, Taiwan
Shingo Otsuka	Kanagawa Institute of Technology, Japan
Tomoki Yoshihisa	Osaka University, Japan
Pruet Boonma	Chiang Mai University, Thailand
Izwan Nizal Mohd Shaharanee	University Utara Malaysia
Youngseop Kim	Dankook University, Korea
Cheong-Ghil Kim	Namseoul University, Korea

Fumitaka Ono	Tokyo Polytechnic University, Japan
Schelkens Peter	Vrije Universiteit Brussel, Belgium
You-Seok Won	Uzbrainnet Inc., Korea
Zhenhua Tan	Northeastern University, China
Tianhan Gao	Northeastern University, China
Nan Guo	Northeastern University, China
Yung-Fa Huang Huang	Chaoyang University of Technology, Taiwan
Tzu-Liang Kung	Asia University, Taiwan

9. Cloud Computing and Service-Oriented Applications

Track Co-chairs

Baojiang Ciu	Beijing University of Posts and Telecommunications, China
Francesco Palmieri	University of Salerno, Italy

PC Members

Aniello Castiglione	University of Salerno, Italy
Ashiq Anjum	University of Derby, UK
Beniamino Di Martino	University of Campania Luigi Vanvitelli, Italy
Gang Wang	Nankai University, China
Shaozhang Niu	Beijing University of Posts and Telecommunications, China
Jianxin Wang	Beijing Forestry University, China
Jie Cheng	Shandong University, China
Shaoyin Cheng	University of Science and Technology of China, China
Jingling Zhao	Beijing University of Posts and Telecommunications, China
Qing Liao	Beijing University of Posts and Telecommunications, China
Xiaohui Li	Wuhan University of Science and Technology, China
Chunhong Liu	Hainan Normal University, China
Yan Zhang	Yan Hubei University, China

10. Ontology and Semantic Web

Track Co-chairs

Alba Amato	Italian National Research Center (CNR), Italy
Fong-Hao Liu	National Defense University, Taiwan
Farookh Hussain	University of Technology Sydney, Australia

PC Members

Flora Amato	University of Naples Federico II, Italy
Claudia Di Napoli	Italian National Research Center (CNR), Italy
Salvatore Venticinque	University of Campania Luigi Vanvitelli, Italy
Marco Scialdone	University of Campania Luigi Vanvitelli, Italy
Wei-Tsong Lee	Tamkang University, Taiwan, ROC
Tin-Yu Wu	National Ilan University, Taiwan, ROC
Liang-Chu Chen	National Defense University, Taiwan, ROC
Omar Khadeer Hussain	University of New South Wales, Canberra, Australia
Salem Alkhalaf	Qassim University, Saudi Arabia
Omar Khadeer Hussain	University of New South Wales, Canberra, Australia
Salem Alkhalaf	Qassim University, Saudi Arabia
Osama Alfarraj	King Saud University, Saudi Arabia
Thamer AlHussain	Saudi Electronic University, Saudi Arabia
Mukesh Prasad	University of Technology, Sydney, Australia

11. IoT and Opportunistic Networking

Track Co-chairs

Antonio Jara	HES-SO, Switzerland
Isaac Woungang	Ryerson University, Canada

PC Members

Gianluca Rizzo	HES-SO, Switzerland
Marcin Pawlowski	Warsaw University of Technology, Poland
Gaetano Manzo	HES-SO, Switzerland
Dominique Gabioud	HES-SO, Switzerland
Markus Jung	Osram, Germany
Sebastien Ziegler	Mandat International, Switzerland

Pierre-André Mudry	HES-SO, Switzerland
Leire Bastida	Tecnalia, Spain
Charalampos Doukas	CREATE-NET, Italy
Marco Fiore	CNR, Italy
Diego Gachet	Universidad Europea de Madrid, Spain
Dominique Genoud	HES-SO, Switzerland
Dimosthenis Ioannidis	CERTH-ITI, Greece
Ved Kafle	NICT, Japan
Daeyoung Kim	KAIST, Korea
Lambros Lambrinos	Cyprus University of Technology, Cyprus
Pedro Malo	Universidade Nova de Lisboa, Portugal
Septimiu Nechifor	Siemens, Romania
Alexis Olivereau	CEA, France
Marcin Piotr	Jagiellonian University, Poland
Mirko Presser	Alexandra Institute, Denmark
Martin Brynskov	OASC, Denmark
Dario Russo	National Research Council of Italy, Italy
Martin Serrano	NUIG, Ireland
Felix Villanueva	UCLM, Spain

12. Embedded Systems and Wearable Computers

Track Co-chairs

Jiankang Ren	Dalian University of Technology, China
Keita Matsuo	Fukuoka Institute of Technology, Japan
Kangbin Yim	SCH University, Korea

PC Members

Xiulong Liu	The Hong Kong Polytechnic University, Hong Kong
Yong Xie	Xiamen University of Technology, Xiamen, China
Yang Zhao	Tianjin University, Tianjin, China
Fangmin Sun	Shenzhen Institutes of Advanced Technology, Chinese Academy of Sciences, China
Shaobo Zhang	Central South University, Changsha, China
Kun Wang	Liaoning Police Academy, Dalian, China

IMIS-2018 Reviewers

Leonard Barolli
Makoto Takizawa
Fatos Xhafa
Isaac Woungang
Hae-Duck Joshua Jeong
Fang-Yie Leu
Kangbin Yim
Marek Ogiela
Makoto Ikeda
Keita Matsuo
Francesco Palmieri
Massimo Ficco
Salvatore Venticinque
Mauro Migliardi
Noriki Uchida
Nik Bessis
Yoshitaka Shibata
Kaoru Sugita
Antonio J. Jara
Admir Barolli
Elis Kulla
Evjola Spaho
Arjan Durresi
Bhed Bista
Muhammad Younas
Hsing-Chung Chen
Kin Fun Li
Hiroaki Kikuchi
Lidia Ogiela
Nan Guo
Xiaofeng Chen
Hwamin Lee
Tetsuya Shigeyasu
Fumiaki Sato
Kosuke Takano
Florin Pop
Flora Amato
Tomoya Enokido
Minoru Uehara
Santi Caballé
Tomoyuki Ishida
Hwa Min Lee
Jiyoung Lim
Tianhan Gao
Danda Rawat
Farookh Hussain
JongSuk Lee
Omar Hussain
Wang Xu An

Welcome Message from WAIS-2018 International Workshop Organizers

It is our great pleasure to welcome you to the 12th International Workshop on Advances in Information Security (WAIS-2018). The workshop is held in conjunction with 12th International Conference on Innovative Mobile and Internet Services in Ubiquitous Computing (IMIS-2018) from July 4th to July 6th, 2018, at Kunibiki Messe, Matsue, Japan.

As computing systems have begun to pervade every aspect of daily life, people need to be able to trust them—so much of their lives depend on them. Today, many of these systems are far too vulnerable to cyber attacks that can inhibit their operation, corrupt valuable data, or expose private information. Future systems will include sensors and computers everywhere, exacerbating the attainment of security and privacy. Current security practices largely address current and known threats, but there is a need for research to take account of future threats too.

The goal of WAIS-2018 is to bring together computer scientists, industrial engineers, and researchers to discuss and exchange experimental or theoretical results, novel designs, work-in-progress, experience, case studies, and trend-setting ideas in the area of Information Security.

We would like to thank all authors for submitting their research works to the workshop and the PC Members for checking carefully the papers.

We look forward to meet all of you in Matsue, Japan.

WAIS-2018 Workshop Co-chairs

Leonard Barolli	Fukuoka Institute of Technology (FIT), Japan
Arjan Durresi	Indiana University–Purdue University at Indianapolis (IUPUI), USA
Hiroaki Kikuchi	Meiji University, Japan

WAIS-2018 Organizing Committee

Workshop Co-chairs

Leonard Barolli	Fukuoka Institute of Technology (FIT), Japan
Arjan Durresi	Indiana University–Purdue University at Indianapolis (IUPUI), USA
Hiroaki Kikuchi	Meiji University, Japan

Advisory Co-chairs

Makoto Takizawa	Hosei University, Japan
Raj Jain	Washington University in St. Louis, USA

Program Committee Members

Sriram Chellappan	Missouri University of Science and Technology, USA
Koji Chida	NTT, Japan
Tesuya Izu	Fujitsu Laboratories, Japan
Qijun Gu	Texas State University, San Marcos, USA
Youki Kadobayashi	Nara Institute of Science and Technology, Japan
Akio Koyama	Yamagata University, Japan
Michiharu Kudo	IBM Japan, Japan
Sanjay Kumar Madria	Missouri University of Science and Technology, USA
Masakatsu Morii	Kobe University, Japan
Masakatsu Nishigaki	Shizuoka University, Japan
Vamsi Paruchuri	University of Central Arkansas, USA
Hiroshi Shigeno	Keio University, Japan
Yuji Suga	Internet Initiative Japan Inc., Japan
Keisuke Takemori	KDDI Co., Japan
Ryuya Uda	Tokyo University of Technology, Japan
Xukai Zou	Indiana University–Purdue University Indianapolis, USA
Wenye Wang	North Carolina State University, USA
Hiroshi Yoshiura	University of Electro-Communications, Japan

Welcome Message from MCNCS-2018 International Workshop Chair

Welcome to the 8th International Workshop on Mobile Commerce, Cloud Computing, Network and Communication and their Securities (MCNCS-2018) which will be in conjunction with the 12th International Conference on Innovative Mobile and Internet Services in Ubiquitous Computing (IMIS-2018) from July 4th to July 6th, 2018, at Kunibiki Messe, Matsue, Japan.

Computer network and communication have been a part of our everyday life. People use them to contact others almost anytime anywhere. On the other hand, hackers due to business benefits, enjoying their skill/professional achievement or some other reasons very often attack, intrude, or penetrate our systems. This is the key reason why computer/network and their securities have been the important issues in computer research. Many researchers have tried to do their best to develop system network, security techniques, and the methods to protect a system. But system attacks still occur worldwide every day. In fact, current network techniques and system security technology are still far away from convenient to use and completely secure and should be further improved.

This workshop aims to present the innovative researches, methods and for mobile commerce, cloud computing, network and communication and their securities. The workshop contains high-quality research papers, which were selected carefully by Program Committee Members.

It is impossible to organize such a successful program without the help of many individuals. We would like to express our appreciation to the authors of the submitted papers and to the program committee members, who provided timely and significant reviews.

We hope all of you will enjoy MCNCS-2018 and find this a productive opportunity to exchange ideas with many researchers.

MCNCS-2018 International Workshop Organizers

MCNCS-2018 Workshop Co-chairs

Fang-Yie Leu	Tunghai University, Taiwan
Aniello Castiglione	University of Salerno, Italy
Chu-Hsing Lin	Tunghai University, Taiwan

MCNCS-2018 Advisory Co-chairs

Yi-Li Huang	Tunghai University, Taiwan
Jung-Chun Liu	Tunghai University, Taiwan
Kun-Lin Tsai	Tunghai University, Taiwan

MCNCS-2018 Organizing Committee

Workshop Co-chairs

Fang-Yie Leu	Tunghai University, Taiwan
Aniello Castiglione	University of Salerno, Italy
Chu-Hsing Lin	Tunghai University, Taiwan

Advisory Co-chairs

Yi-Li Huang	Tunghai University, Taiwan
Jung-Chun Liu	Tunghai University, Taiwan
Kun-Lin Tsai	Tunghai University, Taiwan

Program Committee Members

Alessandra Sala	University of California Santa Barbara, USA
Antonio Colella	Italian Army, Italy
Chin-Cheng Lien	Soochow University, Taiwan
Chin-Ling Chen	Chaoyang University of Technology, Taiwan
Chiu-Ching Tuan	National Taipei University of Technology, Taiwan
Claudio Soriente	Universitat Politecnica de Madrid, Spain
Francesco Palmieri	University of Salerno, Italy
Fuw-Yi Yang	Chaoyang University of Technology, Taiwan
I-Long Lin	Central Police University, Taiwan
Ilsun You	Korean Bible University, Korea
Jason Ernst	University of Guelph, Canada
Jinn-Ke Jan	National Chung Hsing University, Taiwan
Jung-Chun Liu	TungHai University, Taiwan
Lein Harn	University of Missouri Kansas City, USA
Sen-Tang Lai	Shih Chien University, Taiwan
Sergio Ricciardi	Universitat Politècnica de Catalunya, Spain
Shyhtsun Felix Wu	University of California, Davis, USA
Tzung-HerChen	National Chiayi University, Taiwan
Ugo Fiore	University of Naples, Italy
Heru Susanto	University of Brunei, Brunei

Welcome Message from ITACU 2018 International Workshop Co-chairs

Welcome to the 8th International Workshop on Intelligent Techniques and Algorithms for Ubiquitous Computing (ITAUC-2018) which will be held in conjunction with the 12th International Conference on Innovative Mobile and Internet Services in Ubiquitous Computing (IMIS-2018) at Kunibiki Messe, Matsue, Japan, from July 4th to July 6th, 2018.

The aim of this workshop is to present the innovative researches, methods, and algorithms for wireless networks, sensor networks, and ubiquitous computing. It is intended to facilitate exchange of ideas and collaborations among researchers from computer science, network computing, mathematics, statistics, intelligent computing and such related sciences, to discuss various aspects of innovative intelligent algorithms and networks security, intelligent techniques for sensor networks and radio networks, and their applications.

Many people have kindly helped us to prepare and organize the ITAUC-2018 workshop. First, we highly thank the authors who submitted high-quality papers and reviewers who carefully evaluated these submissions. We thank Honorary Co-chair, General Co-chairs, PC Co-chairs, and Workshops Co-chairs of IMIS-2018 for their advice and support to make possible organization of ITAUC-2018.

We hope you will enjoy the conference and have a great time in Torino, Italy.

ITAUC-2018 Co-chairs

Leonard Barolli	Fukuoka Institute of Technology, Japan
Hsing-Chung Chen	Asia University, Taiwan
Tzu-Liang Kung	Asia University, Taiwan

ITAUC-2018 Organizing Committee

Workshop Co-chairs

Leonard Barolli	Fukuoka Institute of Technology, Japan
Hsing-Chung Chen	Asia University, Taiwan
Tzu-Liang Kung	Asia University, Taiwan

Program Committee Members

Arjan Durresi	Indiana University–Purdue University at Indianapolis (IUPUI), USA
Makoto Ikeda	Fukuoka Institute of Technology, Japan
Akio Koyama	Yamagata University, Japan
Neng-Yih Shih	Asia University, Taiwan
Yeong-Chin Chen	Asia University, Taiwan
Timothy K. Shih	National Central University, Taiwan
Hsi-Chin Hsin	National United University, Taiwan
Ming-Shiang Huang	Asia University, Taiwan
Chia-Cheng Liu	Asia University, Taiwan
Chia-Hsin Cheng	National Formosa University Yunlin County, Taiwan
Tzu-Liang Kung	Asia University, Taiwan
Gene Shen	Asia University, Taiwan
Jim-Min Lin	Feng Chia University, Taiwan
Chia-Cheng Liu	Asia University, Taiwan
Yen-Ching Chang	Chung Shan Medical University, Taiwan
Shu-Hong Lee	Chienkuo Technology University, Taiwan
Ho-Lung Hung	Chienkuo Technology University, Taiwan
Gwo-Ruey Lee	Lung-Yuan Research Park, Taiwan
Li-Shan Ma	Chienkuo Technology University, Taiwan
Chung-Wen Hung	National Yunlin University of Science & Technology University, Taiwan
Yung-Chen Chou	Asia University, Taiwan
Chen-Hung Chuang	Asia University, Taiwan
Jing-Doo Wang	Asia University, Taiwan
Jui-Chi Chen	Asia University, Taiwan
Young-Long Chen	National Taichung University of Science and Technology, Taiwan

Web Administrators

Donald Elmazi	Fukuoka Institute of Technology, Japan
Miralda Cuka	Fukuoka Institute of Technology, Japan

Welcome Message from FINGNet-2018 International Workshop Organizers

Welcome to the 8th International Workshop on Future Internet and Next Generation Networks (FINGNet-2018), which is held in conjunction with the 12th International Conference on Innovative Mobile and Internet Services in Ubiquitous Computing (IMIS-2018) from July 4th to July 6th, 2018, at Kunibiki Messe, Matsue, Japan.

Over the last years, new paradigms and concepts have emerged in telecommunication networks that are currently being realized in the Internet. Among those are the social networking, the peer-to-peer, and the quality of experience paradigms. Nevertheless, it is difficult or even impossible to predict how the network of the future will emerge. However, it is quite clear that some major issues have to be addressed for the future Internet or next-generation networks. This includes solutions, e.g., for security and privacy issues, such as spam or service-oriented approaches to enable flexible networking. In addition, new applications or services emerge such as social media networks or crowdsourcing platforms enabling collaborative networking of Internet users. To address these issues, there is an increased interest in the scientific community to propose and design new algorithms and methodologies as well as to understand and model new applications and services.

The aim of this workshop is to present such innovative research, methods, and numerical analysis related to advanced Internet and network technologies. The workshop contains high-quality research papers, which were carefully selected by the technical program committee members.

It is impossible to organize such a successful program without the help of many authors, program committee members, and organizers. We also would like to deeply express our appreciation to the authors of the submitted papers and to the program committee members, who provided timely and significant reviews.

We hope all of you will enjoy FINGNet-2018 and find this a productive opportunity to exchange ideas with many researchers. Finally, we would like to thank everyone for participating in FINGNet-2018 workshop. We hope that you will find the workshop along with other joint workshops and conferences stimulating.

FINGNet-2018 Workshop Co-chairs

Tobias Hossfeld	University of Wuerzburg, Germany
Inshil Do	Ewha Womans University, Korea
Hae-Duck Joshua Jeong	Korean Bible University, Korea

FINGNet-2018 Workshop Advisory Co-chairs

Kijoon Chae	Ewha Womans University, Korea
Akihiro Nakao	University of Tokyo, Japan
Jiyoung Lim	Korean Bible University, Korea

FINGNet-2018 Organizing Committee

Workshop Co-chairs

Tobias Hossfeld	University of Wuerzburg, Germany
Inshil Do	Ewha Womans University, Korea
Hae-Duck Joshua Jeong	Korean Bible University, Korea

Advisory Co-chairs

Kijoon Chae	Ewha Womans University, Korea
Akihiro Nakao	University of Tokyo, Japan
Jiyoung Lim	Korean Bible University, Korea

Program Committee Members

Kento Aida	National Institute of Informatics, Japan
Xiangjiu Che	Jilin University, China
Kyo-Il Chung	ETRI, Korea
Cuong Dinh	University of Science Ho Chi Minh City, Vietnam
Greg Ewing	University of Canterbury, New Zealand
Shanmugasundaram Hariharan	B.S. Abdur Rahman University, India
P. T. Ho	University of Hong Kong, Hong Kong
Ching-Hsien Hsu	Chung Hua University, Taiwan
Wen-Tzeng Huang	Minghsin University of Science and Technology, Taiwan
Hai Jin	Huazhong University of Science and Technology, China
Hyung Chan Kim	ETRI, Korea
Joo-Man Kim	Pusan National University, Korea
Wing-Keung Kwan	University of Hong Kong, Hong Kong
Chong Deuk Lee	Chonbuk National University, Korea
JongSuk Ruth Lee	KISTI, Korea
Hing Yan Lee	IDA, Singapore
Fang-Pang Lin	NCHC, Taiwan
Te-Lung Liu	NCHC, Taiwan
Nobutaka Matsumoto	KDDI R&D Laboratories Inc., Japan
Don McNickle	University of Canterbury, New Zealand
Nam Ng	University of Hong Kong, Hong Kong
Seungjin Park	University of Southern Indiana, USA
Anan Phonphoem	Kasetsart University, Thailand
Zhuzhong Qian	Nanjing University, China

Kiwook Sohn	ETRI, Korea
Torab Torabi	La Trobe University, Australia
Putchong Uthayopas	Kasetsart University, Thailand
Xiaohui Wei	Jilin University, China
Lawrence WC Wong	National University of Singapore, Singapore
Kun-Ming Yu	Chung Hua University, Taiwan

Welcome Message from FIMIS-2018 International Workshop Chair

It is our great pleasure to welcome you the 7th International Workshop on Frontiers in Innovative Mobile and Internet Services (FIMIS-2018), which will be held in conjunction with the 12th International Conference on Innovative Mobile and Internet Services in Ubiquitous Computing (IMIS-2018) from July 4th to July 6th, 2018, at Kunibiki Messe, Matsue, Japan.

The workshop focuses on challenges and solutions for Ubiquitous and Pervasive Computing (UPC) with an emphasis on innovative, mobile and Internet services. Especially, the main goal of FIMIS-2018 is to bring together researchers, practitioners, and decision-makers to demonstrate the state of the art as well as future directions for UPC.

The organization of a workshop needs the help of many people. We would like to express our special thanks to the authors and reviewers whose contribution makes this workshop a reality.

Hopefully, you will enjoy the workshop and have a great time in Matsue, Japan.

FIMIS-2018 Workshop Chair

Leonard Barolli Fukuoka Institute of Technology, Japan

FIMIS-2018 Organizing Committee

Workshop Co-chairs

Leonard Barolli	Fukuoka Institute of Technology, Japan

Program Committee Members

Feilong Tang	Shanghai Jiao Tong University, China
Arjan Durresi	IUPUI, USA
Hae-Duck Joshua Jeong	Korean Bible University, Korea
Hiroaki Kikuchi	Meiji University, Japan
Chu-Hsing Lin	Tunghai University, Taiwan
Jinshu Su	National University of Defense Technology, China
Tomoya Enokido	Rissho University
Fang-Yie Leu	Tunghai University, Taiwan
Chunqing Wu	National University of Defense Technology, China
Makoto Ikeda	Fukuoka Institute of Technology, Japan
Marek Ogiela	AGH University of Science and Technology, Poland
Kin Fun Li	University of Victoria, Canada
Joan Arnedo	Open University of Catalonia, Spain
Tetsuya Shigeyasu	Prefectural University of Hiroshima, Japan
Kangbin Yim	Soonchunhyang University, Korea

Web Administrator

Donald Elmazi	Fukuoka Institute of Technology, Japan

Welcome Message from SMEUCE-2018 International Workshop Co-chairs

We welcome you to the 7th International Workshop on Sustainability Management of e-Business and Ubiquitous Commerce Engineering (SMEUCE-2018), which will be held in conjunction with the 12th International Conference on Innovative Mobile and Internet Services in Ubiquitous Computing (IMIS-2018) at Kunibiki Messe, Matsue, Japan, from July 4th to July 6th, 2018.

The aim objective of SMEUCE-2018 is to provide a platform for academics, business leaders, consultants, and other professionals from all over the world to exchange the latest research findings in the field of sustainability management of e-business, e-business, and u-business (ubiquitous business) engineering, Cloud commerce and their relative applications. This conference provides opportunities for the delegates to exchange new ideas and application experiences face to face, to establish business or research relations and to find global partners for future collaboration.

Many people have kindly helped us to prepare and organize the SMEUCE-2018 workshop. First, we highly thank the authors who submitted high-quality papers and reviewers who carefully evaluated these submissions. We are thankful to Prof. Makoto Takizawa, Hosei University, Japan, as Honorary Chair of IMIS-2018. We would like to give our special thanks to Prof. Leonard Barolli as General Chair of IMIS-2018 for their strong encouragement and guidance to organize this workshop. We would like to thank the PC Co-chairs and Workshops Co-chairs for their advice and support to make possible organization of SMEUCE-2018.

We hope you will enjoy the conference and have a great time in Matsue, Japan.

SMEUCE-2018 Co-chairs

Kuei-Yuan Wang — Asia University, Taiwan
Ying-Li Lin — Asia University, Taiwan
Mei-Hua Liao — Asia University, Taiwan
Hsing-Chung Chen — Asia University, Taiwan
Hidekazu Sone — Shizuoka University of Art and Culture, Japan

SMEUCE-2018 Organizing Committee

Workshop Co-chairs

Kuei-Yuan Wang	Asia University, Taiwan
Ying-Li Lin	Asia University, Taiwan
Mei-Hua Liao	Asia University, Taiwan
Hsing-Chung Chen	Asia University, Taiwan
Hidekazu Sone	Shizuoka University of Art and Culture, Japan

Program Committee Members

Ching-Hui Shih	Asia University, Taiwan
Shu-Hui Chuang	Asia University, Taiwan
Ting-Chang Chang	Asia University, Taiwan
Ya-Wen Yu	Asia University, Taiwan
Shyh-Weir Tzang	Asia University, Taiwan
Chia-Hsin Cheng	National Formosa University Yunlin County, Taiwan
Shu-Hong Lee	Chienkuo Technology University, Taiwan
Ho-Lung Hung	Chienkuo Technology University, Taiwan
Chung-Wen Hung	National Yunlin University of Science & Technology University, Taiwan
Horimoto Saburo	Shiga University, Japan
Pulukkuttige Don Nimal	University of Peradeniya, Sri Lanka
Mei Hua Huang	Asia University, Taiwan
Young-Long Chen	National Taichung University of Science and Technology, Taiwan

Web Administrator Chair

Donald Elmazi	Fukuoka Institute of Technology, Japan

Welcome Message from BDITS-2018 International Workshop Co-chairs

Welcome to the 4th International Workshop on Big Data and IoT Security (BDITS-2018), which is held in conjunction with the 12th International Conference on Innovative Mobile and Internet Services in Ubiquitous Computing (IMIS-2018), from July 4th to July 6th, 2018, at Kunibiki Messe, Matsue, Japan.

The aim objective of BDITS-2018 is to provide a platform for academics, the leaders of network security, consultants and other professionals from all over the world to exchange the latest research findings in the field of Security Technologies of Big Data and Internet of Things (IoT), cloud and ubiquitous computing and their applications. This workshop provides opportunities for the delegates to exchange new ideas and application experiences face to face, to establish research relations and to find global partners for future collaboration.

Many people have kindly helped us to prepare and organize the BDITS-2018 workshop. First, we would like to thank the authors who submitted high-quality papers and reviewers who carefully evaluated the submitted papers.

We would like to give our special thanks to Prof. Leonard Barolli for his strong encouragement and guidance to organize this workshop. We would like to thank PC Co-chairs and Workshop Co-chairs of IMIS-2018 for their kind support. We would like to thank all PC members for their serious review works in order to make successful organization of BDITS-2018.

We hope you will enjoy the conference and have a great time in Matsue, Japan.

BDITS-2018 Co-chairs

Baojiang Cui	Beijing University of Posts and Telecommunications, China
Zheli Liu	Nankai University, China
Tainhan Gao	National Pilot Software College, China
Hsing-Chung Chen	Asia University, Taiwan

BDITS-2018 Organizing Committee

Workshop Co-chairs

Baojiang Cui	Beijing University of Posts and Telecommunications, China
Zheli Liu	Nankai University, China
Tainhan Gao	National Pilot Software College, China
Hsing-Chung Chen	Asia University, Taiwan

Program Committee Members

Nan Guo	National Pilot Software College, China
Makoto Ikeda	Fukuoka Institute of Technology, Japan
Neng-Yih Shih	Asia University, Taiwan
Timothy K. Shih	National Central University, Taiwan
Yung-Fa Huang	Chaoyang University of Technology, Taiwan
Chia-Hsin Cheng	National Formosa University Yunlin County, Taiwan
Tzu-Liang Kung	Asia University, Taiwan
Shu-Hong Lee	Chienkuo Technology University, Taiwan
Ho-Lung Hung	Chienkuo Technology University, Taiwan
Gwo-Ruey Lee	Lung-Yuan Research Park, Taiwan
Li-Shan Ma	Chienkuo Technology University, Taiwan
Chung-Wen Hung	National Yunlin University of Science & Technology University, Taiwan
Jui-Chi Chen	Asia University, Taiwan

Web Administrator

Donald Elmazi	Fukuoka Institute of Technology, Japan

CISIS-2018 Keynote Talks

SNS as Social Sensors: Technologies for Extracting Knowledge from SNSs

Takahiro Hara
Osaka University, Osaka, Japan

Abstract. Due to surprisingly rapid popularization of smartphones and Social Network Services (SNSs), ordinary people generate and share a large amount of data using SNSs on smartphones. Recent studies have revealed that messages posted on SNSs such as Twitter can be used for detecting various kinds of facts in the real world such as events, trends, and user sentiment, which can be considered as kinds of social sensor data. Social sensor data are very useful for Big Data analysis because these tell many things representing the real world, which cannot be known by only analyzing traditional Big Data.

In this talk, I will present our recent studies on social sensing based on data mining on SNSs such as Twitter. We will start with our approaches for knowledge extraction from Twitter and then move on to some fundamental techniques which can be used commonly for such knowledge extraction. I will also present our ongoing work for building a framework to share not only social sensor data (i.e., SNS analytical results) but also definitions of social sensor data and analytical procedures. I will conclude this talk with some discussion on future research opportunities for social sensing.

Fog Computing: A New Research Direction in Distributed Computing—Applications, Issues and Challenges

Farookh Khadeer Hussain
University of Technology Sydney (UTS), Sydney, Australia

Abstract. The objective of this talk is to introduce the emerging research area of 'Fog Computing' or 'Edge Computing' and discuss the research opportunities and research challenges in this area. In order to do so, firstly, this talk presents the transition in the discipline of Distributed Computing from 'Grid Computing' to 'Cloud Computing' to 'Cloud-of-Things' and then to a 'Fog computing' environment. It then highlights clearly the role of 'Internet of Things' and 'Cloud Computing' in realizing 'Fog Computing.' Secondly, it briefly discusses some research issues and challenges associated with realizing and developing practical Fog Computing-driven solutions. Finally, it outlines and discusses the future role of 'Cloud-of-Things' in achieving business efficiencies in both developed and developing economies.

Contents

The 4th International Workshop on Big Data and IoT Security (BDITS-2018)

The 12th International Conference on Innovative Mobile and Internet Services in Ubiquitous Computing (IMIS-2018)

An Energy Efficient Scheduling of a Smart Home Based on Optimization Techniques

Aqib Jamil[1], Nadeem Javaid[1(✉)], Muhammad Usman Khalid[1], Muhammad Nadeem Iqbal[2], Saad Rashid[2], and Naveed Anwar[3]

[1] COMSATS Institute of Information Technology, Islamabad 44000, Pakistan
nadeemjavaidqau@gmail.com

[2] COMSATS Institute of Information Technology, Wah Cantt 47040, Pakistan

[3] University of Wah, Wah Cantt 47040, Pakistan

Abstract. After the introduction of smart grid in power system, two-way communication is now possible which helps in optimizing the energy consumption of consumers. To optimize the energy consumption on the consumer side, demand side management is used. In this paper, we focused on the optimization of smart home appliances with the help of optimization techniques. Cuckoo search algorithm, earthworm optimization and a hybrid technique cuckoo-earthworm optimization are used for scheduling the smart home appliances. Home appliances are classified into three groups and real-time pricing scheme is used. Techniques are evaluated and a performance comparison is performed. Results show that the proposed hybrid technique has decreased the electricity cost by 49% as compared to unscheduled cost and a trade-off exists between electricity cost and user comfort.

Keywords: Cuckoo search algorithm · Earthworm optimization
Smart grid · Demand side management · Heuristic techniques

1 Introduction

In last few decades, there is a huge increase in electricity consumption whereas, traditional grids are insufficient to meet such immense demand and they had lot of problems such as unidirectional power flow, manual-monitoring, manual-healing and incompatibility with new technologies. Also, 65% of produced electricity is wasted during production, transmission and distribution [1]. For this purpose, the concept of smart grid (SG) is evolved, which includes two-way power flow and communication, self-monitoring, self-healing and compatibility with new technologies in the power system. SG is considered to be the future of power system because it is an energy efficient system. To solve the energy supply and demand difference problem, supply side management (SSM) or demand side management (DSM) is used in power system. SSM includes installation of new power plants however, installation of these power plants requires a lot of time to complete. In DSM, consumer energy demand can be adjusted by load strategies:

L. Barolli et al. (Eds.): IMIS 2018, AISC 773, pp. 3–14, 2019.
https://doi.org/10.1007/978-3-319-93554-6_1

load shifting, valley filling and flexible load. Load shifting is considered as the most efficient load management strategy [8].

Most usually load shifting strategy is implemented by demand response (DR) program. DR program is the changes in consumer electricity usage patterns in response to price signal sent by utility. DR is considered as a most efficient and reliable program [2]. DR program has two types: incentive-based DR program (IDR) and priced-based DR (PDR) program. IDR program gives consumers some incentives on using more electricity during off-peak hours in order to decrease some stress from the power system during on-peak hours. In PDR program, price of electricity is set for each hour that encourages the users to shift their load to low price hours in order to decrease their electricity cost. Basically, DR program aims in educating the consumers to use most of the electricity during the off-peak hours.

Most common price signals used by utilities are real-time pricing (RTP), day ahead pricing (DAP), time of use (TOU) and inclined block rate (IBR). RTP is considered as most efficient pricing scheme for scheduling in DSM [3].

In this paper, a load shifting strategy is implemented while considering a RTP scheme. In our scenario, a single home is considered which consists of different appliances which are divided into three groups. Meta-heuristic techniques: cuckoo search algorithm (CSA), earthworm optimization (EWA) and a hybrid technique cuckoo-earthworm optimization (HCEO) are implemented to schedule the household appliances to evaluate the performance of these three optimization techniques. The objective is to decrease the electricity cost and results indicate that hybrid technique reduced the cost by 49%.

The remainder of the paper is described as: in Sect. 2, literature review of related papers is discussed. Section 3 explains the working of proposed system model. Section 4 gives information about the implemented optimization techniques. Section 5 demonstrates the simulations results and conclusion is given in Sect. 6.

2 Related Work

Some literature review about the home energy management system (HEMS) is presented in this section. Pedram Samadi *et al.* considered a system model for scheduling the home appliances and they integrated the renewable energy sources (RES) [4]. For this purpose, dynamic programming is implemented to shift home appliances to certain time intervals and game theory is used for the coordination of extra produced energy with neighbors and utility. In their proposed system model, consumers generate electricity from RES and they can sell extra electricity to neighbors and utility.

A HEMS is considered by Sheraz *et al.* to decrease the electricity cost and peak average ratio (PAR) using genetic algorithm (GA), CSA and crow search algorithm [5]. Their proposed system model consists of a smart building and authors integrated energy storage system (ESS) to reduce the cost and PAR. Simulations demonstrate that they achieved their objectives.

In [6], a HEMS is proposed where appliances are scheduled by using heuristic techniques. They used GA and particle swarm optimization (PSO) to decrease the total electricity cost and PAR. To achieve their objectives, multiple knapsack problem (MKP) is implemented with three price signals i.e., RTP, TOU and CPP. GA performed better than PSO in case of cost.

An expert energy management system while focusing on the supply side management is proposed in [7]. They used modified bacterial foraging algorithm to schedule the different distributed energy resources to get an optimal point. Results show that, their proposed system has reduced the operational cost and net emissions. However, they ignored the demand side management. In [8], residential, industrial and commercial users are considered for solving optimization problem, to decrease the electricity cost and PAR by using heuristic based evolutionary algorithm. Results demonstrate that they handled the large number of devices and achieved a considerable amount of reduction in cost and PAR.

Muralitharan *et al.* [9] considered a model to decrease the electricity cost and waiting time using multi-objective evolutionary algorithm. Authors implemented this algorithm with threshold limit to balance the load and to avoid peaks. If consumer's load exceeds from utility's defined threshold limit then consumer will pay extra amount in the form of penalty. From simulations it can be concluded that desired objectives are achieved. In [10], authors proposed a system model in which each house has two types of loads: flexible and essential loads. Flexible loads are divided into delay sensitive and delay insensitive loads. On the bases of this division, authors formulated an optimization problem to reduce the cost and delay of sensitive loads.

In [11], authors proposed a robust-index method to handle the uncertainties associated with the consumers behavior to increase the user comfort in home appliance scheduling. Simulations represent that proposed model succeeded in increasing user comfort. A residential DR is studied in [12] using multi-integer non-linear programming (MINLP). Results demonstrates that more than 25% cost is reduced and most consumers shifted their load to low price hours to gain some incentives.

In [13], a mixed integer linear programming (MILP) technique is considered to optimize the household appliances while integrating the RESs, ESSs and electric vehicles (EV) in a smart home. A heuristic technique is used for scheduling and to reduce consumers electricity cost. Three case studies for different time horizons are performed by varying few factors by using taguchi method.

In this work, we used meta-heuristics techniques CSA, EWA and a hybrid technique HCEO to schedule the home appliances.

3 System Model

System model consists of advance metering infrastructure (AMI), smart meter (SM), scheduling unit (SU) and home appliances. This system model is proposed to decrease the electricity expenses of users in a smart home. Smart home is equipped with SM, SU, household appliances and each SM is connected to

AMI. AMI makes a bridge between SM and utility. AMI and SM make the two-way communication possible between utility and consumer. Further, SM informs the utility about the consumption patterns of consumers. With the help of information received from SM, utility calculates the energy demand for the upcoming hours and informs the generation station about the power requirement for the upcoming hours. On the basis of this information, generation station changes the power generation to fulfill the utility requirements.

SU schedules the appliances according to user's defined hours by using the meta-heuristic techniques EWA, CSA and HCEO. System model of this work is given in Fig. 1.

In this work, a residential area where each home contains appliances, SM and SU is considered. In each home, appliances, SU and SM communicate with each other through home area network (HAN). Moreover, communication between SM and utility takes place through wireless networks. Price scheme used in this scenario is RTP. Utility sends the RTP signal to SM through wireless communication and then, SM sends the pricing signal to SU. In SU, CSA, EWA and proposed hybrid scheme schedule the appliances on the basis of RTP signal for each hour of a day to decrease cost. These appliances are scheduled for 24 time slots.

3.1 Load Classification

In this work, home devices are classified into following groups: Interruptible appliances, non-interruptible appliances and base appliances. This classification helps in scheduling the home appliances more efficiently and whole day scheduling of household appliances is performed as explained below

$t \in T, \forall, T = \{t_1, t_2, t_3, t_4, ...t_{24}\}$

A residential area is considered where each smart home consists of fifteen appliances. These appliances have different power ratings and operational time.

3.1.1 Interruptible Appliances

Interruptible appliances considered in this scenario are water heater, water pump, air conditioner, refrigerator, iron, dish washer and vacuum cleaner. These appliances are shiftable and can be shifted to any time slot during their operational time.

Let A_I, represents the set of interruptible appliances and $a_I \in A_I$ is an appliance from the interruptible group and $\wp_I$ is the power rating of every interruptible appliance. Then, electricity consumed by interruptible appliances for 24 h can be given as

$$\varepsilon_I = \sum_{a_I \epsilon A_n} \left(\sum_{t=1}^{24} \wp_I \times \Im_I(t) \right) \tag{1}$$

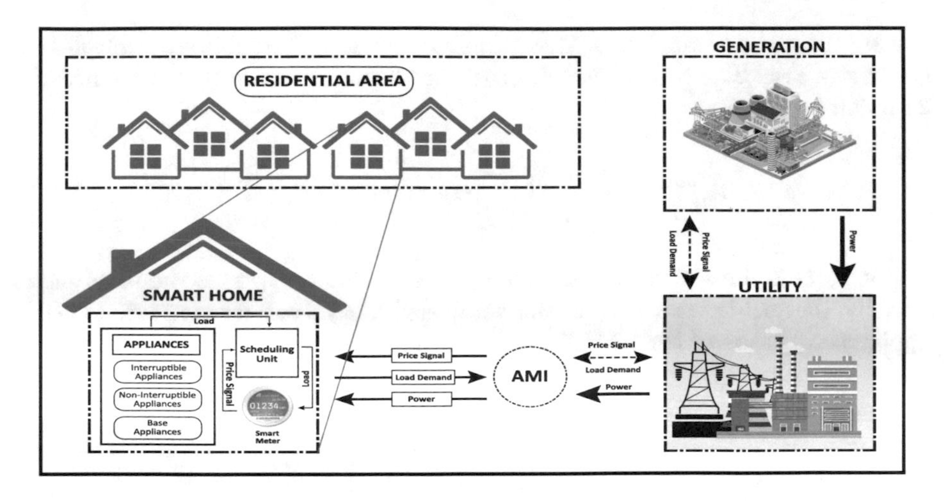

Fig. 1. Proposed system model

The total electricity cost of 24 h in a day which will be paid to the utility by the consumer against all interruptible appliances is given by

$$\aleph_{a_I}^{Total} = \sum_{a_I \epsilon A_n} \left(\sum_{t=1}^{24} \wp_I \times \rho(t) \times \Im_I(t) \right) \tag{2}$$

here $\Im_I(t)$ shows the status of interruptible appliances which can be 0 or 1, $\rho(t)$ is the price signal sent by the utility and $\aleph_I$ is the total electricity cost for interruptible appliances. Status of interruptible appliances is given as

$$\Im_I(t) = \begin{cases} 0 & \text{if appliance is OFF} \\ 1 & \text{if appliance is ON} \end{cases} \tag{3}$$

3.1.2 Non-interruptible Appliances

These appliances are washing machine and cloth dryer. Due to their operation, these appliances can not be interrupted during their operational time. As cloth dryer is needed after the Washing machine operation therefore, and cloth dryer is scheduled immediately after the washing machine.

Let $a_{NI} \in A_{NI}$, represents each non-interruptible appliance where, A_{NI} is the set of non-interruptible appliances. $\wp_{NI}$ is the power rating then, total electricity ε_{NI} consumed by these appliances can be explained as

$$\varepsilon_{NI} = \sum_{a_{NI} \epsilon A_n} \left(\sum_{t=1}^{24} \wp_{NI} \times \Im_{NI}(t) \right) \tag{4}$$

During their operational time, these appliances will not be shifted, regardless of electricity cost. The total electricity cost for all non-interruptible appliances for 24 h can be given as

$$\aleph_{a_{NI}}^{Total} = \sum_{a_{NI} \epsilon A_n} \left(\sum_{t=1}^{24} \wp_{NI} \times \rho(t) \times \Im_{NI}(t) \right) \tag{5}$$

where $\Im_I(t)$ is the status of non-interruptible appliances, $\rho(t)$ is the price signal sent by the utility and $\aleph_{NI}$ is the total electricity cost for non-interruptible appliances. Status of these appliances is given by following equation

$$\Im_I(t) = \begin{cases} 0 & \text{if appliance is OFF} \\ 1 & \text{if appliance is ON} \end{cases} \tag{6}$$

3.1.3 Base Appliances

Base appliances or fixed appliance are oven, blender and four lights of different power rating. These appliances are not scheduled and consumer can switch ON these appliances any time according to desired requirements.

Let $a_B \in A_B$, represents each base appliance where, A_B is the set of base appliances and $\wp_B$ is the power rating of each appliance then, total energy ε_B consumed during the whole day can be explained as

$$\varepsilon_B = \sum_{a_B \epsilon A_n} \left(\sum_{t=1}^{24} \wp_B \times \Im_B(t) \right) \tag{7}$$

The cost of electricity consumed by base appliances for 24 h can be found as

$$\aleph_{a_B}^{Total} = \sum_{a_B \epsilon A_n} \left(\sum_{t=1}^{24} \wp_B \times \rho(t) \times \Im_B(t) \right) \tag{8}$$

where $\Im_I(t)$ shows the status of interruptible appliances which can be 0 or 1, $\rho(t)$ is the price signal sent by the utility and $\aleph_B$ is the total electricity cost for base appliances. Status of base appliances is given as

$$\Im_I(t) = \begin{cases} 0 & \text{if appliance is OFF} \\ 1 & \text{if appliance is ON} \end{cases} \tag{9}$$

To find the total cost of electricity consumed during the whole day can be found by Eqs. 2, 5 and 8 as

$$\top_{Cost} = \left(\aleph_{a_I}^{Total} + \aleph_{a_{NI}}^{Total} + \aleph_{a_B}^{Total} \right) \tag{10}$$

where, $\top_{Cost}$ is the total cost of all appliances in a day.

4 Optimization Techniques

In SG, each smart home contains a SM which makes the communication between consumer and the utility. Consumers are able to send their energy usage patterns to utilities and receive price signals from the utility using SM. To optimize the consumer load, optimization techniques are implemented in this paper which are explained below:

4.1 CSA

The CSA is a nature-inspired algorithm which is based upon the natural behavior of some special kinds of cuckoo. Some species of cuckoo such as *Ani* and *Guira*, tries to lay eggs in other birds' nests and it removes other eggs from host nests to increase its child's food share in the nest. If host bird founds the alien eggs it will either throw them out or leave the nest and make a new nest [14].

For optimization problem, each egg represents a solution. Cuckoo performs *Levy flight* to find the host nest and lays a egg which can be explained as

$$X_a^{(t+1)} = X_a^{(t)} + \gamma \oplus Levy(\lambda) \tag{11}$$

where, γ is step size and mostly it is taken as 1. Levy flight is used for random walk and its step length is random and it only depends upon its current location. It is an efficient way to explore the search space.

4.2 EWA

EWA is a bio-inspired optimization technique, inspired from the reproduction system of earthworms. In earthworms, two types of reproduction exists. In Reproduction-1, off-springs are produced by single parent and in reproduction-2 off-springs are produced by two earthworm parents [15].

Reproduction-1 is formulated as

$$x_{i1,k} = x_{max,k} + x_{min,k} - \sigma x_{i,k} \tag{12}$$

where $x_{i,k}$ is the kth element of x_i that shows the position of the earthworm i. Similarly, $x_{i1,k}$ is the kth element of x_{i1} which is the new position of the earthworm $i1$ and σ is the similarity factor. $x_{max,k}$ and $x_{min,k}$ are upper and lower limits respectively.

Reproduction-2 is an improved form of crossover operator. Crossover can be single point, multipoints or uniform. In this paper, uniform crossover is used and an off-spring $x_{i,2}$ is produced. After implementing above reproductions, both off-springs are combine by following equation

$$x_i^{'} = \delta x_{i1} + (1 - \delta) x_{i2} \tag{13}$$

where δ is a proportional factor which adjusts the proportion of off-springs. At the end *Cauchy mutation* is applied to increase the search ability and moves it towards the global optimal point [15].

4.3 Hybrid CSA and EWA

Hybrid optimization technique (HCEO) is proposed in this paper, which is hybridized from CSA and EWA. Hybridization is done in order to enhance the performance of an optimization technique as explained in [16]. Hybridization can be loosely coupled or strongly coupled, in strongly coupled hybridization, flow of a technique is followed and in loosely coupled hybridization internal steps of techniques are interchanged. Our proposed hybrid technique is hybrid in such a way that steps of EWA are fully followed whereas, only few steps from EWA are opted in order get an optimal solution.

5 Simulations and Discussions

To show the effectiveness of our scheme in HEMS, simulations are performed in MATLAB 2017a and RTP scheme is used. Algorithms: EWA, CSA and proposed hybrid scheme HCEO are evaluated and performance is compared with each other on the basis of PAR, electricity cost, consumed electricity and waiting time of appliances. Simulations are conducted for a smart home with fifteen appliances.

Figure 2 illustrates the hourly electricity cost of single home with RTP signals. Results show that most of the load is shifted to off-peak hours and only necessary load is scheduled during high price hours therefore, very low cost is paid during on-peak hours. Figure 3 indicates the total cost comparison with each other and with unscheduled cost. It is clear from figure that proposed hybrid scheme has outperformed CSA in case of cost. Further, all implemented techniques have decreased the total electricity cost as compared to unscheduled cost. From Fig. 3 it is cleared that EWA, CSA and HCEO have reduced cost by 49%, 25% and 49% respectively by using RTP scheme.

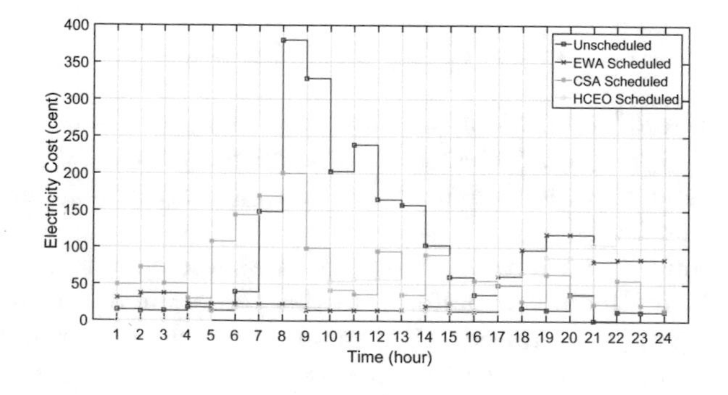

Fig. 2. Per hour electricity cost of scheduling techniques

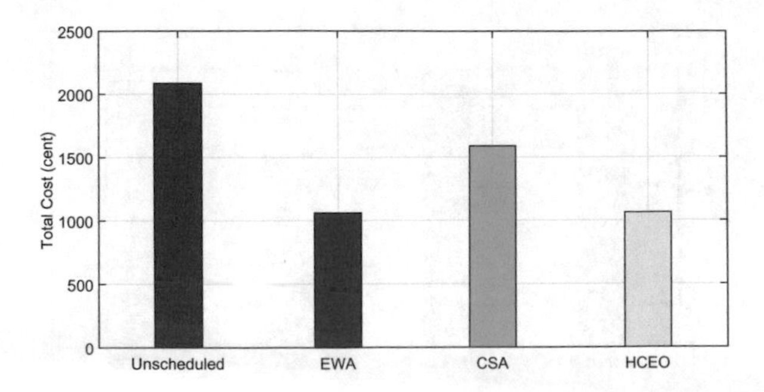

Fig. 3. Total electricity cost in a day

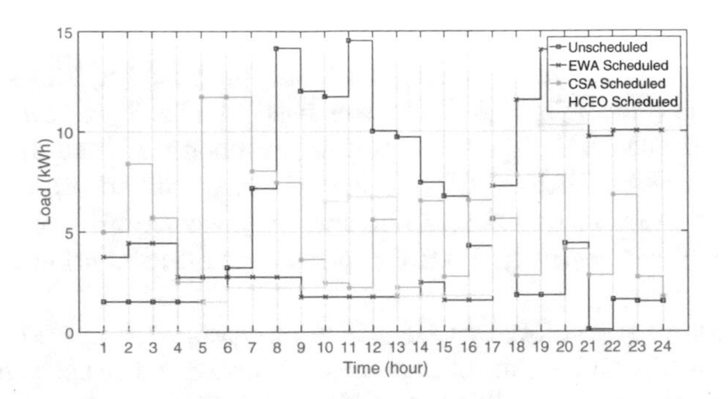

Fig. 4. Hourly electricity consumption of household appliances

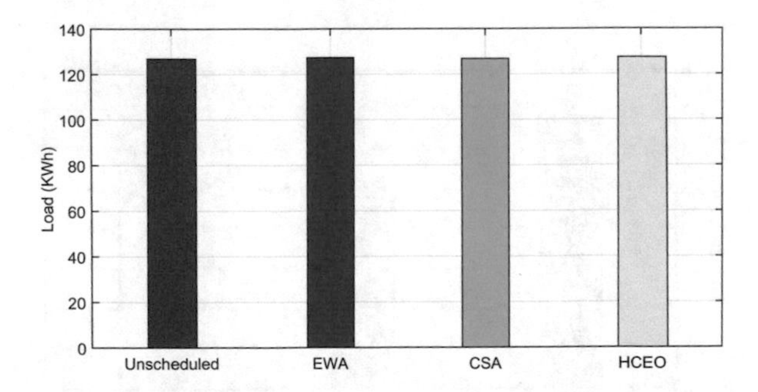

Fig. 5. Total load consumed by implemented optimization techniques

Proposed hybrid scheme, in case of per hour electricity consumption, schedules most of the load during off-peak hours and schedules very small load during on-peak hours thus, reducing cost as shown in Fig. 4. Moreover, Fig. 5 illustrates that total load before and after scheduling remains same because load is only shifted not terminated. As we have only scheduled the load optimally and not implemented any load shedding strategy therefore, total load with and without scheduling remains same.

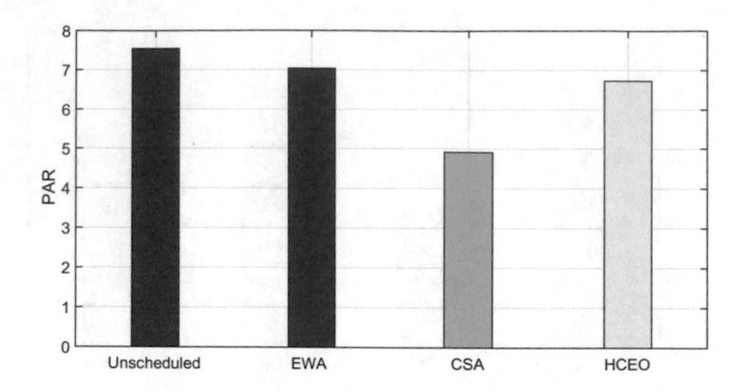

Fig. 6. PAR of implemented optimization techniques

Performance comparison of CSA, EWA and proposed hybrid scheme on the basis of PAR is shown in Fig. 6. On the base of Fig. 6, it can be concluded that PAR of all used techniques is less than unscheduled load and also in case of scheduled schemes, PAR of CSA is less as compared to other two schemes. Further PAR reduction by CSA, EWA and proposed hybrid scheme is 34.8%, 6.72% and 10.8% respectively. PAR comparison of all used schemes is shown in Fig. 6.

In this work, user comfort by CSA, EWA and proposed hybrid scheme is calculated in terms of waiting time of appliances. It is defined as the time interval for which a user waits for an appliance to switch ON. User comfort of implemented schemes is given in Fig. 7.

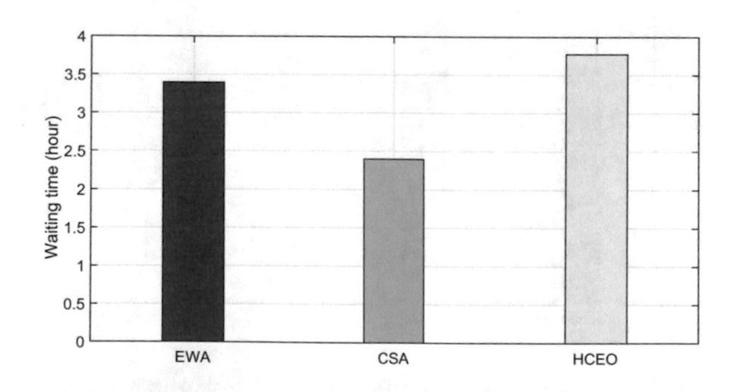

Fig. 7. Waiting time of home appliances using EWA, CSA and HCEO

From simulation findings, it is concluded that, hybrid technique performs better in case of electricity cost and it surpassed the EWA and CSA in case of cost. A trade-off exists between electricity cost and user comfort. To reduce the electricity bill, HCEO shifts most of the home appliances from high price hours to low price hours as a result, waiting time of appliances is increased. So if consumer wants to reduce the waiting time of household appliances then, electricity cost will be high.

6 Conclusion

An optimization based HEMS is proposed, in which appliance scheduling is performed using load shifting strategy. For scheduling the home appliances, optimization algorithms (CSA, EWA and HCEO) are used and their performance is calculated using four parameters: electricity cost, electricity consumed, PAR and waiting time of appliances using RTP scheme sent by the utility. Objective of this study is to decrease the electricity cost by shifting the household load. The hybrid technique HCEO decreased the electricity cost by 49% with respect to unscheduled electricity cost because it has scheduled the most of the appliances during the low price hours. Thus, it has reduced the electricity consumption during on-peak hours. From simulations, it is observed that there is a trade-off between electricity cost and user comfort. So, if consumer wants to decrease the waiting time of home electrical devices then, consumer needs to compromise on electrical cost.

In future, we will integrate the pumped storage system to generate the electricity when demand and prices are low and will use it during high price and demand hours. In addition, we will focus on the coordination of smart homes to sell extra electricity produced by consumers.

References

1. Evangelisti, S., Lettieri, P., Clift, R., Borello, D.: Distributed generation by energy from waste technology: a life cycle perspective. Process Safety Environ. Protect. **93**, 161–172 (2015)
2. Vardakas, J.S., Zorba, N., Verikoukis, C.V.: A survey on demand response programs in smart grids: pricing methods and optimization algorithms. IEEE Commun. Surv. Tutor. **17**(1), 152–178 (2015)
3. Yi, P., Dong, X., Iwayemi, A., Zhou, C., Li, S.: Real-time opportunistic scheduling for residential demand response. IEEE Trans. Smart Grid **4**(1), 227–234 (2013)
4. Samadi, P., Wong, V.W.S., Schober, R.: Load scheduling and power trading in systems with high penetration of renewable energy resources. IEEE Trans. Smart Grid **7**(4), 1802–1812 (2016)
5. Aslam, S., Iqbal, Z., Javaid, N., Khan, Z.A., Aurangzeb, K., Haider, S.I.: Towards efficient energy management of smart buildings exploiting heuristic optimization with real time and critical peak pricing schemes. Energies **10**(12), 2065 (2017)
6. Mahmood, A., Javaid, N., Khan, N.A., Razzaq, S.: An optimized approach for home appliances scheduling in smart grid. In: 19th International Multi-topic Conference (INMIC), Islamabad, pp. 1–5 (2016)
7. Motevasel, M., Seifi, A.R.: Expert energy management of a micro-grid considering wind energy uncertainty. Energy Convers. Manage. **83**, 58–72 (2014)
8. Logenthiran, T., Srinivasan, D., Shun, T.Z.: Demand side management in smart grid using heuristic optimization. IEEE Trans. Smart Grid **3**(3), 1244–1252 (2012)
9. Muralitharan, K., Sakthivel, R., Shi, Y.: Multiobjective optimization technique for demand side management with load balancing approach in smart grid. Neurocomputing **177**, 110–119 (2016)

10. Liu, Y., Yuen, C., Yu, R., Zhang, Y., Xie, S.: Queuing-based energy consumption management for heterogeneous residential demands in smart grid. IEEE Trans. Smart Grid **7**(3), 1650–1659 (2016)
11. Wang, C., Zhou, Y., Wu, J., Wang, J., Zhang, Y., Wang, D.: Robust-index method for household load scheduling considering uncertainties of customer behavior. IEEE Trans. Smart Grid **6**(4), 1806–1818 (2015)
12. Setlhaolo, D., Xia, X., Zhang, J.: Optimal scheduling of household appliances for demand response. Electr. Power Syst. Res. **116**, 24–28 (2014)
13. Melhem, F.Y., Grunder, O., Hammoudan, Z., Moubayed, N.: Optimization and energy management in smart home considering photovoltaic, wind, and battery storage system with integration of electric vehicles. Canad. J. Electr. Comput. Eng. **40**(2), 128–138 (2017)
14. Yang, X.S., Deb, S.: Cuckoo search via levy flights. In: World Congress on Nature and Biologically Inspired Computing (NaBIC), Coimbatore, pp. 210–214 (2009)
15. Wang, G.G., Deb, S., Coelho, L.D.S.: Earthworm optimization algorithm: a bio-inspired metaheuristic algorithm for global optimization problems. Int. J. Bio Inspired Comput. **7**, 1–23 (2015)
16. Xhafa, F., Gonzalez, J.A., Dahal, K.P., Abraham, A.: A GA (TS) hybrid algorithm for scheduling in computational grids. In: International Conference on Hybrid Artificial Intelligence Systems (HAIS), Berlin, pp. 285–292 (2009)

Differential-Evolution-Earthworm Hybrid Meta-heuristic Optimization Technique for Home Energy Management System in Smart Grid

Nadeem Javaid(✉), Ihtisham Ullah, Syed Shahab Zarin, Mohsin Kamal, Babatunji Omoniwa, and Abdul Mateen

COMSATS Institute of Information Technology, Islamabad 44000, Pakistan
nadeemjavaidqau@gmail.com
http://www.njavaid.com

Abstract. In recent years, advanced technology is increasing rapidly, especially in the field of smart grids. A home energy management systems are implemented in homes for scheduling of power for cost minimization. In this paper, for management of home energy we propose a meta-heuristic technique which is hybrid of existing techniques enhanced differential evolution (EDE) and earthworm optimization algorithm (EWA) and it is named as earthworm EWA (EEDE). Simulations show that EWA performed better in term of reducing cost and EDE performed better in reducing peak to average ratio (PAR). However proposed scheme outperformed in terms of both cost and PAR. For evaluating the performance of proposed technique a home energy system proposed by us. In our work we are considering a single home, consists of many appliances. Appliances are categorized into two groups: Interruptible and uninterruptible. Simulations and results show that both algorithms performed well in terms of reducing costs and PAR. We also measured waiting time to find out user comfort and energy consumption.

Keywords: EDE algorithm · EWA algorithm · User comfort
Hybrid meta-heuristic technique

1 Introduction

Population of the world is increasing day by day and due to increase in population demands also increasing. The use of electricity is also increasing. The demand is increasing for electricity, in order to provide reliable electricity the concept of smart grid (SG) is implemented everywhere. The SG acts as a medium between the utility and consumer for exchange of information about price signals and energy demand. Demand side management (DSM) is very important for consumers to reduce electricity cost. Some strategies are discussed in the literature to reduce electricity cost, PAR and energy consumption. For optimization of

L. Barolli et al. (Eds.): IMIS 2018, AISC 773, pp. 15–31, 2019.
https://doi.org/10.1007/978-3-319-93554-6_2

energy consumption and by applying various methods these strategies are used. To balance supply from the utility and demand from consumers, DSM strategies are very useful.

Demand and supply are balanced by using dynamic pricing schemes. These schemes are more effective, especially for electricity management to reduce costs and PAR rapidly. Some of dynamic pricing schemes like Real time pricing (RTP), time of use (TOU), critical peak pricing (CPP) and inclined block rate (IBR) are very useful. Smart grid provides consumers a safe way to reduce cost, PAR, energy consumption and to maximize user comfort. Demand and response is the best technique to achieve all objectives like in [1], authors using DR technique to reduce the cost and PAR and achieving maximum user comfort. Electricity consumption is increasing because demand is increasing due to high number of consumers, so to fulfill this problem we use different techniques in home energy management to reduce energy consumption, cost and PAR and to increase user comfort.

Demand side management strategy has been used in [2], saving cost and reducing PAR. Energy consumption when increases, so load on utility side also increases, which results in decreasing performance of utility. The increase in demand can create a shortage of electricity. When in peak hours electricity demand is high, so utility have to use extra energy resources to cover that space, which results in creating an imbalance between supply and demand. For the management of demand and response issue in DSM demand response (DR) is used. DR provides sufficient knowledge to users to use minimum electricity during peak hours, in order that energy loss can be controlled.

Two algorithms teaching and learning based optimization (TLBO), shuffled the frog leaping (SFL) are used in [3]. Scheduling different appliances in smart homes in smart grid (SG). The main objective is to reduce consumer cost. Results showed that the program we used demand response DR by shifting the load minimizing cost successfully. The Large number of appliances is aggregated which are based on DR and its decomposition it non-convex. Heuristic algorithm is applied in [4], using an intelligent decision support system (IDSS) for cost minimizing costs of appliances. Two models for cost generic and flexible are utilized in this model. Real time pricing (RTP), RTP/2-RTP schemes is being used, but the performance of combination works well than RTP alone.

Energy is consumed in such way in [5], that there is the utilization of appliances and minimization of cost is balanced. Realistic scheduling algorithm (RSM) is used to achieve user comfort, scheduling appliances into time slots sub slots effectively. The results shows that it maximize appliance utilization and cost is reduced. RSM is working along with BPSO. In this paper we are using meta-heuristic techniques EDE and EWA and hybrid meta-heuristic technique EEDE, our aim is to minimize cost, PAR, energy consumption to increase user comfort.

Rest of paper is organized as follows: Sect. 2 contains the Related Work. The Problem statement is presented in Sect. 3. Proposed solution is discussed in Sect. 4. Section 5 consists of the system model. Optimization techniques are explained in Sect. 6. Simulation and the results are presented in Sect. 6 and conclusion of the paper is in Sect. 8.

2 Related Work

Many researchers around the world are working in the smart grid, scheduling home appliances using different algorithms. In this regard, some of the papers are discussed as follows:

The results shows that it maximize appliance utilization and cost is reduced. RSM is working along with BPSO. In [6], authors are using renewable energy for battery optimal control and management. The neural network is used to apply adaptive dynamic programming (ADP). In each iteration discount factor is chosen appropriately, the simulations show the effectiveness of developed algorithm. In [7], the authors presented a new HEM model, they have worked on using the TOU pricing scheme in two conditions. In the first condition with the addition of RES energy and in second condition without renewable energy resource (RES) energy. They have evaluated performance on a base of three algorithms, genetic algorithm (GA), binary particle swarm optimization (BPSO) and Cuckoo, minimizing the user bill and achieving low peaks. The result shows that the cuckoo has shown better results in comparison to GA and BPSO, reducing cost and high peaks.

The authors presented the energy management model to schedule in [8], the resident load by optimization and also to reduce electricity bills, carbon emissions. Another objective is to achieve high user comfort in zero energy buildings. The algorithms used are GA, TLBO and EDE. With the use of RES after performing simulations the electricity cost, carbon emissions and PAR are reduced up to 67%, 55% and 29%.

The authors used GA, BPSO and ant colony optimization (ACO) techniques [9], to minimize cost, PAR and achieving user comfort. The price schemes used are TOU and IBR. In this paper Hybrid of GA, BPSO and ACO are used to combine with the energy management controller (EMC) to reduce costs, PAR. The results show that GA-EMC performed well. Proposed optimal and dynamic energy control flow strategies in [10]. In this paper residential energy local network (RELN) combined with DSM and using RTP scheme. The results show that cost is minimized. Several conclusions are made on the basis of RES. DR is combined with a DG program in [11]. The results show that PAR and cost are reduced for multiple homes with the least effect on user comfort. Several issues are needed to be addressed in the future like securing and reliable home energy management system.

In [12], using different pricing schemes like RTP and TOU, the author indicates that the home energy management system over a varying household set may not perform equally with respect to the load. Shifting and reducing the load are two factors. The results show that the home energy management system is not only based on DR but also we have to look after individual preferences. A multi-period artificial bee colony (MABC) is proposed to achieve maximum operational efficiency and minimum cost in home micro-grids by authors in [13], the proposed energy management scheme (EMS) performs well in increasing convergence speed also in combination mixed integer nonlinear programming (MINLP) algorithm is also used. An optimal algorithm is used for minimization of electricity tariffs from the user and demand power on the supply side in [14], also solar energy is used to minimize demand load in distribution network in order to increase stability of the grid. In both user mode and manually an adaptable autonomous thermostat used for residential energy management in order to save more energy and reduce costs, Adaptive Fuzzy Logic Model (AFLM) was also used to adapt itself to new user preferences, the results shows that proposed scheme performed well but users opts maximum in automatic mode in [15]. In near future smart home technology will be preferred everywhere for energy management in homes and also for better performance of appliances, users want to reduce cost and on demand side stability is important. Electricity utilization in a better way is important both for users and utility, so authors in [16] have focused on smart technologies, giving brief knowledge about different technologies and also predicting about future aspects.

Awareness on energy management is very important which type of appliances should be used and how to control cost, authors in [17] worked on increasing life cycle cost of different appliance in reducing expenses of organization in term of cost and operation. Using inventor in AC can reduce energy consumption by 38%. The main idea is to manage energy. A DSM technique consists of a smart grid is used in [18], smart consumers connected to the grid. A distributed algorithm micro-grid energy management, distributed optimization Algorithm (MEM-DOA) designed for customers of each type, installed on smart meters is used for simulations, results show that PAR and peak demand reduced. This algorithm also shows that active consumers benefited more than passive.

EWA method performed well in finding values of better function as compare to other algorithms in [19], it is also very useful for parallel computation. A hybrid teacher learning, TLBO is used in [20] to minimize cost-user dis-comfort and PAR, TLBO performed well in reducing costs while GA performed well in term of reducing cost.

3 Problem Statement

In SG the main issues which need to be solved are cost minimization, to reduce PAR, to balance load and increasing user comfort. In [7], the authors HEM model, using the TOU pricing scheme in two conditions. The algorithms used are GA, TLBO and EDE. With the use of RES after performing simulations the electricity cost, carbon emissions and PAR are reduced. However the user comfort is not considered. Demand side management strategy has been used in [2]. The results show that proposed algorithm can handle a large number of controllable devices, saving cost and reducing PAR but again user comfort is not considered in this work. To overcome the problem of waiting time we propose a useful home energy model, we have used EDE and EWA heuristic techniques to overcome the problem of user comfort. In terms of cost our proposed technique performed as well as maximizing user comfort. Energy is consumed in [5], in such way that there is the utilization of appliances and minimization of cost is balanced. However, PAR reduction is not considered in this paper.

In our work, we apply EDE and EWA and hybrid meta-heuristic technique EEDE in a single home. We schedule different appliances. Appliances are categorized into two groups as discussed in system model. Overall objectives of our proposed work are: Cost reduction PAR reduction Minimizing energy consumption Increasing user comfort.

4 Proposed Solution

The objectives of our work are: cost reduction, minimizing PAR and load balancing along with increasing user comfort. For the solution of our problem we are using EDE and EWA algorithms and hybrid meta-heuristic technique EEDE, using the dynamic CPP pricing scheme. For 24 h schedule, we are taking 96 time slots. We are dealing with seven appliances. Our proposed technique will schedule all appliances to minimize cost, PAR and balancing load. We will also increase user comfort dealing with waiting time.

We will calculate all parameters for different number of operational time intervals (OTI). In our paper we will optimize the problem by taking 15, 30 and 60 min OTI and then comparing them all.

5 System Model

We discussed working of our designed system in this portion. As shown in Fig. 1, we propose a model of HEMS for single homes. In order to schedule appliances of home to minimize cost, PAR and maximizing user comfort.

In our scenario we are scheduling appliances of single smart home. The smart home appliances are connected with the smart meter (SM). An EMC is used in order to receive price signals from SM. SM working jointly and it also takes data about energy consumption from EMC and sends it to the utility. SM and Utility are communicating with each other through a wireless network. All information between SM, utility and smart meter are exchanged in the home area network. Formulas for Cost, load and PAR are mentioned in Eqs. (1), (2) and (3).

$$Cost = \sum_{t=1}^{24} (E_{hourRate} * P_{AppRate}) \tag{1}$$

$$Load = P_{AppRate} * App \tag{2}$$

$$PAR = max \frac{(Loads)}{Avg(Loads)} \tag{3}$$

In our research work we are considering single home with 7 appliances. For electricity bill we are using the CPP pricing scheme. In our scenario a single time slot for one appliance to complete its working is 15 min. OTI in our scenario is 15 min. We have also calculated the cost on the basis of 30 min OTI and 60 min. The length of operational time of appliances is 15 min, so dividing 24 h in 96 slots.

Each hour consist of 4 slots. One slot is about 15 min. Dividing the whole day on the basis of hours, like scheduling an appliance for hours will result in loss of time, because each appliance working time is not more than 15 min. So diving day on the basis of appliance working time will reduce cost and will the whole system stability greater.

For scheduling according to their nature appliances are divided into two categories according to their nature on the basis of power consumption.

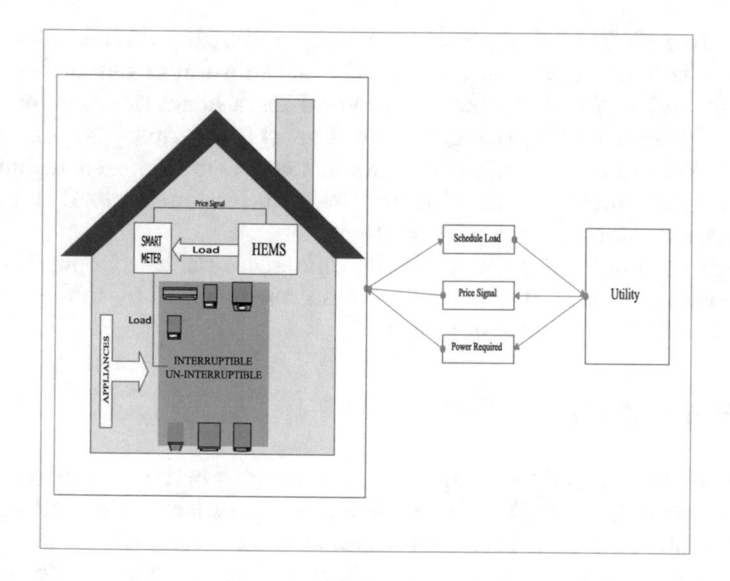

Fig. 1. System model

5.1 Interruptible Appliances

A type of appliances which can be interrupted during operational time and can be shifted. In our case, interruptible appliances are air conditioner and refrigerator, they are also known as heavy loads. Heavy loads operating at the same time will be costly.

5.2 Un-interruptible

Type of appliances can be shifted to any time slots, but once their operation is started they cannot be interrupted until they complete their operation. In our model non-interruptible appliances are Dishwasher, cloth washer and dryer, oven1, oven2, and the electric vehicle.

5.3 Price Model

The price rate set by utility electricity cost is calculated on that basis, different pricing schemes are used to calculate the price. A utility used different tariff schemes. For the reduction of cost and PAR mostly dynamic pricing schemes are used like TOU, IBR, CPP, DAT and RTP. Mostly dynamic pricing scheme is used because the user can shift their appliances from high rate hours to low in order to reduce cost. Mostly RTP is used but in our scenario we have used CPP pricing scheme.

6 Optimization Techniques

We have applied heuristic algorithms for our proposed work. Traditional techniques are known to not good for real time optimization. The techniques we used are the EDE and EWA. EDE and EWA are described in detail in the following sections.

6.1 EDE

Differential evolution (DE) was the first version followed by EDE. EDE is an evolutionary algorithm mostly used for population generation, Storn developed it in 1955. DE algorithm execution is in three steps: mutation, crossover and selection. Initially a random population is generated, then after generation of population mutation is performed. In mutation step three random vectors are selected in order for each target vector, the difference between any two random vectors is added into a third vector to form a new mutant vector. The equation of the mutant vector is:

$$Rj, i, M + 1 = Xa1, M + Q(Xa2, M * Xa3, M) \tag{4}$$

In Eq. (4), where a1, a2 and a3 are distinct vectors and Q is a scaling factor. After generation of mutant vector crossover starts, in a crossover trial vector generation takes place. By combing elements of mutant and the target vector, a trial vector is generated. For the formation of the trial vector, a rate is decided which is known as a crossover rate. The crossover rates decides from target and a mutant vector how much information is collected, a random number is generated to compare with the crossover rate to decide a value for a specific position is chosen either form target or mutant vector. When trial vector is created, then in phase of selection trial and target vectors are compared. A vector having better fitness value, for the next generation that vector is selected.

In DE algorithm we have 3 numbers of parameter used to control: population size (PZ), the factor of scaling Q and crossover (C), in EDE these are reduced to two parameters only PZ and Q. The only difference between EDE and DE is that at crossover phase, five trial vectors are generated in EDE instead of one.

Different crossover rates are used for the generation of first three trial vectors. To increase convergence speed fourth trail vector is used and the fifth one is used only to enhance the diversity of the trial vectors.

A vector with maximum fitness value is selected from the trial vectors finally generated, after the selection of a final trial vector, it is compared with the target vector and then the vector with a fitness value which is better selected for the next generation.

6.2 EWA

EWA is a bio-inspired heuristic optimization algorithm. It consists of two kinds of reproductions of worm's reproduction 2 and reproduction 2. The offspring are generated through reproductions and then we calculate weightage sum of all offspring, so we can get final earthworms for the next generation. One offspring is only generated by reproduction 1 which is special, reproduction two generates more than one offspring whit the help of crosser operators which are nine in numbers. For improvement of searching ability a Cauchy mutation (CM) is added to EWA. The earthworm has the finest fitness permit on straight next generation, and operators can not alter them. The population of earthworm cannot fail in generations in the increment this is the assurance.

6.3 EEDE

Our proposed hybrid technique is EEDE. In EWA in last step best population is generated, in this paper, we assigned best generated population of EWA to EDE to make it hybrid in form of EEDE. EEDE is more better it consists of both EDE and EWA and results in better performance.

7 Simulation and Results

This section composed of simulations and results. In this section we evaluated the performance of EDE and EWA techniques and new hybrid meta-heuristic technique EEDE. All the simulations are conducted by using MATLAB. We evaluated the proposed EDE and EWA and EEDE Techniques on the basis of four parameters: Electricity cost, Energy consumption, PAR and user comfort.

For simulation purpose, we are considering single home, consists of seven appliances using the CPP price scheme. We have categorized appliances into two groups discussed in the Table 2.

We applied both techniques using three different OTI of 15 min, 30 min and 60 min, using a CPP price scheme for 24 h. Simulations and results are discussed in the following sections for four parameters, cost, PAR, energy consumption and user comfort.

7.1 Cost

The overall cost for single home for a day is illustrated in Fig. 2, sub-figures a, b, c. In these figures we have calculated the cost of electricity of single home for one day. Total cost reduction in case of 15 min OTI is 2.8% by using EDE algorithm. Now for OTI of 30 min using EDE the total cost reduction is 1%.

In case of 60 min OTI the cost is using EDE, the total cost is reduced up to 6%. Overall, our proposed technique showed good performance in reducing cost. A comparison of the costs of different OTI. EWA also showed good performance in reducing cost. In case 15 min, 30 min and 60 min EWA reduced costs by 1%, 5%, 8% respectively.EEDE performed better than both techniques, reducing cost by 8%, 9%, 10% in case of 15 min, 30 min and 60 min OTI respectively.

7.2 Energy Consumption

In my paper, I have taken two types of appliances: interruptible and non-interruptible. In case of interruptible appliances, the running time of the appliances is not reduced, but is shifted from a peak to non-peak hours. In Fig. 3, sub-figures a, b, c energy consumption of appliances is shown during different hours. The electrical appliances consume energy in case of 15 min, 30 min and 60 min OTI, using EWA energy consumption is low during (0 to 5) hours for 24 h and EDE algorithm appliances consume more energy for all OTIs. During peak hours (10 to 4) for 24 h in comparison to EEDE, EDE and EWA algorithm appliances consume more energy. The normal consumption of energy for both schedule and unscheduled case are same. During high peak hours (10 to 4) EDE and EWA algorithm appliances consume more energy than EEDE, on the other hand EEDE reduce energy consumption efficiently by scheduling appliances during low and mid peak hours. EDE and EWA do not turn on appliances during peak hours.

7.3 PAR

DSM is good for utility as well as consumers. PAR reduction helps in reducing costs as well as in retaining utility stability. In comparison to the un-scheduled case from Fig. 4, sub-figures a, b, c, it is clear that in case of 15, 30 and 60 min OTI, EEDE algorithm performed well than EDE and EWA in reducing PAR.

In case of 15 min OTI PAR is reduced up to 1% after using EDE. In 30 min OTI PAR is reduced up to 6% using EDE, and for 30 min, reducing it up to 4%. EWA reducing PAR for 15 min, 30 min and 60 min up to 2%, 3%, 4% respectively. EEDE reduced PAR for 15 min, 30 min and 60 min by 13%, 7%, 8%, performed better then EDE and EWA.

7.4 User Comfort

This term is related to cost as well as waiting time. We calculated in terms of waiting time the user comfort that how much the user will wait to turn on the appliance, in order to have minimum cost users must schedule their appliances in low rate hours. For reduction of cost user will compromise on user comfort. User comfort and cost are inversely related to each other. We are calculating waiting time for 15 min, 30 min and 60 min OTI.

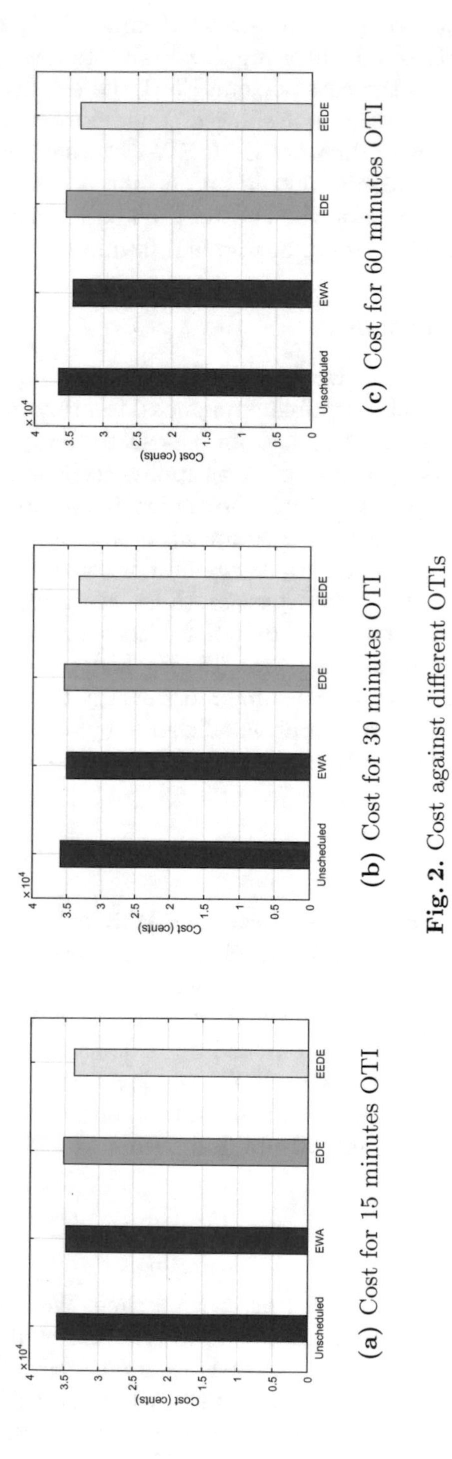

(a) Cost for 15 minutes OTI

(b) Cost for 30 minutes OTI

(c) Cost for 60 minutes OTI

Fig. 2. Cost against different OTIs

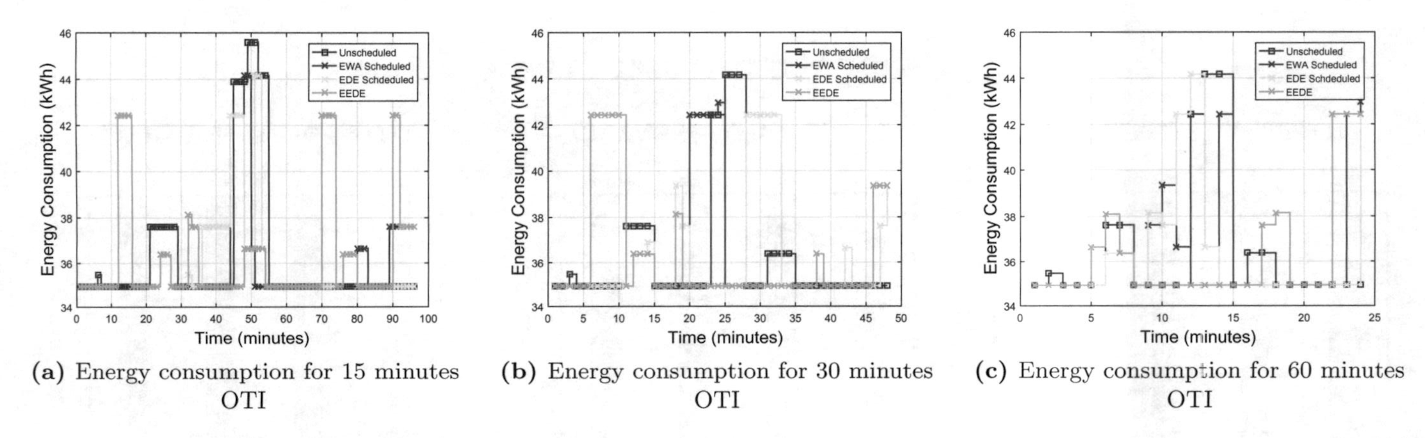

(a) Energy consumption for 15 minutes OTI

(b) Energy consumption for 30 minutes OTI

(c) Energy consumption for 60 minutes OTI

Fig. 3. Energy consumption comparison for different OTIs

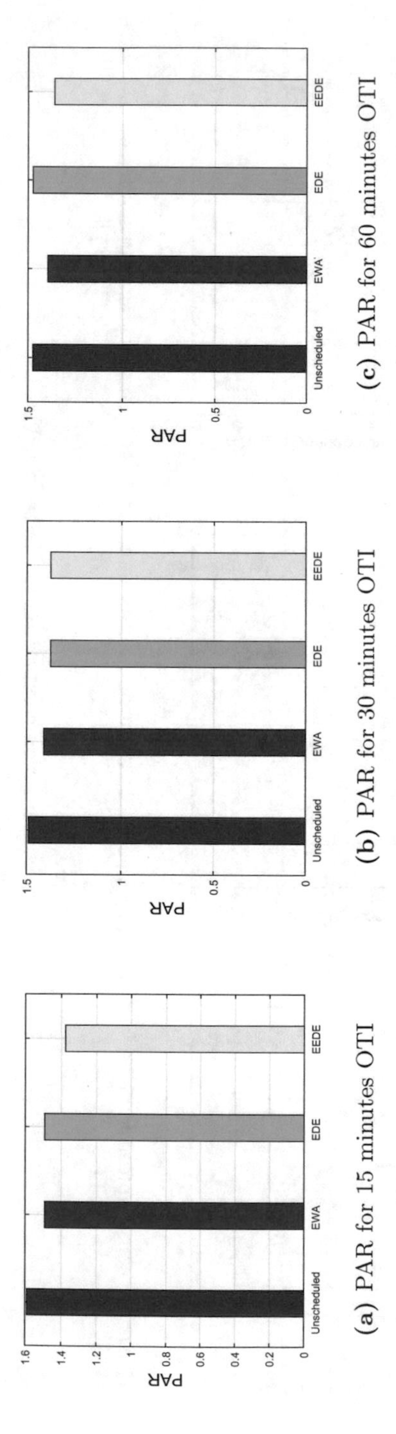

(a) PAR for 15 minutes OTI

(b) PAR for 30 minutes OTI

(c) PAR for 60 minutes OTI

Fig. 4. PAR in case of different OTIs

The results show in Fig. 5, sub-figures a, b, c, that EDE algorithm has maximum waiting time as compared to EWA. There is a tradeoff between cost and user comfort, and other hand EWA has less waiting time as compared to both EDE and EEDE and unscheduled case. It shows that EWA performs well in increasing user comfort, and reducing cost. So EWA is considered good for those consumers who are sensitive about cost and also wants user comfort.

7.5 Feasible Region

A region containing solutions bonded by set of points. In this paper cost and energy consumption depends on price of electricity for different time slots and the power consumption. As all the consumers wants to minimize cost, so they can only rely on power consumption adjustment. All the consumers can not change electricity price. There is indirect relationship between cost and user comfort.

Feasible Region for Energy Consumption and Cost. The Feasible region for energy consumption and cost for 15 min, 30 min and 60 min OTI is shown in Fig. 6, sub-figures a, b, c. A bounded region by P1 (0.1396, 0.0991), P2 (0.1396, 0.3085), P3 (0.7740, 1.7105), P4 (0.7740, 0.5945) in Fig. 6(a) shows an overall cost region for 15 min OTI. Similarly, when minimum load is scheduled using minimum price is shown by P1 (0.1396, 0.0991) and P2 (0.1396, 0.3085) represents the cost at maximum price using minimum load. Coordinates (P3, P4) represents the cost by keeping load maximum using maximum and minimum prices. The maximum cost at any time slot must be less than the maximum cost in unscheduled case which is 1.5 dollars. After applying constraints a feasible region is found which consists of coordinates (P1, P2, P3, P4, P5, P6) where coordinate P5 (0.52, 1.1525) shows that schedule load must not exceed than 0.52 kWh at that time slot where the price is the maximum and maximum load, i.e. 0.7740 kWh is represented by coordinate P6 (0.7740, 1.525), at any time slot price is higher than 1.5 dollars should not be scheduled.

Similarly, we find feasible region for 30 min OTI, a bounded region by P1 (0.8375, 3.5594), P2 (0.8375, 11.0969), P3 (4.6790, 61.9968), P4 (4.6790, 19.8858) in Fig. 6(b) shows an overall cost region for 30 min OTI. Similarly, when minimum load is scheduled using minimum price is shown by P1 and P2 represents the cost at maximum price using minimum load. Coordinates (P3, P4) represent the cost of keeping load maximum using maximum and minimum prices. The maximum cost at any time slot must be less than the maximum cost in unscheduled case which is 58.0969 dollars. After applying constraints a feasible region is found which consists of coordinates (P1, P2, P3, P4, P5, P6) where coordinate P5 (4.4, 58.0969) shows that schedule load must not exceed than 4.4 kWh at that time slot where the price is the maximum and maximum load, i.e. 4.6790 kWh is represented by coordinate P6 (4.6790, 58.0969), at any time slot price is higher than 58.0969 dollars should not be scheduled.

Figure 6(c) shows feasible region for 60 min OTI, where, at any time slot if the price is higher than 159 dollars should not be scheduled, and at any time

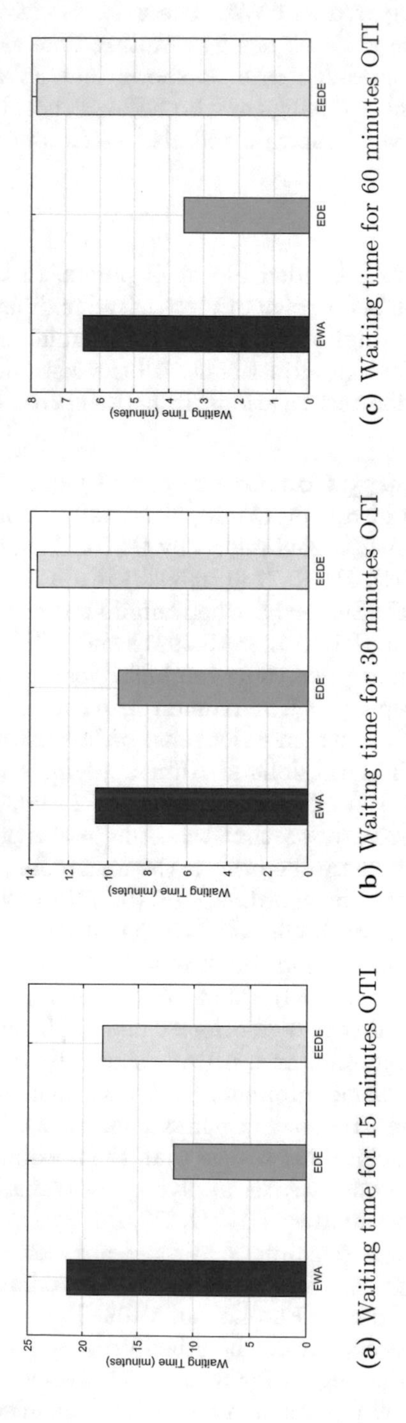

(a) Waiting time for 15 minutes OTI

(b) Waiting time for 30 minutes OTI

(c) Waiting time for 60 minutes OTI

Fig. 5. Waiting time comparison for different OTIs

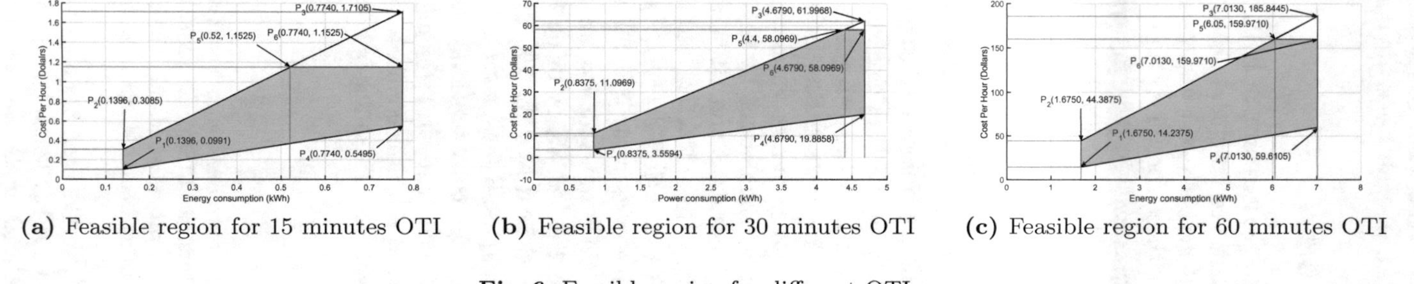

(a) Feasible region for 15 minutes OTI **(b)** Feasible region for 30 minutes OTI **(c)** Feasible region for 60 minutes OTI

Fig. 6. Feasible region for different OTIs

slot P5 shows that schedule load must not be exceed than 6.05 kWh where the price the price is maximum. All possible cases are discussed in Table 5.

7.6 Performance Trade-Off

In this paper, we evaluated the performance of EDE and EWA and EEDE algorithms, the hybrid meta-heuristic technique performed well in minimizing cost, PAR, but there is a trade-off between cost and user-comfort. EWA maximize user-comfort, so it is good for those consumers who not worries about cost. On the other hand EEDE performed well in reducing cost. If a consumer wants to reduce costs, then he must sacrifice user comfort. In short, there is a trade-off between cost and user comfort.

8 Conclusion

In this paper, we evaluated the performance of EDE, EWA and a hybrid meta-heuristic technique EEDE using the CPP pricing scheme. We take 7 appliances on a single home divided them into two categories according to the pattern of their energy consumption. The optimization technique we used to measure performance on the basis of four parameters, energy consumption, cost, PAR and user comfort. We have done simulations by taking three different OTI, in each case our proposed techniques EDE and EWA and EEDE performed well. EDE shows the reduction in cost up to 2%, 1% and 6%, while EWA shows cost reduction up to 1%, 5% and 8% in each OTI. EEDE showed better performance than other techniques, reducing costs up to 1%, 2% and 4%, in case of 15 min, 30 min, and 60 min OTI the PAR was reduced up to 13%, 20% and 14% by using EDE and reduced up to 2%, 3%, 4% by using EWA for 15, 30 and 60 min OTI respectively.EEDE reduced PAR by 13%, 7%, 8% in case of each OTI, performed better than other techniques. Hybrid meta-heuristic technique performed well in reducing cost, PAR, energy consumption. The results show that EEDE performed better than other techniques in reducing cost and PAR. However, there is a trade-off between cost and user comfort. We will study other pricing schemes in future work and will consider multiple homes instead of single home.

References

1. Rasheed, M.B., et al.: Energy optimization in smart homes using customer preference and dynamic pricing. Energies **9**(8), 1–25 (2016)
2. Logenthiran, T., Member, S., Srinivasan, D., Member, S., Shun, T.Z.: Demand side management in smart grid using heuristic optimization. IEEE Trans. Smart Grid **3**(3), 1244–1252 (2015)
3. Mhanna, S., Chapman, A.C., Verbic, G.: A fast distributed algorithm for large-scale demand response aggregation. IEEE Trans. Smart Grid **7**(4), 2094–2107 (2016)
4. Ogwumike, C., Short, M., Abugchem, F.: Heuristic optimization of consumer electricity costs using a generic cost model. Energies **9**(1), 6 (2016)

5. Mahmood, D., et al.: Realistic Scheduling Mechanism for smart homes. Energies **9**(3), 1–28 (2016)
6. Wei, Q., Lewis, F.L., Shi, G., Song, R.: Error-tolerant iterative adaptive dynamic programming for optimal renewable home energy scheduling and battery management. IEEE Trans. Ind. Electron. **64**(12), 1–1 (2017)
7. Javaid, N., et al.: An intelligent load management system with renewable energy integration for smart homes. IEEE Access **5**, 13587–13600 (2017)
8. Javaid, N., et al.: Demand side management in nearly zero energy buildings using heuristic optimizations. Energies **10**(8), 1131 (2017)
9. Rahim, S., et al.: Exploiting heuristic algorithms to efficiently utilize energy management controllers with renewable energy sources. Energ. Build. **129**, 452–470 (2016)
10. Yang, X., Zhang, Y., Zhao, B., Huang, F., Chen, Y., Ren, S.: Optimal energy flow control strategy for a residential energy local network combined with demand-side management and real-time pricing. Energ. Build. **150**, 177–188 (2017)
11. Celik, B., Roche, R., Suryanarayanan, S., Bouquain, D., Miraoui, A.: Electric energy management in residential areas through coordination of multiple smart homes. Renew. Sustain. Energ. Rev. **80**(May), 260–275 (2017)
12. Koolen, D., Sadat-razavi, N., Ketter, W.: Applied sciences machine learning for identifying demand patterns of home energy management systems with dynamic electricity pricing (2017)
13. Marzband, M., Ghazimirsaeid, S.S., Uppal, H., Fernando, T.: A real-time evaluation of energy management systems for smart hybrid home Microgrids. Electr. Power Syst. Res. **143**, 624–633 (2017)
14. Zhou, Y.: The optimal home energy management strategy in smart grid. J. Renew. Sustain. Energ. **8**, 45101 (2016)
15. Keshtkar, A., Arzanpour, S.: An adaptive fuzzy logic system for residential energy management in smart grid environments. Appl. Energ. **186**, 68–81 (2017)
16. Lobaccaro, G., Carlucci, S., Löfström, E.: A review of systems and technologies for smart homes and smart grids. Energies **9**(5), 1–33 (2016)
17. Schulze, M., Heidenreich, S., Spieth, P.: The impact of energy management control systems on energy efficiency in the german manufacturing industry, vol. 0, no. 0, pp. 1–14 (2017)
18. Longe, O.M., Ouahada, K., Rimer, S., Ferreira, H.C., Han Vinck, A.J.: Distributed optimisation algorithm for demand side management in a grid-connected smart microgrid. Sustainability **9**(7), 1088–1104 (2017)
19. Wang, G.G., Deb, S., Coelho, L.D.S.: Earthworm optimization algorithm: a bio-inspired metaheuristic algorithm for global optimization problems. Int. J. Bio Inspired Comput. **1**(1), 1 (2015)
20. Manzoor, A., Javaid, N., Ullah, I., Abdul, W., Almogren, A., Alamri, A.: An intelligent hybrid heuristic scheme for smart metering based demand side management in smart homes. Energies **10**(9), 1–28 (2017)

Performance Analysis of Simulation System Based on Particle Swarm Optimization and Distributed Genetic Algorithm for WMNs Considering Different Distributions of Mesh Clients

Admir Barolli[1], Shinji Sakamoto[2(✉)], Leonard Barolli[3], and Makoto Takizawa[4]

[1] Department of Information Technology, Aleksander Moisiu University of Durres, L.1, Rruga e Currilave, Durres, Albania
admir.barolli@gmail.com

[2] Department of Computer and Information Science, Seikei University, 3-3-1 Kichijoji-Kitamachi, Musashino-shi, Tokyo 180-8633, Japan
shinji.sakamoto@ieee.org

[3] Department of Information and Communication Engineering, Fukuoka Institute of Technology, 3-30-1 Wajiro-Higashi, Higashi-Ku, Fukuoka 811-0295, Japan
barolli@fit.ac.jp

[4] Department of Advanced Sciences, Faculty of Science and Engineering, Hosei University, Kajino-Machi, Koganei-Shi, Tokyo 184-8584, Japan
makoto.takizawa@computer.org

Abstract. The Wireless Mesh Networks (WMNs) are becoming an important networking infrastructure because they have many advantages such as low cost and increased high speed wireless Internet connectivity. In our previous work, we implemented a Particle Swarm Optimization (PSO) based simulation system, called WMN-PSO, and a simulation system based on Genetic Algorithm (GA), called WMN-GA, for solving node placement problem in WMNs. In this paper, we implement a hybrid simulation system based on PSO and distributed GA (DGA), called WMN-PSODGA. We analyze the performance of WMN-PSODGA by computer simulations considering different client distributions. Simulation results show that the WMN-PSODGA has good performance when the client distribution is Normal compared with the case of Exponential distribution.

1 Introduction

The wireless networks and devises are becoming increasingly popular and they provide users access to information and communication anytime and anywhere [11–14, 22, 28–30]. Wireless Mesh Networks (WMNs) are gaining a lot of attention because of their low cost nature that makes them attractive for providing wireless Internet connectivity. A WMN is dynamically self-organized

L. Barolli et al. (Eds.): IMIS 2018, AISC 773, pp. 32–45, 2019.
https://doi.org/10.1007/978-3-319-93554-6_3

and self-configured, with the nodes in the network automatically establishing and maintaining mesh connectivity among them-selves (creating, in effect, an ad hoc network). This feature brings many advantages to WMNs such as low up-front cost, easy network maintenance, robustness and reliable service coverage [1]. Moreover, such infrastructure can be used to deploy community networks, metropolitan area networks, municipal and corporative networks, and to support applications for urban areas, medical, transport and surveillance systems.

Mesh node placement in WMN can be seen as a family of problems, which are shown (through graph theoretic approaches or placement problems, e.g. [8,17]) to be computationally hard to solve for most of the formulations [34]. In fact, the node placement problem considered here is even more challenging due to two additional characteristics:

(a) locations of mesh router nodes are not pre-determined, in other wards, any available position in the considered area can be used for deploying the mesh routers.
(b) routers are assumed to have their own radio coverage area.

Here, we consider the version of the mesh router nodes placement problem in which we are given a grid area where to deploy a number of mesh router nodes and a number of mesh client nodes of fixed positions (of an arbitrary distribution) in the grid area. The objective is to find a location assignment for the mesh routers to the cells of the grid area that maximizes the network connectivity and client coverage. Node placement problems are known to be computationally hard to solve [15,16,35]. In some previous works, intelligent algorithms have been recently investigated [2–4,6,7,9,10,18,20,23–25,36].

In our previous work, we implemented a Particle Swarm Optimization (PSO) based simulation system, called WMN-PSO [26]. Also, we implemented another simulation system based on Genetic Algorithm (GA), called WMN-GA [21], for solving node placement problem in WMNs.

In this paper, we design and implement a hybrid simulation system based on PSO and distributed GA (DGA). We call this system WMN-PSODGA. We analyze the performance of the implemented WMN-PSODGA system by computer simulations considering Normal and Exponential client distributions.

The rest of the paper is organized as follows. The mesh router nodes placement problem is defined in Sect. 2. We present our designed and implemented hybrid simulation system in Sect. 3. The simulation results are given in Sect. 4. Finally, we give conclusions and future work in Sect. 5.

2 Node Placement Problem in WMNs

For this problem, we have a grid area arranged in cells we want to find where to distribute a number of mesh router nodes and a number of mesh client nodes of fixed positions (of an arbitrary distribution) in the considered area. The objective is to find a location assignment for the mesh routers to the area that maximizes the network connectivity and client coverage. Network connectivity is measured by Size of Giant Component (SGC) of the resulting WMN graph, while the user coverage is simply the number of mesh client nodes that fall within the radio coverage of at least one mesh router node and is measured by Number of Covered Mesh Clients (NCMC).

An instance of the problem consists as follows.

- N mesh router nodes, each having its own radio coverage, defining thus a vector of routers.
- An area $W \times H$ where to distribute N mesh routers. Positions of mesh routers are not pre-determined and are to be computed.
- M client mesh nodes located in arbitrary points of the considered area, defining a matrix of clients.

It should be noted that network connectivity and user coverage are among most important metrics in WMNs and directly affect the network performance.

In this work, we have considered a bi-objective optimization in which we first maximize the network connectivity of the WMN (through the maximization of the SGC) and then, the maximization of the NCMC.

In fact, we can formalize an instance of the problem by constructing an adjacency matrix of the WMN graph, whose nodes are router nodes and client nodes and whose edges are links between nodes in the mesh network. Each mesh node in the graph is a triple $\boldsymbol{v} = < x, y, r >$ representing the 2D location point and r is the radius of the transmission range. There is an arc between two nodes $\boldsymbol{u}$ and $\boldsymbol{v}$, if $\boldsymbol{v}$ is within the transmission circular area of $\boldsymbol{u}$.

3 Proposed and Implemented Simulation System

3.1 Particle Swarm Optimization

In PSO a number of simple entities (the particles) are placed in the search space of some problem or function and each evaluates the objective function at its current location. The objective function is often minimized and the exploration of the search space is not through evolution [19]. However, following a widespread practice of borrowing from the evolutionary computation field, in this work, we consider the bi-objective function and fitness function interchangeably. Each particle then determines its movement through the search space by combining some aspect of the history of its own current and best (best-fitness) locations with those of one or more members of the swarm, with some random perturbations. The next iteration takes place after all particles have been moved. Eventually the swarm as a whole, like a flock of birds collectively foraging for food, is likely to move close to an optimum of the fitness function.

Each individual in the particle swarm is composed of three $\mathcal{D}$-dimensional vectors, where $\mathcal{D}$ is the dimensionality of the search space. These are the current position $\mathbf{x}_i$, the previous best position $\mathbf{p}_i$ and the velocity $\mathbf{v}_i$.

The particle swarm is more than just a collection of particles. A particle by itself has almost no power to solve any problem; progress occurs only when the

Algorithm 1. Pseudo code of PSO.

```
/* Initialize all parameters for PSO */
Computation maxtime:= Tp_max, t := 0;
Number of particle-patterns:= m, 2 ≤ m ∈ N^1;
Particle-patterns initial solution:= P_i^0;
Particle-patterns initial position:= x_ij^0;
Particles initial velocity:= v_ij^0;
PSO parameter:= ω, 0 < ω ∈ R^1;
PSO parameter:= C_1, 0 < C_1 ∈ R^1;
PSO parameter:= C_2, 0 < C_2 ∈ R^1;
/* Start PSO */
Evaluate(G^0, P^0);
while t < Tp_max do
   /* Update velocities and positions */
   v_ij^{t+1} = ω · v_ij^t
          +C_1 · rand() · (best(P_ij^t) − x_ij^t)
          +C_2 · rand() · (best(G^t) − x_ij^t);
   x_ij^{t+1} = x_ij^t + v_ij^{t+1};
   /* if fitness value is increased, a new solution will be accepted. */
   Update_Solutions(G^t, P^t);
   t = t + 1;
end while
Update_Solutions(G^t, P^t);
return Best found pattern of particles as solution;
```

particles interact. Problem solving is a population-wide phenomenon, emerging from the individual behaviors of the particles through their interactions. In any case, populations are organized according to some sort of communication structure or topology, often thought of as a social network. The topology typically consists of bidirectional edges connecting pairs of particles, so that if j is in i's neighborhood, i is also in j's. Each particle communicates with some other particles and is affected by the best point found by any member of its topological neighborhood. This is just the vector $\mathbf{p}_i$ for that best neighbor, which we will denote with $\mathbf{p}_g$. The potential kinds of population "social networks" are hugely varied, but in practice certain types have been used more frequently. We show the pseudo code of PSO in Algorithm 1.

In the PSO process, the velocity of each particle is iteratively adjusted so that the particle stochastically oscillates around $\mathbf{p}_i$ and $\mathbf{p}_g$ locations.

3.2 Distributed Genetic Algorithm

Distributed Genetic Algorithm (DGA) has been focused from various fields of science. DGA has shown their usefulness for the resolution of many computationally hard combinatorial optimization problems. We show the pseudo code of DGA in Algorithm 2.

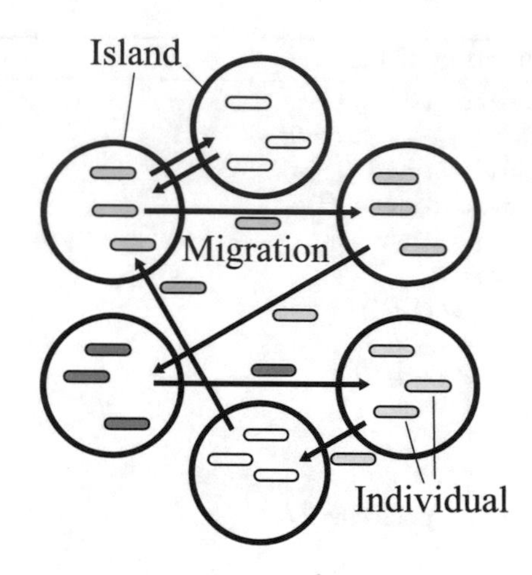

Fig. 1. Model of migration in DGA.

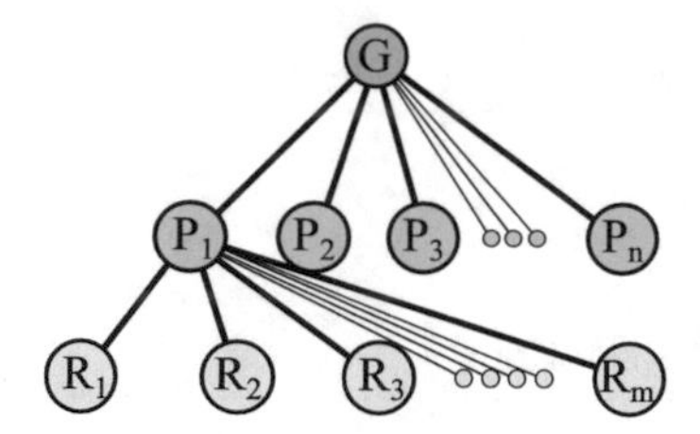

Fig. 2. Relationship among global solution, particle-patterns and mesh routers in PSO part.

Algorithm 2. Pseudo code of DSA.

/* Initialize all parameters for DGA */
Computation maxtime:= Tg_{max}, $t := 0$;
Number of islands:= n, $1 \leq n \in \boldsymbol{N}^1$;
initial solution:= $\boldsymbol{P}_i^0$;
/* Start DGA */
Evaluate($\boldsymbol{G}^0, \boldsymbol{P}^0$);
while $t < Tg_{max}$ **do**
  for all islands **do**
    Selection();
    Crossover();
    Mutation();
  end for
  $t = t + 1$;
end while
Update_Solutions($\boldsymbol{G}^t, \boldsymbol{P}^t$);
return Best found pattern of particles as solution;

Population of individuals: Unlike local search techniques that construct a path in the solution space jumping from one solution to another one through local perturbations, DGA use a population of individuals giving thus the search a larger scope and chances to find better solutions. This feature is also known as "exploration" process in difference to "exploitation" process of local search methods.

Fitness: The determination of an appropriate fitness function, together with the chromosome encoding are crucial to the performance of DGA. Ideally we would construct objective functions with "certain regularities", i.e. objective functions that verify that for any two individuals which are close in the search space, their respective values in the objective functions are similar.

Selection: The selection of individuals to be crossed is another important aspect in DGA as it impacts on the convergence of the algorithm. Several selection schemes have been proposed in the literature for selection operators trying to cope with premature convergence of DGA. There are many selection methods in GA. In our system, we implement 2 selection methods: Random method and Roulette wheel method.

Crossover operators: Use of crossover operators is one of the most important characteristics. Crossover operator is the means of DGA to transmit best genetic features of parents to offsprings during generations of the evolution process. Many methods for crossover operators have been proposed such as Blend Crossover (BLX-α), Unimodal Normal Distribution Crossover (UNDX), Simplex Crossover (SPX).

Mutation operators: These operators intend to improve the individuals of a population by small local perturbations. They aim to provide a component of randomness in the neighborhood of the individuals of the population. In our system, we implemented two mutation methods: uniformly random mutation and boundary mutation.

Escaping from local optima: GA itself has the ability to avoid falling prematurely into local optima and can eventually escape from them during the search process. DGA has one more mechanism to escape from local optima by considering some islands. Each island computes GA for optimizing and they migrate its gene to provide the ability to avoid from local optima (See Fig. 1).

Convergence: The convergence of the algorithm is the mechanism of DGA to reach to good solutions. A premature convergence of the algorithm would cause that all individuals of the population be similar in their genetic features and thus the search would result ineffective and the algorithm getting stuck into local optima. Maintaining the diversity of the population is therefore very important to this family of evolutionary algorithms.

3.3 WMN-PSODGA Hybrid Simulation System

In this subsection, we present the initialization, particle-pattern, fitness function and replacement methods.

Initialization

Our proposed system starts by generating an initial solution randomly, by *ad hoc* methods [36]. We decide the velocity of particles by a random process considering the area size. For instance, when the area size is $W \times H$, the velocity is decided randomly from $-\sqrt{W^2 + H^2}$ to $\sqrt{W^2 + H^2}$.

Particle-Pattern

A particle is a mesh router. A fitness value of a particle-pattern is computed by combination of mesh routers and mesh clients positions. In other words, each particle-pattern is a solution as shown is Fig. 2.

Gene Coding

A gene describes a WMN. Each individual has its own combination of mesh nodes. In other words, each individual has a fitness value. Therefore, the combination of mesh nodes is a solution.

Fitness Function

One of most important thing in PSO algorithm is to decide the determination of an appropriate objective function and its encoding. In our case, each particle-pattern has an own fitness value and compares it with other particle-pattern's fitness value in order to share information of global solution. The fitness function follows a hierarchical approach in which the main objective is to maximize the SGC in WMN. Thus, the fitness function of this scenario is defined as

$$\text{Fitness} = 0.7 \times \text{SGC}(\boldsymbol{x}_{ij}, \boldsymbol{y}_{ij}) + 0.3 \times \text{NCMC}(\boldsymbol{x}_{ij}, \boldsymbol{y}_{ij}).$$

Client Distributions

In our system, many kinds of client distributions are generated. In this paper, we consider Normal and Exponential distributions for mesh clients. Clients are distributed randomly to a considered area for Normal distribution. On the other hand, Exponential distribution is a hot-spot model supposition.

Routers Replacement Method for PSO Part

A mesh router has x, y positions and velocity. Mesh routers are moved based on velocities. There are many moving methods in PSO field, such as:

Constriction Method (CM)
: CM is a method which PSO parameters are set to a week stable region ($\omega = 0.729$, $C_1 = C2 = 1.4955$) based on analysis of PSO by Clerc et al. [5,32].

Random Inertia Weight Method (RIWM)
: In RIWM, the ω parameter is changing randomly from 0.5 to 1.0. The C_1 and C_2 are kept 2.0. The ω can be estimated by the week stable region. The average of ω is 0.75 [32].

Linearly Decreasing Inertia Weight Method (LDIWM)
In LDIWM, C_1 and C_2 are set to 2.0, constantly. On the other hand, the ω parameter is changed linearly from unstable region ($\omega = 0.9$) to stable region ($\omega = 0.4$) with increasing of iterations of computations [32,33].

Linearly Decreasing Vmax Method (LDVM)
In LDVM, PSO parameters are set to unstable region ($\omega = 0.9$, $C_1 = C_2 = 2.0$). A value of V_{max} which is maximum velocity of particles is considered. With increasing of iteration of computations, the V_{max} is kept decreasing linearly [31].

Rational Decrement of Vmax Method (RDVM)
In RDVM, PSO parameters are set to unstable region ($\omega = 0.9$, $C_1 = C_2 = 2.0$). The V_{max} is kept decreasing with the increasing of iterations as

$$V_{max}(x) = \sqrt{W^2 + H^2} \times \frac{T - x}{x}.$$

Where, W and H are the width and the height of the considered area, respectively. Also, T and x are the total number of iterations and a current number of iteration, respectively [27].

3.4 WMN-PSODGA Web GUI Tool and Pseudo Code

The Web application follows a standard Client-Server architecture and is implemented using LAMP (Linux + Apache + MySQL + PHP) technology (see Fig. 4). Remote users (clients) submit their requests by completing first the parameter setting. The parameter values to be provided by the user are classified into three groups, as follows.

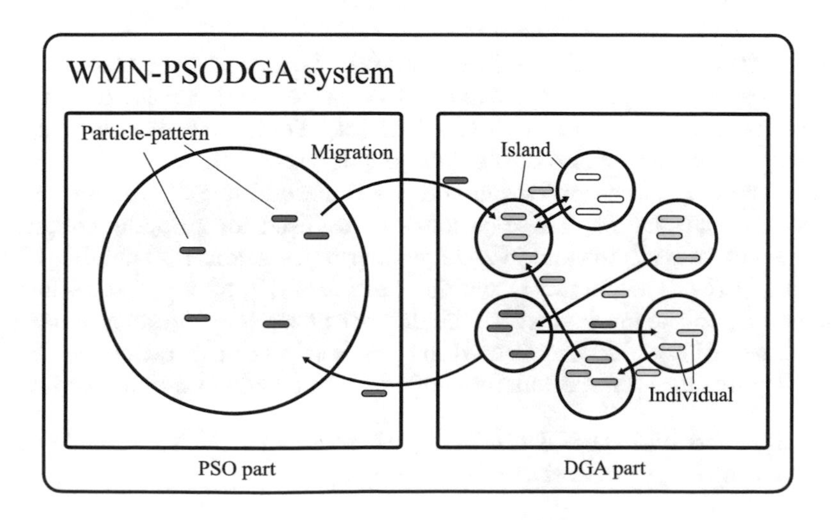

Fig. 3. Model of WMN-PSODGA migration.

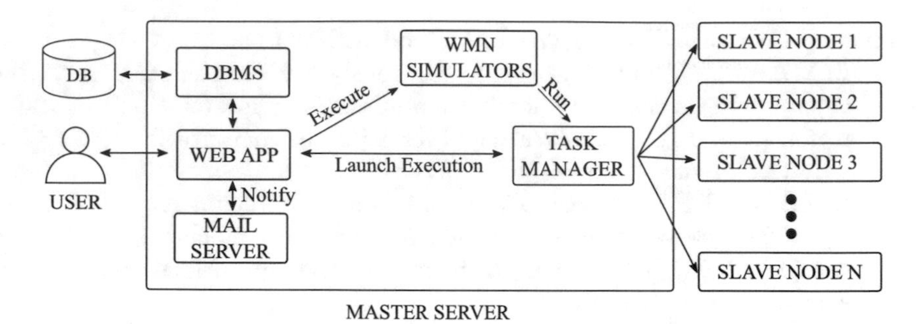

Fig. 4. System structure for web interface.

Algorithm 3. Pseudo code of WMN-PSODGA system.

Computation maxtime:= T_{max}, $t := 0$;
Initial solutions: $\boldsymbol{P}$.
Initial global solutions: $\boldsymbol{G}$.
/* Start PSODGA */
while $t < T_{max}$ **do**
  Subprocess(PSO);
  Subprocess(DGA);
  WaitSubprocesses();
  Evaluate($\boldsymbol{G}^t, \boldsymbol{P}^t$)
  /* Migration() swaps solutions (see Fig. 3). */
  Migration();
  $t = t + 1$;
end while
Update_Solutions($\boldsymbol{G}^t, \boldsymbol{P}^t$);
return Best found pattern of particles as solution;

- Parameters related to the problem instance: These include parameter values that determine a problem instance to be solved and consist of number of router nodes, number of mesh client nodes, client mesh distribution, radio coverage interval and size of the deployment area.
- Parameters of the resolution method: Each method has its own parameters.
- Execution parameters: These parameters are used for stopping condition of the resolution methods and include number of iterations and number of independent runs. The former is provided as a total number of iterations and depending on the method is also divided per phase (e.g., number of iterations in a exploration). The later is used to run the same configuration for the same problem instance and parameter configuration a certain number of times.

We show the WMN-PSODGA Web GUI tool in Fig. 5. The pseudo code of our implemented system is shown in Algorithm 3.

Simulator parameters, Distributed Genetic Algorithm and Particle Swarm Optimization

Distribution	Uniform	
Number of clients	48 (integer)(min:48 max:128)	
Number of routers	16 (integer) (min:16 max:48)	
Area size (WxH)	32 (positive real number)	32 (positive real number)
Radius (Min & Max)	2 (positive real number)	2 (positive real number)
Number of migration	200 (integer)	
Number of islands	200 (integer)	
Populations parameter	1 (integer)	
Independent runs	1 (integer) (min:1 max:100)	
Replacement method	Constriction Method	
Number of evolution steps	10 (integer) (min:1 max:64)	
Crossover rate	0.8 (positive real number)	
Mutation rate	0.2 (positive real number)	
Select method	Random Selection	
Crossover method	BLX-a Method	
Mutation method	Uniform Mutation	
Send by mail		

Run

Fig. 5. WMN-PSODGA Web GUI Tool.

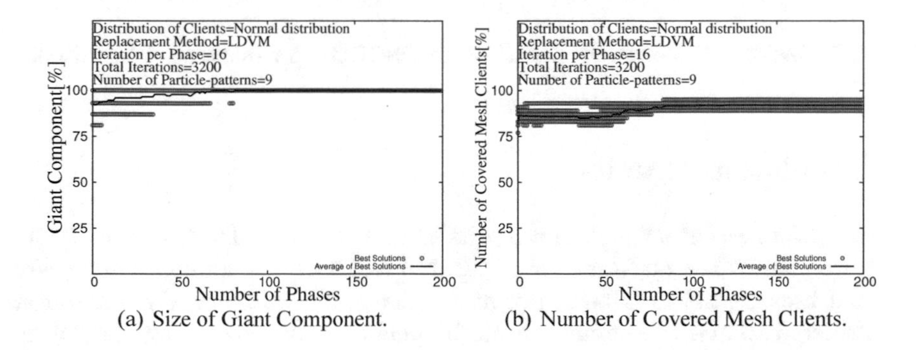

(a) Size of Giant Component.

(b) Number of Covered Mesh Clients.

Fig. 6. Simulation results of WMN-PSODGA for Normal distribution of mesh clients.

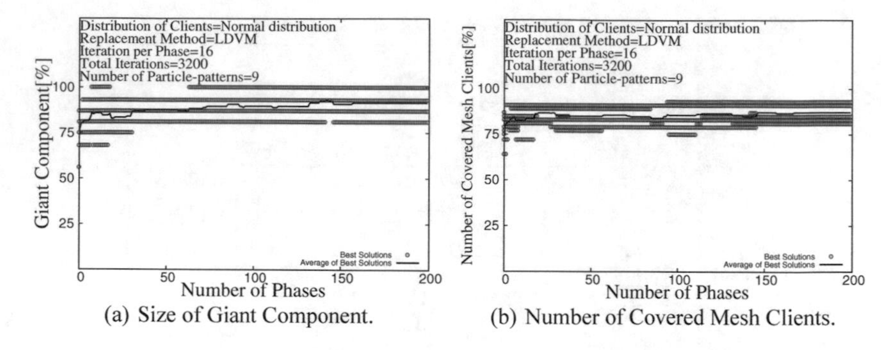

(a) Size of Giant Component. (b) Number of Covered Mesh Clients.

Fig. 7. Simulation results of WMN-PSODGA for Exponential distribution of mesh clients.

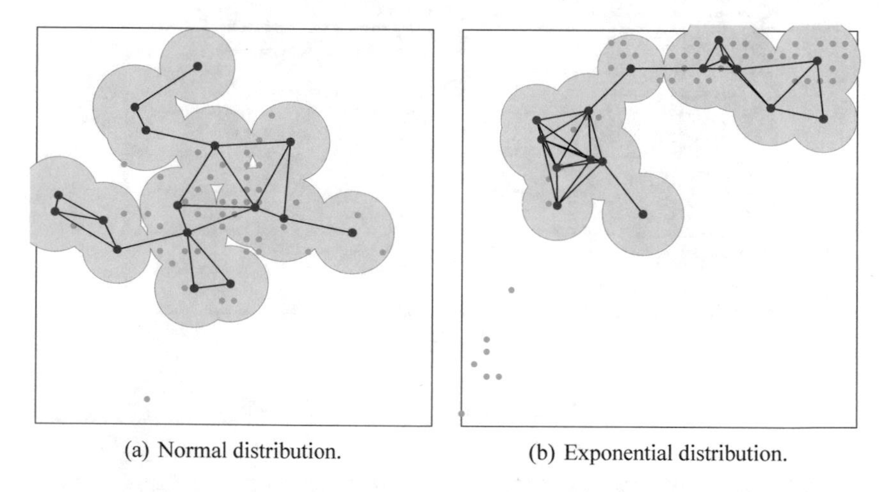

(a) Normal distribution. (b) Exponential distribution.

Fig. 8. Visualized simulation results of WMN-PSODGA for different client distributions.

4 Simulation Results

In this section, we show simulation results using WMN-PSODGA system. In this work, we analyse the performance of WMN-PSODGA system considering Normal and Exponential distribution of mesh clients. The number of mesh routers is considered 16 and the number of mesh clients 48. We conducted simulations 10 times, in order to avoid the effect of randomness and create a general view of results. We show the parameter setting for WM-PSODGA in Table 1.

We show simulation results from Figs. 6, 7 and 8. In Figs. 6a and 7a, we see that WMN-PSODGA performs better when the client distribution is Normal. For Normal distribution, WMN-PSODGA can find the maximum SGC for all cases. However, for Exponential distribution, the average of best solutions does not reach the maximum. Also, in Figs. 6b and 7b, the WMN-PSODGA system perfoms better when the client distribuion is Normal. We show the visualized

Table 1. WMN-PSODGA parameters.

Parameters	Values
Clients distribution	Normal, Exponential
Area size	32.0×32.0
Number of mesh routers	16
Number of mesh clients	48
Number of migrations	200
Evolution steps	9
Number of GA islands	16
Radius of a mesh router	2.5–3.5
Replacement method	LDVM
Selection method	Roulette wheel method
Crossover method	SPX
Mutation method	Boundary mutation
Crossover rate	0.8
Mutation rate	0.2

simulation results in Fig. 8. For Normal distribution, only one mesh client is not covered. However, seven mesh clients are not covered when the client distribution is Exponential.

5 Conclusions

In this work, we evaluated the performance of a hybrid simulation system based on PSO and DGA (called WMN-PSODGA). Simulation results show that the WMN-PSODGA has good performance when the client distribution is Normal compared with the case of Exponential distribution.

In our future work, we would like to evaluate the performance of the proposed system for different parameters and patterns.

References

1. Akyildiz, I.F., Wang, X., Wang, W.: Wireless mesh networks: a survey. Comput. Netw. **47**(4), 445–487 (2005)
2. Amaldi, E., Capone, A., Cesana, M., Filippini, I., Malucelli, F.: Optimization models and methods for planning wireless mesh networks. Comput. Netw. **52**(11), 2159–2171 (2008)
3. Barolli, A., Spaho, E., Barolli, L., Xhafa, F., Takizawa, M.: QoS routing in Ad-hoc networks using GA and multi-objective optimization. Mob. Inf. Syst. **7**(3), 169–188 (2011)

4. Behnamian, J., Ghomi, S.F.: Development of a PSO-SA hybrid metaheuristic for a new comprehensive regression model to time-series forecasting. Expert Syst. Appl. **37**(2), 974–984 (2010)
5. Clerc, M., Kennedy, J.: The particle swarm-explosion, stability, and convergence in a multidimensional complex space. IEEE Trans. Evol. Comput. **6**(1), 58–73 (2002)
6. Cunha, M.d.C., Sousa, J.: Water distribution network design optimization: simulated annealing approach. J. Water Resour. Plann. Manag. **125**(4), 215–221 (1999)
7. Del Valle, Y., Venayagamoorthy, G.K., Mohagheghi, S., Hernandez, J.C., Harley, R.G.: Particle swarm optimization: basic concepts, variants and applications in power systems. IEEE Trans. Evol. Comput. **12**(2), 171–195 (2008)
8. Franklin, A.A., Murthy, C.S.R.: Node placement algorithm for deployment of two-tier wireless mesh networks. In: Proceedings of Global Telecommunications Conference, pp. 4823–4827 (2007)
9. Ge, H., Du, W., Qian, F.: A hybrid algorithm based on particle swarm optimization and simulated annealing for job shop scheduling. In: Third International Conference on Natural Computation (ICNC-2007), vol. 3, pp. 715–719 (2007)
10. Girgis, M.R., Mahmoud, T.M., Abdullatif, B.A., Rabie, A.M.: Solving the wireless mesh network design problem using genetic algorithm and simulated annealing optimization methods. Int. J. Comput. Appl. **96**(11), 1–10 (2014)
11. Goto, K., Sasaki, Y., Hara, T., Nishio, S.: Data gathering using mobile agents for reducing traffic in dense mobile wireless sensor networks. Mob. Inf. Syst. **9**(4), 295–314 (2013)
12. Inaba, T., Elmazi, D., Sakamoto, S., Oda, T., Ikeda, M., Barolli, L.: A secure-aware call admission control scheme for wireless cellular networks using fuzzy logic and its performance evaluation. J. Mob. Multimed. **11**(3&4), 213–222 (2015)
13. Inaba, T., Obukata, R., Sakamoto, S., Oda, T., Ikeda, M., Barolli, L.: Performance evaluation of a QoS-aware fuzzy-based CAC for LAN access. Int. J. Space Based Situated Comput. **6**(4), 228–238 (2016)
14. Inaba, T., Sakamoto, S., Oda, T., Ikeda, M., Barolli, L.: A testbed for admission control in WLAN: a fuzzy approach and its performance evaluation. In: International Conference on Broadband and Wireless Computing, Communication and Applications, pp. 559–571. Springer (2016)
15. Lim, A., Rodrigues, B., Wang, F., Xu, Z.: k-center problems with minimum coverage. In: Computing and Combinatorics, pp. 349–359 (2004)
16. Maolin, T., et al.: Gateways placement in backbone wireless mesh networks. Int. J. Commun. Netw. Syst. Sci. **2**(1), 44 (2009)
17. Muthaiah, S.N., Rosenberg, C.P.: Single gateway placement in wireless mesh networks. In: Proceedings of 8th International IEEE Symposium on Computer Networks, pp. 4754–4759 (2008)
18. Naka, S., Genji, T., Yura, T., Fukuyama, Y.: A hybrid particle swarm optimization for distribution state estimation. IEEE Trans. Power Syst. **18**(1), 60–68 (2003)
19. Poli, R., Kennedy, J., Blackwell, T.: Particle swarm optimization. Swarm Intell. **1**(1), 33–57 (2007)
20. Sakamoto, S., Kulla, E., Oda, T., Ikeda, M., Barolli, L., Xhafa, F.: A comparison study of simulated annealing and genetic algorithm for node placement problem in wireless mesh networks. J. Mob. Multimed. **9**(1–2), 101–110 (2013)
21. Sakamoto, S., Kulla, E., Oda, T., Ikeda, M., Barolli, L., Xhafa, F.: A comparison study of hill climbing, simulated annealing and genetic algorithm for node placement problem in WMNs. J. High Speed Netw. **20**(1), 55–66 (2014)

22. Sakamoto, S., Kulla, E., Oda, T., Ikeda, M., Barolli, L., Xhafa, F.: A simulation system for WMN based on SA: performance evaluation for different instances and starting temperature values. Int. J. Space Based Situated Comput. **4**(3–4), 209–216 (2014)
23. Sakamoto, S., Kulla, E., Oda, T., Ikeda, M., Barolli, L., Xhafa, F.: Performance evaluation considering iterations per phase and SA temperature in WMN-SA system. Mob. Inf. Syst. **10**(3), 321–330 (2014)
24. Sakamoto, S., Lala, A., Oda, T., Kolici, V., Barolli, L., Xhafa, F.: Application of WMN-SA simulation system for node placement in wireless mesh networks: a case study for a realistic scenario. Int. J. Mob. Comput. Multimed. Commun. (IJMCMC) **6**(2), 13–21 (2014)
25. Sakamoto, S., Oda, T., Ikeda, M., Barolli, L., Xhafa, F.: An integrated simulation system considering WMN-PSO simulation system and network simulator 3. In: International Conference on Broadband and Wireless Computing, Communication and Applications, pp. 187–198. Springer (2016)
26. Sakamoto, S., Oda, T., Ikeda, M., Barolli, L., Xhafa, F.: Implementation and evaluation of a simulation system based on particle swarm optimisation for node placement problem in wireless mesh networks. Int. J. Commun. Netw. Distrib. Syst. **17**(1), 1–13 (2016)
27. Sakamoto, S., Oda, T., Ikeda, M., Barolli, L., Xhafa, F.: implementation of a new replacement method in WMN-PSO simulation system and its performance evaluation. In: The 30th IEEE International Conference on Advanced Information Networking and Applications (AINA-2016), pp. 206–211 (2016). https://doi.org/10.1109/AINA.2016.42
28. Sakamoto, S., Obukata, R., Oda, T., Barolli, L., Ikeda, M., Barolli, A.: Performance analysis of two wireless mesh network architectures by WMN-SA and WMN-TS simulation systems. J. High Speed Netw. **23**(4), 311–322 (2017)
29. Sakamoto, S., Ozera, K., Barolli, A., Ikeda, M., Barolli, L., Takizawa, M.: Implementation of an intelligent hybrid simulation systems for WMNs based on particle swarm optimization and simulated annealing: performance evaluation for different replacement methods. Soft Comput. 1–7 (2017)
30. Sakamoto, S., Ozera, K., Ikeda, M., Barolli, L.: Implementation of intelligent hybrid systems for node placement problem in WMNs considering particle swarm optimization, hill climbing and simulated annealing. Mob. Netw. Appl. **23**, 1–7 (2017)
31. Schutte, J.F., Groenwold, A.A.: A study of global optimization using particle swarms. J. Global Optim. **31**(1), 93–108 (2005)
32. Shi, Y.: Particle swarm optimization. IEEE Connections **2**(1), 8–13 (2004)
33. Shi, Y., Eberhart, R.C.: Parameter selection in particle swarm optimization. In: Evolutionary Programming VII, pp. 591–600 (1998)
34. Vanhatupa, T., Hannikainen, M., Hamalainen, T.: Genetic algorithm to optimize node placement and configuration for WLAN planning. In: Proceedings of 4th IEEE International Symposium on Wireless Communication Systems, pp. 612–616 (2007)
35. Wang, J., Xie, B., Cai, K., Agrawal, D.P.: Efficient mesh router placement in wireless mesh networks. In: Proceedings of IEEE International Conference on Mobile Adhoc and Sensor Systems (MASS-2007), pp. 1–9 (2007)
36. Xhafa, F., Sanchez, C., Barolli, L.: Ad hoc and neighborhood search methods for placement of mesh routers in wireless mesh networks. In: Proceedings of 29th IEEE International Conference on Distributed Computing Systems Workshops (ICDCS-2009), pp. 400–405 (2009)

A Hybrid Flower-Grey Wolf Optimizer Based Home Energy Management in Smart Grid

Pamir[1], Nadeem Javaid[1(✉)], Attiq ullah Khan[2], Syed Muhammad Mohsin[1], Yasir Khan Jadoon[1], and Orooj Nazeer[2]

[1] COMSATS Institute of Information Technology, Islamabad 44000, Pakistan
nadeemjavaidqau@gmail.com
[2] Abasyn University, Islamabad 44000, Pakistan
http://www.njavaid.com

Abstract. Demand side management (DSM) in smart grid (SG) makes users able to take informed decisions according to the power usage pattern of the electricity users and assists the utility in minimizing peak power demand in the duration of high energy demand slots. Where, this ultimately leads to carbon emission reduction, total electricity cost minimization and maximization of grid efficiency and sustainability. Nowadays, many DSM strategies are available in existing literature concentrate on house hold appliances scheduling to decrease electricity cost. However, they ignore peak to average ratio (PAR) and consumer's delay minimization. In this paper, a load shifting strategy of DSM is considered, to decrease PAR and waiting time. To gain aforementioned objectives, the flower pollination algorithm (FPA), grey wolf optimizer (GWO) and their hybrid i.e., flower grey wolf optimizer (FGWO) are used. Simulations were conducted for a single home consist of 15 appliances and critical peak pricing (CPP) tariff is used for computing user's electricity payment. The results show and validate that load is successfully transferred to low price rate hours using our proposed FGWO technique, which ultimately leads to 50.425% reduction in PAR, 2.4148 h waiting time and with 54.654% reasonable reduction in cost.

Keywords: Flower pollination algorithm · Grey wolf optimizer
Metaheuristic Techniques · Heuristic techniques
Appliances scheduling · Home energy management
Demand side management · Smart grid

1 Introduction

Conventional power grid is insufficient to mitigate electricity grid challenges; i.e., security, scalability, robustness and reliability [1]. Hence, an advanced and smart infrastructure of the existing power grid is required to satisfy these challenges smartly. So, in this way, smart grid (SG) adds various information and communication technologies (ICTs) to conventional grid. With the emerging of SG,

L. Barolli et al. (Eds.): IMIS 2018, AISC 773, pp. 46–59, 2019.
https://doi.org/10.1007/978-3-319-93554-6_4

the electricity consumers are no more only consumers but they are prosumers as well because now, they have the authority to sell back the surplus energy to the grid. Utility companies are focus on maximizing their profit with minimizing peak to average ratio (PAR). However, customers want to reduce electricity cost without paying cost of their comfort. Demand side management (DSM) is an important element of SG which keeps balance between demand side and supply side. There are two main elements or aspects of DSM [2]; demand response (DR) and load management. The second function of DSM i.e., load management concentrates on energy efficiency improvement to avoid load shedding problem. DR is an action taken by electricity customers against the dynamic price schemes.

The most common goals and objectives of SG are minimization of PAR, total electricity payment and maximization of consumer satisfaction. To gain these objectives, a large number of DSM strategies and techniques are proposed by the researchers. Wang et al. [3,4] use mixed integer nonlinear programming (MINLP) technique to minimize electricity payment and aggregated power usage. In [5], the authors use integer linear programming (ILP) to minimize the load in low price hours, to maximize user satisfaction. On the other hand, these techniques cannot deal with the appliances having nonlinear and complex power usage pattern. In this article, we use meta-heuristic techniques: flower pollination algorithm (FPA) [8], grey wolf optimizer (GWO) [9] and the proposed one which is the hybrid of FPA and GWO i.e., flower-grey wolf optimizer (FGWO) algorithms are used. In literature, a lot of work is available that they have worked to improve the efficiency of DSM using heuristic techniques. For example, in [6], the authors use genetic algorithm (GA), ant colony optimization (ACO) and binary particle swarm optimization (BPSO) to reduce PAR, electricity bill and user comfort maximization. Different models present to minimize electricity cost using GA in [7]. Although, aforementioned work done well for gaining their objectives. However, there is still a big gap for research, as up to now there is no comprehensive solution available to fulfill all energy problems of existing power grid.

Our achievements and contributions are, designing of the smart energy management controller (EMC) based on FPA, GWO and FGWO for a single home and compared their performance in terms of PAR, consumer comfort level, electricity bill and power usage pattern. Results show that our prosed FGWO based EMC outperformed the other schemes. Remainder of this article is arranged as: Sect. 2 describes the background. Section 3 elaborates the problem statement. Next, Sect. 4 illustrates the proposed system model. Section 5 discusses about the optimization algorithms. Simulation results and discussion are elaborated in Sect. 6. Section 7 shows the feasible region. Finally, Sect. 8 conclusion of our paper.

2 Background

Many scientists and researchers in the world have worked to schedule residential appliances in an optimal way. Here, some of the research papers are summarized below.

In [5], authors concentrate on power scheduling. They employed integer linear programming (ILP) algorithm for minimization of energy usage in high price rate hours, to maximize user satisfaction level. The smart meter (SM) is connected with the user interface (UI) and gets the necessary data for scheduling; the priority of appliances, the power consumption related plan. Using the data and information collected, the SM minimizes the hourly usage of energy. Simulations have been carried out and the results presented that hourly based high load is optimized. On the other hand, user comfort and electricity payment are not considered.

In [6], authors targeted renewable energy sources (RES) integration and user satisfaction maximization. In this context, the authors has given solution using three techniques; ACO, BPSO and GA. Where, these heuristic algorithms would be installed into EMC. Simulation results depicted that GA-EMC is more effective in achieving high user satisfaction, low electricity payments and low PAR as compared to ACO based designed EMC and BPSO based designed EMC. However, RES installation and maintenance cost is neglected. On the other hand, high consumer satisfaction level can be achieve in further less payment than this paper achieved. A DSM model is presented in [7] for residential area consumers to decrease PAR and electricity payments. GA is employed for scheduling of appliances in optimal way. A tradeoff is exist between consumer delay and electricity payment. Simulation Results prove the effectiveness of the proposed DSM architecture for both single consumer and multiple consumers.

In [10], authors propose residential appliances scheduling strategy to minimize total electricity payment, user's dissatisfaction and carbon dioxide emissions. A cooperative multi swarm particle swarm optimization (PSO) technique is used to achieve aforementioned objectives. Simulation results shown that the performance convergence of DMS-PSO-CLS is performed well. However, PAR is not considered. An optimal real time schedule controller is presented in [11] for smart home energy management. A binary backtracking search technique (BBSA) is used to reduce energy usage, total electricity payment and PAR. The BBSA is used in two various cases; first, usage of appliances in week days from 4pm to 11pm, and second, usage of appliances at various time of the day in weekend. From simulation results, it is cleared that in terms of, reduction in power usage, total electricity payment and PAR, the binary backtracking search technique based schedule controller outperformed the BPSO based controller. However, the user satisfaction is neglected.

In [12], authors propose an approach to reduce high load along with home area network (HAN) activated inside home. HAN system algorithm ran by the HAN controller. The benefits of the proposed system are the power reduction for achieving a comparatively lower cost. However, the user satisfaction is compromised. In [13], authors address the minimization of electricity usage and PAR. A new smart homes community architecture is proposed. In which, the controller of the community works as virtual energy distribution, and they also, introduced a true RTP between smart houses community, consumers and community controller. Although, these mentioned propositions are best met in addressing of PAR and energy usage reduction. However, user satisfaction is neglected.

In [14], authors worked on home energy management (HEM) approach to investigate two parameters and objectives in a sequential way. The first one is, reducing monetary cost, Then, the second objective is, reducing deviation in consumers load profile. Results revealed minimized energy monetary cost and energy profile deviation. On the other hand, users satisfaction is ignored. In [15], authors focus on PAR and electricity monetary payment minimization. A HEM model is designed. This model considers both power consumption and generation. Although, It is very good option to have such model that at the same time it deals with both energy utilization and generation to achieve PAR and electricity cost reduction. However, the they ignored consumers comfort.

3 Problem Statement

The most targeted issues in SG are; balancing of load, PAR minimization, energy monetary payment reduction and user satisfaction maximization. Some of the optimization algorithms are used in the literature to gain above mentioned objectives. In [10], a cooperative multi swarm PSO technique is used to minimize total electricity payment, user dissatisfaction and carbon dioxide emissions. Results shown that the performance convergence of DMS-PSO-CLS is good under various scenarios. However, PAR is not considered. In [12], authors employed HAN system algorithm ran by the HAN controller for power reduction, to achieve a comparatively small cost. The load and cost are minimized. However, the user satisfaction is compromised. In [14], authors worked on home energy management (HEM) approach to investigate multi objectives. The first one is, reducing monetary cost. The second objective is, reducing deviation in consumers load profile. Simulation results revealed minimized energy monetary cost and energy profile deviation. On the other hand, consumer satisfaction is ignored. In our paper, we employ FPA, GWO and besides, we propose a hybrid FGWO algorithm by considering a single home with 15 appliances scheduling.

4 Proposed System Model

In SG, DSM allows efficient grid operations. In residential sector, every smart house is equipped with SM having EMCs to enable reliable two way communication between end users and utilities. All smart components of the smart home submit their information to the EMC using HAN. And EMC controls the appliances scheduling. After receiving all information, EMC transfers it to the utility via SM and the communication is performed among EMC, SM and utility is based on wide area network (WAN) as shown in the proposed system architecture in Fig. 1. An EMC receives information from all smart elements of smart home. A SM receives the information from EMC which can be about the load demand, extra load demand for the next day, or other demands of a specific smart home where the EMC is operating. This information sends from EMC to SM via WAN and SM after receiving, sends these information to the utility by WAN. And the utility may tell to SM about pricing information or about a new policy if exists. Then, SM communicates and shares these informaion to the EMC. Now, EMC has pricing information from utility and also has appliances and other smart home elements information like, appliances power rating information, length of operation time (LoT) information, operational time interval (OTI) information. Finally, based on the power rating, pricing signal and other required information, the EMC based on FPA, GWO or FGWO makes an optimal power scheduling for PAR reduction, consumer satisfaction maximization and cost reduction.

In this paper, we perform scheduling for a single home with 15 appliances of various classes. The appliances are categorized into two main classes; first class is, non-schedule able and the second class is, Schedulable appliances. Schedulable appliances category is further divided into two types which are, interruptible and non-interruptible appliances. Non interruptible appliances are the appliance that can not disturb during its functioning or working time. And non-interruptible appliances cannot be switch on and scheduled on the basis of demand of the consumer. Where, interruptible appliances are the appliances that can be scheduled at various time slots. There is only one constraint that cloth dryer will always start its working after completion of washing machine working time. Appliances with their corresponding category, power rating and length of operation time (LoT) are shown in Table 1 below. Our main objectives of this work are: to reduce PAR, increase consumer satisfaction level by decreasing their waiting time for an appliance to operate and to decrease energy consumption during high price time intervals to reduce cost. The total energy utilization in 24 h is computed in Eq. 1.

$$Load = \sum_{t=1}^{24} \rho * S(t), \quad S(t) = [1/0] \tag{1}$$

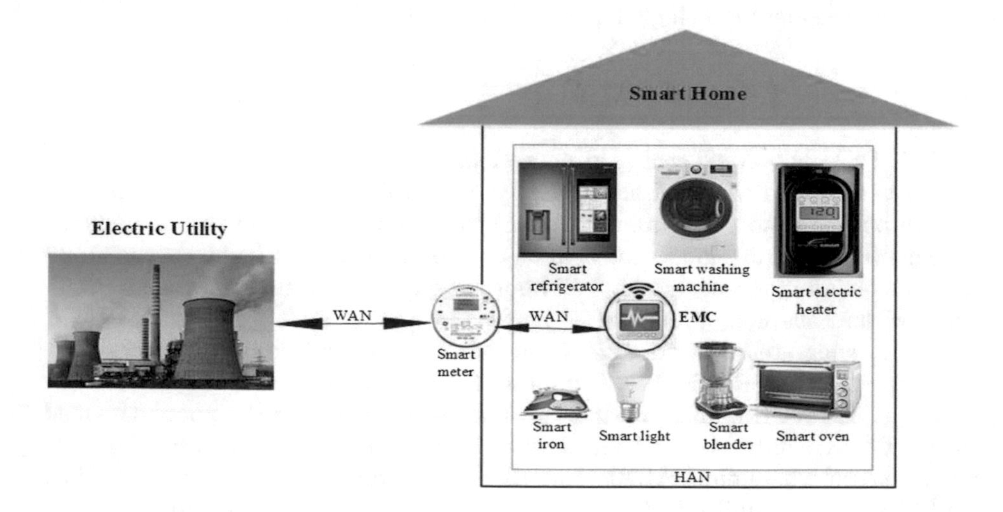

Fig. 1. Proposed System Model

PAR is computed in Eq. 2 and Total cost for 24 h is calculated in Eq. 3.

$$PAR = Max(Load)/Avg(Load) \tag{2}$$

$$Cost = \sum_{t=1}^{24}(EP * \rho * S(t)), \quad S(t) = [1/0] \tag{3}$$

where, in Eq. 1, ρ shows the power rating of appliances, and S(t) is the (on or off) status of the appliances where 1 means that appliance is on at a specific time slot and 0 means the appliance is off. In Eq. 2 PAR is calculated as, take maximum load from 24 h energy usage pattern and then divide it by the average of the same 24 h energy usage pattern. And then, in Eq. 3, EP shows the electric price for 24 h.

Table 1. Appliances employed in simulations

Appliance group	Appliances	Power ratings (kWh)	LoT (h)
Interruptible load appliances	AC	1.5	14
	Iron	1	3
	Water heater	5	8
	Dish washer	1.8	4
	Water pump	1	8
	Vacuum cleaner	0.7	6
	Refrigerator	0.225	20
Non-interruptible load appliances	Washing machine	0.7	5
	Cloth dryer	5	4
Non-schedule able load appliances	Light1	0.03	12
	Light2	0.03	10
	Light3	0.011	9
	Light4	0.18	8
	Blender	0.3	2
	Oven	2.15	4

5 Optimization Algorithms

Conventional optimization algorithms which are mostly mathematical techniques such as, MINLP, ILP are not working satisfactory in case of large number of electronic devices and also they are computationally slow. That is why, we employ meta heuristic techniques FPA and GWO to gain our objectives. We prposed a hybrid of FPA and GWO i.e., FGWO. Selected algorithms are discussed in detail in the subsections below.

5.1 FPA

In this section, we are discussing about FPA which is inspired from the flowers' pollination process. FPA is proposed by Yang [8] in 2013. Flower pollination is basically related with the transfer of pollens by different pollinators such as, birds, insects etc. Flower pollination can be performed using two main forms; abiotic or local or self pollination and biotic or cross pollination or global pollination. The wind blowing and water diffusion help abiotic pollination of flowers to take place. The local or self pollination is occurred from pollen of the same flower or from different flowers of the same plant. Whereas, biotic pollination can occur at huge distance and pollinators like birds can fly a huge distance and they may act as Levy flight.

Now we can represent the flower pollination characteristics:

1. Biotic pollination is considered as the global pollination.
2. Abiotic pollination is recognized as the local pollination.
3. Another characteristics of flower pollination process flower constancy is considered as the probability of reproduction is related with the two flowers' similarity that are involved.
4. global pollination and local pollination or biotic pollination and abiotic pollination is controlled and directed by probability p ϵ [0, 1].

5.2 GWO

GWO [9], is a new meta-heuristic algorithm inspired by the leadership chain and hunting behaviour of grey wolves. The leadership chain of grey wolves consist of four types of wolves; alpha (α), beta (β), delta (δ) and omega (ω). Moreover, there are three main phases of hunting that are, searching for the prey, encircling the prey and attacking over the prey. According to the social hierarchy of grey wolves, the alpha is considered as the most fittest solution, then the beta, delta and omega comes respectively. The main steps involved in GWO are:

1. Encircling the prey.
2. Hunting: Grey wolves have the capability of prey location recognition and encircle them.
3. Exploitation: It is also called attacking the prey. Grey wolves end the hunt while starting attack over the prey when the prey stops movement.
4. Exploration: It is also known as search for prey. The searching is mostly performed in grey wolves based the location of α, β and δ. The grey wolves diverge for "search for prey" and converge in order to attack prey or exploitation.

5.3 FGWO

In this section, we discusses about our proposed hybrid algorithm. In FPA, the updating of the population is performed using the comparison of a random number with the probability (p) value which is equal to 0.9 in our scenario, and then, the p value decides either to perform global or local pollination. In addition, in

GWO the population update is performed, in GWO, the update of the population is fully dependent on the location of the first three best candidates that are, alpha, beta and delta. So in GWO, we record first three best solutions and by force, oblige the other wolves or search agents to modify and update their locations on the basis of the location of the best search agent. Here, in GWO the best search agent location is calculated by Eq. 4 which is the average location of the top three best solutions. Hence, we can say that FPA based population update strategy is better as compared to the GWO based population update strategy, because here, our goal is to find the more optimized result as possible and the more optimized result can be possible by having a huge random and diverge population. And in FPA, the population is updated under the condition i.e. comparison of the random number with the p which is probability switch value then based on this comparison it decides either global or local pollination will be performed. Where, in GWO, update of the population is performed according to the position of the best search agent. Where, GWO based initialization strategy is better than FPA. So, we selected FPA based population update strategy and GWO based initialization strategy, so that, we proposed a new hybrid algorithm FGWO.

$$\overrightarrow{X}(t+1) = \frac{\overrightarrow{X_1} + \overrightarrow{X_2} + \overrightarrow{X_3}}{3} \tag{4}$$

6 Simulation Results and Discussion

In this section, simulation results are shown and discussed in detail. Using simulations, algorithms are validated on the bases of PAR, waiting time, electricity cost and energy consumption. For the simulation, we consider single home with 15 appliances. CPP pricing scheme is used here for the computation of electricity bill. CPP pricing signal is depicted in Fig. 2 below.

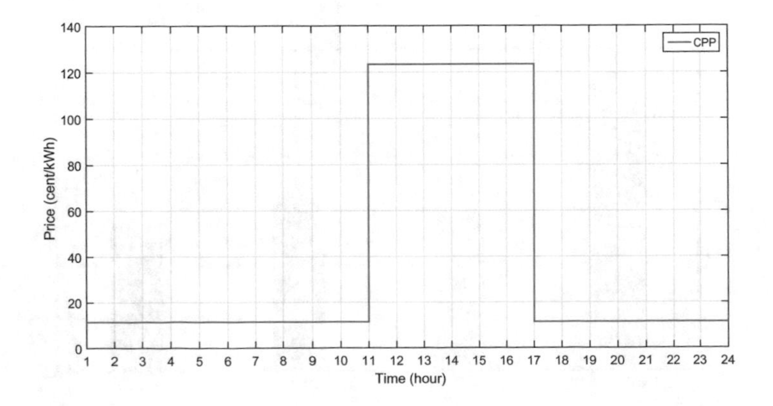

Fig. 2. Pricing signal

Figure 3 depicts the power consumption pattern for each slot in unscheduled as well as scheduled scenarios. The results for energy consumption show that the proposed FGWO technique schedules the power in efficient way. FGWO overall energy consumption peak is less than the other techniques because in order to avoid peak creation at any slot of the day, this technique is proposed. It is noteworthy that the energy consumption pattern of all scheduling algorithms is quite better as compared to unscheduled power consumption pattern. It is finalized that the proposed FGWO optimization technique optimizes daily load consumption pattern in a better way through shifting the load from peak time intervals to off peak time intervals. Shifting of load affects the user satisfaction level, on the other hand, it gives benefit to the user in terms of cost reduction. DSM is not only useful for consumers but it is also beneficial for utility. Decrease in PAR assists the utility to keep its stability and it finally, leads to cost reduction. The performance of various optimization algorithms in the form of PAR is illustrated in Fig. 4. The figure is clearly shows that, proposed FGWO outper-

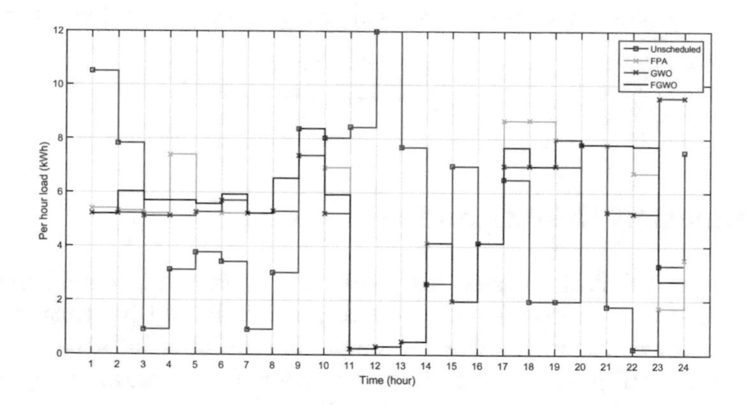

Fig. 3. Hourly load

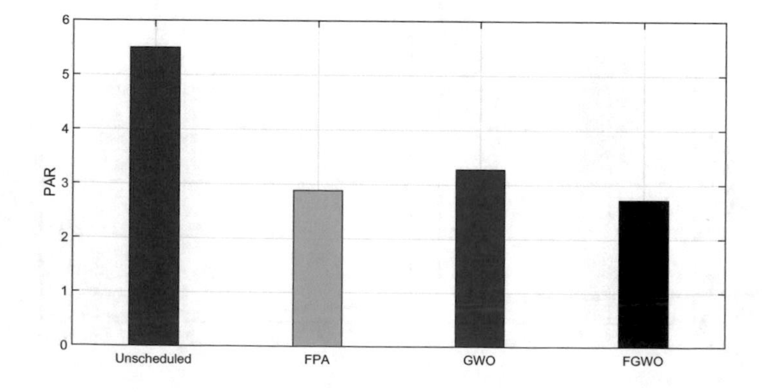

Fig. 4. PAR

formed the other techniques in terms of PAR. The PAR is minimized 50.425%, 40.266% and 47.448% by FGWO, GWO and FPA respectively in comparison with unscheduled scenario.

Figure 5 shows per hour cost for FPA, GWO, and Proposed FGWO. Plot shows that cost of the aforementioned scheduling techniques in peak hours is lower than unscheduled scenario the reason is that the load is transferred from high price hours to low price hours. Figure 6 indicates the performance ratio between unscheduled case and scheduled algorithms with respect to total cost. It is clearly visible in the plot that our proposed FGWO technique outperformed the GWO and unscheduled in overall cost. However, FPA beats our proposed FGWA as well as GWO and unscheduled. It is worth mentioning that the cost is minimized at all aforementioned scheduling algorithms in comparison with unscheduled scenario. As we can say that total cost is minimized 54.654%, 48.699% and 58.026% in cases of FGWO, GWO and FPA respectively in difference with unscheduled case.

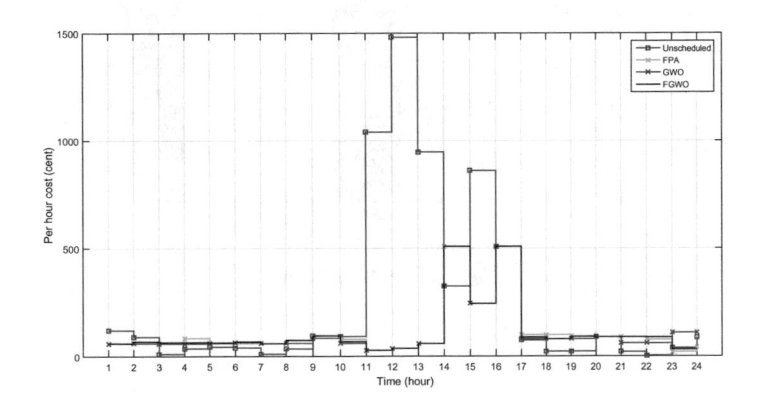

Fig. 5. Hourly cost

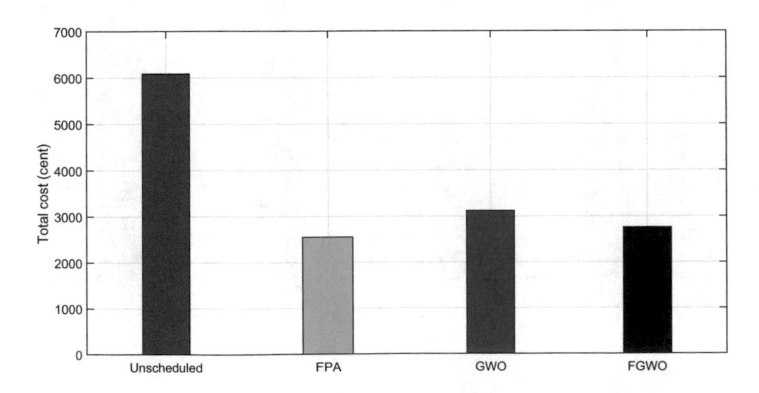

Fig. 6. Total cost

In Fig. 7, the waiting is shown. The user or consumer comfort is computed in the form of waiting time. In our scenario, the user comfort is the amount of time a user waits for a particular appliance to switch on. So, the user comfort is inversely proportional to the delay time or waiting time. The delay time is 2.4148 h, 3.3854 h and 2.4527 h by FGWO, GWO and FPA algorithms respectively. Hence, it is noteworthy, that our proposed FGWO technique outperformed all other techniques in terms of delay time. Figure 8 depicts the total power consumption which is 122.799 kWh for a single home consists of 15 appliances. The total energy consumption is equal in all the scenarios; in unscheduled case as well as in scheduling techniques.

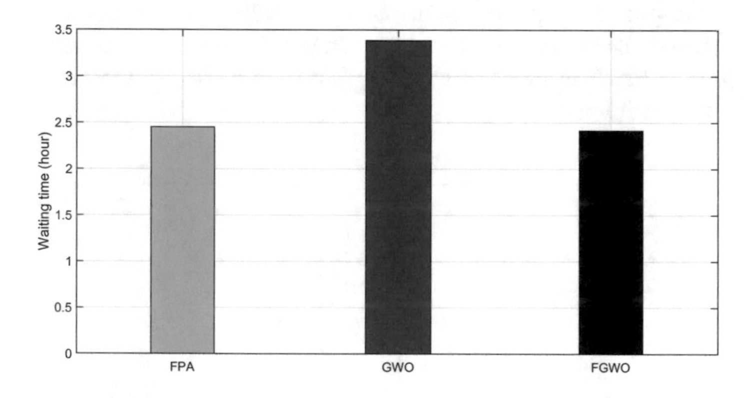

Fig. 7. Waiting time

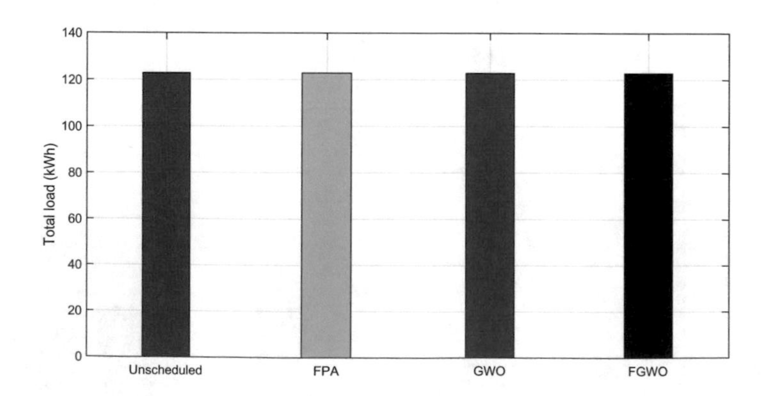

Fig. 8. Total load

7 Feasible Region

Feasible region is an area, which is covered by a set of points in which the fitness function satisfies its results. In this work, we discuss the feasible region of cost and energy utilization. The feasible region of cost reduction through reducing the hourly energy usage is shown in Fig. 9 below. This set of points which will be draw a complete cost region are the constraints to the problem. In order to find the constraints or sets of points use to draw overall cost region, we employ CPP scheme ranges from 11.4 to 123.4 cent/kWh. Furthermore, the energy utilization profile in unscheduled case is ranges from 0.225 to 13.375 kWh is also employed. Finally, we calculate all the possible combinations of unscheduled power utilization with CPP as shown in Table 2.

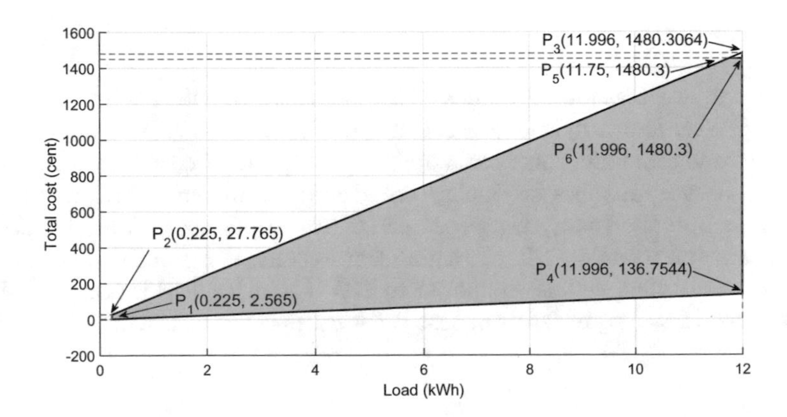

Fig. 9. Feasible Region

Table 2. All possible combinations

Combinations	Unscheduled load (kWh)	CPP price (cent)	Cost (cent)
Minimum load, Minimum price	0.225	11.4	2.565
Minimum load, Maximum price	0.225	123.4	27.765
Maximum load, Minimum price	11.996	11.4	136.7544
Maximum load, Maximum price	11.996	123.4	1480.3064

The overall cost region is shown using these points: P1(0.225, 2.565), P2(0.225, 27.765), P3(11.996, 1480.3064) and P4(11.996, 136.7544). Where, P1 demonstrates the electricity payment if the minimum energy is scheduled at minimum price. Furthermore, P2 shows the electricity cost if minimum power is used at maximum price. In the same way, P3 represents the electricity payment when maximum load is utilized at maximum price. Finally, P4 shows the electricity cost when maximum energy is consumed at minimum price. Now, we have an another core constraint that maximum electricity cost at any time slot must be smaller than maximum electricity cost in unscheduled case i.e., 1480.3 cents. After the consideration of all the constraints, the feasible region is drawn which is depicted and demonstrated by the shaded region covered by points (P1, P2, P4, P5 and P6). Where, P5(11.75, 1480.3) depicts that energy usage in scheduled case must not cross 11.75 kWh at hour where the electricity cost is maximum. Furthermore, point P6(11.996, 1480.3) shows that maximum possible energy consumption i.e., 11.996 kWh must never be scheduled at a time slot where electricity payment is more than 1480.3 cents.

8 Conclusion

In this work, we considered load shifting strategy of DSM with EMC based on FPA, GWO, and a hybrid of FPA and GWO i.e., FGWO to decrease PAR and increase user comfort with an affordable amount of cost. A various sets of residential area appliances are incorporated and appliances scheduling are performed based on the energy usage pattern of the appliances. From performance evaluation of the techniques in the simulation results, it is finalized, that FGWO technique performed well as compared to GWO and FPA in terms of PAR and waiting time. However, in terms of cost, FPA outperformed the other techniques.

References

1. Gungor, V.C., Sahin, D., Kocak, T., Ergut, S., Buccella, C., Cecati, C., Hancke, G.P.: Smart grid technologies: communication technologies and standards. IEEE Trans. Ind. Inf. **7**(4), 529–539 (2011)
2. Rahimi, F., Ipakchi, A.: Demand response as a market resource under the smart grid paradigm. IEEE Trans. Smart Grid **1**(1), 82–88 (2010)
3. Wang, J., Sun, Z., Zhou, Y., Dai, J.: Optimal dispatching model of smart home energy management system. In: 2012 IEEE Innovative Smart Grid Technologies-Asia (ISGT Asia), pp. 1-5. IEEE (2012)
4. Ahmad, A., Javaid, N., Alrajeh, N., Khan, Z.A., Qasim, U., Khan, A.: A modified feature selection and artificial neural network-based day-ahead load forecasting model for a smart grid. Appl. Sci. **5**(4), 1756–1772 (2015)
5. Zhu, Z., Tang, J., Lambotharan, S., Chin, W.H., Fan, Z.: An integer linear programming based optimization for home demand-side management in smart grid. In: 2012 IEEE PES Innovative Smart Grid Technologies (ISGT), pp. 1-5. IEEE (2012)
6. Rahim, S., Javaid, N., Ahmad, A., Khan, S.A., Khan, Z.A., Alrajeh, N., Qasim, U.: Exploiting heuristic algorithms to efficiently utilize energy management controllers with renewable energy sources. Energy Buildings **129**, 452–470 (2016)
7. Khan, M.A., Javaid, N., Mahmood, A., Khan, Z.A., Alrajeh, N.: A generic demand-side management model for smart grid. Int. J. Energy Res. **39**(7), 954–964 (2015)

8. Yang, X.-S.: Flower pollination algorithm for global optimization. In: UCNC, pp. 240–249 (2012)
9. Mirjalili, S., Mirjalili, S.M., Lewis, A.: Grey wolf optimizer. Adv. Eng. Softw. **69**, 46–61 (2014)
10. Ma, K., Shubing, H., Yang, J., Xia, X., Guan, X.: Appliances scheduling via cooperative multi-swarm PSO under day-ahead prices and photovoltaic generation. Appl. Soft Comput. **62**, 504–513 (2018)
11. Ahmed, M.S., Mohamed, A., Khatib, T., Shareef, H., Homod, R.Z., Ali, J.A.: Real time optimal schedule controller for home energy management system using new binary backtracking search algorithm. Energy Buildings **138**, 215–227 (2017)
12. Bazydło, G., Wermiński, S.: Demand side management through home area network systems. Int. J. Electr. Power Energy Syst. **97**, 174–185 (2018)
13. Anees, A., Chen, Y.-P.P.: True real time pricing and combined power scheduling of electric appliances in residential energy management system. Appl. Energy **165**, 592–600 (2016)
14. Sattarpour, T., Nazarpour, D., Golshannavaz, S.: A multi-objective HEM strategy for smart home energy scheduling: a collaborative approach to support microgrid operation. Sustainable Cities Soc. **37**, 26–33 (2018)
15. Shirazi, E., Jadid, S.: Cost reduction and peak shaving through domestic load shifting and DERs. Energy **124**, 146–159 (2017)

A Fuzzy-Based Approach for Improving Peer Coordination Quality in MobilePeerDroid Mobile System

Yi Liu[1](✉), Kosuke Ozera[1], Keita Matsuo[2], Makoto Ikeda[2], and Leonard Barolli[2]

[1] Graduate School of Engineering, Fukuoka Institute of Technology (FIT), 3-30-1 Wajiro-Higashi, Fukuoka, Higashi-ku 811-0295, Japan
ryuui1010@gmail.com, kosuke.o.fit@gmail.com

[2] Department of Information and Communication Engineering, Fukuoka Institute of Technology (FIT), 3-30-1 Wajiro-Higashi, Fukuoka, Higashi-ku 811-0295, Japan
{kt-matsuo,barolli}@fit.ac.jp, makoto.ikd@acm.org

Abstract. In this work, we present a distributed event-based awareness approach for P2P groupware systems. In our approach, the awareness of collaboration will be achieved by using primitive operations and services that are integrated into the P2P middleware. We propose an abstract model for achieving these requirements and we discuss how this model can support awareness of collaboration in mobile teams. We present a fuzzy-based system for improving peer coordination quality according to three parameters. This model will be implemented in MobilePeerDroid system to give more realistic view of the collaborative activity and better decisions for the groupwork, while encouraging peers to increase their reliability in order to support awareness of collaboration in MobilePeerDroid Mobile System. We evaluated the performance of proposed system by computer simulations. From the simulations results, we conclude that when GS and SCT are high, the peer coordination quality is high. With increasing of AA, the peer coordination quality is increasing.

1 Introduction

Peer to Peer technologies has been among most disruptive technologies after Internet. Indeed, the emergence of the P2P technologies changed drastically the concepts, paradigms and protocols of sharing and communication in large scale distributed systems. As pointed out since early 2000 years [1–5], the nature of the sharing and the direct communication among peers in the system, being these machines or people, makes possible to overcome the limitations of the flat communications through email, newsgroups and other forum-based communication forms.

The usefulness of P2P technologies on one hand has been shown for the development of stand alone applications. On the other hand, P2P technologies,

L. Barolli et al. (Eds.): IMIS 2018, AISC 773, pp. 60–73, 2019.
https://doi.org/10.1007/978-3-319-93554-6_5

paradigms and protocols have penetrated other large scale distributed systems such as Mobile Ad hoc Networks (MANETs), Groupware systems, Mobile Systems to achieve efficient sharing, communication, coordination, replication, awareness and synchronization. In fact, for every new form of Internet-based distributed systems, we are seeing how P2P concepts and paradigms again play an important role to enhance the efficiency and effectiveness of such systems or to enhance information sharing and online collaborative activities of groups of people. We briefly introduce below some common application scenarios that can benefit from P2P communications.

Awareness is a key feature of groupware systems. In its simplest terms, awareness can be defined as the system's ability to notify the members of a group of changes occurring in the group's workspace. Awareness systems for online collaborative work have been proposed since in early stages of Web technology. Such proposals started by approaching workspace awareness, aiming to inform users about changes occurring in the shared workspace. More recently, research has focussed on using new paradigms, such as P2P systems, to achieve fully decentralized, ubiquitous groupware systems and awareness in such systems. In P2P groupware systems group processes may be more efficient because peers can be aware of the status of other peers in the group, and can interact directly and share resources with peers in order to provide additional scaffolding or social support. Moreover, P2P systems are pervasive and ubiquitous in nature, thus enabling contextualized awareness.

Fuzzy Logic (FL) is the logic underlying modes of reasoning which are approximate rather then exact. The importance of FL derives from the fact that most modes of human reasoning and especially common sense reasoning are approximate in nature [6]. FL uses linguistic variables to describe the control parameters. By using relatively simple linguistic expressions it is possible to describe and grasp very complex problems. A very important property of the linguistic variables is the capability of describing imprecise parameters.

The concept of a fuzzy set deals with the representation of classes whose boundaries are not determined. It uses a characteristic function, taking values usually in the interval [0, 1]. The fuzzy sets are used for representing linguistic labels. This can be viewed as expressing an uncertainty about the clear-cut meaning of the label. But important point is that the valuation set is supposed to be common to the various linguistic labels that are involved in the given problem.

The fuzzy set theory uses the membership function to encode a preference among the possible interpretations of the corresponding label. A fuzzy set can be defined by examplification, ranking elements according to their typicality with respect to the concept underlying the fuzzy set [7].

In this paper, we propose a fuzzy-based system for MobilePeerDroid system considering three parameters: Activity Awareness (AA), Group Synchronization (GS), Sustained Communication Time (SCT) to decide the Peer Coordination Quality (PCQ). We evaluated the proposed system by simulations.

The structure of this paper is as follows. In Sect. 2, we introduce the scenarios of collaborative teamwork. In Sect. 3, we introduce the group activity awareness model. In Sect. 4, we introduce FL used for control. In Sect. 5, we present the proposed fuzzy-based system. In Sect. 6, we discuss the simulation results. Finally, conclusions and future work are given in Sect. 7.

2 Scenarios of Collaborative Teamwork

In this section, we describe and analyse some main scenarios of collaborative teamwork for which P2P technologies can support efficient system design.

2.1 Collaborative Teamwork and Virtual Campuses

Collaborative work through virtual teams is a significant way of collaborating in modern businesses, online learning, etc. [8]. Collaboration in virtual teams requires efficient sharing of information (both data sharing among the group members as well as sharing of group processes) and efficient communication among members of the team. Additionally, coordination and interaction are crucial for accomplishing common tasks through a shared workspace environment. P2P systems can enable fully decentralized collaborative systems by efficiently supporting different forms of collaboration [9]. One such form is using P2P networks, with super-peer structure as show in Fig. 1.

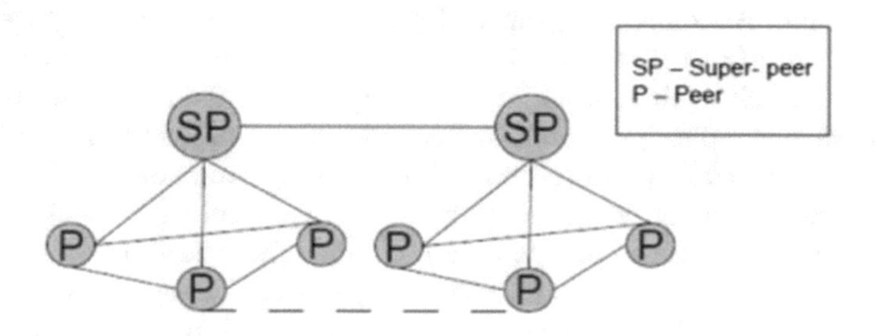

Fig. 1. Super-peer P2P group netwok.

During the last two decades, online learning has become very popular and there is a widespread of virtual campuses or combinations of face-to-face with semi-open teaching and learning. Virtual campuses are now looking at ways to effectively support learners, especially for online courses implemented as PBL-Project Based Learning or SBL Scenario Based Learning there is an increasing need to develop mobile applications that support these online groupwork learning paradigms [10]. In such setting, P2P technologies offer interesting solutions for (a) decentralizing the virtual campuses, which tend to grow and get further centralized with the increase of number of students enrolled, new degrees, and increase in academic activity; (b) in taking advantage of resources of students and developing volunteerbased computing systems as part of virtual campuses and (c) alleviating the communication burden for efficient collaborative teamwork. The use of P2P libraries such as JXTA have been investigated to design P2P middleware for P2P eLearning applications. Also, the use of P2P technologies in such setting is used for P2P video synchronization in a collaborative virtual environment [11]. Recently, virtual campuses are also introducing social networking among their students to enhance the learning activities through social support and scaffolding. Again the P2P solutions are sought in this context [12] in combination with social networking features to enhance especially the interaction among learners sharing similar objectives and interest or accomplishing a common project.

2.2 Mobile Ad Hoc Networks (MANETs)

Mobile ad-hoc networks are among most interesting infrastructureless network of mobile devices connected by wireless having self-configuring properties [13]. The lack of fixed infrastructure and of a centralized administration makes the building and operation in MANETS challenging. P2P networks and mobile ad hoc networks (MANETs) follow the same idea of creating a network without a central entity. All nodes (peers) must collaborate together to make possible the proper functioning of the network by forwarding information on behalf of others in the network [14]. P2P and MANETs share many key characteristics such as self-organization and decentralization due to the common nature of their distributed components. Both MANETs and P2P networks follow a P2P paradigm characterized by the lack of a central node or peer acting as a managing server, all participants having therefore to collaborate in order for the whole system to work. A key issue in both networks is the process of discovering the requested data or route efficiently in a decentralized manner. Recently, new P2P applications which uses wireless communication and integrates mobile devices such as PDA and mobile phones is emerging. Several P2P-based protocols can be used for MANETs such as Mobile P2P Protocol (MPP), which is based on Dynamic Source Routing (DSR), JXTA prtotocols, and MANET Anonymous Peer-to-peer Communication Protocol (MAPCP), which serves as an efficient anonymous communication protocol for P2P applications over MANET.

3 Group Activity Awareness Model

The awareness model considered here focuses on supporting group activities so to accomplish a common group project, although it can also be used in a broader scope of teamwork. The main building blocks of our model (see also [15,16] in the context of web-based groupware) are described below.

Activity awareness: Activity awareness refers to awareness information about the project-related activities of group members. Project-based work is one of the most common methods of group working. Activity awareness aims to provide information about progress on the accomplishment of tasks by both individuals and the group as a whole. It comprises knowing about actions taken by members of the group according to the project schedule, and synchronization of activities with the project schedule. Activity awareness should therefore enable members to know about recent and past actions on the project's work by the group. As part of activity awareness, we also consider information on group artifacts such as documents and actions upon them (uploads, downloads, modifications, reading). Activity awareness is one of most important, and most complex, types of awareness. As well as the direct link to monitoring a group's progress on the work relating to a project, it also supports group communication and coordination processes.

Process awareness: In project-based work, a project typically requires the enactment of a workflow. In such a case, the objective of the awareness is to track the state of the workflow and to inform users accordingly. We term this process awareness. The workflow is defined through a set of tasks and precedence relationships relating to their order of completion. Process awareness targets the information flow of the project, providing individuals and the group with a partial view (what they are each doing individually) and a complete view (what they are doing as a group), thus enabling the identification of past, current and next states of the workflow in order to move the collaboration process forward.

Communication awareness: Another type of awareness considered in this work is that of communication awareness. We consider awareness information relating to message exchange, and synchronous and asynchronous discussion forums. The first is intended to support awareness of peer-to-peer communication (when some peer wants to establish a direct communication with another peer); the second is aimed at supporting awareness about chat room creation and lifetime (so that other peers can be aware of, and possibly eventually join, the chat room); the third refers to awareness of new messages posted at the discussion forum, replies, etc.

Availability awareness: Availability awareness is useful for provide individuals and the group with information on members' and resources' availability. The former is necessary for establishing synchronous collaboration either in peer-to-peer mode or (sub)group mode. The later is useful for supporting members' tasks requiring available resources (e.g. a machine for running a software program). Groupware applications usually monitor availability of group members by simply looking at group workspaces. However, availability awareness encompasses not only knowing who is in the workspace at any given moment but also who is available when, via members' profiles (which include also personal calendars) and information explicitly provided by members. In the case of resources, awareness is achieved via the schedules of resources. Thus, both explicit and implicit forms of gathering availability awareness information should be supported.

4 Application of Fuzzy Logic for Control

The ability of fuzzy sets and possibility theory to model gradual properties or soft constraints whose satisfaction is matter of degree, as well as information pervaded with imprecision and uncertainty, makes them useful in a great variety of applications.

The most popular area of application is Fuzzy Control (FC), since the appearance, especially in Japan, of industrial applications in domestic appliances, process control, and automotive systems, among many other fields.

4.1 FC

In the FC systems, expert knowledge is encoded in the form of fuzzy rules, which describe recommended actions for different classes of situations represented by fuzzy sets.

In fact, any kind of control law can be modeled by the FC methodology, provided that this law is expressible in terms of "if ... then ..." rules, just like in the case of expert systems. However, FL diverges from the standard expert system approach by providing an interpolation mechanism from several rules. In the contents of complex processes, it may turn out to be more practical to get knowledge from an expert operator than to calculate an optimal control, due to modeling costs or because a model is out of reach.

4.2 Linguistic Variables

A concept that plays a central role in the application of FL is that of a linguistic variable. The linguistic variables may be viewed as a form of data compression. One linguistic variable may represent many numerical variables. It is suggestive to refer to this form of data compression as granulation [17].

The same effect can be achieved by conventional quantization, but in the case of quantization, the values are intervals, whereas in the case of granulation the values are overlapping fuzzy sets. The advantages of granulation over quantization are as follows:

- it is more general;
- it mimics the way in which humans interpret linguistic values;
- the transition from one linguistic value to a contiguous linguistic value is gradual rather than abrupt, resulting in continuity and robustness.

4.3 FC Rules

FC describes the algorithm for process control as a fuzzy relation between information about the conditions of the process to be controlled, x and y, and the output for the process z. The control algorithm is given in "if ... then ..." expression, such as:

If x is small and y is big, then z is medium;
If x is big and y is medium, then z is big.

These rules are called *FC rules*. The "if" clause of the rules is called the antecedent and the "then" clause is called consequent. In general, variables x and y are called the input and z the output. The "small" and "big" are fuzzy values for x and y, and they are expressed by fuzzy sets.

Fuzzy controllers are constructed of groups of these FC rules, and when an actual input is given, the output is calculated by means of fuzzy inference.

4.4 Control Knowledge Base

There are two main tasks in designing the control knowledge base. First, a set of linguistic variables must be selected which describe the values of the main control parameters of the process. Both the input and output parameters must be linguistically defined in this stage using proper term sets. The selection of the level of granularity of a term set for an input variable or an output variable plays an important role in the smoothness of control. Second, a control knowledge base must be developed which uses the above linguistic description of the input and output parameters. Four methods [18–21] have been suggested for doing this:

- expert's experience and knowledge;
- modelling the operator's control action;
- modelling a process;
- self organization.

Among the above methods, the first one is the most widely used. In the modeling of the human expert operator's knowledge, fuzzy rules of the form "If Error is small and Change-in-error is small then the Force is small" have been used in several studies [22,23]. This method is effective when expert human operators can express the heuristics or the knowledge that they use in controlling a process in terms of rules of the above form.

4.5 Defuzzification Methods

The defuzzification operation produces a non-FC action that best represent the membership function of an inferred FC action. Several defuzzification methods have been suggested in literature. Among them, four methods which have been applied most often are:

- Tsukamoto's Defuzzification Method;
- The Center of Area (COA) Method;
- The Mean of Maximum (MOM) Method;
- Defuzzification when Output of Rules are Function of Their Inputs.

5 Proposed Fuzzy-Based System for Peer Coordination Quality

The P2P group-based model considered is that of a superpeer model. In this model, the P2P network is fragmented into several disjoint peergroups (see

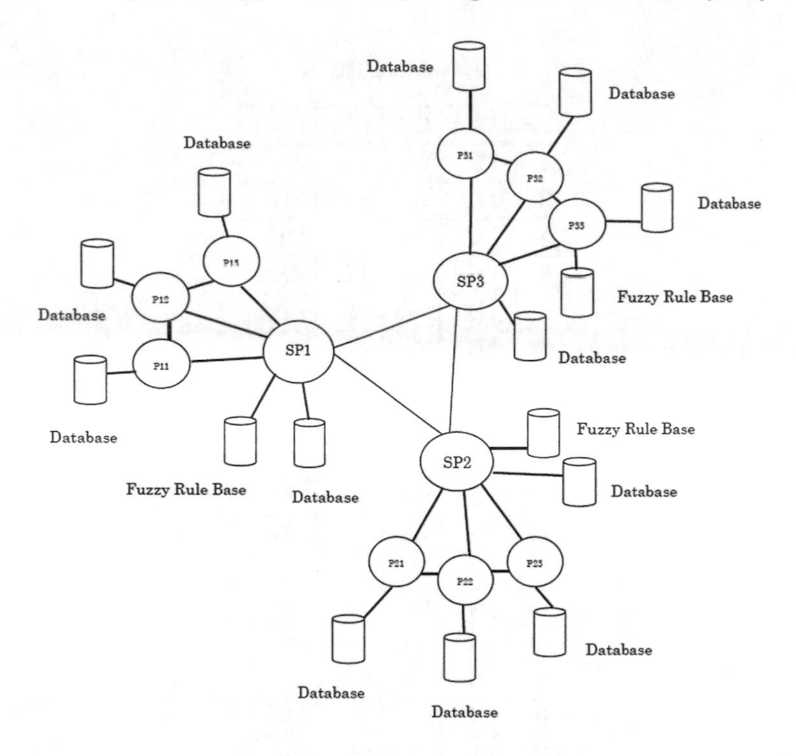

Fig. 2. P2P group-based model.

Fig. 2). The peers of each peergroup are connected to a single superpeer. There is frequent local communication between peers in a peergroup, and less frequent global communication between superpeers.

To complete a certain task in P2P mobile collaborative team work, peers often have to in teract with unknow peers. Thus, it is important that group members must select reliable peers to interact.

In this work, we consider three parameters: Activity Awareness (AA), Group Synchronization (GS), Sustained Communication Time (SCT) to decide the Peer Coordination Quality (PCQ). The structure of this system called Fuzzy-based Coordination Qualityt System (FCQS) is shown in Fig. 3. These three parameters are fuzzified using fuzzy system, and based on the decision of fuzzy system the peer coordination quality is calculated. The membership functions for our system are shown in Fig. 4. In Table 1, we show the Fuzzy Rule Base (FRB) of our proposed system, which consists of 36 rules.

The input parameters for FCQS are: AA, GS and SCT the output linguistic parameter is PCQ. The term sets of AA, GS and SCT are defined respectively as:

Table 1. FRB.

Rule	AA	GS	SCT	PCQ
1	B	Ba	VS	EB
2	B	Ba	S	EB
3	B	Ba	L	MG
4	B	Ba	VL	EB
5	B	Nor	VS	BD
6	B	Nor	S	PG
7	B	Nor	L	BD
8	B	Nor	VL	MG
9	B	Go	VS	VG
10	B	Go	S	EB
11	B	Go	L	BD
12	B	Go	VL	PG
13	N	Ba	VS	BD
14	N	Ba	S	MG
15	N	Ba	L	G
16	N	Ba	VL	MG
17	N	Nor	VS	PG
18	N	Nor	S	VG
19	N	Nor	L	BD
20	N	Nor	VL	MG
21	N	Go	VS	G
22	N	Go	S	MG
23	N	Go	L	PG
24	N	Go	VL	VG
25	G	Ba	VS	PG
26	G	Ba	S	G
27	G	Ba	L	VVG
28	G	Ba	VL	MG
29	G	Nor	VS	PG
30	G	Nor	S	VG
31	G	Nor	L	PG
32	G	Nor	VL	G
33	G	Go	VS	VVG
34	G	Go	S	G
35	G	Go	L	VG
36	G	Go	VL	VVG

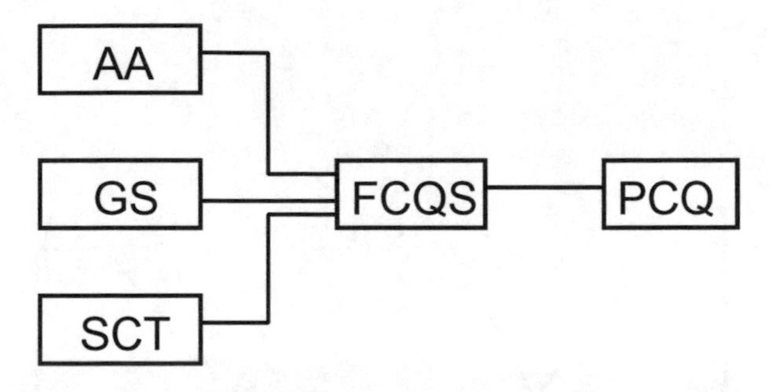

Fig. 3. Proposed sytem of stucture.

$$\begin{aligned} AA &= \{Bad,\ Normal,\ Good\} \\ &= \{B,\ N,\ G\}; \\ GS &= \{Bad,\ Normal,\ Good\} \\ &= \{Ba,\ Nor,\ Go\}; \\ SCT &= \{Very\ Short,\ Short,\ Long,\ Very\ Long\} \\ &= \{VS,\ S,\ L,\ VL\}; \end{aligned}$$

and the term set for the output PCQ is defined as:

$$PCQ = \begin{pmatrix} Extremely\ Bad \\ Bad \\ Minimally\ Good \\ Partially\ Good \\ Good \\ Very\ Good \\ Very\ Very\ Good \end{pmatrix} = \begin{pmatrix} EB \\ BD \\ MG \\ PG \\ G \\ VG \\ VVG \end{pmatrix}$$

6 Simulation Results

In this section, we present the simulation results for our proposed system. In our system, we decided the number of term sets by carrying out many simulations. These simulation results were carried out in MATLAB.

From Fig. 5(a) to (c), we show the relation between AA, GS, SCT and PCQ. In this simulation, we consider the AA as a constant parameter.

In Fig. 5(a), we consider the AA value 10 units. We change the GS value from 0 to 100 units. When the GS increases, the PCQ is increased. Also, when the SCT increases, the PCQ is increased.

In Fig. 5(b) to (c), we increase the AA values to 50 and 90 units, respectively. We see that, when the AA increases, the PCQ is increased.

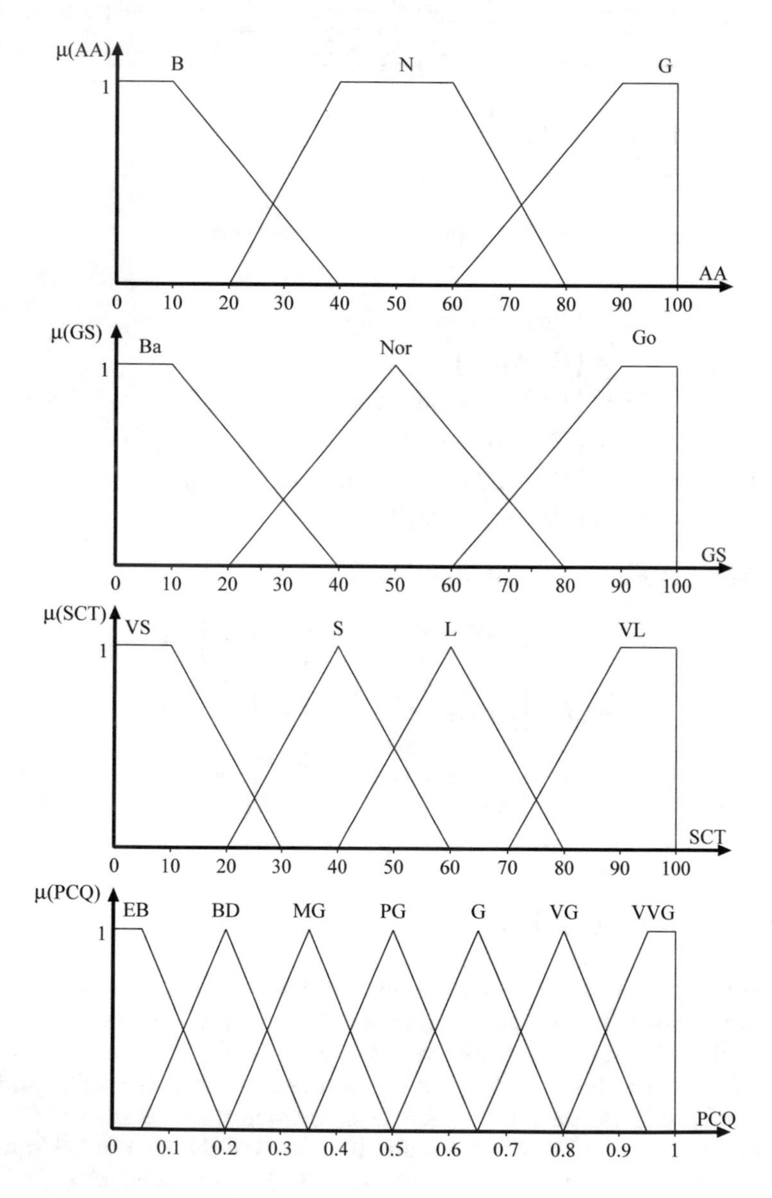

Fig. 4. Membership functions.

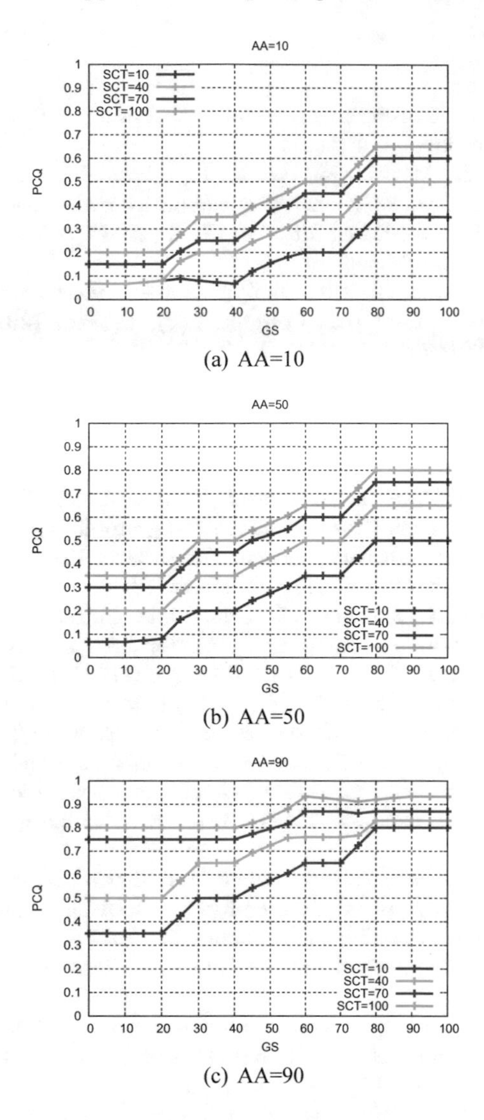

Fig. 5. PCQ for different AA.

7 Conclusions and Future Work

In this paper, we proposed a fuzzy-based system to decide the PCQ. We took into consideration three parameters: AA, GS and SCT. We evaluated the performance of proposed system by computer simulations. From the simulations results, we conclude that when GS and SCT are high, the peer coordination quality is high. With increasing of AA, the peer coordination quality is increased.

In the future, we would like to make extensive simulations to evaluate the proposed systems and compare the performance with other systems.

References

1. Oram, A. (ed.): Peer-to-Peer: Harnessing the Power of Disruptive Technologies. O'Reilly and Associates, CA (2001)
2. Sula, A., Spaho, E., Matsuo, K., Barolli, L., Xhafa, F., Miho, R.: A new system for supporting children with autism spectrum disorder based on IoT and P2P technology. Int. J. Space-Based Situated Comput. **4**(1), 55–64 (2014). https://doi.org/10.1504/IJSSC.2014.060688
3. Di Stefano, A., Morana, G., Zito, D.: QoS-aware services composition in P2PGrid environments. Int. J. Grid Util. Comput. **2**(2), 139–147 (2011). https://doi.org/10.1504/IJGUC.2011.040601
4. Sawamura, S., Barolli, A., Aikebaier, A., Takizawa, M., Enokido, T.: Design and evaluation of algorithms for obtaining objective trustworthiness on acquaintances in P2P overlay networks. Int. J. Grid Util. Comput. **2**(3), 196–203 (2011). https://doi.org/10.1504/IJGUC.2011.042042
5. Higashino, M., Hayakawa, T., Takahashi, K., Kawamura, T., Sugahara, K.: Management of streaming multimedia content using mobile agent technology on pure P2P-based distributed e-learning system. Int. J. Grid Util. Comput. **5**(3), 198–204 (2014). https://doi.org/10.1504/IJGUC.2014.062928
6. Inaba, T., Obukata, R., Sakamoto, S., Oda, T., Ikeda, M., Barolli, L.: Performance evaluation of a QoS-aware fuzzy-based CAC for LAN access. Int. J. Space-Based Situated Comput. **1**(1) (2011). https://doi.org/10.1504/IJSSC.2016.082768
7. Terano, T., Asai, K., Sugeno, M.: Fuzzy Systems Theory and Its Applications. Academic Press Inc. Harcourt Brace Jovanovich, Publishers, SanDiego (1992)
8. Mori, T., Nakashima, M., Ito, T.: SpACCE: a sophisticated ad hoc cloud computing environment built by server migration to facilitate distributed collaboration. Int. J. Space-Based Situated Comput. **1**(1) (2011). https://doi.org/10.1504/IJSSC.2012.050000
9. Xhafa, F., Poulovassilis, A.: Requirements for distributed event-based awareness in P2P groupware systems. In: Proceedings of AINA 2010, pp. 220–225 (2010)
10. Xhafa, F., Barolli, L., Caballé, S., Fernandez, R.: Supporting scenario-based online learning with P2P group-based systems. In: Proceedings of NBiS 2010, pp. 173–180 (2010)
11. Gupta, S., Kaiser, G.: P2P video synchronization in a collaborative virtual environment. In: Proceedings of the 4th International Conference on Advances in Web-Based Learning, ICWL 2005, pp. 86–98 (2005)
12. Martnez-Alemn, A.M., Wartman, K.L.: Online Social Networking on Campus Understanding What Matters in Student Culture. Taylor and Francis, Routledge (2008)
13. Puzar, M., Plagemann, T.: Data sharing in mobile ad-hoc networks – a study of replication and performance in the MIDAS data space. Int. J. Space-Based Situated Comput. **1**(1) (2011). https://doi.org/10.1504/IJSSC.2011.040340
14. Spaho, E., Kulla, E., Xhafa, F., Barolli, L.: P2P solutions to efficient mobile peer collaboration in MANETs. In: Proceedings of 3PGCIC 2012, pp. 379–383, November 2012
15. Gutwin, C., Greenberg, S., Roseman, M.: Workspace awareness in real time distributed groupware: framework, widgets, and evaluation. In: BCS HCI 1996, pp. 281–298
16. You, Y., Pekkola, S.: Meeting others -supporting situation awareness on the WWW. Decis. Support Syst. **32**(1), 71–82 (2001)

17. Kandel, A.: Fuzzy Expert Systems. CRC Press, Boca Raton (1992)
18. Zimmermann, H.J.: Fuzzy Set Theory and Its Applications. Kluwer Academic Publishers (1991). Second Revised Edition
19. McNeill, F.M., Thro, E.: Fuzzy Logic: A Practical Approach. Academic Press Inc., Boston (1994)
20. Zadeh, L.A., Kacprzyk, J.: Fuzzy Logic for the Management of Uncertainty. Wiley, New York (1992)
21. Procyk, T.J., Mamdani, E.H.: A linguistic self-organizing process controller. Automatica **15**(1), 15–30 (1979)
22. Klir, G.J., Folger, T.A.: Fuzzy Sets, Uncertainty, and Information. Prentice Hall, Englewood Cliffs (1988)
23. Munakata, T., Jani, Y.: Fuzzy systems: an overview. Commun. ACM **37**(3), 69–76 (1994)

A Fuzzy-Based System for Selection of IoT Devices in Opportunistic Networks Considering IoT Device Contact Duration, Storage and Remaining Energy

Miralda Cuka[1(✉)], Donald Elmazi[1], Keita Matsuo[2], Makoto Ikeda[2], and Leonard Barolli[2]

[1] Graduate School of Engineering, Fukuoka Institute of Technology (FIT), 3-30-1 Wajiro-Higashi, Higashi-Ku, Fukuoka 811-0295, Japan
mcuka91@gmail.com, donald.elmazi@gmail.com

[2] Department of Information and Communication Engineering, Fukuoka Institute of Technology (FIT), 3-30-1 Wajiro-Higashi, Higashi-Ku, Fukuoka 811-0295, Japan
{kt-matsuo,barolli}@fit.ac.jp, makoto.ikd@acm.org

Abstract. The OppNets are a subclass of delay-tolerant networks where communication opportunities (contacts) are intermittent and there is no need to establish an end-to-end link between the communication nodes. The Internet of Things (IoT) is the network of devices, vehicles, buildings and other items embedded with software, electronics, sensors and network connectivity that enables these objects to collect and exchange data. There are different issues for these networks. One of them is the selection of IoT devices in order to carry out a task in opportunistic networks. In this work, we implement a Fuzzy-Based System for IoT device selection in opportunistic networks. For our system, we use three input parameters: IoT Contact Duration (IDCD), IoT Device Storage (IDST) and IoT Device Remaining Energy (IDRE). The output parameter is IoT Device Selection Decision (IDSD). The simulation results show that the proposed system makes a proper selection decison of IoT-devices in opportunistic networks. The IoT device selection is increased up to 19% and 53% by increasing IDCD and IDRE respectively.

1 Introduction

The OppNets have appeared as an evolution of the MANETs. They are also a wireless based network and hence, they face various issues similar to MANETs such as frequent disconnections, highly variable links and limited bandwidth. In OppNets, nodes are always moving which makes the network easy to deploy and decreases the dependence on infrastructure for communication [1]. The Internet of Things (IoT) can seamlessly connect the real world and cyberspace via physical objects that embed with various types of intelligent sensors. A large number of Internet-connected machines will generate and exchange an enormous amount

L. Barolli et al. (Eds.): IMIS 2018, AISC 773, pp. 74–85, 2019.
https://doi.org/10.1007/978-3-319-93554-6_6

of data that make daily life more convenient, help to make a tough decision and provide beneficial services. The IoT probably becomes one of the most popular networking concepts that has the potential to bring out many benefits [2,3].

OppNets are the variants of Delay Tolerant Networks (DTNs). Owing to the transient and un-connected nature of the nodes, routing becomes a challenging task in these networks. Sparse connectivity, no infrastructure and limited resources further complicate the situation [4,5]. Routing methods for such sparse mobile networks use a different paradigm for message delivery. They utilize node mobility by having nodes carry messages, waiting for an opportunity to transfer messages to the destination or the next relay rather than transmitting them over a path [6]. Hence, the challenges for routing in OppNets are very different from the traditional wireless networks and their utility and potential for scalability makes them a huge success.

However, most of the proposed routing schemes assume long contact durations such that all buffered messages can be transferred within a single contact. For example, when hand-held devices communicate via Bluetooth that has a typical wireless range of about 10 m, the contact duration tends to be as short as several seconds if the users are walking. For high speed vehicles that communicate via WiFi (802.11g), which has a longer range (up to 38 m indoors and 140 m outdoors), the contact duration is still short. In the presence of short contact durations, there are two key issues that must be addressed. First is the relay selection issue. We need to select relay nodes that will contact the message's destination long enough so that the entire message can be successfully transmitted. Second is the message scheduling issue. Since not all messages can be exchanged between nodes within a single contact, it is important to schedule the transmission of messages in such a way that will maximize the network delivery ratio [7].

In an OppNet, when nodes move away or turn off their power to conserve energy, links may be disrupted or shut down periodically. These events result in intermittent connectivity. When there is no path between the source and the destination, the network partition occurs. Therefore, nodes need to communicate with each other via opportunistic contacts through store-carry-forward operation. Since these types of networks require the IoT devices to store some information, storage is an important parameter in evaluation of their performance. However, the storage capacity of the device is limited which makes storage a requirement to be considered.

The Fuzzy Logic (FL) is unique approach that is able to simultaneously handle numerical data and linguistic knowledge. The fuzzy logic works on the levels of possibilities of input to achieve the definite output. Fuzzy set theory and FL establish the specifics of the nonlinear mapping.

In this paper, we propose and implement a simulation system for selection of IoT devices in OppNets. The system is based on fuzzy logic and considers three parameters for IoT device selection. We show the simulation results for different values of parameters.

The remainder of the paper is organized as follows. In the Sect. 2, we present a brief introduction of IoT. In Sect. 3, we describe the basics of OppNets including research challenges and architecture. In Sect. 4, we introduce the proposed system model and its implementation. Simulation results are shown in Sect. 5. Finally, conclusions and future work are given in Sect. 6.

2 IoT

2.1 IoT Architecture

The typical IoT architecture can be divided into five layers as shown in Fig. 1. Each layer is briefly described below.

Perception Layer: The perception layer is similar to physical layer in OSI model which consists of the diffThe OppNets are the variants of Delay Tolerant Networks (DTNs). These networks can be useful for routing in places where there are few base stations and connected routes for long distances. In an OppNets, when nodes move away or turn off their power to conserve energy, links may be disrupted or shut down periodically. These events result in intermittent connectivity. When there is no path existing between the source and the destination, the network partition occurs. Therefore, nodes need to communicate with each other via opportunistic contacts through store-carry-forward operation. This layer generally deals with identification and collection of specific information by each type of sensor devices. The gathered information can be location, wind speed, vibration, pH level, humidity, amount of dust in the air and so on. The gathered information is transmitted through Network layer toward central information processing system.

Network Layer: The Network layer plays an important role in securely transferring and keeping the sensitive information confidential from sensor devices to the central information processing system through 3G, 4G, UMTS, WiFi, WiMAX, RFID, Infrared and Satellite dependent on the type of sensors devices. Thus, this layer is mainly responsible for transfering the information from Perception layer to upper layer.

Middleware Layer: The devices in the IoT system may generate various type of services when they are connected and communicate with others. Middleware layer has two essential functions, including service management and store the lower layer information into the database. Moreover, this layer has capability to retrieve, process, compute information, and then automatically decide based on the computational results.

Application Layer: Application layer is responsible for inclusive applications management based on the processed information in the Middleware layer. The IoT applications can be smart postal, smart health, smart car, smart glasses, smart home, smart independent living, smart transportation, etc.

Business Layer: This layer functions cover the whole IoT applications and services management. It can create practically graphs, business models, flow chart

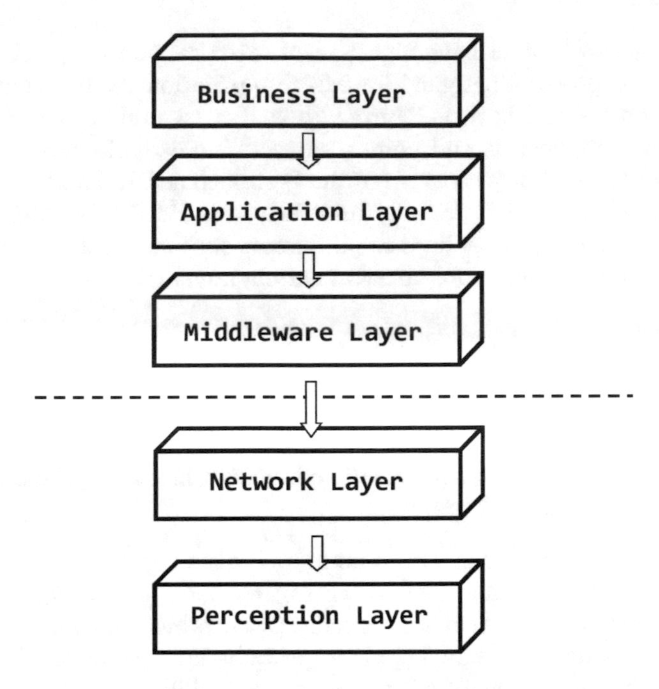

Fig. 1. IoT architecture layers

and executive report based on the amount of accurate data received from lower layer and effective data analysis process. Based on the good analysis results, it will help the functional managers or executives to make more accurate decisions about the business strategies and roadmaps.

2.2 IoT Protocols

In following we will briefly describe about the most frequently used protocols for Machine-to-Machine (M2M) communication.

The Message Queue Telemetry Transport (MQTT) is a Client Server publishes or subscribes messaging transport protocol. It is light weight, open, simple and designed so as to be easy to implement. The protocol runs over TCP/IP or over other network protocols that provide ordered, lossless, bi-directional connections. The MQTT features include the usage of the publish/subscribe message pattern which provides one-to-many message distribution, a messaging transport that is agnostic to the content of the payload. Furthermore, the MQTT protocol has not only minimized transport overhead and protocol exchange to reduce network traffic but also has an extraordinary mechanism to notify interested parties when an abnormal disconnection occurs as well.

The Constraint Application Protocol (CoAP) is a specialized web transfer protocol for use with constrained nodes and constrained networks. The nodes often have 8-bit microcontroller with small amounts of ROM and RAM, while constrained network often have high packet error rate and typical throughput is 10 kbps. This protocol designed for M2M application such as smart city and building automation. The CoAP provides a request and response interaction model between application end points, support build-in discovery services and resources, and includes key concepts of the Web such as URIs and Internet media types. CoAP is designed to friendly interface with HTTP for integration with the Web while meeting specialized requirements such as multicast support, very low overhead and simplicity for constrained environments.

3 OppNets

3.1 OppNets Challenges

In this section, we consider two specific challenges in an OppNets: the contact opportunity and the node storage.

- *Contact Opportunity:* Due to the node mobility or the dynamics of wireless channel, a node can make contact with other nodes at an unpredicted time. Since contacts between nodes are hardly predictable, they must be exploited opportunistically for exchanging messages between some nodes that can move between remote fragments of the network. Mobility increases the chances of communication between nodes. When nodes move randomly around the network, where jamming signals are disrupting the communication, they may pass through unjammed area and hence be able to communicate. In addition, the contact capacity needs to be considered [8,9].
- *Node Storage:* As described above, to avoid dropping packets, the intermediate nodes are required to have enough storage to store all messages for an unpredictable period of time until next contact occurs. In other words, the required storage space increases as a function of the number of messages in the network. Therefore, the routing and replication strategies must take the storage constraint into consideration [10].

3.2 OppNets Architectures

In an OppNet, a network is typically separated into several network partitions called regions. Traditional applications are not suitable for this kind of environment because they normally assume that the end-to-end connection must exist from the source to the destination.

The OppNets enables the devices in different regions to interconnect by operating message in a store-carry-forward fashion. The intermediate nodes implement the store-carry-forward message switching mechanism by overlaying a new protocol layer, called the bundle layer, on top of heterogeneous region-specific lower layers.

In an OppNet, each node is an entity with a bundle layer which can act as a host, a router or a gateway. When the node acts as a router, the bundle layer can store, carry and forward the entire bundles (or bundle fragments) between

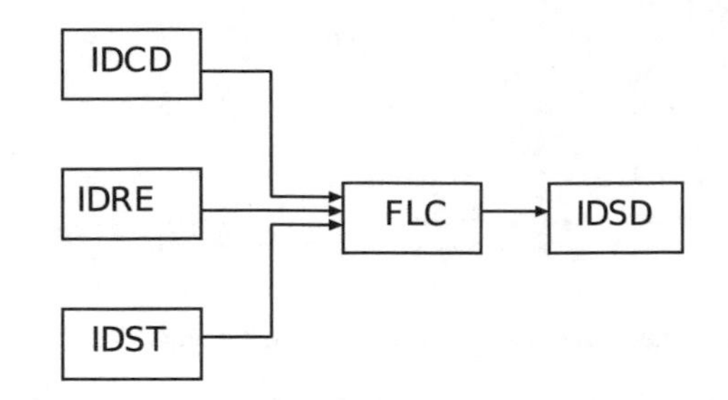

Fig. 2. Proposed system model.

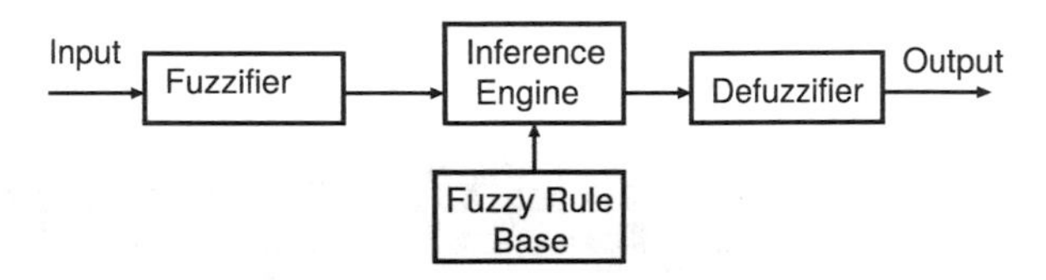

Fig. 3. FLC structure.

the nodes in the same region. On the other hand, the bundle layer of gateway is used to transfer messages across different regions. A gateway can forward bundles between two or more regions and may optionally be a host, so it must have persistent storage and support custody transfers.

4 Proposed System

4.1 System Parameters

Based on OppNets characteristics and challenges, we consider the following parameters for implementation of our proposed system.

IoT Device Contact Duration (IDCD): This is an important parameter in mobility-assisted networks as contact times represent the duration of message communication opportunity upon a contact. Contact durations is the time in which all buffered messages can be transferred within a single contact.

IoT Device Storage (IDST): In delay tolerant networks data is carried by the IoT device until a communication opportunity is available. Considering that different IoT devices have different storage capabilities, the selection desicion should consider the storage capacity.

IoT Device Remaining Energy (IDRE): The IoT devices in OppNets are active and can perform tasks and exchange data in different ways from each other. Consequently, some IoT devices may have a lot of remaining power and other may have very little, when an event occurs.

IoT Device Selection Decision (IDSD): The proposed system considers the following levels for IoT device selection:

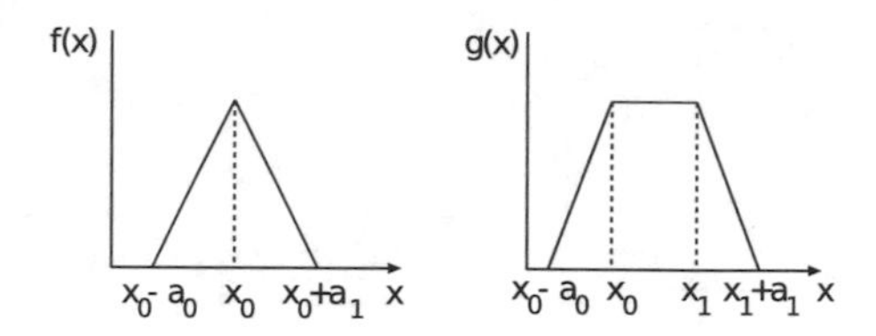

Fig. 4. Triangular and trapezoidal membership functions.

Table 1. Parameters and their term sets for FLC.

Parameters	Term sets
IoT Device Contact Duration (IDCD)	Short (Sho), Medium (Med), Long (Lg)
IoT Device Remaining Energy (IDRE)	Low (Lo), Medium (Mdm), High (Hgh)
IoT Device Storage (IDST)	Small (Sm), Medium (Me), High (Hi)
IoT Device Selection Decision (IDSD)	VLSP, LSP, MSP, HSP, VHSP

- Very Low Selection Possibility (VLSP) - The IoT device will have very low probability to be selected.
- Low Selection Possibility (LSP) - There might be other IoT devices which can do the job better.
- Middle Selection Possibility (MSP) - The IoT device is ready to be assigned a task, but is not the "chosen" one.
- High Selection Possibility (HSP) - The IoT device takes responsibility of completing the task.
- Very High Selection Possibility (VHSP) - The IoT device has almost all the required information and potential to be selected and then allocated in an appropriate position to carry out a job.

4.2 System Implementation

Fuzzy sets and fuzzy logic have been developed to manage vagueness and uncertainty in a reasoning process of an intelligent system such as a knowledge based system, an expert system or a logic control system [11–24]. In this work, we use fuzzy logic to implement the proposed system.

The structure of the proposed system is shown in Fig. 2. It consists of one Fuzzy Logic Controller (FLC), which is the main part of our system and its basic elements are shown in Fig. 3. They are the fuzzifier, inference engine, Fuzzy Rule Base (FRB) and defuzzifier.

As shown in Fig. 4, we use triangular and trapezoidal membership functions for FLC, because they are suitable for real-time operation [25]. The x_0 in $f(x)$ is the center of triangular function, $x_0(x_1)$ in $g(x)$ is the left (right) edge of trapezoidal function, and $a_0(a_1)$ is the left (right) width of the triangular or trapezoidal function. We explain in details the design of FLC in following.

Table 2. FRB of proposed fuzzy-based system.

No.	IDCD	IDRE	IDST	IDSD
1	Sho	Lo	Sm	VLSP
2	Sho	Lo	Me	VLSP
3	Sho	Lo	Hi	LSP
4	Sho	Mdm	Sm	VLSP
5	Sho	Mdm	Me	VLSP
6	Sho	Mdm	Hi	MSP
7	Sho	Hgh	Sm	LSP
8	Sho	Hgh	Me	MSP
9	Sho	Hgh	Hi	VHSP
10	Med	Lo	Sm	VLSP
11	Med	Lo	Me	VLSP
12	Med	Lo	Hi	HSP
13	Med	Mdm	Sm	LSP
14	Med	Mdm	Me	LSP
15	Med	Mdm	Hi	HSP
16	Med	Hgh	Sm	MSP
17	Med	Hgh	Me	HSP
18	Med	Hgh	Hi	VHSP
19	Lg	Lo	Sm	LSP
20	Lg	Lo	Me	MHSP
21	Lg	Lo	Hi	VHSP
22	Lg	Mdm	Sm	MSP
23	Lg	Mdm	Me	HSP
24	Lg	Mdm	Hi	VHSP
25	Lg	Hgh	Sm	HSP
26	Lg	Hgh	Me	VHSP
27	Lg	Hgh	Hi	VHSP

4.3 Description of FLC

We use three input parameters for FLC: IoT Device Contact Duration (IDCD), IoT Device Storage (IDST), IoT Device Remaining Energy (IDRE).

The term sets for each input linguistic parameter are defined respectively as shown in Table 1.

$$T(IDCD) = \{Short(Sho), Medium(Med), Long(Lg)\}$$
$$T(IDRE) = \{Low(Lo), Medium(Mdm), High(Hgh)\}$$
$$T(IDST) = \{Small(Sm), Medium(Me), High(Hi)\}$$

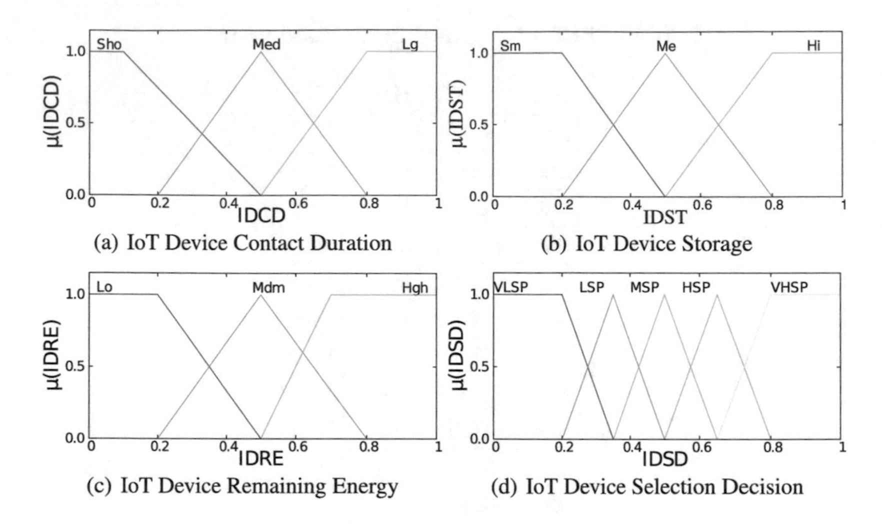

(a) IoT Device Contact Duration (b) IoT Device Storage (c) IoT Device Remaining Energy (d) IoT Device Selection Decision

Fig. 5. Fuzzy membership functions.

The membership functions for input parameters of FLC are defined as:

$$\begin{aligned}
\mu_{Sho}(IDCD) &= g(IDCD; Sho_0, Sho_1, Sho_{w0}, Sho_{w1}) \\
\mu_{Mi}(IDCD) &= f(IDCD; Med_0, Med_{w0}, Med_{w1}) \\
\mu_{Lg}(IDCD) &= g(IDCD; Lg_0, Lg_1, Lg_{w0}, Lg_{w1}) \\
\mu_{We}(IDRE) &= g(IDRE; Lo_0, Lo_1, Lo_{w0}, Lo_{w1}) \\
\mu_{Mo}(IDRE) &= f(IDRE; Mdm_0, Mdm_{w0}, Mdm_{w1}) \\
\mu_{St}(IDRE) &= g(IDRE; Hgh_0, Hgh_1, Hgh_{w0}, Hgh_{w1}) \\
\mu_{Sm}(IDST) &= g(IDST; Sm_0, Sm_1, Sm_{w0}, Sm_{w1}) \\
\mu_{Me}(IDST) &= f(IDST; Me_0, Me_{w0}, Me_{w1}) \\
\mu_{Hi}(IDST) &= g(IDST; Hi_0, Hi_1, Hi_{w0}, Hi_{w1}).
\end{aligned}$$

The small letters *w0* and *w1* mean left width and right width, respectively.

The output linguistic parameter is the IoT device Selection Decision (IDSD). We define the term set of IDSD as:

$$\begin{aligned}
\{&Very\ Low\ Selection\ Possibility\ (VLSP), \\
&Low\ Selection\ Possibility\ (LSP), \\
&Middle\ Selection\ Possibility\ (MSP), \\
&High\ Selection\ Possibility\ (HSP), \\
&Very\ High\ Selection\ Possibility\ (VHSP)\}.
\end{aligned}$$

The membership functions for the output parameter *IDSD* are defined as:

$$\begin{aligned}
\mu_{VLSP}(IDSD) &= g(IDSD; VLSP_0, VLSP_1, VLSP_{w0}, VLSP_{w1}) \\
\mu_{LSP}(IDSD) &= f(IDSD; LSP_0, LSP_{w0}, LSP_{w1}) \\
\mu_{MSP}(IDSD) &= f(IDSD; MSP_0, MSP_{w0}, MSP_{w1}) \\
\mu_{HSP}(IDSD) &= f(IDSD; HSP_0, HSP_{w0}, HSP_{w1}) \\
\mu_{VHSP}(IDSD) &= g(IDSD; VHSP_0, VHSP_1, VHSP_{w0}, VHSP_{w1}).
\end{aligned}$$

The membership functions are shown in Fig. 5 and the Fuzzy Rule Base (FRB) for our system are shown in Table 2. The FRB forms a fuzzy set of dimensions $|T(IDCD)| \times |T(IDST)| \times |T(IDRE)|$, where $|T(x)|$ is the number of terms on $T(x)$. We have three input paramters, so our system has 27 rules. The control rules have the form: IF "conditions" THEN "control action".

5 Simulation Results

We present the simulation results in Fig. 6. In this figure, we show the relation between the probability of an IoT device to be selected (IDSD) to carry out a task, versus IDCD, IDST and IDRE. We consider IDCD as a constant parameter and change the values of IDST and IDRE. We see that IoT devices with more remaining energy, have a higher possibility to be selected for carrying out a job. By increasing the IDST value, the IDSD is also increased because devices with more storage capacity are more likely to carry the message until there is a contact opportunity.

In Fig. 6(a), when IDRE is 0.1 and IDST is 0.7, the IDSD is 0.18. For IDRE 0.5, the IDSD is 0.3 and for IDRE 0.9, IDSD is 0.72, thus the IDSD is increased about 23% and 53%, for IDRE 0.5 and IDRE 0.9, respectively.

In Fig. 6(b) and (c), we increase the IDCD value to 0.5 and 0.9, respectively. The contact duration between two nodes depends on the speed of device. The speed makes the device move faster, making them break contact more often but increases the possibility of having a contact with another node. We see that

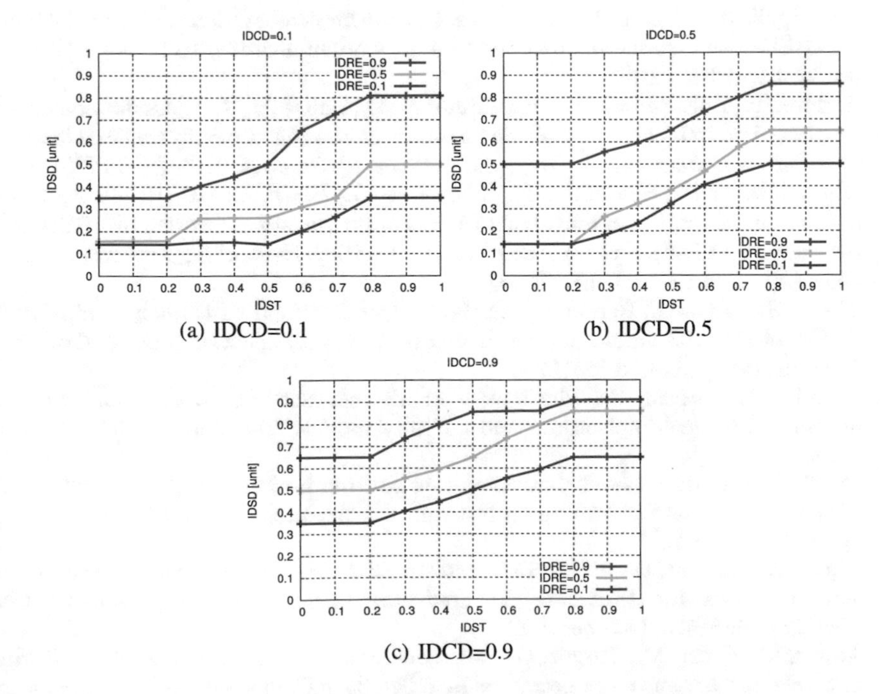

(a) IDCD=0.1

(b) IDCD=0.5

(c) IDCD=0.9

Fig. 6. Results for different values of $IDCD$.

with the increase of the IDCD parameter, the possibility of an IoT device to be selected is increased, since contacts last longer. For IDST = 0.7 and IDRE = 0.9, comparing Fig. 6(b) with (a) and Fig. 6(c) with (a), the IDSD is increased 13% and 19%, respectively.

6 Conclusions and Future Work

In this paper, we proposed and implemented a fuzzy-based IoT device selection system for OppNets, which is used to select an IoT device for a required task.

We evaluated the proposed system by computer simulations. The simulation results show that the devices with high contact opportunity, are more likely to be selected for carrying out a job, so with the increase of IDCD the possibility of an IoT device to be selected increases. We can see that by increasing IDST and IDRE, the IDSD is also increased.

In the future work, we will also consider other parameters for IoT device selection such as Message Time-out Ratio, Node Computational Time and make extensive simulations to evaluate the proposed system.

References

1. Sharma, D.K., Sharma, A., Kumar, J., et al.: KNNR: K-nearest neighbour classification based routing protocol for opportunistic networks. In: 2017 Tenth International Conference on Contemporary Computing (IC3), pp. 1–6. IEEE (2017)
2. Kraijak, S., Tuwanut, P.: A survey on internet of things architecture, protocols, possible applications, security, privacy, real-world implementation and future trends. In: IEEE 16th International Conference on Communication Technology (ICCT), pp. 26–31. IEEE (2015)
3. Arridha, R., Sukaridhoto, S., Pramadihanto, D., Funabiki, N.: Classification extension based on IoT-big data analytic for smart environment monitoring and analytic in real-time system. Int. J. Space-Based Situated Comput. **7**(2), 82–93 (2017)
4. Dhurandher, S.K., Sharma, D.K., Woungang, I., Bhati, S.: HBPR: history based prediction for routing in infrastructure-less opportunistic networks. In: IEEE 27th International Conference on Advanced Information Networking and Applications (AINA), pp. 931–936. IEEE (2013)
5. Spaho, E., Mino, G., Barolli, L., Xhafa, F.: Goodput and PDR analysis of AODV, OLSR and DYMO protocols for vehicular networks using cavenet. Int. J. Grid Util. Comput. **2**(2), 130–138 (2011)
6. Abdulla, M., Simon, R.: The impact of intercontact time within opportunistic networks: protocol implications and mobility models. TechRepublic White Paper (2009)
7. Le, T., Gerla, M.: Contact duration-aware routing in delay tolerant networks. In: 2017 International Conference on Networking, Architecture, and Storage (NAS), pp. 1–8. IEEE (2017)
8. Akbas, M., Turgut, D.: APAWSAN: actor positioning for aerial wireless sensor and actor networks. In: IEEE 36th Conference on Local Computer Networks (LCN-2011), pp. 563–570, October 2011
9. Akbas, M., Brust, M., Turgut, D.: Local positioning for environmental monitoring in wireless sensor and actor networks. In: IEEE 35th Conference on Local Computer Networks (LCN-2010), pp. 806–813, October 2010

10. Melodia, T., Pompili, D., Gungor, V., Akyildiz, I.: Communication and coordination in wireless sensor and actor networks. IEEE Trans. Mob. Comput. **6**(10), 1126–1129 (2007)
11. Inaba, T., Sakamoto, S., Kolici, V., Mino, G., Barolli, L.: A CAC scheme based on fuzzy logic for cellular networks considering security and priority parameters. In: The 9-th International Conference on Broadband and Wireless Computing, Communication and Applications (BWCCA-2014), pp. 340–346 (2014)
12. Spaho, E., Sakamoto, S., Barolli, L., Xhafa, F., Barolli, V., Iwashige, J.: A fuzzy-based system for peer reliability in JXTA-overlay P2P considering number of interactions. In: The 16th International Conference on Network-Based Information Systems (NBiS-2013), pp. 156–161 (2013)
13. Matsuo, K., Elmazi, D., Liu, Y., Sakamoto, S., Mino, G., Barolli, L.: FACS-MP: a fuzzy admission control system with many priorities for wireless cellular networks and its performance evaluation. J. High Speed Netw. **21**(1), 1–14 (2015)
14. Grabisch, M.: The application of fuzzy integrals in multicriteria decision making. Eur. J. Oper. Res. **89**(3), 445–456 (1996)
15. Inaba, T., Elmazi, D., Liu, Y., Sakamoto, S., Barolli, L., Uchida, K.: Integrating wireless cellular and ad-hoc networks using fuzzy logic considering node mobility and security. In: The 29th IEEE International Conference on Advanced Information Networking and Applications Workshops (WAINA-2015), pp. 54–60 (2015)
16. Kulla, E., Mino, G., Sakamoto, S., Ikeda, M., Caballé, S., Barolli, L.: FBMIS: a fuzzy-based multi-interface system for cellular and ad hoc networks. In: International Conference on Advanced Information Networking and Applications (AINA-2014), pp. 180–185 (2014)
17. Elmazi, D., Kulla, E., Oda, T., Spaho, E., Sakamoto, S., Barolli, L.: A comparison study of two fuzzy-based systems for selection of actor node in wireless sensor actor networks. J. Amb. Intell. Hum. Comput. **6**(5), 635–645 (2015)
18. Zadeh, L.: Fuzzy logic, neural networks, and soft computing. ACM Commun. **37**(3), 77–84 (1994)
19. Spaho, E., Sakamoto, S., Barolli, L., Xhafa, F., Ikeda, M.: Trustworthiness in P2P: performance behaviour of two fuzzy-based systems for JXTA-overlay platform. Soft. Comput. **18**(9), 1783–1793 (2014)
20. Inaba, T., Sakamoto, S., Kulla, E., Caballe, S., Ikeda, M., Barolli, L.: An integrated system for wireless cellular and ad-hoc networks using fuzzy logic. In: International Conference on Intelligent Networking and Collaborative Systems (INCoS-2014), pp. 157–162 (2014)
21. Matsuo, K., Elmazi, D., Liu, Y., Sakamoto, S., Barolli, L.: A multi-modal simulation system for wireless sensor networks: a comparison study considering stationary and mobile sink and event. J. Amb. Intell. Hum. Comput. **6**(4), 519–529 (2015)
22. Kolici, V., Inaba, T., Lala, A., Mino, G., Sakamoto, S., Barolli, L.: A fuzzy-based CAC scheme for cellular networks considering security. In: International Conference on Network-Based Information Systems (NBiS-2014), pp. 368–373 (2014)
23. Liu, Y., Sakamoto, S., Matsuo, K., Ikeda, M., Barolli, L., Xhafa, F.: A comparison study for two fuzzy-based systems: improving reliability and security of JXTA-overlay P2P platform. Soft. Comput. **20**(7), 2677–2687 (2015)
24. Matsuo, K., Elmazi, D., Liu, Y., Sakamoto, S., Mino, G., Barolli, L.: FACS-MP: a fuzzy admission control system with many priorities for wireless cellular networks and its perforemance evaluation. J. High Speed Netw. **21**(1), 1–14 (2015)
25. Mendel, J.M.: Fuzzy logic systems for engineering: a tutorial. Proc. IEEE **83**(3), 345–377 (1995)

Efficient Routing in Geographic and Opportunistic Routing for Underwater WSNs

Ghazanfar Latif[1], Nadeem Javaid[1(✉)], Aasma Khan[1], Aisha Fatima[1], Landing Jatta[1], and Wahab Khan[2]

[1] COMSATS Institute of Information Technology, Islamabad 44000, Pakistan
nadeemjavaidqau@gmail.com
[2] Beijing Institute of Technology, Beijing 100000, China
http://www.njavaid.com

Abstract. Underwater wireless sensor networks (UWSNs) are capable of providing facilities for the wide range of aquatic applications. However, due to the adverse environment, UWSNs face huge challenges and issues i.e., limited bandwidth, node mobility, higher propagation delay, high manufacturer and deployment costs etc. In this paper, we propose two techniques: the geographic and opportunistic routing via transmission range (T-GEDAR) and the geographic and opportunistic routing via the backward transmission (B-GEDAR). Firstly, in the absence of forwarder node, we increase the transmission range to determine the forwarder node. Because of this, we can send packets to the sink; Secondly, when the forwarder node is unavailable in adjustable transmission range. Then, the B-GEDAR is used for determining the forwarder node so that the packet delivery ratio (PDR) can be increased effectively. This is because, our simulation results perform better network performance in terms of an energy efficiency, PDR, and the fraction of void nodes.

1 Introduction

Underwater wireless sensor networks (UWSNs) are gaining huge interest due to the demanding oceanic applications of natural disaster prevention, military surveillance, aquatic environmental monitoring and resource investigations etc. UWSNs consist of various sensors are deployed in depth of water. However, sonobuoys are deployed at the water surface to perform the collaborative tasks [1]. The UWSNs use in the acoustic channel, which is five orders of magnitude less than the radio channel. The radio channel cannot perform well in the underwater network. This is because, the high propagation delay, limited bandwidth, and interference are in the channel. However, the acoustic channel has some problems to be solved i.e., long propagation delay and limited bandwidth as compared to radio waves. The speed of the acoustic signal is 1500 m/s which is five orders of magnitude less than the radio signal and their speed is 3 x 10^8 m/s. In UWSNs, the bit error rate (BER) occurs due to the multi-path fading, path loss and low bandwidth [2].

L. Barolli et al. (Eds.): IMIS 2018, AISC 773, pp. 86–95, 2019.
https://doi.org/10.1007/978-3-319-93554-6_7

It is described in the literature that balanced load distribution (BLOAD) [2], an adaptive hop-by-hop vector based forwarding (AHH-VBF) [3], and balanced routing (BR) [14], that provide the efficient and reliable communications. The BLOAD provides the load distribution among different coronas. In BLOAD, the energy hole problem can be reduced and maximized the lifespan of the network. Moreover, the AHH-VBF ensures the reliability and reducing the void nodes via adaptively adjusting the pipeline radius of the virtual pipeline. While BR provides the balanced routing protocol in which the load distribution is balanced among different coronas. However, It is also to increase the lifespan of network. The energy hole problem occurs near the sink due to imbalance load distribution.

The limitation of geographic and opportunistic routing protocol is communication void region. The communication void regions are those regions where the forwarder nodes do not occur in the region. The nodes are located in the void region called the void node. Sometimes the number of the data packet does not transmit due to the void hole problem. The proposed routing protocol transmits the data via the alternative path. Furthermore, the sensor nodes are activated by batteries. After deployment of sensor nodes, it is difficult to recharge because of an adverse environment, especially in the depth of water. So, the energy efficiency is one of the important issues in the deployment of UWSNs.

Taking motivation from the above considerations, we have proposed an efficient communication-based routing protocol over sensor nodes geographic and opportunistic routing via transmission range (T-GEDAR) and geographic and opportunistic routing via the backward transmission (B-GEDAR) for UWSNs. In this paper, our proposed routing protocols prevent void nodes during the data forwarding from source to the sink on the surface of the water. We also to increase the PDR with the increasing number of sensor nodes. In this way, the network lifetime can be increased and energy consumption can be minimized.

Contributions: In this paper, we have proposed two routing protocols named T-GEDAR and B-GEDAR, respectively for UWSNs. The T-GEDAR defines the maximum communication range upon failing the depth adjustment topology. In the topology of T-GEDAR, if the forwarder node is unavailable in the communication range, then the communication range is maximized up to the certain limit and data will be transmitted to sink via adjustable transmission range. However, because of this scheme, the data delivery ratio can be increased but the chance of duplicate packets are generated by the collision. Moreover, it is also to increase the network lifetime by the presence of forwarder node in communication range.

Furthermore, we have determined the B-GEDAR scheme for hole avoidance among sensor nodes during the communication of network nodes. The B-GEDAR is performed upon failing the T-GEDAR routing scheme. Because of this, the void nodes decrease and network lifetime is also increased. However, B-GEDAR can determine the T-GEDAR of forwarder node for data transmission among network nodes. By this process, the nodes dissipate energy, so network lifetime, PDR and other parameters can be increased.

The rest of the paper is organized as follows. Related work is discussed in Sect. 2. The problem statement of our system is given in Sect. 3. The system model is explained in Sect. 4. The simulation and results are presented in Sect. 5. The conclusions are in Sect. 6 and finally, references are listed for related work.

2 Related Work

Recently, researchers have interested in terrestrial wireless sensor networks (WSNs) due to the distinctive characteristics of UWSNs. In this section, we explain some existing literature in this domain.

In [2], authors proposed the routing protocol called the BLOAD. In this paper, there are three types of data fractions i.e., small, medium and large. The advantages of this protocol are avoidance of energy hole problem due to unbalance energy consumption. This work achieves the higher network lifetime and stability period. The limitation of this protocol, the energy consumption is high due to the direct transmission at long distance.

Haitao *et al.* proposed the routing protocol called AHH-VBF [3]. In this work, authors investigate the void node problem and increase the data delivery ratio. This article adaptively adjusts the communication range by maximizing the pipeline radius. The simulation results show that the propagation delay and energy consumption can be reduced effectively. This paper achieves to improve the network lifetime. However, their proposed scheme produces the duplicate packets and high manufacturing costs.

Jarnet *et al.* proposed the routing protocol called focused beam routing (FBR) in UWSNs [4]. This protocol is to minimize the extra flooding. In this strategy, before transmitting the packets, the nodes increase the transmitting range time-by-time according to adjust the flooding angle and the communication power level via power gradient. Moreover, the nodes need to determine the request to send (RTS) message in the sparse network. Due to this scheme, the wastage of energy consumption and propagation delay are high. However, the flooding angle is affected by the network performance.

In [5], authors introduced the routing protocol called vector-based void avoidance (VBVA). This paper investigates the void hole problem in a mobile-based scenario of UWSNs. There are two methods of this protocol i.e., vector shift and back pressure which are to resolve the void nodes. The vector shift procedure is used for sending data along the routing hole boundary. Moreover, the back pressure technique describes the packet forwarding back in routing path and packet move away from the destination. However, this work achieves to minimize the routing hole and high PDR.

Youngtae *et al.* proposed the void-aware pressure routing (VAPR) for UWSNs [6]. This paper knows about the depth knowledge of nodes to forward the packets toward the sink on the sea surface. This protocol just like the geographic and opportunistic routing where the next hop forwarder determines through greedy forwarding approach. In this protocol, every node prevents the void node from sink's reachability data disseminated in the network from periodic beaconing. Every node uses this information to make a directional path towards some surface sonobuoy. The next-hop forwarder node is appointed

Table 1. State of-the-art-work

Technique (s)	Feature (s)	Achievement (s)	Limitation (s)
BLOAD [2]	Balanced load distribution, energy hole avoidance	Higher network lifetime, stability period is high	Higher energy consumption due to long distance
AHH-VBF [3]	Reduces propagation delay, higher network lifetime	Reliability, energy efficiency	Duplicate packets, high manufacturing and deployment cost
FBR [4]	Focused beam routing, minimize the extra flooding	Due to flooding angle the packet delivery ratio is high	Wastage of energy consumption and propagation delay is high
VBVA [5]	Vector-based void avoidance, vector shift and back pressure	Void node decreases, better network performance	Higher propagation delay
VAPR [6]	Void-aware pressure routing, greedy forwarding strategy	Same as geographic and opportunistic routing, high packet delivery ratio	Higher energy consumption, greater propagation delay
GEAR [7]	Transmission based upon cluster head and multi-hop strategy	Higher network lifetime, energy efficiency	Larger end-to-end delay
BECHA [8]	Load distribution is balanced, greater network lifetime	Network lifetime, energy efficiency	Energy hole problem due to load imbalance
EEDBR [9]	Energy efficient depth based routing, forwarder is selected on the basis of residual energy and depth	Network lifetime	Packet is neglected due to low energy
WDFAD-DBR [10]	Forwarder nodes select up to two hop neighbor, avoidance of void nodes	Higher reliability, higher PDR and lower propagation delay	High manufacturer and deployment cost
EEBET [11]	Enhanced efficient and balancing energy technique, solves the deficiencies in balanced transmission mechanism	Energy efficiency, higher network lifetime	Energy hole problem
RDBF [12]	Relative distance based forwarding protocol, fitness factor for appropriate forwarder	Data delivery ratio, low propagation delay and energy efficiency	Load is an imbalance

through neighbor node's direction, which is those paths in which there is the same transmitting path with the current forwarder i.e., upward or downward. However, VAPR achieves the high PDR at the cost of high energy consumption and propagation delay as data packets are routed through more hops for the avoidance of void nodes in the networks (Table 1).

In [7], authors proposed the routing protocol called gateway based energy-efficient routing protocol (GEAR). In this paper, there are four types of regions. Each region performs different approaches. Firstly, two regions use direct transmission, while other two regions are divided on the basis of cluster head (CH) and perform multi-hop transmissions. This protocol achieves to improve the lifespan of network and minimize the energy consumption.

Naeem *et al.* proposed the routing protocol called balanced energy consumption and hole alleviation (BECHA) balances the load distribution among different coronas [8]. The energy balancing is a precious resource for maximizing the lifespan of network. Due to imbalance load distribution, the death of sensor nodes very quickly and it causes the energy hole problem. This scheme resolves, the energy hole problem which is located near the sink due to imbalance load distribution. This strategy is an important to improve the throughput. It is also to balance the load distribution and provide the energy efficiency.

In [9], authors proposed the routing protocol called energy efficient depth based routing (EEDBR) for UWSNs. In this paper, the forwarder node is selected on the basis of depth and residual energy. Because of this selection, the energy can be balanced and improved network lifetime. In this work, the sensor nodes retain the data for some time before transmitting. The holding time depends on the residual energy of sensor nodes. In this way, if the residual energy is high then the data is directly transmitted to sink, otherwise, the packet is discarded. The limitation of this paper, the packet is neglected due to the low energy. This protocol does not need to the localization of the network nodes.

Haitao *et al.* proposed a weighting depth and forwarding area division depth based routing protocol (WDFAD-DBR) [10]. The forwarder node is selected on the basis of two-hop neighbors. This protocol achieves to high reliability, higher PDR and lower propagation delay by the increasing number of nodes in the network. However, their proposed scheme produces the high manufacturing cost.

In [11], authors proposed the routing protocol called enhanced efficient and balancing energy technique (EEBET) to solve the deficiencies in balanced transmission mechanism (BTM). Similarly, the efficient and balancing energy consumption technique (EBET) achieves to improve the energy efficiency and determine the suitable energy level. By this process, the network lifetime can be increased. The limitation of this work is energy hole problem.

In [12], proposed a routing protocol called relative distance based forwarding protocol (RDBF). This paper determines an appropriate forwarder node which acts as a fitness function. It means that the selection of forwarder node via fitness function. This paper achieves the better PDR, an end-to-end delay, and energy efficiency. However, due to the minimum number of hop counts, the load is an imbalance.

Latif *et al.* proposed the routing protocol called the energy hole and coverage hole in terms of network lifetime and throughput. By this protocol, the network lifetime and throughput can be maximized. The energy hole and coverage hole problems resolve in depth based routing and the residual energy of each node act as a forwarding metrics for the data packet. A node selects as a forwarder node for the data packet if it has a smaller holding time than other neighbor nodes in order to suppress the retransmission of duplicate packets. Moreover, a hole repair technique is used to maintain the connectivity among sensor nodes in order to maximize the network lifetime. Moreover, an adaptive transmission power level is introduced to improve the energy efficiency of the network. However, the limitation of this protocol is higher propagation delay of the network [13].

3 Problem Statement

The communication void region and energy consumption degrade the performance of the network. The data transmission failing upon the absence of forwarder node. It means that the void node is available in the void region creates the problem. In geographic and opportunistic for depth adjustment routing protocol (GEDAR) [1], it avoids the void node region via depth adjustment topology. In this paper, we are mitigating the void node region problem and it is also to increase the PDR. The node located in the void region is known as the void node. Moreover, due to the multi-hop scenario, the energy hole problem occurs closer to the sink located on the water surface. In the presence of void node, the packet gets stuck in a void region. The existing routing protocols should determine to transmit the packet via some proposed techniques.

4 Proposed Scheme

In our proposed scheme, the sensor nodes are randomly deployed and the sinks are located on the water surface. The details of network model and proposed schemes are given below:

4.1 Network Model

In this section, the sensor nodes are randomly deployed in an area the size of 1500 m $\times$ 1500 m $\times$ 1500 m and the sonobuoys are 45. In our proposed system, we consider a three-dimensional model, where the sensor nodes are randomly deployed.

4.2 T-GEDAR

We determine the maximum communication range upon failing the depth adjustment technique. In this scheme, if the forwarder node is unavailable in the communication range of sensor nodes for data transmission. Then, the communication range is maximized up to the certain distance. However, the data is transmitted to sink by adjustable range. Moreover, in this scheme the data delivery ratio is high and the probability of duplicate packets generated due to sending the multiple copies of the same data packet from the source node. This is because it also to increase the network lifetime.

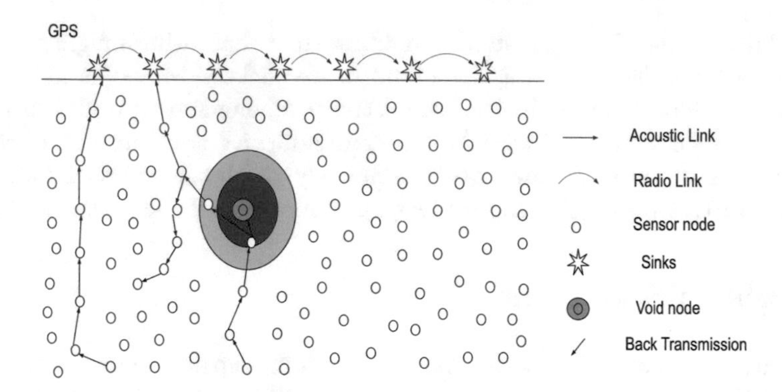

Fig. 1. Network model of proposed scheme

Table 2. Simulation parameters

Control parameters	Values	Units
Area (A)	1500 m × 1500 m × 1500 m	Meters
Transmission power (T_P)	2	Watts
Reception power (R_P)	0.1	Watts
Idle power (I_P)	0.01	Watts
Number of sensors (n)	150	–

4.3 B-GEDAR

We propose the B-GEDAR scheme for the hole avoidance among sensor nodes during the communication of network nodes. The B-GEDAR is performed upon failing the T-GEDAR routing protocol. This is because of the number of void nodes decreases and network lifetime increases. However, in B-GEDAR determines the communication range of forwarder node for data transmission among nodes are shown in Fig. 1 . By this process, the nodes dissipate energy, so network lifetime, PDR and other parameters can be increased.

5 Simulation Results

In this section, we evaluate the performance of our scheme and compare the results with existing protocol (GEDAR) in terms of PDR and energy efficiency of the network. The sensor nodes are randomly deployed. The total network area is 1500 m × 1500 m × 1500 m. We assume that the 45 sinks are located at the surface of the water. The transmission power of sensor node is set to be $T_P = 2W$, the reception power is set to be $R_P = 0.1W$ and idle power is set to be $I_P = 0.01W$. Let us assume that, the sensor nodes are equal to $n = 150$. The control parameters of this paper are given in Table 2.

Figure 2 depicts the fraction of void nodes decreases when the network density increases for both techniques. Similarly, T-GEDAR and B-GEDAR achieve

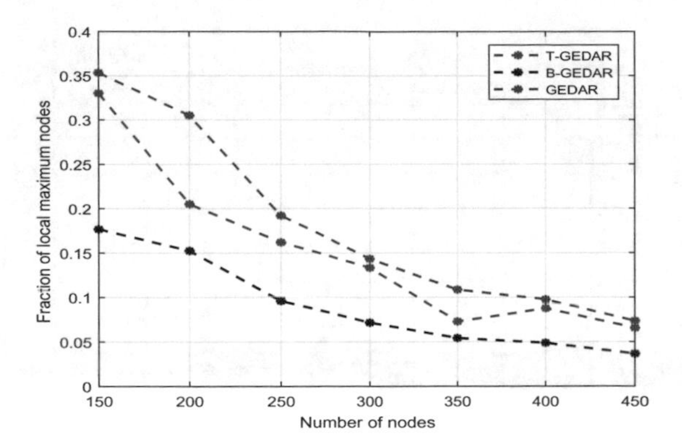

Fig. 2. Fraction of void nodes

the best performance of results are compared with the GEDAR. When GEDAR is used, the proposed schemes like T-GEDAR and B-GEDAR are to minimize the 33% and 50% fraction of void nodes respectively. However, the overall performance comparison is better for B-GEDAR, because via this scheme achieves approximately 34% is compared with the increase of communication range.

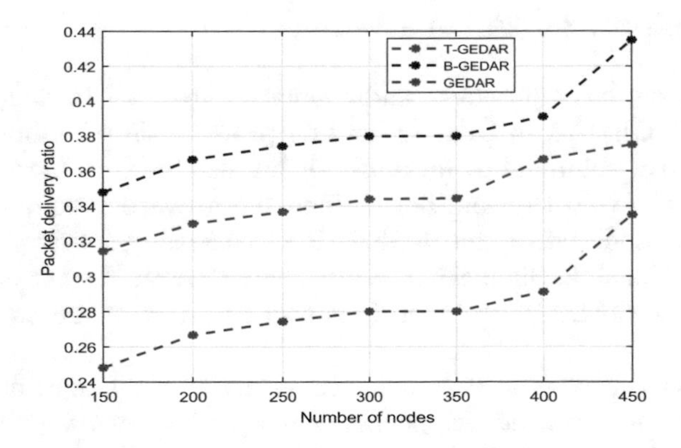

Fig. 3. Packet delivery ratio

Figure 3 depicts the results of PDR. The overall result is to increase in PDR when the number of nodes increases. GEDAR has the high data delivery ratio and better performance due to increase the transmission range of communication and the B-GEDAR. This is because; the T-GEDAR and B-GEDAR are to achieve the 18.6% and 29.1% data delivery ratio among nodes, respectively. However, the overall achievement is 36% through the B-GEDAR.

Figure 4 depicts the results of energy consumption per received message per node. In GEDAR, the energy consumption is high due to the lower node density scenario. However, because of T-GEDAR and B-GEDAR the node density

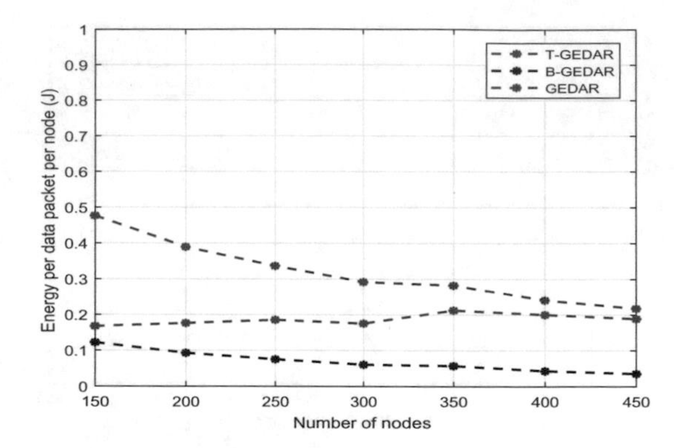

Fig. 4. Energy consumption per message per node

increases then the energy consumption is relatively decreased. Moreover, the proposed schemes are to minimize the 66% and 78% of energy consumption per packet per node respectively. However, the overall performance comparison is better for B-GEDAR because this scheme achieves approximately 15.6% and it is compared to increase the communication range.

6 Conclusion and Future Work

In this paper, we have proposed and evaluated the two techniques called T-GEDAR and B-GEDAR in GEDAR routing protocol. Furthermore, T-GEDAR provides a greater communication range for finding the suitable forwarder node and it is used to avoid the void hole. When the forwarder node is unavailable, then the sensor nodes start the B-GEDAR so that the packets can be received successfully at the sink. Simulation results show that the better performance as compared to the GEDAR Protocol. It is also to improve the energy efficiency and PDR.

As future work, we plan to address these major challenges in the UWSNs, considering the various needs for packet delivery. Moreover, we plan to improve the lifespan of the network, so that we balance the energy consumption among various nodes.

References

1. Coutinho, R.W.L., Azzedine, B., Luiz, F.M.V., Antonio, A.F.L.: Geographic and opportunistic routing for underwater sensor networks. IEEE Trans. Comput. **65**(2), 548–561 (2016)
2. Azam, I., Javaid, N., Ahmad, A., Wadood, A., Almogren, A., Alamri, A.: Balanced load distribution with energy hole avoidance in underwater WSNs. IEEE Access (2017)
3. Yu, H., Yao, N., Liu, J.: An adaptive routing protocol in underwater sparse acoustic sensor networks. Ad Hoc Netw. **34**, 121–143 (2015)

4. Jornet, J.M., Stojanovic, M., Zorzi, M.: Focused beam routing protocol for underwater acoustic networks. In: Proceedings of the Third ACM International Workshop on Underwater Networks, pp. 75–82. ACM (2008)
5. Xie, P., Zhou, Z., Peng, Z., Cui, J.-H., Shi, Z.: Void avoidance in three-dimensional mobile underwater sensor networks. In: International Conference on Wireless Algorithms, Systems, and Applications, pp. 305–314. Springer, Berlin (2009)
6. Noh, Y., Lee, U., Wang, P., Choi, B.S.C., Gerla, M.: VAPR: void-aware pressure routing for underwater sensor networks. IEEE Trans. Mob. Comput. **12**(5), 895–908 (2013)
7. Nadeem, Q., Rasheed, M.B., Javaid, N., Ali Khan, Z., Maqsood, Y., Din, Q.: M-GEAR: gateway-based energy-aware multi-hop routing protocol for WSNs. In: Proceedings of 2013 Eigth International Conference on Broadband and Wireless Computing, Communication and Applications (BWCCA), pp. 164–169. IEEE (2013)
8. Jan, N., Javaid, N., Javaid, Q., Alrajeh, N., Alam, M., Ali Khan, Z., Niaz, I.A.: A balanced energy-consuming and hole-alleviating algorithm for wireless sensor networks. IEEE. Access **5**, 6134–6150 (2017)
9. Wahid, A., Kim, D.: An energy efficient localization-free routing protocol for underwater wireless sensor networks. Int. J. Distrib. Sens. Netw. **8**(4), 307246 (2012)
10. Yu, H., Yao, N., Wang, T., Li, G., Gao, Z., Tan, G.: WDFAD-DBR: weighting depth and forwarding area division DBR routing protocol for UASNs. Ad Hoc Netw. **37**, 256–282 (2016)
11. Javaid, N., Shah, M., Ashfaq, A., Imran, M., Khan, M.I., Vasilakos, A.V.: An enhanced energy balanced data transmission protocol for underwater acoustic sensor networks. Sensors **16**(4), 487 (2016)
12. Li, Z., Yao, N., Gao, Q.: Relative distance based forwarding protocol for underwater wireless networks. Int. J. Distrib. Sens. Netw. **10**(2), 173089 (2014)
13. Latif, K., Javaid, N., Ahmad, A., Khan, Z.A., Alrajeh, N., Khan, M.I.: On energy hole and coverage hole avoidance in underwater wireless sensor networks. IEEE Sens. J. **16**(11), 4431–4442 (2016)
14. Zidi, C., Bouabdallah, F., Boutaba, R.: Routing design avoiding energy holes in underwater acoustic sensor networks. Wirel. Commun. Mob. Comput. **16**(14), 2035–2051 (2016)

Supporting Online/Offline Collaborative Work with WebRTC Application Migration

Fatos Xhafa[1(✉)], David Zaragoza[1], and Santi Caballé[2]

[1] Universitat Politècnica de Catalunya, Barcelona, Spain
fatos@cs.upc.edu
[2] Open University of Catalonia, Barcelona, Spain
scaballe@uoc.edu

Abstract. With the fast development of mobile computing and increasing computing capacities of mobile devices, new collaborative applications and platforms are appearing to support collaboration on the move. Indeed, nowadays, members of a team can be not only geographically distributed but they can also work anytime and anywhere thanks to the use of mobile devices. Often, however, team members would like to work either online or offline on a common project; likewise, they may wish to switch among various devices such as laptops, tablets and mobile phones and still work in the same application environment, sharing the same data, etc. In this paper we present a platform that enables application and services migration at runtime between different platforms using the WebRTC (Web Real-Time Communication) framework. We have studied applications migration both through a central server and through a distributed (Peer-to-Peer) model. Various issues that arise in application migration such as profile matching, application context, data synchronisation and consistency are discussed. The efficiency and scalability of the WebRTC framework and mobile devices (peers) under Android in a real computing infrastructure are studied. Some experimental results on the application migration time according to application state data size are reported.

1 Introduction

Application migration consists in moving an application from one computing device or environment to another one. Application migration can be useful in various contexts. For instance, in Cloud environments, through migration it is possible to move applications from one Cloud environment (e.g. a VM) to another one; or, in a broader context, to move applications across Clouds [13, 15]. It is also useful for legacy application migration in enterprises, businesses, institutions and alike. Other interesting uses of application migration appear in collaborative work in premises of enterprise environments. Indeed, in this case, different users (employees) can share and use the same application and its *state* in different terminals [16].

In this work we are interested on the usefulness of application migration for supporting collaborative team work *on the move* [18, 19]. Collaboratively working on the move means that team members can work not only geographically distributed but also while changing their locations. In the later case, team members can work online (connected

L. Barolli et al. (Eds.): IMIS 2018, AISC 773, pp. 96–104, 2019.
https://doi.org/10.1007/978-3-319-93554-6_8

and synchronised to the rest of the team through groupware system) or offline (disconnected from the groupware system). Using various computing devices (laptops, tablets and smartphones) on the move has become a common practice, which requires, therefore, that users can switch from one computing device to another one. In this context, we are interested to support application migration among various computing devices in *real-time*.

Application migration, in any of its forms, requires addressing several issues due to differences between the original and target computing devices or environments. Generally speaking, applications are developed for concrete computing devices/platforms and are thus bound to operating system, networking, data storage, configurations, etc., of the original computing environment. In our problem setting, additionally, we are interested to migrate not only the application itself but also its *current state* as well as the *data* associated with the application to enable team members to resume and follow up the collaborative work from within another computing devices and complete the migration process in real-time. For instance, we would like to migrate a project management system, which uses a calendar of events or milestones related to a project development and a contact team members lists. To achieve the application migration in real-time, we specify and develop a platform for application migration using WebRTC (Web Real Time Communication) technology [1], which is a recent technology to support real-time communications for a variety of applications such as video and audio streaming, teleconferencing, application window sharing, etc. The advantages of using WebRTC are, on the one hand, that it adds real-time communication to the browser and thus solves the problem that native applications cannot be used across different platforms, and on the other, it has a focus on Peer-to-Peer communication. The later feature is particularly useful in the case of collaborative (*peer-group*) teamwork. Also, WebRTC can be used for data sharing by using computing resources on Web browsers, not server resources, through a Peer-to-Peer network [14].

We have therefore studied applications migration both through a central server and through a distributed (Peer-to-Peer) model. Various issues that arise in application migration such as profile matching, application context, data synchronisation and consistency are considered. The efficiency and scalability of the WebRTC framework and mobile devices (peers) under Android in a real computing infrastructure. Some experimental results on the application migration time according to application state data size are reported.

The rest of the paper is structured as follows. In Sect. 2 we describe main characteristics of WebRTC and the related work using this technology. In Sect. 3 we present the main requirements, design and implementation of the prototype. The experimental study is summarised in Sect. 4. Finally, in Sect. 5 we give conclusions and indications for future work.

2 WebRTC and Related Work

Web Real Time Communication (WebRTC) technology [1] is a recent technology to support real-time communications for a variety of applications such as video and audio streaming, tele-conferencing, application window sharing, etc. by embedding real-time communications capabilities into Web browsers. The advantages of using WebRTC are, on the one hand, that it adds real-time communication to the browser (without the need for browser plugins) and thus solves the problem that native applications cannot be used across different platforms; different browsers can directly exchange media and data between them, and on the other, it has a focus on Peer-to-Peer communication. It is also reported that WebRTC offers secure communications.

There has been several research works in the literature using WebRTC for a variety of applications, all of them having in common the need for a real-time communication and waiving the obstacle of native applications, which cannot be used across different platforms.

Jang-Jaccard *et al.* [11] presented a WebRTC-based video conferencing service for tele-health, which allows online meetings between remotely located care coordinators and patients at their homes. Their system is aimed to overcome some limitations of current solutions in tele-health domain such as high development and maintenance cost, use of proprietary incompatible technologies, etc. There is also recent interest in exploring WebRTC technology for IoT applications (see e.g., Sandholm *et al.* [21]; Janak and Schulzrinne [12]). Edan *et al.* [6] and Wang and Mei [24] presented the design and evaluation of video/multimedia conferencing systems based on WebRTC.

Garcia *et al.* [7] and Gouaillard and Roux [8] analyse some challenges and practical solutions through testing and validating WebRTC-based applications.

In particular, several works address the use of WebRTC in distant learning and virtual campuses (there are many previous works that were based on JXTA library and on Per-to-Peer paradigm, see e.g. [3, 9, 17, 25]). Bandung *et al.* [2] analyse video-conference quality from the implementation of WebRTC based system for supporting distance learning, actually their work showed that WebRTC is a reasonable solution to address the challenges of real-time video-conferencing on bandwidth limited network. WebRTC can also be used for casting screen images and voices from a host PC to client web browsers on many other PCs in real-time, which can be useful for students to attend class sessions and learning activities either on-premises or remotely. This actually falls within sharing features enabled by WebRTC, which could be data sharing, screen sharing, etc. [10]. Using WebRTC it is possible to operate multiple PCs using a single device and share any-application window between PCs.

Finally, WebRTC has been exploited for application migration purposes, which is also the objective of this paper. In such context, migrating the application state among various computing environments or devices is a key component [16, 20]. Schuchardt [23] presented a vision and methodology for moving mobile applications between Android mobile devices seamlessly. Chu *et al.* [5] presented Roam System, a framework for making applications migratable that can be executed on different platforms. Their programming platform is somehow specific by creating "roamlets", which are downloaded from a server and will be adapted to the device. Schmidt and Hauck [22] introduced SAMProc, a middleware for self-adaptive mobile processes in heterogeneous ubiquitous environments. Belluci *et al.* [4] address user interfaces able to migrate across various types of devices while preserving task continuity. They focus on the issue of preserving the JavaScript state while moving from one device to another one.

3 Application Migration: Requirements, Design and Implementation

3.1 Requirements

In order to achieve an application migration from one device to another one, it is necessary to migrate at runtime an application defined by its code, states and data (related information) of the application. From this basic definition we can derive the main requirements on a platform to support migratable applications and application migration process.

Among such requirements we could distinguish the following ones (corresponding use cases diagrams are omitted here):

[R1] The application should be identified by information of its proprietary. Likewise, users allowed to use the application should be identified as well. The application should be able to be used even offline.
[R2] The application should know the list of current available connected devices, where migration could take place.
[R3] The application state should be readable/interpretable, comparable and updateable.
[R4] Associated data with the application should be efficiently stored/saved.
[R5] Migration process should be fast so that users will not perceive any interruption effect while working.
[R6] Migration should be possible among devices running under different operating systems.

The application migrating process usually involves the following three phases:

1. **Suspension phase**: the application is paused and the application state is saved.
2. **Migration phase**: the application of the current device is transferred to the new device where we want to run it.
3. **Resuming phase**: the application is executed in the new device.

Upon successful completion of the above phases, the user can continue the work at the point where he had it left at the former device. A detailed diagram of executing a newly migrated application can be seen in Fig. 1. During the process, the user has the option to select the best fit (in terms of dimensions, battery level, etc.) among connected devices where to migrate the application.

3.2 Communication Models Among Devices: Centralized vs. Peer-to-Peer

The phase 2 above (see Subsect. 3.1), can be implemented either according to a centralized model or a distributed one. In the centralised model, a migration server is used where applications are first uploaded from devices and then are downloaded for use in other devices (Fig. 2).

The migration server is in charge of registering users, applications and devices. Applications are stored in a repository/directory and have associated their configuration parameters and their state. Furthermore, various devices can be connected to the

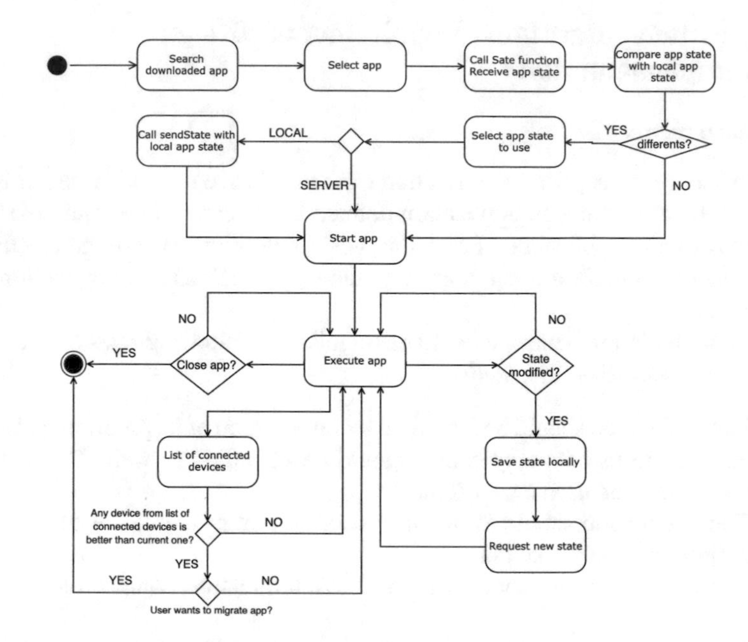

Fig. 1. Diagram of executing a migratable application.

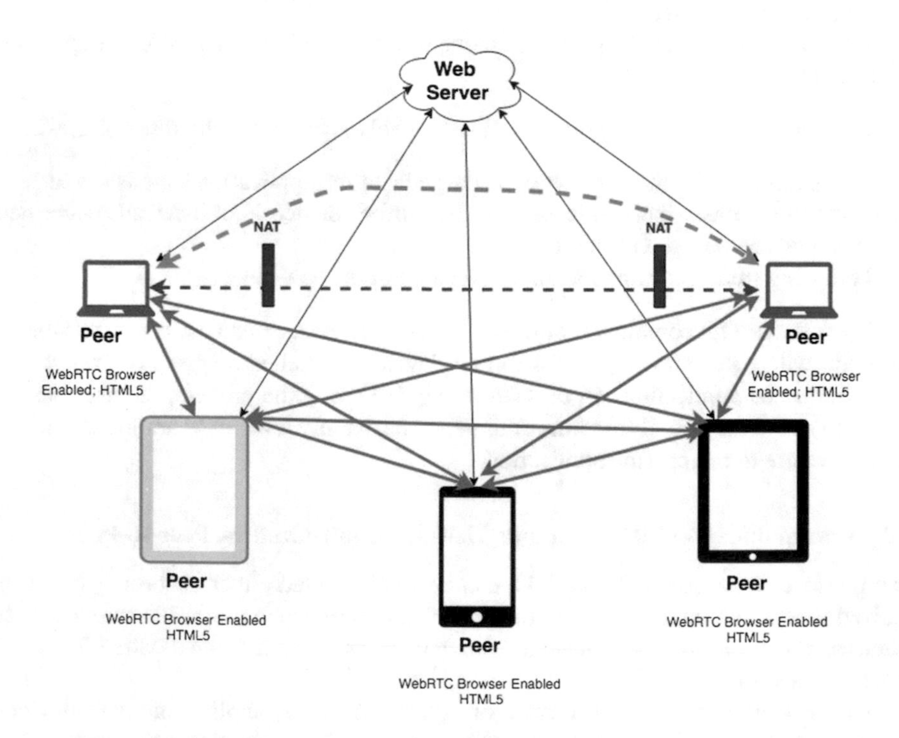

Fig. 2. Communication models among devices: centralized vs. peer-to-peer.

server (a registered user can have several devices registered on the server). The entity-relationship diagram can be see in Fig. 3.

On the other hand, the Peer-to-Peer application migration model is implemented via WebRTC to establish and enable Peer-to-Peer communication among peer Android devices (Crosswalk library and peerjs scripts are used.) As in the case of other Peer-to-Peer protocols, peer communications can be established also for peers in private networks/behind NAT devices.

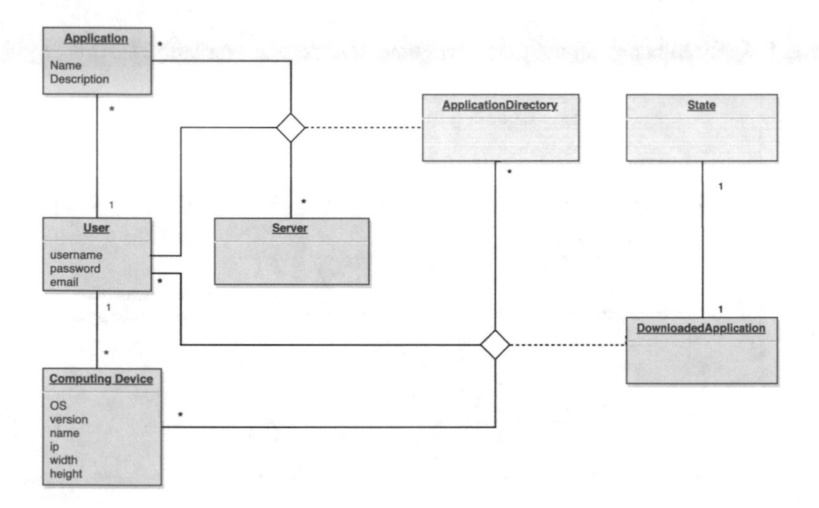

Fig. 3. Entity diagram.

4 Experimental Study

An aspect of interest while implementing platforms for application migration is the migration time among devices. It is therefore interesting to compare the migration time when a migration server is used *vs.* WebRTC-based migration. While measuring the migration time, application state size is taken as a relevant parameter. A simple contacts agenda application (adding, modifying, deleting contacts) was used for the purpose of this empirical study. Various data sizes for application states were generated to measure the migration time.

In Fig. 4 can be seen the migration time (in the left, migration time via a migration server, upload/download time are also given; in the right, WebRTC migration time, Peer-to-Peer connection time and data transmissions are given). As expected, data upload (for the server migration case) and data transfer (for the WebRTC case) were predominant as compared to download time and Peer-to-Peer connection time, respectively.

A comparison of total migration time for server migration and WebRTC Peer-to-Peer migration is shown in Fig. 5. As can be seen, for small size of application state (in this experiment, up to 100000 Bytes), migration is more efficient via a migration

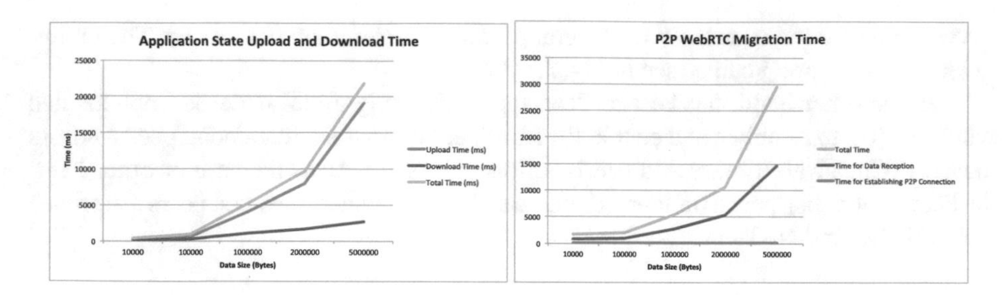

Fig. 4. Application migration time (server time, left; peer-to-peer WebRTC time, right).

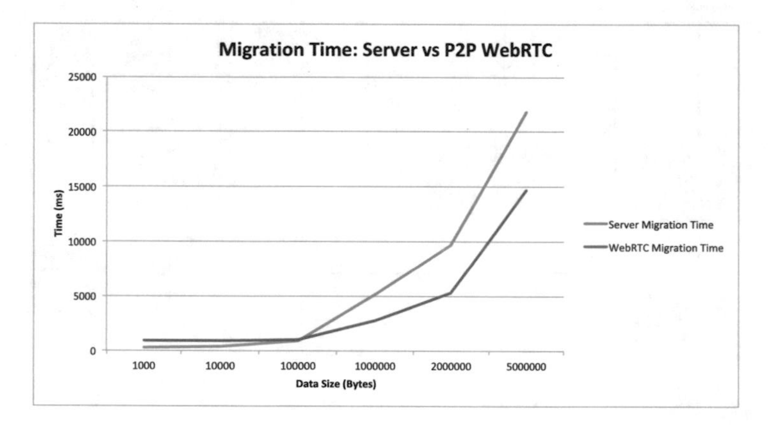

Fig. 5. Comparison of application migration time (server vs. peer-to-peer WebRTC).

server than WebRTC migration. The efficiency of the WebRTC is therefore seen more clearly for complex application of larger size states (larger than 100000 Bytes for this experiment).

5 Conclusions and Future Work

In this paper we have presented the requirements, design and implementation for supporting application migration among various computing devices. The aim is to provide members of a teamwork to be able to work online, while connected with other peer members and to change from one devices to another one without interrupting the collaborative work and seamlessly to other peer members. Likewise, team members can work offline and still change from one devices to another one without interrupting their own work. Such kind of application migration platforms are becoming very important at present with the increasing number of multi-devices (laptops, tablets and smartphones) being used by individuals. In particular, application migration supports working *on the move* either on-premise or off-premise environments, during travelling, etc.

For comparison purpose, we considered two architectures for the application migration platform: a web server application migration and a WebRTC Peer-to-Peer application migration (WebRTC was selected for its interesting features.) A simple empirical study was also conducted, from which we observed that for small application state size, server migration would be the option, while for larger and more complex applications of larger state size, the WebRTC would be the choice.

In our future work, we plan to fully evaluate the proposal in the Virtual Campus of Open University of Catalonia (UOC) for online learning teams of about 5 members working collaboratively. Besides aiming to get more conclusive results, we would like to evaluate the usefulness of the application migration platform to the profile of the students at UOC, where many of them have to switch from on-premise to off-premise collaborative activity during a working day, and quite often have to travel. Also, we would like to extend our platform to enable application migration from one member's device to devices of all members of the team, by taking advantage of WebRTC Peer-to-Peer features such as one-to-many connections. Finally, we are evaluating WebRTC as a solution for delivery of educational content in a virtual classroom of the Virtual Campus.

Acknowledgements. This research was supported by the European Commission through the project "colMOOC: Integrating Conversational Agents and Learning Analytics in MOOCs" (588438-EPP-1-2017-1-EL-EPPKA2-KA).

References

1. Barz, H.W., Gregory, A.: Protocols, Design and Applications Wiley Telecom eBook Chapters, Bassett. WebRTC. Multimedia Networks (2016)
2. Bandung, Y., Subekti, L.B., Tanjung, D.: Chrysostomos chrysostomou. QoS analysis for WebRTC videoconference on bandwidth-limited network. In: Proceedings of the 20th International Symposium on Wireless Personal Multimedia Communications (WPMC), pp. 547–553. IEEE CPS (2017)
3. Barolli, L., Xhafa, F.: JXTA-overlay: a P2P platform for distributed, collaborative, and ubiquitous computing. IEEE Trans. Indus. Electron. **58**(6), 2163–2172 (2011)
4. Bellucci, F., Ghiani, G., Paternò, F., Santoro, C.: Engineering javaScript state persistence of web applications migrating across multiple devices. In: Proceedings of the 3rd ACM SIGCHI Symposium on Engineering Interactive Computing Systems (EICS 2011), pp. 105–110. ACM, New York (2011). https://doi.org/10.1145/1996461.1996502
5. Chu, H., Song, H., Wong, C., Kurakake, S., Katagiri, M.: Roam, a seamless application framework. J. Syst. Softw. **69**(3), 209–226 (2004). https://doi.org/10.1016/S0164-1212(03)00052-9
6. Edan, N.M., Al-Sherbaz, A., Turner, S.: Design and evaluation of browser-to-browser video conferencing in WebRTC. In: Global Information Infrastructure and Networking Symposium (GIIS), pp. 75–78. IEEE CPS (2017)
7. Garcia, B., Gortazar, F., Lopez-Fernandez, L., Gallego, M., Paris, M.: WebRTC testing: challenges and practical solutions. IEEE Commun. Stand. Magaz. **1**(2), 36–42 (2017). https://doi.org/10.1109/MCOMSTD.2017.1700005
8. Gouaillard, A., Roux, L.: Real-time communication testing evolution with WebRTC 1.0. In: Principles, Systems and Applications of IP Telecommunications (IPT Comm), pp. 1–8. IEEE (2017)
9. Higashino, M., Hayakawa, T., Takahashi, K., Kawamura, T., Sugahara, K.: Management of streaming multimedia content using mobile agent technology on pure P2P-based distributed e-learning system. Int. J. Grid Util. Comput. **5**3, 198–204 (2014). https://doi.org/10.1504/IJGUC.2014.062928

10. Iwata, S., Ozono, T., Shintani, T.: Any-application window sharing mechanism based on WebRTC. In: Proceedings of 6th IIAI International Congress on Advanced Applied Informatics (IIAI-AAI), pp. 808–813. IEEE CPS (2017)
11. Jang-Jaccard, J., Nepal, S., Celler, B., Yan, B.: WebRTC-based video conferencing service for telehealth. Computing **98**(1–2), 169–193 (2016). https://doi.org/10.1007/s00607-014-0429-2
12. Janak, J., Schulzrinne, H.: Framework for rapid prototyping of distributed IoT applications powered by WebRTC. In: Principles, Systems and Applications of IP Telecommunications (IPTComm), pp. 1–7. IEEE CPS (2016)
13. Kandil, A., El-Deeb, H.: Exploration of application migration to cloud environment. In: Proceedings of 2016 6th International Conference - Cloud System and Big Data Engineering (Confluence), pp. 109–114. IEEE CPS (2016)
14. Kohana, M., Okamoto, S.: A data sharing method using WebRTC for web-based virtual world. In: Barolli, L., Xhafa, F., Javaid, N., Spaho, E., Kolici, V. (eds.) Advances in Internet, Data & Web Technologies. EIDWT 2018. Lecture Notes on Data Engineering and Communications Technologies, vol. 17. Springer, Cham (2018)
15. Kumar, K.M., Sanil Kumar, D., Sardesai, R.P., Akhil, M.B.S.S., Kumar, N.: Application migration architecture for cross clouds analysis on the strategies methods and frameworks. In: Proceedings of 2017 IEEE International Conference on Cloud Computing in Emerging Markets (CCEM), pp. 107–112. IEEE CPS (2017)
16. Lin, J., Feng, Z., Chen, S., Huang, K., Tan, W., Li, X.: ATM: a framework to support application state migration in multiple terminals environment. In: IEEE International Conference on Services Computing (SCC), pp. 188 - 195. IEEE CPS (2017)
17. Nejdl, W., Wolf, B., Qu, C., Decker, S., Sintek, M., Naeve, A., Nilsson, M., Palmér, M., Risch, T.: EDUTELLA: a P2P networking infrastructure based on RDF. In: Proceedings of the 11th International Conference on World Wide Web (WWW 2002), pp. 604-615. ACM, New York (2002). https://doi.org/10.1145/511446.511525
18. Rivadulla, X., Xhafa, F., Caballé, S.: An event-based approach to support A3: anytime, anywhere, awareness in online learning teams. In: CISIS, pp. 947–952. IEEE CPS (2012)
19. Roig-Torres, J., Xhafa, F., Caballé, S.: Supporting online coordination of learning teams through mobile devices. In: CISIS, pp. 941–946. IEEE CPS (2012)
20. Sato, K., Mouri, K., Saito, S.: Design and implementation of an application state migration mechanism between android devices. In: Fourth International Symposium on Computing and Networking (CANDAR), pp. 696–700. IEEE (2016)
21. Sandholm, T., Magnusson, B., Johnsson, B.A.: An on-demand WebRTC and IoT device tunneling service for hospitals. In: International Conference on Future Internet of Things and Cloud, pp. 53–60. IEEE CPS (2014)
22. Schmidt, H., Hauck, F.J.: SAMProc: middleware for self-adaptive mobile processes in heterogeneous ubiquitous environments. In: Proceedings of the 4th on Middleware Doctoral Symposium (MDS 2007), Article 11 , p. 6. ACM, New York (2007). https://doi.org/10.1145/1377934.1377935
23. Schuchardt, V.: Moving mobile applications between mobile devices seamlessly. In: Proceedings of International Conference on Software Engineering. IEEE CPS (2012). https://doi.org/10.1109/ICSE.2012.6227028
24. Wang, W., Mei, L.: A design of multimedia conferencing system based on WebRTC Technology. In: Proceedings of 8th IEEE Annual Information Technology, Electronics and Mobile Communication Conference (IEMCON), pp. 148–153. IEEE (2017)
25. Noriyasu, Y.: An improved group discussion system for active learning using smartphone and its experimental evaluation. Int. J. Space-Based Situated Comput. **6**(4), 221– 227 (2016). https://doi.org/10.1504/IJSSC.2016.082763

One-to-One Routing Protocols for Wireless Ad-Hoc Networks Considering the Electric Energy Consumption

Emi Ogawa[1(✉)], Shigenari Nakamura[1], Tomoya Enokido[2], and Makoto Takizawa[1]

[1] Hosei University, Tokyo, Japan
emi.ogawa.2q@stu.hosei.ac.jp, nakamura.shigenari@gmail.com, makoto.takizawa@computer.org
[2] Rissho University, Tokyo, Japan
eno@ris.ac.jp

Abstract. In wireless ad-hoc networks, messages have to be energy-efficiently delivered to destination nodes by exchanging the messages among neighboring nodes. In our previous studies, the reactive type EAO, LEU, and IEAO protocols are proposed to unicast messages. In the EAO protocol, the total electric energy of nodes and delay time from a source node to a destination node can be reduced compared with the ESU and AODV protocols. However, a source-to-destination route may not be found if the communication range of each node is shorter. In the IEAO protocol, a route can be found even in short communication range. In the forwarding phase, a node p_i does not necessarily receive an RQ messages which sent by a node p_j whose level parameter is bigger than p_j and keeps the information which is contained in the RQ messages. Here, by neglecting superfluous RQ messages, the electric energy consumption of the forwarding phase can be reduced. In this paper, we propose an IEAO2 protocol by improving the IEAO protocol so that the electric energy consumption is reduced in the forwarding phase. In the evaluation, we show the information which each node keeps can be reduced in the IEAO2 protocol compared with other protocols.

1 Introduction

Wireless ad-hoc networks [4,10,11] are widely used in various types of applications, especially in vehicle-to-vehicle (V2V) communication [15] and delay-tolerant network (DTN) [2]. Nodes forward messages to neighbor nodes in wireless networks. In flooding protocols [12], the number of messages transmitted by nodes exponentially increases as the number of nodes increases. In the multi-point relay (MPR) protocols [12], on receipt of a message, only relay nodes forward the message to their first-neighbor nodes. Protocols for energy-efficiently broadcasting messages in a group of multiple nodes are also proposed [1,13,14].

L. Barolli et al. (Eds.): IMIS 2018, AISC 773, pp. 105–115, 2019.
https://doi.org/10.1007/978-3-319-93554-6_9

In this paper, we would like to discuss a unicast type of energy-efficient ad-hoc routing protocol to energy-efficiently deliver messages from a source node to a destination node in wireless ad-hoc networks. In our previous studies [9], reactive type [10] ad-hoc routing protocols, ESU (Energy-Saving Unicast routing) [8], EAO (Energy-Aware One-to-one routing) [5], LEU (Low-Energy Unicast Ad-hoc routing) [6], and IEAO (Improved EAO) [7] protocols are proposed. In the ESU protocol, the electric energy consumed by each node can be reduced but the length of a route is larger since nearer neighbor nodes are selected in the neighbor nodes. In the EAO protocol, the total electric energy consumed by nodes and the route length can be reduced. However, a source-to-destination route may not be found if the communication range of each node is shorter. In the LEU protocol, a source-to-destination route can be found with the shorter communication range of each node but the electric energy consumed by nodes is bigger than the EAO protocol. In the IEAO (Improved EAO) protocol, not only a source-to-destination route can be found even in the shorter communication range but also the electric energy consumption of nodes can be reduced. However, superfluous RQ messages are transmitted in the forwarding phase. In these protocols, it is assumed each node does not spend electric energy to receive a message. In reality, even if the communication device of each node just consumes small electric energy, the node consumes electric energy to perform the protocol module. In this paper, we propose an IEAO2 protocol by improving the forwarding phase of the IEAO protocol to reduce the number of RQ messages transmitted and received. Here, each node consumes the electric energy to send and receive messages.

We evaluate the IEAO2 protocol compared with the IEAO [7], EAO [5], and AODV [10] protocols in the simulation. We show the electric energy of the forwarding phase consumed by nodes in the IEAO2 protocol is smaller than the IEAO protocol and the other protocols. We also show the information which each node keeps can be reduced in the IEAO2 protocol compared with other protocols.

In Sect. 2, we present a system model. In Sect. 3, we propose the IEAO2 protocol. In Sect. 4, we evaluate the IEAO2 protocol.

2 System Model

A network N is composed of n ($\geq$1) nodes $p_1, \ldots, p_n$ which are cooperating with one another by exchanging messages in wireless networks [3]. Let d_{ij} be the distance between a pair of nodes p_i and p_j. In this paper, we assume the distance d_{ij} between every pair of nodes p_i and p_j is *a priori* known. Each node does not move, i.e. stays at fixed location.

Let $maxSE_i$ show the maximum electric energy [J] consumed by a node p_i to send a message. $wd_i(maxSE_i)$ shows the maximum communication range of a node p_i. A node p_j can receive a message sent by a node p_i if the node p_j is a $first-neighbor$ node of a node p_i, i.e. $d_{ij} \leq wd_i(maxSE_i)$. Otherwise, a node p_j cannot receive the message from a node p_i. We assume the maximum communication range $wd_i(maxSE_i)$ of each node is the same. This means, the maximum electric energy [J] consumed by each node is also the same, i.e. $maxSE_i = maxSE$.

In the forwarding phase, each node is in one state, 1 or 2 on receiving a message. At state 1, a node receives an *RQ* message q and checks the level parameter which is contained in the message q, i.e. $q.l$. Here, the electric energy consumed by a node to receive the message q and check the level parameter $q.l$ is rcE [J]. At state 2, the node finishes the state 1 and stores the data of the *RQ* message into the memory. A node is assumed to consume the electric energy dE [J] to process the data. In the forwarding phase, RE_i shows the electric energy [J] consumed by a node p_i. Therefore, RE_i of a node of state 1 is $RE_i = rcE$. At a node p_i of state 2 is $RE_i = rcE + dE$.

3 IEAO Protocol

In this paper, we newly propose an *IEAO2* (*Improved Energy-Aware One-to-one routing*) protocol for a source node to unicast messages to a destination node so that the electric energy consumed by the forwarding phase can be reduced and the information which each nodes keeps also can be reduced.

3.1 Overview

The IEAO2 protocol is composed of two phases, forwarding and backtracking phases as discussed in the IEAO [7] and EAO [5] protocols. At first, a source node p_s initiates the forwarding phase to find a shortest route to a destination node p_d to exchange messages. Here, each node obtains information of first-neighbor nodes in the network N by flooding *RQ* (request) messages to the destination node in a similar way to the EAO [5] and AODV [10] protocols. A shortest route from the source node p_s to the destination node p_d is found by using the first-neighbor information. Here, a directed link $p_j \rightarrow p_i$ shows that the node p_i is a first neighbor node of the node p_j, i.e. p_j receives an *RQ* message q from p_i. If the destination node p_d receives an *RQ* message q, a shortest route $p_s \rightarrow \ldots \rightarrow p_d$ is obtained as the forwarding route. Here, the forwarding phase terminates.

Then, the destination node p_d initiates the backtracking phase to find a more energy-efficient route by backtracking the forwarding route from the destination node p_d. Here, suppose a directed link $p_j \rightarrow p_i$ from a node p_j to a node p_i is found in the forwarding phase by the forwarding phase. For the node p_i, a common first-neighbor node p_k of the nodes p_i and p_j is checked if a path $p_j \rightarrow p_k \rightarrow p_i$ consumes smaller electric energy than the original path $p_j \rightarrow p_i$ as shown in Fig. 1. If so, the route $p_j \rightarrow p_k \rightarrow p_i$ is newly taken as a new route from p_j to p_i. Then, for the node p_k, it is checked if there is another node p_h such that $p_j \rightarrow p_h \rightarrow p_k$ in the same way. Until the source node p_s is found, the backtracking procedure is iterated. Thus, a new route from the source node p_s to the destination node p_d is found. In our previous studies, only electric energy to send a message is considered. Here, the electric energy to send a message from a node p_i to p_j is given d_{ij}^2. Hence, if $d_{ij}^2 < d_{ik}^2 + d_{kj}^2$, the path $p_i \rightarrow p_k \rightarrow p_j$ spends smaller electric energy than $p_i \rightarrow p_j$.

In our experiment, a node consumes so large to perform the protocol modules that the electric energy consumed by communication devices can be neglected. Since a node p_i selects a first-neighbor node p_j which the level parameter $p_j.l$ is smaller than $p_i.l$ and has an uncover neighbor node is selected as a prior node to male a route, the node p_i does not necessarily keep the information of nodes which the level parameter is bigger than $p_i.l$. This means, the node p_i neglected the RQ message q which is sent from the nodes if $q.l > p_i.l$ in the forwarding phase. Suppose there are three nodes p_i, p_j, and p_k. $p_i.l = 1$, $p_j.l = 1$, and $p_k.l = 2$. Here, we focus on the node p_i. In the EAO protocol, each node receives and keeps all the RQ messages. Nodes are all on the state 2 which the electric energy consumed by nodes is $RE = rcE + dE$. Hence, the node p_i receives and processes the message from the nodes p_j and p_k. The electric energy consumed by the node p_i is $RE_i = 2 \cdot (rcE + dE)$ in the EAO protocol. In the IEAO2 (improved IEAO) protocol, the node neglects the message q which the level parameter $q.l$ is bigger than the level parameter of the node. The node which neglects the RQ message is on the state 1. The electric energy RE of state 1 is rcE. Therefore, the node p_i neglects the message from the node p_k and receives the node p_j. The electric energy consumed by the node p_i is $RE_i = 2 \cdot rcE + dE$ in the IEAO2 protocol.

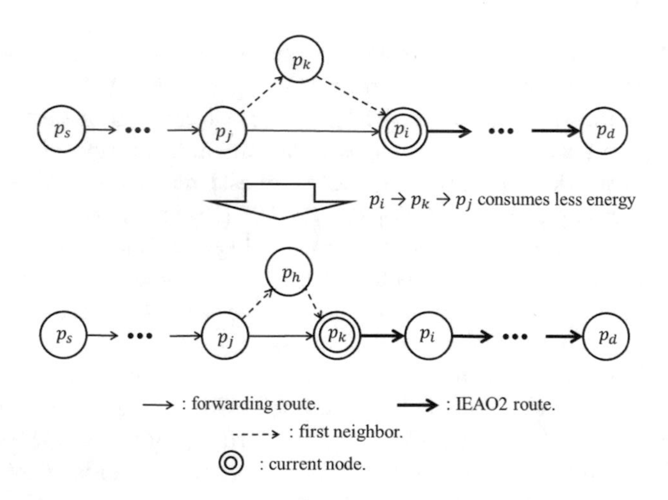

Fig. 1. Backtracking phase.

Each node p_i manipulates the following variables to find a route.

- $p_i.l$ = level parameter of the node p_i, initially 0.
- $p_i.FN$ = set of first-neighbor nodes of the node p_i, initially ϕ.
- $p_i.p_j.FN$ = set of first-neighbor nodes of each first-neighbor node p_j of the node p_i, initially does not exist.

- $p_i.Nfs$ = set of nodes which are not only first-neighbor nodes of the node p_i but also first-neighbor nodes of a first-neighbor node p_j of the node p_i, i.e. $p_i.Nfs = p_i.Nfs \cup (p_i.FN \cap p_i.p_j.FN)$, initially ϕ.

A variable $p_i.p_j.FN$ is created and $p_i.p_j.FN = \phi$ if the node p_i finds a node p_j to be a first-neighbor node of the node p_i, i.e. $p_j \in p_i.FN$ in the forwarding phase.

An *RQ* message q sent by a node p_i is composed of the following fields:

- $q.l$ = level parameter of the source node p_i, i.e. $q.l = p_i.l$.
- $q.FN$ = set of first-neighbor nodes of the source node p_i, i.e. $q.FN = p_i.FN$.
- $q.src$ = source node p_s.
- $q.dst$ = destination node p_d.

3.2 Forwarding Phase

First, a source node p_s would like to deliver messages to a destination node p_d in the network N. Here, $p_s.l = 0$ and $p_s.FN = \phi$. The source node p_s sends an *RQ* message q. In the forwarding phase, the source node p_s first sends an *RQ* message q to every first-neighbor node p_j with the maximum electric energy $SE_i = maxSE$. Here, $q.l = p_s.l$ $(= 0)$ and $q.FN = p_s.FN$ $(= \phi)$. Then, the node p_j sends an *RQ* message q where $q.l = 1$. Thus, *RQ* messages are forwarded to nodes in a flooding manner and the destination node p_d eventually receives an *RQ* message. Here, an *RQ* message q sent by a node p_j carries a level parameter $q.l$ and a set $q.FN$ of first-neighbor nodes which the node p_i knows.

[**Source node** p_s] A source node p_s would like to communicate with a destination node p_d.

1. $p_s.l = 0$; $p_s.FN = \phi$;
2. $q.l = p_s.l$ $(= 0)$;
   $q.FN = p_s.FN$ $(= \phi)$;
   $q.dst = p_d$;
3. **send** an *RQ* message q **to** every first neighbor node p_j with the maximum electric energy $maxSE_s$;

As presented here, a node p_i receives an *RQ* message q from a node p_j. If $p_i.l = 0$ and p_i is not the source node, the level parameter $q.l$ is stored in the variable $p_i.l$ and then the variable $p_i.l$ is incremented by one. The variable $p_i.l$ shows that the node p_i is at one level higher than the node p_j. If the node p_i had already received the *RQ* message q from another node, i.e. $q.l \neq 0$, the node p_i compares the level parameter $p_i.l$ with $q.l$. Here, if the level parameter $p_i.l$ is bigger than $q.l$ $(p_i.l > q.l)$, the variable $p_i.l$ is changed with $q.l + 1$. Otherwise, the variable $p_i.l$ is not changed.

The *RQ* message q carries the field $q.FN$ which is a set of first-neighbor nodes of the node p_j which the node p_j knows. Here, the node p_i knows that the node p_j is a first-neighbor node of the node p_i. If $p_i.l = 0$ or $p_i.l$ $(\neq 0) \geq$

$q.l$, $p_i.FN = p_i.FN \cup \{p_j\}$. Then, the node p_i sends an RQ message q to the first-neighbor nodes where $q.l = p_i.l$ and $q.FN = p_i.FN$.

A node p_i may receive RQ messages from multiple nodes. Each time the node p_i receives an RQ message q from a node p_j, the node p_i knows the node p_j is a first-neighbor node. Then, the node p_i checks the level parameter $q.l$ in the message q. If $p_i.l = 0$ or $p_i.l\ (\neq 0) \geq q.l$, $p_i.FN = p_i.FN \cup \{p_j\}$. The RQ message q also carries the information of first-neighbor nodes of the node p_j in the field $q.FN$. The node p_i keeps in record of $q.FN$ in a set variable $p_i.p_j.FN$. Initially, no variable $p_i.p_j.FN$ exists for any node p_j. On receipt of a message from a node p_j, a variable $p_i.p_j.FN$ is created. Then, $p_i.p_j.FN = \phi$. Nodes in $q.FN$, i.e. a set $p_j.FN$ of first neighbor nodes of the node p_j are stored in the variable $p_i.p_j.FN$, i.e. $p_i.p_j.FN = p_i.p_j.FN \cup q.FN$. Thus, the variable $p_i.p_j.FN$ stands for first-neighbor nodes of a first-neighbor node p_j which a node p_i knows. The variable $p_i.Nfs$ includes the information of the nodes which are not only the first-neighbor nodes but also the second-neighbor nodes of the node p_i, i.e. $p_i.Nfs = p_i.Nfs \cup (p_i.FN \cap p_i.p_j.FN)$.

[**Node** p_i] **on receipt** of an RQ message q **from** a node p_j,
  if $p_i.l = 0$,
  begin
    if p_i is the source node and $p_i.l < q.l$,
    begin
      p_i delete the message q;
      $q.l = p_i.l$; $q.FN = p_i.FN$;
      send an RQ message q **to** every first-neighbor node p_k with the maximum electric energy $maxSE_i$;
    end
    else /*p_i is not the source node*/
      if $p_i \leq q.l$,
      begin
        $p_i.l = q.l + 1$;
        $p_i.FN = p_i.FN \cup \{p_j\}$; /*$p_j$ is a first-neighbor node*/
        $p_i.p_j.FN = q.N \cup p_i.p_j.FN$;
        $p_i.Nfs = p_i.Nfs \cup (p_i.FN \cap p_i.p_j.FN)$;
        $q.l = p_i.l$; $q.FN = p_i.FN$;
      send an RQ message q **to** every first-neighbor node p_k with the maximum electric energy $maxSE_i$;
      end /*$p_i.l \leq q.l$*/
    end /*p_i is not the source node*/
  else /*$p_i.l \neq 0$*/
    if $p_i.l < q.l$,
    begin
      $p_i.l$ does not change and p_i delete the message q;
      $q.l = p_i.l$; $q.FN = p_i.FN$;
      send an RQ message q **to** every first-neighbor node p_k with the maximum electric energy $maxSE_i$;

end /*$p_i.l < q.l$*/
if $p_i.l = q.l$,
begin
$p_i.l$ does not change;
$p_i.FN = p_i.FN \cup \{p_j\}$; /*$p_j$ is a first-neighbor node*/
$p_i.p_j.FN = q.N \cup p_i.p_j.FN$;
$p_i.Nfs = p_i.Nfs \cup (p_i.FN \cap p_i.p_j.FN)$;
$q.l = p_i.l$; $q.FN = p_i.FN$;
send an *RQ* message q **to** every first-neighbor node p_k with the maximum electric energy $maxSE_i$;
end /*$p_i.l = q.l$*/
if $p_i.l > q.l$,
begin
$p_i.l = q.l + 1$;
$p_i.FN = p_i.FN \cup \{p_j\}$; /*$p_j$ is a first-neighbor node*/
$p_i.p_j.FN = q.N \cup p_i.p_j.FN$;
$p_i.Nfs = p_i.Nfs \cup (p_i.FN \cap p_i.p_j.FN)$;
$q.l = p_i.l$; $q.FN = p_i.FN$;
send an *RQ* message q **to** every first-neighbor node p_k with the maximum electric energy $maxSE_i$;
end /*$p_i.l > q.l$*/
end /*$p_i.l \neq 0$*/

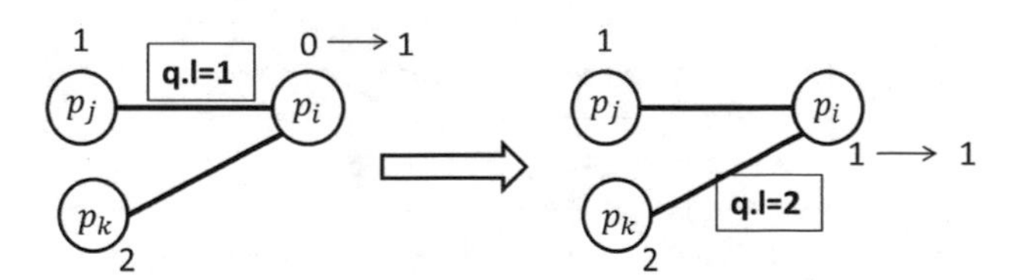

Fig. 2. On receipt of an *RQ* message q.

Suppose there are three nodes p_i, p_j, and p_k as shown in Fig. 2. The level parameter of the nodes p_j and p_k is 1 and 2, respectively. The level parameter of the node p_i is 0 before receiving the *RQ* message q from another node. First, the node p_j sends the *RQ* message q to the node p_i, i.e.$q.l = 1$ and $q.FN = p_j.FN$. The node p_i receives the message q and checks the level parameter $q.l$. Here, $p_i.l$ $(= 0) < q.l = 1$. The node p_i keeps the data which is contained in the message q. $p_i.FN = p_i.FN \cup \{p_j\}$. The *RQ* message q also carries the information of first-neighbor nodes of the node p_j in the field $q.FN$. The node p_i keeps in record of $q.FN$ in a set variable $p_i.p_j.FN$. Then, the node p_i receives the *RQ* message q from the node p_k. The level parameter $q.l$ of the message q is bigger than the node p_i and $p_i.l \neq 0$. Therefore, $p_i.l$ is not changed and the node p_i does not keep the information which is contained in the message q from the node p_k. The node

p_i deletes the message q from the node p_k after checking the level parameter $q.l$ compared with itself.

Eventually, the destination node p_d receives an RQ message q. Suppose a node p_j is a first-neighbor node of a node p_i. Here, the node p_j precedes the node p_i ($p_j \rightarrow p_i$) iff p_j is a first-neighbor node of the node p_i and $p_j.l < p_i.l$.

4 Evaluation

We evaluate the IEAO2 protocol in terms of electric energy consumption of nodes in the forwarding phase compared with the IEAO [7], EAO [5] and AODV [10] protocols. In the evaluation, n (≥ 1) nodes $p_0, p_1, \ldots, p_{n-1}$ are uniformly deployed on an $m \cdot m$ mesh network N. In the evaluation, we consider a 128 times 128 mesh network, i.e. $m = 128$. We randomly select $n \cdot (n-1)$ pairs of a source node p_s and a destination node p_d in the n nodes. Here, the maximum communication range $maxd_i$ of each node p_i is the same $maxd$. Then, IEAO2, IEAO, EAO, and AODV routes are found for each pair of a source node p_s and a destination node p_d on the mesh network in the IEAO2, IEAO, EAO, and AODV protocols, respectively, on each deployment of nodes in the mesh network. We assume the communication range $maxd_i$ of each node p_i is the same $maxd$.

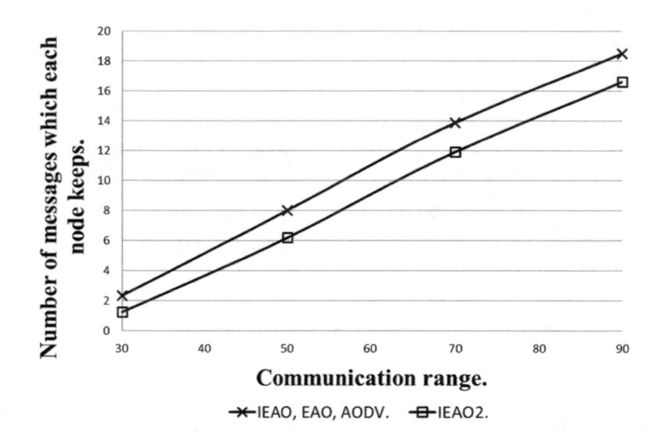

Fig. 3. Number of messages which are kept in each node.

Figure 3 shows the number of messages which are kept in each node in the IEAO2, IEAO, EAO, and AODV protocols for $n = 30$ where $30 \leq maxd \leq 90$. The number of messages kept in each node of the IEAO2 protocol is fewer than the other protocols. The smaller the communication range gets, the fewer number of messages is kept in each node. For example, the number of messages in the IEAO2 protocol is about 90% of the IEAO, EAO, and AODV protocols where the communication range $maxd$ is 30.

Figure 4 shows the electric energy consumption ratios of the forwarding phase in the IEAO2 route to the IEAO, EAO, and AODV routes for $n = 30$ and the

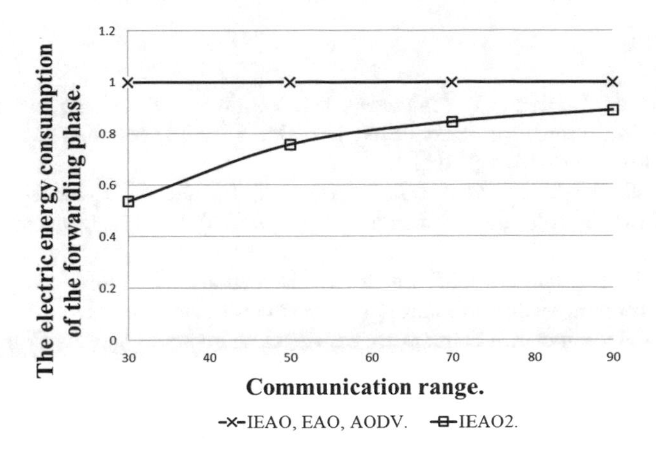

Fig. 4. Electric energy ratio of the forwarding phase.

communication range *maxd* is 30 to 90. The electric energy consumption of the IEAO2 protocol is smaller than the other protocols. The larger the communication range *maxd* gets, the larger electric energy is consumed. For example, the electric energy consumption of the forwarding phase in the IEAO2 protocol is about 54% of the IEAO, EAO, and AODV protocols for $maxd = 90$.

5 Concluding Remarks

In this paper, we newly proposed the forwarding phase of the IEAO2 protocol to deliver messages from a source node to a destination node in a wireless ad-hoc network. In the IEAO2 protocol, information of first-neighbor nodes from a source node p_s to a destination node p_d is collected by flooding *RQ* messages in a similar way to the AODV protocol. Then, starting from the destination node as a current node, a more energy-efficient prior node is tried to be found for each current node. We evaluated the IEAO2 protocol compared with IEAO, EAO, and AODV protocols. We showed the electric energy consumed by nodes in the forwarding phase and the number of messages which are kept in each node can be reduced compared with the IEAO, EAO, and AODV protocols. In the IEAO protocol, the superfluous *RQ* messages are transmitted and processed by nodes in the forwarding phase. In the IEAO2 protocol, the data of the superfluous *RQ* message is not kept in each node. This means, the electric energy consumed by nodes to perform the protocol module can be reduced. We are now designing the backtracking phase of the IEAO2 protocol.

Acknowledgements. This work was supported by JSPS KAKENHI grant number 15H0295.

References

1. Aikebaier, A., Enokido, T., Takizawa, M.: Reliable and efficient way to broadcast messages in a group by Trust-based Broadcast (TBB) scheme. Comput. Inform. (CAI) **30**(6), 1001–1015 (2011)
2. Cerf, V., Burleigh, S., Hooke, A., Torgerson, L., Durst, R., Scott, K., Fall, K., Weiss, H.: Delay-tolerant network architecture (2007). https://tools.ietf.org/html/rfc4838
3. Clausen, T., Hansen, G., Christensen, L., Behrmann, G.: The optimized link state routing protocol, evaluation through experiments and simulation. In: Proceedings of the IEEE Symposium Conference on Wireless Personal Mobile Communications (2001)
4. Jacquet, P., Muhlethaler, P., Clausen, T., Laouiti, A., Qayyum, A., Viennot, L.: Optimized link state routing (OLSR) protocol for ad hoc network (2003). https://www.ietf.org/rfc/rfc3626.txt
5. Ogawa, E., Nakamura, S., Enokido, T., Takizawa, M.: An energy-aware one-to-one routing protocol in wireless ad-hoc network. In: Proceedings of the 20th International Conference on Network-Based Information Systems (NBiS-2017), pp. 102–113 (2017)
6. Ogawa, E., Nakamura, S., Enokido, T., Takizawa, M.: A low-energy unicast ad-hoc routing protocol in wireless networks. In: Proceedings of the 12th International Conference on Broad-Band Wireless Computing, Communication and Applications (BWCCA-2017), pp. 173–184 (2017)
7. Ogawa, E., Nakamura, S., Enokido, T., Takizawa, M.: Unicast routing protocols to reduce electric energy consumption in wireless ad-hoc networks. In: Proc. of IEEE the 32st International Conference on Advanced Information Networking and Applications (AINA-2018), pp. 1162–1168 (2018)
8. Ogawa, E., Nakamura, S., Takizawa, M.: An energy-saving unicast routing protocol in wireless ad-hoc network. In: Proceedings of the 11th International Conference on Innovative Mobile and Internet Services in Ubiquitous Computing (IMIS-2017), pp. 110–120 (2017)
9. Ogawa, E., Nakamura, S., Takizawa, M.: A trustworthiness-based ad-hoc routing protocol in wireless networks. In: Proceedings of IEEE the 31st International Conference on Advanced Information Networking and Applications (AINA-2017), pp. 1162–1168 (2017)
10. Perkins, C., Belding-Royer, E., Das, S.: Ad hoc On-Demand Distance Vector (AODV) routing (2003). https://www.ietf.org/rfc/rfc3561.txt
11. Perkins, C.E., Bhagwat, P.: Highly dynamic Destination Sequenced Distance Vector routing (DSDV) for mobile computers. In: Proceedings of ACM SIGCOMM 1994, pp. 234–244 (1994)
12. Qayyum, A., Viennot, L., Laouiti, A.: Multipoint relaying: an efficient technique for flooding in mobile wireless networks. Technical report 3898, Inria (2000)
13. Sugino, M., Nakamuara, S., Enokido, T., Takizawa, M.: Energy-efficient broadcast protocols in wireless networks. In: Proceedings of the 18th International Conference on Network-Based Information Systems (NBiS-2015), pp. 357–364 (2015)

14. Sugino, M., Nakamuara, S., Enokido, T., Takizawa, M.: Protocols for energy-efficiently broadcasting messages in wireless networks. In: Proceedings of the AINA-2016 Workshops (WAINA-2016), pp. 286–293 (2016)
15. Yang, X., Liu, J., Zhao, F., Vaidya, N.H.: A vehicle-to-vehicle communication protocol for cooperative collision warning. In: Proceedings of the 1st Annual for International Conference on Mobile and Ubiquitous System: Networking and Services, pp. 114–123 (2004)

Virtual Machine Migration Algorithms to Reduce Electric Energy Consumption of a Server Cluster

Ryo Watanabe[1(✉)], Dilawaer Duolikun[1], Tomoya Enokido[2], and Makoto Takizawa[1]

[1] Hosei University, Tokyo, Japan
ryo.watanabe.4h@stu.hosei.ac.jp, dilewerdolkun@gmail.com, makoto.takizawa@computer.org
[2] Rissho University, Tokyo, Japan
eno@ris.ac.jp

Abstract. In this paper, we discuss a virtual machine migration approach to reducing the electric energy consumption of servers. In our previous algorithms, one virtual machine migrates from a host server to a guest server. While the electric energy consumption of servers can be reduced by migrating some number b of processes, there might not be a virtual machine with the same number b of processes on a host server. In this paper, we propose an ISEAM2T algorithm where multiple virtual machines can migrate from a host server to a guest server. Here, multiple virtual machines on a host server are selected so that the total number of processes on the virtual machines can be more easily adjusted to the best number b of processes. In the evaluation, we show the total electric energy consumption and active time of the servers and the average execution time of processes can be reduced in the proposed algorithm.

1 Introduction

We have to reduce electric energy consumption of servers in clusters like cloud computing systems [2] in order to realize eco society [4]. Energy-efficient hardware devices like CPUs [3] are developed in the hardware-oriented approach. On the other hand, we aim at reducing the total electric energy consumed by a server to perform application processes in our macro-level approach [8–10]. Here, types of power consumption and computation models are proposed [8–10].

If an application process is issued to a cluster of servers, one energy-efficient server is selected to perform the process in types of server selection algorithms [10–12]. Furthermore, an application process performed on a host server migrates to a guest server which is expected to consume smaller electric energy than the host server in the migration approach [6,7]. By using virtual machines [5], a cloud computing system supports applications with virtual computation service which is independent of locations and heterogeneity of servers. In addition, a virtual machine with application processes can easily migrate from a host server

L. Barolli et al. (Eds.): IMIS 2018, AISC 773, pp. 116–127, 2019.
https://doi.org/10.1007/978-3-319-93554-6_10

to another guest server without suspending the processes like the live migration [5]. Energy-efficient migration algorithms of virtual machines [15–17,19] are so far proposed to migrate virtual machines on host servers to energy-efficient guest servers. Here, one virtual machine is selected to migrate. In papers [19,20], the mathematical relation among the electric energy consumed by a host server s_t and a guest server s_u and the number of processes to migrate is discussed. By using the relation, we can find such number b of processes to migrate from a host server s_t to a guest server s_u that the total electric energy to be consumed by the servers s_t and s_u is minimized. However, it is not easy to find a virtual machine on a host server by migrating which the electric energy consumption of the host and guest servers can be mostly reduced. In this paper, we proposed an ISEAM2T algorithm where multiple virtual machines can migrate from a host server to a guest server where the total number of processes on the virtual machines is near to the best number b.

We evaluate the ISEAM2T algorithm in terms of the total electric energy consumption and active time of servers and the average execution time of processes compared with other non-migration algorithms and migration algorithms.

In Sect. 2, we present a system model. In Sect. 4, we propose the ISEAM2T algorithm. In Sect. 5, we evaluate the ISEAM2T algorithm.

2 System Model

A cluster S is composed of servers $s_1, \ldots, s_m$ $(m \geq 1)$ and supports applications on clients with virtual service on computation resources by using virtual machines $vm_1, \ldots, vm_v$ $(v \geq 1)$ [5]. If a client issues an application process p_i, one virtual machine vm_h is selected to perform the process p_i. $VCP_h(\tau)$ is a set of resident processes on the virtual machine vm_h at time τ. Here, $|VCP_h(\tau)|$ shows the size nv_h of a virtual machine vm_h. A virtual machine vm_h is *active* if $nv_h \geq 1$, else *idle*. An active server hosts at least one active virtual machine. A server which hosts at least one virtual machine is *engaged*, otherwise *free*. A virtual machine vm_h is *smaller* than a virtual machine vm_k $(vm_h < vm_k)$ iff (if and only if) $nv_h < nv_k$.

In this paper, a *process* means a computation type of application process which uses CPU on a server. In the SPC (Simple Power Consumption) model [10], the electric power consumption $CE_t(n)$ is given as follows:

$$CE_t(n) = \begin{cases} minE_t \text{ if } n = 0. \\ maxE_t \text{ if } n > 0. \end{cases} \tag{1}$$

That is, a server s_t consumes the electric power $maxE_t$ [W] if at least one process is performed, otherwise $minE_t$. The total electric energy consumed by a server s_t from time st to time et is defined to be $\sum_{\tau=st}^{et} CE_t(|CP_t(\tau)|)$ [W tu].

Each process p_i is performed on a host server s_t. It takes T_{ti} time units [tu] to perform a process p_i on a thread of a server s_t. If only a process p_i is performed on a server s_t without any other process, the execution time T_{ti} of the process p_i is shortest, i.e. $T_{ti} = minT_{ti}$. In a cluster S of servers $s_1, \ldots, s_m$ $(m \geq 1)$, $minT_i$

shows a shortest one of $minT_{1i}, \ldots, minT_{mi}$. If $minT_{fi} = minT_i$, a thread of a server s_f is $fastest$ and the server s_f is $fastest$. We assume one virtual computation step [vs] is performed on a thread of a fastest server s_f for one time unit [tu], i.e. the computation rate TCR_f of the thread is one [vs/tu]. The total number VC_i of virtual computation steps of a process p_i is defined to be $minT_i$ [tu] $\cdot$ TCR_f [vs/tu] $= minT_i$ [vs]. The maximum computation rate $maxCR_{ti}$ of a process p_i on a server s_t is VC_i [vs]$/minT_{ti}[tu] = minT_i/minT_{ti}$ [vs/tu] (≤ 1). In the SC (Simple Computation) model [10], the computation rate $NSR_t(n)$ of a server s_t is given as follows:

$$NSR_t(n) = \begin{cases} TCR_t & \text{if } n > 0. \\ 0 & \text{if } n = 0. \end{cases} \tag{2}$$

The server computation rate $NSR_t(n)$ is fairly allocated to each process p_i of n current processes. The process computation rate $NPR_{ti}(n)$ [vs/tu] of a process p_i with $(n-1)$ processes on a server s_t is $NSR_t(n)/n$.

If a process p_i starts at time st and ends at time et, $\sum_{\tau=st}^{et} NPR_{ti}(|CP_t(\tau)|) = VC_i$ [vs]. VC_i shows the total number of computation steps to be performed by a process p_i. At time τ a process p_i starts, $lc_{ti}(\tau) = VC_i$. Then, $lc_{ti}(\tau)$ is decremented by the computation rate $NPR_{ti}(|CP_t(\tau)|)$, i.e. $lc_{ti}(\tau+1) = lc_{ti}(\tau) - NPR_{ti}(|CP_t(\tau)|)$ at each time τ. If $lc_{ti}(\tau) > 0$ and $lc_{ti}(\tau+1) \leq 0$, the process p_i terminates at time τ.

3 Estimation Model

3.1 Expected Electric Energy Consumption

A client issues a process p_i to a set VM of virtual machines $vm_1, \ldots, vm_v$ ($v \geq 1$) in a cluster S of servers $s_1, \cdots s_m$. We assume the total number VC_i of virtual computation steps of each process p_i to be a constant, $VC_i = 1$. Suppose n processes are currently performed on a server s_t and k new processes are issued to the server s_t. The total amount of computation to finish the n current processes is assumed to be $n/2$. The total computation to finish the k new processes is k. The expected termination time $SET_t(n, k)$ [tu] and expected electric energy consumption $SEE_t(n, k)$ [W tu] of a server s_t to perform n current processes and k new processes are given as follows:

$$SET_t(n,\ k) = (n/2 + k)/NSR_t(n+k). \tag{3}$$

$$\begin{aligned} SEE_t(n,\ k) &= SET_t(n,k) \cdot NE_t(n+k) \\ &= (n/2 + k) \cdot NE_t(n+k)/NSR_t(n+k). \end{aligned} \tag{4}$$

Let n_t be the number $|CP_t(\tau)|$ of processes performed on each server s_t. First, suppose no virtual machine on the server s_t migrates to the server s_u. The servers s_t and s_u consume electric energy $EE_t(=SEE_t(n_t, 0))$ by time $ET_t(=SET_t(n_t, 0))$ and $EE_u(=SEE_u(n_u, 0))$ by $ET_u(=SET_u(n_u, 0))$ to perform n_t and n_u current processes, respectively. The servers s_t and s_u totally consume the electric energy EE_{tu} until every current process on s_t and s_u terminates:

$$EE_{tu} = \begin{cases} EE_t + EE_u + (ET_t - ET_u) \cdot minE_u \text{ if } ET_t \geq ET_u. \\ EE_t + EE_u + (ET_u - ET_t) \cdot minE_t \text{ if } ET_t < ET_u. \end{cases} \quad (5)$$

Next, suppose a virtual machine vm_h on the host server s_t migrates to the guest server s_u. Here, $(n_t - nv_h)$ processes are performed on the server s_t while $(n_u + nv_h)$ processes on the server s_u. The servers s_t and s_u totally consume the electric energy $NE_t = SEE_t(n_t - nv_h, 0)$ by time $NT_t = SET_t(n_t - nv_h, 0)$ and the electric energy $NE_u = SEE_u(n_u + nv_h, 0)$ $NT_u = SET_u(n_u + nv_h, 0)$, respectively, to perform every process. That is, the servers s_t and s_u consume electric energy NE_t and NE_u to perform all the current processes, respectively. Hence, the servers s_t and s_u totally consume the electric energy NE_{tu} as discussed in formula (5):

$$NE_{tu} = \begin{cases} NE_t + NE_u + (NT_t - NT_u) \cdot minE_u \text{ if } NT_t \geq NT_u. \\ NE_t + NE_u + (NT_u - NT_t) \cdot minE_t \text{ if } NT_t < NT_u. \end{cases} \quad (6)$$

If $EE_{tu} > NE_{tu}$, the virtual machine vm_h can migrate from the host server s_t to the guest server s_u. If there is no virtual machine on the server s_t such that $EE_{tu} > NE_{tu}$, no virtual machine migrates to the server s_u.

3.2 Selection of a Virtual Machine

If no virtual machine migrates from a host server s_t to a guest server s_u, the expected termination time ET_t is $n_t/maxCR_t$ [tu] and the expected electric energy consumption EE_t is $ET_t \cdot maxE_t = n_t \cdot maxE_t/maxCR_t$ [W tu] for $n_t > 0$ for a host server s_t. $ET_u = n_u/maxCR_u$ and $EE_u = ET_u \cdot maxE_u = n_u \cdot maxE_u/maxCR_u$ for a server s_u.

Next, suppose a virtual machine vm_h with nv_h processes migrates from a host server s_t to a guest server s_u. The expected termination time NT_t is $(n_t - nv_h)/maxCR_t$ and the expected electric energy consumption NE_t is $(n_t - nv_h) \cdot maxE_t/maxCR_t$ for the host server s_t. NT_u is $(n_u + nv_h)/maxCR_u$ and NE_u is $(n_u + nv_h) \cdot maxE_u/maxCR_u$ for a guest server s_u. The expected termination time $NT_t(nv_h)$ and $NT_u(nv_h)$ of the servers s_t and s_u are $(n_t - nv_h)/maxCR_t$ and $(n_t + nv_h)/maxCR_u$, respectively. The total electric energy consumption $EC_{tu}(nv_h)$ of the servers s_t and s_u is given for the size nv_h of the virtual machine vm_h:

$$EC_{tu}(nv_h) = \begin{cases} A_{1tu} \cdot nv_h + C_{1tu} \text{ if } NT_t(nv_h) \geq NT_u(nv_h). \\ A_{2tu} \cdot nv_h + C_{2tu} \text{ if } NT_t(nv_h) < NT_u(nv_h). \end{cases} \quad (7)$$

$$NT_t(nv_h) = (n_t - nv_h)/maxCR_t. \quad (8)$$

$$NT_u(nv_h) = (n_u + nv_h)/maxCR_u. \quad (9)$$

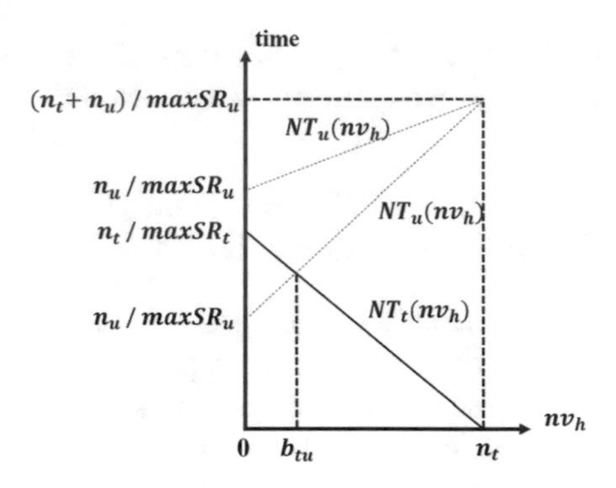

Fig. 1. $NT_t(nv_h)$ and $NT_u(nv_h)$.

$$A_{1tu} = (maxE_u - minE_u)/maxCR_u - (maxE_t + minE_u)/maxCR_t. \tag{10}$$

$$A_{2tu} = (maxE_u + minE_t)/maxCR_u - (maxE_t - minE_t)/maxCR_t. \tag{11}$$

$$C_{1tu} = n_u \cdot (maxE_u - minE_u)/maxCR_u + n_t \cdot (maxE_t + minE_u)/maxCR_t. \tag{12}$$

$$C_{2tu} = n_u \cdot (maxE_u + minE_t)/maxCR_u + n_t \cdot (maxE_t - minE_t)/maxCR_t. \tag{13}$$

In Fig. 1, the straight line shows the expected termination time $NT_t(nv_h)$ and the dotted line indicates $NT_u(nv_h)$. If $n_u/maxCR_u > n_t/maxCR_t$, $NT_u(nv_h) > NT_t(nv_h)$ for every number $nv_h(\geq 0)$ of processes on a virtual machine vm_h. Otherwise, there is some value b_{tu} $(0 \leq b_{tu} \leq n_t)$ such that $NT_t(b_{tu}) = NT_u(b_{tu})$:

$$b_{tu} = (n_t \cdot maxCR_u - n_u \cdot maxCR_t)/(maxCR_t + maxCR_u). \tag{14}$$

For $nv_h < b_{tu}$, $NT_t(nv_h) > NT_u(nv_h)$. Here, the guest server s_u terminates before the host server s_t if a virtual machine vm_h of size nv_h migrates from the host server s_t to the guest server s_u. For $nv_h > b_{tu}$, $NT_t(nv_h) < NT_u(nv_h)$.

If $A_{1tu} \geq 0$ and $A_{2tu} \geq 0$, the total electric energy consumption $EC_{tu}(nv_h)$ linearly increases as the size nv_h of a virtual machine vm_h increases. If $A_{1tu} < 0$ and $A_{2tu} < 0$, $EC_{tu}(nv_h)$ linearly decreases as nv_h increases. If $A_{1tu} < 0$ and $A_{2tu} > 0$, $EC_{tu}(b_{tu})$ is minimum. If a virtual machine vm_h migrates whose size nv_h is b_{tu}, the electric energy consumption of the servers s_t and s_u can be minimized as shown in Fig. 2. If $A_{1tu} > 0$ and $A_{2tu} < 0$, $EC_{tu}(0)$ or $EC_{tu}(n_t)$ is minimum and $EC_{tu}(b_{tu})$ is maximum.

4 Energy-Efficient Selection and Migration Algorithms

4.1 VM Selection Algorithm

First, suppose a client issues a process p_i to a cluster S of servers $s_1, \ldots, s_m$ $(m \geq 1)$ with a set VM of virtual machines $vm_1, \ldots, vm_v$ $(v \geq 1)$ at time τ.

[VM selection] A client issues a process p_i to a cluster S.

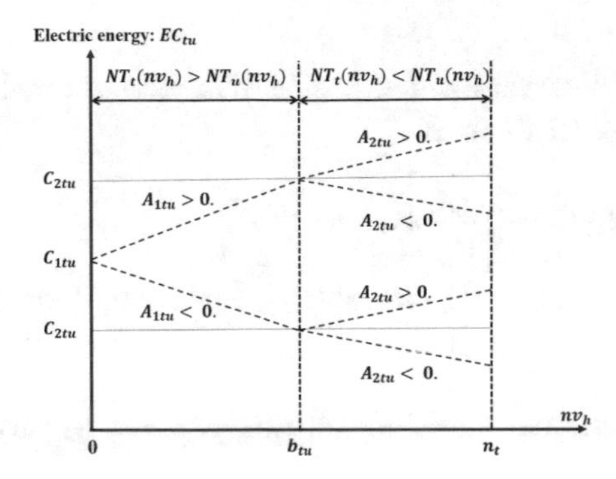

Fig. 2. Energy consumption $EC_{tu}(nv_h)$.

1. **select** an engaged host server s_t where at least one virtual machine resides in a selection algorithm.
2. **select** a smallest virtual machine vm_h on the selected host server s_t.
3. **perform** the process p_i on the virtual machine vm_h of the host server s_t.

As discussed in the SGEA algorithm [14], not only a host server s_t of a new process p_i but also the other servers consume electric energy even if the servers are idle. Here, let NT_t and NE_t be the expected termination time $SET_t(n_t, 1)$ and expected electric energy consumption $SEE_t(n_t, 1)$ of a host server s_t, respectively, where a new process is assumed to be performed. Here, $SET_t(n_t, 1) = (n/2+1)/NSR_t(n_t+1)$ and $SEE_t(n_t, 1) = (n/2+1) \cdot CE_t(n_t+1)/NSR_t(n_t+1)$. Let ET_u and EE_u be the expected execution time $SET_u(n_u, 0)$ and expected electric energy $SEE_u(n_u, 0)$ of a server s_u, respectively, where only n_u current processes are performed. The expected electric energy consumption $NSEE_{tu}$ of another server s_u ($\neq s_t$) for a host server s_t of a process p_i is given by $FE_u(NE_u, NT_t, NT_u)$, which another server s_u is expected to consume by the time NT_t:

$$FE_t(E, T, U) = \begin{cases} E + minE_t \cdot (T - U) & \text{if } T \geq U. \\ E \cdot U/T & \text{if } T < U. \end{cases} \tag{15}$$

The expected total electric energy consumption $SGEE_t$ of the servers $s_1, \ldots, s_m$ for a host server s_t is $NE_t + \sum_{u=1,\ldots,m(\neq t)} NSEE_{tu}$.

[**Server selection algorithm SGEA**]

1. **select** a server s_t **where** $SGEE_t$ is minimum;
2. **select** a smallest virtual machine vm_hon s_t and perform the process p_i;

4.2 VM Migration (VMM) Algorithm

Each engaged server s_t is periodically checked. For each server $s_u(\neq s_t)$, a virtual machine vm_h on the server s_t is selected where the total electric energy expected to be consumed by a pair of the servers s_t and s_u is minimum:

[VMM algorithm]

```
X = a set of engaged servers which host active virtual machines in a cluster S;
for each engaging server s_t in X,
while (|X| > 0) {
    for each engaging server s_u(≠ s_t) in S,
    EE_t = SEE_t(n_t, 0); EE_u = SEE_u(n_u, 0);
    b_tu = (n_t · maxCR_u − n_u · maxCR_t)/(maxCR_t + maxCR_u);
    if b_tu ≤ 0, {
      if A_1tu < 0, {
        select all virtual machines on the server s_t as target ones in TMV_t;
        NE_t = SEE_t(0, 0); NE_u = SEE_u(n_u + n_t, 0);
      };
      else, no virtual machine migrates;
    }; /* if b ≤ 0 end */
    else /* 0 < b_tu < n_t */, {
    if A_1tu < 0 and A_2tu < 0, {
      select all virtual machines on the server s_t as target ones in TMV_t;
      NE_t = SEE_t(0, 0); NE_u = SEE_u(n_u + n_t, 0);
    };
    if A_1tu < 0 and A_2tu ≥ 0, {
      while (b ≥ tnv_tu) {
        select a smallest virtual machine vm_tu on the server s_t as a target
one in TMV_t
        where |b_tu − nv_tu − tnv_tu| is minimum;
        tnv_tu = tnv_tu + nv_tu;
      }; /* while end */
      NE_t = SEE_t(n_t − tnv_tu, 0); NE_u = SEE_u(n_u + tnv_tu, 0);
    };
    if A_1tu ≥ 0 and A_2tu < 0,
      if A_2tu · n_t < C_1(n_t, n_u), {
        select all virtual machines on the server s_t as target ones in TMV_t;
        NE_t = SEE_t(0, 0); NE_u = SEE_u(n_u + n_t, 0);
      };
    }; /* else end */
    NEE_tu = NE_t + NE_u;
    select a server s_u where (EE_t + EE_u) − (NE_t + NE_u) > 0 and NEE_tu
is smallest;
    if s_u is found, {
      migrate the target virtual machines in TMV_t from s_t to s_u;
      if s_u is in X, X = X− {s_u};
    }; /* if end */
    X = X− {s_t};
}; /* while X end */
```

Here, a subset TMV_t of virtual machines on a host server s_t are selected for each guest server s_u, where $nv_{tu} = \sum_{vm_h \in TMV_t} nv_h$ is nearest to b_{tu}. Then, a server s_u is selected where expected electric energy consumption $EC_{tu}(nv_{tu})$ is minimum. If a guest server s_u is found, the virtual machines in TMV_t migrate to the guest server s_u. Otherwise, no virtual machine migrates from the host server s_t.

5 Evaluation

We consider a cluster S of four homogeneous servers s_1, s_2, s_3, and s_4 ($m = 4$), respectively, in our laboratory and 40 virtual machines $vm_1, \ldots, vm_{40}$ ($v = 40$). A virtual machine vm_h is on a host server s_t where $t = (h - 1)$ module 4 +1. Initially, each server s_t hosts ten virtual machines. The performance parameters like TCR_t and electric energy parameters like $maxE_t$ of each server s_t are shown in Table 1. In the simulation, the total electric energy consumption EE_t and total active time AT_t of each server s_t are obtained for each server s_t. The total active time AT_t of a server s_t is time when the server s_t is active, i.e. at least one process is performed.

In the cluster S, n (> 0) processes $p_1, \ldots, p_n$ are performed. The starting time $stime_i$ of each process p_i is randomly taken from time 0 to $xtime - 1$. Here, $xtime$ is 1,000 time units [tu]. In this paper, $3n/4$ processes are randomly taken from time 0 to $xtime - 1$. Then, $stime_i$ of each process p_i of the other $n/4$ processes is randomly taken from $xtime/4 - 10$ to $xtime/4 + 10$. In fact, one time unit [tu] shows 100 [msec] [13]. The minimum execution time $minT_i$ is randomly taken from 5 to 20 [tu]. The number VC_i of virtual computation steps of each process p_i is 5.0 to 20.0 [vs]. The parameters of each process p_i are shown in Table 2. Each process p_i starts at time $stime_i$ and terminates at time $etime_i$ which is obtained in the simulation. The execution time T_i of a process p_i is $etime_i - stime_i + 1$ [tu]. The simulation ends at time $etime$ when every process terminates, i.e. $etime = max(etime_1, \ldots, etime_n)$. The electric energy consumption EE_t of each server s_t is $\sum_{\tau=0}^{etime} CE_t(|CP_t(\tau)|)$.

We consider the random (RD), round robin (RR), SGEA [14], ISEAM [15], ISEAM2 [19], and ISEAM2T algorithms. In the RD, RR, and SGEA algorithms, virtual machines do not migrate. In the ISEAM and ISEAM2 algorithms, a smallest virtual machine migrates to a more energy-efficient server. In the ISEAM algorithm, at most one virtual machine migrates. In the ISEAM2T algorithm, multiple virtual machines migrate. The average execution time AT is $(T_1 + \cdots + T_n)/n$.

The simulation is time-based. Initially, the electric energy consumption variable EE_t and active time variable AT_t are 0 and a variable CP_t is ϕ for each server s_t. At each time τ, if a process p_i states, i.e. $stime_i = \tau$, one server s_t is selected in a selection algorithm and $CP_t = CP_t \cup \{p_i\}$. Then, EE_t is incremented by the electric power $CE_t(n)$ when $n = |CP_t|$. AT_t is also incremented by one if $n \neq 0$. For each process p_i, VC_i is decremented by the computation rate $NSR_t(n)/n$. If $VC_i \leq 0$, p_i terminates, i.e. $CP_t = CP_t - \{p_i\}$ and $etime_i = \tau$. The simulation is implemented in SQL [1] on a database.

Figure 3 shows the total electric energy consumption TEE [W · tu] of the four servers in the cluster S for number n of processes. The total electric energy consumption TEE of the ISEAM2T algorithm is smallest. For example, the TEE of the ISEAM2T is about 35% smaller than the RD and RR algorithms and about 15% smaller than ISEAM and ISEAM2 algorithms for n = 2,000.

Figure 4 shows the total active time TAT [tu] of the four servers. The TAT of the ISEAM2T algorithm is the shortest compared with the other algorithms. Servers are less loaded in the ISEAM2T algorithm than the other algorithms.

Figure 5 shows the average execution time AT [tu] of the n processes. The AT of the ISEAM2T algorithm is shortest compared with the other algorithms.

Table 1. Parameters of servers.

Parameters	DSLab2 (s_1)	DSLab1 (s_2)	Sunny (s_3)	Atira (s_4)
np_t	2	1	1	1
nc_t	8	8	6	4
nt_t	32	16	12	8
CRT_t [vs/tu]	1.0	1.0	0.5	0.7
$maxCR_t$ [vs/tu]	32	16	6	5.6
$minE_t$ [W]	126.1	126.1	87.2	41.3
$maxE_t$ [W]	301.1	207.3	131.2	89.5
bE_t [W]	30	30	16	15
cE_t [W]	5.6	5.6	3.6	4.7
tE_t [W]	0.8	0.8	0.9	1.1

Table 2. Parameters of processes.

Parameters	Values
n	number of processes $p_1, \ldots, p_n$
$minT_i$ [tu]	minimum computation time of a process p_i
VC_i [vs]	5.0 ~ 20.0 $(VC_i = minT_i)$
$stime_i$ [tu]	starting time of p_i $(0 \leq st_i < xtime$ - 1)
$xtime$ [tu]	simulation time (= 1,000 (= 100[sec]))

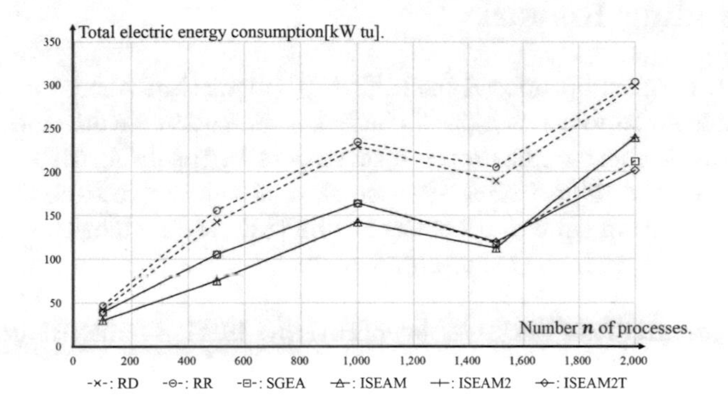

Fig. 3. Total electric energy consumption ($m = 4$, $v = 40$).

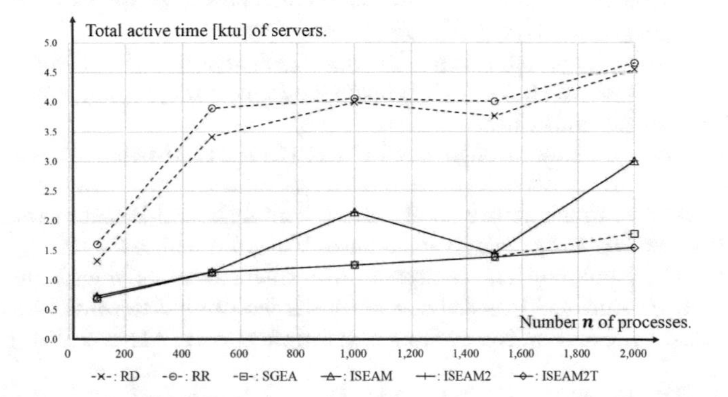

Fig. 4. Total active time ($m = 4$, $v = 40$).

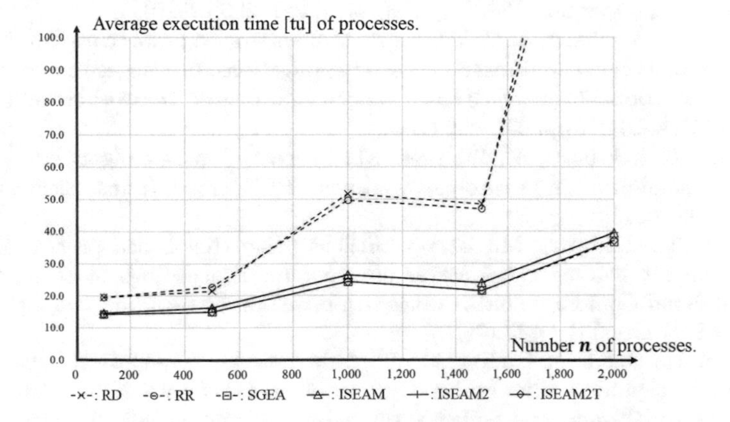

Fig. 5. Average execution time of processes ($m = 4$, $v = 40$).

6 Concluding Remarks

In this paper, we newly proposed the ISEAM2T algorithm where multiple virtual machines migrate to more energy-efficient servers. In the evaluation, we showed the total electric energy consumption of servers in the ISEAM2T algorithm is smallest in the other algorithms. We also showed the active time of servers and the average execution time of processes in the ISEAM2T algorithm can be most reduced compared with the other algorithms.

Acknowledgment. This work was supported by JSPS KAKENHI grant number 15H0295.

References

1. Datebase managiment system sybase. http://infocenter.sybase.com/help/index.jsp
2. Google, google green. http://www.google.com/green/
3. Intel xeon processor 5600 series: The next generation of intelligent server processors white paper. http://www.intel.com/content/www/us/en/processors/xeon/xeon-processor-e5-family.html
4. United nations climate change conference (COP21). https://en.wikipedia.org/wiki/2015
5. A virtualization infrastructure for the linux kernel (kernel-based virtual machine). https://en.wikipedia.org/wiki/Kernel-basedVirtualMachine
6. Duolikun, D., Enokido, T., Takizawa, M.: Asynchronous migration of process replica in a cluster. In: Proceedings of IEEE the 29th International Conference on Advanced Information Networking and Applications (AINA-2015), pp. 271–278 (2015)
7. Duolikun, D., Enokido, T., Takizawa, M.: Asynchronous migration of process replica in a cluster. In: Proceedings of the 9th International Conference on Complex, Intelligent, and Software Intensive Systems (CISIS-2015), pp. 118–125 (2015)
8. Enokido, T., Aikebaier, A., Takizawa, M.: A model for reducing power consumption in peer-to-peer systems. IEEE Syst. J. **4**(2), 221–229 (2010)
9. Enokido, T., Aikebaier, A., Takizawa, M.: An integrated power consumption model for communication and transaction based applications. In: Proceedings of IEEE the 25th International Conference on Advanced Information Networking and Applications (AINA-2011), pp. 627–636 (2011)
10. Enokido, T., Aikebaier, A., Takizawa, M.: Process allocation algorithms for saving power consumption in peer-to-peer systems. IEEE Trans. Indus. Electron. **58**(6), 2097–2105 (2011)
11. Enokido, T., Takizawa, M.: Energy-efficient delay time-based process allocation algorithm for heterogeneous server clusters. In: Proceedings of IEEE the 29th International Conference on Advanced Information Networking and Applications (AINA-2015), pp. 279–286 (2015)
12. Kataoka, H., Duolikun, D., Enokido, T., Takizawa, M.: Evaluation of energy-aware server selection algorithm. In: Proceedings of the 9th International Conference on Complex, Intelligent, and Software Intensive Systems (CISIS-2015), pp. 318–325 (2015)

13. Kataoka, H., Duolikun, D., Enokido, T., Takizawa, M.: Power consumption and computation models of a server with a multi-core CPU and experiments. In: Proceedings of IEEE the 29th International Conference on Advanced Information Networking and Applications Workshops (WAINA-2015), pp. 217–223 (2015)
14. Kataoka, H., Duolikun, D., Enokido, T., Takizawa, M.: Simple energy-aware algorithms for selecting a server in a scalable cluster. In: Proceedings of IEEE the 31st International Conference on Advanced Information Networking and Applications Workshops (WAINA-2017), pp. 146–153 (2017)
15. Watanabe, R., Duolikun, D., Cuiqin, Q., Enokido, T., Takizawa, M.: A simple energy-aware virtual machine migration algorithm in a server cluster. In: Proceedings of the 19th International Conference on Network-Based Information Systems (NBiS -2017), pp. 55–65 (2017)
16. Watanabe, R., Duolikun, D., Enokido, T., Takizawa, M.: An energy-efficient migration algorithm of virtual machines in server clusters. In: Proceedings of the 11th International Conference on Complex, Intelligent, and Software Intensive Systems (CISIS-2017), pp. 94–105 (2017)
17. Watanabe, R., Duolikun, D., Enokido, T., Takizawa, M.: A simply energy-efficient migration algorithm of processes with virtual machines in server clusters. J. Wirel. Mob. Netw. Ubiquitous Comput. Dependable Appl. (JoWUA) **8**(2), 1–18 (2017)
18. Watanabe, R., Duolikun, D., Enokido, T., Takizawa, M.: Simple models of processes migration with virtual machines in a cluster of servers. In: Proceedings of IEEE the 32nd International Conference on Advanced Information Networking and Applications (AINA-2018), pp. CD–ROM (2018)
19. Watanabe, R., Duolikun, D., Qin, C., Enokido, T., Takizawa, M.: A simple migration algorithm of virtual machine in a server cluster. In: Proceedings of the 12th International Conference on Broadband and Wireless Computing, Communication and Applications (BWCCA-2017), pp. 149–160 (2017)
20. Watanabe, R., Duolikun, D., Qin, C., Enokido, T., Takizawa, M.: Eco migration algorithms of processes with virtual machines in a server cluster. In: Proceedings of the 6th International Conference on Emerging Internet, Data and Web Technologies (EIDWT2018), pp. 130–141 (2018)

Evaluating Motion and Heart Rate Sensors to Measure Intensity of Physical Activity

Miguel A. Wister(✉), Pablo Pancardo, and Ivan Rodriguez

Academic Division of Information Technology and Systems,
Juarez Autonomous University of Tabasco, Cunduacan, Tabasco, Mexico
{miguel.wister,pablo.pancardo}@ujat.mx, jesucito.cukin@gmail.com

Abstract. Using a device for measuring the intensity of a physical activity when a person carries out their daily routines is an important support to monitor their health, especially if this person is overweight or obese since it exists risk for their health when demanding a lot of energy while performing physical activities. To confront this problem, there are new generation devices for measuring physical activity, that can be used to know physical intensity levels and consequently, establish exercise programs if were necessary to lose weight or maintain a certain level of training derived from a medical prescription. This paper evaluates the relationship between values of a motion sensor and heart rate sensor for measuring the intensity of physical activity of overweight or obese people. We propose to use these two sensors to determine the correlation between both so that at a given time, the motion sensor can be a useful alternative to measure the intensity of physical activity. This option makes easier for people to measure physical intensity with a conventional device equipped with an accelerometer, many people that use smartphones might avoid going to an expert to keep track of physical exercises.

1 Introduction

Currently, according to the World Health Organization (WHO) [11], overweight and obesity have become one of the main factors of disability and death worldwide. Overweight and obesity are defined as an abnormal or excessive accumulation of fat that can be harmful to health.

Overweight and obesity are defined as abnormal or excessive fat accumulation that may impair health. Body Mass Index (BMI) is a simple index of weight-for-height that is commonly used to classify overweight and obesity in adults. It is defined as a person's weight in kilograms divided by the square of his height in meters (kg/m^2).

World Health Organization [11] continues explaining, Obesity is one of the most pervasive, chronic diseases in need of new strategies for medical treatment and prevention. Obesity is defined as excess adipose tissue. There are several

L. Barolli et al. (Eds.): IMIS 2018, AISC 773, pp. 128–138, 2019.
https://doi.org/10.1007/978-3-319-93554-6_11

different methods for determining excess adipose (fat) tissue; the most common being the Body Mass Index (BMI). Obesity is a disease that affects more than one-third of the U.S. adult population (approximately 78.6 million Americans).

People who have obesity, compared to those with a normal or healthy weight, are at increased risk for many serious diseases and health conditions. The main causes that origin overweight and obesity is due to the consumption of high-calorie foods very rich in fat and to their sedentary lifestyle [11] and in turn, can cause cardiovascular disease, high blood pressure, diabetes mellitus type 2, hyperlipidemia, different types of cancer, degenerative arthropathy and psychological disorders [2].

Most overweight people want to perform physical activities to lose weight and enjoy good health, but due to the sedentary lifestyle that they have been living and the lack of good physical condition, it can be dangerous to start doing certain physical activities suddenly. Given this, the main cause of death can be a heart attack.

In these situations, there are new technology devices to measure physical activity of a person, that can serve to have control over physical intensity levels that have and consequently establish exercise programs when it is necessary to lose weight or maintain a certain level of exercise derived from a medical prescription [3].

The most used devices to measure the intensity of physical activity are heart rate sensors; since heart rate is an adequate indicator to estimate the level of effort that represents a physical activity. One of these devices is BASIS B1. This device belongs to the Intel company and has a highly effective heart rate sensor.

Another way to measure physical activity is by triaxial accelerometers that measure frequency and magnitude of accelerations and decelerations of body movement to monitor the intensity of physical activity of the subject. Triaxial accelerometers determine intensity and number of movements in three axes (x, y, z).

Some examples of these devices are FitbitBlaze and Applewatch, which are wrist-type watches that allow measuring physical exercises such as running, cycling and cardio. These devices are able to monitor the activity when a person sleeps, their diet, in addition to being able to synchronize with smartphones and are currently competing strongly in the market [4].

One more of these new innovative devices and it will work on this research project to measure the acceleration of movements is GENEActiv [4]. GENEActiv is the raw data accelerometer in form of a bracelet, useful for measuring physical activities and frequencies of movement during daily activities.

Some investigations that use GENEActiv device are related to lack of physical activity and obesity and overweight. A sample is shown in the research project "Project Energise: Using participatory approaches and real-time prompts to reduce occupational sitting and increase physical activity in office workers", that conducted a study in 57 office workers (age 47 $\pm$ 11) who have office work where it shows a sedentary lifestyle who they are living due to their kind of work and this opens the door to possible chronic diseases [5].

An opportunity to perform an analysis that shows the relationship between a heart rate sensor (BASIS B1) and a motion sensor (GENEActiv) is this research project.

2 Related Works

In [8], Calibration of the GENEA accelerometer for assessment of physical activity intensity in children, the study established intensity levels of activity using the GENEA accelerometer and by calibration with oxygen consumption (VO2). In this study, forty-four children, aged 8–14 years, were taken into account, who performed eight activities (ranging from the supine position to a medium-speed race) while carrying three GENEA accelerometers (one on each wrist and another on the right side of the hip), an ActiGraph GTIM on the hip and a portable gas analyzer. The GENEA showed good validity in both wrists (right: $r = 0.900$, left: $r = 0.910$, both $p < 0.01$), but the monitor placed on the hip showed significantly greater validity ($r = 0.965$ $p < 0.05$). In conclusion, GENEA can be used in children to accurately assess the intensity of physical activity, either in the wrist or hip.

In [10] is considered an architecture of wearable health monitoring system composed by wearable sensor system, network transmission system and the information processing system on the basis of the three-layer architecture of the internet of things. These data collected are transmitted through Bluetooth technology to a gateway for providing real-time monitoring. All data acquired through network are transmitted in real time to the database server.

Another study [1] entitled: Healthy obesity and objective physical activity, it aimed to examine the differences in moderate to vigorous physical activity between healthy and unhealthy obese groups by using questionnaires answered by the participants themselves, as well as with the evaluations of an accelerometer placed on the wrist.

The analysis was performed on 3,457 adults aged 60–82 years (77% males). The differences between groups in physical activity, based on questionnaires and evaluations with the triaxial wrist accelerometer (GENEActiv), were examined using linear regression. When doing the evaluation with the GENEA ($P = 0.002$) it was determined that the healthy obese adults did a more physical activity than the unhealthy obese adults. In conclusion, it was found that there was greater physical activity in healthy obese than in obese adults not healthy, when measured objectively (accelerometer), which suggests that physical activity has an important role in promoting health in obese populations.

The work [7] presents a design and implementation of a testbed for AmI using Raspberry Pi mounted on Raspbian OS. It is analyzed the optimized link state routing (OLSR) and wired equivalent privacy (WEP) protocol in an indoor scenario, and mean shift clustering algorithm considering sensing data.

In [6], Physical Activity in Hemodialysis Patients Measured by Triaxial Accelerometer, it published in the journal BioMed Research International mentions that there are several factors that can contribute to a sedentary lifestyle among patients on hemodialysis (HD), including the time that they go on dialysis. The objective of this study was to evaluate the characteristics of physical activities in the daily life of hemodialysis patients through the use of a triaxial accelerometer and the correlation of these characteristics with physiological variables.

Nineteen patients in HD were evaluated using the DynaPort accelerometer and compared with nineteen control individuals (not ill), with respect to the time spent in different activities and everyday situations, and the number of steps when walking. Patients with HD were more sedentary than control individuals because they spend less time walking or standing and spend more time in bed.

The sedentary lifestyle was more visible in the days of dialysis. According to the number of steps given per day, 47.4% of hemodialysis patients were classified as sedentary versus 10.5% in the control group. The level of hemoglobin, the muscle strength of the lower limbs, and the physical functioning of the SF-36 questionnaire correlated significantly with walking and active time.

3 Evaluation Overview

Intensity reflects the speed that an activity is performed or the magnitude of the effort required to perform an exercise or activity. It can be estimated by asking how much a person has to strive to perform that activity [11]. The World Health Organization [11] defines the following types of physical activity intensities:

- Mild activity: It refers to natural movements, simple, that do not require much effort (walking, getting up from a chair).
- Moderate activity: Indicates moderate strength exercises (normal running, dancing).
- Vigorous or intense activity Requires a large amount of effort and causes rapid breathing and a substantial increase in heart rate (playing football or basketball match by actively intervening).

ChooseMyPlate.gov website [9] defines physical activity simply means movement of the body that uses energy. For health benefits, physical activity should be moderate or vigorous intensity.

- Moderate physical activities include: Walking briskly (about 3 miles per hour), bicycling (less than 10 miles per hour), dancing, golf (walking and carrying clubs), water aerobics, canoeing, and tennis (doubles).
- Vigorous physical activities include: Running/jogging (5 miles per hour), walking very fast (4 miles per hour), bicycling (more than 10 miles per hour), heavy yard work, such as chopping wood, swimming (freestyle laps), aerobics, basketball (competitive), and tennis (singles).

As instruments to estimate the intensity of physical activity, a BASIS B1 Smartwatch was used, that is a device that calculates the heart rate. This device is a bracelet and is worn on the right wrist. The BASIS B1 is capable of noticing exercises (walking, running, cycling) and keeps track of the heart rate all the time.

We also used a GENEActiv bracelet, that contains a motion sensor. GENEActiv produces raw data, useful for the measurement of physical activities and frequencies of movement during daily activities. This device is scientifically validated by numerous investigations.

GENEActiv device is worn on the hip (GENEActiv is bracelet but was adapted to a belt for wearing on the hip), studies suggest that placing on the hip makes a better estimation of the intensity of movements during physical activities because it is a point close to the center of mass, it also allows to represent values that would be obtained from a smartphone with motion sensor.

Physical activities consisted of performing physical effort, therefore, all participants walked and ran on a treadmill at different speeds and under a certain period of time. In detail, the activities were: walking (speed: 3 Km/h), jogging (speed: 5 Km/h), and running (speed: 7 Km/h). For each speed, the execution time was at least three minutes, that was sought to reflect the physical effort in the motion sensor and heart rate.

Based on the data obtained from both sensors, mathematical methods were applied to show the relationship of the intensity level between both devices. Once different evaluations were made, the model that best reflected the relationship between both devices was selected.

4 Experimental Results

Equipment

To measure the intensity of physical activities, two devices were used: In Fig. 1 (left) BASIS B1 (heart rate sensor) and Fig. 1 (right) GENEActiv (motion sensor).

Fig. 1. Basis 1 device (left) GENEActiv device (right)

As we mentioned, overweight and obesity are defined as an abnormal or excessive accumulation of fat in a person's body that can be harmful to health

[11], while, the BMI (Body Mass Index) is a formula that uses both weight and height to estimate body fat.

All tests were done using a treadmill, the speed in K/h in these experiments was setting on the sensors for collecting values of physical intensity measurement according to each device. Figure 2 (left) shows to the GENEActiv device placed on the hip, it also shows that a belt was used to hold the device. BASIS B1 bracelet device was placed on the wrist as is shown in Fig. 2 (right), each user wears this device on the right wrist.

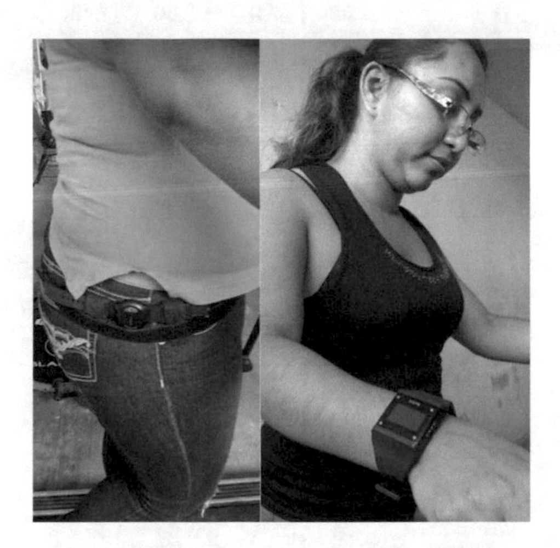

Fig. 2. GENEActiv device placed on hip (left) and BASIS B1 device placed on wrist (right)

This project used a non-probabilistic sampling, where were chosen participants for convenience, in Table 1 we summarize all data of the participants. To measure physical activity, 11 men and 9 women were selected, who had time to perform all the tests. These studies were conducted on people between 20 and 30 years old, regular or average weight, overweight or obesity (determined according to their BMI) and who did not suffer from any type of disease that could be aggravated during the test or affect the readings of the motion and heart rate sensors.

Our tests were carried out at the gym, where a treadmill is located. At the time of the activity, participants wore a BASIS B1 bracelet on their wrist (right hand) and a GENEActiv on their hip, because it is an ideal place to measure physical activity since it is near the center of body mass. See Fig. 3.

GENEActiv device obtained samples at a frequency of 75 Hz generating 75 samples per second. These 75 samples are summarized by the device in such a way that it has only one value per second for each of the axes. Table 2 is a small sample of the x, y, and z values for 10 s. To obtain data from the BASIS B1, it was only necessary to touch the screen of the device and record the amount that reflected the heart rate.

Table 1. Respondent's data

Respondents	Age	Weight	Height	BMI
1	22	52.4	1.47	24.2
2	22	81.7	1.58	32.7
3	21	50.6	1.52	21.9
4	26	51.0	1.60	19.9
5	23	85.2	1.59	33.7
6	28	62.0	1.60	24.2
7	24	54.4	1.59	21.5
8	20	56.3	1.56	23.1
9	20	56.0	1.59	22.2
10	23	70.2	1.65	28.4
11	22	65.2	1.65	23.9
12	22	69.1	1.66	25.1
13	25	74.4	1.69	26.0
14	23	67.6	1.64	25.1
15	22	73.8	1.68	29.7
16	27	82.3	1.70	28.5
17	22	74.0	1.72	25.0
18	21	79.2	1.72	26.8
19	24	78.9	1.72	26.7
20	24	95.4	1.64	35.5

[a] Weight (Kg) and Height (m)

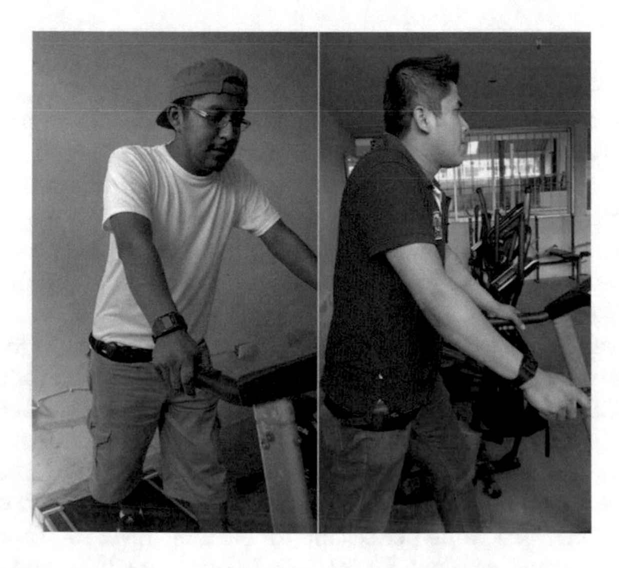

Fig. 3. GENEActiv bracelet and BASIS B1 devices placed on wrist and hip, respectively.

Table 2. Sample of data obtained by the GENEActiv

Second	X	Y	Z
1	−0.1900	−0.9920	0.0271
2	−0.2315	−0.9721	0.0365
3	−0.2108	−1.0024	0.0136
4	−0.1868	−1.0062	0.0175
5	−0.1737	−0.9652	0.0868
6	−0.2350	−1.0150	0.0966
7	−0.2151	−0.9658	0.0265
8	−0.2271	−0.9710	0.0431
9	−0.2070	−0.9962	0.0348
10	−0.1632	−1.0109	−0.0130

Table 3 shows the final results of a female participant, where the intensity levels at the different speeds turned out to be the same with the different devices.

Table 3. Female - intensity level

Speed (Km/h)	Respondent	Gender	GENEActiv	BASIS B1
3	1	Female	Light	Light
5	1	Female	Moderate	Moderate
7	1	Female	Vigorous	Vigorous

Regarding the final results of a male participant are shown in Table 4, where the intensity levels with both devices and at different speeds turned out to be the same.

Table 4. Male - intensity level

Speed (Km/h)	Respondent	Gender	GENEActiv	BASIS B1
3	10	Male	Light	Light
5	10	Male	Moderate	Moderate
7	10	Male	Vigorous	Vigorous

Table 5 lists the final results of a female participant (Number 5), who has the average age corresponding to all the participants. The intensity levels with both devices at the different speeds turned out to be almost the same, the exception was

Table 5. Average age - intensity level

Speed (Km/h)	Respondent	Gender	GENEActiv	BASIS B1
3	5	Female	Light	Light
5	5	Female	Light	Moderate
7	5	Female	Vigorous	Vigorous

in the 5 km/h test, where the results of the participant number 5 are reviewed, here there was a small difference in level but it is not considered relevant.

Table 6 shows the final results of a male participant (Number 12), who has the average weight of all participants, the intensity levels for both devices at different speeds turned out to be very similar, the exception was in the 5 km/h test, where this participant obtained a vigorous level with GENEActiv and Moderate level with BASIS B1, it could be appreciated that the Vigorous level was the result of the intense style of participant number 12, when running freely on the treadmill.

Table 6. Average weight - intensity level

Speed (Km/h)	Respondent	Gender	GENEActiv	BASIS B1
3	12	Male	Light	Light
5	12	Male	Vigorous	Moderate
7	12	Male	Vigorous	Vigorous

In regards the final results of a female participant (Number 4) are shown in Table 7, He has the average height of all the participants, the intensity levels at different speeds for both devices were practically the same.

Table 7. Average height - intensity level

Speed (Km/h)	Respondent	Gender	GENEActiv	BASIS B1
3	4	Female	Light	Light
5	4	Female	Vigorous	Moderate
7	4	Female	Vigorous	Vigorous

Table 8 shows the final results of a male participant, who has a light intensity level at all speeds by the BASIS B1 device. The result was produced because participant 11 is a person who practices a lot of sport, and since the BASIS B1 measures the heart rate, this person showed a low rate of heart rate at performing the activities. It is established that usually, people with the very good physical condition have a lower heart rate compared to those who lack good physical condition.

Table 8. Person with unequal intensity level

Speed (Km/h)	Respondent	Gender	GENEActiv	BASIS B1
3	11	Male	Light	Light
5	11	Male	Vigorous	Light
7	11	Male	Vigorous	Moderate

5 Conclusions

This article evaluated the relationship between values of motion and heart rate to measure the intensity of physical activity of overweight or obese people. For this, a group of people performed different physical activities.

According to the values showed in the results of different physical activities, through a series of mathematical operations and an analysis to locate the results in the different ranges of physical intensity, it was possible to obtain the physical intensity level from each sensor.

Both the BASIS B1 and the GENEActiv showed a very similar level of intensity for each participant, hence means that at a given moment the movement sensor can be used to estimate the physical activity of a person, specifically an activity where is used the entire body to perform it, or else, an activity where there is no significant muscular effort to try to lift or load a heavy object.

It is also concluded that to estimate physical activity you can make use of the motion sensor contained in a smartphone, thus having financial savings if it already has one. That is, an additional device is not necessary.

As shown in Table 4, the use of a motion sensor is not useful if the person in whom it is applied practices a lot of sport, since due to its very good physical condition it will not be reflected by the motion sensor calculation of the intensity level.

References

1. Bell, J.A., Hamer, M., van Hees, V.T., Singh-Manoux, A., Kivimäki, M., Sabia, S.: Healthy obesity and objective physical activity. Am. J. Clin. Nutr. **102**(2), 268–275 (2015). https://doi.org/10.3945/ajcn.115.110924. http://www.ncbi.nlm.nih.gov/pmc/articles/PMC4515867/
2. CDC: The Health Effects of Overweight and Obesity (2015). https://www.cdc.gov/healthyweight/effects/index.html
3. Ciuti, G., Ricotti, L., Menciassi, A., Dario, P.: Mems sensor technologies for human centred applications in healthcare, physical activities, safety and environmental sensing: a review on research activities in Italy. Sensors **15**(3), 6441–6468 (2015). https://doi.org/10.3390/s150306441. http://www.mdpi.com/1424-8220/15/3/6441
4. GENEActiv: Fitbit Blaze ™Smart Fitness Watch (2016). https://www.fitbit.com/es/blaze

5. Gilson, N., Ng, N., Pavey, T., Ryde, G., Straker, L., Brown, W.: Project energise: using participatory approaches and real time prompts to reduce occupational sitting and increase physical activity in office workers (2015)
6. Gomes, E.P., de Moura Reboredo, M., Carvalho, E.V., Teixeira, D.R., d'Ornellas Carvalho, L.F.C., Filho, G.F.F., de Oliveira, J.C.A., Sanders-Pinheiro, H., Chebli, J.M.F., de Paula, R.B., Pinheiro, B.: Physical activity in hemodialysis patients measured by triaxial accelerometer. In: BioMed Research International (2015)
7. Obukata, R., Oda, T., Barolli, L.: Design of an ambient intelligence testbed for improving quality of life. In: 2016 30th International Conference on Advanced Information Networking and Applications Workshops (WAINA), pp. 714–719 (2016). https://doi.org/10.1109/WAINA.2016.148
8. Phillips, L.R., Parfitt, G., Rowlands, A.V.: Calibration of the genea accelerometer for assessment of physical activity intensity in children. J. Sci. Med. Sport **16**(2), 124–128 (2013). https://doi.org/10.1016/j.jsams.2012.05.013. http://www.sciencedirect.com/science/article/pii/S1440244012001120
9. USDA: The USDA Center for Nutrition Policy and Promotion (CNPP) (2015). https://www.choosemyplate.gov/physical-activity-what-is
10. Wang, X.: The architecture design of the wearable health monitoring system based on internet of things technology **6**, 207 (2015)
11. WHO: World Health Organizatiion Official Web. Obesity and overweight (2018). http://www.who.int/mediacentre/factsheets/fs311/en/

Implementation of WiFi P2P Based DTN Routing and Gateway for Disaster Information Networks

Noriki Uchida[1(✉)], Haruki Kuga[1], Tomoyuki Ishida[2], and Yoshitaka Shibata[2]

[1] Fukuoka Institute of Technology, 3-30-1 Wajirohigashi, Higashi-ku, Fukuoka 811-0214, Japan
n-uchida@fit.ac.jp, s14b1024@bene.fit.ac.jp

[2] Iwate Prefectural University, 152-52 Sugo, Takizawa, Iwate 020-0693, Japan
t-ishida@fit.ac.jp, Shibata@iwate-pu.ac.jp

Abstract. If there is an ultra-large scale disaster happened, the important messages such as life safety or rescues would not be transferred because of the serious damages of information networks. It is supposed that the Delay Tolerant Networking (DTN) is one of the effective routing methods for realizing against such robust network conditions, but some problems such as IP configurations or gateway functions have been considered for the realistic mobile networks. Therefore, this paper proposed the layer 2 level DTN routing methods by using WiFi P2P and the gateway functions between the DTN and IP networks. Then, the implementations of the prototype systems by Android smartphones and the field experiments are reported in the paper, and the experimental results are discussed for the effectiveness of the proposed methods and future studies.

1 Introduction

If there is an ultra-large scale disaster happened such as the Great East Japan Earthquake in 2011, the important messages such as life safety or rescues would not be transferred because of the serious damages of information networks. Actually, the paper [1] reported the network analysis during the disaster, it discussed the future subjects of the network requirements for the disaster information networks. When it is considered about the difficulties of the network conditions during the disaster, it is widely known that the Delay Tolerant Networking (DTN) [2] is one of the effective routing methods for realizing against such robust network conditions.

However, some previous papers [3–5] have pointed out the problems of network latency or delivery rate of massages through the DTN routing, and they proposed the enhanced DTN routing such as the Spray and Wait [3], the Maxprop [4], and the PRoPHET [5] method. Moreover, some problems such as IP configurations or gateway functions have been considered for the realistic mobile networks.

Therefore, this paper proposed the layer 2 level DTN routing methods by using the WiFi P2P and the gateway functions between the DTN and IP networks in consideration with the realistic mobile networks. In the proposed methods, these functions are

L. Barolli et al. (Eds.): IMIS 2018, AISC 773, pp. 139–146, 2019.
https://doi.org/10.1007/978-3-319-93554-6_12

additionally introduced to the previously proposed Data Triage Methods [6] that is the queue ordering functions by the user policy, and the node priority function is additionally realized by using WiFi P2P functions. Then, the implementations of the prototype systems by Android smartphones and the field experiments are reported in the paper, and the experimental results are discussed for the effectiveness of the proposed methods and future studies.

In the followings, the system architecture and the requirements for the proposed methods are discussed in Sect. 2, and Sect. 3 introduces the proposed the WiFi P2P and the gateway functions for the realization of the node priority and the realistic mobile networks. Section 4 explains the implementations of the proposed methods. Finally, Sect. 5 introduces the prototype system and the experiments, and the conclusion and future study are presented in Sect. 6.

2 Proposed Networks

The DTN routing is the stored-and-carried typed routing protocol, and so it is supposed to be the effective routing methods for the robust network conditions such as the disaster. However, it has been some enhanced DTN routing methods for improving the efficiency or the data transmissions. For example, the Spray and Wait [2] introduced to reduce the overhead of the data transmissions, the Maxprop [3] proposed the queue ordering methods by using the probability of message delivering by the node encounters, and the PRoPHET [4] proposed to the epidemic routing based on the history of encounters and transitivity.

However, in the case of realistic mobile networks, there are some problems such as the gateway between the DTN and IP networks or the duration of the IP configurations such as DHCP for the new encounters. Figure 1 shows the proposed network architecture of the proposed DTN routing.

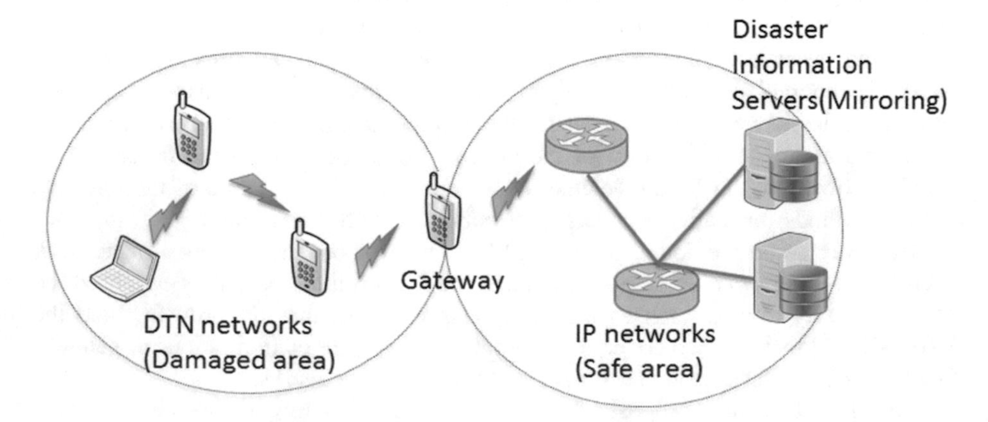

Fig. 1. The network architecture of the proposed methods. The messages are firstly delivered in the DTN networks until the mobile node are going to the inside of the IP networks.

In the proposed network as shown in Fig. 1, there are the DTN networks that consists of the wireless mobile devices in the damaged area, and there are the IP networks that survived from the disaster consists of the wired and wireless network. Then, it is consumed that every wireless mobile device equips the gateway function between the DTN and IP networks, and so the evacuator's messages are automatically transmitted to the disaster information servers if the mobile device reaches to the inside of the IP networks.

Also, it is assumed that the proposed DTN networks supports the Data Triage Methods [6] for the effectivity of the data delivery rate and the latency. The Data Triage Methods for the DTN routing is the queue-order data transmission methods by the user policy [6], and the data priority function is decided by the user policy changes through the time after the disaster happened. That is, if there is the disaster, text data such as rescue, evacuation, and safety status information are decided as the significantly important during a couple of days. Next, bi-directional real-time information such as relief rations based on VoIP becomes important after a couple of days, and then the broadband communication data such as video conference finally afford in the Data Triage Methods after a week. The paper [7] also presents the implementations of the Data Triage Methods by using the Android smartphone.

This paper additionally introduced the WiFi P2P functions for the node priority decisions and the gateway functions between the DTN and IP networks in consideration with the realistic mobile networks in the following sections.

3 The WiFi P2P and the Gateway Functions

The previous study [9] proposed that the necessary of the node selections and the data error corrections except the data priority for the disaster information networks. Especially, the evacuators just after the disaster tends to need the information in the close area because they need such as the evacuation shelters, grocery stores, or gas station. Besides, according to the previous study [9], it is known that there are two major movement's patterns after the disaster as shown in Fig. 2. There, the nodes inside of the one's area is focused in the proposed methods.

Each node continuously observed the location by the GPS or WiFi networks in the methods, and the maximum node territory value is calculated as the following.

$$S_i(t_n) = \sum_{k=1}^{n} \{\alpha_k D_i(t_k)\} \quad (1)$$

Here, t_n is the periodic observation time, S_i is the calculated territory value in the area i, α_k is the weighted values that the total sum is one, and D_i is the observed duration inside of the area i. Then, the nodes that are mostly stayed in the same local area are decided as the priority node, when there are some transmittable mobile nodes in the assumed network.

Then, the Android API of the WiFi P2P is used for the proposed node selection methods. The WiFi P2P framework [10] complies with the WiFi Alliance's Wi-Fi

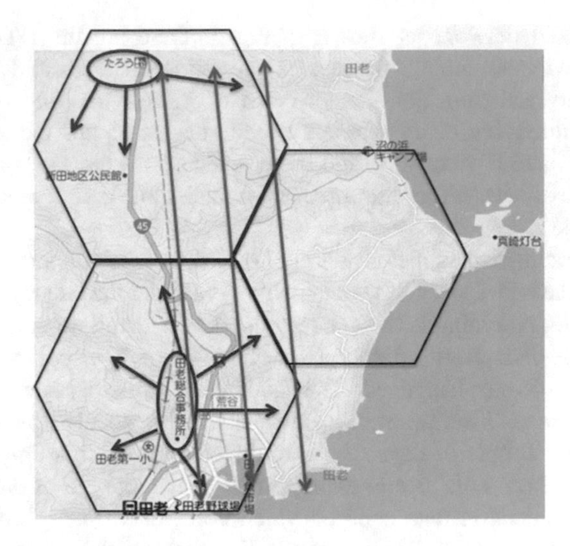

Fig. 2. Two major movements of the evacuators after the Great East Japan Earthquake. One is the movements inside of the near area that is mainly evacuators, and another is the movements across areas that are mainly consists of the rescues.

Direct certification program, and it is provided by the Android 4.0 or later. It allows the direct connection to each other device with the WiFi connections, and it also support additional functions such as the broadcast transmission by the grouping IDs. Therefore, the proposed DTN routing supports both P2P transmission and multicast data transmission by the grouping IDs.

4 The Prototype System

The prototype system is implemented for the effectiveness of the proposed methods as shown in Fig. 3. Nexus 5X (Android OS 6.0.1, IEEE802.11a/b/g/n/ac, 2 GB MEM, 16 GB Storage), Nexus 7 (Android OS 6.0.1, IEEE802.11a/b/g/n/ac, 2 GB MEM, 16 GB Storage), Android Studio 2.1.3, and Java 8.0 are used for the implementations of the node selections and the gateway functions in the prototype system. Also, the Wifip2pManager class is used for the P2P and multicast data transmission in the system.

Then, the flowcharts of the proposed node selections and gateway functions are presented in Fig. 4.

First of all, the right figure is about the gateway functions in the prototype system. After the application is activated in the smartphone, the network connection to the servers of the Disaster Information System. Here, since it is supposed that the server IP address is not changed so frequently, the previously confirmed IP addresses are checked by Ping command periodically. Therefore, if the smartphone start to connect the Internet through WiFi or LTE networks, the stored data in the smartphone is automatically merged to the disaster servers. When the Ping is not worked, the

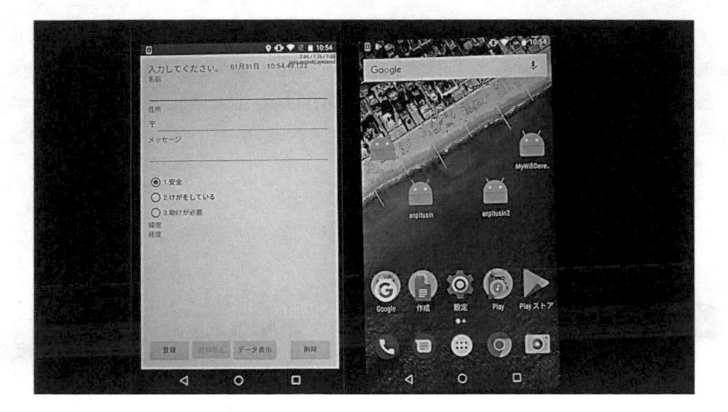

Fig. 3. The prototype system by the Android smartphone.

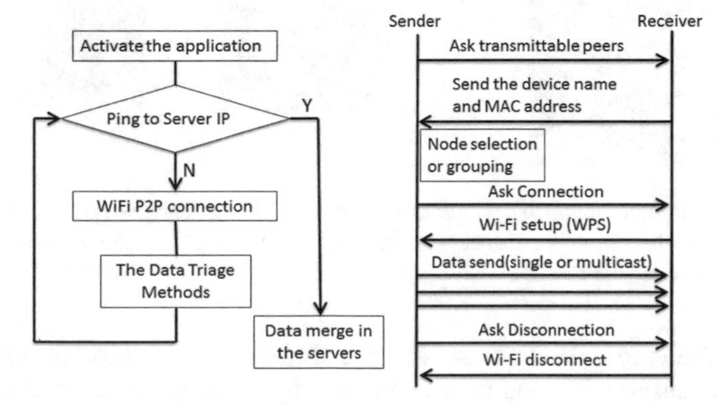

Fig. 4. The flowcharts of the implementations. The left figure is the methods of the gateway between the DTN and IP networks. The right figure is about the WiFi P2P connection by the Wifip2pManager class in Android OS.

smartphone confirm the WiFi P2P connection flow that is in the left figure. Then, there are WiFi devices that are in the transmittable wireless range, the single or multicast data transmissions by the right flow are confirmed without the IP address. Therefore, it is considered that the proposed methods improved the DTN transmission with the IP configuration such as DHCP and the efficiency of the data delivery rate and the latency.

5 Experiments

For the effectiveness of the proposed methods, the field experiments of the data transmissions were measured by the prototype system. The experiments were held in the Fukuoka Institute of Technology, Japan, and the periods and the number of received messages were measured when the 100 messages were transmitted by the DHCP and the proposed methods. As shown in Fig. 5, the sender pretended to fall down under the boxes and chairs, and the receiver pretended to walk through the corridor with the short look to the inside of the laboratory. The 100 messages consisted of 100 Japanese characteristics, and the receiver was shortly stayed in front of the door about five seconds in the experiments. Also, the 100 messages are previously divided as the 1st priority messages, 2nd priority messages, and 3rd priority messages by the Data Triage Methods.

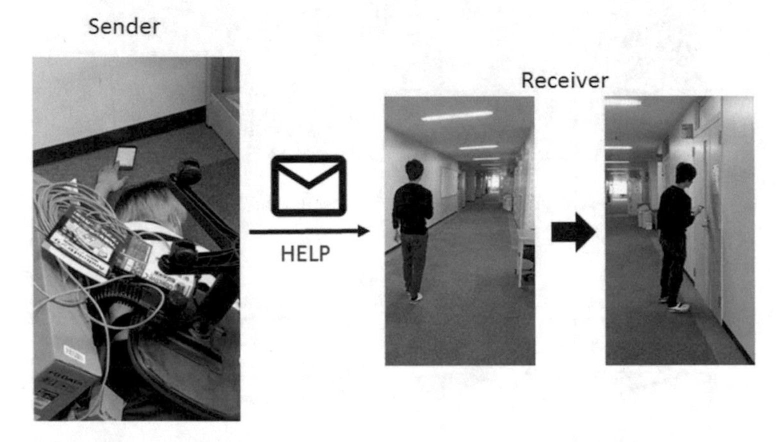

Fig. 5. The field experiments by the prototype system. The sender pretended to fall down the boxes and chairs by assuming the earthquake, and the receiver pretend to walk through the corridor by assuming the rescues.

As the results in Table 1, there were the message transmission errors by DHCP WiFi connections as 97%, and there is no error by the proposed methods. Moreover, the period of data transmission was 8.031 s by the DHCP. However, the proposed methods spent 1.998 s to complete the data transmission.

Table 1. The results of the data transmission under the DHCP and the proposed methods.

	DHCP		Proposed methods	
	Send messages	Received messages	Send messages	Received messages
1st priority	33	31	33	33
2nd priority	33	32	33	33
3rd priority	34	34	34	34
SUM	100	97	100	100
Delivery rate	97%		100%	
Period	8.031 s		1.998 s	

Therefore, it is supposed that the proposed methods by the WiFi P2P connection effectively in comparison with the DHCP WiFi connection by using IP networks.

6 The Conclusion and Future Study

If there is an ultra-large scale disaster happened, the important messages such as life safety or rescues would not be transferred because of the serious damages of information networks. It is supposed that the Delay Tolerant Networking (DTN) is one of the effective routing methods for realizing against such robust network conditions. However, in the case of the realistic IP networks, it is supposed that the usage of the IP configurations or the gateway functions might be problems for the effectiveness of the data delivery and the latency. Therefore, this paper proposed the layer 2 level DTN routing methods by using WiFi P2P and the gateway functions between the DTN and IP networks. Then, the implementations of the prototype systems by Android smartphones and the field experiments are reported in the paper.

As the results of the field experiments by the implemented prototype system, the delivery rate of the data transmission was increased as the comparison with the DHCP WiFi connections. Moreover, the period of the data transmission was increased significantly, and it is supposed that the proposed methods works effectively by these experimental results.

Now, the additional field experiments including the effectiveness of the gateway functions are planning for the future studies, and the error correction methods for DTN routing by FEC (Forward Error Correction) is also considered for the future works.

References

1. Shibata, Y., Uchida, N., Shiratori, N.: Analysis of and proposal for a disaster information network from experience of the Great East Japan Earthquake. IEEE Commun. Mag. **52**, 44–50 (2014)
2. Fall, K., Hooke, A., Torgerson, L., Cerf, V., Durst, B., Scott, K.: Delay-tolerant networking: an approach to interplanetary internet. IEEE Commun. Mag. **41**(6), 128–136 (2006)
3. Spyropoulos, T., Psounis, K., Raghavendra, C.S.: Spray and wait: an efficient routing scheme for intermittently connected mobile networks. In: WDTN 2005 Proceedings of the 2005 ACM SIGCOMM Workshop on Delay-Tolerant Networking, pp. 252–259 (2005)
4. Burgess, J., Gallagher, B., Jensen, D., Levine, B.: MaxProp: routing for vehicle-based disruption-tolerant networks. In: INFOCOM 2006, Proceedings of the 25th IEEE International Conference on Computer Communications, pp. 1–11 (2006)
5. Lindgren, A., Doria, A., Scheln, O.: Probabilistic routing in intermittently connected networks. ACM SIGMOBILE Mob. Comput. Commun. Rev. **7**(3), 19–20 (2003)
6. Uchida, N., Kawamura, N., Shibata, Y., Shiratori, N.: Proposal of data triage methods for disaster information network system based on delay tolerant networking. In: The 7th International Conference on Broadband and Wireless Computing, Communication and Applications (BWCCA 2013), pp. 15–21 (2013)
7. Uchida, N., Kawamura, N., Shibata, Y.: Delay tolerant networks on vehicle-to-vehicle cognitive wireless communication with satellite system for disaster information system in a coastal city. IT CoNvergence PRActice (INPRA) **1**(1), 53–66 (2013). ISSN: 2288–0860 (Online)

8. Uchida, N., Kawamura, N., Sato, G., Shibata, Y.: Delay tolerant networking with data triage method based on emergent user policies for disaster information networks. Mob. Inf. Syst. **10**(4), 347–359 (2014)
9. Uchida, N., Kawamura, N., Takahata, K., Shibata, Y.: Proposal of dynamic FEC controls with population estimation methods for delay tolerant networks. In: The 6th International Workshop on Disaster and Emergency Information Network Systems (IWDENS 2014), pp. 633–638 (2014)
10. Android Developers: WiFi peer-to-peer over overview. https://developer.android.com/guide/topics/connectivity/wifip2p. Accessed 2018

A New Contents Delivery Network Mixing on Static/Dynamic Heterogeneous DTN Environment

Shoko Takabatake[1(✉)] and Tetsuya Shigeyasu[2]

[1] Program in Information and Management Systems, Graduate School of Comprehensive Scientific Research, Prefectural University of Hiroshima, Hiroshima, Japan
q822003fr@ed.pu-hiroshima.ac.jp

[2] Department of Management and Information Systems, Prefectural University of Hiroshima, Hiroshima, Japan
sigeyasu@pu-hiroshima.ac.jp

Abstract. For reducing the damage of disaster, it is needed to correct/deliver disaster information rapidly. However, under the disaster occasion, it is not easy to engage the usual communication due to the lack of perfect operation of communication infrastructure. Hence, in this paper, we propose to construct new contents centric data delivery system over the network consisting of DTN nodes. The performance evaluations confirm that our proposal effectively reducing the cache acquisition delay.

1 Introduction

Recently, a variety of communication infrastructures including smartphones have been widely deployed. In addition to the smartphone used by consumers, WSN (Wireless Sensor Network) consists of a bunch of small unmanned devices has been developed [1]. Such an environment made a major change in our daily life. In addition, it is reported that those IT communication tools are used to obtain emergency information under disaster situation.

However, communication infrastructure built before occasion of disaster, have a high risk for being inactive state. If so, powerful tools mentioned in the above, could not work effectively for the disaster relief activities. Under the situations that fixed infrastructure broke due to the effects of disaster, offices of local government relying on public communication line, can not grasp important information from disaster affected areas, quickly. For enabling to grasp disaster information from such area even if the public communication line has been broken, number of network systems have been proposed in [4–6].

One of the most powerful network system against effects of a disaster, is DTN (Delay Tolerant Network) [7]. By using DTN, any couple of nodes can communicate without fixed communication infrastructure even if these nodes disconnected each other. DTN is used not only in disasters but also in many situations. In the literature [2,3], authors have proposed to use DTN on V2V (Vehicle-to-Vehicle) network.

L. Barolli et al. (Eds.): IMIS 2018, AISC 773, pp. 147–158, 2019.
https://doi.org/10.1007/978-3-319-93554-6_13

In DTN, message transmitted by sender traverses entire a network by relaying among the other nodes. A node received the message forwards it to next node immediately if it has neighbor nodes in its communication range, otherwise, it moves around with holding the message until appropriate neighbor appeared.

This process is called as *move-carry-forward* mechanism well works under disaster effected area. However, performance of message delivery delay is much influenced by the density and moving speed of the relay nodes. In addition, throughput of DTN will be easily to degraded in proportion to in cease the amount of network traffic.

Therefore, in this paper, for improving performance of network system which robust for drastically changing network conditions, we propose a new disaster information delivery system. Our proposal consists of DTN and CCN (Content Centric Network) [9] for reducing network traffic and message delay. Even though the technology of CCN is originally developed for wired network, mechanism of in-network cache of CCN can effectively reduce the data obtaining time, for networks other than wired network.

For clarifying the availability of CCN on DTN situation, we firstly report the results of computer simulation comparing DTN with and without CCN, those destined for disaster information collecting. Next, we also report the performance of CCN on DTN in influenced by the amount network traffic.

2 Communication Risk on Earthquake and Disaster Information System

2.1 Communication Risk

At 14:46 on March 11 in 2011, a large-scale earthquake had happen in Japan and it gave serious damage to communication infrastructure. In such situation, there were number of tremendous serious affected damages: collapse and/or submergence on buildings and loss in communication facilities, damage of underground cables and pipelines, destruction of utility poles, collapse of cellular phone base stations. Hence, victims in disaster affected areas had hard to acquire disaster information.

2.2 Disaster Information System

2.2.1 ADES: Autonomous Data Exchange System

In order to overcome weakness of the existing systems, we have proposed another network system which establishes wireless network among shelters to avoid damage on communication facilities in literatures.

In the literature [4], authors have proposed the system, named, ADES (Autonomous Data Exchange System), establishes a WLAN network based on IEEE802.11 because it is widely used and low cost and many victims can get the network connections by their smartphones. The system overview of the ADES is shown in Fig. 1. In this system, own wireless network with backbone network (Fig. 1(A)) and branch network (Fig. 1(B)) will be established, after disaster.

The backbone network consists of earthquake resistant buildings and it covers entire disaster affected area. The branch network consists of medium and small evacuation shelters around each backbone base stations. The branch base stations collect the disaster information at each shelters and the information is shared with the nearest backbone base stations. In addition, by exchanging information between adjacent backbone base stations, victims can grasp disaster information from the entire disaster area. With this network system, however, it is impossible for victims in the disaster area to get disaster information. Because only information from victims who could evacuate at the shelters and countermeasure office will be shared.

2.3 DTN

DTN has been proposed as a technology that enables multi-hop communication even in an end-to-end communication path are disconnected [7]. In DTN, when the relay node can not establish the route to the next node, the data is temporarily held in its own buffer and wait for opportunity to forward. By the method, data exchange will be engaged even in case of changing dynamically network topology.

In addition, in the routing protocol called Epidemic Routing [8] used mainly in DTN, a node holding data forward its replicated data to all neighbor nodes. This phenomenon improves message acquisition rate though amount of network traffic largely increases.

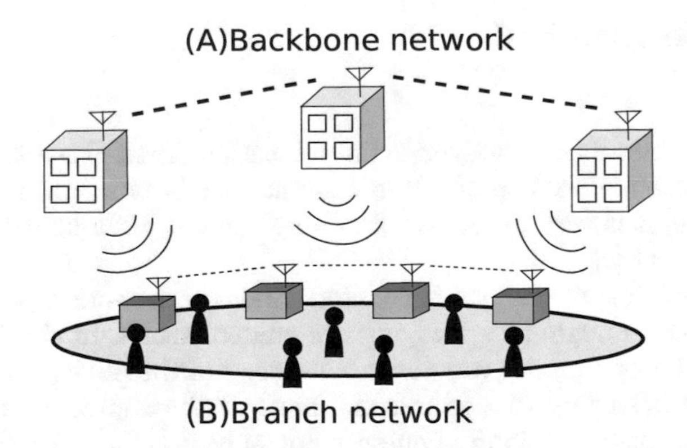

Fig. 1. Overview of ADES

2.4 Related Works Based on DTN

Several studies based on DTN have been conducted [5,6].

Literature [5] has been proposed for improving delay and throughput by using the message ferry method. More concretely, a special mobile node called a message ferry moves within a certain length in times, and plays a role of carrying data between nodes. For ferry nodes, authors supposed to use airplanes and/or vehicles. Authors have reported that the message arrival time can be improved though it is costly because we use multiple ferries.

Although, several disaster information collecting systems have been proposed as described in the above, it is difficult to apply to real environments.

2.5 A Method for Message Relay While Suppressing Unneeded Transmission

As a disaster information collecting method using DTN, we have also proposed a method to suppress redundant transfers by introducing message relay areas [6].

The purpose of this method was to make victims, who have not completed evacuation to shelters, collect information efficiently. In this method, messages will be forwarded if the messages traverse in the field of ideal route. The ideal route is established around the direct line from sender and destination. Messages will be discarded without relaying if those are received at the places away more than given certain distance from ideal route. As a result, the number of unneeded packet transmissions are decreased and the average message arrival can be rate improved.

This method, however, does not consider to deliver collected disaster information from countermeasure office/shelters to each suffers. Hence, as a disaster information distribution method to victims, we will consider a new network system introducing the concept of CCN, which is the content delivery infrastructure technology to DTN, and greatly shorten the communication delay time.

3 Proposed Method

3.1 CCN

The modern IP network is categorized in a location oriented network that relies on which servers are used to obtain a content, but in order to acquire content in a short time, it is very useful for obtaining content without regarding to the location of servers [9].

In CCN, which is representative content oriented network architecture, contents acquisition is made by specifying the content name. In the CCN, when a message is exchanged among any two nodes, copy of the message will be cached at relay nodes belonging to a forwarding path. If the appropriate contents are cached in path, user can obtain contents not only from the original server but also from the relay node, named contents router. In this way, contents acquisition time can be shorten.

More detailed mechanism of the CCN is described in Fig. 2. First, the user 1 transmits Interest as a content request. The server receiving the Interest returns the corresponding content to user 1. If there is no cache in the intermediate contents router, the cache is left in the contents router, and the transfer is continued. Next, when the user 2 request the same content as user 1, user 2 can receive the content from contents router if the cache is left in the contents router (Fig. 2).

3.2 Outline of Proposed Method

In the existing system, there were problems as shown as below.

- It is premised collection of disaster information only from victims staying at shelters
- Setting multiple ferry nodes is costly
- Information distribution to victims is not considered

Therefore, it is necessary to solve these problems in the proposed system.

The outline of the proposed method is shown in Fig. 3. In this system, we suppose that victims staying around disaster affected area are not able to move while staying in the disaster area, and they are requesting the information regarding to the disaster area. Hereafter, we call a victim who requesting disaster information as requester. We suppose that relay nodes are, for example, areas toward shelters, rescue vehicles carrying rescue items. They forward content requests receiving from requesters to the headquarter, and deliver contents from the headquarter to requesters. Moreover, they transfer according to Epidemic Routing.

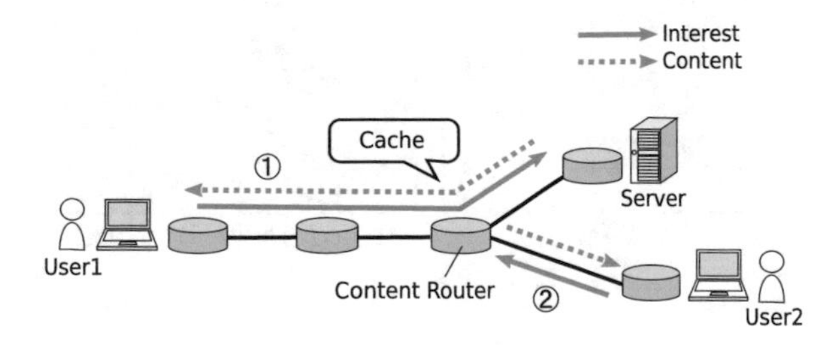

Fig. 2. Contents flow on CCN

In the method only employing DTN (in Fig. 3(a)), a request is transferred to the headquarter via relay nodes. The headquarter returns the content corresponding to requesters request. When relaying these request messages and contents, relay nodes transmit the duplicate data to all encountered nodes. Therefore, there is a possibility that a number of copied contents are generated and diffused with respect to the one request. This phenomenon induces under utilization of network.

In the proposed method, we place contents router on existing infrastructure, or staying node within a certain range (in Fig. 3(b)). When the content traverses over a content router, copies of the content are cached in the content router. As a result, content can be acquired not only from the headquarters but also from the content router. Therefore, in our proposal, the spread of content becomes less compared with DTN, and it is expected that the data acquisition time can be drastically reduced.

4 Evaluation

In this section, by using simulation, we evaluate the effectiveness of the proposed system. In the evaluation, we use ONE (The Opportunistic Network Environment simulator) [10]. The simulation topology used for the evaluation is shown in Fig. 4, and the simulation parameters are shown in Table 1.

As shown in the Fig. 4, we put the contents router, 200 [m] away from the headquarter. In the evaluation, two requesters sequentially request the content which randomly selected one from 100 contents stored at a headquarter. Contents request will be forwarded over the DTN network consists of mobile nodes. Once the request is arrived at the headquarter, corresponding content will be returned toward the requester.

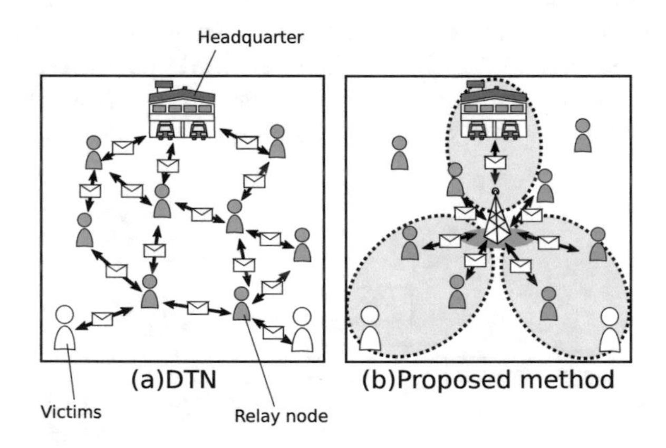

Fig. 3. Overview of message delivery on DTN and proposed method

Returned contents are forwarded by the intermediate nodes including contents router. The contents router copies and stores it if the received content is firstly arrived at the contents router. The stored caches are used to respond for future requests. The contents router returns cache in its buffer if the corresponding request is arrived. By this process, delay for obtaining content can be reduced in our proposed system.

In the evaluations, transmission intervals of content requests are varied from 30, 50, 100, 200, 300 [s].

Table 1. Simulation parameters

Parameter	Value
Simulation period	20,000 [s]
Simulation range	500 × 500 [m]
Relay node speed	9.5–10.5 [m/s]
Transmission range	50 [m]
Transmission speed	2 [Mbps]
Message size	500K-1M [Byte]
Number of relay nodes	100
Number of contents	100
Routing protocol	Epidemic routing

4.1 Evaluation Result

4.1.1 RTT and Number of Hops

Figures 5 and 6 show RTT and Number of Hops, respectively. In these figures, lines shown as DTN/CCN are the results of our proposed system. For the RTT, we can see that the proposed method improves performance except for the case of the content request interval is 30 s in Fig. 6. (The reasons why RTT are not improved in all conditions will be described later.) From these results, it is confirmed that the performance improvement by the proposed method increases in the region where the content request interval is under 100 s. On the other hand, the improvement becomes constant over 100 s of content request interval both RTT and Number of Hops.

These phenomenon can be though as follows: the requests for the same content by the requester 2 firstly arrives at the contents router while the content requested by the requester 1 is not yet cached in case of the content request interval is short. In addition, the improvement became constant after the content request

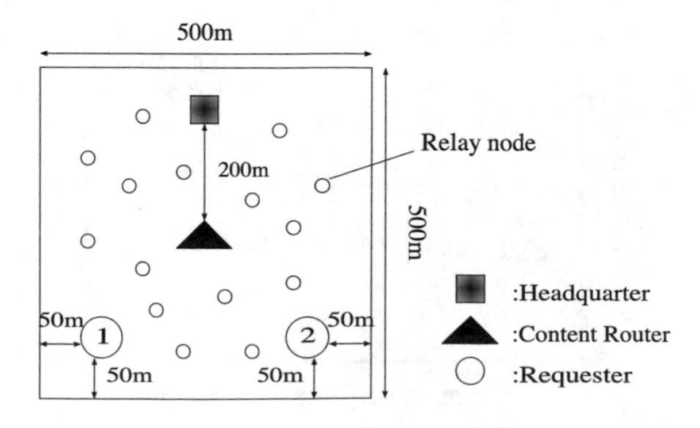

Fig. 4. Simulation topology

interval exceeds a certain length, the content is already cached at the contents router. Therefore, it seems that the number of cache usage has hardly changed even if the content request interval becomes longer.

4.1.2 Content Router Location and Cache Reach

Let the differences of arrival of content request be R, and the cache stored time be T, respectively. Here T is the average time period between transmission tiome of a content request and cached time at the contents router. Figure 7 shows the results of cache arrival time for changing the distance between the contents router and the headquarter.

In the evaluations described at Sect. 4.1.1, the distance between the headquarter and the contents router was fixed as 200 [m]. Moreover, from the result shown in Fig. 7, it is confirmed that the cache has reached the contents router at Approx. 120 s when the distance is 200 [m]. In the results shown in Figs. 5 and 6, this value of 120 s almost corresponds to the value at which the performance improvement of the proposed method becomes constant.

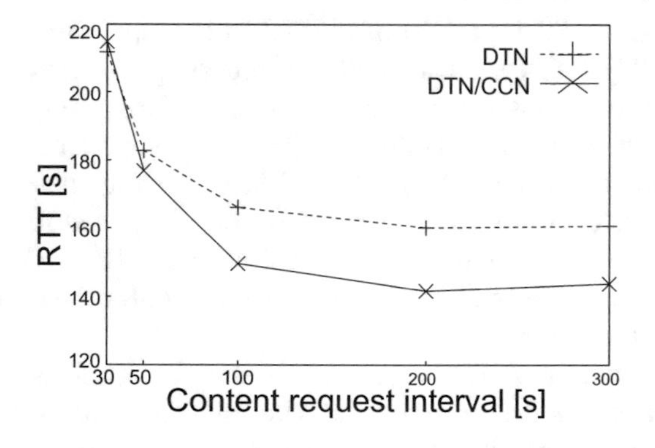

Fig. 5. Contents request interval and RTT

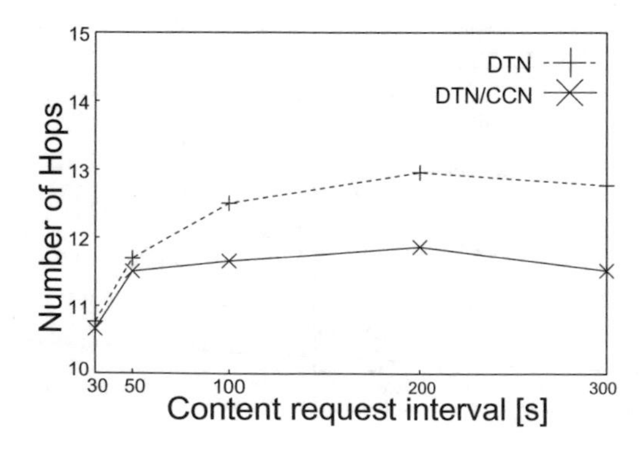

Fig. 6. Contents request interval and number of hops

From the above, it can be said that the number of utilized caches increases when $T < R$, and the number of caches used decreases when $T > R$.

5 Evaluation Under the Condition of Large Population of Requesters

This section reports results of the performance evaluations under the situation that number of requesters is increased. Simulation topology is shown in Fig. 8, and parameters used for the simulation are shown in Table 1.

All requesters are placed to have a same distance from contents router, each other. In the evaluation, requesters request contents in an order of requesters 1 to 4, and they request contents having content name with a number selected randomly in a range of 1 to 100. In this section, time period between two successive content requests of any requester, is called message creation interval. The simulation parameters are the same as in Table 1.

5.1 Evaluation Result

The results are shown in Figs. 9 and 10.

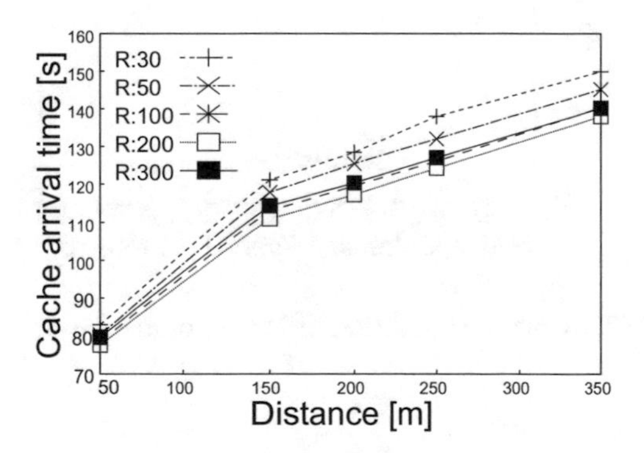

Fig. 7. Location of content router and cache arrival time

As Fig. 10 shows, the performance of the proposed method is improved entire all message creation intervals compared with the DTN. However, Fig. 9 confirms that the RTT of the proposed method increased significantly when the content request interval in less 120 s. In order to clarify the phenomenon, next section discusses further characteristics of the proposed method.

5.2 Evaluation Under the Condition of Varying Message Transmission Speed

This section reports the results of the performances under the transmission speed is varied. Simulation parameters used for the evaluation are same as Table 1,

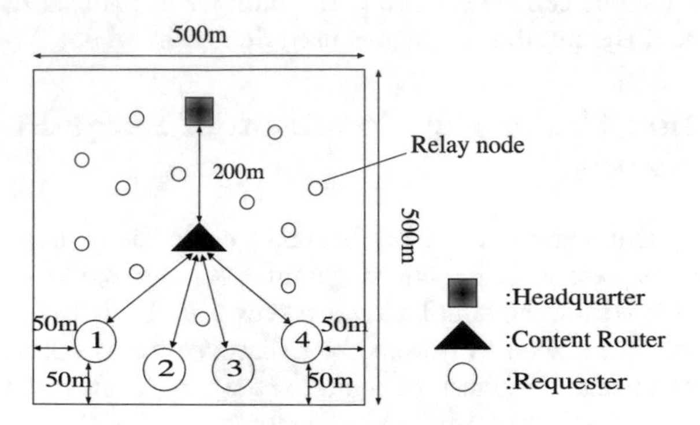

Fig. 8. Simulation topology

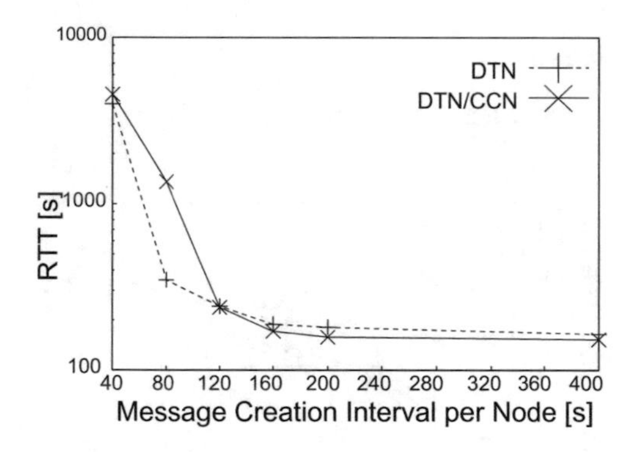

Fig. 9. RTT under the condition of large population of requesters

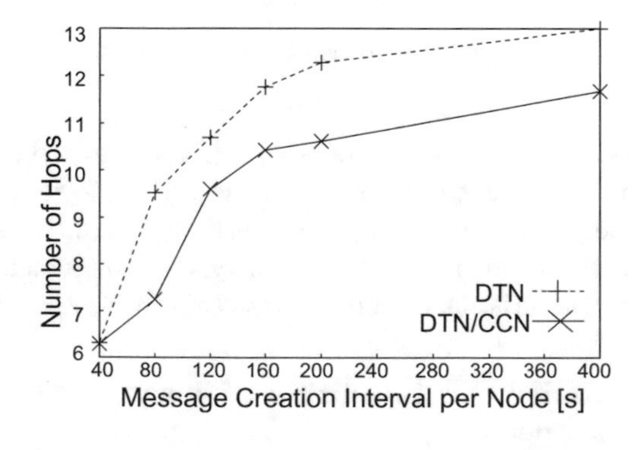

Fig. 10. Number of hops under the condition of large population of requesters

except for the transmission speed. RTT and Number of Hops are evaluated when the message creation interval per requester is fixed to 40 [s], and the message transfer speed is set to 2, 20, 200 [Mbps]. The results are shown in Figs. 11 and 12, respectively.

From these figures, it can be seen that when the message transfer rate increases, advantage of the proposed system is increased. It is obviously that length of RTT is mainly induced by migration of relay nodes and total transmission time in DTN. Here, the arrangement of location of the contents router does not affect migration time of relay nodes. Then, in order to cancel the effects of transmission times, Figs. 11 and 12 show the results of RTT and Number of Hops under the varying transmission speed. These results clarify that advantages of our proposal become high if the total transmission times can be ignored. In addition, because of total transmission times are increased in proportion to increase of the number of packet exchanged, it is needed to prevent unneeded copies of Interest/contents.

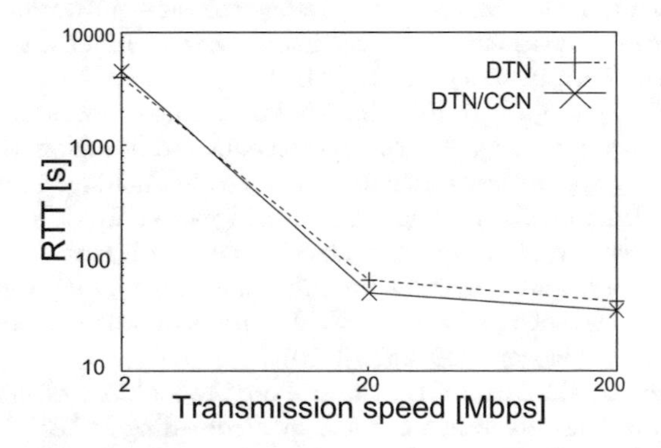

Fig. 11. RTT under the condition of varying message transmission speed

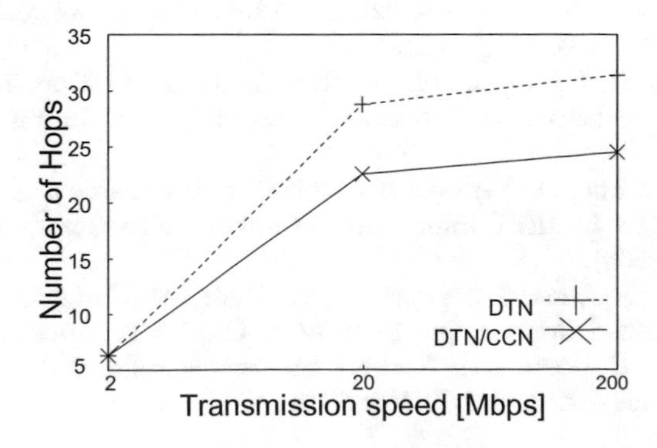

Fig. 12. Number of hops under the condition of varying message transmission speed

6 Conclusion

In this paper, we proposed a new network system employing CCN based on DTN.

The results show that the number of Hop can be reduced regardless of the content request interval, but if the content request interval is short, RTT increases. The cause of the increase in traffic is due to the return of contents from both the original server and the contents router. Therefore, in the future, we will deal with the network congestion by implementing limitations for content forwarding. Such as a restricted angle forwarding.

References

1. Chen, L., Liu, L., Qi, X., Zheng, G.: Cooperation forwarding data gathering strategy of wireless sensor networks. Int. J. Grid Utility Comput. **8**(1), 46–52 (2017)
2. Uchida, N., Ishida, T., Shibata, Y.: Delay tolerant networks-based vehicle-to-vehicle wireless networks for road surveillance systems in local areas. Int. J. Space-Based Situated Comput. **6**(1), 12–20 (2016)
3. Bylykbashi, K., Spaho, E., Barolli, L., Xhafa, F.: Impact of node density and TTL in vehicular delay tolerant networks: performance comparison of different routing protocols. Int. J. Space-Based Situated Comput. **7**(3), 136–144 (2017)
4. Urakami, M., Innami, S., Kamegawa, M., Shigeyasu, T., Matsuno, H.: Wireless distributed network system for relief activities after disasters - a field experiment for channel setting and a system for information exchange among shelters. In: International Conference on Broadband, Wireless Computing, Communication and Applications, Fukuoka, pp. 260–267 (2010)
5. Zhao, W., Ammar, M., Zegura, E.: Controlling the mobility of multiple data transport ferries in a delay-tolerant network. In: Proceedings IEEE 24th Annual Joint Conference of the IEEE Computer and Communications Societies, vol. 2, pp. 1407–1418 (2005)
6. Kawamoto, M., Shigeyasu, T.: Message relay decision algorithm to improve message delivery ratio in DTN-based wireless disaster information systems. In: IEEE 29th International Conference on Advanced Information Networking and Applications, Gwangiu, pp. 822–828 (2015)
7. Farrell, S., Cahill, V., Geraghty, D., Humphreys, I., McDonald, P.: When TCP Breaks: delay- and disruption- tolerant networking. IEEE Internet Comput. **10**4, 72–78 (2006)
8. Ip, Y., Lau, W., Yue, O.: Performance modeling of epidemic routing with heterogeneous node types. In: IEEE International Conference on Communications, Beijing, pp. 219–224 (2008)
9. Jacobson, V., Smetters, D., Thornton, J., Plass, M., Briggs, N., Braynard, R.: Networking Named Content, pp. 1–12. ACM CoNEXT, Rome, Italy (2009)
10. The ONE : The Opportunistic Network Environment simulator. https://akeranen.github.io/the-one/. Accessed 12 Mar 2018

Predicate Clustering-Based Entity-Centered Graph Pattern Recognition for Query Extension on the LOD

Jongmo Kim, Junsik Kong, Daeun Park, and Mye Sohn(✉)

Sungkyunkwan University, Suwon, Korea
{dignityc, jsgong94, eksdl23, myesohn}@skku.edu

Abstract. In this paper, we propose a method to reduce the difficulties of query caused by lack of information about graph patterns even though the graph pattern is one of the important characteristics of the LOD. To do so, we apply the clustering methodology to find the RDF predicates that have similar patterns. In addition, we identify representative graph patterns that imply its characteristics each cluster. The representative graph patterns are used to extend the users' query graphs. To show the difficulties of the query on the LOD, we developed an illustrative example. We propose the novel framework to support query extension using predicate clustering-based entity-centered graph patterns. Through the implementation of this framework, the user can easily query the LOD and at the same time collect appropriate query results.

1 Introduction

To increase the effectiveness of data mining and broaden its scope, many researchers have interested in hybridizing data mining and Linked Open Data (LOD) [1]. Connecting datasets developed for closed domains using the LOD prior to performing data mining, the following effects can be expected. First, the LOD can contribute to automatic discovery or integration of the heterogeneous data sets required to perform data mining [2]. Second, we can acquire additional information or knowledge needed to analyze the dataset in a machine-readable form in the LOD without the help of a domain expert [1]. At this time, the acquired information is used to identify features with good quality. Last but not least, the LOD is the only alternative that allows data mining to be applied to domains where data mining is difficult to be applied due to the burden of data preprocessing. However, despite the obvious benefits of the LOD on data mining, it is not easy to leverage it because users must select the source highly related to the query, have to know the exact vocabulary of RDF, and should be aware of the graph pattern of RDF dataset [3].

In this paper, we propose a method to reduce the difficulties of query caused by lack of information about graph patterns in the LOD. Many researchers have proposed two approaches to address this lack of information about the graph patterns: top-down approach [6–9] and bottom-up approach [10, 11]. Top-down approach relies on predefined knowledge base (KB) to find the graph patterns need for the queries. So, users can perform the queries appropriately and efficiently in the cloud such as dbpedia, which

L. Barolli et al. (Eds.): IMIS 2018, AISC 773, pp. 159–170, 2019.
https://doi.org/10.1007/978-3-319-93554-6_14

features an encyclopedia. However, there is a limit to apply the top- down approach to the entire LOD clouds because the RDFs of the LOD clouds are dynamically changed and the size of them is very large. On the contrary, the bottom-up approach obtains the information needed to extend the query based only on RDF data. So, this approach is suitable for querying the LOD clouds with huge amounts of data. However, most of the bottom-up approaches have limitations in that they do not take into account the inherent characteristics of the graph patterns within the LOD. They simply transform and apply RDF (a.k.a. graph patterns) to general features for data mining.

The graph pattern is one of the important characteristics of the LOD. In particular, the predicates that help to understand the relationships between the resources are the core elements that imply information about the LOD. So, we apply data mining techniques to identify the graph patterns considering the meaning of predicates on the LOD. To do this, we apply the document clustering methodology to perform the clustering for the RDF predicates that have similar patterns. After clustering, we identify representative graph patterns that imply its characteristics for each cluster. The representative graph patterns are used to extend the users' query graphs. As results, this not only overcomes the limitations of the bottom-up approach, but also increases the accuracy of the LOD queries.

This paper is organized as follow. In Sect. 2, we have developed an illustrative example to show the difficulty of performing an LOD query when there is insufficient information about the graph patterns. Section 3 offers detailed descriptions about the overall architecture and the components of the framework. Section 3 performs experiment and evaluation the proposal framework. Section 5 reviews the related research papers. Finally, Sect. 6 presents the conclusions and further research.

2 Illustrative Example

John will make a SPARQL query to find the artworks of his favorite artist, Alberto Giacometti, the galleries that display his artworks, and the location of the galleries in the LOD clouds. At this time, he is having trouble writing a SPARQL query, because the information about Alberto Giacometti is dispersed in on the LOD clouds, which are interconnected but different. Despite these difficulties, he wrote the SPARQL query as the left pane in Fig. 1, and known that "L'Homme qui marche I (The Walking Man I)" was an artwork by Alberto Giacometti and the artwork was on display at the Albright-Knox Art Gallery. As depicted in the right pane of Fig. 1, the LOD cloud contains much more information in graph patterns than John knows. However, until he knows exact structure of the LOD graph patterns and the predicate vocabularies, it is very difficult to perform SPARQL queries that take into account the LOD graph patterns. What makes John more difficult is that it is almost impossible to know the structure of the LOD graph patterns and the predicate vocabularies ahead of time because of hugeness and complexity of the RDF datasets aka LOD clouds.

In this paper, we try the expansion of the user's query using the predicate patterns to provide relevant and sufficient information to the user. As shown in Fig. 1, when the user writes a query with lack of knowledge about the graph patterns, our framework analyses the LOD graph patterns based on the user query. Furthermore, it identifies predicates that can query the most useful information from the analyzed results. Then, it uses the identified predicate to extend the user's query so that the user can obtain useful information from the LOD. As results, it allows users to query useful information without sufficient knowledge of the LOD graph patterns, as well as obtain useful information from a variety of RDF datasets, regardless of the domain.

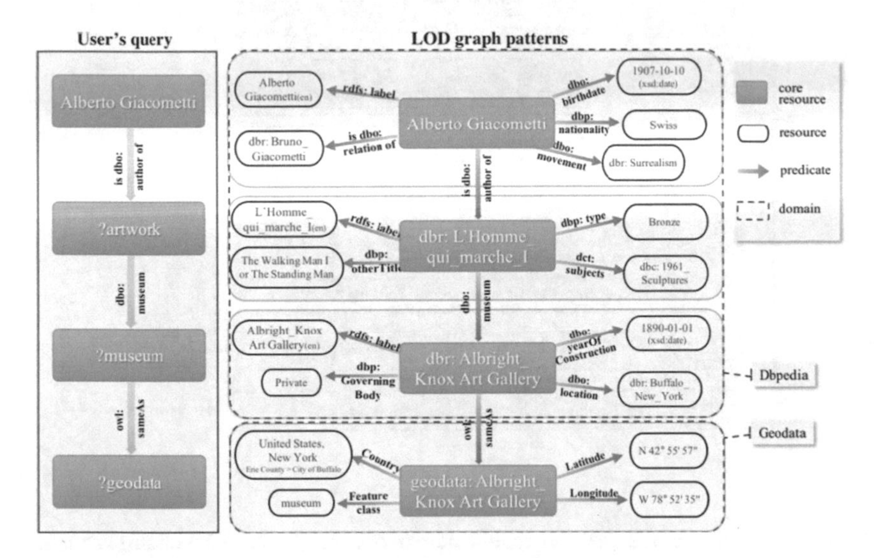

Fig. 1. Illustrative SPARQL query and relevant graph patterns

3 Overall Framework

We propose the ***Qu**ery **E**xtension framework based **E**ntity-centric predicate patterns recognition (hereafter, QUEEST framework). The overall framework of the QUEEST is depicted in Fig. 2. It is composed of four modules. First three modules identify the entity-centered graphs and two kinds of predicate patterns, and the last one combines the two kinds of patterns into one representative predicate pattern to be used for query extension.

3.1 Entity-Centered Graph Identification Module

This module extracts the entities that are core components of the predicate patterns. We use the predicate "rdf: type" to identify entities among the resources. The predicate "rdf: type" associated with the entities has a value that specifies the type of the various resources such as "rdf:Property," "owl:Class," "dbo:person," "dbo:Place," and/or "dbo: album." In order to identify only informative the entities among the resources, we perform filtering on predicates that have "rdf:Property" or "owl:Class" as a value among the resources linked to the predicate "rdf: type." The entities are defined as follows.

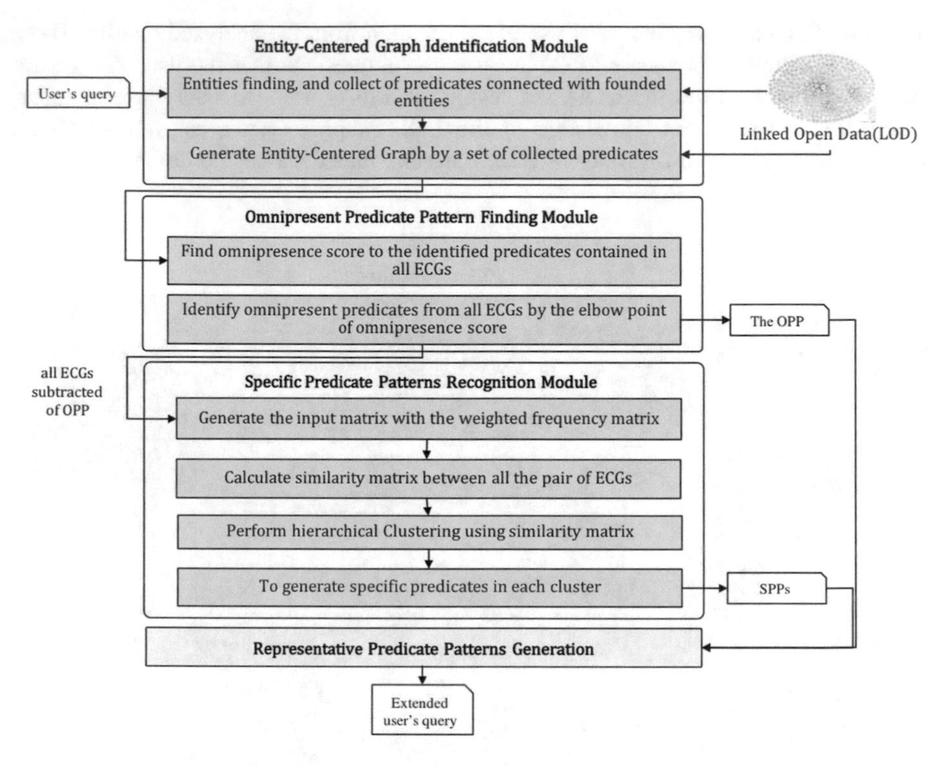

Fig. 2. The overall of the query extension framework based predicate patterns recognition

Definition 1. Entity (E) means a subset of RDF entities whose a value of "rdf:type" is neither "rdf: Property" nor "owl: Class." It is simply represented as

$$E = \{e_1, e_2, \ldots, e_i, \ldots\}, 1 \le i \le n$$

where e_i is the ith entity, not the literal. At this time, the entities in the RDF clouds are linked to a number of predicates to describe their attributes. The RDF predicates associated with e_i are used to generate i^{th} entity-centered graph (ECG_i) that is a multiset. ECG_i is simply represented as

$$ECG_i = \{p_{i1}, p_{i2}, \ldots, p_{ij}, \ldots\}, \quad j = 1, 2, \ldots$$

where p_{ij} is a jth predicate connected with e_i. Like e_i, only p_{ij}, not a literal, becomes an element of the ECG_i.

Through the identification of the ECG_i, the QUEEST framework identifies only those entities that contain information that is closely related to the user query among the LOD resources. So, even if the query of the user becomes complicated or the size of the LOD explosively increases, the QUEEST framework can significantly reduce the search burden by identifying the ECG_i.

In this paper, we only collect and store predicates to find two kinds of predicate patterns. In the next two modules, using all kinds of ECG_i only contained predicates (p_{ij}), our framework generates one omnipresent predicate pattern (*OPP*) and k specific predicate patterns ($SPP_k, k = 1, 2, \ldots$).

3.2 Omnipresent Predicate Pattern Finding Module

OPP is a predicate pattern that appears frequently in most ECGs regardless of the query. The omnipresent predicates are independent of the entity's characteristics, while provides basic information such as a name or comments. The OPP is defined as follows.

Definition 2. The *OPP* is a set of RDF predicates consisting of predicate patterns that are ubiquitous across all ECGs. To identify the elements of the *OPP*, we perform the predicate frequency-inverse ECG frequency (PFIEF), which is a variant of term frequency–inverse document frequency (TFIDF).

$$OPP = \{op_1, op_2, \ldots, op_l \ldots\}, \quad l = 1, 2 \ldots; \ l \leq j$$

where op_l is a lth predicate, which is not duplicated with other predicates in the OPP.

To find OPP, we applied the idea of TFIDF to evaluate the omnipresence of the predicates across all ECGs. In our framework, the Term in TFIDF becomes the predicate (p_{ij}) and the document is replaced by the ECG_i. The PFIEF-based omnipresence score (om_{ij}) of jth predicate in i^{th} ECG is calculated as follows.

$$PF_{ij} = \sum_{i'=1}^{all} \sum_{j'=1}^{all} M\left(p_{ij}, p_{i'j'}\right)$$

$$IEF_{ij} = \log(n / \sum_{s=1}^{n} O\left(p_{ij}, ECG_s\right)$$

where $\mathrm{M}\left(p_{ij}, p_{i'j'}\right) = \begin{cases} 1, p_{ij} = p_{i'j'} \\ 0, p_{ij} \neq p_{i'j'} \end{cases}$, p_{ij} and $p_{i'j'}$ are predicates ($p_{ij}, p_{i'j'} \in OPP$, $i \neq i'$ and $j \neq j'$). And $\mathrm{O}\left(p_{ij}, ECG_{i'}\right) = \begin{cases} 1, p_{ij} \in ECG_{i'} \\ 0, p_{ij} \notin ECG_{i'} \end{cases}$.

Using above two equations, we calculate om_{ij} as follows.

$$\mathrm{om}_{ij} = 1/IEF_{ij}$$

The above equation means that if p_{ij} is an omnipresent predicate appearing in all ECGs, the om_{ij} score becomes larger, whereas if this predicate appears only in some ECGs, the om_{ij} score becomes smaller. This module uses the om_{ij} score to determine whether p_{ij} can be an element of the OPP for all ECGs. The determination procedure is summarized in Fig. 3.

```
Input     ECG_i  for all i
Output    The OPP

1         OM = []
2         For all i,j:
3             Calculate(om_ij)
4             OM.append(om_ij)
5         EndFor
6         sortedOM = sorted(OM, order = descending)
7         diffsVector = []
8         For var in range(0, var-1):
9             diffsVector.append(sortedOM[var]-sortedOM[var+1])
10        EndFor
11        MaxDiffs = max(diffsVector)
12        WhereAtMaxDiffs = diffsVector.index(MaxDiffs)
13        The_Elbowpoint = sortedOM[WhereAtMaxDiffs]
14        OPP = []
15        For all i, j:
16            If o_ij >= The_Elbowpoint:
17                OPP.append (p_ij)
18            EndIf
19        EndFor
```

Fig. 3. Omnipresent predicate pattern finding algorithm

3.3 Specific Predicate Patterns Recognition Module

In general, it is a known that the entities with similar characteristics have similar predicate patterns in whatever ECG they contain. If we are able to identify the similar predicate patterns contained in all ECGs, the completeness of the LOD query will be greatly improved. To achieve this goal, our framework generates $SPP_k(k = 1, 2, \ldots)$ through clustering the predicate pattern contained in all ECGs. As a result of the clustering, *kSPPs* are derived.

Definition 3. The SPP_k a set of predicates that have a significant effect on k^{th} cluster.

$$SPP_k = \{sp_{k1}, sp_{k2}, \ldots, sp_{kr} \ldots\}, 1 \leq r$$

where sp_{kr} is a rth specific predicate in SPP_k.

In order to recognize the SPPs, this module is performed as follows. First, it generates the input matrix for the clustering using the ECGs, which are subtracted from the OPP. The input matrix can be expressed as a $n \times t$ matrix, where n is the number of ECGs and t is the number predicates remaining after eliminating duplicate predicates in the all ECGs. Furthermore, each entry pf_{io} of the input matrix is the occurrence frequency of the oth specific predicate in i^{th} ECG ($o = 1, 2, \ldots, t; o \leq m$). The next step is to calculate the weighted frequency of the predicate (wpf_{io}) that is modified pf_{io}, taking $PFIEF_{io}$ into account. The weighted frequency of the predicate (wpf_{io}) is calculated as follows.

$$wpf_{io} = PFIEF_{io} \times \sum_{j=1}^{all} M(p_o, p_{ij})$$

where p_o is o^{th} predicate on the weighted frequency matrix.

We perform clustering using a matrix with the weighted frequency of the predicates (wpf_{io}) as entries. We filtered out the p_o that did not affect the clustering in order to avoid the "curse of dimension" that may occur during the clustering process and at the same time to increase the accuracy of clustering. At this time, the filtering criterion is whether or not the number of entries whose entry value of p_o is "0" is larger than a predefined threshold (δ). After filtering p_o, we can obtain the reduced input matrix ($r - IM$). Using the $r - IM$, we calculate the similarity among all pair of the ECGs.

To calculate the similarity between two arbitrary ECGs in $r - IM$, we applied the cosine similarity that is suitable for our wpf_{io} which does not have a fixed maximum value [4]. The cosine similarity-based similarity $SIM(ECG_i, ECG_{i'})$ between two arbitrary ECGs is calculated as follows.

$$SIM(ECG_i, ECG_{i'}) = \frac{\sum_{t=1}^{all} (wpf_{io} \times wpf_{i'o})}{\sqrt{\sum_{t=1}^{all} wpf_{io}} \sqrt{\sum_{t=1}^{all} wpf_{i'o}}}, \quad i \neq i'$$

Using the similarity matrix, we perform hierarchical clustering that is suitable for data patterns that have a huge data volume and low sparsity like our $r - IM$ [5]. After clustering is finished, we select multiple $p_{o'}$ which well represents of the characteristics of kth cluster ($p_{o'} \subseteq p_o$), and assign multiple $p_{o'}$ as the elements of SPP_k. The determination procedure of multiple $p_{o'}$ is summarized in Fig. 4.

3.4 Representative Predicate Patterns Generation Module

In the last module, the user's query (q) consisting of p $qECG_p$(p = 1, 2..., t) is extended using the representative predicate pattern (RPP_k). To do this, First, we integrate the elements of OPP with the elements of kth SPP_k to generate RPP_k. This process is repeated for all k. As a next step, we perform the comparison between the elements of the pth $qECG_p$ and the elements of the each RPP_k. This comparison process repeats $lenght(k)$ times for all k. Finally, we extend the pth $qECG_p$ by appending an $RPP_{k_{opt}}$ that shares the most elements with p^{th} $qECG_p$ to the pth $qECG_p(k_{opt} \in k)$. The whole process is summarized in Fig. 5.

```
Input     r − IM
Output    SPP_k

1    SIMTX = [n][n]
2    For all i, i′:
3            If i > i′:
4                    SIMTX[i][i′].append (SIM(ECG_i, ECG_i′))
5            EndIf
6    EndFor
7    PerformClustering (SIMTX, method=hierarchical)
8    For all o:
9        av_o^all = Average (r − IM[1:n][o])
10   EndFor
11   For all i, k :
12       If ECG_i ∈ Cluster_k:
13           C_k.append(ECG_i)
14           CI_k.append(i)
15       EndIf
16   EndFor
17   For all k;
18       For all o:
19           If p_o in C_k:
20               av_o^k = Average (r − IM[CI_k][o])
21               If av_o^k > av_o^all:
22                   p_o′ ← p_o
23                   SPP_k.append(p_o′)
24               EndIf
25           EndIf
26       EndFor
27   EndFor
```

Fig. 4. Clustering-based specific predicate patterns recognition algorithm

```
Input     OPP, SPP_k
          qECG_p: p^th ECG in the user's query (q)
Output    RPP_k

1    For all k:
2        RPP_k ← Merge(OPP, SPP_k)
3    EndFor
4    For all p:
5        temp=[]
6        For all k:
7            temp.append.set(qECG_p ∩ RPP_k)
8            MaxNumber = MaxNumberOf(temp)
9        k^opt = temp.where(MaxNumber)
10       qECG_p ← Merge(qECG_p, RPP_{k^opt})
11       EndFor
12   EndFor
```

Fig. 5. Generation of representative predicate patterns and query extension

4 Experiments and Performance Evaluation

To demonstrate the superiority of the QUEEST framework, we performed experiments based on the scenarios described in Sect. 2. The purpose of the experiment is to evaluate the completeness of the extended query using 'hit ratio' with the infobox of the Wikipedia. To do this, we collected 162 entities from the SPARQL endpoints of the dbpedia, 41 artists, 84 artworks, and 37 museums with collection of 15,994 RDFs associated with them.

The QUEEST framework found one OPP and three SPPs. As shown in Table 1, although OPP contains 15 predicates, it can be seen that these are general predicates that cannot explain the characteristics of the entities included in the user query. On the other hand, the predicates of the SPPs that extend the criteria of the three entities included in the user query consist of those that are highly related to the characteristics of each entity.

Table 1. The a part of predicates in the OPP and SPPs

	OPP	SPP_1	SPP_2	SPP_3
Predicates	dbo:abstract	dbp:caption	dbo:author	geo:geometry
	dbo:wikiPageID	dbo:birthDate	dbp:imageFile	georss:georss
	owl:sameAs	dct:description	dbp:year	foaf:homepage
	rdfs:label	dbo:birthPlace	dbo:museum	dbo:location
	dbo:wikiPageRevisionID	foaf:gender	dbp:metricUnit	dbp:established
	foaf:isPrimaryTopicOf	dbo:deathDate	dbp:City	dbp:publictransit
	rdfs:comment			dbp:director
	rdf:type			
	……			

As a next step, the SPPs are extended to the $qECG_p(\mathrm{p} \leq 3)$ of the artist, artwork and museum using the three RPPs generated by OPP and SPPs. In order to evaluate the completeness of the extended $qECG_p(qECG_p^*)$, we use the following equation.

$$completness = \frac{Num\left(qECG_p^* \cap infobox - key\right)}{Num(infobox - key)}$$

where, infobox-key is a set of keys in Wikipedia infobox and $qECG_p^*$ is an extended $qECG_p$ with RPPs. The reason why we compared $qECG_p^*$ with Wikipedia infobox is a structured document containing a set of key-value pairs, and represents a basic information about an entity.

To do so, we collected 30 Wikipedia infoboxes that are related to $qECG_p^*$ and calculated the average completeness (Fig. 6). As results, for museums with relatively uniform information on Wikipedia, QUEEST framework also has a very high average completeness (95.65%). On the other hand, the artworks that can be explained in various ways exhibit relatively low average completeness (75%). Even if the average completeness is not high enough, the QUEEST framework can return query results that satisfy the user because most of the information in the Wikipedia infobox can be found.

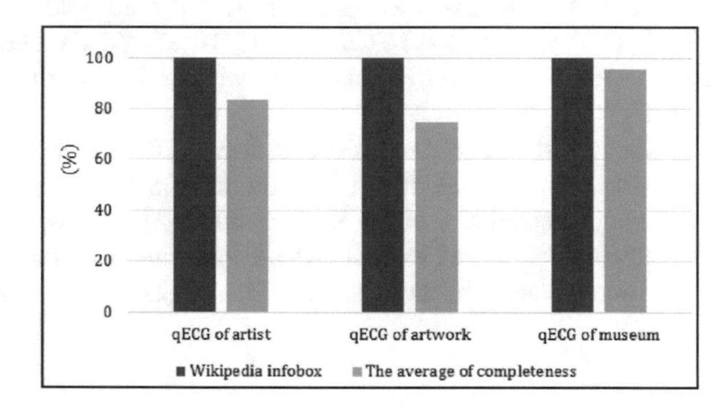

Fig. 6. The average completeness of $qECG_p^*$

5 Related Works

There are three types of research to improve the LOD query. First, some researchers have attempted to improve LOD queries by ensuring interoperability of RDF vocabularies [14]. Second, another researcher selected an RDF dataset that stored the most relevant resources from the LOD cloud and then executed queries against that dataset only [13]. Last but not least, research was carried out utilizing the features of the LOD graph pattern in order to improve the completeness of the query [6]. The last research started recently compared to the previous two research. This is because the LOD has a very complex and variety graph patterns, and the graph patterns of the LOD has a structure different from that of graph theory or network science, which is interested in the graph patterns.

However, in order to improve the LOD query, it is essential to perform a query based on understanding the graph patterns in the LOD. In other words, it is required to write or extend the queries that can collect appropriate data from the LOD., that is, a complete query. To meet the requirement, two approaches were performed by the researchers. The top-down approach assumes and exploits an external knowledge base that can describe the LOD graph pattern for general information. [7] defined the relationship between the elements of the actual application environment and the LOD graph patterns with the predefined concept model. [8] rewrote the query to improve the completeness of the query by adding new inferred relations or SPARQL operators using the two layers developed ontology. [9, 10] proposed the query method on the LOD using knowledge base or knowledge graph. This method relies heavily on knowledge of the graph pattern in the knowledge base or the knowledge graph regardless of the characteristics of the RDF datasets, and only the RDF datasets is used as a database.

To the contrary, the bottom-up approach improves the completeness of the LOD query by relying solely on the pattern or distribution of the RDF datasets, without assuming any knowledge base that may exist in the past. [11] found important graph patterns between RDF datasets to select the sources that can affect the LOD query results. To identify important graph patterns, they redefined LOD graph patterns using predicate "rdf:type" to distinguish important entities and predicates. [12] proposed the framework for finding new graph patterns based on the RDF datasets acquired during the query. The newly discovered graph pattern was added to the user's query result by evaluating the relevance of the query and the efficiency of the information gain.

6 Conclusion and Future Works

We proposed the ***Qu****ery* ***E****xtension framework based* ***E****ntity-centric predicate patterns recognition (hereafter, QUEEST framework) to find two kinds of predicate patterns in the LOD based on user's query and to extend the query by utilizing these predicate patterns. By utilizing this framework, the users are able to collect rich information using simple SPARQL query, without complete knowledge of the LOD. Furthermore, it can dramatically improve the utility of the LOD, which has been criticized so far by researchers.

However, this research has the following limitations. The first limitation is that we consider only the entity-centered graph type in the various graph patterns that the LOD contains. It is also a limitation that cross-domain graph patterns, which are the strengths of LOD, are not identified. To overcome these limitations, we will apply the data mining technique to the RDF dataset in the future to discover various types of graph patterns, and to select the most effective type of graph pattern and apply them to our framework. In addition, we will research to find extraordinary graph pattern occurring between a pair of domains and to propose the LOD query extension methodology that maximizes the serendipity of the LOD.

Acknowledgements. This research is supported by C2 integrating and interfacing technologies laboratory of Agency for Defense Development (UD180014ED).

References

1. Lausch, A., Schmidt, A., Tischendorf, L.: Data mining and linked open data – new perspectives for data analysis in environmental research. Ecol. Model. **295**, 5–17 (2015)
2. Ristoski, P., Paulheim, H.: Semantic web in data mining and knowledge discovery: a comprehensive survey. Web Semant. Sci. Serv. Agents World Wide Web **36**, 1–22 (2016)
3. Ristoski, P., Paulheim, H.: RDF2Vec: RDF graph embeddings for data mining. In: The Semantic Web – ISWC 2016, pp. 498–514 (2016)
4. Steinbach, M., Karypis, G., Kumar, V.: A comparison of document clustering techniques. In: KDD Workshop on Text Mining, vol. 400, no. 1, pp. 525–526, August 2000
5. Hierarchical Clustering, Clustering, pp. 31–62
6. Harth, A., Speiser, S.: On completeness classes for query evaluation on linked data. In: AAAI, July 2012

7. Karagiannis, D., Buchmann, R.A.: Linked open models: extending linked open data with conceptual model information. Inf. Syst. **56**, 174–197 (2016)
8. Makris, K., Bikakis, N., Gioldasis, N., Christodoulakis, S.: SPARQL-RW. In: Proceedings of the 15th International Conference on Extending Database Technology - EDBT 2012 (2012)
9. Yih, W., Chang, M.-W., He, X., Gao, J.: Semantic parsing via staged query graph generation: question answering with knowledge base. In: Proceedings of the 53rd Annual Meeting of the Association for Computational Linguistics and the 7th International Joint Conference on Natural Language Processing (Volume 1: Long Papers) (2015)
10. Zheng, W., Zou, L., Peng, W., Yan, X., Song, S., Zhao, D.: Semantic SPARQL similarity search over RDF knowledge graphs. Proc. VLDB Endowment **9**(11), 840–851 (2016)
11. Li, J., Wang, W.: Graph summarization for source selection of querying over Linked Open Data. In: 2017 IEEE 2nd Information Technology, Networking, Electronic and Automation Control Conference (ITNEC), December 2017
12. Zou, L., Özsu, M.T., Chen, L., Shen, X., Huang, R., Zhao, D.: gStore: a graph-based SPARQL query engine. VLDB J. **23**(4), 565–590 (2013)
13. Vander Sande, M., Verborgh, R., Dimou, A., Colpaert, P., Mannens, E.: Hypermedia-based discovery for source selection using low-cost linked data interfaces. Int. J. Semant. Web Inf. Syst. **12**(3), 79–110 (2016)
14. Vandenbussche, P.-Y., Atemezing, G.A., Poveda-Villalón, M., Vatant, B.: Linked Open Vocabularies (LOV): a gateway to reusable semantic vocabularies on the web. Semant. Web **8**(3), 437–452 (2016)

Discrimination of Eye Blinks and Eye Movements as Features for Image Analysis of the Around Ocular Region for Use as an Input Interface

Shogo Matsuno[1(✉)], Masatoshi Tanaka[2], Keisuke Yoshida[2], Kota Akehi[3], Naoaki Itakura[3], Tota Mizuno[3], and Kazuyuki Mito[3]

[1] Hotto Link Inc., Chiyoda-ku, Tokyo, Japan
matsuno@vpac.cs.tut.ac.jp
[2] Tokyo Denki University, Adachi-ku, Tokyo, Japan
[3] The University of Electro-Communications, Chofu, Tokyo, Japan

Abstract. This paper examines an input method for ocular analysis that incorporates eye-motion and eye-blink features to enable an eye-controlled input interface that functions independent of gaze-position measurement. This was achieved by analyzing the visible light in images captured without using special equipment. We propose applying two methods. One method detects eye motions using optical flow. The other method classifies voluntary eye blinks. The experimental evaluations assessed both identification algorithms simultaneously. Both algorithms were also examined for applicability in an input interface. The results have been consolidated and evaluated. This paper concludes by considering of the future of this topic.

1 Introduction

The eye-gaze input system uses eye information such as eye gaze and eye blink, to direct user interfaces. The system is attracting increasing attention as a hands-free input means because of developments in the miniaturization of information equipment. Many eye-gaze input-driven interfaces rely on algorithms that interpret gazing markers by collecting precise measurements of user gaze positions. The gaze input measures with high accuracy and realizes intuitive operation. However, to perform gaze input, indices for input targets and a screen area for displaying the targets are required. Therefore, the accuracy required of the measurement device depends on the size of the screen. For this we infer that difficulties with input increases as information terminals become more compact. Also, when attempting to control all computer operations using gaze input, developers encounter the Midas Touch problem, in which all markers in the same direction as the line of sight are activated. For this reason, the fixed time, or input time, is a bottleneck for input speed, and complicated input such as character input is slow compared to other input interfaces.

To address these issues, this study proposes a novel method of realizing complicated input by combining analysis of eye motions and blinking, rather than relying only on gaze. Several successful attempts at systems combining eye motions and blinking

L. Barolli et al. (Eds.): IMIS 2018, AISC 773, pp. 171–182, 2019.
https://doi.org/10.1007/978-3-319-93554-6_15

have been reported [1, 2]. These reports show that the combination of eye motions and blinking has a certain effect. However, prior studies obtained measurements using an electro-oculogram or an infrared camera, which required an enhanced and expanded data collection device.

To mitigate the cost and other constraints associated with such cameras, this study attempted to realize an input interface that collected eye motions and blinking using a general visible light camera of the sort common to web cameras and often mounted on a notebook computers and smartphones. If input with a visible light camera becomes possible, then it would follow that gaze input could be realized on compact terminals by introducing software, and without requiring extension or alteration of the hardware. In this paper, we propose an algorithm for measuring and identifying eye motion and blinking by analyzing the moving images captured by a visible light camera. This would be the first step towards realizing an ocular input interface that can be used in small devices. This paper reports on the experimental evaluation of the proposed method.

2 Related Works

Several attempted systems that capture and analyze both eye motion and blinking have been reported [1, 2]. However, most of these systems require expansion of hardware such as the incorporation of an infrared light source. In our previous study [3], we evaluated a communication support device based on blinking and eye motions using electrooculography (EOG). In a different study [4], a wheelchair operates using blinking and laser range finder. Previous studies [5, 6] also consider horizontal eye movement using a pair of EOG electrodes to direct character input using voluntary blinking. Another study [7] proposed a conversation support system that uses a head-mounted display (HMD) and a web camera to input information about eye gaze and blink. In a 2014 study [8], robot arm operation is performed using gaze and blink. However, all these studies and approaches are limited by the common reliance on wearable equipment for measurement, which limits their practical applications. As a non-contact method, one study [9] proposes operating a GUI under natural light using a home video camera and a personal computer and controlling the mouse cursor using eye gaze and blink data. Although this method achieves non-contact input, it is not suitable for the operation of compact information terminals because it requires a specific screen size, measurement accuracy, and measurement device to fix user position and operate the mouse cursor.

In response to the previously identified constraints, the authors exploit the fact that the movement of the eyeball is the earliest indicator of movement in the encompassing facial region, using this phenomenon to quantitatively detect the movement of the line of sight and identify direction by measuring the optical flow. Note that it is difficult to distinguish Eyeblink types even though blink detection is a viable optical flow measurement [10]. Eyeblinks are classified into intentional voluntary blinking and involuntary blinking performed unconsciously or under reflection. To use blinking as an input, it is necessary to distinguish involuntary blinks from voluntary blinks and discard involuntary blink data. By applying the method used in our previous research [11], we identify discretionary blinks. This study combines these blink detection and type identification algorithms to track the movement of the line of sight and measure and identify eyeblinks by analyzing moving images captured by a visible light camera.

3 Method of Classifying Eye Motions and Eye Blinks

In the proposed method, the real-time input detection is achieved by operating the algorithm for detecting the gaze movement direction and the algorithm for discriminating the voluntary blink in parallel. Both methods perform detection and identification by analyzing moving images of the eyeballs and their immediate vicinity.

3.1 Measurement and Classification Method of Round Trip Eye Movements (Eye Glance)

A system that inputs characters by measuring round trip eye movements (Eye Glance) in diagonal four directions using EOG has been proposed [12, 13]. Eye Glance input realizes complicated input by repeating a simple operation (eye line movement in oblique direction). The authors attempted to detect Eye Glance by analyzing moving images. Comparing two temporally consecutive images and expressing the motion of the pixel by a vector is known as optical flow. By using the shading of the pixels representing the black and white portions of the eyeball captured in the moving images, it is possible to express eye movement as an optical flow. Eye Glance will be explained using the optical flow obtained from this technique of image analysis.

Figure 1 depicts an ideal optical flow waveform obtained when Eye Glance is captures movement in the upper right direction. When assessing horizontal direction, the left is reflected with a positive value and the right is recorded with a negative value. Vertical measurements attribute a positive value to the upward direction and a negative value to the downward direction. The waveform of the optical flow is calculated by reducing the average value in the stationary state from the average value of the velocity vector obtained from the measurement point of the rectangular eyeball neighborhood represented in a

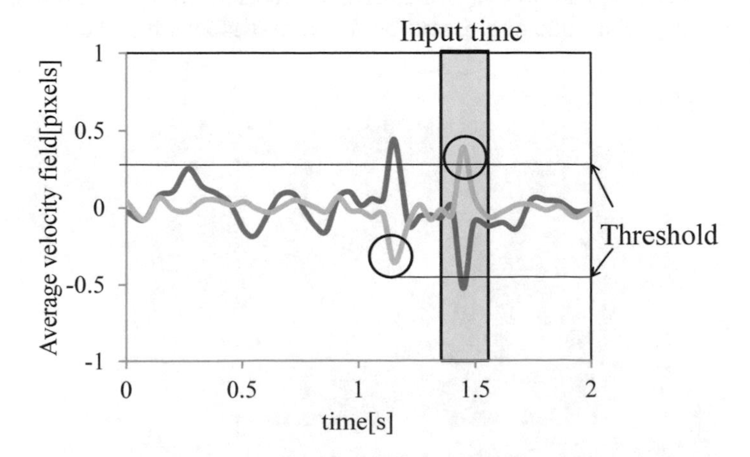

Fig. 1. Threshold and input time for Eye Glance input detection

200 by 80 pixel image area. Currently, an period during which reverse waveforms in horizontal and vertical direction components occur consecutively and a slight dwell time occurs during two eye movements, is regarded as an Eye Glance waveform and is extracted an individual entity. The overall direction is identified based on the order of the mountains and valleys of the horizontal and vertical direction components.

3.2 Method of Detection and Classification of Voluntary Eye Blinks

Blinking is a fast behavior that completes a series of actions in the span of hundreds of milliseconds. For this reason, it is difficult to distinguish voluntary blinks and involuntary blinks by blink detection using optical flow. Therefore, in this paper, we expand upon previous research. The proposed method measures the change in the area of the eyeball opening, detects the blink, and distinguishes the type of blink by temporally integrating the change in the amount of eyeball opening area.

We suggest that if the proposed system can record the changes in eye opening area, this data can be used to construct a blink waveform, as illustrated in Fig. 2. The opening of the eyeball can be defined by extracting pixels dissimilar to the flesh tone color information derived from the moving images that capture the eyeball and its immediate vicinity [14]. Therefore, we binarize the opening of the eyeball based on the skin color information taken from the same image, as demonstrated in Fig. 3. An image of RGB color photographed by a camera is converted into a YCrCb colorimetric system from which color difference ratios for each pixel are calculated and used to create a histogram. In most cases, the histogram forms two mountains, one in the flesh color part and a second in the other part. The numerical value associated with the valley between the two mountains is adopted as the binarization threshold value. Thus, the number of pixels representing the area of the eyeball opening portion of the moving image is obtained. Since blinking is an operation in which eye closure and eye opening occur continuously at high speed, it is possible to identify the time segment when blinking occurs by extracting the fragment where the eyeball opening area abruptly changes. In our previous study [11], we established that voluntary blinks increase the amplitude of the change in blinking duration and eyeball opening area as compared to instances of involuntary blinking. Based on this observation, when the value obtained

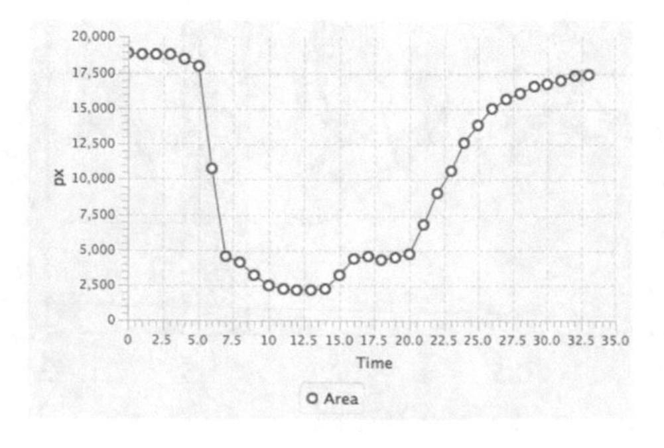

Fig. 2. Eyeblink waveform.

by temporally integrating the change amount of the eyeball opening area exceeds a certain threshold, voluntary blinking is distinguished.

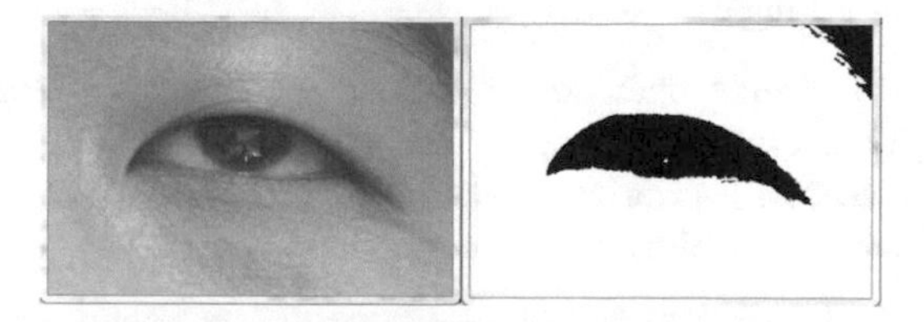

Fig. 3. Result of binarization using color information.

3.3 Parallel Execution of Eye-Glance Detection and Blink Detection

Figure 4 outlines the processing flow. In this scenario, when identification modules are simply executed in parallel, there is concern that blinking could be incorrectly interpreted as eye motion, and, conversely, that eye motions may be mistakenly recognized as blinking. To resolve this problem, during the calibration of the blink and eye movement waveforms, both the number of pixels in the eyeball area and the optical flow are acquired respectively for each algorithm. This protocol enables further study of approaches to reduce false recognition.

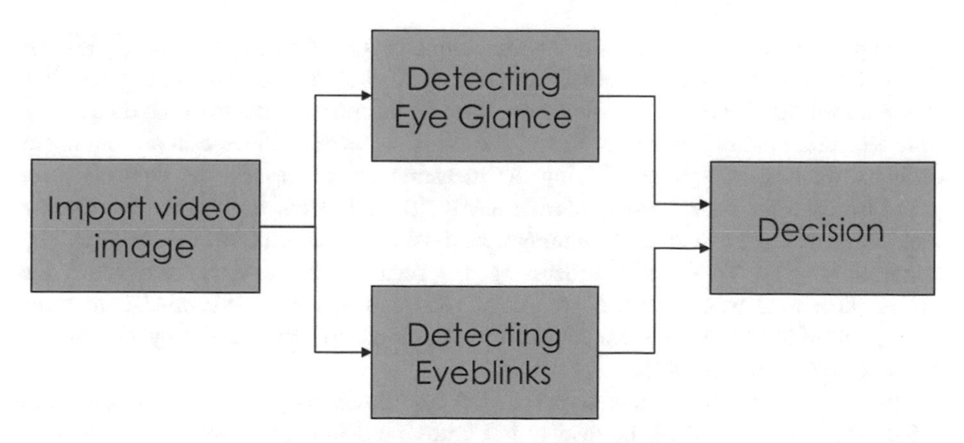

Fig. 4. Outlines the processing flow.

Detection of eye movement and blink detection are processed separately, though in parallel, from the same moving image. Therefore, when operating in real time, each detection result is judged at a unique point in time. That is, when erroneous detection occurs, detection output is obtained from both modules. In this scenario, if the output of detection is reflected as it is, it is always processed as an erroneous detection. Therefore, when there is output from one of the detection modules, the operation judgment module waits for a predetermined period of time. Also, when there are multiple detection outputs with a short time period, only one output is selected to be the final detection result.

When the same moving image is analyzed by the detection modules simultaneously, the computations employed by the blink detection algorithm frequently achieve the fastest output. Therefore, comparisons of the two detection methods were made using the following settings:

(1) Output of Eye Glance detection is not accepted for a specified time period immediately after blink detection has been output.
(2) Eye Glance detection is prioritized when Eye Glance detection is output at a fixed time immediately after blink detection has been output.

4 Experimental Results

This section describes the experimental system that we constructed based on the input method combining input from voluntary blinking and Eye Glance proposed in Sect. 3. We report results for the detection accuracy evaluation experiment conducted on five subjects in their twenties. Experiment participants received sufficient explanations of the method and purpose of the experiment, and agreed to take part in the experiment, allowing disclosure of the experimental data with the provision that the subject's identity is concealed.

4.1 Experimental Environment

The experimental system used a notebook computer (OS X El Capitan: MacBook Air 13-in., Intel Core i7 2 GHz, Intel HD Graphics 4000) with a video camera was installed at the top of the display. The video camera acquired moving images with an average time resolution of approximately 25 fps. Experiments were conducted in a room under ordinary fluorescent lighting. During the measurement processes, the subjects were asked to sit on the chair located approximately 40 cm in front of the monitor. During measurement, the facial area was photographed using the video camera at the top of the notebook display. To simplify processing, the region near the eye, from which the optical flow data was collected, was fixed. The subjects were informed before the experiment that both eyes were captured in the frame to produce the eyeball neighborhood region examined the experiment.

Figure 5 shows the display screen of the experimental system. The purpose and information in each part of the display in Figure are detailed below.

① Original image: The image output from the camera installed in the experimental environment is displayed as captured.
② Eyeball vicinity area image: The image of the eyeball neighborhood region extracted from the original image. Subsequent image processing is performed on the area in this image.
③ Optical flow of the eyeball vicinity area image: This image monitored of the optical flow detected by the method described in Sect. 3.1.

④ Binarized image of the region near the eyeball: This image monitored the binarization performed by the method described in Sect. 3.2. The white slide could be adjusted using the slide at the bottom.

⑤ Feedback display area for gaze movement and blink detection: When a gaze movement or blink were detected in the input image, a feedback image was displayed.

⑥ Detection log of gaze movement: A list of the detected gaze movements in order from the left, classified by color for each direction of movement. The numerical values in the bottom rows indicated the system time at the time of detection.

⑦ Blink detection log: A list of detected blinks, stored sequentially from the left. Optionality and spontaneity are classified by color. The numerical values in the lowest row refer to the system time at the time of detection.

⑧ Shape feature and waveform of specified detection result: This display section exhibited the graphs depicted in Figs. 6 and 7. Figure 6 charts an eye movement waveform by selecting individual log items in the area of display part 6. Figure 7 depicts a blink waveform drawn from the individual log items in the area of display part 7.

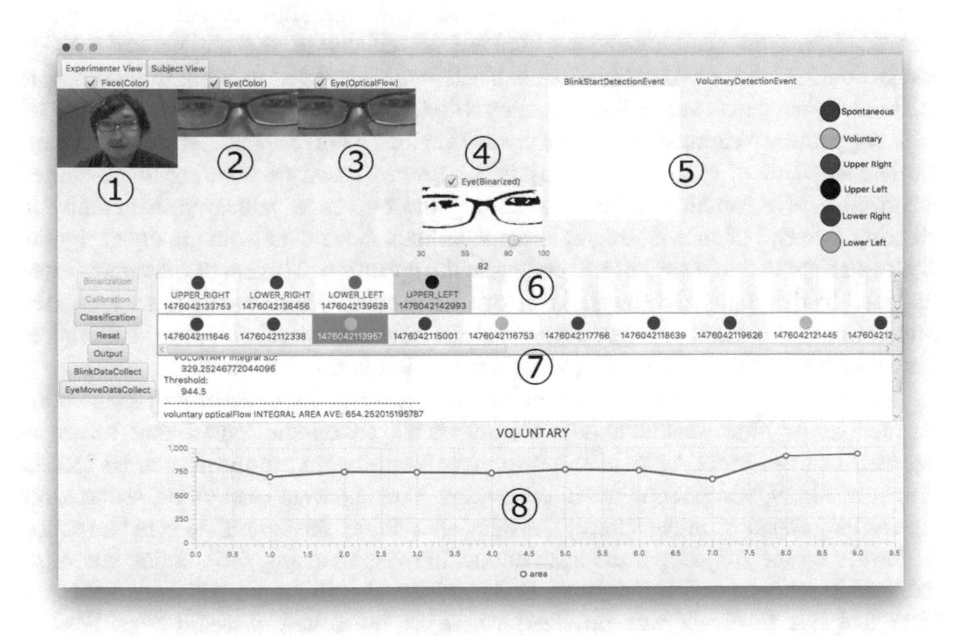

Fig. 5. Display screen of the experimental system.

4.2 Experimental Protocol

Figure 8 summarizes the workflow of the experiment. First, calibration was performed to determine the threshold value for input discrimination. The subjects were taught in

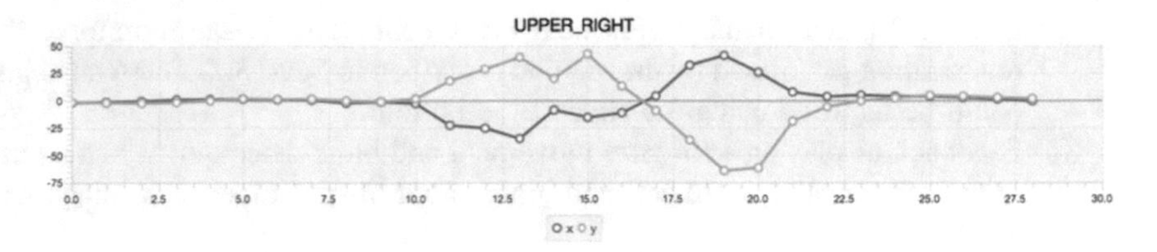

Fig. 6. An eyeblink waveform measuring by optical flow.

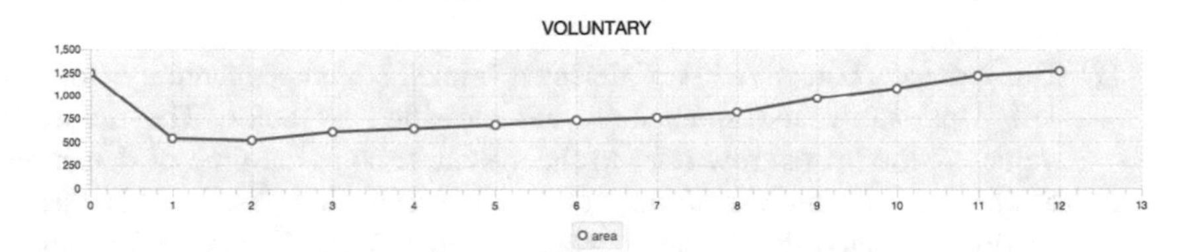

Fig. 7. An eyeblink waveform measuring by optical flow

advance to perform blinks and eye glances according to the instructions simultaneously displayed on the screen and audibly delivered. However, subjects were also advised that there was no need to endure blinking, especially in the absence of explicit instruction. At the time of calibration, subjects were instructed to blink three times, to establish a reference value for voluntary blinking from shape feature quantity of blinking performed immediately after the instructions were received. At the same time, a reference value of spontaneous blink is determined based on blinking that occurred irrespective of instruction. After determining the reference values corresponding to blinking, one Eye Glance input was initiated to track movement from the upper right to the lower right to the lower left and ending in the upper left. This set of movements was analyzed to determine the threshold of eye movement. These calibrations measured waveform, to determine the dwell time and the detection threshold. After calibration, the experiment was executed after an arbitrary intermission period.

In the experiment, the subjects were asked to generate input according to direction instructions or blink instructions randomly displayed on the experiment screen at intervals of three seconds. In addition to visually presenting the instructional indices shown in Fig. 9, instructions for direction and blinking were read aloud at the same time as they appeared in the subject's display area. When the system detected an action performed by the subject per the instructions, the system changed the index color and shape to provide feedback indicating to the subject that the input had been accepted. Note that this feedback was provided whenever input was detected regardless of whether the input was true or false. This operation was performed once for each type of input, a sum of five times, for a total of six enforcements. The order of input instructions for each enforcement was random, and the subjects took arbitrary breaks between each enforcement.

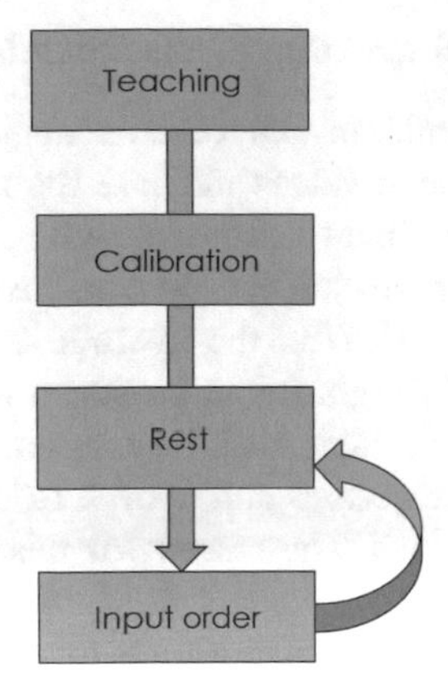

Fig. 8. The experiment workflow.

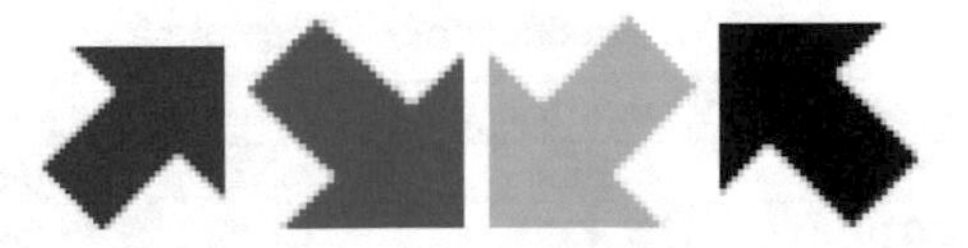

Fig. 9. The instructional indices presented to subjects in the visual display.

4.3 Experimental Results for Setting 1, Prioritizing Blink Detection

Table 1 shows the experimental results of five subjects, respectively. Numeric values indicate the ratio of times that the identification result matched the given input instruction with respect to the total of 30 input instructions administered to each subject. In this experiment, the overall accuracy rate for all subjects and instances was 68.00%. The low concordance rate of gaze movement recorded for subjects A and B may have resulted from gaze movements that were incorrectly interpreted as voluntary blinks. In this experiment, when the detection of a voluntary blink and the detection of movement in the line-of-sight movement were almost simultaneous, the blink detection was prioritized. For this reason, we posit that many erroneous determinations were generated when rapid sequences of waveform features were recorded.

Table 1. Input accuracy rate.

Subjects	Voluntary (%)	Upper right (%)	Lower right (%)	Lower left (%)	Upper left (%)	Average (%)
A	6/6 (100)	1/6 (16.7)	1/6 (16.7)	1/6 (16.7)	3/6 (50.0)	12/30 (40.00)
B	5/6 (83.3)	3/6 (50.0)	1/6 (16.7)	1/6 (16.7)	3/6 (50.0)	13/30 (43.33)
C	6/6 (100)	3/6 (50.0)	6/6 (100)	5/6 (83.3)	3/6 (50.0)	23/30 (76.67)
D	6/6 (100)	5/6 (83.3)	4/6 (66.7)	5/6 (83.3)	6/6 (100)	26/30 (86.67)
E	6/6 (100)	6/6 (100)	5/6 (83.3)	5/6 (83.3)	6/6 (100)	28/30 (93.33)
Average	29/30 (96.67)	18/30 (60.00)	17/30 (56.67)	17/30 (56.67)	21/30 (70.00)	102/150 (68.00)

4.4 Experimental Results for Setting 2, Prioritizing Gaze Movement

Table 2 shows the experimental results of five subjects, respectively. As in the experiment for Setting 1, numeric values indicate the ratio of times that the identification result matched the given input instruction with respect to the total of 30 input instructions administered to each subject. In this experiment, the overall subject accuracy rate was 79.33%. At 96.67%, the average accuracy rate achieved for voluntary blinking was particularly high. It can be stated that eye movements were generally correctly classified under Setting 2. However, although subjects A, D, and E had overall high agreement rates, subjects B and C both exhibited lower accuracy rates for the upper and lower left directions. The reason for this may be that the magnitude of eye movements accompanying movement of the line of sight differed on the left and right. When observing the waveform at the time of eye movements that failed to be correctly identified, in many cases the amplitude was relatively small compared to the waveform produced the same subject at instances of successful identification.

Table 2. Accuracy rate of inputting.

Subjects	Voluntary (%)	Upper right (%)	Lower right (%)	Lower left (%)	Upper left (%)	Average (%)
A	6/6 (100)	5/6 (83.3)	6/6 (100)	5/6 (83.3)	4/6 (66.6)	26/30 (86.67)
B	6/6 (100)	6/6 (100)	5/6 (83.3)	3/6 (50.0)	3/6 (50.0)	23/30 (76.67)
C	5/6 (83.3)	4/6 (66.7)	4/6 (66.7)	2/6 (33.3)	2/6 (33.3)	17/30 (56.67)
D	6/6 (100)	4/6 (66.7)	4/6 (66.7)	6/6 (100)	5/6 (83.3)	25/30 (83.33)
E	6/6 (100)	5/6 (83.3)	6/6 (100)	5/6 (83.3)	6/6 (100)	28/30 (93.33)
Average	29/30 (96.67)	24/30 (80.00)	25/30 (83.33)	21/30 (70.00)	20/30 (66.67)	119/150 (79.33)

5 Discussion

This section discusses the experimental results detailed in Sect. 4, compare our results with the eye motion measurement process in previously proposed eye input systems.

The gaze input system in the previous study [7] used a dedicated eyeglass type device with a small display and a web camera placed in front of one eye, with input that combined nine option indicators. Experiment literature claimed that the input speeds as high as 34.9 characters per minute were achieved, but the key selection shows up to 26 failures for 12 required input activities. In addition, of six subjects, half reported that they could not successfully select a key by blinking. Compared with this, the experimental result of our proposed method demonstrate a high detection rate for blinking and the recognition rate for each direction was also approximately 80%, so without requiring equipment, our results approximate those of the previous study. It can be expected that input precision of about degree can be obtained.

In the previous study [15], a system that used gaze point measurement on the display and voluntary blink detection under natural light was proposed. The system collected input from the images captured by a web camera attached to the eyeglasses. In the literature, the gazing point measurement error and the high detection rate of the voluntary blink are demonstrated, but the detection of the voluntary blink is compiled with the fixed five second threshold. Because the gazing point measurement evaluation method differs from that of this experiment, a simple comparison cannot be made. However, it should be noted that the detection threshold of the voluntary blink, as evaluated in our experiment, was less than two seconds for all subjects. From this, we deduce that the proposed method can select and input at a higher speed than the previous study [15].

In a previous study [6], a character input system using horizontal eye movement and voluntary blinking using AC-EOG was proposed. In this system, the cursor was moved by moving the line of sight from the character displayed on the screen, and it was selected by blinking. A decision is made twice for rows and columns per character. The literature evaluating the experiment reported an average input speed of 6.2 characters per interval during which ten subjects input 100 characters. As it was possible to input approximately 15 characters per minute in the evaluation conducted in previous study [12], assuming that the same input speed can be realized using the proposed method, we expect that the input time can be reduced by 50%.

6 Conclusions

In this paper, to enable the use eyes for the handsfree operation of small devices and compact information terminals, we experimentally evaluated the practicality of input drawn from eye motions and blinking, rather than conventional gaze-based input. Furthermore, this study realized an input interface operated by blink and eye motion that employed the type of visible light camera already commonly found in most small devices, and therefore evaded the requirement of special equipment, such as infrared light cameras, utilized in current conventional methods. The proposed method investigated the analysis of the image of the eyeball and it immediate facial region. The proposed method constructed an automatic detection system that realizes blink measurement and eye motion detection simultaneously in real time and executed experimental evaluations. Our results demonstrated that it is possible to detect and discriminate eye motion and blinking in real time, a task difficult when approached in previous studies. We consider this a valuable contribution to the improvement of a ocular input interface that does not rely solely on gaze position measurement.

In future work, we hope to implement the proposed method as an input interface, by devising an optimized input screen designed to specifically accommodate different usage scenarios. Because the proposed input method features a small number of choices and facilitates rapid input, it is suitable not only for general uses such as character input, but also for macroscopic operations such as environment control and key unlocking. Therefore, we would like to design and build a user interface individually tailored to these varied use cases.

References

1. Bhaskar, T.N., Ranganath, F.T.K., Venkatesh, Y.V.: Blink detection and eye tracking for eye localization. In: IEEE Conference on region 10, TENCON, vol. 2, pp. 821–824 (2003)
2. Lalonde, M., Byrns, D., Gagnon, L., Teasdale, N., Laurendeau, D.: Real-time eye blink detection with GPU-based SIFT tracking. In: Fourth Canadian Conference on Computer and Robot Vision, pp. 481–487 (2007)
3. Hori, J., Chiba, S.: Development of EOG-Based letter input interface on hierarchical screen keyboard considering the characteristics of eye movements. Far East J. Electron. Commun. **14**(1), 53–69 (2015)
4. Okugawa, K., Nakanishi, M., Mitukura, Y., Takahashi, M.: Experimental verification for driving control of a powered wheelchair by voluntary eye blinking and with environment recognition. Trans. JSME **80**(813), 1–15 (2014)
5. Kajiwara, Y., Nakamura, M., Murata, H., Kimura, H.: Human-computer interface controlled by eye movements and voluntary blinks. Trans. Jpn. Soc. Welf. Eng. **14**(1), 22–29 (2012)
6. Kajiware, Y., Murata, H., Kimura, H., Abe, K.: Human-computer interface controlled by horizontal directional eye movements and voluintary blinks using AC EOG signals. Trans. IEEJ. C **132**(4), 555–560 (2012)
7. Arai, K., Yajima, K.: Communication aid with human eyes only. Trans. IEEJ. C **128**(11), 1679–1686 (2008)
8. Sasaki, M., Ito, S., Takeda, K., Okamoto, T., Rusydi, M.I.: Developing a two-link robot arm controller using voluntary blink. Jpn. AEM J. **22**(4), 475–481 (2014)
9. Sato, H., Abe, K., Ohi, S., Ohyama, M.: A text input system based on information of voluntary blink and eye-gaze using an image analysis. Trans. IEEJ. C **137**(4), 584–594 (2017)
10. Matsuno, S., Mizuno, T., Itakura, N.: Method for measuring intentional eye blinks by focusing on momentary movement around the eyes. In: RISP International Workshop on Nonlinear Circuits, Communications and Signal Processing 2016 (NCSP), pp. 137–140, March 2016
11. Matsuno, S., Ohyama, M., Abe, K., Ohi, S., Itakura, N.: Differentiating conscious and unconscious eyeblinks for development of eyeblink computer input system. In: Usability- and Accessibility-Focused Requirements Engineering. LNCS, vol. 9312, pp. 160–174. Springer (2016)
12. Matsuno, S., Ito, Y., Akehi, K., Itakura, N., Mizuno, T., Mito, K.: A multiple-choice input interface using slanting Eye Glance. Trans. IEEJ. C **137**(4), 621–627 (2017)
13. Matsuno, S., Ito, Y., Itakura, N., Mizuno, T., Mito, K.: Study of an intention communication assisting system using eye movement. In: 15th International Conference on Computers Helping People with Special Needs (ICCHP), pp. 495–502, July 2016
14. Sato, H., Abe, K., Ohi, S., Ohyama, M.: Automatic classification between involuntary and two types of voluntary blinks based on an image analysis. In: HCII (2), pp. 140–149 (2015)
15. Chen, B.C., Wu, P.C., Chien, S.Y.: Real-time eye localization, blink detection, and gaze estimation system without infrared illumination. In: IEEE International Conference on Image Processing (ICIP), pp. 715–719 (2015)

Dynamic Group Formation for an Active Learning System Using Smartphone to Improve Learning Motivation

Noriyasu Yamamoto(✉) and Noriki Uchida

Department of Information and Communication Engineering,
Fukuoka Institute of Technology,
3-30-1 Wajiro-Higashi, Higashi-Ku, Fukuoka 811-0295, Japan
{nori,n-uchida}@fit.ac.jp

Abstract. In our previous work, we presented an interactive learning process in order to increase the students learning motivation and the self-learning time. We proposed an Active Learning System (ALS) for student's self-learning. For each level (low, middle and high level) class, we showed that the students could keep concentration using our proposed ALS. However, in the group discussion, some students who understood the lecture could teach other students who did not understand "study points". But, many students complained that they did not feel like a lecture style. In this paper, to solve this problem, we propose a new method for dynamic group formation. After the system decides the level of lecture understanding, the students who did not understand the lecture make the presentation and show the points that they did not understood. Also, the students who understood the lecture explain by presentation the questions or the points that other students did not understand. So, in the group discussion, each group member presents the points that they understood or did not understand. Thus, most of students can study considering their understanding level using above group discussion and they can keep their learning motivation.

1 Introduction

There is a total of 777 universities today in Japan. The "middle-level" universities (which are ranked at around the middle in terms of academic level) form the largest group in Japan and their undergraduates students are the main workforces. Because the group represents the "middle-level", academic capacities of students often vary significantly. For this reason, the teaching speed should be controlled when deciding the understanding level. When a lesson offers advanced contents or is fast-paced, good students find it satisfying, while middle level or low level students fail to catch up. If a lesson is designed too simple or slow, the middle level and low level students find it satisfying, while good students get frustrated. To solve this problem, there are many e-learning systems [1–9]. Also, today's classes increasingly utilize information terminals, such as notebook PCs and projectors. These information tools have enabled lecturers to offer more information to students. But, their effectiveness has no positive impact on students who lack concentration and motivation. The students just look at the

L. Barolli et al. (Eds.): IMIS 2018, AISC 773, pp. 183–189, 2019.
https://doi.org/10.1007/978-3-319-93554-6_16

projector screen and get the information materials. Their satisfaction level for the lecture is increased, but their scores do not improve.

Usage of the desk-top computers and notebook PCs for lecture may be inconvenient and they occupy a lot of space. Therefore, it will be better that students use small and lightweight terminals like Personal Digital Assistant (PDA) devices. Also, because of wireless networks are spread over university campuses, it is easy to connect mobile terminals to the Internet and now the students can use mobile terminals in many lecture rooms, without needing a large-scale network facility. Our idea has been to use various information terminals and tools to boost students' concentration and motivation.

We considered a method of acquiring/utilizing the study record using smartphone in order to improve the students learning motivation [10]. During the lecture the students use the smartphone for learning. The results showed that the proposed study record system has a good effect for improving students' motivation for learning.

For the professors of the university, it is difficult to offer all necessary information to the students. In addition, they cannot provide the information to satisfy all students because the quantity of knowledge of each student attending a lecture is different. Therefore, for the lectures of a higher level than intermediate level, the students should study the learning materials by themselves.

In our previous work, it was presented an interactive learning process in order to increase the students learning motivation and the self-learning time [11]. However, the progress speed of a lecture was not good. To solve this problem, we proposed an Active Learning System (ALS) for student's self-learning [12, 13]. Also, to improve student's self-learning procedure, we proposed some functions for the ALS [14]. We performed experimental evaluation of our ALS and showed that the proposed ALS can improve student self-learning and concentration [15]. Although, the self-learning time and the examination score were increased, the number of student that passed examination didn't increase significantly. To solve this problem, we proposed the group discussion procedure to perform discussion efficiently [16].

The learning system proposed so far only gives a feedback of students' learning record to lecturer. We propose a mechanism to enhance the learning effects by using a record of students' learnings in the whole process of learning system [17].

The previous interfaces for interactive learning had limited adequacy in maintaining participants' concentration when their skill levels vary. Thus, we proposed the interface of the mobile devices (Smartphone/Pad) on our ALS to improve the learning concentration [18]. We presented the performance evaluation of ALS when using the improved human interface on the mobile devices [19]. When the lecture uses ALS, the average dropping out was half compared with the conventional lecture. In addition, we showed the flow of the group discussion and the performance evaluation of ALS for high level class [20]. The evaluation results showed that the interactive lecture by using ALS increases the students' concentration, and the average dropping out for the lecture was half compared with the conventional ALS. Also, the method of group discussion increased the students' concentration for high level class. However, in the group discussion, some students who understood the lecture could teach other students who did not understand "study points". But, many students complained that they did not feel like a lecture style. In this paper, to solve this problem, we propose a new method for dynamic group formation.

The paper structure is as follows. In Sect. 2, we introduce the related work on learning systems and present ALS with Interactive Lecture and Group Discussion. Then, in Sect. 3, we present the dynamic group formation for discussion. Finally, in Sect. 4, we give some conclusions and future work.

2 Active Learning System with Interactive Lecture and Group Discussion Using Smartphone

We developed an Active Learning System (ALS) and show ALC (Active Learning Cycle) of the proposed ALS in Fig. 1. The system facilitates a learning cycle consisting of a lecture, reviews at home and group discussions. The student and the lecturer confirm the movement by setting each cycle and information by the smartphone. At the beginning of ALC, the lecturer performs the interactive lecture by confirming the understanding degree of the student using their smartphone in real time.

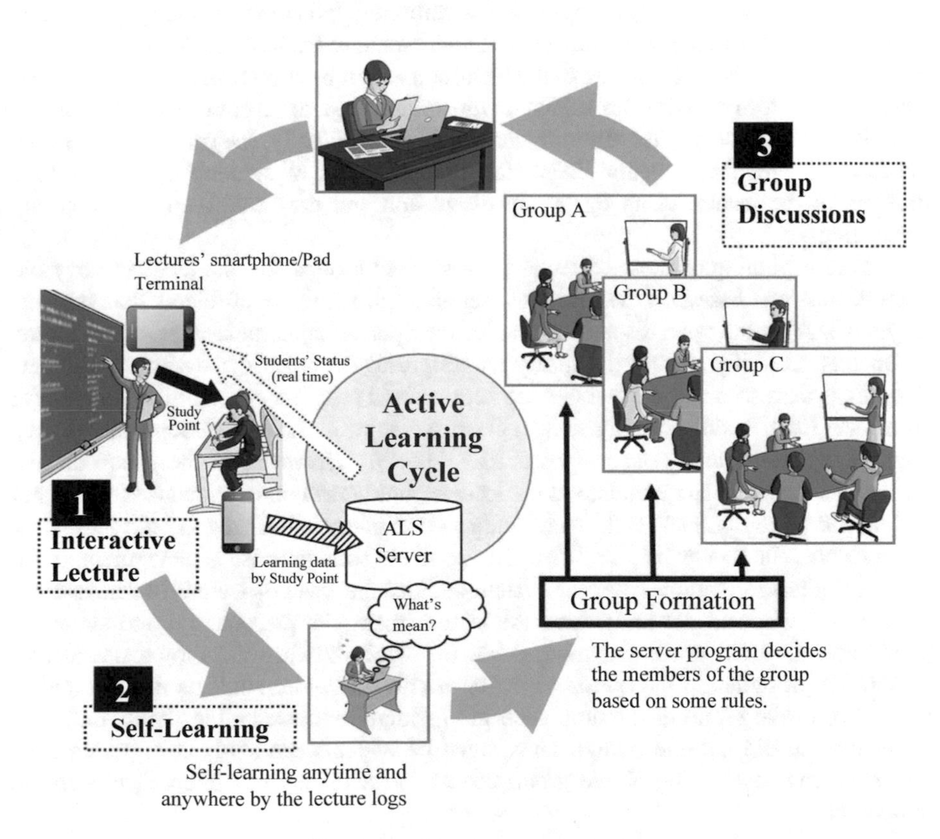

Fig. 1. ALC using a smartphone.

Prior to the lecture, the lecturer prepares "study points". "Small examination" refers to a mini quiz prepared for each study point on students' smartphone. A mini quiz consists of simple multiple-choice questions. "Understanding level" is set by the result of the "Small examination". "Lecture speed" suggests whether students find the lecture progress too fast or too slow. By "Understanding level" and "Lecture speed", students' understanding can be judged. These two functions are used as feedbacks through the application on students' smartphone (students' application). During the lecture, these data are recorded in the database to be reflected in the application for the lecturer.

Based on the record accumulated through button operations, the system creates logs and updates the database on real-time. Furthermore, the log automatically updates the study points in the database to be reflected on the lecturer's screen.

Therefore, a lecturer can transmit the knowledge to the student effectively. After the lecture is finished, the student can read the lecture log by their smartphone. The student can review the lecture content using the lecture log anywhere and anytime.

Then, the student discusses the lecture content in small groups. Firstly, the formation of the group are randomly performed by the ALS server and the students are informed of the group member by their smartphone application. In the group discussion, the students discuss what they did not understand in the lecture. Then, they submit the result of the group discussion to the lecturer as short reports. Secondly, the group is automatically formed with the students who understood the lecture and the students who did not understand the lecture by the ALS server. Finally, the group is formed by the lecturer with the students' discussion reports. Most of students can solve their problem in the lecture using the group discussion and they can keep their learning motivation.

At the beginning of the next lecture, the student groups and the lecturer carry out open discussions based on the submitted reports and solve the problems that students may have. After the open discussion, the lecturer performs the next interactive lecture.

In this ALC, by adding the group learning and the open discussion, the understanding is increased and the lecturer can keep a fixed progress speed of the lecture. For each level (low, middle and high level) class, we showed that the students could keep high concentration using our proposed ALS [18–20]. However, in the group discussion, some students who understood the lecture could teach other students who did not understand "study points". But, many students complained that they did not feel like a lecture style. In this paper, to solve this problem, we propose a new method for dynamic group formation. After the system decides the level of lecture understanding, the students who did not understand the lecture make the presentation and show the points that they did not understood. Also, the students who understood the lecture explain by presentation the questions or the points that other students did not understand. So, in the group discussion, each group member presents the points that they understood or did not understand. Thus, most of students can study considering their understanding level using above group discussion and they can keep their learning motivation.

3 Dynamic Group Formation Method

In this section, we explain the proposed flow of the dynamic group formation after students' self-learning. In the previous system, the member group were formed automatically by selecting the students in random way. The students who understood the "study points" were teaching other students who did not understood the lecture. But, many students complained that they did not feel like a lecture style. In this paper, to solve this problem, we propose a new method for dynamic group formation.

In Fig. 2, we show the flow of group discussion on ALS with the dynamic group formation. It is a two steps discussion and the students send a short report to the lecturer.

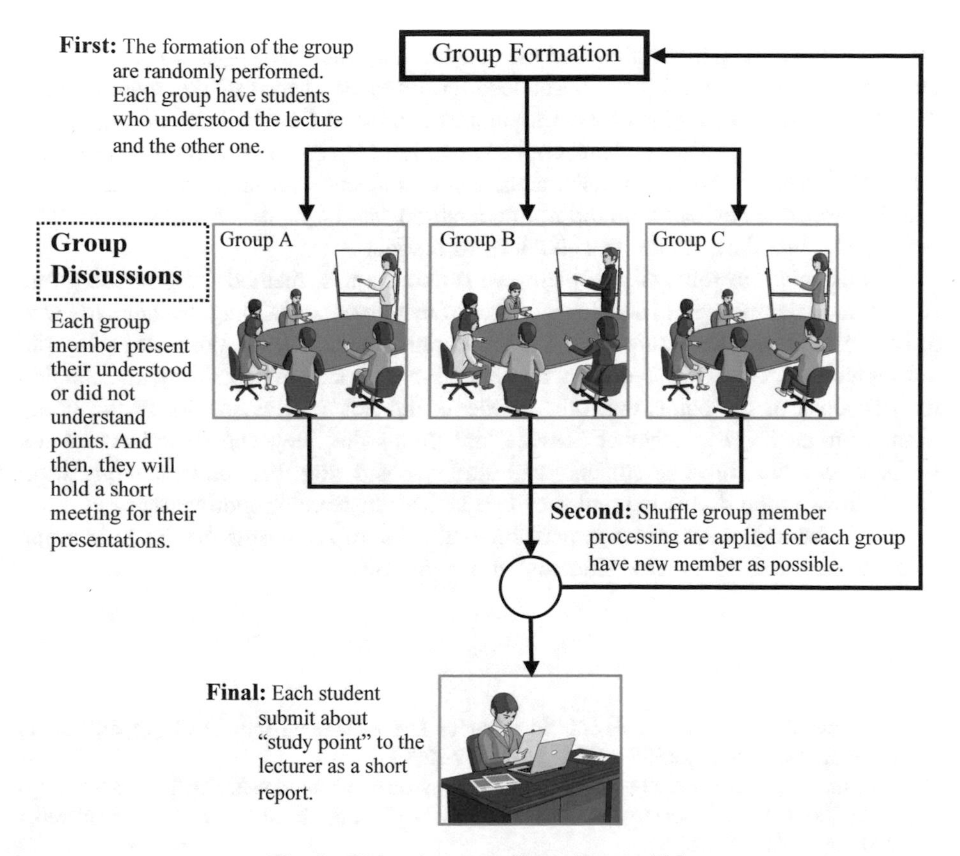

Fig. 2. Dynamic group formation on ALS.

Firstly, the formation of the group are randomly performed by the ALS server and the students are informed of the group members by their smartphone application. Each group is constructed by some students who understood the lecture and other students who did not understand the lecture. In the group discussion, each group member presents the points that understood or did not understand. Then, they will hold a short meeting for their presentations. After that, they submit the "study points" to the lecturer as a short report.

Secondly, shuffle group member processing is applied to each group in order to have new members by the ALS server. Then, the second group discussion is applied the same as first one.

Finally, each student submits the "study points" to the lecturer as a short report. Most of students can study by considering their understanding level using above group discussion and they can keep their learning motivation.

4 Conclusions

In our previous work, we presented an interactive learning process in order to increase the students learning motivation and the self-learning time. We proposed an ALS for student's self-learning. Also, we introduced the interface of the mobile devices on our ALS to improve the learning concentration. For each level (low, middle and high level) class, we showed that the students could keep high concentration using our proposed ALS. However, in the group discussion, some students who understood the lecture could teach other students who did not understand "study points". But, many students complained that they did not feel like a lecture style.

In this paper, to solve this problem, we propose a new method for dynamic group formation. After the system decides the level of lecture understanding, the students who did not understand the lecture make the presentation and show the points that they did not understood. Also, the students who understood the lecture explain by presentation the questions or the points that other students did not understand. So, in the group discussion, each group member presents the points that they understood or did not understand. Thus, most of students can study considering their understanding level using above group discussion and they can keep their learning motivation.

As the future work, we plan to perform evaluation of ALS using the dynamic group formation in order to increase students' concentration.

References

1. Underwood, J., Szabo, A.: Academic offences and e-learning: individual propensities in cheating. Br. J. Edu. Technol. **34**(4), 467–477 (2003)
2. Harashima, H.: Creating a blended learning environment using moodle. In: The Proceedings of the 20th Annual Conference of Japan Society of Educational Technology, September 2004, pp. 241–242 (2004)
3. Brandl, K.: Are you ready to "Moodle"? Lang. Learn. Technol. **9**(2), 16–23 (2005)
4. Dagger, D., Connor, A., Lawless, S., Walsh, E., Wade, V.P.: Service-oriented e-learning platforms: from monolithic systems to flexible services. IEEE Internet Comput. **11**(3), 28–35 (2007)
5. Patcharee, B., Achmad, B., Achmad, H.T., Okawa, K., Murai, J.: Collaborating remote computer laboratory and distance learning approach for hands-on IT education. J. Inf. Process. **22**(1), 67–74 (2013)

6. Emi, K., Okuda, S., Kawachi, Y.: Building of an e-learning system with interactive whiteboard and with smartphones and/or tablets through electronic textbooks. Information Processing Society of Japan (IPSJ), IPSJ SIG Notes 2013-CE-118(3), 01 February 2013, pp. 1–4 (2013)
7. Yamaguchi, S., Ohnichi, Y., Nichino, K.: An efficient high resolution video distribution system for the lecture using blackboard description. Techn. Rep. IEICE **112**(190), 115–119 (2013)
8. Hirayama, Y., Hirayama, S.: An analysis of the two-factor model of learning motivation in university students. Bulletin of Tokyo Kasei University, 1, Cultural and Social Science 41, pp. 101–105 (2001)
9. Ichihara, M., Arai, K.: Moderator effects of meta-cognition: a test in math of a motivational model. Jpn. J. Educ. Psychol. **54**(2), 199–210 (2006)
10. Yamamoto, N., Wakahara, T.: An interactive learning system using smartphone for improving students learning motivation. In: Information Technology Convergence, Lecture Notes in Electrical Engineering, vol. 253, pp. 305–310 (2013)
11. Yamamoto, N.: An interactive learning system using smartphone: improving students' learning motivation and self-learning. In: Proceeding of the 9th International Conference on Broadband and Wireless Computing, Communication and Applications (BWCCA-2014), November 2014, pp. 428–431 (2014)
12. Yamamoto, N.: An active learning system using smartphone for improving students learning concentration. In: Proceeding of International Conference on Advanced Information Networking and Applications Workshops (WAINA-2015), March 2015, pp. 199–203 (2015)
13. Yamamoto, N.: An interactive e-learning system for improving students motivation and self-learning by using smartphones. J. Mob. Multimedia (JMM) **11**(1&2), 67–75 (2015)
14. Yamamoto, N.: New functions for an active learning system to improve students self-learning and concentration. In: Proceeding of the 18th International Conference on Network-Based Information Systems (NBIS-2015), September 2015, pp. 573–576 (2015)
15. Yamamoto, N.: Performance evaluation of an active learning system to improve students self-learning and concentration. In: Proceeding of the 10th International Conference on Broadband and Wireless Computing, Communication and Applications (BWCCA-2015), November 2015, pp. 497–500 (2015)
16. Yamamoto, N.: Improvement of group discussion system for active learning using smartphone. In: Proceeding of the 10th International Conference on Innovative Mobile and Internet Services in Ubiquitous Computing (IMIS-2016), July 2016, pp. 143–148 (2016)
17. Yamamoto, N.: Improvement of study logging system for active learning using smartphone. Proceedings of the 11th International Conference on P2P, Parallel, Grid, Cloud and Internet Computing (3PGCIC–2016), November 2016, pp. 845–851 (2016)
18. Yamamoto, N., Uchida, N.: Improvement of the interface of smartphone for an active learning with high learning concentration. In: Proceeding of the 31st International Conference on Advanced Information Networking and Applications Workshops (AINA-2017), March 2017, pp. 531–534 (2017)
19. Yamamoto, N., Uchida, N.: Performance evaluation of a learning logger system for active learning using smartphone. In: Proceeding of the 20th International Conference on Network-Based Information Systems (NBiS-2017), August 2017, pp. 443–452 (2017)
20. Yamamoto, N., Uchida, N.: Performance evaluation of an active learning system using smartphone: a case study for high level class. In: Proceeding of the 6th International Conference on Emerging Internet, Data and Web Technologies (EIDWT-2018), March 2018, pp. 152–160 (2018)

Evaluation of 13.56 MHz RFID System Considering Communication Distance Between Reader and Tag

Kiyotaka Fujisaki(✉)

Fukuoka Institute of Technology, 3-30-1 Wajiro-higashi, Higashi-ku, Fukuoka 811-0295, Japan
fujisaki@fit.ac.jp

Abstract. RFID system becomes one of the very useful tools for the management of the library. Using electromagnetic coupling, an RFID tag can get power supplier by a reader and communicate with it for data exchange. Because the RFID system enables non-contact communication, various services and applications including the management of a library catalogue are possible. However, because the system is affected easily by neighboring environment, the communication performance is low. In this paper, by using 13.56 MHz RFID system, we evaluate the resonant frequency of RFID tag and the communication distance between the reader and the target tag when some tags becoming as interference sources come close to each other, and show the possibility to expand the communication distance by using tags near the target tag.

1 Introduction

The progress of the radio technology provide us with many wireless services such as: TV, Radio, mobile phone, and so on. Recently, using a radio wave technique to get the information from goods without direct contact is very important. This technique is called Radio Frequency Identification (RFID) [3].

The RFID technique uses electromagnetic coupling for data exchange between the reader/writer and the tag. The RFID system using this technique can be applied in the case when a large quantity of goods is managed. For example, in the library, by using RFID system, we expect the efficiency of the following services: (1) rental of book and the return, (2) collection inventory, (3) search of the book, (4) access control of users [4,15]. Furthermore, if RFID system is integrated with smart phones and sensor networks as new services for the library system to trace books and users, the system may send useful information to users.

A lot of applications using RFID are proposed [7,13,16,18]. Moreover, the development of RFID devices and performance evaluation using RFID system are performed to realize reliable RFID systems [1,2,5,6,10–12,14,17]. For example, at Kyushu University Library, a RFID system was implemented and the usefulness of RFID system was evaluated as a joint study with Mitsubishi Materials

L. Barolli et al. (Eds.): IMIS 2018, AISC 773, pp. 190–200, 2019.
https://doi.org/10.1007/978-3-319-93554-6_17

Corp. and Checkpoint Systems Inc. [4]. Furthermore, in [5], using 13.56 MHz RFID system based on the international standard of ISO15693, the influence that papers or another RFID tag give to the resonant frequency of an RFID tag have been evaluated. In [6], the performance of a table type RFID reader using 13.56 MHz have been evaluated and the relation between the reading rate and the distance between metallic plate and RFID reader was investigated. In these experiments, it is reported that the performance depends on the reading range of RFID system. Especially, when tags get too close, the performance largely decreased. The performance of the tag system changed by the distance of the metal bookshelf or the desk including the metal. The main reason of this problem is the shift of the resonant frequency by the influence of other things.

In this paper, using 13.56 MHz RFID system based on the international standard of ISO15693, we evaluate the influence that other RFID tags, which become as the interference sources, give to the resonant frequency of the RFID tag and the communication distance between the reader and the target tag.

The paper structure is as follows. In the next section, we present the RFID system. Then, we give the evaluations of the resonant frequency and the communication distance RFID tag system. Finally, we conclude the paper.

2 RFID System

An RFID system is one of the technique used for the automatic identification. The automatic identification means to "automatically input bar-code, magnetic-card, RFID data, etc. with the use of hardware and software and not human intervention in order to recognise the content of the data". Also, the biometrics, OCR, the machine vision are included in this technique. Because RFID system uses the wireless communication is able to get the ID from the tag without touching the tag.

An RFID system is made up of two components as shown in Fig. 1 [3]. One is the RFID tag, which is located on the object to be identified, and another is the

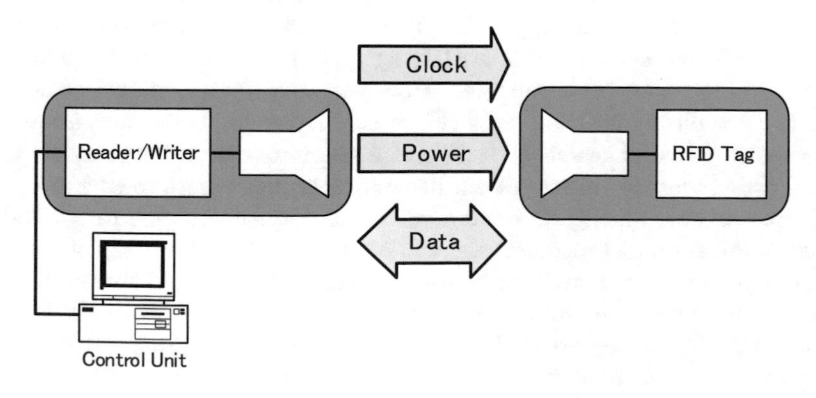

Fig. 1. Basic concept of RFID system

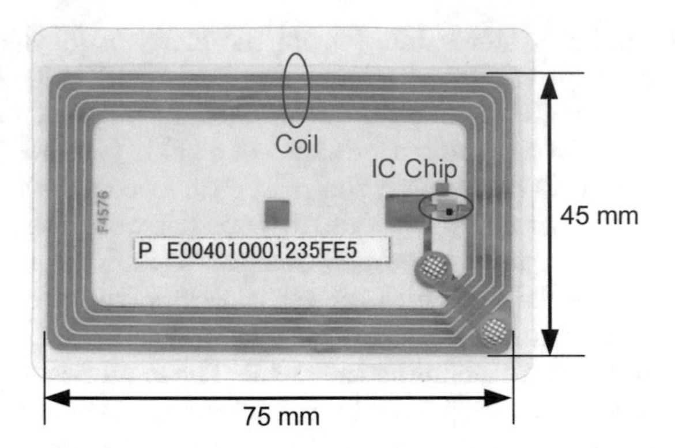

Fig. 2. Example of 13.56 MHz RFID tag in practical use

reader/writer. The RFID tag normally does not have the power supply to work, so the reader/writer not only exchange the data, but also supply the power and clock signal to the RFID tag. To do this, the RFID tag has an antenna or a coil as a coupling element for the communication. Figure 2 shows a practical example of label type RFID tag using 13.56 MHz. The small black box inside the coil in this figure is an IC chip. The RFID tag has various shapes, for example, label, card, coin, and stick.

RFID systems are classified according to the operating frequency, the physical coupling method and the communication range. For example, available operating frequency is 135 kHz, 13.56 MHz, 900 MHz, and 2.45 GHz. The operating frequency is used properly depending on a purpose.

3 Evaluation of 13.56 MHz RFID System

13.56 MHz RFID systems are widely known and used for the management of a library catalogue at various libraries. For example, in Kyushu University, using 13.56 MHz RFID system, the effects of the introduction of the RFID system to the library was evaluated [4]. In [5], using 13.56 MHz RFID system, the influence that papers or another RFID tag give to the resonant frequency of a target RFID tag have been evaluated. However, in a library, many books with a RFID tag are placed on the bookshelf. In this situation, some tags are located in the access range of reader and affect the communication between the target tag and the reader. In this situation, because tags come close to each other, the interference occurs among tags and a resonant frequency of the target tag and the communication distance between the reader and the target tag change.

In this paper, we assume the situation mentioned above and the resonant frequency of the target tag and the communication distance between RFID reader and the target RFID tag are evaluated when there are some tags, which become as the interference sources, stand in a single line with the target tag. 13.56 MHz RFID system based on the international standard of ISO15693 is used in this experiment.

3.1 Resonant Frequency of RFID Tag

13.56 MHz RFID system is connected by the inductive coupling. In this case a tag is combined with a reader in a resonant frequency that corresponds with the transmission frequency of the reader. Generally, RFID tag has the characteristic between induced voltage at a tag coil and frequency [8,9]. If the induced voltage to occur in the transmission frequency of the reader is lower than the minimum of the voltage necessary to drive a tag, the tag does not work. On the other hand, in the case that the resonant frequency of the tag becomes the transmission frequency of the reader, the induced voltage becomes maximum and maximum power is supplied to the tag. In this case, because the induced voltage of the transmission frequency of the reader is larger than the minimum of the voltage necessary to drive a tag, the tag works. The resonant frequency of the tag is affected easy by neighboring environment and a resonant frequency becomes lower. To know the characteristics of the resonant frequency of the tag in real situation is very important to develop the reliable RFID systems.

For this experiment is prepared a testing bench as shown in Fig. 3. Using the grid dip oscillator DELICA DMS-230S2 as shown in Fig. 4, the resonant frequency of the target RFID tag is observed when some tags are in a single line with the target tag as the interference sources. The number of tags becoming

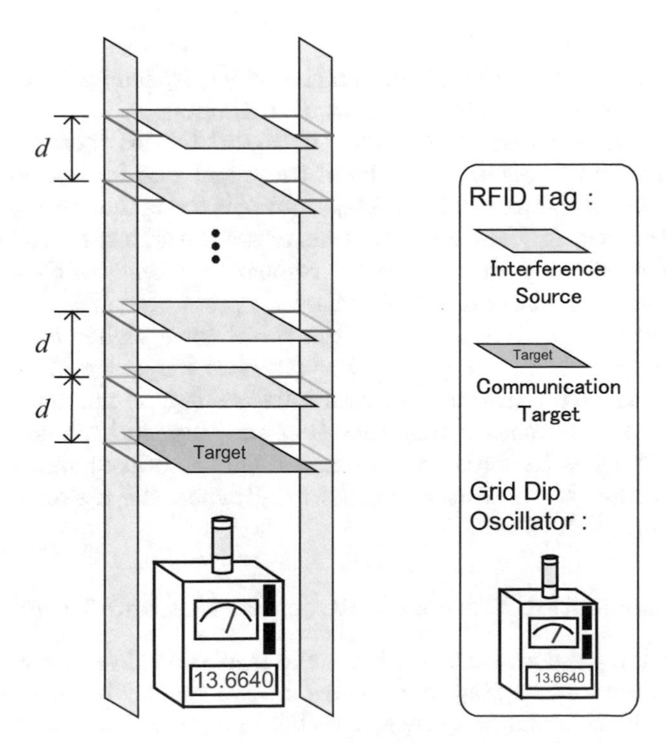

Fig. 3. Measurement image of the resonant frequency of RFID tag affected by another RFID tags

Fig. 4. Photo of Grid dip oscillator DMC-230S2

the interference sources, which is put on the target, is changed from 1 to 6 and the distance between adjacent tags is set to d [mm].

The experimental results are shown in Fig. 5. In this figure, vertical axis shows the resonant frequency f [MHz] of the target tag, horizontal axis shows the distance d [mm] between adjacent tags and N is the number of tags becoming the interference sources. The measurements of each situation are carried out five times. The dotted line in Fig. 5 shows the resonant frequency of the target RFID tag when observed alone and is 13.87 MHz.

These results show that tags near the target have higher influence to the resonant frequency of the target tag. The resonant frequency of the target tag becomes low when the distance between adjacent tags is narrow. In addition, the variation of the resonant frequency becomes large in accordance with the increment of tags as the interference sources. On the other hand, the coupling among the target tag and other tags is small when the distance of these is separated more than 50 mm.

3.2 Communication Distance Between Reader and Target Tag

As shown in the previous section, when some tags come close to each other, the interference occurs among tags and the resonant frequency becomes low. In this situation, if the resonant frequency have big difference from the transmission frequency of the reader, because the induced voltage becomes low, the tag may not sufficiently work. In order to communicate with the tag under this condition

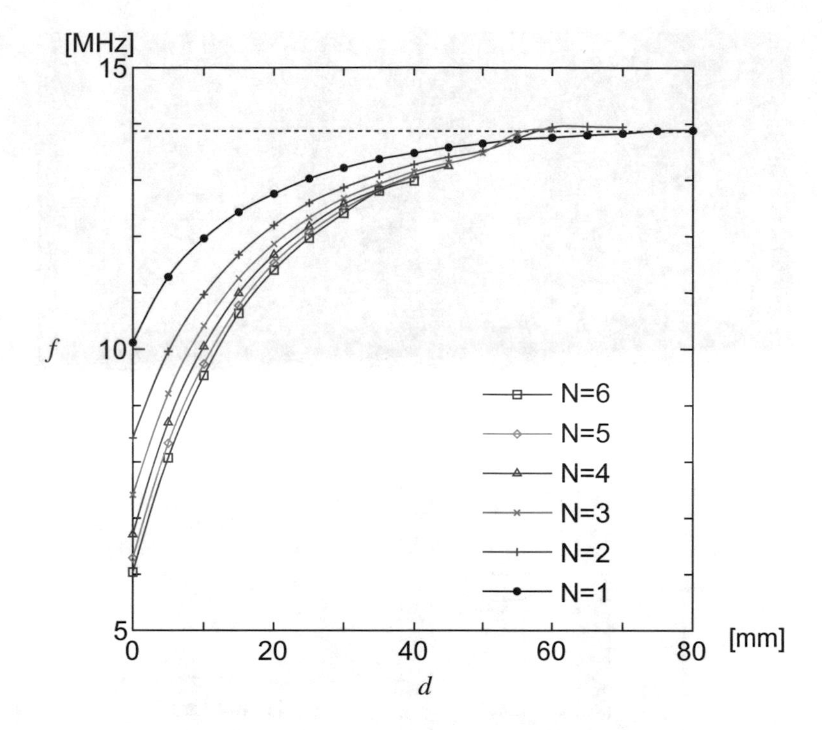

Fig. 5. Resonant frequency f of RFID tag vs distance d between adjacent tags

is needed to give a higher magnetic flux density to the tag, and we may consider the necessity to short the communication distance. In this section, we evaluate the communication distance between the reader and the target tag when the tags becoming as the interference sources are placed between the reader and the target tag.

In this experiment, we use a handmade type RFID reader as shown in Fig. 6. As the reader is used ISO15693 IC tag kit sold by SOFEL, and has an approximately 45 mm in diameter loop coil as the communication antenna of the reader.

Figure 7 shows the measurement image of the communication distance between the reader and the target tag. Some tags becoming as the interference sources are put between the target tag and the reader. The number of tags is changed from 1 to 6 and the distance between adjacent tags is set to d [mm].

The experimental results are shown in Fig. 8. In this figure, vertical axis shows the communication distance L [mm] between the target tag and the reader, horizontal axis shows the distance d [mm] between adjacent tags and N is the number of tags becoming as the interference sources. The measurements of each situation are carried out five times and the average values are shown in this figure. The dotted line in Fig. 8 shows the communication distance between the target RFID tag and the reader when observed alone and is 101 mm.

Before carrying out this experiment, because the resonant frequency becomes low by the interference from other tags, we assumed that the communication

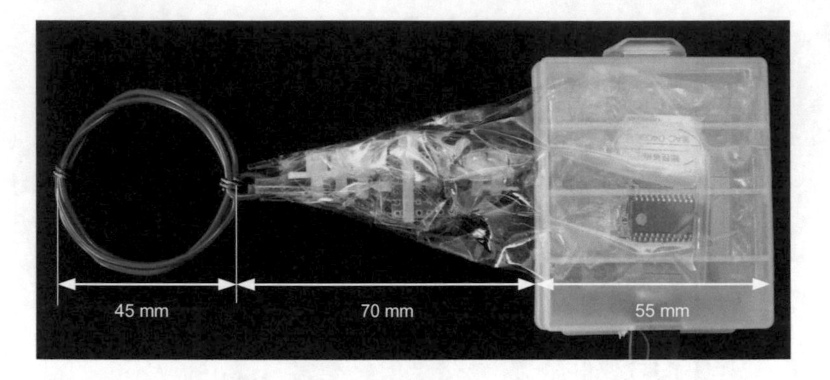

Fig. 6. Photo of Handmade type RFID reader

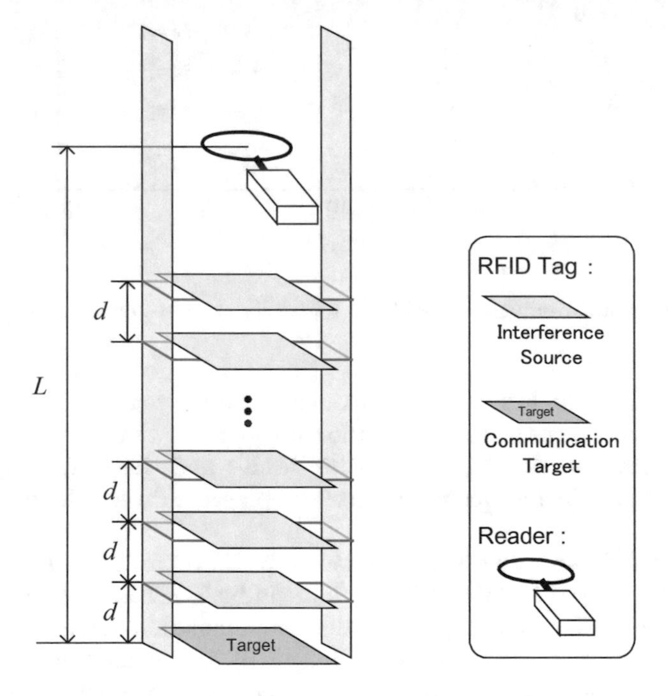

Fig. 7. Measurement image of the communication distance between reader and target tag

distance decreased. In fact, when the distance between adjacent tags is less than 10 mm, in all cases the communication distance was decreased compared with the distance of a single tag case. However, when the distance between adjacent tags was more than 10 mm, it is possible to have a long communication distance compared with the single tag case. For example, when there were 5 or 6 pieces of tags as the interference source and the distance between adjacent tags was about 30 mm, the communication distance was increased by about double. Even in the

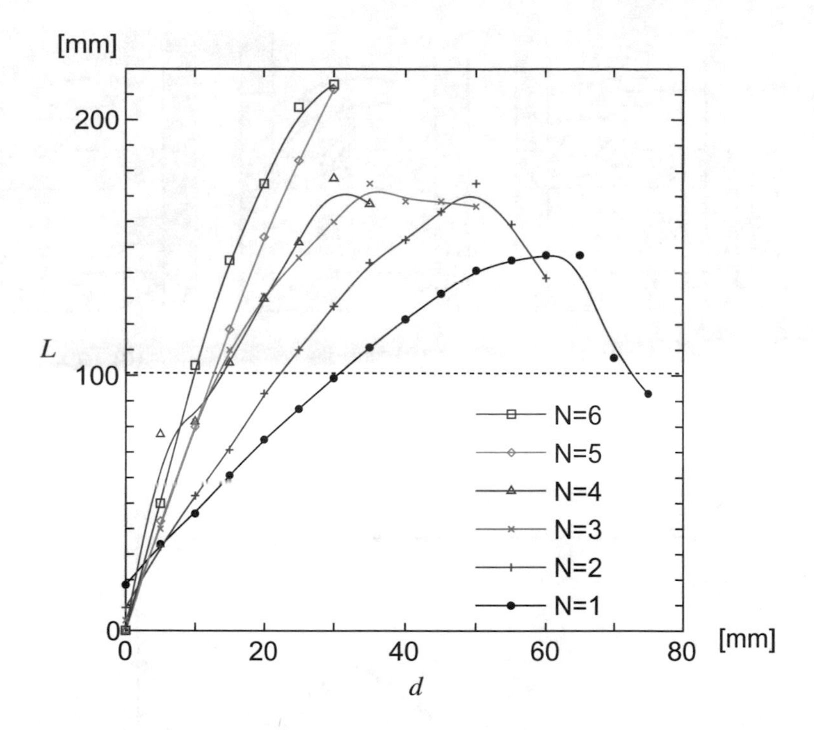

Fig. 8. Communication distance L between RFID tag and reader vs distance d between adjacent tags

case of one piece of the interference source could expand the communication distance to 1.5 times in the case when the distance between adjacent tags was about 60 mm.

From these results, we conclude that when there are some tags becoming as the interference sources and stand in a single line with the target tag, the tags sandwiched between the reader and the target tag collect the magnetic flux around the tag, and relay it to the target tag or/and the reader.

Finally, we change the position of the tags becoming as the interference sources and evaluate the communication distance between the reader and the target tag. Figure 9 shows the measurement image of the communication distance between the reader and the target tag. In this experiment, 2 pieces of tags are used as the interference sources and change the position of the target tag as shown in Fig. 9. The experimental results are shown in Fig. 10.

In the case of Fig. 9(a), the communication distance between the reader and the target tag is expanded about 1.5 times by the effect of the interference tags, and this result is the same as the result shown when $N = 2$ in Fig. 8. On the other hand, in the case of Fig. 9(c), the communication distance between the reader and the target tag slightly increased, when the distance between adjacent tags became bigger than 60 mm. In the case of Fig. 9(b), the communication distance is expanded by the interference tags. Furthermore, this result is similar with the

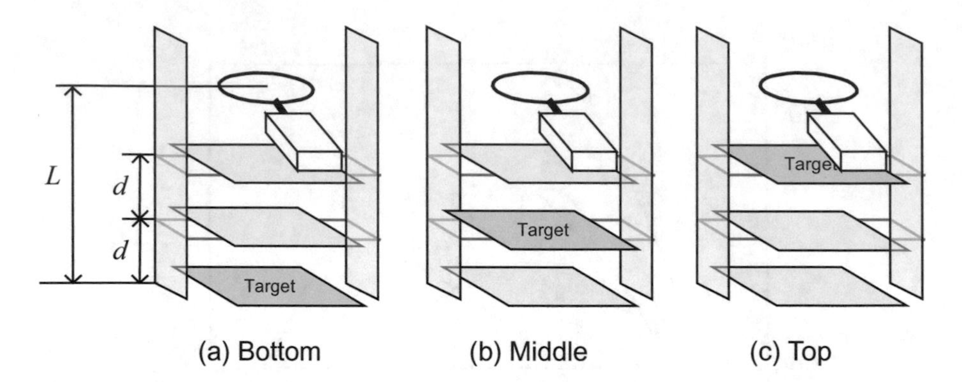

Fig. 9. Measurement image of the communication distance between reader and target tag

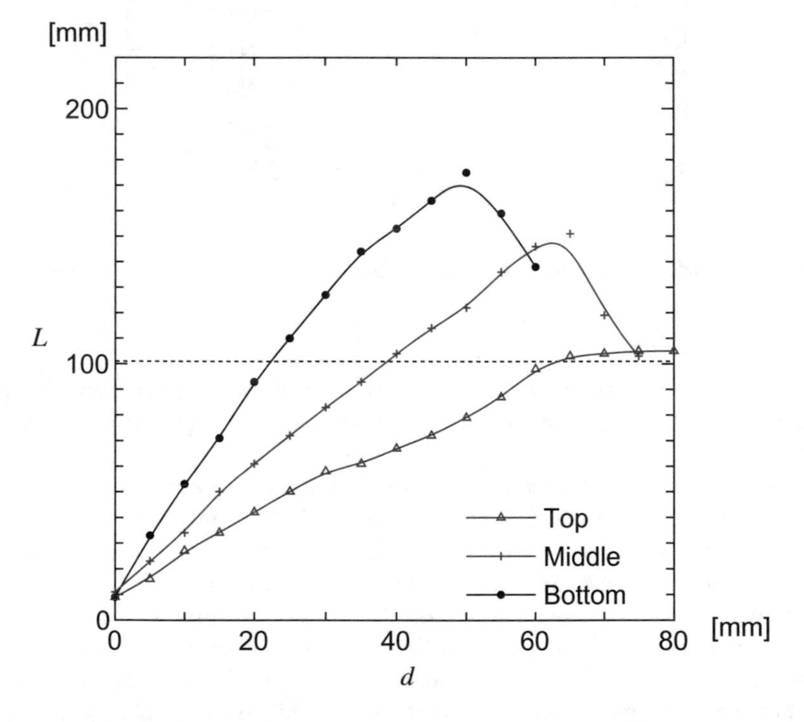

Fig. 10. Communication distance L between RFID tag and reader vs distance d between adjacent tags

result of Fig. 8 when $N = 1$. From these results, it is clearly shown that when there are tags becoming as the interference sources between the reader and the target tag, the tags have the possibility to expand the communication distance between the reader and the target tag.

4 Conclusions

In this paper, we evaluated the resonant frequency of RFID tag and the communication distance between the reader and the target tag when some tags becoming as the interference sources came close to each other.

Using a grid dip oscillator, we measured the resonant frequency of RFID tag under various conditions and showed that the tags near the target have a big influence to the resonant frequency of the target tag. Specifically, the resonant frequency of the target tag becomes low when the distance between adjacent tags is narrow and the variation of the resonant frequency becomes large in accordance with the increment of tags becoming as the interference sources. On the other hand, the coupling among the target tag and other tags is small when the distance of these is separated more than 50 mm.

Furthermore, using the handmade type reader, we evaluated the communication distance between the reader and the target tag and showed the possibility to have a long communication distance compared with the single tag case when the tags becoming as the interference sources sandwiched between the reader and the target tag.

In the future work, we want to evaluate in details the influence that many other RFID tags becoming as the interference sources gives to the communication performance of RFID system for various situation and to suggest methods to improve the performance.

Acknowledgements. I am indebted to Hirotaka Kato for his assistance in experiments.

References

1. Bolomey, J.C, Capdevila, S., Jofre, L., Romeu, J.: Electromagnetic modeling of RFID-modulated scattering mechanism. Application to tag performance evaluation. Proc. IEEE (2010). https://doi.org/10.1109/JPROC.2010.2053332
2. Cantatore, E., Geuns, T.C.T., Gelinck, G.H., et al.: A 13.56-MHz RFID system based on organic transponders. IEEE J. Solid-State Circ. (2007). https://doi.org/10.1109/JSSC.2006.886556
3. Finkenzeller, K.: RFID Handbook, 3rd edn. Wiley (2010). https://doi.org/10.1002/9780470665121
4. Fujisaki, K.: The implementation of the RFID technology in the library, and electromagnetic compatibility (in Japanese). Monthly EMC **183**, 86–94 (2003)
5. Fujisaki, K.: Implementation of a RFID-based system for library management. Int. J. Distrib. Syst. Technol. (2015). https://doi.org/10.4018/IJDST.2015070101
6. Fujisaki, K.: Evaluation and measurements of main features of a table type RFID reader. J. Mob. Multimed. **11**, 21–33 (2015)
7. Ha, O.K., Song, Y.S., Chung, K.Y., et al.: Relation model describing the effects of introducing RFID in the supply chain: evidence from the food and beverage industry in South Korea. Pers. Ubiquitous Comput. Arch. (2014). https://doi.org/10.1007/s00779-013-0675-x
8. Iga, T., Morise, H.: How to Use IC Tag. Nikkan Kogyo Shinbun (2005) (in Japanese)

9. Karibe, H.: Introduction to Contactless Smart Card Design. Nikkan Kogyo Shinbun (2005) (in Japanese)
10. Kuoa, S.K., Hsub, J.Y., Hungb, Y.H.: A performance evaluation method for EMI sheet of metal mountable HF RFID tag. Measurement (2011). https://doi.org/10.1016/j.measurement.2011.02.018
11. Li, N., Gerber, B.B.: Performance-based evaluation of RFID-based indoor location sensing solutions for the built environment. Adv. Eng. Inf. (2011). https://doi.org/10.1016/j.aei.2011.02.004
12. Potyrailo, R.A., Morris, W.G., Sivavec, T., et al.: RFID sensors based on ubiquitous passive 13.56-MHz RFID tags and complex impedance detection. Wirel. Commun. Mob. Comput. (2009). https://doi.org/10.1002/wcm.711
13. Prasad, N.R.K., Rajesh, A.: RFID-based hospital real time patient management system. Int. J. Comput. Trends Technol. **3**, 1011–1016 (2012)
14. Basat, S.S., Kyutae, L., Laskar, J., Tentzeris, M.M.: Design and modeling of embedded 13.56 MHz RFID antennas. In: Proceedings of IEEE International Symposium on Antenna Propagation (2005). https://doi.org/10.1109/APS.2005.1552740
15. Sing, J., Brar, N., Fong, C.: The State of RFID applications in libraries. Inf. Technol. Libr. (2006). https://doi.org/10.6017/ital.v25i1.3326
16. Symonds, J., Seet, B.C., Xiong, J.: Activity inference for RFID-based assisted living applications. J. Mob. Multimed. **6**, 15–25 (2010)
17. Uysal, D.D., Gainesville, F., Emond, J., Engles, D.W.: Evaluation of RFID performance for a pharmaceutical distribution chain: HF vs. UHF. In: Proceedings of 2008 IEEE International Conference on RFID (2008). https://doi.org/10.1109/RFID.2008.4519382
18. Zhonga, R.Y., Dai, Q.Y., Qu, T., et al.: RFID-enabled real-time manufacturing execution system for mass-customization production. Robot. Comput. Integr. Manuf. (2013). https://doi.org/10.1016/j.rcim.2012.08.001

An Efficient Routing Protocol via Depth Adjustment and Energy Gradation in Underwater Wireless Sensor Networks

Ghazanfar Latif[1], Nadeem Javaid[1](✉), Arshad Iqbal[2], Javed Ahmad[3], Ather Abdul Jabbar[1], and Muhammad Imran[4]

[1] COMSATS Institute of Information Technology, Islamabad 44000, Pakistan
nadeemjavaidqau@gmail.com
[2] Bahauddin Zakariya University, Multan 66000, Pakistan
[3] Virtual Univeristy of Pakistan, Lahore 54000, Pakistan
[4] King Saud University, Riyadh, Saudi Arabia
http://www.njavaid.com

Abstract. Underwater wireless sensor networks (UWSNs) provide the wide range of aquatic applications. The limited bandwidth, long propagation delay, energy consumption, high manufacturing, and deployment costs are many challenges in the domain of UWSNs. In this paper, we present the two techniques i.e., energy gradation (EG) and depth adjustment (DA) in without the number of coronas. Firstly, the forwarder node determines the higher energy node and it is directly transmitted to sink; secondly, if the forwarder node occurs in transmission void region then the node moves to the new depth so that the data delivery ratio can be ensured effectively. Simulation results define that our proposed schemes show better performance in terms of energy efficiency, packet delivery ratio (PDR) and network lifetime etc.

1 Introduction

Almost 75% of the land is covered with water. With the decrease of land resource, human being pays more focus to the investigation of ocean resource. UWSNs are used in the depth of the sea and their in-depth deployment leads to high pressure and less visibility. UWSNs are widely used for disaster prevention, military defense, and pollution monitoring etc. [1]. The radio signal is important for terrestrial communication and acoustic signal for the underwater communication. However, the limited bandwidth, high propagation delay, energy consumption, high manufacturing, and deployment costs are many challenges in the domain of UWSNs. The sound wave is equal to 1500 m/s. However, the speed of radio signal is 3×10^8 m/s [3]. The bit error rate (BER) is caused by the temporary transmission loss and greater interference.

The limitation of this protocol is transmission void region problem. The transmission void region problem degrades the network performance [2]. Due to the void node degrades an energy efficiency of the network. The node located in a

L. Barolli et al. (Eds.): IMIS 2018, AISC 773, pp. 201–211, 2019.
https://doi.org/10.1007/978-3-319-93554-6_18

transmission void region is known as the void node. The transmission void region is also known as the void hole. Moreover, a data gets fail in a restricted area, the routing scheme should determine to transmit the data via DA or neglected it.

In this work, we have proposed two routing schemes i.e., the EG and DA of the void node. The EG scheme divides the initial energy into the number of chunks. The EG determines the higher energy node then data is directly transmitted to sink. The advantages of EG are to improve the energy efficiency, PDR and network lifetime, respectively. Moreover, we have considered the DA technique of void node. In DA scheme, data is forwarded in the presence of void node. When a node is available in a transmission void region then the void node moves to new depth via greedily forwarding strategy. Because of this, the data delivery ratio, reliability, and energy efficiency can be increased.

Contributions: In this paper, we are mitigating the void nodes and improving the energy efficiency via EG and DA. Firstly, we have proposed an EG routing scheme, it divides the initial energy of sensor nodes into the number of chunks. However, the forwarder node finds the higher energy node and it is directly transmitted to sink. During data transmission the energy consumption of sensor nodes can be improved effectively; Secondly, we have proposed the DA technique, data is forwarded in the presence of void node. When a data gets fail in a void node, the routing scheme should calculate to forward the data via DA or neglected it. In DA technique, when the void node occurs then the data is transmitted via DA topology. The void node moves to the new depth via greedy forwarding strategy. However, in EG the data is forwarded on the basis of energy comparison among sensor nodes. The advantages of these schemes are to increase the energy efficiency, and network lifetime, respectively. The performance evaluation explains that our schemes show better results than the current scheme like energy efficiency, PDR, and lifespan of the network.

The rest of the paper is organized as follows. Related work is discussed in Sect. 2. The problem statement of our system is given in Sect. 3. System Model explains in Sect. 4. Simulation and Results are presented in Sect. 5 and conclusion is in Sect. 6 and finally, references are listed of related work.

2 Related Work

To understand UWSNs, we have discussed various past routing methodologies like geographic and opportunistic routing strategies.

Authors in [2] proposed the routing scheme called DA in order to overcome the void hole issue of geographic and opportunistic routing. Because of DA technique, it is to minimize the void node. The limitation of this protocol, the data transmission is failing upon the void node region. This routing strategy achieves to improve the lifespan of the network, PDR due to increasing number of sensors (Table 1).

In [3], authors proposed the network protocol called the adaptive hop-by-hop vector based forwarding routing protocol (AHH-VBF). This work explains to adaptively adjust the forwarding range of data packets. Because of this, the

Table 1. Related work

Technique (s)	Feature (s)	Achievement (s)	Limitation (s)
GEDAR [2]	Geographic and opportunistic routing	Higher network lifetime, PDR is high	High propagation delay
AHH-VBF [3]	Higher network lifetime	Reliability, energy efficiency	High propagation delay due to holding time
VBF [4]	Vector based forwarding	Average energy via transmission range, higher network lifetime	Increases the void nodes due to the void hole issue
CH [5]	Clustering, cluster head, setup phase and communication phase	Energy efficiency, time-critical applications	High propagation delay
MPT [6]	Multi-path routing protocol, time critical application	Network lifetime	Higher energy consumption
VBVA [7]	Vector-based void avoidance, vector shift and back pressure	Void node decreases, better network performance	Higher energy consumption than VBF
BECHA [8]	The energy consumption can be balanced	Network lifetime	Low throughput and high packet drop
EEDBR [9]	Energy efficient depth based routing, forwarder selects based upon the residual energy and depth	Network lifetime	Packet is neglected because of low energy
ELBAR [13]	Efficient load balanced routing	Energy efficiency, higher network lifetime	High propagation delay

reliability can be ensured in the sparse network. Moreover, this paper can be reduced duplicate packet in the dense network. It can also be increased energy efficiency. The limitation of this protocol is the high propagation delay due to holding time. It is also to increase the manufacturing and deployment cost [3].

Authors in [4] proposed the routing scheme called vector based forwarding (VBF). This protocol explains the transmitting range of data is restricted to be a virtual pipeline with the direction from sender to the destination. This paper achieves to increase the efficiency of energy and reliability. The limitation of this paper is not essential for the sparse network. However, the pipeline radius has a greater effect on the network efficiency.

In [5], authors proposed the routing scheme called clustering. The clustering is based upon the cluster heads (CHs). The CHs are important for UWSNs. There are two phases of clustering technique i.e., setup phase and communication phase. In the setup phase, the CHs are selected on the basis of residual energy or location of sensor nodes. However, in communication phase, the member nodes are collecting data and transmitted to the correspondence CHs. This work achieves to improve the energy efficiency, PDR, respectively. However, the limitation of this paper is the higher propagation delay.

In [6], authors proposed a routing scheme is known as the multipath power control transmission (MPT) for time-critical applications in UWSNs. This work is used for the cross-layer approach. The goal of this paper is the low propagation delay and greater reliability. Simulation results show the better performance than the current scheme like the efficiency of energy, PDR. The limitation of this work is to increase the collision due to the increasing number of node density.

Xie *et al.* proposed the routing scheme called vector-based void avoidance (VBVA). The sensor nodes are deployed in the three-dimensional environment. This protocol is also called the void avoidance problem. VBVA defines two schemes i.e., vector shift and back pressure. These schemes resolve the void node problem. The vector shift scheme defines to forward the data via routing hole boundary. However, the pressure shift defines the route data via the backward transmission. This paper can avoid the void nodes in mobile UWSNs. The limitation of this work is more energy consumption than VBF [7].

Naeem *et al.* proposed the routing protocol called balanced energy consumption and hole alleviation (BECHA). The sensor nodes are deployed among various coronas. Because of this, the load distribution is balanced among different coronas. This paper achieves to improve the network lifetime. This work is also to improve the energy efficiency via balanced load distribution. The limitation of this paper is to increase the packet drop ratio. However, it is also to decrease the throughput [8].

In [9], authors proposed the network protocol called energy efficient depth based routing (EEDBR) for UWSNs. In this paper, the forwarder node is selected on the basis of depth and residual energy. Because of this selection, the energy can be balanced and improved network lifetime. In this work, the sensor nodes keep the data for some time before transmitting. The holding time depends on the residual energy of sensor nodes. In this way, the data is directly transmitted to sink on the basis of higher residual energy. The limitation of this paper, the packet is neglected due to the low energy. This protocol does not need the localization of the network nodes.

In [10], authors proposed the routing protocol called the balanced routing (BR). The objective of this work is balancing the energy consumption among sensor nodes. This protocol determines the load weight for each appropriate forwarder nodes. It is also to lead the fair energy consumption throughout the network. Because of this, the energy hole problem can be avoided. This work achieves to improve the network lifetime. The limitation of this work is more energy consumption due to long distance.

Authors in [11] proposed the routing protocol for data gathering problem with data importance consideration in UWSNs. The sensor nodes are located near the sink deplete their energy very quickly due to imbalance load distribution. Because of this, the energy hole problem occurs closer to the sink. In this paper, the unbalanced energy consumption of sensor nodes due to multi-hop communication in deep water is effectively reduced by introducing autonomous underwater vehicles (AUVs) for collecting the data from deep underwater nodes. In this paper, introducing AUVs for data gathering from sensor nodes at deep water improves the network performance via maximizing the network lifetime and PDR with the reduced propagation delay in the network.

Nadeem *et al.* proposed the gateway based energy-efficient routing protocol (M-GEAR) for WSNs [12]. In this strategy, the sensor nodes divide into four logical regions. In each region perform different scenarios. Two regions perform direct transmission and two regions further sub-divided on the basis of cluster head and perform multi-hop transmission. This protocol achieves to improve the network lifetime and energy efficiency.

Nguyen *et al.* proposed the routing protocol called efficient load balanced routing (ELBAR) for WSNs. In this work, the sensor nodes divide into three regions. This protocol performs data transmission in the presence of the void hole. Data transmits along the boundary of void hole via default and escape mode. The achievement of this paper is to improve the energy efficiency, higher network lifetime. The limitation of this protocol is higher propagation delay [13].

3 Problem Statement

The energy hole problem, energy consumption, and void nodes degrade the network performance in terms of network lifetime, PDR, etc. In [1], the sensor nodes are transmitted data via three transmission ranges i.e., one hop, two hops, and direct transmission, respectively. Moreover, the energy consumption is high at direct transmission due to the long distance. The other problem is communication void region occurs due to the void node. The node is located in the void region called the void node. In the case of the void node, the data gets fail in the prohibited area then the current scheme should calculate to forward the data via some routing strategies. In this paper, we are mitigating the energy consumption and number of void nodes, respectively.

4 Proposed Scheme

In this section, the details of network model and proposed schemes are given below:

4.1 Network Model

we suppose that the N number of sensor nodes are uniformly deployed in a circular network area. While the sinks are deployed on the surface of the water.

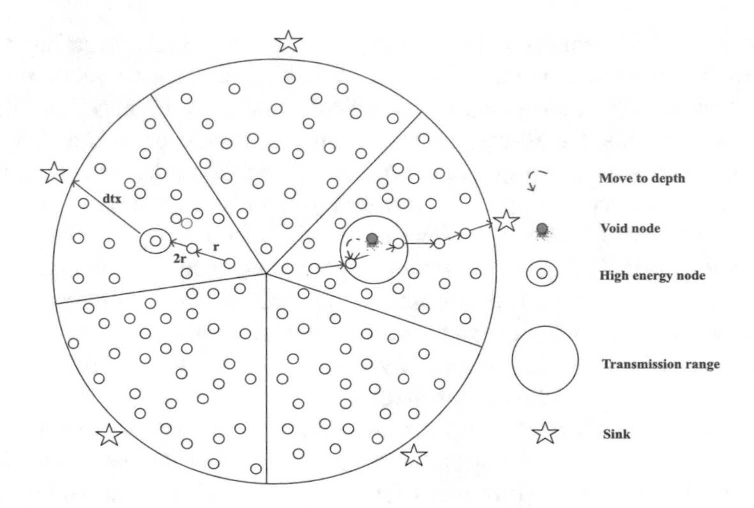

Fig. 1. Network model of proposed scheme

The circular network field divides into equal wedges W. Because of this, the sensor nodes are equally divided into five wedges. The sensor nodes are equal to $N = 50$, respectively. The data transmission is always done by sensor nodes. However, data transmission stopped due to the occurrence of the void node and it can be resumed through DA and EG. These transmissions are given in Fig. 1. The transmission and reception energies are the same when the distance between two nodes is the same. Moreover, every node is equipped with acoustic modem while sinks have both modems radio and acoustic to communicate with underwater nodes and with the offshore data centers. When nodes have the data packet, they forward it through multi-hopping by detecting and avoiding the void node at each hop. Further details are given in the following subsections.

4.2 EG

We have proposed the EG scheme to reduce the energy and the void node during communication of network nodes. The gradation is performed by fragmenting the node battery into small chunks called in this paper m. However, the gradation can be calculated by equation $1-2$. Now, if segments are not introduced then in direct transmission more energy is consumed. Basically, the segmentation helps to set out a threshold to control the dissipation of the nodes. For instance, nodes in the direct vicinity of the sink are only allowed to deplete one grade of the battery. The first node is larger than the other node then the first node is directly transmitted to sink. However, if energy grade is small then they transmit to multi-hop and find the higher energy node for data transmission as shown in Fig. 1. By this process, the nodes dissipate energy evenly, so network lifetime, energy efficiency, and other parameters can be increased.

$$m = min(\sqrt{\frac{N * E}{Etx}}, \frac{E}{Etx}) \tag{1}$$

where, N represents the sensor nodes, E represents the energy of sensor nodes, Etx represents the transmission energy, respectively.

$$div - Eo = \frac{Eo}{m} \tag{2}$$

where, $div - Eo$ represents the dividing initial energy, E_o is set to be initial energy and m represents the total number of grades, respectively.

4.3 DA

When a void node occurred then the data is transmitted via DA technique. Because of this technique, we can calculate and manage the routing path through void regions. When the packet cannot forward to next neighbor because of the void node, then the void node moves to new depth for data transmission. The DA technique is the more effective technique for data forwarding and energy efficiency. In this technique, the data transmission is performed via greedy forwarding strategy. The DA technique is also called the data recovery procedure. In data recovery procedure, when the source node comes in the void area then the void node moves to new depth for data forwarding. In geographic and opportunistic for depth adjustment routing (GEDAR), the DA is used for geographic and opportunistic routing [2]. Here, the DA technique is used for circular network region. This technique achieves to improve the network performance. The data transmission is shown in Fig. 1. However, the process of DA via without corona network is shown in an Algorithm 1.

5 Simulation Results

In this work, we have evaluated the efficiency of our scheme and compared the results with the current protocol i.e., balanced load distribution (BLOAD) [1] in terms of PDR and energy consumption. The network nodes are uniformly deployed in a circular area with radius $r = 100$ m. The sensor nodes are equal to $N = 50$, respectively. We assume that the five sinks are located on the water surface. The initial energy of a sensor node is set to be $Eo = 1$ J, the transmission energy is set to be $Etx = 0.0005$ J and the reception energy is $Erx = 0.00005$ J. The circular monitoring area is $R = 1000$ m, respectively. The control parameters used in this paper are given in Table 2.

Figure 2 depicts the number of alive nodes per unit time. The alive nodes increase when the time of transmissions are successfully scheduled in the network. Due to the higher number of alive nodes, the stability period becomes higher. The DA scheme achieves the better result than EG and BLOAD because the node comes in communication void region then the void node moves to new depth for data transmission. According to Fig. 2, the proposed scheme performs 70.5% better than the BLOAD. The overall achievement is 5.53% than BLOAD scheme. The reason is that we are applying the DA and EG proposed schemes. These schemes are compared with the BLOAD. By the process of DA and EG,

Algorithm 1. DA

```
begin
    Initialize parameters
    Initial energy := Ei
    Assume EnergyForDepthAdjustment
    return true
    Procedure RecoveryProcess()
    Start BeaconMessage
    Determine VoidNode
    VoidNode := true
    Stop BeaconMessage
    E_i := E_1, E_2,...,E_n
    SetNeigh(σ) := c_1, c_2,...,c_n
    DataFractions := small, medium, large
    UpdateNeighList := (SetNeigh, E_i, DataFractions)
    return true
    end Procedure
    c_v n := void node
    σ := set of neighbor nodes to act as a next forwaders
    D := collection of depth nodes to the void node c_v n
    Procedure DetermineNewDepth()
    Perform DA
    Data forwarding on the basis of depth adjustment
    for i = 1 to n do
        Calculate the distance of neighbour nodes
        TransmissionRange :=r
        if distNeigh¡= r then
            Data forwards to neighbor nodes or sink
            Select σ act as a next hop forwarder
        else
            c_v n the void node is occur
            Start RecoveryModeProcedure
            EnergyConsumedOnDepthAdjustment
            DepthAdjustmentInitialize
        end if
        Calculate the distance with void node
        if dist(c_v n)¡= r then
            Void node becomes the neighbor node
            Data Forwards to next forwarder node
        else
            c_v n the void node is occur
            Repeat RecoveryModeProcedure
        end if
    end for
    ControlVoidNode()
    n_v n moves to newDepth
    VoidNode := false
    Data Transmission on RecoveryMode Process()
    return true
    Data packet is transmitted by DA
end begin
```

Table 2. Simulation parameters

Control parameter	Value	Unit
Initial energy (E*o*)	1	Joule
Transmission energy (Etx)	0.0005	Joule
Reception energy (Erx)	0.00005	Joule
Circular area (A)	1000	Meters
Radius of circular area (r)	100	Meters
Sensor nodes (N)	50	—

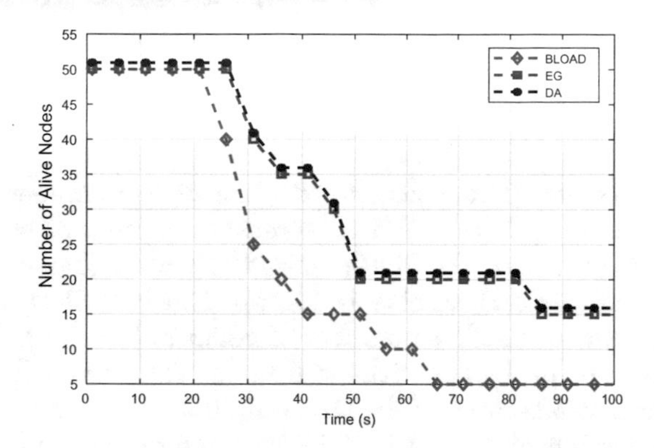

Fig. 2. Number of alive nodes

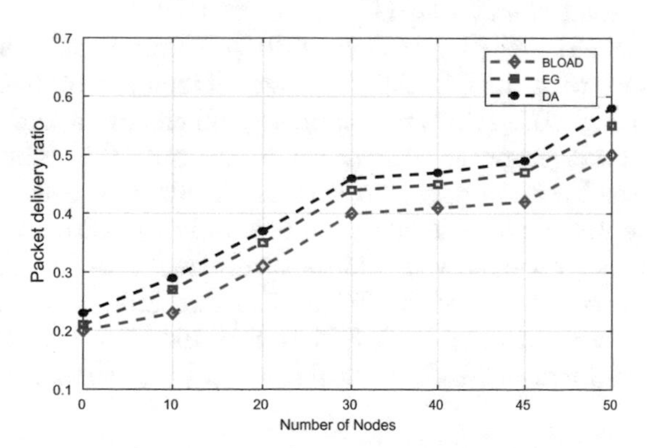

Fig. 3. Packet delivery ratio

fewer numbers of nodes are dead in the network. However, the large numbers of nodes are alive. The simulation results show that the nodes are totally dead in BLOAD. Moreover, in EG and DA schemes are not completely dead. The alive nodes represent the higher network lifetime. Therefore, the network is stable up to 35 s.

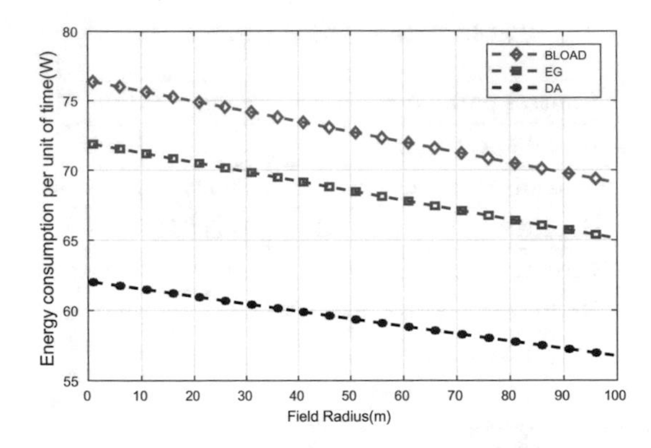

Fig. 4. Energy consumption per unit time

Figure 3 depicts the PDR increases with the higher number of nodes. The PDR of DA and EG increases with the increasing number of nodes among the network. According to figure, the PDR of DA is 20% more than EG and BLOAD. While the PDR of BLOAD scheme is 11.1%. The overall performance is to improve the PDR when the higher number of nodes in the network. The number of nodes increases due to the DA of the void node. The proposed scheme has the best PDR performance because of the DA technique in without corona network. Therefore, the overall achievement is 70.3% of the network performance via DA and EG in without the number of coronas. Moreover, our proposed protocol performs better result than BLOAD in terms of PDR.

Figure 4 depicts the energy consumption per unit time. These results are better performance with the BLOAD. The overall trend is minimizing the energy consumption through DA scheme. The energy efficiency of the current scheme is greater than both proposed schemes like EG and DA. Therefore, the EG scheme divides the initial energy into various chunks and compare them. when the energy of the first node is higher then the node will directly be transmitted to the sink. However, the energy should be minimized than BLOAD scheme. The overall performance is done by the EG and DA technique in without coronas. The proposed scheme achieves 22.86% better result than the BLOAD scheme. The overall achievement is 85.6% using DA and EG in without corona network.

6 Conclusion and Future Work

The energy efficiency and void nodes are the important issues in improving the lifespan of the network. In UWSNs, because of imbalanced energy consumption, the network nodes are expired very quickly. Because of this, the network performance will be degraded. The energy consumption can be balanced and avoidance of void node issues, we invented the EG and DA in without corona network. The EG scheme is explained to improve the PDR, reduce the energy of sensor nodes and increase the lifespan of the network. However, the DA scheme is described to avoid the void node and forward the packets to sink successfully. In EG, if the node consumes more energy then they will directly be transmitted to sink. However, if the node comes in transmission void region then the void node moves to new depth via DA. Simulation results show better performance as compared to BLOAD technique. It is also to improve the stability period, energy efficiency, lifespan of the network and PDR. Further, we will work on various combinations of data. However, we will resolve the major problems with different scenarios in the domain of UWSNs.

References

1. Azam, I., Javaid, N., Ahmad, A., Wadood, A., Almogren, A., Alamri, A.: Balanced load distribution with energy hole avoidance in underwater WSNs. IEEE Access (2017)
2. Coutinho, R.W.L., Boukerche, A., Vieira, L.F.M., Loureiro, A.A.F.: Geographic and opportunistic routing for underwater sensor networks. IEEE Trans. Comput. **65**(2), 548–561 (2016)
3. Yu, H., Yao, N., Liu, J.: An adaptive routing protocol in underwater sparse acoustic sensor networks. Ad Hoc Netw. **34**, 121–143 (2015)
4. Xie, P., Cui, J.-H., Lao, L.: VBF: vector-based forwarding protocol for underwater sensor networks. In: Networking, vol. 3976, pp. 12161221 (2006)
5. Liu, G., Wei, C.: A new multi-path routing protocol based on cluster for underwater acoustic sensor networks. In: 2011 International Conference on Multimedia Technology (ICMT), pp. 91–94. IEEE (2011)
6. Zhou, Z., Peng, Z., Cui, J.-H., Shi, Z.: Efficient multipath communication for time-critical applications in underwater acoustic sensor networks. IEEE/ACM Trans. Netw. **19**(1), 28–41 (2011)
7. Xie, P., Zhou, Z., Peng, Z., Cui, J.-H., Shi, Z.: Void avoidance in three-dimensional mobile underwater sensor networks. In: International Conference on Wireless Algorithms, Systems, and Applications, pp. 305-314. Springer, Heidelberg (2009)
8. Jan, N., Javaid, N., Javaid, Q., Alrajeh, N., Alam, M., Khan, Z.A., Niaz, I.A.: A balanced energy-consuming and hole-alleviating algorithm for wireless sensor networks. IEEE Access **5**, 6134–6150 (2017)
9. Wahid, A., Kim, D.: An energy efficient localization-free routing protocol for underwater wireless sensor networks. Int. J. Distrib. Sens. Netw. **8**(4), 307246 (2012)
10. Zidi, C., Bouabdallah, F., Boutaba, R.: Routing design avoiding energy holes in underwater acoustic sensor networks. Wirel. Commun. Mob. Comput. **16**(14), 2035–2051 (2016)
11. Cheng, C.-F., Li, L.-H.: Data gathering problem with the data importance consideration in underwater wireless sensor networks. J. Netw. Comput. Appl. **78**, 300–312 (2017)
12. Qureshi, N., Rasheed, M.B., Javaid, N., Khan, Z.A., Maqsood, Y., Din, A.: M-GEAR: gateway-based energy-aware multi-hop routing protocol for WSNs. In: 2013 Eighth International Conference on Broadband and Wireless Computing, Communication and Applications (BWCCA), pp. 164–169. IEEE (2013)
13. Nguyen, K.-V., Nguyen, P.L., Vu, Q.H., Do, T.V.: An energy efficient and load balanced distributed routing scheme for wireless sensor networks with holes. J. Syst. Softw. **123**, 92–105 (2017)

Vulnerability Analysis on the Image-Based Authentication Through the PS/2 Interface

Insu Oh[1], Kyungroul Lee[2], Sun-Young Lee[1], Kyunghwa Do[3], Hyo beom Ahn[4], and Kangbin Yim[1(✉)]

[1] Department of Information Security Engineering, Soonchunhyang University, Asan, South Korea
{catalyst32, sunlee, yim}@sch.ac.kr
[2] R&BD Center for Security and Safety Industries (SSI), Soonchunhyang University, Asan, South Korea
carpedm@sch.ac.kr
[3] Department of Software, Konkuk University, Seoul, South Korea
doda@konkuk.ac.kr
[4] Division of Information and Telecommunication Engineering, Kongju National University, Chonan, South Korea
hbahn@kongju.ac.kr

Abstract. The mouse is one of the most widely used I/O devices on a computer. Most user authentication methods are password-based through the keyboard, but there exists a vulnerability through which passwords are exposed through data input, such as keyloggers. Thus, image-based authentication, which authenticates through data input from a mouse, has been discovered. Image-based authentication method is widely used in various Web sites and Internet banking services. This paper analyzes the vulnerability of image-based authentication, which is based on the input data through the mouse. This paper also analyzes an experiment where passwords are exposed by taking mouse data through the PS/2 controller, and we also implemented the proof-of-concept tool and confirm the result of mouse data exposure in the image-based authentication applied in the Internet banking service.

1 Introduction

Mouse is one of the popular peripherals I/O devices with a keyboard, and it performs various features such as moving mouse cursor and clicking buttons. Inputted mouse information is utilized for various areas. For example, mouse can not only sophisticate control that is hard to perform with the keyboard, but it also inputs private information such as password instead of using an exposed keyboard.

Currently, most user authentication methods require ID and password in the string form. In case of the string form, passwords are inputted from the keyboard, and the secret information is exposed to third parties and not authorized users. For this reason, studies have been conducted to improve keyboard security, and secure software have been used to prevent key-loggers. Nevertheless, this approach is practically insufficient. Therefore, novel authentication method has been developed using a mouse. This

L. Barolli et al. (Eds.): IMIS 2018, AISC 773, pp. 212–219, 2019.
https://doi.org/10.1007/978-3-319-93554-6_19

method is known as the image-based authentication. This authentication method displays images such as key pad or number pad, and mouse information clicked by users who used a password. Most banking sites supporting Internet banking service utilize image-based authentication for user authentication [1–3].

Unfortunately, mouse data is not protected, and this is problem that cases security threat. For instance, if clicked mouse data is exposed to third parties such as an attacker, he or she can steal authentication information hence enabling them to bypass the authentication.

Consequently, we analyzed vulnerabilities that occur when mouse data is exposed while in service. We also verified exposure possibility of mouse data, as one of the key data, for image-based authentication.

2 Related Work

2.1 Service Utilizing Mouse Data

Services utilizing mouse data include the image-based authentication and a service using mouse movement pattern. Image-based authentication involves inputting password using a mouse to prevent security threats that resulted from the use of a keyboard. Service using mouse movement pattern is utilized by applying various services and analyzing the correlation between mouse cursors; clicking and moving patterns.

2.2 Image-Based Authentication

Image-based authentication utilizes mouse data as a password. To determine the password, an image such as virtual keyboard or number pad is displayed, and the user clicks keys or numbers on the image. For this reason, this authentication does not expose the keyboard data, hence, it is safe from keyboard hooking attacks.

Image-based authentication is mainly used when authentication using mouse data is required for services such as Internet banking and e-commerce. Nevertheless, mouse data is exposed by mouse loggers [4–7].

2.3 Mouse Logger

Mouse logger is a program that collects and records mouse data such as coordinates of the cursor coordinates is based on the event generated by the mouse. In addition, mouse movements can trace using replay feature based on recorded logs. Representative loggers are Ghost mouse, Axife Mouse Recorder Automatic, Mouse Recorder Pro 2, and Automatic Mouse and Keyboard.

We researched exposure possibility of mouse data using the mentioned mouse loggers in Internet banking and e-commerce sites. To verify the safety of mouse data, we experimented five sites that incorporate the use of Internet banking services. Results from the experiment clearly depicted that, Ghost mouse logger did not steal mouse data while Automatic Mouse and Keyboard steals mouse data in almost sites.

Likewise, mouse data does not protect by just mouse loggers which can download easily from online. Therefore, attackers can steal mouse data using downloaded mouse loggers [8].

2.4 Existing Attack and Defense Techniques of Mouse Data Exposure

An attacker can trace mouse movements through mouse loggers, so exposure of mouse data has been researched using mouse loggers easily obtained from the Internet. The research verified that passwords can be stolen by exposing mouse data to various Internet banking services [8].

The Microsoft Windows operating system provides various APIs to manage and support the mouse position, and the GetCursorPos() function provides the current mouse position in the form of x and y coordinates. Hence, if an attacker collects x and y coordinates by calling the GetCursorPos() function periodically, he or she can trace the mouse movement inputted from the user. This study verified password exposure by collecting mouse data in real e-commerce sites in Fig. 1 [12].

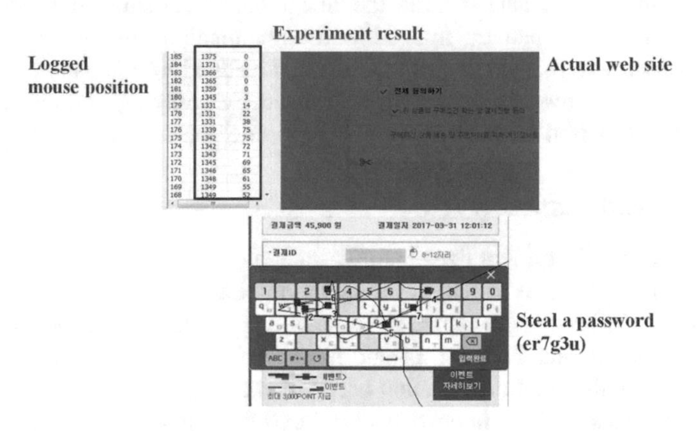

Fig. 1. Attack scenario using the GetCursorPos() function

To prevent this attack technique, the Windows operating system generates mouse position data, which makes it possible to generate random irrelevant mouse position data through the SetCursorPos() function in Fig. 2 [12].

Lee et al. verified the exposure of mouse data using the WM_INPUT message handler that extracts mouse data inputted from the mouse to analyze the vulnerability of image-based authentication. As a result, mouse data was exposed on most banking and payment sites in South Korea in Fig. 3 [13].

Lee et al. mingles mouse movements generated using WM_INPUT message handler to prevent the exposure of mouse position. The experiment result shows in Fig. 4 [14].

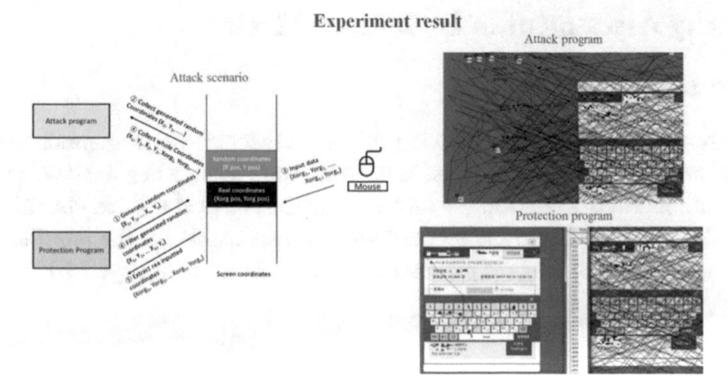

Fig. 2. The operation process of the prevention technique using SetcursorPos() function

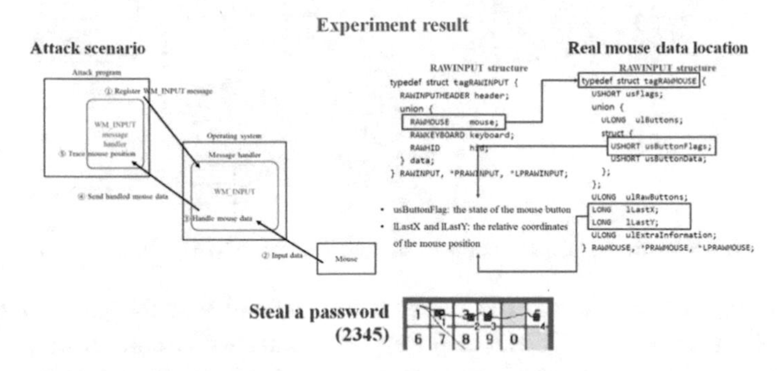

Fig. 3. Attack scenario using the WM_INPUT message handler

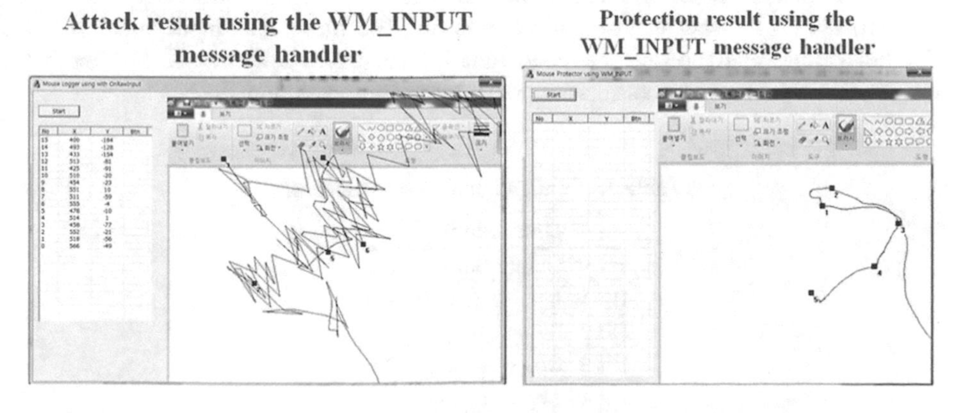

Fig. 4. Experiment results of protection technique using the WM_INPUT message handler

3 Security Assessment of PS/2 Mouse Data

3.1 PS/2 Data Transmission

Figure 5 shows mouse data transmission of the PS/2 interface. Data input from mouse and keyboard is transmitted to the host in the scan code form, when the data has to move via PS/2 controller. PS/2 controller which is sometimes called PS/2 keyboard controller (PS/2 KBC), is located in the main board and its responsible for the interface between the device processor and the host processor when PS/2 mouse or keyboard is connected.

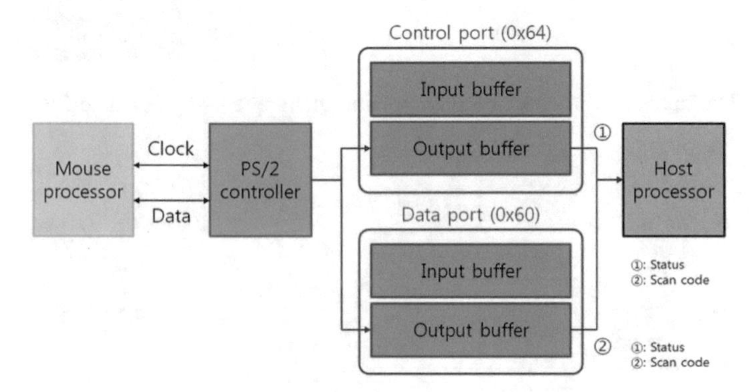

Fig. 5. Experiment results of protection technique using the WM_INPUT message handler

PS/2 controller connects to I/O ports as 0×60 and 0×64 ports. Data port as 0×60 collects mouse scan codes and keyboard data, while control port as 0×64 collects status data of status register. Status register information impacts the current state of the PS/2 controller. Table 1 shows configuration of status register [9–11].

To preempt accessing scan code, we checked whether mouse data was inputted or not. As shown in Table 1, 0 and 1 bits of status register denotes OBF and IBF, respectively, OBF means that output buffer is full, IBF means that input buffer is full. For this reason, when mouse or keyboard data is transmitted, OBF is set to 1. After that, we obtained keyboard or mouse scan codes by reading data port [9–11].

Table 1. Configuration of status register

Bit	Description
Bit 0	Output buffer status
Bit 1	Input buffer status
Bit 2	System flag
Bit 3	C/D (Command/Data)
Bit 4	Keyboard lock
Bit 5	Receive time-out
Bit 6	Time-out error
Bit 7	Parity error

3.2 Structure of PS/2 Mouse Data

The most basic mouse features include mouse movement and mouse click. For these features, mouse scan code consists of three bytes, and Table 2 shows structure of PS/2 mouse data packet [6].

Table 2. Structure of PS/2 mouse data packet

Byte	Bit 7	Bit 6	Bit 5	Bit 4	Bit 3	Bit 2	Bit 1	Bit 0
Byte 1	Y overflow	X overflow	Y sign	X sign	Always 1	Middle Btn	Right Btn	Left Btn
Byte 2	X movement							
Byte 3	Y movement							

From 0 to 2 bits of first byte denotes flag bits for the left, right, and middle buttons. From 4 to 5 bits denotes sign bit of X and Y movements, so these bits set 1 when negative whereas 0 when positive. These bits are important for trace because mouse movements have both positive and negative movements.

Second and third bytes denote X and Y movements. These movements indicate two's complement of nine bits, so movement is included from −255 to +255. If the boundary is exceeded, overflow bits are set. The movement denotes relative distance of previous movement. For example, if mouse is moved to right, X movement is increased, and if mouse is moved to left, X movement is decreased. Likewise, if mouse is moved up, Y movement is increased, and if mouse is moved down, Y movement is decreased. Consequently, we assume that mouse movement can trace by collecting X and Y movements with X and Y sign bits. In addition, we can obtain clicked information by collecting flag bits from the left, right, and middle buttons [9–11].

3.3 Experiment Result

To assess the safety of mouse data, we implemented proof-of-concept tool utilizing drawn vulnerability. First, we implement a device driver to monitor 0×60 and 0×64 ports. Secondly, when OBF was set after obtaining status register from 0×64 port, the tool obtained scan codes from 0×60 port. Thirdly, scan codes as shown in Table 2 are parsed bit by bit to analyze packet. Fourth, increment and decrement values of X and Y coordinates are displayed to a list with flag of button, and the captured screens are sent to an attacker. Finally, the attacker draws mouse movements by a line based on the received mouse data. Moreover, when buttons are clicked, it is indicated by a dot with the sequence number. Accordingly, the attacker can steal the victim's password when victim inputs image-based password. Figure 6 shows experiment result, and Table 3 shows exposure result.

The experiment result provided that, mouse data in five internet banking sites were exposed by proof-of-concept tool utilizing vulnerability.

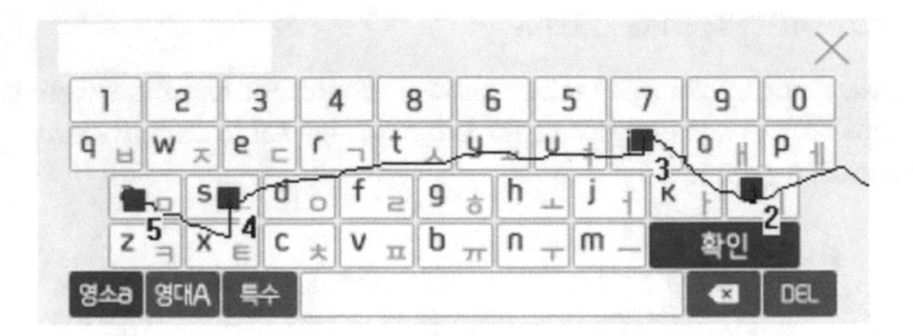

Fig. 6. Experiment results

Table 3. Structure of PS/2 mouse data packet

Internet banking company	Coordinate position	Cursor position	Input window	Exposure result
Company A	O	O	O	O
Company B	O	O	O	O
Company C	O	O	O	O
Company D	O	O	O	O
Company E	O	O	X	O

4 Conclusions

To assess safety of PS/2 mouse data, we experimented the exposure results by implementing proof-of-concept tool, in result, and most sites exposed mouse data. Image-based authentication to prevent security threats of the keyboard, are exposed inputted key data by key loggers, is used to input password from the mouse on the displayed image. Despite the mouse data being protected by various security software, the password is still exposed by hardware vulnerability.

If an attacker steals mouse data, it results to a serious damage. For example, mouse data and status information are exposed, and this depends on hardware vulnerability, so there is an urgent countermeasure against this vulnerability. For this reason, the study was mainly oriented towards preventing exposure of mouse data.

Acknowledgments. This research was supported by the Basic Science Research Program through the National Research Foundation of Korea (NRF) that is funded by the Ministry of Education (NRF-2015R1D1A1A01057300).

References

1. Lee, K., Bae, K., Yim, K.: Hardware approach to solving password exposure problem through keyboard sniff. In: Proceedings of the Academic Science Research, WASET, Singapore, 26–28 August 2009
2. Lee, S., Lee, K., Yim, K.: Security assessment of keyboard data based on Kaspersky product. In: Proceedings of the International Conference on Broadband and Wireless Computing, Communication (2016)
3. Sagiroglu, S., Canbek, G.: Keyloggers. IEEE Technol. Soc. Mag. **28**(3) (2009)
4. Oh, I., Lee, K., Yim, K.: Security assessment of the image-based authentication using screen-capture tools. In: Proceedings of the International Conference on Innovative Mobile and Internet Services in Ubiquitous Computing (IMIS), Torino, Italy, 10–12 July 2017
5. Akula, S., Devisetty, V.: Image based registration and authentication system. In: Proceedings of Midwest Instruction and Computing Symposium, Morris, USA, 16–17 April 2004
6. Almuairfi, S., Veeraraghavan, P., Chilamkurti, N.: A novel image-based implicit password authentication system (IPAS) for mobile and non-mobile devices. Math. Comput. Model. **58**, 1 (2013)
7. Eljetlawi, A.M., Ithnin, N.: Graphical password: comprehensive study of the usability features of the recognition base graphical password methods. In: Proceedings of the IEEE International Conference on Convergence and Hybrid Information Technology (ICCIT), Busan, South Korea, 11–13 November 2008
8. Lee, H., Lee, Y., Lee, K., Yim, K.: Security assessment on the mouse data using mouse loggers. In: Proceedings of the International Conference on Broadband and Wireless Computing, Communication and Applications (BWCCA), Asan, South Korea, 5–7 November 2016
9. Chapweske, A.: Computer-engineering (2003)
10. Chapweske, A.: Computer-engineering (1999)
11. Chen, X.: Analysis and application of PS/2 device interface protocol. J. Int. Electron. Elem. **4** (2004)
12. Lee, K., Oh, I., Yim, K.: A protection technique for screen image-based authentication protocols utilizing the SetCursorPos function. In: Proceedings of the World conference on Information Security Applications (WISA), Jeju Island, Korea, 24–26 August 2017
13. Lee, K., Yim, K.: Vulnerability analysis on the image-based authentication: through the WM_INPUT message. In: Proceedings of the International Workshop on Convergence Information Technology (IWCIT), Busan, Korea, 21–23 December 2017
14. Lee, K., Yim, K.: A protection technique for screen image-based authentication utilizing the WM_INPUT message. In: Proceedings of the Korea Society of Computer Information (KSCI) Conference, Busan, South Korea, 11–13 January 2018

A Spacecraft AIT Visualization Control System Based on VR Technology

Wei Peng[1(✉)], Zhang Liwei[1], Wu Qiong[2,3], Wang Miaoxin[2,3], and Li Jian[2,3]

[1] Beijing Institute of Spacecraft Environment Engineering, Beijing, China
Weip2.13@sem.tsinghua.edu.cn
[2] Beihang University, Beijing, China
wuqbuaa@buaa.edu.cn
[3] Beijing Spacenovo Technologies Co., Ltd., Beijing, China
teksail@foxmail.com

Abstract. This article proposes to construct a spacecraft AIT (Assembly Integration Test) visualization control system based on VR technology, which can further enhance the digitization level of spacecraft manufacturing. The article first outlines the research and development background of the system, and then elaborates on the key technologies and software architecture used in the construction of the system, and gives the actual operation situation of the system, and finally summarizes the article.

1 Introduction

In recent years, with the introduction and development of advanced manufacturing theories such as Internet of Things, Cloud Manufacturing, Internet plus manufacturing, Industry 4.0, and intelligent manufacturing, more and more industrial manufacturers have begun to pay attention to how to integrate advanced information technologies with the traditional manufacturing mode to improve capabilities in real-time sensing, optimization control, and intelligent decision in the manufacturing process. For aerospace equipment manufacturers, the strategic planning for aerospace equipment manufacturing development is to promote digitalization, networking and intelligence, and to establish a digital and network-based manufacturing integration system that is suitable for the characteristics of multi-species and small-batch production of aerospace products [1].

China Academy of Space Technology has been focusing on optimizing and innovating the AIT production and management for many years and has obtained a series of results. MES (Manufacturing Execution System) oriented to the production execution layer has been established, laying the foundation for realizing the digitalization, networking, and intelligence of the AIT's whole process [2]. The management information system represented by MES, although initially solved the requirements for the acquisition, storage and management of spacecraft manufacturing data, there are still many problems that need to be solved urgently. For example, how to make managers use these data more effectively, how to identify problems in the production

L. Barolli et al. (Eds.): IMIS 2018, AISC 773, pp. 220–230, 2019.
https://doi.org/10.1007/978-3-319-93554-6_20

process in real time and gain insight into the lack of management behind the data, and how to obtain more scientific and intelligent visual aided decision-making information.

This article proposes to establish a spacecraft AIT production visualization management and control system, based on the data of real factory buildings, workstations, equipment, process flows. The system combines real-time production process data of the MES system with advanced VR technology for rendering and display. The system can provide managers with visual aided decision support for the whole AIT process, further improving the digital, visual and intelligent levels of spaccecraft manufacturing.

To build a spacecraft AIT visualization and control system based on VR technology, we must first solve the problem of lightweight assembly of aerospace complex assembly model. Secondly, in the aspect of VR rendering, the relationship between management requirements, visual effects, and rendering efficiency needs to be balanced based on a comprehensive consideration of the functional structure of the spacecraft. In addition, considering the management requirements of the centralized information system of aerospace manufacturing enterprises, it is necessary to comprehensively optimize the problems of network data transmission efficiency and user experience. Finally, the data interface issues involved in the integration with existing production MES systems need to be considered.

2 The Key Technologies

2.1 Lightweight for Complex Assembly Models

Currently, aerospace manufacturers use digital modeling tools such as Pro/E or CATIA for digital modeling and design [3]. The capacity of 3D model data files stored by such 3D modeling software is often several GB. As the complexity of the model increases, the capacity of the data file increases exponentially, which brings great inconvenience to copy, download, open, and render the model [4]. Therefore, it is necessary to reduce the weight of complex models. The lightweight model has the advantages of small file size, fast reading, memory saving. Lightweight mainly includes the following methods:

- **Delete non-critical information of the model file**
  Study on the model data file shows that the file contains geometric information and non-geometric information of the 3D model. Geometry information includes: coordinate system definition, geometry description, history record, feature parameters, etc. Non-geometric information includes manufacturing information and attribute information. Under the premise of retaining the original design intent, the non-critical information that is not related to the model structure, such as historical record information, feature parameter information, and annotation information, can be deleted, which can greatly reduce the capacity of the model file.
- **Polygonalization of geometric parameters**
  Pro/E, CATIA and other commercial 3D modeling software use parametric curves/faces to describe the geometry of objects. This method can accurately define the geometric information of objects to meet the precision requirements of spacecraft component manufacturing. However, from the perspective of AIT visualization and control, the description accuracy is too high, which results in a significant

decrease in the speed of model loading and rendering. Therefore, a mesh model which easy to render can be used to describe the geometry of the object instead of the parametric model. In this way, the number of vertices and polygons in the 3D model can be greatly reduced, and the loading speed and rendering efficiency of the 3D model can be improved.

- **Lossless compression coding**
  After eliminating key information and polygonizing geometry parameters, the data capacity is greatly reduced relative to the original model file. The lossless compression coding algorithm is used to encode and compress the mesh model to obtain a lightweight model with a smaller capacity. Common lossless compression algorithms include run-length coding, LZW coding, Huffman coding, arithmetic coding (Fig. 1).

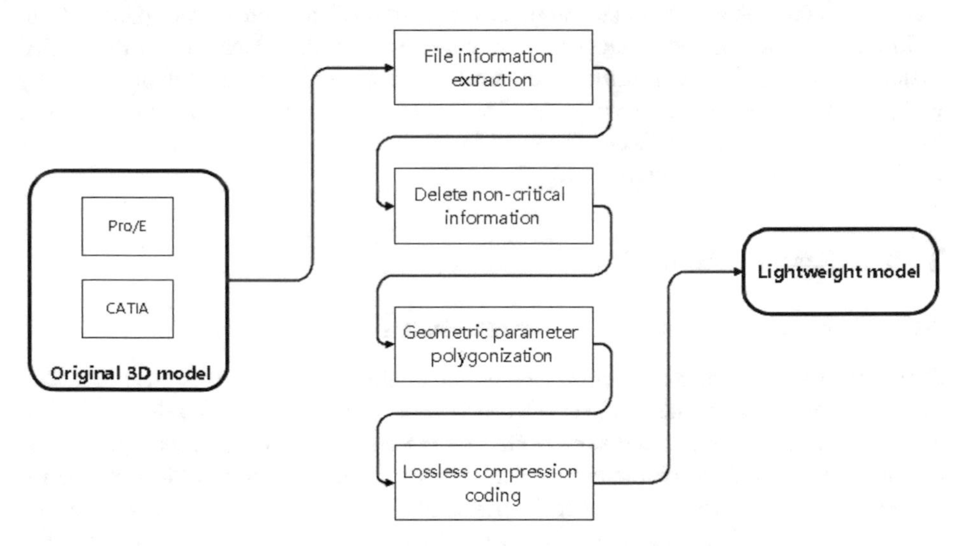

Fig. 1. The lightweight process of the model

Taking a certain type of satellite as an example, we compared the number of graphics before and after weight reduction. The results are shown in the following Table 1.

Table 1. Model complexity comparison table

	Number of parts	Number of vertices	Number of triangles
Before	752	1103245	1015381
After	327	775269	827537

From the above table, it can be seen that the lightweighting of the model greatly reduces the number of components in the model, and greatly reduces the number of model vertices and triangular patches. Of course, the lightweight effect is also closely related to the structural characteristics of the model itself.

2.2 VR Rendering Technology Based on Unity3D Web Player

Building a spacecraft AIT visualization and control system based on VR technology requires selecting a suitable graphics rendering engine to load and render various scenes and models. At the same time, it is necessary to combine the actual requirements of the management of the internal information system of the aerospace enterprise, and consider the system compatibility, operating efficiency, experience fluency, visualization effects and other factors. The current major 3D graphics rendering engines are compared in the following Table 2.

Table 2. 3D graphics engine function comparison table

Compare items	OGRE	Unity3D	OSG	Vega Prime
Operating system	Windows/Linux/iOS/Android	Windows/Linux/iOS/Android	Windows/Linux/iOS/Android	Windows/Linux
Model support	3dmax	3dmax	3dmax/Creator	Creator
Development language	C++	C++/Javascript	C++	C++
Physical system	Basic physical systems, collision detection, and rigid bodies	Built-in PhysX physics engine	Bullet, ODE physics engine	Basic physical system collision detection, seamlessly connectable with other ODE physics engines
Shadow system	Shadow mapping and projection	Shadow mapping	OSG shadow toolbox	Cloud shadows, dynamic shadows
Shader model	4.0	4.0	4.0	2.0
Network	Unsupported	Single-user connection and full real-time multi-user connection	Client/server, P2P network, master server mode	Internet mode
Scene editor	The open source Ogitor is very immature, only the export tool of modeling software	Powerful and easy-to-use graphical user interface	Lynx Prime GUI configuration tool	Lynx Prime GUI configuration tool

Due to the different architectures of OGRE, OSG, Unity3D, and Vega Prime, as well as different third-party API support capabilities, the simulation program running under the same scenario complexity has different frame rates. On Windows platform, OGRE and Unity3D have basically the same efficiency, but have higher rendering efficiency than Vega Prime and OSG, due to the support of D3D rendering. On iOS platform, Unity3D has the best performance and the highest efficiency. On Android platform, Unity3D is also the best performer. On Linux platform, the operating efficiency of the four engines are basically the same.

In theory, OGRE, OSG, and Unity3D can achieve high-level effects of commercial engines, but they need to be studied on the algorithm and it is difficult to develop. For the spacecraft AIT visualization control system, the three rendering effects can meet the requirements. However, due to the stagnation of its own development, Vega Prime has a certain gap compared to the other three engines.

Unity3D makes it easier to get started with C# or JS and a powerful editor, which is undoubtedly the shortest for the development cycle. However, if there are some mature projects developed by other engines which also has mature third-party plug-ins, libraries, and platforms (such as OGRE's SkyX, OSG's osgEarth), it is also a good choice to develop with these engines. The development methods used by OSG and OGRE to write 3D-rendered programs sometimes seem very complicated. But as open source engine, there are a lot of resources on the network that can be used in development. Therefore, the development process with OSG and OGRE is fast, and the rendering effect is also better. Vega Prime is basically the same as OGS and OGRE in terms of development mode. But since all the effects rely on Vega Prime's internal functional modules, it has the lowest development efficiency. What's more, it badly relies on the developer's familiarity with Vega Prime.

At present, the environment of the intranet IT system is that various types of information systems are deployed in a centralized manner, and the entrances to the various systems are linked to the intranet portal and accessed through the browser. The operating system is based on Windows XP/7, and the browser version is mainly based on IE 6/8. After comprehensively evaluating the functions, performance, compatibility, and rendering capabilities of each engine, Unity3D was selected as the 3D graphics rendering engine.

2.3 Asset Bundle

In spacecraft AIT visualization and control system, a large number of scenes and spacecraft models need to be loaded. If waiting for all scenes, models, and textures to be loaded before entering the system, it will inevitably lead to a long system response time and poor user experience. Therefore, it is necessary to develop resource dynamic loading technology, that is, to first load the resources most urgently needed by the user, and dynamically load the remaining resources in the background without affecting the user operations until all the resources are completely loaded, which can significantly improve the loading speed and reduce memory resource consumption.

Asset Bundle technology can be used to achieve dynamic loading of system resources. The Asset Bundle can store any resource storage file that Unity3D can recognize. Developers can first pack and upload potentially available resources to the server and selectively download them according to actual needs. The program loads them dynamically through the load module, which facilitates the user to download and update resources on the server (Fig. 2).

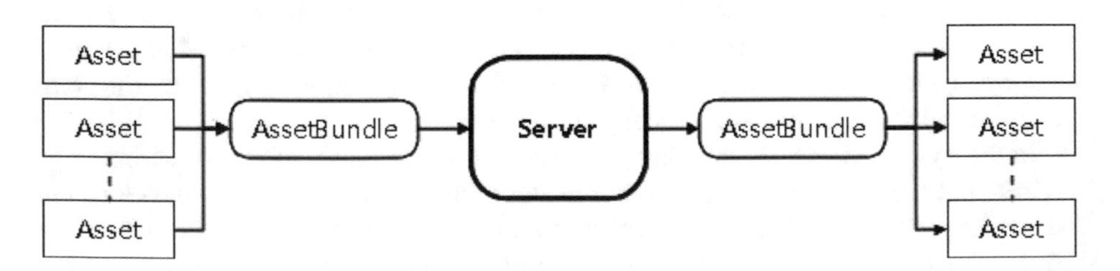

Fig. 2. Asset bundle upload/download process

Assets that can be packaged using the Asset Bundle in Unity3D are divided into several categories, which are scenarios, models, and special effects.

Scene resources usually include surface models, instantiated objects, and scene objects. The instantiated object includes camera, light source, and three-dimensional object model required by an actual scene. The scene object refers to factory buildings, work stations, devices, tools, and the like. In each scenario, when resources that are reused in multiple scenarios appear, they can be extracted as common resources and packaged separately. Then private resources related to each scenario are packaged into various scenario packages. In this way, when each scene is loaded, the private resources of each scene can be loaded first, and then the common resource package stored in the memory is loaded according to the configuration information xml file. Resources such as model files, animations, materials, special effects, sounds, and texts are all packaged in asset bundle binary format file using the default Prefab (Fig. 3).

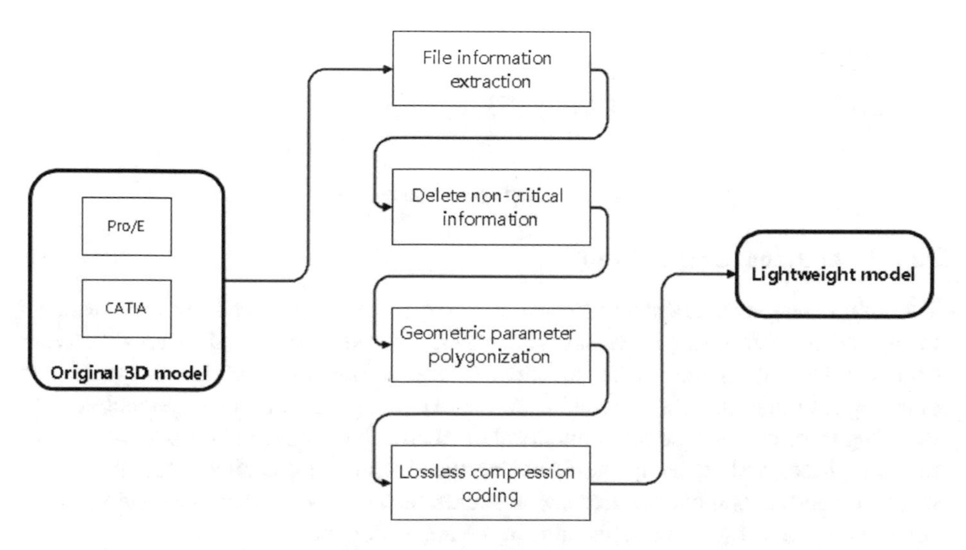

Fig. 3. Asset bundle process

The professional performance testing tool Profiler is used to test the system. The memory resource data without Asset Bundle is 1.27 GB, and the memory resource data with Asset Bundle is 736.4 MB. This shows that the use of Asset Bundles significantly reduces system memory usage. The system runs more efficiently on the client with Asset Bundle.

3 Software Architecture

The spacecraft AIT visualization control system consists of a data management module, a visualization resource management module, and an information query and positioning module. Among them, the data management module and the visualized resource management module are background configuration modules, and the information query and positioning module is foreground display module (Fig. 4).

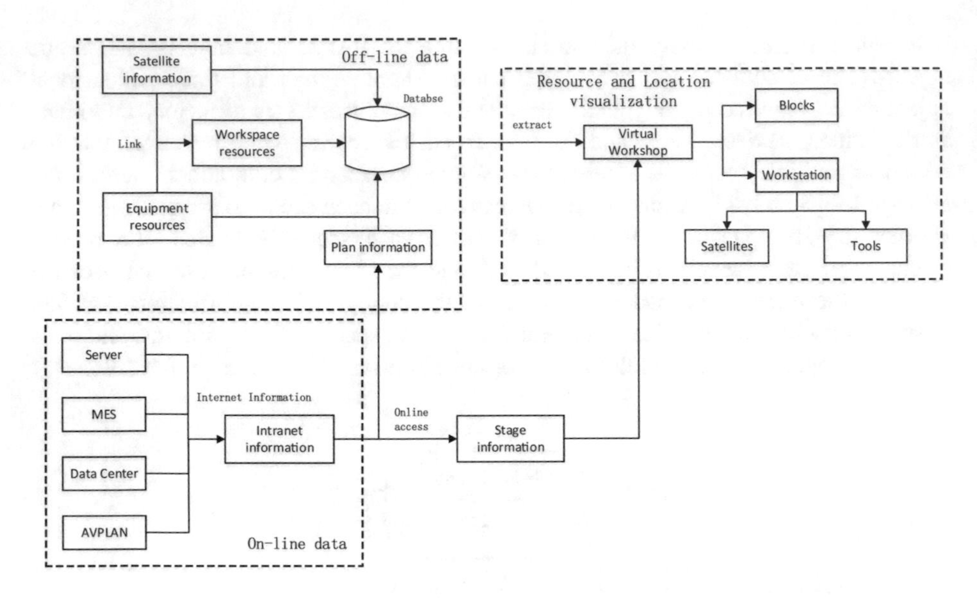

Fig. 4. Software architecture

3.1 Data Management Module

The data management module provides data support for visualized resource management, and updates status according to spacecraft resource data and external network data. On-line data mainly refers to satellite stage information, production plans, and planning information. It is imported from the existing production management system. Planning information is stored in the local database. Stage information is used to drive satellite phases and status updates. Off-line data includes all available satellite status, workshop and equipment resources, workstation resources, and plan information which edited and stored by the administrator in advance (Fig. 5).

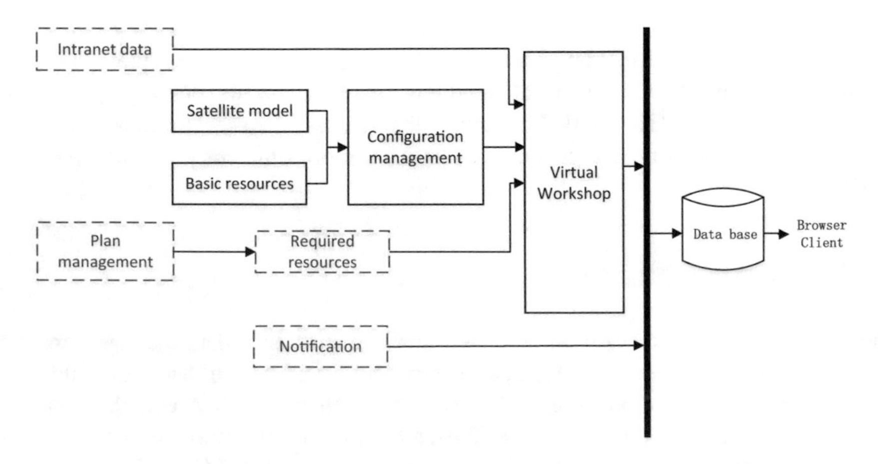

Fig. 5. Data management architecture

3.2 Visualization Resource Management Module

Visualization resource management module realizes the management and configuration of visualization resources such as satellite models and basic resources. It associates different levels of visual models to form a complete factory model. Functions of creation, modification, and updating of model resources such as workshops, workstations, satellites, and tools are supported. Functions of maintenance of related resources in the device resource maintenance module to implement functions such as adding, modifying, deleting, exporting, and querying are supported. Satellite resource maintenance operations include adding, modifying, deleting, and querying satellites (Fig. 6).

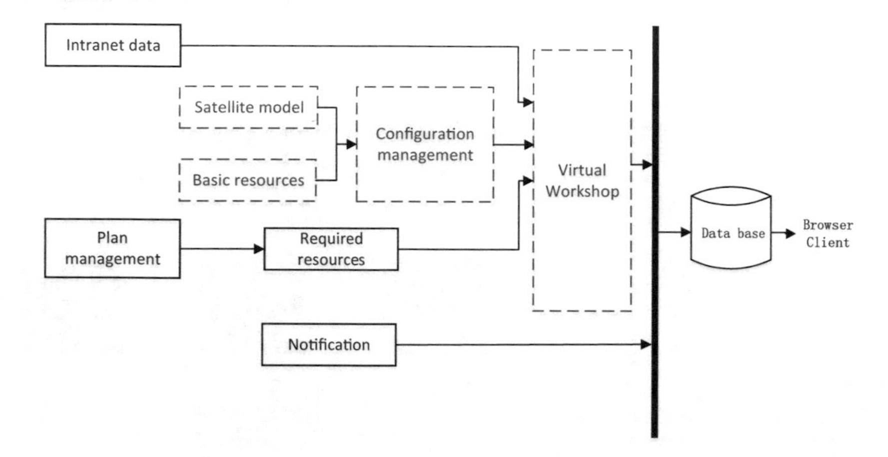

Fig. 6. Visualization resource management architecture

3.3 Information Query Module

Information query module realizes free browsing and information inquiry and display functions of workshops, tools and spacecraft which based on 3D models. The virtual workshop browsing includes fixed route view mode and free view mode which can be freely switched. Information that can be queried on workstation includes plan

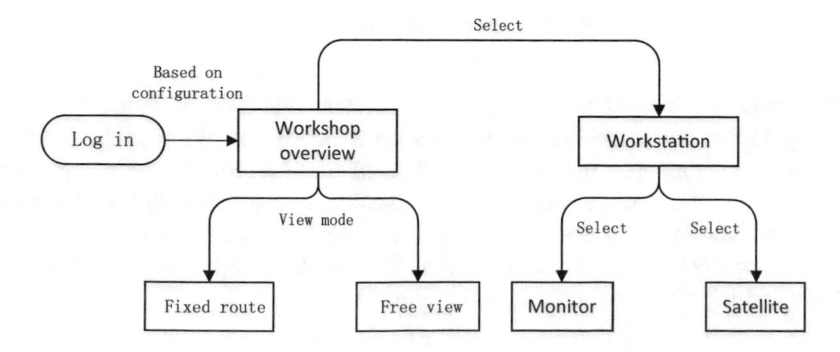

Fig. 7. Information query module architecture

information, stage information and satellite assembly status information. Take Beidou-3 as an example, satellite status query and browsing can be divided into 12 stages and 16 types of display status (Fig. 7).

4 System Operation Situation

The visual control system is designed and deployed with B/S architecture. The system accesses the internal network through web server, obtains formatted data file and parses it. The working phase and status are stored in the local database. The visualization module drives the virtual workshop and updates information through database. It provides users browsing functions on workshop, workstation, satellite status and production information with browser access. The management and maintenance module realizes functions such as visualized resource management, plan management, workshop resource management, and notification publishing (Fig. 8).

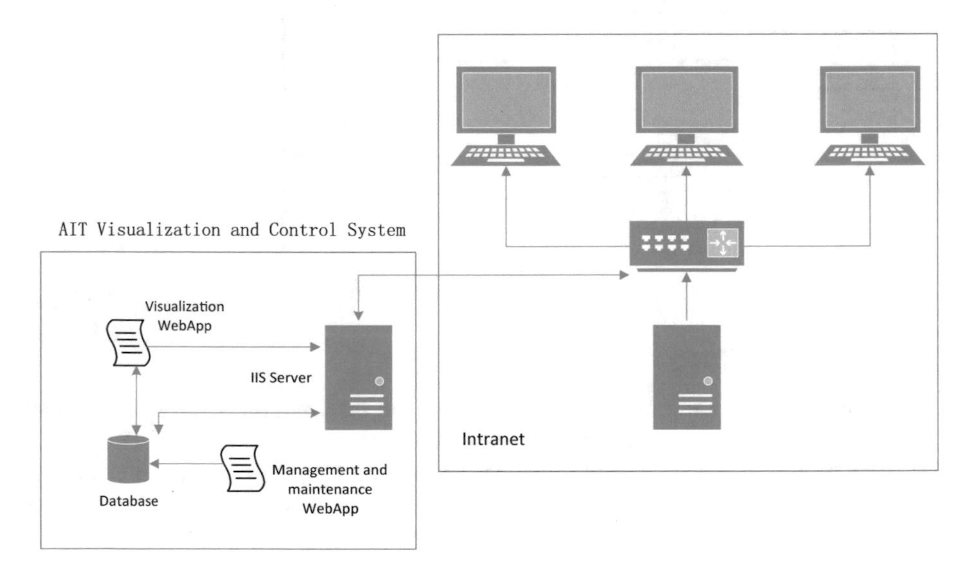

Fig. 8. System deployment structure

The screen capture of the visual control system interface is shown below. The workshop resources include the assembly hall, welding room and assembly room. The content of management is mainly divided into plan management, workshop resource management, notification publishing, basic resource management and model resource management. The function that roaming in workshop is supported by the system. At the same time, registered user can use the system to view the current progress of spacecraft assembly and regularly maintain updates (Fig. 9).

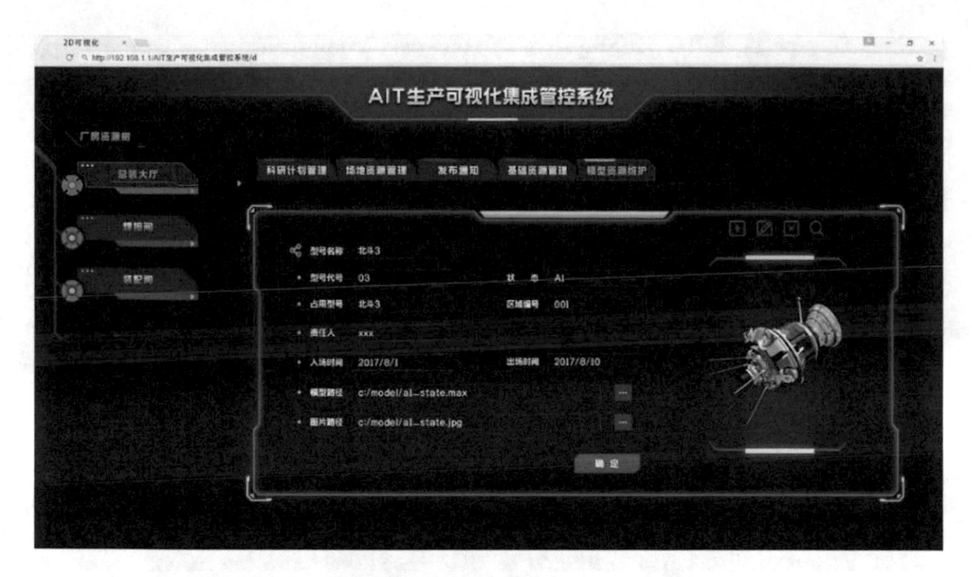

Fig. 9. System screenshot

5 Summary

This article proposes to establish a spacecraft AIT visualization control system based on VR technology for further improving the digitization, visualization and intelligence of spacecraft AIT process. The system solves the technical problems of complex model, large capacity, slow resource loading and low rendering efficiency in the traditional 3D visualization system, and realizes seamless connection with the existing MES system. The system makes up for the insufficiency of real-time visualization and

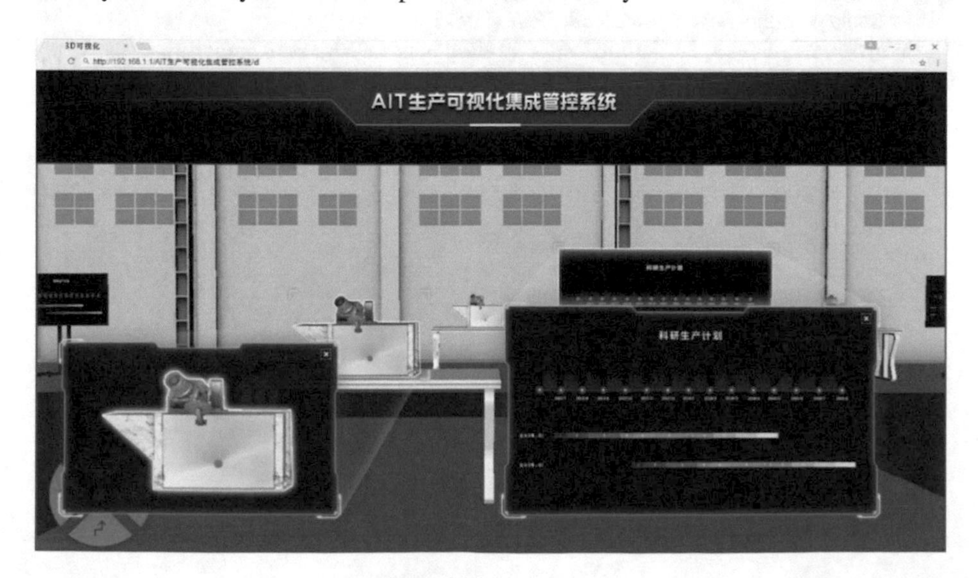

Fig. 10. System screenshot

perceivable aspects of the MES system of aerospace equipment manufacturing companies, and provides policy makers with more intuitive and lively AIT real-time production process support decision-making information (Fig. 10).

References

1. Zhao, H., Co, H.C.: Adoption and implementation of advanced manufacturing technology in Singapore. Int. J. Prod. Econ. **48**, 7–19 (1997)
2. Shi, Y., Xu, N.: Digital factory and its implementation techniques. Program. Controllers Factory Autom. **13**(11), 37–39 (2011)
3. Yuan, F., Xie, L.: Research on the modeling and simulation system of digital production line manufacturing process. Comput. Eng. Appl. **32**, 18–21 (2004)
4. Wu, Y., Qiu, H.: Digital factory platform for design and manufacturing. Chin. Manuf. Inf. **40** (1), 1–5 (2011)
5. Bracht, U., Masurat, T.: The digital factory between vision and reality. Comput. Ind. **56**(4), 325–333 (2005)
6. Liu, J.: Research and implementation of SOA-based enterprise integration system. In: Proceedings of 2013 3rd International Conference on Education and Education Management (EEM 2013), vol. 26(C) (2013)
7. Chlebus, E., Krot, K.: CAD 3D models decomposition in manufacturing processes. Arch. Civil Mech. Eng. **16**(1), 20–29 (2016)
8. Yap, H.-J., Taha, Z., Md Dawal, S.Z.: A generic approach of integrating 3D models into virtual manufacturing. J. Zhejiang Univ. Sci. C Comput. Electr. **13**(01), 20–28 (2012)
9. Novak-Marcincin, J.: Selected applications of virtual reality in manufacturing. J. Technol. Plast. **36**(1), 26–33 (2011)
10. Bamodu, O., Ye, X.M.: Virtual manufacturing and components of virtual reality. Appl. Mech. Mater. **20**(5), 318 (2005)
11. Zhang, X.Y., Xiong, Y., Li, X.J., Wang, D.: The development of virtual manufacturing technology. Appl. Mech. Mater. **58**, 1287 (2011)
12. Weyrich, M., Drews, P.: An interactive environment for virtual manufacturing: the virtual workbench. Comput. Ind. **38**(1), 5–15 (1999)

An Efficient Home Energy Management and Power Trading in Smart Grid

Sheraz Aslam[1], Sakeena Javaid[1], Nadeem Javaid[1](✉),
Syed Muhammad Mohsin[1], Saad Sulman Khan[2], and Mariam Akbar[1]

[1] COMSATS Institute of Information Technology, Islamabad 44000, Pakistan
nadeemjavaidqau@gmail.com
[2] COMSATS Institute of Information Technology, Wah Cantt 47040, Pakistan
http://www.njavaid.com

Abstract. In this work, we propose a DSM scheme for electricity expenses and peak to average ratio (PAR) reduction using two well-known heuristic approaches: the cuckoo search algorithm (CSA) and strawberry algorithm (SA). In our proposed scheme, a smart home decides to buy or sell electricity from/to the commercial grid for minimizing electricity costs and PAR with earning maximization. It makes a decision on the basis of electricity prices, demand and generation from its own microgrid. The microgrid consists of a wind turbine and solar panel. Electricity generation from the solar panel and wind turbine is intermittent in nature. Therefore, an energy storage system (ESS) is also considered for stable and reliable power system operation. We test our proposed scheme on a set of different case studies. The simulation results affirm our proposed scheme in terms of electricity cost and PAR reduction with profit maximization.

Keywords: Smart grid · Heuristic algorithms · Energy management
Power trading

1 Introduction

World increasing population, global warming, the rise in carbon emissions and increasing electricity demand create an alarming situation for electricity producing and distributing companies as well as for governments to take any strong action against these alarming situations. The electricity producing companies are detained from integrating renewable energy sources (RESs) to overcome global warming and carbon emissions [1]. The present fossil fuel based electric grid is working on the centralized approach: only a few large electricity producing plants are operating at 50 Hz or more. High-power electricity plants are operating at very high voltage (i.e., 400 kV or more). Then, the produced electricity from large plants is distributed to the electricity consumers. A large number of supply lines supply the high voltage load to heavy industries and low voltage load to residential consumers and small-scale industries.

L. Barolli et al. (Eds.): IMIS 2018, AISC 773, pp. 231–241, 2019.
https://doi.org/10.1007/978-3-319-93554-6_21

The power flow in the present power system is unidirectional due to the centralized approach. Electricity consumers are considered just passive users; they cannot play any role in the stability and reliability of the electric grid. According to [2], more than 65% of total electricity is wasted during generation, transmission and distribution of electricity. The basic reasons behind electricity wastage are that the present electricity system has unidirectional communication and a lack of monitoring technologies. The novel approach is distributed based on bidirectional communication. It provides the widely and higher distributed intelligence in electric power generation, distribution and flow of information. Furthermore, the novel approach provides the multiple opportunities for electricity consumers to manage their electricity consumption for bill reduction and reliable grid operation.

The novel approach is a smart grid and integration of cost-efficient RESs hold promise to tackle the above-discussed problems in the traditional electric grid. In the smart grid, electricity is generated via cheaper and efficient resources and then distributed to electricity consumers through smart transmission lines. Electricity prosumers are the consumers they can utilize as well as produce the electricity from their own local microgrid, which consists of multiple RESs, i.e., solar panel, wind turbine, hydro power plant, etc. They utilize electricity from their own generation and are also interconnected with the commercial grid. In case of less electricity generation, as compared to load demand, they purchase electricity from utilities or neighbors. If the electricity production from their own microgrid is more than the load demand, then the excess electricity is sold back to the commercial grid or stored in batteries for future use when electricity generation is low. The batteries may discharge only when electricity production from the microgrid is low or per unit electricity price is high.

In smart grids, two-way communication provides an opportunity to optimize consumption costs along with peak to average ratio (PAR) minimization. Due to the advent of a smart grid, a lot of studies have focused in regard to cost and PAR reduction via DSM [3–6]. However, none of this work has included the capability to generate and store electricity for future use. The authors in [7] have proposed a DSM scheme by considering different types of electricity consumers. Optimum electricity consumption with maximum user comfort is determined in [8] within a smart home containing different types of smart appliances. They also investigate their proposed scheme on smart buildings, comprised of multiple smart homes with different living patterns (power rating and load demand). A cost efficient home energy management scheme has been proposed in [9]. The authors of [9] also integrate the RESs to minimize the electricity cost and carbon emissions. Furthermore, the consumers are able to store excess electricity in batteries for future use; when electricity rates are high, stored electricity is consumed. The authors of [10] have proposed a new smart grid architecture for electricity consumers. They also integrate the RESs for electricity generation. According to their work, the consumers are able to consume, generate, store and sell excess electricity. The excess electricity is sold back to the electric grid for earnings maximization. Most of the recently proposed schemes used look at the problem from the grid or electricity consumer's perspective.

In this work, a DSM scheme has been proposed for electricity cost and PAR reduction via integrating the RESs and energy storage system (ESS) in a residential area. We consider a smart home that not only utilizes electricity from the commercial grid but is also capable of generating electricity from its own microgrid and storing electricity for future use. Furthermore, the smart home is independent from making autonomous decisions for electricity cost and PAR reduction with revenue maximization in each hour. Therefore, maximizing the revenue generated from power trading is also the objective of this work. A smart home makes decisions according to electricity tariffs. When prices are high, the home tries to reduce load demand and excess electricity is sold back to the commercial grid. The home purchases electricity in low price time slots while minimizing the PAR. Furthermore, two different RTP schemes are considered for electricity cost calculation in terms of purchasing and selling electricity. At the end, a comparative analysis is performed to show the performance of the cuckoo search algorithm (CSA) and strawberry algorithm (SA).

The remainder of the document is organized as follows: a literature review is presented in Sect. 2. Section 3 explains the problem statement and proposed system model. The simulation results are presented in Sect. 4. Finally, paper findings and future work are explained in Sect. 5.

2 Related Work

Electricity cost reduction and load equilibrium between demand and supply are the interesting and challenging research problems that have been tackled by researchers in the last few decades. Many DSM strategies have been presented in the last few years for electricity costs and PAR minimization while maximizing user comfort. Some of the existing work is presented below.

In [11], a home appliances scheduling scheme was proposed to reduce the total electricity cost and balance the load demand by mixed integer linear programming (MILP) in residential areas. The experimental results present that their proposed scheme rapidly obtains the desired targets, i.e., reduction in peak load and electricity cost. An integer linear programming (ILP) based strategy is proposed in [12]. The basic objective of this study is to find equilibrium between electricity supply and demand in the residential area. Their proposed strategy efficiently shifts optimal operation time and optimal power for time-shiftable and power-shiftable appliances, respectively. Experimental results present that their proposed technique sharply archived the claimed objectives.

A MILP based model is presented in [13] for PAR and electricity cost reduction along with RESs integration. The simulation results validate this proposed model for electricity cost and PAR alleviation with efficient integration of RESs. Another optimization technique is presented by Mohamed et al., in [14] using a genetic algorithm (GA). The objectives of this technique are energy cost reduction and load balancing between electricity supply and load demand. The authors of [15] further proposed a heuristic based technique using particle swarm optimization (PSO) and GA. The optimization problem is formulated through a multiple knapsack problem (MKP) along with three different pricing signals: ToU,

RTP and CPP. Electricity bill and peak load reduction are the main objectives of this work. Experimental analysis shows the efficacy of proposed scheme using PSO and GA. Moreover, they also provide the comparison between GA and PSO, which shows that GA outperforms PSO. In [16], a dynamic programming (DP) based electricity cost reduction scheme was proposed. Electricity costs are reduced through scheduling of home appliances from ON-peak to OFF-peak time intervals. Furthermore, the game theory based approach is adopted to interact with the electricity consumers with extra electricity generation. In [17], the electricity cost optimization problem under the ToU environment is presented. Furthermore, they categorize the total load into three categories known as: interruptible load, shiftable load and weather based load. The authors of [18] considered three main objectives for the optimization problem, which are scheduling preference optimization, cost minimization and climatic comfort maximization. The demand response (DR) policy is presented in [19] and intends to obtain PAR reduction and electricity cost savings via scheduling of smart appliances according to hourly electricity prices. The home energy management system is presented in [20] using MILP to adjust the load demand among the PV panel, electric grid, ESS and electric vehicles. An electricity cost minimization scheme using GA was presented in [21], with RES and ESS integration. The ESS is used to balance the electricity supply and load demand. Experimental analysis shows the efficacy of their proposed scheme.

An intelligent home energy management scheme was presented in [22] for PAR and electricity cost minimization with the integration of RESs. In [9], an MILP based cost reduction and load balancing scheme is presented in a residential area. A MILP based home energy management system (HEMS) [10] considers electricity users as prosumers. They proposed HEMS for a single smart home and for a community of 39 prosumers. Each smart home has its own PV panel for electricity generation and is also connected with a commercial grid to meet the load demand. The prosumers store electricity in batteries when they have more generation than the requirements. However, they are not able to export electricity in high price hours for maximizing profit. In [11], a scheme is presented for cost minimization via integrating distributed energy sources. Moreover, each consumer has its own solar panel for electricity generation along with battery storage. Electricity consumers are able to sell or purchase electricity according to price signals. A RTP scheme is used in their work for electricity cost calculation. However, this work only considers non-interruptible appliances, which is not realistic for a smart home because there are appliances that can be interrupted, for instance, TV, etc. The authors of [23] studied the trading problem in smart grids. Electricity consumers are able to generate, purchase and sell electricity. However, they put their excessive amount of electricity on an auction market for bidding instead of selling to the main grid or single utility. Nevertheless, the ability of electricity consumers to generate electricity, store electricity and adjust their demand according to electricity tariffs, and selling excess electricity back to the electric grid simultaneously has not been considered in [9–11,23]. Here, we propose a novel energy management scheme using CSA and SA to tackle the above-mentioned problems.

3 Proposed System Model

This work investigates a design of future smart grid that targets the reduction of electricity costs for consumers in a residential area, make electric grid stable and minimize the overall peak load during electric grid operation. The proposed optimization model, which is presented in Fig. 1, is explained below.

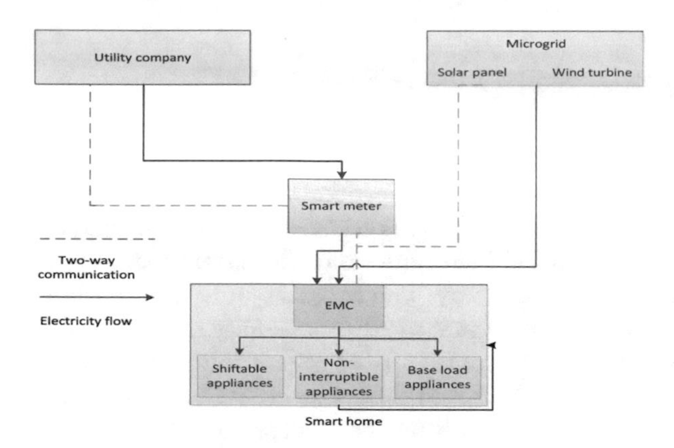

Fig. 1. Proposed system model.

3.1 Microgrid

We consider the local microgrid with m renewable energy sources, i.e., solar panel and wind turbine, which is explained in the next sub-sections.

3.1.1 Solar Panel

The proposed energy management scheme tries to maximize user comfort in terms of cost minimization via utilizing electricity in high price hours, which is generated from the microgrid. Solar energy is converted into electricity via photovoltaic cells and converting the direct current (DC) to alternating current (AC) through the converter. The performance of photovoltaic cells is calculated by the following relationship:

$$i_L - i_S exp[\alpha(v_{pv} + R_S i - pv)] - 1 v_{pv} + R_S i_{pv} / R_{Sh} - i_{pv} = 0, \;\; p = V_{PV} I_{PV}, \quad (1)$$

where $'p'$ is the power generation by solar panel, i_L is the light current, i_s is the diode saturation current, and series resistance and shunt resistance are represented by R_s and R_{sh}, respectively. The ideality factor $\alpha = q/n_s kT$, where $k = 1.38 \times 10^{23} j/K$, $q = 1.60 \times 10^{-19}$, C is the electronic charge, $T = 298\,\text{K}$ temperature and n_s shows the number of solar cells.

3.1.2 Wind Turbine

The kinetic energy is converted into electric energy via a wind turbine. The electric power that is produced by the wind turbine ${p_t}^{wt}$ in time t is explained by Eq. (2). This equation calculates the electric energy on the basis of the following parameters: wind turbine blade area, the wind speed and wind turbine efficiency.

$$P_t^{wt} = 1/2.C_p.(\lambda).\rho.A.({V_t}^{wt})^3, \tag{2}$$

3.2 ESS

ESS is also installed for storage of electricity and the basic purpose is to exploit the efficiency of the proposed home energy management scheme. ESS stores electricity from the commercial grid when the price is low and also stores from the microgrid in high electricity generation hours. ESS is considered as a shiftable load in this work, and ESS charging and discharging can be scheduled in any time interval in an adaptive way. The capacity of ESS is considered 5 kW in this work and ESS stores a maximum of 90% electricity instead of 100% due to safety reasons. ESS supports the home in the limit of minimum and maximum energy storage levels which are 10% and 90%, respectively. ESS discharges only when the rates of electricity are high or the microgrid is unable to meet electricity demand. When the ESS has maximum storage, the excess electricity sells back to the commercial grid.

3.3 Household Electricity Load

In this work, we consider 24 h for implementation, 1 h for each time slot. Furthermore, the household consists of multiple appliances that are further categorized into three main categories: shiftable appliances (a_s), non-interruptible appliances (a_{ni}) and base-load appliances (a_b), where a_s, a_{ni} and $a_b \epsilon A_n$ (A_n is the combination of all appliances). All appliances and their parameters are taken from [8], which are explained in next section.

3.3.1 Shiftable Appliances

In this section, shiftable appliances are defined and these types of appliances can be shifted or interrupted in any of the given time period depending upon their necessity. Shiftable appliances class includes the electrical car, vacuum cleaner, laptop, etc. A set of all shiftable appliances are demonstrated by A_s and $a_s \in A_s$ demonstrates each appliance in the shiftable category. In Eq. (3), the power rating for shiftable smart appliances is shown by λ_s. ε_s shows the total electricity consumption from the commercial grid or microgrid against shiftable appliances in a day, and is calculated by the equation given below:

$$\varepsilon_s = \sum_{t=1}^{T} \Bigg(\sum_{a_s \epsilon A_s} \lambda_s \times \alpha_s(t) \Bigg). \tag{3}$$

3.3.2 Non-interruptible Appliances

In this section, we define the second category of appliances named non-interruptible appliances. This type of appliance may not be interrupted when execution starts but shifted to any time slot before starting their execution. The operation time of non-interruptible appliances cannot be changed. However, this type of appliance may be scheduled between possible earliest starting and possible least ending time. Let $a_{ni} \in A_{ni}$ represent each appliance in this category. The λ_{ni} and ε_{ni} express the power rating and electricity consumption of these types of appliances, respectively. The electricity consumption per day against these types of appliances is calculated in Eq. (4):

$$\varepsilon_{ni} = \sum_{t=1}^{T} \left(\sum_{a_{ni} \epsilon A_{n} i} (\lambda_{ni} \times \alpha_{ni}(t)) \right). \tag{4}$$

3.3.3 Base-Load Appliances

The base-load appliances A_b are such types of appliances that cannot be shifted or interrupted while performing their operations. Generally, these appliances considered the main load of any household; these appliances are also called non-shiftable and non-interruptible appliances. We consider interior lighting and refrigerators as base load appliances. Let $a_b \in A_b$ represent a single appliance from the base-load appliances' category. The λ_b presents the power rating of each appliance in this category and total consumed electricity ε_b in a day is calculated as:

$$\varepsilon_b = \sum_{t=1}^{T} \left(\sum_{a_b \epsilon A_b} (\lambda_b \times \alpha_b(t)) \right). \tag{5}$$

3.4 Electricity Tariff and Bill Calculation

An Electricity tariff is another dynamic attribute of our proposed system model. The utility company provides different electricity tariffs for consumers' motivation to manage their load requirements. In this work, we consider the RTP signals to calculate the electricity consumption cost. There are two different electricity rates for each time slot (hour); one rate is for electricity purchasing EP^{pur} and the other rate for excess electricity selling EP^{sell} to the commercial grid. However, the electricity selling rate is 90% of the purchasing rate in each hour [24], calculated in Eq. (6).

$$EP^{sell} = EP^{pur} \times 0.90. \tag{6}$$

4 Simulation Results and Discussion

In this section, to evaluate and demonstrate the versatility of our proposed scheme, simulations are carried out. Then, simulation results are shown to uncover an optimal scheduling and power trading for the household using CSA and SA. The procedure of CSA and SA is presented in [25].

4.1 Case 1: Home Without Energy Management and Microgrid

In this case, we have studied the operation of a conventional home without a microgrid and ESS integration. This conventional home is not able to manage its electricity consumption and does not have any excess electricity to sell back to the electric grid. In addition, the home cannot make any decisions about electricity usage. It blindly purchases and utilizes electricity while ignoring the electricity tariff or any other parameters.

Figure 2 depicts the electricity bought by conventional homes from electric grid. The results clearly demonstrate that the conventional home ignores the pricing signals and blindly utilizes the electricity. The conventional home purchases electricity in time slots 9 and 10 when the price is at a maximum and also creates peaks due to maximum electricity consumption. PAR of conventional home is presented in Fig. 3. Therefore, the conventional home would pay maximum electricity costs against blind utilization of electricity. The hourly and total electricity cost against unscheduled consumption are presented in Figs. 4 and 5, respectively. However, the results presented in this section consider a baseline for our later comparisons.

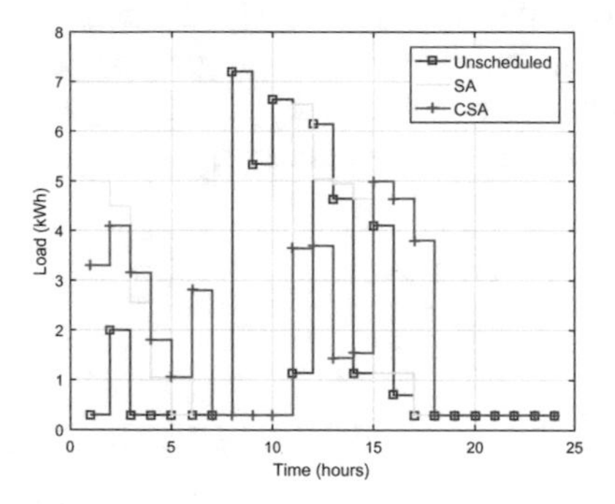

Fig. 2. Hourly electricity consumption.

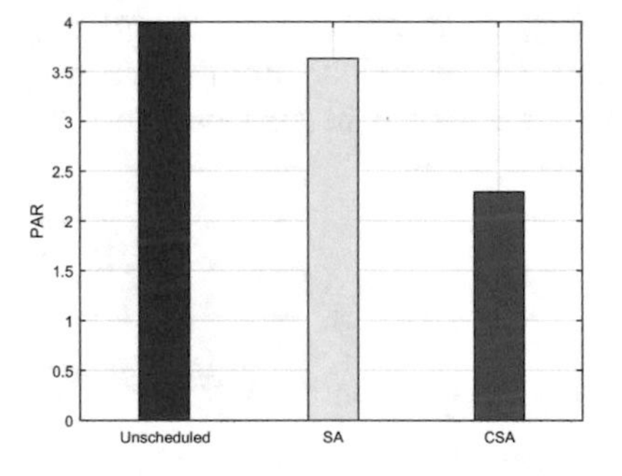

Fig. 3. PAR.

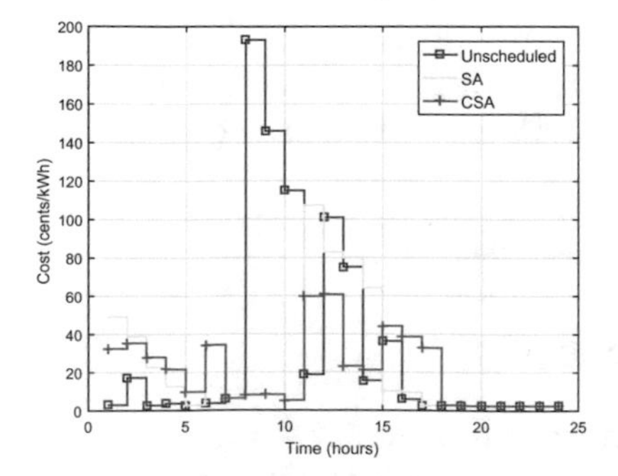

Fig. 4. Hourly electricity cost.

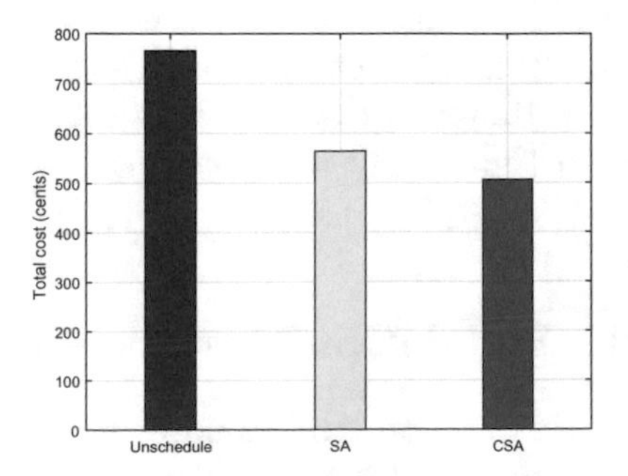

Fig. 5. Total cost.

4.2 Case 2: Home with Energy Management

Energy management of a home is considered in this case and the smart home is able to manage its electricity consumption. The shiftable appliances may have shifted from ON-peak to OFF-peak hours. EMC is installed in this case and EMC shifts the load according to price and load information. Figure 2 represents the hourly electricity consumption without energy management and with energy management using SA and CSA. The result demonstrates that electricity consumption is high without energy management in ON-peak hours; however, electricity consumption is low in ON-peak hours using SA and CSA. Our proposed algorithms efficiently shift the load while minimizing the PAR, which is presented in Fig. 3. CSA shows high performance for PAR reduction as compared to SA and an unscheduled case. The PAR reduction is 15% and 45% using SA and CSA, respectively. Furthermore, when microgrid is installed, more electricity cost is reduced. The electricity generation from microgrid is depicted in Fig. 6. Furthermore, the Fig. 7 shows the electricity cost and earning against imported and sold electricity, respectively.

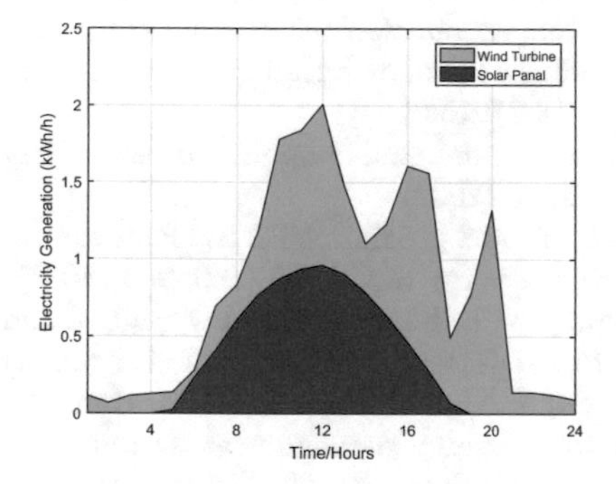

Fig. 6. Electricity generation from wind turbines and solar panels.

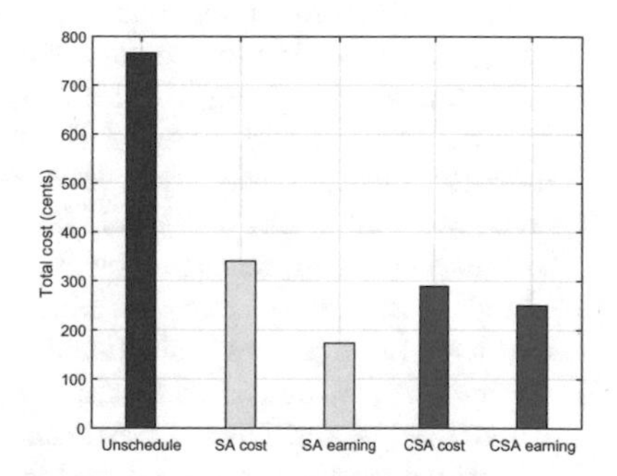

Fig. 7. Cost and earnings against imported and sold electricity.

5 Conclusion and Future Work

In this paper, an electricity load management scheme is developed using SA and CSA along with an RTP scheme for DSM. We considered the load scheduling and power trading problem simultaneously in a smart home that has a grid-connected microgrid. An ESS is also installed to enhance the microgrid performance and reliable operation. The smart home makes autonomous decisions against selling, purchasing or storing electricity according to electricity generation and pricing signals. Furthermore, a comparison of different case studies has been provided to investigate the effectiveness of our proposed scheme. Simulation results demonstrate that our proposed scheme outperforms in terms of electricity cost and PAR reduction with maximizing the earnings. It has been observed from simulation results that the proposed scheme using SA and CSA minimized the electricity cost by 26.63% and 36.42%, respectively, in case 2, and by 56.26% and 62.83%, respectively, in case 3. In addition, our proposed scheme using SA and CSA earns 173.39 and 249.39 (cents), respectively. The overall performance of CSA is superior to SA because the CSA spends a maximum amount of time on global search instead of local search.

In the future, we will focus our research on implementing the mathematical techniques for cost and PAR alleviation with profit maximization in residential and commercial areas.

References

1. Benzi, F., Anglani, N., Bassi, E., Frosini, L.: Electricity smart meters interfacing the households. IEEE Trans. Ind. Electron. **58**, 4487–4494 (2011)
2. Evangelisti, S., Lettieri, P., Clift, R., Borello, D.: Distributed generation by energy from waste technology: a life cycle perspective. Process Saf. Environ. Prot. **93**, 161–172 (2015)
3. Khalid, A., Javaid, N., Guizani, M., Alhussein, M., Aurangzeb, K., Ilahi, M.: Towards dynamic coordination among home appliances using multi-objective energy optimization for demand side management in smart buildings. IEEE Access (2018). https://doi.org/10.1109/ACCESS.2018.2791546
4. Albadi, M.H., El-Saadany, E.F.: A summary of demand response in electricity markets. Electr. Power Syst. Res. **78**, 1989–1996 (2008)
5. Avci, M., Erkoc, M., Rahmani, A., Asfour, S.: Model predictive HVAC load control in buildings using real-time electricity pricing. Energy Build. **60**, 199–209 (2013)
6. Yang, J., Zhang, G., Ma, K.: Matching supply with demand: a power control and real time pricing approach. Int. J. Electr. Power Energy Syst. **61**, 111–117 (2014)
7. Ahmad, A., Khan, A., Javaid, N., Hussain, H.M., Abdul, W., Almogren, A., Alamri, A., Azim Niaz, I.: An optimized home energy management system with integrated renewable energy and storage resources. Energies **10**, 549 (2017)
8. Aslam, S., Iqbal, Z., Javaid, N., Khan, Z.A., Aurangzeb, K., Haider, S.I.: Towards efficient energy management of smart buildings exploiting heuristic optimization with real time and critical peak pricing schemes. Energies **10**, 2065 (2017)
9. Van der Stelt, S., AlSkaif, T., van Sark, W.: Techno-economic analysis of household and community energy storage for residential prosumers with smart appliances. Appl. Energy **209**, 266–276 (2018)
10. Liu, R.S., Hsu, Y.F.: A scalable and robust approach to demand side management for smart grids with uncertain renewable power generation and bi-directional energy trading. Int. J. Electr. Power Energy Syst. **97**, 396–407 (2018)
11. Bradac, Z., Kaczmarczyk, V., Fiedler, P.: Optimal scheduling of domestic appliances via MILP. Energies **8**, 217–232 (2014)
12. Zhu, Z., Tang, J., Lambotharan, S., Chin, W.H., Fan, Z.: An integer linear programming based optimization for home demand-side management in smart grid. In: Proceedings of the 2012 IEEE PES Innovative Smart Grid Technologies (ISGT), Washington, DC, USA, 16–20 January 2012, pp. 1–5 (2012)
13. Zhang, D., Evangelisti, S., Lettieri, P., Papageorgiou, L.G.: Economic and environmental scheduling of smart homes with microgrid: DER operation and electrical tasks. Energy Convers. Manag. **110**, 113–124 (2016)

14. Mohamed, F.A., Koivo, H.N.: Online management genetic algorithms of microgrid for residential application. Energy Convers. Manag. **64**, 562–568 (2012)
15. Hafeez, G., Javaid, N., Iqbal, S., Ali Khan, F.: Optimal residential load scheduling under utility and rooftop photovoltaic units. Energies **11**(3), 611 (2018)
16. Samadi, P., Wong, V.W., Schober, R.: Load scheduling and power trading in systems with high penetration of renewable energy resources. IEEE Trans. Smart Grid **7**, 1802–1812 (2016)
17. Qayyum, F.A., Naeem, M., Khwaja, A.S., Anpalagan, A., Guan, L., Venkatesh, B.: Appliance scheduling optimization in smart home networks. IEEE Access **3**, 2176–2190 (2015)
18. Agnetis, A., de Pascale, G., Detti, P., Vicino, A.: Load scheduling for household energy consumption optimization. IEEE Trans. Smart Grid **4**, 2364–2373 (2013)
19. Tushar, M.H.K., Assi, C., Maier, M., Uddin, M.F.: Smart microgrids: optimal joint scheduling for electric vehicles and home appliances. IEEE Trans. Smart Grid **5**, 239–250 (2014)
20. Erdinc, O.: Economic impacts of small-scale own generating and storage units, and electric vehicles under different demand response strategies for smart households. Appl. Energy **126**, 142–150 (2014)
21. Mary, G.A., Rajarajeswari, R.: Smart grid cost optimization using genetic algorithm. Int. J. Res. Eng. Technol. **3**, 282–287 (2014)
22. Javaid, N., Ullah, I., Akbar, M., Iqbal, Z., Khan, F.A., Alrajeh, N., Alabed, M.S.: An intelligent load management system with renewable energy integration for smart homes. IEEE Access **5**, 13587–13600 (2017)
23. Wang, Y., Saad, W., Han, Z., Poor, H.V., Baar, T.: A game-theoretic approach to energy trading in the smart grid. IEEE Trans. Smart Grid **5**, 1439–1450 (2014). **3**, 282–287 (2014)
24. Ding, Y.M., Hong, S.H., Li, X.H.: A demand response energy management scheme for industrial facilities in smart grid. IEEE Trans. Ind. Inform. **10**, 2257–2269 (2014)
25. Aslam, S., Javaid, N., Ali Khan, F., Alamri, A., Almogren, A., Abdul, W.: Towards efficient energy management and power trading in a residential area via integrating grid-connected microgrid. Sustainability **10**(4), 1245 (2018). https://doi.org/10.3390/su10041245

Hierarchical Based Coordination Strategy to Efficiently Exchange the Power Among Micro-grids

Aqib Jamil[1], Nadeem Javaid[1(✉)], Zafar Iqbal[2], Muhammad Abdullah[1], Muhammad Zaid Riaz[1], and Mariam Akbar[1]

[1] COMSATS Institute of Information Technology, Islamabad 44000, Pakistan
nadeemjavaidqau@gmail.com

[2] PMAS, Arid Agriculture University, Rawalpindi 46000, Pakistan

Abstract. Micro-grid (MG) is an emerging component of a smart grid and it is increasing the efficiency and reliability of the power system with the passage of time. MGs often need power in order to fulfill its load requirements, which is transmitted form macro station (MS). Transmission of power from MS cause power line losses. To decrease these power line losses, a hierarchical based coordination (HBC) strategy is proposed for efficiently exchanging the power among MGs. HBC aims to decrease power line losses by making hierarchical coalitions. Results are evaluated and compared with conventional non-coordination model (NCM). This comparison shows the effectiveness of proposed HBC strategy. Results indicate that HBC has decreased the power line losses by 40.1% as compared to conventional NCM.

Keywords: Smart grid · Micro grid · Energy management Coordination strategy · Hierarchical coalitions

1 Introduction

In past few years, electricity consumption is increasing day by day with the increase in the population and technology growth. Due to this immense increase in electricity consumption, existing power system is becoming inefficient and it is not possible to meet such demand of electricity with existing generation plants. In addition, most of the electricity is generated from natural sources, which are also depleting with time. Therefore, there is a need for traditional power system to move on to renewable energy resources (RESs) [1]. However, traditional grid cannot integrate RESs in the power system and it has a lot of other challenges: reliability, sustainability, one-way communication and self-monitoring. To overcome the aforementioned challenges, concept of smart grid (SG) is evolved. SG solved the most of the challenges of traditional grid including the challenge of RESs' integration in power system. However, RESs are highly intermittent in nature, they can not guarantee the required output all the time. Therefore,

L. Barolli et al. (Eds.): IMIS 2018, AISC 773, pp. 242–251, 2019.
https://doi.org/10.1007/978-3-319-93554-6_22

depending upon the RESs only, is not a feasible solution [2]. Moreover, new generation plants' installation take a lot of time and money. Therefore, instead of increasing the power generation, existing power system can be optimized by managing the power on the supply and demand side.

Micro grid (MG) is a small component of SG, which comprises of RESs and distributed generators (DGs) with its own load. MGs are operated in two modes: grid connected mode (GCM) and islanded mode (IM) [3]. In GCM, MGs are connected to main grid and during surplus power hours, these MGs inject surplus power into main grid. Similarly, during deficient power hours, they receive power from main grid. In IM, MGs are operated as standalone and during deficient power hours, they use load curtailment strategy in order to balance the load and supply. MGs exchange power with the macro station (MS) in GCM to meet the power requirements, however, due to long distance between MG and MS, power line losses are high. Therefore, in order to optimize the power on supply side, power line losses can be minimized using efficient power sharing strategy among MGs.

For this purpose, we have proposed a hierarchical based coordination (HBC) strategy to efficiently exchange the power among the MGs. In HBC, MGs coordinate with MG coordination center (MCC) to send their location, generation and load details of each hour. As distance between MG and MS is long therefore, power line losses during power exchange are high. Whereas, HBC makes hierarchical coalitions (HCs) among MGs located near to each other. HCs are formed depending upon the power line losses, which directly relate with the distance among MGs. Results are computed and compared with conventional non-coordination model (NCM). In NCM, MGs exchange power only with MS. Results demonstrate that HBC has minimized the power line losses efficiently as compared to NCM.

The rest of the paper is organized as follows: back ground and literature of paper is described in Sect. 2. The system model along with coordination strategy is presented in Sect. 3. Results and comparison of proposed HBC strategy with conventional NCM are discussed in Sect. 4. Finally, conclusion of proposed strategy is given in Sect. 5.

2 Related Work

MG is an emerging concept and it is getting a lot of interest from research community. Many approaches have been presented for balancing the supply and load gap. Some researchers have focused on the supply side whereas, some have focused on the demand side. In this section, approaches which are related to our work are described. Coalitional game theory (CGT) is implemented in [4,5] to reduce the power line losses during power sharing. Authors in [4] have developed a three stage algorithm: request, merge-split and transaction stages, based on CGT. They also considered the load shedding strategy in their scenario. In [5], disjoint coalitions of MGs are formed for reducing power line losses and they have studied the effect of increase in the number of MGs on the average power line losses.

Cooperative energy management [6] is proposed in MGs based on multi agent system (MAS). MAS is used for control and dispatch of power flows among MGs. Cooperative game is implemented for the formation of coalitions and MGs coordinate with each other to decrease the power line losses by sharing power among neighbor MGs. In [7], authors used optimization techniques to optimize the household load. They have considered RESs and energy storage system (ESS) in their model. Results indicate that 47.7% and 49.2% cost is reduced using RESs and ESS respectively.

Rahbar *et al.* [8] have studied the energy cooperation problem of two MGs. Each MG is equipped with RESs and ESS, and two algorithms are proposed for energy management in order to study the effect of EES on energy cost. Evolutionary game theory is proposed in [9] for energy cooperation between MG and traditional grid. Authors have concluded that cooperation cost directly relates with direct benefits and indirect benefits of MGs and traditional grid.

A hierarchical coordination [10] is implemented for optimal power sharing among MGs. Both alternating current (AC) and direct current (DC) MGs with IM and GCM are considered. Authors have exchanged the power among AC and DC MGs while keeping the rated frequency and voltage of MGs constant. In [11], two-way power flow is considered among smart homes, which contain electric vehicles, ESSs and DGs. Moreover, authors have considered the transformer load limit in order to reduce the stress on the transformer during peak hours.

A game theoretic approach is presented in [12] for energy trading among MGs. MGs with surplus power, sell power to neighboring MGs and MGs with deficient power, buy from neighboring MGs. Results represent that their proposed approach has performed well in decreasing the energy bill. An optimal coalition mechanism is proposed in [13] with an aim to decrease the power line losses and stress on the main grid.

After analyzing the related literature, we proposed a power sharing strategy among MGs in order to decrease the power line losses and stress on the main grid. For this purpose, a hierarchical coordination is proposed and compared with conventional power sharing model. It is further explained in next section.

3 System Model and Coordination Strategy

This section describes the proposed system model and its components. A power system model is considered, which consists of n MGs, randomly distributed in an area k. These MGs are connected to each other and with the MS. Each MG has different RESs or DGs and they supply power to n number of users: residential, industrial, hospitals and companies.

The system model is illustrated in Fig. 1. There is a two-way communication line between MGs and MS through which MGs communicate with the MCC. Two-way power flow is considered among MGs and with MS as shown in system model. System model can be divided into three levels: utility grid, MS and MGs level. At MGs level, power is supplied to end-users and power exchange is performed among MGs using proposed HBC. In HBC, each MG at the lower

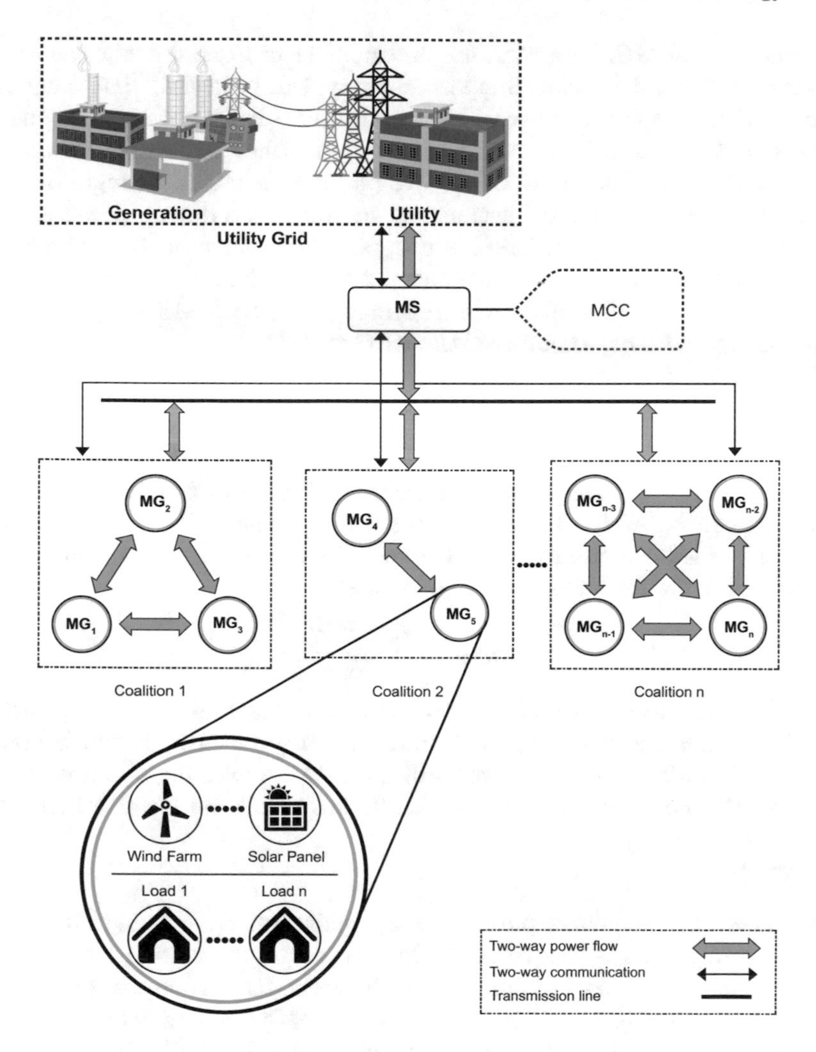

Fig. 1. Proposed system model

level, sends information about its location, generation and load of each time slot (1 h) to the MCC which is the upper level of hierarchy. On the basis of this information, MCC calculates the surplus or deficient power of each MG and makes HCs of nearby located MGs. HCs are formed on the basis of power losses, which depend upon the distance among the MGs.

Utility grid and MS levels are responsible for receiving power from MGs or transmitting it to MGs. However, in our case a MG transmits power to MS only when no other MG with deficient power is left within that coalition. Similarly, it receives power from MS only when no more MGs with surplus power are left within that coalition. Remember that each coalition must contains at least one MG with surplus power and a MG with deficient power. As shown in system model, number of MGs in a coalition and number of formed coalitions are kept changing after each time slot. Suppose, for a typical time slot, HBC may makes five HCs with different number of MGs in it and a HC contains only one MG, it shows that this MG is located very far from other MGs as a result power line losses are high. Therefore, HBC have placed this MG in a single HC and it exchanges power with the MS because power line losses during power exchange with MS is less than the line losses during power exchange in HCs. There is also a possibility that for a typical time slot, all of the MGs have surplus power or deficient power only, then MGs will exchange power with MS.

In this paper, we have compared the HBC model with conventional NCM and results are computed, which are explained and discussed in simulation section.

3.1 NCM

In this case, MGs exchange power with the MS only and no power exchange takes place among MGs. As the distance between the MG and MS is long, as compared to distance among MGs, therefore power line losses are high. Power line losses of MG_k for NCM can be given as:

$$P_{NCM} = \frac{R \times EP_k^2}{V_k}, \tag{1}$$

where R is the resistance of the transmission line and it is considered as $0.20\,\Omega/\text{km}$. V_k is the voltage of the transmission line through which power is transmitted to MS or received from MS. EP_k is the exchanged power between MG_k and MS. Power line losses of each MG are calculated for all 24 time slots.

3.2 HBC

In this case, MGs exchange power among themselves by forming HCs. As the distance among MGs is less than the MS, therefore power line losses are low. In order to exchange power among MGs efficiently, HBC is implemented among MGs, where each MG sends its information to MCC, where MCC makes HCs among nearby located MGs. The power line losses for MG_k during HBC are calculated as:

$$P_{HBC} = \frac{R \times EP_{jk}^2}{V_{jk}}, \tag{2}$$

here P_{HBC} is the power line losses during HBC. R is resistance of transmission line and V_{jk} shows the line voltage, which is 11 kV. EP_{jk} is the exchanged power between MG_j and MG_k. HBC makes HCs using following rules:

- Each HC must has at least one MG with surplus power and a MG with deficient power.
- Each HC contains MS as a imaginary member because after exchanging all power among MGs within a coalition, if a MG still has surplus or deficient power then it will exchange power with MS. Remember that MGs will exchange power with MS only when no other MG is left to exchange power within a coalition.

Table 1. Power requirements of MGs

MGs	Power exchanged (MW)
MG-1	0.7353
MG-2	0.2503
MG-3	−2.1804
MG-4	−1.0445
MG-5	−1.2678
MG-6	−1.7093
MG-7	1.9139
MG-8	2.3741
MG-9	−0.1928
MG-10	−0.2428
MG-11	−1.5423
MG-12	0.8555

4 Simulations and Discussions

To evaluate the performance of proposed strategy, simulations are performed in MATLAB 2017*a*. Twelve MGs are randomly deployed in an area of 14 km^2 as illustrated in Fig. 2. Each MG has different RESs (solar panels or wind farms) or DGs with its own individual loads: residential consumers and industrial consumers. For simulation purposes, MS is assumed to be placed at the coordinates (0,0) and MGs are randomly placed around the MS. The generation and demand are randomly generated for each MG and theses values are kept constant for 1 h

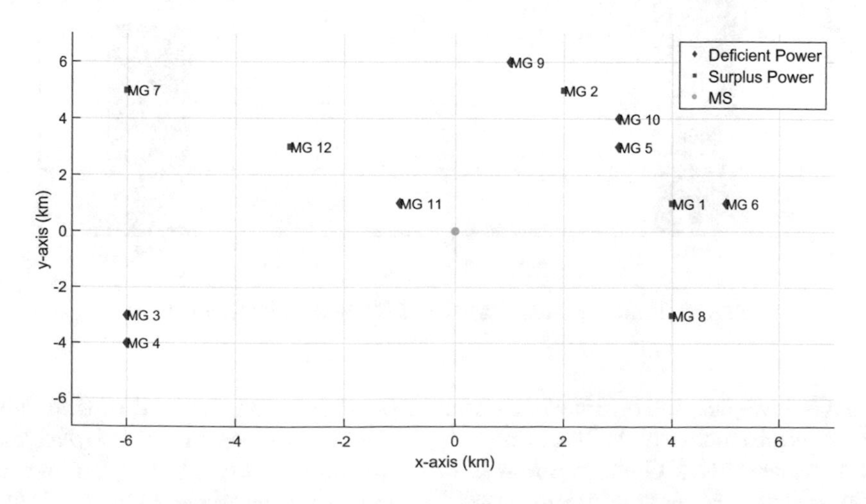

Fig. 2. MGs and MS deployment in an $7 \times 7\,\mathrm{km}^2$ area

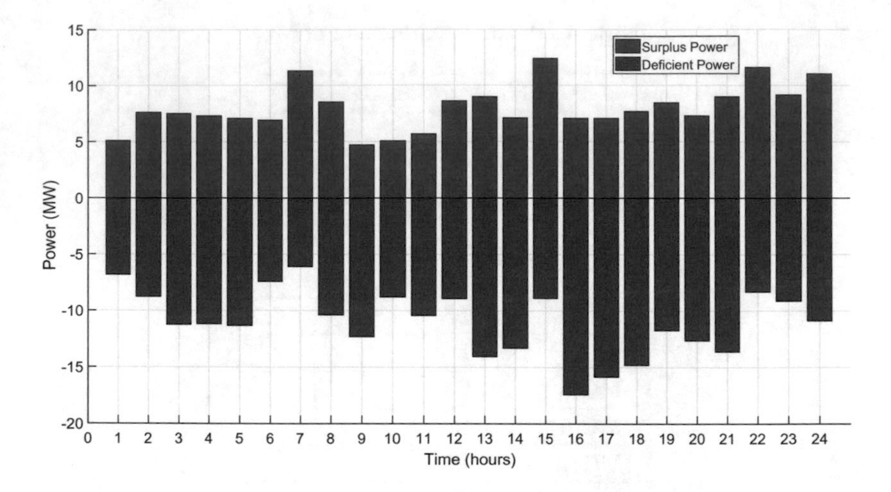

Fig. 3. Surplus and deficient power of proposed system during each hour of a day

time slot. Demand and supply for each MG is distributed from 0 to 4 MW randomly for each hour of a day. Remember that these assumptions can be easily replaced by real-time energy data for implementing this system in real-time scenario.

Fig. 4. Total power line losses of NCM and HBC strategy

MGs have surplus or deficient power depending upon the load of that hour. Power requirements of MGs during a typical hour is given in Table 1. Negative power shows that MG needs power in order to fulfill its load. Figure 2 shows the MGs having surplus or deficient power in a hour of a day. Power status of MGs, keeps changing after each hour because of variable load and generation. Figure 3 demonstrates the total surplus and deficient power of proposed system during each hour. Maximum surplus power is 12.40 MW at time slot 15 and minimum surplus power is 4.72 MW, which occurs at 9th time slot. Similarly, maximum and minimum deficient power occurs at 16th and 7th time slots respectively.

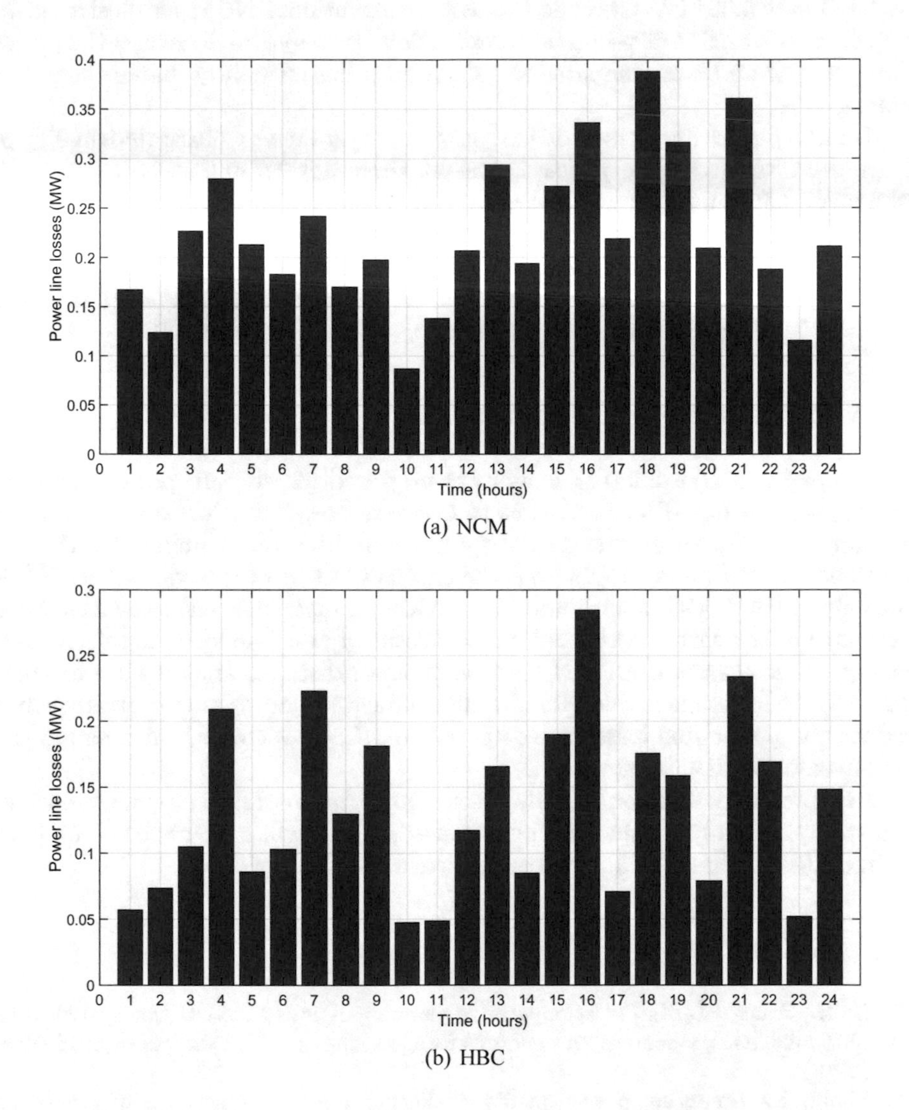

Fig. 5. Hourly power line losses of proposed system during a day.

Transmission line resistance is set as 0.20 Ω/km and voltage level among the MGs and MS is set as 11 kV. These values are set in order to calculate the power line losses for both NCM and HBC scenarios. Figure 4 depicts the power line losses of both cases. As during conventional NCM, MGs fulfill their power requirements directly form MS, which is usually located at longer distance. Therefore, it has high power line losses as compared to HBC where MGs, which are close to each other, make coalitions and exchange power with each other. Results show that 3.19 MW power is lost during power exchange in case of HBC and 5.33 MW power is lost with conventional NCM as illustrated in Fig. 4. It is cleared that proposed coordination strategy has decreased the power line losses by 40.1% as compared to NCM and it has efficiently shared the power among MGs.

Hourly power line losses of the proposed system are illustrated in Fig. 5. During NCM, maximum power is lost at time slot 18 and it is 0.3878 MW. Moreover, minimum power is lost at 10th time slot which is 0.0870 MW. In case of HBC, maximum power lost is 0.2845 MW, at time slot 16. Minimum power is lost during 10th time slot and it is 0.0475 MW. In addition, if each time slot of both cases are compared with each other. We can be concluded that power losses of HBC during each time slot of a day, are less than the NCM.

5 Conclusion and Future Work

In this paper, a coordination strategy is proposed for efficient power exchange among MGs using HBC. HBC aims to decrease power line losses during power exchange. Results of proposed strategy are computed and compared with conventional NCM in order to find out the effectiveness of proposed strategy. With the help of HBC, MGs coordinate with MCC and make efficient coalitions. MGs exchange power within each coalition as a result power line losses are decreased. Power line losses are decreased because of small distance among MGs as compared to the distance from MS. Results indicate that proposed strategy has reduce the power line losses as compared to NCM by 40.1% which shows the effectiveness of HBC strategy.

In future, we will determine the effect of integration of ESS on power sharing. Moreover, a pricing scheme will be proposed, which will encourage the MGs to form coalitions and it will decrease the stress on main grid.

References

1. Li, R., Wang, W., Xia, M.: Cooperative planning of active distribution system with renewable energy sources and energy storage systems. IEEE Access **6**, 5916–5926 (2018)
2. Liang, X.: Emerging power quality challenges due to integration of renewable energy sources. IEEE Trans. Ind. Appl. **53**(2), 855–866 (2017)
3. Han, Y., Li, H., Shen, P., Coelho, E.A.A., Guerrero, J.M.: Review of active and reactive power sharing strategies in hierarchical controlled microgrids. IEEE Trans. Power Electron. **32**(3), 2427–2451 (2017)
4. Ni, J., Ai, Q.: Economic power transaction using coalitional game strategy in microgrids. IET Gener. Transm. Distrib. **10**(1), 10–18 (2016)

5. Wei, C., Fadlullah, Z.M., Kato, N., Takeuchi, A.: GT-CFS: a game theoretic coalition formulation strategy for reducing power loss in micro grids. IEEE Trans. Parallel Distrib. Syst. **25**(9), 2307–2317 (2014)
6. Mangiatordi, F., Pallotti, E., Panzieri, D., Capodiferro, L.: Multi agent system for cooperative energy management in micro-grids. In: 2016 IEEE 16th International Conference on Environment and Electrical Engineering (EEEIC), Florence, pp. 1–5 (2016)
7. Ghulam, H., Javaid, N., Iqbal, S., Khan, F.A.: Optimal residential load scheduling under utility and rooftop photovoltaic units. Energies **11**(3), 1–27 (2018)
8. Rahbar, K., Chai, C.C., Zhang, R.: Energy cooperation optimization in microgrids with renewable energy integration. IEEE Trans. Smart Grid **9**(2), 1482–1493 (2018)
9. Pan, C., Long, Y.: Evolutionary game analysis of cooperation between micro-grid and conventional grid. Math. Prob. Eng. **2015**, 1–10 (2015)
10. Che, L., Shahidehpour, M., Alabdulwahab, A., Al-Turki, Y.: Hierarchical coordination of a community microgrid with AC and DC microgrids. IEEE Trans. Smart Grid **6**(6), 3042–3051 (2015)
11. Paterakis, N.G., Erdin, O., Pappi, I.N., Bakirtzis, A.G., Catalao, J.P.S.: Coordinated operation of a neighborhood of smart households comprising electric vehicles, energy storage and distributed generation. IEEE Trans. Smart Grid **7**(6), 2736–2747 (2016)
12. Yaagoubi, N., Mouftah, H.T.: Energy trading in the smart grid: a distributed game-theoretic approach. Can. J. Electr. Comput. Eng. **40**(2), 57–65 (2017)
13. Chakraborty, S., Nakamura, S., Okabe, T.: Real-time energy exchange strategy of optimally cooperative microgrids for scale-flexible distribution system. Expert Syst. Appl. **42**(10), 4643–4652 (2015)

Weighted Cuckoo Search Based Load Balanced Cloud for Green Smart Grids

Muhammad Hassan Rahim[1], Nadeem Javaid[1(✉)], Sahar Rahim[2], Muqaddas Naz[1], Mariam Akbar[1], and Farhana Javed[3]

[1] COMSATS Institute of Information Technology, Islamabad 44000, Pakistan
nadeemjavaidqau@gmail.com
[2] COMSATS Institute of Information Technology, Wah Cantt 47040, Pakistan
[3] Huazhong University of Science and Technology, Wuhan, China

Abstract. The concept of cloud computing is becoming popular with each passing day. Clouds provide virtual environment for computation and storage. Number of cloud users is increasing drastically which may cause network congestion problem. To avoid such situation, fog computing is used along with cloud computing. Cloud act as a global system and fog works locally. As the requests from users are increasing so load balancing is also required on fog side. In this paper, a three layered cloud and fog based architecture is proposed. Fog computing acts as a middle layer between users and the cloud. Users' requests are handled at fog layer and filtered data is forwarded to cloud. A single fog has multiple virtual machines (VMs) that are assigned to the users' requests. The load balancing problem of these requests is managed by proposed weighted cuckoo search (WCS) algorithm. Simulations are carried out to evaluate the performance of proposed model. Results are presented in the form of bar graphs for comparison and detailed values of each parameter are presented in tables. Results show the effectiveness of proposed technique.

Keywords: Cloud computing · Fog computing
Demand request time · Demand response time
Demand processing time · Energy management

1 Introduction

In the information technology world, cloud computing is one of the recent paradigms. Cloud computing is an integrated concept of parallel and distributed computing which shares resources like hardware, software, and information with computers or other devices on demand [1]. Moreover, it provides three major type of services: (i) software as a service (SaaS), (ii) infrastructure as a service (IaaS), and (iii) platform as a service (PaaS) to the users [2]. The users have to pay certain amount to use these services. However, limited network bandwidth, increased response time, delays in data transfer, and data transfer cost are some of the challenging issues in cloud computing environment. Moreover, the load

L. Barolli et al. (Eds.): IMIS 2018, AISC 773, pp. 252–264, 2019.
https://doi.org/10.1007/978-3-319-93554-6_23

balancing in cloud computing systems is one of a major challenge [4]. Cisco introduced a new concept of fog computing in 2014 to extend and tackle the aforementioned challenges of the cloud computing. Fog computing provides the same services as cloud computing. The concept of fog computing can be implemented locally within a region on the basis of user's demand. A cloud may have n number of fog servers responding and processing the users' request more efficiently. The communication among cloud, fog, and the users is establish by different wireless communication protocols. The fog servers have limited amount of storage memory and store the data temporarily. To permanently store the data of users, it communicate with cloud and send it to the cloud for storage.

Virtual machine (VM) act as an execution unit and provide the foundation for fog and cloud computing technology. Visualization consists of creation, execution, and management of a hosting environment for various applications and resources. The VMs in fog and cloud environment share resources like system bus and processing cores. Each VM is constrained by the limited processing power of computing resources [5]. The users send requests in a uncertain way continuously. The poor management of critical load requests lead to poor system performance and under utilization of resources. Hence, load balancing has become a critical task. Thus, to address the issue of load balancing many load balancing algorithms are introduced such as round robin, weighted round robin, dynamic load balancing, Throttled algorithm, etc.

In this paper, an integrated fog and cloud based model for load management is proposed. Fogs are integrated with the cloud to tackle the challenges of cloud. They manage the load requests generated by the users. The world is divided into six regions based on the six continents. Six fogs are considers in this model which are allocated to the users' requests on the basis of response time, data center, processing time, etc. A single group of building contains multiple smart buildings and users. Additionally, there is a microgrid (MG) near each building and the fog has the detailed information about of the MG, utility, and the users' demand. Fog fulfills the users' load demand by analyzing the available data of MG and utility. If MG generates enough energy to fulfill the load requests of user it establishes a connection between MG and user to meet the user's request. A load balancing algorithm is also proposed in this paper named as weighted cuckoo search (WCS). The proposed algorithm handles the requests more effectively and able to minimize the response time and processing time of the requests with different service broker policies. The main contributions of this work are:

- The world is divided into six region based on the continent.
- Fog is integrated with the model to improve the efficiency of cloud.
- Three layered architecture is proposed for energy exchange between user and MG.
- A new VM load balancing technique WCS is proposed for load balancing.

The rest of the paper is organized as follow: Sect. 2 presents an overview of current literature work. System model is given in Sect. 3. In Sect. 4, simulation and results are discussed. Section 5, concludes the paper.

2 Literature Review

Extensive research is going on to cope the challenges of fog and cloud computing paradigms around the world. Some of the existing literature is discussed in this section. Authors in [1] discuss the problem of efficient resource allocation in cloud computing from both user and utility's prospective. For the scheduling of computational resources, first come first serve (FCFS) and round robin scheduling techniques are implemented. Performance of both algorithms, in terms of cost and waiting time, is evaluated. To mimic the real cloud environment, CloudSim simulation toolkit has been utilized. This toolkit provides real cloud like environment for the performance evaluation of algorithms. Two new algorithms are also proposed. The first one is FCFS integrated with shortest job first and second one is round robin based on average burst time of task. It is concluded that the later two algorithms are more efficient for CPU allocation and scheduling purposes.

Authors proposed an integrated fog and cloud based model in [3] to evaluate the energy consumption of the residential sectors in all the regions. Moreover, the authors proposed a energy management model and a new service broker policy to efficiently select the nearby fogs to minimize the delay and response time. Comparative results show the effectiveness of the proposed algorithm.

In [4], another load balancing algorithm for clod computing environment is proposed. The proposed algorithm successfully balances the load by using scheduling algorithm and resource allocation techniques. In this system, the requests are received by proposed throttled algorithm. These requests are then forwarded to the load balancer which allocates the suitable VM to the job. The performance of throttled load balancing algorithm is compared with already existing algorithms: round robin and equally spread current executed algorithms. The performance of the proposed algorithm is best as compared to other two already existing algorithms.

The load balancing problem of VMs is discussed in [5]. It is stated that without proper scheduling and optimization, VMs are under utilized. The VMs should be assigned to each request according to the nature of its requirements. For example, for non-preemptive tasks, suitable VMs should be allocated before the initialization of tasks. Moreover, some jobs consist of multiple independent tasks that can be executed on multiple VMs or different cores of same VMs simultaneously. To accommodate all these heterogeneous jobs efficiently, a suitable load balancing scheduling of jobs is required. In this regard, authors have proposed improved weighted round robin algorithm. This algorithm assigns the VM to a job on the basis of in formation like: processing capacity, current load on VM, length of job and its priority among other available jobs. Multiple simulations are conducted to evaluate the performance of proposed algorithm. The proposed algorithm outperforms than already existing round robin and weighted round robin algorithms.

A load balancing algorithm known as particle swarm optimization with simulated annealing (PSOSA) is proposed by Anila *et al.* in [6]. Additionally, they also proposed a three layer model for cloud and fog computing environment.

Moreover, they compare the proposed algorithm with the existing Throttled and round robin algorithms on the basis of performance parameters: average response and processing times. The effectiveness of the proposed technique is shown by the simulation results.

In [7], it is highlighted that the integration of EVs has positive affect on power system and environment but they can have adverse effects on smart grid. EVs consume a large amount of energy and their irregular charging and discharging patterns can cause peaks in power system. So, there is a need of a system to efficiently control the changing and discharging process of EV's. In this regard, authors have proposed a cloud based architecture for EVs public supply stations. The proposed algorithm schedules these stations and optimizes the waiting times of EVs. The performance of proposed algorithms is evaluated through extensive simulations. Results of these simulations validate the performance of these algorithms. The peaks in power consumptions are significantly reduced which prove the efficiency of proposed work.

The problem of poor utilization of cloud resources for demand side management is discussed in [8]. The authors have highlighted the issue of under utilization problem of resources in a cloud environment. It is stated that in centralized resources environment, one to one allocation is implemented where one computational resource is allocated to only one application. In this scenario, a single application cannot fully utilize the computational power of a source which results in under utilization of resource and increases the cost. To tackle such problems in centralized environment, authors have proposed an effective cost oriented model for allocation of computational resources for demand side management. An optimization algorithm is also proposed to solve the optimization problem during allocation and load profile is also taken into account. The performance of the proposed system is evaluated using extensive simulations. The results of these simulations depict the effectiveness of the proposed system. Moreover, the validity of the system is also proved by numerical studies.

Authors in [9], proposed a fog to cloud based frame work considering residential sector to enhance the service of cloud in term of latency. Moreover, the authors compare the result of shortest job first with the existing algorithms: round robin and equally spread current execution. Simulation results show that the proposed techniques outperform the existing technique.

In [10], it is stated that concept of smart cities is becoming more popular with the emergence of smart devices and communication technologies. It has made resource utilization and management more efficient and also plays a very important role in automating and improving the quality of life. A large number of devices are connected together for information exchange, these communication and information exchange require a platform. Such platform is provided by cloud and fog computing. Cloud provides the services globally and fog computing is designed to provide services locally. To make the operation more effective these two technologies need to be integrated together, which is not a simple task. For this reason, authors have proposed a service oriented middle ware named Smart City Ware. The implementation details of the proposed architecture along with simulation results are presented which validate the performance of this architecture.

In [11], an extended version of round robin algorithm is proposed for the load balancing in cloud computing environment. Number of cloud users is increasing day by day and instead of purchasing the required resources, users prefer pay as you go services of cloud. The increasing popularity of clouds can cause the load congestion problems. This problem makes the users to wait in queues to get the required resources. To avoid such situations load balancing techniques play a very important role. In this study, authors have proposed a system which balances the load of data centers. It receives the requests and allocates the suitable data center to each request. The proposed allocation algorithm is the extension of two existing algorithms: round robin and randomized scheduling. Simulations are carried out to validate the performance of proposed algorithm. Results show that the proposed algorithm successfully schedules the requests and balances the load in an efficient manner.

Authors in [12] have proposed a cloud based architecture for demand side management. In this architecture, there are multiple regions and every region has its own edge cloud. A centralized cloud is also present named as core cloud. In this study, demand side management problem is addressed where, day ahead optimization is performed. Two types of pricing tariffs are used: inclined block rate and real time pricing. Users are also categorized into three categories according to their participation in demand response programs. The type 0 users do not take part in demand side management program, the type 1 users take part in these programs, whereas, the type 2 users take part is demand side management programs and they also have their own local power generation resources. The optimization problem of the users living in a region is run on its edge cloud. This information is then sent to the core cloud which also requires the energy generation and consumption patterns of MG. The core cloud uses this information and generates the appropriate power consumption patterns for MG. The simulation results show the effectiveness of the proposed system.

3 System Model

A three layer cloud and fog based architecture is proposed in this study. The pictorial representation of the system model is shown in Fig. 1. The first and last layers of the system represent the cloud and users, respectively, whereas, fog computing act as an intermediate layer between cloud and users. The world is divided into six regions on the basis of different continents. Each region is allocated a suitable fog to facilitate the users' requests and balance load. Cloud computing provide infrastructure, platform and software as a service to facilitates the users. The main challenges in cloud computing environment are less network bandwidth, response time, data transfer cost and delays in data transfer. To tackle the cloud computing issues, fog computing is integrated in the system. Fog is considered much closer and easily accessible to the users. It provides the same services as the cloud. However, It has limited computational power and

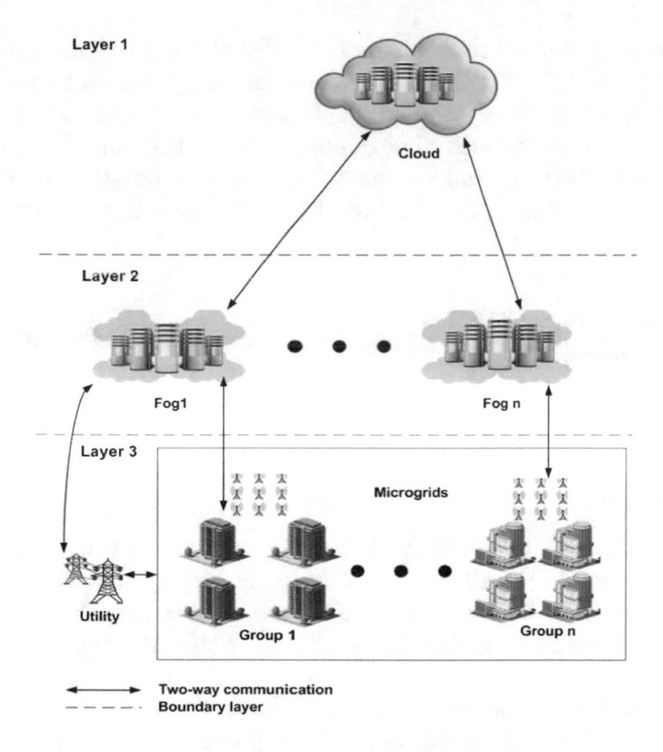

Fig. 1. System model

storage capacity. Fog computing can be implemented locally with in a region to fulfill the demand of users. It acts as a hop between the cloud and the users.

In this study, we assume that there are total six fog servers connecting all six regions of the world. In each region there are n number of users in the form of a group. Additionally, each group of buildings is also connected to a MG. The users send the load requests to fog having fifty VMs. The fog allocates the task to different VMs on the basis of the schedule generated by the proposed WCS VM load balancing technique to balance the load effectively with minimum latency and delay. Different wireless communication protocols are used to establish the connection among cloud, fog and the users. The requests made by the cluster of buildings will be handled by the fog server and cloud server. Each smart building's information is stored at the data center in the fog or the cloud environment, where resource manager is used for the resources' allocation and management. Fog allocate VMs to each building's request which is monitored and controlled by the hypervisor virtually.

3.1 Scheme

In this study, a heuristic technique WCS is proposed. It is inspired by the behaviour of cuckoo bird. The proposed algorithm explores the search space by using the Levy step taken from the Levy distribution proposed by Mantegna algorithm [13]. The WCS follow three basic rules: (i) cuckoo lays one egg at a time, and drop its egg in a randomly selected nest, (ii) the next generation is carried by the best nests with high quality of eggs, and (iii) The total number of the host nests is fixed, and the host bird can discover the cuckoo egg with the probability between [0,1]. The host bird will leave the nest on discovering cuckoo's egg.

3.2 Service Broker Policies

A service broker policy decides that which fog should provide services in response to the requests coming from clusters of smart buildings [14]. Service broker controls the traffic of requests. It uses a set of rules or policies on the basis of which the fog is selected for the processing of user's requests. Some of the policies are described below;

- Optimise Response Time Policy: This policy maintains the index of all available fogs located in the all regions. It keep the history that which fog provide best response time in the history. The requests coming from the group of buildings are assigned to the fog available within the region with optimal response time.
- Dynamically Reconfigure with Latency: This policy manages the fogs on currently basis and selects the fog with best response time which is closed to the group of buildings. If current response time is maximum than the best one then it divert the traffic to next available fog. Moreover, it also maintains the index of all fogs available with minimum response time.

4 Simulation Results

In this work, a new heuristic swarm based technique (WCS) is proposed. A fog based model is designed to simulate the proposed technique. Simulations are implemented on cloud analyst using different service broker policies such as closest data center, optimised response time, configure dynamically with latency and advanced service broker policies. For simulation purpose, the world is divided into six regions with multiple fogs and group of buildings. The allocation and division of fogs and regions is based on the number of users in each region while also considering surface area of the region. Additionally, each fog has 50 VMs to facilitates the users' requests.

Tables 1, 2, 3 and 4 represents the response time for each group of buildings using WCS VM load balancing technique on the basis of aforementioned service broker policies. Additionally, the processing time of the requests from fog is shown in Tables 5, 6, 7 and 8. Moreover, each fog response and process the requests of single group of buildings. These tables show the response and processing time of requests for each group of buildings and fogs as average, minimum and maximum computed time for a day using the proposed VM load balancing technique under all policies.

Table 1. Closest data center

Group of buildings	Avg (ms)	Min (ms)	Max (ms)
G1	94.33	46.56	171.34
G2	66.85	45.22	100.2
G3	95.63	45.66	153.4
G4	97.81	42.23	176.35
G5	118.26	47.1	225.73
G6	119.41	53	198.54

Table 2. Optimal response time

Group of buildings	Avg (ms)	Min (ms)	Max (ms)
G1	104.17	47.06	322.71
G2	66.52	42.46	96.93
G3	95.43	46.67	158.42
G4	97.23	44.06	165.11
G5	121.24	44.05	395.5
G6	133.7	45.88	424.73

Table 3. Configure dynamically with latency

Group of buildings	Avg (ms)	Min (ms)	Max (ms)
G1	121.1	50.89	239.58
G2	83.65	46.01	139.7
G3	123.35	47.3	218.09
G4	127.01	43.91	259.13
G5	118.33	47.1	225.73
G6	119.38	53	198.54

Table 4. Advanced service broker policy

Group of buildings	Avg (ms)	Min (ms)	Max (ms)
G1	121.13	50.89	239.58
G2	88	46.31	148.27
G3	123.28	47.3	218.09
G4	127.02	43.91	259.13
G5	118.27	47.1	225.73
G6	119.35	53	198.54

Table 5. Closest data center

Data center	Avg (ms)	Min (ms)	Max (ms)
Fog1	44.6	2.19	122.45
Fog2	17.26	0.6	47.01
Fog3	46.03	0.98	105.64
Fog4	48.1	1.01	120.08
Fog5	68.4	3.03	171.14
Fog6	69.35	2.06	151.07

Table 6. Optimal response time

Data center	Avg (ms)	Min (ms)	Max (ms)
Fog1	44.57	1	127.4
Fog2	16.91	0.41	49.33
Fog3	45.61	0.98	106.9
Fog4	47.63	1.02	115.01
Fog5	68.03	1.79	165.95
Fog6	65.19	2.25	162.22

Table 7. Configure dynamically with latency

Data center	Avg (ms)	Min (ms)	Max (ms)
Fog1	71.41	3.32	194.06
Fog2	34.06	1.39	88.17
Fog3	73.73	1.55	170.33
Fog4	77.29	1.57	202.85
Fog5	68.48	3.03	171.14
Fog6	69.33	2.06	151.07

Table 8. Advanced service broker policy

Data center	Avg (ms)	Min (ms)	Max (ms)
Fog1	71.41	3.32	194.06
Fog2	38.41	1.44	96.74
Fog3	73.67	1.55	170.33
Fog4	77.30	1.57	202.85
Fog5	68.41	3.03	171.14
Fog6	69.32	2.06	151.07

Table 9. Response time

Group of buildings	ORT	CDL	CDC	ASB
C1	104.17	121.10	94.33	121.13
C2	66.52	83.65	66.85	88.00
C3	95.43	123.35	95.63	123.28
C4	97.23	127.01	97.81	127.02
C5	121.24	118.33	118.26	118.27
C6	133.70	119.38	119.41	119.35

Table 10. Processing time

Data center	ORT	CDL	CDC	ASB
Fog1	44.57	71.41	94.33	121.13
Fog2	16.91	34.06	66.85	88.00
Fog3	45.61	73.73	95.63	123.28
Fog4	47.63	77.29	97.81	127.02
Fog5	68.03	68.48	118.26	118.27
Fog6	65.19	69.33	119.41	119.35

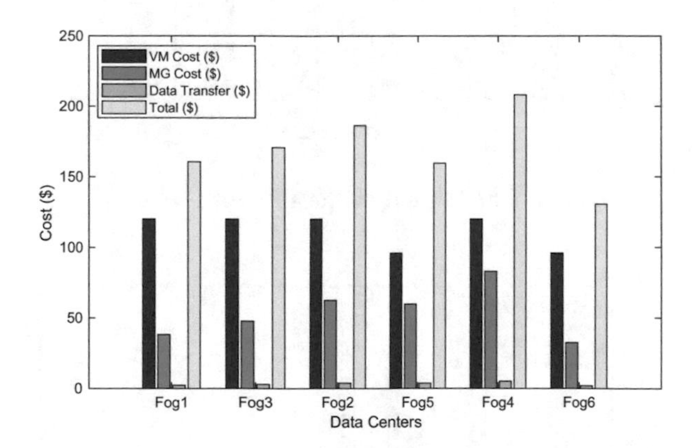

Fig. 2. Closest data center

The bar plots of computed cost for VMs, MGs, data transfer and the aggregated cost, for all aforementioned parameters, is shown in Figs. 2, 3, 4 and 5 for each service broker policy while integrating the proposed VM load balancing algorithm (WCS), respectively.

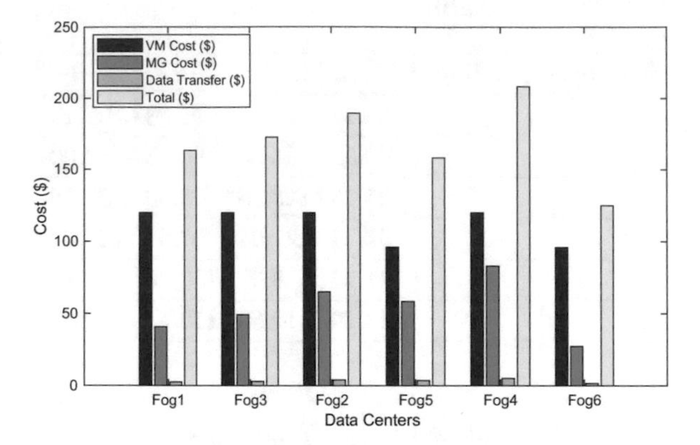

Fig. 3. Optimal response time

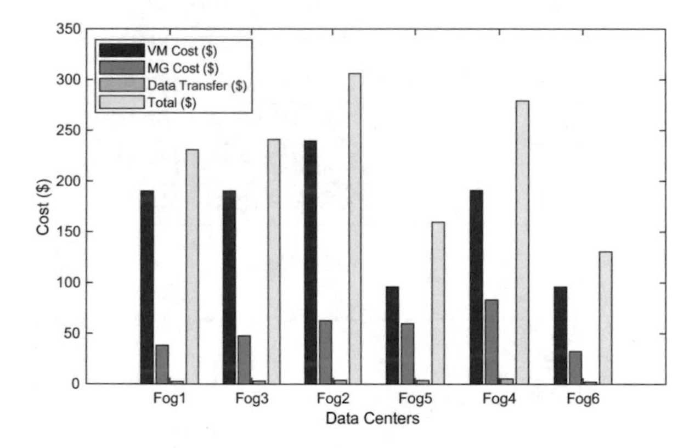

Fig. 4. Configure dynamically with latency

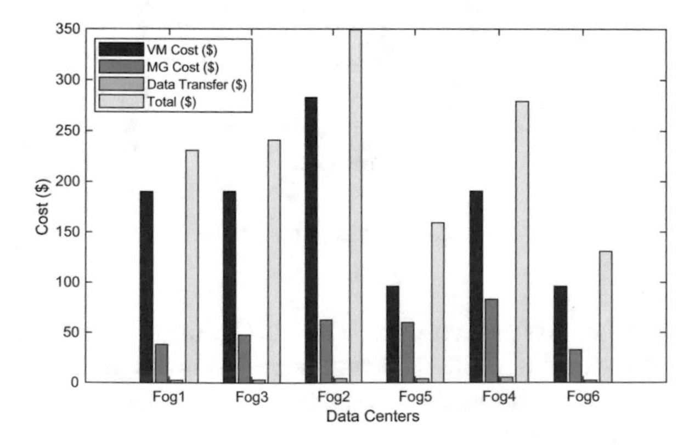

Fig. 5. Advanced service broker policy

Moreover, in Table 9 a comparison in average response time is provided using the proposed technique for all used service broker policies. Additionally, a comparison of average processing time of the load requests is also provided in Table 10. For a fair comparison among different service broker policies all input parameters and the proposed WCS VM load balancing technique are kept same.

5 Conclusion

In this paper, a three-layered architecture of cloud and fog based system is proposed to entertain users' requests. Fog act as an intermediate between the cloud and users. Load balancing and management is one of the major challenges in fog and cloud computing paradigm. It causes the problem of congestion and under utilization of available resources. Without proper scheduling and optimization, VMs are under utilized. The VMs should be assigned to each request according to the nature of its requirements. Hence, WCS load balancing algorithm is proposed and implemented while considering the different service broker policies to efficiently manage the load on VMs of a fog. Eclipse JAVA platform is used for the simulations. Results are presented in the simulation and result section that depict the effectiveness of the proposed model. In future, we are planning to manage and optimize the load on users' side which will contribute in stable and sustainable smart grid. Moreover, we can use the historical data stored on cloud for prediction of users' demand in a particular interval and schedule the power generation of MG accordingly.

References

1. Tani, H.G., El Amrani, C.: Cloud computing CPU allocation and scheduling algorithms using CloudSim simulator. Int. J. Electr. Comput. Eng. **6**(4), 1866 (2016)
2. Manasrah, A.M., Ba Ali, H.: Workflow scheduling using hybrid GA-PSO algorithm in cloud computing. Wirel. Commun. Mob. Comput. (2018)
3. Fatima, I., Javaid, N., Iqbal, M.N., Shafi, I., Anjum, A., Memon, U.: Integration of cloud and fog based environment for effective resource distribution in smart buildings. In: 14th IEEE International Wireless Communications and Mobile Computing Conference (IWCMC-2018), pp. 2–6 (2018)
4. Patel, D., Rajawat, A.S.: Efficient throttled load balancing algorithm in cloud environment. Int. J. Mordern Trends Eng. Res. **2**(3) (2015)
5. Devi, D.C., Uthariaraj, V.R.: Load balancing in cloud computing environment using improved weighted round robin algorithm for nonpreemptive dependent tasks. Sci. World J. (2016)
6. Yasmeen, A., Javaid, N., Iftkhar, H., Rehman, O., Malik, M.F.: Efficient resource provisioning for smart buildings utilizing fog and cloud based environment. In: 14th IEEE International Wireless Communications and Mobile Computing Conference (IWCMC-2018), pp. 1–6 (2018)
7. Chekired, D.A., Khoukhi, L.: Smart grid solution for charging and discharging services based on cloud computing scheduling. IEEE Trans. Industr. Inf. **13**(6), 3312–3321 (2017)

8. Cao, Z., Lin, J., Wan, C., Song, Y., Zhang, Y., Wang, X.: Optimal cloud computing resource allocation for demand side management in smart grid. IEEE Trans. Smart Grid **8**(4), 1943–1955 (2017)
9. Javaid, S., Javaid, N., Tayyaba, S., Sattar, N.A., Ruqia, B., Zahid, M.: Resource allocation using Fog-2-Cloud based environment for smart buildings. In: 14th IEEE International Wireless Communications and Mobile Computing Conference (IWCMC-2018), pp. 1–6 (2018)
10. Mohamed, N., Al-Jaroodi, J., Jawhar, I., Lazarova-Molnar, S., Mahmoud, S.: SmartCityWare: a service-oriented middleware for cloud and fog enabled smart city services. IEEE Access **5**, 17576–17588 (2017)
11. Gautam, P., Bansal, R.: Extended round robin load balancing in cloud computing. Int. J. Eng. Comput. Sci. **3**(8), 7926–31 (2014)
12. Yaghmaee, M.H., Moghaddassian, M., Leon-Garcia, A.: Autonomous two-tier cloud-based demand side management approach with microgrid. IEEE Trans. Industr. Inf. **13**(3), 1109–1120 (2017)
13. Naik, M., Nath, M.R., Wunnava, A., Sahany, S., Panda, R.: A new adaptive cuckoo search algorithm. In: 2015 IEEE 2nd International Conference on Recent Trends in Information Systems (ReTIS), pp. 1–5. IEEE, July 2015
14. Patel, H., Patel, R.: Cloud analyst: an insight of service broker policy. Int. J. Adv. Res. Comput. Commun. Eng. **4**(1), 122–127 (2015)

Foged Energy Optimization in Smart Homes

Ayesha Anjum Butt[1], Nadeem Javaid[1(✉)], Sana Mujeeb[1], Salman Ahmed[2], Malik Muhammad Shahid Ali[2], and Waqar Ali[1]

[1] COMSATS Institute of Information Technology, Islamabad 44000, Pakistan
nadeemjavaidqau@gmail.com
[2] International Islamic University, Islamabad 44000, Pakistan
http://www.njavaid.com/

Abstract. In this paper, Smart Grid (SG) efficiency is improved by introducing Cloud-based environment. To access the services and hostage of cloud large number of requests are entertained from Smart Homes (SHs). These SHs exists in clusters of smart buildings. When the number of requests increase, delay, latency and response time also increase. To overcome these issues, Fog is introduced, which act as an intermediate layer between the cloud and consumer. Five Micro Grids (MGs) are attached to each cluster of the smart building to manage its requests. By using Fog base environment, the delay and latency decreases. The response time also increases with less processing time. To handle the load on cloud different load balancing algorithms and service broker policies exist. In order to manage the load, Honey Bee (HB) is implemented. HB is compared with existing algorithm Round Robin (RR). It gives better results than RR.

Keywords: Fog · Smart Homes · Load balancing
Service Broker Policy · Round Robin · Honey Bee

1 Introduction

To overcome the limitations of traditional grid stations, Smart Grid (SG) is introduced. To use the electricity efficiently and effectively, SG is used. SG performs work by using the Smart Meter (SM), smart appliances, and renewable energy resources in an efficient manner. It helps to manage a load of consumers and utility through Demand Response (DR), which balances the load from on-peak hours to off-peak hours without affecting the users' comfort level. The growth in Information and Communication Technologies become the revolutionary change in the power sector which realizes the importance of SG. Demand Side Management (DSM) is the important key factor of the SG because its scenario is implemented on the consumer side. Residential, industrial and commercial side lie on the consumer side.

L. Barolli et al. (Eds.): IMIS 2018, AISC 773, pp. 265–275, 2019.
https://doi.org/10.1007/978-3-319-93554-6_24

According to proposed scheme, residential side is discussed as Smart Homes (SHs). The main aim of the SG is to manage a load of SHs during peak hours, save energy, cost minimization, and less delay. SG effectively handles the load balancing problem on the consumer side by deploying the energy management system. When the discussion is about SHs of a large number of area and six regions of the world; it becomes difficult to manage the load in the SHs. It also becomes difficult to communicate the grid of other regions when consumers demand is not fulfilled in their own region. It is because, consumer wants more resources and a quick response from the grid in less time without considering maximum cost and without any delay. To overcome these issues in the SG field and also to enhance the ability and efficiency of SG, the concept of Cloud Computing is introduced. It provides the global solution of the DR on the cloud [1] and the services of sharing and storage of electricity from SH to cloud and SH to SH. It also plays an impactful role in load balancing. The cloud which handles the requests from consumers' side becomes more efficient and reliable when it handles a couple of requests from the consumers side.

By using Fog in their system, the efficiency of the cloud is also increased. The latency and the range of bandwidth are also improved. Fog acts as an intermediate layer between cloud and cluster of SHs at consumer side [2]. Fog is used as a distributed, according to geographical nature. SM is also introduced to enhance the efficiency of Fog. By using SM, data of each cluster can easily monitor or store according to the demand; Then this data can easily transmit to the cloud by using Fog. It reduces the latency, delay of the requests and increases the processing time. The response is generated from Fog to SH is minimum then the response which is generated from cloud to SH. Because of its closeness to the consumer side, it can give direct access to the consumer to the source. Fog's distributive nature also helps to improve the privacy of the consumer's data in two categories, private and public data. It also improves the reliability of SG which is directly connected to the Fog and actually entertains the requests of the Fog.

To manage a load of SHs, by using Fog environment, different load balancing algorithms with different service broker policies are used. These broker policies with load balancing algorithms manage the electric load of SHs of different cluster size, and the number of requests on FOG. Different Virtual Machines (VMs) and Fog are used to manage these loads. For this purpose, six regions of the world are used. In this paper, to balance the load using Fog environment, Honey Bee (HB) algorithm is implemented and compared with existing algorithm Round Robin (RR). After evaluating the results of HB and RR, it has been seen that our implemented HB has minimum average time and less response time than RR. It is also compared HB on the basis of existing four services broker policies: Closet DC (CDC), Optimize Response Time (ORT), Reconfigure Dynamically with Load (RDL) and Advance Service Broker Policy (ASP) in [3].

The rest of the paper is organized as follow, Sect. 2 defines the related work of the paper, Sect. 3 defines the proposed system model, then implemented algorithm HB, the last Sect. 4 discusses the results and simulations of the paper after performing simulations by using cloud analyst simulator. Section 5 defines the conclusions of this paper.

1.1 Motivation

In [1], the cloud base environment is deployed to manage the communication between the consumers and SG. It improves the efficiency in SG environment. On the other hand, it fails to handle the issue of delay and latency. Because, when bandwidth is low then all cloud based system becomes slow. When there is On-peak hours, the delay increases due to large number of requests. In [2], energy management for home and MG is described by using two prototypes. HEM prototype and MG level prototype. These are implemented by using Fog based environment. They give the reference of cloud, however, cloud in not integrated with their Fog based environment. When their proposed system will be implemented for large number of homes, in that case the Fog services become limited and cannot be able to fulfill the consumers' demands. Load Balancing and assigning of Fog among the SHs are the required steps to manage the energy efficiency. When we implement load balancing algorithms and policies with different scenarios in order to achieve the energy efficiency and cost minimization [3]. In this paper, ASP is proposed and implemented which results in minimum delay. The reasons behind this are: it is implemented on large scale (six regions), and because of maximum number of requests, limited number of Fog are assigned for processing the requests. So, it is difficult to manage that which Fog is assigned for entertaining the requests.

1.2 Contribution

The contributions of existing work are defined as follows:

- In this paper, two regions are taken to simulate the results, and these regions are North America, South America.
- To improve the efficiency of cloud, we introduce Fog in our proposed work.
- Ten Fogs are taken in account for the selected regions and clusters of smart buildings in order to improve the efficiency of the Fog.
- We have selected 10 clusters in each region.
- MG is attached to the clusters, the size of MG depends on the size of the cluster which increases the efficiency of Fog based environment.
- Population based optimization technique HB is implemented to check the effectiveness of this system.
- After implementation, this technique gives better results than the existing technique RR.

By applying different load balancing algorithms on different VMs; they find that load balancing in cloud can be improved by using these algorithms. HB has less processing time and maximum response time rather than RR.

2 Related Work

Milani *et al.* [4] perform survey on different techniques of load balancing algorithm then describe the future trend of those algorithms and techniques. To perform the survey for the load balancing; different VMs are used to evaluate the results. By applying different load balancing algorithms on different VMs; they find that load balancing in cloud can be improved by using these algorithms. If hybrid load balancing scheme is applied, then it gives more efficient results than other load balancing techniques. After adopting these techniques, carbon dioxide emission is decreased and efficiency is improved. On the other hand, implementation cost may be high than the expected cost.

Cloud Load Balancing (CLB) novel based load balancing algorithm is proposed by Chen *et al.* [5] they also compared their proposed algorithm with the previous existing algorithms like RR, Min-Max, etc., algorithms. In their work, they check the efficiency, reliability, and scalability of the cloud based system on the basis of their proposed CLB algorithm. Then they compare their work with previous existing load balancing algorithm. They check the results of their defined parameters on the basis of load on web server and load on cloud. The implemented algorithms were applied on physical systems and web servers. After experiments, they achieve that physical system are efficient and scalable than web servers on the basis of their defined parameters. They achieve that physical system are scalable then web servers. However, authors can calculate the response time, average time, and cost of both systems.

A survey performed on different techniques of Particle Swarm Optimization (PSO) is described by Masdari and Salehi in [6]. They define that which type of PSO technique can perform better on which type of VM scenario. They elaborate the limitations of these techniques in different scenarios on the basis of cost and other parameters. The measuring objectives that used for their experiments are scheduling Parameters: time base parameters, and cost based parameters. After implementation, different techniques achieve efficient results on different load balancing scenarios. However, it is not defined by them what they will do when load increases.

Manasrah and Smadi proposed a Variable Service Broker Policy (VSBP) in [7]. This policy based on meta heuristic technique. By using communication channel bandwidth, latency, and job size; they aimed to reduce the response time of the DCs that occur because of the load on DCs. To achieve the results, according to their technique; they consider response time and processing time to show the efficiency of their proposed broker policy. They compare their results with previous existing policies to show the differences. For this purpose, they compare their proposed policy on the basis of different scenarios which contain various service broker policies. After implementation, they achieve minimum response time by DCs through considering the size of requests. Their policy also shows less processing time then existing techniques. On the other hand, they fail to describe the cost consumed by their implemented scheme.

New Service Broker Policy (SBP) on RR was proposed by Radi in [8]. In this work, author wants to propose SBP for large scale; then implemented a policy and compare it with three existing policies on the basis of various VMs and load balancing algorithms. For this implementation and evaluation purpose cloud analyst simulator is used. Author wants to improve the overall average response time then existing policies. For achieving better results author places various number of DCs at different regions with numerous VMs to calculate the results. These results were conducted in different time period like on peak and off peak hours. After implementation, author achieves improvement in overall average response time. However, author does not discuss the processing time and cost consumed by proposed policy.

K-Nearest-Neighbor Based service policy is implemented by Irbid and Jordanin [9]. In this paper, they implement this policy on the basis of DC characteristics to know that how DC responds the user's requests. By using the policy; they want to evaluate the average processing and response time. For this purpose, they used five DCs in closed environment and placed these DCs at different regions to check the average response and processing time of that DCs. To run this scenario, they use cloud analyst tool to get the results. After run the simulations, they achieve better results on the basis of proposed modification of Routing policy. This is because that DCs are placed near to each other. However, cost consumed by their implemented modified policy was not defined by them. They also does not get the value of K means during their work.

In [10], Anila *et al.* proposed a load balancing algorithm known as particle swarm optimization with simulated annealing (PSOSA). The proposed algorithm is then compare with existing algorithms RR and Throttled (TH). In this paper authors also proposed a model which is known as three layered model. This model contains cloud layer, fog layer, and end user layer. For implementation purposes, they are considering main input parameters: average processing and response time, cost on the basis of 24 h slot. They considered buildings to calculate the simulations. After implementation, they achieved that their proposed algorithm performs better than existing RR and TH. However, they can consider numerous buildings by using two or more regions.

Fog 2 cloud based frame work is presented by Sakina *et al.* in [11]. In this paper, they considered residential side to improve the services of cloud in terms of latency. They compare the results of 24 h time slot by using Shortest Job First (SJF) algorithm. The input parameters are: response time, request per hour, and cost. Results are compared to the existing two algorithms RR and Equally Spread Current Execution (ESCE). After implementation, their techniques achieve outperform results than existing techniques. However, they can also implement broker policies to improve their results more.

3 Proposed System Model

In this paper, Fog based environment is proposed. Cisco has introduced a concept in January, 2014 which is known as Fog. To overcome the deficiencies of Cloud, in recent years, Fog concept was introduced. The proposed system model is shown in Fig. 1, which describes the key features and functionality of our Fog based environment. In this system, we used ten clusters with ten Fogs. Our clusters contain multiple number of smart buildings in the range of (10–100). These buildings also contain numerous SHs or consumers in the range of (100 to 1000).

In this paper, a Fog based environment is proposed which consists of multiple numbers of Fog and VMs. How buildings connect with Fog and how VM are allocated to consumers' according to their requirements. When consumer send the request to the Fog the Fog assigns VM and Fog for users according to their requirement. However, users can send their requests or can communicate only through Fogs.

The proposed model is basically Fog based environment, so the components of Fog are defined in Table 1, "Components of Fog". This table defines all the components of the Fog which are used in this system model. It describes that total ten number of Fogs exists in this implemented system. Five Fog lies in region 0 and other five lies in region 1. MG also sends request to the Fog according to the consumer's requirement. Each region contains multiple number of Fogs, whose ranges lie between 1 to 5. These MGs send request to the regional Fog, and the other end they fulfill the user's requirement through Fog.

Table 2 defines the specification of the Fog. It defines that in this model, there are ten number of Fogs, each Fog has two VMs, which is connected with Fog to fulfill the consumers' requirement. The memory used by VM of the Fog is 512, and the BW and image size is 1000. The Xen is the VM Manager (VMM) of the Fog, the Operating System (OS) used by Fog VM is Linux and the architecture is X86. To implement this proposed system model, two load balancing algorithm RR and implemented HB are used with four service broker policies that are CDC, ORT, RDL, and ASP. The purpose of service broker policies is to manage the Fog, means that which Fog is nearer to which cluster or SH, or when will it assign to which one. In this proposed system model, two regions are considered region 0 and 1; North America and South America. These two regions is taken in this proposed system because they have large population. That's why they also have large number of buildings and SHs.

4 Results and Simulations

In this section, algorithms, policies, scenario, and scenario of this paper are defined which are done by using the Cloud Analyst tool; which also helps to compare the results of four service broker policies; CDC, ORT, RDL, and ASP in order to check the correctness of aforementioned parameters. To evaluate the results of service broker policies; two load balancing algorithms are compared. To describe the results of service broker policy, ten clusters with the same number of Fogs are used which are placed in two different regions of the world. These clusters are connected to Fog and by using this Fog platform. The division of regions are shown in Fig. 2.

In Fig. 3a, average processing time of implemented load balancing algorithm HB is compared with existing load balancing algorithm RR on the basis of service broker policies. However, the average processing time is calculated on the basis of CDC, ORT, RDL, and ASP policies. It is concluded that, on CDC, HB has 5.70% and RR has 2.77% average response time. Furthermore, on ORT, HB has 5.74% and RR has 2.78% response time. HB has response time 44.32% and RR

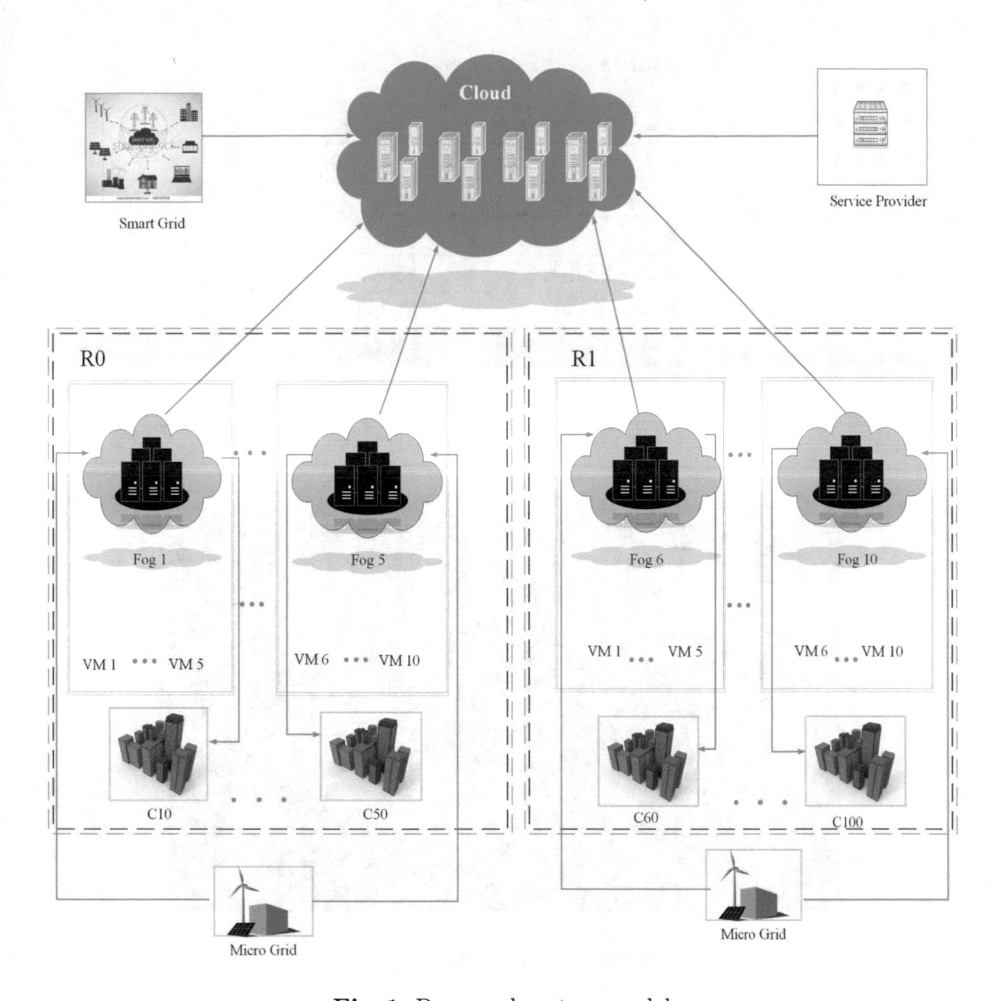

Fig. 1. Proposed system model

Table 1. Components of Fog

Number of Fog	Regions	MG
F1	0	2
F2	1	4
F3	0	2
F4	1	4
F5	0	5
F6	1	4
F7	0	2
F8	1	4
F9	0	3
F10	1	1

Table 2. Specification of Fog

Specification of Fog	Range
Fog	1 to 10
No. of VMs	2
Memory	512
Band Width (Bw)	1000
Image Size	1000
VMM	Xen
Operating System (OS)	Linux
Architecture (Arch.)	X86

Fig. 2. Division of world map

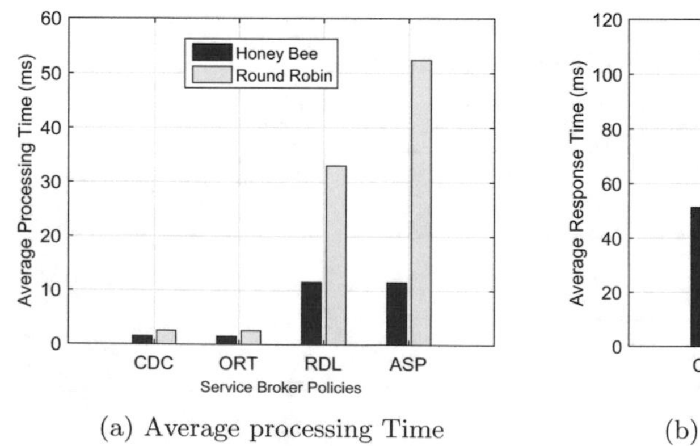

(a) Average processing Time

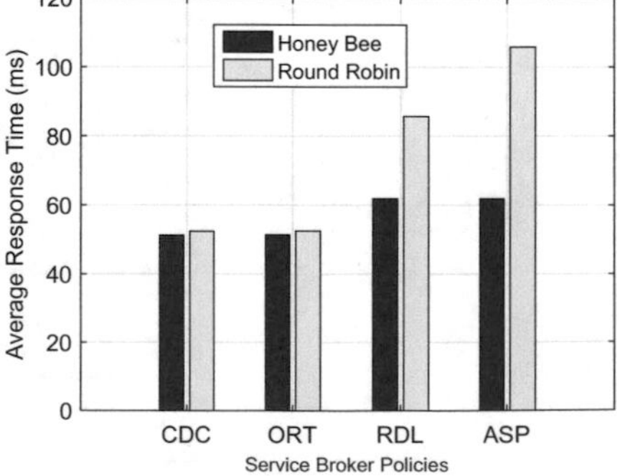

(b) Average Response Time

Fig. 3. Average processing and response time

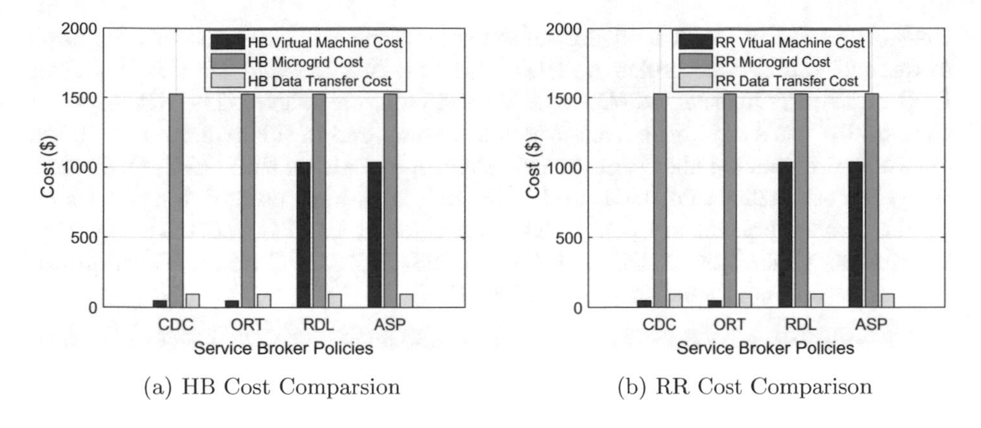

(a) HB Cost Comparsion (b) RR Cost Comparison

Fig. 4. HB and RR cost comparison

has 36.52% on RDL. On the basis of ASP 44.21% and RR has 57.90% average response time. In Fig. 3b, the HB load-balancing algorithm is compared with existing RR in order to evaluate the average response time. After evaluating the results, it is concluded that on the basis of CDC, HB has 51.24% and RR has a 17.65% average response time. On ORT, HB has 51.35% and RR has 17.70%. Besides, on the topic of RDL, HB has 27.34% and RR has 28.93%. Moreover, the HB has 27.327% and RR has 17.65% average response time. It is clearly shown that on the basis of CDC and ORT, HB has better response time than RR while RR has better response time on the basis of RDL and ASP.

In Fig. 4a, cost of HB is compared by using three parameters: VM cost, MG cost, data transfer cost (DT). VM cost of CDC and ORT is less than RDL and ASP. DT cost is less than VM and GM on all aforementioned policies. The overall data transfer cost of HB is minimum because it selects best local and global path to transfer the data. That's why it reduces the cost. In Fig. 4b,

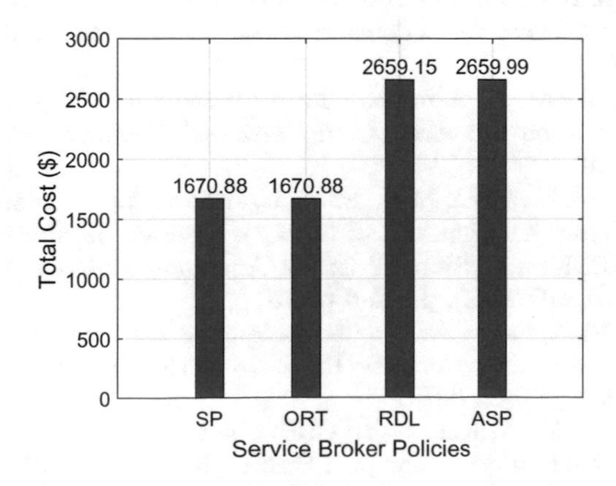

Fig. 5. Total cost

Cost of RR is compared by using three parameters: VM, MG, DT on the basis of defined four service broker policies. VM cost of CDC and ORT is less than DLR and ASP. On other side DT cost is less in all four policies. The RR response time on the basis of four policies, which are compared in this paper on the basis of two load balancing algorithm, in this Figure, we can see that CDC, ORT have less VM cost rather than RDL and ASP. In Fig. 5, total cost of described four service broker policies are compared. The total of CDC is 16070.8%, ORT is 16070.88%, RDL is 2659.15% and ASP is 2659.99%. CDC and ORT minimize total cost % as compared to DLR and ASP.

5 Conclusion and Futurework

In this paper, load balancing with HB is implemented to compute the energy consumption of SHs in the cluster of smart buildings. The implemented HB is compared with existing RR load balancing algorithm. These load balancing algorithms are compared on the basis of aforementioned policies: CDC, ORT, RDL, and ASP. For the simulations, three input parameters are used: average response time, average processing time, and cost. To calculate the results of SHs on the basis of their requests; two VMs are assigned to the Fog which SHs exist in ten clusters. After implementation, it has been concluded that on the basis of ASP, HB has 44.21% better average processing time than RR. When average response time is calculated, it is shown that on CDC, HB has 51.24%, on ORT and ASP, HB has 51.35% and 27.327% average response time. Overall, HB has better results than RR on the basis of CDC, ORT, and ASP. In future, we will implement this work for multiple regions to minimize latency and affordable delay by applying different load balancing algorithms.

References

1. Moghaddam, M.H.Y., Leon-Garcia, A., Moghaddassian, M.: On the performance of distributed and cloud-based demand response in smart grid. IEEE Trans. Smart Grid 1–15 (2017)
2. Okay, F.Y., Ozdemir, S.: A fog computing based smart grid model. In: International Symposium on Networks, Computers and Communications (ISNCC), pp. 1–6. IEEE (2016)
3. Fatima, I., Javaid, N., Iqbal, M.N., Shafi, I., Anjum, A., Memon, U.: Integration of cloud and fog based environment for effective resource distribution in smart buildings. In: 14th IEEE International Wireless Communications and Mobile Computing Conference (IWCMC-2018), pp. 2–6 (2018)
4. Milani, A.S., Navimipour, N.J.: Load balancing mechanisms and techniques in the cloud environments: systematic literature review and future trends. J. Netw. Comput. Appl. **71**, 86–98 (2016)
5. Chen, S.L., Chen, Y.-Y., Kuo, S.-H.: CLB: a novel load balancing architecture and algorithm for cloud services. Comput. Electr. Eng. **58**, 154–160 (2017)
6. Masdari, M., Salehi, F., Jalali, M., Bidaki, M.: A survey of PSO-based scheduling algorithms in cloud computing. J. Netw. Syst. Manag. **25**(1), 122–158 (2017)

7. Manasrah, A.M., Smadi, T., ALmomani, A.: A variable service broker routing policy for data center selection in cloud analyst. J. King Saud Univ. Comput. Inf. Sci. **29**(3), 365–377 (2017)
8. Radi, M.: Efficient service broker policy for large-scale cloud environments. arXiv preprint arXiv:1503.03460 (2015)
9. Al Sukhni, E.: K-nearest-neighbor-based service broker policy for data center selection in cloud computing environment. Int. Res. J. Electron. Comput. Eng. **2**(3), 5–9 (2016)
10. Yasmeen, A., Javaid, N., Iftkhar, H., Rehman, O., Malik, M.F.: Efficient resource provisioning for smart buildings utilizing fog and cloud based environment. In: 14th IEEE International Wireless Communications and Mobile Computing Conference (IWCMC-2018), pp. 1–6 (2018)
11. Javaid, S., Javaid, N., Tayyaba, S., Sattar, N.A., Ruqia, B., Zahid, M.: Resource allocation using Fog-2-Cloud based environment for smart buildings. In: 14th IEEE International Wireless Communications and Mobile Computing Conference (IWCMC-2018), pp. 1–6 (2018)
12. Wickremasinghe, B., Buyya, R.: CloudAnalyst: a CloudSim-based tool for modelling and analysis of large scale cloud computing environments. MEDC Proj. Rep. **22**(6), 433–659 (2009)

Short Term Load Forecasting based on Deep Learning for Smart Grid Applications

Ghulam Hafeez[1], Nadeem Javaid[1](✉), Safeer Ullah[1], Zafar Iqbal[2], Mahnoor Khan[1], Aziz Ur Rehman[1,3], and Ziaullah[4]

[1] COMSATS Institute of Information Technology, Islamabad 44000, Pakistan
nadeemjavaidqau@gmail.com
[2] PMAS Arid Agriculture University, Rawalpindi 46000, Pakistan
[3] The University of Lahore, Gujrat Campus, Gujrat, Pakistan
[4] Huazhong University of Science and Technology, Wuhan, China
http://www.njavaid.com

Abstract. Short term load forecasting is indispensable for industrial, commercial, and residential smart grid (SG) applications. In this regard, a large variety of short term load forecasting models have been proposed in literature spaning from legacy time series models to contemporary data analytic models. Some of these models have either better performance in terms of accuracy while others perform well in convergence rate. In this paper, a fast and accurate short term load forecasting framework based on stacked factored conditional restricted boltzmann machine (FCRBM) and conditional restricted boltzmann machine (CRBM) is presented. The stacked FCRBM and CRBM are trained using rectified linear unit (RelU) and sigmoid functions, respectively. The proposed framework is applied to offline demand side load data of US utility. Load forecasts decide weather to increase or decrease the generation of an already running generator or to add extra units or exchange power with neighboring systems. Three performance metrics i.e., mean absolute percentage error (MAPE), normalized root mean square (NRMSE), and correlation coefficient are used to validate the proposed framework. The results show that stacked FCRBM and CRBM are accurate and robust as compared to artificial neural network (ANN) and convolutional neural network (CNN).

1 Introduction

The energy demand is increasing rapidly with the increase in population. The current net energy demand of commercial and residential sector is about 40%–50% in developed countries of the world [1]. In order, to meet this increasing energy demand the entire globe is transitioning from traditional grid to smart grid (SG). The SG has the ability to forecast, monitor, plan, schedule, and make real time decisions regarding generation and consumption of energy.

L. Barolli et al. (Eds.): IMIS 2018, AISC 773, pp. 276–288, 2019.
https://doi.org/10.1007/978-3-319-93554-6_25

One of the features of SG is advanced metering infrastructure which realized the active participation of both utility and consumers in electricity market. The goals and objectives of SG are to improve efficiency and effective utilization of electricity framework. Modeling and forecasting load realized the achievement these objectives.

For SG distribution side performance optimization a proper decision making is necessary which leads to reduce cost, alleviate peaks, and power losses. The current research keeping the aforementioned objectives in mind perform power scheduling using optimization techniques [2,3]. However, prior to optimization techniques based scheduling, load forecasting is necessary for efficient energy management. Typically, load forecasting have three categories: (a) short term load forecast, ranging from hours to week, (b) medium term forecast, for a duration of weeks to year, and (c) long term forecast, for a duration more than one year. Especially, the focus of this work is on short term load forecasting, it is a challenging task due to stochastic and non-linear consumption behavior of consumers. Many short term load forecasting models have been proposed in [4–11]. However, these models have either accuracy or convergence rate problems. In [12], authors used ANN based forecasting model to reduce the error performance and improve the forecast accuracy. However, while improving forecast accuracy the computational complexity and execution time is compromised due to tradeoff between accuracy and convergence rate.

The contribution of this paper is twofold, first, we propose a new way to adopt stacked factored conditional restricted boltzmann machine (FCRBM) model with aim to learn the non-linear and stochastic electrical patterns from offline data. Second, a short term load forecasting model is proposed based on the stacked FCRBM and CRBM to improve the relative forecast accuracy. The proposed model is validated by comparing with existing short term load forecasting models based on ANN and CNN in terms in terms of mean absolute percentage error (MAPE), normalized root mean square error (NRMSE), and correlation coefficient. We adopt a moduler strategy for short term load forecasting in which the output of each former module is fed into the later module. In short, our system model comprises of three modules data processing and feature extraction module, deep learning-based training module, and deep learning-based forecasting module.

The organisation of the paper is as follows: In Sect. 2, we introduce proposed methods. Section 3 demonstrates proposed architecture. In Sect. 4, simulation results are provided. Finally, the paper is concluded in Sect. 5.

2 Proposed Methods

In this section, we introduce deep learning techniques CRBM and stacked FCRBM for short term load forecasting. For the aforementioned deep learning techniques, we describe three ingredients i.e., error function, conditional probability, and update/learning rules. Error function of a given network provides scalar values that are essential for the configuration. Conditional probability calculates the probability of an event over specific condition. Update/learning rules are required for tuning of the free parameters.

2.1 CRBM

CRBM [13] is an amendment in the RBM [14]. It is a probabilistic model used to model human activities, weather data, collaborative filtering, classification, and time series data [15]. In this paper, we use CRBM having three layers visible layer, hidden layer, and conditional history layer. The generic infrastructure of CRBM is shown in Fig. 1. The detailed description of CRBM three ingredients such as error function, conditional probability, and learning rules are as follows.

Error Function: The error function express the possible correlations between input, conditional history layer, hidden layer, and output. The error function is calculated as:

$$E(v, u, h; w) = -(v^T w^{vh} h + u^T w^{uv} v + u^T w^{uh} h + u^T a + h^T b), \quad (1)$$

where $v = [v_1, v_2, ..., v_n]$ is the real valued vector having visible unit neurons from 1 to n neuron, $u = [u_1, u_2, ..., u_n]$ shows the real valued vector having history neurons from 1 to n, $h = [h_1, h_2, ..., h_n]$ denotes the binary vector having hidden neurons from 1 to n, w is the weight matrix, a is the visible layer bias, and b is the hidden layer bias. The weight matrix w^{vh} is bidirectional while the weight matrices w^{uh} and w^{uv} are unidirectional.

Conditional Probability: Conditional probability in CRBM determines the probability distribution over two inferences. First inference $p(h/v, u)$, is to determine the probability of hidden layer inferenced on all the layers, while the second inference $p(v/h, u)$, is to determine the probability of visible layer conditioned on all the other layers. The two inferences are leading to:

$$p(h/v, u) = sigmoid(u^T w^{uh} + v^T w^{vh} + b) \quad (2)$$

$$p(v/h, u) = sigmoid(w^{uv} u^T + w^{vh} h + a) \quad (3)$$

Weights and Biases Learning and Update Rules: We use a stochastic gradient decent method for learning and updating the weights and biases of the layers because other methods some time have the problem of vanishing gradient which made the network hard to train. The parameters are fine tuned by maximizing the probability function and the weights and biases metrics are updated to minimize the gap between real and forecasted value.

The weights are updated as follows:

$$\begin{aligned} w_{t+1}^{uh} &= w_t^{uh} - \eta \frac{\partial E}{\partial w^{uh}} \\ w_{t+1}^{uv} &= w_t^{uv} - \eta \frac{\partial E}{\partial w^{uv}} \\ w_{t+1}^{vh} &= w_t^{vh} - \eta \frac{\partial E}{\partial w^{vh}}. \end{aligned} \tag{4}$$

The biases are updated by the following Equation:

$$\begin{aligned} a_{t+1} &= a_t - \eta \frac{\partial E}{\partial a^v} \\ b_{t+1} &= b_t - \eta \frac{\partial E}{\partial b^h} \end{aligned} \tag{5}$$

where η is the learning rate and t is the iteration number. The aforementioned procedure is repeated for the number of epochs until to converge the model.

2.2 Stacked FCRBM

FCRBM is an extension of the CRBM introduced by Taylor and Hinton in [15]. In FCRBM [16], they add the concept of factor and styles to mimic multiple human actions. We propose a new way to adopt deep learning technique i.e., stacked FCRBM for short term load forecasting, where the successive layers take output from the previous trained layers to overcome the problem of overfitting and to improve the forecast accuracy. The stacked FCRBM comprised of four layers as shown in Fig. 2: (a) visible layer v, (b) history layer u, (c) hidden layer h, and (d) style layer y.

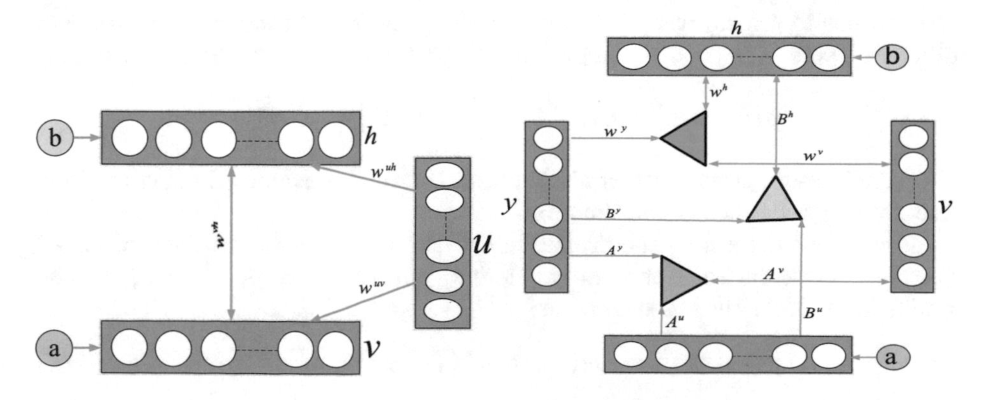

Fig. 1. Generic infrastructure of CRBM

Fig. 2. Generic infrastructure of stacked FCRBM

The visible and history layers are real valued while the hidden layer is binary. The visible layer is responsible for encoding the present time series data to forecast the future value, while the history layer will encode historical time series data. Hidden layer is responsible for the discovery of significant features required for analysis. The different styles and parameters essential for forecasting are embedded into the style layer. The relation and interaction between the layers, weights, and factors is expressed by an error function as:

$$E(v, u, h; w) = -v^T \hat{a} - h^T \hat{b} - \sum \{(v^T w^v) \circ (y^T w^y) \circ (h^T w^h)\} \quad (6)$$

where E is the error function, $v^T w^v$ is the visible factored, $y^T w^y$ is the style factored, and $h^T w^h$ is the hidden factored. It $() \circ ()$ is the product known as hadamard product, in which the product operation is element wise. The elements $\hat{a}$ and $\hat{b}$ represent dynamic biases for visible and hidden layers, respectively, which are defined as follows:

$$\begin{aligned} \hat{a} &= a + A^v \{ (u^T A^u) \circ (y^T A^y) \}^T \\ \hat{b} &= b + B^h \{ (u^T B^u) \circ (y^T B^y) \}^T \end{aligned} \quad (7)$$

where w^v, w^y, w^h are weights of the corresponding layers and A^v, A^u, A^y, B^h, B^u, B^y are the connections of the corresponding layers to factors, are known as model free parameters.

2.3 Conditional Probability

In case of stacked FCRBM, conditional probability determines probability distribution of one layer conditioned over all the remaining layers. In first case, we define probability distribution of hidden layer conditioned over all the remaining layers $p(h|v, u, y)$. The restriction is that there is no intra-connection between the neurons in any layer while there is inter-connection between the neurons of different layers. The conditional probability of hidden layer can be calculated as:

$$p(h|v, u, y) = \text{RelU}\left[\hat{b} + w^h \{ (v^T w^v) \circ (y^T w^y) \}\right] \quad (8)$$

For all inputs probability of hidden layer neurons is evaluated using rectified linear unit (RelU) activation function.

Finally, we determine the probability of the visible layer conditioned on all the reaming layers such as history, hidden, and style layers $p(v|h, u, y)$. The visible layer probability is defined as:

$$p(v|h, u, y) = \text{RelU}\left[\hat{a} + w^v \{ (h^T w^h) \circ (y^T w^y) \}\right] \quad (9)$$

2.4 Stacked FCRBM Weights and Biases Learning Rules

We adopt stochastic gradient decent for learning and updating rules because to overcome the problem of vanishing gradient. Moreover, the stochastic gradient decent on a large dataset converge faster and avoid overfitting as compared to mini-batch.

The weights of corresponding layers are updated as:

$$w_{t+1}^h = w_t^h - \eta \frac{\partial E}{\partial w^h}$$
$$w_{t+1}^y = w_t^y - \eta \frac{\partial E}{\partial w^y} \tag{10}$$
$$w_{t+1}^v = w_t^v - \eta \frac{\partial E}{\partial w^v}$$

The connections and biases are updated as follows:

$$A_{t+1}^u = A_t^u - \eta \frac{\partial E}{\partial A^u}$$
$$A_{t+1}^v = A_t^v - \eta \frac{\partial E}{\partial A^v} \tag{11}$$
$$A_{t+1}^y = A_t^y - \eta \frac{\partial E}{\partial A^y}$$
$$B_{t+1}^u = B_t^u - \eta \frac{\partial E}{\partial B^u}$$
$$B_{t+1}^h = B_t^h - \eta \frac{\partial E}{\partial B^h} \tag{12}$$
$$B_{t+1}^y = B_t^y - \eta \frac{\partial E}{\partial B^y}$$
$$\hat{a}_{t+1} = \hat{a}_t - \eta \frac{\partial E}{\partial v}$$
$$\hat{b}_{t+1} = \hat{b}_t - \eta \frac{\partial E}{\partial h} \tag{13}$$

3 Proposed Architecture

In this section, we introduce our proposed fast and accurate short term load forecasting model based on deep learning techniques such as stacked FCRBM and CRBM as shown in Fig. 3. We adopt the modular strategy for short term load forecasting based on deep learning techniques, in which the output of each former module is fed into the later module. In short, our system model consists of three modules data processing and feature extraction module, deep learning-based training module, and deep learning-based forecasting module. The detailed description is as follows:

3.1 Data Processing and Feature Extraction Module

First the twenty zones historical data of US utility consisting of hourly load and weather data is taken from the Kaggle repository. This data is given as an input to the data processing and feature extraction module. The three data operations: cleansing, normalization, and structuring are performed on the received data. The cleansing operation is performed in order to replace the missing and defective values by the mean of the previous values. After cleansing, the data is normalized in order to reduce and eliminate the redundancy. Moreover, the

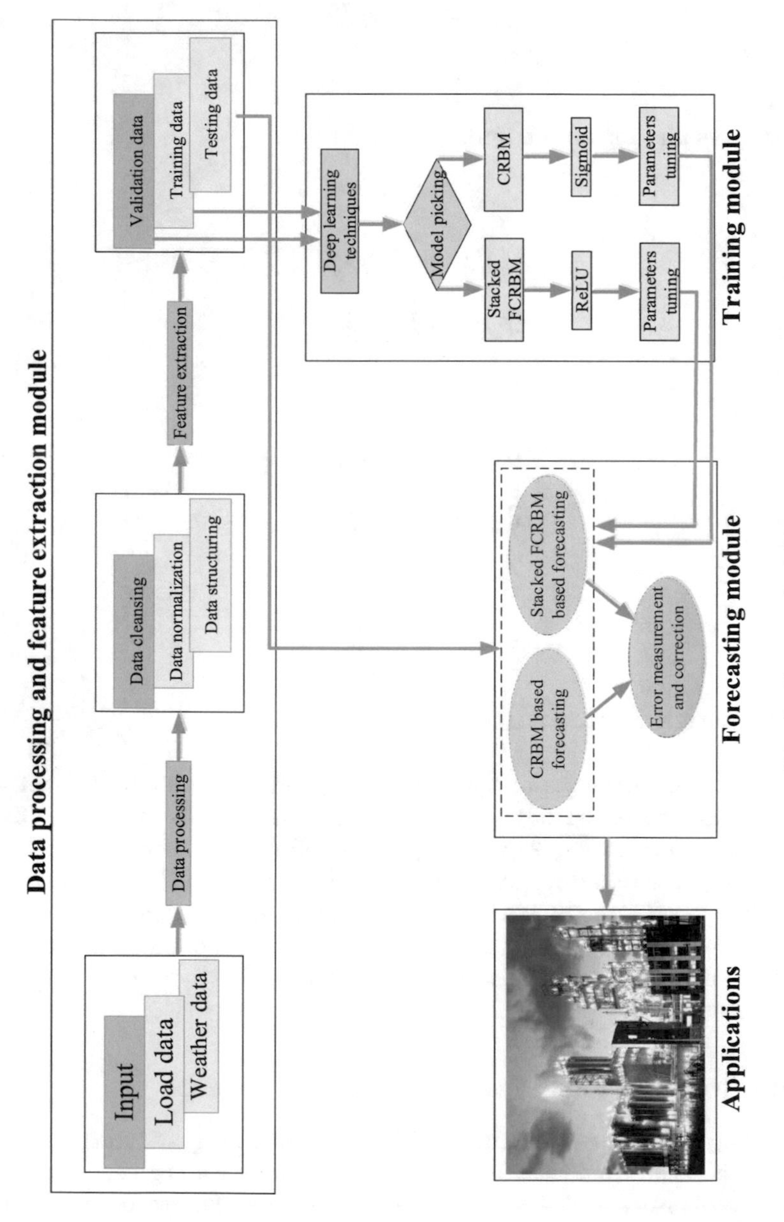

Fig. 3. Proposed architecture

data has large values, the normalization is performed to make the weighted sum within in the limits of the sigmoid function. At the end, we denormalize the data to achieve the desired load predictions. After cleansing and normalization, the data is structured in ascending or descending order. The desired features from the dataset is extracted by feature extraction process and finally the data is split into training and testing dataset. The training data have hourly load and weather data to train the deep learning techniques such as CRBM and stacked FCRBM. The testing data is used to evaluate the accuracy of the proposed deep learning techniques in terms of MAPE, NRMSE, and correlation coefficient. For the proper parameters tuning, we construct validation dataset.

3.2 Training Module

Deep learning techniques such as CRBM and stacked FCRBM based training is the main part of this architecture. These techniques are trained with the training data, they learn non-linear relationship between demand load profile and historical observations. The output of the data processing and feature extraction module is given as an input to the training module. This module takes the training data and chose one of the models CRBM or stacked FCRBM for training. If the chosen model is stacked FCRBM, training module will train the stacked FCRBM using RelU activation function because it overcomes the problems of vanishing gradient and curse of dimensionality. If the model chosen for training is CRBM training module will train it with sigmoidal activation function. In this way, the deep learning-based training module is enabled via learning to forecasts the future load.

3.3 Forecasting Module

The output of the training module is fed into the forecasting module on the basis of trained deep learning models, the forecasting module forecasts the future load. The accuracy of the proposed deep learning techniques is evaluated in terms of MAPE, NRMSE, and correlation coefficient using the testing data. The forecasted results are used for SG applications such as power generation planning, economic operation and unit commitment, power system maintenance and planning, load switching, power purchasing, demand side management, and contract evaluation.

4 Simulation Results

In this section, simulation results of our proposed fast and accurate short term load forecasting model based on deep learning techniques such as stacked FCRBM and CRBM are presented. To validate accuracy of the proposed short term load forecasting model based on deep learning technique i.e., stacked FCRBM is compared with CRBM, ANN, and CNN in terms of performance metrics MAPE, NRMSE, and correlation coefficient. The detailed description is as follows: First, we define the forecast accuracy, in terms of NRMSE as:

$$RMSE = \sqrt{\frac{1}{\tau}\sum_{t=1}^{\tau}(R_t - F_t)^2}$$
$$NRMSE = \frac{RMSE}{\max(R_t - F_t)} \qquad (14)$$

where τ is the number of steps forecasted in future, R_t represents the real value, and F_t is the forecasted value. Second, we present MAPE performance metric to get statistical significance for accuracy assessment. For minimum MAPE accuracy will be maximum and vice versa. The MAPE is calculated as:

$$MAPE = \left(\frac{1}{\tau}\sum_{t=1}^{\tau}\frac{|R_t - F_t|}{|R_t|}\right) * 100 \qquad (15)$$

Finally, the correlation coefficient is implemented to check the accuracy that how close is the forecasted value to real value and is defined by the Eq. 16. The correlation coefficient returns a value between −1 and 1. If the returned value is close to 1 the real and forecasted values are positively correlated, if the returned value is closed to −1 shows that the real and forecasted values are negatively correlated, and if zero is returned then real and forecasted values are not correlated.

$$r = \frac{\mathrm{E}\{(R_t - \mu_R)(F_t - \mu_F)\}}{\sqrt{\sum(R_t - \mu_R)^2 * \sum(F_t - \mu_F)^2}} \qquad (16)$$

where E is the expected value operator, μ_R is the mean of real values, and μ_F is the mean of forecasted value.

4.1 Historical Load Data

The historical data is taken from publicly available repository Kaggle of global energy forecasting competition 2012 [17]. The dataset consists of hourly load (kW) of twenty zones of US utility and temperature of eleven stations. This historical dataset (load and weather) is of four years ranges from 1st hour of 1/1/2004 to 6th hour of 30/6/2008. The dataset is divided into training and testing data. The three years data is used to train the network and one year data is used to test the network. In summer during daytime there is a significant load increase of consumers as compared to night time. In winter during daytime there is a slight increase in the load as compared to the night. The electricity consumption from 2004 to 2008 maximum and minimum 540393 kWh and 149 kWh.

4.2 Learning Curve

Learning curve describes the error rate across the number of epochs. It examines the difference between the training and testing data, when the gap between the training and testing data is more the forecasting results will be in accurate and vice versa. Generally, at some point where test error starts to increase as compared to training error, this simply means that overfitting is occurred. In such situation the model memorizes the given training data rather than learning and the forecasted results will be inaccurate. This problem is solved by several methods such as drop out and early stopping. However, we observed learning curves of both deep learning techniques stacked FCRBM and CRBM, we did not notice overfitting because test error deceases as training error does as shown in Fig. 4. In this situation, the network is learning rather than memorizing and the forecasted results will be accurate.

4.3 Cumulative Distribution Function of Errors

The NRMSE is expressed int terms of cumulative distribution function (CDF) as shown in Fig. 5. The stacked FCRBM has 50% better CDF of NRMSE as compared to ANN, CNN, and CRBM because the stacked FCRBM has more computational capability. Stacked FCRBM based prediction is reliable, even if the load is uncertain with high error of predication. The results show that deep learning technique stacked FCRBM is robust and accurate as compared to CRBM, ANN, and CNN. Moreover, the stacked FCRBM would be the best choice for the consumers load forecasting as compared to CRBM because it has more computational power to capture the highly abstract data.

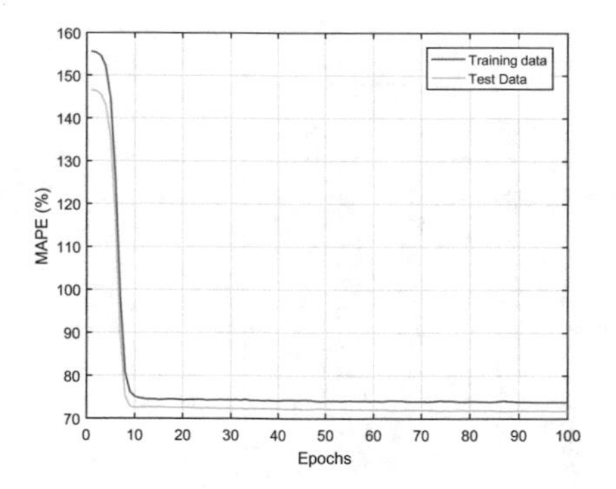

Fig. 4. Learning curve for both training and testing data

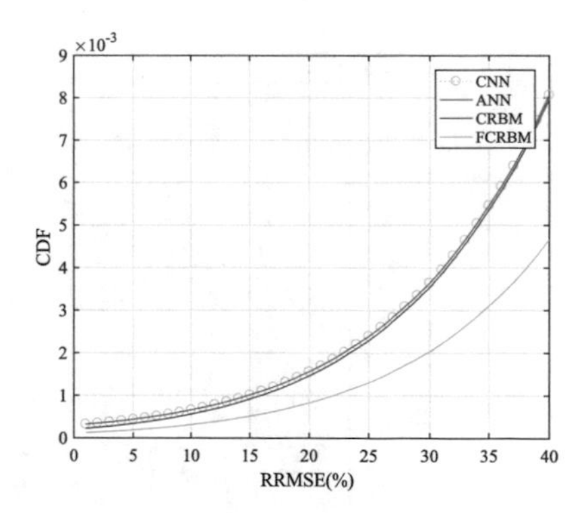

Fig. 5. CDF analysis interms of error performance

4.4 Deep Learning Based Short Term Load Forecasting

The short term load forecasting for one week time horizon with hourly resolution is described in Fig. 6. We forecast load of one week in middle of each season. It is observed that FCRBM model closely follow the real curve as compared to the

CRBM, ANN, and CNN. Moreover, the stacked FCRBM has more stable and accurate forecasted curve as compared to ANN, CNN, and CRBM. The observation numerical values in terms of performance metrics i.e., NRMSE, MAPE, and correlation coefficient are listed in Table 1. It is ensured from the results listed in Table 1 that stacked FCRBM forecasted are more accurate as compared to other techniques in terms of NRMSE, MAPE, and correlation coefficient. Moreover, the adopted stacked FCRBM have 99.62% accuracy with affordable execution time and complexity. Also, we have observed that deep learning techniques stacked FCRBM and CRBM performance are dependent on the type, number, and size of dataset chosen. The performance of deep learning techniques improves as the size of the training data increases and vice versa.

Table 1. Accuracy analysis in terms of performance metrics

Techniques	MAPE (%)	NRMSE	Correlation coefficient
ANN	3.500	2.048	0.4189
CNN	1.700	1.325	0.5300
CRBM	1.225	1.032	0.7287
Stacked FCRBM	0.377	0.6570	0.9987

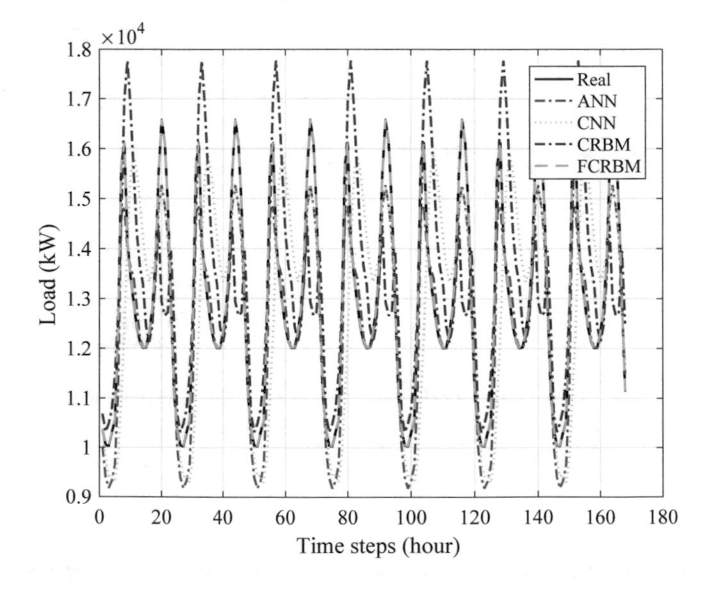

Fig. 6. Short term energy forecasting based on deep learning technique stacked FCRBM

5 Conclusion

A short term load forecasting model based on deep learning techniques i.e., stacked FCRBM and CRBM is proposed in this paper. The proposed model consists of data processing and feature extraction module, deep learning-based training module, deep learning-based forecasting module, and forecast weakly load profile on the basis of past energy consumption. We found that the stacked FCRBM and CRBM are effective to learn from past energy consumption and exhibit better performance compared to literature forecasting models. The performance evaluation of the proposed model is performed in terms MAPE, NRMSE, and correlation coefficient. Simulation results demonstrate that stacked FCRBM and CRBM are accurate and robust as compared to ANN and CNN. Moreover, the adopted stacked FCRBM achieved 99.62% accuracy with affordable execution time and complexity. In future, we will work in medium term forecasting and long term forecasting.

References

1. https://www.eia.gov/tools/faqs/faq.php?id=86&t=1. Accessed 19 Apr 2018
2. Hafeez, G., Javaid, N., Iqbal, S., Ali Khan, F.: Optimal residential load scheduling under utility and rooftop photovoltaic units. Energies **11**(3), 611 (2018)
3. Hafeez, G., Javaid, N., Zahoor, S., Fatima, I., Ali Khan, Z.: Energy efficient integration of renewable energy sources in smart grid. In: International Conference on Emerging Internetworking, Data and Web Technologies, pp. 553–562. Springer, Cham (2017)
4. Marino, D.L., Amarasinghe, K., Manic, M.: Building energy load forecasting using deep neural networks. In: IECON 2016-42nd Annual Conference of the IEEE Industrial Electronics Society, pp. 7046-7051. IEEE (2016)
5. Zeng, N., Zhang, H., Liu, W., Liang, J., Alsaadi, F.E.: A switching delayed PSO optimized extreme learning machine for short-term load forecasting. Neurocomputing **240**, 175–182 (2017)
6. Cecati, C., Kolbusz, J., Różycki, P., Siano, P., Wilamowski, B.M.: A novel RBF training algorithm for short-term electric load forecasting and comparative studies. IEEE Trans. Ind. Electron. **62**(10), 6519–6529 (2015)
7. Dedinec, A., Filiposka, S., Dedinec, A., Kocarev, L.: Deep belief network based electricity load forecasting: an analysis of Macedonian case. Energy **115**, 1688–1700 (2016)
8. Fan, C., Xiao, F., Zhao, Y.: A short-term building cooling load prediction method using deep learning algorithms. Appl. Energy **195**, 222–233 (2017)
9. Amjady, N., Keynia, F., Zareipour, H.: Short-term load forecast of microgrids by a new bilevel prediction strategy. IEEE Trans. Smart Grid **1**(3), 286–294 (2010)
10. Mocanu, E., Nguyen, P.H., Gibescu, M., Kling, W.L.: Deep learning for estimating building energy consumption. Sustain. Energy Grids Netw. **6**, 91–99 (2016)
11. Ryu, S., Noh, J., Kim, H.: Deep neural network based demand side short term load forecasting. Energies **10**(1), 3 (2016)
12. Ahmad, A., Javaid, N., Guizani, M., Alrajeh, N., Ali Khan, Z.: An accurate and fast converging short-term load forecasting model for industrial applications in a smart grid. IEEE Trans. Ind. Inf. **13**(5), 2587–2596 (2017)

13. Mnih, V., Larochelle, H., Hinton, G.: Conditional restricted Boltzmann machines for structured output prediction. In: Proceedings of the International Conference on Uncertainty in Artificial Intelligence (2011)
14. Hinton, G.E.: A practical guide to training restricted Boltzmann machines. In: Neural Networks: Tricks of the Trade, pp. 599–619. Springer, Heidelberg (2012)
15. Taylor, G.W., Hinton, G.E., Roweis, S.T.: Two distributed-state models for generating high-dimensional time series. J. Mach. Learn. Res. **12**, 1025–1068 (2011)
16. Mocanu, D.C., Ammar, H.B., Lowet, D., Driessens, K., Liotta, A., Weiss, G., Tuyls, K.: Factored four way conditional restricted boltzmann machines for activity recognition. Pattern Recogn. Lett. **66**, 100–108 (2015)
17. https://www.kaggle.com/c/global-energy-forecasting-competition-2012-load-forecasting. Accessed 26 Mar 2018

Efficient Resource Allocation Model for Residential Buildings in Smart Grid Using Fog and Cloud Computing

Aisha Fatima[1], Nadeem Javaid[1](✉), Momina Waheed[1], Tooba Nazar[1], Shaista Shabbir[2], and Tanzeela Sultana[3]

[1] COMSATS Institute of Information Technology, Islamabad 44000, Pakistan
nadeemjavaidqau@gmail.com
[2] Virtual University of Pakistan, Kotli Campus, Lahore 11100, Azad Kashmir, Pakistan
[3] University of Azad Jammu and Kashmir, Kotli 11100, Azad Kashmir, Pakistan
http://www.njavaid.com

Abstract. In this article, a resource allocation model is presented in order to optimize the resources in residential buildings. The whole world is categorized into six regions depending on its continents. The fog helps cloud computing connectivity on the edge network. It also saves data temporarily and sends to the cloud for permanent storage. Each continent has one fog which deals with three clusters having 100 buildings. Microgrids (MGs) are used for the effective electricity distribution among the consumers. The control parameters considered in this paper are: clusters, number of buildings, number of homes and load requests whereas the performance parameters are: cost, Response Time (RT) and Processing Time (PT). Particle Swarm Optimization with Simulated Annealing (PSOSA) is used for load balancing of Virtual Machines (VMs) using multiple service broker policies. Service broker policies in this paper are: new dynamic service proximity, new dynamic response time and enhanced new response time. The results of proposed service broker policies with PSOSA are compared with the existing policy: new dynamic service proximity. New dynamic response time and enhanced new dynamic response time performs better than the existing policy in terms of cost, RT and PT. However, the maximum RT and PT of proposed policies is more than the existing policy. We have used Cloud-Analyst for conducting simulations for the proposed scheme.

Keywords: Smart rid · Cloud computing
Particle Swarm Optimization · Simulated Annealing

1 Introduction

Utilization of advanced Information and Communication Technology (ICT) in Demand Side Management (DSM) has been considered as one of the main

L. Barolli et al. (Eds.): IMIS 2018, AISC 773, pp. 289–298, 2019.
https://doi.org/10.1007/978-3-319-93554-6_26

characteristics of Smart Grids (SGs) [1]. Bi-directional flow of energy and communication has been done by SG to get information of users and to distribute energy between consumers. The traditional grid is converted into a SG to reduce the Carbon Dioxide (CO_2). The number of devices have been utilized on the demand side. Many new concepts including Electric Vehicles (EVs) charging and discharging, intelligent home appliances, smart meters and so on have been used in DSM in the SG environment [1].

Cloud computing is generally associated with the services of the internet. The internet has connected to the world. Users can transfer a large amount of data and can also enjoy new technologies and services provided by the cloud at any time and any place [2]. Cloud computing has provided various facilities including minimum cost, maximum speed, high performance, and elasticity. Cloud can be public, private or hybrid. Netflix, skype, emails, microsoft office 365 and so on are the examples of cloud computing. However, there are some issues in cloud computing: latency, and less security. To tackle aforementioned issues, the concept of fog computing was introduced.

Fog computing concept is introduced by Computer Information System Company (CISCO) in 2014. Fog computing has emerged as a promising infrastructure to provide elastic resources at the network edge to minimize latency and to increase security. Fog computing is used to reduce the burden on cloud and for direct communication with consumers. Communication between fog and consumer is done through some communication medium (wireless, i.e., wi-fi). Fog provides local services and can be accessed without internet.

The integrated cloud-fog based environment is three-layered architecture. Fog is an intermittent layer between cloud and end user layer. The concept of cloud and fog is almost same. The differences are: size, distance from user, memory, and processing. The distance of cloud from ground level is thousands of kilometer whereas the fog must be on ground level. The services provided by cloud computing and fog computing are: Software as a Service (SaaS), Platform as a Service (PaaS), and Infrastructure as a Service (IaaS) [2].

SaaS:

- SaaS is the highest level of abstraction.
- It is accessed by the users through a web browser.
- SaaS provides an access to the licensed software.

PaaS:

- PaaS provides simplicity and convenience for consumers.
- A user can access PaaS services anywhere through a web browser.
- PaaS then charges the users for that access.

IaaS:

- It is a fundamental building block for cloud services.
- A cloud service provider provides the infrastructure components like data centers, servers, storage, and networking hardware.

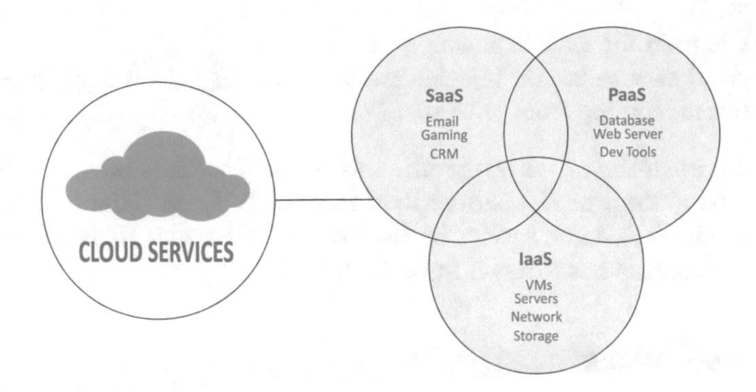

Fig. 1. Cloud and fog services

- The main use of IaaS includes the actual development and deployment of PaaS and SaaS.

Different service providers provide different services as shown in Fig. 1. SG with cloud and fog based environment is considered. The proposed scenario is divided into three layers: SG based layer, fog layer, and the cloud layer. In each cluster, hundred buildings are considered, a controller is used in SG layer to communicate with fog layer. Clusters are connected to fog in the same region. The data on fog is stored temporarily and fog sends data to the cloud for permanent storage. Consumers have to make a profile to communicate with fog. A profile contains information about consumers' location and daily electricity usage. These profiles help the fog and the cloud to maintain its data accordingly.

1.1 Motivation

A cloud-fog based platform is presented in [4,5], where fog devices are installed in a region between the end user layer and the cloud layer to minimize latency. Six regions are considered on the basis of six continents to cover the whole world [4]. One fog in each region is used rather than two to minimize the cost [4]. Fifty VMs are used instead of twenty-five to increase the efficiency in terms of PT. Five MGs are placed in each region to fulfill consumers' requests as much as needed to minimize RT. Three clusters instead of two [4] with hundred buildings in each are considered to achieve the results closer to real-time scenario.

1.2 Contributions

In this paper, SG application is integrated with the cloud-fog based environment, which covers a large area based on six continents of the world. It provides numerous benefits for SG applications such as;

- Low latency services are provided, as fog devices are placed near the end user.
- MGs are used to fulfill the electricity requirements of consumers.

- PSOSA is used for load balancing.
- Two hybrid service broker policies are used for the selection of fogs to entertain requests coming from the user.

Remaining part of the paper is organized as: related work is presented in Sect. 2. The proposed system model is described in Sect. 3. Load balancing algorithms are discussed in Sect. 4 and service broker policies in Sect. 5. However, simulation results and conclusion are drawn in Sects. 6 and 7.

2 Related Work

Fog computing is used as an intermittent layer between end user layer and the cloud layer. Fog computing is used to manage renewable energy resources and is accessible without internet. It provides true support for mobility and the Internet of Things (IoT) devices. Fog computing brings data closer to the end user layer. Cloud computing has some limitations for the SG, a huge number of SG devices need enormous data storage, networking, and processing. So fog is used near the end user layer to manage the SG resources.

Cao *et al.* [1] have proposed a cost-oriented optimization model. A Modified Priority List (MPL) and Simulated Annealing (SA) algorithms have been used to solve the proposed optimization model efficiently. Computing instance is a minimal unit that a user can take from the cloud. On-Demand Instances (ODI) and Reserved Instances (RI) are considered in this paper. ODI idea is like pay as you go while in a RI; users have a relatively long-term computing demands. RI has been declared better than ODI. However, a user has to pay the upfront payment in RI.

In [2], PSO based on Service Cost Optimization (PSOSC) scheduling algorithm has been proposed to schedule the tasks coming from users. PSOSC balance a load of VMs to minimize the cost and shortens the completion time. Task scheduling of workflow in the cloud is very important. However, RT has increased.

The authors in [4] have proposed a new dynamic service proximity policy for the selection of VMs. A VM having minimum latency is allocated to fulfill the consumers' need. The communication has been performed between the end user, fog, and the cloud. However, using two fogs in the same region is quite expensive.

Simulation technology has become a powerful and useful tool in cloud computing for research community [6]. The authors have compared the two cloud simulation tools CloudSim and CloudAnalyst. CloudAnalyst is declared the best option if anyone wants to work particularly on service broker policy or on load balancing algorithm as compare to CloudSim. However, CloudAnalyst is not a comprehensive solution for all complex tasks.

The authors in [7] have found some common security gaps of existing fog computing applications. Some impacts of security issues and possible solutions have been discussed in this paper. The detailed comparison between edge computing, cloudlet, and micro-data center has been given. However, the security issue is still there for a huge number of IOT devices.

Yasmeen *et al.* [9] have used cloud-fog based environment for efficient resource allocation. The author has proposed PSOSA and Cuckoo Search (CS) for balancing the load of VMs. The proposed service broker policy has been used for the selection of fog to entertain the requests coming from consumers. However, the RT and the PT is increased with the proposed service broker policy.

The authors have proposed PSO scheduling based algorithm for workflow scheduling. Workflow scheduling is a complicated scheduling containing a set of dependent tasks communicating with each other. Masdari *et al.* [15] have discussed the types of PSO algorithm, their objectives, and properties. However, load balancing of VMs is still a big problem and must be considered for efficient resource allocation.

3 Proposed System Model

In this study, an efficient resource allocation model is presented to address the following issues: minimization of PT, RT and the overall cost of VMs, MG, and total data transfer. The proposed structure has three layers: layer 1 (SG layer), layer 2 (fog layer) and the layer 3 (cloud layer). The centralized cloud platform is used for data storage and macrogrid availability. The world is divided into six regions based on the continents [1], as graphically shown in Fig. 2. Each region contains one fog that minimizes the RT and PT, three clusters and five MG. There are 100 buildings in one cluster and each building comprises of 50 to 80 apartments. A smart meter is appended to the all apartments.

MG incorporates with renewable energy. It has it's own power generation resources and have small-scale power. Macrogrid produces a large amount of electricity. Windmills, fossil fuels, water turbine, etc are the source of electricity for macrogrids. Fog in a region is able to respond the requests of three clusters and based on the energy demand, forward these requests to the cloud server. MGs are situated near the clusters of buildings. However, consumers are not permitted to communicate directly with the MGs. The requests for electricity from clusters are sent to the fog through the controller. The fog communicates with the MGs in the same region to fulfill the consumers' need. MGs send back an acknowledgement of the power they have. On the other hand, if they do not have adequate power, then the fog communicates with the cloud to provide the macrogrid facility. Proposed system model is shown in Fig. 3.

4 Load Balancing Algorithms

Load balancing algorithms are used for the distribution of the workload to achieve minimum RT and PT. Round robin and throttled algorithms were used in [4] to balance the load of VMs; a new load balancing algorithm (PSOSA) is used in this scenario. A number of particles form a swarm. These particles

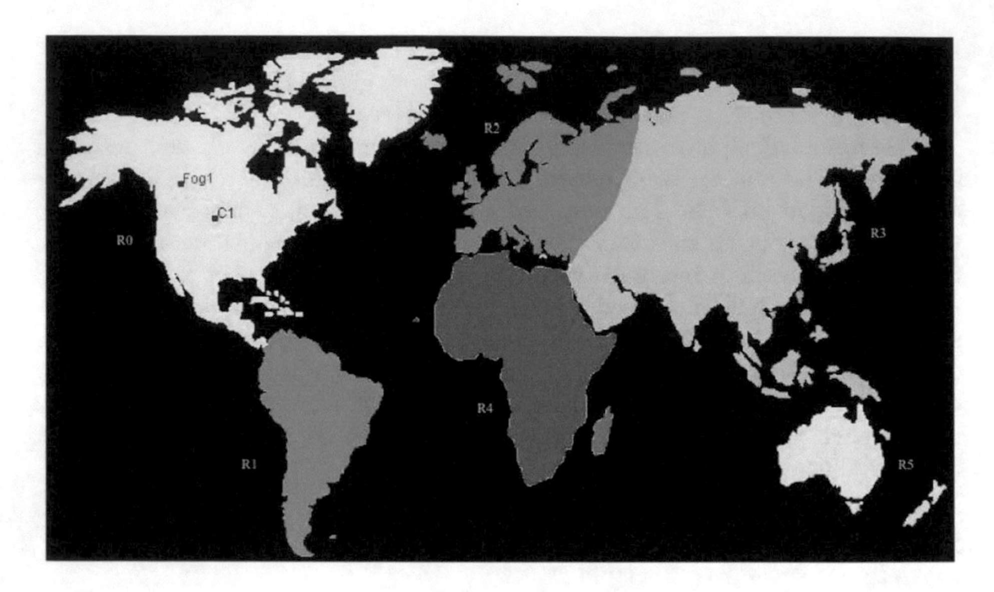

Fig. 2. Regions

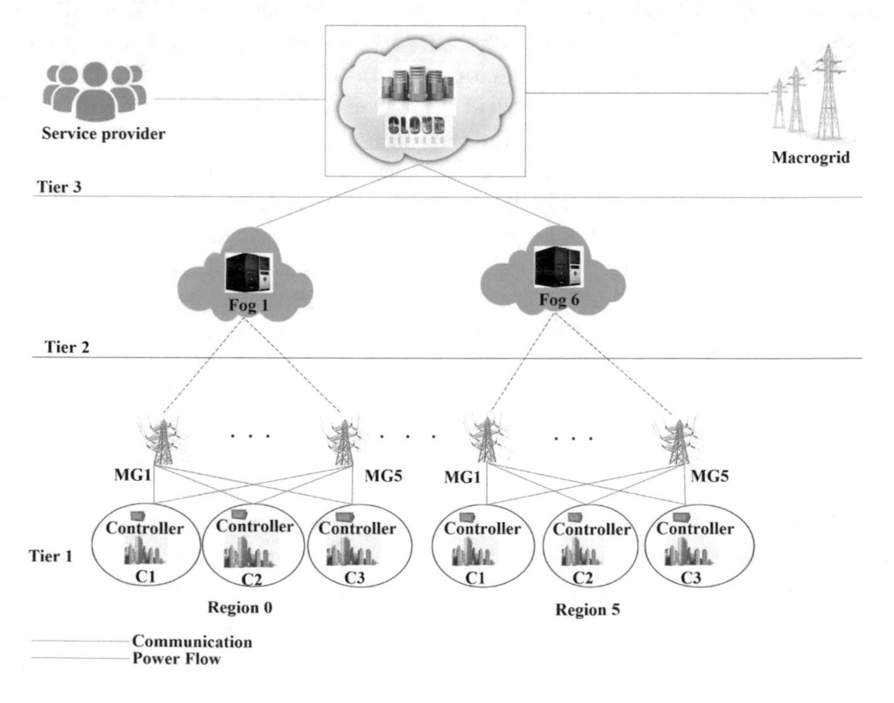

Fig. 3. Proposed system model

communicate with each other. A particle is composed of 3 vectors (x-vector, p-vector, and v-vector). These vectors record the current location of a particle, the best solution found so far and a direction for which particle will travel. Following steps are performed in PSOSA load balancing algorithm [2].

1. Initialize number of particle swarms, a number of tasks and a number of VMs.
2. Initialize velocity and positions of the particles.
3. Definition of adaptive functions, which includes tasks allocation strategy and fitness value to measure the merits of the allocation strategy. *f(i)* = fitness function and the *SumCost(i)* = total cost of the ith particle.
4. Compare fitness value with individual extremum and global extremum.
5. Update particle's speed and position.

5 Service Broker Policies

Resources are little bit complicated to manage. Cloud computing creates a set of virtual resources, i.e., VMs. Service broker policies are used to route the traffic coming from the end user to the fog. These policies decide which fog should deal with consumers' request. Following policies were used in [4].

A. Service Proximity Policy

- Service Proximity policy is easy to implement.
- It maintains the index table of all fogs in each region.
- The fog is selected which has minimum latency and closed to the cluster located in the same region.
- The fog is selected randomly if all fogs in the same region have minimum latency.

B. Optimize Response Time Policy

- It maintains the index table of all available fogs located in all regions.
- It checks the history that which fog provides best RT.
- The fog in the same region with best RT is assigned to the consumer.

C. Dynamically Reconfigure with Load

- This is the hybrid of service proximity policy and optimize response time.
- The fog is selected which is closed to the cluster of the same region with best RT.
- It also provides a facility for scalability.

D. New Dynamic Service Proximity

- New dynamic service proximity policy is the extension of dynamically reconfigure with load and service proximity policy.
- The fogs are allocated on the basis of minimum latency and already existing traffic load on fog and predicts next fog to be selected.

Following are the proposed policies in this paper.

5.1 New Dynamic Response Time

- It is the extension of dynamically reconfigure with load and optimize response time.
- The history of all fogs is sustained in the form of an index table.
- The fog is assigned on the basis of best RT in the same region by checking the history of all fogs.

5.2 Enhanced New Dynamic Response Time

- This is the extension of new dynamic response time and service proximity policy.
- The RT of all fogs is maintained in a table.
- The fog having best RT and minimum latency is allocated to the request coming from the cluster in the same region.

6 Simulations and Discussion

In this paper, CloudAnalyst tool is used for simulations. CloudAnalyst is used to work specifically on service broker policies and load balancing algorithms. The simulation results using PSOSA load balancing algorithm with three service broker policies are discussed. For the experimental purpose first PSOSA with new dynamic service proximity [1] is considered and than compared it with two proposed policies: new dynamic response time and enhanced new dynamic response time.

Table 1. Overall RT and PT

New dynamic proximity policy	Avg (ms)	Min (ms)	Max (ms)
RT	111.71	37.91	31729
PT	61.67	0.12	31682
New dynamic response time			
RT	98.4	36.71	33175
PT	44.29	0.05	33124
Enhanced new dynamic response time			
RT	99.05	37.92	33888
PT	44.94	0.05	33837

Minimum requests are serviced to minimize the cost in on-peak hours. RT is the time interval between the time, when the request is sent to fog and the response received against that request. The total time to process a request is known as PT. The overall RT and PT for PSOSA and service broker policies: new dynamic service proximity, new dynamic response time, and enhanced new dynamic response time is shown in Table 1. Each fog has some VM cost, MG cost, and data transfer cost. The grand total cost using three different policies with PSOSA is shown in Table 2.

Table 2. Cost comparison

	New dynamic proximity policy	New dynamic response time	Enhanced new dynamic response time
Total VM cost ($)	1334.25	816.01	816.01
Total MG cost ($)	266.85	163.2	163.2
Total data transfer cost ($)	289.11	289.11	289.1
Grand total ($)	1890.21	1268.32	1268.31

7 Conclusion

In this paper, an integrated fog and cloud based model is proposed to manage the SG resources optimally. It is analyzed that energy management is very important for both demand side and the supply side. Some service broker policies are also used for efficient selection of fog. PSOSA algorithm along with two hybrid service broker policies is implemented. Furthermore, we observed that the overall cost of PSOSA with new dynamic response time and enhanced new dynamic response time is approximately 20% better as compared to the existing policy. However the maximum RT and PT of new dynamic service proximity is approximately 3% better than the proposed policies. Simulations are performed on JAVA platform using CloudAnalyst. In future, we will extend this study for five clusters and elaborate system model.

References

1. Cao, Z., Lin, J., Wan, C., Song, Y., Zhang, Y., Wang, X.: Optimal cloud computing resource allocation for demand side management in smart grid. IEEE Trans. Smart Grid **8**(4), 1943–1955 (2017)
2. Xue, S., Shi, W., Xu, X.: A heuristic scheduling algorithm based on PSO in the cloud computing environment. Int. J. U- E-Serv. Sci. Technol. **9**(1), 349–362 (2016)
3. Chekired, D.A., Khoukhi, L.: Smart grid solution for charging and discharging services based on cloud computing scheduling. IEEE Trans. Ind. Inform. **13**(6), 3312–3321 (2017)
4. Fatima, I., Javaid, N., Iqbal, M.N., Shafi, I., Anjum, A., Memon, U.: Integration of cloud and fog based environment for effective resource distribution in smart buildings. In: 14th IEEE International Wireless Communications and Mobile Computing Conference (IWCMC-2018) (2018)
5. Zahoor, S., Javaid, N., Khan, A., Muhammad, F.J., Zahid, M., Guizani, M.: A cloud-fog-based smart grid model for efficient resource utilization. In: 14th IEEE International Wireless Communications and Mobile Computing Conference (IWCMC-2018) (2018)
6. Patel, H., Patel, R.: Cloud analyst: an insight of service broker policy. Int. J. Adv. Res. Comput. Commun. Eng. **4**(1), 122–127 (2015)
7. Khan, S., Parkinson, S., Qin, Y.: Fog computing security: a review of current applications and security solutions. J. Cloud Comput. **6**(1), 19 (2017)

8. Okay, F.Y., Ozdemir, S.: A fog computing based smart grid model. In: 2016 International Symposium on Networks, Computers and Communications (ISNCC), pp. 1–6. IEEE (2016)
9. Yasmeen, A., Javaid, N., Rehman, O.U., Iftikhar, H., Malik, M.F., Muhammad, F.J.: Efficient resource provisioning for smart buildings utilizing fog and cloud based environment. In: 14th IEEE International Wireless Communications and Mobile Computing Conference (IWCMC-2018) (2018)
10. Javaid, S., Javaid, N., Tayyaba, S.K., Sattar, N.A., Ruqia, B., Zahid, M.: Resource allocation using fog-2-cloud based environment for smart buildings. In: 14th IEEE International Wireless Communications and Mobile Computing Conference (IWCMC-2018) (2018)
11. Moghaddam, M.H.Y., Leon-Garcia, A., Moghaddassian, M.: On the performance of distributed and cloud-based demand response in smart grid. IEEE Trans. Smart Grid (2017)
12. Chekired, D.A., Khoukhi, L., Mouftah, H.T.: Decentralized cloud-SDN architecture in smart grid: a dynamic pricing model. IEEE Trans. Ind. Inform. **14**(3), 1220–1231 (2018)
13. Gai, K., Qiu, M., Zhao, H., Tao, L., Zong, Z.: Dynamic energy-aware cloudlet-based mobile cloud computing model for green computing. J. Netw. Comput. Appl. **59**, 46–54 (2016)
14. Chen, S.L., Chen, Y.-Y., Kuo, S.-H.: CLB: a novel load balancing architecture and algorithm for cloud services. Comput. Electr. Eng. **58**, 154–160 (2017)
15. Masdari, M., Salehi, F., Jalali, M., Bidaki, M.: A survey of PSO-based scheduling algorithms in cloud computing. J. Netw. Syst. Manage. **25**(1), 122–158 (2017)

Feature Selection and Extraction Along with Electricity Price Forecasting Using Big Data Analytics

Isra Shafi[1], Nadeem Javaid[2](✉), Aqdas Naz[2], Yasir Amir[1], Israr Ishaq[1], and Kashif Naseem[1]

[1] Department of Computing and Technology, Abasyn University, Islamabad 44000, Pakistan
[2] COMSATS Institute of Information Technology, Islamabad 44000, Pakistan
nadeemjavaidqau@gmail.com
http://www.njavaid.com

Abstract. The most important part of the smart grid (SG) is prediction of electricity price and by this prediction SG becomes cost efficient. To tackle with large amount of data in SG, it is a challenging task for existing techniques to accurately predict the electricity price. So, to handle the above mentioned problem, a framework has been proposed with three different steps: feature selection, feature extraction and classification. The purpose of feature selection is to remove irrelevant data by using extra tree classifier on the basis of pearson correlation coefficient. Feature extraction is performed using t-distributed stochastic neighbor embedding method to reduce redundancy from the selected data. For accurate electricity price forecasting, support vector machine classifier is used. Simulation results show that the proposed framework outperforms than the other methods.

Keywords: Forecasting · Electricity price · Support vector machine Extra tree classifier

1 Introduction

The industry is facing different type of issues while handling the large data sets [1]. The volume of the global data from 2005 to 2020 is growing by the factor of 300 and this growing factor shows that after every two years our growth will be double. Data is produced in different volume, velocity and variety and to cope with this different types of data "big data" term was introduced and this term helps to understand the meaning of the evolving data. In science and engineering domain big data is becoming very popular technology. Big data used a set of tools to store, acquire and for processing the huge amount of data. Data is collecting from multiple sources and this data is growing day by data so, to apply a big data framework on a specific task is not an easy method.

L. Barolli et al. (Eds.): IMIS 2018, AISC 773, pp. 299–309, 2019.
https://doi.org/10.1007/978-3-319-93554-6_27

To improve the accuracy of power supply utilities are transferring to smart grids (SGs) and smart meters (SMs). Utilities are doing this to integrate distributed generation resources, establish solutions of storage, efficiently use the power plant, and give the chance to the end users to cooperate in supervising their use of energy. Many suppliers at a first step deploying SM to achieving the aforementioned goal. SM producing a massive amount of data that read by the meter in a month. Huge amount of data is generated via smart meter and to efficiently manage that huge data can make it easy to understand the behavior of end users and economies the electric tariffs.

Now a days, from generation to transmission and from transmission to distribution technology is changing. Meters are receiving different data from different areas and this data has to be correlated to control, to monitor and also for different research purposes. However, the huge amount of data generated by SM has to be managed efficiently to increase the grid accuracy, capability and sustainability. This is a challenge for big data to handle a massive amount of data by using advanced technologies of big data. The thing which truly reflects the smart grid is big data [2]. Big data deals with "4Vs" that are:

- **Variety:** It shows that data comes from different sources and of different formats such as structured, semi-structured. The data can be in the form of videos, photos and logs, etc. This is the big challenge to store and analyze this type of data.
- **Volume:** Different machines, social media etc. are generated a huge amount of data and to analyze this massive data is challenge.
- **Veracity:** Veracity shows the quality and to manage the veracity of big data's different tools are used to discard noise from the data.
- **Velocity:** The rate at which data arrives is known as velocity. The time which it take to process and understand the data is also included in this time.

To handle the "4Vs" by using traditional resources is not possible in smart grid. Different challenges in smart grid encouraged the developers to move towards the new paradigm that is big data. For numerous phenomena's, big data is considered as a powerful data driven tool which is used in different fields e.g., biological systems, financial systems and also in wireless communication network. If big data will be used in power systems then it will produce a fruitful results. Data is a vital resource and it should be efficiently used in power systems. To handle the huge amount of data such as cloud computing, database parallel processing and scalable storage systems different techniques are discussed in papers.

The SM data is increasing day by day and this huge data is collected by different wireless sensors and consumers are also interacted by communication objects. To handle such a huge amount of data with irrelevant and redundant features which are generated by multiple sources is challenging with existing methods. From a huge amount of data only partial set of data is meaningful to perform any task in SG. Without prepossessing of the data no one can get accurate results. To increase the accuracy and efficiency of the data different steps has to be performed and on the basis of these steps, good results can been achieved [3].

Balancing of gap between user demand and power supply and to shift the load from on peak hours to off peak hours is the main objective of SM. End users are eager to know about the pricing scheme because according to those pricing schemes they schedule their load to minimize the energy cost. The requirement of the industry and the economy is to predict the accurate electricity price. The end users are eager to know about the electricity price that whether it cross the predefined threshold or not. On the basis of predefined threshold they manage their load. Due to this, end user needs classification of electricity price. Different types of classifiers are used to classify the electricity price. In price forecasting, the electricity price forecasting becomes challenge. Different factors influenced the electricity price such as requirement of electricity, energy resources etc. and on the hourly basis it varies [4].

1.1 Contributions

The contributions of this research work are summed up as follows:

- To achieve the accurate big data price forecasting for SM, authors incorporate the electricity price forecasting framework.
- To tackle the problem of large amount of data in SM, feature selection, extraction and classification are combined in proposed frame work.
- For selection, extraction and classification ETC, t-SNE and SVM has been used.

2 Related Work

To balance the power supply and demand side authors in [3], proposed framework based on cloud to computing the big data. In this paper, historical weather data has been used to predict the production of energy from different sources and also they analyze the behavior of end users to capture their demand. Authors in [5], explains the two methods for electricity price forecasting i.e, time series and machine learning. On the basis of information of load and temperature a hybrid model forecast day ahead electricity.

On the basis of harmonic state estimation authors proposed a practical data utilization technique for the harmonic analysis of power system. In any specific field it was considered as method of data processing and only available when engineering model is perfectly accurate [6].

In [7], to forecast the price of per hour electricity, probability methodology has been used. To study the uncertainty, bootstrapping method has been used and for the calculation of wavelet neural networks, extreme machine learning methods has been used.

For electricity forecasting time series model are also used and auto regressive integrated moving average (ARIMA) performed good for the market stable electricity. Raw data of market makes the accuracy of forecasting unstable so by this ARIMA has some outliers [8]. In functional time series to calculate the mobile average values the autoregressive moving average Hilbertian has been used to forecast the price of electricity [9].

Feature engineering is the important method and feature selection and extraction are its two main operations. For feature engineering different methods have been used. In [10], authors discussed that in price forecasting how accuracy can be achieved by selecting relevant features.

In paper [11], support vector machine (SVM) is used as a classifier and for SVM its not east to tackle the uncertain information. In paper [12], authors discussed about principle component analysis (PCA) that for high dimensional computations the performance of PCA is not so good.

[13] explains in order to analyze load utilization efficiently system load analysis is not sufficient, therefore author focus on analyzing user load consumption. OS-ELM technique is used for load forecasting after applying k means.

3 System Model

Figure 1 elaborates the structure of the proposed model. In this model, first step is the selection of data and the second step is the data preprocessing. After this, the data flows into the feature selection step, where irrelevant features are removed from the data. Feature extraction is used to remove redundancy from the selected features. Finally, the classifier is applied on the processed data.

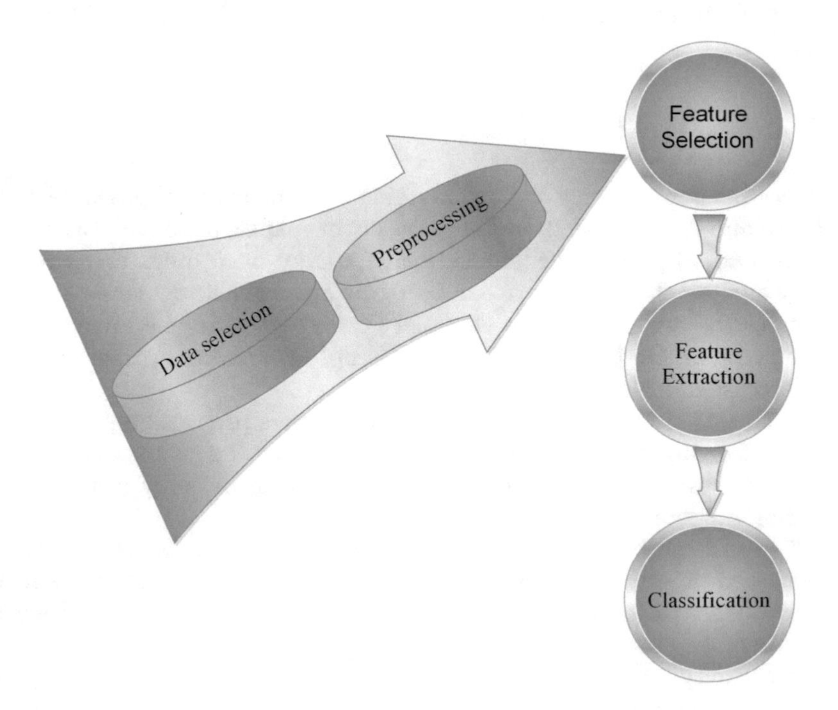

Fig. 1. System model

3.1 Feature Selection

Feature selection is also known as attribute selection or variable selection. Selection of features are different from dimensionality reduction. The main purpose of both methods are to reduce the number of variables. The feature selection method reserve or delete the variables without changing them. Feature selection is very useful, it act as a filter. Feature selection helps to choses those features which performs good in prediction. Selection of feature are having three folds: Improvement in the performance of prediction, predictors can be more cost effective and fast, a good understanding of the data generated by underlying process.

Electricity price data is denoted by a matrix as given in Eq. 1. In matrix, rows represent the time stamps and columns represent the index of features.

$$M = \begin{bmatrix} X_{11} & X_{12} & \dots & X_{1n} \\ X_{21} & X_{22} & \dots & X_{2n} \\ \dots & \dots & \dots & \dots \\ X_{m1} & X_{m2} & \dots & X_{mn} \end{bmatrix} \tag{1}$$

Here, X_{ij} is the j-th data component of the i hour ahead of the predicted class. Normalization has been performed on data to reshaped the selected data and to improve integrity. The data can be normalized by using Eq. 2.

$$X_{new} = \frac{X - X_{min}}{X_{max} - X_{min}} \tag{2}$$

After normalization, the statistical term, correlation has to be applied for showing that how strongly the features are related. Prediction is influenced by different features having different degree of influence. So, pearson correlated coefficient (PCC) has been used to measure the influence of different features. PCC measures the relationship among different features by using Eq. 3 [14].

$$P_{X,Y} = \frac{cov(X,Y)}{\sigma X \sigma Y} \tag{3}$$

where cov is covariance, it tells us that is there any relationship between two variables or not. σX and σY, shows the standard deviation of X and Y.

Features are ranked according to the relationship measured by PCC. There are many features having weak relationship for removing that irrelevant features ETC has been used. Purpose of feature selection is to remove those features which are irrelevant. Feature selection is a funnel like approach as shown in Fig. 2.

On results of PCC, extra tree classifier (ETC) has been applied as a feature selector technique and ETC belongs from the extremely randomized trees. ETC is the variant of Random Forest. In ETC, the entire data is used and according to this, decision boundaries are randomly selected. On the basis of PCC Ranking, we set the threshold and according to that threshold, the features having less value will be dropped from the data.

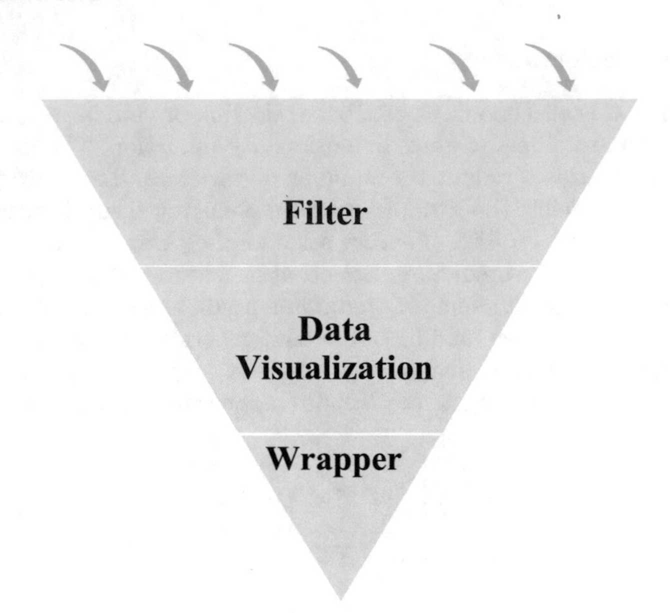

Fig. 2. Funnel approach

Selection of feature is based on ETC, and these features are normalized by Eq. 4.

$$Z = Z/max(ETC) \tag{4}$$

where Z gives the feature importance. The feature selection is based on the value of threshold. If the value of feature is greater than the threshold it will be the part of the data otherwise, the feature will be dropped from the data.

3.2 Feature Extraction

In this section, feature extraction has been proposed which removed redundancy among the features. However, it reduces the computational time for learning and also increased the accuracy of forecasting [15]. The paper [16], states that the crowding problem of SNE has been removed by t-SNE and now its easy to map a high dimensional visualization in a single map. This technique represents the points with high probability in such a way that the similar objects are represented by neighboring points and contradictory objects are placed by distant points. In this section, for dimensionality reduction the input comes from feature selection and t-SNE has been applied on that input.

3.3 Classification

Classification is used to build a model on the basis of trained data to assign a class label to independent data samples. Classification is of two types: binary and multi class classification. In machine learning different types of classifier has been used and SVM is more popular due to its robustness, adaptability and generalization capability. Classification error is minimized by estimating the optimal hyper plane in SVM. The margin of the closest data points are estimated by these hyper plane. On the basis of these closest points, the margin has been defined to separate the hyper plane and known as support vector. In SVM, Kernel tricks are used to transform the data and on the basis of these transformations, SVM finds the boundaries that are optimal for available outputs.

The aim of this paper is to utilize the data in such a way that it become affective for prediction. After the above two sections, the irrelevant and redundant information has been dropped from the data. In this section, final electricity price has been predict via SVM as a classifier. For higher accuracy in SVM, parameters of SVM are tuned by using grid search.

4 Simulation Results

To forecast the electricity price, three different steps are performed. The simulations are performed in python to examine the efficiency of proposed framework. For simulations, the platform is used with Intel core i5 and 4 GB RAM. For this framework, the hourly base electricity dataset of ISO new England control area (ISO-NE-CA) is used. The dataset have different attributes and on the basis of those attributes forecasting of electricity price has been done.

4.1 Feature Selection Performance of PCC Based ETC

Time series dataset depict different behaviors, on the basis of time stamps. The main purpose is to predict the electricity price. Features having minimum impact on price has been removed. Before applying ETC, PCC calculates correlation between target and features as shown in Fig. 3. PCC defines the correlation among different features. Nevertheless, if feature are closer to each other, than the correlation is greater among the features.

The significant features are selected on the basis of the results that are generated by the PCC. In this study, the threshold selected value is 0.6 in order to control the selection of features. The Fig. 4 shows the ranking of features on the basis of PCC.

The Fig. 4 specifies that most of the features' grade is above 0.6. ETC is applied to select the values of important features from the electricity price data. The features are dropped that have low grades, i.e., 0.58 and 0.57 as shown in Fig. 5.

4.2 Feature Extraction by t-SNE

The features selected by ETC can be supposed that they have no irrelevant features. However, the feature selected is redundant to some extent. Moreover, the redundant information from the features is removed by t-SNE. The features having negative value are extracted from the data as shown in Fig. 6. Nonetheless, the feature with negative value shows that they have redundancy in them.

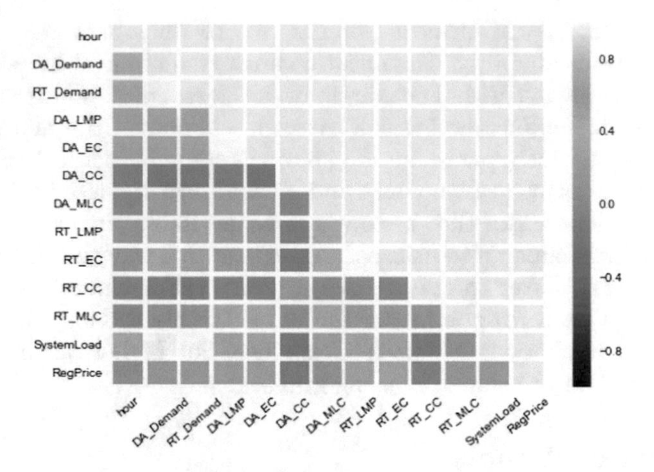

Fig. 3. Correlation between features and target values

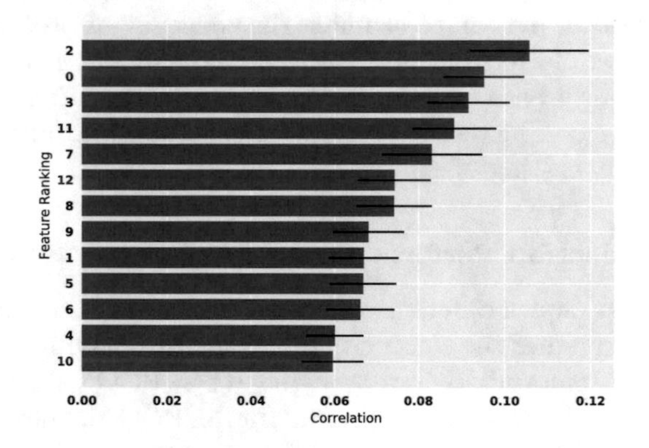

Fig. 4. Ranking of features

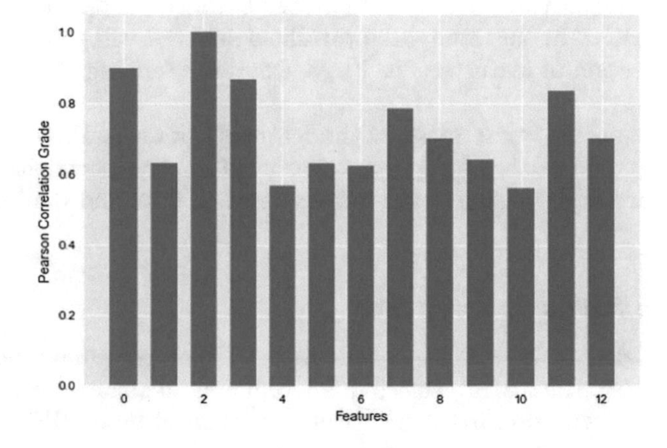

Fig. 5. Feature importance on the basis of PCC

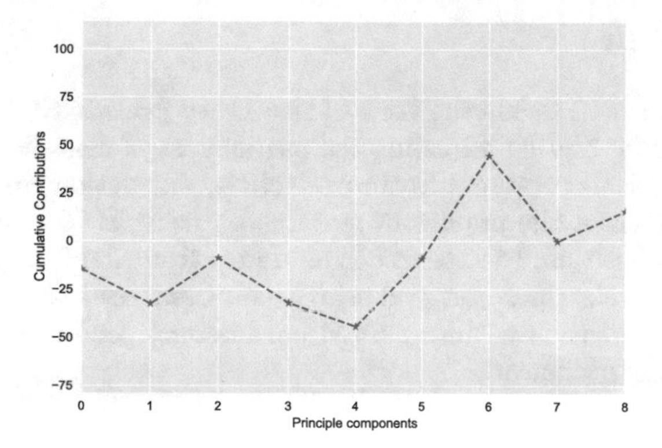

Fig. 6. Performance of t-SNE

4.3 Performance of SVM

After feature selection and extraction, the irrelevant and redundant information have been removed. SVM is applied for electricity price forecasting through the processed data and then the model trained by SVM. Moreover, the model can predict the electricity price that fits well with real time value and also with Decision tree (DT). The resultant value is compared with the real values. Tuning of parameters are done by grid search. For price forecasting, the SVM performs well as shown in Fig. 7. SVM has some outliers when time series is 9, 11 and 12.

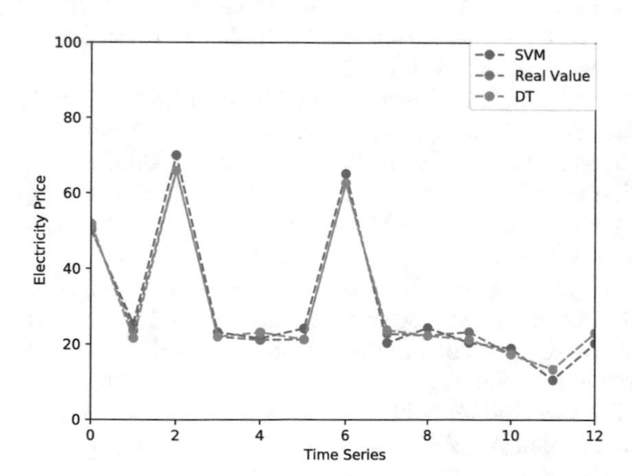

Fig. 7. Comparison on price forecasting

5 Conclusion

In this paper, authors tackle the issue of electricity price in SG with the help of feature engineering and by adjusting the parameters of classifier. In this paper, authors proposed three steps: feature selection, extraction and classification. Firstly, authors select the important features by using ETC from the data and then t-SNE has been used to remove redundancy from the features. Finally, the authors use SVM classifier and grid search has been used to tune the parameters of SVM. In future, for tuning SVM parameters, genetic algorithm will be considered in this framework.

References

1. Kezunovic, M., Xie, L., Grijalva, S.: The role of big data in improving power system operation and protection. In: 2013 IREP Symposium on Bulk Power System Dynamics and Control-IX Optimization, Security and Control of the Emerging Power Grid (IREP), pp. 1–9. IEEE (2013)
2. Munshi, A.A., Yasser, A.R.M.: Big data framework for analytics in smart grids. Electr. Power Syst. Res. **151**, 369–380 (2017)
3. Hou, W., Ning, Z., Guo, L., Zhang, X.: Temporal, functional and spatial big data computing framework for large-scale smart grid. IEEE Trans. Emerg. Topics Comput. (2017)
4. Wang, K., Xu, C., Zhang, Y., Guo, S., Zomaya, A.: Robust big data analytics for electricity price forecasting in the smart grid. IEEE Trans. Big Data (2017)
5. Varshney, H., Sharma, A., Kumar, R.: A hybrid approach to price forecasting incorporating exogenous variables for a day ahead electricity market. In: IEEE International Conference on Power Electronics, Intelligent Control and Energy Systems (ICPEICES), pp. 1–6. IEEE, July 2016
6. Kanao, N., Yamashita, M., Yanagida, H., Mizukami, M., Hayashi, Y., Matsuki, J.: Power system harmonic analysis using state-estimation method for Japanese field data. IEEE Trans. Power Delivery **20**(2), 970–977 (2005)
7. Rafiei, M., Niknam, T., Khooban, M.H.: Probabilistic forecasting of hourly electricity price by generalization of ELM for usage in improved wavelet neural network. IEEE Trans. Ind. Inform. **13**(1), 71–79 (2017)
8. Ozozen, A., Kayakutlu, G., Ketterer, M., Kayalica, O.: A combined seasonal ARIMA and ANN model for improved results in electricity spot price forecasting: case study in Turkey. In: 2016 Portland International Conference on Management of Engineering and Technology (PICMET), pp. 2681–2690. IEEE, September 2016
9. González, J.P., San Roque, A.M., Pérez, E.A.: Forecasting functional time series with a new Hilbertian ARMAX model: application to electricity price forecasting. IEEE Trans. Power Syst. **33**(1), 545–556 (2018)
10. Zhao, J., Dong, Z., Li, X.: Electricity price forecasting with effective feature preprocessing. In: IEEE Power Engineering Society General Meeting, p. 8-pp. IEEE, January 2006
11. Han, H., Dang, J., Ren, E.: Comparative study of two uncertain support vector machines. In: 2012 IEEE Fifth International Conference on Advanced Computational Intelligence (ICACI), pp. 388–390. IEEE, October 2012
12. Li, Y., Guo, P., Li, X.: Short-term load forecasting based on the analysis of user electricity behavior. Algorithms **9**(4), 80 (2016)

13. Wang, K., Hu, X., Li, H., Li, P., Zeng, D., Guo, S.: A survey on energy internet communications for sustainability. IEEE Trans. Sustain. Comput. **2**(3), 231–254 (2017)
14. Bakar, N.A., Rosbi, S.: Robust statistical pearson correlation diagnostics for bitcoin exchange rate with trading volume: an analysis of high frequency data in high volatility environment (2017)
15. Ramadevi, G.N., Rani, K.U., Lavanya, D.: Importance of feature extraction for classification of breast cancer datasets-a study. Int. J. Sci. Innov. Math. Res. **3**(2), 763–768 (2015)
16. Abdelmoula, W.M., Pezzotti, N., Hölt, T., Dijkstra, J., Vilanova, A., McDonnell, L.A., Lelieveldt, B.P.: Interactive visual exploration of 3D mass spectrometry imaging data using hierarchical stochastic neighbor embedding reveals spatiomolecular structures at full data resolution. J. Proteome Res. **17**(3), 1054–1064 (2018)

Proposal of a Disaster Support Expert System Using Accumulated Empirical Data

Tatsuya Ohyanagi[1(✉)], Tomoyuki Ishida[2], Noriki Uchida[2], Yoshitaka Shibata[3], and Hiromasa Habuchi[1]

[1] Ibaraki University, Hitachi, Ibaraki 316-8511, Japan
{17nm704r,hiromasa.habuchi.hiro}@vc.ibaraki.ac.jp
[2] Fukuoka Institute of Technology, Fukuoka, Fukuoka 811-0295, Japan
{t-ishida,n-uchida}@fit.ac.jp
[3] Iwate Prefectural University, Takizawa, Iwate 020-0693, Japan
shibata@iwate-pu.ac.jp

Abstract. In this paper, we implemented a disaster support expert system for emergency response headquarters. This system consists of the disaster information storage system and the disaster information visualization system. The disaster information storage system stores disaster case, disaster response record, and local disaster prevention plan. And, the disaster visualization system visualizes past disaster information and correspondence records accumulated in the disaster information storage system. By using this system, the emergency response headquarters can promptly and appropriately disaster response through accumulated past disaster cases and disaster response records.

1 Introduction

According to World Risk Report 2016 [1], the risk of occurrence of natural disasters (earthquakes, typhoons, floods, drought, sea level rise) in Japan is fourth in the world. In other words, Japan is a natural disaster nation. Protecting citizens' lives and property from natural disasters is a national priority. Under such circumstances, research and development of resilient disaster prevention/reduction technologies by disaster prediction, disaster prevention, disaster response, and information sharing using the latest technology is an important research theme. The "Emergency Response Headquarters" play a central role of disaster countermeasures in emergencies. The role of emergency response headquarters is as follows.

- Collection, processing, transmission of various disaster information
- Discuss and decide disaster countermeasures
- Directing various disaster emergency activities

In order to conduct timely and appropriate disaster measures at the time of large-scale natural disaster, it is necessary for the emergency response headquarters to function effectively. However, as Numata et al. [2] pointed out, the emergency response headquarters will be pressed by various disaster response such as evacuation center management, relief supplies management, incident investigation, rubble investigation, and injured person correspondence at the time of large-scale natural disaster.

L. Barolli et al. (Eds.): IMIS 2018, AISC 773, pp. 310–319, 2019.
https://doi.org/10.1007/978-3-319-93554-6_28

Moreover, it is difficult to prompt disaster correspondence because the local government office buildings and municipal officials may be affected by the disaster. In fact, the affliction of local government office buildings and municipal officials caused a major obstacle to the collection, sharing and dissemination of disaster information at time of the Great East Japan Earthquake [3]. On the other hand, although the communication line was largely restricted, communication restrictions on packet communications and Internet lines were small at the time of the Great East Japan Earthquake. Therefore, information sharing utilizing social media was effective [4]. Rescue cases triggered by social media information have also been reported. However, social media tends to transmit many misinformation and hoax information. Therefore, when emergency response headquarters use information on social media, it is necessary to determine the authenticity of information.

2 Related Works

Hirohara et al. [5–8] constructed a decision-support cloud system for emergency response headquarters using an interactive large-scale ultra-high definition display. This system consists of disaster information input function, disaster information output function, disaster information transmitting function, and disaster information interactive sharing function. Disaster information input by the disaster information input function is visualized in the form of a timeline by the disaster information output function. And this system visualizes disaster information and big data at the same time, and supports emergency response headquarters decision making by using the large-scale ultra-high definition display. Disaster information transmitting function transmits real-time disaster information to residents. And disaster information interactive sharing function reflects disaster information on large-scale ultra-high definition display with one touch/one flick from tablet terminal. However, this decision-support cloud system has not realized smooth data update by manual update of disaster data. In addition, this system has not realized data linkage with other disaster prevention systems.

Hamamura et al. [9] developed an “Akari Map” which is evacuation support system for everyday use in offline environment. The Akari Map has a notification function of evacuation support information, a disaster mode function that can experience the disaster situation, and a flood area display function. Moreover, this Akari Map realizes the use in offline environment by caching data in the terminal, and realizes provide of evacuation support information without launching applications by the widget function. However, this application only supports the Android platform. In addition, the user must install the application beforehand in order to use this system.

Nonaka et al. [10] developed a decision support system for emergency operation to support disaster countermeasures operations quickly and accurately. This decision support system realizes the provide of active information by the if-then rule by digitizing the disaster prevention plan in advance. Moreover, this system has an aspect as an expert system that provides active information and an aspect as a groupware system that shares information with multiple users. However, this system has not realized support for unexpected disaster events.

3 Research Objective

In this paper, we implement a disaster support expert system capable of dealing with various kinds of natural disasters. This system aims to support disaster response of emergency response headquarters at the time of large-scale natural disaster, and consists of disaster information storage system and disaster information visualization system. The disaster information storage system realizes registration of disaster information such as disaster case, disaster response record, and local disaster prevention plan etc. On the other hand, the disaster visualization system realizes visualization of past disaster information and realizes past disaster correspondence record retrieval by using disaster information stored in the database. The emergency response headquarters can make prompt decision making by visualizing past disaster information. Also, the emergency response headquarters can refer to appropriate correspondence records according to the scale and type of natural disaster by retrieving for past disaster response records. In addition, we realize real-time tweet presentation function, and support information gathering at the time of disaster by using the Twitter API [11].

4 System Configuration

The system configuration of the Disaster Support Expert System is shown in Fig. 1. The roles of each system are described below.

- Staff Agent
  The staff agent is municipal officials of the emergency response headquarters or the disaster sites. The staff agent registers the disaster data by using the information storage system.
- Emergency Response Headquarters
  The emergency response headquarters is an organization that handles disaster response and decision making. The emergency response headquarters shares disaster information by using disaster information visualization system.
- Disaster Information Storage System
  The disaster information storage system provides functions for registering disaster cases, disaster response records, local disaster prevention plans in the database.
- Disaster Information Visualization System
  The disaster information visualization system provides disaster information arrangement and sharing function, disaster information visualization function, and disaster information analysis function stored by the disaster information storage system.
- Application Server/Database Server
  The application server manipulates the disaster data stored in the database, and various disaster information is stored in the database server.
- Twitter API
  The Twitter API is an API for acquiring tweets, and in this research, we acquire tweets related to disaster information in real-time.

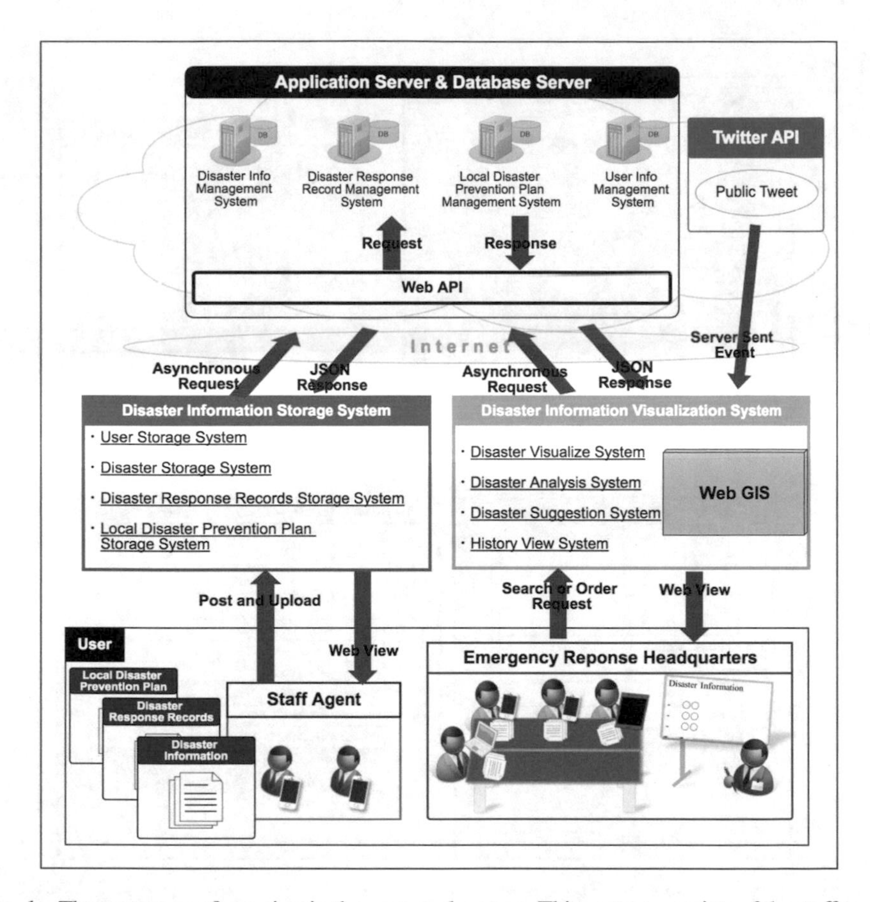

Fig. 1. The system configuration in the proposed system. This system consists of the staff agent, the emergency response headquarters agent, the disaster information storage system, the disaster information visualization system, the application server & the database server, and the Twitter API.

The user can operate the system by executing instructions such as data input and retrieval. These instructions are requested to the Web API using asynchronous communication technology. The Web API executes data editing processing and acquisition processing according to the requested instruction. When the process is completed, the execution result is transmitted to the Web API as a response. The Web API transmits the received response to each function in JSON format. Each function that receives the JSON file visualizes the data on the Web View.

5 System Architecture

The system architecture of the Disaster Support Expert System is shown in Fig. 2.

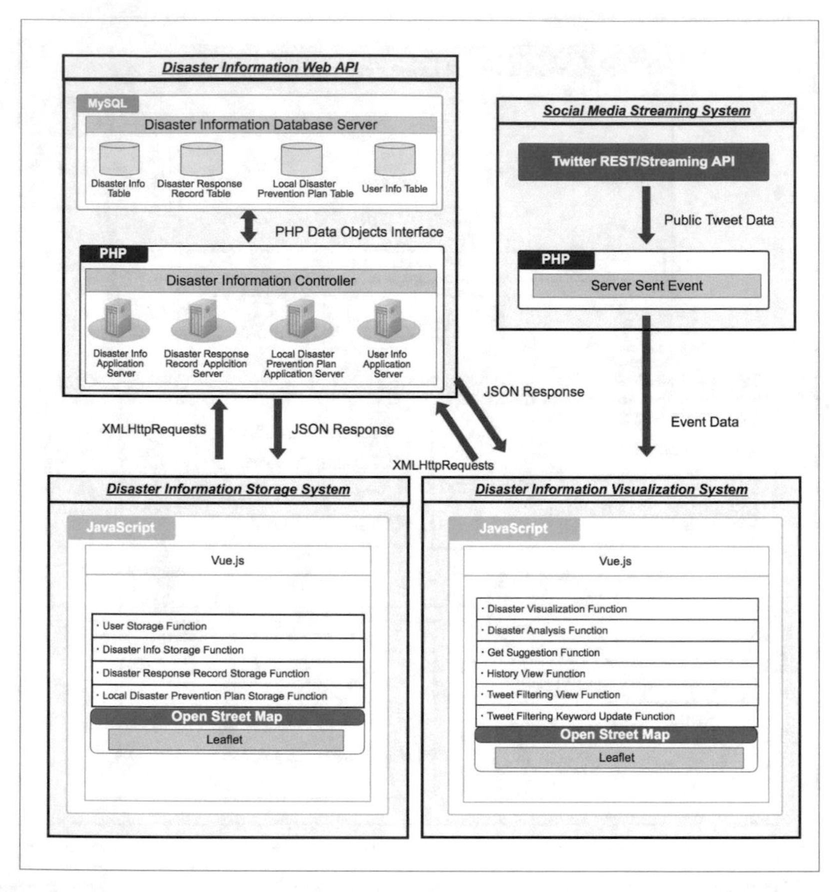

Fig. 2. The system architecture in the proposed system. This system is roughly classified in four systems: the disaster information storage system, the disaster information visualization system, the disaster information Web API, and the social media streaming system.

6 Prototype System

Our prototype system consists of the disaster information storage system and the disaster information visualization system.

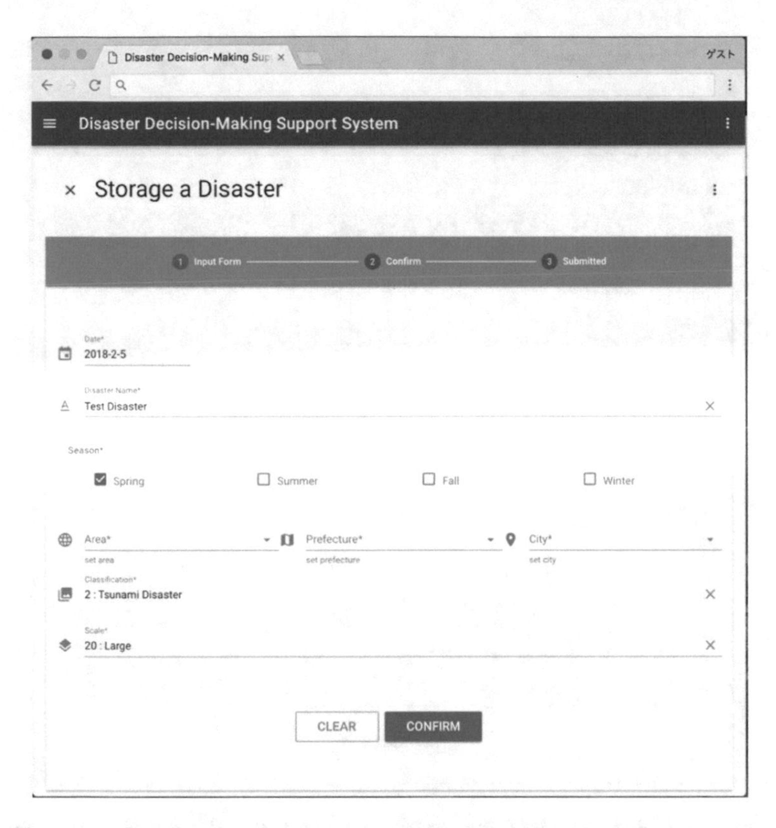

Fig. 3. The date and time, name, season, and area are displayed on the registration screen. When the user inputs data and presses the "SUBMIT" button, data is transmitted to the Disaster Information Web API by the asynchronous communication technology.

6.1 Prototype of Disaster Information Storage System

The disaster information storage system consists of the disaster case storage system, the disaster response record storage system, and the local disaster prevention plan storage system. The registration screen of the Disaster Information Storage System is shown in Fig. 3. By registering various data in the database via this system, it becomes possible to visualize enormous data on the disaster information visualization system.

The disaster response record storage system registers disaster response for each disaster case. The disaster response record storage system is shown in Fig. 4.

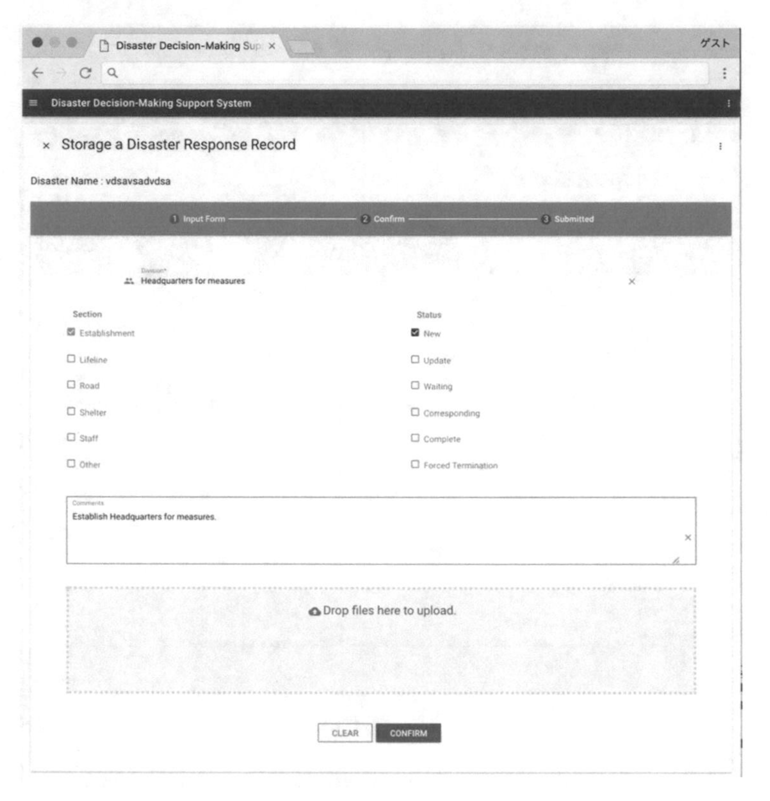

Fig. 4. The registration screen of the disaster response record storage system consists of corresponding division, classification of disaster response record, contents of disaster response, current disaster response status, and remarks.

6.2 Prototype of Disaster Information Visualization System

The disaster information visualization system performs data stored in the database by the disaster information storage system. This system consists of the disaster response suggestion system and the social media system. The disaster information visualization system is shown in Fig. 5. Emergency Response Headquarters can make prompt decision-making for disaster response by using disaster information visualization system.

The disaster response suggestion system is shown in Fig. 6. When the user inputs information into the search field at the top of the screen, the disaster response record matching the search condition is displayed.

The social media system is shown in Fig. 7. Tweets are acquired in real-time by using Twitter Streaming API. At this time, tweets can be filtered by setting keywords in advance.

Fig. 5. The user can confirm detailed information on disasters by using this system. A disaster response record related to the selected disaster is displayed at the bottom of the page. Moreover, the damage area of the disaster is overlap displayed on the Open Street Map [12].

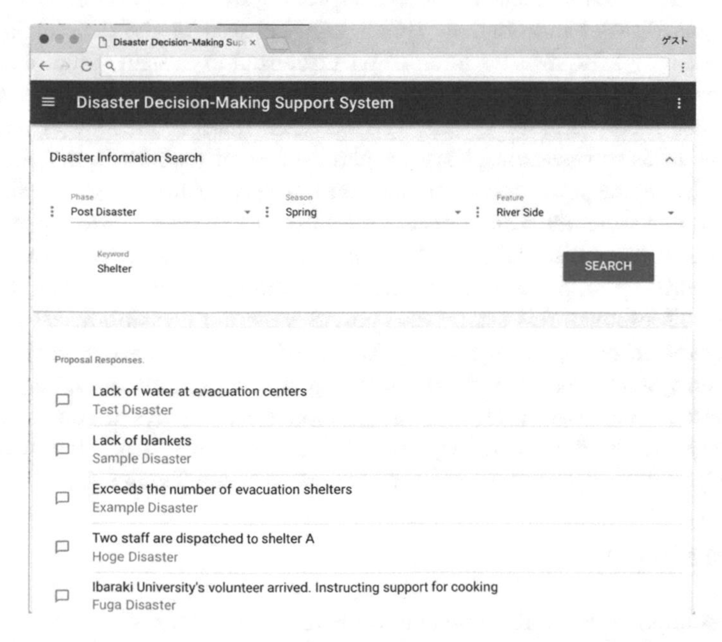

Fig. 6. The search field consists of disaster response phase, season, topography, and keywords. When the user executes the search, the search result is displayed in the middle part of the screen.

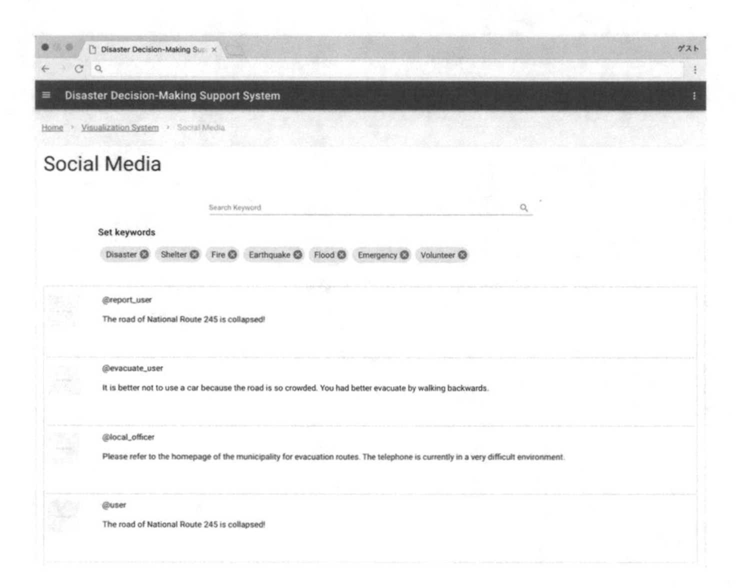

Fig. 7. A keyword is registered by the user inputting a keyword in "Search Keyword" at the top of the screen. The registered keywords are displayed in "Set keywords".

7 Conclusion

In this research, we implemented a disaster support expert system for emergency response headquarters of local governments. This system supports disaster decisions based on past disaster information. This system consists of the disaster information storage system and the disaster information visualization system. Moreover, the disaster information storage system consists of the disaster case storage system, the disaster response record storage system, and the local disaster prevention plan storage system. The disaster case storage system plays a role of registering disaster information. The disaster response record storage system plays a role of registering disaster response records. And, the local disaster prevention plan storage system plays a role of registering a regional disaster prevention plan. The disaster information visualization system provides prompt disaster information sharing to the emergency response headquarters. Moreover, this system also provides prompt decision-making support to the emergency response headquarters. On the other hand, the disaster information visualization system consists of the disaster response suggestion system and the social media system. The disaster response suggestion system plays a role of proposing disaster response to the emergency response headquarters. And, the social media system plays a role of acquiring real-time tweets by using Twitter API.

8 Future Works

In order to improve the disaster support expert system, we will construct a disaster information analysis function and further improve the user interface. The disaster information analysis function visualizes the disaster occurrence frequency of each year and the disaster characteristics of each area by referring to the accumulated disaster information. Moreover, this function also compares local disaster prevention plans with disaster response records. By realizing the disaster information analysis function, local governments can improve disaster prevention measures according to regional characteristics. On the other hand, currently we are conducting interview survey on user interfaces to municipalities in Ibaraki prefecture. In the future, we plan to improve the user interface based on the results of the interview survey. Moreover, we plan to evaluate the usability and effectiveness of the new disaster support expert system through evaluation experiments with municipal officials.

Acknowledgments. This research was supported by JSPS KAKENHI Grant Number JP16K00119.

References

1. United Nations University: World Risk Report 2016. http://collections.unu.edu/eserv/UNU:5763/WorldRiskReport2016_small.pdf. Accessed Apr 2018
2. Numada, M., Meguro, K.: Development of disaster process system case for Yabuki town in Fukushima prefecture, Seisan-kenkyu, vol. 67, no. 2, pp. 227–231 (2015)
3. Cabinet Office: The major issues of emergency response measures in the Great East Japan Earthquake, http://www.bousai.go.jp/jishin/syuto/taisaku_wg/5/pdf/3.pdf. Accessed Apr 2018
4. The council on possibility of utilization of emergency call by social networking service in a large-scale disaster, The report of the council on possibility of utilization of emergency call by social networking service in a large-scale disaster, http://www.fdma.go.jp/neuter/topics/houdou/h25/2503/250327_1houdou/02_houkokusho.pdf. Accessed Apr 2018
5. Hirohara, Y., Ishida, T.: Proposal of a cloud disaster information sharing system for disaster headquarters. In: Proceedings of the Visualization Society of Japan Visualization Conference (Hitachi 2016), vol. 36, no. 2, p. B106 (2016)
6. Ishida, T., Hirohara, Y., Kukimoto, N., Shibata, Y.: Proposal of a decision support system for the local government's disaster control headquarters. In: Proceedings of the 22nd International Symposium on Artificial Life and Robotics, pp. 649–652 (2017)
7. Ishida, T., Hirohara, Y., Kukimoto, N., Shibata, Y.: Implementation of a decision support system using an interactive large-scale high-resolution display. J. Artif. Life Robot. **22**(3), 385–390 (2017)
8. Ishida, T., Hirohara, Y., Uchida, N., Shibata, Y.: Implementation of an integrated disaster information cloud system for disaster control. J. Internet Serv. Inf. Secur. (JISIS) **7**(4), 1–20 (2017)
9. Hamamura, A., Fukushima, T., Yoshino, T., Esuga, N.: AkariMap: evacuation support system for everyday use in offline environment. In: Proceedings of the Multimedia, Distributed, Cooperative, and Mobile Symposium (DICOMO2014) Symposium, pp. 2070–2078 (2014)
10. Nonaka, H., Shoujima, H.: A decision support system for emergency operation. J. Jpn. Soc. Artif. Intell. **15**(3), 469–476 (2000)
11. Twitter API. https://developer.twitter.com/. Accessed Apr 2018
12. Open Street Map. http://openstreetmap.org/. Accessed Apr 2018

Proposal of a Regional Knowledge Inheritance System Using Location-Based AR and Historical Maps

Hayato Ito[1(✉)], Tatsuya Ohyanagi[1], Tomoyuki Ishida[2], and Tatsuhiro Yonekura[1]

[1] Ibaraki University, Hitachi, Ibaraki 316-8511, Japan
{17nm702f,17nm704r, tatsuhiro.yonekura.z}@vc.ibaraki.ac.jp
[2] Fukuoka Institute of Technology, Fukuoka, Fukuoka 811-0295, Japan
t-ishida@fit.ac.jp

Abstract. In this paper, we propose collecting personal culture of local residents to collective culture, and returning collective culture to local residents as personal culture by using AR technology and historical map. And, we propose a prototype system of regional knowledge inheritance system in order to support the protection and inheritance of regional knowledge. The prototype system has the function of overlaying the historical map on the basic map and the function of presenting the regional knowledge related to the place as an AR.

1 Introduction

Currently, the importance of a mechanism to inherit history and culture to the next generation is increasing. In history and culture, personal memory, experience, and tradition etc. are called personal culture. On the other hand, the accumulated personal culture is called collective culture [1]. Personal culture and collective culture bring mutual agglomeration and reduction action by “narrative” [2]. Narrative is event reproduction and information with stories. Regional unique narratives contribute to the maturation of community awareness among local residents and have the effect of sustaining the region. However, regional unique narratives and regional knowledge are fragmentary, and there is no system for inheriting to the next generation [3]. In order to inherit regional knowledge to the next generation, it is important to leave regional knowledge as a culture.

On the other hand, the historical map is a tool for examining past information. The historical map finds the relevance of past things and contributes to finding a new viewpoint from the data on the map [4]. Moreover, by comparing the historical map with the current map, it is expected to discover new relevance to the time-axis and the subject-axis. At this point, taking a look at the latest Information and communication technology, the augmented reality technology attracts attention. Among the AR technology, the location-based AR presents the information associating the location information with the content. By using this technology, the user can browse the AR contents with high presence. Like the historical map, the AR technology belongs to

L. Barolli et al. (Eds.): IMIS 2018, AISC 773, pp. 320–328, 2019.
https://doi.org/10.1007/978-3-319-93554-6_29

narrative because it has aspects to reproduce the event. In other words, by implementing the regional knowledge inheritance system using AR technology and historical maps, it is possible to protect the culture and contribute to the maturity of community awareness of local residents. In addition, the "Basic Plan of History and Culture" formulated by the Agency for Cultural Affairs defines a policy to accurately grasp the local cultural properties and to preserve and utilize local cultural properties comprehensively [5]. Therefore, utilization of local cultural properties by the latest ICT is in conformity with the purpose of this plan.

2 Related Works

There are many studies being performed in order to inherit regional knowledge of individuals.

Suda et al. [6] developed a "GBVoice" to accumulate the experience and knowledge of local residents as voice data and to inherit it. This system provides people with the state of life and appeal of the old days by inheriting the experience and knowledge of local residents as voice data. However, Suda et al. [6] point out that records of places without abstract events and landmarks are difficult to inherit.

Nakahara et al. [7] developed Web-GIS that enables efficient communication among local residents based on the accumulation and sharing of regional knowledge. And, Yanagisawa et al. [8] integrated the three Web applications of Web-GIS, Wiki, SNS and developed an information sharing GIS that is optimal for accumulating regional knowledge. However, the issue of both systems is to improve the significance of using the system.

Ŝaito et al. [2] developed "KACHINA CUBE" which provides a deep understanding of specific regional culture by accumulating and sharing a large number of narratives. This system realizes easy recognition of personal culture and manifestation of collective culture. However, the issues of this system are to visualize a large amount of narrative easily and to visualize the relation of narratives.

From the above, we believe that the following elements are important for inheriting regional knowledge.

- Easy to handle information of various subjects
- Does not depend on existing buildings and landmarks
- Presenting a large number of information and relationships readily

3 Research Objective

In this research, we implement a regional knowledge inheritance system using AR technology and Historical Map. By collecting personal culture of local residents to collective culture and conversely returning collective culture to local residents as personal culture, this system supports protection and inheritance of regional knowledge. This system presents various information on the historical map by operate of the time-axis, the space-axis, and the map-axis. And this system realizes complementing of the landscape and the association of the position information by extending past information using location-based AR technology. Moreover, since AR technology and historical map have sides as a narrative, this research contributes to effective inheritance of regional knowledge and maturity of community awareness.

4 System Configuration

The system configuration in this research is shown in Fig. 1. The roles of each system are described below.

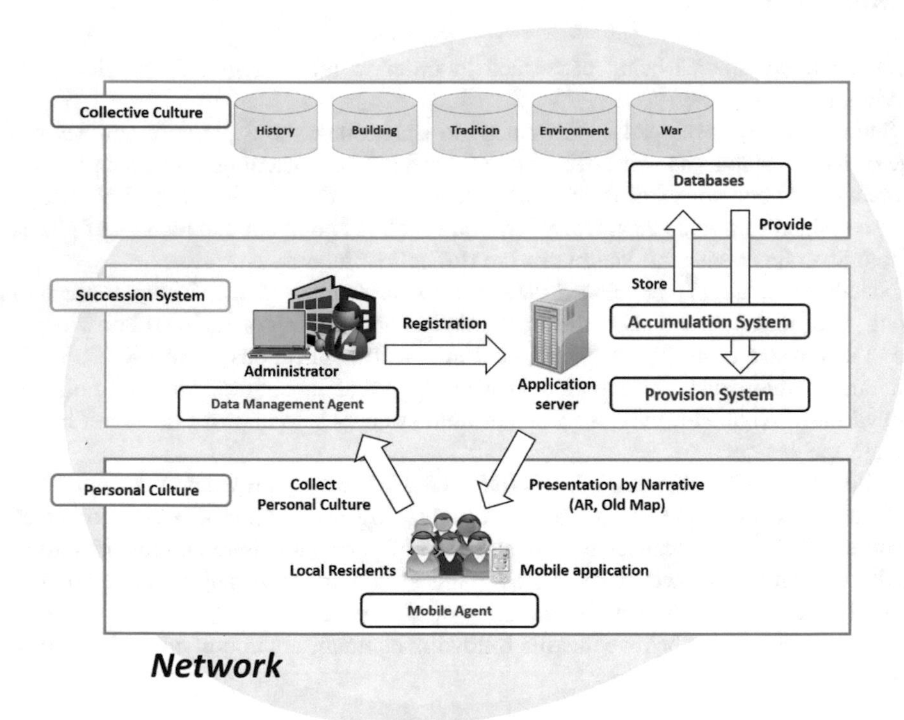

Fig. 1. The system configuration in the proposed system. This system consists of the mobile agent, the data management agent, the regional knowledge storage system, the regional knowledge distribution system, and the database server.

- Mobile Agent
  The mobile agent assumes local residents with regional knowledge. The personal culture of the mobile agent is accumulated into the collective culture by the data management agent. Users of this system can browse collective cultures expressed by AR and historical maps through mobile applications.
- Databa Management Agent
  The data management agent assumes museum curators. The data management agent accumulates the personal culture collected from the mobile agent in the collective

culture through the regional knowledge inheritance system on the application server.

- Regional Knowledge Distribution System
  The regional knowledge inheritance system operates on the application server. The regional knowledge inheritance system stores the personal culture received from the data management agent in the database server.
- Database Server
  The personal culture of various fields is stored as collective culture in the database server. The data in the database server is edited through the regional knowledge inheritance system on the application server.

5 System Architecture

The system architecture of the regional knowledge inheritance system is shown in Fig. 2.

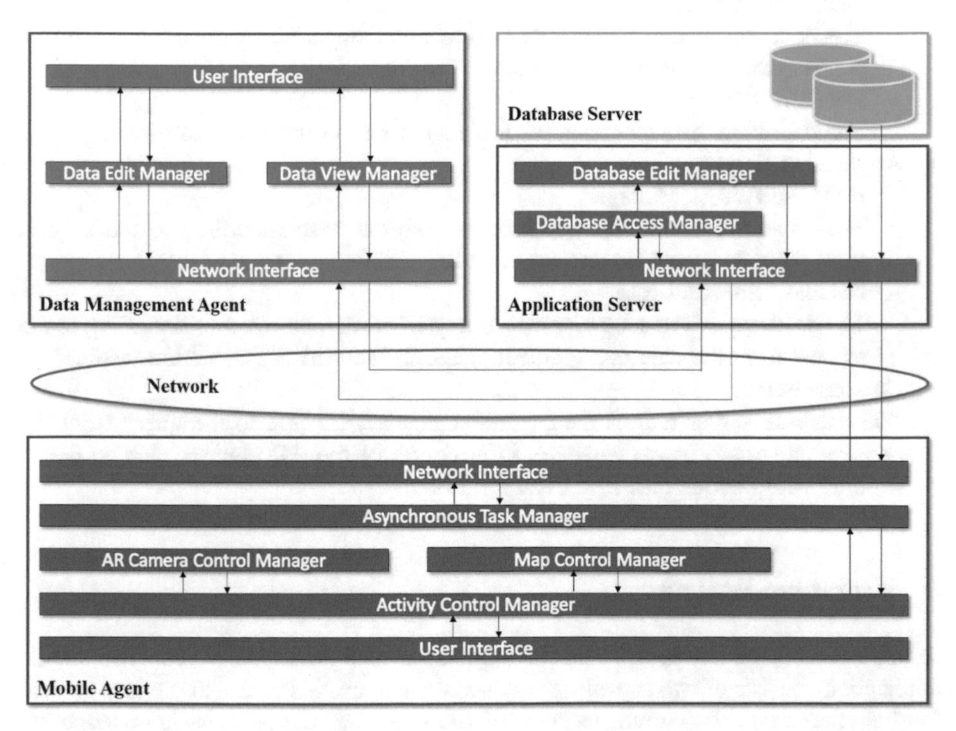

Fig. 2. The architecture of this system consists of the mobile agent, the data management agent, the application server, and the database server.

- Mobile Agent
  - User Interface
    The user interface provides contents related to the old map expressed by AR technology.

  - Activity Control Manager
    The activity control manager manages the activity of the mobile application.
  - AR Camera Control Manager
    The AR camera control manager controls the camera view of the mobile terminal and superimposes the AR content.
  - Map Control Manager
    The map control manager displays a historical map on the mobile application, and controls map switching, transmitting, and scaling.
  - Asynchronous Task Manager
    The asynchronous task manager controls asynchronous communication. This manager transmits a data acquisition request to the application server and receives data from the application server.
- Data Management Agent
  - User Interface
    The user interface provides an operation screen of the database server to the administrator user.
  - Data Edit Manager
    The data edit manager transmits data to the application server in response to a request from the administrator user.
  - Data View Manager
    The data view manager acquires the data stored in the database server.
- Application Server
  - Database Access Manager
    The database access manager accesses the database according to the request from the data management agent and the mobile agent.
  - Database Edit Manager
    The database edit manager issues a query to the database server according to the request from the data management agent and returns the execution result.
- Database Server
  The database server returns the execution result according to the query from the application server. In the database, information of the AR content such as name, caption, latitude/longitude and URL are stored.

6 Prototype System

In this research, we developed a mobile application as a prototype system. The mobile application consists of two screens, a map screen and an AR screen. In the map screen, the historical map, the current location of the user, and the regional knowledge are displayed. And in the AR screen, the regional knowledge is superimposed on the real space as an AR.

6.1 Map Screen

When the user starts the mobile application, a map screen as shown in Fig. 3 is displayed. The Open Street Map [9] is displayed in the center of the map screen, and the historical map is overlaid on the Open Street Map. In addition, the registered regional knowledge is displayed as a marker. The menu at the bottom of the map screen consists of transition button to AR screen, historical map switching spinner, transparency switching slide bar, and time-axis switching slide bar. The user can expansion/reduction the basic map by pinch-in/pinch-out operation. At this time, the overlaid historical map is also expansion/reduction interlocking with the size of the basic map. And, the current location of the user acquired from the GPS satellite is displayed as a marker on the map. Moreover, the regional knowledge related to the set time-axis is displayed as a marker on the map. The user can browse the historical map and the regional knowledge corresponding to the age by operating the time-axis switching slide bar.

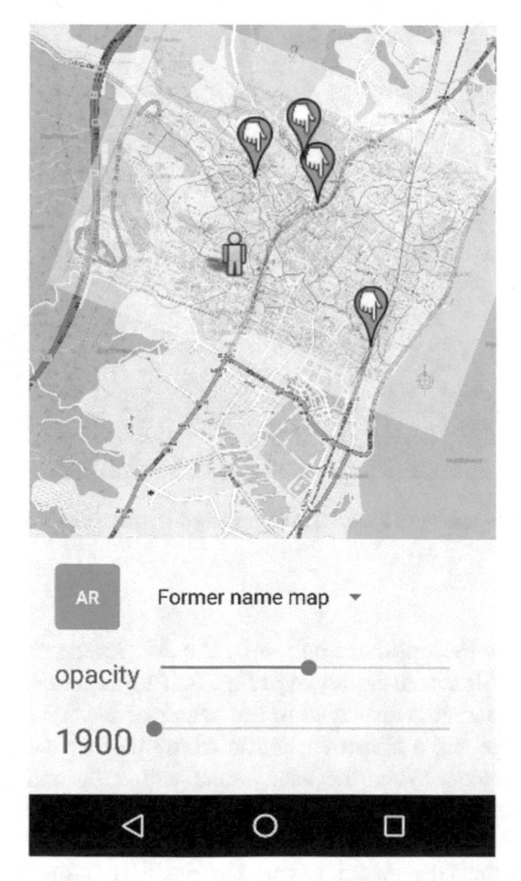

Fig. 3. The user can browse the historical map and regional knowledge on the map screen.

The type of the historical map to be overlaid on the base map can be switched by operating the historical map switching spinner. The operation of the historical map switching spinner is shown in Fig. 4.

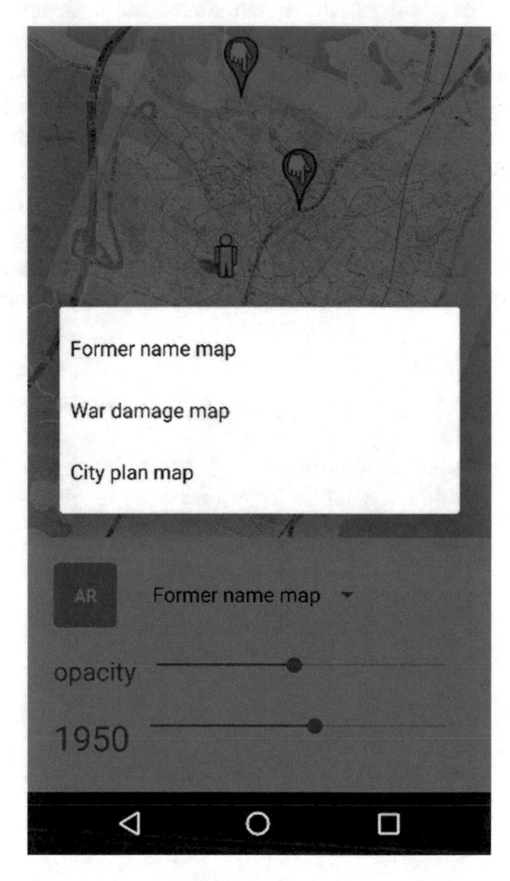

Fig. 4. The user can switch the historical map using the historical map switching slide bar.

6.2 AR Screen

When the user selects the transition button to the AR screen on the map screen, the screen transits to the AR screen as shown in Fig. 5. The camera view acquired from the rear camera of the terminal is displayed in the center of the AR screen. An annotation of regional knowledge and a transition button to the map screen are displayed at the bottom of the AR screen. When the user enters within the radius 30 meters of the registered regional knowledge, the regional knowledge is displayed as the AR content on the AR screen. At this time, the system compares the position information of the user acquired from the GPS satellite with the position information of the regional knowledge registered in the database.

Fig. 5. The user can browse the regional knowledge displayed as AR and related annotations on the AR screen.

7 Evaluation

In the future, we plan to evaluate this prototype system. Perform a questionnaire of effectiveness and satisfaction by five-point scale. Users browse contents linked to the space of the city in advance and fill the questionnaire. Target users assume the local residents of Hitachi City and the Hitachi City museum's users. Then, aggregate results, and shows whether the constructed system is effective.

8 Conclusion

In this research, we proposed the regional knowledge inheritance system for supporting regional knowledge protection and inheritance by using historical map and AR technology, and constructed a prototype system. In the prototype system, we implemented the overlaid function of the historical map and the presentation function of the regional knowledge related to the historical map. And, the prototype system has a map-axis and a time-axis in addition to the space-axis originally owned by the map. By presenting information using three axes, the user can browse the regional knowledge from several points of view. Moreover, the user can browse the regional knowledge related to the location as an AR by using the location-based AR technology.

9 Future Works

As a future development, we will implement the function to visually present the relationship between regional knowledges on the map. The following information will be used for the relationship between regional knowledges.

- Information on the time-axis representing the passage of time and aging
- Information on the space-axis based on landmarks and various position information
- Information on the map-axis between events with different fields

By this, we find relationships between knowledge in three axes of space-axis, time-axis, and map-axis, and try to create new viewpoints that cannot be found with one axis. Moreover, we plan to implement a system which enables users to capture the regional history and culture more stereoscopically.

Acknowledgments. The authors would like to thank J. Ohmori of Hitachi City Museum for great assistance with the system construction.

References

1. Valsiner, J.: Personal culture and conduct of value. J. Soc. Evol. Cult./ Psychol. **1**(2), 59–65 (2007)
2. Saito, S., Inaba, M.: Archiving knowledge in regions - development of Japanese cultural archives through collaborative construction of narratives. IPSJ SIG technical report, 2008-CH-78, pp. 61–68 (2008)
3. Science Council of Japan, Regional Research Committee: Efforts for Accumulation and Utilization of "Regional Knowledge". http://www.scj.go.jp/ja/info/kohyo/pdf/kohyo-21-h-1-7.pdf. Accessed Apr 2018
4. Gifu Prefectural Library: Use Historical Map. https://www.library.pref.gifu.lg.jp/reference/shirabekata/72_kochizu.pdf. Accessed Apr 2018
5. The Agency for Cultural Affairs, Basic Plan of History and Culture. http://www.bunka.go.jp/seisaku/bunkazai/rekishibunka/index.html. Accessed Apr 2018
6. Suda, M., Hiroi, K., Yamanouchi, M., Kato, A., Sunahara.: the development of audio contents of conveying characteristics of local regions and a proposal of distribution system for them. IPSJ SIG technical report, vol. 2013-IOT-20, no. 3, pp. 13–18 (2013)
7. Nakahara, H., Yanagisawa, T., Yamamoto, K.: Study on a web-GIS to support the communication of regional knowledge in regional communities: focusing on regional residents' experiential knowledge. Soc. Soc. Inf. **1**(2), 77–92 (2012)
8. Yanagisawa, T., Yamamoto, K.: A study on information sharing GIS to accumulate local knowledge in local communities. Theor. Appl. GIS **20**(1), 61–70 (2012)
9. Open Street Map. http://openstreetmap.org/. Accessed Apr 2018

QoS Management for Virtual Reality Contents

Ko Takayama[1] and Kaoru Sugita[2(✉)]

[1] Graduate School of Fukuoka Institute of Technology, Fukuoka, Japan
mgm18102@bene.fit.ac.jp
[2] Fukuoka Institute of Technology, Fukuoka, Japan
sugita@fit.ac.jp

Abstract. In the VR (Virtual Reality) content, the immersion is important sensibility in order to be alive in the VR space. Also, the vision should be improved by following the sight of view and high quality of videos. For this reason, the VR content streaming uses a high-speed network for sending continuously high frame rate and high frame size videos. There are many studies about video streaming technologies and QoS (Quality of Service) control mechanisms. However, they can't be used for VR streaming technology in case of limited computer network resources. In this paper, we introduce a QoS Management for Virtual Reality Contents by controlling QoS parameters according to user's requests to keep the immersive experience quality in case of limited computer network resources.

1 Introduction

The Virtual Reality (VR) contents are popular on the Internet and published on online video sites such as the Youtube. The content includes: VR video, VR communication tools and VR games. Many VR video and VR communication tools are played on VR goggles attached to a smartphone. Also, most of VR games are played on VR-HMD such as HTC vive, PS VR and so on. These devices are used for higher immersive experiences in a virtual space and they can be divided as follows:

(D1) HMD (Head Mounted Display) : Used for only viewing a virtual space,
(D2) Acceleration Sensor : Mounted for a head tracking in a virtual space,
(D3) 3D Display : Feeling a 3D effect expressed by a parallax on a display.

The quality of immersive experience can be improved by changing the frame size, frame rate and the omnidirectional viewing synchronized with head direction in the VR contents. These parameters should have higher values, but they are limitations on the performances in the Internet environment. For this reason, the VR content should keep the QoS and these parameters should be changed simultaneously without being noticed by the user [1].

There are some studies proposed for streaming of virtual reality content to mobile users [2]. In this study, VR content can be reduced when there is a deterioration of quality of the content which is played at high bit rate for ROI (Region of Intensity) and

L. Barolli et al. (Eds.): IMIS 2018, AISC 773, pp. 329–335, 2019.
https://doi.org/10.1007/978-3-319-93554-6_30

at low bit rate for other regions in an immersive omnidirectional content provided to control a viewpoint. Also, a framework is introduced for a resource allocation to construct a VR model and optimize QoS parameters in the VR contents [3]. However, these studies have not investigated the effects of immersive experience in streaming contents. Therefore, we focus on the streaming framework of VR contents for keeping the immersive experience reflected by QoS parameters in the Internet environment.

In this paper, we introduce the QoS Management for VR Contents by controlling the bit rate, the frame rate and the frame size according to user's requests to keep the immersion quality in case of limited computer network resources.

This paper is organized as follows. The QoS management for VR contents is introduces in Sect. 2. Our implementation is presented in Sect. 3. The preliminary evaluation results are discussed in Sect. 4. Finally, Sect. 4 concludes the paper.

2 QoS Management for Virtual Reality Contents

By the QoS management, the VR content is suspended when there is a low throughput on a streaming service. In this case, the streaming software does not receive frames of the VR content during the playing time. The QoS management gives priorities and changes the QoS parameters according to the limitation of available resources and the user's requests as shown in Fig. 1.

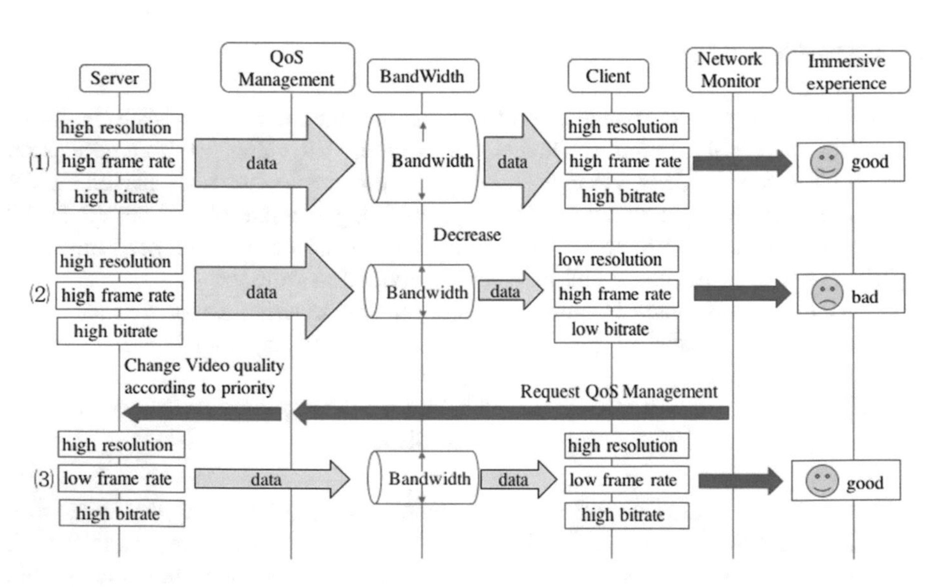

Fig. 1. QoS control for keeping immersive experience on virtual reality contents.

Using the HMDs and the VR goggles, the immersive experience is affected by following elements:

(IE1) Interactivity : The controlling response on a view point synchronized with the head direction.

(IE2) Content Quality : The visual performance in a real time and a precision of images.
(IE3) 3D Expression : The parallax on images for recognizing the depth of content.

The interactivity can be given by both the omnidirectional video and the acceleration sensor. It shows the feeling for the expansion of the virtual space. In the interactivity, the high-quality content is set at high frame size to spread all the VR space. The content quality can be provided by the frame size, frame rate and bit rate for the visual performance of the virtual space. In the content quality, a high-quality content is used for high CPU and network load to play a high precision image in a real time. For realizing the 3D expression is needed the parallax of 2 videos for recognizing the depth of a virtual space.

Table 1. Examples of QoS parameters.

QoS parameter	High	Normal	Low	None
Frame rate	60 fps	30 fps	15 fps	Still Frame
Frame size	Over FHD (Ex. UHD and QHD)	FHD	HD	Invalid
Bit rate	15 Mbps	8 Mbps	4 Mbps	Invalid
Stereo video	Always	Active	Passive	Invalid
Omnidirectional video	Always	Active	Passive	Invalid

The QoS parameters of VR content are shown in Table 1. Especially, the QoS parameter 'None' meaning is as follows:
(QP1) Frame Rate : Displaying a still frame (invalid updating a video streaming),
(QP2) Frame Size : Invalid playing a video streaming,
(QP3) Bit Rate : Invalid playing a video streaming,
(QP4) Stereo Video : Displaying a general video (invalid displaying a parallax video),
(QP5) Omnidirectional Video : Invalid supporting head-tracking or interactive operation for viewing direction.

These parameters are selected by a preset considering user's requests as shown in Table 2. The preset is a set of priority values given by the QoS parameters when the other parameters are degraded according to the available throughput.

In the VR content, the QoS is managed by the above mentioned QoS parameters for keeping the immersive experience reflecting both user's request and performances of a computer network environments as shown in Fig. 1. In a normal condition, the VR content is played as an original content by keeping the QoS using the available bandwidth. In a bad condition, the VR content is changed to low using QoS parameters by applying a short transmission rate and retaining the immersive experience. In this procedure, new QoS parameters are used for the VR content after checking requested QoS parameters and the available transmission rate when a server receives a notification message for a short transmission rate of QoS. The priority values of QoS parameters at starting point of playing of the VR content are set by default values and then modified by considering repeated user's requests and evaluations.

Table 2. Examples of presets for user's requests in the VR content.

Preset	Preferential in 3D expression	Preferential in smooth motion	Preferential in precision draw	Preferential in interactivity
Frame rate	Normal	High	Normal	High
Frame size	High	Normal	High	Normal
Bit rate	Normal	High	High	Normal
Stereo video	High	Low	Low	Low
Omnidirectional video	Low	Normal	Normal	High

3 System Implementation

Our system is implemented as a video streaming Web application running on a WebRTC. In the application, a VR content is an SBS (Side by Side) video jumping and running a 3D character as shown in Fig. 2. A smartphone is used as a client software which performs both a video sender and a video receiver as shown in Table 3. The default frame size of SBS video is QHD (2560 $\times$ 1440 [pixel]) for a MacBook and FHD (1920 $\times$ 1080 [pixel]) for a smartphone. The default frame rate was set to 60 [fps]. A Macintosh Mini is used as a server software for establishment of P2P

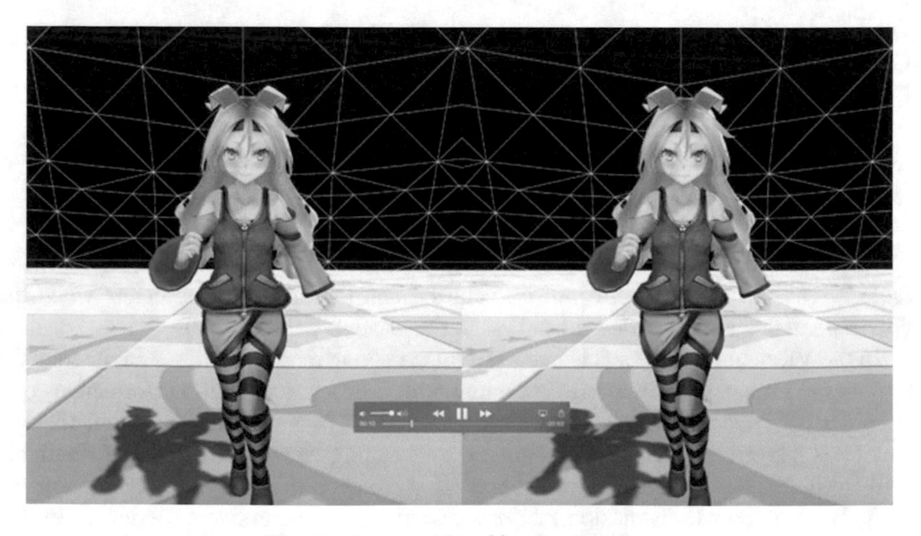

Fig. 2. An example of implementation.

connection constructed on a virtual desktop environment and developed by a Node.js and an Express as shown in Tables 4 and 5, respectively. The starting process of video streaming is shown in Fig. 3. This process is carried out by (1) Accessing a Web Server, (2) Displaying a Web page, (3) Loading a video at a video sender, (4) Selecting a Peer for a destination of video streaming, (5) Signaling the Web server (as a Web RTC signaling server), (6) Establishment of P2P connection with a Peer and (7) Starting to play a video streaming in order.

Table 3. Client environment.

Types of client	Video sender	Video receiver
Hardware	MacBook (Retina, 13-in., Late2013)	AQUOS SERIE mini SHV38
OS	mac OS Sierra	Android7.0
CPU	Intel Core i5	Snapdragon617
GPU	Intel Iris 1536 MB	Adreno 405
RAM	DDR3 16 GB 1600 MHz	16 GB
Display	2560 × 1600 pixel	1080 × 1920 pixel

Table 4. Server environment.

Hardware	Mac mini (Late 2014)
OS	macOS Sierra
Processor	Intel Core i5
Memory	DDR3 16 GB 1600 MHz

Table 5. Virtual server on specifications.

Type of software	Product
Virtualization software	Virtual Box 5.0.40r115130
Management software for guest OSs	Vagrant 1.7.4
Guest OS	Ubuntu 16.04.3 LTS
Java script environment on the server	Node.js v6.11.1
Web application frame work	Express 4.13.0

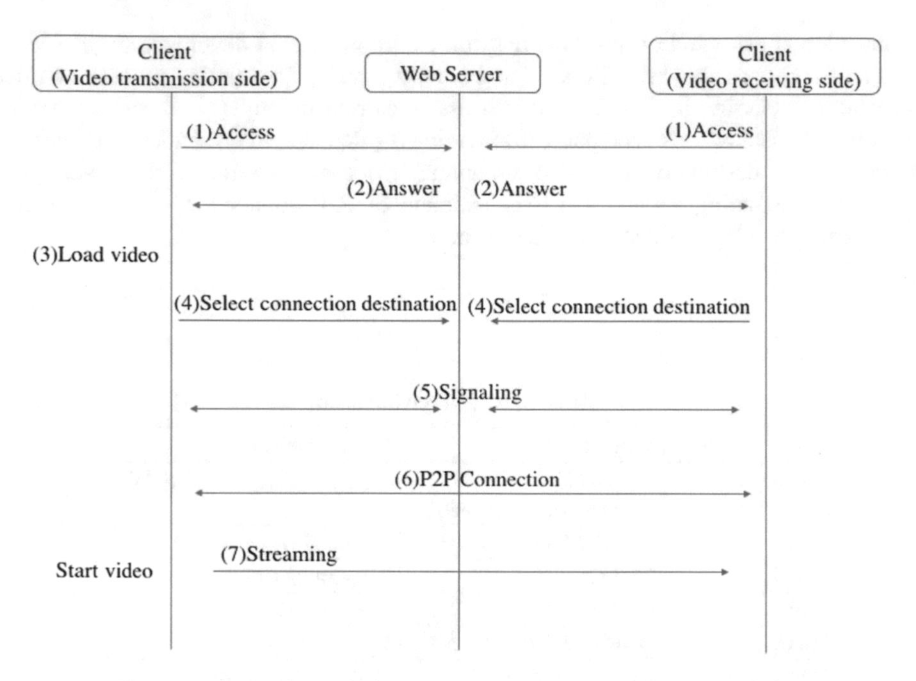

Fig. 3. Starting process of live video streaming over the Web system.

4 Conclusions

In this paper, we introduced a QoS Management for Virtual Reality Contents by controlling QoS parameters according to user's requests to keep the immersive experience quality in the case of limited computer network resources.

We discussed about the influence on immersive experience related with some elements of VR device and VR content. We also considered QoS parameters of VR content and its QoS management. The VR content quality of the immersive experience is decreased by suspending the play in the case of low throughput for transmitting a frame in the available computer network environment. In this case, the VR content should be degrading by considering QoS parameters priority and according to the user's requests for the streaming service.

Currently, we are carrying out some preliminary evaluation to find the available reduction data size for QoS parameters in some type of video formats.

In the future, we will propose a management framework for controlling QoS in VR content. Also, we would like to develop an immersive experience software supporting the QoS controlling function in the VR content.

References

1. Ejder, B., Mehdi, B., Muriel, M., Mérouane, D.: Toward interconnected virtual reality: opportunities, challenges, and enablers. IEEE Commun. Mag. **55**(6), 110–117 (2017)
2. Hamed, A., Omar, E., Mohamed, H.: Adaptive multicast streaming of virtual reality content to mobile users. In: Proceedings of the on Thematic Workshops of ACM Multimedia, pp. 170–178 (2017)
3. Mingzhe, C., Walid, S., Changchuan, Y.: Resource management for wireless virtual reality: machine learning meets multi-attribute utility. In: IEEE Global Communications Conference, pp. 4–8 (2017)
4. Unity Version 2017.3.0f3 (a9f86dcd79df). (c) 2017 Unity Technologies ApS. All rights reserved

A Webshell Detection Technology Based on HTTP Traffic Analysis

Wenchuan Yang[1], Bang Sun[2](✉), and Baojiang Cui[2]

[1] School of Cyberspace Security, Beijing University of Posts and Telecommunications, Beijing, China
yangwenchuan@bupt.edu.cn

[2] Nation Engineering Laboratory for Mobile Network Security, Beijing University of Post and Telecommunications, Beijing, China
{confuoco, cuibj}@bupt.edu.cn

Abstract. Webshell is a common backdoor program of web applications. After an attacker uploads Webshell successfully by using a vulnerability. Attacker can get a command execution environment to control the web server by access Webshell. In this paper, an attack detection technology based on SVM algorithm is proposed by analyzing the network traffic of attackers accessing Webshell. This technology realizes the detection of Webshell attack traffic in HTTP traffic by means of the method of supervised machine learning model. And this technology achieves high accuracy and recall rate. After detecting abnormal traffic, the system can locate the Webshell according to traffic information. And eliminate the backdoor in time to ensuring the security and stability of the web server. So it also can help to monitor the trend of intrusion and network security.

1 Introduction

With the continuous development of mobile Internet technology, the Web application of B/S architecture are becoming more and more popular. Web applications are also expose various types of vulnerabilities, such as SQL injection, XSS, and uploading malicious Webshell. Uploading Webshell is a serious vulnerability. Attackers can get higher permissions once the Webshell executed by web server, such as download or upload any files in server, read the database and command execution [1].

Webshell is a common backdoor file, which is often used by attackers as backdoor tools for Web server operation management. It is also an auxiliary tool to promote web permissions to system permissions in penetration testing.

Webshell is very harmful. It shows that the attacker had used Webshell to obtaining higher privileges if the Webshell is found in the Web server. Therefore, it is very important that accurately detect the attack traffic of Webshell from the HTTP traffic for locating Webshell malicious files and ensure the safety of Web servers [2].

L. Barolli et al. (Eds.): IMIS 2018, AISC 773, pp. 336–342, 2019.
https://doi.org/10.1007/978-3-319-93554-6_31

2 Related Work

The existing Webshell detection technologies are all white box detection. Detect the source code of Webshell script file or detect based on the host network traffic. It is divided into two categories, host based detection and network based detection.

2.1 Host Based Detection

The commonly used detection method in industry is to use the known keywords directly as a feature. Use 'grep' sentences to search for suspicious files after manual analysis, or use programs to check regularly the MD5 values of existing files and check whether new files are generated. This intuitionistic detection method can easily evaded by attacker using confusing means.

Emposha's WebShelDetector is a PHP script. It uses a large Wehshel feature library, which can be used to identify Webshell of php/cgi/asp/aspx, and the recognition rate is up to 99% [3].

2.2 Network Based Detection

The current research focuses on configuring an intrusion detection system or WAF to detect Webshell at the entrance of the network. Fireeye proposes to use Snort configuration feature rules to detect a sentence of "Chinese kitchen knife" in a sentence of Trojan horse. Yang detect uploading Webshell behavior by configuring the ModSecurity core rules.

The above two methods determine whether an attacker is uploading a malicious script file by analyzing of whether the HTTP request contains special keywords, such as <form, <php, <?. So that all upload HTML or script behavior will be considered a uploaded Webshell attack. This method requires a lot of memory overhead and there is a possibility of false positives. And it only can detect the behavior of uploading Webshell. It is helpless for detecting the existing Webshell in Web server [4].

This paper presents a Webshell detection technology based on HTTP traffic analysis. Training the HTTP traffic detection model by machine learning. The model can detect the malicious Webshell traffic in HTTP traffic and then locate the attacker ip address and Webshell in the Web server. And then the hidden danger of Web server will clear.

3 Definition and Classification of Webshell

Webshell is a collection of Web and shell, and Web represents a server that opens Web services, and shell means access to server operation permissions. Webshell gets the management permissions of the Web server through the Web service, so as to penetrate and control the Web server. From the attacker's point of view, Webshell is the back door of a scripted Trojan written by PHP, JSP, or ASP. Attacker uploaded Webshell to the Web server directory by exploiting upload vulnerabilities or system vulnerabilities is found on the site. By accessing script files, it sends instructions to script files, which can control Web servers, such as reading website database, downloading arbitrary files on the server, executing commands, etc. [5].

The attacker usually divides Webshell into three categories: "great Trojan", "Trojan", "one sentence Trojan" according to the size and function of Webshell.

Great Trojan has the most comprehensive function, usually including the operation interface, and can perform file operation, command execution and database operation under the graphical interface. It often has large files, including all the characteristics of the Trojan, at the same time, use the confusing means to prevent the detection, and the concealment is better.

Trojan usually contains only a single function, including file upload or database operation. Attackers will use Trojan as an upload springboard when Web website has a limitation on the size of uploaded files.

One sentence Trojan usually refers to a code containing a script that can execute system commands, such as the 'eval ()' function, and so on. Because there is only one line of code, one sentence Trojan has better hiding characteristics and can be inserted into the original code of Web website. And it is hard for administrators to find it. One sentence Trojan is the most common Trojan used by the attackers at present [6].

4 Approach

The Webshell detection technology based on HTTP traffic analysis uses a supervised machine learning algorithm. The Webshell detection technology based on HTTP traffic analysis uses a supervised machine learning algorithm. The training set contains the normal flow and abnormal flow of the web access logs. The SVM algorithm classification model is obtained by learning and training on the training set. Then, according to the classification model, we determine whether the HTTP requests to be detected are the attack traffic. The threat index is 1 if it is not attack traffic. If it is attack traffic, then the threat index is 1, if the corresponding HTTP response state code is 200, and the response body length is greater than 0, the threat index is 2. The system flowchart is shown in Fig. 1.

The Webshell detection system based on SVM algorithm first preprocessed the labeled HTTP request training set. The parameter part and post form part of URL are extracted from HTTP request and stitching together. Then 4-gram algorithm is used to segment the preprocessed samples and form bag of words. After the bag of words is formed, all the samples in the sample are vectorized. And then trained the classification model by the samples after preprocessing.

The initial data is a pair of HTTP request and response packages. First, preprocess HTTP request. And then vectorize the sample by using bag of words. Using the classification model to judge the vectorized HTTP requests. The threat index is 0 if it is not malicious requests. If it is malicious requests, but the corresponding HTTP response is not included in the package such as the status code is 404, the threat index is 1. If the HTTP response corresponding to the content contained in the package, then web server successfully returned the information. So attacker attacked success that the threat index is 2.

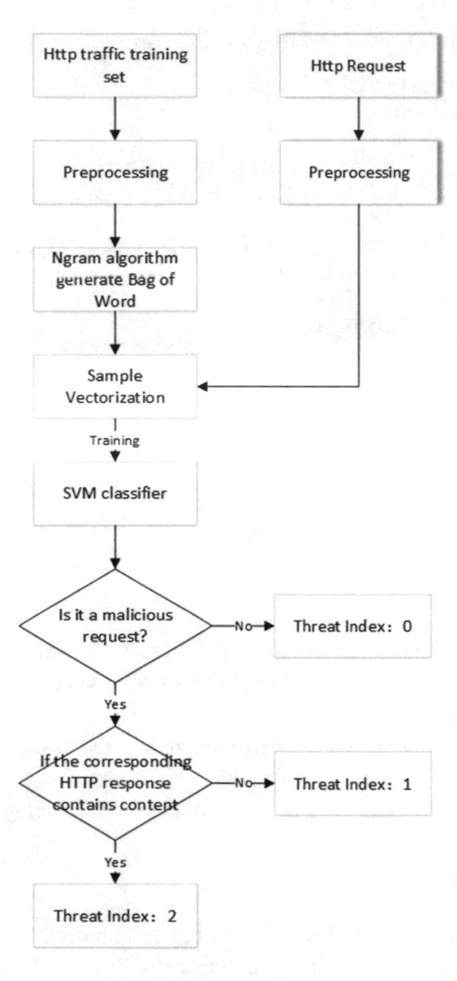

Fig. 1. System flowchart

4.1 Sample Vectorization

For each sample of training set, extract the parameters of get request part and the parameters of post request part. Decode them and merge them together. Decoding includes Base64 decoding, URL decoding, etc.

Divide every sample in the sample set to generate bag of words by using 4-gram algorithm and filter all words that are spaces, numbers or symbols in the bag. And convert all words into lowercase. Each sample in the sample set generate a text vector based on the generated word bag.

The following algorithm is used to slice the sample

```
def ngram(str):
    lexicon = []
    n = 4
    for i in range(len(str)):
        if len(str[i:i + n]) < n:
            break
        if str[i + 1:i + n].isdigit():
            continue
        if str[i:i + n] == ' '*n:
            continue
        if str[i:i + n] == '.'*n:
            continue
        lexicon.append(str[i:i + n].lower())

    return ' '.join(lexicon)
```

5 Implementation and Experiments

5.1 Experimental Data

The training set used in the experiment is mostly from the data after desensitizated of a competition. A total of 1743 Webshell malicious traffic samples and 3500 normal flow samples were included.

In order to evaluate the effect of the experiment, the training sample and the test sample need to be extracted first. In this paper, 80% samples were selected as training data by random sampling, and 20% of the samples were testing data.

5.2 Experimental Method

First, preprocess all training samples. And the Ngram algorithm is used to generate bag of words. And then vectorizing training samples while save the bag of words which is convenient for the samples of the test set to be vectorized.

In this paper, we use 3-gram algorithm and 4-gram algorithm to generate bag of words respectively. And use naive bayes, random forest, and SVM three kinds of classification algorithms to do comparative tests. Finally resulting in 6 sets of experimental results, and then choose the best algorithm.

5.3 Experimental Result

In this experiment, the Webshell malicious traffic sample is represented as 1, the normal flow sample is 0, and the classification is shown in Table 1.

Table 1. Classification.

		Forecast results	
		1	0
Actual results	1	True positive (TP)	False negative (FN)
	0	False positive (FP)	True negative (TN)

Use three evaluation indexes: accuracy rate, recall rate, F1-score. Accuracy rate defined as Precision $= \frac{TP}{TP+FP}$. Recall rate defined as Recall $= \frac{TP}{TP+FN}$. F1-score defined as F1 $= \frac{2}{1/Precision + 1/Recall}$.

For the two kinds of bag of words generation algorithm mentioned in Sect. 5.2 and three classification algorithms, 6 groups of classification results are produced which is showed in Table 2.

Table 2. Classification results.

Bag of words generation algorithm	Classification algorithm	Accuracy rate	Recall rate	F1-score
3-gram	Navie Bayes	95.4%	95.4%	95.4%
4-gram	Navie Bayes	94.8%	94.7%	94.7%
3-gram	SVM	98.2%	98.2%	98.2%
4-gram	SVM	98.9%	98.9%	98.9%
3-gram	Random Forest	97.3%	97.3%	97.3%
4-gram	Random Forest	97.6%	97.5%	97.5%

It can be seen from the table that the results of the 4 algorithms all have good performance. And the accuracy rate, recall rate and F1-score are above 90%. The highest evaluation index is obtained by using the 4-gram algorithm to generate the bag of words and the SVM algorithm to classification. This method is also used in real system testing to identify whether a HTTP request is malicious.

After the system is actually deployed, the real attack traffic is captured as shown in Fig. 2. This is a typical attacker who sends instructions to a Trojan and executes the command. But the HTTP status code returned by the server was 503 and did not return the content, so the attack did not attack successfully, and the threat index was 1. But there may still be potential risks, and the network administrator should locate the file in time and see if it is a Trojan.

Fig. 2. One of system detection result

6 Conclusion

From the perspective of network traffic, this paper analyzes the traffic characteristics of attackers attacking by Webshell and obtains a detection model based on 4-gram algorithm and SVM algorithm. A trained classification model is used to detect malicious requests of HTTP and combined with the characteristics of HTTP response packets. Each HTTP session can be evaluated by threat. This technology has practical application value.

References

1. Duan, L.H., Zhang, Z.F., Chen, et al.: Survey of web backdoor detection and protection. Institute of Computer Science and Technology of Peking University, pp. 893–899 (2014)
2. Jian, K., Ma, D., et al.: Research on Webshell detection method based on decision tree. Netw. New Media Technol. **6**, 15–19 (2012). https://doi.org/10.3969/j.issn.2095-347X.2012.06.004
3. Biwei, H.: Research on Webshell detection method based on Bayesian theory. Sci. Technol. Square **6**, 66–70 (2016). https://doi.org/10.3969/j.issn.1671-4792.2016.06.016
4. Chen, J.-X.: A Method for Detecting Webshell on Windows Platform, pp. 533–535 (2011)
5. Du, H., Fang, Y.: PHP real time dynamic detection of Webshell. Netw. Secur. Technol. Appl. **12,** 120–121, 125 (2014). https://doi.org/10.3969/j.issn.1009-6833.2014.12.074.4. Michalewi
6. Tu, R.D., Guang, C., Xiaojun, G., et al.: Webshell detection techniques in web applications, pp. 1–7. School of Computer Science & Engineering, Southeast University, Nanjing, China (2014)

Enhanced Secure ZigBee Light Link Protocol Based on Network Key Update Mechanism

Jun Yang[1(✉)], Ruiqing Liu[2], and Baojiang Cui[3]

[1] School of Computer Science and Technology, Beijing University of Posts and Telecommunications, Beijing 100876, China
junyang@bupt.edu.cn
[2] National Engineering Laboratory for Mobile Network Security, Beijing University of Posts and Telecommunications, Beijing 100876, China
liurq@bupt.edu.cn
[3] School of Cyberspace Security, Beijing University of Posts and Telecommunications, Beijing 100876, China
cuibj@bupt.edu.cn

Abstract. In recent years, the market demand for smart devices continues to increase. As a widely used communication technology, applications of ZigBee Light Link protocol and its security have received extensive attention in recent years. This article discusses the security of the ZigBee Light Link commissioning protocol. Based on the hash chain technology, this paper proposes a network key update protocol based on the security-enhanced ZLL protocol. In the end, the performance evaluation and comparison are given.

Keywords: ZigBee Light Link · Hash chain · Key update

1 Introduction

IoT (Internet of things) technology is an important part of the new generation of information technology. ZigBee, one of the communication protocols, plays a vital role in the development of IoT technology. ZigBee protocol specification is created by the ZigBee Alliance based on the IEEE 802.15.4 standard specification. ZigBee Alliance is formed of a set of companies, providing an open-source wireless networking standard aimed at monitoring, and controlling services applications where responsiveness is more important for real-time operations than high bandwidth [1]. Compared with other traditional communication protocols, ZigBee has many advantages like low cost, low energy and high fault tolerance.

The ZigBee protocol has a wide range of practical applications, such as health care [2], voice communication [3], wireless sensor networks in agriculture [4], home automation networks [5] and smart lighting [6]. In lighting industry, ZigBee technology defines a global communication standard for interoperable and very easy-to-use consumer lighting and control products.

Since more and more smart lighting devices are equipped with ZigBee as their communication protocol, more security flaws and problems will also emerge. This paper points out that ZLL touchlink commissioning protocol is vulnerable to attack due

L. Barolli et al. (Eds.): IMIS 2018, AISC 773, pp. 343–353, 2019.
https://doi.org/10.1007/978-3-319-93554-6_32

to lack of identity authentication and network key update mechanism. Therefore, a security enhancement protocol that incorporates a network key update mechanism has been proposed to address security flaws.

2 Related Works

ZigBee Light Link standard describes ZLL stack profile and defines the protocol procedure. It was published in 2012 by ZigBee Alliance [7].

In order to research the vulnerabilities of ZigBee systems, many experts have made great contributions. In 2009, some researchers developed a ZigBee Exploitation framework [8]. Radmand and Domingo pointed out how to exploit ZigBee protocol based on compromised cryptographic keys in 2010 [9]. In 2014, Olawumi with others sum up the attack and defense experience and put forward practical attacking scenarios on ZigBee, plus with some schemes about how to patch those vulnerabilities [10]. In Black Hat conference 2015, Zillner showed how to exploit ZigBee vulnerabilities using keys defects [11].

Some experts proposed enhanced ZigBee systems by deploying secure mechanisms. Son Thanh and NguyenChunming Rong researched on ZigBee Security using identity-based cryptography [12]. Radmand researched on ZigBee PRO security assessment based on compromised cryptographic keys and provided some countermeasures in 2010 [9]. In 2016, Frederik Armknecht transformed ZLL commissioning procedure to a formal security model for better protocol designing [13].

3 Prerequisite Knowledge for ZLL

Generally, ZigBee protocol stack contains four layers: Physical layer, Medium Access Control layer, Network layer and Application Security layer. Physical layers and Medium Access Control layer are included in IEEE 802.15.4 standard to provide basic communication support.

ZigBee protocol defines three types of logical devices and each type has a specific role:

1. Coordinators:
   It is responsible for setting up the network parameters, e.g. selecting the network encryption key and creating the network.
2. Routers:
   Routers are usually the intermediates nodes and route the information sent by end devices to the coordinator; in some cases routers could also act as an end device.
3. End Devices:
   End devices are usually the sensor nodes that collect environment data, e.g. thermostat, motion detectors, and infrared devices.

3.1 ZLL Touchlink Commissioning

End devices joins into networks by means of “commissioning”. Procedure “commissioning” consists of three processes: “Device discovery”, “Identify”, and “Start a new network”.

“Device discovery” is initiated by the initiator to find devices that are willing to join the network. After it, a list of detailed information of potential devices is generated. In the “Identify” session, the target will allow users to identify themselves in some way. In the “Start a new network” session, the initiator will exchange network keys with the target to complete the process of adding new devices to the network.

The key bitmask field (KBM) is 16-bits in length and specifies which keys (and hence which encryption 15 algorithms) are supported by the device. The appropriate key shall be present on the device only if its 16 corresponding bit is set to 1 otherwise the key is not supported. Bit i of the key bitmask field shall 17 correspond to key index i, where $0 \leq i \leq 15$.

The detailed interaction process is shown in the Fig. 1, where the nouns can be found in Table 1. Here only the calculation process of the network key is specifically introduced.

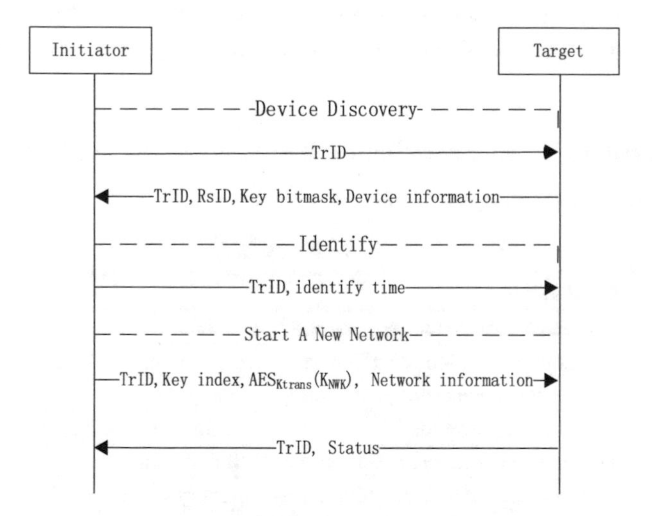

Fig. 1. ZLL commissioning flow

(1) Concatenate TrID and RsID twice to a 128 bits string.
(2) Select the key in the key bitmask KI as encryption key and AES as encryption method to encrypt the input in step.1 thus generate the transport key Ktrans.

$$\text{Ktrans} = \text{AES}_{\text{KI}}(\text{TrID}\| \text{TrID} \|\text{RsID}\|\text{RsID}) \quad (1)$$

(3) The initiator randomly generate a 32-bit network key K_{NWK} and use the Ktrans to encrypt K_{NWK}.

$$\text{Encrypted}(K_{NWK}) = AES_{Ktrans}(K_{NWK}) \tag{2}$$

(4) Send TrID || Key index || $AES_{Ktrans}(K_{NWK})$ || Network as ***network start request frame*** to the target without encryption.

3.2 Security Assessment

In the ZLL touchlink commissioning protocol, there are mainly three concerns. First, the communication between targets and the initiator are transparent easily leads to information theft. Second, without identity identification of both initiator and end targets, attackers can easily steal information. Third, without a network key update mechanism, it is easy to cause known key attacks

4 Enhanced ZLL Commissioning Protocol Design

Aiming at the defections mentioned above, we propose an enhanced protocol by adding some protection mechanisms: identity authentication, key updating and timestamps technique.

4.1 Enhanced ZLL Touchlink Commissioning Based on Hash Chain

Table 1 lists the notations used below

Table 1. Notations

Notation	Explanation
TrID	The transaction identifier generated by the initiator
NTrID	The new transaction identifier generated in the network key update section
RsID	The response identifier generated by the target
NRsID	The new response identifier generated in the network key update section
ks	The fixed ZLL key stored in every ZigBee device
h	A strong secure hash function stored in every ZigBee device
Si	A secret seed generated by the initiator
St	A secret seed generated by the target
Ti(1,2,3……)	Timestamps generated by the initiator
Tt(1,2,3……)	Timestamps generated by the target
KBM	Key bitmask, which contains the ID and content of the available keys
KI	Key index, the ID of selected key from KBM
||	Concatenation of expressions
()ks	Use ks as key and user-selected symmetric encryption algorithm to encrypt
Ki	The number of times of hash function for the initiator
Kt	The number of times of hash function for the target
K_{NWK}	The negotiated network key
NK_{NWK}	The new network key

The concrete interactive process are just like Fig. 2.

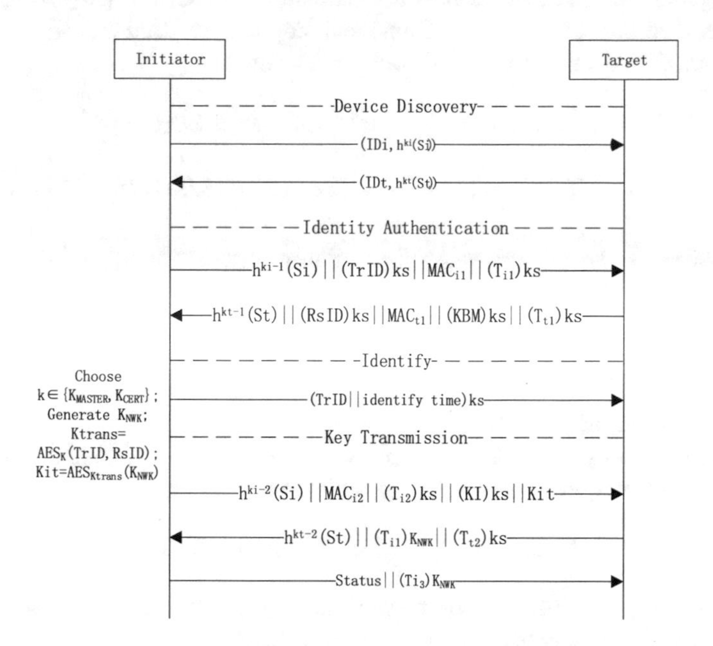

Fig. 2. Enhanced ZLL commissioning flow

4.1.1 Device Discovery

1. First, the initiator will generate the initial parameters: ***Si***, ***Ki*** and ***h^{Ki}(Si)***. Then broadcast ***scan request frame*** including its identification ***(ID, h^{Ki}(Si))***.
2. Targets willing to join in the network will generate: ***St, Kt and h^{Kt}(St)*** then unicast a ***scan response frame*** to the initiator.

4.1.2 Identity Authentication

1. Initiator → Target

The initiator calculate MACi1 and ß as below then send ß to target.

$$\mathrm{MACi1} = \mathrm{h}(\mathrm{h}^{\mathrm{ki}-2}(\mathrm{Si})\|\mathrm{Ti1}) \tag{3}$$

$$ß = \mathrm{h}^{\mathrm{ki}-1}(\mathrm{Si})\ \|\ (\mathrm{TrID})\mathrm{ks}\ \|\mathrm{MACi1}\|[\mathrm{Ti1}]\mathrm{ks} \tag{4}$$

2. Target → initiator

(1) Decrypt Ti1 to verify the time to ensure the freshness of messages.

(2) Compute $h(h^{ki-1}(Si))$ to compare it with stored $h^{ki}(Si)$. If they are equal, initiator is legal. Then target stores initiator's identity as (IDi, $h^{ki-1}(Si)$). If they are not same, then target discards the frame and stop this transaction.
(3) Calculate MACt1 and ß,send ß back to the initiator.

$$MACt1 = h(Tt1 || h^{kt-2}(St) || MACi1) \tag{5}$$

$$ß = h^{kt-1}(St) \,||\, (RsID)ks \,||\, MACt1 \,||\, (KBM)ks \,||\, (Tt1)ks \tag{6}$$

Encrypt TrID and identify time together to the target so that target can identify itself.

4.1.3 Key Transmission

1. Initiator → Target
   (1) Select the key in the key bitmask KI as encryption key. Calculate Ktrans and use it to encrypt stochastically generated network key.

$$Ktrans = AES_{KI}(TrID \,||\, TrID \,||\, RsID \,||\, RsID) \tag{7}$$

   (2) Randomly generate a 32-bit network key K_{NWK} and use the Ktrans to encrypt K_{NWK}. Kit means the encrypted network key.

$$Kit = Encrypted(K_{NWK}) = AES_{Ktrans}(K_{NWK}) \tag{8}$$

   (3) Calculate MACi2 and ß, then unicaste ß to the target.

$$MACi2 = h(Ti2 \,||\, K_{NWK}) \tag{9}$$

$$ß = h^{ki-2}(Si) \,||\, MACi2 \,||\, (Ti2)ks \,||\, (KI)ks \,||\, Kit \tag{10}$$

2. Target → Initiator
   (1) Decrypt Ti2 to verify the message's effectiveness.
   (2) Utilize $h^{ki-2}(Si)$ received this time and Ti1 stored to calculate MACi1 and compare it with MACi1 received before. Close the transaction if they are different.
   (3) Decrypt Kit using TrID and RsID received. Decrypt K_{NWK} and Ti2 to compute MACi2 to make sure the K_{NWK} is correct. If it is, then successfully establish network key else stop the transaction.
   (4) Send initiator the ß.

$$ß = h^{kt-2}(St) \,||\, (Ti1)K_{NWK} \,||\, (Tt2)ks \tag{11}$$

3. Initiator → Target
   (1) Decipher Tt2 to ensure the freshness of that message.

(2) Use Tt1, h^{kt-2}(St) with MACi1 received before to compute MACt1. If this one is same with MACt1 stored. Stop the transaction if any bit is different.
(3) Use K_{NWK} to decipher Ti1 to ensure the target have the correct network key. Send ß back including the status of network key negotiation.

$$ß = \text{Status} \,||\, (Ti3)\, K_{NWK} \tag{12}$$

5 Network Key Update Mechanism of ZLL

The network key is used to protect the messages transmitted after. It doesn't change in the lifespan of the network in the old protocol, which will lead to known key attack and other security problems. As Fig. 3 shows, we design a network key update mechanism based on hash chain for ZLL to strengthen network security. Since network key is shared between initiator and every target, it may lead to single point of failure if the initiator have much computing task. Thus the new network key is calculated by end devices in our design.

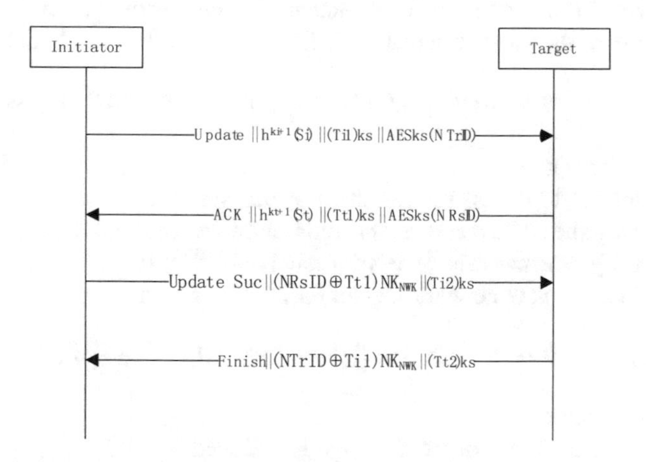

Fig. 3. Network key update protocol

Network key update interval is determined by the initiator and the surrounding environment of the network.

The new network key is calculated based on the new transaction identifier "NTrID" new response identifier "NRsID", which are randomly generated in the network key section, plus with the old network key K_{NWK} and the dependable primary key Ks. The detailed algorithm just like Fig. 4 illustrates.

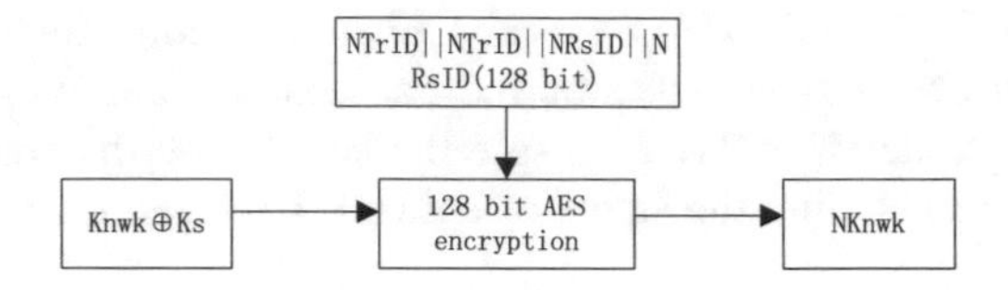

Fig. 4. Network key calculating procedure

1. Initiator → Target
The initiator will generate a timestamp Ti and new transaction identifier NTrID, then concatenate them as ß. Send ß to the target.

$$ß = \text{Update} \; || \; h^{ki+1}(Si) \; || \; AESks(NTrID) \; || \; (Ti1)ks \tag{13}$$

2. Target → Initiator
   (1) Decrypt Ti1 to ensure the freshness of the message.
   (2) On reception of the update message, the target will deploy hash function on stored $h^{ki}(Si)$ and verify whether it is same with received $h^{ki+1}(Si)$. This step aims at making sure the updating request comes from the trusted initiator.
   (3) Decrypt NTrID and store it to calculate the new network key.
   (4) In the end, the target calculate NRsID, then send ß to the initiator.

$$ß = ACK \; || \; h^{kt+1}(St) \; || \; AESks(NRsID) \; || \; (Tt1)ks \tag{14}$$

3. Initiator → Target
   (1) Decrypt Tt1 to ensure the freshness of the message.
   (2) Receiving the ACK message, the initiator deploy hash function on stored $h^{kt}(St)$ and verify whether it is same with received $h^{kt+1}(St)$.
   (3) Calculate the new network key and send ß to target.

$$ß = \text{Update Suc} \; || \; (NRsID \oplus Tt1)NK_{NWK} || \; (Ti2)ks \tag{15}$$

4. Target → Initiator
   (1) Decrypt Ti2 in case of the message is replaced.
   (2) Since target has known NTrID and NRsID, therefore it knows NK_{NWK}. Decipher ß and compare the result with calculated NRsID ⊕ Tt1 using stored data. No mistake means the initiator has correctly updated the network key.
   (3) Now tell the initiator that the target has updated successfully also. ß is transmitted to the other as follows.

$$ß = \text{Finish} \; || \; (NTrID \oplus Ti1)NK_{NWK} || \; (Tt2)ks \tag{16}$$

6 Performance Evaluation

6.1 Security Analysis

Table 2. Performance Comparison

Properties	ZLL touchlink commissioning	Improved ZLL touchlink commissioning
Identity authentication	×	✓
Anti-replay attacks	×	✓
Known key attack	×	✓
Privacy protection	×	✓
Integrity	×	✓

1. Identity authentication
   The ZLL touchlink commissioning does not apply trusted identity detection, which will leads to identity forge attacks and other malicious attacks. The enhanced protocol proposed in this paper can protect the protocol from this attack (Table 2).
2. Anti-replay attacks
   Adding timestamp in each data interaction ensures the freshness and effectiveness of the message. It effectively protect the data transmission form anti-reply attacks.
3. Known key attack
   If the network key is leaked, it will not lead to severe results. Since the network key is updated timely and parameters related is randomly generated by the trusted initiator and target, malicious third party cannot get the new key.
4. Privacy Protection
   In the original protocol, the privacy is transparent to attackers. In contrast with it, all of the privacy messages is transmitted encrypted in the new protocol. Thus privacy protection is ensured.
5. Integrity
   When attackers change the data, the data frame will be distorted and begin a new transaction. That's how the data integrity is protected in the enhanced protocol.

6.2 Storage Complexity

In the enhanced mechanism of the ZLL touchlink commissioning, the initiator both save the safe hash function and some temporary variables. So compared to before, the storage complexity has no big change.

6.3 Computational Cost

The original protocol's computation cost focus on the encrypting process of the network key. But the enchanced mechanism includes hash computation and some symmetric encryption. Thus the enhanced touchlink commissioning is both efficient and secure.

7 Conclusion

With the rapid update of wireless network technology and hardware devices, intelligent lighting equipment has become the mainstream of the market. Since hardware always carries key data of research and industry value, protecting them from malicious attack is of importance.

At the beginning, this article introduced the relevant research of the ZLL protocol and pointed out that there is a lack of a secure ZLL protocol. Then research on the defects of ZigBee Light Link protocol and propose a safety-enhanced protocol based on network key updating mechanism. In the end, this paper gives some analyzation on the performance of the new protocol.

Acknowledgement. This work is supported by National Natural Science Foundation of China (No. U1536122, No. 61502536).

References

1. ZigBee Alliance: Base Device Behavior Specification. Accessed 24 Feb 2016
2. Xi, X., Cheng, T., Xingyuan, F.: A health care system based on PLC and ZigBee. In: International Conference on Wireless Communications, Networking and Mobile Computing, pp. 3063–3k066
3. Wang, C., Kazem, S., Rittwik, J., Lusheng, J., Mahmoud, D.: Voice communications over ZigBee networks. IEEE Commun. Mag. **46**(1), 121–127 (2008)
4. Osipov, M.: A survey on Zigbee based wireless sensor networks in agriculture. In: Proceedings of the 3rd International Conference on Trendz in Information Sciences and Computing(TISC), pp. 85–89 (2011)
5. Muhammad, A., Ahsan, R.: An overview of home automation systems. In: International Conference on Robotics and Artificial Intelligence (ICRAI), pp. 27–31, 19 December 2016
6. Wang, J.: ZigBee light link and its applications. IEEE Wirel. Commun. **20**(4), 6–7 (2013)
7. ZigBee Alliance: ZigBee Light Link Standard (2012)
8. Wright, J.: KillerBee: practical ZigBee exploitation framework. In: ToorCon (2009)
9. Radmand, P., Domingo, M., Singh, J., Arnedo, J., Talevski, A., Petersen, S., Carlsen, S.: ZigBee/ZigBee PRO security assessment based on compromised cryptographic keys. Digital Ecosystem and Business Intelligence Institute, Curtin University of Technology, Perth, Australia (2010)
10. Olawumi, O., Haataja, K., Asikainen, M., Vidgren, N., Toivanen, P.: Three practical attacks against ZigBee security: attack scenario definitions, practical experiments, countermeasures, and lessons learned. In: IEEE 14th International Conference on Hybrid Intelligent Systems (HIS2014), at Kuwai. https://doi.org/10.1109/his.2014.7086198
11. Zillner, T.: ZigBee exploited: the good, the bad and the ugly. In: Black Hat (2015)

12. Nguyen, S.T., Rong, C.: ZigBee security using identity-based cryptography. In: Xiao, B., Yang, L.T., Ma, J., Muller-Schloer, C., Hua, Y. (eds.) ATC 2007. LNCS, vol. 4610, pp. 3–12. Springer, Heidelberg (2007). https://doi.org/10.1007/978-3-540-73547-2_3
13. Armknecht, F., Benenson, Z., Morgner, P., M¨uller, C.: On the security of the ZigBee light link touchlink commissioning procedure in Sicherheit. In: Lecture Notes in Informatics (LNI), Gesellschaft für Informatik, Bonn (2016)

Intrusion Detection Model Based on GA Dimension Reduction and MEA-Elman Neural Network

Ze Zhang(✉), Guidong Zhang, Yongjun Shen, and Yan Zhu

School of Information Science and Engineering, Lanzhou University, Lanzhou, Gansu, China
zhangz2016@foxmail.com

Abstract. Now people are using the network all the time, but the ensuing network attacks are constantly threatening people's lives, so information security is becoming more and more important. In this paper, an intrusion detection model based on the MEA-Elman neural network is proposed. Firstly, GA algorithm is used to reduce the dimension of the dataset, and then verified by the MEA-Elman network model. The experiment results show that the detection model has high accuracy, which can meet the basic requirements of intrusion detection.

1 Introduction

Today's network is more and more developed, and also greatly facilitate people's lives, but people also suffer from network attacks, and it maybe lead to information leakage and loss of property. As these security threats continue to emerge, people are placing more emphasis on information security. In response to this situation, the research of machine learning method and intrusion detection came into being. However, we found that not all research methods are so perfect.

The intrusion detection model needs to be evaluated and tested to verify whether it meets the design goals and requirements after the design is completed. The structure of the general intrusion detection system is shown in Fig. 1.

Among them, data analysis is the most important process. After the information extraction, the data is very huge. In those data, the information of intrusion is only a small part. Therefore, we need to use data analysis to find abnormal data from a large number of data.

[1] review the research status of deep learning in the field of intrusion detection. In [2], an intrusion detection method based on incremental GHSOM neural network model was proposed, incremental learning is carried out for new attack types in the detection process, and it can complete the dynamic expansion of the intrusion detection model. The [3] preprocessed the data firstly, and then combined PCNN (Pulse Coupled Neural Networks) and SVM algorithm to identify and classify attacks in dataset. The [4] applied GRNN (Generalized Regression Neural Network) and MPNN (Multilayer Perceptron Neural Network) in the host-based intrusion detection environment, and improved the detection accuracy. The [5] objectively tests and evaluates the

L. Barolli et al. (Eds.): IMIS 2018, AISC 773, pp. 354–365, 2019.
https://doi.org/10.1007/978-3-319-93554-6_33

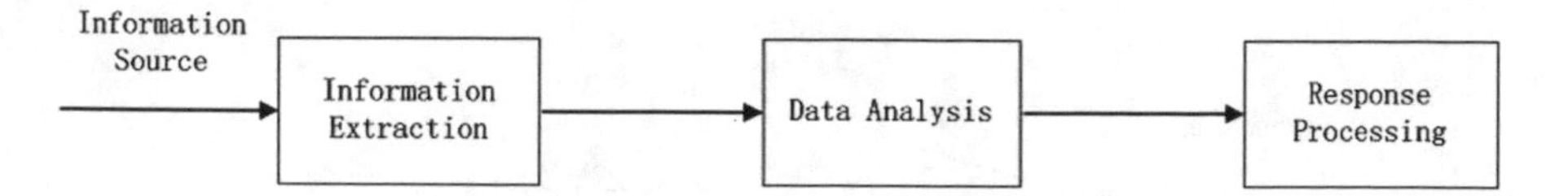

Fig. 1. Structure of intrusion detection system.

performance of artificial neural networks in dealing with intrusion detection data from various perspectives. The [6] combines artificial neural networks with EM clustering and applied intelligent network technology to identify intrusion detection; the [7] adopted accelerated neural network architecture to design intrusion detection model and identify abnormal behavior in the network.

Through the above analysis, this paper selects Elman neural network as the main algorithm to design intrusion detection model, Elman neural network is a dynamic system with feedback capability which has a context layer, and it can directly reflect the changes of dynamic system and have excellent computing performance. Using KDDCUP99 dataset, the data are preprocessed firstly, and then the GA algorithm is used to reduce the dimension of the data. Finally, the Mind Evolutionary Algorithm (MEA) is used to optimize the Elman network. The experiment results show that the accuracy of U2R and R2L classification is very low when using the original data directly. However, the classification accuracy of U2R and R2L is significantly improved after using GA to reduce the dimension of the data, indicating that the algorithm has some significance for the research of intrusion detection model.

2 Elman Neural Network

Elman neural network is divided into 4 layers: input layer, hidden layer, context layer and output layer. The structure of network model is shown as Fig. 2. Three layers of Elman network, including input layer, hidden layer and output layer, have similar connection mode to feed forward neural network, and both have same function too. The context layer can be considered as a one-step time-lapse operator, mainly used for memorizing.

In non-linear solution space, formulas of Elman network are:

$$y(k) = g(\omega^3 x(k)) \tag{1}$$

$$x(k) = f(\omega^1 x_c(k) + \omega^2 (u(k-1))) \tag{2}$$

$$x_c(k) = x(k-1) \tag{3}$$

u is r-dimensional input vector, x is vector of n-dimensional hidden layer, y is m-dimensional output node vector, x_c is n-dimensional feedback vector, ω^3 is the connection weight from hidden layer to output layer, ω^2 is the connection weight from

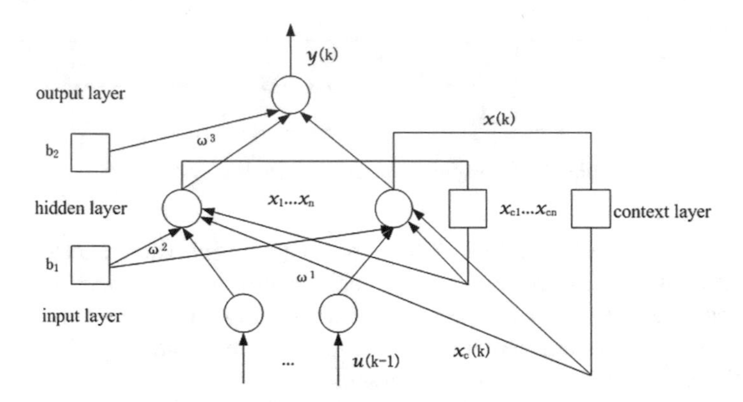

Fig. 2. Structure of Elman neural network.

input layer to hidden layer, ω^1 is the connection weight from context layer to hidden layer; $g(*)$ is the transfer function of output neurons, it is linear combination of output from hidden layer, and $f(*)$ is the transfer function of neurons hidden layer neurons, generally, Sigmoid function is used [8].

Elman neural network is significantly different from static feed-forward neural network. The output of hidden layer of Elman network is sent back to the input of hidden layer via the delay and storage of the context layer, which heightened the ability of handling dynamic information after add internal feedback network [9].

3 Mind Evolutionary Algorithm

Compared with traditional algorithms, Evolutionary Computation (EC) is characterized by group search. EC have successfully solved many problems. However, the defects of algorithms cannot be ignored, such as premature and slow convergence speed. In response to the existing problems of EC, Sun et al. proposed the Mind Evolutionary Algorithm (MEA) [10] in 1998. The main system framework of MEA is shown in Fig. 3.

Different from genetic algorithm, some concepts of MEA are explained as follows:

(1) Population and Group. MEA is an iterative algorithm, the collection of all individuals in the process of evolution is called population, and the population is divided into several groups, including two categories: superior group and temporary group. The superior group records the winner's information in the global competition and the temporary group records the process of global competition.
(2) Billboard. The billboard is equivalent to an information platform, which provides opportunities for the exchange of information between individuals and groups. The billboard records three valid messages: the serial number of individual or group, action, and score. Using the serial numbers, different individuals or groups can be easily distinguished. The descriptions of action are different according to the actual situation. For example, in this paper, we use MEA to optimize

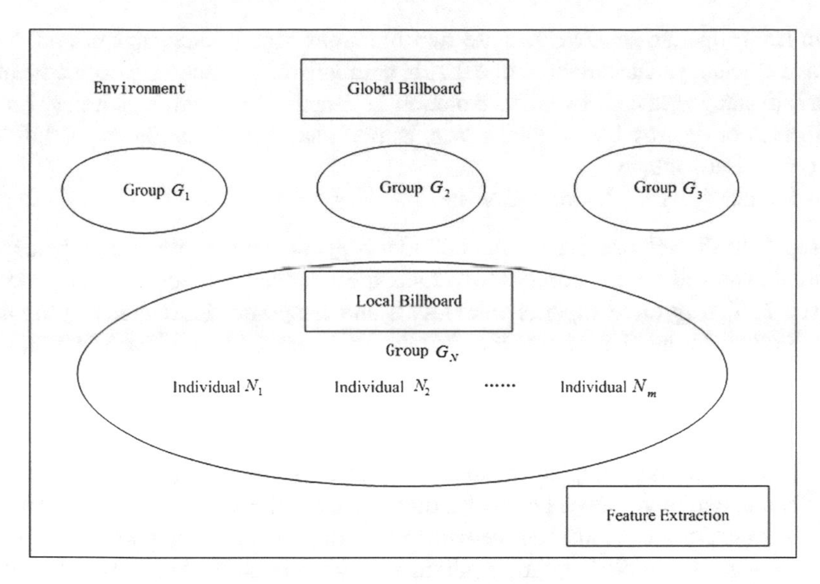

Fig. 3. Framework of MEA.

parameters, so records of action are the specific positions of individuals and groups. Score is the evaluation of individual actions by the environment. In the process of using the MEA, the optimized individuals and groups can be found quickly only by recording the scores of each individuals and groups at all times. Individuals post their own information on local billboards, and global billboards are used to post information about each groups.

(3) Similartaxis.

Definition 1: in the group, the individual becomes the winner and the competition is called similartaxis.

Definition 2: In the process of similartaxis, if a group don't have new winner anymore, it is called the group is already mature. When the group is mature, the similartaxis process is over.

(4) Dissimilation.

Definition 3: In the whole solution space, each group is competing for the winner to explore new points in the solution space. This process is called dissimilation.

The meaning of dissimilation is that all groups compete globally, if the score of a temporary group is higher than the score of a superior group, the superior group is replaced by this temporary group, temporary group becomes a new superior group, and the original superior group is released; if score of a mature temporary group is lower than any of the superior group, the temporary group is discarded, the individuals in it are released, and the released individuals search again globally to form a new temporary group.

Compared with genetic algorithm, MEA has its own characteristics: the population is divided into the superior group and temporary group, the similartaxis and dissimilation operate separately, these two functions remain independent and easy to improve efficiency. The structure of MEA is parallelism, and there are both advantages and disadvantages of crossover and mutation in genetic algorithm, which may produce good genes or destroy the original genes. Similartaxis and dissimilation in MEA can avoid this disadvantage.

In summary, basic idea of MEA is:

Step 1: In the solution space, the individuals generated randomly, and the superior individuals and the temporary individuals are searched according to the score.
Step 2: Taking these superior individuals and temporary individuals as the center respectively, a number of new individuals are generated around each individual, so as to obtain a number of superior groups and temporary groups.
Step 3: The similartaxis is performed within the groups until the group is mature, and the score of the center of the group is taken as the score of this subgroup.
Step 4: After the group is mature, each group's score is posted on the global billboard, and then groups begin dissimilation operations directly, and complete the replacement and abandonment between superior group and temporary group, so as to calculate the global optimal individual and the score. After the dissimilation operation is completed, a new temporary group is needed in the solution space to ensure that the number of the temporary groups keep unchanged.

4 Build MEA-Elman Network Intrusion Detection Model

Using MEA to optimize the initial weights and thresholds of Elman neural network, the flow chart of the algorithm design step is shown in Fig. 4.

Step 1: According to the topological structure of Elman neural network, the solution space is mapped to the encoding space, and each encoding corresponds to a solution (i.e., individual) of the problem;
Step 2: After several experiments, the number of hidden layer neurons in Elman network is determined to be 10, the length of encoding is chosen to be 21, the population size is set to 200, and the number of superior groups is set to 5, the number of temporary groups is set to 5;
Step 3: The reciprocal of the mean square error of the training set is selected as the score function of individual and population. After iteratively iterating, the optimal individual is output by using the MEA, and the initial weights and thresholds of the neural network are obtained;
Step 4: Experiments with trained model to verify whether the accuracy of the model is required.

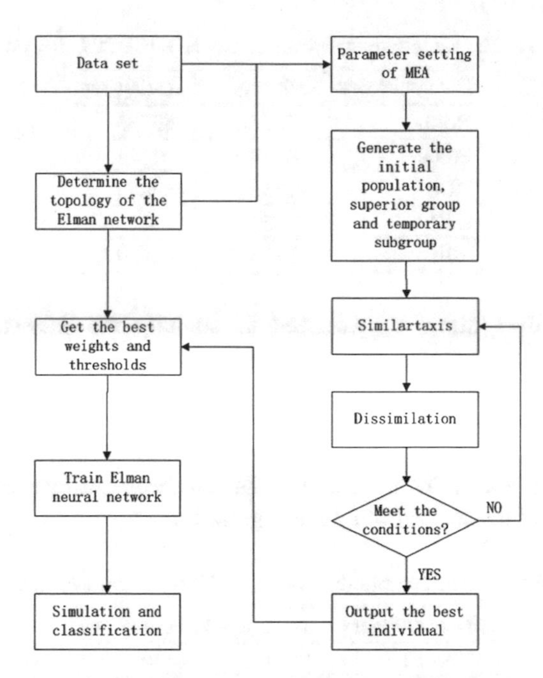

Fig. 4. The flow chart of building MEA-Elman network intrusion detection model.

5 Data Processing and Dimension Reduction by Genetic Algorithm

In this paper, KDDCUP99_10% dataset is selected. KDDCUP99 dataset has 41 features and 1 class label, 41 features of the data type is not uniform, they contain continuous types and discrete types, of which discrete types include 2, 3, 4, 7, 12, 14, 15, 21, 22, total of nine, consecutive types include 1, 5, 6, 8, 9, 10, 11, 13, 16, 17, 18, 19, 20, 23, 24, 25, 26, 27, 28, 29, 30, 31, 32, 33, 34, 35, 36, 37, 38, 39, 40, 41, total of thirty-two, non-uniform data types cannot be identified and processed by the neural network, we have found second dimension, third dimension, fourth dimension data type is discrete, and the protocol types TCP, UDP and ICMP appearing in the second dimension are mapped by number 1, 2 and 3 respectively. Due to there are too 70 state values in the third dimension, they are not mapped one by one, and these state values are sorted according to the degree of importance, the common and more important states are mapped one by one, and all the unimportant states are mapped to a value. The fourth dimension is similar to the third dimension, and treated it in the same way.

The forty-second dimension of KDDCUP99_10% data set is class label. It can be classified into two categories according to attack mode: normal and attack. Attack mode can be further divided into four categories: Probe, DOS, U2R and R2L. The sample distribution of the KDDCUP99_10% dataset is shown in Table 1.

Although the KDDCUP99_10% dataset is a streamlined dataset, it can be seen from Table 1 that the data of the Normal type and the DOS type is huge, but the data of the U2R and R2L types is too small. Training model with this dataset directly, the neural

Table 1. The sample distribution of the KDDCUP99_10% dataset.

Data category	Training set	Testing set
Normal	97278	60593
Probe	4107	4166
DOS	391458	229853
U2R	52	228
R2L	1126	16189

network will be over-fitting, so we need to simplify the dataset, eliminate some redundant data, keep the number of each type balanced, and the sample distribution of the streamlined dataset is shown in Table 2.

As can be seen from Table 2, U2R has a small amount of data and does not streamline it, but other types of data are streamlined, streamlined datasets can represent feature of each type, although the amount of data has been greatly reduced, after the test comparison, the test results will not have a great impact.

Table 2. The sample distribution of the streamlined dataset.

Data category	Training set	Testing set
Normal	9861	4460
Probe	3250	2268
DOS	11084	6350
U2R	52	228
R2L	1126	2797

The MEA-Elman network model was used to experiment with the processed dataset. The training dataset was used to train the network and the testing dataset was used to test the trained model. The experiment results are shown in Table 3:

Table 3. The experiment results with original dataset.

Data category	Total number of samples	Correct number of classification	Error number of classification	Accuracy of classification
Normal	4460	4405	55	98.7668%
Probe	2268	2192	76	96.6490%
DOS	6350	5801	549	91.3543%
U2R	228	35	193	15.3508%
R2L	2797	605	2192	21.6303%

As can be seen from Table 3, MEA-Elman network model has a high accuracy of Normal, Probe and DOS classification of more than 90%, but the accuracy of the U2R and R2L classification is too low, it can not meet the requirements, so it is necessary to find a data processing algorithm. In this paper, the genetic algorithm is used to reduce the dimension of the dataset. After the experimental results show that using the dimension-reduced dataset can effectively improve the accuracy of U2R and R2L classification.

Using genetic algorithm to optimize the calculation, it needs to map the solution space to the encoding space, and each encoding corresponds to a solution of the problem [11]. Because of the features of data are consist of 41 dimensions, so encoding length is designed as 41, each of the chromosomes corresponds to an input independent variable, and each gene value can only be "0" or "1". If one of the chromosomes has a value of 1, it means this input independent variable participates in the final modelling, otherwise, the input independent variable corresponding to the value 0 is not used as the final modelling. Select the reciprocal of the mean square error of the test set data as the fitness function, so that after continuous iterative evolution, representative input variables are selected to participate in the modelling. According to the above ideas, the main steps of the design are as follows:

Step 1: First, the MEA-Elman model is built by using all 41 input independent variables;
Step 2: N string structure data are generated randomly, each string structure data is an individual, and the N individuals constitute a population. The genetic algorithm uses the N string structure data as starting points for iteration. As mentioned above, each string structure data has only two values of 0 or 1;
Step 3: The reciprocal of the mean square error of the training set is selected as the fitness function:

$$f(X) = \frac{1}{SE} = \frac{1}{sse(\hat{T} - T)} = \frac{1}{\sum\limits_{i=1}^{n} (\hat{t}_i - t_i)^2} \tag{4}$$

In the formula, $\hat{T} = \{\hat{t}_1, \hat{t}_2, \ldots, \hat{t}_n\}$ is the prediction value of the test set, $T = \{t_1, t_2, \ldots, t_n\}$ is the true value of the test set, n is the total number of samples in the test set.
Step 4: Selection. The probability of an individual is selected and inherited into a next generation population is proportional to the individual's fitness. First calculate the sum of the fitness of all the individuals in the population:

$$F = \sum_{k=1}^{n_r} f(X_k) \tag{5}$$

then using the formula (6) to calculate the relative fitness of each individual in a population as the probability of this individual being selected and inherited in the next generation:

$$p_k = \frac{f(X_k)}{F} \quad k = 1, 2, \ldots, n_r \tag{6}$$

Step 5: Crossover. Each pair of pair-wise individuals selects a point randomly as a crossover point. According to the location of the crossover point, two parts of the chromosome are exchanged, and two new individuals are generated according to the Formula (7) and (8)

$$c_1 = p_1 \times a + p_2 \times (1 - a) \tag{7}$$

$$c_2 = p_1 \times (1 - a) + p_2 \times a \tag{8}$$

in the formula, p_1, p_2 are two individuals, c_1, c_2 are new individuals after crossover, a is a random number located in the interval (0, 1), that is the crossover probability;

Step 6: Mutation. First, select the mutation point randomly, and then change the corresponding value according to the position of the mutation point, that is, the result of the mutation operation is that 1 becomes 0 or 0 becomes 1;

Step 7: When the experiment satisfies the condition of termination of iteration, the output of the last generation population is the optimal approximate solution of the problem, that is, the most representative combination of input variables is selected;

Step 8: The MEA-Elman network model is rebuilt. The input independent variable selected by the GA algorithm is extracted from the training set and test set, and then re-simulate with MEA-Elman network model.

The flow chart of optimization steps is shown in Fig. 5.

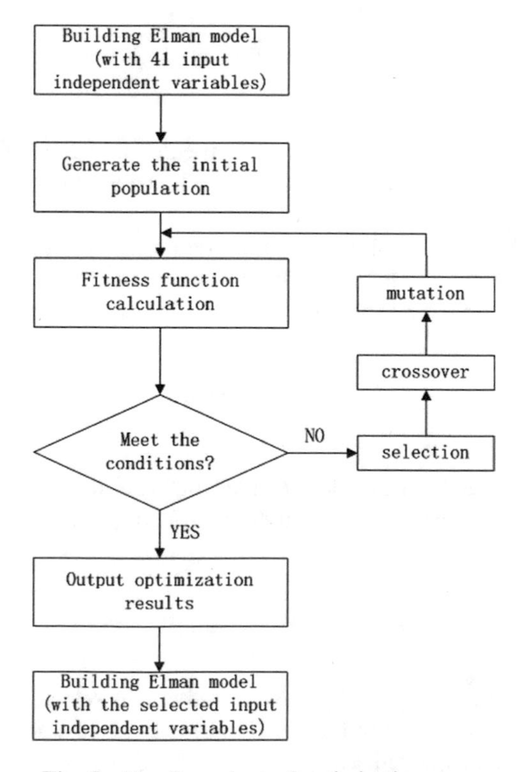

Fig. 5. The flow chart of optimization steps.

6 Experiment and Result

The data set is processed by the GA algorithm. The evolution curve of the fitness function of the population is shown in Fig. 6.

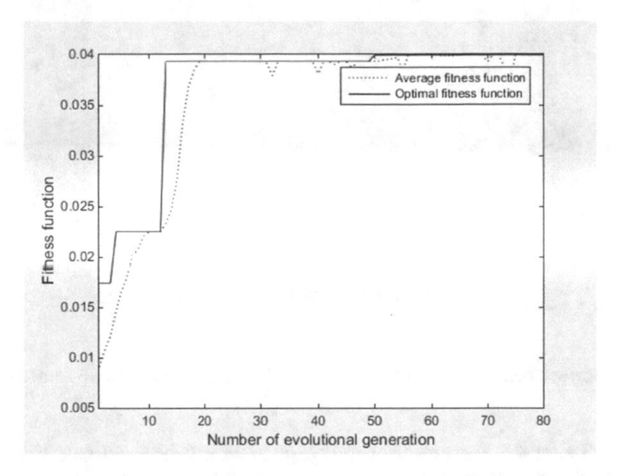

Fig. 6. The evolution curve of the fitness function.

The number of the selected input independent variables is: 1, 2, 5, 7, 8, 9, 10, 14, 18, 20, 21, 26, 31, 33, 34, 35, 36, 37 and 40. Obviously, after GA algorithm selecting, the number of input independent variables involved in the modeling is about half of the number of total input variables.

The MEA-Elman network model is built by using the processed data. After iteration, the similartaxis of the superior group is shown in Fig. 7, the similartaxis of the temporary group is shown in Fig. 8.

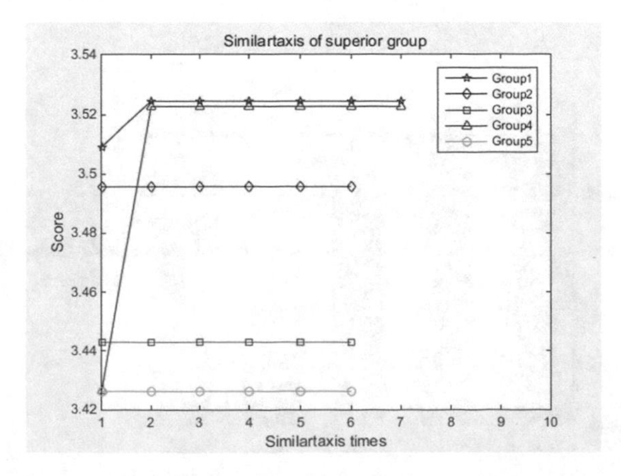

Fig. 7. Similartaxis of the superior group.

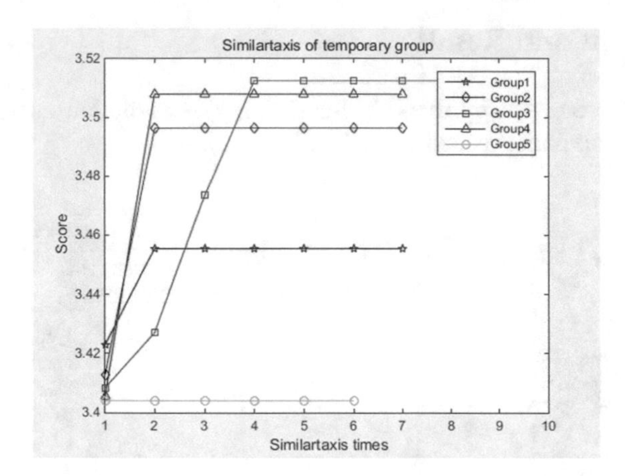

Fig. 8. Similartaxis of the temporary group.

The experimental results of the intrusion detection model are shown in Table 4:

Table 4. The experimental results with processed dataset.

Data category	Total number of samples	Correct number of classification	Error number of classification	Accuracy of classification
Normal	4460	4133	327	92.6682%
Probe	2268	2122	146	93.5626%
DOS	6350	5460	890	85.9842%
U2R	228	123	105	53.9473%
R2L	2797	1761	1036	62.9603%

The experimental results using the original dataset and the results using the processed dataset are compared, and the histogram is shown in Fig. 9:

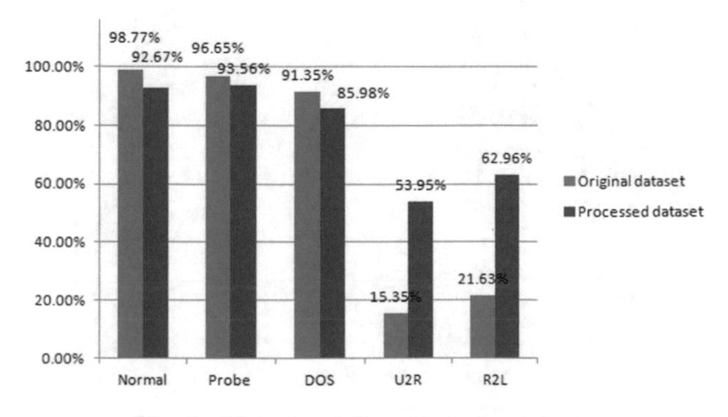

Fig. 9. Histogram of accuracy comparison.

From Fig. 9, we can find that the detection accuracy of U2R and R2L has been greatly improved by using the processed data set, and the detection accuracy of Normal, Probe and DOS has decreased slightly, but it still keeps around 90%. The MEA-Elman intrusion detection model can basically meet the requirements.

7 Conclusion

In this paper, the GA algorithm is used to reduce the dimension of input variables, and then the MEA-Elman model is used to verify it. The experiment results show that the detection accuracy of U2R and R2L has been greatly improved by using the GA algorithm. Detection accuracy of Normal, Probe and DOS has decreased slightly, but it still keeps around 90%. After reducing the dimension of dataset, it has impact on the detection of Normal, Probe and DOS. But in general, the MEA-Elman intrusion detection model based on GA dimension reduction can basically meet the requirements. In the future research, we can explore a better algorithm to improve the accuracy of the intrusion detection model.

References

1. Aminanto, E.M., Kim, K.: Deep learning in intrusion detection system: an overview. In: International Research Conference on Engineering and Technology (2016)
2. Yang, Y., Huang, H.: Research on intrusion detection based on incremental GHSOM. Chin. J. Comput. **37**(5), 1216–1224 (2014)
3. Aditya Shrivastava, M.B., Gupta, H.: A novel hybrid feature selection and intrusion detection based on PCNN and support vector machine. Comput. Technol. Appl. **4**, 922–927 (2013)
4. Gautam, S.K., Om, H.: Computational neural network regression model for host based intrusion detection system. Perspect. Sci. **8**, 93–95 (2016)
5. El Farissi, I., Saber, M., Chadli, S., Emharraf, M., Belkasmi, M.G.: The analysis performance of an intrusion detection systems based on neural network
6. Kosek, A.M., Gehrke, O.: Ensemble regression model-based anomaly detection for cyber-physical intrusion detection in smart grids. In: 2016 IEEE Electrical Power and Energy Conference (EPEC) (2016)
7. Sasanka Potluri, C.D.: Accelerated deep neural networks for enhanced intrusion detection system (2016)
8. Tang, J., Cao, Y., Xiao, J., et al.: Predication of plasma concentration of remifentanil based on Elman neural network. J. Central South Univ. **20**(11), 3187–3192 (2013)
9. Zhang, Q., Xu, Z., Zhao, K.: Prediction of data from pollution sources based on Elman neural network. J. South Chin. Univ. Technol. (Nat. Sci. Edn.) **37**(5), 135–138 (2009)
10. Chengyi, S., Keming, X., Mingqi, C.: Mind-evolution-based machine learning framework and new development. J. Taiyuan Univ. Technol. **5**, 453–457 (1999)
11. Karegowda, A.G., Jayaram, M.A., Manjunath, A.S., et al.: GA based Dimension Reduction for enhancing performance of k-means and fuzzy k-means: a case study for categorization of medical dataset. Adv. Intell. Syst. Comput. **201**, 169–180 (2013)

Location Privacy Protection Scheme Based on Random Encryption Period in VANETs

Tianhan Gao and Xin Xin(✉)

Northeastern University, Shenyang, China
gaoth@mail.neu.edu.cn, xinxin@stumail.neu.edu.cn

Abstract. How to ensure location privacy has become an important issue for VANETs' security. The more effective mechanism is that the vehicles can not be associated by replacing the pseudonyms to protect their location privacy. This paper proposes a novel location privacy protection scheme for VANETs. When vehicles need to change the pseudonyms, they will cooperate with the neighbor nodes to create the encrypted area through group key encryption, so that the external adversary cannot crack the message in this area. During this period, some vehicles change the pseudonyms jointly so that the external adversary can not associate the pseudonyms before and after. Thus the location privacy protection of the vehicle nodes is achieved.

1 Introduction

In recent years, as mobile ad hoc network, VANETs has gradually become a research hot spot. VANETs uses wireless access technologies to connect road entities to form an intelligent network. VANETs allow the vehicle to communicate with other vehicles (V2V) or roadside infrastructure (V2I). According to the Dedicated Short Range Communication protocol(DSRC [1]), any vehicle equipped with OBU will periodically broadcast the location,speed, and other information, which can provide users with real-time traffic and neighbor vehicle's information to warn the accident early.

VANETs confront unique security threats which contain location privacy, identity privacy, and other security threats [2]. Any external eavesdropper should not obtain the driver's true identity or track a particular vehicle. One solution to protecting the driver's identity privacy is to provide a series of anonymous certificates for each vehicle [7,8]. However, these anonymous certificates can not guarantee the location privacy of VANETs, even if the vehicle regularly changes pseudonyms, it is still easy to follow [3].

A good way to protect location privacy is to create Mix-Zone [4–6]. Mix-Zone is a non-monitorable area. The vehicles update the pseudonym in Mix-Zone, making it difficult to correlate messages from the same node, in order to prevent the adversary from obtaining the vehicle's position, speed, etc, from

L. Barolli et al. (Eds.): IMIS 2018, AISC 773, pp. 366–374, 2019.
https://doi.org/10.1007/978-3-319-93554-6_34

the multicast information of the vehicle. However, there are deficiencies in these schemes: the area is fixed, and if the node fails to reach the Mix-Zone, the node's identity is completely exposed.

In this paper, we propose a location privacy protection scheme based on REP: When the vehicle exchanges pseudonym, the surrounding vehicles use the group key to encrypt all the information to establish a random encrypted area, the external attacker is unable to obtain the relevant information of the vehicles, so that it is impossible to correlate the running track of the vehicles and realize the privacy protection. The main work of this paper is as follows:

1. According to the space-time factors in the network to establish a location privacy model and attack model.
2. Proposed location privacy protection scheme based on the random encryption period, the vehicle cooperates with other neighbor nodes to establish a random Mix-Zone, and replaces the pseudonym to ensure location privacy.
3. The security analysis of the program.

In the Sect. 2, we discuss the related work of this paper. Section 3 defines the system model and introduces the location privacy protection scheme. Section 4 is about the security analysis of the scheme. Section 5 summarizes the full text and gives the next research work.

2 Related Work

2.1 Mix-Zone

Freudiger et al. [4] presented the encrypted Mix-Zone (CMIX) to provide location privacy for VANETs. In CMIX, the RSU of the intersection generates a group key, and all the traffic between the intersections encrypts all communication with a shared group key to create the CMIX and change the certificate. Albert et al. [6] proposed a scheme based on random encryption period (REP) in which each vehicle node obtains a series of anonymous certificates, a series of symmetric keys and the initial group key when register with the TA, the node triggers a random encryption period when changing the pseudonym, and the group member encrypts all the information with the group key to interfere with the external eavesdropper. However, the initial load of the TA is heavy. In addition, when the group key is updated, each OBU performs a large number of calculations to update all symmetric keys.

On the basis of [6], this paper uses the RSU instead of part of the TA function, the vehicle generates a pseudonym under the joint action of TA and RSU, changing pseudonym can protect location privacy. In addition, the RSU directly generates and replaces the group key, which improves the working efficiency of the scheme.

2.2 Bilinear Pairing

Bilinear mapping: Given two groups $G1$ and $G2$ with same order p, where $p = q^n$, q is prime and $n \in Z^*$, $G1$ is an additive group, and $G2$ is a multiplicative group, and assume that the discrete logarithm problem on the above two groups is difficult. If the mapping $e : G1 \times G1 \rightarrow G2$ satisfies the following three properties, then say e is a bilinear pairing:

(1) Bilinear: for any $P \in G1$, $Q \in G1$ and $a, b \in Z_q*$, there are $e(aP, bQ) = e(P, Q)ab$.

(2) Non-degeneracy: there must be a certain $P \in G1$, $Q \in G1$ satisfy $e(P, Q) = 1$.

(3) Computability: An efficient algorithm can calculate $e(P, Q) \in G2$, where $P \in G1$, $Q \in G1$.

3 The Proposed Scheme

3.1 System Model

The system model of the program includes the following three entities: Trusted Authority (TA), Roadside Units (RSUs), On-Board Units (OBUs). (As shown in Fig. 1)

- Trusted Authority (TA): TA is usually a trusted third party such as traffic management. It is mainly responsible for the identity authentication of RSUs and OBUs.
- Roadside Units (RSUs): RSUs are managed and regularly monitored by traffic management. The RSUs and TA are connected via the Internet which for secure communication with each other. The default RSU is credible and has strong computing power.
- On-Board Units (OBUs): Each vehicle is equipped with an OBU, including tamper-proof devices (TPD) that store secret information, event data recorder (EDRs) and global positioning systems. Vehicles should regularly broadcast safety information.

Table 1 illustrates the notations used in the presentation of our scheme.

3.2 Description of the Scheme

3.2.1 Registration of Vehicles

When the vehicle registers with TA, the TA generates the ticket δ_a according to the real ID of the vehicle node and produces the corresponding private key S_a. TA loads δ_a and system parameters into TPD of the vehicle. The process is as shown in (Fig. 2).

TA generates a ticket δ_a and a private key S_a for the vehicle based on its real identity ID, then TA stores the mapping of real ID and ticket into the database, meanwhile save the signature of the ticket, private key in the OBU of vehicle.

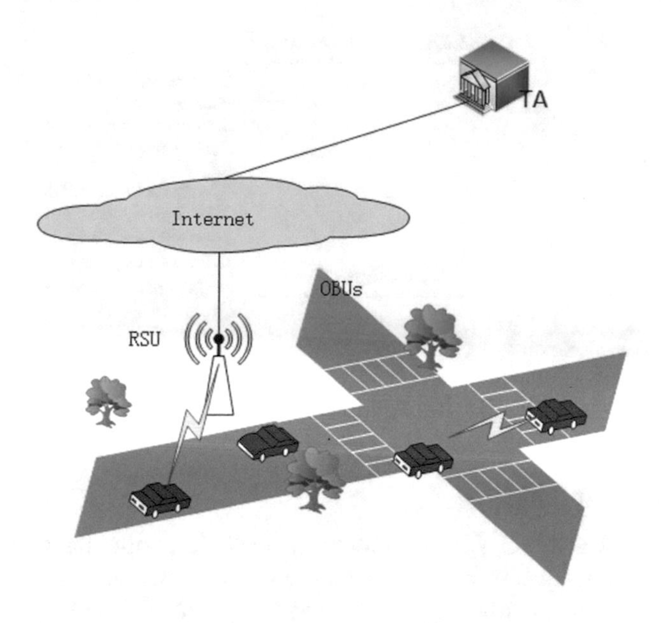

Fig. 1. System model.

Table 1. Explanation of nouns

SYMBOL	Explanation
ID_a	The real ID of the vehicle node a
v_a	The vehicle a
P_{R_i}	The public key of the roadside unit R_i
S_{R_i}	The private key of the roadside unit R_i
$Cert_{R_i}$	The identity certificate of roadside unit R_i issued by TA
$SIG(M;K)$	signature of the message M by using the key K
S_{TA}	The private key of TA
P_{TA}	The public key of TA
k_g	The group key shared by all vehicle nodes in the same RSU
δ_a	The ticket of v_a issued by TA as its public key
S_a	OBU's private key issued by TA
k_a	The shared key between vehicle a and RSU

In this part we can see that the ticket δ_a does not contain the real identity information of the vehicle v_a, but if the vehicle is misbehavior, TA can find the true identity of the vehicle based on the mapping relationship and revoke it. $<\delta_a$, $SIG(\delta_a; S_TA)$, $S_a>$ as the private information of vehicle v_a is stored in TPD, and only the TA can modify the information.

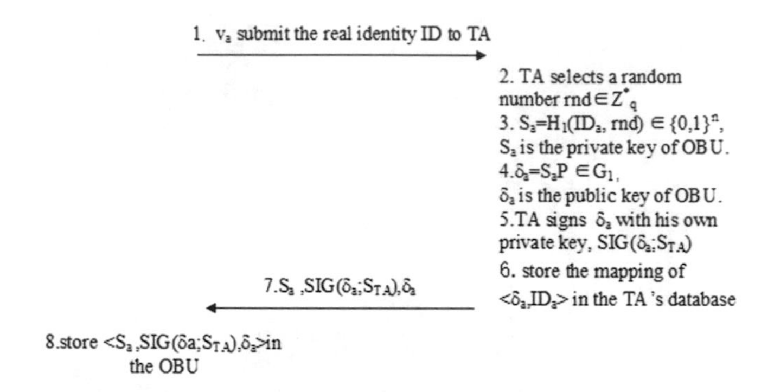

Fig. 2. Registration of vehicles.

3.2.2 Access of Vehicles

After acquiring the ticket δ_a, the vehicle v_a needs to cooperate with the nearby roadside unit R_i to obtain its own pseudonym token. v_a consults shared key k_a with R_i. Meanwhile v_a submits its own ticket δ_a to R_i. After verifying that v_a is legally valid, R_i generate the pseudonym token $T_{(a,i)}$ and the token's valid time $t_{(a,i)}$, then send them with the current group key to v_a. The main algorithm is shown in Fig. 3.

In this scenario, R_i stores a mapping between the token and the ticket δ_a, which helps to determine its corresponding ticket based on the pseudonym of the vehicle during the revocation phase. Our scheme allows v_a to obtain multiple token from a RSU by using the same ticket δ_a. $T^k_{(a,i)}$ represents the k_th token issued by R_i to vehicle v_a who uses the token $T^k_{(a,i)}$ in the pseudonym generation scheme to generate the k_th pseudonym. The information stored in M includes token $T_{(a,i)}$, the token's expiration time $t_{(a,i)}$, R_i's signature $SIG(T_{(a,i)}, t_{(a,i)}; S_{R_i})$, the random number $\gamma_{(a,i)}$ selected by the RSU and the certificate of the R_i, $Cert_{R_i}$, and the current group key k_g. It is to be noted that the group key k_g is a symmetric key generated and managed by the RSU, which can be used to establish a random encrypted area, without affecting the communication between the internal vehicles. The external attacker is prevented from obtaining the safety information of the vehicle. The group key is not static, and in some cases the RSU updates it.

3.2.3 Location Privacy Protection

The vehicle v_a obtains a number of pseudonymous tokens from the RSU, selecting a $T^j_{(a,i)}$ to generate the pseudonym, and using the random number $\gamma^j_{(a,i)}$ obtained from the RSU to generate the corresponding private key. The specific algorithm is shown in Fig. 4:

After obtaining a large number of pseudonym and the current group key, vehicles establish a random encryption period, and change the pseudonyms to protect their own location privacy. In our paper, the external attacker can determine

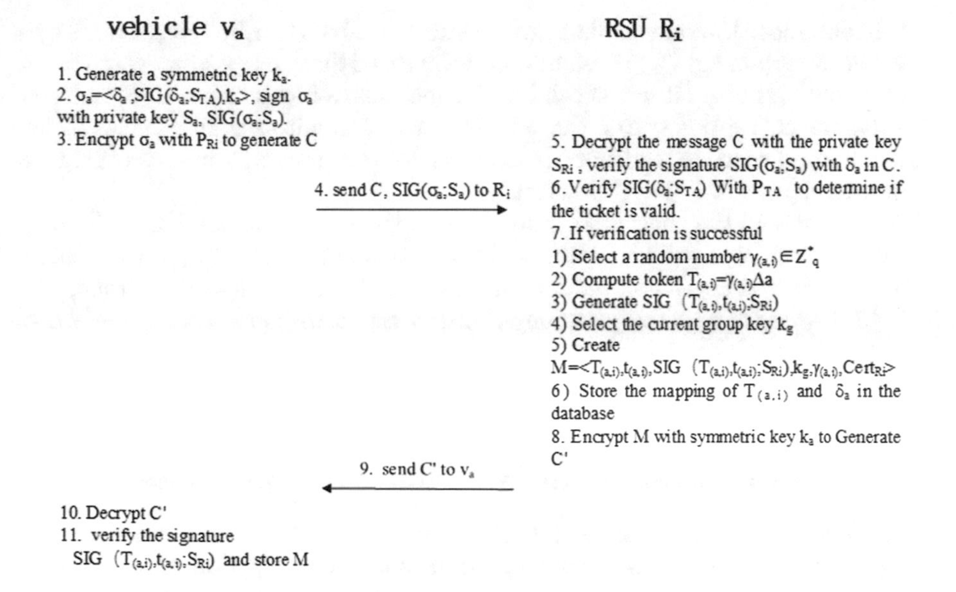

Fig. 3. Access of vehicles.

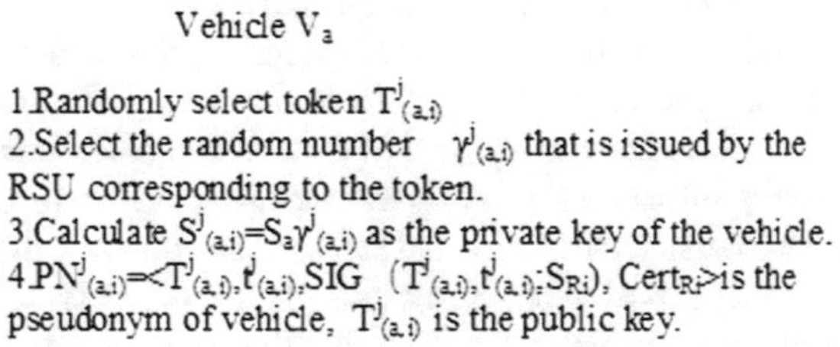

Vehicle V_a

1 Randomly select token $T^j_{(a,i)}$
2.Select the random number $\gamma^j_{(a,i)}$ that is issued by the RSU corresponding to the token.
3.Calculate $S^j_{(a,i)}=S_a\gamma^j_{(a,i)}$ as the private key of the vehicle.
4 $PN^j_{(a,i)}$=<$T^j_{(a,i)}$,$t^j_{(a,i)}$,SIG ($T^j_{(a,i)}$,$t^j_{(a,i)}$;S_{Ri}), $Cert_{Ri}$>is the pseudonym of vehicle, $T^j_{(a,i)}$ is the public key.

Fig. 4. Generate pseudonyms

its trajectory based on the speed and location of the vehicle. If the vehicle only replaces its pseudonym, the external attacker can still associate the two pseudonym. In our scheme, when the vehicle needs to change its pseudonym, it will trigger a random encryption period, the vehicle replaces the pseudonym and changes the speed or trajectory to disturb the external attacker. The specific scenario is described as follows:

1. The vehicle v_i who needs to change the pseudonym sends the message $msg = \{request_{REP}, PN^j_{(a,i)}, T_{REP}\}$ to the neighbor under the same RSU before changing its pseudonym. The v_i encrypts the msg with the group key k_g. The $request_{REP}$ is the request to start the random encryption period, The $PN^j_{(a,i)}$ is the pseudonym used by the vehicle, and T_{REP} is the duration of the random encryption period.
2. All vehicles that have received msg decrypt it with the current group key k_g. And if the decryption is successful, judgment request is valid. All the

vehicles under the same RSU begin to encrypt all of their broadcast messages using the group key k_g. We name the vehicles of the encrypted message as an encrypted group. If it is not valid, the remaining vehicles refuse to cooperate.
3. After the encryption starts, the vehicle v_i starts monitoring all vehicles in the encrypted group. In addition, it changes its pseudonym, while changing its own speed or track (lane/direction).
4. Any vehicle in the encrypted group checks the remaining validity of its current certificate. If the remaining valid period is lower than T_{REP}, the vehicle immediately changes its pseudonym and also changes its speed or track.
5. Vehicle who changes the pseudonym broadcasts a response.
6. V_i monitors all vehicles in the encrypted group if the following two conditions are completed during the encryption time:
   a. The number of vehicles in the group who change the pseudonyms are not less than 2;
   b. The vehicle who changes the pseudonym also changes its speed or track.

If the above conditions are met at the end of T_{REP}, then v_i stops the encryption period by broadcasting a message informing the encryption group to stop encrypting its message. If the condition for terminating the encryption period is not met before T_{REP}, the v_i will broadcast another request to open a new encryption period to protect its own location privacy.

In this process, with the replacement of pseudonyms, the more vehicles, the location privacy of the system is stronger. Because the legitimate members of the group have group key k_g, encryption period does not affect their communication and access of security information. But the external adversaries do not have a group key, which prevents them from listening for messages during the time of certificate replacement, thereby reducing the likelihood of tracking vehicles.

This article proposes a scheme for establishing an encrypted zone to protect the location privacy of the vehicles. This scheme mainly analyzes the intensity of the location privacy protection of the vehicles, and can be calculated using entropy, it is assumed that the number of vehicles changing pseudonyms is $k+1$. The privacy of this program can be expressed as follows:

$$H(x) = -\sum_{i=1}^{k+1} p(x_i) log_2 p(x_i) \tag{1}$$

In this scheme, when $p(x_i) = 1/(k+1)$, the intensity of the location privacy is maximum. In terms of location privacy, the privacy intensity of this scheme and REP [6] scheme is higher than CMIX [4] scheme who has same k. Since the cooperative vehicles in the two schemes jointly replaced the pseudonym and the trajectory, there are still some vehicles running in the original trajectory in the CMIX scheme, which are more easily tracked by the attackers.

In this scenario, in the process of updating the pseudonym, the vehicle only needs to submit a request to the RSU for obtaining the group key, and the RSU can send the new group key to the vehicle. Group key update overhead is only one encryption and decryption overhead, which reduces the overhead of intermediate keys compared to REP schemes. It can be seen that our scheme has higher location privacy and lower group key update overhead.

4 Security Analysis

- Identity privacy protection: In this scheme, only the TA knows the true identity of the vehicle. Though the vehicle uses the ticket to authenticate with RSU, the RSU can't break the true identity of vehicle, for ECDLP is difficult to crack in the calculation, which protects the identity privacy of the vehicle. At the same time, the vehicle through the frequent replacement of pseudonyms, to prevent the external rivals found the true identity of the vehicle. The vehicle then generates its own pseudonyms based on the token and communicates with other vehicles using the pseudonyms. The remaining vehicles can obtain only the vehicle's pseudonym information, which achieves the identity privacy of the system requirements.
- Confidentiality: The vehicle negotiates the shared key at the same time as the RSU authentication. During the group key update, the RSU encrypts the new group key with the shared key and sends it to the vehicle to prevent external rivals from acquiring new group keys to listen for messages in encrypted areas, to achieve the confidentiality requirements.
- Location privacy protection: The external passive global observer can obtain the location from the message released by the vehicle node and estimate the next step of the vehicle node based on the current speed and traffic lane. In this scheme, the vehicle node constructs an encrypted area with the neighbors through the cooperative way. Because the external rivals can't obtain the group key, it cannot crack the information of the encryption period. The vehicle replaces the pseudonym with multiple vehicles during this encryption time to confuse the external adversaries, protect their own trajectories and location privacy.

5 Conclusion and Future Work

This paper describes a scheme to protect the location privacy of vehicles in VANETs. In this paper, the pseudonym authentication scheme is combined with the location privacy protection scheme based on the random encryption period. The vehicle obtains the current group key when authenticating with the RSU. For the vehicle that needs to change the certificate, it will trigger the surrounding vehicle node to build the encrypted area and make the external attacker cannot observe, to achieve the purpose of confusing external attackers. Through the security analysis, the reliability of the scheme is proved. In the future research, we will consider how to implement location privacy protection in the area where there is no RSU.

References

1. Dedicated short range communications (2006). http://www.gmuper.ieee.ors/groups/scc32/dsrc/index.html
2. Wang, J., Wang, Y., Geng, J., et al.: Secure and privacy-preserving scheme based on pseudonyms exchanges for VANET. J. Tsinghua Univ. (Sci & Tech) **52**(5), 592–597 (2012)
3. Wiedersheim, B., Ma, Z., Kargl, F., et al.: Privacy in inter-vehicular networks: why simple pseudonym change is not enough. In: Seventh International Conference on Wireless On-Demand Network Systems and Services, pp. 176–183. IEEE (2010)
4. Freudiger, J., Raya, M., et al.: Mix-zones for location privacy in vehicular networks. In: WiN-ITS 2007 (2007)
5. Lu, R., Lin, X., Luan, T.H., et al.: Pseudonym changing at social spots: an effective strategy for location privacy in VANETs. IEEE Trans. Veh. Technol. **61**(1), 86–96 (2012)
6. Wasef, A., Shen, X.: REP: location privacy for VANETs using random encryption periods. Mob. Netw. Appl. **15**(1), 172–185 (2010)
7. Misra, S., Verma, M.: PACP: an efficient pseudonymous authentication-based conditional privacy protocol for VANETs. IEEE Trans. Intell. Transp. Syst. **12**(3), 736–746 (2011)
8. He, D., Zeadally, S., Kumar, N., et al.: Efficient and anonymous mobile user authentication protocol using self-certified public key cryptography for multi-server architectures. IEEE Trans. Inf. Forensics Secur. **11**(9), 2052–2064 (2016)

An Improvement RFID Security Authentication Protocol Based on Hash Function

Haowen Sun[1], Peng Li[1,2(✉)], He Xu[1,2], and Feng Zhu[1,2]

[1] School of Computer Science,
Nanjing University of Posts and Telecommunications, Nanjing 210003, China
{1217043119, lipeng, xuhe, zhufeng}@njupt.edu.cn
[2] Jiangsu High Technology Research Key Laboratory for Wireless Sensor Networks, Nanjing 210003, Jiangsu, China

Abstract. Radio frequency identification (RFID) technology faces many security issues, and a detailed analysis of these security issues is conducted. It analyzes the existing RFID security authentication protocol and proposes a security authentication protocol that can effectively solve the problems of fraud, retransmission, tracking, and synchronization. In the continuous conversation mode of RFID, it uses the powerful computing power of the reader, the random number identification and comparison between the reader and the tag can resist the denial of service attack. To be able to resist desynchronization attacks, the back-end database stores tag identifiers and dynamically updates the data so that the tags and the background database maintain data synchronization. It utilizes the reader's computing and storage functions to reduce the cost of tags, makes the agreement meet the requirements of low-cost, ensures two-way authentication, and improves the efficiency of the security authentication protocol.

1 Introduction

Radio frequency identification [1] is a non-contact automatic identification and information collection technology of objects using wireless radio frequency. It has the characteristics of fast, real-time and high-efficiency. It is one of the key technologies of the Internet of Things. RFID has been successfully applied to logistics warehouse management, smart medical care, product safety traceability, personnel information management, and national defense. It has gradually become the core competitiveness of various companies. With the wide application of RFID technology in various industries, its security issues have also received attention from all sectors of society. The security and privacy [2] threats in RFID systems are mainly transmitted data security threats, user personal privacy threats, and cloning threats. These problems not only cause damage to the RFID system itself, but also endanger the privacy and life security of individual users. The insufficiency of security and privacy has seriously affected the promotion of RFID worldwide. It is still a recognized problem to improve the security mechanism of RFID systems.

L. Barolli et al. (Eds.): IMIS 2018, AISC 773, pp. 375–384, 2019.
https://doi.org/10.1007/978-3-319-93554-6_35

1.1 RFID System Composition

RFID technology is supported by wireless radio frequency technology, computer software technology, hardware integrated circuit technology and database technology. Figure 1 below shows the basic composition of the RFID system [3].

Figure 1 shows the most basic RFID hardware system [4] consists of a tag, a reader, and a background database.

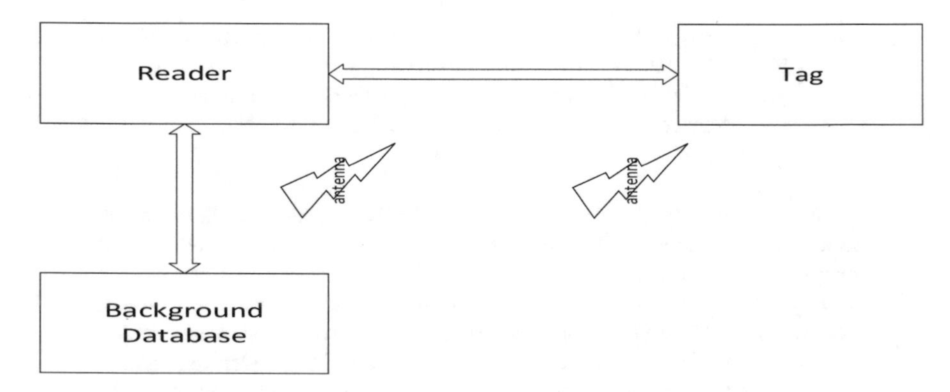

Fig. 1. Composition of the RFID system

1.2 Security Issues Faced by RFID

The communication channel (forward channel) [5] between the RFID reader and the tag is an insecure channel, and the attacker uses a different attack method to attack the channel. The specific security issues [6] faced by an RFID system are mainly the following:

(1) Retransmission attack [7]: When the reader and the tag communicate, the attacker intercepts the response signal returned by the tag to the reader by some means, and then uses the tag's response signal to communicate with the RFID system again, making the information data. Retransmission to achieve the purpose of the attack. For retransmission attacks, challenge-response mechanisms can be used to solve them. Some protocols also use readers and tag random numbers for authentication and identification to resist this attack.

(2) Counterfeit attack [8]: When the reader and the tag communicate, the attacker fakes the reader to intercept the response signal returned by the tag to the reader. At this point, the attacker uses the intercepted response signal to respond to the original reader, causing the reader to mistakenly believe that it is a response signal returned by the tag, and the attacker can thus move or even destroy the tag to achieve the purpose of the attack. Encrypting data and adding security protocols can solve this type of attack.

(3) Tracking [9]: This kind of attack is different from the previous two attacks. It does not directly attack the communication of the system, but it poses a threat to the

security and privacy of the label user. When the attacker again intercepts the response signal returned by the tag, the tag is tracked, including the personal privacy information and history information of the tag itself. For this type of attack, many protocols use the Hash function for encryption, using its randomness and unidirectionality to solve the problem.

(4) Desynchronization [10]: This kind of attack will generally appear in the security protocol of the dynamic ID. After the RFID completes a communication, the attacker actively updates the ID of the tag. During the next communication, the tag information stored in the database and the tag information are not synchronized, which results in failure to authenticate the tag and achieve the purpose of the attack. The protocol in this article uses the background database to store the historical information of all tags, so that the database will not lack the certification of tag history information and solve the problem of desynchronization.

(5) Denial-of-service attacks [11]: Flooding attacks initiated by illegal tags against background databases. This article also studies the method of resisting denial of service attack, using the reader's computing power and storage capacity to process the information of the tag.

(6) Eavesdropping [12]: The attacker uses a certain device to eavesdrop on the electromagnetic characteristics, signal characteristics, and energy performance of the RFID hardware system during operation. The side channel attacks belong to this type of attack. Since this kind of attack belongs to the physical layer, it is not in the scope of this article.

2 Typical RFID Security Authentication Protocol Analysis

There are two main types of security solutions for FID systems: physical security mechanisms [13] and encryption-based technology mechanisms. Physical security mechanisms require the use of additional hardware equipment, which brings with it cost and means limitations. The security mechanism based on encryption technology has attracted more attention. The security authentication protocol of RFID system must provide authentication and encryption functions. The medium-level authentication protocol for message authentication and encryption using Hash function not only provides sufficient security, but also controls the cost requirements of the device and becomes the RFID security authentication protocol [14]. The research focuses and hotspots. The protocol research in this paper is also based on the Hash function.

2.1 Hash-Lock Protocol

Hash-Lock was proposed by Sarma et al. The main method is to use the tag's virtual ID (metaID) instead of the real ID, which can effectively avoid the leakage and tracking of tag information. The specific protocol flow is shown in Fig. 2.

Hash-Lock protocol [15] achieves the reliability of RFID system communication while ensuring less computation and lower resource consumption. However, as can be seen from Fig. 3, in the final stage of authentication, the background database still uses

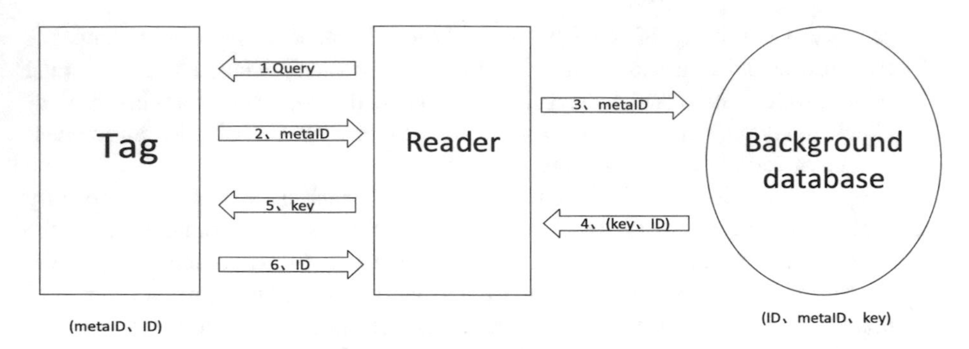

Fig. 2. Hash-Lock protocol

a real tag ID, which is transmitted in clear text on the insecure channel between the reader and the tag, and the protocol has no ID dynamics. Refresh mechanism, metaID remains unchanged, so Hash-Lock protocol is vulnerable to retransmission attacks and counterfeit attacks, attackers can easily trace the label.

2.2 Random Hash-Lock Protocol

The random Hash-Lock protocol was proposed by Weis et al. The goal is to solve the problem of the tag location being tracked in the Hash-Lock protocol. The specific protocol flow is shown in Fig. 3.

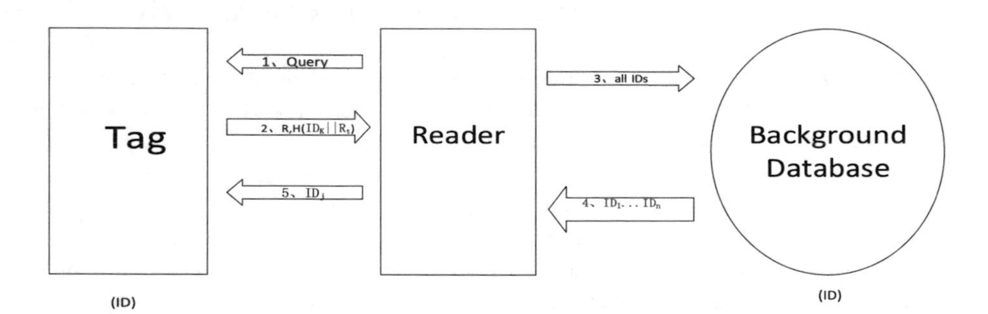

Fig. 3. Random Hash-Lock protocol

However, because the protocol lacks updates to the tag information, each time a fixed ID, the protocol cannot avoid replay attacks. After the background database completes the authentication of the tag information, it still transmits the tag ID important information on the insecure channel in the form of plain text, and is easily traced by the attacker. The security mechanism of the agreement is also not perfect. However, because the protocol lacks updates to the tag information, each time a fixed ID, the protocol cannot avoid replay attacks. After the background database completes the authentication of the tag information, it still transmits the tag ID important information on the insecure channel in the form of plain text, and is easily traced by the attacker. The security mechanism of the agreement is also not perfect. In addition, the protocol requires the tag to have the function of hash function calculation and random number generation, which increases the traffic and tag cost, so this protocol is also not practical.

2.3 Hash-Chain Protocol

The Hash-Chain protocol is based on a shared-key challenge-response mechanism with good forward security. Compared with the random Hash-Lock protocol, the Hash-Chain protocol uses two different Hash functions, H and G. When the reader sends a query and authentication request, the tag will return different response information [16]. Another feature of the protocol is that both the tag and the back-end database share an initial key value. Similarly, the back-end database stores the IDs of all tags. The specific protocol flow is shown in Fig. 4.

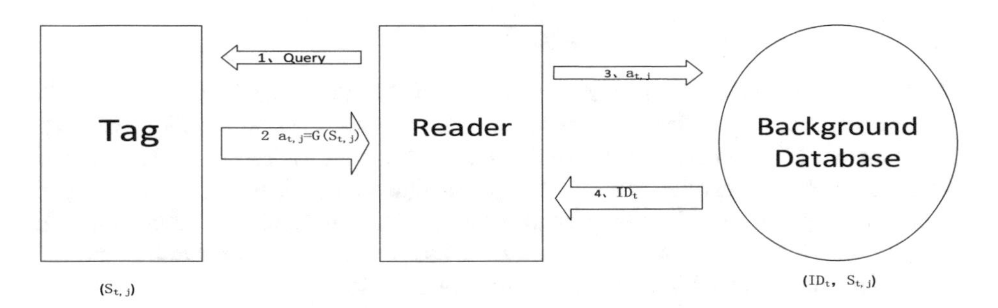

Fig. 4. Hash-Chain protocol

Unlike the other two authentication protocols, the Hash-Chain protocol [17] requires the tag to have an autonomous update ID and be an active tag. Whenever a device in an RFID system completes an authentication communication, the tag is automatically updated, which avoids the leakage of important tag information.

3 Improved Authentication Protocol Based on Hash Function

Based on the above three existing typical security authentication protocol analysis and comparison, the three security authentication protocol fails to meet the security requirements, there are replay attacks, counterfeit attacks, one-way authentication and other issues [18]. In view of the above problems, this paper presents an improved authentication protocol, which is also based on the Hash function. But the difference is that the reader participates in the authentication process and makes full use of the powerful computing and storage functions of the reader. In a communication process, the reader sends a random number and query information to the tag, and calculates the response information returned by the tag to obtain the calculation result, and compares the result with the random number generated by the reader itself. If they are equal, it indicates that the tag is a valid tag; if not, it indicates that the tag is illegitimate, interrupts the transmission of information to the background database, completes the authentication of the reader on the tag's legal identity, completes bidirectional authentication between the reader and the tag, At the same time, the reader filters the tags to prevent denial of service attacks. Dynamic update background database and tag identification, background database storage a large number of tag information, to prevent the synchronization to attack. This protocol also uses a random number generator, the generator exists in the reader, and other protocols is different, each communication process, the reader will produce two different random numbers Nr and Nt.

These two random numbers perform their own duties, Nr used to complete the reader screening of tags to prevent flooding denial-of-service attacks; Nt is used to prevent the attacker from intercepting tag information tag tracking. The improved protocol in this paper uses a reader instead of a tag to generate random numbers, reducing the cost of tags. This shows that the improved protocol in this paper not only meets the security requirements but also has high efficiency.

3.1 Initial Conditions and Symbols

Under the initial conditions, the tag stores the metaID of the dynamic identifier and the identifier RID of the reader, and the tag in this protocol has the functions of Hash operation and XOR operation [19]. Unlike other protocols, the reader in this protocol stores the reader's identifier, RID, with a pseudo-random number generator that performs Hash operations. The background database stores not only the identifiers TID and RID of all valid and authorized readers, but also the tag's dynamic identifiers newmetaID and oldmetaID. Specific settings are as follows:

The improved protocol is shown in Fig. 5.

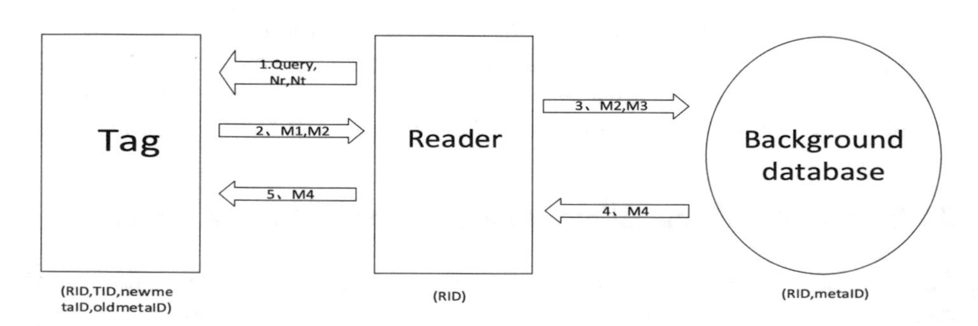

Fig. 5. Improved authentication protocol

H: Hash function is one-way hash, and metaID = H (TID) stored at the tag.
Nr and Nt: are random numbers generated by the reader, assigned to readers and tags.
||: concatenation operation symbol.
⊕: XOR operation symbol.
? =: Compare the operation symbol.

3.2 Improved Protocol Specific Process and Description

The following improvements to the agreement for a specific process analysis [20]:

(1) The reader generates two random numbers Nr and Nt and sends a Query authentication request to the tag.
(2) After receiving the authentication request from the reader, the tag calculates N = Nr ⊕ Nt, M1 = N||Nt, and M2 = mataID ⊕ H (RID ⊕ N), and sends the calculated M1 and M2 as response information To the reader.
(3) The reader receives M1 and M2, separates N and Nt from M1, calculates N ⊕ Nt to obtain Nr′, and then compares Nr′ with the random number Nr generated by the reader itself. If they are equal, M3 = RID||N is calculated, and M2 and M3 are sent together to the background database; if they are not equal, the tag is an illegal tag, and the authentication fails.
(4) The background database receives M2 and M3, separates RID and N from M3, and finds out from the database whether RID′ is equal to the received RID. If it exists, then calculate H (RID′ ⊕ N) ⊕ M2 to get mataID′. And then according to the database stored in the tag identifier to find whether there is a tag newmetaID or oldmetaID and metaID′ equal. If it exists, the background database successfully authenticates the tag. If it does not exist, the authentication fails. After the background database successfully authenticates the tag, M4 = RID′ ⊕ H (metaID′ ⊕ N) is calculated, and M4 is sent to the reader. After the background database need to dynamically update the identifier, according to the background database stored newmetaID and oldmetaID. If metaID′ = newmetaID, then calculate oldmetaID = newmetaID, update newmetaID = metaID′ ⊕ N; if metaID′ = oldmetaID, update newmetaID = metaID′ ⊕ N.
(5) The reader receives M4 and forwards it to the tag.
(6) The tag receives M4, calculates H (metaID ⊕ N) ⊕ M4 gets RID′, and compares with the tag stored RID for equality. If they are equal, the tag completes the authentication of the reader and the background database; otherwise, the authentication fails.

3.3 Improved Protocol Security Analysis

There are many differences between the improved protocol and the three typical Hash authentication protocols in this paper, but whether this protocol meets the actual needs and whether it can withstand common attacks needs to be further analyzed [21].

(1) Anti-retransmission attacks and counterfeit attacks: Based on the continuous conversation mode, the reader generates two different random numbers during each communication process, so that the information transmitted is random. Even after a certain communication, the attacker intercepts the response message from the tag, but due to the unpredictability of the random number, the attacker can not continue to authenticate communication with the reader according to the information and resist the retransmission attack. Coupled with the random nature of the attacker can not be disguised as a legitimate reader, the response information returned by the tag can not be completed certification against counterfeit attacks.
(2) Confidentiality: In this improved protocol, the tag ID is not transmitted directly in the insecure channel. Instead, the metaID is used instead of the real identity. In addition, Hash is one-way secure, Each communication becomes randomized, an attacker can not infer historical information based on current information or predict future information, and the tag is hidden (not traceable).
(3) To resist desynchronization attack: In the continuous conversation mode, the background database stores the information of the tags newmetaID and oldmetaID in two different time periods. Even if the attacker uses some means to cut off the communication in step 5 of the protocol, the background database has completed updating the dynamic identifier newmetaID, and the tag fails to update its dynamic identifier metaID. In this case, the background database uses the oldmetaID, which succeeds in the previous authentication of the tag, to ensure the validity of the tag and update the dynamic identifier accordingly [22].
(4) Resistance to denial of service attacks: In continuous conversation mode, a reader may communicate with multiple tags. In order to resist flood attacks and denial of service attacks, this protocol performs the following two stages of precautionary measures: (1) Reader After receiving the tag M1, M1 is calculated, and the calculated result is compared with the random number generated by the reader itself. If they are not equal, the tag is illegitimate, and the reader will not pass M2 and M3 to the background database. (2) This protocol step 4 for metaID′ comparison, if there is no newmetaID or oldmetaID equal to it, the agreement is interrupted to avoid illegal tags denial of service attacks on the database.
(5) Forward security: in each communication process, the reader will generate two random numbers, to ensure random reader and tag information, coupled with Hash is one-way security, the attacker can not be based on the Times of information for the next communication certification.

In addition to meeting the security requirements of RFID systems, this protocol is optimized in terms of algorithm complexity and algorithm space. During a complete authentication, the tag stores two sets of data and performs two Hash operations. The reader stores a set of data and generates a random number. The background database stores four sets of data and performs two hash operations. The agreement not only can resist various attacks, but also in time and space can be a good compromise.

4 Summary

This paper mainly analyzes and researches the security of RFID system, and proposes an improved Hash-based efficient security authentication protocol based on the existing three typical Hash function security authentication protocols. The reader generates two random numbers to make the information random, the tag has a Hash encryption operation function, the tag and the background database have a dynamic ID refresh mechanism, etc. are all important features of the protocol, can resist multiple attacks, and in time and space Made an eclectic.

With the popularization of RFID technology [23], the security mechanism of this system is constantly being improved. Cryptographic technology-based RFID security authentication protocol is the focus of people's research. However, how to combine high-strength encryption technology with the low-cost requirements of labels is a problem that all walks of life will face. While improving the security of RFID systems, it needs Efforts are made to make full use of RFID technology in life practices.

Acknowledgments. The subject is sponsored by the National Natural Science Foundation of P. R. China (No. 61672296, No. 61602261), the Natural Science Foundation of Jiangsu Province (No. BK20160089), Scientific & Technological Support Project of Jiangsu Province (No. BE2016777, BE2016185).

References

1. Want, R.: An introduction to RFID technology. IEEE Pervasive Comput. **5**(1), 25–33 (2006)
2. Juels, A.: RFID security and privacy: a research survey. IEEE J. Sel. Areas Commun. **24**(2), 381–394 (2006)
3. Lakafosis, V., Traille, A., Lee, H., et al.: An RFID system with enhanced hardware-enabled authentication and anti-counterfeiting capabilities. In: Microwave Symposium Digest, pp. 840–843. IEEE (2010)
4. Bonuccelli, M.A., Martelli, F.: A very fast tags polling protocol for single and multiple readers RFID systems, and its applications. Ad Hoc Netw. **71**, 14–30 (2018)
5. Shen, Z., Zeng, P., Qian, Y., Choo, K.-K.R.: A secure and practical RFID ownership transfer protocol based on chebyshev polynomials. IEEE Access **6**, 14560–14566 (2018)
6. Rieback, M.R., Crispo, B., Tanenbaum, A.S.: The evolution of RFID security. IEEE Pervasive Comput. **5**(1), 62–69 (2006)
7. Schantin, A., Ruland, C.: Retransmission strategies for RFID systems using the EPC protocol. In: IEEE International Conference on RFID-Technologies and Applications, pp. 1–6. IEEE (2013)
8. Li, H., Hou, Y.B., Huang, Z.Q., et al.: Research on the attack model for RFID anti-counterfeit protocol. Acta Electronica Sinica **37**(11), 2565–2573 (2009)
9. Kodialam, M., Nandagopal, T., Lau, W.C.: Anonymous tracking using RFID tags. In: IEEE International Conference on Computer Communications, INFOCOM 2007, pp. 1217–1225. IEEE (2007)
10. Zhou, S., Zhang, Z., Luo, Z., et al.: A lightweight anti-desynchronization RFID authentication protocol. Inf. Syst. Front. **12**(5), 521–528 (2010)

11. Fu, Y., Zhang, C., Wang, J.: A research on Denial of Service attack in passive RFID system. In: International Conference on Anti-Counterfeiting Security and Identification in Communication, pp. 24–28. IEEE (2010)
12. Xie, R., Jian, B., Liu, D.: An improved ownership transfer for RFID protocol. Int. J. Netw. Secur. **20**(1), 149–156 (2018)
13. Zhang, Y., Yang, F., Wang, Q., He, Q., Li, J., Yang, Y.: An anti-collision algorithm for RFID-based robots based on dynamic grouping binary trees. Comput. Electr. Eng. **63**, 91–98 (2017)
14. Cai, Y., Zhang, J.: Comparison and research of RFID security authentication protocol. Digit. Commun. World. **12**, 62–66 (2015)
15. Ding, Z., Li, J., Feng, B.: Research on hash-based RFID security authentication protocol. J. Comput. Res. Dev. **46**(4), 583–592 (2009)
16. Zeng, L., Zhang, X., Zhang, T.: Key value renewal random hash lock for security and privacy enhancement of RFID. Comput. Eng. **33**(3), 151–153 (2007)
17. Zhong, J.Z.: Design and research on RFID security protocol based on hash-chain. Modern Comput. **8**, 139–141 (2010)
18. Zhang W.: Analysis and improvement of RFID security authentication protocol HSAP. Intell. Comput. Appl. **5**, 78–80 (2012)
19. Benssalah, M., Djeddou, M., Drouiche, K.: Security analysis and enhancement of the most recent RFID authentication protocol for telecare medicine information system. Wirel. Pers. Commun. **96**(4), 6221–6238 (2017)
20. Wang, K.-H., Chen, C.-M., Fang, W., Tsu-Yang, W.: On the security of a new ultra-lightweight authentication protocol in IoT environment for RFID tags. J. Supercomput. **74** (1), 65–70 (2018)
21. Gandino, F., Montrucchio, B., Rebaudengo, M.: A security protocol for RFID traceability. Int. J. Commun. Syst. **30**(6), e3109 (2017)
22. Sarker, J.H., Nahhas, A.M.: Mobile RFID system in the presence of denial-of-service attacking signals. IEEE Trans. Autom. Sci. Eng. **14**(2), 955–967 (2017)
23. Wang, C., Shi, Z., Fei, W.: An improved particle swarm optimization-based feed-forward neural network combined with RFID sensors to indoor localization. Information **8**(1), 9 (2017)

Classifying Malicious URLs Using Gated Recurrent Neural Networks

Jingling Zhao[1,3], Nan Wang[1,3](✉), Qian Ma[2,3], and Zishuai Cheng[1,3]

[1] School of Computer Science,
Beijing University of Posts and Telecommunications, Beijing, China
nanwang@bupt.edu.cn
[2] School of Cyberspace Security,
Beijing University of Posts and Telecommunications, Beijing, China
[3] National Engineering Laboratory for Mobile Network Security, Beijing, China

Abstract. The past decade has witnessed a rapidly developing Internet, which consequently brings about devastating web attacks of various types. The popularity of automated web attack tools also pushes the need for better methods to proactively detect the huge amounts of evolutionary web attacks. In this work, large quantities of URLs were used for detecting web attacks using machine learning models. Based on the dataset and feature selection methods of [1], multi-classification of six types of URLs was explored using the random forest method, which was later compared against the gated recurrent neural networks. Even without the need of manual feature creation, the gated recurrent neural networks consistently outperformed the random forest method with well-selected features. Therefore, we determine it is an efficient and adaptive proactive detection system, which is more advanced in the ever-changing cyberspace environment.

1 Introduction

The connection between the Web and our daily activities is getting increasingly close nowadays. Convenience as the Web may bring about, it also poses threats. According to [2], approximately 611,141 web attacks were blocked by Symantec per day in the year of 2017. Enterprises and individuals suffer from service breakdown, information leakage, data hijacking and other incidents due to various types of Web attacks, such as SQL injection and XSS injection attacks. As the attack techniques develop quickly with the expansion of the Internet, it is vital for us to develop reliable and adaptive defense technologies.

Web applications are usually accessed via URLs, which can be manually typed in the browser's address bar or clicked as a link. However, URLs are often embedded with executable codes and used as a main container for attack payloads by attackers. In 2017, the number of malicious URLs grew by 2.8%, with 7.8% of all URLs identified as malicious [2]. What's worse, attackers have developed various disguise techniques to bypass detection. Therefore, a solid step towards detecting Web attacks is to identify malicious URLs.

L. Barolli et al. (Eds.): IMIS 2018, AISC 773, pp. 385–394, 2019.
https://doi.org/10.1007/978-3-319-93554-6_36

In previous studies, malicious keyword databases and matching regulations are applied to block malicious URLs. While such methods can recognize well-known attack patterns efficiently, they are not flexible enough to adapt to more complicated attack behaviors or fuzzy characteristics.

For this work, we focused on using machine learning technique for the multi-classification of malicious URLs, which is a continuation of [1] and followed where the datasets and most of feature selection methods. Specifically, we compared the combination of lexical and statistical analysis of URLs as input for a random forest (RF) classifier against a novel approach that employs gated recurrent neural networks (GRU). RFs, with well-selected features, have been successfully applied to identifying malicious URLs and performed well in [1]. On the other hand, GRU models are competent at detecting long patterns in sequences and well-known for solving text analysis problems. Additionally, the GRU does not require manual feature extraction, since it directly learns a representation from the URL's sequence of characters. The RF model achieved an accuracy rate of 96.4%, and the GRU model 98.5%, according to the experimental results, with GRU model consistently outperformed the RF model on different training sizes.

The remaining of this paper is organized as follows. Section 2 will describe related work on malicious URL detection. Section 3 will introduce the feature extraction method and random forest method. The gated recurrent unit method is presented in Sect. 4. Then, the data and methodology are illustrated in Sect. 5, and the experimental results are shown in Sect. 6. Finally, conclusions are drawn in Sect. 7.

2 Related Work

The studies on malicious URL detection can be mainly divided into two categories, which include blacklisting and machine learning. For blacklisting methods, Jian Zhang focused on the blacklisting strategy and introduced a blacklisting system based on a relevance ranking scheme borrowed from the link-analysis community [4], while Prakash [5] proposed an approach that mainly matched IP addresses, hostnames, directory structures, and brand names to detect phishing URLs. Hegarty detected the transmission of sensitive files based on file signature within the cloud network environment [6].

Although such methods can discover well-known attack patterns efficiently, they are not competent at detecting new malicious URLs and have poor generalization ability. In recent years, a variety of machine learning approaches have been proposed for malicious URL detection. For instance, Garera [7] detected and intercepted malicious URLs by analyzing the URL structure of the phishing site and training the classifier based on the logistic model.

Feature extraction is a crucial part of the machine learning method for detecting malicious URLs. In [3] both heuristic-based and feature-based methods were applied in feature extraction process, which proved to work efficiently.

The work by Bahnsen is most closely related to our study with regard to experimental setup [8]. The paper compared the combination of lexical and statistical analysis of phishing URLs as input for a random forest classifier against LSTM networks. It extracted 14 features for the RF model and compared both methods in terms of accuracy and training times.

Although similar in motivation and methodology to the preceding papers, this paper proposes a new method to identify various types of malicious URLs from benign ones. By comparing different methods based on experimental results, we can see the GRU method has shown great promise for the information security industry.

3 Classifying Malicious URLs Using Random Forest

In this section, we will illustrate our feature extraction strategy and how to apply it in a random forest classifier.

The feature extraction strategy we adopted is basically the same as that in [1], for the reason that we share the same dataset and the strategy has proved to work well in binary classification experiments. To better distinguish malicious URLs among different attacks, we added three heuristic-based features as is presented in [3], where they are applied in semi-supervised multi-classification tasks.

Specifically, the 21 features can be divided into five categories, which include the length features, the number features, the type features, the risk level features and other advanced features.

The Length Features

The length features contain length measurements that describe different features from various viewpoints, as is listed in Table 1.

Table 1. The length features.

Length features			
subPathAvgLen	pathLen	paraNameMaxLen	paraValueMaxLen
paraAvgLen	paraLen	subPathMaxLen	directoryMaxLen

The Number Features

The number features include the numbers of parameters, abnormal characters and sub-paths in the URL.

The Type Features

The type features represent the composition of a URL string, like pure-digital type and pure-letter type. Character types of the parameter value and the whole path are included.

The Risk Level Features

Risk level features include SQL risk level, XSS risk level, sensitive risk level, and other risk level. These values are the result of a comprehensive calculation. As is stated in [3], heuristic-based method is applied with ModSecurity rules for calculating the level value, and a blacklist of risk keywords was used in the process.

Other Advanced Features

Other advanced features include two percentage features, which calculate the percentages of pure digits and pure letters in the entire URL. Additionally, the feature "Value contains IP" utilizes regular expressions to detect whether an IP address appears in a value field and the feature "Nginx test" detects whether a URL is an Nginx attack.

Once the features were extracted, a multiple classifier was trained using the random forest method. The random forest is an ensemble classifier consisting of a collection of tree-structured base classifiers. In the classifier, the best parameter at each node in a decision tree is made from a randomly selected number of features, which helps to not only scale multiple-feature vectors, but also reduce the interdependence between feature attributes [9]. What's more, in [1] where the same dataset was first applied, comprehensive performances of naïve Bayes, decision tree and support vector machine were compared in binary classification tasks. The decision tree model, which can be assembled as random forest, outperformed the other models in both experiment results and practical application, achieving an accuracy above 98.7%.

4 Modeling Malicious URLs Using GRU Networks

In the previous section, we designed a set of features extracted form a URL and fed them into a classification model to predict whether a URL is a case of web attack. We now approach the problem in a different way. Instead of manually extracting the features, we directly learn a representation from the URL's character sequence.

Each character sequence exhibits correlations, that is, nearby characters in a URL are likely to be related to each other. These sequential patterns are important because they can be exploited to improve the performance of predictors [10].

RNN including gated recurrent unit is related to our work. As stated in [11], RNN are able to memorize arbitrary-length sequences of input patterns by building connections between units from a directed cycle. However, the drawback of RNN is that they are unable to learn the correlation between elements more than 5 or 10 time steps apart [12]. To alleviate this issue, long short-term memory networks were first presented by introducing gates function in the design of transition function [13]. In our paper, we adopt another RNN variant: gated recurrent neural networks (GRUs) that can be regarded as a simpler version of LSTMs.

In this work, we used GRU units to build a model that receives as input a URL as character sequence and predict which type of web attacks it belongs to. The architecture is illustrated in Fig. 1. Each input is translated by one-hot embedding, and the 129-dimension vectors are fed into a GRU layer as a 200-step sequence. Finally the classification is performed using a softmax function.

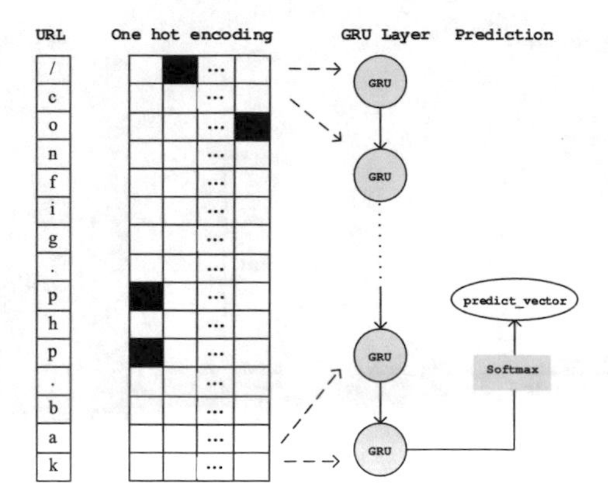

Fig. 1. Architecture of gated recurrent neural networks.

5 Experimental Setup

We used part of the dataset from [1, 3] as the training dataset, which comes from a well-known Chinese Internet security company. The dataset includes more than 150,000 legitimate URLs and 240,000 malicious ones, with both ModSecurity rules and manual labeling applied to get the training set classified. In Table 2, samples of six URL types are shown.

Table 2. Samples of six URL types.

No.	URL	Type
1	/modules/article/addbookcase.php?bid=43489	Legitimate
2	/member.php?mod=-1 OR 1=1&action=login	SQL injection
3	/admin/main.do?%22%20onmousemove='alert(42873)'wb=%22	XSS attack
4	/config/config_ucenter.php.bak	Sensitive file attack
5	/list/list.php?p=../../../../WEB-INF/web.xml?	Directory traversal
6	/ebook/yp/product.php?pagesize=${@phpinfo()}	Other attack

Moreover, in Fig. 2, a comparison of URL length distribution is presented. It can be observed that the URL distribution of SQL injection type is obviously different from other URL types, it tends to contain more longer URLs (measured in number of characters).

In the training dataset, the number of six types of URLs is roughly equal. Specifically, we used a 3-fold cross-validation strategy, which consists of splitting data into 3 folds, and use one fold for model validation each time. The ultimate accuracy of both models is calculated based on the averaged results of each fold. The performance evaluation is done through comparing accuracies.

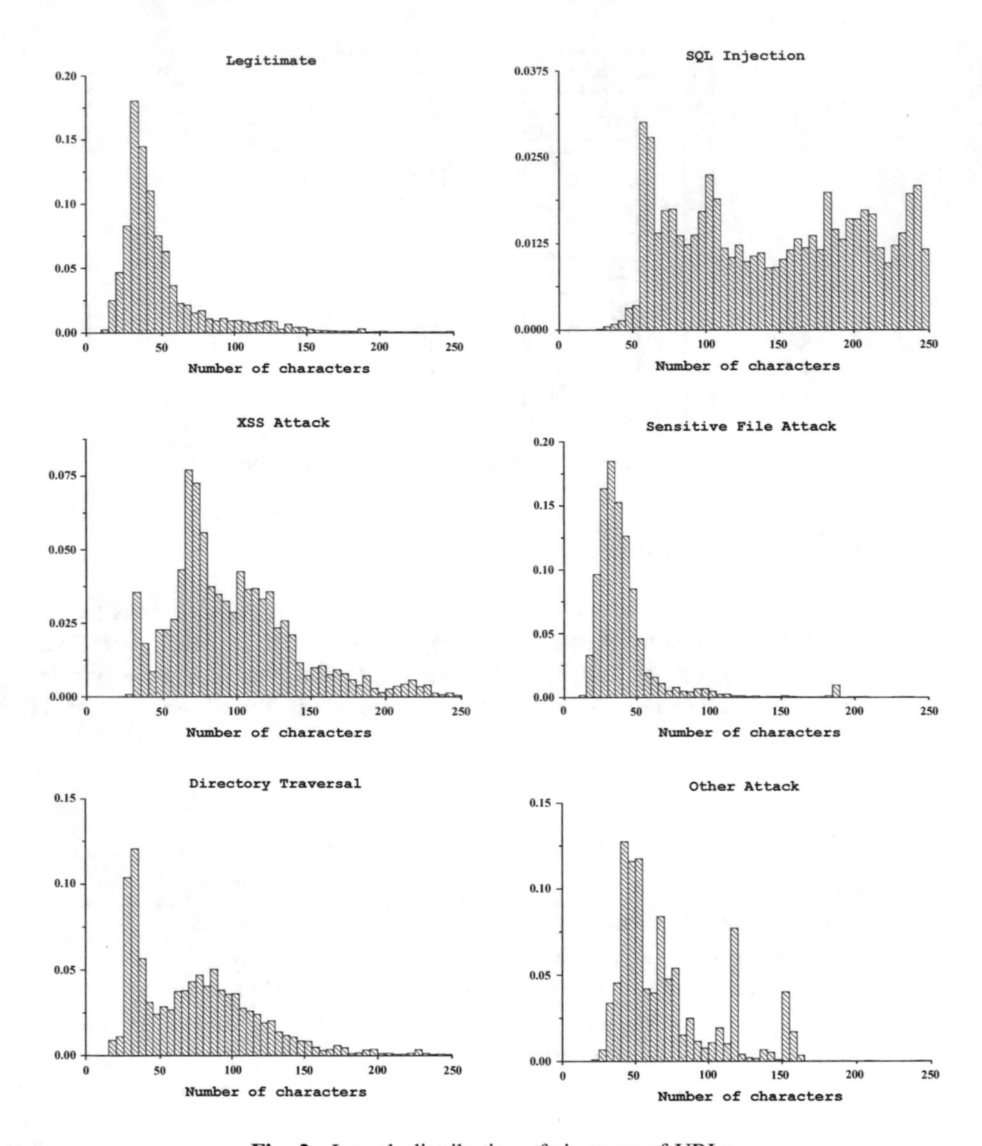

Fig. 2. Length distribution of six types of URLs.

6 Results

In this section we present the experimental results. First, we evaluated the performance of the traditional feature extraction plus the classification-algorithm methodology presented in Sect. 3. We extracted 21 features based on the URL's lexical and statistical analysis. Then we trained a random forest classifier with 20 decision trees. We used the random forest implementation of the Scikit-Learn library [14].

Using 240,000 URLs from the datasets described above, we tested the accuracy of the model using a 3-fold cross validation strategy. The results are shown in Table 3. The average accuracy of the random forest model is 96.4%.

Table 3. Accuracy of RF model.

Folder	1	2	3	Average
Accuracy	0.969	0.963	0.961	0.964

Furthermore, we analyzed the feature importance of the random forest classifier. This is done by counting the number of times each feature was selected in the different decision trees inside the random forest [8]. Feature importance is shown in Fig. 3. The most important features are the risk levels of sensitive file attack, SQL injection and XSS attack, all summarized based on heuristics. To be more specific, this suggests that the high accuracy should be mainly credit to the well-designed risk level evaluation process applied in feature extraction. Additionally, it is observed that features representing length characteristics play an important role in telling different URLs apart.

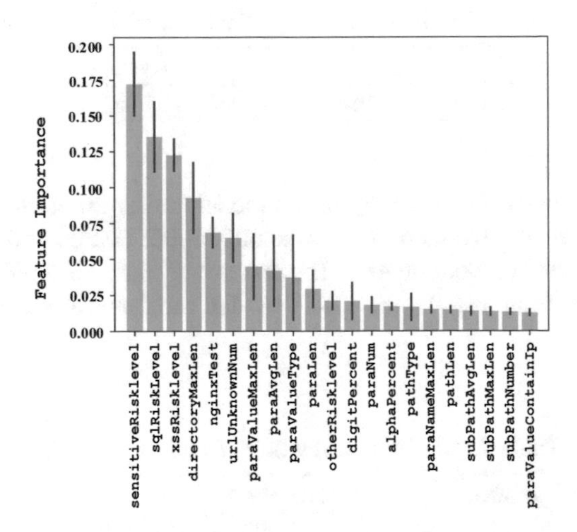

Fig. 3. Feature importance of the random forest classifier.

Subsequently, we trained the GRU network as described in Sect. 4. The experiments are based on the Theano – a python library, which supports efficient symbolic differentiation and transparent use of a GPU [8]. We defined the max epoch number to be 20, and extract 10% of the URLs as validation set. The validation accuracy converged within 13 epochs, increasing to over 98% from epoch 9 upwards (Fig. 4).

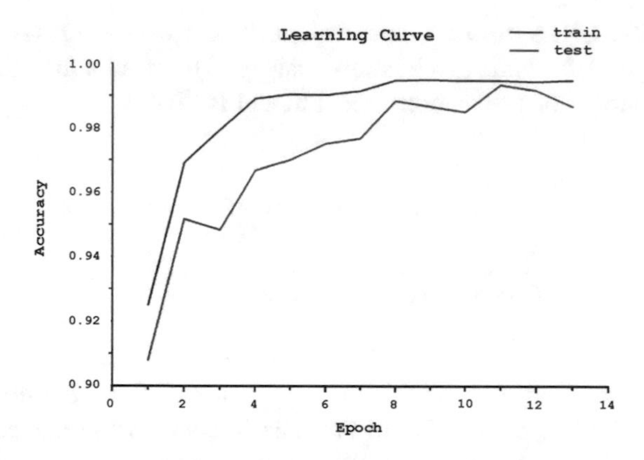

Fig. 4. Learning curve of the GRU model.

As is shown in Table 4, the GRU model has an average accuracy of 98.5%, which is 2.1% higher than that of random forest model.

Table 4. Accuracy of GRU model.

Folder	1	2	3	Average
Accuracy	0.986	0.985	0.983	0.985

Lastly, we compared the accuracy of the models using different numbers of URLs for training. Specifically, we randomly chose 600, 3000, 60,000 and 240,000 URLs as training sets and tested on both models. The results are shown in Table 5 and Fig. 5. As can be concluded from Fig. 5, the GRU model consistently outperformed the RF model.

Table 5. Model accuracy on different training sizes.

Models	URL numbers			
	600	3,000	60,000	240,000
Random forest	0.878	0.917	0.945	0.964
GRU network	0.893	0.956	0.976	0.985

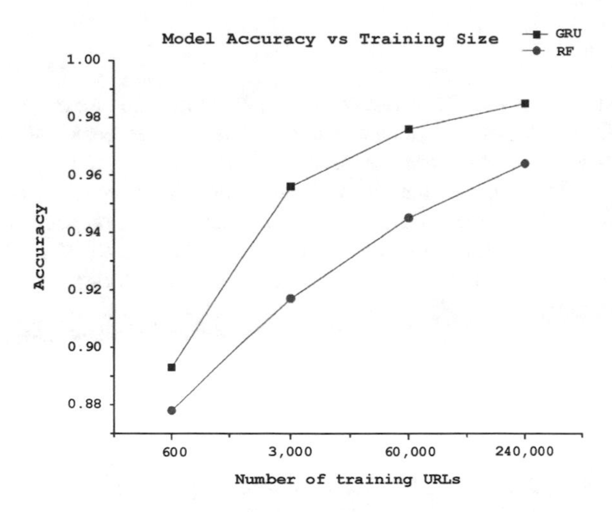

Fig. 5. Model accuracy on different training sizes.

7 Conclusion and Discussion

We explored how well we can classify different malicious URLs using two methodologies: a random forest classifier with well-designed feature extraction, and a gated recurrent neural network method. The former has been widely used since the 1990s, and the latter is a new method within recurrent neural networks. In order to evaluate the approaches, we used approximately 240,000 URLs of six types, which include legitimate, SQL injection, XSS injection, directory traversal, sensitive file attack and other attacks. The results of the 3-fold cross validation showed that the average accuracy of RF model is 96.4% while that of the GRU model is 98.5%.

As can be seen from the analysis above, a high-performance random forest classifier for malicious URL multi-classification largely relies on heuristic-based features which usually consist of malicious keyword blacklists. And domain knowledge is crucial for summarizing such malicious keywords and attack patterns. On the other hand, the GRU model showed an overall higher prediction performance without the need of expert knowledge to extract features. All we need to do is transform all the characters into vectors, and the network will learn hidden patterns by itself. Therefore, the GRU method is a promising new direction for malicious URL multi-classification problems.

However, there is also downside of the GRU method. Just like the other deep learning models, the GRU model needs much more time to train, which requires us to optimize the system architecture so that it could become more applicable and efficient in practice.

Acknowledgments. This work was supported by National Natural Science Foundation of China (No. U1536122).

References

1. Cui, B., et al.: Malicious URL detection with feature extraction based on machine learning
2. Cleary, G., Corpin, M., et al.: Symantec internet security threat report 2017. Symantec Corp., Mountain View, CA, USA, Technical report (2018)
3. Yang, J., et al.: Multi-classification for malicious URL based on improved semi-supervised algorithm. In: 2017 IEEE International Conference on Computational Science and Engineering (CSE) and Embedded and Ubiquitous Computing (EUC), vol. 1. IEEE (2017)
4. Zhang, J., Porras, P.A., Ullrich, J.: Highly predictive blacklisting. In: USENIX Security Symposium (2008)
5. Prakash, P., et al.: PhishNet: predictive blacklisting to detect phishing attacks. In: INFOCOM, 2010 Proceedings IEEE. IEEE (2010)
6. Hegarty, R., Haggerty, J.: Extrusion detection of illegal files in cloud-based systems. Int. J. Space Based Situated Comput. **5**(3), 150–158 (2015)
7. Garera, S., et al.: A framework for detection and measurement of phishing attacks. In: Proceedings of the 2007 ACM Workshop on Recurring Malcode. ACM (2007)
8. Bahnsen, A.C., et al.: Classifying phishing URLs using recurrent neural networks. In: 2017 APWG Symposium on Electronic Crime Research (eCrime). IEEE (2017)
9. Wu, Q., et al.: ForesTexter: an efficient random forest algorithm for imbalanced text categorization. Knowl. Based Syst. **67**, 105–116 (2014)
10. Dietterich, T.G.: Machine learning for sequential data: a review. In: Joint IAPR International Workshops on Statistical Techniques in Pattern Recognition (SPR) and Structural and Syntactic Pattern Recognition (SSPR). Springer, Heidelberg (2002)
11. Schmidhuber, J.: Deep learning in neural networks: an overview. Neural Netw. **61**, 85–117 (2015)
12. Gers, F.A., Schmidhuber, J., Cummins, F.: Learning to forget: continual prediction with LSTM. Neural Comput. **12**, 850–855 (1999)
13. Hochreiter, S., Schmidhuber, J.: Long short-term memory. Neural Comput. **9**(8), 1735–1780 (1997)
14. Pedregosa, F., et al.: Scikit-learn: machine learning in Python. J. Mach. Learn. Res. **12**, 2825–2830 (2011)
15. Zhao, R., et al.: Machine health monitoring using local feature-based gated recurrent unit networks. IEEE Trans. Ind. Electron. **65**(2), 1539–1548 (2018)

A Fast PQ Hash Code Indexing

Jingsong Shan(✉), Yongjun Zhang, Mingxin Jiang, Chunhua Jin, and Zhengwei Zhang

Huaiyin Institute of Technology,
Meicheng Rd, Huaian, Jiangsu, People's Republic of China
shanjings@hyit.edu.cn

Abstract. This paper presents a Compressed PQ Indexing (CPQI) data structure, which realizes the further compression of sparse entries, requires only sub-linear search time, and the sparse entries are no longer stored. The proposed CPQI saves storage space and is suitable for in-memory computing for large-scale data. The CPQI employs the Minimal Perfect Hash to hash the PQ code, preserve non-null entries, and store the structure very compactly; the compressed PQ hash code index no longer stores PQ code. A sub-linear time search is implemented by combining Bloom filtering with a minimum perfect hash function.

1 Introduction

The retrieval of nearest neighbors from huge high-dimensional dataset has become more and more important and has played a fundamental role in machine vision, image processing, and machine learning [1]. The goal of nearest neighbors search is to find the k most similar objects for a given query.

Retrieval of similar items in high-dimensional unstructured data usually uses content-based retrieval techniques [2] (Content Based Image Retrieval: CBIR). (1) Extract feature descriptions of unstructured data, such as SIFT [3], GIST features [4]. These description features are usually represented as high dimensional vectors (hundreds of dimensions or even higher); (2) Given query objects The feature description is to search for similar features in the feature database. The essence of this feature is the Nearest Neighbor Search (NNS) problem in high-dimensional feature space [5, 6].

A commonly used approximate neighbor search method is the tree indexing mechanism, including K-D-tree [7], R-tree, R* tree, VP tree [8]. The tree model index divides the data space in different dimensions, and each node of the tree establishes an index for the corresponding subspace. Approximate nearest neighbor search based on this type of model requires only logarithmic time complexity (O(log(N))). Although these methods have high query efficiency when dealing with low-dimensional data, when the dimensions are high, they will inevitably encounter the "dimensional disaster" bottleneck, and the search time complexity and data dimension will increase exponentially when the dimensions are high. Its performance may even be lower than linear scanning [9, 10].

L. Barolli et al. (Eds.): IMIS 2018, AISC 773, pp. 395–402, 2019.
https://doi.org/10.1007/978-3-319-93554-6_37

In recent years, Product Quantization (PQ) [11] and its variants [12, 13] have attracted widespread attention. Its main idea is to use the Cartesian product to combine the center points of the subspaces. It is very easy to construct a very large number of data partitions (up to 2^{64}) so that the data space is divided very finely; the high dimensional vectors are represented as short The integer hash code, compared with the original vector, greatly reduces the memory consumption and can even store tens of millions of data points in memory; the distance between the original data points can be approximated by an asymmetric distance approximation. Data structures and encoding algorithms are easy to implement. By previously quantizing the database vector into short codes, ANNs for a given vector can be found by linear comparisons from the hash code database using Asymmetric Distance Computation (ADC) and Lookup Table.

Although PQ is easy to use look-up tables and ADC linear scans, the computational cost of N PQ codes is at least O(N). This exhaustive linear scan search over asymmetric distances is only suitable for small data sets. For small-scale datasets, you can simply calculate the distance from each query's data to the hash code to get the most similar objects. However, because of the low efficiency of the linear query and the search time complexity of O(N), linear search becomes infeasible even for medium-scale (10^6-order) data sets.

Intuitively, each center point can be used as an index entry to construct a structure similar to a hash table. In the retrieval phase, the query vector is first PQ-encoded, and then the corresponding item of the PQ code is searched. This seems to be very simple and intuitive, but using the PQ codebook as an index to achieve a fast nearest neighbor search requires solving the problem:

(1) PQ codebooks assembled using Cartesian products are very large. (2) Long code problem. Each entry is represented using a 64-bit integer code. For small-scale PQ hash tables, this code can be used directly, such as millions. However, it is difficult for large-scale entries to be stored in memory.

To address above problems, this paper presents a Compressed PQ Indexing (CPQI) data structure, which realizes the further compression of sparse entries, requires only sub-linear search time, and is no longer stored. PQ encoding saves storage space and is suitable for in-memory computing of large-scale data. The idea of compressing the PQ hash code index is to use the Minimal Perfect Hash to hash the PQ code, preserve non-null entries, and store the structure very compactly; the compressed PQ hash code index no longer stores PQ code A sub-linear time search is implemented by combining Bloom filtering with a minimum perfect hash function.

2 Related Works

PQ quantizes a high-dimensional vector as a PQ code which consists of subspace codeword indexes. For a vector **x**, **x** is equally divided into m disjoint sub-vectors. The PQ quantizer pq(.) quantizes x as:

$$pq(x) = [q_1(x_1), q_2(x_1), \cdots, q_m(x_m)] \tag{1}$$

It is the quantizer of the subspace i which maps the subvector $\mathbf{x}_i$ to the index value of the subcodebook. The quantized value of x is called PQ code. Given a query vector $\mathbf{y}$, it is easy to construct a distance from y to any vector x using the ADC:

$$d_{ADC}^2(x,y) = \sum_{j=1}^{m} d(x_m, q_m(y_m))^2 \tag{2}$$

The PQ linear search method is as follows:

(1) Calculate the ADC distance from y to each sub-word, and construct the distance table for Mdis with size M × K. (2) It is easy to look up the distance from x_i to y_i by using the PQ code of the vector x to calculate the distance from the vector x to the query vector y.

The linear search PQ method is simple and easy to implement, but its search time complexity is O(N) and it cannot adapt to large-scale data neighbor search. In order to avoid the linear search, [11] proposed to use Inverted Indexing File (IVF) to speed up PQ query. The main idea of IVF is to establish an inverted index on the PQ code to avoid exhaustive linear search.

3 Compressing PQ Hash Code Indexes

This article uses two methods to achieve fast PQ search: (1) use a minimal perfect hash to establish a hash index on the PQ encoding, store the PQ code in the storage structure, suitable for medium-sized data sets; (2) store PQ directly for large-scale data sets The code (usually 64 bits) consumes storage space. For example, if there are 109 non-repeating PQ codes (that is, 10^9 PQ blocks are non-empty), the index structure requires 10^9 64-bit index entries. Then, only the index entries need to consume 4 GB of memory space. Need to store in the index structure does not store PQ code, use Bloom filter to solve the problem of the minimum perfect hash.

3.1 Constructing Compressed PQ Hash Code

The process of constructing a compressed PQ hash code mainly consists of three main steps: (1) Utilizing a PQ quantizer to quantize the database vector to a short PQ code; (2) Mapping the PQ code to a corresponding value using a minimal perfect hash For the index entries, this paper uses the minimal perfect hash [15] implementation in [14]; (3) Build the hash table.

The PQ quantizer includes m sub-quantizers $\{q_1, q_2, \cdots, q_m\}$. PQ maps ∀x in the database to a sub-center point index value, which is the index value of the center point $\{i_1^{k_1}, i_2^{k_2}, \cdots, i_m^{k_m}\}$, $i_j^{k_j}$ is called a PQ indexing code. Each x is mapped to an integer value of m, making up the PQ index library. If there is the same PQ index code, i.e. $PQ(x_i) = PQ(x_j)$, x_i, x_j belongs to the same block.

The PQ index is coded to [1 … N] using the least perfect hash. That is, the PQ index code is mapped to an integer value. According to the nature of the minimum perfect hash, different PQ index codes are mapped to unrelated terms. And there is no conflict. Each item of the hash table corresponds to a block center point, and all data points in the center point are hash mapped to the corresponding items of the center point.

3.2 Query Scheme Based on Bloom Filter

The above design scheme needs to store the PQ hash code, because when the given query data y does not belong to the database set X (that is, the minimum perfect hash will still map the x hash to the domain [0 … n], Causes a query error. Therefore, store the hash code in the index item and compare it with the hash code of the query data to avoid query errors. However, storing the hash in the index structure consumes a lot of storage space. The idea of the query scheme based on Bloom filter is to establish Bloom filtering on a vector database, as shown in Fig. 1. When a neighbor search is performed, a Bloom filter is first used to determine whether the PQ hash code of the query data is in the PQ hash database, and if so, then Execute the query, otherwise do not need not execute the query (because the block corresponding to the hash code is empty). The query construction and query filtering of the query filtering scheme based on Bloom Filter [16, 17].

Algorithm 1. Query and searching

---

input :

dataset $X \in R^d$; input query y

output:

T neighbors

1 for $j = 0; i < m; j++$ **do**

2 **for** $i = 0; i < K; i++$ **do**

3 $disTab[i][j] = d(y_j, C[i][j])^2$;

4 **endfor**

5 endfor

6 sorting disTab[][m] in increasing order;

7 generating candidates codeword *canTab*[] of y by *disTab*[][];

8 for j=1;j<K_{cand};j++ **do**

9 *ind*=mphf(canTab[j]);

10 Obtaining ids between hashIndex[*ind*].start and hashIndex[*ind*] and stored in NNTab[];

11 endfor

12 return NNTab[];

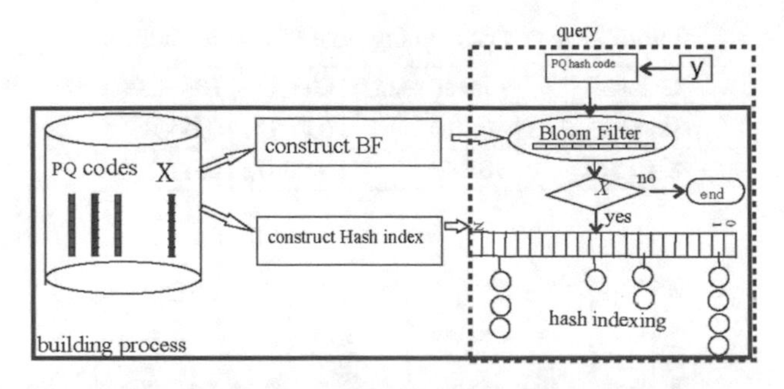

Fig. 1. Process of constructing compressed PQ hash index

4 Experimental Results

The parameters of the product-quantization hash linear search, the compression product-quantization hash index, and the compressive-product-quantization hash index based on the bloom filter are: the number of subspace divisions m = 8, the number of center points in each subspace k = 256; Bloom filter hash function number and bit vector length settings are shown in Table 1. According to the principle of the Bloom filter analyzed in [16, 17], the Bloom filter parameters are set in this experiment so that the proportion of false positives is less than 1%.

Table 1. Dataset and parameters

Dataset	Length of bit (bit)	Number of hash
random dataset	1000000 (10^6)	7
SIFT1M	10000000 (10^7)	7

(1) Search Time Comparison

The average time spent in one lookup of the three search methods is shown in Table 2. The linear search method takes the most time, approximately 150 times the compressed hash index and 120 times the BF-compressed hash index. This is because PQ's search method is to scan the entire PQ hash code library, calculate the ADC distance, and then sort it. Obviously this process is very time consuming. The latter two methods do not need to scan the entire PQ code base, as long as according to the candidate nearest set, the minimal perfect hash is directly mapped to the corresponding block, and the search efficiency is greatly improved.

As shown in Table 2, the PQ linear search is directly proportional to the size of the data set, which is consistent with the algorithm analysis, while the other two methods hardly grow with the increase in data size. The compression hash index and BF-compress hash index are related to the size of the nearest neighbor candidate list. As shown in Fig. 2, when the candidate set's size is fixed, its running time is almost constant.

Table 2. Average running time of search (unit: s)

Dataset	Linear search	CPQI	BF- CPQI
Random data	0.719	0.0052	0.0063
SIFT1M	70.742	0.00672	0.0071

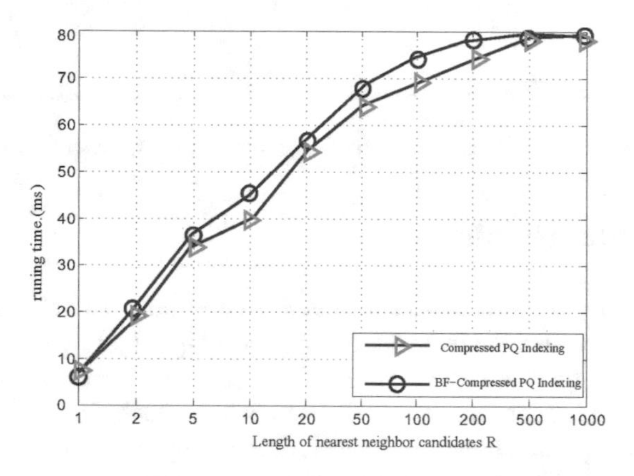

Fig. 2. The relation of running time with candidates

From Table 2 and Fig. 2, it can be seen that the Bloom filtering scheme takes approximately 0.001 ms compared to the compression hash index because the time it takes for the former to perform Bloom filtering before the search. Increasing the Bloom filter did not significantly increase the algorithm runtime.

(2) Comparison of Storage Space

Because all three methods need to store a vector ID, the required storage space is the same, and the required storage space is not counted in the comparison. The PQ linear search stores the full hash code for each vector, each hash code takes up 64 bits, and the storage hash needs to be stored as 64 * N bits. In addition, PQ searches linearly for the stored distance table. The required memory space is 64 * 8 bits. Compressing the hash index also requires 64 * N bits of storage to store the hash code in the index structure. The Bloom filter hash hash index does not need to store the PQ hash code, but it needs to construct the Bloom filter bit vector. Three methods for Bloom filtering compress the hash index to save space.

5 Summary

Although similarity hashing maps large-scale, high-dimensional data into hash hashes, it reduces search computations, reduces memory requirements, and improves search efficiency. However, linear comparison of hash codes still takes a long time. Search time. On the other hand, because the hash map space is very large, it is usually not feasible to use the hash code directly as an index item. This leads to the current hash method cannot be applied to larger-scale search problems.

This paper focuses on the above problems, proposes a perfect hash based compression quantized indexing mechanism, and further compresses the sparse hash map space to a dense hash code, which greatly improves the search speed on the premise of guaranteeing the search accuracy. In order to further reduce memory consumption, this paper proposes a compression hash index mechanism combined with Bloom filtering. This method does not need to store the hash code directly, further reducing the memory space. Although the design presented in this chapter is based on an ensemble hash, in fact, the design idea can also be applied to other hash methods such as LSH.

Acknowledgments. This work was supported by the National Natural Science Foundation of China under Grant 61403060, Huaian Natural Science Foundation HAB201704, Six Talent Peaks project in Jiangsu Province under Grant 2016XYDXXJS-012, the Natural Science Foundation of Jiangsu Province under Grant BK20171267, 533 talents engineering project in Huaian under Grant HAA201738.

References

1. Wang, J., Zhang, T., Song, J., et al.: A Survey on Learning to Hash. CoRR, abs/1606.00185 (2016)
2. Torralba, A., Murphy, K.P., Freeman, W.T., et al.: Context-based vision system for place and objectrecognition. In: Proceedings of 9th IEEE International Conference on Computer Vision (ICCV2003), Nice, France, 14–17 October 2003, pp. 273–280 (2003)
3. Lowe, D.G.: Distinctive image features from scale-invariant keypoints. Int. J. Comput. Vis. **60**(2), 91–110 (2004)
4. Oliva, A., Torralba, A.: Modeling the shape of the scene: a holistic representation of the spatial envelope. Int. J. Comput. Vis. **42**(3), 145–175 (2001)
5. Seidl, T., Kriegel, H.: Optimal multi-step k-Nearest neighbor search. In: Proceedings of SIGMOD 1998, Proceedings ACM SIGMOD International Conference on Management of Data, Seattle, Washington, USA, 2–4 June 1998, pp. 154–165 (1998)
6. Xu, H., Wang, J., Li, Z., et al.: Complementary hashing for approximate nearest neighbor search. In: Proceedings of IEEE International Conference on Computer Vision, ICCV 2011, Barcelona, Spain, 6–13 November 2011, pp. 1631–1638 (2011)
7. Bentley, J.L.: Multidimensional binary search trees used for associative searching. Commun. ACM **18**(9), 509–517 (1975)
8. Fu, A.W., Chan, P.M., Cheung, Y., et al.: Dynamic vp-Tree indexing for n-Nearest neighbor search given pair-wise distances. VLDB J. **9**(2), 154–173 (2000)
9. Indyk, P.: Nearest neighbors in high-dimensional spaces. In: Proceedings of Handbook of Discrete and Computational Geometry, pp. 877–892, 2nd edn. (2004)
10. Datar, M., Immorlica, N., Indyk, P., et al.: Locality-sensitive hashing scheme based on p-stable distributions. In: Proceedings of the 20th ACM Symposium on Computational Geometry, Brooklyn, New York, USA, 8–11 June 2004, pp. 253–262 (2004)
11. Jégou, H., Douze, M., Schmid, C.: Product quantization for nearest neighbor search. IEEE Trans. Pattern Anal. Mach. Intell. **33**(1), 117–128 (2011)
12. Ge, T., He, K., Ke, Q., et al.: Optimized product quantization. IEEE Trans. Pattern Anal. Mach. Intell. **36**(4), 744–755 (2014)

13. Kalantidis, Y., Avrithis, Y.S.: Locally optimized product quantization for approximate nearest neighbor search. In: Proceedings of 2014 IEEE Conference on Computer Vision and Pattern Recognition, CVPR 2014, Columbus, OH, USA, 23–28 June 2014, pp. 2329–2336 (2014)
14. Fox, E.A., Heath, L.S., Chen, Q.F., et al.: Practical minimal perfect hash functions for large databases. Commun. ACM **35**(1), 105–121 (1992)
15. Limasset, A., Rizk, G., Chikhi, R., et al.: Fast and scalable minimal perfect hashing for massive keysets. CoRR, abs/1702.03154 (2017)
16. Mitzenmacher, M., Vadhan, S.P.: Why simple hash functions work: exploiting the entropy in a data stream. In: Proceedings of the Nineteenth Annual ACM-SIAM Symposium on Discrete Algorithms, SODA 2008, San Francisco, California, USA, 20–22 January 2008, pp. 746–755 (2008)
17. Bloom, B.H.: Space/Time trade-offs in hash coding with allowable errors. Commun. ACM **13**(7), 422–426 (1970)

Dynamic Incentive Mechanism for Direct Energy Trading

Nan Zhao(✉), Pengfei Fan, Minghu Wu, Xiao He, Menglin Fan, and Chao Tian

Hubei University of Technology, Wuhan 430068, China
{nzhao,wuxx1005}@mail.hbut.edu.cn,
fanpengfeifmt@163.com, zxad-005@163.com,
Fanmenglin0415@163.com, tianchaotcl123@163.com

Abstract. Direct Energy trading is a promising approach to simultaneously achieve trading benefits and reduce transmission line losses. Due to the characteristics of selfish requirement and asymmetric information, how to provide proper incentives for the electricity consumer (EC) and small-scale electricity supplier (SES) to take part in direct energy trading is an essential issue. Considering the variable characteristic of requirements and environment in direct energy trading, a two-period dynamic contract incentive mechanism is introduced into the long-term direct energy trading. The optimal contract is designed to obtain the maximum expected utility of the EC based on the individually rational and incentive compatible conditions. Simulation result shows that the optimal dynamic contract is efficient to improve the performance of direct energy trading.

1 Introduction

Direct energy trading is considered as a promising method to improve energy efficiency [1]. By building small wind power platforms or solar panels, small-scale electricity suppliers (SESs) can sell surplus power to electricity consumer (EC), which can simultaneously achieve trading benefits and reduce transmission line losses. However, due to the randomness of power resources and the selfish nature of SESs, the potential SESs may be unwilling to participate in direct energy trading without any incentive. In addition, various incentive methods need almost complete information to select SESs. However, due to differences of SES power generation equipment and storage equipment, renewable energy may have random and intermittent characteristics [2, 3]. Such complete information may not be available to all ECs [4]. Furthermore, the information may belong to SESs privately, leading to information asymmetry between EC and SESs. In this paper, we will focus on designing an effective incentive method for direct energy trading to resolve these difficult problems.

The above incentive problems of energy trading have just been researched, using mostly game theory [5, 6]. However, most existing game theory-based methods are based on complete information and symmetry hypothesis. Players know all the necessary information. However, such assumption is impractical in reality. Therefore, in this paper, we focus on contract theory-based incentive approach to direct energy trading under asymmetric information scenario.

L. Barolli et al. (Eds.): IMIS 2018, AISC 773, pp. 403–411, 2019.
https://doi.org/10.1007/978-3-319-93554-6_38

Contract theory [7] investigates the mutually agreeable contract among the economic players under asymmetric or incomplete information scenarios. Recently, contract based solutions have been suggested for direct energy trading [8]. In contract theory, a principal-agent model is utilized, where EC acts as the principal and each SES is an agent. However, most existing works considered the static contract design, where the relationship between EC and SESs is static in direct energy trading. However, in the practical direct energy trading, the electricity load of residents and the transmission cost of electricity during peak hours change over time, and the static incentive mechanism appears to lack flexibility in direct energy trading [9]. Therefore, EC needs to design a long-term dynamic incentive mechanism to avoid renegotiation.

In this paper, to solve the problem of information asymmetry and SESs' selfishness in direct energy trading, dynamic contract design for direct energy trading is proposed [10]. A two-period dynamic contract mechanism is investigated with independently and identically distributed SESs' private types. The optimal contract is designed to obtain the maximum expected utility of the EC based on the individually rational and incentive compatible conditions [11, 12]. A sequential optimization algorithm is proposed to obtain the optimal trading strategy. Simulations demonstrate the performance of the optimal contract strategy in direct energy trading.

2 System Model and Problem Formulation

In this paper, a typical direct energy trading scenario is considered with one EC, and N SESs. The EC recruits some SESs to participate in direct energy trading. The EC wants to get the SESs' help as much as possible [13]. However, due to the selfish nature of SESs, the SESs want to obtain the large reward with a little effort they spend. Therefore, in this study, we will design contract incentive mechanism to deal with the above conflicting demands between EC and SESs.

2.1 Electricity Consumer Model

In general, when the EC needs electricity, it purchases from a power retailer in a traditional electricity market. However, if the price of SES is lower than that of a traditional electricity market, then the EC can buy electricity from the SES. Assume that the EC pays the SES reward π to obtain the amount of electricity q. Then, the EC's utility is the achievable profit gained from the direct energy trading minus the reward to SES, which is given by

$$U = V(q) - \pi. \tag{1}$$

2.2 Small-Scale Electricity Supplier Model

Suppose that one SES obtains the reward π by selling the amount of electricity q to the EC. Taking into account the cost of generation and transmission, the utility of the SES is defined as the difference of the reward π and its own cost, that is,

$$\omega = \pi - \theta q, \tag{2}$$

where θ is the cost factor per unit of electricity, which is considered as the private information of the SES. A higher θ_i indicates that the SES has a worse power generation condition or it has a higher cost.

2.3 Contract Formulation

This paper propose a two-period dynamic contract model to address the long-term goal of contradictions between SESs and EC. Here, N SES types are considered, which are denoted by the set $\Theta = \{\theta_1, \theta_2, \theta_3 ..., \theta_N\}$ with $0 < \theta_1 < \theta_2 < ... < \theta_N$. Assume that the EC has some statistical information about the probability p_i of type-θ_i SES. Obviously $p_i \in [0, 1]$ and $\sum_{i=1}^{N} p_i = 1$ [14]. Moreover, consider that the private information may have some correlation across periods, The ith SES's type in periods 1 (θ_i^1) and period 2 (θ_i^2) are independently drawn from the same support Θ with identical probabilities p_i. Then, to incentivize the SES to reveal their types, a contract needs to be comprised of N contract items [4], which can be written as $\Phi = \{q_i, \pi_i, \ i = 1, 2, ..., N\}$. Figure 1 describes the timing of contracting. At the beginning stage of contracting, the EC offers a long term contract $\{\pi_i^1(q_i^1);\ \pi_i^2(q_i^1, q_i^2)\}$ to the SESs. After evaluating the contract items, the SES will inform the EC its choice if willing to accept certain contract item. At the end of each period, if the trading is successful, the EC will reward the SESs according to their contracts.

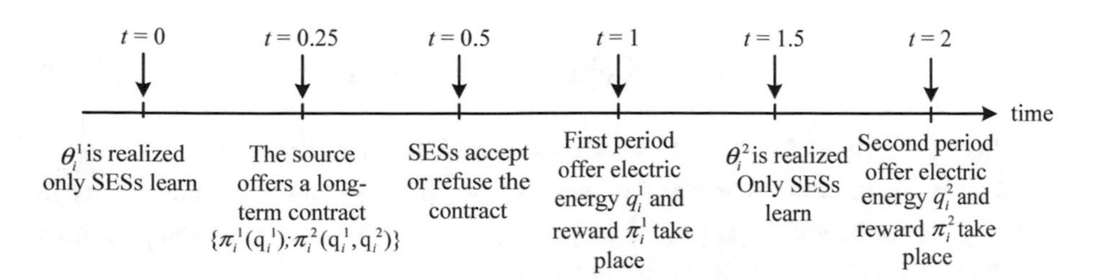

Fig. 1. Two-stage direct energy trading timing

3 Dynamic Incentive Design for Direct Energy Trading

In this section, a dynamic contract model is considered in direct energy trading. The EC expected utility in the two periods can be written as

$$U_{EC_i} = \sum_{i=1}^{N} p_i[v(q_i^1) - \pi_i^1] + \delta \sum_{i=1}^{N} p_i[v(q_i^2) - \pi_i^2], \tag{3}$$

where $q_i^t(resp. \pi_i^t)$ is the ith SES's power generation capacity (resp. reward) in the period $t = 1, 2$, and a discount factor is $\delta > 0$. $\delta > 1$ means that period 2 lasts much longer than period 1.

Assuming that the SESs have the same discount factor as the EC, he i^{th} SES's utility in the two periods can be written as

$$U_{SES_i} = [\pi_i^1 - \theta_i^1 q_i^1] + \delta[\pi_i^2 - \theta_i^2 q_i^2]. \tag{4}$$

At the time of signing the long-term contract with the EC, the ith SES only knows its first period type θ_i^1. After the first period of trading is realized, its second period type θ_i^2 can be learned by the ith SES. Therefore, the optimal long-term contract is obtained by putting together the optimal contract with interim contracting for the first period and the optimal contract with ex ante contracting for the second period. Interim contracting is a commercial process that the EC offers the contract to the SESs once the SESs have learned their hidden types. And ex ante contracting is a commercial process before the SESs learn their hidden types.

3.1 Contracting Design in Period 2

In period 2, to ensure that each type-θ_k^2 SES obtains a non-negative utility by accepting the contract item $(\pi_k^2(\tilde{\theta}_i), q_k^2(\tilde{\theta}_i))$, the following individually rational (IR) constraint should be satisfied,

$$\pi_k^2(\tilde{\theta}_i) - \theta_k^2 q_k^2(\tilde{\theta}_i) \geq 0, \qquad \forall \in \Omega \tag{5}$$

where $\tilde{\theta}_i$ is the ith SES's first-period announcement about his type.

In addition, in order to ensure that each type-θ_k^2 SES gets the maximum utility by choosing the contract item $(\pi_k^2(\tilde{\theta}_i), q_k^2(\tilde{\theta}_i))$ designed for its type, the following incentive compatible (IC) constraint should be satisfied,

$$\pi_k^2(\tilde{\theta}_i) - \theta_k^2 q_k^2(\tilde{\theta}_i) \geq \pi_j^2(\tilde{\theta}_i) - \theta_k^2 q_j^2(\tilde{\theta}_i), \quad \forall k, j \in \Omega \tag{6}$$

3.2 Contracting Design in Period 1

Considering that the ith SES's utility in period 1 can be written as $U^1_{SES_i} = \pi^1_i - \theta^1_i q^1_i$, the inter-temporal IC constraint is written as

$$(\pi^1_i - \theta^1_i q^1_i) + \delta \sum_{k=1}^{N} p_k U^2_{SES_k}(\theta^1_i) \geq (\pi^1_j - \theta^1_i q^1_j) + \delta \sum_{k=1}^{N} p_k U^2_{SES_k}(\theta^1_j), \tag{7}$$

where $\sum_{k=1}^{N} p_k U^2_{SES_k}(\theta^1_i)$ is the expected continuation utility for period 2.

Since the SESs' utility $U^2_{SES_k}(\theta^1_i)$ in period 2 is independent of θ^1_i, that is, $\sum_{k=1}^{N} p_k U^2_{SES_k}(\theta^1_i) = \sum_{k=1}^{N} p_k U^2_{SES_k}(\theta^1_j)$. Then, the above IC constraint can be simplified as

$$\pi^1_i - \theta^1_i q^1_i \geq \pi^1_i - \theta^1_i q^1_j, \qquad \forall i,j \in \Omega. \tag{8}$$

Moreover, considering the expected continuation utility of period 2, the ith SES's intertemporal IR constraint can be written as

$$\pi^1_i - \theta^1_i q^1_i + \delta \sum_{i=1}^{N} q_k U^2_{SES_k}(\theta^1_i) \geq 0, \quad \forall k, \mathrm{j} \in \Omega. \tag{9}$$

Thus, the intertemporal contract optimization problem in the two periods is to obtain the maximum expected utility of EC subject to the above IC and IR constraints, that is,

$$\begin{aligned} &\max_{\pi^1_i, q^1_i, \pi^2_k(\theta^1_i), q^2_k(\theta^1_i)} \quad \mathrm{U}^1_{EC} + \delta U^2_{EC}(\theta^1_i) \\ &\text{s.t. } (5), (6), (8), (9) \end{aligned} \tag{10}$$

where $U^1_{EC} = \sum_{i=1}^{N} q_i[v(p^1_i) - \pi^1_i]$ and $U^2_{EC}(\theta^1_i) = \sum_{k=1}^{N} q_k[v(p^2_k(\theta^1_i)) - \pi^1_k(\theta^1_i)]$.

3.3 Optimal Contract Design

Considering that the optimization problem (10) is generally non-convex, a sequential optimization approach is adopted in this paper.

Since the ith SES's type in periods 1 (θ^1_i) and period 2 (θ^2_i) are independently, we consider the contract $\Phi^2 = \{(\pi^2_k(\theta^1_i), q^2_k(\theta^1_i)), \ \forall k\}$ of period 2 and the contract $\Phi^1 = \{(\pi^1_i, q^1_i), \ \forall i\}$ of period 1 separately. In period 2 and period 1, both the sequential optimization approaches are investigated. Firstly, the best reward is derived given fixed amount of electricity, then, the best amount of electricity for the optimal contract is obtained.

The optimal rewards in period 2 satisfy

$$\pi_k^2(\theta_i^1) = \begin{cases} \theta_N^2 q_N^2(\theta_i^1), & k = \mathrm{N} \\ \theta_N^2 q_N^2(\theta_i^1) + \sum\limits_{j=k}^{N-1} \theta_j^2 \Delta q_j^2, & k \neq \mathrm{N} \end{cases} \tag{11}$$

where $\Delta q_k^2 = q_k^2(\tilde{\theta}_i) - q_{k+1}^2(\tilde{\theta}_i)$.

Then, the type-θ_k^2 SES's utility can be obtained

$$U_{SES_k}^2(\theta_i^1) = \begin{cases} 0, & k = \mathrm{N} \\ \sum\limits_{j=k}^{N-1} (\theta_{j+1}^2 - \theta_j^2) q_{j+1}^2(\theta_i^1), & k \neq N. \end{cases} \tag{12}$$

In period 1, similar to the case of period 2, the optimal rewards in period 1 satisfy

$$\pi_i^1 = \begin{cases} \theta_N^1 q_N^1 - \delta \sum\limits_{k=1}^{N} p_k U_{SES_k}^2(\theta_i^1), & i = \mathrm{N} \\ \theta_N^1 q_N^1 - \delta \sum\limits_{k=1}^{N} p_k U_{SES_k}^2(\theta_i^1) + \sum\limits_{j=1}^{N-1} \theta_j^1 \Delta q_j^1, & i \neq \mathrm{N}, \end{cases} \tag{13}$$

where $\Delta q_j^1 = q_j^1 - q_{j+1}^1$.

Then, the intertemporal contract design optimization problem (10) can be simplified as,

$$\begin{aligned} \max_{\{q_i^1, q_k^2(\theta_i^1)\}} \quad & \mathrm{U}_{EC}^1 + \delta U_{EC}^2(\theta_i^1), \\ s.t. \quad & (11) \text{ and } (13), \end{aligned} \tag{14}$$

where $U_{EC}^1 = \sum\limits_{i=1}^{N} p_i[v(q) - \pi_i^1(q_i^1)]$ and $U_{EC}^2(\theta_i^1) = \sum\limits_{k=1}^{N} p_k[v(q_k^2(\theta_i^1)) - \pi_k^2(q_k^2(\theta_i^1))$.

Therefore, the optimal amount of electricity is obtained from the following formulation

$$\begin{aligned} v'(q_i^{1*}) &= \begin{cases} \theta_1^1, & i = \mathrm{N}, \\ \dfrac{(\theta_i^1 - \theta_{i-1}^1) \sum\limits_{j=1}^{i-1} p_j}{p_i}, & i \neq \mathrm{N}, \end{cases} \\ v'(q_k^{2*}(\theta_i^1)) &= \theta_k^2, \qquad k = 1, \ldots, N. \end{aligned} \tag{15}$$

4 Numerical Results and Discussion

In this section, numerical results are given to evaluate the performance of the proposed dynamic contract design. Experiment parameters are set to $V(q) = -\frac{a}{2}q^2 + bq + c(b << a)$, a = 0.01, b = 1.5 and c = 20. Two private types of SESs are considered with $\theta_1 = 0.2/\text{KWh}$ and $\theta_2 = 0.4/\text{KWh}$. Assume that each SES has 10 KWh – 300 KWh remaining power for trading.

We first evaluate the performance of dynamic contract incentive mechanism. Assume that δ are uniformly distributed in $[0, 2]$. Figure 2 shows that the optimal expected utility of EC increases in the discount factor δ and the probability p_1 of type-θ_1 SES. As δ increases, EC obtain more amount of electricity from the SESs, leading to more expected utility from the SESs' direct energy trading. When the discount factor $\delta = 0$, the two-period contract between EC and SESs is reduced to the one-period contract. Even with one-period direct energy trading, the EC's expected utility is also enhanced compared with non-trading strategy. Moreover, considering that the type-θ_1 SES has a least trading cost, as p_1 increases, the proportion of the low type-θ_1 SES is enhanced, leading to the higher expected utility of EC.

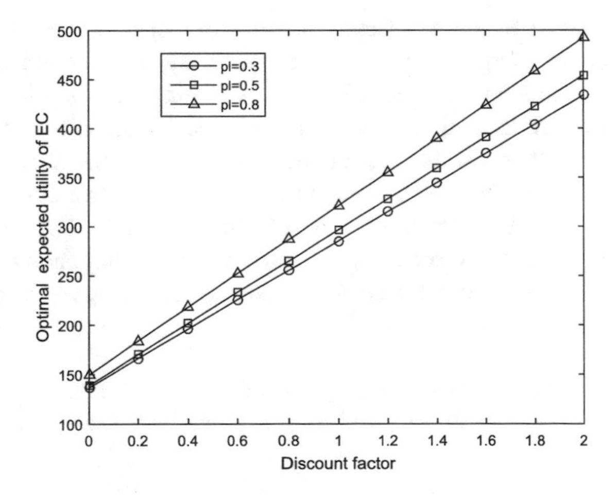

Fig. 2. The optimal expected utility of EC vs. δ and p_1.

The second evaluation method is to analyze the performance of energy suppliers election mechanism. Assume that θ_i^1 and θ_i^2 are uniformly distributed with N = 2. Figure 3 demonstrates that the EC's optimal utility is increasing in the discount factor δ and decreasing in the SESs' types (θ_i^1/θ_i^2). As δ increases, the SESs have more incentive to provide energy trading with the EC, and the EC obtains more utility from the SESs' trading. When the discount factor δ = 0, the contract between the EC and SESs is designed in the single period. However, even with one-period energy trading, the EC's expected utility is also enhanced compared with the non-energy trading. Moreover, as (θ_i^1/θ_i^2) increases, the SES of that type has a higher trading cost for energy trading, thus the EC obtains the less utility by trading with SESs.

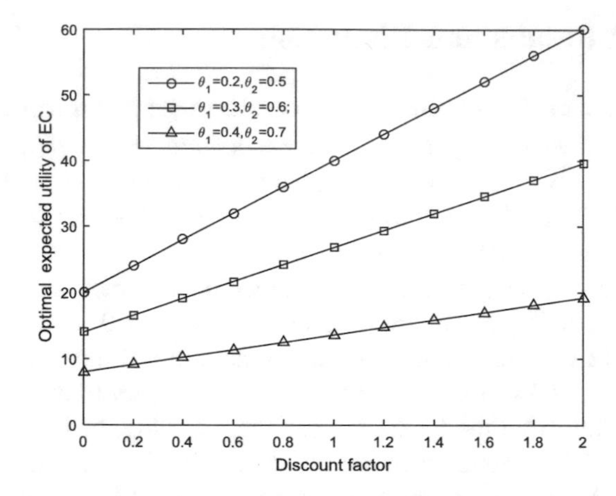

Fig. 3. The EC's optimal expected utility vs. the discount factor δ

5 Conclusion

In this paper, dynamic incentive mechanism between one EC and multiple SESs is investigated in direct energy trading. A two-period dynamic contract mechanism is proposed to obtain the long-term incentive energy trading. Considering that the SESs' types are independent in both periods with identical probability distribution, the contract-theoretic model for ability discrimination trading incentive is formulated. The optimization problem is proposed to obtain the maximum expected utility of EC with the necessary and sufficient constraints. A sequential optimization algorithm is proposed to obtain the optimal direct energy trading strategy. Simulation results indicate that the proposed dynamic contract-theoretic scheme can improve the performance of direct energy trading.

Acknowledgements. This work was supported by the National Natural Science Foundation of China (No. 61501178, No. 61471162) and Project Funded by China Postdoctoral Science Foundation (2017M623004).

References

1. Rahimi, F., Ipakchi, A.: Demand response as a market resource under the smart grid paradigm. IEEE Trans. Smart Grid **1**, 82–88 (2010)
2. Li, Y., Ng, B.L., Trayer, M., et al.: Automated residential demand response: algorithmic implications of pricing models. IEEE Trans. Smart Grid **3**, 1712–1721 (2012)
3. Lee, W., Xiang, L., Schober, R., et al.: Direct electricity trading in smart grid: a coalitional game analysis. IEEE J. Sel. Areas Commun. **32**, 1398–1411 (2014)
4. Zeng, M., Cheng, J., Qian, X., et al.: Research on the trading mechanism of distributed generation bidding online market. East China Electr. Power. **40**, 1–4 (2012)
5. Mei, S., Wei, W.: Master-slave game model and application examples in smart grid environment. Syst. Sci. Math. **34**, 1331–1344 (2014)

6. Dou, C., Jia, X., Li, H.: Multi-agent-based distributed generation of micro-grid for market power generation based on competitive bidding. Power Syst. Technol. **40**, 579–586 (2016)
7. Zhao, N., Chen, Y., Liu, R., Wu, M., Xiong, W.: Monitoring strategy for relay incentive mechanism. In: Cooperative Communications Networks. Comput. Electr. Eng., 14–29 (2017)
8. Zhang, B., Jiang, C., Yu, J.L., et al.: A contract game for direct energy trading in smart grid. IEEE Trans. Smart Grid **99**, 1 (2016)
9. Bolton, P., Dewatripont, M.: Contract Theory. MIT Press, Cambridge (2005)
10. Gibbons, R.: Game Theory for Applied Economists. Princeton University Press, Princeton (1992)
11. Zhao, N., Liang, Y., Pei, Y.: Dynamic contract design for cooperative wireless networks. In: IEEE Globe Communication Conference (GLOBECOM), Singapore, pp. 1–6 (2017)
12. Zhao, N., Liu, R., Chen, Y., Wu, M., Jiang, Y., Xiong, W., Liu, C.: Contract design for relay incentive mechanism under dual asymmetric information in cooperative networks. Wirel. Netw. https://doi.org/10.1007/s11276-017-1518-x
13. Zhao, N., Chen, Y., Liu, R., Wu, M., Xiong, W.: Monitoring strategy for relay incentive mechanism in cooperative communications networks. Comput. Electr. Eng. **60**, 14–29 (2017)
14. Zhao, N., Wu, M., Chen, J.: Android-based mobile educational platform for speech signal processing. Int. J. Electr. Eng. Educ. **54**, 3–16 (2017)

Improve Memory for Alzheimer Patient by Employing Mind Wave on Virtual Reality with Deep Learning

Marwan Kadhim Mohammed Al-shammari(✉) and Gao Tian Han

Northeastern University, Shenyang, China
alkaseralshamary@gmail.com, gaoth@mail.neu.edu.cn

Abstract. Alzheimer disease is associated with many risks, including the destruction of family morale and the loss of experience of many scientists in different areas. However, little research depending on computer science has been conducted to explore this disease. The purpose of this study is trying to find the possibility of using computer techniques to improve the therapeutic methods of Alzheimer disease. This paper elaborates the approach of using EEG signals on virtual reality environment and introducing them as a patient's therapeutic program to improve temporary memory. The patient's memory is rearranging based on a suitable brain signal through the theory of artificial neural network and deep learning technique so that the memory is able to be gradually improved.

Keywords: EEG · VR · Alzheimer · Artificial neural network
Deep learning

1 Introduction

Alzheimer is a disease that is previously known as dementia. A person is sitting in a place who and does not remember all the details and not remember most of the people around him. The disease is surrounded by a kind of shyness. Families are trying to hide the patient since they think it is a shame that people discover the condition of the patient. Today, the situation has been developed by dint of medical associations and humanitarian organizations. These institutions demystify the disease and help families deal with this disease [1].

Disease name came from the name of the discoverer, German physician Alois Alzheimer. In 1907, Alois studied a study case similar to schizophrenia. But after unexpected death of the patient, Alois decided to autopsy patient brain to find out the reason of death. Alois discovered the existence of unknown harmful proteins which created inside the brain, Alois noticed that the disease destroy the cells responsible for memory and the first attack of the disease effect on the temporary memory cells. Here, patient was forgetting essentially information about his family and private life. This information should not be forgotten because it represents his daily routine life [2]. The disease does not lead directly to the death but, due to death if the disease is accompanied by another disease. In addition, the death maybe came if the patient forgets one

L. Barolli et al. (Eds.): IMIS 2018, AISC 773, pp. 412–419, 2019.
https://doi.org/10.1007/978-3-319-93554-6_39

of the most vital things. Such as forget to chew food. The psychological effects on Alzheimer's patient were link events from past with the events from present, return back to childhood stage, or feel embarrass to get hand help [3]. Since the early 1930s, Alzheimer's disease has been the focus of study. The disease was classified as a chronic disease such as cancer. There is no definitive treatment for this disease, but there are preventative steps to reduce disease progression in the early stages of the disease [4]

A group of scientists have focused on the chemical and pharmacological side to reduce the effects of Alzheimer's disease especially in the early stages of the disease. Where, the patient is given a specific set of medicines and foods to reduce the generation of these toxic proteins in the brain [5, 6]. Another group of scientists focused on the clinical side of Alzheimer's patients. They set up medical sessions for patients to activate memory in the patient's mind. During the session, the therapist displays a set of memories for the patient and monitors the patient's response to these memories. The patient produces a range of positive and negative feelings depend on the type of memories. The task of the therapist is to arrange the set of memories in the way that improves the patient's memory [7]. Various approaches are harness the latest advances in virtual reality technology to strengthen the medical aspect. Virtual reality with the aid of bio-sensor to evaluate the effectiveness of therapy has been used. The imaginary technique which includes Electroencephalography (EEG) is named Virtual Reality Therapy (VRT). In the recent years, the VRT has been applied as a sole treatment for attention deficit hyperactivity disorder disease (ADHD). ADHD is a childhood syndrome characterized by short attention span [8]. On the other hand, EEG biofeedback treatment based on virtual reality was studied for fear of flying disease on 36 patients [9]. In terms of psychotherapy, finding best meditation and worst stress has been studied by VRT [10]. Cerebral palsy is a disease caused by decreased brain oxygenation and the conventional treatments for this disease are expensive since the VRT was suggested as an alternative treatment for this disease [11].

The purpose of this study is to find out the possibility of employing technology to help Alzheimer's patient. To achieve the work of the clinical therapist in the medical sessions of the Alzheimer patient, three main factors are necessary. First, create a patient suitable environment. Second, choose device to read feelings of the patient. And finally, employ an appropriate theory to arrange the patient's memory based on the patient's feelings. The study hypothesizes that the appropriate environment for an Alzheimer's patient that can isolate Alzheimer's patient from the work environment is to use virtual reality. The approach of using Electroencephalography (EEG) to read patient brain is suggested. The employment of deep learning technology and artificial neural network theory in particular is hypothesized to arrange patient's memory.

2 Preliminaries

2.1 Electroencephalography (EEG)

Electroencephalography is a medical technique that read electrical activity generated by brain. When brain cells are activated, local current flows are produced. The differences of electrical potentials which caused are measured by clinical neurophysiology tools. The brainwaves have been categorized into four essentially groups, Beta (>13 Hz), Alpha (8–13 Hz), Theta (4–8 Hz) and Delta (0.5–4 Hz). The clearest signal is Alpha, can be usually observed better. Alpha activity is induced by closing the eyes and relaxation. Our paper study focused on relaxation issue. Brain computer interface (BCI) is a communication system that recognizes patient brain waves and sends these signals as readable values to the computer [12].

2.2 Virtual Reality (VR)

Virtual reality is a high end user interface that involves real time interaction through multiple sensorial channels like sound, touch, vision, smell and taste. A computer generate multisensory information program which tracks a user in real time. The medical application of VR is depend on the medical staff and therapy needs to analyze and virtualizes complex medical data during surgery, education or training sessions. Diseases have been diagnosed and treat better with using 3D models representation [13].

2.3 Deep Learning Technique and Neural Network Method

Deep learning agile approach is a strategy to use computational models that are composed from multiple processing and harnesses these computational models to represent multiple levels data. This approach emerged from human mind thinking strategy where human uses fast thinking system to behave in daily routine activities. Feeling detection has been dramatically improved by used deep learning methods. In addition, deep learning methods used as intricate structure in the large data representation [14]. Theoretical studies suggested that, in order to solve the complicated functions which represent high level abstractions, the system may need to build a deep learning architecture which content from multi-processing levels. These levels are non-linear operations, such as in human neural network, with many hidden layers which proposed to reduce complicated of formulae. It's difficult to achieve suitable deep learning architecture, but with the aid of artificial neural network algorithm, which has recently been proposed, this problem has tackled with notable success [15].

Artificial neural network is a network of simple element called nodes (neurons) which receive input and change the nodes internal weights activation according to the required neural network output [16]. At each node of artificial neural network, outputs were calculated according to general equation, as shown in Eq. 1. Where, the output of each node depends on the input multiplied by the weight of the previous level until the end of the desired output. The most influential factor in the equation is the weight, which is changed several times until achieved the desired purpose. The layers in the artificial neural network divided into input layer, hidden layer and output layer. Input data multiply by suitable weights were been received in input layer. Result's function from the input layer multiply with stage weight was received in hidden layer. The artificial neural network may have more than one hidden layer depends on the desired structure. Finally, the output layers received function from the last hidden layer and

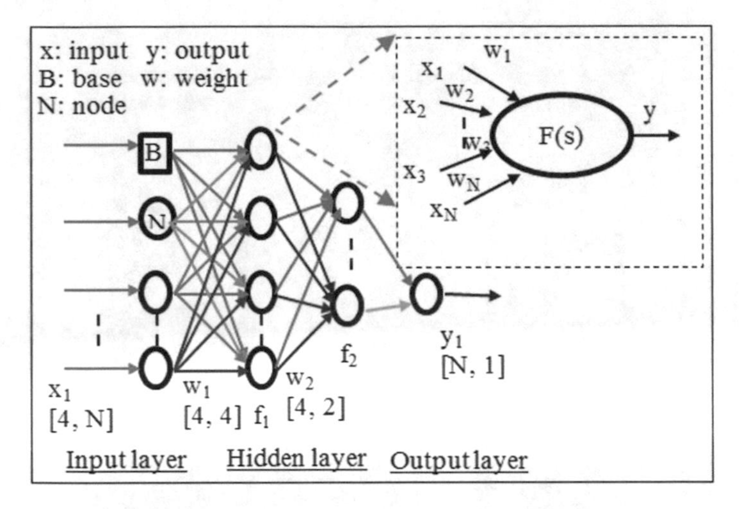

Fig. 1. Artificial neural network structure and function for each node.

multiply this function with stage weight. This procedure called learning artificial neural network to achieve specific output, as Fig. 1.

$$f(s) = \sum_{j=1}^{N} w_j x_j \tag{1}$$

3 System Design and Implementation

3.1 Overall Design

Alzheimer patient system has three main objects. The first element is the inputs of the neural brain where the signals represent the feelings of the patient. These signals are transferred to the system in the form of positive and negative ones. The brain signals are analog signals which are extracted and filtered then decoded to be suitable for later use. The second element is the theory that facilitated patient data regulation to be accurate. The third component is the environment designed to suit the condition of the Alzheimer patient which is more effective and isolated, as shown in Fig. 2. The treatment for objects in this system is how to arrange family pictures of the Alzheimer patient to activate and train memory cells in the early stages of the disease. The images are arranged based on the patient's reactions where negative emotions are excluded and the images are re-indexed based on the positive sensation of the patient and periodically based on a suitable theory.

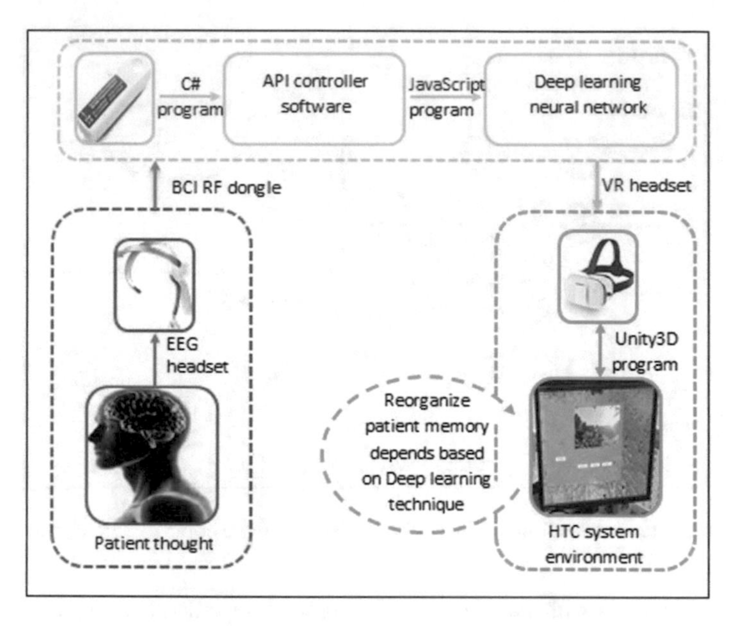

Fig. 2. Block diagram for Alzheimer patient system.

3.2 System Implementation

There is a rapidly varied in the signal of the EEG headset depending on the mental state for patient. Whereas, the results tended to be more accurate by use the method that is suggested in this paper. The sinusoidal signal of the EEG headset which represents the positive and negative feelings of the patient is sent to the computer. The signal is

Fig. 3. An illustration shows the components of the Alzheimer patient system.

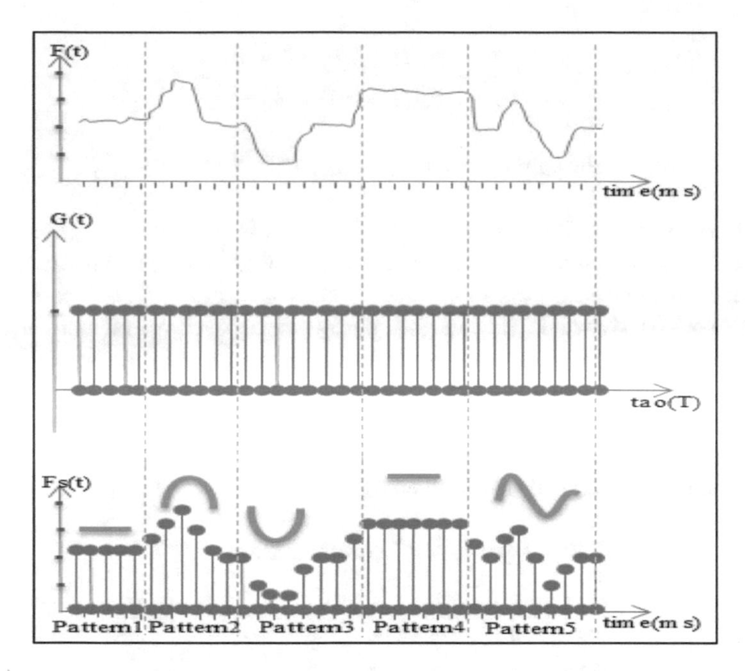

Fig. 4. Analysis signal sent from EEG to Alzheimer disease.

received and converted to the numerical ratio by the BCI dingle. The API plug-in is used as instruction set library to import BCI data. The memories of the patient have been recorded and recognized in the first learning level, as shown in Fig. 3.

The sinusoidal signal from the EEG device is processed, as shown in Eq. 2. In addition, the sinusoidal signal is discrete by multiply the EEG signal with impulse signal, as shown in Eq. 3. In other words, the signal is collected at specific time, as shown in Eq. 4. This signal is received from Alzheimer program by imported think-gear library through API sub routine called think-gear plug-in. It is noted that the signal is divided into five patterns. There is a significant increment in the second and third patterns. On the other hand, there is a marginal increment in the first and fourth patterns. Despite this, there is moderate increment in the fifth pattern, as shown Fig. 4. For more accuracy, the input values are converted from decimal to binary numerical, as shown in Eq. 5. To that end, the input is prepared according to the law of probability, as shown in Eq. 6.

$$F(t) = A1.\sin(\alpha) + A2.\sin(\beta) + A3.\sin(\theta) + A4.\sin(\Delta) \quad (2)$$

$$G(T_n)^n = \sum_{n=0}^{\infty} A\delta(t - nT_s). \quad (3)$$

$$F_s(t) = G(T_n) \times F(t) \quad (4)$$

$$I_n = \begin{cases} F_s(t) \leq 50 \rightarrow I = 0 \\ F_s(t) > 50 \rightarrow I = 1 \end{cases} \quad (5)$$

$$\boldsymbol{Probability} = \boldsymbol{NumaricBase}^{\boldsymbol{rank(Inputnumber)}} \quad (6)$$

The feed forward artificial neural network model is presented to achieve required output from prepared input. Two hidden layers are considered as maximum numbers of hidden layer to reduce network error rate, as shown in Eq. 7. The superiority of the neural network is shown in the weights performance over neural network through learning progresses. The input layer of the neural network has four binary inputs, and the 4 × 4 array of weights have been learned to be from −2 to 2 ranges, as shown in Eq. 8. It is also followed by the first hidden layer multiply by 2 × 2 array of weights and one vector array for hidden layer two, as shown in Eqs. 9, and 10. Finally, the threshold function is applied on all layers.

$$\text{Hidden layer} = \log^{-1}\text{Numaric Base}^{\text{Inputnumber}} \quad (7)$$

$$\begin{bmatrix} \text{H4} \\ \text{H3} \\ \text{H2} \\ \text{H1} \end{bmatrix} = \begin{bmatrix} -2 & 1 & 1 & 1 & 1 \\ -1 & 2 & 2 & 2 & 2 \\ -1 & 1 & 1 & 1 & 1 \\ 1 & -1 & -1 & -1 & -1 \end{bmatrix} \cdot \begin{bmatrix} \text{I4} \\ \text{I3} \\ \text{I2} \\ \text{I1} \end{bmatrix} \quad (8)$$

$$\begin{bmatrix} \text{H5} \\ \text{H6} \end{bmatrix} = \begin{bmatrix} 1 & -1 & 2 & -3 \\ 1 & 2 & -2 & -1 \end{bmatrix} \cdot \begin{bmatrix} \text{H4} \\ \text{H3} \\ \text{H2} \\ \text{H1} \end{bmatrix} \quad (9)$$

$$\text{O} = [2 \quad -1] \cdot \begin{bmatrix} \text{H5} \\ \text{H6} \end{bmatrix} \quad (10)$$

4 Conclusions

Overall, this study suggests that the use of EEG headset to extract brain signal from Alzheimer's patient, is an effective process in the development of patient's mental state, especially after applying this reading signal to the patient's rehabilitation programs. This was considered an interesting result which reduced the feeling of embarrassment for the Alzheimer's patient during the sessions of the treatment, which needed direct communication with the therapist and this eventually led to positive results. It should be borne in mind that this study was conducted a small group of patients over a short period of time. Thus, further studies are hence needed to determine the signal effects of brain stimulation on a larger number of Alzheimer's patients before generalized conclusion can be adopted.

Acknowledgments. Northeastern university, China. Support this project with Neurosky and Emotiv headsets, to read EEG signal.

References

1. Baldonado, M., Chang, C.-C.K., Gravano, L., Paepcke, A.: The Stanford digital library metadata architecture. Int. J. Digit. Libr. **1**, 108–121 (1997)
2. Bruce, K.B., Cardelli, L., Pierce, B.C.: Comparing object encodings. In: Abadi, M., Ito, T. (eds.) Theoretical Aspects of Computer Software. Lecture Notes in Computer Science, vol. 1281, pp. 415–438. Springer, New York (1997)
3. van Leeuwen, J. (ed.): Computer Science Today. Recent Trends and Developments. Lecture Notes in Computer Science, vol. 1000. Springer, New York (1995)
4. Michalewicz, Z.: Genetic Algorithms + Data Structures = Evolution Programs, 3rd edn. Springer-Verlag, New York (1996)
5. Mackenzie, I.R., Munoz, D.G.: Nonsteroidal anti-inflammatory drug use and Alzheimer-type pathology in aging. Neurology **50**(4), 986–990 (1998)
6. Bar-On, P., Millard, C.B., Harel, M., Dvir, H., Enz, A., Sussman, J.L., Silman, I.: Kinetic and structural studies on the interaction of cholinesterases with the anti-Alzheimer drug rivastigmine. Biochemistry **41**(11), 3555–3564 (2002)
7. McKhann, G.M., Knopman, D.S., Chertkow, H., Hyman, B.T., Jack, C.R., Kawas, C.H., Klunk, W.E., Koroshetz, W.J., Manly, J.J., Mayeux, R., Mohs, R.C.: The diagnosis of dementia due to Alzheimer's disease: recommendations from the National Institute on Aging-Alzheimer's Association workgroups on diagnostic guidelines for Alzheimer's disease. Alzheimer's Dement. **7**(3), 263–269 (2011)
8. Cho, B.H., Lee, J.M., Ku, J.H., Jang, D.P., Kim, J.S., Kim, I.Y., Lee, J.H., Kim, S.I.: Attention enhancement system using virtual reality and EEG biofeedback. In: 2002 Proceedings of IEEE Virtual Reality, pp. 156–163. IEEE (2002)
9. Wiederhold, B.K., Jang, D.P., Kim, S.I., Wiederhold, M.D.: Physiological monitoring as an objective tool in virtual reality therapy. CyberPsychology Behav. **5**(1), 77–82 (2002)
10. Perhakaran, G., Yusof, A.M., Rusli, M.E., Yusoff, M.Z.M., Mahalil, I., Zainuddin, A.R.R.: A study of meditation effectiveness for virtual reality based stress therapy using EEG measurement and questionnaire approaches. In: Innovation in Medicine and Healthcare 2015, pp. 365–373. Springer, Cham (2016)
11. de Oliveira, J.M., Fernandes, R.C.G., Pinto, C.S., Pinheiro, P.R., Ribeiro, S., de Albuquerque, V.H.C.: Novel virtual environment for alternative treatment of children with cerebral palsy. Comput. Intell. Neurosci. (2016)
12. Teplan, M.: Fundamentals of EEG measurement. Measur. Sci. Rev. **2**(2), 1–11 (2002)
13. Burdea Grigore, C., Coiffet, P.: Virtual Reality Technology. Wiley-Interscience, London (1994)
14. LeCun, Y., Bengio, Y., Hinton, G.: Deep learning. Nature **521**(7553), 436–444 (2015)
15. Bengio, Y.: Learning deep architectures for AI. Found. Trends® Mach. Learn. **2**(1), 1–127 (2009)
16. Schalkoff, R.J.: Artificial Neural Networks, vol. 1. McGraw-Hill, New York (1997)

Fast FFT-Based Inference in 3D Convolutional Neural Networks

Bo Xie(✉), Guidong Zhang, Yongjun Shen, Shun Liu, and Yabin Ge

Lanzhou University, Chengguan District, Lanzhou, Gansu, China
504577818@qq.com

Abstract. Recognizing real world objects based on their 3D shapes is an important problem in robotics, computational medicine, and the internet of things (IoT) applications. In the recent years, deep learning has emerged as the foremost tool for a wide range of recognition and classification problems. However, the main problem of convolutional neural networks, which are the primary deep learning systems for such tasks, lies in the high computational cost required to train and use them, even for 2D problems. For 3D problems, the problem becomes even more pressing, and requires new methods to keep up with the further increase in computational cost. One such method is the use of Fast Fourier Transforms to reduce the computational cost by performing convolution operations in the Fourier domain. Recently, this method has seen widespread use for 2D problems. In this paper, we implement and test the method for 2D and 3D object recognition problems and compare it to the traditional convolution methods. We test our network on the ShapeNet 3D object library, achieving superior performance without any loss in accuracy compared to conventional methods.

1 Introduction

Convolutional neural networks have proven advantages over traditional machine learning methods on applications such as image classification [1], tracking [2], detection [3], and many others. However, the primary downside of convolutional neural networks is the increased computational cost. This becomes especially challenging for 3D convolution, where handling even the smallest instances requires substantial resources.

3D convolutional neural networks have recently come to the attention of the scientific community. In a database for 3D object recognition named ObjectNet3D was presented [4]. The database focusses on the problem of recognizing the 3D pose and shape of objects from 2D images. Another repository of 3D CAD models of objects is ShapeNet [5], which we use as test and training data in this work. In [6], the authors propose VoxNet, a 3D convolutional neural network to solve the robust object recognition task with the help of 3D information.

In the light of these successful applications, it is worthwhile to explore new ways to speed up the 3D convolution operation. In this paper we do so by extending FFT-based methods suggested by [7] from 2D to 3D, and studying the gain in performance. The remainder of this paper is structured as follows: We first discuss related work in Sect. 2

L. Barolli et al. (Eds.): IMIS 2018, AISC 773, pp. 420–431, 2019.
https://doi.org/10.1007/978-3-319-93554-6_40

and introduce convolutional neural networks and FFT-based methods in Sect. 3, where we also compare the theoretical properties of both methods. We then give a description of our neural network which is designed to solve the 3D object recognition task in Sect. 4. In Sect. 5, we first compare the performance of a traditional convolution method and FFT-based method for both 2D and 3D experimentally, and then test the performance of our neural network.

In this paper, we make four main contributions. Firstly we analyze the memory requirement and the computation of FFT-based method and traditional convolution method in theory. Secondly, we design an efficient convolutional neural network to recognize 3D objects. Thirdly, through our experiments, we prove that FFT-based method can have an obvious advantage in performance over the traditional method for 3D inputs. Finally, we present a system with good performance in 3D object recognition tasks using the FFT-based method.

2 Related Work

Performance is a fundamental problem for the use of convolutional neural networks in a wide range of applications. Consequently, many recent works have been devoted to reducing their computation and memory requirements. There is a great redundancy in the parameters of deep learning models, which allows pruning of many connections from the network. Parsimony expression reduces the number of parameters that need to be stored, and thus allow us to solve a bigger problem within limited memory. Similar work shows how to reduce the redundancy in parameters using a sparse decomposition. In [8], the network is pruned using learned information of important connections, whose weights are then quantized to enforce weight sharing. Huffman encoding is then applied to save memory. Similarly, in [9] the neural network is compressed to reduce memory usage.

Although these methods can reduce the number of parameters significantly, this might still not be enough when running a CNN on an embedded device. To reduce the size further, a common trend in recent literature is to quantize parameters to 16 or 8 bits rather than the standard 32 bits to save memory. In [10], the authors propose a framework named Quantized CNN, which uses this idea to speed up the computation and reduce the memory consumption of CNN models. The quantization is applied to the filter kernels in the convolutional layer and the weight matrices in fully-connected layers. Similarly, in [11] the authors propose a quantizer designed for CNNs that use fixed-point arithmetic, which means that the floating point variables are converted to integers. A similar technique is used in [12], where the authors adopt a stochastic rounding scheme to convert float parameters to a lower precision fixed-point representation. XNOR-Networks extend this idea even further by using binary variables. In [13] both the filters and the input to the convolutional layers are binary. This result in 58x faster convolutional operations and 32x reductions in memory space, at the cost of a considerable decrease in accuracy of the neural network in some non-trival cases.

Other approaches aim directly at reducing the computational cost within CNN. In [14], the authors analyze the algebraic properties of CNNs and propose an algorithmic improvement to reduce the computational workload. They achieve a 47% reduction in computation without affecting the accuracy. In [7], convolution operations are replaced with pointwise products in the Fourier domain, which can reduce the amount of computation significantly. Finally, evaluates two Fast Fourier Transform convolution implementations, one based on Nvidia's cuFFT and the other based on Facebook's FFT implementation. Both methods achieve good performance, which demonstrates the efficacy of the idea. Following this idea, we apply similar methods to the 3D domain.

3 Convolutional Neural Networks

A convolutional neural network is composed of multiple layers of different types. The most frequently used layers are convolution layers, fully connected layers, rectified linear unit (ReLU) layers, normalization layers, and pooling layers. However, by far the most computational work is performed by the convolution layers. Therefore, we only focus on optimized implementations of these layers. In a 2D convolutional layer, 2D kernels are applied to the inputs. Assume there are C input channels and K output feature maps. For ease of notation, we assume that all feature maps and kernels are quadratic. The size of an input channel is N^2, and the kernel size is k^2, which results in an output feature map of size M^2 where $M = N - k + 1$. We can easily calculate that the total number of FLOAT operations required to compute the convolution operation directly is $2CKM^2k^2$. All of these operations are multiplications paired with additions, and can thus benefit from fused multiply-add architectures. In practice, the direct convolution is usually converted to matrix multiplication, which has the same computational cost but can be implemented very efficiently.

3.1 3D Convolutional Network Layers

For 2D convolution, kernels have fixed width and height and they are slid along the width and height of the input feature maps. For 3D convolution, both feature maps and kernels also have depth, which increases the number of required operations to

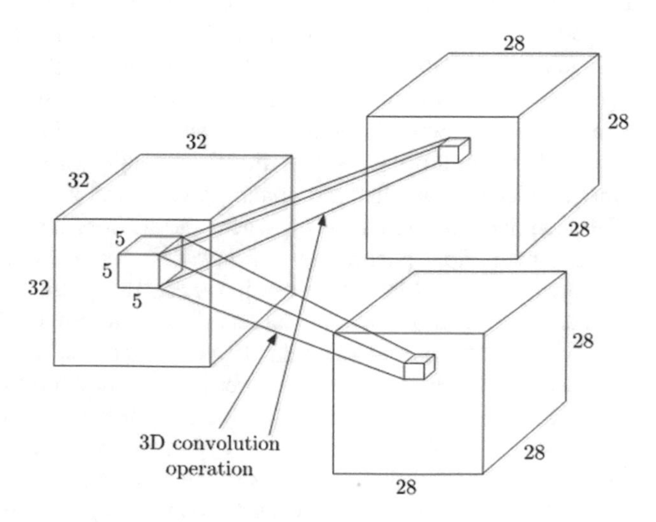

Fig. 1. Example of a 3D convolution operation.

$2CKM^3k^3$, assuming cubic feature maps and kernels. Algorithm 1 shows the code for the 3D convolution. Its output is also a volumetric feature map. An example of a 3D convolution operation is shown in Fig. 1.

Algorithm 1 Direct Convolution Implementation in 3D

```
1: for α ∈ {1, ..., K} do
2:   for β ∈ {1, ..., K} do
3:     for γ ∈ {1, ..., K} do
4:       for δ ∈ {1, ..., K} do
5:         for ϵ ∈ {1, ..., K} do
6:           for φ ∈ {1, ..., K} do
7:             for λ ∈ {1, ..., K} do
8:               for μ ∈ {1, ..., K} do
9:                 Y[α][γ][δ][ϵ]+          =
10:                W[α][β][φ][λ][μ]
11:                *X[β][γ + φ][δ + λ][ϵ + μ];
12:            end for
13:          end for
14:        end for
15:      end for
16:    end for
17: end for
```

3.2 Fast Fourier Transform

It is widely known that convolutions in the spatial domain are equivalent to pointwise products in the Fourier domain. As shown in [7], convolutions between functions f and g can be computed via Eq. 1:

$$f * g = F^{-1}(F(f) \cdot F(g)) \tag{1}$$

F denotes the Fourier transform and F^{-1} denotes the inverse Fourier transform, f represents the input of a convolutional layer and g represents a kernel. We always use Fast Fourier Transform to implement the Fourier transform. Thus we can calculate the number of operations of the 2D and 3D FFT-based method, which amounts to $2\alpha(\mathrm{K}+\mathrm{C}+\mathrm{KC})N^2 \log N + 4KCN^2$ for 2D and to $3\alpha(\mathrm{K}+\mathrm{C}+\mathrm{KC})N^3 \log N + 4KCN^3$ for 3D for some small constant α. Thus, we can denote the running times as $O(\mathrm{KC}N^2 \log N)$ and $O(\mathrm{KC}N^3 \log N)$ respectively. We can thus compare the

Table 1. The computational complexity of convolutions different implementations.

Method	Computational Complexity
2D direct convolution	$2CKM^2k^2$
2D FFT	$O(CKN^2 \log N)$
3D direct convolution	$2CKM^3k^3$
3D FFT	$O(CKN^3 \log N)$

complexity of the direct convolution method to that of the FFT-based method. Table 1 shows the difference.

We can easily calculate the FLOAT operations given those parameters for a convolutional layer. Note that unless the feature maps are very small or the kernels are very large, M and N are almost of the same size. And while the FFT method includes a constant greater than two, we can clearly see from the table that it becomes far superior for larger kernels in 3D, since in that case the running time of the direct method grows cubically in k. The FFT-based method shows a more obvious advantage over the direct convolution method with the increasing size of the kernel. However, for the 2D FFT-based method, the advantage is not obvious when kernel size is small, which limits the wide use of 2D FFT-based method as the kernel size of most convolution layers is smaller than 6. For 3D FFT-based method however, we can achieve some speedup even for small kernel sizes. We will show the benefit of the FFT-based method in the 2D and 3D convolutional neural network in our experiments.

Furthermore, the main problem of using the FFT-based method is its memory requirement. We compare the memory usage of the direct convolution method and the FFT-based method. Table 2 shows the memory required for a single convolution layer. Even though the memory usage of the FFT-based method is very high, the reduction in computation time can make them worthwhile since convolution operations are extremely computationally intensive.

Table 2. Memory usage of the different convolution methods. Values are given as numbers to be stored. Actual memory consumption depends on the data type.

Method	Input	Weights
2D direct convolution	SCN^2	CKk^2
2D FFT	SCN^2	$2SCKN^2$
3D direct convolution	SCN^3	CKk^3
3D FFT	SCN^3	$2SCKN^3$

4 Implementation

Our convolutional neural network for object recognition is based on VoxNet [6], a 3D CNN for the same problem. VoxNet contains two convolution layers, and six additional layers. Since by far the largest part of the computation is performed by the convolutional layers, we only transfer these from the spatial domain to the pointwise product in the Fourier domain, and leave the other layers unchanged. In contrast to VoxNet, which uses the direct convolution method, we apply the FFT-based method in the convolution layers. The other layers work in the same way as in VoxNet. The structure of our network is shown in Fig. 2. The 3D FFT convolution is implemented in Theano, which contains an optimized implementation of the FFT method, including fast routines for the inverse FFT and the element-wise products required by Eq. 1.

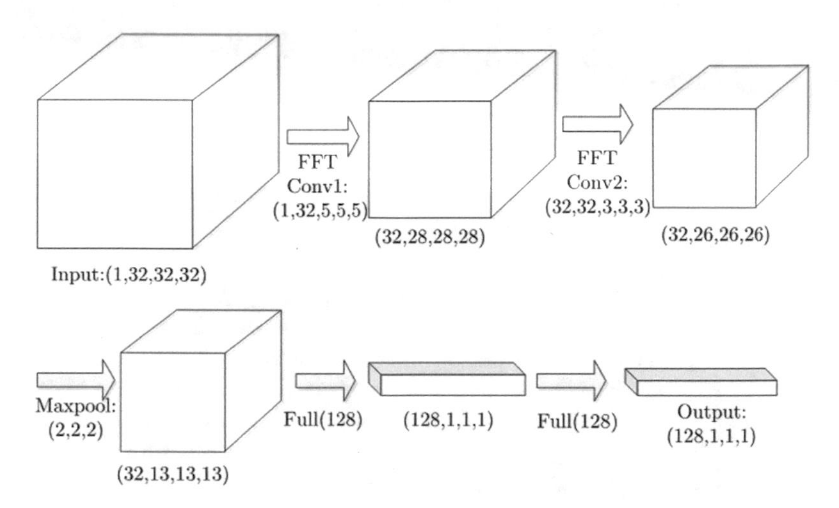

Fig. 2. Structure of our CNN.

5 Experiments

In this section, we present experimental results on the performance of FFT-based methods applied to convolutional layers, in comparison to the direct method. We then test the network described in Sect. 4 for a 3D object recognition task. Data for that task is taken from ShapeNet [5]. All the experiments are performed on an Nvidia Tesla K40 GPU running CUDA 7.0. The GPU is equipped with 12 GB of memory with a memory bandwidth of 288 GB/s. Its 2880 CUDA cores provide it with 4.29 Tflops/s of single-precision performance. All tests codes are implemented using Theano 0.8.

5.1 Performance of the 2D and 3D FFT-Based Method on the Convolution Layers

In our first experiment, we compare the performance of the two implementations under varying kernel size while the number of input channels, output channels, and the input size remains fixed.

We set C = 32 and K = 32. We use an 256×256 input for 2D and an $64 \times 64 \times 64$ input for 3D, while k ranges from 3 to 11. For all kernels the values of k are equal for all dimensions. Results can be found in Figs. 3 and 4, which show the computation time versus kernel size.

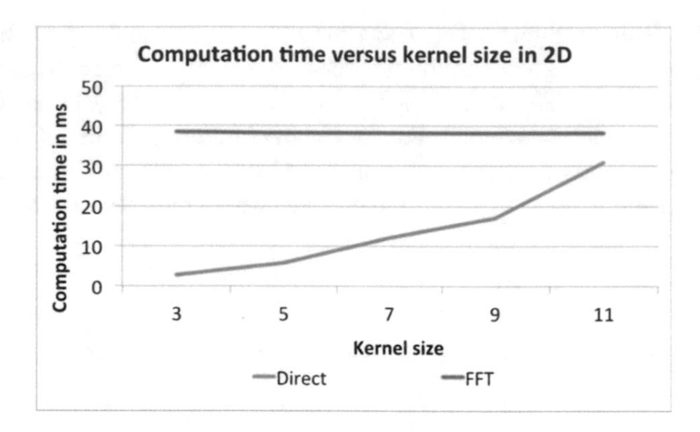

Fig. 3. Computation time with varied kernel size in 2D.

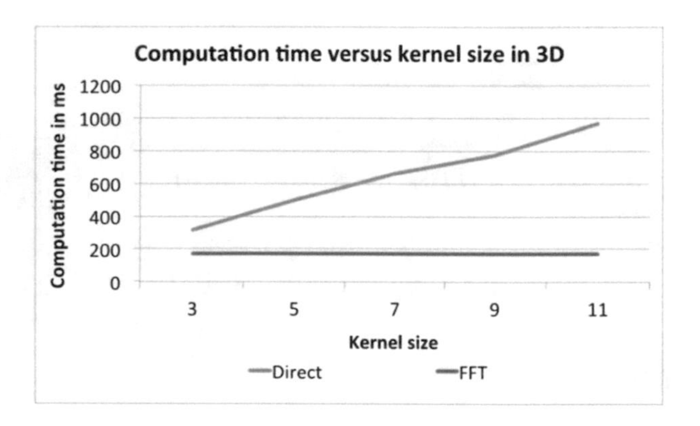

Fig. 4. Computation time with varied kernel size in 3D.

For the 2D convolution, we can see that the performance of the direct convolution is higher than that of the FFT-based method. On the other hand, in the 3D case, we observe a significant performance gain of the FFT-based method over the direct method, which confirms our theoretical analysis of the FFT method in such applications. Note that in this experiment, we process only a single input at a time, even though the FFT method also benefits from larger batch sizes, as shown below.

The second experiment measures the computation time under varying batch size while the input size, number of input channels, number of output channels, and kernel size fixed. We set C = 32 and K = 32. We use an 256×256 input for 2D and an $32 \times 32 \times 32$ input for 3D. We test both for a small kernel with k = 3 and a large kernel with k = 7, while the batch size S ranges from 1 to 64 (32 in the 3D case). Since the performance of the FFT-based method is independent of the kernel size, we plot only one line for FFT. The results are shown in Figs. 5 and 6.

For the 2D case, we observe that for the large kernel, the FFT method becomes competitive as the batch size increases, beating the direct method by a large margin at S = 64. For the smaller kernels however, the direct method remains faster for all

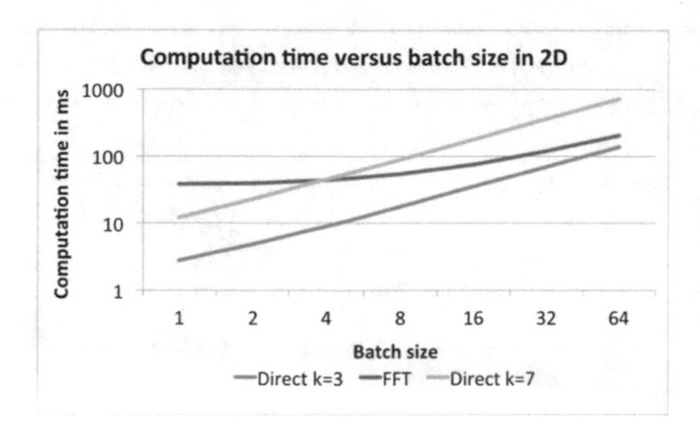

Fig. 5. Computation time with varied batch size in 2D.

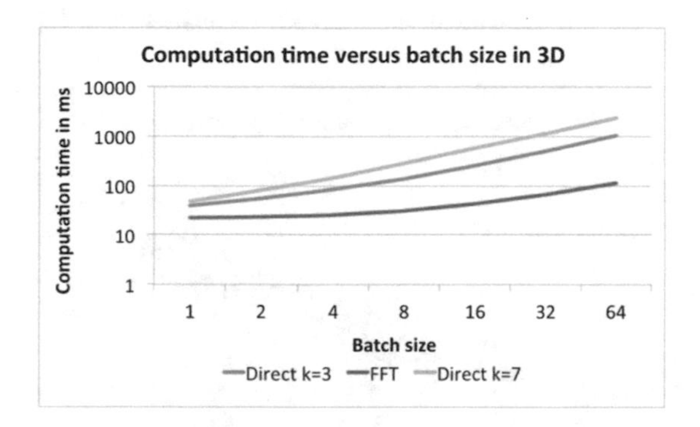

Fig. 6. Computation time with varied batch size in 3D.

batch sizes. On the other hand, in the 3D case, the FFT-based method is faster for all batch sizes, even for $3 \times 3 \times 3$ kernels.

For the third experiment, we test the influence of the input size on the relative performance of the methods. Again, we test both for a small kernel with k = 3 and a large kernel with k = 7. The image size varies from 16 to 128 in each dimension (256 for 2D). We set C = 1 and K = 32. Based on the results of the previous experiment, we set the batch size to 64 for 2D and 32 for 3D. Figure 7 and Fig. 8 show the results.

We observe that in the 2D the relative performance of the FFT-based method depends again on the kernel size. For the large kernels, FFT is significantly faster for all input sizes, and somewhat slower for the small kernel. On the other hand, the relative performance of FFT in the 3D case improves with growing image size, outperforming the direct method even for small kernels at input sizes $64 \times 64 \times 64$ and above. This is consistent with the behaviour for large kernel and batch sizes, making the method well suited for large convolution layers and networks. The downside is that in large networks, the memory requirements can soon become prohibitive. In the cases where it can be used however, the FFT method provides a very substantial benefit, especially for 3D applications. Based on this, we test our FFT-based CNNs for object recognition on a 3D object library.

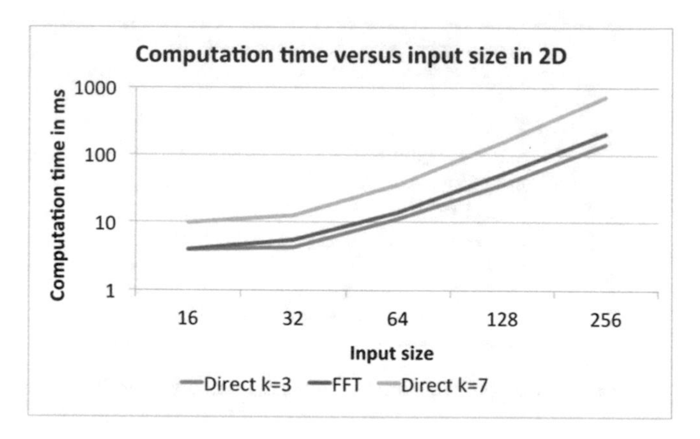

Fig. 7. Computation time with varied input size in 2D.

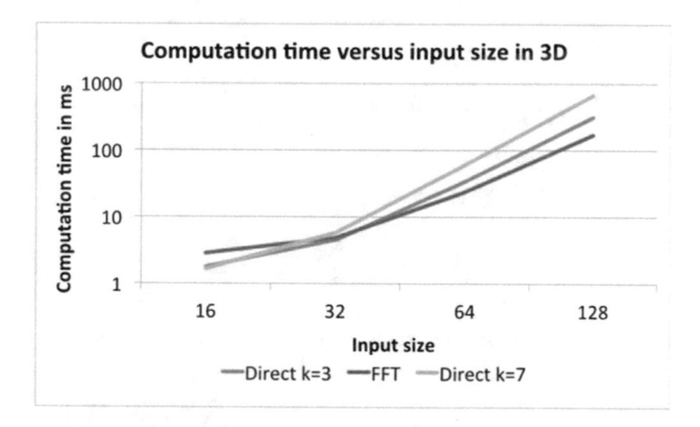

Fig. 8. Computation time with varied input size in 3D.

5.2 Experiments on ShapeNet Using 3D FFT

In this section, we test the performance and accuracy of our FFT-based CNNs described in Sect. 4. First we compare the performance of the traditional 3D convolution method with the 3D FFT-based method on the two convolutional layers. We achieve a 3.29x and a 6.38x speedup respectively. The results are shown in Table 3.

Table 3. The computation time of different implementation methods in 3D convolutional neural network.

Input: S, C, N Kernel: C, K, k	Time(seconds)		Speedup
	Direct-Conv3D	FFT-3D	
32,1,32 1,32,5	0.120	0.036	3.29
32,32,28 32,32,3	0.320	0.050	6.38

While testing the accuracy, we obtained 92.73% accuracy for 3D object classification. Figure 9 shows some classification results. Note that the 3D FFT-based method does not affect the classification accuracy of VoxNet, which means that accuracy results for VoxNet extend to our network. The only problem of the FFT-based method is that it consumes more memory. We calculated that the memory required for the weights of the direct method of the first convolution layer is about 15.6 KB, while the memory used for FFT-based method is about 171 MB.

Fig. 9. Four 3D objects which are labeled correctly using FFT-based method.

Due to the heavy computation and large memory requirement of the 3D convolutional network, the input data sets are necessarily small and simple. Just like the inputs to VoxNet, the 3D objects only have one channel and a small size of 32 per dimension. For widespread use of 3D convolutional neural networks on much more complex data sets, more computational resources and faster implementations are required, as discussed in Sect. 5.1. The large memory requirement is a significant disadvantage of the FFT-based method, which restricts its use, especially on GPUs where memory is limited. However, the new Pascal generation of NVIDIA GPUs is capable of seamlessly transferring data between CPU and GPU, which allows FFT-based methods to make use of the much larger CPU memory rather than the GPU memory.

6 Conclusion

3D CNNs gained a lot attention due to their ability to handle general multi-dimensional data. However, they require a high computational cost, which makes the 3D CNNs not available on limited resource devices. There is a promising way to use the FFT method to implement convolutional neural networks, and recent literature demonstrates the efficacy in 2D cases.

In this paper, we propose a method for solving the 3D object recognition problem, which utilizes Fourier transforms to compute the convolution operation in 3D CNNs. The experiments demonstrated that FFT-based methods can significantly improve the performance of convolutional neural networks, especially in the 3D case. However, the gains in performance come at the cost of a considerably larger memory requirement.

This poses a dilemma for the use of object recognition in embedded systems, which posses neither high computational performance nor large memory. Furthermore, neither can currently be improved without a significant increase in energy consumption, which is another common restriction for embedded devices. Therefore, further research is needed to either improve the performance of the direct method, or to reduce the memory consumption of FFT methods to make 3D object recognition available for embedded systems. In the future, NVRAM will allow embedded systems to be equipped with large amounts of memory without the immense energy cost required for doing so today. At that point, FFT-based methods are likely to become crucial.

References

1. Krizhevsky, A., Sutskever, I., Hinton, G.E.: ImageNet classification with deep convolutional neural networks. In: Advances in Neural Information Processing Systems, pp. 1097–1105 (2012)
2. Fan, J., Xu, W., Wu, Y., Gong, Y.: Human tracking using convolutional neural networks. IEEE Trans. Neural Netw. **21**(10), 1610–1623 (2010)
3. Redmon, J., Divvala, S., Girshick, R., Farhadi, A.: You only look once: unified, real-time object detection, arXiv preprint arXiv:1506.02640
4. Xiang, Y., Kim, W., Chen, W., Ji, J., Choy, C., Su, H., Mottaghi, R., Guibas, L., Savarese, S.: ObjectNet3d: a large scale database for 3D object recognition. In: European Conference on Computer Vision, pp. 160–176. Springer, (2016)
5. Chang, A.X., Funkhouser, T., Guibas, L., Hanrahan, P., Huang, Q., Li, Z., Savarese, S., Savva, M., Song, S., Su, H., et al.: ShapeNet: an information-rich 3D model repository, arXiv preprint arXiv:1512.03012
6. Maturana, D., Scherer, S.: VoxNet: a 3D convolutional neural network for real-time object recognition. In: 2015 IEEE/RSJ International Conference on Intelligent Robots and Systems (IROS), pp. 922–928. IEEE (2015)
7. Mathieu, M., Henaff, M., LeCun, Y.: Fast training of convolutional networks through FFTs, arXiv preprint arXiv:1312.5851
8. Han, S., Mao, H., Dally, W.J.: A deep neural network compression pipeline: pruning, quantization, huffman encoding, arXiv preprint arXiv:1510.00149
9. Chen, W., Wilson, J.T., Tyree, S., Weinberger, K.Q., Chen, Y.: Compressing neural networks with the hashing trick, CoRR, abs/1504.04788
10. Wu, J., Leng, C., Wang, Y., Hu, Q., Cheng, J.: Quantized convolutional neural networks for mobile devices, arXiv preprint arXiv:1512.06473
11. Lin, D.D., Talathi, S.S., Annapureddy, V.S.: Fixed point quantization of deep convolutional networks, arXiv preprint arXiv:1511.06393
12. Gupta, S., Agrawal, A., Gopalakrishnan, K., Narayanan, P.: Deep learning with limited numerical precision, CoRR, abs/1502.02551 392

13. Rastegari, M., Ordonez, V., Redmon, J., Farhadi, A.: XNOR-Net: ImageNet classification using binary convolutional neural networks, arXiv preprint arXiv:1603.05279
14. Cong, J., Xiao, B.: Minimizing computation in convolutional neural networks. In: International Conference on Artificial Neural Networks, pp. 281–290. Springer (2014)

Cognitive Informatics Approaches for Data Sharing and Management in Cloud Computing

Marek R. Ogiela(✉) and Lidia Ogiela

Cryptography and Cognitive Informatics Research Group,
AGH University of Science and Technology,
30 Mickiewicza Ave., 30-059 Krakow, Poland
{mogiela, logiela}@agh.edu.pl

Abstract. In this paper will be described possible applications of cognitive information systems to intelligent and secure information management tasks in Fog and Cloud computing. In particular will be presented the ways of using some semantic descriptors and personal characteristics for creation of protocols dedicated to confidential distribution and secure data management in different distributed environments. The new paradigm of cognitive cryptography will be also described.

Keywords: Cognitive systems · Cryptographic techniques
Cognitive cryptography

1 Introduction

Advanced security solutions are developing rapidly due to the application of cognitive information systems, based on biologically-inspired computational approaches. It seems that such systems will also become more important in future research, as a result of the very dynamic development of new generation information systems, dedicated for ubiquitous computing in cyberspace, smart cities or ambient world. Areas connected with semantic data exploration and information fusion are broadly developed [1–3], especially researches which are based on application of deep learning approaches [2]. Also security areas may be connected with application of cognitive approaches in developing new protocols for authentication, or services management in Cloud computing [4–7].

In this paper will be described new solutions connected with application of cognitive information systems used for semantic pattern evaluation and security applications. Such procedures can be used for evaluation of the meaning of selected patterns or data records, and involve semantic meaning into the classification task or encryption process. Such general protocols may play important role in modern, and efficient information management protocols oriented on information and services management in Cloud or Fog environments.

In particular application of cognitive information systems for semantic analysis of different patterns will be presented. Such cognitive information systems are based on important computational paradigms called cognitive resonance and connected with computer understanding techniques proposed for the semantic interpretation of visual patterns [8, 9].

L. Barolli et al. (Eds.): IMIS 2018, AISC 773, pp. 432–437, 2019.
https://doi.org/10.1007/978-3-319-93554-6_41

In conducted research we proposed several different classes of cognitive information systems, especially dedicated to visual pattern semantic evaluation, scene understanding and also financial data analysis. All such systems try to imitate the natural way of human thinking and simulate cognitive processes, which are based on the model of human visual perception. In such processes some hypothesis can be generated, and next verified with real data or measured features.

2 Cognitive Approaches for Data Semantic Evaluation and Cloud Applications

As mentioned before in our research we proposed several classes of cognitive systems dedicated to image analysis and supporting decision making processes [9]. As a result of application of such systems it was possible to obtain semantic description of complex patterns or analysed information records. Such semantic description can contain both local and global parameters describing single objects as well as some relation between objects and parameters. Comparing cognitive systems with pattern classification methods we can noting that traditional methods of pattern recognition are usually not allow to extract or describe the semantic content. Cognitive systems allow

Fig. 1. Examples of images with different content but having similar semantic meaning. Both of these images presents boats, but in different conditions.

to evaluate the most important semantic features (semantic meaning), which usually is hidden in the whole pattern. Such cognitive and pattern understanding approaches allow to classify into the same semantic class different patterns, which present different objects, but have similar meaning or semantic content. Example of such images are presented in Fig. 1.

Application of cognitive information systems allow to evaluate semantic meaning of analysed patterns or data records, and next apply such semantic information in security procedures implemented in Cloud or Fog computing.

Cloud computing provides computing and storage possibilities, which considerably facilitate data acquisition and enhanced processing. To avoid some transmission limitation in Cloud computation, the data exploration and fusion tasks can be moved closer to the original data sources, to the layer existed between the nodes and Cloud units i.e. Fog or Edge infrastructure. It is very important especially in processing a great amount of signals originating from IoT infrastructures. In such distributed and hierarchical processing structures Fog computing is able to perform initial data processing, and analytics tasks. Edge and Fog layers participate in connecting data coming from multiple sources, and determining if data should be analyzed immediately or may be transferred to higher level i.e. to the Cloud. Finally Cloud and Fog levels create a hierarchical architecture for data processing.

The lower level is connected with sensors and devices, registering different signals or information. The second level (Fog processing units) are distant from real sensors, and consist more powerful computing services, which can perform data acquisition, fusion and analytics tasks. On the highest level it is possible to run Cloud services and application, which can be managed by different computation and storage providers.

3 Cognitive Systems for Encryption and Secure Information Management

Cognitive systems may be applied for semantic meaning evaluation of visual patterns or data records. In such cases, we can try to extract some specific data (semantic descriptors) for security and cryptography purposes [10–13].

Creation of new cryptographic procedures, which are based on cognitive information systems allows to introduce a new security paradigm called cognitive cryptography [14, 15]. This new computational area allow to join techniques, which are usually used for data security with special characteristics describing semantic information extracted from encrypted data.

Application of cognitive systems in cryptographic procedures allows to make encryption process more oriented and dependent from the encrypted information. Cryptographic protocol allow to encode semantic meaning or use it as a special key associated with particular encrypted data record.

The example of application cognitive systems for security purposes is connected with threshold schemes used for secret sharing [16, 17]. In traditional sharing protocols it is no connections between generated parts of shared data and its semantic meaning so we proposed a new class of semantic threshold procedures, which allow to generate shadows depending on the meaning of shared information.

The concept of semantic threshold procedure is based on using cognitive systems to evaluate semantic meaning of processed information, and convert it into the special encryption key (with particular length) using one-way hash functions. Such semantic key allow later to perform sharing procedure which is fully dependent on the semantic meaning. In such cases each secret part may have included some semantic features associated with original information. Example of such protocol is presented in Fig. 2.

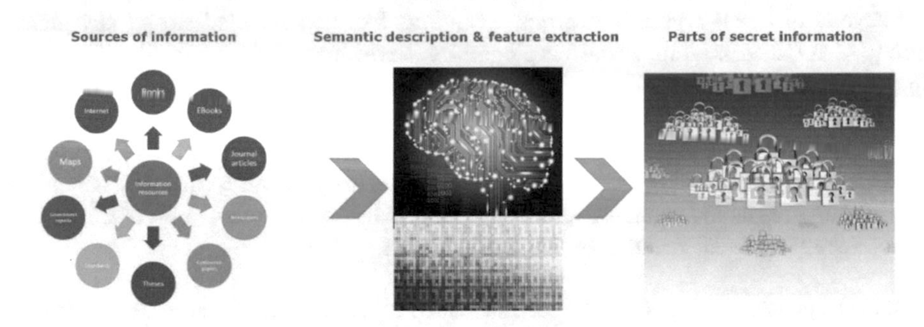

Fig. 2. An idea of semantic threshold procedures.

In this place it is also worth underline that next possible application of cognitive systems in security areas may be connected with crypto-biometric solutions. For such solutions personal data can be evaluated using cognitive systems. Such personal information may be used for authentication purposes, and for creation personalized encryption keys, assigned to particular user. The personal feature vector can be used also in digital watermarking and multi-secret steganography based on fuzzy vaults. Cognitive systems allow to evaluate also behavioral features with possible application in creation of behavioral lock. For such purposes cognitive systems can analyse fingers or hand movements which have the high level of complication, sufficient for obtaining the distinctive features. During such analysis it is possible to consider different types of gestures like fixed patterns, imitating handwritten signatures, personal natural gestures, and special user-defined patterns [15].

Application of cognitive systems is also possible in visual cryptography. In this area such systems allow to determine different perception threshold for participants while restoring visual information. In such procedure it is possible to evaluate individual perception threshold for particular user at which he is able to properly recognize the original visual object, but other participants cannot to do it in the same manner [14].

4 Conclusions

In this paper an idea of application of cognitive information systems for security purposes was described [17, 18]. In particular the way of semantic meaning evaluation was presented and application of such information in modern cryptographic procedures were described. It should be noted that semantic evaluation of many kind of information records, allow to perform a computer understanding tasks, which enables to classify in the same manner different patterns, which have the same semantic meaning [14, 19]. Application of cognitive information systems to semantic pattern analysis allow to extract some semantic descriptors, which may be later used for security purposes. For security protocols we can apply unique features originating from encoded patterns, what finally results that we can even define a new computational paradigm called cognitive cryptography [15]. Beside semantic information, also personal features extracted from different biometric or behavioural patterns may be applied for several different crypto-biometrics operations like personalized strong key generation, secret distribution, visual cryptography, and creation of personal behavioural lock.

Application of such features in security procedures, seems to be very important in future IT solutions, dedicated for Cloud and Fog computing as well IoT systems. Presented solutions have also some important security features like resistance for known cryptographic attacks, efficiency and low complexity, universality, and dependence from real features or semantic content.

Acknowledgments. This work has been supported by the National Science Centre, Poland, under project number DEC-2016/23/B/HS4/00616.

References

1. Lyon, R.F.: Human and Machine Hearing: Extracting Meaning from Sound. Cambridge University Press (2017)
2. LeCun, Y., Bengio, Y., Hinton, G.: Deep learning. Nature **521**, 436–444 (2015)
3. Hahn, Ch., Hur, J.: Efficient and privacy-preserving biometric identification in cloud. ICT Express **2**, 135–139 (2016)
4. Arigbabu, O.A., Ahmad, S.M.S., Adnan, W.A.W., Yussof, S.: Integration of multiple soft biometrics for human identification. Pattern Recogn. Lett. **68**, 278–287 (2015)
5. Gomez-Barrero, M., Maiorana, E., Galbally, J., et al.: Multi-biometric template protection based on Homomorphic Encryption. Pattern Recogn. **67**, 149–163 (2017)
6. Ogiela, L.: Cognitive Information Systems in Management Sciences. Elsevier, Academic Press (2017)
7. Ogiela, L., Ogiela, M.R.: Management information systems. LNEE, vol. 331, pp. 449–456 (2015)
8. Bisantz, A.M., Burns, C.M., Fairbanks, R.J. (eds.): Cognitive Systems Engineering in Health Care. CRC Press, Boca Raton (2014)
9. Ogiela, L., Ogiela, M.R.: Data mining and semantic inference in cognitive systems. In: Xhafa, F., Barolli, L., Palmieri, F., et al. (eds.) 2014 International Conference on Intelligent Networking and Collaborative Systems (IEEE INCoS 2014) Salerno, Italy, pp. 257–261 (2014)
10. Li, L., Yan, Ch.C, Chen, X., et al.: Distributed image understanding with semantic dictionary, and semantic expansion. Neurocomputing **174**, 384–392 (2016)
11. Cox, I.J., Miller, M.L., Bloom, J., Fridrich, J., Kalker, J.: Digital Watermarking and Steganography. Morgan Kaufmann Publishers, Burlington (2008)
12. Han, L., Xie, Q., Liu, W.: An improved biometric based authentication scheme with user anonymity using elliptic curve cryptosystem. Int. J. Netw. Secur. **19**(3), 469–478 (2017)

13. Jin, Z., Teoh, A.B.J., Goi, B.-M., Tay, Y.-H.: Biometric cryptosystems: a new biometric key binding and its implementation for fingerprint minutiae-based representation. Pattern Recogn. **56**, 50–62 (2016)
14. Ogiela, M.R., Ogiela, L.: On using cognitive models in cryptography. In: IEEE AINA 2016 - The IEEE 30th International Conference on Advanced Information Networking and Applications, Crans-Montana, Switzerland, 23–25 March 2016, pp. 1055–1058 (2016)
15. Ogiela, M.R., Ogiela, L.: Cognitive keys in personalized cryptography. In: IEEE AINA 2017 - The 31st IEEE International Conference on Advanced Information Networking and Applications, Taipei, Taiwan, 27–29 March 2017, pp. 1050–1054 (2017)
16. Ogiela, M.R., Ogiela, U.: Linguistic approach to cryptographic data sharing. In: FGCN 2008 – The 2nd International Conference on Future Generation Communication and Networking, Hainan Island, China, vol. 1, pp. 377–380 (2008)
17. Ogiela, M.R., Ogiela, U.: Grammar encoding in DNA-like secret sharing infrastructure. Lecture Notes in Computer Science, vol. 6059, pp. 175–182 (2010)
18. Ogiela, L., Ogiela, M.R.: Insider threats and cryptographic techniques in secure information Manage. IEEE Syst. J. **11**, 405–414 (2017)
19. Vernon, D.: Artificial Cognitive Systems: A Primer. MIT Press, Cambridge (2014)

Implementation of Lane Detection Algorithm for Self-driving Vehicles Using Tensor Flow

Hyunhee Park(✉)

Department of Computer Software, Korean Bible University,
32, Dongil-ro 214gil, Nowon-gu, Seoul, Korea
parkhyunhee@gmail.com

Abstract. Recently, systems for detecting and tracking moving objects from video are gaining research interest in the field of image processing, owing to their applications in fields such as security, observation, and military, and considerable research is being conducted to develop high-accuracy and high-speed processing systems. In particular, as interest in autonomous driving has increased rapidly, various algorithms for lane keeping assistance devices have been developed. This study proposes a lane detection algorithm by comparing color-based lane detection algorithms and using a lane detection algorithm based on representative line extraction. Edge extraction and Gaussian filters are applied for lane detection and a Median filter is applied for image noise reduction. The detection accuracy is improved by extracting the region of interest for the image based on four pointers. Finally, a Hough transform is applied to improve the accuracy of straight line detection, and an algorithm to extract representative lines is applied to increase the detection rate in shadow regions and dark areas. Experimental results show that the proposed algorithm can detect lanes with high accuracy. The application of this algorithm to vehicle black boxes or autonomous driving will help prevent lane departure and reduce accident rates.

1 Introduction

Research on intelligent automobile technologies combined with IT technology is being actively conducted [1,2]. In particular, the advanced safety vehicle (ASV) technology that reduces the risk of accidents is studied [3]. This technology includes lane departure alarms, lane keeping, pedestrian collision warnings, and front collision warnings, and is the basis of intelligent vehicle technology [4]. In the development of lane detection technology, a vision-based method using an image sensor is widely used as the core technology of ASVs. Vision-based lane detection is performed in four steps: feature extraction, outlier removal and pre-processing, post-processing, and lane tracking. The performance and characteristics of the lane detection algorithm are determined based on the algorithm used in each step.

L. Barolli et al. (Eds.): IMIS 2018, AISC 773, pp. 438–447, 2019.
https://doi.org/10.1007/978-3-319-93554-6_42

Features used for lane detection can be divided into boundary line and color line [5]. However, features obtained using the existing feature detection methods depend on changes in illumination [6]. This problem can be resolved using a sensor capable of acquiring high-resolution data or combining a heterogeneous sensor having a long detection distance. However, it increases cost and difficulty in real-time processing [7]. To solve the problem of the existing feature detection method, this study proposes a method to adaptively detect the boundary of the environment [8]. In this paper, we propose a new lane detection algorithm that can effectively cope with various lane shapes and road environments, improve lane detection performance for practical use, and reduce the amount of computation and real-time processing. In this paper, the problems of conventional image-based lane detection methods are discussed in Sect. 2 and the proposed lane detection algorithm is described in Sect. 3. Section 4 shows the experimental results of lane detection using gray color conversion, HSV color conversion, and representative line extraction. Conclusions are drawn from the results of the analysis in Sect. 5.

2 Related Work

Lane and line detection is frequently used in computer vision, e.g. autonomous vehicles and intelligent robots, so there are many methods described in literature. The biggest question mark that we faced at the beginning of the implementation was to decide which lane detection method we should select out of the many available ones. Features used for lane detection can be divided into boundary-line and color-line features. Color-based methods are still in the early stages of research because they exhibit very poor performance when experiencing changes in illumination. Therefore, edge-detection-based methods are widely studied. Wang proposed a method using B-snake and Canny Hough estimation vanishing point algorithms for boundary detection [9]. Many lane boundary detection algorithms in [10] were developed to locate the positions of lane boundaries through a Canny edge detector or a Hough transformation. However, the lack of geometric constraints is a common defect in these methods. For vehicle, obstacle, and pedestrian detection tasks used a method to generate a bounding box for the description of the objects coordinates. However, in a typical self-driving task, the precise location of the bounding box is useless. These methods of lane detection convert the acquired color image into a monochrome image to reduce the complexity and improve the performance of the boundary detection algorithm [11]. They also focus on eliminating outliers, removing inadequate boundaries during post-processing, and compensating for undetected areas. However, if boundary detection does not occur well, performance degradation of the entire system cannot be avoided [12,13].

3 Proposed Lane Detection Algorithm

To output the image, we use VideoCapture of Open CV to import the video (e.g., block box in a car). In this paper, we change the RGB (red, green, and blue) value of the output image to the gray or HSV (Hue, Saturation, Value) color code. In addition, the result of lane detection by extracting representative lines is compared. In this section, we describe the proposed lane detection algorithm of each step.

Color Conversion Technique

If Gray = TRUE, a gray color code performs color conversion to obtain a grayscale image and then perform binarization to remove shadows. Binarization is a task to clarify black and white classification. When the gray color code is used, it is possible to extract the shadow region by the road.

If Gray = FALSE, an HSV color code performs color tracking. For this purpose, only the white color is detected using a color mask. Based on the number of lanes on a road, multiple vehicles can travel simultaneously and shadows from these vehicles can occur. In addition, the lane may be erased due to aging of the road surface. Therefore, it is possible to distinguish kinds of color by the characteristic of the lane. The edge is detected using the HSV color code, and the accuracy of detection is improved by comparing the results with the Canny edge algorithm. The image input in RGB format is converted to HSV color space. By setting the color space area using the HSV color space, only white, yellow, and blue colors are detected. Applying the binary method, these three colors are represented in white while the remaining colors are represented in black. Therefore, using HSV color codes that are insensitive to shadow effects in the side lanes at high speeds, higher lane detection accuracy than using the gray color code can be achieved.

Edge Detection Technique

The most basic method to detect lanes in an image is by detecting the edges of color area because the lane on the road surface forms a boundary with other parts [14]. In this paper, the Canny algorithm is used for lane edge detection. In addition, we use a 5 x 5 Gaussian filter to remove noise and confirm the direction and intensity of the image gradient. The derivative value is used to determine the boundary value, which considers the rapidly changing differential value due to the different color from the surroundings.

Application of Gaussian Filter and Median Filter

gaussian filter removes noise as described above. The principle of the gaussian filter is to convolve each point of the input array into a Gaussian kernel, combine all of them to form an output array, and proceed with the filter operation. After applying the gaussian filter, the Median filter is applied to smoothing the image.

Region of Interest Extraction Technique

In this paper, the region of interest for lane boundary (ROI-LB) is first obtained for lane detection to reduce the number of pixels of the image and the amount of computation required [15,16]. Moreover, since Hough transform is applied only within the region of interest, the lines only in the region of interest are computed and the accuracy is high because other backgrounds are not recognized [17]. In other to extract region of interest, four points are set up to set the region of interest. If the region of interest is extracted and only the edge information is extracted from the color image, lane detection performance can be remarkably improved by eliminating the information that is mistaken for the lane.

Hough Transform Technique

After detecting the candidate images, a straight line must be detected to find the lane. This is achieved by applying Hough transform to the edge extracted by the Canny edge algorithm. The Hough transform uses a linear equation expressed as $y = mx + c$. This equation can be expressed as a trigonometric function as $r = x\cos\theta + y\sin\theta$. minLineLength specifies the minimum length. maxLineGap is the maximum allowable distance between lines. Using the polar coordinate system, the Hough transform can be expressed as a straight line at a vertical distance from the origin and an angle θ between the vertical line and the x-axis and is detected precisely at a constant interval.

Representative Line Detection Technique

For lane detection, it is important to consider not only the straight line but also the curved section [18]. For this purpose, the slope of the straight line is considered. Using the coordinates of the starting and ending points of all the straight lines through the Hough transform, the slope of each straight line can be obtained. In this paper, the horizontal slope is limited to 160° and the vertical slope is limited to 90° (this is the heuristic result, which is determined after applying various values). Then, the filtered straight line is removed and the representative lines for each of the left and right sides are obtained.

4 Experimental Results

Experiments on lane detection were performed on daytime running image. The input image used in the experiment is the color image extracted from a vehicle black box video [19].

(a) Original image of interest region extraction

(b) Color area image of interest region extraction

(c) Edge area image of interest region extraction

Fig. 1. Result image of region of interest extraction from daytime running image in block box of the car.

Since the processing area is reduced by ROI-LB extraction, the amount of calculation required for lane detection is drastically reduced. Although the overall amount of computation varies according to the optimization of the processor or program used and the image of the environment, it is confirmed that the proposed algorithm can reduce the computation time by eight as compared to the conventional Hough transform method. These results are analyzed using ROI-LB optimal extraction and block direction edge direction information. In this paper, four points are set up to set the region of interest from original image in Fig. 1(a). In addition, if the region of interest is extracted as a color image, the result shown in Fig. 1(b) can be obtained. If only the edge is extracted in the region of interest, the result shown in Fig. 1(c) can be obtained. Thus, if the region of interest is extracted and only the edge information is extracted from the color image, lane detection performance can be remarkably improved by eliminating the information that is mistaken for the lane.

In this paper, we compare the results of lane detection obtained by applying gray color conversion for low-speed city driving with those obtained by applying the HSV color code for high-speed driving. Figure 2 shows the result of applying the gray color code to the image. As a result of applying the gray color code, it can be seen that the lanes of the straight road (Fig. 2(a)) and the curved

(a) Gray color code result image of straight road

(b) Gray color code result image of curved road

(c) Gray color code result image of shadow generating road

Fig. 2. Result image of gray color code from daytime running image in block box of the car.

road (Fig. 2(b)) are extracted smoothly. Especially, Fig. 2(c) shows that the gray color code extracts the shadow region from the left. Figure 3 shows the result of applying the HSV color code to the image. As a result of applying the HSV color code, it can be seen that the lanes of the straight road and the curved road are extracted without any large error detection. In contrast to Figs. 2(c), 3(c) shows that the HSV color code does not extract shadow regions from the left. Finally, Fig. 4 shows the result of extracting representative lines according to the slope value and vertical slope value by calculating the slope value after the additional Hough transform is applied. Because Hough transform is additionally applied to the result of applying the color code, the accuracy may be reduced if the operation speed is low, vehicle speed is very high, or slope of the curved section is abrupt.

Table 1 shows the lane detection results of the proposed algorithm. The resulting image includes a straight section, a curved section, a low-speed section, and a high-speed section. The detection rate obtained using a 1252 image of 960×540 pixel size by Open CV is 94.83% at the average 24 fps image, as shown in Table 1. The result of the non-detection was evaluated as the case where detection failed for one or both lanes and the case where the lane was not detected in the case of false detection. According to this criterion, the evaluation results show that lane

(a) HSV color code result image of straight road (b) HSV color code result image of curved road

(c) HSV color code result image of shadow generating road

Fig. 3. Result image of HSV color code from daytime running image in block box of the car.

detection using the proposed algorithm is accurate in straight line and curved sections and under various road environments. In particular, the representative line detection technique predicted the direction of the vehicle based on the vertex at the front of the road. In addition, the result of the representative line rarely occurred when the error was detected, and it was found that the non-detection was slightly occurred in the curved section of the high-speed running. The gray color code was not detected due to the shadow region.

Table 1. Detection rate of experimental results

Category	Total	Correct	Not detected	False detection	Detection rate
Gray code	418	389	22	7	93.06%
HSV code	430	407	19	4	94.65%
Rep. line	404	391	10	3	96.78%
Total result	1252	1187	51	14	94.83%

(a) Representative line extraction result image of straight road

(b) Representative line extraction result image of curved road

(c) Representative line extraction result image of shadow generating road

Fig. 4. Result image of region of Representative line extraction from daytime running image in block box of the car.

5 Conclusion

In this paper, we propose a lane detection algorithm that guarantees both high accuracy and real-time performance. Using the proposed algorithm, lanes can be accurately detected irrespective of the changes in the environment of the input image. In particular, it is confirmed that the proposed algorithm shows high detection rate by evaluating the straight line and curved sections as well as the low-speed and high-speed sections. The average detection rate was 94.83% during daytime. Based on the detected lane and control point, it is possible to determine whether the vehicle has departed from the lane. When the vehicle runs normally, the value of the control point is updated to a new value as the lane changes. If the control point values are not updated and the existing value is maintained for more than ten frames, it is determined that the vehicle has left the lane, and a warning message is output to inform the driver of the situation.

Acknowledgements. This work was supported by the National Research Foundation of Korea (NRF) grant funded by the Korea government (MSIT) (No. 2017R1C1B5017556).

The simulation results of this paper are performed by Hwan Kim (Seoul Sanggye High School, rlaghks1103@gmail.com) and Yonghee Lee (Seoul Sanggye High School, Information Instructor, L7419@naver.com).

References

1. Aziz, M., Prihatmanto, A., Hindersah, H.: Implementation of lane detection algorithm for self-driving car on toll road cipularang using Python language. In: International Conference on Electric Vehicular Technology, Paris, pp. 144–148 (2017)
2. Park, H., Park, S., Kim, E.: A road condition-based routing and greedy data forwarding algorithm for VANETs. Ad Hoc Sens. Wirel. Netw. **33**, 301–3019 (2016)
3. Paula, M., Jung, C.: Automatic detection and classification of road lane markings using onboard vehicular cameras. IEEE Trans. Neural Netw. Learn. Syst. **28**, 690–703 (2017)
4. Uchida, N., Ishida, T., Shibata, T.: Delay tolerant networks-based vehicle-to-vehicle wireless networks for road surveillance systems in local areas. Int. J. Space-Based Situated Comput. **6**, 12–20 (2016)
5. Bente, T., Szaghalmy, S., Fazekas, A.: Detection of lanes and traffic signs painted on road using on-board camera. In: International Conference on Future IoT Technologies, Eger Hungary, pp. 1–7 (2018)
6. Ito, K., Hirakawa, G., Arai, Y., Shibata, Y.: A road condition monitoring system using various sensor data in vehicle-to-vehicle communication environment. Int. J. Space-Based Situated Comput. **6**, 21–30 (2016)
7. Zhang, Y., Wang, J., Wang, X., Dolan, J.: Road-segmentation-based curb detection method for self-driving via a 3D-LiDAR sensor. IEEE Trans. Intell. Transp. Syst. **99**, 1–11 (2018)
8. Li, X., Liu, J., Li, X., Li, H.: A reputation-based secure scheme in vehicular ad hoc networks. Int. J. Grid Util. Comput. **6**, 83–90 (2015)
9. Wang, C., Huang, S., Fu, L.: Driver assistance system for lane detection and vehicle recognition with night vision. In: International Conference on Intelligent Robots and Systems, Edmonton, Canada, pp. 3530–3535 (2005)
10. Satzoda, R., Sathyanarayana, S., Srikanthan, T., Sathyanarayana, S.: Hierarchical additive hough transform for lane detection. IEEE Embed. Syst. Lett. **2**, 23–26 (2010)
11. Li, J., Mei, X., Prokhorov, D., Tao, D.: Deep neural network for structural prediction and lane detection in traffic scene. IEEE Trans. Neural Netw. Learn. Syst. **28**, 690–703 (2017)
12. Hanna, M., Kimmel, S.: Current US federal policy framework for self-driving vehicles: opportunities and challenges. Computer **50**, 32–40 (2017)
13. Cheng, H., Yu, C., Tseng, C., Fan, K., Hwang, J., Jeng, B.: Environment classification and hierarchical lane detection for structured and unstructured roads. IET Comput. Vis. **4**, 37–49 (2010)
14. Huang, R., Chang, B., Tsai, Y., Liang, Y.: Mobile edge computing-based vehicular cloud of cooperative adaptive driving for platooning autonomous self driving. In: International Symposium on Cloud and Service Computing, Kanazawa, Japan, pp. 32–39 (2017)
15. Ozgunalp, U., Xiao, R., Dahnoun, A.: Multiple lane detection algorithm based on novel dense vanishing point estimation. IEEE Trans. Intell. Transp. Syst. **18**, 621–632 (2017)

16. Yoo, J., Lee, S., Park, S., Kim, D.: A robust lane detection method based on vanishing point estimation using the relevance of line segments. IEEE Trans. Intell. Transp. Syst. **18**, 3254–3266 (2017)
17. Chen, S., Shang, J., Zhang, S., Zheng, N.: Cognitive map-based model: toward a developmental framework for self-driving cars. In: International Conference on Intelligent Transportation Systems, pp. 1–8. MIT, Cambridge (2017)
18. Paula, M., Jung, C.: Automatic detection and classification of road lane markings using onboard vehicular cameras. IEEE Trans. Intell. Transp. Syst. **16**, 3160–3169 (2015)
19. Nugraha, B., Fahmizal S.: Towards self-driving car using convolutional neural network and road lane detector. In: International Conference on Automation, Cognitive Science, Optics, Micro Electro–Mechanical System, and Information Technology, Jakarta, Indonesia, pp. 65–69 (2017)

Design and Implementation of Cognitive System of Children's Education Based on RFID

Hongyu Gan[1], Chenghao Wu[1], Jie Xu[1], Peng Li[2,3], and He Xu[2,3](✉)

[1] Bell Honors School, Nanjing University of Posts and Telecommunications, Nanjing, China
{q16010109,q16010217,q16010304}@njupt.edu.cn
[2] School of Computer Science, Nanjing University of Posts and Telecommunications, Nanjing, China
{lipeng,xuhe}@njupt.edu.cn
[3] Jiangsu High Technology Reserarch Key Laboratory for Wireless Sensor Networks, Nanjing, China

Abstract. This paper mainly introduces the creative design of a children's cognitive system based on RFID. The electronic tag is affixed to the real learning object, and the card reader is carried by the children. The main functions to be realized are vivid images, which give children quick and vivid knowledge of objects. Through the education of physical objects that can be touched, they can be vividly visualized, and children can be made to know objects vividly. In addition, based on our design concept, which connects cognition with material objects, it can actively mobilize children's various sensory systems, so that children can have a deeper understanding of the learning objects and reduce the cost of education.

1 Introduction

Vygoski, a famous cognitive educator, has raised the question of the best time limit for children's learning. Teaching before or after this time will be blind and futile, and will have a negative impact on children's intellectual development. So the cognitive education of children is very important.

With the development of modern society, people are becoming more and more aware of the importance of children's cognitive education. At present, the cognitive education of children is mainly realized by ordinary family education or inviting family education to send children to educational institutions. Parents do not have enough time to educate their children, and the other method (such as finding home teacher) is expensive, and it is only a bit stiff about children's cognitive education. If an unprofessional teacher or an educational institution comes, perhaps it will have a negative impact on children's cognitive education. The dot-reading machine on the market only allows children to recognize objects on paper, and cannot teach children in daily life. 3–4 years old children have not yet mastered a certain memory method. You cannot expect to read books to have a profound understanding of things, and it is also not

L. Barolli et al. (Eds.): IMIS 2018, AISC 773, pp. 448–455, 2019.
https://doi.org/10.1007/978-3-319-93554-6_43

going to arouse children's interests. Compared with reading machines and adult dictation, The tangible and vivid things that can be touched are more easily remembered by children naturally. Therefore, we intend to combine cognitive education with real objects to study a cognitive education system for children based on RFID.

This system of children's cognitive education is very convenient, not only to save a lot of money, but also to do it at home, so that children can get cognitive education in their daily lives. It can also protect children's safety to a certain extent, and this kind of education based on family background is helpful to increase parent-child interaction. There are no similar products in the market at present. Stick the chip to a child's cognitive object such as an apple, just put the chip close to the chip on the apple, which can automatically make a sound such as "this is an apple!". This teaching method can not only improve children's cognitive ability, but also actively mobilize the children's various sensory systems, so that children have a deeper understanding of the learning object.

In the field of RFID, most people study the location of RFID and the security of RFID [1, 2]. While in the field of children's cognitive education, the combination of RFID and children's cognitive education is rare and few. RFID is used to record attendance in many applications [3–5]. In this paper, we mainly introduce the creative design of a children's cognitive system based on RFID. The electronic tag is affixed to the real learning object, and the card reader is carried by the children. The main functions to be realized are vivid images, which give children quick and vivid knowledge of objects. Through the education of physical objects that can be touched, they can be vividly visualized, and children can be made to know objects vividly. In addition, based on our design concept, which connects cognition with material objects, it can actively mobilize children's various sensory systems, so that children can have a deeper understanding of the learning objects and reduce the cost of education.

2 RFID Systems

RFID is an acronym of Radio Frequency Identification. It is a contactless automatic identification technology. It can identify objects automatically by means of radio frequency signals, which can quickly track objects and exchange data, and can identify multiple tags at the same time. The operation is fast and convenient. RFID is the successor of the traditional barcode technology. It is called "electronic tag" or "radio frequency tag". Figure 1 shows the RFID systems modules.

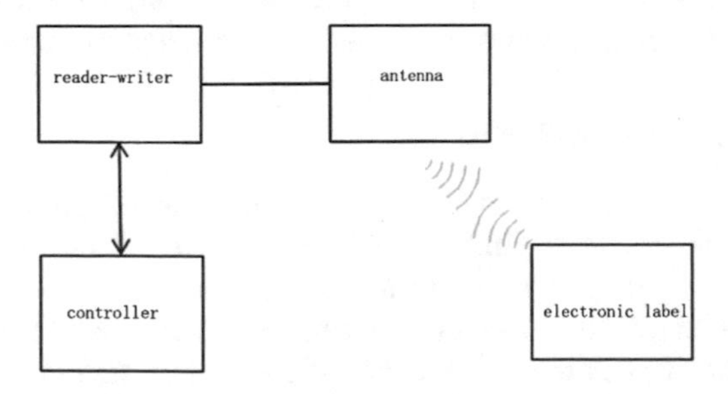

Fig. 1. RFID systems

The most basic RFID system consists of three parts: tag, which consists of coupling elements and chips, each tag has a unique electronic code, attached to the object to identify the target object; A device for reading (and sometimes writing) tag information, designed to be hand-held and fixed; and antenna, which transmits radio frequency signals between tags and readers.

The work flow of RFID system is as follows: the reader transmits the RF signal of a certain frequency by transmitting antenna, when the RF card enters the working area of the transmitting antenna, the inductive current is generated, and the RF card obtains the energy to be activated; The RF card transmits information such as its own code through the built-in transmitting antenna; the system receives the carrier signal from the antenna and transmits it to the reader through the antenna regulator. The reader demodulates and decodes the received signals and sends it to the backstage system for related processing; the main system judges the legitimacy of the card according to logical operations and makes corresponding processing and control for different settings. Then it issues an instruction signal to control the action of the actuator.

3 Overall Design Scheme

The cognitive system of children's education based on RFID is divided into two parts: one is the hand-held part of the card reader, which is held by the child, the other part is the electronic label. The electronic tag on the object is identified by a reader in the hands of a child through radio frequency identification, which reads and feeds back to the development board and calls its corresponding information. When a child puts the reader near an actual object with an electronic tag, such as an apple, when the label has been set in advance, The electronic tag makes the following sound: "this is an apple." Children can pick up the apple's feelings, or they can smell them on their noses. They can use all kinds of feelings to find out the sound of the apple. The reader can make the sounds of Chinese characters. A reader can read out information from multiple tags. At the same time, the system is designed for the convenience of users, and also includes the ability of parents to modify and expand the voice content on their own. Parents only need to open the mobile phone app, change the corresponding content of the label, can use the label on other objects to achieve the effect of reuse.

The device of children's cognitive education system based on RFID includes plastic shell, speech module and RFID identification module, development board, voltage stabilized power supply module, label. Each module is connected with the Arduino development board. Figure 2 shows the system block diagram of the device.

3.1 Arduino Development Board

Arduino is a convenient and easy to use open source electronic prototype platform. It is built on the open source code simple I/O interface. And it has a processing/wiring development environment similar to Java C language. It mainly includes two main parts: the hardware part is the Arduino circuit board which can be used to make circuit connection; The other is Arduino IDE, the programming environment on computer. The code can be written in IDE and the program is uploaded to the Arduino circuit

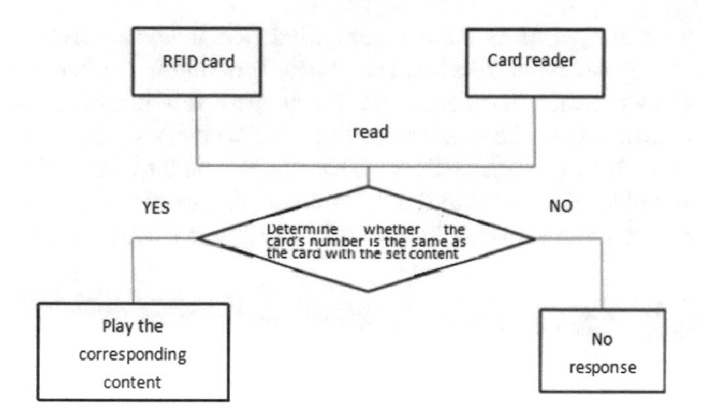

Fig. 2. Block diagram of cognitive system for children's education

board, and the program tells Arduino what to do. Arduino can sense the environment through a variety of sensors, feedback by controlling lights, motors and other devices, and affect the environment. The microcontrollers on the board can be programmed and compiled into binary files through the Arduino programming language. Arduino is programmed through the Arduino programming language (wiring-based) and the Arduino development environment (Processing-based). A Arduino based project can contain only Arduino and other software running on a PC. They communicate with each other (such as Flash, processing, MaxMSPs).

3.2 RFID Module

In order to realize the social function of the design, the RFID module and tag. RFID module are directly connected to the Arduino development board. The tag posted on

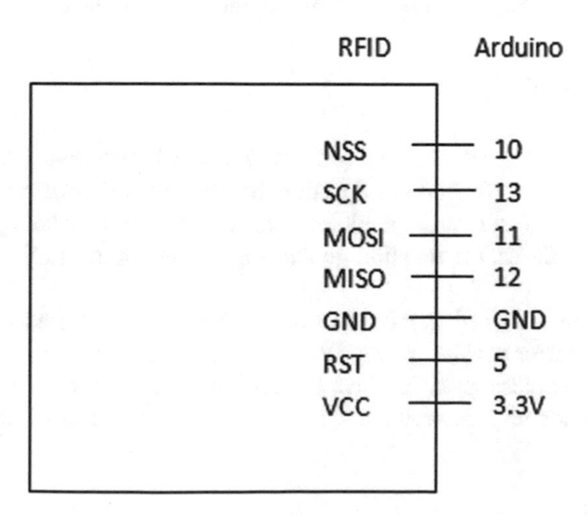

Fig. 3. Connection diagram between RFID card reader and Arduino development board

the object to be recognized is mainly composed of wireless receiver, antenna, data processing module and other functional modules. Working in the 2.4 GHz – 2.5 GHz microwave segment with built-in antennas, it can effectively identify RFID tags with a radius of less than 4 cm. The tags are coded, The wireless signal sent by the tag is received and decoded by the reader through the antenna on the card reader, and then the received information can be sent to the Arduino development board for processing via the connection. Figure 3 shows RFID module is connected with Arduino module.

3.3 Speech Module

In order to realize the speech output function of the system, the speech module is adopted. The speech module is composed of TTS serial port SYN6288 speech synthesis module and speaker. Arduino development board calls the corresponding contents according to the encoding read by RFID to the TTS serial port SYN6288 speech synthesis module. The synthesis module synthesizes the sound output from the loudspeaker. Figure 4 shows the connections between speech module and Arduino.

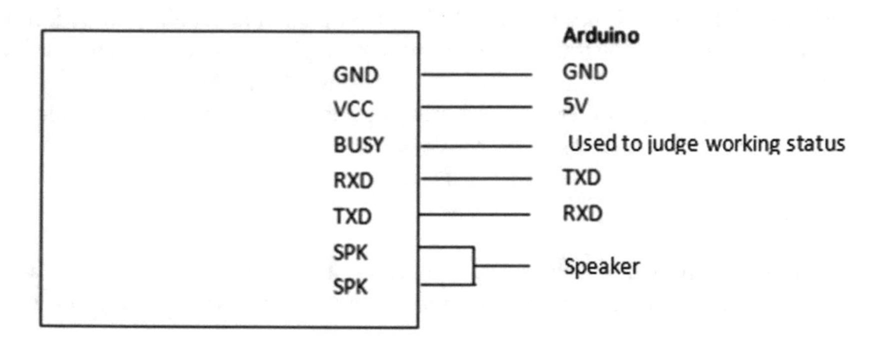

Fig. 4. Speech synthesis module with Arduino

4 Experiment Tests

The function generating system can convert the word to output into the form of a function in the code. If the code in Arduino has a function corresponding to the tag encoding, the speech synthesis module outputs the corresponding speech of the function. Figure 5 shows the function generating system and Fig. 6 show the realized hardware.

The RFID chip is affixed to the cognitive object, and the corresponding content is set on the computer or mobile phone. When the card reader is placed near the RFID chip, the speech synthesis module plays the content which is set in advance. Figure 7 shows that it can identify umbrella and Fig. 8 shows that it can identify water bottle.

Fig. 5. Function generating system

Fig. 6. The realized hardware

Fig. 7. Identify umbrella experiment

Fig. 8. Identify water bottle experiment

5 Conclusions

There are a lot of products on the market today about children's cognition, such as story books, picture books, reading pens, storytelling robots, and better intelligent robots are expensive. However, there are no related products that directly put the actual objects into the education market. Our designed system is small and light, and children can learn in their own hands without feeling burdensome. The shell is mainly round in shape to avoid the damage caused by the corner shape, and the products on the market are of good quality and low price. Our educational cognitive system has a low production cost, but it can also give children tangible and vivid teaching that can be touched, mobilize children's various senses, and enable children to have a deeper understanding of what they are learning about; convenience is also a major advantage. For parents who want to modify the content of voice, who can directly input voice content on the mobile phone APP; the children's education cognitive system caters to the concept of green market. RFID tags can effectively avoid external pollution, which can work in all kinds of harsh environment, long shelf life, and the label can be reused. As long as we modify the voice content, we can attach labels to other objects. The cognitive system of children's education based on RFID takes the combination of actual objects and cognitive education as the design idea, and fully calls on the various senses of children, so that children can learn different aspects of learning objects.

Acknowledgments. This work is financially supported by the National Natural Science Foundation of P. R. China (No.: 61672296, No.: 61602261), Scientific & Technological Support Project of Jiangsu Province (No.: BE2015702, BE2016185, No.: BE2016777), China Postdoctoral Science Foundation (No.: 2014M561696), Jiangsu Planned Projects for Postdoctoral Research Funds (No.: 1401005B), Postgraduate Research and Practice Innovation Program of Jiangsu Province (No.: KYCX17_0798), and NUPT STITP (No.: SYB2017019).

References

1. Xu, H., Ding, Y., Li, P., Wang, R., Li, Y.: An RFID indoor positioning algorithm based on bayesian probability and k-nearest neighbor. Sensors **17**(1806), 1–17 (2017)
2. Xu, H., Ding, J., Li, P., Zhu, F., Wang, R.: A Lightweight RFID mutual authentication protocol based on physical unclonable function. Sensors **18**, 760 (2018)
3. Tan, P., Wu, H., Li, P., Xu, H.: Teaching management system with applications of RFID and IoT Technology. Educ. Sci. **8**, 26 (2018)
4. Liu, C., Fan, K., Pan, Y., et al.: Design of and research on an attendance system based on RFID and WSN technologies for the rail transportation industry. In: International Conference on Electrical and Information Technologies for Rail Transportation, pp. 1013–1024. Springer, Singapore (2017)
5. Zaman, H.U., Hossain, J.S., Anika, T.T., et al.: RFID based attendance system. In: Proceedings of 2017 8th International Conference on Computing, Communication and Networking Technologies (ICCCNT), pp. 1–5. IEEE (2017)

Delay Optimization for Mobile Cloud Computing Application Offloading in Smart Cities

Shan Guo(✉), Ying Wang, Sachula Meng, and Nan Ma

State Key Laboratory of Networking and Switching Technology,
Beijing University of Posts and Telecommunications,
Beijing 100876, People's Republic of China
{guoshan33,wangying}@bupt.edu.cn

Abstract. In smart cites, more and more smart mobile devices (SMDs) have many computation-intensive applications to be processed. Mobile cloud computing (MCC) as an effective technology can help SMDs reduce their energy consumption and processing delay by offloading the tasks on the distributed cloudlet. However, due to long transmission delay resulting from the unstable wireless environment, the SMD may be out of the serving area before the cloudlet transmits responses to the user. Thus, delay is a crucial problem for the MCC offloading. In this paper, we consider a multi-SMDs MCC system, where each SMD having an application to be offloaded asks for computation offloading to a cloudlet. In order to minimize the total delay of the SMDs in the system, we jointly take the offloading cloudlet selection, wireless access selection, and computation resource allocation into consideration. We formulate the total delay minimization problem as a mixed integer nonlinear programming (MINLP) problem which is NP-hard. We propose an improved genetic algorithm to obtain a local optimal result. Simulation results demonstrated that our proposal could effectively reduce the system delay.

1 Introduction

With the continuous growth of the population, the world urbanization level has also been continuously increasing [1]. In order to deal with such problems as traffic congestion and resource shortage in cities due to high population density, the city should become smarter to increase safety, efficiency, productivity and quality of citizens life. This goal can be realized by using information and communication technology (ICT). This implies that a great number of smart mobile devices (SMDs) will be used in the city environment and generate massive data to be processed by SMDs [2]. However, SMDs have resource limitations, such as battery limitation and other hardware constraints. If all the data are processed on the SMDs, they will consume a great deal of energy and storage. To solve this problem, we can rely on mobile cloud computing (MCC) as one of the communication technologies that enable smart cities.

L. Barolli et al. (Eds.): IMIS 2018, AISC 773, pp. 456–466, 2019.
https://doi.org/10.1007/978-3-319-93554-6_44

MCC provides abundant computation resource and storage resource in cloud comparing the resource-limited SMDs to improve the user experience. One of the techniques adopted in MCC is application offloading [3] that SMDs transmit the task to the cloud via wireless networks for data gathering, processing, and storage. It can effectively reduce energy consumption and processing delay of SMDs.

Different offloading schemes have been proposed in [4–7]. They either optimize the energy consumption of SMDs or minimize the bandwidth cost of offloading applications to remote cloud. Some of them only use a deadline to limit the execution delay, i.e., each mobile application must be executed by a given delay deadline. A few works, such as [8–11], consider the tradeoff between energy consumption and execution delay by taking the different factors into account. In [10,11], execution delay includes the uplink transmission time and the processing waiting time. However, due to large size of returned data of some services in smart cities, downlink transmission time cannot be ignored so that we take it into account. Besides, on account of the unstable wireless environment and the mobility of SMDs in smart cities, the response of some computation-intensive and delay-sensitive applications may not return to the users in time so that it will have negative impact on quality of experiment (QoE). Thus, minimizing the execution delay itself is particularly important to improve the quality of services, e.g., interaction gaming, augmented reality (AR), virtual reality (VR), vehicle-mounted devices, etc. In this paper, we optimize the system delay in order to ensure that users can receive the offloading result as soon as possible in case of moving out of the cloudlet service area. Wherein, where to offload and how to offload are the crucial problems to find the shortest system delay.

How to offload means that which kind of wireless access methods the SMD will choose. In order to increase the peak transmission speed, system connection capacity and enlarge coverage area in smart cities, a SMD usually has multiple wireless access methods, such as WiFi and cellular networks, and each connection performs differently in terms of speed and energy consumption.

Where to offloading means which cloudlets the SMD choose to offload the task. Cloudlets which decrease the network latency and accelerate the computing can be considered as a powerless public cloud and reduce the impact of resource constraints by allocating resources reasonably. The amount of allocated resources and the distance from cloudlets to SMDs will affect the selection of the cloud.

Therefore, appropriate transmission method and offloading place can effectively reduce the delay. In this paper, we consider a MCC system consisting of several cloudlets and multiple mobile users, where each SMD having an application to be offloaded asks for computation offloading to a cloudlet. In order to minimize the total delay of the SMDs in the system, we jointly take the offloading cloudlet selection, wireless access selection, and computational resource allocation into the consideration. We formulate the total delay minimization problem as a mixed integer nonlinear programming (MINLP) problem, which is subject to specific SMD energy consumption constraints. While the MINLP problem is proved to have a relatively high complexity to deal with, we proposed an

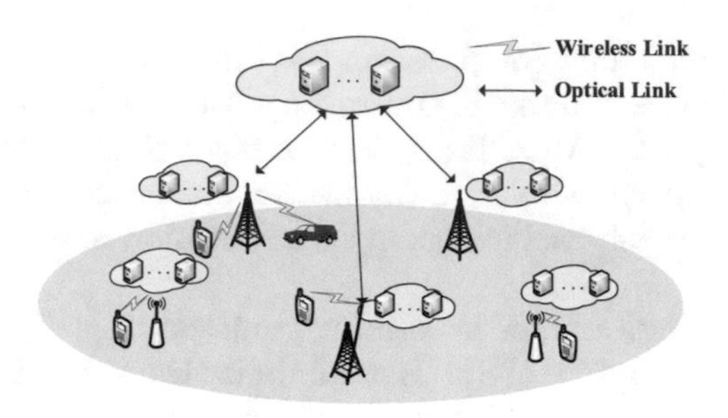

Fig. 1. Mobile cloud computing offloading system model in heterogeneous network

improved genetic algorithm (GA) to obtain a local optimal result. Simulation results demonstrated that our proposal could effectively reduce the system delay.

The rest of the paper is organized as follows. In Sect. 2, we describe the system model and present the problem formulation. In Sect. 3, we provide details of the improved genetic algorithm. We present simulation results in Sect. 4 and conclude in Sect. 5.

2 System Model and Problem Formulation

We assume that there are N SMDs in a region as shown in Fig. 1, denoted as $\mathscr{N} = \{1, 2, \cdots, N\}$. For each user, there is one mobile application to be executed. We use A_i to denote the set of mobile application at device i.

$$A_i = \left(D_{in}^i, S_i, D_{out}^i, E_i^{th}, L_i(m,n)\right), \tag{1}$$

where D_{in}^i is the number of exchanged data for executing the mobile application, S_i is the number of requesting CPU cycles to execute the application. Therein S_i can be calculated based on D_{in}^i, according to [?]. D_{out}^i denotes the returned file size of the offloading application. E_i^{th} denotes the corresponding energy constraint due to the limited energy of SMDs. If executing energy consumption is beyond the E_i^{th}, the SMD should change an executing way. $L_i(m,n)$ denotes the location of the SMD i.

There are K cloudlets which can be used to process the offloading data in our system. All of K cloudlets are homogeneous. Due to the limited processing capacity for every cloudlet, the maximum CPU frequency of the kth cloudlet is F_{CL}^k, $k = (1, 2 \cdots \cdots K)$.

We use α_i to indicate whether the application on the SMD i is offloaded. If the SMD decides to offload the application to the cloudlet, $\alpha_i = 1$. Otherwise, $\alpha_i = 0$. The final decision is

$$\alpha = \{\alpha_1, \alpha_2, \cdots \cdots, \alpha_N\}. \tag{2}$$

2.1 Local Execution Model

We assume that the CPU frequency of the SMD i is f_{li}. The local executing latency can be expressed as:

$$T_{li} = \frac{S_i}{f_{li}}. \tag{3}$$

According to [8], the energy consumption of local processing is

$$E_{li} = \kappa(f_{li})^2 S_i, \tag{4}$$

where κ is the effective switched capacitance depending on the chip architecture. We set $\kappa = 10^{-11}$ so that energy consumption is consistent with the measurements in [12].

2.2 Cloud Offloading Model

In this paper, communication resources include WiFi and cellular access. And we assume that the cellular network is seamless covered, but users cannot connect with WiFi all the time. Different access bandwidth and various distance from user to access point (AP) or base station (BS) will lead to diverse throughout. The throughout of the i-th user can be expressed as:

$$\omega_i = \frac{BW_i}{n} \log_2 \left(1 + \frac{SNR}{{d_i}^2}\right), \tag{5}$$

where BW_i is channel bandwidth of the access method the ith user choose, n denotes the number of SMD that the base station (BS) or the access point (AP) connect. SNR is a reference Signal-to-Noise Ratio value and d_i is the distance between the SMD i and the access node. For simplicity, we do not consider the effects of channel shadow fading, which has no effect on our simulation.

If the application is executed on the cloudlet, the ${T_C}^i$ can be defined as the sum of the wireless transmission time to the cloudlet, the processing time on the cloudlet and the response reception time by the user.

The uplink and downlink transmission time between SMD i and the cloudlet k can be respectively denoted as:

$${T_{tr}}^i = \frac{D_{in}^i}{{\omega_i}^{up}}, \tag{6}$$

$${T_{re}}^i = \frac{D_{out}^i}{{\omega_i}^{down}}. \tag{7}$$

The processing time on the cloudlet can be expressed as:

$${T_{execution}}^i = \frac{S_i}{f_{CL}^k}, \tag{8}$$

where f_{CL}^k is the computation resource allocated by the cloudlet k to the ith user. Though the cloudlet has a relatively strong computing ability, it still has a limited resource. Therefore, f_{CL}^k can not exceed the maximum CPU frequency F_{CL}^k.

On the basis of three parts time cost above, we can get the total amount of time and energy consumption spent.

$$T_C{}^i = \frac{D_{in}^i}{\omega_i{}^{up}} + \frac{S_i}{f_{CL_k}^i} + \frac{D_{out}^i}{\omega_i{}^{down}}. \tag{9}$$

$$E_C{}^i = P_{tr}\frac{D_{in}^i}{\omega_i{}^{up}} + P_{execution}\frac{S_i}{f_{CL_k}^i} + P_{re}\frac{D_{out}^i}{\omega_i{}^{down}}, \tag{10}$$

where P_{tr}, $P_{execution}$, P_{re} are SMD power consumption of transmission, waiting and receive data, respectively.

2.3 Problem Formulation

The objective of the paper is to minimize the total system delay under specified energy consumption constraints. To this end, the problem can be formulated as

$$T = \min \sum_{i=1}^{N} (\alpha_i T_C{}^i + (1-\alpha_i)T_{li}). \tag{11}$$

$$s.t. \quad C1: 0 \le f_{CL_k}^i \le F_{CL}^k, \tag{12}$$

$$C2: \sum_{i=1}^{N} f_{CL_k}^i \le F_{CL}^k, \tag{13}$$

$$C3: \alpha_i E_C{}^i + (1-\alpha_i)E_{li} \le E_i^{th}, \tag{14}$$

$$C4: \alpha_i \in \{0,1\}. \tag{15}$$

$C1$ is the constraint of the available computing resource to be allocated for user i.

$C2$ represents that the total allocated computing resource cannot exceed F_{CL}^k at the cloudlet.

$C3$ indicates that each task A_i must meet the specified energy constraint E_i^{th}.

$C4$ specifies that each SMD processes the task either by local execution or by offloading execution.

3 Genetic Algorithm Based on MCC Offloading

GA is a metaheuristic inspired by the process of natural selection. GA are commonly used to generate high-quality solutions to optimization and search problems by relying on bio-inspired operators such as mutation, crossover and selection [13]. The proposed offloading decision problem has been proved a NP-hard problem. GA can obtain near optimal solution in a short time for such a problem [14]. When comparing with other traditional heuristic algorithms, GA can accelerate the convergence speed, enhance global search capabilities and robustness to prevent trapping into local optimum [15]. Therefore, we use the GA to solve the optimization problem. The genetic algorithm flow chart has displayed as shown in Fig. 2.

In this paper, every chromosome which is encoded in a string of integers and real number corresponds an offloading strategy. The population can be defined as

$$P = \{C_1, C_2, \cdots, C_j, \cdots, C_S\}, \tag{16}$$

where S is the population size. C_j is one of the chromosome correspond one of the solutions. It can be denoted as

$$\mathrm{C}_j = \{x_{j1}, x_{j2}, ..., x_{juser}, y_{j1}, y_{j2}, ..., y_{juser}\}, \tag{17}$$

where x_{ji} is the cloudlet selection of the i-user, y_{ji} is the computing resource allocation for the i-user.

3.1 Fitness Evaluation

Fitness evaluation is calculated according to the objective function which is the optimal objective. The fitness values are computed for the C_j chromosome by substituting selected cloudlet and allocated computing resource into the capacity in (11). Meanwhile, we also normalize the computing resource to ensure that they satisfy the constraints (12) and (13), respectively.

$$C_R{}'(i) = C_R(i) \times \frac{C_R(i)}{\sum\limits_{j=1}^{U} C_R(j)}. \tag{18}$$

The $C_R{}'(i)$ determined above is also checked whether it has satisfied the constraint (14) or not. If not, the corresponding chromosome will not be selected into the mating pool.

3.2 Selection

Fitness of chromosomes in the population are calculated and chromosomes with higher fitness have more probability to be selected. Then, we employ a roulette wheel function to select chromosomes for generating offspring of new generation randomly.

3.3 Reproduction

The crossover step begins by randomly generating a $(User + 1) \times 1$ crossover mask sequence composed by 1 and 0. The first $User$ elements of the mask are for the cloudlet selection, and the remaining element is for the computation resource allocation. To satisfy the constraint (15), when the cloudlet selection is $\{1, \cdots\cdots CL\}$, the decision is offloading, $\alpha_i = 1$. When the cloudlet selection is

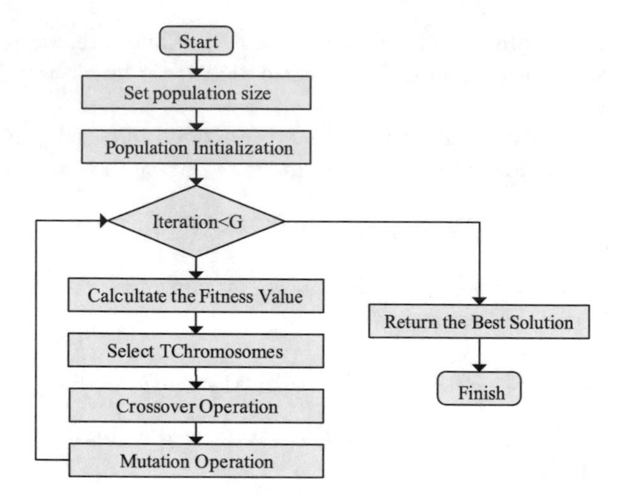

Fig. 2. Improved genetic algorithm flow chart

0, the decision is located processing, $\alpha_i = 0$. This crossover operation will repeat until the size of the population reaches S.

The mutation operation is also divided into the integer mutation for the cloudlet selection and the mutation for the computation resource allocation. First, for each chromosome, we create a 2×1 mutation mask sequence comprising of 1 and 0 generated according to the mutation probability. The mutation probability is not always constant. A higher quality individual will be assigned with a higher mutation probability.

4 Simulation Results

In order to study the performance gain of the proposed solution, we carried out a set of simulation studies using MATLAB under different parameter settings. The improved GA is implemented and compared with Truncated GA. In the selection of Truncated GA, which only reserves T parent chromosomes with higher fitness values in order to preserve better chromosomes to produce better offsprings. T parent chromosomes are put in a mating pool and two parent chromosomes will be randomly selected. However, in our improved GA, fitness of chromosomes in the population are calculated and chromosomes with higher fitness have more probability to be selected. It will prevent premature convergence.

4.1 Simulation Setup

We adopt the mobile device parameters from [12], which is based on a Nokia smart device. We assume that all the SMDs in the system have the same CPU frequency and processing power. We consider the x264 CBR encode application

Table 1. Simulation parameters

Parameters	Values
Number of users	10
Number of cloudlets	5
SMD CPU frequency	500M cycles/s
SNR	30 dB
SMD transmission/receiving power	1.42×10^{-7} J/bit
Cloudlet CPU frequency	10G cycle/s
Input data size	10–30 MB
Output data size	1–3 MB

Table 2. Genetic algorithm settings

Parameters	Values
Population	100
Crossover probability	0.7

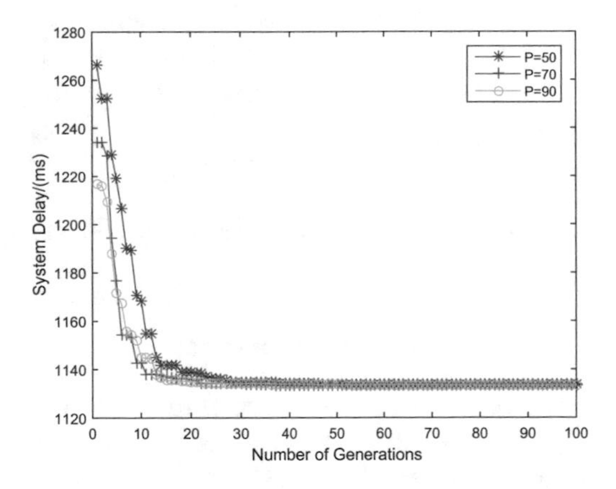

Fig. 3. System delay against the number of generations in the improved GA

[12], which requires 1900 cycles/byte, i.e., $S_i = 1900 \times C_i$. Simulation parameters and genetic algorithm settings are summarized in Tables 1 and 2, respectively.

The total transmission bandwidth between the mobile users and the cloudlet is set according to the wireless access mode, with no additional limit on the uplink or downlink.

Since the size of each task is generated randomly, and the genetic algorithm can only find the local optimal solution, the result will have some differences every time. Thus, in the performance comparison, we compare the average of the obtained 100 results.

4.2 Impact of Iteration Number

From Figs. 3 and 4, we can see that the system delay of three iterations tend to a fixed value with the increase of generations. In other words, the minimum system delay value continues to converge with the increase of generation. It is stabilized in the about 30th generation. The improved genetic algorithm has better convergence by comparing the two figures. When the population is different, the result basically converged to the same optimal value. However, compared with the convergence of the Truncated GA is not stable, when the population is different, there is a certain error in the local optimal solution.

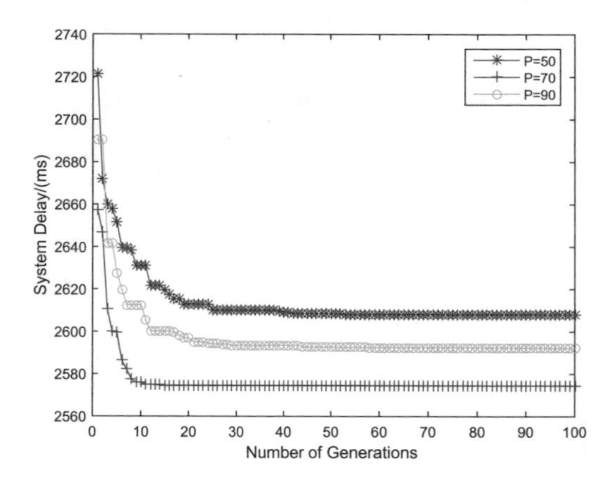

Fig. 4. System delay against the number of generations in the Truncated GA

4.3 User Number Impact on Performance

In Fig. 5, we compare the performance of the improved genetic algorithm and the Truncated GA by changing the number of users in the system from 0 to 20. As the number of users increases, the system delay also increases continuously. However, from the curves in the figure, it clearly shows that when the number of users is constant, the system delay of the improved algorithm obtained is shorter than the Truncated GA system delay. As users grow, the system delay of the improved GA increases more slowly. This shows that the improved algorithm can find a better local optimal solution than the Truncated GA.

4.4 Impact of Iteration Number

In Fig. 6, we compare the improved performance of the improved algorithm and the Truncated GA by changing the cellular network bandwidth. As the bandwidth of the cellular network increases, the system delay also decreases. However, as we can clearly see from the two curves in the figure, when the bandwidth increases, the system delay of the improved algorithm is shorter than the system

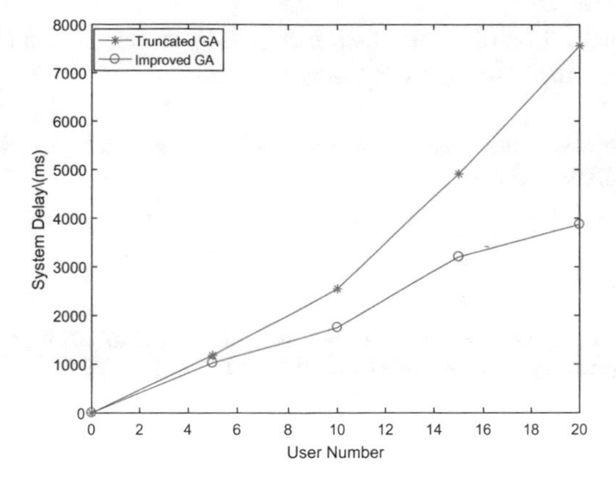

Fig. 5. System delay against number of users

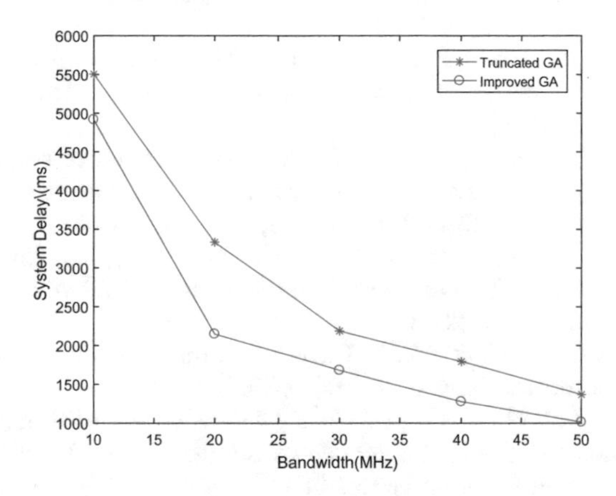

Fig. 6. System delay against bandwidth

delay of Truncated GA and decreasing speed is more obvious. From this point of view, the local optimal solution obtained by the improved algorithm is better than the Truncated GA with the changing bandwidth.

5 Conclusion

In this paper, we introduce a multi-SMDs MCC system, where each SMD having an application to be offloaded asks for computation offloading to a cloudlet. In order to minimize the total delay of the SMDs in the system, we jointly take the offloading cloudlet selection, wireless access selection, and computational resource allocation into consideration. We formulate the total delay minimization problem as a MINLP problem, which is subject to specific SMD energy consumption constraints. While the MINLP problem is proved to have a relatively high complexity to deal with, we propose an improved genetic algorithm to obtain a local optimal result. Simulation results demonstrated that our proposal could effectively reduce the system delay.

Acknowledgements. This paper is supported by the National Key Project under Grant NO. 2017 ZX03001009.

References

1. Mazza, D., Tarchi, D., Corazza, G.E.: A unified urban mobile cloud computing offloading mechanism for smart cities. IEEE Commun. Mag. **55**(3), 30–37 (2017)
2. Gharaibeh, A., Salahuddin, M.A., Hussini, S.J., Khreishah, A., Khalil, I., Guizani, M., Al-Fuqaha, A.: Smart cities: a survey on data management, security, and enabling technologies. IEEE Commun. Surv. Tutorials **19**(4), 2456–2501 (2017)
3. Zhou, B., Dastjerdi, A.V., Calheiros, R.N., Srirama, S.N., Buyya, R.: mCoud: a context-aware offloading framework for heterogeneous mobile cloud. IEEE Trans. Serv. Comput. **10**(5), 797–810 (2017)
4. Zhang, W., Wen, Y., Guan, K., Dan, K., Luo, H., Wu, D.O.: Energy-optimal mobile cloud computing under stochastic wireless channel. IEEE Trans. Wirel. Commun. **12**(9), 4569–4581 (2013)
5. Mukherjee, A., Gupta, P., De, D.: Mobile cloud computing based energy efficient offloading strategies for femtocell network. In: 2014 Applications and Innovations in Mobile Computing (AIMoC), pp. 28–35, February 2014
6. Shu, P., Liu, F., Jin, H., Chen, M., Wen, F., Qu, Y., Li, B.: eTime: energy-efficient transmission between cloud and mobile devices. In: 2013 Proceedings IEEE INFOCOM, pp. 195–199, April 2013
7. Liu, D., Khoukhi, L., Hafid, A.S.: Data offloading in mobile cloud computing: a Markov decision process approach. In: IEEE ICC (2017)
8. Wang, J., Peng, J., Wei, Y., Liu, D., Fu, J.: Adaptive application offloading decision and transmission scheduling for mobile cloud computing. In: 2016 IEEE International Conference on Communications (ICC), pp. 1–7, May 2016
9. Mazza, D., Tarchi, D., Corazza, G.E.: A cluster based computation offloading technique for mobile cloud computing in smart cities. In: 2016 IEEE International Conference on Communications (ICC), pp. 1–6, May 2016
10. Chen, M.H., Liang, B., Dong, M.: Joint offloading and resource allocation for computation and communication in mobile cloud with computing access point. In: IEEE INFOCOM 2017 - IEEE Conference on Computer Communications, pp. 1–9 (2017)
11. Guo, S., Xiao, B., Yang, Y., Yang, Y.: Energy-efficient dynamic offloading and resource scheduling in mobile cloud computing. In: IEEE INFOCOM 2016 - the IEEE International Conference on Computer Communications, pp. 1–9 (2016)
12. Miettinen, A.P., Nurminen, J.K.: Energy efficiency of mobile clients in cloud computing. In: Usenix Conference on Hot Topics in Cloud Computing, p. 4 (2010)
13. Genetic Algorithm. https://en.wikipedia.org/wiki/Genetic_algorithm. Accessed 30 Nov 2017
14. Rai, A., Bhagwan, R., Guha, S.: Generalized resource allocation for the cloud. In: Proceedings of 3rd ACM Symposium on Cloud Computing, San Jose, CA, USA, pp. 1–12, October 2012
15. Weise, T.: Global Optimization Algorithms Theory and Application. http://www.it-weise.de/projects/book.pdf. Accessed 30 Nov 2017

Joint User Association and Power Allocation for Minimizing Multi-bitrate Video Transmission Delay in Mobile-Edge Computing Networks

Hong Wang(✉), Ying Wang, Ruijin Sun, Runcong Su, and Baoling Liu

State Key Laboratory of Networking and Switching Technology,
Beijing University of Posts and Telecommunications, Beijing 100876,
People's Republic of China
{hwang,wangying}@bupt.edu.cn

Abstract. Fast-growing video services place higher demands on network performance especially in terms of latency, but the traditional networks architecture with congested backhaul link can no longer meet the requirement. Recently, mobile edge computing (MEC) has become a promising paradigm to achieve low latency performance and can provide multi-bitrate video streaming at the edge of radio access networks (RAN) with the ability of caching and transcoding. In this paper, we consider the scenario of multi-cell MEC networks, where each BS deployed with one MEC server is connected to the core network through the limited-capacity backhaul link. Our goal is to minimize the system delay which includes backhaul transmission delay and wireless side transmission delay. To this end, we propose a collaborative optimization of user-BS association and power allocation strategy with the given cache status. This is a mixed-integer nonlinear programming (MINLP) problem which is NP-hard. Thus we propose an improved genetic algorithm to solve this problem based on the traditional genetic algorithm. Simulation results demonstrate that our proposed algorithm performs better in terms of convergence and can get better solution as compared with traditional genetic algorithm.

1 Introduction

Driven by the rapid increasing of mobile data especially the explosion of on-demand mobile video requirements, both the wired backhaul and the wireless networks have experienced a heavy burden in the last few years. Mobile video service is predicted to account for 72% of the total mobile data traffic by 2020 [1]. Recently, the need for low latency and high definition video is significantly increased. However, the existing network architecture can not meet the demand for delay anymore. On one hand, the bandwidths of backhaul links connecting base stations and the core network are limited [2]. In some scenarios of massive video deliveries, for example, in intensive areas or during peak traffic hours, a lot

L. Barolli et al. (Eds.): IMIS 2018, AISC 773, pp. 467–478, 2019.
https://doi.org/10.1007/978-3-319-93554-6_45

of contents are transmitted through backhaul links at the same time. Therefore, users may perceive long delay due to the congestion in backhaul links [3] and thus the quality of experience (QoE) of users is degraded badly. On the other hand, the time-varying wireless channel has a decisive impact on the wireless transmission rate. So the delay caused by the wireless side can not be ignored. In order to meet the requirements of low latency, mobile edge computing (MEC) is proposed by the standards organization European Telecommunications Standards Institute which is considered as a promising technique by providing ability of context awareness, caching and computing if deployed at the edge of the radio access networks (RAN) [4,5]. It overcomes the drawbacks of mobile cloud computing which could cause long latency and backhaul bandwidth congestion [6]. The benefits of MEC in reducing latency mainly own to its ability of caching and computing.

Firstly, by deploying the content caching capacity in close proximity to mobile users, MEC servers can store popular videos in advance to reduce the redupli-cate videos transmission in peak period if users' requests hit the ratio [7]. In general, the current researches on caching can be clustered into two categories: cache placement strategy in a relatively long time and scheduling strategy in physical layer transmission slot with a given cache state. For the first category, the authors in [8] consider the distributed caching in densely-deployed small base stations (SBSs) and formulate the optimal cache allocation policy. [9] studies the optimization for cache content placement to minimize the backhaul load subject to cache capacity constraints for caching enabled small cell networks. For the second category, the works in [10] study dynamic content centric BS clustering and multicast beamforming with respect to caching status to minimize the weighted sum of backhaul cost and transmit power. As mentioned in [11], the authors study the joint transmission clustering problem to minimize the backhaul data traffic while each users data rate requirement can be satisfied under the limited radio resource blocks.

Secondly, due to the existence of user heterogeneity, different users may request different bitrate versions of the same video. For example, users prefer low-bitrate-version videos to ensure the fluency of playback while watching a sport event video, but for romantic or documentary films, high-bitrate-version videos are often the choices. With the cloud computing capacity, MEC can deliver multi-bitrate video streaming by means of transferring high-definition videos to low-definition videos in the case that a user requests a low-definition video uncached while its corresponding high-definition video is cached in the MEC server, or a user requests high-definition but the wireless channel condition is not good enough [12]. In this case, the video download delay can be significantly decreased so that QoE of users can be significantly improved. With the ability of caching and transcoding, MEC servers can bring great gains to mobile networks, especially in terms of delay. For example, in the investigation of [6], cache placement policy and resources allocation to different BSs are determined to maximize the utility functions of MEC, BS and RAN. In [13], the authors envision a collaborative joint caching and processing strategy for on-demand video streaming in MEC networks, but they do not consider power allocation so as to take wireless transmission delay into consideration.

Motivated by the advantages of MEC, we consider the system delay in the scenario of multi-cell MEC networks, where each BS is deployed with one MEC server. Since the cache contents are usually updated in several hours or even several days, we can assume it remains the same in the period of physical layer transmission slot which is usually on the millisecond scale. In this paper, we assume a cache-enabled system that the video files have already been stored in MEC server according to popularity of zipf distribution [14] and our scheme aims to minimize global system delay which considers both backhaul delay and wireless transmission delay by investigating user association and power allocation strategy.

The rest of this paper is organized as follows: the system model is presented in the aspects of network model, cache model and radio link model in Sect. 2. In Sect. 3, the network transmission delay function and the optimization problem are formulated. In Sect. 4, the formulated problem is analysed to be NP-hard. Then we propose our improved genetic algorithm to solve the problem. The simulation results are presented and discussed in Sect. 5. Finally, Sect. 6 concludes the paper.

2 System Model

2.1 Network Model

In this paper, we assume a mobile edge computing (MEC) assisted video downlink transmission scenario which is illustrated as Fig. 1. The number of BS is N and each BS is equipped with a MEC server which can be denoted as $\mathscr{N} = \{1, 2, ...N\}$. K users to be severed are denoted as $\mathscr{K} = \{1, 2, ...K\}$. Each BS is connected to the core network through a wired backhaul link with limited capacity C_n. The transcoding resource of each MEC server is limited to T and caching capacity is denoted as M.

Since the cache contents are usually updated in several hours or even several days, we can assume it remains the same in the physical layer transmission slot which is usually on the millisecond scale. Let $q_{k,n}$ be a binary decision variable which represents whether the v-th video with bitrate l is requested by user k. We have $q_{k,v_l} = 1$ if the v-th video with bitrate l is requested by user k and $q_{k,v_l} = 0$ otherwise. We assume that each user can only be entirely served by one BS each time. Let $a_{k,n}$ be a binary decision variable which represents whether the k-th user is connected to the n-th base station or not. We have $a_{k,n} = 1$ when the k-th user associates with n-th BS, and $a_{k,n} = 0$ otherwise.

2.2 Cache Model

We set the videos requested by mobile users as $\mathscr{V} = \{1, 2, ...V\}$ and the popularity of all videos follows the Zipf distribution. Without loss of generality, we

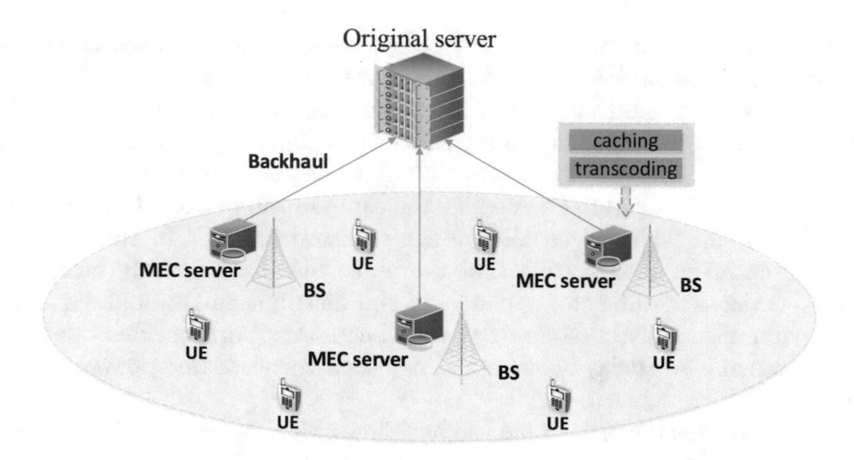

Fig. 1. Network model for video delivery with MEC server.

assume that all videos have the same length. Hence, the size of each video is proportional to its bitrate rate l which is denoted as s_l. For simplicity, we assume only two bitrate versions: low-definition (LD) of which $l = 1$ and high-definition (HD) of which $l = 2$. The MEC can only transcode high-definition videos to low-definition videos. We consider that the cost of transcoding a HD video to LD video each time is denoted as t_l. We have $x_{k,n} = 1$ indicates that a transcoding occurred when k-th user request arrives at n-th MEC server and $x_{k,n} = 1$ otherwise. Without loss of generality, the probability that users request video v follows the Zipf distribution. In addition to the popularity of the videos, users usually have different preference for diverse-class videos. Taking both popularity and user preference into consideration, the probability of user k requesting for video v with version l is expressed as follows:

$$b_{k,v_l} = \frac{v^{-\alpha}}{\sum_{v=1}^{V} v^{-\alpha}} g_{k,v_l}, \tag{1}$$

where $\alpha > 0$ is the Zipf parameter and g_{k,v_l} denotes the k-th user's preference of video v with bitrate l. The final request vector can be generated based on the requesting probability.

2.3 Radio Link Model

The signal-to-interference-plus-noise ratio (SINR) of user k is calculated as follows:

$$SINR_k = \frac{p_{k,n} {h_{k,n}}^2}{\sum_{j \neq n}^{N} p_{k,j} {h_{k,j}}^2 + \sigma^2}, \tag{2}$$

where $h_{k,n}$ is the wireless channel gain between BS n and user k. Let $p_{k,n}$ denote the power that BS n allocate to user k. We assume orthogonal frequency division technology within the same cell, so the formula above only considers the interference between cells. The noise at a user follows Gaussian distribution with a variance σ^2. Then according to the Shannon capacity formula, the transmission rate of user k can be written as :

$$r_k = w \log(1 + SINR_k). \tag{3}$$

The bandwidth of each RB allocated to each user is assumed to be w.

3 Problem Formulation

In this section, we need to calculate the delay for UE u_k to download a specific video. To describe the possible events that need to occupy the backhaul link when a request for video v_l arriving at server n, we introduce the binary variables $\{y_{k,n}^{v_l}, z_{k,n}^{v_l}\} \in \{0, 1\}$, which are explained as follows:

1. The MEC server does not cache the same bitrate video or higher bitrate video requested by the user. In this case, we set $y_{k,n}^{v_l} = 1$.
2. The MEC server does not cache the same quality video but caches higher bitrate version of the same video, whereas the MEC server has no remaining transcoding resources if too many transcoding tasks are proceeded at the same time. In this case, we set $z_{k,n}^{v_l} = 1$.

We denote the backhaul transmission delay for UE k as $D_{k,n}$, relating to average link distance, average traffic load and average number of BSs connecting to a single cell gateway. It can be modeled to be an exponentially distributed random variable with a mean value of D_B [15]. The backhaul delay can be expressed as:

$$d_k^1 = (y_{k,n}^{v_l} + z_{k,n}^{v_l}) D_{k,n}. \tag{4}$$

The second part of network delay is wireless transmission delay between UE k and BS n, which can be calculated as:

$$d_k^2 = \frac{s_l}{r_k}. \tag{5}$$

Consequently, the delay for UE k to download video v_l can be denoted as:

$$d_k^{v_l} = d_k^1 + d_k^2 = (y_{k,n}^{v_l} + z_{k,n}^{v_l}) D_{k,n} + \frac{s_l}{w \log(1 + SINR_k)}. \tag{6}$$

Then, the average network delay of all users is:

$$\bar{t} = \frac{1}{|K|} \sum_{k \in K} \sum_{n \in N} \sum_{v_l \in V} q_{k,v_l} a_{k,n} d_k^{v_l}. \tag{7}$$

Considering that our objective is to make decisions of the BS selection and power allocation to minimize average network transmission delay, then the problem formulation is as follows:

$$\min_{a,p} \bar{t}, \tag{8}$$

$$s.t. \quad r_k \geq R, \tag{9}$$

$$\sum_{n=1}^{N} a_{k,n} = 1, \tag{10}$$

$$\sum_{k=1}^{K} p_{k,n} \leq P_n^{\max}, \tag{11}$$

$$\sum_{k=1}^{K} a_{k,n} x_{k,n} t_l \leq T_n, \tag{12}$$

$$\sum_{k=1}^{K} a_{k,n} (y_{k,n}^{v_l} + z_{k,n}^{v_l}) r_l \leq C_n. \tag{13}$$

The constraints in the formulation above can be explained as follows: (8b) indicate the transmission rate should meet the user's rate requirement denoted as R. Constraints (8c) forces each user only associates with one BS each time. (8d) reveals the transmission power constraint of each BS. Constraint (8e) implies transcoding tasks for each MEC can not exceed its own transcoding capability. (8f) enforces that total video transmitted through backhaul can not exceed its limit.

4 Genetic Algorithm Design

The problem proposed in (8a)–(8f) is proved to be NP-hard and the complexity is pretty high, so no algorithms can provide a guaranteed optimal solution in polynomial time. Genetic algorithm is known to be very successful in such cases [16]. Therefore, we use the genetic algorithm (GA) based algorithm and make improvements on it to ensure near-optimal solutions. We remap the possible association and power allocation solutions into some chromosomes. Each chromosome consists of two parts: the BS selection part and the power allocation part. The genes in BS selection part are formed by an integer string B, whose k-th gene value $B(k)$ indicates the BS selected for user k. Similarly, the k-th gene value $p(k)$ in power distribution part denotes the power allocated to user k at the selected BS. Then, we describe the steps of GA algorithm.

4.1 Fitness Function

Calculate the objective function and use it as a fitness function. In each generation, the fitness values of all chromosomes are calculated. If $\sum_{k=1}^{K} p_{k,n} > P_n^{\max}$, then we need to normalize the power by:

$$p'(k) = p(k) \times \frac{P_n^{\max}}{\sum_{k=1}^{K} p_{k,n}}, \tag{14}$$

to ensure that the constraints (8d) is satisfied.

4.2 Initialization

Firstly, P parent chromosomes are created where P is the population size. The BS selection part comprises K users, and each of them is randomly distributed in $\{1,...,N\}$. The power distribution part consists of K genes, each of which is randomly distributed in $(0, \frac{p_{k,n}}{P_n^{\max}})$.

4.3 Crossover

The proposed new crossover begins by generating a $(K+1) \times 1$ crossover mask sequence which consists of 1 and 0 with equal probability. The first K elements correspond to the BS selection, whereas the remaining element is for the power distribution. For the BS selection part, if the elements in the crossover mask are equal to 1, the genes of the two parent chromosomes in the corresponding positions will be exchanged; whereas, if the elements are equal to 0, they will remain unchanged. The crossover operations in the power distribution parts are based on the arithmetic crossover operation. Note that based on this crossover, the genes at the BS selection part will always lie in $\{1, ... |K|\}$, so the BS selection can be easily made, and the constraint (8c) is always satisfied throughout the evaluation process. This crossover operation will repeat until the size of the population reaches P.

4.4 Mutation

The new mutation operation is also divided into two operations: the integer mutation for the BS selection part and the uniform mutation for the power distribution parts. First, for each chromosome, we create a 2×1 mutation mask sequence comprising of 1 and 0 generated according to the mutation probability p_m. In the BS selection part of the chromosome, if the first element of the mutation mask is 1, two gene values which are randomly selected by the mating point will be exchanged; whereas, if this element is 0, the corresponding genes remain the same. For the genes in the power distribution part, if the corresponding element in the mutation mask is 0, then all of the genes will remain the same; whereas, if it is 1, then a randomly selected mutation point, say i, of the offspring will change as in [17].

5 Performance Evaluation

In this section, we present numerical results to evaluate the performance of the proposed algorithm. We compare our proposed algorithm with traditional genetic algorithm which select offspring randomly. More details of simulation settings are set in Table 1.

Table 1. Simulation settings.

Parameters	Values
Transmit power of BS	46 (dBm)
High-definition video size	2 (Mbits)
Low-definition video size	1 (Mbits)
Number of files	100
MEC server cache capacity	40
MEC server transcoding capacity	10
Number of users	10
Number of BSs (MEC servers)	5
Backhaul delay D_B	150 ms
Zipf parameter	0.95
σ^2	−83.98 dBm
Population size	50
MAX-iteration	100
p_c	0.8
p_m	0.2

Our algorithms ensure that chromosomes with higher fitness values are more likely to be selected when selecting parents to be mated, and offsprings of good fitness are selected for entry into the next generation in improved GA. While in random GA both of all are randomly selected.

First, we investigate the relationship between the network average delay and the number of generations. The user number K is set as 10. As shown in Fig. 2, the average delay decreases as G increases and the average delay remains about the same when $G > 25$. We can note that $P = 50$, $P = 70$, and $P = 90$ eventually reach the same value which means that the improved genetic algorithm has good convergence.

As shown in Fig. 3, traditional GA which randomly choose parent chromosomes and offspring chromosomes does not show a satisfactory convergence compared to our proposed algorithm.

Intuitively, Fig. 4 shows the improved GA performs better than traditional random GA, proving that our algorithm can find a relatively better solution. The improved GA performs better than traditional GA in finding the optimal

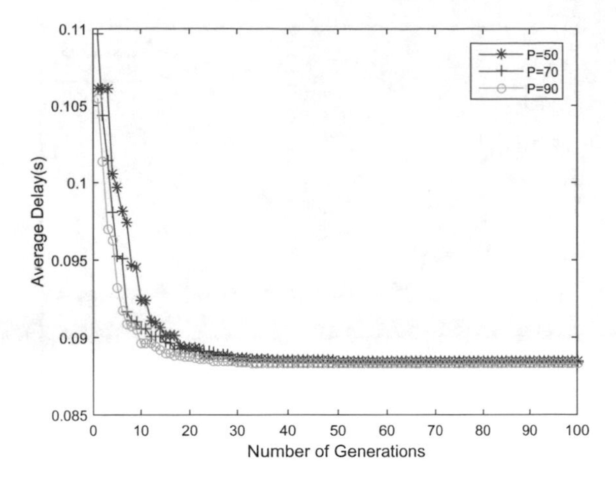

Fig. 2. Average delay against the number of generations for improved algorithm.

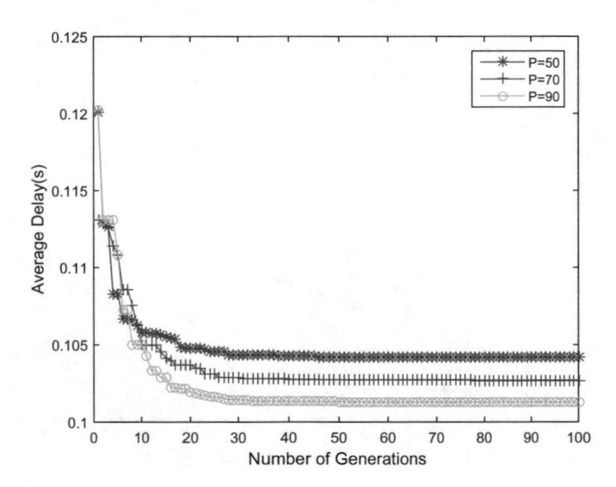

Fig. 3. Average delay against the number of generations for traditional algorithm.

solution of average delay. The reason is that traditional GA repeatedly appears the problem of premature convergence. We can note that our proposed improved GA can achieve lower latency and achieve better global convergence. Also, we can see that improved GA can achieve global convergence in a shorter period of time as compared with traditional GA.

The numerical results in Fig. 5 demonstrate that our improved algorithm outperforms the traditional GA in the aspect of average delay. Also, we can note that average delay increases as the number of users increases. The reasons can be analysed in the following two aspects: on the one hand, when more users request videos at the same time, the transcoding resources of the MEC server may be exhausted, thus the MEC server can not provide transcoding resources anymore.

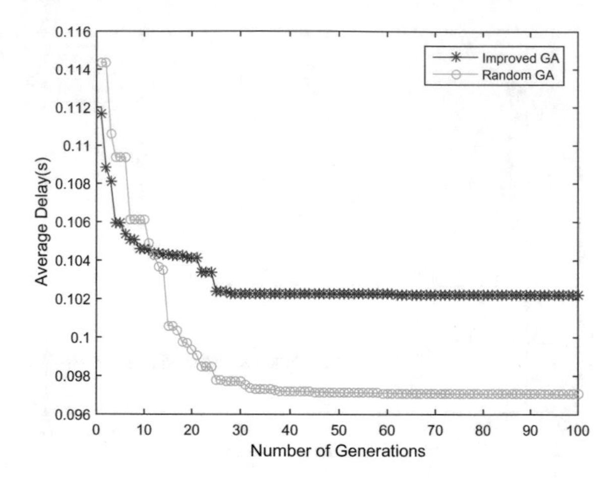

Fig. 4. Performance comparison of different algorithm.

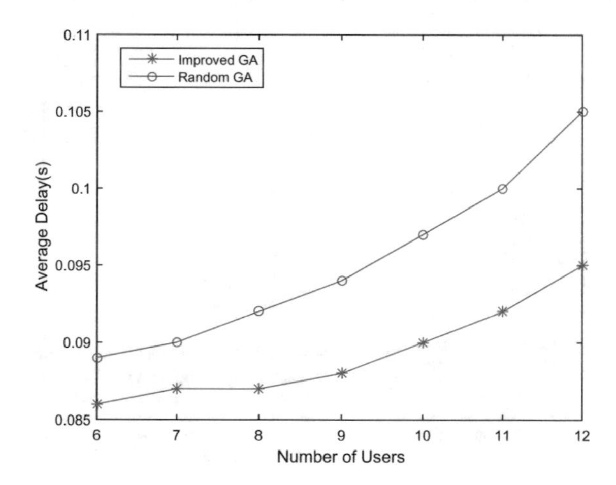

Fig. 5. Average delay versus number of users.

On the other hand, as more users request, the power allocated to each user by the base station will be reduced, resulting in reduced wireless transmission rate, so the wireless transmission delay increases.

6 Conclusion

This paper studies the optimization of minimizing the network delay which takes backhaul transmission delay and the wireless transmission delay into consideration. Then we design the joint BS-user association and power distribution strategy in multi-cell MEC networks where each BS is deployed with one MEC server which can provide caching and transcoding ability at the radio access network.

The joint optimization problem is analyzed to be NP-hard so we propose an improved GA algorithm. The simulation results verify the performance gain of the proposed algorithm as compared with traditional genetic algorithm.

Acknowledgements. This paper is supported by the National Key Project under Grant NO. 2017 ZX03001009.

References

1. Networking, C.V.: Ciscoglobal cloud index: forecast and methodology, 2015–2020. white paper (2017)
2. Peng, X., Shen, J.C., Zhang, J., Letaief, K.B.: Joint data assignment and beamforming for backhaul limited caching networks. In: 2014 IEEE 25th Annual International Symposium on Personal, Indoor, and Mobile Radio Communication (PIMRC), pp. 1370–1374, September 2014
3. Zhou, B., Cui, Y., Tao, M.: Stochastic content-centric multicast scheduling for cache-enabled heterogeneous cellular networks. IEEE Trans. Wirel. Commun. **15**(9), 6284–6297 (2016)
4. Tham, C.K., Chattopadhyay, R.: A load balancing scheme for sensing and analytics on a mobile edge computing network. In: 2017 IEEE 18th International Symposium on A World of Wireless, Mobile and Multimedia Networks (WoWMoM), pp. 1–9, June 2017
5. Abbas, N., Zhang, Y., Taherkordi, A., Skeie, T.: Mobile edge computing: a survey. IEEE Int. Things J. **PP**(99), 1 (2017)
6. Xu, X., Liu, J., Tao, X.: Mobile edge computing enhanced adaptive bitrate video delivery with joint cache and radio resource allocation. IEEE Access **5**, 16406–16415 (2017)
7. Bastug, E., Bennis, M., Debbah, M.: Living on the edge: the role of proactive caching in 5G wireless networks. IEEE Commun. Mag. **52**(8), 82–89 (2014)
8. Abboud, A., Baştuğ, E., Hamidouche, K., Debbah, M.: Distributed caching in 5G networks: an alternating direction method of multipliers approach. In: 2015 IEEE 16th International Workshop on Signal Processing Advances in Wireless Communications (SPAWC), pp. 171–175, June 2015
9. Liao, J., Wong, K.K., Khandaker, M.R.A., Zheng, Z.: Optimizing cache placement for heterogeneous small cell networks. IEEE Commun. Lett. **21**(1), 120–123 (2017)
10. Tao, M., Chen, E., Zhou, H., Yu, W.: Content-centric sparse multicast beamforming for cache-enabled cloud ran. IEEE Trans. Wirel. Commun. **15**(9), 6118–6131 (2016)
11. Yu, Y.J., Tsai, W.C., Pang, A.C.: Backhaul traffic minimization under cache-enabled comp transmissions over 5G cellular systems. In: 2016 IEEE Global Communications Conference (GLOBECOM), pp. 1–7, December 2016
12. Wang, C.C., Lin, Z.N., Yang, S.R., Lin, P.: Mobile edge computing-enabled channel-aware video streaming for 4G LTE. In: 2017 13th International Wireless Communications and Mobile Computing Conference (IWCMC), pp. 564–569, June 2017
13. Tran, T.X., Pandey, P., Hajisami, A., Pompili, D.: Collaborative multi-bitrate video caching and processing in mobile-edge computing networks. In: 2017 13th Annual Conference on Wireless On-demand Network Systems and Services (WONS), pp. 165–172, February 2017

14. Amentie, M.D., Sheng, M., Song, J., Liu, J.: Minimum delay guaranteed cooperative device-to-device caching in 5G wireless networks. In: 2016 8th International Conference on Wireless Communications Signal Processing (WCSP), pp. 1–5, October 2016
15. Chen, D.C., Quek, T.Q.S., Kountouris, M.: Backhauling in heterogeneous cellular networks: modeling and tradeoffs. IEEE Trans. Wirel. Commun. **14**(6), 3194–3206 (2015)
16. Islam, M., Razzaque, A., Islam, J.: A genetic algorithm for virtual machine migration in heterogeneous mobile cloud computing. In: 2016 International Conference on Networking Systems and Security (NSysS), pp. 1–6, January 2016
17. Lai, T.I., Fang, W.H., Lin, S.C.: Efficient subcarrier pairing and power allocation in multi-relay cognitive networks. In: 2016 2nd International Conference on Intelligent Green Building and Smart Grid (IGBSG), pp. 1–5, June 2016

Cyber-Physical Sensors and Devices for the Provision of Next-Generation Personalized Services

Borja Bordel(✉), Teresa Iturrioz, Ramón Alcarria, and Diego Sánchez-de-Rivera

Universidad Politécnica de Madrid, Madrid, España
{bbordel,diego.sanchez}@dit.upm.es,
{teresa.iturrioz,ramon.alcarria}@upm.es

Abstract. Cyber-Physical Systems (CPS) are set to radically transform the world we live in. Prototypes for very different domains have been reported, from Industry 4.0 to Ambient Intelligence and the Internet of Things. Several research works have shown the good performance of these systems, which could be useful for everyday living once they become commercial products. However, no complete application for cyber-physical devices has been reported yet. Thus, the large amount of benefits this new paradigm may push remains very difficult to envision by general society and companies. In any case, personalized services rank among the most direct and interesting applications for cyber-physical devices. So far, no work on this topic has been reported, but the implementation of this new generation of services is a key area for the advancement toward the CPS era. Therefore, in this paper we will explore the concept of cyber-physical personalized services and propose a first example of these new services based on cyber-physical sensors and a cyber-physical device: a smart table. Finally, in order to evaluate the performance of the proposed solution, we will carry out an experimental validation.

Keywords: Cyber-Physical Systems · RFID · Humanized computing
Smart object · Personalized services

1 Introduction

Cyber-Physical Systems (CPS) [1] are new engineered systems widely regarded as the basis of future digital society for the fourth industrial revolution [2]. They integrate and relate the physical and cyber worlds in a seamless way; therefore, the behavior of the resulting networks and environments depends on the programming, and also on natural laws [3].

The earliest CPSs were based on RFID tags, which provided daily living objects with a unique identity in the cyber world [4]. Later, and until today, new miniaturized electronic components such as System-on-Chip enabled the creation of smart objects and smart environments where embedded computing devices become the fundamental element. Unlike the Internet-of-Things [5], CPS does not refer to an infrastructure which expands and/or improves the current capabilities of technological systems (thus allowing for the deployment of new services such as Smart Home applications).

L. Barolli et al. (Eds.): IMIS 2018, AISC 773, pp. 479–490, 2019.
https://doi.org/10.1007/978-3-319-93554-6_46

On the contrary, CPS implies the definition and deployment of a new paradigm where the limits between the cyber and the physical world get fuzzy through the use of implicit interfaces [6].

Implicit human computer interactions [7] are based on actions performed by users; these actions are not primarily aimed at interacting with a computing device, being instead regarded by the system as an input. In order to reach this purpose, CPS must have a certain understanding of human behavior (some new ideas such as the Humanized Cyber-Physical Systems [8] have been defined in relation to this topic), and they must also consider the evolution of nature, since natural phenomena (such as temperature variations) can be also understood as implicit inputs.

Although this proposal seems difficult to implement by using real technologies and devices, many different prototypes of Cyber-Physical devices and networked CPS have been reported: from RFID-based devices [9] to machine learning and artificial intelligence solutions. Besides, these prototypes have performed well, and in most cases the reported results have been beyond the success thresholds.

However, these prototypes are rarely integrated into a complete application for users or companies. In that way, the benefits, synergies and advantages of using this new paradigm may be difficult to identify for users lacking the required technical knowledge. Among the possible applications cyber-physical devices could be integrated into, the greatest revolution can be caused in the personalized services. Personalized services feature the ability to change their configuration, content and/or operation depending on the users' preferences [10]. In traditional approaches, personalization is based on the value of some relevant (but also obvious) variables: location, previous selections, etc. [11] This approach is problematic, since quite often the information is only provided after the users have detected their own needs and made a decision about them. The next generation of personalized services, however, should be able to infer the users' needs before they are completely aware of them. In that way, personalized services may improve their experience, as users will be provided with valuable information at the right time [12]. On the other hand, as we said before, cyber-physical elements must be implicit, and the difference between standard and cyber-physical objects or services should not be noticeable (it has been said that cyber-physical devices must be unobtrusive [13, 14]). Nevertheless, traditional hardware sensors and devices can only acquire clear signals from the physical world; therefore, personalized services will only receive information once it becomes evident. In order to sense hidden signals, an inference and data analysis must be carried out. In this context, cyber-physical sensors are the adequate solution.

Therefore, the aim of this paper is to investigate the concept of cyber-physical personalized services and to propose a first example of these new services, based on a cyber-physical table and on physiological and emotional signals obtained through cyber-physical sensors. The information captured by the cyber-physical devices will be employed to automatically personalize services according to the user's situation. In order to evaluate the performance of the proposed system, different experiments were carried out.

The rest of the paper is organized as follows: Sect. 2 will describe the state of the art on cyber-physical sensors. Section 3 will present a proposed technology. Section 4 will provide an experimental validation of the proposal. Finally, Sects. 5 and 6 will explain some results of this experimental validation and the conclusions of our work.

2 State of the Art

Research on cyber-physical sensors is a very promising field. So far, widely different works have been reported on this topic.

Some authors view cyber-physical sensors as instruments capable of passively accumulating information [15] while the system is, for example, running secondary algorithms. Other proposals have focused on the possibility of implementing sensors with processing capabilities in relation to personal data protection, secure access to private signals (such as biological measures), etc. [16].

A small group of papers also consider cyber-physical sensors as a term that refers to the embedded devices making up a CPS [17]. Other studies, however, differentiate cyber-physical sensors (and actuators) from CPS, identifying some specific particularities and features. For example, cyber-physical sensors may be distributed and could be integrated into pervasive systems; they may also monitor worldwide phenomena (for example, global communities of users) [18]. Following this line of thought, some proposals employ cyber-physical sensors in order to support context-aware applications, as these devices may dynamically change their behavior by means of their software components. Cyber-physical sensors have been conversely applied to power management [19], traffic [20] and decision making [21].

Finally, some works have been published on partial or complete implementations of cyber-physical sensors. In particular, different reports can be found on hybrid microelectromechanical systems with a view to creating cyber-physical inertial sensors [22]. These systems include a statistical framework acting as a computation module, which enables the system to eliminate the noise, thus improving the quality of the measurements. Other contributions have consisted in mere data analysis frameworks to be integrated into cyber-physical sensors [23].

As for cyber-physical devices, various research proposals have emerged since 2000 for the integration of electronic systems at home or in the workplace: [24] presents the concept of "Smart Home"; a "Smart Office" is designed in [25]; an "Intelligent Home" is planned in [26]. The concept of "smart furniture" became very popular in research between 2003 and 2004. Various research papers appeared on this subject, such as the general framework presented in [27]. Furthermore, patents such as [28], in which an RFID smart chair is described, were also given. Finally, in relation to wearable technologies, several papers describe their use in traceability systems, especially cybernetic gloves. Thus, a first prototype for a cybernetic glove appeared in 2005 [29], and recently, in February 2014, Fujitsu presented its own design for industrial environments [30]. In parallel to this, textile industry is advancing toward the seamless integration of electronics with textiles, which will enable a qualitative leap in wearable products [31].

On the other hand, while not being considered integrated systems, several commercial "smart products" have been developed recently for the purposes of stock control, the adaptation of the environment to the user and traceability [32]. In 2011, a smart poker table was presented in Italy by GTI Gaming [33]. Based on RFID technology, this table implements a real-time technology capable of calculating the amount of pot and rake, reporting defects, identifying the dealer, establishing a network of tables and so on. Later, between December 2013 and February 2015, no less than three smart clinical medicine dispensers were presented in Europe. The first, in Portugal [34], was developed in collaboration with Fujitsu. The other two [35], developed in Spain in June 2015, are currently undergoing tests. Nevertheless, the most successful product in the market so far is SATO's VINICITY technology [36]. This technology can read various RFID tags at once, having been employed in several "smart products" such as trays, tables and medicine dispensers.

3 Provision of Personalized Cyber-Physical Services

In this section we will propose a model for cyber-physical services, as well as a possible cyber-physical device to support such services.

3.1 Service Model and the Definition of Cyber-Physical Services

In order to improve their efficiency, next-generation personalized services should follow an event-oriented design. Basically, the operation flow consists of six steps. First, the user must select the service to be executed by his/her personal device. This device sends a service request to the server, which locates the requested application in the service repository. Then, if available, the service is downloaded onto the personal device through the central server. In this server, a record is kept of the services being executed and their logical locations, with the aim of supporting mobility and acquiring information from cyber-physical sensors based on pervasive and/or distributed infrastructures. Once the service is being executed in the personal device, data from cyber-physical sensors are collected. These data must be uploaded onto the server, where personalization events are triggered. Depending on the implementation of the hardware components of the cyber-physical sensors, these data can be sent to the server either directly or with the personal device being used as gateway. After considering the received information, the server will create and send the corresponding personalization events to the personal devices. Personalization events are described in a catalogue stored in the server, which must be taken into account by service developers in order to make their applications interoperable with the proposed provision platform. Considering these personalization events, and the possible personalization rules to be defined by the service provider, the personal device provides the user with custom information. Data formats and implementation technologies strongly rely on their underlying software platforms (web servers, mobile operating systems, etc.), but most common solutions are equipped to support our proposal.

The defined variability model for next-generation personalized services based on cyber-physical sensors and devices is presented in the class diagram (UML) of Fig. 1. In order to support scalability and future designs it is considered that a service must be composed of other previously defined services. The service we are proposing in this first work, however, acts as an atomic unit. For services to be personalized, this model must contain a series of variability points. Each point will represent one variable behavior of the service depending on the physical world, and will be modelled as a decision point controlled by a set of rules and personalization events. These points contain a set of two or more variability options representing either the choices made by the user (e.g. through a list) or the ones made automatically depending on the context (for example, a value provided by cyber-physical sensors). Variability options enhance the personalization of the service; they can be predefined (the option value is provided) or free (with customers entering arbitrary values), and it is also possible to maintain specific default values. Each variability point presents several features. These include certain constraints, which restrict the variability of the behavior depending on the configuration of the point or the contextual conditions (i.e. the measures obtained from cyber-physical sensors).

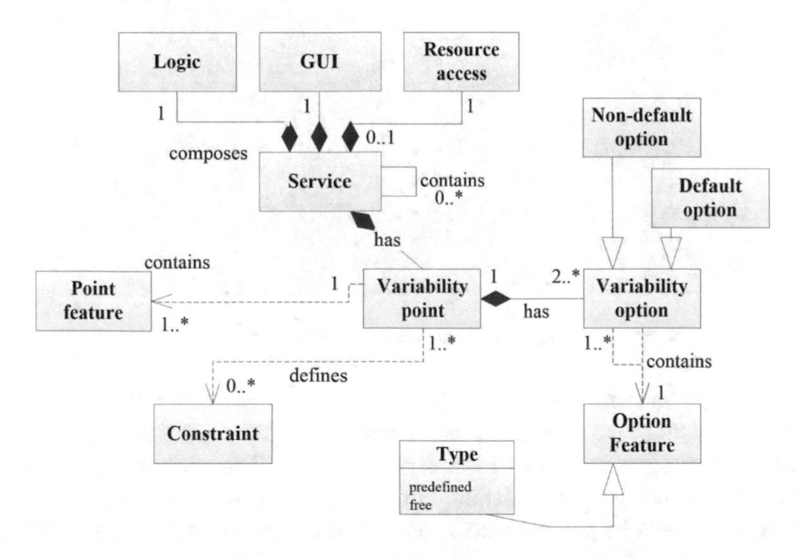

Fig. 1. Variability model for a personalized service

3.2 Cyber-Physical Sensors and Devices: The Smart Table

As we said before, cyber-physical elements must be totally unobtrusive and identical to the corresponding standard objects. Many different devices and implementations could be developed following this paradigm: motivational sensors [37], cybernetic gloves [4], etc. However, in this case, we have proposed an advanced version of one of the first researched cyber-physical devices: the smart table. In Fig. 2 we present our prototype for a RFID-based cybernetic table. It is made of a plastic-based material (which does not interfere with electromagnetic fields) and on its surface we have defined four detector areas. In Fig. 3, an electronic scheme of the table is presented. As can be seen, there are four dedicated microcontrollers (one for each detector area), and one manager microcontroller. The first ones were based on the Arduino Nano architecture and the last one, on Arduino Mega256 architecture. When a dedicated microcontroller must transmit data, it requests permission from the manager and sends the message by using a physical protocol. This physical protocol consists of two lines: the permission line and the request line. The former acts as an output for the manager and as an input for the dedicated microcontrollers. The request line acts as an input for the manager microcontroller and as an output for the dedicated units.

Fig. 2. Enhanced cybernetic table

The manager microcontroller, which employs a Bluetooth interface, finally sends the information about the objects placed or removed from the table to an access point, from which messages are sent to the service server. The user service will change depending on the received information. For example, when employed in a warehouse, this technology could help workers to locate specific items or to determine whether any given item is sold out.

In order to detect the objects on the table, all elements in the system must include an RFID tag. As can be seen in Fig. 3, the RFID readers taken into consideration are based on RFID-RC-522 architecture. These readers include in the same element both the controller and the resonant printed inductive antenna. Besides, they can detect the presence of several elements stacked on the same reader, unlike other technologies such as RDM8800 NFC readers. LEDs indicate whether any object is detected on the corresponding detection area, as a visual element that helps improve the user experience.

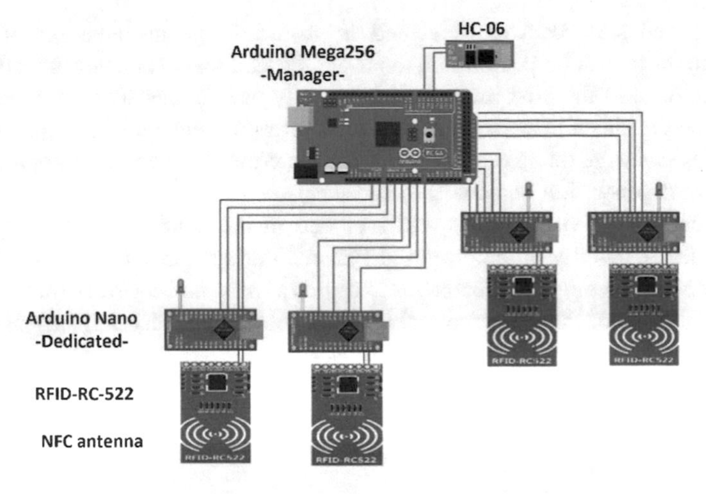

Fig. 3. Electronic scheme of the proposed smart table

4 Experimental Validation

In order to evaluate the performance of the proposed personalized services, a first real implementation has been designed and deployed. Besides, with a view to evaluating more complex features such as scalability, various virtual users were introduced in the system during experiments. Using the infrastructures of the Universidad Politécnica de Madrid, a remote server and a service repository were deployed. Both components were based on standard web servers provided with a REST interface.

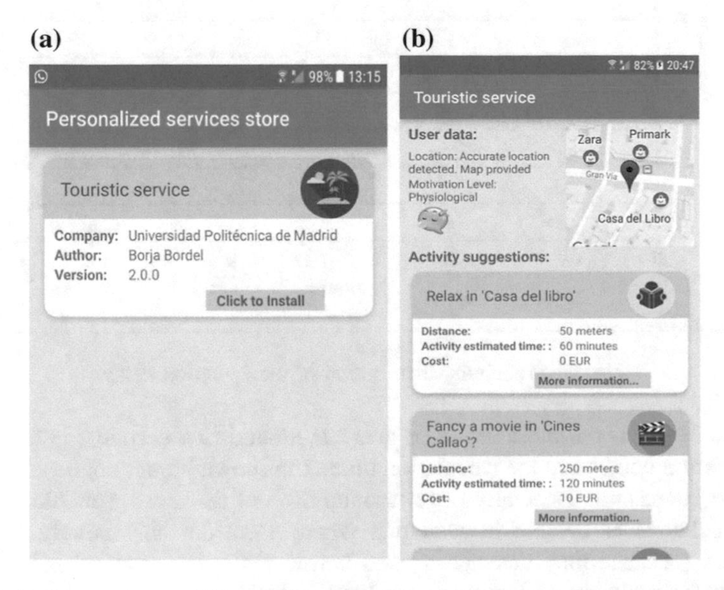

Fig. 4. Screen shots of the mobile applications (a) App store (b) Service interface

Personalized services were executed in Android systems (see Fig. 4), such as smartphones or tablets (depending on the user). Fifteen real users were asked to use the created platform. In this first implementation, only one service was available. In order to download and install this service (i.e. the corresponding mobile application, since Android devices were being employed), users were provided with the appropriate App store (for next-generation personalized services).

A community of virtual users was included in the platform with the purpose of evaluating the scalability of the proposed system. Various virtual machines (VM) were deployed emulating the users' behavior and providing a correlated output information flow about the use of the cybernetic table. These VMs have also implemented and run the mobile application, so each one represents a real service user. The number of users in the community was progressively increased, and the evolution of the platform's technical features was monitored.

5 Results

As we said in Sect. 4, some important technical parameters of the system were monitored during the experiments. The most interesting one is the operation delay. Figure 5 shows the probability distribution of this parameter.

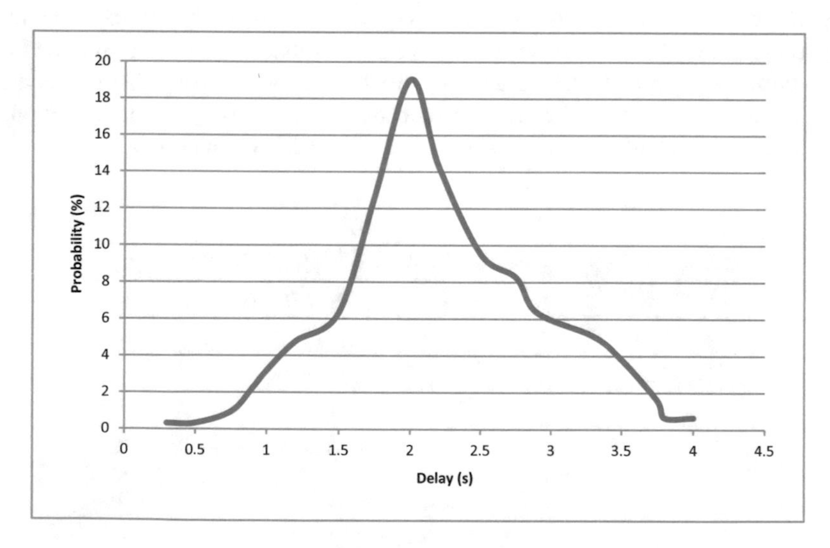

Fig. 5. Probability distribution of the operation delay

As can be seen, the most probable delay is around two seconds, considering the communication delays and the calculation time. The showed function, besides, is pretty narrow, so jitter is not especially huge (around 50% of the most probable delay). The obtained delay is acceptable in scenarios where users are not actively looking for information (because they detected a need before the cyber-physical sensor did). In order to do an adequate comparison, we should consider that the maximum speed for an LTE connection is around 100 Mbps, so regular mobile Internet connections presents a latency of 20 ms. Deployment with more resources and a more efficient programming may improve these results.

Delays could increase if cyber-physical devices do not generate data properly or if a communication error occurs. In this case, traditional recommendation techniques based on the historical record will be triggered to maintain the personalized service available.

Finally, the second experiment evaluated the scalability of the system. The number of users in the system was increased, until the first failures due to system saturation appeared. The failure probability depending on the number of users in the system is evaluated in Fig. 6.

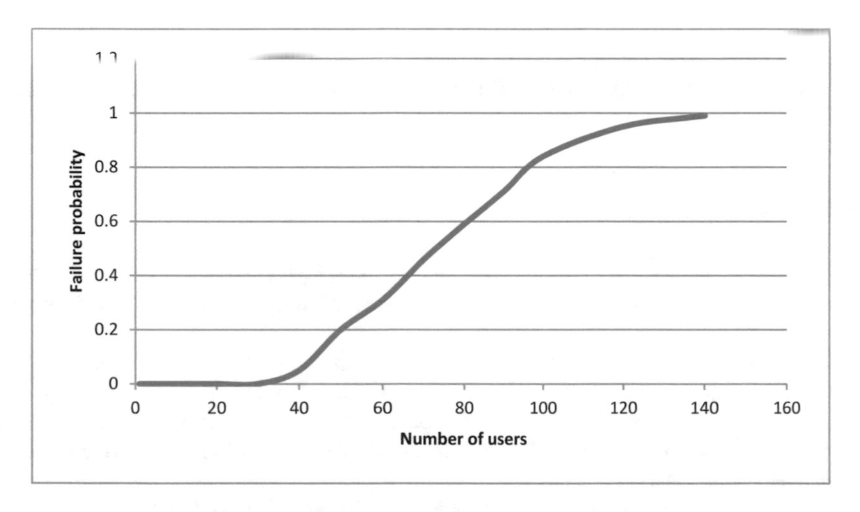

Fig. 6. Failure probability depending on the number of users

The system supports around forty (40) users without fails. This limit is due, overall, to the computing power and communication delay required to acquire the user information from the cyber-physical devices. If the number of users is above 65 people (approximately), the deployed infrastructure fails in 50% of the computed cases. The natural limit of the system is around 140 users; when this number of users is reached, the next-generation personalized service provision platform gets totally saturated and stops operating normally.

It is important to note that this first deployment is only a proof of concept, which has proved to be scalable (both the infrastructure and the cyber-physical sensors). For commercial or real implementations, a server should be considered with more capacity as well as more complex engineering techniques.

6 Conclusions

In this paper we are exploring the concept of personalized services in the context of cyber-physical systems. We have proposed a service model and a cyber-physical device to support the creation of the personalization events, based on a RFID-based cybernetic table. The proposed experimental validation showed that personalized services are scalable and reliable, although enhanced scenarios (including future high-density scenarios) should also be considered before formulating any definitive conclusions. In any case, first evidences are provided.

Acknowledgments. Borja Bordel has received funding from the Ministry of Education through the FPU program (grant number FPU15/03977). Additionally, the research leading to these results has received funding from the Ministry of Economy and Competitiveness through SEMOLA project (TEC2015-68284-R) and from the Autonomous Region of Madrid through MOSI-AGIL-CM project (grant P2013/ICE-3019, co-funded by EU Structural Funds FSE and FEDER).

References

1. Bordel, B., Alcarria, R., Robles, T., Martín, D.: Cyber–physical systems: extending pervasive sensing from control theory to the internet of things. Pervasive Mob. Comput. **40**, 156–184 (2017)
2. Sánchez, B.B., Alcarria, R., de Rivera, D.S., Sánchez-Picot, Á.: Enhancing process control in industry 4.0 scenarios using cyber-physical systems. JoWUA **7**(4), 41–64 (2016)
3. Lee, E.A.: Cyber-physical systems-are computing foundations adequate. In: Position Paper for NSF Workshop on Cyber-Physical Systems: Research Motivation, Techniques and Roadmap, vol. 2, October 2006
4. Bordel Sánchez, B., Alcarria, R., Martín, D., Robles, T.: TF4SM: a framework for developing traceability solutions in small manufacturing companies. Sensors **15**(11), 29478–29510 (2015)
5. Robles, T., Alcarria, R., Martín, D., Morales, A., Navarro, M., Calero, R., López, M., et al.: An internet of things-based model for smart water management. In: Proceedings of 2014 28th International Conference on Advanced Information Networking and Applications Workshops (WAINA), pp. 821–826. IEEE, May 2014
6. Pérez-Jiménez, M., Sánchez, B.B., Alcarria, R.: T4AI: a system for monitoring people based on improved wearable devices. Res. Briefs Inf. Commun. Technol. Evol. (ReBICTE) **2**, 1–16 (2016)
7. Schmidt, A.: Implicit human computer interaction through context. Pers. Technol. **4**(2–3), 191–199 (2000)
8. Bordel, B., Alcarria, R., Martín, D., Robles, T., de Rivera, D.S.: Self-configuration in humanized cyber-physical systems. J. Ambient Intell. Human. Comput. **8**(4), 485–496 (2017)
9. Hong, J., Suh, E.H., Kim, J., Kim, S.: Context-aware system for proactive personalized service based on context history. Expert Syst. Appl. **36**(4), 7448–7457 (2009)
10. Morales, A., Alcarria, R., Martín, D., Robles, T.: Enhancing evacuation plans with a situation awareness system based on end-user knowledge provision. Sensors **14**(6), 11153–11178 (2014). https://doi.org/10.3390/s140611153

11. Chang, W.L., Jung, C.F.: A hybrid approach for personalized service staff recommendation. Inf. Syst. Front. **19**(1), 149–163 (2017)
12. Sánchez, B.B., de Rivera, D.S., Sánchez-Picot, A.: Building unobtrusive wearable devices: an ergonomic cybernetic glove. J. Internet Serv. Inf. Secur. **6**(2), 37–52 (2016)
13. Martín, D., Bordel, B., Alcarria, R., Sánchez-Picot, Á., de Rivera, D.S., Robles, T.: Improving learning tasks for mentally handicapped people using AmI environments based on cyber-physical systems. In: International Conference on Ubiquitous Computing and Ambient Intelligence, pp. 166–177. Springer, Cham (2016)
14. Stehr, M.O., Talcott, C., Rushby, J., Lincoln, P., Kim, M., Cheung, S., Poggio, A.: Fractionated software for networked cyber-physical systems: research directions and long-term vision. In: Formal Modeling: Actors, Open Systems, Biological Systems, pp. 110–143. Springer, Heidelberg (2011)
15. Herlihy, L., Golen, E., Reznik, L., Lyshevski, S.E.: Secure communication and signal processing in inertial navigation systems. In: Proceedings of 2017 IEEE 37th International Conference on Electronics and Nanotechnology (ELNANO), pp. 414–419. IEEE, April 2017
16. Wu, L.L.: Improving system reliability for cyber-physical systems (Doctoral dissertation, Columbia University) (2015)
17. Bellavista, P., Giannelli, C.: Cyber physical sensors and actuators for privacy-and cost-aware optimization of user-generated content provisioning. Int. J. Distrib. Sens. Netw. **15**, 7172–7205 (2015)
18. Javed, F., Ali, U., Nabeel, M., Khalid, Q., Arshad, N., Ikram, J.: SmartDSM: a layered model for development of demand side management in smart grids. In: Proceedings of the 3rd International Workshop on Software Engineering Challenges for the Smart Grid, pp. 15–20. ACM, June 2014
19. Stergiopoulos, G., Valvis, E., Anagnou-Misyris, F., Bozovic, N., Gritzalis, D.: Interdependency analysis of junctions for congestion mitigation in transportation infrastructures (2017)
20. Kureshi, I., Theodoropoulos, G., Mangina, E., O'Hare, G., Roche, J.: Towards an info-symbiotic decision support system for disaster risk management. In: Proceedings of the 19th International Symposium on Distributed Simulation and Real Time Applications, pp. 85–91. IEEE Press, October 2015
21. Lyshevski, S.E.: Signal processing in cyber-physical MEMS sensors: inertial measurement and navigation systems. IEEE Trans. Indus. Electron. (2017)
22. Basta, S., Manco, G., Masciari, E., Pontieri, L.: 20+ years of analytics on complex data: impact, issues, challenges and contributions. In: Flesca, S., Greco, S., Masciari, E., Saccà, D. (eds.) A Comprehensive Guide Through the Italian Database Research Over the Last 25 Years. Studies in Big Data, vol. 31, pp. 353–374. Springer, Cham (2018). https://doi.org/10.1007/978-3-319-61893-7_21
23. Das, S.K., Cook, D.J.: Agent based health monitoring in smart homes. In: Proceedings of the International Conference on Smart Homes and Health Telematics (ICOST), Singapore, pp. 3–14, 15–17 September 2004
24. Le Gal, C., Martin, J., Lux, A., Crowley, J.L.: Smart office: design of an intelligent environment. IEEE Intell. Syst. **4**, 60–66 (2001)
25. Lesser, V., Atighetchi, M., Benyo, B., Horling, B., Raja, A., Wagner, T., Xuan, P., Zhang, S.: The intelligent home testbed. In: Proceedings of the Autonomy Control Software Workshop, Seattle, WA, USA, 29 January 1999
26. Tokuda, H., Takashio, K., Nakazawa, J., Matsumiya, K., Ito, M., Saito, M.: SF2: Smart furniture for creating ubiquitous applications. In: Proceedings of the International Symposium on Applications and the Internet, Tokyo, Japan, pp. 423–429, 26–30 January 2004. Sensors 2015, 15 29509 59

27. Hagale, A.R., Kelley, J.E., Rozich, R.: RFID smart office chair. US Patent 6,964,370, 15 Nov 2005. 61
28. Fishkin, K.P., Philipose, M., Rea, A.: Hands-on RFID: wireless wearables for detecting use of objects. In: Proceedings of the 9th International Symposium on Wearable Computers, Osaka, Japan, pp. 38–43, 18–21 October 2005
29. Fujitsu's Smart Glove. http://www.qore.com/articulos/17335/Fujitsu-desarrolla-unguante-de-realidad-aumentada. Accessed 26 Oct 2015
30. Marculescu, D.: Electronic textiles: a platform for pervasive computing. IEEE Proc. **91**, 1995–2018 (2003)
31. Sakurai, S.: Prediction of sales volume based on the RFID data collected from apparel shops. Int. J. Space-Based Situated Comput. **1**(2–3), 174–182 (2011)
32. Giochi SmartPoker. http://www.gtigaming.com/en/product-service/smartpoker-2/. Accessed 18 Feb 2018
33. Argos Smart Clinical Cupboard. http://sicolareshigia.com/armario-de-control-destock/. Accessed 18 Feb 2018
34. Izco Smart Clinical Cupboard. http://www.farodevigo.es/economia/2014/09/28/viguesa-izco-lanza-fabricacion-armarios/1102202.html. Accessed 18 Feb 2018
35. SATO's VINICITY Technology. http://www.satovicinity.com/sp/products_magellan_pjm_rfid_smart_readers.asp. Accessed 18 Feb 2018
36. Bordel, B., Alcarria, R.: Assessment of human motivation through analysis of physiological and emotional signals in industry 4.0 scenarios. J. Ambient Intell. Humaniz. Comput. **4**(6), 1–21 (2017)

Device Stand-by Management of IoT: A Framework for Dealing with Real-World Device Fault in City Platform as a Service

Toshihiko Yamakami(✉)

ACCESS, 3 Kanda-Neribei-cho, Chiyoda-ku, Tokyo 101-0022, Japan
Toshihiko.Yamakami@access-company.com

Abstract. Expansion of IoT and increasing computing resources provides opportunities in edge computing. There are two types of edge computing: heavy edge computing and lightweight edge computing. The author discusses lightweight edge computing with resource-constraint from real-world use cases from a smart city project. The author proposes device management framework with stand by mechanism and device characteristics from data mining of past device behavior.

1 Introduction

The increase of number of smart devices attract attention in edge computing. Efficient use of resources is an important topic in edge computing. When the number of nodes in edge computing increases, task allocation emerges as an issue.

IoT engines are low-resource standardized IoT platform developed in Japan [6]. In order to leverage potentials of IoT engines, we study a standby task initiation in edge computing.

There are two types of edge computing, heavy-edge computing and lightweight-edge computing. IoT is expected to continue to grow. And the growth will present opportunities of lightweight edge computing because a large part of IoT growth will be attributed to limited resource devices. The future of IoT will accommodate rich sets of devices that are always on and transmit information constantly such as media streaming devices. In order to make use of full potential of rich IoT paradigm in the future, it is important to bridge the current real-world IoT and the resource-abundant cloud computing. Limited resource device-based edge computing requires reconsideration of migration in edge computing.

We are engaged in CPaaS.io [3], an EU-Japan collaboration project for a city-wide IoT and open data platform. A smart city platform is no exception to the challenges of resource-constraint edge computing. In this paper, the author describes the requirements of IoT, and presents a task standby initiation scheme for fault-tolerant IoT computing.

L. Barolli et al. (Eds.): IMIS 2018, AISC 773, pp. 491–502, 2019.
https://doi.org/10.1007/978-3-319-93554-6_47

2 Background

2.1 Purpose of Research

The aim of this research is to develop a framework to enable stand by mechanism to ensure service availability at a device fault.

2.2 Related Work

Research on task migration in IoT consist of three areas: (a) task migration algorithms, (b) migration programming models and middleware, and, (c) collaborative processing and system issues.

First, in regards to algorithms of task migration, Dziurzanski et al. discussed a resource allocation approach for a hard real-time guarantee automotive system [5]. Dwarakanath et al. discussed complex event processing in operator migration of a device-to-device system [4]. Yang et al. discussed real time task migration in Apache Storm [17]. Quan et al. presented a task-migration simulation framework [16]. Ottenwalder et al. discussed task migration for network utilization [14]. Gaspar et al. discussed a fine-grained application-aware task management in heterogeneous embedded platforms [8]. Zhang et al. performed a survey and presented challenges in task migration in a mobile cloud environment [18].

Second, in regard to migration programming models and middleware, Gascon-Samson et al. discussed ThingsJS, a Javascript-based middleware platform [7]. Azzarà et al. presented a macro programming scheme for IoT to hide low-level communication details [1]. Jalali et al. discussed cognitive computing at edge nodes for machine learning at IoT gateways [12]. Cozzolino et al. presented a fine-grained edge off-loading architecture FADES [2].

Third, in regard to collaborative processing and system issues, Kumar et al. discussed a clustering algorithm for collaborative processing in IoT network [13]. Giang et al. discussed distribute data flow for collaborative processing of devices with different owners [9]. Hussain et al. discussed BLE-based handover [10,11]. Pourmohseni et al. discussed predictable reconfiguration of real time applications [15].

The originality of this paper lies in its development of a stand-by-based migration scheme for IoT.

3 Method

The author performs the following steps to lay the foundation of a cost-effective framework for standby mechanism using low-resource devices:

- Identifying real-world IoT migration contexts,
- Developing a framework of lightweight edge computing,
- Developing a stand-by-based migration scheme for edge computing

4 Observation

Applicability of task migration of u2 architecture (ucode version 2) is summarized in Table 1.

When the edge nodes have resources to spare, various task migration can be applied. Resource-constrained environments such as envisioned by u2 architecture need to deal with static and planned task migration. Full flexible task migration is difficult to deploy in this context.

Table 1. Applicability of task migration of u2 architecture

	Dynamic (real-time, automatic)	Static (planned, semi-automatic)
Heavy-edge module Migration	N.A.	N.A.
Lightweight edge module Migration	N.A.	u2

Requirements of edge computing are depicted in Table 2.

Table 2. Requirements of edge computing

Requirement	Description
Load balancing	Computational load is distributed among multiple edge nodes to ensure throughput
Low latency	Computational load is performed at edge nodes not in the cloud, to ensure low latency
Privacy preservation	Sensitive information is maintained in devices and edges to ensure protected access
Off-loading	Computational load is distributed to another available edge node to ensure service availability of front-end node
Fault-tolerance	Alternate device is provided to ensure continued service at a fault in a device

Characteristics of lightweight edge computing are depicted in Table 3.

In the future, many always-on data collection/dissemination service will be important in the IoT paradigm. One example is a rich streaming media service. However, in real-world use cases, it has a higher priority to accommodate fault tolerance of end devices. In CPaaS.io, many end devices consist of single failure points in the Yokosuka medical emergency project.

Table 3. Characteristics of lightweight edge computing

Item	Description
Binary migration	Migration is done by binary code
State-less migration	Limited or no capability of processing states among lightweight edge nodes
Front-end availability	it is crucial to maintain front-end availability
Data loss tolerance	Data loss is prohibiting, however, it is better than no front-end availability

Rare failure take place more than original designers took into consideration. For example, when a fault take places at random and the average interval is 90 days, more than 1 % fault occurs within 1 day from the previous fault in a Poisson distribution case. In the real-world deployment, it is important to take care of rare events, which happens more frequently than they were thought in advance.

Sending service personnel for each such breakdown as the fault occurs will be a nightmare from the viewpoint of service providers. In order to deal with this requirement, the author proposes a managed standby mechanism. It provides a flexible and proactive standby mechanism to balance battery usage and availability. It uses a modified definition of migration in edge computing as shown in Table 4.

Table 4. Modified definition of task migration of edge computing

Migration	Load-balancing or off-loading of computational requirements to multiple available edge nodes
Modified migration	Remove the computational requirements from original device and relocate them to a stand-by device in order to secure the continuity of a service

This modified definition eliminates the computational load of complicated task management from the edge computing. Task management at each device needs to deal with simple on and off. The edge and/or cloud node will take care of the stand-by planning to compensate the lack of complicated coordination at the device side.

Stand-by mechanism is depicted in Fig. 1.

An example of control flow is depicted in Fig. 2.

Requirements of device description language (DDL) are depicted in Table 5.

Task and device descriptions are depicted in Fig. 3.

Device status is depicted in Table 6.

Binary code allocation is depicted in Table 7. In this paper, the on-peer-device mode is considered.

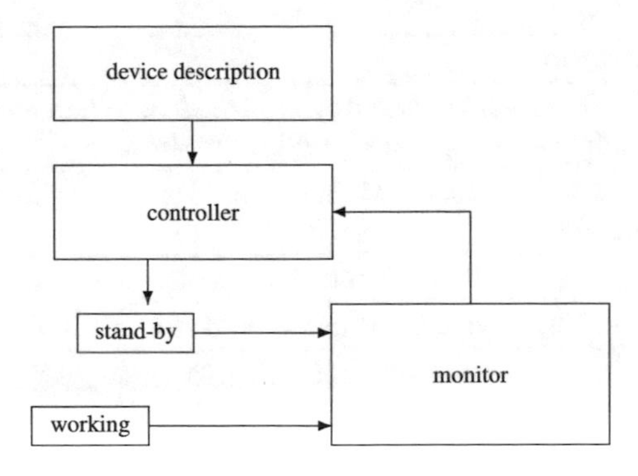

Fig. 1. Stand-by mechanism

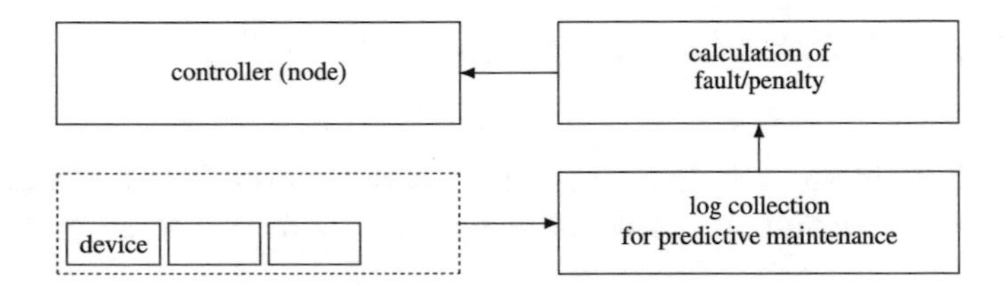

Fig. 2. Example of control flow

Table 5. Requirements of device description language

Item	Description
Availability target	Required range of availability ratio as a service system
Stand-by costs	Costs (hardware, software, networking, battery) for stand-by
Dependency	External dependency to execute a task
Fault characteristics	Characteristics (severity, frequency) of faults

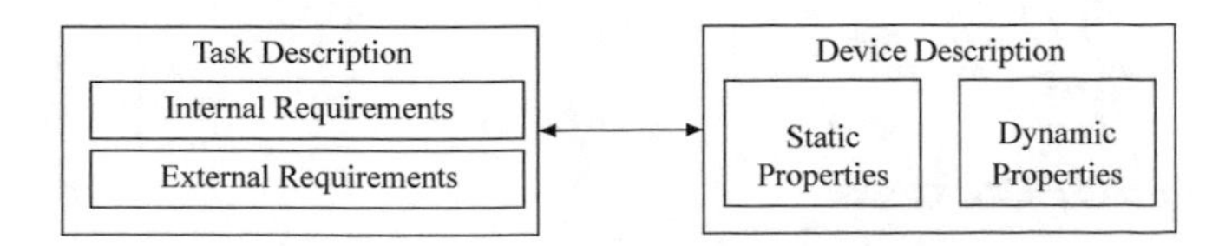

Fig. 3. Task and device descriptions

Table 6. Device status

Status	Description
Normal	Normal status. No standby is provided. When at fault, normal reconfiguration operation will be performed
Dual	Dual working status. Multiple devices are working to ensure fault tolerance
Cold standby	A stand-by device is available, but it is not ready for instant switch
Hot standby	A stand-by device is available, and it is ready for instant switch

Table 7. Binary code allocation

Status	Description
On device	Multiple binary images are maintained on a device
On cloud	A binary image is maintained in cloud, and transferred on demand
On peer device	A binary image is maintained on a peer device
On edge	A binary image is maintained in an edge node

The fault ratio and required maximum down time varies from applications. The examples of maximum down time requirements are depicted in Table 8.

Table 8. Examples of down time requirements

Application type	Description
Chemical plant sensors	Industrial chemical plant sensors require a down time less than 60 s
Real-time traffic monitors	Real-time traffic monitors require a down time less than 5 min
Climate logging sensors	Climate logging sensors require a down time less than an hour

The requirements vary from application to application. The precision requirements and influence from down times will determine the minimum downtime requirements.

5 Evaluation

5.1 Simulation Contexts

Simulation parameters are depicted in Table 9.

Battery consumption, switch time, and average fault intervals are ad hoc values.

Table 9. Simulation parameters

Item	Description
Stand-by number	1
Number of trial	10000
Battery consumption	normal: cold-start: hot-start: dual = 1.0 : 1.1 : 1.6 : 2.8
Switch time	normal 15 min, cold standby 1 min
Average fault interval	30 days (Poisson distribution)

Table 10. Output of simulation

Output	Description
Latency	Average latency to respond a service request
Availability	Average ratio to be ready to serve a request
Costs	Costs of devices and operation to serve a request

There are three outputs in the simulation. Outputs are depicted in Table 10.

In this paper, costs are not considered because emergency medical treatment cannot compromise availability by costs.

In addition to the normal mode, there are four additional modes to deal with faults. Comparison is done in five cases as depicted in Table 11.

Table 11. Five cases in comparison

Case	Description
Regular	No stand by. When a device fails, an alternative device is invoked after the detection
Cold standby	An alternate device is usable, but not ready for working
Managed standby	A stand-by is initiated according to observed device characteristics
Hot standby	An alternate device is always in standby to replace a failed device
Dual	Always two devices are working and synchronized. One fault does not disrupt continuous operation

5.2 Simulation Results

The managed standby is done as follows:

- after initiation or a fault, a standby device is in a cold standby status.
- according to the previous fault records, a transition time is set.
- when the transition time passes, the standby devices becomes a hot standby status.

Table 12. Simulation conditions

Conditions	Description
Switch time	The switch times for the hot standby mode and the dual mode are negligible. t_n for the normal mode and t_c for the cold standby mode
Battery-based defect	Battery status is not considered to influence defect ratios
Fault ratio	Fault ratios are assumed to be random (Poisson distribution arrival) The same in all modes (No impact from standby or dual modes) Average fault interval is 30 days
Operation complexity	Operation complexity of the dual mode is negligible
Battery overhead	Percentage to the normal mode, $1 + b_c$ for the cold standby mode, $1 + b_h$ for the hot standby mode, $1 + b_d$ for the dual mode

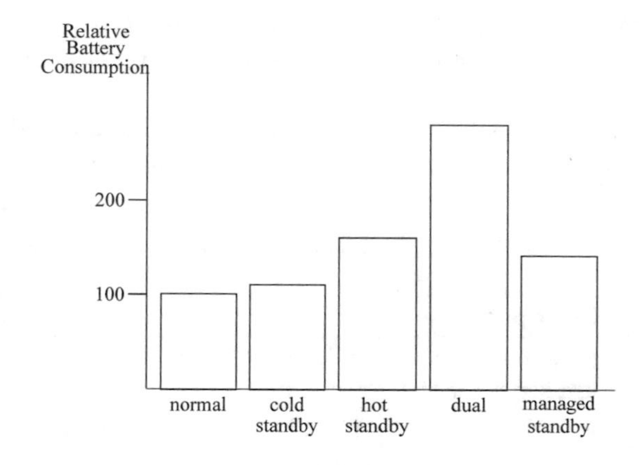

Fig. 4. Comparison of battery consumption (normal mode as 100)

- when a fault occurs before the transition time, the standby device will be activated from the cold standby status.

The simulation conditions are depicted in Table 12.

The simulation is done in R-3.3.3 for Windows (64bit).

The comparison of battery consumption is shown in Fig. 4.

Battery consumption is crucial in the real-world IoT environment. The managed standby mode provides a reasonable alternative for hot-standby and dual modes. It is close to the cold standby mode using a simple threshold method. When a more sophisticated estimate of fault time is possible, it will have a further improvement opportunity.

A comparison of relative downtime is shown in Fig. 5.

For a mission-critical application, hot-standby and dual modes are more suitable than the managed standby mode. The maximum tolerance of down time differs from case to case. When there is some tolerance of down time, the managed standby mode is suitable to balance the downtime and battery efficiency.

6 Discussion

6.1 Advantages of the Proposed Method

Many real-world IoT applications have challenges of hardware faults. In the Cyber-Physical system perspective, it is a challenge to deal with.

The author proposes a hybrid approach of managed standby. The approach uses a combination of cold-standby and hot-standby using data-based proactive management of devices.

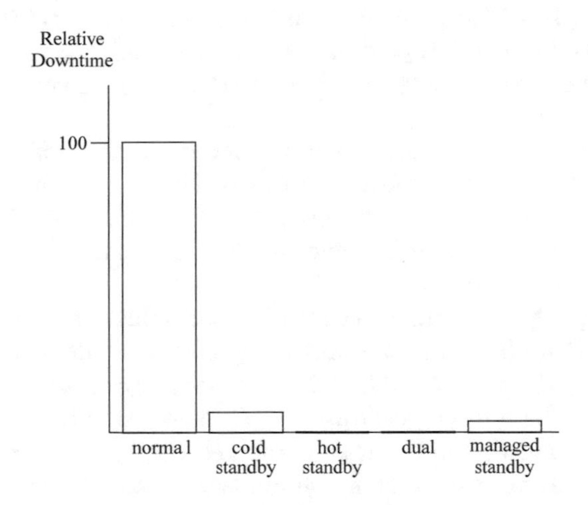

Fig. 5. Comparison of relative downtime (normal mode as 100)

The author performs a preliminary simulation to compare the five modes, normal, cold-standby, hot-standby, dual, and managed standby.

Initial simulation results are promising to pursue proactive management of IoT devices. The faulty devices are difficult to manage by cloud services. The use of relatively inexpensive common CPU boards as a hot-standby or cold-standby for fault situations may be an important consideration in the future where the IoT computing is indispensable for the society. It is conceptually similar to RAID (Redundant Array of Inexpensive Disks) in hard disk.

Depending on a stringent requirement for continuity hot-standby or even a dual configuration may be requested. But our simulation shows that a proactive managed standby may bring in a cost-effective solution.

6.2 Limitations

This paper is exploratory and at its early stage.

Quantitative evaluation of the proposed method is not enough and requires further examination. The proposed control mechanism requires further validation with real-world test data in multiple use cases.

Requirements of maximum downtime vary and require case-by-case examination.

The parameters of simulation are ad hoc, and not based on the real-world data.

Penalties of managed standby is context-dependent and requires further consideration at deployment.

Software-managed fault tolerance of IoT devices is a relatively new field and requires future evaluation.

7 Conclusion

IoT is expanding its coverage in the real world. Increasing computing resources have lead to attention on edge computing. In examining the real-world IoT requirements, there are two types of edge computing: heavy edge computing and lightweight edge computing.

In this paper, the author focuses in the latter, lightweight edge computing. For lightweight edge, it is important to deal with device failures rather than management of complicated computing resources with off-loading. The author defines the binary-level stand-by management is a part of task migration in lightweight edge computing.

The author proposes a managed standby mechanism to make hybrid use of multiple standby mechanisms to dynamically adapt the device management. It facilitates the use of accumulated fault history data to provide proactive management of devices. The author performs a preliminary simulation to evaluate the advantages of managed standby with combined cold standby and hot standby modes. The simulation results show the managed standby method is promising considering the hybrid characteristics of cold-standby and hot-standby. The further research will include proactive management of IoT devices using historical data gathered by IoT monitoring.

This is an encouraging result for future exploration of proactive managed standby in smart city services. The future research include data-mining-based proactive management and deep learning of long-term fault characteristics using aggregated use of IoT maintenance history data.

Acknowledgments. The research results have been achieved by "EUJ-02-2016: IoT/Cloud/Big Data platforms in social application contexts," the Commissioned Research of National Institute of Information and Communications Technology (NICT), JAPAN.

References

1. Azzarà, A., Alessandrelli, D., Petracca, M., Pagano, P.: Demonstration abstract: Pyot, a macroprogramming framework for the IoT. In: Proceedings of the 13th International Symposium on Information Processing in Sensor Networks, IPSN 2014, pp. 315–316. IEEE Press, Piscataway (2014)
2. Cozzolino, V., Ding, A.Y., Ott, J.: FADES: fine-grained edge offloading with unikernels. In: Proceedings of the Workshop on Hot Topics in Container Networking and Networked Systems, HotConNet 2017, pp. 36–41. ACM, New York (2017)
3. CPaaS.io: CPaaS.io – City Platform as a Service (2016). http://www.cpaas.io
4. Dwarakanath, R., Koldehofe, B., Steinmetz, R.: Operator migration for distributed complex event processing in device-to-device based networks. In: Proceedings of the 3rd Workshop on Middleware for Context-Aware Applications in the IoT, M4IoT 2016, pp. 13–18. ACM, New York (2016)
5. Dziurzanski, P., Singh, A.K., Indrusiak, L.S., Saballus, B.: Hard real-time guarantee of automotive applications during mode changes. In: Proceedings of the 23rd International Conference on Real Time and Networks Systems, RTNS 2015, pp. 161–170. ACM, New York (2015)
6. Forum, T.: IoT-engine hardware specification white paper, May 2016. https://www.tron.org/wp-content/uploads/2015/03/TEB061-S101-01.00.00.B0_en.pdf
7. Gascon-Samson, J., Rafiuzzaman, M., Pattabiraman, K.: ThingsJS: towards a flexible and self-adaptable middleware for dynamic and heterogeneous IoT environments. In: Proceedings of the 4th Workshop on Middleware and Applications for the Internet of Things, M4IoT 2017, pp. 11–16. ACM, New York (2017)
8. Gaspar, F., Taniça, L., Tomás, P., Ilic, A., Sousa, L.: A framework for application-guided task management on heterogeneous embedded systems. ACM Trans. Archit. Code Optim. **12**(4), 42:1–42:25 (2015)
9. Giang, N.K., Blackstock, M., Lea, R., Leung, V.C.M.: Distributed data flow: a programming model for the crowdsourced internet of things. In: Proceedings of the Doctoral Symposium of the 16th International Middleware Conference, Middleware Doct Symposium 2015, pp. 4:1–4:4. ACM, New York (2015)
10. Hussain, S.R., Mehnaz, S., Nirjon, S., Bertino, E.: SeamBlue: seamless bluetooth low energy connection migration for unmodified IoT devices. In: Proceedings of the 2017 International Conference on Embedded Wireless Systems and Networks, EWSN 2017, pp. 132–143. Junction Publishing, USA (2017)
11. Hussain, S.R., Mehnaz, S., Nirjon, S., Bertino, E.: Seamless and secure bluetooth LE connection migration. In: Proceedings of the Seventh ACM on Conference on Data and Application Security and Privacy, CODASPY 2017, pp. 147–149. ACM, New York (2017)
12. Jalali, F., Smith, O.J., Lynar, T., Suits, F.: Cognitive IoT gateways: automatic task sharing and switching between cloud and edge/fog computing. In: Proceedings of the SIGCOMM Posters and Demos, SIGCOMM Posters and Demos 2017, pp. 121–123. ACM, New York (2017)
13. Kumar, J.S., Zaveri, M.A.: Clustering for collaborative processing in IoT network. In: Proceedings of the Second International Conference on IoT in Urban Space, Urb-IoT 2016, pp. 95–97. ACM, New York (2016)
14. Ottenwälder, B., Koldehofe, B., Rothermel, K., Ramachandran, U.: MigCEP: operator migration for mobility driven distributed complex event processing. In: Proceedings of the 7th ACM International Conference on Distributed Event-based Systems, DEBS 2013, pp. 183–194. ACM, New York (2013)

15. Pourmohseni, B., Wildermann, S., Glaß, M., Teich, J.: Predictable run-time mapping reconfiguration for real-time applications on many-core systems. In: Proceedings of the 25th International Conference on Real-Time Networks and Systems, RTNS 2017, pp. 148–157. ACM, New York (2017)
16. Quan, W., Pimentel, A.D.: A system-level simulation framework for evaluating task migration in MPSoCs. In: Proceedings of the 2014 International Conference on Compilers, Architecture and Synthesis for Embedded Systems, CASES 2014, pp. 13:1–13:9. ACM, New York (2014)
17. Yang, M., Ma, R.T.: Smooth task migration in apache storm. In: Proceedings of the 2015 ACM SIGMOD International Conference on Management of Data, SIGMOD 2015, pp. 2067–2068. ACM, New York (2015)
18. Zhang, W., Tan, S., Xia, F., Chen, X., Li, Z., Lu, Q., Yang, S.: A survey on decision making for task migration in mobile cloud environments. Pers. Ubiquit. Comput. **20**(3), 295–309 (2016)

The 12th International Workshop on Advances in Information Security (WAIS-2018)

A Security-Aware Fuzzy-Based Cluster Head Selection System for VANETs

Kosuke Ozera[1(✉)], Kevin Bylykbashi[2], Yi Liu[1], and Leonard Barolli[3]

[1] Graduate School of Engineering, Fukuoka Institute of Technology (FIT), 3-30-1 Wajiro-Higashi, Higashi-Ku, Fukuoka 811–0295, Japan
kosuke.o.fit@gmail.com, ryuui1010@gmail.com

[2] Faculty of Information Technologies, Polytechnic University of Tirana, Bul. "Dëshmorët e Kombit", "Mother Theresa" Square, Nr. 4, Tirana, Albania
kevini_95@hotmail.com

[3] Department of Information and Communication Engineering, Fukuoka Institute of Technology (FIT), 3-30-1 Wajiro-Higashi, Higashi-Ku, Fukuoka 811–0295, Japan
barolli@fit.ac.jp

Abstract. In recent years, inter-vehicle communication has attracted attention because it can be applicable not only to alternative networks but also to various communication systems. In this paper, we propose a security-aware Fuzzy-based cluster head selection system for VANETs. We evaluate the proposed system by simulations. From the simulation results, we found that when GS, DC and RA are high, the CHS is high. The simulation result show that the performance of the system is increased when security parameter value is increased.

Keywords: Inter-vehicle communication · VANET · Fuzzy · Cluster

1 Introduction

In recent years, a number of disasters have been occurred around the world. The technologies of disaster management system are improved, however the communication system does not work well in disaster area due to the traffic concentration, device failure, and so on. A key for creating a valuable disaster rescue plan is to prepare alternative communication systems. In disaster area, mobile devices are often disconnected in the network due to network traffic congestion and access point failure. Inter-vehicle communication has attracted attention as an alternative network in disaster situations. In this case, Delay/Disruption/Disconnection Tolerant Networking (DTN) are used as one of a key alternative option to provide the network services [37].

The DTN aims to provide seamless communications with a wide range of network, which have not good performance characteristics [4]. DTN has the potential to interconnect vehicles in regions that current networking protocol

L. Barolli et al. (Eds.): IMIS 2018, AISC 773, pp. 505–516, 2019.
https://doi.org/10.1007/978-3-319-93554-6_48

cannot reach the destination. For inter-vehicle communications, there are different types of communication such as Vehicle-to-Vehicle (V2V), Vehicle-to-Infrastructure (V2I), Vehicle-to-Pedestrian (V2P) and Vehicle-to-X (V2X) communications [2,7,10,14,31]. IEEE 802.11p supports these communications in outdoor environments. It defines enhancements to 802.11 required to support Intelligent Transport System (ITS) applications. The technology operates at 5.9 GHz in various propagation environments to high-speed moving vehicles.

There are different works for Vehicular Ad-Hoc Networks (VANETs). In [16], the authors proposed a Message Suppression Controller (MSC) for V2V and V2I communications. They considered some parameters to control the message suppression dynamically. However, a fixed parameter still is used to calculate the duration of message suppression. To solve this problem, the authors proposed an Enhanced Message Suppression Controller (EMSC) [18] for Vehicular-DTN (V-DTN). The EMSC is an expanded version of MSC [16] and can be used for various network conditions. But, many control packets were delivered in the network.

Security and trust in VANETs is essential in order to prevent malicious agents sabotaging road safety systems built upon the VANET framework, potentially causing serious disruption to traffic flows or safety hazards. Several authors have proposed cluster head metrics which can assist in identifying malicious vehicles and mitigating their impact by denying them access to cluster resources [8].

Security of the safety messages can be achieved by authentication [38]. To make the process of authentication faster [17], vehicles in the communication range of an Road Side Unit (RSU) can be grouped to be in one cluster and a cluster head is elected to authenticate all the vehicles available in the cluster. Formation of clusters in a dynamic VANET and selection of cluster head plays a major role is selected.

In [9] computes a cluster head selection metric based on vehicle direction, degree of connectivity, an entropy value calculated from the mobility of nodes in the network, and a distrust level based on the reliability of a node's packet relaying. Vehicles are assigned verifiers, which are neighbors with a lower distrust value. Verifiers monitor the network behavior of a vehicle, and confirm whether it is routing packets and advertising mobility and traffic information that is consistent with the verifier's own view of the neighborhood. The distrust value for nodes which behave abnormally is then automatically increased, while it is decreased for nodes which perform reliably. In this way, the trustworthiness of a node is accounted for the cluster head selection process.

Fuzzy Logic (FL) is the logic underlying modes of reasoning which are approximate rather then exact. The importance of FL derives from the fact that most modes of human reasoning and especially common sense reasoning are approximate in nature [19]. FL uses linguistic variables to describe the control parameters. By using relatively simple linguistic expressions it is possible to describe and grasp very complex problems. A very important property of the linguistic variables is the capability of describing imprecise parameters.

In this paper, we propose a security-aware Fuzzy-based cluster head selection system for VANETs. We evaluate the proposed system by simulation. The simulation results show that the performance of the system is increased when security parameter value is increased.

The structure of the paper is as follows. In Sect. 2, we present VANETs and DTNs. We give application of Fuzzy Logic for control in Sect. 3. In Sect. 4, we present our proposed systems. In Sect. 5, we show simulation results. Finally, conclusions and future work are given in Sect. 6.

2 VANETs and DTNs

VANETs are considered to have an enormous potential in enhancing road traffic safety and traffic efficiency. Therefore various governments have launched programs dedicated to the development and consolidation of vehicular communications and networking and both industrial and academic researchers are addressing many related challenges, including socio-economic ones, which are among the most important [15,33].

VANET technology uses moving vehicle as nodes to form a wireless mobile network. It aims to provide fast and cost-efficient data transfer for the advantage of passenger safety and comfort. To improve road safety and travel comfort of voyagers and drivers, Intelligent Transport Systems (ITS) are developed. ITS proposes to manage vehicle traffic, support drivers with safety and other information, and provide some services such as automated toll collection and driver assist systems [22].

In essence, VANETs provide new prospects to improve advanced solutions for making reliable communication between vehicles. VANETs can be defined as a part of ITS which aims to make transportation systems faster and smarter in which vehicles are equipped with some short-range and medium-range wireless communication [3]. In a VANET, wireless vehicles are able to communicate directly with each other (i.e., emergency vehicle warning, stationary vehicle warning) and also served various services (i.e., video streaming, internet) from access points (i.e., 3G or 4G) through roadside units [5,22].

The DTN are occasionally connected networks, characterized by the absence of a continuous path between the source and destination [1,13]. The data can be transmitted by storing them at nodes and forwarding them later when there is a working link. This technique is called message switching. Eventually the data will be relayed to the destination. The inspiration for DTNs came from an unlikely source: efforts to send packets in space. Space networks must deal with intermittent communication and very long delays [36]. In [13], the author observed the possibility to apply these ideas for other applications.

The main assumption in the Internet that DTNs seek to relax is that an End-to-End (E2E) path between a source and a destination exists for the entire duration of a communication session. When this is not the case, the normal Internet protocols fail. The DTN architecture is based on message switching. It is also intended to tolerate links with low reliability and large delays. The architecture is specified in RFC 4838 [6].

Bundle protocol has been designed as an implementation of the DTN architecture. A bundle is a basic data unit of the DTN bundle protocol. Each bundle comprises a sequence of two or more blocks of protocol data, which serve for various purposes. In poor conditions, bundle protocol works on the application layer of some number of constituent Internet, forming a store-and-forward overlay network to provide its services. The bundle protocol is specified in RFC 5050 [34]. It is responsible for accepting messages from the application and sending them as one or more bundles via store-carry-forward operations to the destination DTN node. The bundle protocol provides a transport service for many different applications.

3 Application of Fuzzy Logic for Control

The ability of fuzzy sets and possibility theory to model gradual properties or soft constraints whose satisfaction is matter of degree, as well as information pervaded with imprecision and uncertainty, makes them useful in a great variety of applications.

The most popular area of application is Fuzzy Control (FC), since the appearance, especially in Japan, of industrial applications in domestic appliances, process control, and automotive systems, among many other fields [11,12,20,24–28,35].

3.1 FC

In the FC systems, expert knowledge is encoded in the form of fuzzy rules, which describe recommended actions for different classes of situations represented by fuzzy sets.

In fact, any kind of control law can be modeled by the FC methodology, provided that this law is expressible in terms of "if ... then ..." rules, just like in the case of expert systems. However, FL diverges from the standard expert system approach by providing an interpolation mechanism from several rules. In the contents of complex processes, it may turn out to be more practical to get knowledge from an expert operator than to calculate an optimal control, due to modeling costs or because a model is out of reach.

3.2 Linguistic Variables

A concept that plays a central role in the application of FL is that of a linguistic variable. The linguistic variables may be viewed as a form of data compression. One linguistic variable may represent many numerical variables. It is suggestive to refer to this form of data compression as granulation [21].

The same effect can be achieved by conventional quantization, but in the case of quantization, the values are intervals, whereas in the case of granulation the values are overlapping fuzzy sets. The advantages of granulation over quantization are as follows:

- it is more general;
- it mimics the way in which humans interpret linguistic values;
- the transition from one linguistic value to a contiguous linguistic value is gradual rather than abrupt, resulting in continuity and robustness.

3.3 FC Rules

FC describes the algorithm for process control as a fuzzy relation between information about the conditions of the process to be controlled, x and y, and the output for the process z. The control algorithm is given in "if ... then ..." expression, such as:

$$\text{If } x \text{ is small and } y \text{ is big, then } z \text{ is medium;}$$
$$\text{If } x \text{ is big and } y \text{ is medium, then } z \text{ is big.}$$

These rules are called *FC rules.* The "if" clause of the rules is called the antecedent and the "then" clause is called consequent. In general, variables x and y are called the input and z the output. The "small" and "big" are fuzzy values for x and y, and they are expressed by fuzzy sets.

Fuzzy controllers are constructed of groups of these FC rules, and when an actual input is given, the output is calculated by means of fuzzy inference.

3.4 Control Knowledge Base

There are two main tasks in designing the control knowledge base. First, a set of linguistic variables must be selected which describe the values of the main control parameters of the process. Both the input and output parameters must be linguistically defined in this stage using proper term sets. The selection of the level of granularity of a term set for an input variable or an output variable plays an important role in the smoothness of control. Second, a control knowledge base must be developed which uses the above linguistic description of the input and output parameters. Four methods [29,32,39,40] have been suggested for doing this:

- expert's experience and knowledge;
- modelling the operator's control action;
- modelling a process;
- self organization.

Among the above methods, the first one is the most widely used. In the modeling of the human expert operator's knowledge, fuzzy rules of the form "If Error is small and Change-in-error is small then the Force is small" have been used in several studies [23,30]. This method is effective when expert human operators can express the heuristics or the knowledge that they use in controlling a process in terms of rules of the above form.

3.5 Defuzzification Methods

The defuzzification operation produces a non-FC action that best represent the membership function of an inferred FC action. Several defuzzification methods have been suggested in literature. Among them, four methods which have been applied most often are:

- Tsukamoto's Defuzzification Method;
- The Center of Area (COA) Method;
- The Mean of Maximum (MOM) Method;
- Defuzzification when Output of Rules are Function of Their Inputs.

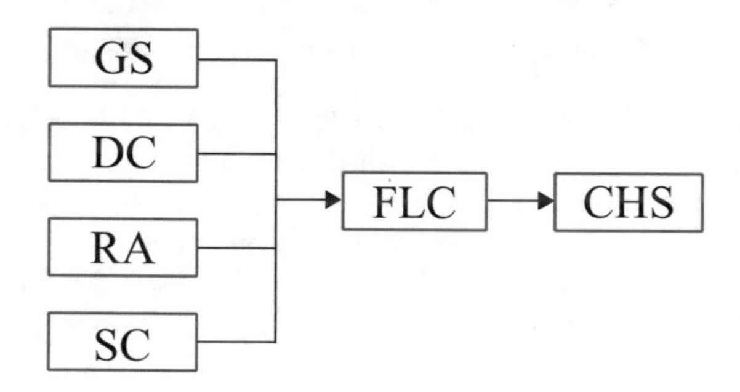

Fig. 1. Proposed system model.

4 Proposed System

The proposed model is shown in Fig. 1. We consider four parameters: Group Speed (GS), Degree of Connectivity (DC), Relative Acceleration (RA) and Security (SC) to decide the Cluster Head Selection (CHS) output parameter. These four parameters are not correlated with each other, for this reason we use fuzzy system. The membership functions are shown in Fig. 2. In Table 1, we show the Fuzzy Rule Base (FRB), which consists of 81 rules.

The term sets of *GS*, *DC*, *RA* and *SC* are defined respectively as:

$$\begin{aligned} GS &= \{Slow,\ Middle,\ Fast\} \\ &= \{Sl,\ Mi,\ Fa\}; \\ DC &= \{Low,\ Middle,\ High\} \\ &= \{L,\ M,\ H\}; \\ RA &= \{Decelerate,\ Same,\ Accelerate\} \\ &= \{Dec,\ Sam,\ Acc\}; \\ SC &= \{Weak,\ Middle,\ Strong\} \\ &= \{We,\ Mi,\ St\}. \end{aligned}$$

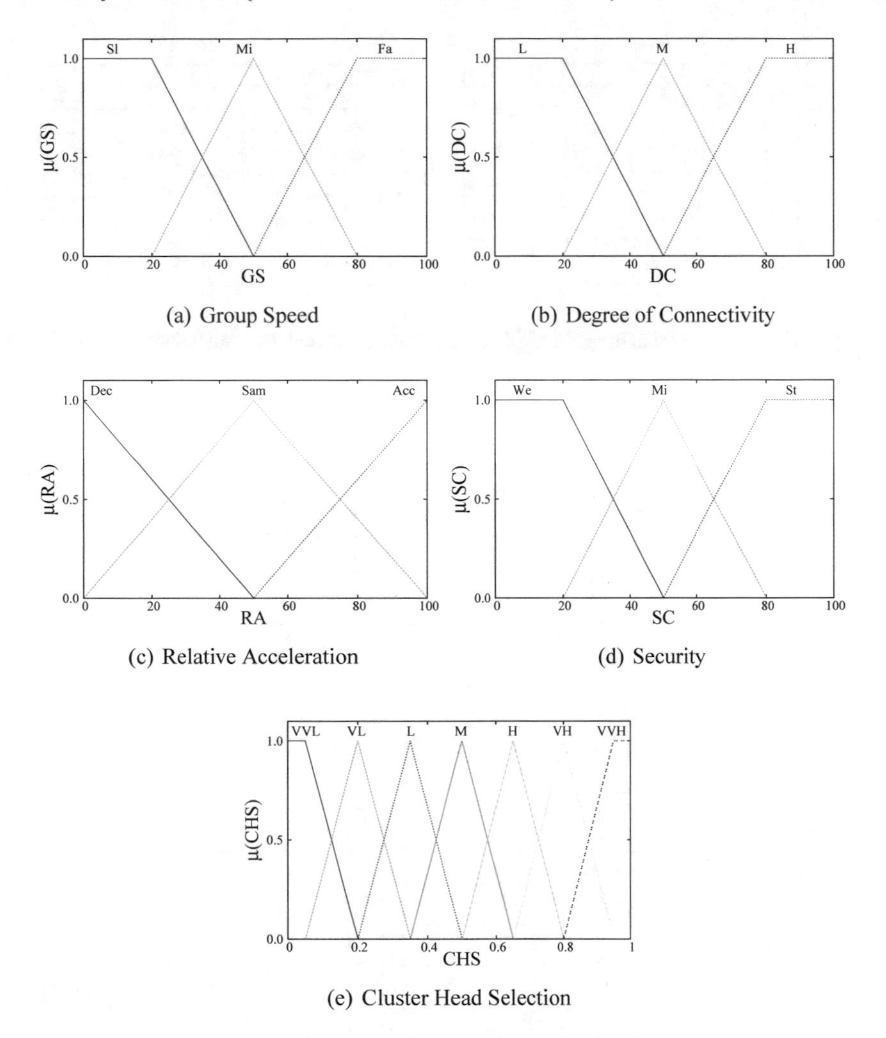

(a) Group Speed

(b) Degree of Connectivity

(c) Relative Acceleration

(d) Security

(e) Cluster Head Selection

Fig. 2. Membership functions.

and the term set for the output CHS is defined as:

$$CHS = \begin{pmatrix} VeryVeryLow \\ VeryLow \\ Low \\ Middle \\ High \\ VeryHigh \\ VeryVeryHigh \end{pmatrix} = \begin{pmatrix} VVL \\ VL \\ L \\ M \\ H \\ VH \\ VVH \end{pmatrix}.$$

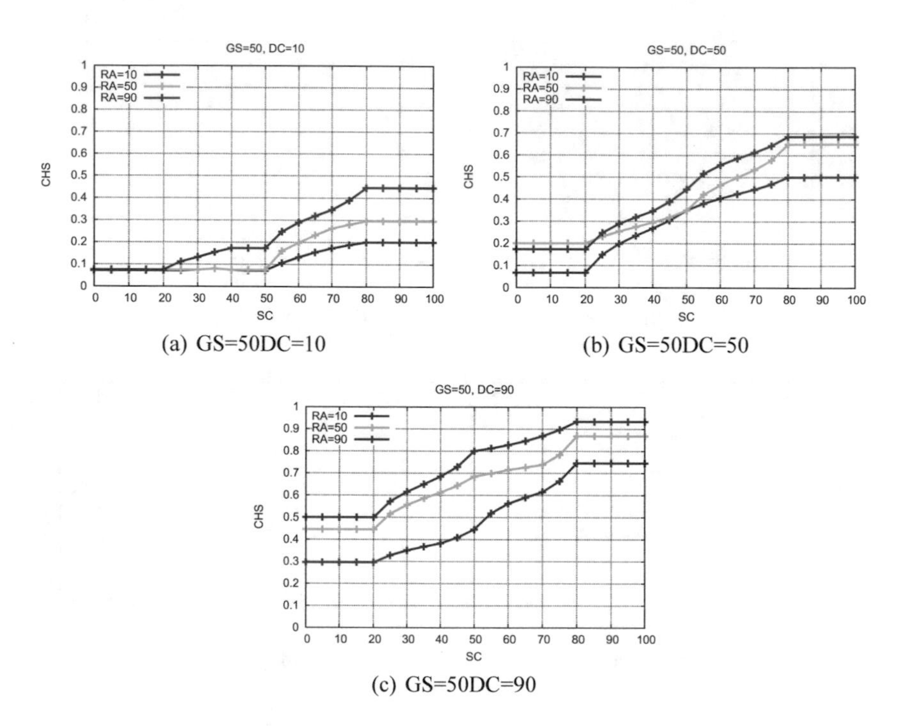

(a) GS=50DC=10

(b) GS=50DC=50

(c) GS=50DC=90

Fig. 3. Simulation results when GS is 50.

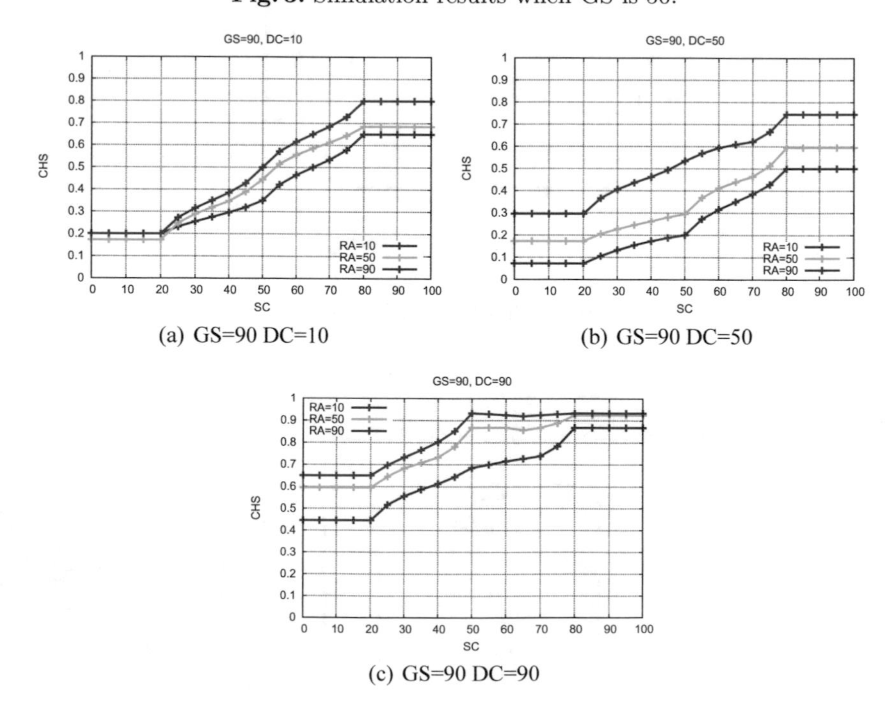

(a) GS=90 DC=10

(b) GS=90 DC=50

(c) GS=90 DC=90

Fig. 4. Simulation results when GS is 90.

Table 1. FRB.

Rule No.	GS	DC	RA	SC	CHS	Rule No.	GS	DC	RA	SC	CHS	Rule No.	GS	DC	RA	SC	CHS
1	Sl	L	Dec	We	VVL	28	Mi	L	Dec	We	VVL	55	Fa	L	Dec	We	VVL
2	Sl	L	Dec	Mi	VVL	29	Mi	L	Dec	Mi	VVL	56	Fa	L	Dec	Mi	VL
3	Sl	L	Dec	St	VVL	30	Mi	L	Dec	St	VL	57	Fa	L	Dec	St	L
4	Sl	L	Sam	We	VVL	31	Mi	L	Sam	We	VL	58	Fa	L	Sam	We	L
5	Sl	L	Sam	Mi	VVL	32	Mi	L	Sam	Mi	L	59	Fa	L	Sam	Mi	M
6	Sl	L	Sam	St	VL	33	Mi	L	Sam	St	M	60	Fa	L	Sam	St	H
7	Sl	L	Acc	We	VL	34	Mi	L	Acc	We	L	61	Fa	L	Acc	We	H
8	Sl	L	Acc	Mi	L	35	Mi	L	Acc	Mi	H	62	Fa	L	Acc	Mi	VH
9	Sl	L	Acc	St	M	36	Mi	L	Acc	St	VH	63	Fa	L	Acc	St	VVH
10	Sl	M	Dec	We	VVL	37	Mi	M	Dec	We	VVL	64	Fa	M	Dec	We	VL
11	Sl	M	Dec	Mi	VVL	38	Mi	M	Dec	Mi	VL	65	Fa	M	Dec	Mi	M
12	Sl	M	Dec	St	VL	39	Mi	M	Dec	St	L	66	Fa	M	Dec	St	H
13	Sl	M	Sam	We	VVL	40	Mi	M	Sam	We	L	67	Fa	M	Sam	We	M
14	Sl	M	Sam	Mi	VL	41	Mi	M	Sam	Mi	M	68	Fa	M	Sam	Mi	VH
15	Sl	M	Sam	St	L	42	Mi	M	Sam	St	H	69	Fa	M	Sam	St	VH
16	Sl	M	Acc	We	L	43	Mi	M	Acc	We	H	70	Fa	M	Acc	We	VH
17	Sl	M	Acc	Mi	M	44	Mi	M	Acc	Mi	VH	71	Fa	M	Acc	Mi	VVH
18	Sl	M	Acc	St	H	45	Mi	M	Acc	St	VVH	72	Fa	M	Acc	St	VVH
19	Sl	H	Dec	We	VVL	46	Mi	H	Dec	We	L	73	Fa	H	Dec	We	M
20	Sl	H	Dec	Mi	VL	47	Mi	H	Dec	Mi	M	74	Fa	H	Dec	Mi	VH
21	Sl	H	Dec	St	L	48	Mi	H	Dec	St	H	75	Fa	H	Dec	St	VH
22	Sl	H	Sam	We	L	49	Mi	H	Sam	We	H	76	Fa	H	Sam	We	VH
23	Sl	H	Sam	Mi	M	50	Mi	H	Sam	Mi	VH	77	Fa	H	Sam	Mi	VVH
24	Sl	H	Sam	St	H	51	Mi	H	Sam	St	VVH	78	Fa	H	Sam	St	VVH
25	Sl	H	Acc	We	H	52	Mi	H	Acc	We	VH	79	Fa	H	Acc	We	VVH
26	Sl	H	Acc	Mi	VH	53	Mi	H	Acc	Mi	VVH	80	Fa	H	Acc	Mi	VVH
27	Sl	H	Acc	St	VVH	54	Mi	H	Acc	St	VVH	81	Fa	H	Acc	St	VVH

5 Simulation Results

In this section, we present the simulation results for our proposed system. In our system, we decided the membership functions by carrying out many simulations. The simulation results were carried out in MATLAB.

In Figs. 3 and 4, we show the relation between GS, DC, RA and SC. We consider the GS and DC as constant parameters.

In Figs. 3(a) and 4(a), we consider the DC value 10. We change the SC value from 0 to 100. When SC increases, the CHS is increased. Also, when the RA increases, the CHS is increased.

In Figs. 3(b) and 4(b), we increase the DC value to 50. We can see that by increasing DC value, the CHS is increased. Then in Figs. 3(c) and 4(c), we increase the DC value to 90. It can be seen that the CHS value is increased much more.

6 Conclusions

In this paper, we proposed a security-aware fuzzy-based cluster head selection system for VANETs. We considering four parameters: GS, DC, RA and SC. We evaluated the performance of proposed system by computer simulations. The simulation results show that the performance of the system is increased when all four parameters have the highest value.

In the future work, we will consider other parameters for the simulation system and carry out extensive simulations.

References

1. Delay- and disruption-tolerant networks (DTNs) tutorial. NASA/JPL's Interplanetary Internet (IPN) Project (2012). http://www.warthman.com/images/DTN_Tutorial_v2.0.pdf
2. Araniti, G., Campolo, C., Condoluci, M., Iera, A., Molinaro, A.: Lte for vehicular networking: a survey. IEEE Commun. Mag. **21**(5), 148–157 (2013)
3. Booysen, M.J., Zeadally, S., van Rooyen, G.J.: Performance comparison of media access control protocols for vehicular ad hoc networks. IET Netw. **1**(1), 10–19 (2012)
4. Burleigh, S., Hooke, A., Torgerson, L., Fall, K., Cerf, V., Durst, B., Scott, K., Weiss, H.: Delay-tolerant networking: an approach to interplanetary internet. IEEE Commun. Mag. **41**(6), 128–136 (2003)
5. Calhan, A.: A fuzzy logic based clustering strategy for improving vehicular ad-hoc network performance **40**(2), 351–367 (2015)
6. Cerf, V., Burleigh, S., Hooke, A., Torgerson, L., Durst, R., Scott, K., Fall, K., Weiss, H.: Delay-tolerant networking architecture. IETF RFC 4838 (Informational), April 2007
7. Cheng, X., Yao, Q., Wen, M., Wang, C.X., Song, L.Y., Jiao, B.L.: Wideband channel modeling and intercarrier interference cancellation for vehicle-to-vehicle communication systems. IEEE J. Sel. Areas Commun. **31**(9), 434–448 (2013)
8. Cooper, C., Franklin, D., Ros, M., Safaei, F., Abolhasan, M.: A comparative survey of vanet clustering techniques. IEEE Commun. Surv. Tutorials **19**(1), 657–681 (2017)
9. Daeinabi, A., Rahbar, A.G.P., Khademzadeh, A.: VWCA: an efficient clustering algorithm in vehicular ad hoc networks. J. Netw. Comput. Appl. **34**(1), 207–222 (2011)
10. Dias, J.A.F.F., Rodrigues, J.J.P.C., Xia, F., Mavromoustakis, C.X.: A cooperative watchdog system to detect misbehavior nodes in vehicular delay-tolerant networks. IEEE Trans. Industr. Electron. **62**(12), 7929–7937 (2015)
11. Elmazi, D., Kulla, E., Oda, T., Spaho, E., Sakamoto, S., Barolli, L.: A comparison study of two fuzzy-based systems for selection of actor node in wireless sensor actor networks. J. Ambient Intell. Humaniz. Comput. **6**(5), 635–645 (2015)
12. Elmazi, D., Sakamoto, S., Oda, T., Kulla, E., Spaho, E., Barolli, L.: Two fuzzy-based systems for selection of actor nodes in wireless sensor and actor networks: a comparison study considering security parameter effect. Mob. Netw. Appl. 21(1), 1–12 (2016)
13. Fall, K.: A delay-tolerant network architecture for challenged Internets. In: Proceedings of the International Conference on Applications, Technologies, Architectures, and Protocols for Computer Communications SIGCOMM 2003, pp. 27–34 (2003)
14. Grassi, G., Pesavento, D., Pau, G., Vuyyuru, R., Wakikawa, R., Zhang, L.: VANET via named data networking. In: Proceedings of the IEEE Conference on Computer Communications Workshops (INFOCOM WKSHPS 2014), pp. 410–415, April 2014

15. Hartenstein, H., Laberteaux, L.: A tutorial survey on vehicular ad hoc networks. IEEE Commun. Mag. **46**(6), 164–171 (2008)
16. Honda, T., Ikeda, M., Ishikawa, S., Barolli, L.: A message suppression controller for vehicular delay tolerant networking. In: Proceedings of the 29th IEEE International Conference on Advanced Information Networking and Applications (IEEE AINA-2015), pp. 754–760, March 2015
17. Huang, J.L., Yeh, L.Y., Chien, H.Y.: Abaka: an anonymous batch authenticated and key agreement scheme for value-added services in vehicular ad hoc networks. IEEE Trans. Veh. Technol. **60**(1), 248–262 (2011)
18. Ikeda, M., Ishikawa, S., Barolli, L.: An enhanced message suppression controller for vehicular-delay tolerant networks. In: Proceedings of the 30th IEEE International Conference on Advanced Information Networking and Applications (IEEE AINA-2016), pp. 573–579, March 2016
19. Inaba, T., Obukata, R., Sakamoto, S., Oda, T., Ikeda, M., Barolli, L.: Performance evaluation of a qos-aware fuzzy-based cac for lan access. Int. J. Space Based Situ. Comput. **6**(4), 228–238 (2016)
20. Inaba, T., Sakamoto, S., Oda, T., Ikeda, M., Barolli, L.: A secure-aware call admission control scheme for wireless cellular networks using fuzzy logic and its performance evaluation. J. Mob. Multimedia **11**(3&4), 213–222 (2015)
21. Kandel, A.: Fuzzy Expert Systems. CRC press, Boca Raton (1991)
22. Karagiannis, G., Altintas, O., Ekici, E., Heijenk, G., Jarupan, B., Lin, K., Weil, T.: Vehicular networking: a survey and tutorial on requirements, architectures, challenges, standards and solutions. IEEE Commun. Surv. Tutorials **13**(4), 584–616 (2011)
23. Klir, G.J., Folger, T.A.: Fuzzy Sets, Uncertainty, and Information. Prentice Hall, Upper Saddle River (1988)
24. Kolici, V., Inaba, T., Lala, A., Mino, G., Sakamoto, S., Barolli, L.: A fuzzy-based CAC scheme for cellular networks considering security. In: Proceedings of the 17th International Conference on Network-Based Information Systems (NBiS-2014), pp. 368–373 (2014)
25. Liu, Y., Sakamoto, S., Matsuo, K., Ikeda, M., Barolli, L.: Improving reliability of JXTA-overlay platform: evaluation for E-learning and trustworthiness. J. Mob. Multimedia **11**(2), 34–50 (2015)
26. Liu, Y., Sakamoto, S., Matsuo, K., Ikeda, M., Barolli, L., Xhafa, F.: A comparison study for two fuzzy-based systems: improving reliability and security of JXTAoverlay P2P platform. Soft Comput. **20**(7), 2677–2687 (2016)
27. Liu, Y., Sakamoto, S., Matsuo, K., Ikeda, M., Barolli, L., Xhafa, F.: Improvement of JXTA-overlay P2P Platform: evaluation for medical application and reliability. Int. J. Distrib. Syst. Technol. (IJDST) **6**(2), 45–62 (2015)
28. Matsuo, K., Elmazi, D., Liu, Y., Sakamoto, S., Mino, G., Barolli, L.: FACS-MP: a fuzzy admission control system with many priorities for wireless cellular networks and its performance evaluation. J. High Speed Netw. **21**(1), 1–14 (2015)
29. McNeill, F.M., Thro, E.: Fuzzy Logic: A Practical Approach. Academic Press, Cambridge (1994)
30. Munakata, T., Jani, Y.: Fuzzy systems: an overview. Commun. ACM **37**(3), 68–76 (1994)
31. Ohn-Bar, E., Trivedi, M.M.: Learning to detect vehicles by clustering appearance patterns. IEEE Trans. Intell. Transp. Syst. **16**(5), 2511–2521 (2015)
32. Procyk, T.J., Mamdani, E.H.: A linguistic self-organizing process controller. Automatica **15**(1), 15–30 (1979)

33. Santi, P.: Mobility Models for Next Generation Wireless Networks: Ad hoc, Vehicular and Mesh Networks. Wiley, Hoboken (2012)
34. Scott, K., Burleigh, S.: Bundle protocol specification. IETF RFC 5050 (Experimental), November 2007
35. Spaho, E., Sakamoto, S., Barolli, L., Xhafa, F., Ikeda, M.: Trustworthiness in P2P: performance behaviour of two fuzzy-based systems for JXTA-overlay platform. Soft. Comput. **18**(9), 1783–1793 (2014)
36. Tanenbaum, A.S., Wetherall, D.J.: Computer Networks, 5th edn. Pearson Education Inc., Prentice Hall, Upper Saddle River (2011)
37. Uchida, N., Ishida, T., Shibata, Y.: Delay tolerant networks-based vehicle-to-vehicle wireless networks for road surveillance systems in local areas. Int. J. Space-Based Situ. Comput. **6**(1), 12–20 (2016)
38. Wen, H., Ho, P.H., Gong, G.: A novel framework for message authentication in vehicular communication networks. In: Global Telecommunications Conference GLOBECOM 2009, pp. 1–6. IEEE (2009)
39. Zadeh, L.A., Kacprzyk, J.: Fuzzy Logic for the Management of Uncertainty. Wiley, Hoboken (1992)
40. Zimmermann, H.J.: Fuzzy Set Theory and Its Applications. Springer Science & Business Media, New York (1991)

Research on Food Safety Traceability Technology Based on RFID Security Authentication and 2-Dimensional Code

Jie Ding[1,2], He Xu[1,2(✉)], Peng Li[1,2], and Feng Zhu[1,2]

[1] School of Computer, Nanjing University of Posts and Telecommunications, Nanjing, China
1248395233@qq.com, {xuhe, lipeng, zhufeng}@njupt.edu.cn
[2] Jiangsu High Technology Research Key Laboratory for Wireless Sensor Networks, Nanjing, China

Abstract. With the development of society, there are food safety issues. From Sanlu's "melamine incident", the chemical composition of American McLean chickens to "false zisha" and "clenbuterol", these hot topics have made people more and more concerned about food safety issues. In order to improve the quality and safety of food, and also to meet the transparent demand of producers and consumers for food production, it is particularly urgent to build a set of standardized, intelligent anti-counterfeiting traceability systems. Nowadays, thanks to the rapid development of Internet of Things technology, the realization of anti-counterfeiting traceability systems has become possible. This paper researches on RFID technology, combined with anti-counterfeiting technology and Internet technology, puts forward a realization project of food security anti-counterfeiting tracing system based on RFID security authentication and 2-dimensional code. The purpose of this system is to allow consumers to reassure about the products they purchase. The realized system can quickly and efficiently inquire into raw materials or processing problems when there are product quality problems. It can contribute to quality control and recalls products when necessary, so as to improve the competitiveness of the company.

1 Introduction

Food safety issues concern everyone's health and affect everyone's brain nerves. In particular, the frequent occurrence of food safety incidents such as waste oil, rubber injection, and genetically modified oil in recent years has directly affected the lives safety of people. The issue of food safety has become the issue most concerned by the public and all over the world. This shows that China's food safety supervision is being strengthened, and people's food safety awareness is also improving. However, there are still many problems in China's food safety. The phobia of food safety in the whole society has been triggered, and solving food safety problems has become an urgent task. In recent years, China has been actively investigating food safety issues. However, it has been repeatedly prohibited. In fact, food safety issues not only originate from the integrity and ethics of food production and processing companies, but also have problems in production and processing technologies. Therefore, we must attach

L. Barolli et al. (Eds.): IMIS 2018, AISC 773, pp. 517–526, 2019.
https://doi.org/10.1007/978-3-319-93554-6_49

importance to the problems of food safety and analyze the causes of the problems before we can address the problem, effectively solve food safety problems and ensure consumer safety.

2 Relevant Technology

2.1 Radio Frequency Identification Technology

Radio frequency identification technology is able to recognize objects and people automatically, and it can also automatically obtain related data of recognized objects, which is the non-contract recognition technique [1]. The recognition of RFID does not only need the artificial interference, but also work well in the severe environment. By now, most RFID system is based on the electric induction [2]. Attaching an RFID tag on an object, which involving the information of this object, the dedicated recognition terminal can recognize this attached object. Also, since RFID products read data which has no contact and can pass through external material, the service life is durable. Compared with bar code, RFID products have more advantages [3].

RFID has been used into many areas, such as supply chain management, electric passport, credit card, driving license, vehicle system (charging system, keyless entry systems), entrance guard card (building gate, public transport) and health care [4, 5]. Particularly, in the USA, Japan, and other developed countries, they have equipped with advanced and mature RFID systems [6]. Some retailers have also invested RFID technology, and also authorized RFID producers to attach tags on their goods, so that the low-cost RFID tags are pervasively produced. Wal-Mart passed a resolution, which producers must sufficiently take advantage of the RFID, and attach RFID tags on all products, in order to reduce manpower and material resources [7].

2.2 Spring Framework

Spring is an open source framework, which was a lightweight Java development framework that emerged in 2003 [8]. It was derived from some of the ideas and prototypes that Rod Johnson elaborated in his book "Expert One-On-One J2EE Development and Design". It was created to solve the complexity of enterprise application development. Spring uses basic javaBeans. However, the use of Spring is not limited to server-side development. Any Java application can benefit from Spring for the standpoint of simplicity, testability, and coupling. In simple terms, Spring is a lightweight container control framework for Inversion of Control (IoC) and Aspect Oriented Programming (AOP).

2.3 Spring MVC

Spring MVC is a follow-up product of SpringFrameWork and has been integrated into Spring Web Flow [9]. Spring MVC separates the controllers, model objects, dispatchers, and handler objects. This separation makes them easier to customize.

2.4 Mybatis

MyBatis is an open source project (ibatis) of apache [10]. In 2010, the project was migrated from Apache software foundation to google code, and changed its name to MyBatis. In essence, Mybatis made some improvements to ibatis. MyBatis is an excellent persistence framework that encapsulates the process of JDBC's operations on the database, so that developers only need to pay attention to the SQL language itself, and do not need to spend effort to deal with JDBC complicated process code, such as registration driver, create connection, create statement, manually set parameters, result set retrieval.

Mybatis configures the various statements (Statement, PreparedStatement, and CallableStatement) through xml or annotation. And it generates the final executed sql statement by mapping the SQL in the java object and the statement. Finally, the SQL is executed by the mybatis framework and the result is mapped to the Java object and returned.

3 RFID Anti-counterfeiting Technology

3.1 The Basic Principle of RFID Anti-counterfeiting

RFID tags are divided into five categories according to their functions. This paper uses Class1 tags which only have information storage capabilities and do not have cryptographic capabilities. Electronic Product Code (EPC) is an electronic coding method commonly used in Europe and the United States. Its purpose is to provide unique electronic identifiers for each product, including information on manufacturers, product categories, etc. The product uses the EPC code to uniquely identify the product and meets the current standards for the development and construction of the Internet of Things.

Due to the limitations of RFID readers and ultra-high frequency tags, the concepts of "random tail" and "number of reads" are adopted in this paper to prevent forgery. In the initial stage of the tag, each tag can be assigned an ID and a random number. Every time a label passes through a place, the reader adds a random mark after the initial random number of the label, and the updated random number is also kept in the background database, so that these random signs constitute a "tail" in the whole supply chain. As time goes on, the "tail" of the actual tag will be distinguished from the "tail" of the clone tag, so the clone tag can be detected by the inconsistency of their "tails".

When the false label and the certified label are in the same reader reading range, the method of anti-counterfeiting called "read number" is adopted. It is to use the reader's bottom to read the RFID tag information to optimize the anti-counterfeit. There are multiple labels to be read in the scope of the reader, so the reader displays the label information in a round of labels. In addition. The reading order of the labels in each round is random. If the reader reads 200 times and there are 4 tags to be certified for anti-counterfeiting within the scope of the reader, the 4 tags are only read 50 times. By comparing the number of times each label is read and the average number of times each tag is read, we find that the label is not valid because it is out of range. If there are 4 tags (marked as A tag, B tag, C tag and D tag) to be authenticated in the reading range of the reader and the B tag is the clone a tag, which means that the A tag with the same

EPC code of the B tag. When the reader reads 200 times, each label is read 50 times according to the above principle but the A tag and B tag are the same as the EPC code, so the system adds the number of reads of A tag and B tag. So A tag or B tag was read 100 times, C tag or D tag was 50 times, and the average number of reads per tag was 50 at the end of the show. At this time, the system will detect that the number of A tag or B tag is 100 and the average number of tags read is 50, which is far beyond the scope. So the system regards that A tag or B tag suffers from the attack of clone.

3.2 The Implementation of RFID Anti-counterfeiting Principle

(1) The initialization of tags

Each RFID tag has a unique identification number EPC code and uses this EPC code to identify the corresponding tag. The administrator initialized the user area of the tag to zero, and synchronize the read information with the database in the background server.

(2) The reading and verifying of tag information

When the tag enters the read range of the reader, the reader reads the tag information and matches the background database. If the information of the tag can be found in the database, the tag is considered to be authenticated; otherwise, the tag is considered to be fake.

(3) Add the tag's random number tail

When the tag is authenticated, a random number is added to the end of the user area of the tag to be authenticated as a random number tail.

(4) Label anti-counterfeiting certification

The reader uses the above basic principles of anti-counterfeiting for authentication.

4 Anti-counterfeit Traceability System

4.1 The Actual Link Design of Anti-counterfeit Traceability System

Our concern for food is preservation, deterioration, and timeliness. When these problems arise, traceability is necessary to every aspects of food: production, procurement, processing, storage, transportation, and sales, in order to achieve the purpose of tracking product information.

The food safety anti-counterfeiting traceability system records the information of each traceability link. It is a necessary tool for achieving full food quality control, and is an important window for consumers to enjoy the right to know about food quality. The purpose of the food safety anti-counterfeiting traceability system is that to provide a solution to food safety problems. When a problematic food is found, the food safety anti-counterfeit traceability system can be used to find out where the problem is, so as to solve the problem at the root of the problem, and can timely recover the unsold food on the market, minimize the harm and loss.

In practice, it is mainly divided into production link, processing link, warehousing link, transport link and sales link. The following gives a case of milk production.

(1) **Production Link**

In a relatively large scale dairy farms, because dairy farms generally carry out large-scale dairy farming and intensive management, they have the conditions of using RFID technology. When applying RFID tags, the base management staff can bind a label for each cow. During the each link of from birth to milk production, the base management staff can write the information of the cow to each tag through RFID equipment. A unique code for the milk produced by each cow is set. Thus, in the first step (growth phase) of the milk tracking traceability system, the RFID tag has stored the basic information of the milk.

(2) **Processing Link**

Because RFID tags have the advantages of convenient information reading and storage, milk processing companies can store the necessary information for processing in accordance with their own needs and regulations of the regulatory authorities. The information data of this link includes: the name of the company, the processing date, the packaging materials used in the processing, the additives used in the deep processing, and the weight. Once the processing enterprise input the information of this link to the electronic label, the retailer in the wholesale market inquires the milk product information through data fusion.

(3) **Warehousing Link**

Milk as a product with high preservation requirements has a high storage environment for the warehouse. If the storage environment does not meet the requirements, warehouse storage time should be reduced as much as possible. The RFID devices used in the warehousing link include handheld devices and fixed RFID devices. For the problem of collision and interference of tag signals that occur when multiple electronic tags interact with the reader, there is no further study in this paper.

(4) **Transport Link**

The application of RFID technology in the transport link is mainly reflected in logistics tracking. Combining RFID technology and GPS global positioning system can provide logistics companies with real-time tracking and inquiry services. On the other hand, for customers, they can log on to the agricultural product e-commerce platform through the Internet to understand the real-time location of orders and whether they have been dropped. The transport supervision unit does not need to unpack the goods during the inspection at the crossing. The reader can read the electronic tags on the packages to understand the cargo information, which greatly improves the speed of the goods during the transportation.

(5) **Sales Link**

Providing customer or consumer milk agricultural product information inquiry service, retailers reading electronic labels on agricultural products packaging through readers, and printing detailed information on emerging technology labels such as two-dimensional codes. The product details can be found by scanning the QR code. Figure 1 shows the frame diagram for milk prodcution.

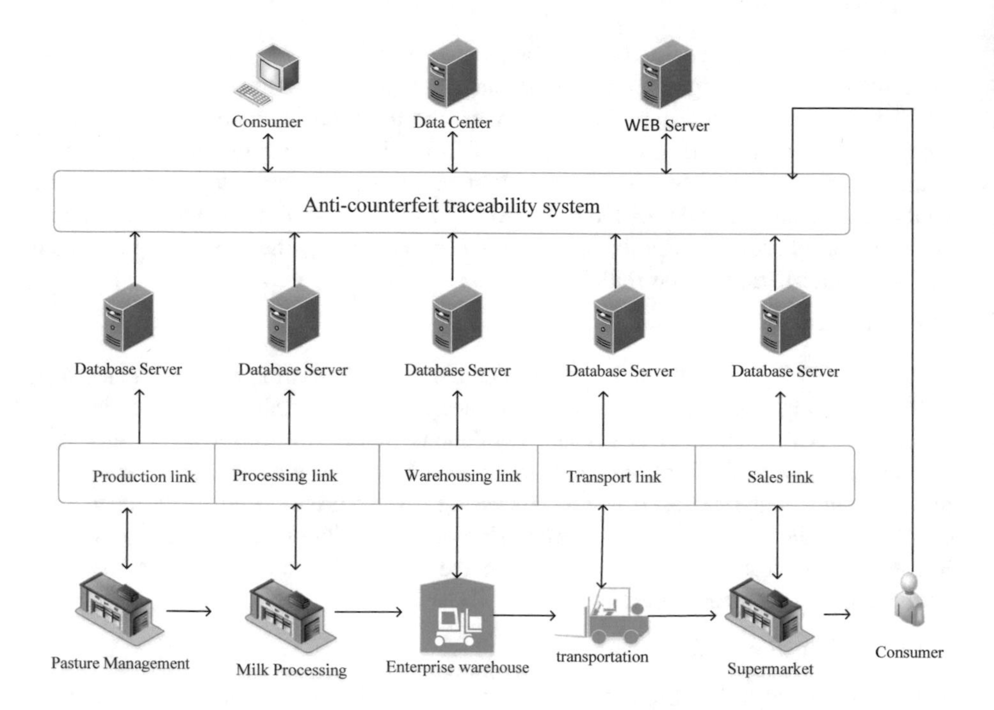

Fig. 1. Frame diagram of the actual link

4.2 The Implementation of Anti-counterfeit Traceability System

Anti-Counterfeit Tracing System is developed using SSM framework. SSM is an integration framework using Spring + SpringMVC + MyBatis. It is a popular open source framework for Web applications.

The flow of the SSM framework is: First, the Controller in SpringMVC receives the client's request, and then the Controller invokes the method in the service layer. Next, the service queries the database via MyBatis, and returns the result to the controller. Finally, the controller returns the result to the interface to display. Throughout the process, spring acts as a container to manage the SpringMVC framework and the MyBatis framework. Next, we will elaborate on the various framework processes (Fig. 2).

(1) **Presentation Layer SpringMVC Architecture Flow**

1) The user sends a request to the front controller DispatcherServlet.
2) DispatcherServlet received a request to call the HandlerMapping.
3) The HandlerMapping finds a specific processor according to the request URL, generates a processor object and a processor interceptor (if generated), and returns it to the DispatcherServlet.

4) DispatcherServlet calls the handler through the HandlerAdapter.
5) The implementation of the processor (Controller, also known as the back-end controller).
6) Controller finishes to return ModelAndView.
7) The HandlerAdapter returns execution result ModelAndView to the DispatcherServlet.
8) The DispatcherServlet passes the ModelAndView to the ViewReslover.
9) ViewReslover returns a specific View after parsing.
10) DispatcherServlet renders a view (Filling the model data into the view).
11) DispatcherServlet responds to the user.

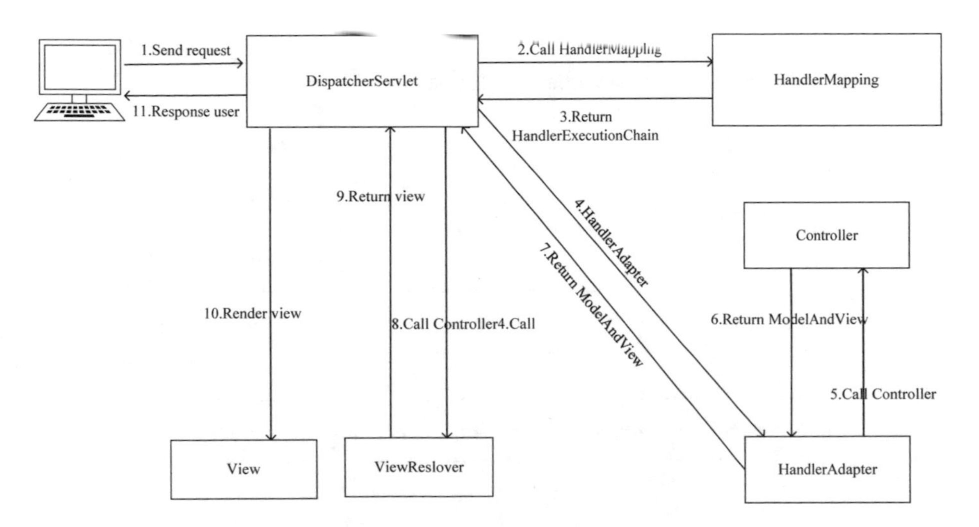

Fig. 2. The workflow flowchart of SpingMVC

(2) **Persistence Layer Mybatis Architecture**

1) Mybatis configure SqlMapConfig.xml. This file as the mybatis global configuration file that configure the mybatis operating environment and other information. The mapper.xml file is the SQL mapping file. The SQL statement that manipulates the database is configured in the file. This file needs to be loaded in SqlMapConfig.xml.
2) It constructes SqlSessionFactory that is the session factory through the mybatis environment and other configuration information.
3) The sqlSession session is created by the session factory. The operation of the database needs to be performed through sqlSession.
4) The bottom of mybatis custom executor interface operation database, Executor interface has two implementations(one is the basic executor, the other is the cache executor).

5) Mapped Statement is also a bottom package object of mybatis, it wraps mybatis configuration information and SQL mapping information. A sql in the mapper.xml file corresponds to a Mapped Statement object. The id of sql is the id of the Mapped statement.
6) Mapped Statement defines the sql execution input parameters, including HashMap, basic types, and pojo. Executor maps the input java object to sql through Mapped Statement before executing sql. The input parameter mapping is the parameter setting of preparedStatement in jdbc programming.
7) Mapped Statement sql implementation of the output of the definition of the results, including HashMap, basic types, pojo, Executor through the Mapped Statement in the implementation of sql after the output is mapped to the java object. The output of the mapping process is equivalent to the analysis of the results of the JDBC programming process.

Figure 3 shows the workflow chart of Mybatis.

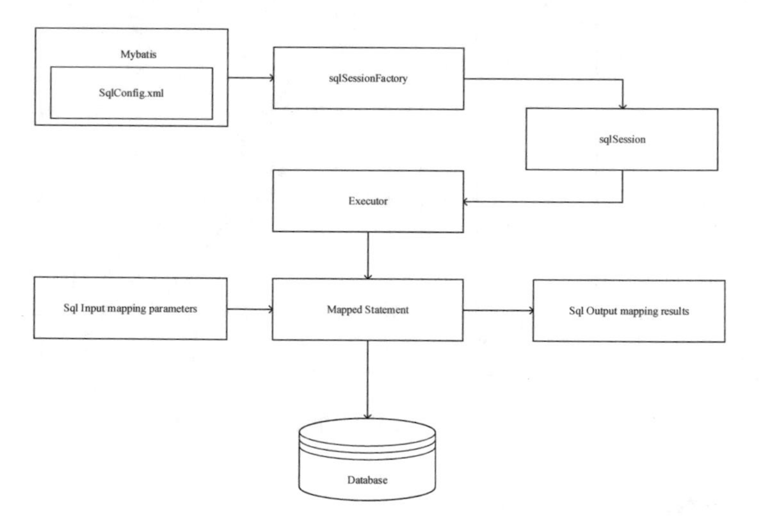

Fig. 3. The workflow chart of Mybatis

5 Implementation

The system uses SSM framework, Mysql5.5.7.20 is used as a back-end database, tomcat9.0 is used as a software server, JDK8.0 and eclipse Mars are used as a corresponding development environment.

Figures 4 and 5 experimental results for detecting cloned labels and an interface diagram of anti-counterfeiting traceability system. The query code can be obtained from the 2-dimensional code.

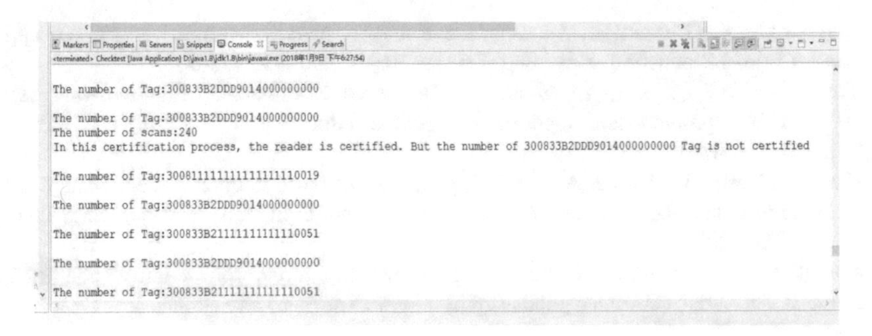

Fig. 4. The result of security certification

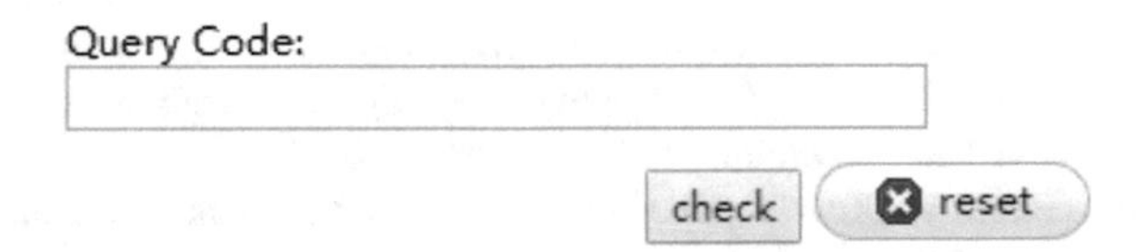

Fig. 5. The login main interface

6 Conclusion

Aimed at the security issue of the food industry, this paper researches on RFID technology, which is used in food anti-counterfeiting, puts forward a realization project of food security anti-counterfeiting tracing system based on RFID security authentication and 2-dimensional code. The purpose of this system is to allow consumers to reassure about the products they purchase. The realized system can quickly and efficiently inquire into raw materials or processing problems when there are product quality problems. It can contribute to quality control and recalls products when necessary, so as to improve the competitiveness of the company. The future work is to design a cloud based system to test our realized system.

Acknowledgments. This work is financially supported by the National Natural Science Foundation of P. R. China (No.: 61672296, No.: 61602261), Scientific & Technological Support Project of Jiangsu Province (No.: BE2015702, BE2016185, No.: BE2016777), China Postdoctoral Science Foundation (No.: 2014M561696), Jiangsu Planned Projects for Postdoctoral Research Funds (No.: 1401005B), Postgraduate Research and Practice Innovation Program of Jiangsu Province (No.: KYCX17_0798).

References

1. Gu, C.: Fast discrepancy identification for RFID-enabled IoT networks. IEEE Access, **PP** (99), 1 (2017)
2. Jayadi, R., Lai, Y.C., Lin, C.C.: Efficient time-oriented anti-collision protocol for RFID tag identification. Comput. Commun. **112**, 141–153 (2017)

3. Domdouzis, K., Kumar, B., Anumba, C.: Radio-frequency identification (RFID) applications: a brief introduction. Adv. Eng. Inform. **21**(4), 350–355 (2007)
4. Yang, L., Wu, Q., Bai, Y., et al.: An improved hash-based RFID two-way security authentication protocol and application in remote education. J. Intell. Fuzzy Syst. **31**(5), 2713–2720 (2016)
5. Kour, R., Karim, R., Parida, A., et al.: Applications of radio frequency identification (RFID) technology with eMaintenance cloud for railway system. Int. J. Syst. Assur. Eng. Manag. **5** (1), 99–106 (2014)
6. Miragliotta, G., Perego, A., Tumino, A.: A quantitative model for the introduction of RFID in the fast moving consumer goods supply chain: Are there any profits. Int. J. Oper. Prod. Manag. **29**(10), 1049–1082 (2009)
7. Coltman, T., Gadh, R., Michael, K.: RFID and supply chain management: introduction to the special issue. J. Theor. Appl. Electron. Commer. Res. **3**(4), 5 (2008)
8. Hu, Q., Xue, J., Zhong, L.: Lightweight J2EE architecture based on spring framework and its application. Comput. Eng. Appl. **44**(5), 115–118 (2008)
9. Xue, F., Liang, F., Xu, S., Wang, B.: Research on spring MVC framework based web and its application. J. Hefei Univ. Technol. **35**(3), 337–340 (2012)
10. Xu, W., Gao, J.: Research of web application framework based on spring MVC and mybatis. Microcomput. Appl. **28**(7), 1–4 (2012)

DoS Attack Pattern Mining Based on Association Rule Approach for Web Server

Hsing-Chung Chen[1,2(✉)] and Shyi-Shiun Kuo[1,3]

[1] Department of Computer Science and Information Engineering, Asia University, Taichung, Taiwan
shin8409@ms6.hinet.net, cdma2000@asia.edu.tw
[2] Department of Medical Research, China Medical University Hospital, China Medical University, Taichung, Taiwan
[3] Department of Multimedia Animation and Application, Nan Kai University of Technology, Nantou County, Taiwan

Abstract. In recent years, lots of web servers increasingly often suffer from Denial of Service (DoS) attacks within application layer. Many approaches provide abnormal traffic detecting in order to prevent any malicious traffic. However, the attack features or patens of the malicious traffics did not addressed clearly. Thus, the aim of this paper is to provide an approach based on the association rule mining technique for traffics appeared in the integrated web services, such as HTTP, HTTPS, and FTP traffic, in order to discover the strong attack features or patens of DoS attacks. Association rule mining is employed in this paper to deal with the DoS patens and then find out the strong relations among features of DoS attacks in large well-known dataset, e.g. NSL-KDD. The strong relations which are determined on when the major attack features are discovered from the open dataset would be considered as the strong patterns of DoS attacks. Finally, the outputted strong patterns could be used in the intrusion detection system (IDS) to enhance the effects of detecting application layer DoS attacks.

1 Introduction

Because of the popularization of the Internet applications, more and more commercial services are built on the Internet. The threats on the network are becoming more frequent and complicated. In recent year, the increasing amount of cyber-attacks has caused unavailability of website, even considerable commercial losses. There have been a variety of threats emerged, in the past years, due to the vulnerability of original design of the Internet. One of the most common threats is Denial of Service (DoS) attacks. Moreover, the traffic volume caused by modern DoS attacks have been more than 500 Gbps [1] and have become a major threat to network security [2], in the recent years. Typically, DoS attack is intended to disrupt the availability of web service to legitimate users by exhausting the network or computing resources of their victims. The kind of attack trying to exhaust the network resources, such as bandwidth or router processing capacity, could be referred to a network layer flooding attack. On the other hand, the kind of attack trying to exhaust the computing resources, such as sockets,

L. Barolli et al. (Eds.): IMIS 2018, AISC 773, pp. 527–536, 2019.
https://doi.org/10.1007/978-3-319-93554-6_50

CPU time, memory, and disk bandwidth, is known as an application layer flooding attack. The network layer flooding attack disrupts the legitimate user's connectivity, and the application layer flooding attack disrupts the legitimate user's service requests. An attacker may launch one or both attacks to block the victim's services. Due to the vulnerabilities of application layer protocol, application layer, DoS attacks are able to cause a same level of impact but much lower cost, if they are compared to network layer DoS attacks. Application layer DoS attack is a rapidly growing category of network threat in past few years. The duration, frequency and size of attacks are represented to be increasing every year [3]. Although a lot of literatures had been designed on the detection and prevention of network layer DoS attack, the researchers focused on detection and prevention of application layer DoS attack is more urgent and important for modern network.

Because of the attack packets are similar to legitimate traffic, the attacks are difficult to detect. Nevertheless, there are many features could be observed in the network traffics. Each type of attack may represent strong relationship among features, that is, some feature has high intensity is possibly related to another high intensity feature. Thus, if a relationship among features is identified, the relationship could be considered as a rule to filter the attack. Therefore, this paper presents an approach to discover the strong attack pattern of application layer DoS attack by using association rule mining technique [4]. Association rule technique is intended to mine the potential relationship among features based on two measured metrics consisting of support and confidence. The features of network traffic are evaluated on the well-known dataset known as NSL-KDD [5] which consists of the 41 features of network traffics. The NSL-KDD dataset is a refined version of its predecessor KDDcup'99 excluding redundant instances [6]. It includes two main classes of network traffic. One is t is normal class and another one is attack class. The attack class is further classified into four categories: DoS, Probing, users-to-root (U2R), and remote-to-local (R2L) [6]. The DoS attack disrupts the availability of victim's service to make it unable to handle legitimate requests. The DoS category includes different types of attacks, such as Apache2, Back, Land, Neptune, and Smurf [7]. The probing attack attempts to scan the network to gather information about the victim in order to find its vulnerabilities. The probing category includes attacks issued by Satan, Ipsweep, Nmap, Portsweep, Mscan, and Saint [7]. The U2R attacker uses a normal user's account to exploit the vulnerabilities of the victim's system by trying access to the root of the system. The examples of U2R attack include Buffer_overflow, Loadmodule, Rootkit, Sqlattack. In R2L attack, an attacker intrudes into a remote machine and gain local access to it. The R2L category includes the attacks such as Guess_Password, Ftp_write, Imap, Multihop [7].

Due to its high rank among various types of attack, the application layer DoS attacks to web servers are focused to explore their attack patterns. Therefore, the three attack types, Apache2, Back, and Neptune, are considered in this paper, which they are the popular application layer flooding attacks issued toward a web server. These three attack types are illustrated in Table 1.

The rest of this paper is organized as follows. The related work for the detection of application layer DoS attacks is described in Sect. 2. Section 3 presents the proposed approach which consists of two phases. In first phase, the necessary features are selected from the 41 features of NSL-KDD in which represents application layer attack.

In second phase, association rule mining method is applied to discover strong relationships among those selected features. Section 4, features selection is discussed. Finally, the conclusions are drawn in Sect. 5.

Table 1. The description of application layer DoS attack types.

Attack types	Description
Apache2	An attacker sends requests with many http headers to an apache web server. The server will slow down and may eventually crash if it receives too many this type of requests
Back	An attacker issues requests with URL containing many front-slashes. The server tries to process these requests and it will slow down and becomes unable to process other requests
Neptune	An attacker generates a large number of SYN packets to a target sever to request session establishment. The server waits for session establishment and its buffer will exhaust

2 Related Work

Detection of DoS attacks is a challenging issue because it has high rank among various types of attacks. Especially, the increasing resources and techniques available to attackers in modern network, the detection of DoS is more difficult in the application level. In recent years, many approaches [8–11] have been proposed to detect anomaly on the web based services.

The defense mechanisms against DoS flooding attacks in application level could be classified into two categories: server side and distributed mechanisms [8]. The classification is based on the deployment location of defense mechanisms. The server side application level DoS attacks are focused in this paper. The application layer DoS flooding attacks usually target the specific characteristics of application services, such as HTTP, DNS, and SIP. In recent year, the HTTP based, i.e., web based, DoS flooding attacks have become the major threats in the Internet. Ajagekar and Jadhav [9] propose a classifier based system to extract the important fields from captured application layer packets, and then to apply classifier to detect attacks. The aim of this method is to improve the accuracy of DDoS attack detection at application layer. Adi *et al.* [10] conduct DoS attack traffic models against HTTP/2 services and present a novel and stealthy DoS attack variant. Then, they conduct the attack traffic analysis which employs four machine learning techniques, namely Naïve Bayes, decision tree, JRip and support vector machines, to show the stealthy attacks through higher percentage of false alarms. Jazi *et al.* [11] focused on the impact of sampling techniques on detection of application layer DoS attacks. They demonstrated that the most sampling techniques introduce significant distortion in the traffic. This type of distortion minimizes the ability of a detection algorithm to capture the traces of stealthy attacks.

A number of studies discussed the detection strategies for application layer DoS attacks. There are few researches focus on the pattern of DoS attack. A pattern consists of several essential features captured from a network connection, which could be represented the characteristics of the connection. The same type of application layer DoS attacks has probably similar patterns. If the pattern of a certain type of attack is determined, the attack connection will be identified. In this paper, the association rule mining technique will be employed to explore the relationships among features of application layer DoS attacks.

3 Proposed Approach

This paper attempts to find possible patterns of application layer DoS attacks by analyzing the features in NSL-KDD dataset. When an attack occurs, some features' values arise significantly in contrast with those values in normal connections. For example, when an attack of Apache2 occurs, the time duration of a connection becomes high and the number of bytes transferred from source to destination also turns into high, thus, the duration and the number of bytes have an association with each other. This feature association may represents one pattern of Apache2 attack, which could be used as a rule to identify this type of attack. In this work, there are three phases to complete the feature association rule mining.

Phase 1: *Feature selection*. It is not necessary to mine the association rules with all features in dataset, since there is no correlation between the number of features and attack detection rate. The essential features will be selected in this phase which is explained in Subsect. 3.1.

Phase 2: *Feature value normalization*. There are two types of feature value in the dataset: continue and discrete. The type of 'continue' is an integer or real number. The type of 'discrete' is an enumerated value in a finite set. The association rule mining technique works only on discrete type of values. Thus, the feature values of continue type need to be converted into discrete type, then they will be normalized into a set of finite numbers. First, all the instances of attack types listed in Table 1 are screened out as a dataset for mining. Then, the values of each feature in this attack dataset will be normalized. The normalization is described in Subsect. 3.2.

Phase 3: *Feature association rule mining*. The relationships of the selected features will be discovered in this phase by using association rule mining method. After applying association rule mining method on the attack dataset, the resulting rules will be listed in descending order depending on their confidence metric. The details of this phase is illustrated in Subsect. 3.3.

3.1 Features Selection

Each record in NSL-KDD dataset represents the instance of a network connection, and each of which consists of 41 features. All features are grouped into three categories: basic features, content related features, and traffic related features [7, 12]. There are 9 features in basic category, 13 features in content related category, and 19 features in traffic category, respectively. In this work, the application layer DoS attack traffic is focused. The traffic related features have intrinsic relation to DoS attack. The basic features contain the basic information of network connection, such as duration, protocol type, number of data bytes transferred between client and host. The content related features have strong relation to the R2L and U2R attacks, e.g., number of login attempts, number of root accesses. They have weak relation to DoS traffic. Therefore, the basic features and traffic related features are selected to be analyzed the difference between normal connections and DoS attacks.

There is no direct correlation between the number of features and attack detection rate, that is, more number of features not necessarily has higher attack detection rate. This irrelevance has been proven in some researches [12–14]. Table 2 lists the detection rate with regard to the number of features. The classification algorithms, such as Adaboost and Naïve Bayes, are employed to select the essential features.

Table 2. The performance of DoS attack detection [12, 13].

References	Number of features	Percentage of detection rate
Yi and Phyu [12]	41	95
	7	100
Natesan and Balasubramanie [13]	41	97.8
	15	98.9

The goal of this work aims to discover the association rules among features to identify the DoS attack patterns. According to the proven results in Table 2, it is not necessary to select all 41 features to mine their relationships. If too many features are selected, there will be too many association rules to be discovered and the result could not enhance the performance of attack detection. On the other hand, the attack patterns could not be comprehensively mined with too less features, and the discovered association rules will be too less and simple that could not identify the real attacks and may cause a large number of false alarms. Therefore, after carefully examining on 28 features in both basic and traffic related categories as well as referring to the essential features mentioned in [12–14], nine features are selected in this work for further mining their relationships. The descriptions of these selected features are listed in Table 3.

Table 3. The selected features and their descriptions.

Selected features	Label	Value types	Category	Description
duration	f_1	Continue	Basic	Time duration of the connection in seconds
src_bytes	f_2	Continue	Basic	Number of data bytes from source to destination in the connection
count	f_3	Continue	Traffic-related	Number of connections to the same destination host as the current connection in the past two seconds
dst_host_count	f_4	Continue	Traffic-related	Number of connections that have the same destination host IP address in the past two seconds
service	f_5	Discrete	Basic	Network service on the destination host, e.g., HTTP, ftp_data
flag	f_6	Discrete	Basic	Status of the connection representing normal or error
srv_serror_rate	f_7	Continue	Traffic-related	Percentage of connections that have 'SYN' errors to the same service
srv_rerror_rate	f_8	Continue	Traffic-related	Percentage of connections that have 'REJ' errors to the same service
dst_host_srv_rerror_rate	f_9	Continue	Traffic-related	Percentage of connections that have 'SYN' errors from same service to the destination host

3.2 Feature Value Normalization

In order to discovery the feature association rules of attacks listed in Table 1, all the instances of these attack types are screened out as an attack dataset for mining. Then, the values of each feature in this attack dataset will be normalized. Since the association rule mining technique works only on discrete type of values, the feature values of continue type need to be converted into discrete type. The values of each feature is normalized as the values between [0, 1] with fixed decimal precision as defined in Eq. (1).

$$R\left(\frac{f_{ij}}{\max f_i}, d_i\right), \quad i = 2, 3, 4, \tag{1}$$

where R is a rounding function, f_{ij} is the j-th instance value of the i-th feature, and d_i is the precision of decimals. The precision d_i will restrict the maximum cardinal number of a finite set. If $d_i = 2$, the f_i is normalized into [0, 1] with 2 decimals precision in which there are at most 101 cardinalities in this feature. The continue type of feature is converted to discrete type with a finite set by using Eq. (1). In addition, each feature could have its own precision by determining d_i.

For the duration feature (f_1), after observing all instances in dataset, the feature values almost are larger than 200 s in attack instances. Thus, f_1 is normalized into binary type of $\{0, 1\}$ by setting a threshold (t) of 200. For features f_5 and f_6, they are intrinsic discrete type with a finite set. For other features f_7, f_8, and f_9, they are already in two decimals precision of percentage. For example, Table 4 is the 10 attack instances of NSL-KDDTest+ dataset, and Table 5 is the normalization version of Table 4.

Table 4. Ten attack instances of NSL-KDDTest+ dataset with selected features as an example.

f_1	f_2	f_3	f_4	f_5	f_6	f_7	f_8	f_9
902	57964	42	241	http	RSTR	0.02	0.95	0.16
0	0	111	255	http_443	REJ	0	1	1
2061	72564	11	255	http	RSTR	0	1	0.3
904	75484	21	255	http	RSTR	0.05	0.9	0.07
781	79864	1	230	http	RSTR	0	1	0.02
0	0	133	255	ftp_data	S0	1	0	0
0	0	76	255	http	S0	1	0	0.63
2068	93004	7	255	http	RSTR	0	0.86	0.03
2100	68184	8	255	http	RSTR	0	0.88	0.03
2069	101764	3	255	http	RSTR	0	0.67	0.01

Table 5. Feature value normalization with precision of two decimals.

f_1	f_2	f_3	f_4	f_5	f_6	f_7	f_8	f_9
1	0.57	0.32	0.95	http	RSTR	0.02	0.95	0.16
0	0	0.83	1	http_443	REJ	0	1	1
1	0.71	0.08	1	http	RSTR	0	1	0.3
1	0.74	0.16	1	http	RSTR	0.05	0.9	0.07
1	0.78	0.01	0.9	http	RSTR	0	1	0.02
0	0	1	1	ftp_data	S0	1	0	0
0	0	0.57	1	http	S0	1	0	0.63
1	0.91	0.05	1	http	RSTR	0	0.86	0.03
1	0.67	0.06	1	http	RSTR	0	0.88	0.03
1	1	0.02	1	http	RSTR	0	0.67	0.01

The procedure for feature normalization is described as follows.

Procedure 1. Normalization.

Input: Original dataset $F=\{f_{ij}\}$ *with features selected in Phase 1.*

Output: Normalized dataset $F'=\{f'_{ij}\}$.

Step 1: *If* $f_{1j} < t$ *then* $f'_{1j} = 0$ *else* $f'_{1j} = 1$, *for every instance j. Let t be a predefined threshold of time duration.*

Step 2: $f'_{ij} = R\left(\frac{f_{ij}}{\max f_i}, d_i\right)$, *for i = 2, 3, 4, and for every instance j. Let* d_i *be a specific precision of decimals.*

Step 3: $f'_{ij} = f_{ij}$, *for i = 5, ..., 9, and for every instance j.*

3.3 Feature Association Rule Mining

Let X and Y be the sets of features, the implication $X \Rightarrow Y$ is called an association rule which represents the relationship of X and Y. There are two measures for association rules, i.e., support and confidence, which indicate the strength of the implication $X \Rightarrow Y$. They are defined as Eqs. (2) and (3), respectively. The support of the association rule $X \Rightarrow Y$ is the probability of $X \cup Y$ in dataset. An association rule could be regarded as an strong attack pattern.

$$support(X \Rightarrow Y) = P(X \cup Y) \tag{2}$$

$$confidence(X \Rightarrow Y) = \frac{support(X \cup Y)}{support(X)} = \frac{P(X \cup Y)}{P(X)} \tag{3}$$

Table 6. The top 10 association rules with minimum support of 0.4 and minimum confidence of 0.8.

Association rules	Confidence
{flag=RSTR} ⇒ {service=http}	1
{duration=0, service=http} ⇒ {dst_host_count=1}	0.9985
{duration=0} ⇒ {dst_host_count=1}	0.9975
{service=http} ⇒ {dst_host_count=1}	0.9796
{srv_serror_rate=0} ⇒ {dst_host_count=1}	0.9669
{srv_serror_rate=0, service=http} ⇒ {dst_host_count=1}	0.9642
{flag=RSTR, service=http} ⇒ {dst_host_count=1}	0.9568
{flag=RSTR} ⇒ {dst_host_count=1}	0.9568
{flag=RSTR, service=http} ⇒ {duration=1}	0.906
{flag=RSTR} ⇒ {duration=1}	0.906

In order to discover the appropriate strong rules, the minimum support and the minimum confidence should be determined before association rule mining. A famous algorithm implementing association rule mining is Apriori [15], which is used in this work to mine the association rules in the attack dataset. After applying Apriori algorithm on attack dataset with selected features, top 10 association rules based on confidence value are picked up. In this paper, let the support value be 0.4 and the confidence value be 0.8, resulting rules after Apriori execution are shown in Table 6.

For more explanation, consider rule 1 and rule 2. Rule 1 indicates that if a connection have been established but aborted by responder (RSTR), and also this connection uses http service, then this connection has high confidence to be regarded as an application layer DoS strong attack pattern. Rule 2 shows that if a connection uses http service and its time duration less than threshold (t = 200 s), but also the number of connections like this one having the same destination IP address (dst_host_count) is almost the maximum number (255 in the attack dataset), then rule 2 has high confidence to be regarded as an strong attack pattern.

The generated association rules will be used in an IDS to enhance its attack detection rate.

4 Discussions

The strong attack pattern is discovered based on the association rule mining method. In this paper, there are some parameters could be adjusted to adapt the strength and precision of association rules. For phase 1, the different set of selected feature will result in different set of strong attack patterns which will lead the IDS to make an appropriate decision. In phase 2, the normalization scheme will affect the complexity of generating combined feature-set. For the time duration feature, different attack type may have different threshold. For continue type features, the normalization function R in Eq. (1) could be replaced by any other normalization method which is able to significantly discriminate attack instances and normal ones. For phase 3, the minimum support and minimum confidence are thresholds to decide the precision and number of association rules. In general, if too many feature selected or too high cardinality of a normalized feature, there will generate too many association rules and may cause a large number of false alarms in IDS.

5 Conclusions

The aim of this paper is to introduce an approach based on association rule mining for web server to discover the pattern of application layer DoS attacks. The features of network traffic are evaluated on NSL-KDD dataset. The proposed approach could be adaptable to any dataset, which could rely on the characteristics of each feature to adjust the parameters for features value normalization. The adaptability of the proposed approach could generate more appropriate association rules. To improve the effects of association rule mining, the machine learning techniques, such as particle swarm optimization and deep learning, could be employed for enhancing the tool of the association rule mining.

Acknowledgement. This work was supported by the Ministry of Science and Technology (MOST), Taiwan, Republic of China, under Grant MOST 106-2632-E-468-003.

References

1. Akamai: State of the Internet Security Report (2017). https://www.akamai.com/
2. Ponemon Institute: Global Report on the Cost of Cyber Crime. HP Enterprise Security (2014)
3. Arbor Networks: Worldwide Infrastructure Security Report (2016). https://www.arbornetworks.com/
4. Agrawal, R., Imielinski, T., Swami, A.: Mining association rules between sets of items in large databases. In: Proceedings of 1993 ACM SIGMOD International Conference on Management of Data, Washington, DC, pp. 207–216 (1993)
5. Canadian Institute for Cybersecurity: NSL-KDD. University of New Brunswick. http://www.unb.ca/cic/datasets/
6. Tavallaee, M., Bagheri, E., Lu, W., Ghorbani, A.: A detailed analysis of the KDD CUP 99 data set. In: Proceedings of IEEE Symposium on Computational Intelligence for Security and Defense Applications (2009)
7. Dhanabal, L., Shantharajah, S.P.: A study on NSL-KDD dataset for intrusion detection system based on classification algorithms. Int. J. Adv. Res. Comput. Commun. Eng. **4**(6), 446–452 (2015)
8. Zargar, S.T., Joshi, J., Tipper, D.: A survey of defense mechanisms against distributed denial of service (DDoS) flooding attacks. IEEE Commun. Surv. Tutor. **15**(4), 2046–2069 (2013)
9. Ajagekar, S.K., Jadhav, V.: Study on web DDOS attacks detection using multinomial classifier. In: Proceedings of 2016 IEEE International Conference on Computational Intelligence and Computing Research, pp. 1–5 (2016)
10. Adi, E., Baig, Z., Hingston, P.: Stealthy denial of service (DoS) attack modelling and detection for HTTP/2 services. J. Netw. Comput. Appl. **91**, 1–13 (2017)
11. Jazi, H.H., Gonzalez, H., Stakhanova, N., Ghorbani, A.A.: Detecting HTTP-based application layer DoS attacks on web servers in the presence of sampling. Comput. Netw. **121**, 25–36 (2017)
12. Yi, M.A., Phyu, T.: Layering based network intrusion detection system to enhance network attacks detection. Int. J. Sci. Res. **2**(9), 499–508 (2013)
13. Natesan, P., Balasubramanie, P.: Multi stage filter using enhanced AdaBoost for network intrusion detection. Int. J. Netw. Secur. Appl. **4**(3), 121–135 (2012)
14. Khan, S., Gani, A., Wahab, A.W.A., Singh, P.K.: Feature selection of denial-of-service attacks using entropy and granular computing. Arab J. Sci. Eng. **43**(2), 499–508 (2017)
15. Agarwal, R., Srikant, R.: Fast algorithms for mining association rules. In: Proceedings of the 20th International Conference on Very Large Databases, vol. 1215, pp. 487–499 (1994)

Current Status on Elliptic Curve Discrete Logarithm Problem
(Extended Abstract)

Maki Inui and Tetsuya Izu(✉)

FUJITSU Laboratories Ltd., Kawasaki, Japan
izu@jp.fujitsu.com

Abstract. This paper reports the current status and records on elliptic curve discrete logarithm problem (ECDLP), which is tightly connected to the security of elliptic curve cryptography (ECC) such as ECDSA and ECDH.

1 Introduction

Elliptic curve cyrptograph (ECC) is one of the standardized public-key cryptography with smaller key size compared to other standardized public-key cryptography such as RSA. The security of ECC depends on the hardness of elliptic curve discrete logarithm problem (ECDLP), so that, analyzing previous records of solving ECDLP and predicting which parameter will be solved at when are essential in order to choose the proper ECC parameter, especially the key size, and use ECC securely.

This article reports the current records on solving ECDLP in three categories, namely, ECDLP over prime field, ECDLP over characteristics 2 field, and Koblitz curve in Tables 1, 2, and 3. Here, Size represents the size of the group where target ECDLP problem is defined rounded up to integers, and Curve Name corresponds to elliptic curves defined in Certicom ECC Challenge [C] for curves begining with ECC, or defined in SEC 2 [SECG10] published by SECG (Standards for Efficient Cryptography Group) for curves begining with `sec`. All records in these tables were established by ρ-method or λ-method. From the theoretical point of view, there has been little progress on the algorithm, however, it should be noted that these methods are energetically implemented on FPGA. On November 2016, Bernstein et al. solved 118-bit ECDLP over characteristics 2 field, which is the record on solving ECDLP. According to the report [BEL+16], the experiment spent about 6 months with 576 FPGAs (Spartan 6 XC6SLX150).

According to the estimation by Yasuda et al., which shows the equivalently secure key sizes between RSA and ECC [YSKI16], the record by Bernstein et al. [BEL+16] corresponds to 118-bit ECDLP over prime field and 124-bit ECDLP over Koblitz curve. In addition, Wenger and Wolfger evaluated required resources to solve 129-bit ECDLP over Koblitz curve in 1 year as 72 FPGAs (122,400 USD) [WW16], which is roughly identical to Yasuda et al.'s estimation. Consequently, these ECDLPs can be solved in a few years.

L. Barolli et al. (Eds.): IMIS 2018, AISC 773, pp. 537–539, 2019.
https://doi.org/10.1007/978-3-319-93554-6_51

Table 1. ECDLP records over prime field

Size	Curve name	Year	Author(s)	Computing resource
79	ECCp-79	Dec 06, 1997	Harley et al. [C]	CPU
89	ECCp-89	Jan 12, 1998	Harley et al. [C]	CPU
97	ECCp-97	Mar 18, 1998	Tsapakidis et al. [C]	CPU
109	ECCp-109	Oct 15, 2002	Monico et al. [C]	CPU
112	`secp112r1`	Jul 08, 2009	Bos et al. [BKK+12]	CPU (PS3)

Table 2. ECDLP records over characteristics 2 field

Size	Curve name	Year	Author(s)	Computing resource
79	ECC2-79	Dec 16, 1997	Harley et al. [C]	CPU
89	ECC2-89	Feb 09, 1998	Harley et al. [C]	CPU
97	ECC2-97	Sep 22, 1999	Harley et al. [C]	CPU
109	ECC2-109	Apr 08, 2004	Monico et al. [C]	CPU
113	`sect113r1`	Jan, 2015	Wenger, Wolfger [WW16]	FPGA
113	`sect113r2`	Apr 17, 2016	Bernstein et al. [BEL+16]	FPGA
118	`target117`	Nov 29, 2016	Bernstein et al. [BEL+16]	FPGA

Table 3. ECDLP records over Koblitz curve

Size	Curve name	Year	Author(s)	Computing resource
95	ECC2K-95	May 21, 1998	Harley et al. [C]	CPU
108	ECC2K-108	Apr 04, 2000	Harley et al. [C]	CPU
113		Apr, 2014	Wenger, Wolfger [WW14]	FPGA
130	ECC2K-130	(Unsolved)	Bailey et al. [BBB+09]	CPU

References

[BBB+09] Bailey, D.V., Batina, L., Bernstein, D.J., et al.: Breaking ECC2K-130, IACR Cryptology ePrint Archive, 2009/541 (2009)

[BEL+16] Bernstein, D.J., Engels, S., Lange, T., et al.: Faster discrete logarithms on FPGAs, IACR Cryptology ePrint Archive, 2016/382 (2016)

[BKK+12] Bos, J.W., Kaihara, M.E., Kleinjung, T., et al.: Solving a 112-bit prime elliptic curve discrete logarithm problem on game consoles using sloppy reduction. IJACT **2**(3), 212–228 (2012)

[C] Certicom, Certicom ECC Challenge, 10 November 2009. https://www.certicom.com/images/pdfs/challenge-2009.pdf

[SECG10] Standards for Efficient Cryptography Group, SEC 2: Recommended Elliptic Curve Domain Parameters (2010). http://www.secg.org/sec2-v2.pdf

[WW14] Wenger, E., Wolfger, P.: Solving the discrete logarithm of a 113-bit Kobilitz curve with an FPGA cluster. In: SAC 2014, LNCS 8781, pp. 363–379. Springer, Berlin (2014)

[WW16] Wenger, E., Wolfger, P.: Harder, better, faster, stronger - elliptic curve discrete logarithm computation on FPGAs. J. Cryptographic Eng. **6**(4), 287–297 (2016)

[YSKI16] Yasuda, M., Shimoyama, T., Kogure, J., Izu, T.: Computational hardness of IFP and ECDLP. In: Applicable Algebra in Engineering, Communication and Computing (AAECC), 09 April 2016

Study on Signature Verification Process for the Firmware of an Android Platform

Eunseon Jeong, Junyoung Park, Byeonggeun Son, Myoungsu Kim, and Kangbin Yim(✉)

Department of Information Security Engineering,
Soonchunhyang University, Asan, Korea
{gomdoli10, apple, ios, brightprice, yim}@sch.ac.kr

Abstract. Recently, Android is expanding its application area including vehicle infotainment system, smart TV, AI speaker as IoT devices as well as mobile terminals. To maintain and support these systems, the manufacturer distributes the firmware through the Android firmware build framework and updates after evaluating firmware integrity by a signature verification process. However, attackers potentially falsify the firmware and raise critical security problems on mobile terminals in cases that developers use the public test key or SDK key to sign the firmware for release, due to lack of security readiness of mobile manufacturers. This paper analyzes the firmware signing, verification and update process of the Android platform, introduces vulnerabilities invented when an unsafe key is used and implements an evaluation tool for signature verification to find the firmware features if signed by an unsafe key.

1 Introduction

As Smartphone market is growing, Google Android and Apple iOS have become two dominant leaders for mobile operating systems. Between them, Android as an open source platform touched 74.24% of global smartphone market revenue in 2017 and now is still continuously gaining the rest [1]. Recently, Android is installed in various embedded system products such as tablet PC, smart watch, AI speaker, in-vehicle telematics system and etc., all of which are incorporated with the firmware update facility for convenient maintenance purpose.

Manufacturers release their renewed firmware of the products after signing it with their private key. During when the pre-installed firmware updates with the signed renewed firmware on Android, verifies the signature of the firmware with the public key stored in the pre-installed certificate and replaces with new one when the verification is successful [2].

Android recommends manufacturers sign the private key to sign the renewed firmware by a unique key that can be acquired by only themselves when they publically release or update the firmware and also the renewed firmware is also possible to be updated only when it is signed by the same private key [2].

If the firmware is signed by a publically known key such as test key or SDK key, an attacker can generate a fake renewed firmware by replacing the system app incorporated in the OS image and proceed a normal update by passing through all of the signature

L. Barolli et al. (Eds.): IMIS 2018, AISC 773, pp. 540–545, 2019.
https://doi.org/10.1007/978-3-319-93554-6_52

verification steps during the firmware update process. This paper analyzes over-the-air firmware update and signature verification process on Android and implements an evaluation tool for firmware signature verification to identify the firmware sighed by an illegal key.

2 Firmware Update Process on Android

2.1 Android Firmware Update and Signature Verification Process

Figure 1 depicts the firmware update process on an Android-based mobile terminal in which each step is executed on Android system through the recovery console as the following [3–5].

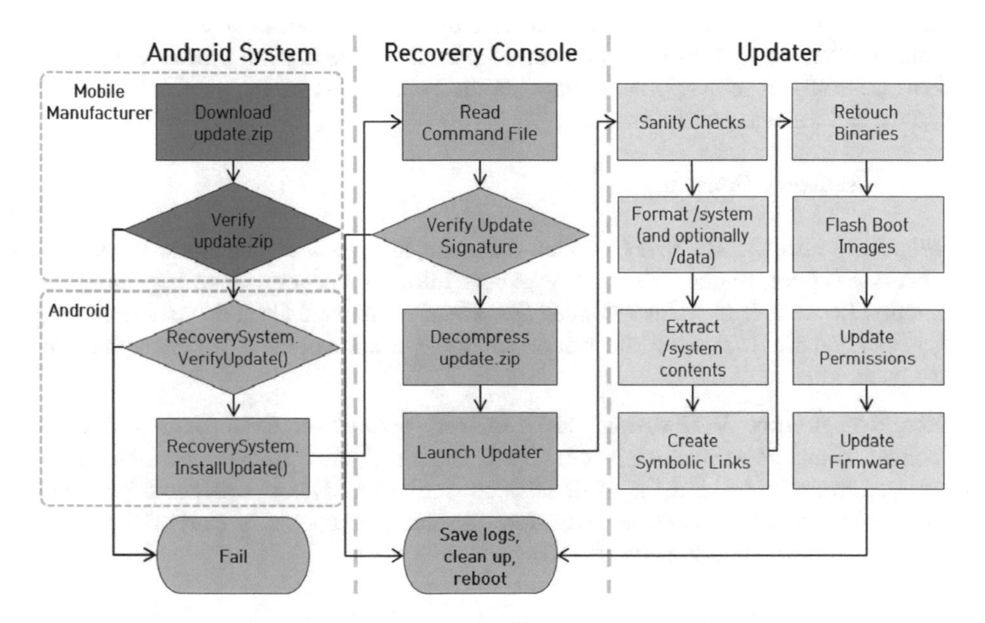

Fig. 1. Firmware update process on Android

2.1.1 Android System

Downloading a Renewed Firmware: The application for the update process is developed by the manufacturer and installed inside the Android device, which checks periodically if a renewed firmware for update is released from the update server. If a new update has become necessary, the mobile terminal is delivered with a download URL and the description for the firmware update and the renewed firmware is downloaded into the cache or a data partition only when the user accepts the download. When the firmware download process has completely done, the firmware goes into the integrity checking sequence, which is MD5 verification or signature verification by a public key according to the manufacturers' design concepts.

First Signature Verification for Renewed Firmware: When the firmware has passed the integrity checking sequence at the update application as described, Android also starts another signature verification process using its function RecoverSystem.VerifyUpdate().

otacerts.zip	이름	원본 크기	압축 크기	압축률	종류
tmp	..				
4173189	releasekey.x509.pem	1,736	1,736	0%	PEM 파일

Fig. 2. Certificate for signature verification

The signature value of the firmware is attached at the end of the firmware image file and the certificate to be used for the signature verification is located at the /system/etc./ security/directory, of which file is named otacerts.zip and includes the public key as the pair of the private key used when the Android system of the terminal was signed by the manufacturer before release as shown in Fig. 2. Therefore, for the firmware update to be successfully done the signature keys should be the same for the presented Android system and the renewed firmware.

2.1.2 Recovery Console

Rebooting into the Recovery Mode: After the first firmware signature verification process has been passed, the system reboots into the recovery mode by calling the function RecoverySystem.InstallUpdate() and reads/command file at the/cache/recovery directory and tries to find the information about the location of the downloaded renewed firmware.

Second Signature Verification for Renewed Firmware: Even though the first firmware signature verification process has been passed, there might be possible for the renewed firmware to be replaced or falsified during the reboot sequence of recovery mode and the second firmware signature verification process is started right after getting into the recovery mode (Fig. 3).

Fig. 3. Certificate for signature verification in recovery mode

The second firmware signature verification process uses the public key file “keys” that is located at the “/res” directory in the recovery partition. This file includes the C language structure converted from the public key data in the public certificate used during the first signature verification process and it is used to public key signature verification at the recovery console. Only when the second firmware signature verification is passed using this public key, the system decompresses the renewed firmware and starts the update.

Running Updater: The updater reads the 'updater-script' file located at the path 'META-INF/com/google/android' in the firmware file and executes the script step-by-step to update the firmware. As shown in Fig. 4, the 'updater-script' file is a text script in which the sequence of the firmware update is described and installs the renewed firmware after formatting the system partition.

```
updater-script
1  ui_print("Removing unneeded files...");
2  delete(
3          " ");
4  ui_print("Patching system files...");
5  ui_print("Unpacking new files...");
6  assert(package_extract_dir("system", "/system"));
7  package_extract_dir("modem", "/modem");
8  package_extract_dir("3rdmodem", "/3rdmodem");
9  assert(package_extract_dir("cust", "/cust"));
10 ui_print("Removing empty directorys...");
11 delete_recursive(
12         " ");
13 ui_print("add link type file...");
14 set_file_attr();
15 setprop("ro.modem_update", "false");
16 delete("/modem/modem_image/balong/modem_updated");
17 delete_recursive("/data/package_cache");
18 unmount("/system");
19 unmount("/cust");
20 hw_ui_print("fullpkg");
21 ui_print("write radio image...");
22 assert(update_huawei_pkg_from_ota_zip(UPDATE.APP));
```

Fig. 4. Updater script for firmware update

Storing the Log and Rebooting: When the firmware update is finished finally, the related log is saved and the terminal reboots again.

3 Identification of the Firmware Sighed by an Unexpected Key

As stated in Sect. 2, during the process of updating Android firmware, signature verification is performed twice, and in the case that a manufacturer signs the firmware with an unsafe key such as a test key or an SDK key, an attacker can generate and distribute malicious firmware that can pass the signature verification procedure, thereby altering the signature system app for the purpose of identity theft or inducing the abnormal behavior of a terminal [6–9]. Therefore, it is necessary to verify whether the firmware was correctly signed with a safe key. This paper implemented a program that checks the update firmware of an Android-based terminal and inspects whether the firmware was signed with a safe key.

As for the signature verification tool, as shown in Fig. 5, there are the test key list and the firmware file list. The test key list is a set of publicly-known public key certificates, and public keys are extracted from certificates and are stored in a buffer. The firmware file list is a set of firmware files whose signature keys are to be inspected. And a public key is extracted from the CERT.RSA of a firmware file, and then is compared with the public key data extracted from a test key in order to inspect whether the same key is used.

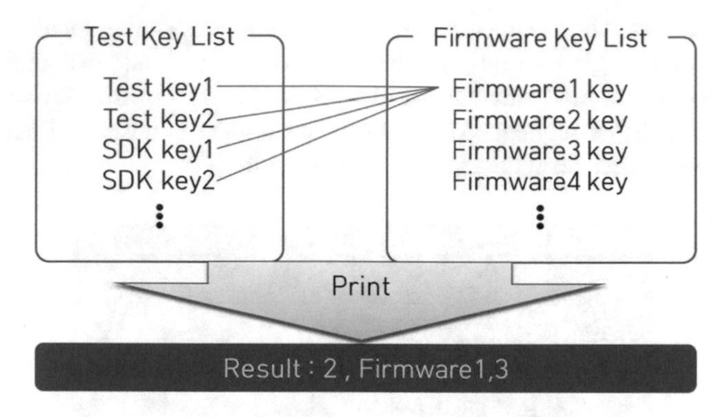

Fig. 5. Schematic diagram of the signature verification tool

From the schematic diagram of the signature verification procedure in Fig. 5, a tool to automatically verify the signature key of Android firmware was implemented, and the result is shown in Fig. 6.

The implemented tool was designed based on the open-source "ASN.1 Editor", which analyzes the firmware signed with the keys same to the public keys in the test key list, and shows their number and detailed information about the firmware [10].

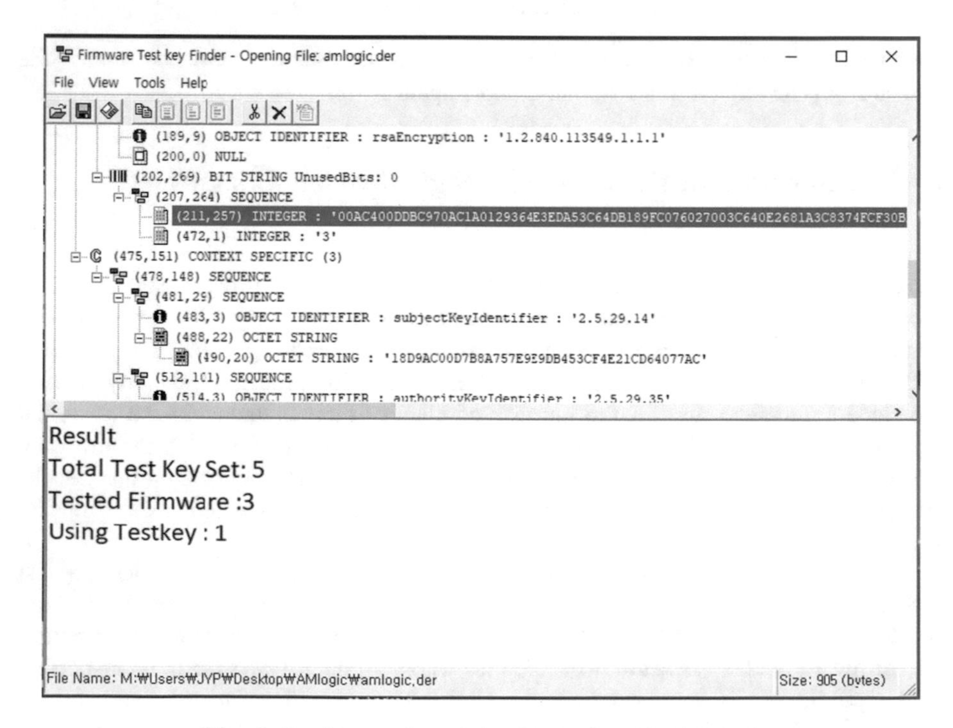

Fig. 6. Implementation of the signature verification tool

4 Conclusions

This paper analyzes the process of the firmware signature verification and update on the Android platform. During the firmware update process, two consequent signature verification processes are executed. The first signature verification is done on the Android system and the second verification is done at the recovery mode. However, if the manufacturer signed the firmware using the unsafe test key or SDK key, the attacker may falsify the firmware, sign it again to generate a malicious firmware that can be updated as the normal one through the update process.

Android still doesn't present any countermeasure against the above problem. Based on the result of this paper, a new secure protocol to exchange the signature key for the firmware update should be researched to recover the situation when the manufacturer released the firmware signed by an unsafe signature key.

Acknowledgments. This research was supported by the MSIT (Ministry of Science, ICT), Korea, under the ITRC (Information Technology Research Center) support program (IITP-2018-2015-0-00403) supervised by the IITP (Institute for Information & communications Technology Promotion).

References

1. STATECOUNTER, Mobile Operating System Market Share Worldwide (2017). http://gs.statcounter.com/os-market-share/mobile/worldwide
2. Android Open Source Project. https://source.android.com/devices/tech/ota/sign_builds#sign-ota-packages
3. Boie, A.: Android Software Updates, Embedded Linux Conference (2015)
4. Bisson, G.: Android OTA update, NXP FTF (2016)
5. Jeong, E.: FOTA Vulnerability Analysis on Android-based Mobile Devices (2017)
6. Ferreira, A.M.: Android OTA Update Another Security Evaluation. University of Systems and Networks (2013)
7. Zheng, M., Sun, M., Lui, J.C.S.: DroidRay: a security evaluation system for customized android firmwares. In: ACM Symposium on Information, Computer and Communications Security, pp. 471–482 (2014)
8. Jo, H.J., Choi, W., Na, S.Y., Woo, S., Lee, D.H.: Vulnerabilities of Android OS-based telematics system, pp. 1511–1530. Springer, New York (2016)
9. Zetter, K.: Attackers stole certificate from foxconn to hack Kaspersky With Duqu 2.0, Wired (2015). https://www.Wired.Com/2015/06/Foxconn-Hack-Kaspersky-Duqu-2
10. Dai, L.: ASN.1 Editor, Code Project (2008). https://www.codeproject.com/Articles/4910/ASN-Editor

Detecting and Extracting Hidden Information from Stego-Images Based on JPG File on StegoMagic

Kyungroul Lee[1], Sun-Young Lee[2], Kyunghwa Do[3], and Kangbin Yim[2(✉)]

[1] R&BD Center for Security and Safety Industries (SSI), Soonchunhyang University, Asan, South Korea
carpedm@sch.ac.kr

[2] Department of Information Security Engineering, Soonchunhyang University, Asan, South Korea
{sunlee,yim}@sch.ac.kr

[3] Department of Software Engineering, Konkuk University, Seoul, South Korea
doda@konkuk.ac.kr

Abstract. Steganography is a technique used for concealing information by embedding messages or data into other data. However, problems can arise when this technique is used to steal confidential information, for unlawful purposes such as spying, terrorist attacks, and so on. Moreover, when the information is hidden, serious problems are caused if the evidence is not obtained. Therefore, we propose a detection method, based on StegoMagic-a steganography tool-that can be used on hidden information within a JPG file. We also propose an extraction method for the hidden information.

1 Introduction

Cryptography technologies have emerged based on mathematical algorithms to transfer information securely. These technologies are used to ensure security of the data by taking a lot of time to calculate decrypting cipher text to plain text, even though the cipher text is exposed to anyone. Unlike these technologies, another technology has emerged in which information is transmitted safely by hiding the existence of the information itself. This technology is called steganography [1]. Steganography uses a technique to conceal the information by embedding a message into other data. In order to conceal the information, the message is hidden inside a file such as an image, an audio, or a video, making it more difficult for a human to recognize than a message contained within a document file such as a text file.

At first, steganography was utilized for positive purposes, to send confidential information securely to third parties. However, some people use this technique to steal confidential information for the purposes of spying, terrorist attacks, and so on. Moreover, when the information is hidden, serious problems can occur if the data is not obtained. To solve these problems, several countermeasures for the detection and extraction of the information hidden by steganography tools have been researched.

L. Barolli et al. (Eds.): IMIS 2018, AISC 773, pp. 546–553, 2019.
https://doi.org/10.1007/978-3-319-93554-6_53

Unfortunately, most of these methods are focused only on the detection; and these methods of detection have practical limitations.

In this paper, we propose a detection method based on StegoMagic-a Steganography tool- that can be used on hidden information within a JPG file. We also propose an extraction method for the hidden information.

2 Related Work

2.1 Steganography

Steganography conceals information by hiding the message itself, unlike cryptography, which transforms data to prevent third parties from decoding the contents of a message. For this reason, when a file such as an image or a video that is embedded with a message is transmitted to third parties, only the receiver can extract the hidden message. Other entities cannot extract the hidden message because they do not know the message exists within the file.

As shown in Fig. 1, Steganography creates a stego image with the message "lisa.sch.ac.kr" inserted into the original image. The existence of the hidden message is undetectable within the generated stego image.

Fig. 1. Example of steganography

2.2 Steganography Tools

According to Wikipedia, about twenty steganography tools have been introduced. Most tools support hiding a message in an image file, and several tools also support hiding a message in an audio file. However, these tools have limitations, in that only a few tools support using many types of files, including video files, document files, executable files, and so on. In addition, these tools conceal and protect the inserted message by using cryptography, compression, and different hiding methods. Table 1 shows the file formats supported by steganography tools [4].

Table 1. Supported file formats of steganography tools

Supported file	Number of supported tools	Supported file formats
Image file	20	BMP, JPG, TIFF, GIF, PNG, PSD, TGA, MNG, JPEG
Audio file	6	WAV, Audio CD, APE tag, FLAC, MP3, WMA
Video file	1	3gp, MP4, MPEG-1, MPEG-2, VOB, SWF, FLV
Document file	2	TXT, HTML, XML, ODT, PDF
Other files	3	EXE, DLL, NTFS streams, Unused floppy disk space, steganographic file system for Linux

2.3 StegoMagic

As described above, most steganography tools support an image file. However, the tools do not support an audio file, a video file, a document file, or other files such as an executable file. Moreover, there is a limitation in that the size of the message is fixed due to the limitation of the file size. StegoMagic is a tool used to overcome these limitations [5]. This tool is able to hide text and files of any size, regardless of file format and file size, and the hidden data is transformed based on a key. Hence, only entities which know the key can recover the transformed hidden data.

2.4 Existing Detection Methods of Hidden Information

Statistics and probabilistic detection methods have been studied [2, 3]. These methods detect the hidden information based on the changing difference according to the information to be inserted. Namely, the stego image is detected by analyzing the difference caused by the fuzzy hashing of image files and text files of various sizes in StegoMagic.

Unfortunately, these methods have a high probability of false positives and false negatives because the methods are dependent upon statistics, probability, and hashing information. In addition, it is difficult to recover the hidden information because the hidden information can only be decoded entities which know the key. Therefore, we have researched a detection method of stego files by analyzing the embedding logic of StegoMagic; we also propose an extraction method for the hidden information in the stego files.

3 Detection Method of Hidden Information

3.1 Analysis of Operation and Concealment Process

StegoMagic is one of the steganography tools used to insert text or a file that a person wants to hide into supporting various file formats. StegoMagic conceals the text or files as follows.

In order to hide text or files, StegoMagic needs an image. After selecting the image file, the text to be hidden is automatically inserted into the image file once the image is selected. This tool displays a key for extracting the hidden text after the concealment process is completed. The key is delivered to the other parties who want to extract the hidden message, and the receivers obtain the hidden text by inputting the key received into StegoMagic. Figure 2 shows the concealment process of StegoMagic.

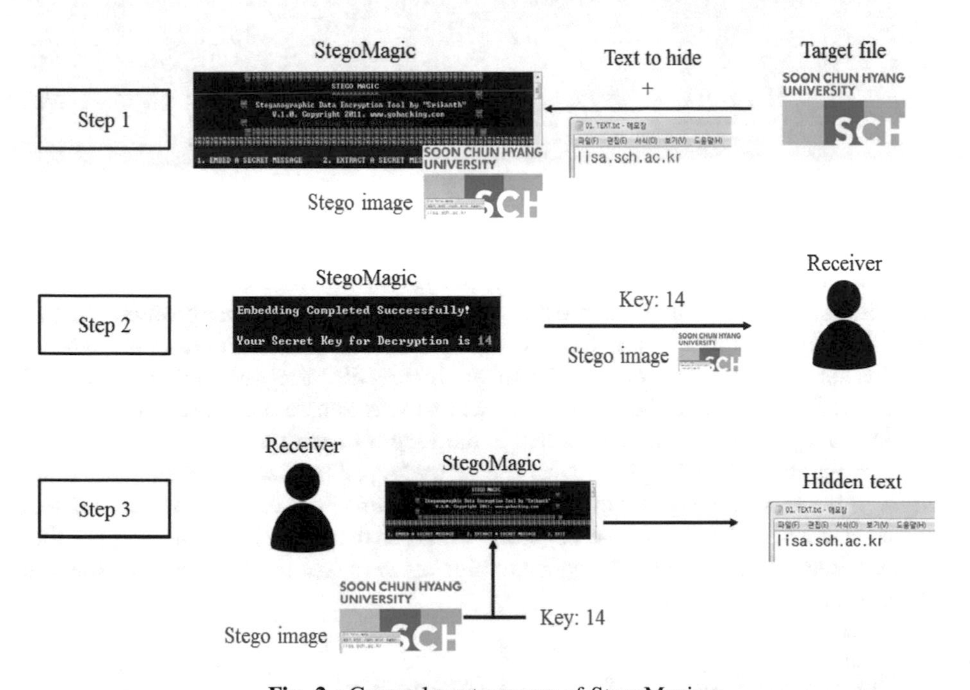

Fig. 2. Concealment process of StegoMagic

In Step 1, a user inputs both an image and a text to be hidden in StegoMagic; the tool creates a stego image. In Step 2, StegoMagic prints a key to extract the hidden text from the stego image, and the user sends the created stego image and the key to the other entities. In Step 3, StegoMagic extracts the hidden text when the receiver inputs the received stego image and the key.

3.2 Correlation Analysis Between Hidden Information and Key

As described above, StegoMagic embeds the text into the target image and then displays a key for decoding. In this paper, we analyze the correlation between the stego image, the key, and the inserted text to detect the existence of hidden information within the stego image. For the correlation analysis, we compare an original file and the stego file. Analyzed result is shown in Fig. 3.

Analyzed result show the data is the same up to 0xFF 0xD9, and then 14 bytes were added in the stego image. According to the result, the added data is related to the text to be hidden because a steganography tool has to embed the information in the target file. Therefore, we analyzed the correlation with the key based on this result, have concluded that the changed text size is the same as the key.

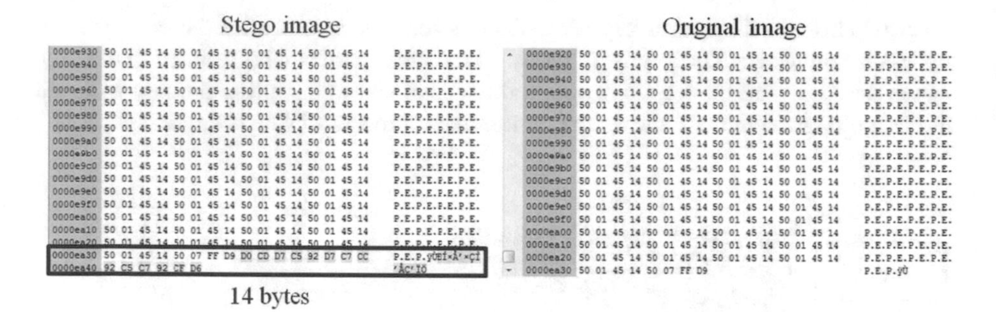

Fig. 3. Analysis of the correlation between the stego image and the original image

In other words, in order to insert text, StegoMagic finds the end position of the target image, and the transformed data is then inserted into the last position. At this point, the length of the text is used as a key for the transformation. For example, when a user inserts the text "lisa.sch.ac.kr" into an image file, StegoMagic finds the end position of the image, and the 14 bytes of text that was transformed based on the key is then inserted. The key used is 14 which is the length of the text.

As mentioned, StegoMagic extracts and recovers just the data corresponding to the length of the text from the end of the target file. For this reason, it is possible to detect the stego image if the size of the target file is larger than the size of the original file. Because StegoMagic inserts information to hide at the end of the target file, the file size increases accordingly.

4 Extraction Method of Hidden Information

4.1 Analysis of Key Extraction Method

As in Sect. 3, if the size of the target file is larger than the size of the original file, or if data is added to the last position of the target file, the target file can be identified as the stego file. When the stego file is found, the hidden information can be extracted by using the length of the data added from the end position of the file as a key. In this paper, we analyze the key extraction method of StegoMagic based on a JPG file. To analyze the method, it is necessary to acquire the size and end position of the target JPG file. Table 2 shows the JPF file format [6, 7].

As shown in the above table, each marker must be tracked to calculate the file size because JPG files use a compression algorithm. For this reason, we derive the key candidates by finding the end of image (EOI), rather than calculating the file size. That is, after finding 0xFF 0xD9 which means EOI in the JPG file, the length of subsequent data is derived as a key candidate. We searched 0xFF 0xD9 in the original JPG file. As a result, a total of four EOIs were found as shown in Table 3.

As shown in the table, EOIs exist at offset 0x6258, 0xCAA9, 0xEA2D, and 0xEA36. Based on the searched results, 34796, 8091, 1559, and 14 are obtained as key candidates.

Table 2. JPG file format structure

Segment	Code	Description
SOI	FF D8	Start of image
JFIF-APP0 JFXX-APP0	FF E0 *s1 s2* 4A 46 49 46 00 … FF E0 *s1 s2* 4A 46 58 58 00…	JFIF APP0 marker segment Optional
Additional marker segments (for example SOF, DHT, COM)		
SOS	FF DA	Start of scan
Compressed image data		
EOI	FF D9	End of image

Table 3. Searched results of EOI

Offset	Data	EOI offset
0x00006250	AE 33 DD 0A F9 E3 E3 06 8F FD 99 E3 97 B9 8D 76	0x6258
0x0000caa0	0B A9 42 B9 96 24 18 FB 42 8F FD 98 7E BD 2B C1	0xCAA9
0x0000e420	C9 74 8F 0A CD 92 72 93 DF 2F 41 EA B1 9F FD 9B	0xE42D
0x0000ea30	50 01 45 14 50 07 FF D9 D0 CD D7 C5 92 D7 C7 CC	0xEA36

4.2 Extraction of Hidden Information

We extract the hidden text from the stego image based on the derived key candidates. The extracted results are shown in Table 4.

Table 4. Extraction results of the hidden information

EIO offset	Extracted hidden text	Key candidate
0x6258	5•3U)•a•?•뽄톾쿂=?뙤•?QAH?狼 8NdM#d?p?파?lh ??2?+5 뵉 Z 쨜{•r 둗•;?: ?•?•?헒슈적맏 e	34796
0xCAA9	4•Y?+ # 뗼?•샖?꼍 n 늦축 O[•귐*?E 삮? W⒀댱?覇?_꺆•8 奬켮•l?캉繡?Z~	8091
0xE42D	7 랹 J•b 붑뛿&밟?q"8•갹?簡 w••릣??뱜뱺•{"	1559
0xEA36	lisa.sch.ac.kr	14

Key candidates 34796, 8091, and 1559 were extracted as the unknown information. Key candidate 14 was extracted the hidden information and was used to obtain the inserted text "lisa.sch.ac.kr".

Finally, we implemented a proof-of-concept tool that extracts key candidates based on the detection and extraction methods analyzed in this paper. Figure 4 shows the implemented result.

As shown in Fig. 4, a user selects the stego file by clicking the "Choose File" button and extracts key candidates by clicking the "Extract Text" button. In the message box below the figure, the size and file format of the stego file are printed, and key candidates are then extracted. Thus, we can obtain the hidden text using the key candidates extracted from this tool.

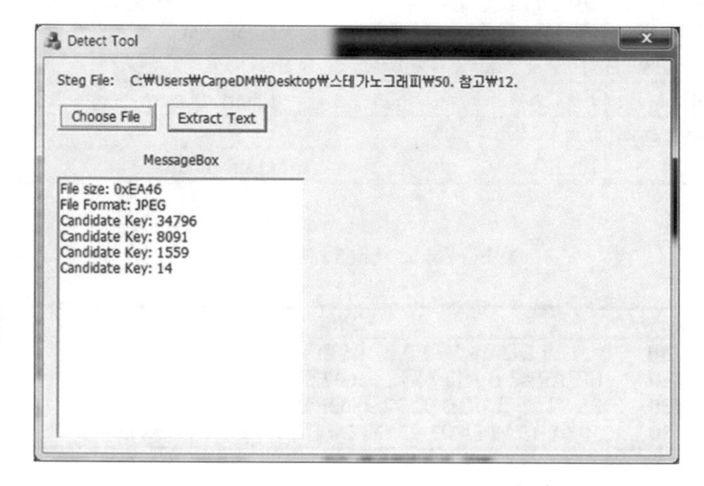

Fig. 4. Implement result of proof-of-concept tool

5 Conclusion

In this paper, we analyzed detection and extraction methods of hidden information based on StegoMagic, a steganography tool. StegoMagic inserts the information to be hidden at the end of the target file and utilizes the length of the information to be hidden as a key. We derived an extraction method of the key based on a JPG file and verified it by implementing a proof-of-concept tool. In the future, we will analyze an extraction method of the stego file and hidden information based on image files such as PNG and GIF, audio files, video files, etc.

Acknowledgments. This research was supported by the Basic Science Research Program through the National Research Foundation of Korea (NRF) that is funded by the Ministry of Education (NRF-2015R1D1A1A01057300).

References

1. Wikipedia: Steganography. https://en.wikipedia.org/wiki/Steganography
2. Dodson, J.D., Siraj, A.: Applying fuzzy hashing to steganography. J. Future Comput. Commun. **4**, 421–425 (2015)
3. Beh, D., Yin, M.: Data hiding for image using sequential colour cycle algorithm. Ph. D. Thesis. University Malaya (2010)
4. Wikipedia: Steganography tools. https://en.wikipedia.org/wiki/Steganography_tools

5. Ramesh, S.: Hide data in image, audio and video files: steganography. https://www.gohacking.com/hide-data-in-image-audio-video-files-steganography
6. Wikipedia: JPEG. https://en.wikipedia.org/wiki/JPEG
7. Wikipedia: JPEF File Interchange Format. https://en.wikipedia.org/wiki/JPEG_File_Interchange_Format

The 8th International Workshop on Mobile Commerce, Cloud Computing, Network and Communication Security (MCNCS-2018)

Clickbait Detection Based on Word Embedding Models

Vorakit Vorakitphan[1], Fang-Yie Leu[2], and Yao-Chung Fan[1(✉)]

[1] National Chung Hsing University, Taichung, Taiwan
bberin@outlook.com, yfan@nchu.edu.tw
[2] Tung-Hai University, Taichung, Taiwan
leufy@thu.edu.tw

Abstract. In recent years, social networking platform serves as a new media of news sharing and information diffusion. Social networking platform has become a part of our daily life. As such, social media advertising budgets have explosively expanded worldwide over the past few years. Due to the huge commercial interest, *clickbait* behaviors are commonly observed, which use attractive headlines and sensationalized textual description to bait users to visit websites. Clickbaits mainly exploit the users' curiosity's gap by interesting headlines to entice its readers to click an accompanying link to articles often with poor contents. Clickbaits are bothersome either to social media users or platform site owners. In this paper, we propose an approach called Ontology-based LSTM Model (OLSTM) to detect clickbaits. Compared with the existing solutions for clickbait detection, our approach is characterized by the following three components: word embedding model, Recurrent Neural Networks (RNN), and word ontology information. The observation is that preserving semantic relationships is significantly an important factor to be considered in detecting clickbaits. Therefore, we propose to capture semantic relationships between words by word embedding models. In addition, we adopted RNN as our classification models to consider word orders in a sentence. Furthermore, we consider the word ontology relation as another feature set for clickbait classification, as clickbaits often uses words with generalized concepts to induce curiosity. We conduct experiments with real data from Twitter and news websites to validate the effectiveness of the proposed approach, which demonstrates that the employment of the proposed method improves clickbait detection accuracy from 80% to 90% compared with the existing solutions.

1 Introduction

In recent years, social networking platform serves as a new media of news sharing and information diffusion. Social networking platform has become a part of our daily life. As such, social media advertising budgets have explosively expanded worldwide over the past few years. The advertising budget goes from $16 billion U.S. in 2014 to $31 billion in 2016 [10]. Due to the huge commercial interest,

L. Barolli et al. (Eds.): IMIS 2018, AISC 773, pp. 557–564, 2019.
https://doi.org/10.1007/978-3-319-93554-6_54

clickbaits [7] that use attractive headlines and sensationalized textual description to bait users to visit websites are commonly observed on social media platforms. In Fig. 1, we show clickbait examples, from which we can see that clickbaits on social platforms mainly exploit the users' curiosity gap by interesting headlines to entice its readers to click an accompanying link to articles with poor contents.

The motivation for clickbaits is that each click produces advertisement-based incomes for the content providers. The more clicks the sites received, the more revenue gained. Clickbaits are bothersome either to social media users or platform site owners. Detecting clickbaits therefore has received significant attention in related research communities [2,3]. A common paradigm for clickbait detection is to cast the problem as a binary classification task and then employing well-designed classification algorithms e.g., Support Vector Machine or Naive Bayesian Classifier, to automatically detect clickbaits based on predefined features. For example, in [2] the authors proposed a clickbait detection model based supervised learning linear classification by extracting the part-of-speech information, the average length of words, the length of headlines, and the number stop words as classification features. In [3], the main idea is to apply association rule mining to discover word relations within clickbait headlines and employ the relations as a feature set to proceed binary classification to detect clickbaits.

Fig. 1. Clickbait examples

The key in modeling clickbait detection problem as a classification task lies in the feature extraction. In this paper, we report a study of extracting the following three feature sets for clickbait detection. First, we propose to employ word embedding models [6,11] as building blocks for feature representation, as preserving semantic relationships is significantly an important factor to be considered in detecting clickbaits. Second, we propose to capture the sequential order of words in a sentence for clickbait detection; we propose to adopt RNN as our classification models to consider word orders in a sentence. Third, we propose to capture ontology information as another feature set for clickbait detection, as clickbaits often uses words with generalized concepts to induce users' curiosity.

The contributions of this paper are as follows.

- For modeling clickbait detection as a classification task, we conduct a data observation study by collecting real clickbait and non-clickbait headlines from various social media platforms to observe the characteristics of clickbait and non-clickbait headlines.
- Second, we propose a framework called Ontology-based LSTM (OLSTM) for clickbait detection problem. The OLSTM framework employs word embedding techniques and ontology information as building blocks for feature representation, and further models the clickbait detection problem as a sequence classification task to capture the word order information in a given sentence.
- Extensive experiments with real datasets are conducted. We also implement the state-of-the-art methods for clickbait detection and compare our proposal with the existing methods. The performance evaluation result demonstrates that the effectiveness of the proposed clickbait detection framework.

The rest section of this paper is organized as follows. In Sect. 2, we introduce the proposed clickbait detection framework, and In Sect. 3 we present the results of performance evaluation. Finally, Sect. 4 concludes this paper and discusses the future directions.

2 Methodology

2.1 Word Embedding Vectors

In this paper, we propose to apply word embedding information to boost the effectiveness of clickbait detection. Word embedding is one of the linguistic feature learning techniques. According to [8], word embedding plays an important role in text classification due to its similarity measurement based on text semantics. The word embedding model takes a large collection of plain text as input and produces a vector space, with each unique word in the collection being assigned a vector in the vector space. An important feature for embedding words as vectors is that the words having similar semantics in the collection are located in close proximity in the vector space. A serial of word embedding techniques has been proposed since 2000, such as [6,11]. The research on developing techniques for word embedding models has drawn significant attention due to its importance to many computational linguistics applications.

The idea of employing the word embedding model for feature representation is as follows. First, for a given sentence, we tokenize the given sentence S into a set of separated words $S = w_1, w_2, w_3, .., w_n$. Second, for every word w_i in S, we obtain the word vector $\mathbf{w_i}$ of w_i through a word embedding model. Then, we aggregate all $\mathbf{w_i}$ to form a vector representative $\mathbf{S}$ for the sentence. Formally, we have

$$\mathbf{S} = \sum_{i=1}^{n} \mathbf{v_i} \tag{1}$$

2.2 Ontology Information

According to [2], clickbaits have some sentence characteristics, as they attempt to catch users' attention and curiosity. One common characteristic of clickbait headline is that some words in a clickbait statement is not specifically referred to deep meaning words but instead generalized words. For example, "Car" in "This Car Will Help You To Get Rid Of Your Dad!", "Technology" in "One and Only Technology You'll Need Right Now In Your Love Life", and "Fruits" in "10 Fruits You'd Never Try After Your Night Out". Those words (i.e., car, technology and fruits) are general concepts. On the other hand, the words in non-clickbait headlines are often specialized words and semantically related. Instead of referring only to a car, technology or fruit, non-clickbait statements usually use more specific words, such as Toyota Prius, iPhone 8 or grape wine to state the real-world facts.

For employing this observation for detecting clickbaits, we propose to employ "Ontology [4,5,9]" structure to capture word meaning hierarchies. Mainly, we consider "Hypernym and Hyponym" information. Hypernym generally refers to a word with a broad meaning constituting a category into which words with more specific meanings fall; a superordinate. For example, color is a hypernym of red. Hyponym refers to a word of more specific meaning than a general or superordinate term applicable to it. For example, a spoon is a hyponym of a container. Our idea is that the number of the hypernyms and the hyponyms of a word can be used as a measurement to determine whether a given word has a general meaning or deep meaning within its domain. The observation is that clickbaits have simple and generalized meaning rather than detailed meaning. On the other hand, non-clickbaits sometimes refer specifically to a fact or details about a thoughtful topic. Normally, if a targeted word has many hypernyms and fewer hyponyms, the targeted word is likely to be a part in a non-clickbait since the targeted word seems to be in a deep level word hierarchy. On the other hand, if a targeted word has many hyponyms and fewer hypernyms, the targeted word has higher possibility to be observed in a clickbait statement. Therefore, we propose to count the number of the hypernyms and the hyponyms of a word as another set of features for clickbait detection. We employ the information from DBpedia to obtain the number of the hypernyms and the hyponyms of a given word.

2.3 Word Order Information

Our third idea for boosting the effectiveness of clickbait detection is to consider the word order in a given headline, as the word order naturally captures the meaning of a sentence. We propose to use Recurrent Neural Networks (RNN) to capture word relations. In this paper, we adopt Long Short-Term Memory networks (LSTM) [12] which extends RNN by adding forget/remember gates in each hidden layer to remember information from in previous stages from the beginning till the end of each learning procedure. LSTM is able to actively maintain self-connecting layers. At this moment, important information of new inputs will be preserved by the time the learning goes to a deeper layer.

2.4 Joint Features

By considering the above-mentioned features, we use LSTM as the fundamental classification model. Our LSTM takes a sequence of $n+2$-dimensional vectors as inputs. Among the $n+2$ dimensions, there are n dimensions for word embedding features and two dimensions for the number of hypernyms and hyponyms of a given word. In Fig. 2, we show an example of using $n = 400$ word embedding techniques. In the experiment, we set a maximum length of each sentence S to be 40 due to every input sequence has to be equal size in LSTM.

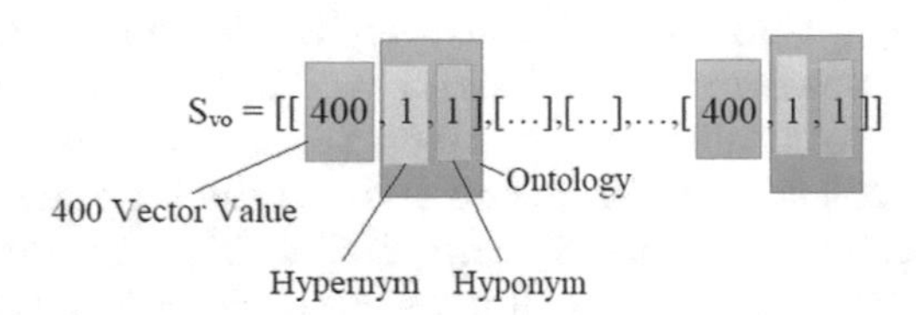

Fig. 2. Joint features for clickbait detection

3 Performance Evaluation

In this section, we evaluate clickbait detection performance based on both clickbait and non-clickbait d from Twitter and news headlines from official news sites. in order to train and validate the proposed models in precision, recall, and f1-score of each proposed model. In the following, we first describe the training set and testing set. We then focus on the performance of the mentioned clickbait detection methods (Table 1).

3.1 Data Collection

In this study, clickbait and non-clickbait posts are collected from Twitter and official news sites [1]. In total, we have 54,033 clickbait and non-clickbait posts. As a training data set, we randomly select 10,000 clickbait posts and 10,000

Table 1. Data set

	Clickbait	Non-Clickbait	Total Instances
Training Set	10,000	10,000	20,000
Testing Set	5,000	5,000	10,000

non-clickbait posts to train classification models. As a testing data set, we picked another set of random instances (not included in the training dataset). In the testing data set, there are 5000 clickbaits and 5000 non-clickbaits. The details of the data collection are as follows.

Data Collection from Social Platform. We collected 22,033 clickbait and non-clickbait tweets from Twitter. Clickbait tweets are collected from social news channels (e.g., @BuzzFeed, @Forbes, @Indy100). Non-clickbait tweets are collected from official news channels (e.g., @BusinessInsider, @CNN, @Independent). After that, all tweets are judged by five graduate students to label the collected tweets (as a clickbait or not).

Data Collection from News Sites. We divide clickbait and non-clickbait news sites into two parts. For clickbait headlines, we obtain the article headline from 'BuzzFeed', 'Upworthy', 'ViralNova', 'Scoopwhoop', and 'ViralStories.' These sites are well-recognized clickbait sites. All the headlines are judged by six graduates to give a label to a headline (as a clickbait or not). For non-clickbaits, we use articles from Wikinews and use the headline of the articles as non-clickbait statement. Within Wikinews site, articles are produced by a community by their outsourcing from real news channels in English. The Wikinews articles are verified by the community before publication. Therefore, we consider them as a source for non-clickbait posts. We in total collected 32,000 headlines as our data set.

3.2 Experimental Settings

Compared Methods. We implement the following methods for performance comparison.

- **NBC-LF** (Naive-Bayes with Lexical Features). With the collected data set, in this implementation, we extract lexical-based vocabulary features from the sentences in the training data set to learn a Naive Bayes classifier.
- **SVM-SF** (Support Vector Machine with Syntactic Features). With the collect data set, we extract the following syntactic features (the number of determiners, the number of stop words, the number of words contractions, the number of hyperbolic, sentence length, Common bait phrases, sentence subjects, punctuation patterns, Part of Speech Tags, and vocabulary statistics) for a training sentence to learn a Support Vector Machine classification model.

- **LR-WEF** (Logistic Regression with Word Embedding Features). In this method, we employ the word embedding model to translate words into their word embedding vectors and aggregate all the word vectors in a given sentence as feature representation (as proposed in Subsect. 2.1) to learn a Logistic Regression classification model.
- **RF-WEF** (Random Forrest with Word Embedding Features). With the same word embedding feature representation as LR-WEF, in this method, we employ Random Forrest model (using Gini index as the measurement for constructing decision trees) to learn a clickbait classifier.
- **OLSTM** (LSTM with Word Embedding and Ontology Features). In this method, we implement the proposed ideas by employing word embedding and ontology features and modeling the detection problem as a sequence classification task.

3.3 Performance Evaluation

In this section, we show the performances of detecting clickbaits by the compared methods. In Table 2, we show the Precision, the Recall, F-1 Score, and Running time for the compared methods. We have the following observations for the experiment results. First, the employment of word embedding techniques indeed improves the effectiveness of detecting clickbaits. From the experiment results, LR-WEF, RF-WEF, and OLSTM significantly outperform the methods without the word embedding features, i.e., NBC-LF and SVM-SF. Second, from the experiment results, we can see that the proposed OLSTM framework has the best performance for clickbait detection, which again validates the superiority of the proposed ideas.

Table 2. The performance overview for the compared methods

	Precision	Recall	F1-Score
NBC-LF	0.8086	0.8069	0.8021
SVM-SF	0.8026	0.7960	0.7967
LR-WEF	0.8517	0.8479	0.8475
RF-WEF	0.8587	0.8533	0.8529
OLSTM	0.8994	0.8990	0.8990

4 Conclusion

In this paper, we report a study of employing word embedding models, recurrent neural network, and word ontology information for clickbait detection. Extensive experiments with real datasets are conducted. We also implement the state-of-the-art methods for clickbait detection and compare our proposal with the existing methods. The performance evaluation result demonstrates that the effectiveness of the proposed clickbait detection framework.

Acknowledgement. This research was supported by the Ministry of Science and Technology Taiwan R.O.C. under grant number 106-2221-E-005-082-, and also partially supported by the Project H367B83300 conducted by ITRI under sponsorship of the Ministry of Economic Affairs, Taiwan, R.O.C.

References

1. Potthast, M., Gollub, T., Komlossy, K., Schuster, S., Wiegmann, M., Garces, E., Hagen, M., Stein, B.: Crowdsourcing a Large Corpus of Clickbait on Twitter. arXiv: 1710.08721v1 (2017)
2. Chakraborty, A., Paranjape, B., Kakarla, S., Ganguly, N.: Stop Clickbait: detecting and preventing clickbaits in online news media. In: IEEE/ACM International Conference on Advances in Social Networks Analysis and Mining (ASONAM) (2016)
3. Heartfield, R., Loukas, G., Gan, D.: You are probably not the weakest link: towards practical prediction of susceptibility to semantic social engineering attacks. In: IEEE Access, vol. 4 (2016)
4. Yang, S., Chen, H., Vorakitphan, V., Fan, Y.: Learning term taxonomy relationship from a large collection of plain text. In: Computer Symposium (ICS) (2016)
5. Arnold, P., Rahm, E.: Extracting semantic concept relations from wikipedia. In: WIMS 2014 Proceedings of the 4th International Conference on Web Intelligence, Mining and Semantics (WIMS14), Article No. 26 (2014)
6. Bengio, Y., Ducharme, R., Vincent, P., Jauvin, C.: A neural probabilistic language model. J. Mach. Learn. Res. **3**, 1137–1155 (2003)
7. Ahmed, S., Monzur, R., Palit, R.: Development of a rumor and spam reporting and removal tool for social media. In: 3rd Asia-Pacific World Congress on Computer Science and Engineering (APWC on CSE) (2016)
8. Sang, L., Xie, F., Liu, X.: WEFEST: word embedding feature extension for short text classification. In: IEEE 16th International Conference on Data Mining Workshops (ICDMW) (2016)
9. Wong, W., Lui, W., Bennamoun, M.: Ontology learning from text: a look back and into the future. In: ACM Computing Surveys CSUR, pp. 1–36 (2011)
10. Fuller, S.: U.S. Social Media Marketing - Statistics Facts (2016)
11. Mikolov, T., Sutskever, I., Chen, K., Corrado, G.S., Dean, J.: Distributed representations of words and phrases and their compositionality. In: Advances in Neural Information Processing Systems, pp. 3111–3119 (2013)
12. Graves, A., Mohamed, A., Hinton, G.: Speech recognition with deep recurrent neural networks. In: 2013 IEEE International Conference on Acoustics, speech and signal processing (ICASSP), pp. 6645–6649 (2013)

A Micro Services Quality Measurement Model for Improving the Efficiency and Quality of DevOps

Sen-Tarng Lai[1(✉)] and Fang-Yie Leu[2]

[1] Department of Information Technology and Management, Shih Chien University, Taipei 10462, Taiwan
stlai@mail.usc.edu.tw

[2] Department of Computer Science, Tunghai University, Taichung 40704, Taiwan
leufy@thu.edu.tw

Abstract. DevOps is an important practice environment and future operation trend for software development and maintenance. DevOps has important features that are continuous integration, continuous delivery, automation and high efficiency to increase enterprise market competition. Micro services architecture is a critical item for keeping the advantages of DevOps environment. In addition, micro services have many advantages than monolithic applications in software development and maintenance. However, many challenges of micro services architecture need to be overcome. Quantified quality characteristics can identify and assist to improve the defects of micro services that affect the efficiency and quality of DevOps operation. In this paper, the authors propose the Micro Services Quality Measurement (MSQM) Model to evaluate and identify work process defects of micro services. Based on MSQM model, the quality improvement procedure is designed for improving micro services work process defects and increasing the overall DevOps operation efficiency and application quality.

1 Introduction

In early 1990s, software engineering scholar Pressman once speculated that software maintenance costs would account for 70–80% of the total software cost [1]. Schach also mentioned in his book Classical and Object-Oriented Software Engineering that two-thirds (about 67%) of the total cost of software is used in software maintenance [2]. Many enterprises and organizations even spend 80% of their time and cost in software maintenance [3]. Under the pressure of fierce market competition, the maintenance operation of continuous changes is the most effort and cost expensive stage in the SDLC, and shows the importance of software maintenance. For this, how to improve the efficiency of software maintenance to reduce maintenance costs has become a topic worthy in depth exploration in software maintenance operations.

DevOps aims to promote collaboration between development and operations teams [4–6]. DevOps uses agile methods [7] to help enterprises and organizations respond quickly and easily to changing needs, advocating collaboration between development and operations, and working to increase the market competitive advantage of customers.

L. Barolli et al. (Eds.): IMIS 2018, AISC 773, pp. 565–575, 2019.
https://doi.org/10.1007/978-3-319-93554-6_55

DevOps provides major advantages such as collaboration, automation, cloud monitoring, business agility, productivity, continuous integration (CI) [8], and instant delivery deployment [9, 10]. In addition, DevOps replace the monolithic application architecture with the micro services architecture [11]. Micro services architecture has become the key of the effective operation of DevOps [12]. However, many problems of micro services architecture may affect the operational effect of DevOps [11–13]. In this paper, the authors proposes the Micro Services Quality Measurement (MSQM) model to resolve the problems and improve the operational efficiency and application quality of DevOps. The paper divides into six sections. Section 2 discusses the DevOps operating environment and the trends of micro services. Section 3 describes the advantages and facing challenges of micro services and discusses the key quality characteristics of micro services. In Sect. 4, the MSQM model and quality improvement procedure is proposed. In Sect. 5, evaluates the effectiveness of the MSQI (Micro Services Quality Improvement) procedure. Section 6 once again emphasizes the importance and contributions of MSQM model to DevOps and concludes on this topic.

2 DevOps Operating Environment and Micro Services Trends

2.1 Discussion on the Operation of DevOps

With rapid environmental changes and continuous evolution of information technology, software system must accept change and expansion requirements to meet the requirements of the current situation for continuing their life cycle. However, software development and maintenance operations belong to different teams, respectively. There are often many misunderstandings and frictions that make software maintenance more difficult and maintenance costs are increasing continuously. DevOps extends the method of agile software development into the maintenance phase, prompting the software system to stretch CI and continuous deployment capabilities into the software operating, significantly increasing the market competitive advantage of enterprises and organizations [5]. Lwakatare et al. [14] believe that there is currently no common understanding between the academic and practitioners regarding the structure of DevOps. It is necessary to study and investigate how DevOps can improve the development and maintenance of software. For this, they defined the four major elements of DevOps:

- Collaboration: Collaboration and information sharing can expand technology and capabilities so that the software system can be successfully delivered.
- Automation: In order to continue agile software development, focus on product reconfiguration and continuous delivery of software operating, and with TDD (Test Driven Development) [15] and CI technology [8], DevOps operation process needs to have fully automation.
- Measurement: Quantitative indicators can identify the defects of DevOps operational processes and products, and help improve product development efficiency. Quality measurement process can identify issues related to DevOps operation and product quality. It is a necessary step to improve operational efficiency and software quality.

- Monitoring: DevOps addresses the challenge of effective monitoring by emphasizing collaboration. Collaboration between developers and operators allows the system to share and control relevant information.

2.2 Micro Services Architecture and DevOps Relations

DevOps supports the business changes proposed by users and improves the efficiency of product completion or service delivery. The main purpose and advantage of DevOps is to reduce the obstacles to the continuous delivery process and improve the efficiency and quality of product development and operations. However, in order to achieve the benefits of DevOps, software applications must turn the development of the monolithic architecture into a micro services architecture. Several advantages of the micro services architecture are important mechanisms for achieving the purpose and expected benefits of DevOps, as described below [15, 16] (showed as Table 1):

- Decompose monolithic application into a micro services architecture. Each service uses its own container and operating environment and is highly independent.
- Micro services designed as one simple function. A micro service can only implement one service function, control its own data and reduce complexity.
- Micro services architecture is connected via REST APIs or message interfaces to avoid proprietary protocols such as shared databases or JMS and improve security.
- Micro Services support CI and Continuous Delivery (CD), allowing the system to continue to evolve with automation capabilities and without sacrificing business.

In micro services architecture, each service has a high degree of reusability and flexibility to adjust, fully utilize and share existing resources, and maintainability. in addition, each service can be developed by an independent team, but necessary to effectively control the phenomenon of inconsistencies in the interface and provide monitoring capabilities. Monolithic applications and application continuous changes take many operational challenges of enterprise and organization:

- Updating and repairing a monolithic application becomes more and more difficult.
- Enterprises are forced to shift their applications to a modern use micro services architecture to withstand changes in the environment.
- Many companies are limited by the need for rapid change and are limited by a monolithic application.
- With the popularization of cloud computing applications, the development and application of the cloud adopts different technologies and methods of development and operations.

Enterprise and organization introduced the micro services architecture to solve the facing problems.

Table 1. Comparison of micro service architecture and monolithic architecture

Key features	Micro services architecture	Monolithic architecture
Independence of components	High	Low
Application complexity	Low	High
Processing efficiency	High	Low
Expansibility	High	Low
Flexible modification	High	Low
Overcome personnel turnover	High	Low

3 The Importance and Influence of Micro Services

3.1 Advantages and Challenges of Micro Services

Many features of micro services are important elements that assist DevOps in achieving cultural integration, coordinating cooperation, sharing resources, continuous integration, continuous delivery, and rapid deployment [15, 16]. However, there are many challenges to the micro services architecture that must be overcome. The problems that may be encountered for micro services are described as follows [13, 17]:

- Size problems: Micro services should provide several services that are most appropriate, and some developers advocate minimal service (approximately 10–100 strokes). Although small services are easier to manage and maintain, in order to achieve major goals, small services must undergo frequent integration actions. The goal of micro services is to fully decompose the application to facilitate the development and deployment of agile applications.
- Integration problems: Another challenge for micro services is the decentralized server structure. Micro services-based applications need to combine multiple servers with different services. Micro services integration without a complete or automated integration tools, is highly challenging for developers.
- Change problems: Another major challenge of the micro services architecture is the implementation of changes across multiple servers. Unlike the use of a monolith application, changes of micro services must be carefully planned and coordinated to management each service. Micro services architecture needs the management features such as configuration management (CM) system, version management mechanisms, and interdependency relationship, which can improve the problems of change operations.
- Deployment problems: The deployment of micro services applications is more complex than the monolithic application. A monolithic application just need to be deployed on the servers with traditional load balancer mechanisms. The process of micro services applications deployment is usually consist of a large number of services. The process requires configuration, deployment, monitoring, and expansion of more service items. Therefore, the successful deployment of micro services applications requires developers to have better technologies, deployment methods, and more automation.

- Automation problems: DevOps attaches great importance to the efficiency of software development and maintenance, and special emphasis on CI, CD and automation. Therefore, micro services apply TDD of agile software development and the appropriate unit and integration test tools. Automation is one of major elements of DevOps. Micro services must enhance automation capability to achieve the efficiency of CI, CD and rapid deployment.

The problems of micro services cannot be concretely improved which must affect the efficiency and effect of DevOps. Software quality characteristics and quantification mechanism can assist to timely identify the problems and improve the defects of micro services architecture.

3.2 Quality Characteristics of Micro Services

In order to overcome the problems that micro services may encounter, the following five quality characteristics to improve the defects of micro services work process:

(1) Modularity: High modularity micro services provide only a single service and must have high cohesion (e.g., information cohesion) and low coupling (e.g., data coupling) [2]. This can not only effectively improve the size of micro services, but also solve part of the integration problems.
(2) Perfect Interface Quality: CI is a key and important activity of DevOps. Micro services must have clear interface definition, low coupling interface, and correct, complete, and consistent interface checking capabilities to improve the quality of the interface. Micro services with the perfect interface quality can achieve the integration capabilities of micro services automation and help overcome the issues of integration and change.
(3) Manageability: Micro services must possess manageability qualities such as construction management system, version control mechanism, and interdependence relationship between micro services and micro services, and can be managed when changes or revisions are made to application software. Sexuality can identify the affected micro services (placed on different servers) in a timely manner and assist in subsequent adjustments and revisions.
(4) Object-oriented quality: Security is one of the important indicators of DevOps. Micro services must have object hiding characteristics such as information hiding, data protection, and Exception handling mechanisms, and information hiding. Data protection can effectively control the access of data and avoid unnecessary data access. The exception handling mechanism can prevent unexpected situations and prevent the leakage of important data. Integrating object-oriented qualities can help overcome the security problems of micro services.
(5) Automation quality: Automation is one of the important advantages of DevOps. Micro services must possess the automated features of agile software development TDD features, unit test case generation, automatic unit testing, etc. DevOps pays special attention to the efficiency of the process, and micro services can integrate multiple. The automated quality can achieve automated unit testing activities in the way of TDD and assist in overcoming the integration and change issues of continuous integration and continuous delivery.

This study integrates key quality features into micro services to effectively improve the predicament of micro services. In addition to overcoming most issues of micro services, it can also specifically improve information sharing, CI, CD, automation, security, and high efficiency of DevOps. The relationship between disadvantages of micro services and micro services quality characteristics (shown as Table 2).

Table 2. Relationship between micro services problems and improvable quality characteristics

Micro services problems	Improvable quality	Quality factors
Service size	Modularity quality	Unit function
		High cohesion
		Low coupling
Service integration	Perfect interface quality	Clear interface definition
		Interface checking
		Interface simplification
Service change and extension	Manageability quality	CM system
		Version control mechanism
		Interdependence relationship
Service automation	Automation testing quality	TDD-based testing
		Automatic unit testing
		Automatic integration
Service security	Object-oriented quality	Information hiding
		Private data protected
		Exception handling

4 MSQM Model and Improvement Procedure

4.1 Micro Services Quality Measurement Model

In this paper, LCM (linear combination model) is applied to the micro services quality measurement [18]. Different levels of quality activities will show different quantified values. Therefore, the quality factor must be normalized before using the LCM in combination with the formula. The normalized quantified value is between 0 and 1, the best quality quantified value is close to 1, and the worst quality quantified value is close to zero. Improving the quality of micro services dilemmas should consider five quantitative quality features (shown as Table 1). Using LCM, relevant quality factors can be incorporated into quality metrics, then related quality metrics can be incorporated into quality metrics, and finally, quality metrics for micro services can be generated:

(1) Modularity Quality Measurement (MoQM) is combined unit service, high cohesion, and low coupling three quality metrics. Combination formula shows as Eq. (1):

MoQM : Modularity Quality Measurement
USM : Unit Service Metric W_1 *: Weight of USM*
HCM : High Cohesion Metric W_2 *: Weight of HCM*
LSM : Low Cohesion Metric W_3 *: Weight of LCM*

$$MoQM = W_1^* USM + W_2^* HCM + W_3^* LCM \quad W_1 + W_2 + W_3 = 1 \tag{1}$$

(2) Perfect Interface Quality Measurement (PIQM) is combined clear interface definition, interface checking and interface simplification three quality metrics. Combination formula shows as Eq. (2):
PIQM : Perfect Interface Quality Measurement
CIDM : Clear Interface Definition Metric W_1 *: Weight of CIDM*
ICM : Interface Checking Metric W_2 *: Weight of ICM*
ISM : Interface Simplification Metric W_3 *: Weight of ISM*

$$PIQM = W_1^* CIDM + W_2^* ICM + W_3^* ISM \quad W_1 + W_2 + W_3 = 1 \tag{2}$$

(3) Manageability Quality Measurement (MaQM) is combined CM System, version control mechanism and interdependence relationship three quality metrics. Combination formula shows as Eq. (3):
MaQM : Manageability Quality Measurement
CMsM : CM system Metric W_1 *: Weight of CMsM*
VCM : Version Control mechanism Metric W_2 *: Weight of VCM*
IRM : Interdependence Relationship Metric W_3 *: Weight of IRM*

$$MaQM = W_1^* CMsM + W_2^* VCM + W_3^* IRM \quad W_1 + W_2 + W_3 = 1 \tag{3}$$

(4) Object-Oriented Quality Measurement (OOQM) is combined information hiding, data protection and exception handling three quality metrics. Combination formula shows as Eq. (4):
OOQM : Object-Oriented Quality Measurement
IHM : Information Hiding Metric W_1 *: Weight of IHM*
DPM : Data Protection Metric W_2 *: Weight of DPM*
EHM : Exception Handling Metric W_3 *: Weight of EHM*

$$OOQM = W_1^* IHM + W_2^* DPM + W_3^* EHM \quad W_1 + W_2 + W_3 = 1 \tag{4}$$

(5) Automation Testing Quality Measurement (ATQM) is combined TDD-based, automated unit testing and automated integration three quality metrics. Combination formula shows as Eq. (5):
ATQM : Automation Testing Quality Measurement
TDDM : TDD – based Metric W_1 *: Weight of TDDM*
AuTM : Automated Unit testing Metric W_2 *: Weight of AuTM*
AIM : Automated Integration Metric W_3 *: Weight of AIM*

$$ATQM = W_1^* TDDM + W_2^* AuTM + W_3^* AIM \quad W_1 + W_2 + W_3 = 1 \tag{5}$$

(6) Finally, combining MoQM, PIQM, MaQM OOQM and ATQM into a Micro Services Quality Indictor (MSQI). Combination formula shows as Eq. (6):
$MSQI$: *Micro Services Quality Indicator*
$MoQM$: *Modularity Quality Measurement* W_1 : *Weight of MoQM*
$PIQM$: *Perfect Interface Quality Measurement* W_2 : *Weight of PIQM*
$MaQM$: *Manageability Quality Measurement* W_3 : *Weight of MaQM*
$OOQM$: *Object – Oriented Quality Measurement* W_4 : *Weight of OOQM*
$ATQM$: *Automation Testing Quality Measurement* W_5 : *Weight of ATQM*

$$QMSQI = W_1^* MoQM + W_2^* PIQM + W_3^* MaQM + W_4^* OOQM + W_5^* ATQM \quad W1 + W2 + W3 + W4 + W5 = 1 \tag{6}$$

Quality measurement model is divided into three layers. First layer, based on the primitive quality factors, combines with quantified quality factors into the data quality metrics. Second layer, based on the data quality metrics, combines with the quality metrics into the data quality measurement. Finally, the third layer, combines with the quality measurements into a Micro Services Quality Indicator (MSQI). We call the quality measurement combination process is a Micro Services Quality Measurement (MSQM) Model.

4.2 Micro Services Quality Improvement Procedure

The MSQI generated through the MSQM model can be cooperated with rule-based inspection activities. The rule-based inspection activities identify quality defects, and can be further determined the work process problems of micro services. Then propose

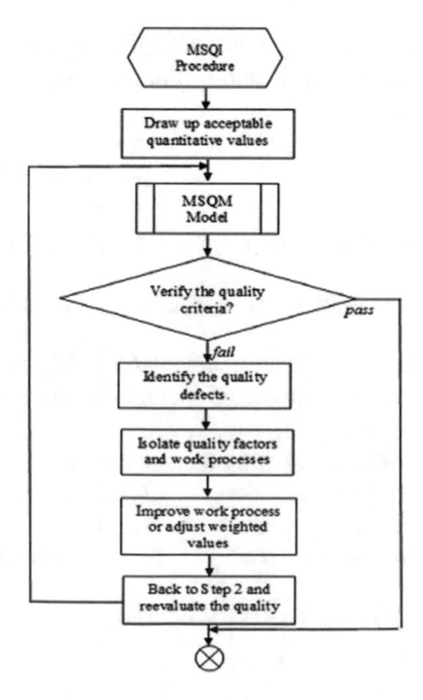

Fig. 1. Flowchart of MSQI procedure

substantial improvement and remedial measures, with specific revisions and correct the work process that is not applicable. Micro Services Quality Improvement (MSQI) Procedure includes five steps (shown as Fig. 1):

Step 1. The DevOps project manager must negotiate with the client and stakeholder to formulate the quality criteria for the micro services. And, set the acceptable quantified value for each measurements of the improved quality characteristic of the micro services.

Step 2. According to the formula (6) of the MSQM mode, determine whether the MSQI satisfies the quality criteria of the micro services. If it can be satisfied, the improvement procedure will end the operation; otherwise, the improvement procedure will judge the quality characteristics that does not reach the quantified value and enter Step 3.

Step 3. According to the quality characteristic measurement value, identify the corresponding work items of the MSQM model that does not reach the qualified value (quality defects).

Step 4. Revise and correct the work process of micro services based on the improvement and corrective measures, and must pass the improvement inspection activities.

Step 5. Collect the quantified quality factors that have been revised work process, and the process returns to Step 2

5 Efficiency Evaluation of MSQI Procedure

Based on MSQM model, this paper designs MSQI procedure to identify the defects of work process and improve the quality of micro services. MSQI procedure combines the quantified measurements to monitor and control the quality of micro services work process timely and effectively enhance quality of micro services. For evaluating the advantages of MSQI procedure, five impact items include efficiency (automation), quality Assurance, CI, CD, sharing to evaluate the MSQI procedure (shown as Table 3).

Table 3. MSQI procedure evaluation table

Advantages of DevOps	Use MSQI procedure	Use micro services	Use monolith
Efficiency	More certain	Certain	Uncertain
Quality	High	Improvable	Unexpected
CI	High	Middle	Low
CD	High	Middle	Low
Sharing	High	Middle	Low

6 Conclusions

DevOps is an important trend that makes software development and operations teams' collaboration. High efficiency and quality, sharing, cooperation, CI, CD and automation are major features of DevOps. Micro services architecture is a critical item that assist DevOps has these advantage features. However, micro services architecture facing many challenges that need to be urgently overcome. The problems of micro services cannot be concretely improved which must affect the efficiency and effect of DevOps. Software quality characteristics and quantification mechanism can assist to timely identify the problems and improve the defects of micro services architecture. For this, in this paper, a MSQM Model is proposed to identify micro services work process defects. Based on the MSQM model, the MSQI procedure can timely identify and modify the problems of micro services architecture to increase the micro services effectiveness and quality. In DevOps operation, micro services architecture with high quality can improve DevOps operation efficiency and products quality. The concrete contributions of the MSQI procedure that based on MSQM Model described as follows:

- Applying quantified quality activities to timely identify work process problems and quality defects of micro services architecture.
- Timely modify work process defects of micro services and concretely increase DevOps operation efficiency and products quality.
- Quality improvement of micro services directly makes DevOps operation to hold the original objectives and advantages.

References

1. Pressman, R.S.: Software Engineering: A Practitioner's Approach. Palgrave Macmillan, Basingstoke (2010)
2. Schach, S.R.: Object-Oriented Software Engineering. McGraw-Hill, New York (2008)
3. Yourdon, E.: Rise and Resurrection of the American Programmer. Yourdon Press, Upper Saddle River (1996)
4. Httermann, M.: DevOps for Developers. Apress, Berkely (2012)
5. Loukides, M.: What is DevOps? O'Reilly Media, Inc., Sebastopol (2012)
6. Erich, F., Amrit, C., Daneva, M.: Cooperation between information system development and operations. In: 8th International Symposium on Empirical Software Engineering and Measurement, p. 1. ACM Press, New York (2014)
7. Szalvay, V.: An Introduction to Agile Software Development, CollabNet, Inc. (2004)
8. Duvall, P., Matyas, S., Glover, A.: Continuous Integration: Improving Software Quality and Reducing Risk. Pearson Education, Inc., Boston (2007)
9. Roche, J.: Adopting DevOps practices in quality assurance. Commun. ACM **56**(11), 38–43 (2013)
10. Schaefer, A., Reichenbach, M., Fey, D.: Continuous integration and automation for DevOps. In: IAENG Transactions on Engineering Technologies, pp. 345–358. Springer, Dordrecht (2013)

11. Bass, L., Weber, I., Zhu, L.: DevOps: A Software Architect's Perspective. Addison-Wesley Professional, Boston (2015)
12. Balalaie, A., Heydarnoori, A., Jamshidi, P.: Microservices architecture enables DevOps: migration to a cloud-native architecture. IEEE Softw. **33**(3), 42–52 (2016)
13. Richardson, C., Smith, F.: Microservices from Design to Deployment. NGINX (2016)
14. Lwakatare, L.E., Kuvaja, P., Oivo, M.: Dimensions of DevOps. In: International Conference on Agile Software Development. Springer, Cham (2015)
15. Beck, K.: Test-Driven Development: By Example, Addison-Wesley, New York (2003)
16. Rahman, M., Gao, J.: A reusable automated acceptance testing architecture for microservices in behavior-driven development. In: 2015 IEEE Symposium on Service-Oriented System Engineering (SOSE), pp. 321–325 (2015)
17. Dmitry, N., Manfred, S.S.: On micro-services architecture. Int. J. Open Inf. Technol. **2**(9), 39 (2014)
18. Fenton, N.E.: Software Metrics - A Rigorous Approach. Chapman & Hall, London (1991)

A Study on Firewall Services with Edge Computers of vEPC in 5G Networks

Fang-Yie Leu(✉) and Ping-Hung Chou

Department of Computer Science, Tunghai University, Taichung, Taiwan
{leufy,s04351038}@thu.edu.tw

Abstract. Following the fast development of 5G networks and Internet of Things (IoT) techniques, in the future, billions of User Equipment (UE) and IoT devices will connect to networks and send data to backend servers for situation monitoring or required processes. Since mobile devices roam to everywhere in the world, firewalls which are designed to protect fixed-position devices cannot be used to secure mobile users. In this study, we would like to use firewall to protect mobile devices, even though they are roaming in the networks other than their home networks. UEs are secured in data layer with firewalls, distributed topology storage (DT storage) and SDN controller, in which firewalls are installed in the vEPC's edge computers. We also develop the corresponding processing algorithms.

Keywords: Firewall · EPS-AKA · SDN controller · SDN · NFV
OpenFlow · FlowTable · Fault tolerant mechanism

1 Introduction

Up to present, firewalls are used to secure those network entities that are geographically fixed. They are hard to protect mobile devices (in the following, we use UEs as examples.), except delivering packets to a firewall website before sending them to their destinations. This is not a good solution since it will worsen network traffic and delay packet delivery. Furthermore, edge computing has attracted researchers' attentions since they process data for users with nearby computers. For example, two people talk with each other with different languages. An edge computer interprets the sentences they speak before these sentences arrive at their destinations. So, even though the two people do not speak the same language, they can communicate with each other under the help of a language interpreter run on an edge computer. In fact, edge computing is one of the key functions of 5G networks since it effectively decreases network traffic and packet-delivery delay.

Of course, firewall can be run on a mobile phone. But this is currently inappropriate due to mobile phone's limited CPU capability, memory size and battery capacity. Someday when these problems are solved, mobile phone is one of the best candidates of network entities acting as a firewall. In this study, we set up firewall in an edge computer attached to vEPC [1] and transfer the firewall following the handover of UE.

L. Barolli et al. (Eds.): IMIS 2018, AISC 773, pp. 576–584, 2019.
https://doi.org/10.1007/978-3-319-93554-6_56

Our firewall is installed in vEPC's edge computer at data layer. The advantage is neither increasing a heavy burden for mobile phones, nor requesting additional network facilities, since firewall is provided by an existing subsystem of 5G networks, i.e., edge computer. Its cost is lower than that when firewall is placed at a dedicated fixed-position server.

The rest of this paper is organized as follows. Section 2 briefly reviews literature related to this study. Section 3 introduces the proposed system. Section 4 concludes this paper.

2 Literature Review

Nguyen *et al.* [2] studied the feasibilities of SDN-based and NFV-based Mobile Packet Core (MPC) from different viewpoints, and defined 4 dimensional classification criteria, including the weights of different methodologies, techniques, implementation approaches and deployment policies. The potential research topics of them were also studied.

Arins in [3] proposed firewalls as one of the security services to protect Internet Service Provider (ISP), allowing users to install matching rules on the ISP's edge computers. OpenFlow is employed as the protocol so that firewall can protect remote users. However, this research only supports three simple APIs and the protected targets should be geographically fixed, i.e., if UE moves to somewhere, it will be protected by local firewalls. In a 5G network, we need powerful network functions and the APIs providing UEs with firewalls, by which system developers can implement many more security functions to effectively protect UEs.

Cliou *et al.* [4] claimed that Mobile-edge computer (MEC, now it is Multi-access Edge Computer) should be those with internal features, neighborhood, and low delay, and then provide the infrastructure of MEC implemented with network function virtualization (NFV). Authors also introduced an Infrastructure as a Service (IAAS) to integrate their system, and described related applications. However, their system can be further enhanced, particularly on system integration.

Zope *et al.* in [5] presented that virtualized networking, which utilizes OpenFlow and SDN Controller and adopts load balance policy, can substitute for the firewall hardware in developing applications. The SDN Controller they used is floorlight. Authors also described how to help users to develop and test their applications so as to gradually deploy the network required by production lines. But this article did not deal with UE handover and its authentication.

Gray *et al.* in [6] tried to establish a firewall with a cluster environment. The server connecting to firewall's element manager (EM) is Commodity off the Shelf (COTS) hardware server. Its operations are monitored by the EM. But this system needs to have parallel capability for processing network traffic.

Guoa *et al.* in [7] established an SDN Controller, which has the features of low burden, short packet-delivery delay and high transmission capability, with three network mechanisms, including Path-set Database Generation, Flow-table Management and Routing Decision. Authors also compared the performance of LRU+OSPF, AC +OSPF and STAR, nevertheless missing ONOS and OpenDayLight which are popular tools of SDN controller.

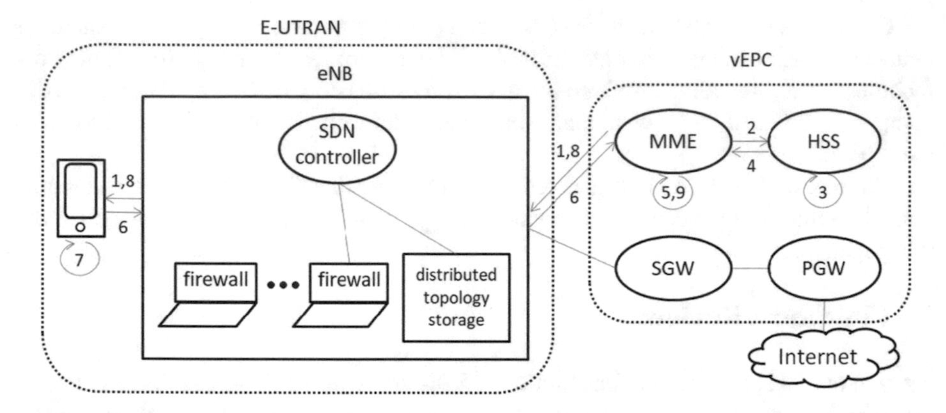

Fig. 1. The firewall in an eNB and the 9 steps of EPS-AKA protocol.

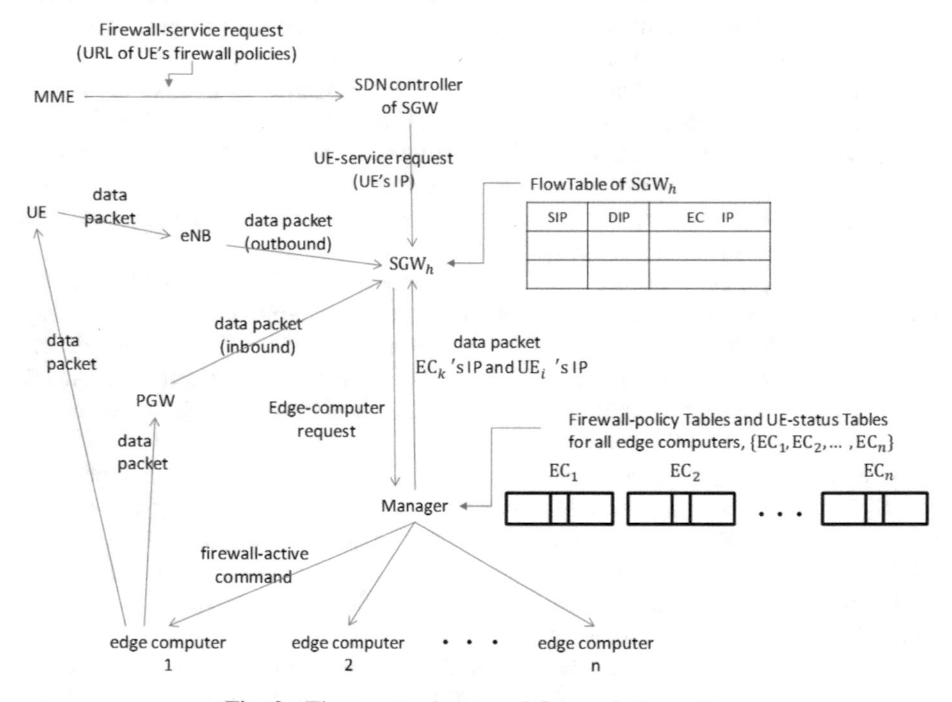

Fig. 2. The message/command flows of our system.

3 The Proposed System

As mentioned above, we will develop a firewall system with the edge computers attached to vEPC. When mobile phone is powered on, it will contact nearby base stations to request authentication and network services. Currently, LTE-4G adopts EPS-AKA protocol [8] which as shown in Fig. 1 consists of 9 steps. At first, UE sends its International Mobile Subscriber Identity (IMSI) to HSS through eNB (Step 1) and

MME (Step 2). After successfully authenticating UE (Step 3), HSS in Step 4 delivers n Authentication Vectors (AVs) to MME. MME in Step 5 chooses an AV containing 6 authentication parameters and keeps two of them, i.e., XRES and K_{ASME}. Other four (including RAND and AUTH where AUTH contains 3 parameters) in Step 6 are sent to UE via eNB. UE in Step 7 authenticates MME and HSS with the parameters received and its own. If it successes, it transmits parameter RES to MME in Step 8. In Step 9, MME checks to see whether RES is equal to the parameter it kept, i.e., XRES, to authenticate UE.

3.1 Firewall on vEPC's Edge Computer

In this study, we modify Step 3, in which, besides the original activities performed by HSS, we request HSS to search its firewall-database based on the UE's IMSI. It then retrieves the URL of this firewall and sends it to MME with the mentioned n AVs. In other words, a total of 7 parameters will be there in an AV. On receiving the n AVs, MME chooses one, keeps two parameters as usual and then sends the remaining 4 parameters to UE through eNB. A Firewall-service request which carries the URL is also delivered to SGW's SDN Controller, as shown in Fig. 2. The controller then selects a SGW, e.g., SGW_h, to serve the UE, and delivers an UE-service request containing the URL and UE_i's IP to SGW_h which will request Manager to choose one of its edge computers with light load to serve as the UE's firewall server. The algorithm of SDN Controller is listed in Algorithm 1.

Algorithm 1: SDN controller of SGW

Input: a packet Q

Output: a selected SGW, e.g., SGW_h

{If (Q is firewall-service request message, carrying an UE, e.g., UE_i's firewall-policy URL, sent by MME)

select a SGW, e.g., SGW_h, to serve UE_i;

sent an UE-service request to SGW_h;}

Table 1. FlowTable of SGW_h

SIP	DIP	EC IP
UE_i's IP	UE_i's IP	**EC_k's IP**

Algorithm 2: SGW, e.g., SGW_h

Input: a packet Q with source IP, e.g., SIP, and destination IP, e.g., DIP

Output: the results of corresponding activities

{If (Q is an UE-service request carrying an UE, e.g., UE_i's IP and firewall-policy URL)

/*sent by SDN controller of SGW*/

{Insert (UE_i's SIP, UE_i's DIP, nil) to SGW_h's FlowTable;

Send an edge-computer request to Manager;}

else if (Q is a packet carrying UE_i's IP and an edge computer, e.g., EC_k's IP from Manager)

Insert EC_k's IP into the EC IP field of UE_i's tuple in SGW_h's FlowTable;

/* the third field of SGW_h's FlowTable*/

else if (Q is a data-packet sent to UE_i by PGW and DIP is in the second field of SGW_h's FlowTable)

Send Q to Manager;

else if (Q is a data-packet sent to SGW by UE_i and SIP is in the first field of SGW_h's FlowTable)

Send Q to Manager;}

On receiving the UE-service Request, SGW_h inserts a tuple (UE_i's IP, UE_i's IP, nil) into its FlowTable for UE_i. The schema of this table is shown in Table 1. After that, it will receive the IP of the chosen edge computer, e.g., EC_k, from Manager. The third field of UE_i's tuple in SGW_h's FlowTable will be inserted with EC_k's IP. On the other hand, when a packet Q related to this UE arrives at this SGW, no matter whether it is an outbound one sent by or inbound one delivered to UE, SGW_h will check its FlowTable and then send Q to Manager which will transmit the packet to the corresponding firewall, i.e., EC_k. The algorithm of SGW_h is shown in Algorithm 2. We will describe the roles of Manager in the following. If Q cannot pass the checking, it will be dropped by the firewall. Otherwise, Q will be sent to its destination (UE or PGW) depending on the fact that it is an inbound or outbound packet.

3.2 Manager

Manager is a Virtual Machine (VM). When it is woken up from virtual VM to an edge computer, two tables will be established in DT Storage for it. The first is called Firewall-policy Table, the schema of which as illustrated in Table 2 is UE-IP, https and Firewall Policies, saving UE's IP, firewall URL and firewall policies, respectively.

The second table is named UE-status Table, the schema of which as shown in Table 3 contains UE-IP, Packet Data and Step, recording UE's IP, packet itself and UE's current authentication step, respectively. The purpose is keeping the status of UE authentication so that when the edge computer fails, takeover edge computer can follow the status to continue the UE's authentication, rather than from the very beginning of EPS-AKA.

Table 2. Schema of firewall-policy table

UE-IP	https	Firewall policies

Table 3. Schema of UE-status table

UE-IP	Packet data	Step

On receiving an edge-computer request from SGW_h, Manager selects an edge computer, e.g., EC_k, from its edge-computer pool, sends a Firewall-active command carrying UE_i's firewall-policy URL to EC_k, and requests EC_k to download UE_i's firewall policies based on the URL and serve as UE_i's firewall server. Next, it replies (EC_k's IP and UE_i's IP) to SGW_h which will be filled in the third field of UE_i's tuple in SGW_h's FlowTable. Manager further individually creates a tuple in EC_k's Firewall-policy Table and UE-status Table for UE_i, inserts (UE_i's IP, URL, nil) into UE_i's tuple in EC_k's Firewall-policy Table, and inserts (UE_i IP, nil, 0) into UE_i's tuple in EC_k's UE-status Table.

On the other hand, when Manager receives an authentication packet Q, which carries EC_k's IP, from SGW_h, meaning UE_i is still in its authentication stage, Manager retrieves UE_i's IP from Q, Calculates authentication-step value, inserts (UE_i's IP, Q, new authentication-step value) into UE_i's tuple in EC_k's UE-status Table and then sends Q to EC_k for firewall checking.

The way Manager calculates the authentication-step value is that when Manager receives a data packet Q, besides saving Q in the Packet Data field of UE_i's tuple in EC_k's UE-status Table, it also stores Step = 1 in the third field, i.e., Step, of this table. After successfully saving Q, Step will be increased to 2 so that when the edge computer fails, a takeover firewall can know which step it should continue. After UE_i's authentication, UE_i's tuple in EC_k's UE-status Table will be deleted. If Q is a data packet, Q is sent to EC_k for firewall checking. The algorithm of Manager is listed in Algorithm 3.

Algorithm 3：Manager

Input：a packet Q

Output：the results of corresponding activities

{If (Q is an edge-computer request, carrying UE_i's IP and UE_i's firewall-policy URL, sent by SGW_h)

{Select an edge computer, e.g., EC_k, from its edge-computer pool;

Send a Firewall-active command carrying UE_i's firewall-policy URL to EC_k.

Request EC_k downloading UE_i's firewall policies based on the URL and serve as UE_i's firewall server;

Reply (EC_k's IP and UE_i's IP) to SGW_h; /* to fill in the third field of UE_i's tuple in SGW_h's FlowTable;

Create a tuple in EC_k's Firewall-policy Table and UE-status Table for UE_i;

Insert (UE_i's IP, URL, nil) into UE_i's tuple in EC_k's Firewall-policy Table;

Insert (UE_i's IP, nil, 0) into UE_i's tuple in EC_k's UE-status Table;}

else if (Q is an authentication packet sent by SGW_h through Manager)

{Retrieve UE_i's IP from Q;

Calculate authentication step;

Insert (UE_i's IP, Q, new authentication step) into UE_i's tuple in EC_k's UE-status Table;

Send Q to EC_k; /* for firewall checking*/

If (this is the end of UE_i's authentication)

Delete UE_i's tuple in EC_k's UE-status Table; /* no longer needed*/}

Else if (Q is a data packet)

Send Q to EC_k; /* for firewall filtering*/}

3.3 Edge Computer

Edge computer's processing algorithm is shown in Algorithm 4. When an edge computer, e.g., EC_k, receives a Firewall-active command from Manager, it retrieves UE_i's firewall policies based on the URL carried in this command, stores these policies into UE_i's tuple in EC_k's Firewall-policy Table, i.e., the third field. However, on receiving an authentication packet, EC_k checks to see whether Q can pass UE_i's firewall policies or not. If not, Q will be dropped. Otherwise, Q will be forwarded toward its destination, i.e., UE_i or PGW.

However, if Q is a data packet Q sent by Manager, it first checks to see whether this is a Real Time Protocol packet. If yes, it sends Q to UE_i (inbound) or PGW (outbound) directly without checking it with firewall policies. The purpose is accelerating packet processing speed. Other types of data packets will be checked by firewall. After that, Q will be dropped or forwarded toward its destination. Of course, whether Q is an attacking packet is detected by Intrusion Detection/Prevention Systems (IDS/IPS) installed somewhere in the network system.

Algorithm 4：edge computer, e.g., EC_k

Input: a packet Q

Output: the results of corresponding activities

{If (Q is Firewall-active command issued by Manager)

{Retrieve UE_i's firewall policies based on URL retrieved from Q;

Store these policies into UE_i's tuple in EC_k's Firewall-policy Table;

/*the third field*/

Else if (Q is an authentication packet sent by Manager)

{Check Q with UE_i's firewall policies;

If (Q cannot pass the checking)

Drop Q;

Else Forward Q to its destination;}

Else if (Q is a data packet sent by Manager)

If (Q is a RTP packet or a voice packet)

Send Q to UE_i (inbound) or PGW (outbound);

/*without firewall checking*/

Else If (Q cannot pass UE_i's firewall policies filtering)

Drop Q;

Else if (Q's destination is UE_i's IP)

Send Q to UE_i; /* inbound packet*/

Else Send Q to PGW;} /* outbound packet*/

4 Conclusions

In this study, we propose a fault tolerant approach in which when an edge computer, no matter it belongs to a eNB or vEPC, that serves as a firewall in a 5G network fails, SDN Controller assigns another edge computer to take over for it so that the firewall service can be continued. When an eNB fails, we do the similar procedure. In this study, we also develop the algorithms with which when UE hands over, the destination eNB can smoothly continue the firewall service for this UE. We are still working on this project. The proposed system will be simulated and its performance will be measured as our future work.

References

1. W.K., The 5G Core Network Revolution When SDN Meet NFV, March 2016 (in Chinese). http://www.2cm.com.tw/technologyshow_content.asp?sn=1603030019
2. Nguyen, V.-G., Brunstrom, A., Grinnemo, K.-J.: SDN/NFV-based mobile packet core network architectures: a survey. IEEE Commun. Surv. Tutor. **19**, 1567–1602 (2017)
3. Arins, A.: Firewall as a service in SDN OpenFlow network. In: IEEE Workshop on Advances Information, Electronic and Electrical Engineering, Riga, Latvia, pp. 1–5 (2015)
4. Cliou, Y.C., Tsai, I.S., Yang, R.S.: Edge computing based on device relay communication, August 2015 (in Chinese). https://ictjournal.itri.org.tw/content/Messagess/contents.aspx?&MmmID=654304432061644411&CatID=654313611231473607&MSID=654526020703022212
5. Zope, N., Pawar, S., Saquib, Z.: Firewall and load balancing as an application of SDN. In: Conference on Advances in Signal Processing, Pune, India, pp. 354–359, June 2016
6. Gray, N., Lorenz, C., Müssig, A.: A priori state synchronization for fast failover of stateful firewall VNFs. In: International Conference on Networked Systems, Gottingen, Germany, pp. 1–6 (2017)
7. Guoa, Z., Liub, R., Xuc, Y., Gushchind, A., Walide, A., Chaoc, H.J.: STAR: preventing flow-table overflow in software-Defined networks. Comput. Netw. **125**, 15–25 (2017)
8. Abdrabou, M.A., Eldien Elbayoumy, A.D., Abd El-Wanis, E.: LTE authentication protocol (EPS-AKA) weaknesses solution. In: IEEE Seventh International Conference on Intelligent Computing and Information Systems, Cairo, Egypt, pp. 434–441, December 2015

The Study of MME Pool Management and Fault Tolerance in 5G Networks with SDN Controllers

Fang-Yie Leu(✉) and Cheng-Yan Gu

Department of Computer Science, Tunghai University, Taichung, Taiwan
{leufy,s04351025}@thu.edu.tw

Abstract. In this study, we would like to deal with two topics. The first one is that we add a machine, named Mediator, to SDN Controller for managing and keeping track of the data generated by MME during UE authentication. When a VM fails, other MMEs can successfully take over its authentication tasks. The second is that when an MME fails, other MMEs can know this immediately and response properly. The purpose of these is to increase the QoS an UE can receives.

Keywords: SDN · MME · vEPC · VM · EPS-AKA · Fault tolerance

1 Introduction

According to mobility report by Ericsson [1], before 2022, the number of 5G users will achieve 500 million, occupying 15% of global population. The number of devices connecting to 5G will be more than 25 billion, in which 18 billion are IoT devoices. Since the complex of IoT device functions and their working models are both higher, and 5G services are versatile, wireless security, transmission speed and capacity are the key issues to be managed in the near future.

Basically, the functions of 5G Evolved Packet Core (EPC) are divided into control plane and data plane. SDN Controller is responsible for the former. It submits commands to switches and VMs of data plane requesting them to provide users with specific functions so that UE can receive its requested services [2].

Currently, seldom literature studied the possible problems when Software Defined Networking (SDN) Controller manages MME-VM. Also due to business secret, we cannot find any detection and take-over procedure when MME which is now serving UE fails. Of course, if virtual EPC (vEPC) does not provide the management mechanism, when UE authentication fails. it will scan nearby stations, trying to re-enter the network. But this is not the case we like to see since it prolongs the time from the time point when UE is powered on to the time when UE successfully connects to the network. Basically, when an IoT device is powered on, it will connect to the network intermittently. After transmitting the data that has been so far collected, it will disconnect itself from the network. If the connection is frequently performed, authentication burden of MME will be higher. Hence, in this study, we solve this problem by effectively managing MME pool. To mitigate the burden of SDN Controller, especially

L. Barolli et al. (Eds.): IMIS 2018, AISC 773, pp. 585–595, 2019.
https://doi.org/10.1007/978-3-319-93554-6_57

in a crowded city or area, we add an assisting machine, named Mediator, to a vEPC. The Mediator is VM used to keep track of data and status of UE authentication and look up related data for SDN Controller. Furthermore, if an MME fails when it is authenticating an UE, the Mediator can help other MMEs to take over for the failed smoothly.

5G networks not only have very low transmission delays, but also support the operation of IoT systems. To ensure the availability of 5G networks and their high performance, network facilities need the functions of fault tolerance.

The rest of this paper is organized as follows. Section 2 introduces literature review and background of this study. Section 3 describes our system and its algorithms. Section 4 concludes this paper.

2 Literature Review and Background

Li *et al.* [2] introduced the applications of service chaining, including firewall, load balancer, and IDS. Due to the complexity, time consuming and high costs of these facilities' deployment and installation, it is not easy for users to employ them. But using Network Function Virtualization (NFV) and SDN can make such deployment and installation easier.

When an UE is handing over to the next base station, [3] delivered UE's authentication data to the next MME through Gprs Transfer Protocol version 2 (GTPv2), and adopted UE offload to balance the burden of MMEs. However, they focused on the case when an UE newly joins a network, rather than the case of load balance when an MME fails. They also did not deal with the procedure for the situation when SDN Controller takes over for a failed MME.

Aujla *et al.* [4] provides entertainment and navigation services for mobile users with different vehicular network techniques. Users can connect to networks via IEEE 802.11p/WAVE protocol to increase the security and comfortability of their network services. Authors adopted data offloading to solve several 5G challenges, particularly on its low delivery delay, high transmission speed, heavy traffic, load balance and service QoS. They also adopted SDN techniques to design an intelligent Controller which is used to determine the time, sequence and management of data offloading. However, the study did not consider the takeover algorithm for the case when MME fails.

4G employed EPS-AKA protocol to authenticate UE. Yuan et al. in [5] presented the mechanism of this protocol and pointed out its possible security vulnerability, e.g., LTE parameter K may be exposed, or there is a faked UE. For this, they proposed a more energy-saving and safer EPS-AKA protocol. But our approach establishes a more completed storage method to adopt itself to the case when the EPC faces a huge amount of IoT devices, and an MME fails, how can other MMEs smoothly take over for the failed to continue authenticating UEs that the failed MME originally handles?

Garzon *et al.* in [6] proposed a queue model for LTE Virtual MME (vMME) with the architecture of network virtualization, and evaluated the cost of signaling workload. They also suggested a data traffic model with which to establish applications for UEs and predict the time for traffic offload. At last, the delay time due to data transmission for vMME was measured given some amount of UEs to this system. Their conclusion is that when the number of UEs is higher, the delay time will exponentially increase.

Tanabe *et al.* [7] pointed out the possible problems that an vEPC has. For example, the transmission delay of an M2M communication is often limited. But owing to the virtulization of EPC, the tasks of controlling control plane and data plane are performed by VMs, meaning that M2M devices will share MME resources with other mobile devices. So when many UEs send authentication requests to MME, heavy traffic may lead to the delay time of M2M over its guaranteed limits, consequently being dropped by the receiving end. If many more resources are provided to MME, the resources offered to SGW/PGW will be less. Hence, authors proposed the concepts of vEPC-ORA which based on resource granularity effect can lower the probability that M2M packets to be dropped and reduce delay time that M2M packets are delivered so as to optimize resource distribution and usage.

Hu in [8] introduced the mechanism used by OpenFlow to process Flow Table, including checking packets and three mechanisms utilized to delete Flow Entries. We employ a similar approach to process VMs when these VMs are in idle state.

Wang in [9] compared performance, energy saving, operation stability and resource availability of two SDN Controllers, i.e., ONOS and OpenDayLight, and concluded that ONOS has better usability, and OpenDayLight outperforms ONOS in system performance. The conclusion is also the criteria with which we choose SDN Controllers.

3 Our System

In a 5G network, EPC entities have been virtualized to virtual EPC (vEPC) by using SDN and NFV. In vEPC, MME pool is managed by SDN Controller. It also provides programmable APIs with which users can monitor the operations of MME pool. In an MME pool, the interface between two arbitrary MMEs is S10 [3]. All MMEs are tightly connected to a complete graph. In other words, if there are n MMEs, the number of employed S10 links is $n(n-1)/2$.

As shown in Fig. 1, the EPS-AKA protocol has 9 Steps. Its original design is that when a step is completed, the security system records the result of this step. In 5G networks, the reliabilities of VM's hardware and software are lower than that of dedicated entities in 4G/3G network. When an MME, e.g., MME_f, fails, the following problems need to be addressed.

(1) How to detect the failure of MME_f by its SDN Controller or other MMEs?
(2) How can other MMEs smoothly take over for MME_f to continue authenticating those UEs originally authenticated by MME_f?
(3) When a VM is woken up and requested to serve as an MME, e.g., MMEn, how can MMEn continue the authentication tasks originally performed by MME_f? What is the takeover procedure?

Due to high transmission speed, 5G users request high QoS services. Service interruption is then a serious problem. In fact, MME records the results of a step when the step is finished. This cannot effectively reflect the details of a step. For example, in

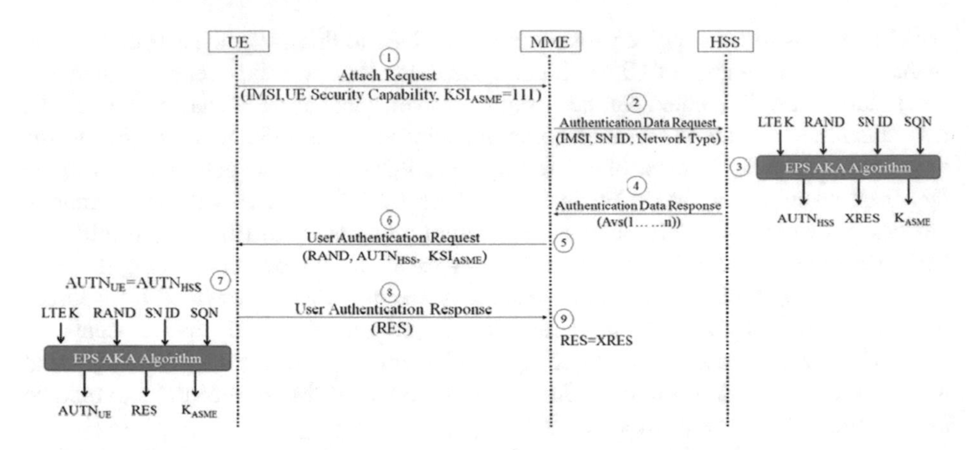

Fig. 1. The sequence chart of EPS-AKA [10].

Step 1 of EPS-AKA, Attach Request is sent to MME from UE, if MME fails before it sends Authentication Data Request to HSS, due to incompleteness of this step, no record will be kept. Hence, UE has to resend Attach Request to MME. On the other hand, after receiving Attach Request, if MME keeps the message, produces Authentication Data Request and then MME fails, the takeover MME can realize the situation. It then sends the Authentication Data Request to HSS and records that Step 1 is finished. UE does not need to send Attach Request again. Our opinion is that the granularity of original steps can be refined, meaning a step can be further divided into sub-steps.

In 4G networks, once the dedicated entity, e.g., MME, fails, except there is a standby MME, the system will be out of work since no MME will take over for the failed. Consequently, no UE can be authenticated. In this research, we divide a step of EPS-AKA into three sub-steps. The purpose is to reduce unnecessary duplicated sub-steps.

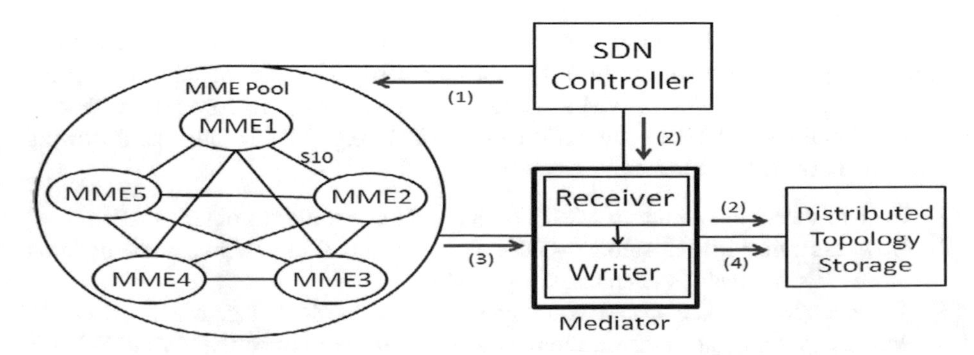

Fig. 2. The relationship among MME Pool, SDN Controller, Mediator and DT Storage. (1) SDN Controller starts up an VM and requests it serving as an MME; (2) SDN Controller requests Mediator to build a Data Table in DT Storage for this MME; (3) MME sends the data generated for authenticating UE to Receiver of Mediator; (4) Writer of Mediator writes the data into the MME's Data Table.

3.1 Mediator

The relationship among Mediator and network entities, including MME Pool, SDN Controller and Distributed Topology Storage (DT Storage for short) is shown in Fig. 2. Mediator consists of Receiver and Writer, in which Receiver is responsible for receiving messages from other entities, e.g., the commands submitted by SDN Controller and the data generated by MME when authenticating UE. Writer takes charge of writing the data, that Receiver receives, to DT storage. Also, when Mediator detects the failure of an MME, e.g., MME*i*, $1 \leqq i \leqq$ n, where *n* is the number of MMEs currently in MME Pool, it notifies SDN Controller. SDN Controller then redistributes those UEs currently being authenticated by MME*i* to other *n*–1 MMEs.

3.2 Starting up MME-VM and Establishing StatusTable

Once an UE is powered on, its authentication process will be started. Basically, SDN Controller will assign the authentication task of this UE to an MME that has lighter load. If all MMEs have reached their full load, SDN Controller will trigger an MME-VM, requesting this VM to act as an MME, e.g., MME*i*, and authenticate this UE. After that, SDN Controller needs to finish three tasks.

(1) Requesting Mediator to establish a Data Table in DT Storage, denoted by StatusTable*i*, for MME*i*. The purpose is to keep all the data generated by MME*i* during UE's authentication.
(2) Allowing Mediator to access StatusTable*i*.
(3) On receiving data packets, e.g., P, from MME*i*, Mediator retrieves UE's International Mobile Subscriber Identity (IMSI) from P, with which to differentiate which UE that P belongs to. It then writes the data carried in P to the UE's tuple in StatusTable*i*.

3.3 MME to DT Storage

When MME*i* fails, to smoothly take over for it, we further divide an authentication step into three sub-steps, namely receiving, processing and sending a message. The symbol X-Y represents the sub-step Y of step X, X = 2, 5 or 9, Y = 1, 2, 3, since the steps related to MME are Steps 2, 5 and 9. But 9–3 does not exist. We will describe this later.

As shown in Fig. 3, we adopt receiving/storage, processing/storage and sending message/storage as our authentication procedure for the 3 authentication steps related to MME. For example, on receiving message 1, i.e., Attach Request, MME*i* saves it in StatusTable*i* via Mediator. If MME*i* fails before requesting Mediator to save the data, since StatusTable*i* does not keep any data, i.e., Step = 0, the takeover MME, e.g., MME*j*, needs to authenticate this UE from very beginning. That is, UE has to resend Attach Request to initiate the authentication process. Step 1 is only related to UE. UE does not send its status to Mediator. So in a StatusTable, there is no Step 1-Y, Y = 1, 2, 3. On the other hand, if MME*i* fails after Mediator successfully records the Attach Request in StatusTable*i*, since the recorded status is 2-1, rather than 0, MME*j* will retrieve the Attach Request from StatusTable*i* and generate Authentication Data Request message. If Mediator successfully records the generated message in StatusTable*i,* then Step will be 2–2. After that, if MME*j* fails, the takeover MME, e.g., MME*k*, knows the status, it retrieves Authentication Data Request from StatusTable*i*, sends it to HSS and records the transmitted message. Now the status is 2–3. On receiving this message, HSS starts authenticate UE (Step3) and then sends *n* Authentication vectors (AVs) to MME*k* (Step 4).

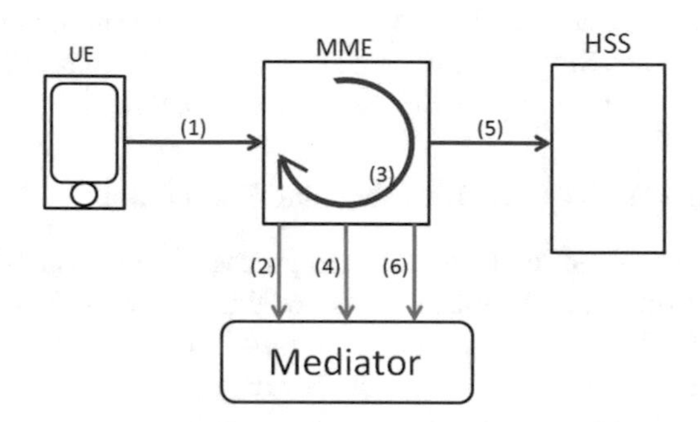

Fig. 3. Following Action-Storage procedure ((1) MME receives data; (2) it sends the data to Mediator (status is now 2-1); (3) MME processes the data; (4) MME sends the processed data to Mediator (the status is now 2–2); (5) MME delivers the processed data to HSS; (6) MME sends the data to Mediator (status is now 2–3)).

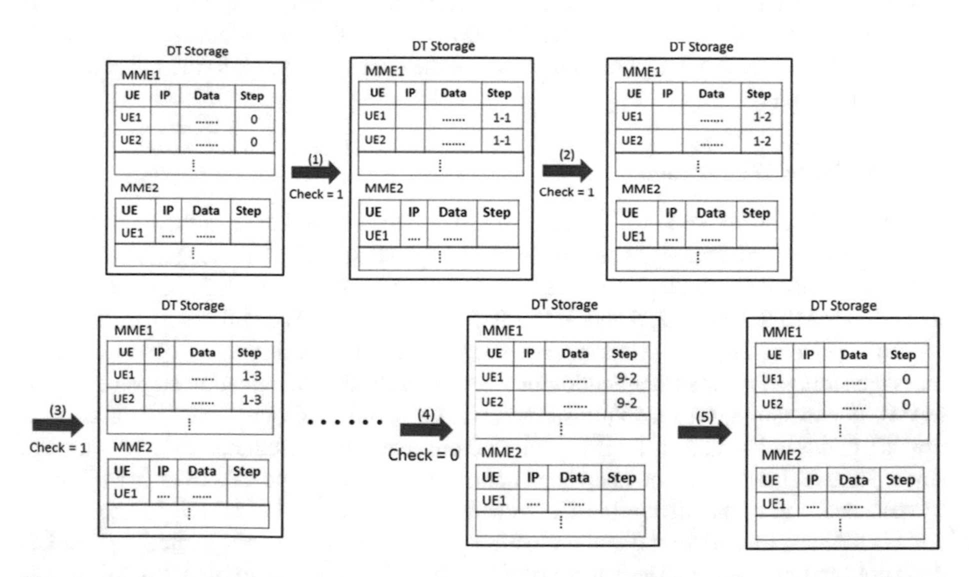

Fig. 4. Mediator and DT Storage operation procedure, ((1) MME receives Attach Request (Check = 1), Writer records Check = 1 and Step = 2-1; (2) MME processes Attach Request (Check = 1), then Check = 1 and Step = 2–2; (3) MME delivers Authentication Data Request (Check = 1), Check = 1 and Step = 2–3; (4) MME verifies whether RES = XRES (Check = 1), Check = 1 and Step = 9-2; (5) UE1 (UE2) finishes its authentication, MME notifies eNB to serve UE1 (UE2) and Step = 0).

Algorithm 1: Mediator records the steps when authenticating UE

```
  Input: MMEi sends an authentication packet P to Mediator
         for UE
  Output: UEj's authentication status recorded in DT
          Storage
{On receiving a packet P, according to the IMSI carried in
     P, Mediator knows the UE to which P belongs, e.g., UEj.
if(P == P0) then//P is the Attach Request sent by UEj
      Creating a tuple in MMEi's status table, i.e.,
        StatusTablei, in DT Storage for UEj;
Switch(Check){
    CASE 1: call Auth_begin (P);
    CASE 0: call Auth_end (P);
    otherwise: Printf("error packet");
               Discard it;}}

Auth_begin (P)  //UEj is under authentication
    /*structure Step*/ {
    int step_value; //the value can only be 0, 2, 5 or 9
    int status_value;}//0, 1 or 2*/
    {Writing the authenticating data and Check value to the
       tuple created for UEj in StatusTablei according to the
       IMSI carried in P;
    Step.status_value++;
    If (Step.status_value > 2)  then {
        Switch (Step. step_value)
        {CASE 0:Step.step_value = 2;
        CASE 2:Step.step_value = 5;
        CASE 5:Step.step_value = 9;
        CASE 9:Step.step_value = 0;}
        Step.status_value ← 0;}}


Auth_end (P):  //UEj has been authenticated
    {Notify eNB to start serving UEj;
    Writing the authenticating data and Check value to the
    tuple created for UEj in StatusTablei accroding to the
    IMSI carried in P;
    Step.step_value ← 0;
    Step.status_value ← 0;}
```

After that, if MME*k* fails, the takeover MME, e.g., MMEl, will wait for HSS to reply *n* AVs carried in Authentication Request Response message, i.e., Step 4 of EPS-AKA protocol. However, HSS sends Authentication vectors (AVs) to MME*k* since HSS receives Authentication Data Request from MME*k*. In this study, we assign IP of MME*i* to MME*j* when MME*j* takes over for MME*i* and allow MME*j* to access StatusTable*i*. In other words, MME*j* now has two IPs, one of its own and the other coming from taking over for MME*i*. It means MME*k* will has three IPs, containing those of MME*i*'s IP and MME*j*'s IP and its own IP. Of course, MME*k* can access StatusTable*i*, StatusTable*j* and StatusTable*k*. MMEl will have 4 IPs and it is allowed to access 4 StatusTables. We call this phenomena, Accumulating takeover IPs (ATIP for short) due to taking over.

When MME*i* sends a message to Mediator, a flag, named Check, will be added. The value of Check can only be 0 and 1. Its initial value is 0. When MME is authenticating UE, the value is 1. When Receiver receives the data sent by MME*i*, if Check = 1, representing that UE is under authentication, Mediator's Writer will write the data into StatusTable*i* as shown in Fig. 4.

In a StatusTable, there is a field named Step used to record the status of UE authentication. When Check = 1 and an authentication step is finished, the value of Step will be increased by one. For example, when MME1 receives Attach Request from UE and delivers the request to Mediator, Writer will save the request in StatusTable1 with Check = 1 and Step = 2-1. When receiving the second message generated by MME, i.e., Authentication Data Request, Writer increases status from 2–1 to 2–2. As MME sends Authentication Data Request to HSS, Writer changes the UE's authentication status to 2–3. But when MME receives Authentication Data Response, since Check = 1 and it is the beginning of Step 5, Writer keeps the message, and Step = 5-1. After MME processes and sends User Authentication Request to UE, Step = 5-3. On receiving RES from UE, MME sends RES to Mediator, Step = 9-1. After verifying whether or not RES = XRES, Step = 9.2. EPS-AKA protocol will be terminated. Following that, MME notifies eNB to serve underlying UE and the authentication is completed. In other words, there is no 9-3 since both Check flag and Step will be 0. Mediator deletes the corresponding tuple concerning this UE from this MME's StatusTable.

The procedure is shown in Algorithm 1.

3.4 Fault Recognition and Process

Generally, MME can finish an UE's authentication in a few miliseconds. Therefore, each step in the authentication stage has its maximum delay time, e.g., D. If the time duration from the time point when receiving the last message to current time point is over D, Check = 1 and Mediator has not received the next message, it is possible that the underlying MME, e.g., MME*i*, fails, the connection between UE and eNB is disconnected or UE is switched off. Our procedure is as follows.

(1) Mediator first sends a heartbeat to MME*i* to see whether MME*i* still works properly or not. If before timer times out, Mediator has not received reply from MME*i*, Mediator will send a heartbeat again.
(2) if MME*i* has not answered the heartbeats for three times, Mediator informs SDN Controller of this event. SDN Controller distributes those UEs currently authenticated by MME*i*, e.g., U = {UE1, UE2, ……, UE*m* }, to other MMEs, e.g., M = {MME1, MME2, ……, MME*n*}, $m \geqq n$, following load balance policy. If

UE*j* is distributed to MME*k*, SDN Controller will authorize MME*k* to access UE*j* and its data in StatusTable*i*.

(3) On receiving a takeover command from SDN Controller, MME*k* retrieves UE*j*'s last authentication status, i.e., Step, from the tuple concerning UE*j* in StatusTable*i* via Mediator, $1 \leqq k \leqq n$, $1 \leqq j \leqq m$, and continue UE*j*'s authentication.

(4) However, after sending a heartbeat if Mediator receives an ACK from MME*i*, Mediator continues waiting for the following message delivered by MME*i*. If it times out again, Mediator will redo the above process.

(5) On the other hand, after the authentication on an UE, e.g., UE1, Check = 0 and the corresponding tuple of UE1 in StatusTable*i* is then deleted. If StatusTable*i* is now empty, Mediator informs SDN Controller of this event. SDN Controller then commands MME*i* to be VM again. The purpose is saving energy and following VM' working policies. Mediator also deletes StatusTable*i* from DT storage. The following algorithm, i.e., Algorithm 2, shows the procedure.

Algorithm 2: Fault recognition and process

```
    Input: Trigger message from Mediator //when Mediator does
           not receive UEj's authenticating packet for a
           while where UEj is served by MMEi
   Output: MMEi Fault Recognition
{Counter ← 0;
While (Counter < 3) do {//sending heartbeat up to 3 times
          Mediator sends a Heartbeat message to MMEi;
    TM = 3;     //down timer, TM = 3 sec
    TM starts;  //waiting for 3 sec
    If (TM times out but does not receive an ACK from
          MMEi) then
      Counter++;
    Else {
      Printf ("MMEi works properly");
      return (normal);}}
Printf ("MMEi fails");
M ← M - {MMEi};
return (IP address of MMEi);}//Notifying maintenance staff
                       of the fact that MMEi fails
```

Algorithm 3: SDN Controller distributes those UEs authenticated by the fault MME, e.g., MME*f*

```
  Input: IP address of fault MMEf, The set of UEs, U={UE1,
         UE2, UE3,……, UEn}, authenticated by MMEf, Current
         MME Pool, M={MME1, MME2,……, MMEq},
 Output: All UEs in U are assigned to other available
         MMEs
{Switch(|U|)
    CASE 0: Printf("No UE is required to be passed to other
                 MME");
            return 0;
    otherwise: call Distribute (U, M);}

Distribute (U, M):
{for k:= 1 to n do{
    min_MME ← Ø; // The set of MMEs with minimum load,
    U' ← Ø;// The set of UEs authenticated by one of MME in
                       min_MME
    min_MME  =  min_{1≦i≦q} {load(MMEi)}, MMEi ∈ M;//all MMEs with
                                           minimum load
    if (|min_MME| > 1)  then
        Randomly choosing an MME, e.g., MMEj, MMEj ∈ M,
          1≤j≤q;
    else
        assuming that the one in min_MME is MMEj;
    load(MMEj)++;
    Printf("UEk is now handled by MMEj ") , 1 ≤ j ≤ q;
    Authorizing MMEj to access UEk's data in the
           StatusTablei;}}
```

Generally, SDN has three policies to recover MME to VM. One is Controller commands MME to be VM. The second is that there is a timeout mechanism. After timing out, MME will be VM. The third is that an Eviction Mechanism [8] provided by SDN system requests MME to be VM. In this study, we only choose the first one since we would like to monitor the operation of all MMEs.

4 Conclusions

In this paper, we propose an authentication tracking system that records the status of authentication on UE by an MME so that when MME fails, takeover MME can successfully continue the authentication without redo those finished steps. The service is provided by requesting Mediator to keep an UE's authentication status. We also establish a system with which we can know the failure of an MME by monitoring its authentication steps. On discovering such a failure, Mediator sends heartbeats to the

MME and waits for receiving its reply. If it fails to receive the heartbeat reply three times, Mediator notifies SDN Controller to remove the VM from MME pool and VM pool. We are still working on this study which will be mature in the coming days.

References

1. Ericsson Mobility Report, June 2017. https://www.ericsson.com/assets/local/mobility-report/documents/2017/ericsson-mobility-report-june-2017.pdf
2. Li, Y., Chen, M.: Software-defined network function virtualization: a survey. IEEE Access **3**, 2542–2553 (2015)
3. Cisco, Load Balance MME in Pool, 19 June 2015. https://www.cisco.com/c/en/us/support/docs/wireless/mme-mobility-management-entity/119021-config-mme-00.html
4. Aujla, G.S., Chaudhary, R., Kumar, N., Rodriques, J.J., VinelM, A.: Data offloading in 5G-enabled software-defined vehicular networks: a Stackelberg game-based approach. IEEE Commun. Mag. **55**(8), 100–108 (2017)
5. Bai, Y., Wang, Q., Jia, Q., Zhang, H.: An efficient and secured AKA for EPS networks. J. Beijing Univ. Posts Telecommun. **38**, 10–14 (2015). (in Chinese)
6. Garzon, J.P., Ramos-Munoz, J.J., Ameigeiras, P., Maldonado, P.A., Lopez-Soler, J.M.: Latency evaluation of a virtualized MME. In: IEEE Wireless Days, pp. 1–3, May 2016
7. Tanabe, K., Nakayama, H., Hayashi, T., Yamaoka, K.: A study on resource granularity of vEPC optimal resource assignment. In: IEEE IWQoS, pp. 1–2, June 2017
8. Hu, K.C.: The study of OpenFlow Protocol: Detailed Comparison Mechanism in Flow Table, October 2016. http://www.netadmin.com.tw/article_content.aspx?sn=1610070003. (in Chinese)
9. Wang, H.C., Lin, I.D.: Comparison on two SDNs: ONOS vs OpenDayLight. CIS. National Chiao Tung University Hsin-Chu, 23 September 2016 (2016). (in Chinese)
10. Leu, F.Y., You, I., Huang, Y.L., Yim, K., Dai, C.R.: Improving security level of LTE authentication and key agreement procedure. In: IEEE Globecom Workshops, pp. 1032–1036, December 2012

The 8th International Workshop on Intelligent Techniques and Algorithms for Ubiquitous Computing (ITAUC-2018)

Home Energy Management Using Hybrid Meta-heuristic Optimization Technique

Orooj Nazeer[1], Nadeem Javaid[2](✉), Adnan Ahmed Rafique[3], Sajid Kiani[4], Yasir Javaid[5], and Zeeshan Khurshid[1]

[1] Department of Computing and Technology, Abasyn University, Islamabad 44000, Pakistan
[2] COMSATS Institute of Information Technology, Islamabad 44000, Pakistan
nadeemjavaidqau@gmail.com
[3] University of Poonch, Rawalakot 12350, Azad Kashmir, Pakistan
[4] Allama Iqbal Open University, Islamabad 44000, Pakistan
[5] Government College of Technology (TEVTA), Rawalakot 12350, Azad Kashmir, Pakistan
http://www.njavaid.com

Abstract. Home energy management systems have been widely used for energy management in smart homes. Management of energy in smart home is a difficult task and requires efficient scheduling for smart appliances in a home. A meta-heuristic optimization technique is proposed in this paper. The proposed Harmony Search Gray Wolf Optimization (HSGWO) is a hybrid of Harmony Search Algorithm (HSA) and Gray Wolf Optimization (GWO). The pricing signal used for the calculation of electricity cost is Real Time Pricing (RTP). The basic aim of this paper is to reduce electricity cost, Peak to Average Ratio (PAR) and maximization of user comfort. Simulation results show that HSGWO performs better as compared to HSA and GWO. The findings demonstrate that there is a trade off between electricity cost and user comfort.

1 Introduction

The growing demand of electricity expands the infrastructure of power grid which increases the energy cost consequently. In conventional grid, transmission lines are used to deliver electric power to consumers. Traditional grid integrated with Information and Communication technology (ICT) makes a new concept, Smart Grid (SG). It uses Advance Metering Infrastructure (AMI), intelligent softwares, and hardware resources for energy management. It uses two-way communication for delivering energy to consumers in an efficient way. The basic aim of the SG is to improve the efficacy and reliability of the power system. Demand Response (DR) is an important aspect of SG which includes Demand Side Management (DSM) and Supply Side Management (SSM). DSM instigates the consumers to maintain their energy consumption pattern so that electricity consumption is mitigated during on peak hours to reduce electricity cost.

L. Barolli et al. (Eds.): IMIS 2018, AISC 773, pp. 599–609, 2019.
https://doi.org/10.1007/978-3-319-93554-6_58

Major objectives of energy management are reduction of electricity cost, Peak to Average Ratio (PAR) and user comfort maximization. Many heuristic algorithms are applied to manage energy consumption patterns in order to reduce electricity cost. From the last century, meta-heuristic optimization techniques have grown to be incredibly famous. In various fields of study, meta-heuristic optimization techniques have been used. They have been generally inspired by natural phenomenon. Normally, evolutionary concepts, physical phenomena are sources of these techniques. Meta-heuristic optimization techniques such as Particle Swarm Optimization (PSO), Ant Colony Optimization (ACO) and Genetic Algorithm (GA) have become very famous while solving many optimization problem. Meta heuristics techniques are simple so we can easily apply them on a problem and get quick results. A large amount of electricity is consumed by a residential area and its consumption is increasing day by day. This fact attracts researchers towards scheduling of appliances in a home. In this paper, a residential area is considered. The basic purpose is to optimize the consumption load for reduction of electricity cost, PAR and user comfort maximization. We have evaluated the performance of three meta-heuristic optimization techniques: Harmony Search Algorithm (HSA), Gray Wolf Optimization (GWO) and the proposed Harmony Search Gray Wolf Optimization (HSGWO).

Paper is organized as follows: Sect. 2 presents the related work. System model is presented in Sect. 3. Heuristic algorithms are discussed in Sect. 4. Simulation results are presented in Sect. 5 and Sect. 6 concludes the paper.

2 Related Work

In paper [1], combinations of optimization, machine learning and structural designing is being developed to fulfill the requirement of real life in a Home Energy Management System (HEMS). Three types of smart appliances are considered. Heating, Ventilation and Air Conditioners (HVACs) are developed for the regularization of load. HVACs is attained through prediction of weather and electricity. We have learned real measured to obtain energy consumption model of a house. Proposed demand response and HEMS is more competent as compared to others systems. Consumers are participating in demand response programs more actively in a HEMS in [2,3]. Therefore, there exists ambiguity in renewable energy resources. A stochastic model is proposed for the optimization of customers cost. Probabilistic model performing well only in planning studies, however, it is not good in actual time processes. Authors proposed a strategy which is based on decision making for managing Plug-in Electric Vehicles (PEVs).

There exists many swarm intelligence techniques which are inspired by hunting. However, no technique exists for imitating the gray wolf leadership hierarchy. For solving this problem, a new technique inspired by gray wolves is proposed in this [4]. GWO shows virtuous performance on both constraints and un-constraints problems. A Hybridized Gray Wolf (HGW) technique is used for solving financial dispatch complications in paper, [5] however, HGW perform efficiently as compared to others techniques. Many optimization techniques like SI techniques, hybrid techniques is used for solving generation gap of electricity nevertheless, HSA shows efficient performance in optimizing of Energy Storage System (ESS) by considering Time of Use (ToU) in [6].

The limitation of PSO is failed to tune the solution in finding global solution. However, a hybrid of Global Dimension with Particle Swarm Optimization (GDPSO) is proposed in paper [7], which performs efficiently as compared to other solutions. In [8], authors control actual and volatile power of grid and island using HSA. Proposed HSA obtains optimum solution as compared to GA and Gradient Method (GM). Scheduling of residential load is done by using a hybrid of Teaching Learning Based Optimization (TLBO) and GA in [9]. Proposed technique outperformed in terms of cost reduction and user comfort maximization.

Cost minimization and user comfort have non-flexible nature, it is hard to tackle these problems while using existing techniques. Therefore, a hybrid Genetic Wing Driven (HGW) optimization is used in paper, [10]. HGW have flexible nature and efficient performance as compared to existing techniques. Power distribution hub for social welfare is proposed in [11], which optimize energy consumption for a smart community. Proposed power distribution hub (PDH) offers a local switching mechanism that take decision of whether to buy electricity from utility. The objective of PDH is apply limit on electricity. In [12], by using Real Time Pricing (RTP) in a home area network, the design of energy management system is introduced for scheduling of appliances. However, basic objectives have been discussed is minimization of cost and waiting time of appliances. Logenthiran presented in [13], response surface method based scheme and HSA is used.

Authors in [14], for minimization of cost and PAR, HEMS is proposed. There are smart appliances in a smart home. In paper [15], energy expenditure in houses have been a emergent problem. Although, in a home, HEMS is very important for scheduling the appliances. Binary Backtracking Search Algorithm (BBSA) is proposed for management of energy consumption. Scheduling of home appliances at exact time is done by BBSA (Table 1).

3 System Model

In this section, the proposed system model is described in detail as shown in Fig. 1, a HEMS is proposed for scheduling the smart appliances. The main objective of scheduling the appliances is to mitigate the electricity cost, PAR and maximize user comfort. To calculate the cost of electricity consumption, RTP signal is used. A smart home includes Smart Meter (SM), smart appliances and an Energy Management Controller (EMC). SM acts as a bridge between power utility and users. The patterns of energy consumption is sent to EMC. EMC schedules appliances according to the price signal which is sent by the utility. SM receives pricing signals from utility and sends it to EMC. Consequently, it takes information of energy consumption from EMC and sends it to utility. Utility and SM communicate with each other through wireless networks

Table 1. Summarized related work

Techniques	Features	Objectives	Limitations
Optimal HVAC demand response mechanism [1]	Learning based demand response strategy	Minimization of cost	User comfort not considered
Stochastic model for HEMS [2]	Uncertainties of EV availability and renewable energy generation	Optimizes electricity cost	PAR is not considered
Proposed SCR coordination algorithm [3]	An online intelligent demand coordination of PEVs in distribution system	Maximize the users' satisfaction in terms of availability of energy	Increased system complexity
GWO [4]	A new meta heuristic algorithm	Outperforms in unconstrained problems as well as constrained problems	Pollutant emission not considered
hybrid GWO [5]	For solving economic dispatch problems	Solving economic dispatch problem	PAR is neglected
HSA [6]	Scheduling of ESS	Reduction in electricity bills of customers	User comfort is neglected
HHSPSO-GDS [7]	A hybrid HSPSO with global dimension selection	Decreased inertia weight, balanced the global exploration and local exploration	Security and privacy issues
HSA, GA [8]	The utilization of HSA of optimal design of PI controller	Control of power converter and grid resynchronization operations	System complexity is increased
TLGO [9]	SG is used to fulfill increasing demand of power	Minimization of cost and maximization of user comfort	PAR is not considered
Hybrid GWD [10]	DSM techniques are presented	cost and PAR reduction, and user comfort maximization	Security issues
PDH [11]	Multi-agent-based locally administrated PDH for social welfare	Minimization of cost	Apply limit on cost
Proposed power scheduling method [12]	An efficient scheduling method for EMS	Reducing electricity expenses and PAR	High PAR
EA [13]	Heuristic based EA is discussed	Reduced overall operational cost and carbon emission	Ignore user comfort
Proposed power scheduling method [14]	An efficient scheduling method for EMS	Electricity cost reduction and manage the problem of high peaks	Possibility of power quality disturbance
Proposed BBSA [15]	Reducing and scheduling energy usage is discussed	Reduction in energy consumption	Scheduling of home appliances only at specific time of the day

such as Z-wave. A single home equipped with 12 appliances is considered in this system model. One day is divided into 120 equal time slots and each time slot is considered for 12 min. For this reason, chosen time interval of 12 min would minimize the electricity cost and make the system more vigorous. Smart appliances are divided into different categories considering their power utilization patterns which are given below.

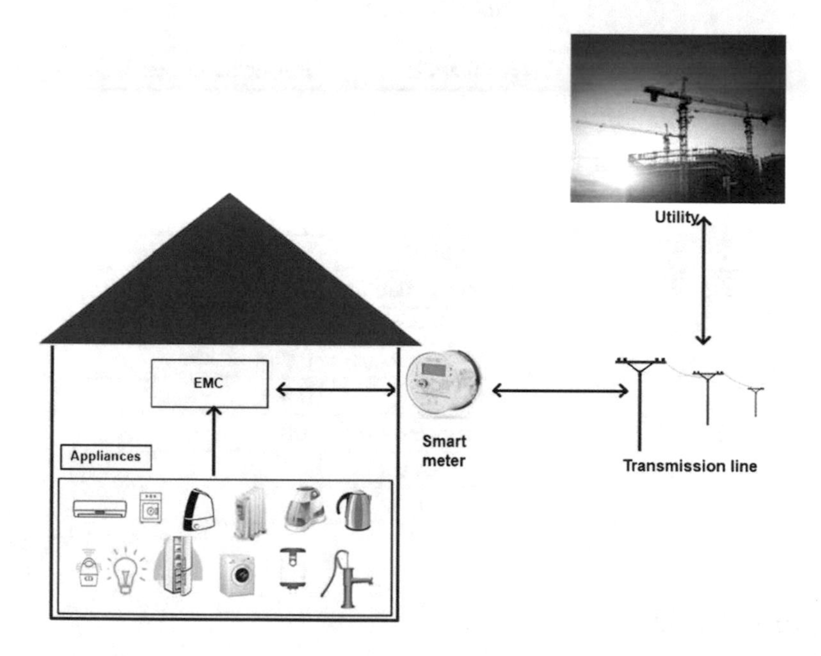

Fig. 1. System model

3.1 Load Categorization

According to the power utilization pattern, appliances are categorized into three extensive classes: base appliances, interruptible and non-interruptible appliances as shown in Table 2. Each category is explained as follows:

Interruptible Appliances. Interruptable appliances can be shifted to any time slot and can be interrupted during their operation. Interruptible are further categorized into heavy loads. Appliances which have power rating more than or equal to 1.5 kWh are considered to be heavy loads.

Non-interruptible Appliances. These appliances can be shifted to any time slot though, once their operation is executed, they must complete their operation without any interruption.

Base Appliances. These appliances are also known as fixed appliances. Base appliances are those appliances that cannot be manageable. Energy utilization pattern and time of operation while considering base appliances cannot be changed. These appliances must be on whenever client wants to turn on these appliances.

Table 2. Appliances

Category	Appliances	Power rating (kWh)
Interruptible	Air conditioner	1
	Dish washer	0.6
	Electric kettle	1.5
	Humidifier 1	0.05
	Humidifier 2	0.05
Non interruptible	Washing machine	0.38
	Cloth dryer	0.8
	Refrigerator	0.24
	Light	0.3
Base	Electric radiator	1.8
	Water heater	1.5
	Water pump	1.7

3.2 Price Tariff

For calculation of electricity price different dynamic pricing schemes are used. We can shift our load from peak hours to off peak hours for minimization of electricity cost and PAR. In this scenario we have used RTP for electricity cost calculation. RTP means tariffed retail charges for delivered electric power and energy that vary hour-to-hour and are determined from wholesale market prices. RTP lets consumers adjust their electricity usage accordingly.

4 Scheduling Technique

Traditional optimization techniques are not enough to handle large amount of appliances. As a result, we apply metaheuristic algorithms HSA and GWO. A hybrid of HSGWO is proposed and then compare the result with traditional HSA and GWO. Meta heuristic algorithm HSA and GWO are describe in detail in the following sub section.

4.1 HSA

Zoong Woo Geem introduces harmony search in 2001. It is a spontaneity procedure of players. In music, best player is selected from harmony. Following are three steps for generating a harmony: memory is considered, pitch is adjusted and harmony is selected randomly. Initially harmony memory is randomly initialized. After initialization of harmony memory, harmony improvisation process start. When new harmony is generated, two control parameters are considered which are Harmony Memory Consideration Rate (HMCR) and pitch adjustment rate. Two control parameters Harmony memory Consideration rate (HMCR) and pitch adjustment rate are considered when new harmony is generated.

4.2 GWO

GWO is a nature inspired optimization technique. The working of gray wolves is on hunting behaviours. ω, δ, β and α are four types of gray wolves. GWO pretend the leadership hierarchy and hunting mechanism of gray wolves in nature proposed by Mirjalili in 2014. Four types of gray wolves such as alpha, beta, delta and omega are employed for simulating the leadership hierarchy. In addition three main steps of hunting: searching for prey, encircling prey and attacking prey. Gray wolves have ability to recognize the location of the prey and encircle them. The hunt is usually guided by the alpha, beta and delta might also participate in hunting occasionally. In smart appliance we initialize the α, β, δ and ω with some value and generate the population Np randomly.

4.3 HSGWO

Latest harmony will be generated by using three steps: random selection, memory contemplation, pitch adjustment. Initially harmony memory is randomly initialize. After initialization of harmony memory, harmony improvisation process starts. We generate new harmony in the harmony search algorithm. GWO works on the hunting behavior of gray wolves. Gray wolves have the following categories beta, alpha, omega and delta. The steps for hunting: searching for victim, surrounding victum and attacking on a victum. Alpha is considered as leader and beta helps alphas in decision making. In smart appliance we initialize the α, β, δ and ω with some value and generate the population Np randomly. We proposed HSGWO for better result as compared to others techniques. The proposed technique performed well as compared to HSA and GWO. PAR and UC of our technique is best and there exist a tradeoff between cost and UC. We initialized our population using HSA and rest of the steps of GWO. The steps of HSA and GWO are mentioned earlier.

5 Simulation and Reasoning

Proposed architecture performance is appraised and results shows proposed approach is optimal as compared to existing approaches. 12 appliances are used

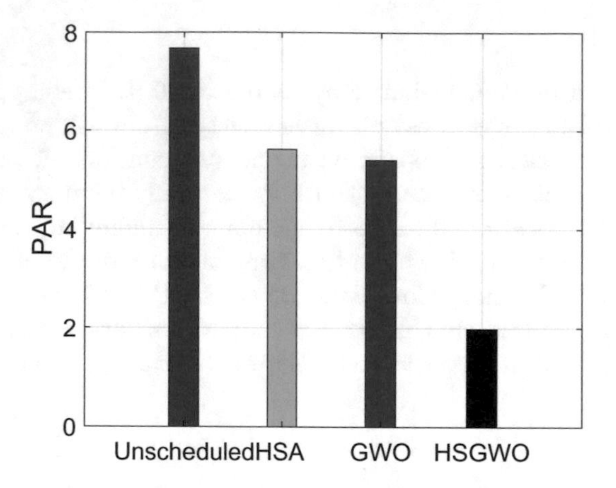

Fig. 2. PAR

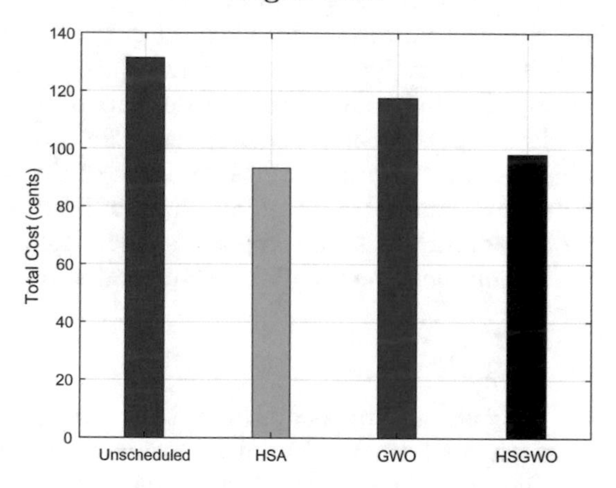

Fig. 3. Total cost

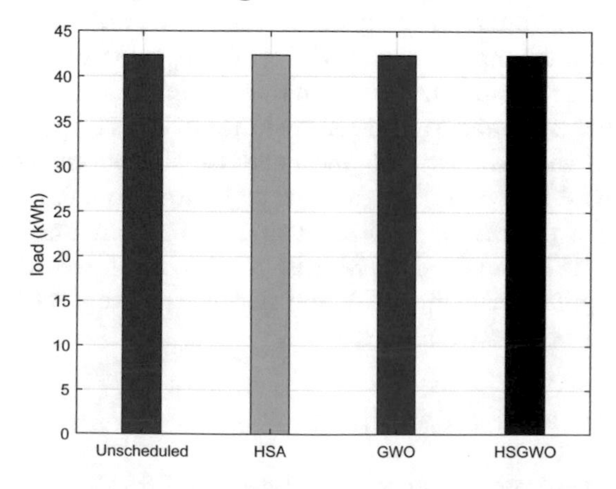

Fig. 4. Load

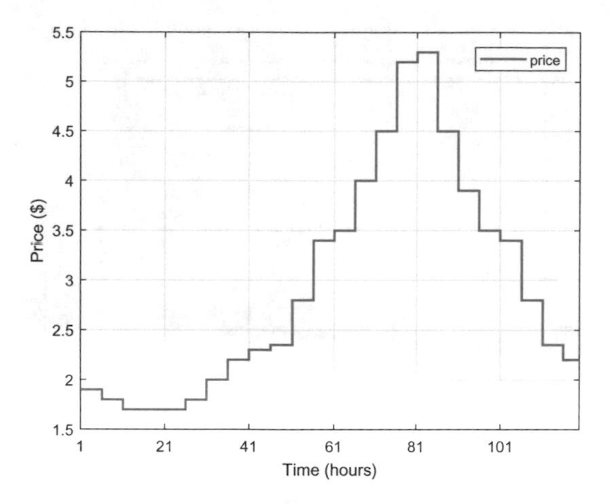

Fig. 5. Price

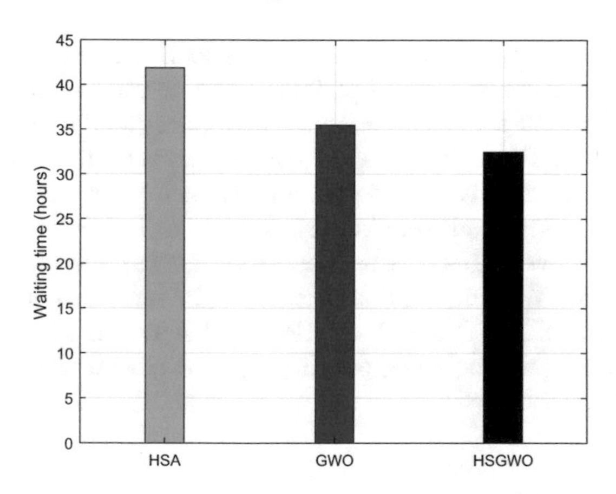

Fig. 6. Waiting time

in this scenario that are further classified into interruptible, non-interruptible, heavy and hard. To estimate performance of approaches, PAR, cost and waiting time are used as performance metrics. Cost and PAR are minimized and user comfort is maximized moreover, RTP signal is used. Unit of cost is cent, waiting-time is measured in hours and load is in kWh. Performance in PAR of proposed HSGWO is shown in Fig. 2, compared with HSA and GWO. HSGWO compares with existing and unscheduled case however, hybrid shows efficient results. HSGWO effectively reduced PAR in a HEMS, the value of HSGWO is 1.9, which is less in comparison of HSA and GWO. Hybrid approach consists of best features of HSA and GWO, it shows efficient results. In un-scheduling electricity cost is 134 cents as shows in Fig. 3, which reduces 97 cents in hybrid optimizer. There exists a tradeoff among user comfort and electricity.

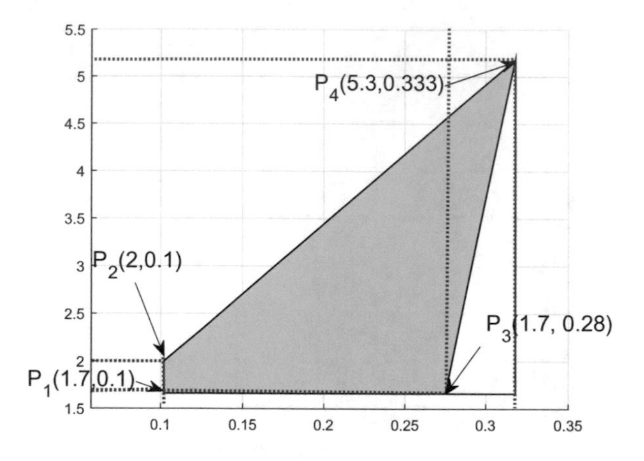

Fig. 7. Feasible region

Consumption of electric power is known as load. Battery and generator are good source of power generation. Appliances are considered as load in circuits. The term may also refer to the power consumed. Scheduled and unscheduled load shows in Fig. 4. As shown in Fig. 5, RTP price signal is used, customers notifies of RTP prices afortime of delivery. Customers and utility companies benefits using RTP. Customers shifted load from highest peak hour to lowest peak hours to get reduction in bills. Low PAR helps in reduction of cost generation and reliability of grid is improved. Waiting time is calculated in term of comfort of users. In Fig. 6, waiting time of HSA, GWO and HSGWO is shown which is 503, 494 and 401 respectively. The waiting time of our proposed HSGWO is low as compared to HSA and GWO.

5.1 Feasible Region

A feasible region is the set of all possible points of an optimization problem that satisfy the problem's constraints. The feasible region of the objective function minimize cost by controlling energy consumption as shown Fig. 7. To find constraints we use RTP pricing tariff ranging. The overall cost region is shown by points P1 (1.7, 0.1), P2 (2, 0.1), P3 (1.7, 0.28) and P4 (5.3, 0.333).

6 Conclusion

In this paper, a HEMS is proposed in which smart appliances are scheduled using meta-heuristic algorithms while RTP pricing signal is used. A hybrid of HSA and GWO, HSGWO is proposed in this paper. HSGWO is compared with two algorithms: HSA and GWO. Simulations are conducted for scheduling of appliances in MATLAB on the basis of power utilization pattern. Simulations results shows that proposed HSGWO performs better as compared to GWO and HSA. There is a trade off between user comfort and cost.

References

1. Zhang, D., Li, S., Sun, M., O'Neill, Z.: An optimal and learning-based demand response and home energy management system. IEEE Trans. Smart Grid **7**(4), 1790–1801 (2016)
2. Shafie-khah, M., Siano, P.: A stochastic home energy management system considering satisfaction cost and response fatigue. IEEE Trans. Ind. Inform. (2017)
3. Akhavan-Rezai, E., Shaaban, M.F., El-Saadany, E.F., Karray, F.: Online intelligent demand management of plug-in electric vehicles in future smart parking lots. IEEE Syst. J. **10**(2), 483–494 (2016)
4. Mirjalili, S., Mirjalili, S.M., Lewis, A.: Grey wolf optimizer. Adv. Eng. Softw. **69**, 46–61 (2014)
5. Jayabarathi, T., Raghunathan, T., Adarsh, B.R., Suganthan, P.N.: Economic dispatch using hybrid grey wolf optimizer. Energy **111**, 630–641 (2016)
6. Geem, Z.W., Yoon, Y.: Harmony search optimization of renewable energy charging with energy storage system. Int. J. Electr. Power Energy Syst. **86**, 120–126 (2017)
7. Ouyang, H.B., Gao, L.Q., Kong, X.Y., Li, S., Zou, D.X.: Hybrid harmony search particle swarm optimization with global dimension selection. Inf. Sci. **346**, 318–337 (2016)
8. Ambia, M.N., Hasanien, H.M., Al-Durra, A., Muyeen, S.M.: Harmony search algorithm-based controller parameters optimization for a distributed-generation system. IEEE Trans. Power Delivery **30**(1), 246–255 (2015)
9. Manzoor, A., Javaid, N., Ullah, I., Abdul, W., Almogren, A., Alamri, A.: An intelligent hybrid heuristic scheme for smart metering based demand side management in smart homes. Energies **10**(9), 1258 (2017)
10. Javaid, N., Javaid, S., Abdul, W., Ahmed, I., Almogren, A., Alamri, A., Niaz, I.A.: A hybrid genetic wind driven heuristic optimization algorithm for demand side management in smart grid. Energies **10**(3), 319 (2017)
11. Mahmood, D., Javaid, N., Ahmed, I., Alrajeh, N., Niaz, I.A. Khan, Z.A.: Multi-agent-based sharing power economy for a smart community. Int. J. Energy Res. **41**, 2074–2090 (2017)
12. Zhao, Z., Lee, W.C., Shin, Y., Song, K.B.: An optimal power scheduling method for demand response in home energy management system. IEEE Trans. Smart Grid **4**(3), 1391–1400 (2013)
13. Logenthiran, T., Srinivasan, D., Shun, T.Z.: Demand side management in smart grid using heuristic optimization. IEEE Trans. smart grid **3**(3), 1244–1252 (2012)
14. Rajalingam, S., Malathi, V.: HEM algorithm based smart controller for home power management system. Energy Build. **131**, 184–192 (2016)
15. Ahmed, M.S., Mohamed, A., Khatib, T., Shareef, H., Homod, R.Z., Ali, J.A.: Real time optimal schedule controller for home energy management system using new binary backtracking search algorithm. Energy Build. **138**, 215–227 (2017)

On the Channel Congestion of the Shortest-Path Routing for Unidirectional Hypercube Networks

Tzu-Liang Kung[1](✉), Chun-Nan Hung[2], and Yuan-Hsiang Teng[3]

[1] Department of Computer Science and Information Engineering, Asia University, Taichung, Taiwan
tlkung@asia.edu.tw

[2] Department of Information Management, Da-Yeh University, Changhua, Taiwan
chunnan.hung@gmail.com

[3] Department of Computer Science and Information Engineering, Hungkuang University, Taichung, Taiwan
yhteng@sunrise.hk.edu.tw

Abstract. Interconnection networks are emerging as an approach to solving system-level communication problems. An interconnection network is a programmable system that serves to transport data or messages between components/terminals in a network system. The hypercube is one of the most widely studied network structures for interconnecting a huge number of network components so that it is usually considered as the fundamental principle and method of network design. The unidirectional hypercube, which was proposed by Chou and Du (1990), is obtained by orienting the direction of each edge in the hypercube. Routing is crucial for almost all aspects of network functionalities. In this paper, we propose a dimension-ordered shortest-path routing scheme for unidirectional hypercubes and then analyze the incurred channel congestion from a worst-case point of view.

1 Introduction

Interconnection networks are emerging as a solution to the system-level communication problems. In many distributed-memory multiprocessor computers, processing and memory units are structured using various types of interconnected systems, and the design of interconnection mechanisms affects the practical efficiency of computation significantly. An interconnection network is a programmable system that serves to transport data or messages between components/terminals in a network system [11]. A common challenge for network designers is to match the data communication scheme of the problem at hand to the topology of the network. In literature, many interconnection networks are developed based on some well-known undirected graphs such as meshes, torus, hypercubes, crossed cubes, exchanged hypercubes, butterfly graphs, star graphs,

L. Barolli et al. (Eds.): IMIS 2018, AISC 773, pp. 610–619, 2019.
https://doi.org/10.1007/978-3-319-93554-6_59

arrangement graphs, Gaussian graphs, etc. [1,12,14,16,17,19,20]. Among the different kinds of network topologies, the hypercube network [22] has been one of the most attractive candidates for parallel and distributed computing [17,23] due to its promising advantages, such as regularity, vertex/edge symmetry, maximum connectivity, optimal fault-tolerance, Hamiltonicity, etc.

In addition to interconnection networks based on undirected graphs, directed interconnection networks have been studied as well. Some research papers in this area include [6–10,13] and these references include many additional ones. In particular, Ref. [9] gave an architectural model for the studies of unidirectional graph topologies coupled with a comparison of the diameters among many known unidirectional interconnection networks. Ref. [10] proposed unidirectional hypercubes as the basis for high speed networking. However, researchers know less about directed interconnection networks than their undirected counterparts. Maybe the main reason is that the directed version is usually more difficult and complicated.

Once a network topology has been already determined, there usually exist multiple paths (ordered sequences of nodes and channels) on which a message could take to travel through the network from the start to destination. The aim of routing is to determine which of these possible paths should be chosen. Routing algorithms can be classified into two categories: deterministic and nondeterministic. Given two nodes x and y, a deterministic routing algorithm always chooses the same path between x and y, whereas a nondeterministic routing algorithm may choose a different path occasionally between x and y. In addition, routing methods can be either oblivious or adaptive. An oblivious routing method chooses a path without considering any information about the present status of the network. Every deterministic routing algorithm is also an oblivious routing method; an adaptive routing method adapts to the state of the network in making routing decisions. The state information includes the status of each terminal and channel, the queues for network resources, and historical channel load records.

The load on a resource, such as communication port and channel, is a metric of how often the associated resource is being utilized. A good routing algorithm balances load across the network channels and/or communication ports. An empirical result indicates that the traffic between each pair of nodes in a network usually follows a predetermined path until this path is unavailable, so as to result in severe load imbalance [11]. On the other hand, a good choice of routes/paths also has to minimize the corresponding hop counts, measured as the number of nodes visited (or the number of channels traversed). The length of a route/path definitely influences communication latency while a message travels from its start to destination through the network. Routing is almost the most important task to achieve all aspects of network functionalities.

The channel congestion of an interconnection network is an important indicator for performance evaluation. Because channels with heavy congestion may lead to communication bottleneck and transmission delay, routing algorithms are required to balance communication loads on them. Motivated by this application, the following concepts related to edge congestion are introduced [15]. Suppose that the probability that information will exchanged between a given pair of nodes is the same as it is for any other pair of nodes. Based on this assumption, we consider all pairs for routing when calculating the level of edge congestion. For a specific routing algorithm in a undirected network, for each edge e, the congestion of e is the number of pairs of vertices that will be routed through e. The worst-case channel congestion of a network under a specific routing algorithm is the maximum value over the congestions of all edges. Then, the optimized channel congestion of a network is the minimum value over all routing algorithms. Obviously, a lower level of channel congestion is preferred for networks. Furthermore, the channel congestion can also be used to determine the lower bound on the area and the longest wire length required by VLSI network layouts. In recent years, many studies have focused on the topic of edge congestion with respect to some specific classes of interconnection networks, such as hypercubes [15], crossed cubes [4], twisted cubes [18], and other instances of networks [2,3,5,15,21]. As these references address only undirected networks, in this paper we propose a dimension-ordered shortest-path routing scheme for unidirectional hypercubes and then analyze the incurred channel congestion from a theoretical point of view.

The rest of this paper is structured as follows. Section 2 introduces the topological properties of unidirectional hypercubes. Section 3 studies the channel congestion for the proposed shortest-path routing algorithm. Finally, some concluding remarks are given in Sect. 4.

2 Unidirectional Hypercubes

We introduce some fundamental graph-theory definitions and notations in advance. Let $G = (V, A)$ be a directed graph, which consists of a nonempty node set and an arc set. A *path* P of length k in G is an ordered sequence of $k + 1$ distinct vertices $\langle v_0, v_1, \cdots, v_k\rangle$ such that $(v_i, v_{i+1}) \in A$ for every $0 \le i \le k - 1$ if $k \ge 1$; otherwise, a path of length 0 consists of a single vertex. We write P as $\langle v_0, v_1, \cdots, v_i, Q, v_j, \cdots, v_k\rangle$, where $Q = \langle v_i, \cdots, v_j\rangle$ is a segment of P whenever $i \le j$. For convenience we denote by $\ell(P)$ and $P[i]$ the length of P and the $(i + 1)$th vertex v_i of P for any $0 \le i \le k$, respectively. A *cycle* C of length $k \ge 3$ in G is a sequence of k distinct vertices $v_1, v_2, \cdots, v_k$ such that $\{(v_i, v_{i+1}) \mid 1 \le i \le k - 1\} \cup \{(v_k, v_1)\} \subseteq A$, and C is represented as $\langle v_1, v_2, v_3, \cdots, v_k, v_1\rangle$.

An n-dimensional hypercube Q_n has 2^n nodes, each of which has exactly n bidirectional communication ports and is labelled by a binary string of length n. It is customary to denote a binary string of length n by $\mathbf{v} = v_{n-1}v_{n-2}\cdots v_0$, and the *Hamming weight* of $\mathbf{v}$ is defined by $h(\mathbf{v}) = v_{n-1} + v_{n-2} + \cdots + v_0$. Two adjacent nodes $\mathbf{u}$ and $\mathbf{v}$ of Q_n are linked by an *i-edge*, $0 \le i \le n - 1$, if $u_i \ne v_i$ and $u_j = v_j$ for every $j \in \{0, 1, \cdots, n - 1\} - \{i\}$. From the perspective of communication, we can imagine that $\mathbf{u}$ and $\mathbf{v}$ communicate with each other through their *i-port*. Instead of the undirected interconnection networks, some research papers [6–10] take into account their directed counterparts as the basis of high-speed networking. On these grounds, Chou and Du [10] proposed the n-dimensional unidirectional hypercube UQ_n by orienting every edge of Q_n. The orientation is regularized as follows: For any vertex $\mathbf{v}$ in Q_n and any integer $0 \le i \le n-1$, the polarity function $Polarity(\mathbf{v}, i) = (-1)^{h(\mathbf{v})+i}$ denotes the polarity of $\mathbf{v}$'s i-port. Given an i-edge linking $\mathbf{u}$ and $\mathbf{v}$, then exactly one of $Polarity(\mathbf{u}, i)$ and $Polarity(\mathbf{v}, i)$ is positive, and this i-edge is oriented to be an *i-arc* from $\mathbf{u}$ to $\mathbf{v}$ if $Polarity(\mathbf{u}, i)$ is positive. In this scenario, $\mathbf{u}$ and $\mathbf{v}$ are *i-neighbor* of each other. For the sake of clarity, Fig. 1 illustrates UQ_2, UQ_3, and UQ_4. Obviously, we have $\rho^-(\mathbf{v}) = \lfloor n/2 \rfloor$ and $\rho^+(\mathbf{v}) = \lceil n/2 \rceil$ (respectively, $\rho^-(\mathbf{v}) = \lceil n/2 \rceil$ and $\rho^+(\mathbf{v}) = \lfloor n/2 \rfloor$) if $h(\mathbf{v})$ is even (respectively, odd), where $\rho^-(\mathbf{v})$ and $\rho^+(\mathbf{v})$ are the in-degree and out-degree of $\mathbf{v}$.

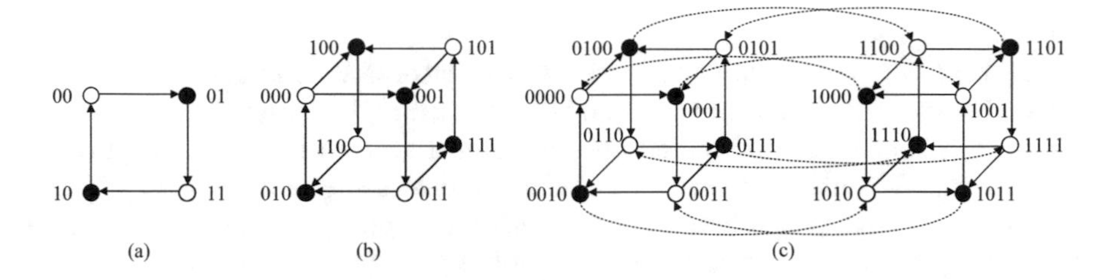

Fig. 1. Illustration of (a) UQ_2, (b) UQ_3, and (c) UQ_4.

For any $i \in \{0, 1\}$, let UQ^i_{n-1} be the subgraph of UQ_n induced by the set of nodes $\{ia_{n-2}a_{n-3} \cdots a_1a_0 \mid a_j \in \{0, 1\}$ for $0 \le j \le n-2\}$. Then both UQ^0_{n-1} and UQ^1_{n-1} are isomorphic to UQ_{n-1}. Let $\mathbf{v}$ be any node of UQ_n. The i-neighbor of $\mathbf{v}$ is denoted by $(\mathbf{v})^i$ for any $i \in \{0, 1, \cdots, n-1\}$.

As in conventional hypercubes, shortest-path routing from node $\mathbf{s} = s_{n-1}s_{n-2} \cdots s_0$ to node $\mathbf{t} = t_{n-1}t_{n-2} \cdots t_0$ can be determined by the distance pattern between $\mathbf{s}$ and $\mathbf{t}$, which is a binary number denoted by $DP(\mathbf{s}, \mathbf{t}) = DP_{n-1}DP_{n-2} \cdots DP_0$ with polarity assigned to every bit DP_i, $0 \le i \le n-1$, determined by the following rules [10]:

1. DP_i is equal to $s_i \oplus t_i$, where $\oplus$ denotes the exclusive-or operation.
2. DP_i has the same polarity as port i of s.
3. DP_i is denotes by d or $\bar{d}$, $d \in \{0, 1\}$, if the polarity of DP_i is positive or negative, respectively.

In short, node $\mathbf{s}$ can evaluate $DP(\mathbf{s}, \mathbf{t})$ for routing packets to the destination node $\mathbf{t}$.

Example 1. The distance pattern between the source node 000000 and the destination node 111111 in UQ_6 is $\bar{1}1\bar{1}1\bar{1}1$; that is, $DP(000000, 111111) = \bar{1}1\bar{1}1\bar{1}1$. A shortest path between 000000 and 111111 is

$$\langle 000000, 010000, 110000, 110100, 111100, 111101, 111111 \rangle.$$

Because the shortest path between any two nodes **s** and **t** corresponds to the distance pattern between them, the length of the shortest path is equal to $\delta(pos, neg)$ defined as follows [10]:

$$\delta(pos, neg) = \begin{cases} 2pos - (pos - neg) \bmod 2 & \text{if } pos > neg \\ 2neg + (neg - pos) \bmod 2 & \text{otherwise,} \end{cases} \tag{1}$$

where pos and neg are number of 1's with positive and negative polarity in $DP(\mathbf{s}, \mathbf{t})$, respectively.

Theorem 1. *[10] The diameter of UQ_n is $n+1$ (respectively, $n+2$) if n is even (respectively, odd).*

3 Channel Congestion

The channel congestion can serve as one indicator for comparing the performances between different routing algorithms. Although many advanced results have been obtained for undirected network, much less is known for the directed counterparts. A routing algorithm α for an undirected graph $G = (V, E)$ or a directed graph $G = (V, A)$ is seen as a function mapping any pair s, t of distinct nodes to a path that connects s to t, denoted by $\alpha(s, t)$. Since an undirected graph can be seen as a graph whose links are all bidirectional, below we define the notion of channel congestion with respect to directed graphs.

For any arc $a \in A$, its channel congestion under a given shortest-path routing algorithm α, denoted by $c(a|\alpha)$, is defined as follows:

$$c(a|\alpha) = |\{\langle u, v\rangle \mid u, v \in V, a \in \alpha(u, v)\}|.$$

The worst-case channel congestion of the routing algorithm α in the graph G is formulated by

$$c^*(\alpha) = \max\{c(a|\alpha) \mid \forall a \in A\},$$

which is the maximum channel congestion over all arcs of G.

Example 2. Given three routing methods $\alpha_1, \alpha_2, \alpha_3$ for the bidirectional 4-cycle C_4, which is a cycle of 4 nodes. See Fig. 2 for illustration, in which each number along an arc denotes the channel congestion. According to the definition presented above, we have $c^*(\alpha_1) = 4$, $c^*(\alpha_2) = 3$, and $c^*(\alpha_3) = 2$.

Theorem 2. *Let n be an even number. For any shortest-path routing algorithm α of UQ_n, its worst-case channel congestion $c^*(\alpha)$ is lower-bounded by $2^n + \binom{n}{n/2}$.*

Proof. For any pair $\mathbf{s}, \mathbf{t}$ of distinct nodes of UQ_n, $\alpha(\mathbf{s}, \mathbf{t})$ is a shortest path connecting $\mathbf{s}$ to $\mathbf{t}$. Clearly, α incurs a smaller channel congestion if it can balance as well as possible all-pair shortest paths over all of those $n \times 2^{n-1}$ arcs in UQ_n. Let $V_0(UQ_n)$ be the set of nodes in UQ_n that have even Hamming weights, and

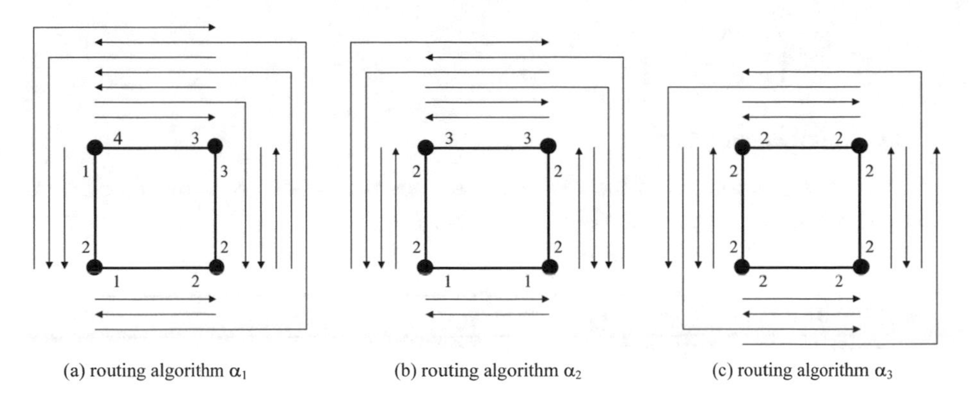

(a) routing algorithm α_1 (b) routing algorithm α_2 (c) routing algorithm α_3

Fig. 2. Illustrating the channel congestion of the given routing algorithms for a bidirectional 4-cycle.

let $V_1(UQ_n)$ be the set of nodes in UQ_n that have odd Hamming weights. We therefore estimate the lower bound on the channel congestion as follows:

$$
\begin{aligned}
c^*(\alpha) &\geq \frac{\sum_{\mathbf{s}\in V_0(UQ_n)\cup V_1(UQ_n)} \sum_{\mathbf{t}\in V_0(UQ_n)\cup V_1(UQ_n)\setminus\{\mathbf{s}\}} \ell(\alpha(\mathbf{s},\mathbf{t}))}{n\times 2^{n-1}} \\
&= \frac{\sum_{\mathbf{s}\in V_0(UQ_n)} \sum_{\mathbf{t}\in V_0(UQ_n)\cup V_1(UQ_n)\setminus\{\mathbf{s}\}} \ell(\alpha(\mathbf{s},\mathbf{t}))}{n\times 2^{n-1}} \\
&\quad + \frac{\sum_{\mathbf{s}\in V_1(UQ_n)} \sum_{\mathbf{t}\in V_0(UQ_n)\cup V_1(UQ_n)\setminus\{\mathbf{s}\}} \ell(\alpha(\mathbf{s},\mathbf{t}))}{n\times 2^{n-1}} \\
&= \frac{\sum_{\mathbf{s}\in V_0(UQ_n)} \left(\sum_{pos=0}^{n/2}\sum_{neg=0}^{n/2}\binom{n/2}{pos}\binom{n/2}{neg}\delta(pos,neg)\right)}{n\times 2^{n-1}} \\
&\quad + \frac{\sum_{\mathbf{t}\in V_1(UQ_n)} \left(\sum_{pos=0}^{n/2}\sum_{neg=0}^{n/2}\binom{n/2}{pos}\binom{n/2}{neg}\delta(pos,neg)\right)}{n\times 2^{n-1}} \\
&= \frac{\left(\sum_{pos=0}^{n/2}\sum_{neg=0}^{n/2}\binom{n/2}{pos}\binom{n/2}{neg}\delta(pos,neg)\right)\times 2^{n-1}}{n\times 2^{n-1}} \\
&\quad + \frac{\left(\sum_{pos=0}^{n/2}\sum_{neg=0}^{n/2}\binom{n/2}{pos}\binom{n/2}{neg}\delta(pos,neg)\right)\times 2^{n-1}}{n\times 2^{n-1}} \\
&= \frac{\sum_{pos=0}^{n/2}\sum_{neg=0}^{n/2}\binom{n/2}{pos}\binom{n/2}{neg}\left(\delta(pos,neg)+\delta(neg,pos)\right)}{n} \\
&= \frac{\sum_{pos=0}^{n/2}\sum_{neg=0}^{n/2}\binom{n/2}{pos}\binom{n/2}{neg}\times 4\max(pos,neg)}{n} \\
&= \frac{4\times\left(\sum_{pos=0}^{n/2}\sum_{neg=1}^{n/2} neg\binom{n/2}{pos}\binom{n/2}{neg} + \sum_{pos=1}^{n/2}\sum_{neg=0}^{n/2-pos} pos\binom{n/2}{pos+neg}\binom{n/2}{neg}\right)}{n} \\
&= \frac{4\times\left(n2^{n-2}+\frac{n}{4}\binom{n}{n/2}\right)}{n}
\end{aligned}
$$

$$= 2^n + \binom{n}{n/2}.$$

The proof is completed.

Here we propose a dimension-ordered shortest-path routing algorithm, namely SPR_{UQ_n}, as presented in Algorithm 1.

Algorithm 1. $SPR_{UQ_n}(\mathbf{s}, \mathbf{t})$

```
Input: A pair of distinct nodes s and t of UQ_n.
Output: A shortest path P from s to t in UQ_n.
begin
    DP_{n-1} DP_{n-2} ··· DP_0 ← DP(s, t) ;
    v ← s ;
    length ← 0 ;
    P[length] ← v ;
    length ← length + 1 ;
    while v ≠ t do
        if there exists an i-port (0 ≤ i ≤ n − 1) with DP_i = 1 and
        Polarity(v, i) = 1 then                          /* Regular moves. */
            find the most significant r-port such that DP_r = 1 and
            Polarity(v, r) = 1 ;
            P[length] ← (v)^r ;
            DP_r ← 0 ;
        else                                             /* Extra moves. */
            if there exists an r-port with DP_r = 1̃ and Polarity(v, r) = 1 then
                P[length] ← (v)^r ;
                DP_r ← 0 ;
            else
                if Polarity(v, 0) = 1 then
                    r ← 0 ;
                else
                    r ← 1 ;
                end
                P[length] ← (v)^r ;
                DP_r ← 1̃ ;
            end
        end
        v ← (v)^r ;
        length ← length + 1 ;
    end
    return P ;
end
```

Theorem 3. *Let n be an even number. The worst-case channel congestion of SPR_{UQ_n} is upper-bounded by $2^n + \frac{n}{2}\binom{n}{n/2}$.*

Proof. For any pair of two distinct nodes $\mathbf{s}, \mathbf{t}$ of UQ_n, let *pos* and *neg* are number of 1's with positive and negative polarity in $DP(\mathbf{s}, \mathbf{t})$, respectively. Then the shortest path $SPR_{UQ_n}(\mathbf{s}, \mathbf{t})$ is composed of $\gamma(pos, neg)$ regular moves and $\varepsilon(pos, neg)$ extra moves, where $\gamma(pos, neg)$ and $\varepsilon(pos, neg)$ are defined as follows:

$$\gamma(pos, neg) = pos + neg \tag{2}$$

and

$$\varepsilon(pos, neg) = \begin{cases} 2 \times \lceil \frac{pos-neg-1}{2} \rceil & \text{if } pos > neg \\ 2 \times \lceil \frac{neg-pos}{2} \rceil & \text{if } pos \leq neg. \end{cases} \tag{3}$$

According to the proposed shortest-path routing algorithm SPR_{UQ_n}, extra moves go through only 0- and 1-arcs. Therefore, both 0- and 1-arcs have higher channel congestion than arcs of the other dimensions. Obviously, the number of regular moves on $SPR_{UQ_n}(\mathbf{s}, \mathbf{t})$ is equal to the Hamming distance between $\mathbf{s}$ and $\mathbf{t}$. Let γ_i, $0 \leq i \leq n-1$, denote the total amount of i-dimensional regular moves in shortest-path routing for all combinations of distance patterns. Then we have

$$\gamma_i = \sum_{pos=0}^{n/2-1} \sum_{neg=0}^{n/2} \binom{n/2-1}{pos} \binom{n/2}{neg} + \sum_{pos=0}^{n/2} \sum_{neg=0}^{n/2-1} \binom{n/2}{pos} \binom{n/2-1}{neg} = 2^n.$$

- If $\mathbf{s}$ has even Hamming weight, its even ports have positive polarity. Let ε_0 and ε_1 denote the numbers of the 0- and 1-dimensional extra moves in all combination of distance patterns, respectively. Consequently, the total number of extra moves is

$$\varepsilon_0 + \varepsilon_1 = \sum_{pos=0}^{n/2} \sum_{neg=0}^{n/2} \binom{n/2}{pos} \binom{n/2}{neg} \varepsilon(pos, neg) = \frac{n}{2} \binom{n}{n/2}.$$

- If $\mathbf{s}$ has odd Hamming weight, its odd ports have positive polarity. Let ε'_0 and ε'_1 denote the numbers of the 0- and 1-dimensional extra moves in all combinations of distance patterns, respectively. Clearly, we have $\varepsilon'_0 = \varepsilon_1$ and $\varepsilon'_1 = \varepsilon_0$. Thus,

$$\varepsilon'_0 + \varepsilon'_1 = \varepsilon_1 + \varepsilon_0 = \frac{n}{2} \binom{n}{n/2}.$$

As a consequence, the overall number of the 0-dimensional extra moves in SPR_{UQ_n} for all combinations of distance patterns is $\varepsilon_0 + \varepsilon'_0 = \frac{n}{2}\binom{n}{n/2}$. Similarly, the overall number of the 1-dimensional extra moves is $\varepsilon_1 + \varepsilon'_1 = \frac{n}{2}\binom{n}{n/2}$.

Suppose that $(\mathbf{u}, \mathbf{v})$ is a 0- or 1-arc. The upper bound on the worst-case channel congestion of SPR_{UQ_n} is given below:

$$c^*(SPR_{UQ_n}) \leq c((\mathbf{u}, \mathbf{v}) \mid SPR_{UQ_n}) = \gamma_0 + \varepsilon_0 + \varepsilon'_0 = 2^n + \frac{n}{2} \binom{n}{n/2}.$$

To summarize the derived formulations, Table 1 shows some numerical results of the lower and upper bounds on the channel congestion for the proposed shortest-path routing algorithm.

Table 1. The lower and upper bounds on the channel congestion of SPR_{UQ_n}.

n	Lower bound	Upper bound	Difference between the lower and upper bounds
4	22	28	6
6	84	124	40
8	326	536	210
10	1,276	2,284	1,008
12	5,020	9,640	4,620

4 Concluding Remarks

In this paper we investigate the channel congestion of the shortest-path routing for unidirectional hypercube networks. We propose a dimension-ordered shortest-path routing scheme for unidirectional hypercubes and then analyze both the lower and upper bounds on the incurred channel congestion from a theoretical point of view.

Acknowledgements. This work is supported in part by the Ministry of Science and Technology, Taiwan, under Grands No: MOST 105-2221-E-468-015 and MOST 106-2221-E-468-003. The authors also gratefully acknowledge the helpful comments and suggestions of the reviewers, which have improved the quality of this paper significantly.

References

1. Akers, S.B., Krishnamurthy, B.: A group theoretic model for symmetric interconnection networks. IEEE Trans. Comput. **38**(4), 555–566 (1989)
2. Andrews, M., Chuzhoy, J., Guruswami, V., Khanna, S., Talwar, K., Zhang, L.: Inapproximability of edge-disjoint paths and low congestion routing on undirected graphs. Combinatorica **30**(5), 485–520 (2010)
3. Aroca, J.A., Anta, A.F.: Bisection (band)width of product networks with application to data centers. IEEE Trans. Parallel Distrib. Syst. **25**(3), 570–580 (2014)
4. Chang, C.-P., Sung, T.-Y., Hsu, L.-H.: Edge congestion and topological properties of crossed cubes. IEEE Trans. Parallel Distrib. Syst. **11**(1), 64–80 (2000)
5. Chekuri, C., Khanna, S., Shepherd, F.B.: Edge-disjoint paths in planar graphs with constant congestion. SIAM J. Comput. **39**(1), 281–301 (2009)
6. Cheng, E., Lindsey, W.A., Stey, D.E.: Maximal vertex-connectivity of $S_{n,k}$. Networks **46**, 154–162 (2005)
7. Cheng, E., Lipman, M.J.: On the Day-Tripathi orientation of the star graphs: connectivity. Inf. Process. Lett. **73**, 5–10 (2000)
8. Cheng, E., Lipman, M.J.: Orienting split-stars and alternating group graphs. Networks **5**, 139–144 (2000)
9. Chern, S.C., Jwo, J.S., Tuan, T.C.: Uni-directional alternating group graphs. In: Lecture Notes in Computer Science, vol. 959, pp. 490–495 (1995)
10. Chou, C.H., Du, D.H.C.: Unidirectional hypercubes. In: Proceedings of the Supercomputing 1990, pp. 254–263 (1990)

11. Dally, W.J., Towles, B.: Principles and Practices of Interconnection Networks. Morgan Kaufmann, San Francisco (2004)
12. Day, K., Tripathi, A.: Arrangement graphs: a class of generalized star graphs. Inf. Process. Lett. **42**(5), 235–241 (1992)
13. Day, K., Tripathi, A.: Unidirectional star graphs. Inf. Process. Lett. **45**, 123–129 (1993)
14. Efe, K.: The crossed cube architecture for parallel computing. IEEE Trans. Parallel Distrib. Syst. **3**, 513–524 (1992)
15. Fiduccia, C.M., Hedrick, P.J.: Edge congestion of shortest path systems for all-to-all communication. IEEE Trans. Parallel Distrib. Syst. **8**(10), 1043–1054 (1997)
16. Flahive, M., Bose, B.: The topology of Gaussian and Eisenstein-Jacobi interconnection networks. IEEE Trans. Parallel Distrib. Syst. **21**(8), 1132–1142 (2010)
17. Leighton, F.T.: Introduction to Parallel Algorithms and Architectures: Arrays · Trees · Hypercubes. Morgan Kaufmann, San Mateo (1992)
18. Li, T.-K., Tan, J.J.M., Hsu, L.-H., Sung, T.-Y.: Optimum congested routing strategy on twisted cubes. J. Interconnection Netw. **1**(2), 115–134 (2000)
19. Loh, P.K.K., Hsu, W.J., Pan, Y.: The exchanged hypercube. IEEE Trans. Parallel Distrib. Syst. **16**(9), 866–874 (2005)
20. Martínez, C., Beivide, R., Stafford, E., Moretó, M., Gabidulin, E.M.: Modeling toroidal networks with the Gaussian integers. IEEE Trans. Comput. **57**(8), 1046–1056 (2008)
21. Ostrovskii, M.I.: Minimal congestion trees. Discrete Math. **285**, 219–226 (2004)
22. Saad, Y., Shultz, M.H.: Topological properties of hypercubes. IEEE Trans. Comput. **37**, 867–872 (1988)
23. Xu, J.-M.: Topological Structure and Analysis of Interconnection Networks. Kluwer Academic Publishers, Dordrecht/Boston/London (2001)

A Diagnosis Algorithm on the 2D-torus Network

Lidan Wang[1], Ningning Liu[1], Cheng-Kuan Lin[1], Tzu-Liang Kung[2], and Yuan-Hsiang Teng[3(✉)]

[1] School of Computer Science and Technology, Soochow University, Suzhou 215006, China
[2] Department of Computer Science and Information Engineering, Asia University, Taichung City 413, Taiwan R.O.C.
[3] Department of Computer Science and Information Engineering, Hungkuang University, Taichung City 433, Taiwan R.O.C.
yhteng@sunrise.hk.edu.tw

Abstract. In this article, we design a three test rounds diagnosis algorithm for a 2D-torus network. Suppose that $\mathrm{TDT}(n,m)$ is a 2D-torus network with $n \geq 4$, $m \geq 6$ and m being even. Let F be the faulty set in $\mathrm{TDT}(n,m)$. With our algorithm, a diagnosis on $\mathrm{TDT}(n,m)$ is completed in three test rounds if $|F| \leq 4$.

1 Introduction

High-speed multiprocessor systems have become more eye-catching in computer technology in recent years. In the multiprocess system, more than one processor operate many programs at the same time. The reliability is crucial since even a tiny malfunction would disable the service, and thus, the processors or links in a multiprocessor system will fail to work. Whenever a fault is found, the device should be replaced with fault-free one immediately to guarantee the effect. Thus, it is important to identify every fault device in a multiprocessor system. This is known as *system diagnosis*. The *diagnosability* is the maximum of faulty devices that can be identified in a correct way. A system is *t-diagnosable* if we can correctly find out the faulty processors with the ones being t at most. Many results about the system diagnosis and the diagnosability have been proposed in literature [7,8,10]. Quite a few famous approaches on diagnosis have been developed before. One classic approach, named the PMC diagnosis model, was proposed by Preparata, Metze and Chien in [9]. The PMC model is the tested-based diagnosis with a processor diagnosing by testing on the neighboring ones via the links between them. Hakimi and Amin [4] proved that a system is t-diagnosable if it is t-connected with at least $2t+1$ vertices under the PMC model.

A diagnosis testing signal should be delivered from one processor to another through the communication bus at one time. Under the PMC model, only processors with direct connections are allowed to test each other. If all the neighboring

L. Barolli et al. (Eds.): IMIS 2018, AISC 773, pp. 620–625, 2019.
https://doi.org/10.1007/978-3-319-93554-6_60

processors v are faulty simultaneously, it is unlikely to determine whether v is faulty or not. Thus the diagnosability is bounded by the minimum degree. Fujita and Araki [3] proposed a scheme that completes a diagnosis in at most three test rounds under the PMC model. It is optimal that three rounds are needed for the adaptive diagnosis. In practice, processors in many systems are connected sparsely. Thus, some research concerns with the measure that can better reflect fault patterns in real systems [1,6]. The 2D-torus network is the graph with degree being 4. Suppose that F is the faulty set in a 2D-torus. In this article, we propose the three test rounds local diagnosis algorithm for the 2D-tours network if $|F| \leq 4$.

2 Preliminaries

For the graph definitions and notations, we follow [5]. Let $G = (V, E)$ be a *graph* if V is a finite set and E is a subset of $\{\{u, v\} \mid \{u, v\}$ is an unordered pair of $V\}$. Supposedly, V is the *vertex set* and E is the *edge set* of G. Two vertices u and v are *adjacent* if $\{u, v\} \in E$; u is a *neighbor* of v, and vice versa. We use $N_G(u)$ to denote the neighborhood set $\{v \mid \{u, v\} \in E(G)\}$. The *degree* of a vertex v in a graph G, denoted by $\deg_G(v)$, is the number of edges incident to v. In the article, suppose that $m \geq 6$ and $n \geq 4$ with m being an even integer. For any two positive integers r and s, we use $[r]_s$ to denote $r(\text{mod} s)$. The 2D-torus network $\text{TDT}(n, m)$ is a graph with the vertex set $V = \{(x, y) \mid 0 \leq x \leq n-1$ and $0 \leq y \leq m-1\}$ and the edge set $E = \{((x_1, y_1), (x_2, y_2)) \mid (x_1, y_1) \in V, (x_2, y_2) \in V$ and $|x_1 - x_2| + |y_1 - y_2| = 1\}$. Figure 1 shows a 2D-torus network $\text{TDT}(6, 6)$.

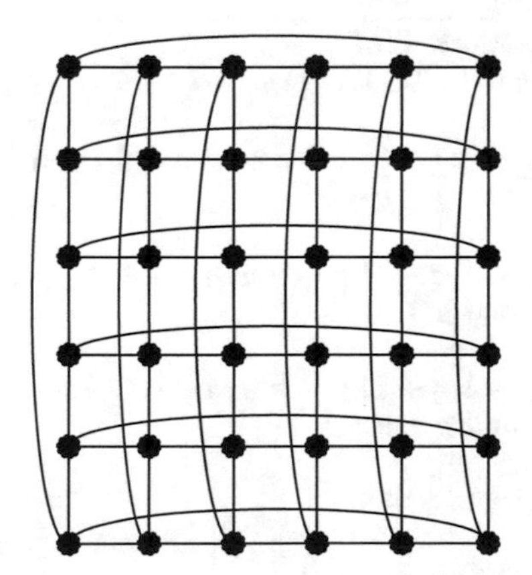

Fig. 1. A 2D-torus network $\text{TDT}(6, 6)$.

Under the PMC diagnosis model [9], assuming that adjacent processors are able to perform tests on each other. Let $G = (V, E)$ denote the underlying topology of a multiprocessor system. For any two of the adjacent vertices $u, v \in V(G)$,

Algorithm 1. AL(TDT(n, m))

Input: A 2D-torus network TDT(n, m).
Output: Fault elements in TDT(n, m).
begin
  for each i with $0 \leq i \leq n-1$ and for each j with $0 \leq j \leq \frac{m}{2} - 1$, $([i]_n, [2j+1]_m)$ tests $([i]_n, [2j]_m)$, and $([i]_n, [2j]_m)$ tests $([i]_n, [2j-1]_m)$; /* Test round 1. */
  set $C_i = \sum_{j \in \{0,1,\ldots,m-1\}} \sigma(([i]_n, [j]_m), ([i]_n, [j-1]_m))$ and set $C = \{i \mid C_i > 0$ for $0 \leq i \leq n-1\}$;
  set
  $F = \{([i]_n, [j]_m) \mid \sigma(([i]_n, [j+1]_m), ([i]_n, [j]_m)) = 1, \sigma(([i]_n, [j+2]_m), ([i]_n, [j+1]_m)) = 0$ and $\sigma(([i]_n, [j+3]_m), ([i]_n, [j+2]_m)) = 0\}$;
  if $|C| == 0$ **then return** $\emptyset$
  else if $|C| == 1$ **then**
    set $C = \{p\}$;
    for each j with $0 \leq j \leq m-1$, $([p+1]_n, [j]_m)$ tests $([p]_n, [j]_m)$;
    return $\{([p]_n, [j]_m) \mid \sigma(([p+1]_n, [j]_m), ([p]_n, [j]_m)) = 1\}$;
  end
  else if $|C| == 2$ **then return AL2**(TDT(n, m))
  else if $|C| == 3$ **then return AL3**(TDT(n, m))
  else return F
end

Algorithm 2. AL2(TDT(n, m))

Input: A 2D-torus network TDT(n, m).
Output: Fault elements in TDT(n, m).
begin
  set $C = \{p, q\}$; **if** $\{p, q\} \cap \{p+1, q+1\} == \emptyset$ **then**
    set $x = p+1$ and $y = q+1$
  end
  else if $\{p, q\} \cap \{p+1, q-1\} == \emptyset$ *and* $p+1 \neq q-1$ **then**
    set $x = p+1$ and $y = q-1$
  end
  else if $\{p, q\} \cap \{p-1, q+1\} == \emptyset$ *and* $p-1 \neq q+1$ **then**
    set $x = p-1$ and $y = q+1$
  end
  else set $x = p-1$ and $y = q-1$
  for each j with $0 \leq j \leq m-1$, $([x]_n, [j]_m)$ tests $([p]_n, [j]_m)$, and $([y]_n, [j]_m)$ tests $([q]_n, [j]_m)$; /* Test round 2. */
  return $\{([p]_n, [j]_m) \mid \sigma(([x]_n, [j]_m), ([p]_n, [j]_m)) = 1\} \cup \{([q]_n, [j]_m) \mid \sigma(([y]_n, [j]_m), ([q]_n, [j]_m)) = 1\}$;
end

Algorithm 3. AL3(TDT(n, m))

```
Input: A 2D-torus network TDT(n, m).
Output: Fault elements in TDT(n, m).
begin
    set C = {p, q, r};
    if max{C_i} - 1 then
        for each k ∈ C, ([k]_n, [j]_m) tests ([k]_n, [j+1]_m);
        if σ(([k]_n, [j+2]_m), ([k]_n, [j+1]_m)) =
        0 and σ(([k]_n, [j+3]_m), ([k]_n, [j+2]_m)) = 1
        then
            return {([k]_n, [j+1]_m) | σ(([k]_n, [j]_m), ([k]_n, [j+1]_m)) = 1} ∪ F;
        end
    end
    else if max{C_i} = 4 then
        Let j be the index such that C_j = 4;
        if there is a node (a, b) in C_j such that σ((a, b+t), (a, b+t+1)) = 1 for
        each t ∈ {0, 1, 2, 3} then
            return {(a, b+1), (a, b+3)} ∪ F;
        end
        else
            Let (a, b) and (c, d) be two distinct nodes in C_j where
            σ((a, b+t), (a, b+t+1)) = 1 and σ((c, d+t), (c, d+t+1)) = 1 for
            each t ∈ {0, 1} return {(a, b+1), (c, d+1)} ∪ F;
        end
    end
    else
        if Σ_{t=0,1,2} σ(([i]_n, [j+t+1]_m), ([i]_n, [j+t]_m)) == 1 for each j then
            return {([i]_n, [j]_m) | σ(([i]_n, [j+1]_m), ([i]_n, [j]_m)) = 1};
        end
        else if there is a vertex (a, b) such that σ(([a]_n, [b+1]_m), ([a]_n, [b]_m)) = 1
        and σ(([a]_n, [b+3]_m), ([a]_n, [b+2]_m)) = 1 then
            ([a]_n, [b-1]_m) tests ([a]_n, [b]_m);
            if σ(([a]_n, [b-1]_m), ([a]_n, [b]_m)) == 1 then return
            {([a]_n, [b]_m)} ∪ F; else return {([a]_n, [b+1]_m)} ∪ F;
        end
        else
            for each k ∈ C, ([k]_n, [j]_m) tests ([k]_n, [j+1]_m);
            if σ(([k]_n, [j+2]_m), ([k]_n, [j+1]_m)) = 1 and
            σ(([k]_n, [j+3]_m), ([k]_n, [j+2]_m)) = 1 then
                return {([k]_n, [j+1]_m) | σ(([k]_n, [j]_m), ([k]_n, [j+1]_m)) = 1} ∪ F;
            end
        end
    end
end
```

the ordered pair (u, v) represents a *test* that processor u has ability to diagnose processor v. Under the circumstance, u is a *tester*, and v is a *testee*. The outcome of a test (u, v) is 1 (respectively, 0) if u evaluates v to be faulty (respectively, fault-free). Since the faults identified here are existing, the outcome is *reliable* if and only if the tester is fault-free. A *test assignment* for system G is a collection of tests, modeled as a directed graph $T = (V, L)$, where $(u, v) \in L$ means that u and v are adjacent in G. A *syndrome* refers to a collection of all test results from the assignment T. Formally, a syndrome of T is a mapping $\sigma : L \rightarrow \{0, 1\}$. A *faulty set* F is the set of all faulty processors in G. It is obvious that F can be any subset of V. The process of identifying all faulty vertices is the system diagnosis. Besides, the maximum of faulty vertices correctly identified in a system G is called the diagnosability of G, denoted by $\tau(G)$. Supposedly, a system G is t-diagnosable if we correctly found out all faulty vertices in G with the total number being at most t. For any given syndrome σ resulting from a test assignment $T = (V, L)$, a subset of vertices $F \subseteq V$ is *consistent* with σ if for a $(u, v) \in L$ such that $u \in V - F$, then $\sigma(u, v) = 1$ if and only if $v \in F$. This corresponds to the assumption that fault-free testers are always correct, but the results from faulty testers may be unreliable. Hence, a given set F of faulty vertices may be consistent with different syndromes. Let $\sigma(F)$ denote the set of all possible syndromes with which the faulty set F can be consistent. Then two distinct faulty sets F_1 and F_2 of V are *distinguishable* if $\sigma(F_1) \cap \sigma(F_2) = \emptyset$; otherwise, F_1 and F_2 are *indistinguishable*. In other words, (F_1, F_2) is a *distinguishable pair* of faulty sets if $\sigma(F_1) \cap \sigma(F_2) = \emptyset$, else, (F_1, F_2) is an *indistinguishable pair.* For any two distinct faulty sets F_1 and F_2 of G with $|F_1| \leq t$ and $|F_2| \leq t$, a system G is t-diagnosable if and only if (F_1, F_2) is a distinguishable pair [2].

3 A Diagnosis Algorithm on the 2D-torus Network

In this section, we propose a three test rounds diagnosis algorithm **AL** on the 2D-torus network TDT(n, m) under the PMC model in Algorithm 1. For the case $|C| = 2$ and $|C| = 3$ in **AL1**, we give the algorithm **AL2** (Algorithm 2) and **AL3** (Algorithm 3), respectively.

4 Conclusions

In this article, we propose an algorithm for determining the faults in a 2D-torus network under the PMC model. Future research will develop some other efficient diagnosis algorithms for other networks, and prove the diagnosability of these networks under the PMC diagnosis model.

Acknowledgments. This work was supported in part by the Ministry of Science and Technology of the Republic of China under Contract MOST 106-2221-E-241-001.

References

1. Das, A., Thulasiraman, K., Agarwal, V.K., Lakshmanan, K.B.: Multiprocessor fault diagnosis under local constraints. IEEE Trans. Computers **42**(8), 984–988 (1993)
2. Dahbura, A.T., Masson, G.M.: An $O(n^{2.5})$ fault identification algorithm for diagnosable systems. IEEE Trans. Comput. **33**(6), 486–492 (1984)
3. Fujita, S., Araki, T.: Three-round adaptive diagnosis in binary n-cubes. In: Lecture Notes in Computer Science, vol. 3341, pp. 442–451 (2005)
4. Hakimi, S.L., Amin, A.T.: Characterization of connection assignment of diagnosable systems. IEEE Trans. Comput. **23**(1), 86–88 (1974)
5. Hsu, L.H., Lin, C.K.: Graph Theory and Interconnection Networks. CRC Press, Boca Raton (2008)
6. Lai, P.L., Tan, J.J.M., Chang, C.P., Hsu, L.H.: Conditional diagnosability measures for large multiprocessor systems. IEEE Trans. Comput. **54**(2), 165–175 (2005)
7. Lin, C.K., Teng, Y.H.: The diagnosability of triangle-free graphs. Theor. Comput. Sci. **530**, 58–65 (2014)
8. Lin, L., Zhou, S., Xu, L., Wang, D.: The extra connectivity and conditional diagnosability of alternating group networks. IEEE Trans. Parallel Distrib. Syst. **26**, 2352–2362 (2015)
9. Preparata, F.P., Metze, G., Chien, R.T.: On the connection assignment problem of diagnosis systems. IEEE Trans. Electron. Comput. **16**(12), 848–854 (1967)
10. Teng, Y.H., Lin, C.K.: A test round controllable local diagnosis algorithm under the PMC diagnosis model. Appl. Math. Comput. **244**, 613–623 (2014)

Construction Schemes for Edge Fault-Tolerance of Ring Networks

Chun-Nan Hung[1(✉)], Tzu-Liang Kung[2], and En-Cheng Zhang[1]

[1] Department of Information Management, Da-Yeh University, Dacun Township 515, Changhua County, Taiwan, R.O.C.
spring@mail.dyu.edu.tw, R0521017@cloud.dyu.edu.tw

[2] Department of Computer Science and Information Engineering, Asia University, Taichung City 413, Taiwan, R.O.C.
tlkung@asia.edu.tw

Abstract. The k-edges fault-tolerance-Hamiltonian graphs have been studied by many researchers. In this paper, we introduce the 2-path-required Hamiltonian graphs. We will show that the complete bipartite graph $K_{n,n}$ is $(n-3)$-edges fault-tolerance 2-path-required Hamiltonian graphs. We also prove the relationship between hyper-Hamiltonian laceability and 2-path-required Hamiltonian property. Moreover, we present the construction scheme for 2-path-required Hamiltonian graphs, named vertex join. Applying this scheme, we can construct many new 2-path-required Hamiltonian graphs with edges fault-tolerant property.

1 Introduction

An interconnection network is usually represented by a graph. Let $G = (V, E)$ be an undirected *graph*, where V is the node set and E is the edge set of G. The *degree* of vertex v in G, denoted by $d_G(v)$, is the number of edges incident to v. A *path* is a sequence of nodes in which every two consecutive nodes are adjacent. A *Hamiltonian path* of G is a path V if every node of V is exactly visited once. A *cycle* is a path if the first node and the last node are the same node. A *Hamiltonian cycle* is a cycle which traverses every node of V exactly once. A *Hamiltonian graph* is a graph that contains a Hamiltonian cycle.

Fault tolerance is desirable in the design of interconnection networks. The faults of a network correspond to removing edges and/or nodes from the graph. In this paper, we will concentrate the edges fault-tolerance of bipartite graphs. Let $F \subseteq E$ denote the set of faulty edges of graph G. Let $G - F = (V, E - F)$ denote the subgraph induced G. A graph G is a *k-edges fault-tolerance Hamiltonian graph* if $G - F$ is Hamiltonian for all $F \subset E$ and $|F| = k$.

A graph $G = (V_b \cup V_w, E)$ is *bipartite* if $V_b \cap V_w = \emptyset$ and $E \subseteq \{(u, v)|u \in V_b$ and $v \in V_w\}$. The author in [8] introduced the concept of Hamiltonian laceability of bipartite graphs. A bipartite graph $G = (V_w \cup V_b, E)$ is Hamiltonian laceable if there exists a Hamiltonian path between every pair vertices u, v for $u \in V_w$, $v \in V_b$. The concept of *hyper-Hamiltonian laceability* is proposed in [5].

L. Barolli et al. (Eds.): IMIS 2018, AISC 773, pp. 626–631, 2019.
https://doi.org/10.1007/978-3-319-93554-6_61

A bipartite graph $G = (V_w \cup V_b, E)$ is hyper-Hamiltonian laceable if there exists a Hamiltonian path between every pair vertices x, y in $G - \{z\}$ for any $x, y \in V_i$, $z \in V_j$ with $\{i, j\} = \{b, w\}$. Many researchers [1,2,4,6] introduced the edges fault-tolerance for Hamiltonian laceability and hyper-Hamiltonian laceability.

In this paper, we propose a new concept, called 2-path-required Hamiltonian graphs, for the fault tolerance of Hamiltonian graphs. A graph G is *2-path-required Hamiltonian* if for every path of G with length 2, denoted p_2, there exists a Hamiltonian cycle passing through p_2. A graph G is a *k-edges fault-tolerance 2-path-required Hamiltonian graph* if the graph $G - F$ is 2-path-required Hamiltonian, for all $F \subset E$ and $|F| = k$. We will show the relationship between hyper-Hamiltonian laceability and 2-path-required Hamiltonian property. Furthermore, we will propose the construction schemes, called vertex join, for k-edges fault-tolerance 2-path-required Hamiltonian graphs.

2 The Preliminary Properties

In this section, we will introduce some basic properties for 2-path-required Hamiltonian graphs. We first show the following lemma.

Lemma 1. *Let G be a graph with $\delta(G) = k + 3$. If the graph G is k-edges fault-tolerance 2-path-required Hamiltonian, G is $(k + 1)$-edges fault-tolerance Hamiltonian.*

Proof. Let F be a set of arbitrary faulty edges with $|F| = k + 1$. We will construct a Hamiltonian cycle of $G - F$. Let $f = (u, v) \in F$ be a faulty edge. Since $\delta(G) = k+3$, there exist two fault-free edges $(x, u), (u, y)$ in $G-(F-\{f\})$. Thus, we can construct a Hamiltonian cycle C_1 passing through the path $\langle x, u, y \rangle$ in $G-(F-\{f\})$ since G is k-edges fault-tolerance 2-path-required Hamiltonian. Therefore, the C_1 is a Hamiltonian cycle in $G - F$. □

In the follows, we will prove the bipartite graphs $K_{n,n}$ is $(n - 3)$-edges fault-tolerance 2-path-required Hamiltonian.

Lemma 2. *The complete bipartite graphs $K_{n,n}$ is $(n - 3)$-edges fault-tolerance 2-path-required Hamiltonian for $n \geq 3$.*

Proof. We will prove this lemma by induction on n. We can verify that $K_{3,3}$ is 2-path-required Hamiltonian by brute-force methods. By the induction hypothesis, we can assume that $K_{n-1,n-1}$ is $(n - 4)$-edges fault-tolerance 2-path-required Hamiltonian for $n \geq 4$. Let F denote an arbitrary set of faulty edges of $K_{n,n} = (V_b \cup V_w, E)$ with $|F| = n - 3$. Let p_2 denote an arbitrary fault-free path with length 2 and $E(p_2)$ be the set of edges in p_2, $V(p_2)$ be the set of vertices in p_2. Since $|F| = n - 3$, there exists some vertices $s \in V_b, t \in V_w$ such that the degrees of s and t are both n of $K_{n,n} - F$ and $s, t \notin V(p_2)$. Let the graph $J = K_{n,n} - \{s, t\}$. We will show that $K_{n,n}$ is $(n - 3)$-edges fault-tolerance 2-path-required Hamiltonian for $n \geq 4$.

Let $e = (v_b, v_w) \in F$ be a faulty edge. Since $K_{n-1,n-1}$ is $(n - 4)$-edges fault-tolerance 2-path-required Hamiltonian, we can obtain a Hamiltonian cycle

C_1 containing p_2 in $J - (F - \{e\})$. We choose an edge of $C_1 - E(p_2)$, denoted as (x_b, x_w). Moreover, when C_1 containing the faulty edge e, we will denote $e = (x_b, x_w)$. We can denote the cycle C_1 as $\langle x_b, x_w \rightarrow P_1 \rightarrow x_b \rangle$ since $n \geq 4$. Therefore, $\langle x_b, t, s, x_w \rightarrow P_1 \rightarrow x_b \rangle$ is the Hamiltonian cycle containing p_2 of $K_{n,n} - F$. □

3 The Hyper-Hamiltonian Laceability

In this section, we will show that a $(k-1)$-edges fault-tolerance hyper-Hamiltonian laceable graph is $(k-1)$-edges fault-tolerance 2-path-required Hamiltonian.

Theorem 3. *Let $G = (V_b \cup V_w, E)$ be a bipartite $(k-1)$-edges fault-tolerance hyper-Hamiltonian laceable graph. Thus, G is $(k-1)$-edges fault-tolerance 2-path-required Hamiltonian.*

Proof. Let F be the set of faulty edges of G with $|F| = k - 1$. Let p_2 be an arbitrary fault-free path with length 2. Without loss of generality, we can assume that $p_2 = \langle a, b, c \rangle$ for $a, c \in V_w$ and $b \in V_b$. We will construct a Hamiltonian path passing through the path p_2 in $G - F$. Since G is $(k-1)$-edges fault-tolerance hyper-Hamiltonian laceable, there exists a Hamiltonian path $P(c, a)$ between c and a in $G - F - \{b\}$. Thus, $\langle a, b, c \rightarrow P(c, a) \rightarrow a \rangle$ is the Hamiltonian cycle passing through p_2 in $G - F$. □

In [2], the author proved that the hypercube Q_n is $(n-3)$-edges fault-tolerance hyper-Hamiltonian laceable. Thus, Q_n is also $(n-3)$-edges fault-tolerance 2-path-required Hamiltonian. The bubblesort graphs B_n and star graphs S_n are both showed $(n-4)$-edges fault-tolerance hyper-Hamiltonian laceable in [1,6]. Applying Theorem 3, we can obtain that both B_n and S_n are also $(n-4)$-edges fault-tolerance 2-path-required Hamiltonian.

4 The Construction Scheme: Vertex Join

In this section, we will introduce the vertex join operation and show that the vertex join of two $(k+2)$-regular and $(k-1)$-edges fault-tolerance 2-path-required Hamiltonian graphs is also a $(k-1)$-edges fault-tolerance 2-path-required Hamiltonian graphs.

In [7], we introduced a construction scheme, named 3-join, for 3-regular 1-fault-tolerance Hamiltonian graph. We also proposed the generalized of 3-join, named $(k+2)$-join, for $(k+2)$-regular, k-fault-tolerance Hamiltonian graph in [3]. We will define the *vertex join* as follows.

Definition 4. *Let x be a node of the $(k+2)$-regular graph $G = (V_g, E_g)$ and y be a node of the $(k+2)$-regular graph $H = (V_h, E_h)$ with $V_g \cap V_h = \emptyset$. Let $N(x) = \{x_1, x_2, \ldots, x_{k+2}\}$ be the neighbor of x and $N(y) = \{y_1, y_2, \ldots, y_{k+2}\}$ be the neighbor of y. The vertex join of G and H, denoted by $G \sqcup H$, at x and y is described as follows.*

$V(G \sqcup H) = V_g \cup V_h - \{x, y\}$, *and*
$E(G \sqcup H) = E_g \cup E_h \cup \{(x_i, y_i)|1 \leq i \leq k+2\} - (\{(x, x_i)|x_i \in N(x)\} \cup \{(y, y_i)|y_i \in N(y)\})$.

We can obtain that the vertex join of two $(k+2)$-regular graphs is $(k+2)$-regular.

We will prove that the vertex join of two $(k+2)$-regular and $(k-1)$-edges fault-tolerance 2-path-required Hamiltonian graphs is also a $(k-1)$-edges fault-tolerance 2-path-required Hamiltonian graphs in the following theorem.

Theorem 5. *A vertex join of two $(k+2)$-regular and $(k-1)$-edges fault-tolerance 2-path-required Hamiltonian graphs is also a $(k-1)$-edges fault-tolerance 2-path-required Hamiltonian graph.*

Proof. Let x be a node of the $(k+2)$-regular graph $G = (V_g, E_g)$ and y be a node of the $(k+2)$-regular graph $H = (V_h, E_h)$ with $V_g \cap V_h = \emptyset$. Let $N(x) = \{x_1, x_2, \ldots, x_{k+2}\}$ be the neighbor of x and $N(y) = \{y_1, y_2, \ldots, y_{k+2}\}$ be the neighbor of y. The $G \sqcup H$ is the vertex join at x and y. Let $E_c = \{(x_i, y_i)|1 \leq i \leq k+2\}$. Let F be the set of faulty edges in $G \sqcup H$ with $|F| = k-1$ and $p_2 = \langle u, v, w \rangle$ be an arbitrary fault-free 2-path of $G \sqcup H$. Let $F_g = F \cap E_g$, $F_h = F \cap E_h$ and $F_c = F \cap E_c$. Let $F_{cg} = \{(x_i, x)|$ for $(x_i, y_i) \in F_c\}$ and $F_{ch} = \{(y_i, y)|$ for $(x_i, y_i) \in F_c\}$. We will prove that there exists a Hamiltonian cycle in $(G \sqcup H) - F$ containing the path p_2 in the following cases.

Case 1 $E_c \cap \{(u, v), (v, w)\} = \emptyset$.
Without loss of generality, we can assume that $\{(u, v), (v, w)\} \subset E_g$. Since the graph G is $(k-1)$-edges fault-tolerance 2-path-required Hamiltonian, we can obtain a Hamiltonian cycle C_1 in $G-(F_g \cup F_{cg})$ containing the path p_2. We can denote C_1 as $\langle x_s, x, x_t \rightarrow P_1 \rightarrow x_s \rangle$. We can construct a Hamiltonian cycle of the graph $H - (F_h \cup F_{ch})$ containing the path $\langle y_t, y, y_s \rangle$ since the graph H is $(k-1)$-edges fault-tolerance 2-path-required Hamiltonian. We can denote this cycle as $\langle y_t, y, y_s \rightarrow P_2 \rightarrow y_t \rangle$. Thus, $\langle x_t \rightarrow P_1 \rightarrow x_s, y_s \rightarrow P_2 \rightarrow y_t, x_t \rangle$ is the Hamiltonian cycle containing the path p_2.

Case 2 $E_c \cap \{(u, v), (v, w)\} \neq \emptyset$.
Without loss of generality, we can assume that $p_2 = \langle u, x_a, y_a \rangle$ for some $1 \leq a \leq k+2$. Since the graph G is $(k-1)$-edges fault-tolerance 2-path-required Hamiltonian, we can obtain a Hamiltonian cycle in $G - (F_g \cup F_{cg})$ containing the path $\langle u, x_a, x \rangle$. We can denote this cycle as $\langle u, x_a, x, x_b \rightarrow P_3 \rightarrow u \rangle$. We can also construct a Hamiltonian cycle of the graph $H - (F_h \cup F_{ch})$ containing the path $\langle y_b, y, y_a \rangle$ since the graph H is $(k-1)$-edges fault-tolerance 2-path-required Hamiltonian. We can denote this cycle as $\langle y_b, y, y_a \rightarrow P_4 \rightarrow y_b \rangle$. Therefore, $\langle u, x_a, y_a \rightarrow P_4 \rightarrow y_b, x_b \rightarrow P_3 \rightarrow u \rangle$ is the Hamiltonian cycle containing the path p_2. □

Applying Theorem 5, we can construct more 2-path-required Hamiltonian graphs with edges fault-tolerant property. For example, we can construct a new 1-edge fault-tolerance 2-path-required Hamiltonian graph by vertex-join of Q_4 and $K_{4,4}$, as illustrated in Fig. 1.

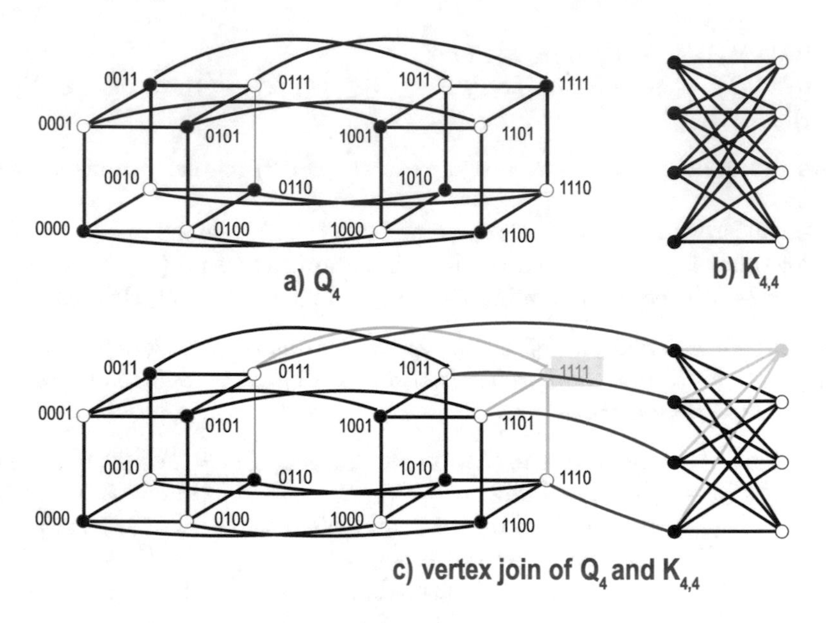

Fig. 1. The example of vertex join

5 Conclusion

In this paper, we introduce 2-path-required Hamiltonian bipartite graphs. We also investigate the edges fault-tolerance about 2-path-required Hamiltonian graphs. We further propose the construction scheme for 2-path-required Hamiltonian graphs. We hope to apply 2-path-required Hamiltonian concept for directed graphs. We will study the arcs fault-tolerance 2-path-required Hamiltonian directed bipartite graphs in the future.

Acknowledgement. This work is supported in part by the Ministry of Science and Technology, Taiwan, under Grands No: MOST 105-2115-M-212-001 and MOST 106-2115-M-212-001.

References

1. Araki, T., Kikuchi, Y.: Hamiltonian laceability of bubble-sort graphs with edge faults. Inf. Sci. **177**, 2679–2691 (2007)
2. Chang, C.-H., Lin, C.-K., Huang, H.-M., Hsu, L.-H.: The super laceability of the hypercubes. Inf. Process. Lett. **92**, 15–21 (2004)
3. Hung, C.-N., Zhu, S.-S.: Construction for strongly k-Hamiltonian graphs. In: Proceedings of the 19th Workshop on Combinatorial Mathematics and Computation Theory, pp. 17–22 (2002)
4. Hsu, L.-H., Lin, C.-K.: Graph Theory and Interconnection Networks. CRC Press, Boca Raton (2008)
5. Lewinter, M., Widulski, W.: Hyper-Hamilton laceable and caterpillar-spannable product graphs. Comput. Math. Appl. **34**, 99–104 (1997)

6. Li, T.-K., Tan, J.J.M., Hsu, L.-H.: Hyper Hamiltonian laceability on edge fault star graph. Inf. Sci. **165**, 59–71 (2004)
7. Wang, J.J., Hung, C.-N., Tan, J.M., Hsu, L.-H., Sung, T.-Y.: Construction schemes for fault-tolerant Hamiltonian graphs. Networks **35**, 233–245 (2000)
8. Wong, S.A.: Hamilton cycles and paths in butterfly graphs. Networks **26**, 145–150 (1995)

Review of RFID-Based Indoor Positioning Technology

Jingkai Zhu[1] and He Xu[2(✉)]

[1] Bell Honors School, Nanjing University of Posts and Telecommunications, Nanjing 210023, China
Q16010211@njupt.edu.cn

[2] School of Computer Science, Nanjing University of Posts and Telecommunications, Nanjing 210023, China
xuhe@njupt.edu.cn

Abstract. Traditional GPS location technology cannot work in indoor environment. In order to sum up the positioning theory of RFID positioning method and find an indoor location algorithm suitable for an indoor environment, this paper reviews the composition of RFID indoor positioning system and the location algorithms of RFID indoor positioning system. And more comprehensive study and a systematic summary are carried out. The paper provides an important basis for the selection of RFID location algorithm and positioning system under different conditions.

1 Introduction

Nowadays, the commonly used technologies for outdoor location are global positioning system (GPS) and cellular base station wireless positioning system. However, because of the limitations of the indoor environment (such as electromagnetic wave shielding), the above two methods cannot meet the user's indoor positioning requirements. In addition, the application of indoor positioning technology has a broad prospect, and the significance of the research is very important. Thus, it becomes a very hot topic. Radio Frequency Identification (RFID) technology is widely favored by researchers because of its high precision, low cost and so on.

In this paper, indoor location technology based on RFID is selected, and several commonly used location algorithms are introduced in detail. Combined with the latest research, several improved RFID positioning systems are summarized, which improve positioning accuracy and efficiency. The paper is organized as the following: Sect. 2 introduces the components of RFID system. Section 3 summarizes some commonly used indoor location algorithms, and Sect. 4 summarizes several ways to improve the accuracy of indoor location. Section 5 concludes the paper.

L. Barolli et al. (Eds.): IMIS 2018, AISC 773, pp. 632–641, 2019.
https://doi.org/10.1007/978-3-319-93554-6_62

2 Components of the RFID System

The basic structure of the RFID system consists of a sensor network and a data transmission network. RFID tags and reader devices make up a sensor network. According to the different ways of obtaining energy, the electronic tags can be divided into three types: active, passive and semi active. Passive tags do not require a power supply, an active tag has a battery, and a semi-active tag can use a battery or not. RFID tags also can be classified into microwave, ultra high frequency, high frequency and low frequency electronic tags, which reading range from large to small according to frequency. The RFID Reader can be classified into two kinds: read-only and read-write, which is the center of data processing and control of the positioning system. In the area covered by the RFID reader, the reader launches the radio frequency signal and activates an RFID tag in the area so that the tag can return data. The basic structure of RFID indoor positioning system, as shown in Fig. 1:

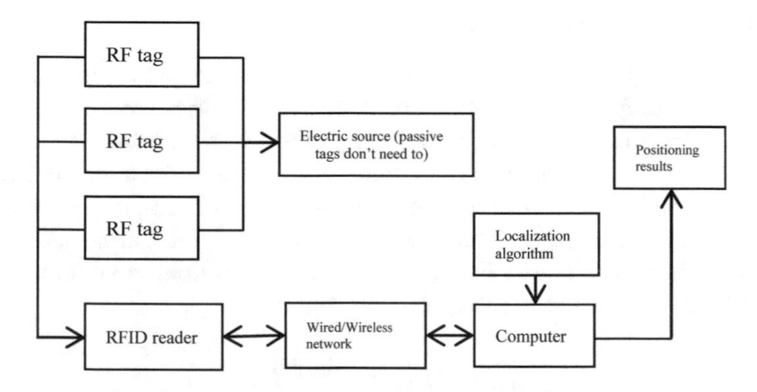

Fig. 1. RFID system fundamental structure

The data transmission network can be divided into two parts: the server and its connection to the reader. The transmission of data contains two aspects: One is to send the user's instructions to the transmission network through the server, the other is to return the received data to the server by the RFID reader. After then the position of the object is calculated through the positioning algorithm. There are also two ways of connecting the server and the reader, that is, the wireless connection and the wired connection.

3 Basic Algorithms for RFID Indoor Location

The location algorithm of the indoor positioning system is usually classified into distance-based location algorithm and distance-independent location algorithm. Distance-based localization algorithm is mainly through measuring the actual distance or orientation information between nodes, and then uses three edge measurement, triangulation or maximum likelihood estimation to calculate the target's location.

Distance-independent location algorithm mainly uses the connectivity between nodes, relative position or specific distance of protocol estimation to calculate the node's position. In the distance-based location algorithms, there are mainly include Time of Arrival method (ToA), Time Different of Arrival method (TDoA), Angle of Arrival method (AoA) and Received Signal Strength Indicator method (RSSI) to measure node spacing and orientation. In addition to these methods, there is a widely used location method called LANDMARC, which is based on RSSI method. VIRE and LEMT are the similar positioning methods [1].

Most of the existing indoor location algorithms are based on the theory of triangulation. The theory is generally realized by measuring the distance or measuring the angle of arrival. Measuring distance, is to measure the distance from the antenna of the reader to the tag. By measuring the angle of arrival, the relative direction between the reader's antenna and the tag is obtained. In the following several indoor location algorithms based on triangulation are reviewed and comparisons.

3.1 Time of Arrival Method (ToA)

The working principle of ToA is that because the speed of radio frequency signal is known, by measuring the time from the reference point to the moving point of the RF signal, the distance is determined between the two points based on the time [1]. As shown in Fig. 2, by collecting data from two or more base stations, the location can be reduced to a circular range. Then the data is gotten from a third-party base station and the current location can be calculated, so that a single point can be identified. This positioning method is also used in the global satellite positioning system.

ToA calculation method use the following formula:

$$d = \frac{T_{total} - T_{sa} - T_{sys} - T_{cab}}{C_o} \tag{1}$$

In the formula (1), T_{sa} is the delay value of tag, T_{sys} is system delay, and T_{cab} is the delay caused by receiving antenna cable. The algorithm requires at least three reference points to locate one two-dimensional target. If more than 3 reference points are available, the least square method can be used to improve the positioning accuracy.

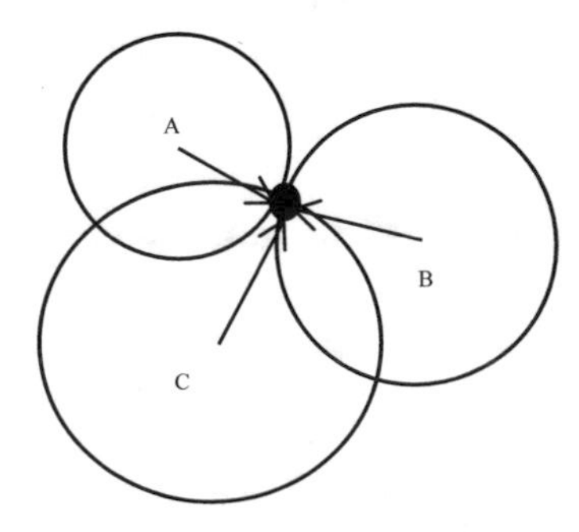

Fig. 2. Time of arrival

The location accuracy of the arrival time positioning method is high, but the algorithm requires that the tag and reader should keep synchronization in time. Because the complexity of the positioning environment leads to the multipath effect, the positioning accuracy of the system is reduced.

3.2 Time Different of Arrival Method (TDoA)

TDoA method synchronously sends signals at different measuring points, and calculates the distance between them by measuring the transmission time difference between signals from different nodes to target objects. The location accuracy of this method is as high as that of the Time of Arrival method, and it requires all the readers to have strict synchronization in time. Otherwise, because of many obstacles, coupled with the complexity of the space, it makes possible for readers to receive signals but which is not sent to tags. In addition, the algorithm is affected by multipath and noise. These factors will reduce the accuracy of RFID positioning system.

TDoA use the following formula:

$$c_o t_d(R) = c_o(t_T - t_{RT}) + ||T - R|| - ||RT - R|| \tag{2}$$

In the formula (2), T is the selected measurement tag, t_T is the response time, t_d is the time differences between the received signal, RT is the known reference tag of the fixed position, t_{RT} is the corresponding time of the reference tag, c_o is the constant. Finally, the estimation of the position of tag is carried out by the weighted mean method.

3.3 Angle of Arrival Method (AoA)

AoA is a positioning method based on angle measurement. Its basic principle is shown in Fig. 3. Two antenna arrays are used to measure the tag signals of the target node A to the angle of arrival of the respective launch points, and the azimuth lines of the two emitter points are obtained [8]. The intersection point of the two azimuth lines is the position of the target node A, and then it calculates its specific coordinates using the triangulation method.

AoA uses the following formula:

$$(x, y) = [h \cdot \frac{\tan(\omega_1)\tan(\omega_2)}{\tan(\omega_1) + \tan(\omega_2)}, \frac{h}{2}\frac{\tan(\omega_1) - \tan(\omega_2)}{\tan(\omega_1) + \tan(\omega_2)}] \tag{3}$$

In the formula (3), h is the center spacing of the two sets of reader connections, ω_1 and ω_2 are the arrival angles estimated by the two lines and the tag.

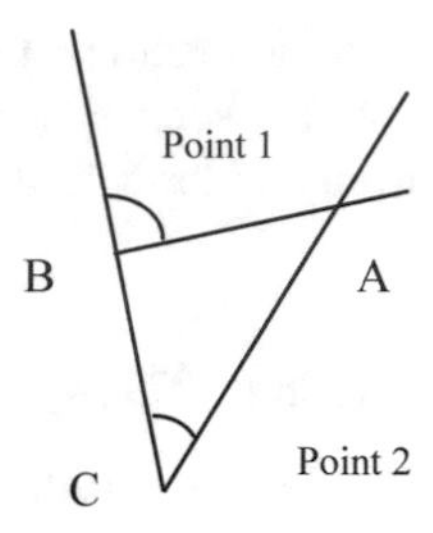

Fig. 3. Angle of arrival

3.4 Received Signal Strength Indicator Method (RSSI)

The RSSI method uses the empirical model to simulate the signal path loss and calculate the target distance, and locates the target coordinates using the location algorithm. The attenuation empirical model of the signal needed can be obtained experimentally or theoretically. It is proved by the experiment that the RSSI method can counteract the effect of multipath to some extent, and it is more suitable for indoor environment. At present, many RFID systems use the RSSI method. The triangulation method based on RSSI requires at least three transmitters to locate objects. These transmitters need to maintain the same angle and distance [1], but also there are many interference factors in indoor radio propagation, such as attenuation, diffraction and so on. These factors affect the accuracy of the positioning of the RSSI method.

RSSI method has the following loss model formula:

$$P_D = P_{D_0} - 10\mu \lg\left(\frac{D}{D_0}\right) + x \tag{4}$$

In the formula (4), D is the distance between locating tag and reference tag, and the loss index is μ, x is a normal distribution random variable with σ as the standard deviation of zero mean, P_D is the RSSI value of distance D, D_0 is the reference distance.

3.5 Location Algorithm Based on RSSI

The LANDMARC location method is based on the RSSI method, which uses the "nearest neighbor distance" (KNN) algorithm, and the fixed reference tags of the known location information is deployed in the monitoring area. The reader has eight different energy levels. K nearest reference label is obtained near the location label by the following formula:

$$E_j = \sqrt{\sum\nolimits_{i=1}^{n} (\theta_{j,i} - S_i)^2} \tag{5}$$

In the formula (5), n is the number of readers; S_i is the signal strength to position the tag reader i obtained; $\theta_{j,i}$ is the signal intensity of the reference label j read by the reader i; E_j is the Euclidean distance between the reference tag j and unknown label. After the K reference tag is obtained, the coordinates of the fixed label can be calculated through the following formula:

$$(x,y) = \sum\nolimits_{t=1}^{k} \omega_t (x_t, y_t) \quad (6)$$

In the formula (6), $\omega_t = \frac{1/E_t^2}{\sum_{t=1}^{k}(1/E_t^2)}$. It is known that a reference label with a minimum E value has the maximum weight. LANDMARC system has low cost, and can adapt to the changing environment. The location information is more accurate and reliable when it is used in indoor environment. However, because of the long computation time, the real-time performance of the system is not good.

4 Review of Improvement of RFID Indoor Positioning System Based on Basic Positioning Algorithm

Although indoor location with RFID technology has many advantages, such as low cost and wide transmission range, in some cases, RFID signal will be interfered by the wireless signal and it will be restricted, which will affect the accuracy of location. In view of this problem, this paper combines the latest research, and summarizes some methods to improve the positioning accuracy and reduce the time of location consumption.

4.1 Improve the Accuracy of the Location by Eliminating the Reflection of the Object

In the indoor positioning system, the specific location of the target object is obtained by the computer based on the tracking data collected by the reader. The effective management of the data collected by the tracker can be used in many indoor tracking applications. Therefore, it is essential to ensure the accuracy of the data collected by the reader.

In order to better describe the location of objects, a data model is usually set up to locate objects in the target location. The specific location is changed into symbolic space, and the symbolic coordinates are used to determine the specific location of objects. Indoor environment is much more complicated, with restrictions on doors, corridors, stairs and other objects. Outdoor space has both geometric and topological characteristics, while indoor environment is only topological. Therefore, the topological properties of indoor location need to be specific. Meanwhile, the distance between the two symbolic coordinates also needs to be specified precisely, just like geometric coordinates [3].

Indoor location, whether using Wi-Fi, Bluetooth or RFID, most of technologies rely on similarity analysis. At the same time, mobile objects cannot be continuously positioned. In actual operation, RFID devices can produce some errors, resulting in the inaccuracy of RFID tracking data. These errors are caused by many different reasons, and one of the important reasons for the tracking error is the object reflection. When RF signals are transmitted indoors, they will be affected by many entities, such as walls, which will reflect or block RF signals, which will disturb the normal transmission of RF signals, and then affect the data acquisition of receiving devices [2]. For example, when an object with RFID tag is detected by two or more RFID readers at the same time or in a very short time, the problem of object reflection will happen. Once such a problem is created, the location of the object being tracked will change constantly and it is difficult to determine a specific location. In addition, when the frequency of RF signals is not consistent, RFID reader devices need to constantly adjust the frequency and scope of monitoring. In the process, there may be wrong location information. These errors prevent advanced applications from making meaningful decisions based on wrong data. Therefore, it is important to remove the error data before it is put into practical application. The data processing can greatly improve the accuracy of the data and make the location more accurate by processing the collected original tracking data.

In order to solve the problem of object reflection, reachability constraints are used and effective device is checked to remove the reflection records left in the reflection period. Asif lqbal Baba et al. designed a three phase bounce handing module, which is generally called an intermediate device. This device is specially designed for indoor positioning and only includes off-line data, as shown in Fig. 4. This component outputs the clean data, which can more accurately determine the location of the object [3].

The first stage: Aggregation Phase.

First, the raw data is processed, the unprocessed reader data is integrated into a mechanism, these data is gathered on time, which become meaningful tracking data. This mechanism sets a threshold τ(limit value), if the time between two consecutive reader data is no more thanτ, these data can be combined as a track record. On the other hand, if the time interval is more than τ, these data will be divided into two track [2].

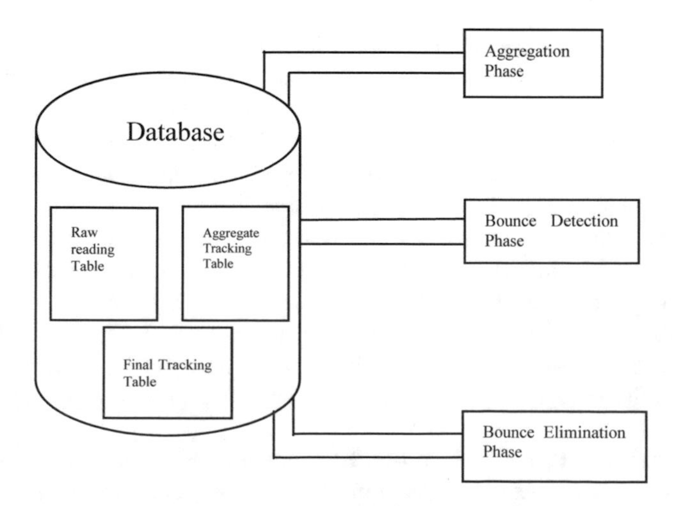

Fig. 4. Three phase bounce handling module

The second stage: Bounce Detection Phase.

The data gathered by the first stage is transmitted to the second stage for processing, and each detection time is compared for tracking data and the one for the track record check time before. If the two time intervals are overlapping, it can be assumed that in the crossover process the two devices have overlapping range check. The overlapped time interval is regarded as a checking interval for a new virtual device to track objects, and at the same time, the tracking information is added to a separate tracking record. In this way, the data of the aggregation phase can be processed, the overlaps that cause the reflection problem are checked, and then the data is processed.

The third stages: Bounce Elimination Phase.

According to the data tables obtained after the second stage processing, the virtual tracking records are searched, which represent two or more equipment inspection overlaps. A method of removing redundant parts based on continuity is needed to remove these overlapped data, because it is impossible for some devices to contain overlapping parts according to the continuous information of the devices deployed according to the deployed RF signals. In addition, in each overlapping period, according to some criteria, it is possible to determine which device is truly monitoring the target object. In this step, the overlap of data combines into a new data set, with longer inspection intervals.

In the RFID indoor tracking error process data, the research is focused on the inspection for the object's reflection and the removal. Through the deployment of RFID reader and the indoor topology, a reflex and removal algorithm is designed which is based on the accessibility of the time limit, to reduce the effect of reflected problems on the RFID data. A large number of experiments have proved that this method is effective [3].

4.2 Indoor Positioning System Based on RFID and Kinect

In recent years, more and more attention has been paid to the research on the combination of RFID and image tracking cameras. Microsoft has developed a device called Kinect that uses an infrared camera to track the image and combines the motion sensing function to track the people in the room. Based on this study, Schindhelm has proposed that Kinect has a practical value for indoor person [4]. Nakano et al. has also proposed to use Kinect for indoor positioning [5]. But similar image tracking technology can only realize the location of the object, but cannot identify the identity. Ching-Sheng Wang et al. put forward a method based on RFID and Kinect technology to locate indoor people [6], which combines the RFID's object recognition function and the Kinect efficient target location acquisition capability. The system can accurately and efficiently determine the location of the indoor person.

The positioning system is consist of two parts, including an active RFID positioning mechanism and a Kinect positioning mechanism. In the RFID mechanism, the reader needs to put in advance, then the RFID tag is associated with the human body. When someone goes to the scope of the reader coverage, the reader can read the label signal, transmit it to the server for analysis, processing and determine the identity of the person. For the Kinect part, Kinect needs to be installed at a higher location. So when Kinect is moving, the motion sensing ability of the sensor can be better positioned. Finally, the location results of RFID and Kinect will be collected to the same terminal, and the final results will be obtained by combining the identification information obtained by RFID and location information obtained by Kinect. However, the location of Kinect is limited by the range of infrared monitoring. Therefore, by adding RGB color camera in Kinect part, people can be tracked far away. It can also be combined with the RFID identity function to form an accurate positioning system for indoor person [6].

4.3 Some Other Improved Positioning Methods

In a variety of indoor positioning systems, active RFID is more applied than infrared, ultrasonic and Wi-Fi. In addition, inertial location system based on inertial measurement unit (IMU) is another better indoor location system, because it enables the radio indoor location infrastructure with high cost to work independently. Therefore, Chian C. Ho et al. proposed a real-time indoor location system [7], which combines the RFID Heron-bilateration judgment based on the external RFID infrastructure and the IMU inertial-navigation based on the internal IMU module, and combines them to improve the accuracy and efficiency of location.

In addition, Anvar Narzullaev et al. put forward a method [9], which can combine Wi-Fi positioning and RFID positioning method together. Caifeng Liu et al. proposed a system which can be applied to real life. It can be used for fire positioning system in the public place. The system based on RFID and Wi-Fi, which is not only ensures the real-time of the location, but also makes use of the WiFi to make the coverage larger, thus making the system more practical [10].

5 Conclusions

In this paper, the positioning principle and location method of RFID system are summarized in detail. The research of RFID location technology is systematically sorted out, and the current popular location algorithms of RFID location technology are summarized and compared. Compared with AoA, ToA, TDoA and other positioning methods, the localization method based on RSSI is relatively high in anti-interference and is easier to be realized. The LANDMARC system has simple structure and high positioning precision. It is the first choice for RSSI positioning. At the same time, this paper summarizes several methods which can improve the accuracy and efficiency of RFID positioning with the latest research. Considering the actual application needs, only a monotonic algorithm usually fails to achieve the desired effect. When using multiple algorithms for joint location, we need to take full account of different conflicting problems when using different algorithms to locate results, as well as the problems of different algorithms and different location patterns. With the development of RFID technology towards miniaturization, integration and intellectualization, RFID indoor positioning will be applied to a more demanding and more complex environment. Because of its low cost and easy implementation of the system, improving the real-time and positioning accuracy of the system is the research trend of the RFID indoor positioning technology in the future.

Acknowledgments. This work is financially supported by the National Natural Science Foundation of P. R. China (No.: 61672296, No.: 61602261), Scientific & Technological Support Project of Jiangsu Province (No.: BE2015702, BE2016185, No.: BE2016777), China Postdoctoral Science Foundation (No.: 2014M561696), Jiangsu Planned Projects for Postdoctoral Research Funds (No.: 1401005B), Postgraduate Research and Practice Innovation Program of Jiangsu Province (No.: KYCX17_0798), and NUPT STITP (No.: XYB2017556).

References

1. Ab Razak, A.A.W., Samsuri, F.: Active RFID-based Indoor Positioning System (IPS) for industrial environment. In: 2015 IEEE International RF and Microwave Conference (RFM 2015), pp. 89–91 (2015)
2. Baba, A.I., Jaeger, M., Lu, H., Pedersen, T.B., Xie, X.: Learning-based cleansing for indoor RFID data. In: SIGMOD, pp. 925–936 (2016)
3. Baba, A.l.: Removing object bouncing from indoor tracking data. In: 2017 ACM SE, 13–15 April 2017, Kennesaw, GA, USA, pp. 1–8 (2017)
4. Schindhelm, C.K.: Evaluating slam approaches for Microsoft Kinect. In: The Eighth International Conference on Wireless and Mobile Communications, pp. 402–407, June 2012
5. Nakano, Y., Kai, K., Izutsu, K., Tatsumi, T., Tajitsu, K.: Kinect positioning system (kps) and its potential applications. In: International Conference on Indoor Positioning and Indoor Navigation (2012)
6. Wang, C.-S., Chen, C.-L.: RFID-based and kinect-based indoor positioning system. IEEE (2014)
7. Ho, C.C., Lee, R.: Real-time indoor positioning system based on RFID Heron-Bilateration location estimation and IMU inertial-navigation location estimation. In: 2015 IEEE 39th Annual International Computers, Software and Applications Conference, pp. 481–486 (2015)
8. Kim, S.-C., Jeong, Y.-S., Park, S.-O.: RFID-Based Indoor Location Tracking to Ensure the Safety of the Elderly in Smart Home Environments. Springer, London (2012)
9. Narzullaev, A., Mohd, H.S.: Wi-Fi signal strengths database construction for indoor positioning systems using Wi-Fi RFID. In: Proceedings of the 2013 IEEE International Conference on RFID Technologies and Applications, 4–5 September 2013, Johor Bahru, Malaysia (2013)
10. Liu, C., Gu, Y.: Research on indoor fire location scheme of RFID based on WiFi. In: 2016 Nicograph International, pp. 116–119 (2016)

Designing a Cybersecurity Board Game Based on Design Thinking Approach

Shian-Shyong Tseng[1(✉)], Tsung-Yu Yang[2], Yuh-Jye Wang[2], and Ai-Chin Lu[3]

[1] Department of M-Commerce and Multimedia Applications, Asia University, Taichung City, Taiwan ROC
sstseng@asia.edu.tw

[2] Department of Computer Science and Information Engineering, Asia University, Taichung City, Taiwan ROC
zongyu212@gmail.com, yjwang@vghtc.gov.tw

[3] Taiwan Network Information Center, Taipei, Taiwan
aclu@twnic.net.tw

Abstract. In this paper, we propose an innovative board game design process to help students to design a cybersecurity board game with a pre-designed board game tool kit, and help them to further learn cybersecurity knowledge by using the design thinking and learning-by-doing strategies. In the process, the board game design course will firstly be given, the CBR-based learning by doing scheme will then be provided for helping the students to develop a similar game by themselves, and finally a preliminary assessment including the questionnaire and the concept map testing will be conducted. The experimental results showed that the given appropriate learning scaffolding can guide students to stimulate the creativity and complete their own cybersecurity board game in a short period of time. Besides, the questionnaire survey result also showed that about 80% of students are very interested in the board game design course, and that they can be able to understand the most frequent attacking techniques.

Keywords: Board game · Board game tool kit · Design thinking
Board game assisted design · Game learning

1 Introduction

Due to the rapid changes in cyber-attacks, cybersecurity researchers have developed various solutions for the cybersecurity [1–3], and a lot of efforts have been devoted to develop the cyber-security learning content. For example, Paypal and VeriSign have provided user quizzes to enhance learning motivation [4, 5], and some e-commerce companies built their own e-book for cybersecurity [6–9]. However, the static content cannot stimulate the motivation of learning [10]. So, the game-based learning, which usually uses an interesting narrative and competitive exercises to motivate students learning according to specific designed learning objectives, has been extensively studied and shown to enhance learner motivation, increase participation, and thus can enhance the effectiveness of learning [11–13]. CyberCIEGE is a security awareness tool which offers a realistic virtual world in which players have to operate and defend a computer network [6, 7, 14].

L. Barolli et al. (Eds.): IMIS 2018, AISC 773, pp. 642–650, 2019.
https://doi.org/10.1007/978-3-319-93554-6_63

In recent years, the board game has been booming, because it is not just for entertainment, but is a social and learn activity. In addition, "learning by doing", focused on experiential learning, receives a lot of attention [15]. Applying "Learning by doing", students can learn to develop students' professional knowledge and problem solving abilities. Therefore, this paper shows how to help students to design a cyber-security board game to learn cybersecurity knowledge.

To design a cybersecurity board game, the student needs to combine the security knowledge with the creativity and problem solving abilities. It is usually an iterative way to refine the current prototype of the game, including the creative brainstorming activity, the evaluation, and the refinement according to the feedback.

Therefore, we propose the idea of the maker's design thinking to design the board games. In other words, students are encouraged to learn knowledge from the process of board game design, to stimulate creativity, and to achieve entertaining learning effectiveness. The web security board game called iMonsters including monsters role attack cards, virus cards, and defense cards which was developed by us several years ago can be used as an example to show the design and the scenario of defense against cyber-attacks during the game playing [16–18]. However, in this paper, the students are asked to design the cyber security board game with the board game tool kit, where the board games cards are composed of three different types of cards, including roles cards, attack cards, and attack target cards. The role cards are mainly world-renowned hacker using some common techniques for cyberattacks. After designing the board game, the students can be able to understand the most frequent attacking techniques through the study of the most famous attacking cases.

2 Related Works

2.1 Game-Based Learning

As we know, game-based learning has been extensively studied to be able to enhance learner motivation, participation and learning outcomes. Game-based learning which is mainly divided into competitions and cooperation provides the appropriate challenges, education, and fun to the users. Therefore, the students can learn the domain knowledge effectively and joyfully through the game playing. Jayakanthan and Wiebenga reported that GBL scenarios are being used for education and training purposes [19, 20]. A number of researchers have applied GBL for expertise training in various domains. For example, Mooney and Bligh applied GBL (CyberIST) to medical education [21]. In addition, Huang and Cappel applied Web-based GBL to information system training [22]. All of these applications showed that GBL can result in superior learning performance. Thus, the game-based learning has been extensively used to enhance learner motivation.

3 Cybersecurity Board Game Design Process

Recently, the science education is emphasized not only on enhancing student's understanding of the domain knowledge but also on enhancing their creativity. The maker movement initiated in 2005 encourages novel applications of technologies and the exploration by practical working among traditionally separate domains, such as metal working, wood working, computer programing, etc. Our idea is to encourage students to act as makers to utilize design thinking approach to bring their thinking and knowledge of cybersecurity into real world. We believe that according to the hands-on design process using our pre-designed tool kit, the students can be able to make an innovative cybersecurity board game and learn the knowledge of cybersecurity.

3.1 Our Cybersecurity Design Process

In this section, we propose a new cybersecurity design process together with a pre-designed board game tool kit to help students to design their own cybersecurity board game. Besides, this approach have been applied to help14 undergraduate students to develop a new cybersecurity board game in 30 min in the Teenager Hackers camp. Based on the design thinking, the Cybersecurity board game design process is shown as follows.

Step1. The teachers teach the basic knowledge of cybersecurity.

Step2. Help students to choose the appropriate type of board game.

Step3. Design the roles and the rules of the board game.

Step4. Make a prototype of the entire Board Game with the given Tool kit.

Step5. Check the consistency and the completeness of the Game Rules.

Example

Through the survey of the well-known role of hackers, we provide ten common attacking scenario as the cybersecurity story background to introduce common Internet security technologies. Therefore, students can learn cyber security knowledge through the board game design process, the steps are as follows:

Step1. The teachers teach the cybersecurity background.

Firstly, the teachers introduce the general network attack methods, the network attack events, and concepts. And then the teachers teach the cybersecurity board game ontology as shown in Fig. 1, including three different types of cards to facilitate the follow-up game rules design and testing.

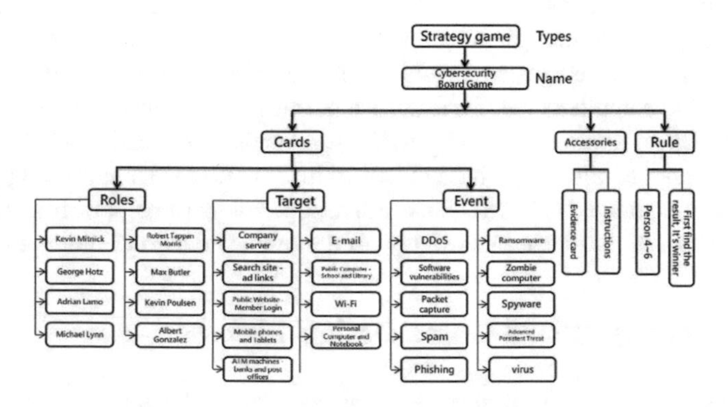

Fig. 1. Cybersecurity board game ontology

Step 2. Choose the appropriate type of board game.

Since the card game is very popular, choosing the appropriate card game can not only make students easier in the design process, but also lower their costs. Thus, the cybersecurity board game is designed with three different types of cards, including role cards, target cards and event cards as shown in Fig. 2. Accordingly, the knowledge structure of cards can be designed based on the stories of the cybersecurity.

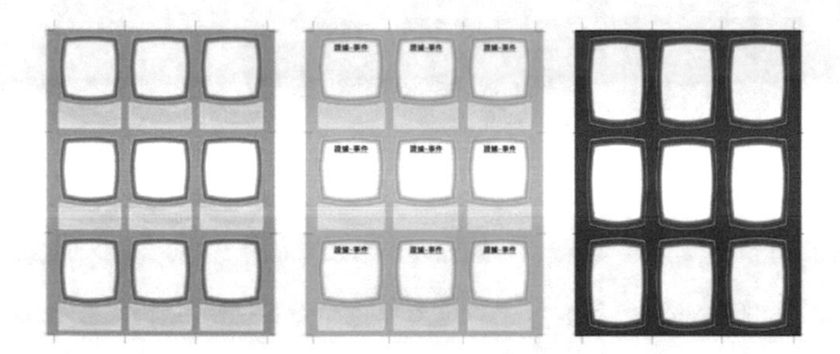

Fig. 2. Three different types of card designs

Step 3. Design the rules of the game: Accordingly, the CBR-based learning by doing scheme for developing a similar game is shown in Fig. 3.

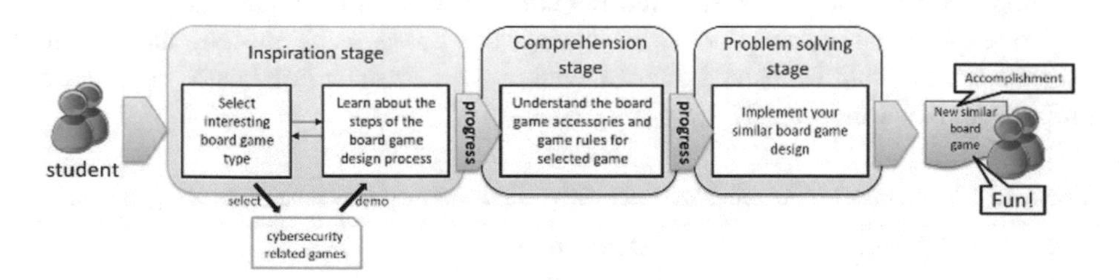

Fig. 3. The CBR-based learning by doing scheme for developing a similar game

In the design of cybersecurity board game, the student may find out one of her/his most interesting game through iteratively selecting different game type. Next, s/he can proceed to the design of cybersecurity board game and download the scenario of the selected game. Once it is well learned, s/he can further proceed to the problem solving stage to design her/his similar game. Through the demonstration of the game and the implementation of a similar game, the students can not only practice the basic skills, but also can explore the embedded knowledge of sample game, the "CLUEDO" board game, to set the rules of the initial game. For example, as the game begins, draw a card from three different types of cards and cover them on the desktop. At this point the first story has appeared on the desktop, such as the well-known hacker attacks using the attack on the target.

Step 4. Make a prototype with the Board Game Tool kit.

Create a board game prototype, and use different styles of accessories, including attack target cards, attack method cards, role cards, as shown in Figs. 4, 5 and 6 respectively.

Fig. 4. Attack target cards

Fig. 5. Attack method cards

Fig. 6. Role cards

Through discussion, card game design is performed with the board game tool kit. The initial planning uses different color card types to distinguish the cards. The role card is red consisting of the current world-famous hackers. The attack target card is orange representing the vulnerable to be attacked. The attack method card is blue, representing such as zombie computers, ransomware, and Phishing, etc.

Step 5. Check the consistency and the completeness of the Game Rules.

To explain the learning activities during the board game design, the learning subjects which could be learned after practicing the corresponding board game design process are listed in Table 1.

Table 1. The mapping of learning subjects with objects of the cybersecurity board game design

Cards type	Attributes	Description
Role	The most famous hackers in the world	By knowing the role of hackers, you can learn about the stories background of hackers
Method	Common cybersecurity attack method	Understand the most common cybersecurity attack method
Target	Common cybersecurity attack target	Understand the most common cybersecurity attack target
Game rule	Instructions	Test and modify game rules with game testing

While evaluating the game, check the balance between education, games, and knowledge through three key features of an educational board game design. The first is to check the number of cards sent to the players, and to check the elapse time is not too long to reach the goal of the game. The results may help the students to adjust the card type or the game rules.

4 Experiment and Findings

Based on the design thinking approach and convenient board game tool kits, we can carry out board game design courses. We further conducted the cybersecurity course in the Teenager Hackers camp.

The cybersecurity board game simulates the role of Internet police to help students understand cybersecurity attacks. There are three different types of cards, including role, attack goals and attack methods. The role cards are mainly world-renowned hacker; attack methods is a common technique for cyber attacks; the attack target is an object that is often attacked by hackers. One of the three cards was randomly selected, the table was covered, and the rest was mixed and distributed to everyone. The story, world-renowned hackers are committing cybercrime on the global network. Internet policemen everywhere have issued countermeasures and all hold different evidence. To prevent hackers from stealing information, they use unilateral data transmission methods to exchange evidence and find out the ultimate hackers, attack methods and targets, and use the narrative methods to discuss with the players. The game is over.

4.1 Evaluation of Concept Map

In addition to playing the board game, this cybersecurity board game can train students to observe any evidence in the game, where the cybersecurity board game concept maps describe the cybersecurity attack event in a storytelling manner. As shown in Table 2, the attack methods consists of three parts, vulnerability attacks, link attacks and denial of service attacks. The connection between two concept map nodes represents the correct attack method and attack target. Current board game attack targets fall into three categories, personal devices, public facilities, and corporate devices as shown in Fig. 7.

Table 2. Cybersecurity board game attack methods and attack targets table

Events	Target
Vulnerability attack	**Personal device**
Software vulnerabilities	Mobile phones and Tablets
Spyware	E-mail
virus	Computer and Notebook
Advanced Persistent Threat	**public utilities**
Link attack	ATM machines - banks
Packet capture	Public Computer – School or Library
Spam	Wi-Fi
Phishing	**company device**
Denial of service attack	Company server
Ransomware	Search site - ad links
Zombie computer	Website - Member Login
DDoS	

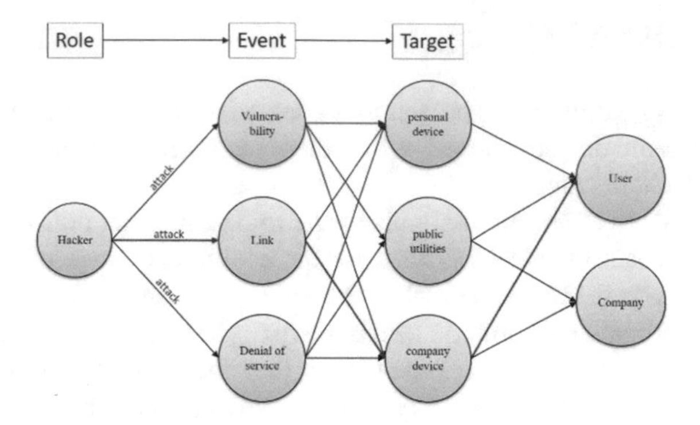

Fig. 7. Concept map of the cybersecurity board game

If the player successfully finds the answer and explains that the hackers use the phishing attack method to perform data theft on the member login system at the end of the game, the attack may lead to user data leakage. Therefore, according to the concept map of cybersecurity, member login system belongs to the company device type (server). The connection is correct, indicating that the player has understood the Internet hacking attack strategy and Attack targets.

For example, the three different types of cards on the table, the role is Max Butler, the event is Phishing, and the target is public website member login. If the player succeeds in finding the attack evidence, s/he must say "Max Butler used phishing attack to the public website member login system.", and explain the situation with other players. For example, the three different types of cards covered in the table, the role is Albert Gonzalez, the event is virus, and the target is ATM machines banks. If the player succeeds in finding out the attack evidence, s/he must say "Albert Gonzalez used virus attack to the bank ATM machines.", and explain the situation with other players (Fig. 8).

Role	Event	Target	Storytelling
Kevin Mitnick	Ransomware	Personal Computer	Kevin Mitnick used ransomware attack to the personal computer.
George Hotz	Zombie computer	Public Computer - School	George Hotz used zombie computer attack to the school computers.
Adrian Lamo	Spyware	Wi-Fi	Adrian Lamo used spyware attack to the wi-fi.
Michael Lynn	Spam	E-mail	Michael Lynn used spam attack to the e-mail.
Max Butler	Phishing	Public Website - Member Login	Max Butler used phishing attack to the public website member login system.
Kevin Poulsen	DDoS	Company server	Kevin Poulsen used DDoS attack to the company server.
Albert Gonzalez	virus	ATM machines banks or post offices	Albert Gonzalez used virus attack to the bank ATM machines.

Fig. 8. Storytelling of the cybersecurity board game

4.2 Questionnaire

After the completion of the board game design course, we conduct a preliminary assessment according to the questionnaire. The first question wants to know the satisfaction rate of the board game design content. The second one is to find out the relevance between the content and the cybersecurity related knowledge. The third one is to know whether the board game design course is interesting or not? The analysis of the survey found that in the board game design teaching course, the good and very satisfactory is about 86%, indicating that the students were very interested in the learning process. According to cybersecurity-related knowledge content, the good and very satisfactory is about 79%, indicating that students can learn cybersecurity-related knowledge in game playing. As to the interestingness of the board game design course, the good and very satisfactory is about 79%, indicating that students are very supportive of board game design courses. In addition, students said that such design courses are very interesting and they can design their own board games by themselves (Figs. 9, 10 and 11).

Fig. 9. The semi-finished board game design teaching content

Fig. 10. This board game attached to cybersecurity related knowledge

Fig. 11. Did you find this board game design course interesting?

5 Conclusion

In this paper, we proposed an innovative board game design process to help students to design a cybersecurity board game with a pre-designed board game tool kit by using the design thinking and learning-by-doing strategies. According to the students' characteristics, exploring the appropriate cybersecurity cases as the learning scaffolding to trigger students' creative thinking can achieve the goal of learning the knowledge of cybersecurity. To evaluate the performance of the design course, 14 students have participated in the experiment. The experimental results showed that the given appropriate learning scaffolding can guide students to stimulate the creativity and complete their own cybersecurity board game in a short period of time.

Acknowledgment. This search was partially supported by the National Science Council of the Republic of China under grants MOST 106-2511-S-468-004-MY2 and MOST 106-2511-S-468-002-MY3.

References

1. Qi, M., Zou, C.-Y.: A study of anti-phishing strategies based on TRIZ. In: International Conference on Networks Security, Wireless Communications and Trusted Computing, pp. 536–538 (2009)
2. Wilson, C., Argles, D.: The fight against phishing: technology, the end user and legislation. In: Proceedings of IEEE, pp. 501–503 (2011)
3. Tseng, S.-S., et al.: Building a frame-based anti-phishing model based on phishing ontology. In: AIT (2013)
4. Sheng, S., Magnien, B., Kumaraguru, P., Acquisti, A., Cranor, L.F., Hong, J., Nunge, E.: Anti-phishing phil: the design and evaluation of a game that teaches people not to fall for phish. In: Proceedings of the 3rd Symposium on Usable Privacy and Security, pp. 88–99. ACM, July 2007
5. Paypal (2011). https://www.paypal.com/au/cgibin/webscrcmd = xpt/Marketing/securitycenter/antiphishing/CanYouSpotPhishing-outside
6. Bay (2011). http://pages.ebay.com/education/spooftutorial
7. Microsoft (2011). http://www.microsoft.com/athome/security/email/phishing.mspx
8. Dennis, D., William, R.: "Games are made for fun": lessons on the effects of concept maps in the classroom use of computer games. Comput. Educ. **56**(3), 604–615 (2011)
9. OnGuard Online (2011). http://www.onguardonline.gov/games/phishing-scams.aspx
10. Gorling, S.: The myth of user education. In: Proceedings of the 16th Virus Bulletin International Conference (2006)
11. Admiraal, W., et al.: The concept of flow in collaborative game-based learning. Comput. Hum. Behav. **27**(3), 1185–1194 (2011)
12. Papastergiou, M.: Digital game-based learning in high school computer science education: impact on educational effectiveness and student motivation. Comput. Educ. **52**, 1–12 (2009)
13. Tseng, S.-S., et al.: Building a game-based internet security learning system by ontology crystallization approach. In: Proceedings of the International Conference on e-Learning, e-Business, Enterprise Information Systems, and e-Government (EEE), p. 98. The Steering Committee of the World Congress in Computer Science, Computer Engineering and Applied Computing (WorldComp), January 2015
14. Wilson, C., Argles, D.: The fight against phishing: technology, the end user and legislation. In: 2011 International Conference on Information Society (i-Society), pp. 501–504. IEEE, June 2011
15. AEE: Association for Experiential Education. AEE definition of experiential education. AEE Horizen **15**(1), 21 (1995)
16. Yang, T.-Y., Tseng, S.-S., Wang, Y.-J., Liao, W.-L.: iMonsters gaming portfolio analysis. In: Global Chinese Conference on Computers in Education 2017 (2017)
17. Wang, Y.-J., Tseng, S.-S., Yang, T.-Y., Weng, J.-F.: Building a frame-based cyber security learning game. In: MobiSec 2016 (2016)
18. Lee, T.-J., et al.: Game-based anti-phishing training. In: TWELF 2010 (2010)
19. Wiebenga, S.R.: Guidelines for selecting, using, and evaluating games in corporate training. Perform. Improve. Q. **18**(4), 19–36 (2005)
20. Jayakanthan, R.: Application of computer games in the field of education. Electr. Libr. **20**(2), 98–102 (2002)
21. Mooney, G.A., Bligh, J.G.: CyberIST: a virtual game for medical education. Med. Teach. **20** (3), 212–216 (1998)
22. Huang, Z., Cappel, J.J.: Assessment of a web-based learning game in an information systems course. J. Comput. Inf. Syst. **45**(4), 42–49 (2005)

The 8th International Workshop on Future Internet and Next Generation Networks (FINGNet-2018)

A Localization Algorithm Based on Availability of Direct Signal from Neighbor Anchor Nodes in a Sensor Networks

Megumi Yamamoto[1(✉)] and Shigetomo Kimura[2]

[1] Graduate School of Systems and Information Engineering, University of Tsukuba, Tsukuba, Japan
s1720744@s.tsukuba.ac.jp

[2] Faculty of Systems and Information Engineering, University of Tsukuba, Tsukuba, Japan
kimura@netlab.cs.tsukuba.ac.jp

Abstract. When using information observed by sensor nodes in a sensor network, it is important in many cases to know the positions of the nodes. One of the well-known methods of estimating node positions is APIT (Approximate Point-in-Triangulation Test). To improve APIT, the authors previously proposed a localization algorithm based on the positions and received signal strengths of neighbor nodes. However, this algorithm leaves the positions of many nodes unknown. To solve this problem, this paper proposes a new localization algorithm which is based on the availability of direct signals from neighbor anchor nodes. Our simulation experiments have confirmed the effectiveness of the proposed algorithm.

1 Introduction

As wireless terminals have become smaller and cheaper in recent years, many sensor nodes have been installed to build sensor networks in order to monitor the status of environments and facilities. It is important in most cases to identify the observation positions, but generally sensor nodes' positions are not input in advance. If a sensor node is to be equipped with GPS (Global Positioning System) capability, it becomes costly and consumes a lot of power [1]. It is desirable that the position of each sensor node can be estimated without using GPS [2]. One of the well-known range-free localization algorithms that is relatively accurate and stable is APIT (Approximate Point-in-Triangulation Test) [3]. This algorithm assumes that a certain percentage of nodes are special nodes, called "anchor nodes," and that they are dispersed in the sensor network. Anchor nodes are those equipped with GPS and have a wider signal transmission range than normal nodes. A normal node estimates its position by receiving signals from anchor nodes and building triangles with anchor nodes as their vertices. However, if a normal node is found to be outside of all the constructed triangles, it cannot estimate its position correctly. The authors extended APIT and proposed an improved localization algorithm based on the positions and received signal strength indicators of surrounding nodes that have already estimated their positions

L. Barolli et al. (Eds.): IMIS 2018, AISC 773, pp. 653–664, 2019.
https://doi.org/10.1007/978-3-319-93554-6_64

correctly. This algorithm is called "the previous algorithm" in this paper. However, this algorithm leaves the positions of many nodes unknown [4]. It has another problem in that the surrounding nodes consume a lot of power because they must temporarily send signals at higher strength than usual for other nodes. To solve these problems, this paper proposes a new localization algorithm that is based on the availability of direct signals from neighbor anchor nodes [5].

This paper is constructed as follows. Section 2 introduces localization algorithms for sensor networks and describes APIT and the previous algorithm. Section 3 proposes an improved localization algorithm. Section 4 presents simulation experiments that have confirmed the effectiveness of the proposed algorithm. Finally, Sect. 5 gives conclusions and indicates future research.

2 Localization Algorithms for Sensor Networks

Localization algorithms for sensor networks can be classified into range-based and range-free algorithms. Those in the former category estimate the positions of nodes by measuring absolute distances between pairs of nodes. Well-known algorithms in this category include the Time of Arrival (TOA) method, which estimates node positions by estimating distances from signal propagation time between pairs of nodes, and the Arrival of Angle (AOA) method, which estimates node positions by measuring signal arrival angles. These methods achieve a relatively high degree of estimation accuracy but require expensive hardware to be installed in each node. Thus, they are too costly for large-scale sensor networks. There are methods in this category that do not require additional hardware. For example, the Received Signal Strength Indicator (RSSI) scheme converts the strength of the signal received from a sender into the distance from that sender, but their degree of accuracy depends on the particular network environment.

The algorithms in the latter category estimate the relative positions of nodes from connectivity between nodes. They require anchor nodes, which are nodes that know their positions, to be distributed in the network. Normal nodes estimate their absolute positions from the positions of the anchor nodes. Several range-free algorithms have been proposed. For example, the Centroid algorithm [6] estimates that a normal node's position $\left(X_{e,} Y_{e}\right)$ is the median point $\left(\frac{\sum_{i=1}^{A} x_i}{A}, \frac{\sum_{i=1}^{A} y_i}{A}\right)$ of the coordinates of all the observed anchor nodes, where A is the number of observed anchor nodes and $\left(x_{i,} y_i\right)$ is the coordinates of the i - th observed anchor node. The calculation load is extremely light because all that is required is to add coordinates and make two divisions. However, the estimation error can be large if the number of observed anchor nodes is small. In the DV-Hop algorithm [7], normal nodes forward the signals they have received from anchor nodes to other normal nodes. Each normal node determines the minimum hop count from each anchor node and estimates its position from the hop counts from three anchor nodes. The estimation accuracy is low if normal nodes are randomly distributed. In addition, nodes consume a lot of power because they must transmit many signals.

The next section describes the two algorithms with which the algorithm proposed in this paper will be compared: APIT and the previous algorithm.

2.1 APIT

APIT (Approximate Point-in-Triangulation Test) [3] is a range-free algorithm with a relatively high degree of accuracy and stability. It assumes that a certain percentage of nodes are special nodes, called "anchor nodes," and that they are dispersed in the sensor network. Anchor nodes are those equipped with GPS and have a wider signal transmission range than normal nodes. A normal node receives a beacon signal from each anchor node, and constructs triangles with anchor nodes as their vertices, as shown in Fig. 1.

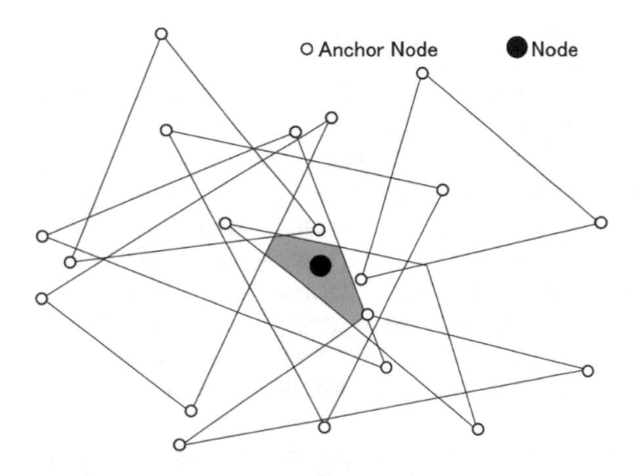

Fig. 1. Example of position estimation by APIT.

APIT also assumes that the node density is relatively high. Normal nodes exchange information about the signal strengths of the beacons from anchor nodes with neighbor nodes. A normal node determines that it is outside a triangle if there exists a neighbor normal node for which the distances from all the anchor nodes that make up the triangle are shorter or the distances from the all the anchor nodes are longer than the corresponding distances from these anchor nodes to the original normal node. Otherwise, it determines that it is inside the triangle. Consider all the triangles where the normal node is outside of these triangles. The area where the regions of these triangles overlap is removed from the region where the regions of all the triangles, inside of which is the node, overlap. Among the resulting regions, select the one with the most overlaps. The position of the node is assumed to be at the center of gravity of this selected region. If a node is outside all of the constructed triangles, its position cannot be estimated. Such a node is called an "unknown node" in this paper. Nodes tend to become unknown nodes if they are located on the periphery of the sensor network or if the density of the neighbor nodes is low.

2.2 Previous Algorithm

To solve APIT's problem, namely, that the positions of many nodes remain unknown, the authors extended APIT and proposed a localization algorithm based on the positions and received signal strength indicators of surrounding nodes that have already estimated their positions correctly. This algorithm assumes that normal nodes can temporally transmit a signal with higher power than usual [4].

After applying APIT, the previous algorithm carries out the following for each unknown node, a:

1. a broadcasts an auxiliary beacon with higher power.
2. A normal node that has already estimated its position (called an "identified node" in this paper) sends back its coordinates and the signal strength of the auxiliary beacon from a.
3. a draws lines through all pairs of identified nodes that replied in 2. Let $v = (x_v, y_v)$ be the intersecting point of two lines.
4. Let n_{far} and n_{max} be, respectively, the node with the minimum received signal strength and the node with the maximum received signal strength among the nodes that replied in 2. Let d be the distance between these two nodes. If the vertex obtained in 3 is outside of the square formed with $n_{max} = (x_{max}, y_{max})$ at its center, i.e., if it does not satisfy $x_{max} - d < x_v < x_{max} + d$ and $y_{max} - d < y_v < y_{max} + d$, it is eliminated from the following process because it is deemed too far from a and thus unreliable as the source of position information. Consider $v = (x_v, y_v)$ which is the intersecting point of the line drawn through nodes m and n and the line drawn through nodes o and p. Let $c(v)$ be the sum of the difference between the signal strength at m and that at n and the difference between the signal strength at o and that at p.
5. Let V be a set of all vertices obtained in 4. Then, the estimated position of a is:

$$(X_e, Y_e) = \left(\frac{\sum_{v \in V} c(v) x_v}{\sum_{v \in V} c(v)}, \frac{\sum_{v \in V} c(v) y_v}{\sum_{v \in V} c(v)} \right).$$

3 Proposed Algorithm

The previous algorithm has two problems. First, there are unknown nodes if V is empty. Our simulation experiments indicated that there are many unknown nodes. Second, identified nodes consume a lot of power because they must send auxiliary beacons with higher power than usual for unknown nodes. This section proposes an improved algorithm in which the number of messages that nodes send is the same as that in APIT and nodes do not need to have an auxiliary beacon.

First of all, APIT is applied as follows:

1. Just as normal nodes do, each anchor node receives beacons from other anchor nodes and saves the coordinates contained in the beacons.
2. The beacon broadcast by an anchor node contains the anchor nodes coordinates as well as the saved coordinates of other anchor nodes.

An anchor node from which a normal node can receive a beacon is called a "neighbor anchor node" for that normal node. An anchor node the coordinates of which are contained in the beacon from a neighbor anchor node but which itself is not a neighbor anchor node is called a "remote anchor node." For example, in Fig. 2, each circle represents the cell of an anchor node (i.e., area within which the beacon from the anchor node can reach). Normal node n in the figure is within the cells of anchor nodes a and b. Thus, these two anchor nodes are neighbor anchor nodes for n. Anchor node a is within the cells of anchor nodes c and d, and b is within the cell of anchor nodes a and e. Thus, anchor nodes c, d, and e. are remote anchor nodes for n.

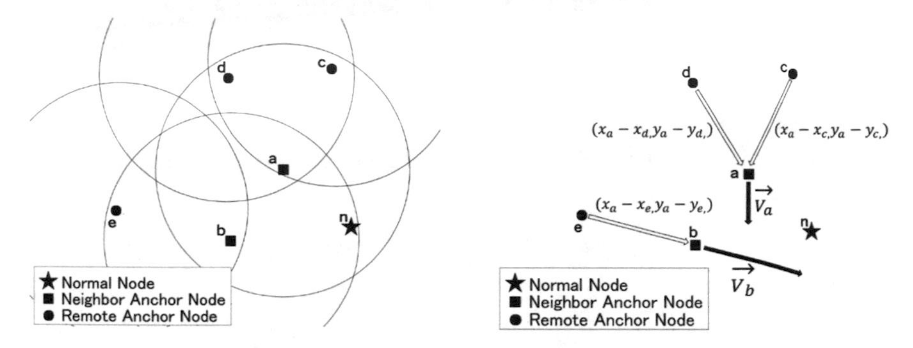

Fig. 2. Example of nodes in the proposed method.

Fig. 3. Example vectors in (a).

After APIT has been applied, there will be some unknown nodes. The proposed algorithm carries out the following for each unknown node n in order to estimate its position.

3. Let A be a set of neighbor anchor nodes for n. For $a \in A$, let $F(a)$ be a set of all remote anchor nodes contained in the beacon from a.
   (a) Calculate the composition vector $\overrightarrow{V_a}$ of the vectors from all remote anchor nodes to a. Let (x_a, y_a) be the coordinates of a and (x_f, y_f) be the coordinates of $f \in F(a)$. Then,

$$\overrightarrow{V_a} = \sum_{f \in F(a)} (x_a - x_f, y_a - y_f).$$

   Neighbor anchor node a in Fig. 2 calculates the composition vector $\overrightarrow{V_a}$ of vectors from anchor nodes c and d to itself, as shown in Fig. 3 (arrows are vectors in the figure). Neighbor anchor node b also calculates $\overrightarrow{V_b}$ using the same process.
   (b) Let $r_s(a)$ be the signal strength of the beacon sent from neighbor anchor node a, and $r_r(a)$ be the signal strength of the beacon received by n. For each $\overrightarrow{V_a}$ in (a), calculate the composition vector $\overrightarrow{V} = (V_x, V_y)$ as follows, where λ is the wavelength of the signal.

$$d(a) = \sqrt{\frac{\lambda^2}{10^{\frac{\log(r_s(a) - r_r(a))}{-10}} 16\pi^2}}$$

$$\overrightarrow{V} = \sum_{a \in A} d(a) \overrightarrow{V_a} / \left| \overrightarrow{V_a} \right|$$

As shown in Fig. 4, $\overrightarrow{V_a}$ and $\overrightarrow{V_b}$ in Fig. 3 are converted into vectors, lengths of which represent signal strengths $d(a)$ and $d(b)$, respectively. Then, the composition vector $\overrightarrow{V}$ is calculated.

(c) Let $a_{\max}$ be the neighbor anchor node the beacon from which has the strongest received signal strength $d(a_{\max})$. Then, convert $\overrightarrow{V}$ into a vector the origin point of which is the coordinates of $a_{\max}$ and for which the length is $d(a_{\max})$. The estimated position (X_e, Y_e) of n is the terminal point of this vector, i.e.,

$$(X_e, Y_e) = (x_{a_{\max}}, y_{a_{\max}}) + d(a_{\max,}) \overrightarrow{V} / \left| \overrightarrow{V} \right|.$$

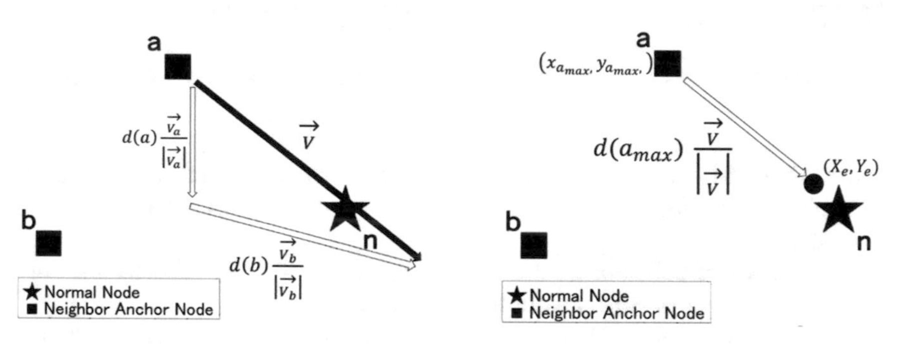

Fig. 4. Example vectors in (b).

Fig. 5. Example vector in (c).

Since a is nearer to n than b is in Fig. 4, the stronger signal strength $d(a_{\max})$ is $d(a)$. The estimated position of n is obtained by converting the length of $\overrightarrow{V}$ to $d(a_{\max})$, as shown in Fig. 5.

4 Simulation Experiments and Results

Simulation experiments were carried out to verify the effectiveness of the proposed method. Normal nodes and anchor nodes were randomly distributed in a 250 m $\times$ 250 m area. In the experiments, the proposed method was compared with APIT and the previous method in terms of position estimation accuracy and the number of unknown nodes. The simulation parameters used are shown in Table 1. The grid interval is used by the grid

Table 1. Simulation parameters.

Parameters	Values
The number of normal nodes	330
The number of anchor nodes	Varied (30, 40, and 50)
Receiving sensitivity	–76.066597 dBm
Signal strength from normal nodes	–10 dBm (for a cell radius of 20 m)
Signal strength from anchor nodes	4 dBm (for a cell radius of 100 m)
Frequency	2.4 GHz
Grid interval	2 m (= 10% of normal node's cell radius)
DOI	Varied (0, 0.3, 0.6, and 0.9%)

scan algorithm in APIT to determine whether a normal node is inside or outside a triangle of anchor nodes.

4.1 Simulation Conditions

The free-space path loss model was used to calculate the received signal strength. The received signal strength was calculated using the following equation [8], where λ is a signal's wavelength and d is the distance between the sender and receiver nodes.

$$(\text{Received signal strength}) = (\text{Sent signal strength}) - (-10\log_{10}\frac{\lambda^2}{16\pi^2 d^2})$$

As shown in Table 1, the cell radius of a normal node was about 20 m and that of an anchor node was about 100 m. It was assumed that signals from a node can reach any node within its cell without errors.

Although a cell is usually approximated by a circle, it is not exactly a circle in real environments [9]. DOI (Degree of Irregularity) was defined to model the cell's shape. DOI is the rate of change in the maximum path loss with respect to a unit change in the angle of the signal propagation direction [10]. Figure 6 shows examples of cells modified by DOI. DOI = 0 means that the cell is an exact circle. If DOI is bigger, the cell's shape becomes more irregular.

In the experiments, a DOI was set for each node. $K_i (i = 0, 1, \cdots, 359)$ was generated for each angle between 0° and 359° using the following equation, where Rand is a uniform random number between 0 and 1, and N is a set of natural numbers. K_i must be such that $|K_0 - K_{359}| \leq \text{DOI}$ is satisfied. If angle i° is not a natural number, K_i is linearly interpolated from the previous and next angles, K_t and K_s.

$$K_i = \begin{cases} 1 \text{ if } i = 0 \\ K_{i-1} \pm \text{Rand} \times \text{DOI} \text{ if } 0 < i < 360 \wedge i \in N \\ K_s + (i - s) \times (K_t - K_s) \text{ if } s = \lfloor i \rfloor \wedge t = \lceil i \rceil \text{mod}\, 360 \\ \wedge 0 < i < 360 \wedge i \notin N \end{cases}$$

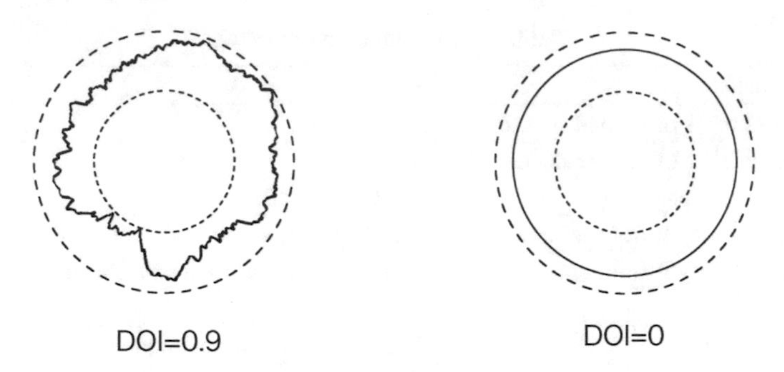

Fig. 6. Examples of cells modified by DOI.

The received signal strength is calculated as:

$$(\text{Received signal strength}) = (\text{Sent signal strength}) - K_i\left(-10\log_{10}\frac{\lambda^2}{16\pi^2 e^2}\right)$$

An anchor node identifies its position using GPS. This position may have some error. To reflect this, some experiments introduced a GPS error as shown in the following equations, where (x_g, y_g) is the anchor node's position identified using GPS, and (x, y) is the correct position, Rand is a random number between 0and 1, e is a constant indicating how large the GPS error generated by the environment is, and r is the cell radius of an anchor node. If the identified position was outside of the simulation area, (x_g, y_g) was generated again until the identified position fell inside the simulation area.

$$x_g = x \pm \text{Rand} \times (e \times r),$$
$$y_g = y \pm \text{Rand} \times (e \times r).$$

In cases where no GPS error was considered, all anchor nodes were assumed to be able to detect their correct positions.

Each normal node considers that it cannot estimate its position if it is outside all of the observed triangles in APIT, if no vertex remains after removal of unreliable nodes in the previous algorithm, and if there is no neighbor anchor node or no remote anchor node in the proposed algorithm.

4.2 Simulation Results

Under the conditions described in the previous section, each simulation experiment was executed twelve times to calculate the average estimation error ε between the correct position (x, y) and the estimated position (x_e, y_e). ε is expressed as $\varepsilon = \frac{\sqrt{(x-x_e)^2 + (y-y_e)^2}}{r}$, where r is the cell radius of a normal node. The average estimation error was calculated for APIT, the previous method, and the proposed method for cases where the average number of unknown nodes was m and the confidence interval was 90%.

Figure 7 shows the average estimation error and the number of unknown nodes for cases where DOI was 0 and the number of anchor nodes was varied from 30 to 50. The average estimation error of the proposed method was greater than that of APIT by about 0.1. The average estimation error of the proposed algorithm was greater because the estimation errors of normal nodes that can find only few anchor nodes are large. In contrast, the average number of unknown nodes when the proposed method was used was almost 0 compared to 64.1 to 89.9 in APIT and 31.5 to 68.4 in the previous algorithm. The reason is that the proposed method can estimate the position of a normal node provided the node finds at least one neighbor anchor node.

Figure 8 shows the average estimation error and the number of unknown nodes when the number of anchor nodes was fixed to 30 and the DOI was varied from 0.3 to 0.9. As the DOI was increased, the average estimation error also increased. When DOI = 0.9, the average estimation error of the proposed method was smaller than that of APIT by 0.026, and smaller than that of the previous algorithm by 0.122. The proposed method is robust against DOI because it does not use received signal strengths to determine how close a node is to neighbor nodes. In the proposed method, the average number of unknown nodes was 0. Nevertheless, the average estimation error was smaller than those of the other methods. This is because APIT and the previous algorithm are more susceptible to variations in DOI.

Figure 9 shows the average estimation error and the number of unknown nodes when constant *e* (indication of the degree of the GPS error) was varied from 0 to 1. The number of anchor nodes was 30 and DOI was 0. The average estimation error remained more or less constant in all the algorithms when *e* was 0 and 0.5. But it increased when *e* was 1. The average estimation error of the proposed method was greater than that of APIT by 0.1 to 0.13. This means that the proposed algorithm is more susceptible to GPS errors and thus requires use of highly accurate GPS sensors. The average number of unknown nodes remained more or less constant in all algorithms even when *e* was varied. The average number of unknown nodes in Fig. 9 corresponds to that for the case where the number of anchor nodes was 30 in Fig. 7. The average number of unknown nodes was relatively unaffected by GPS errors because the decision as to

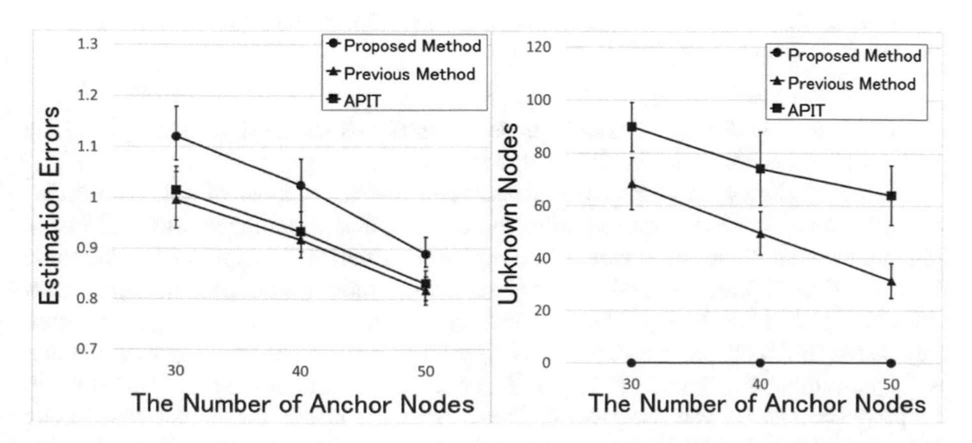

Fig. 7. Average estimation error and average number of unknown nodes when the number of anchor nodes was varied.

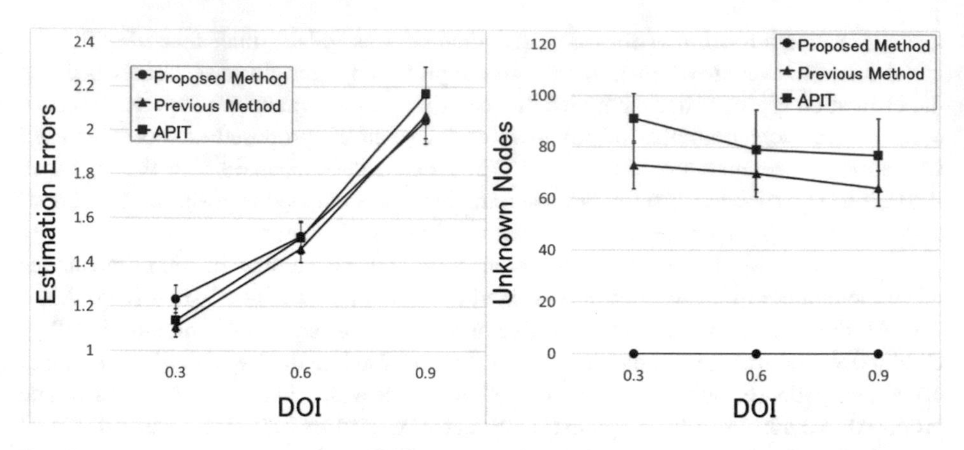

Fig. 8. Average estimation error and average number of unknown nodes when DOI was varied.

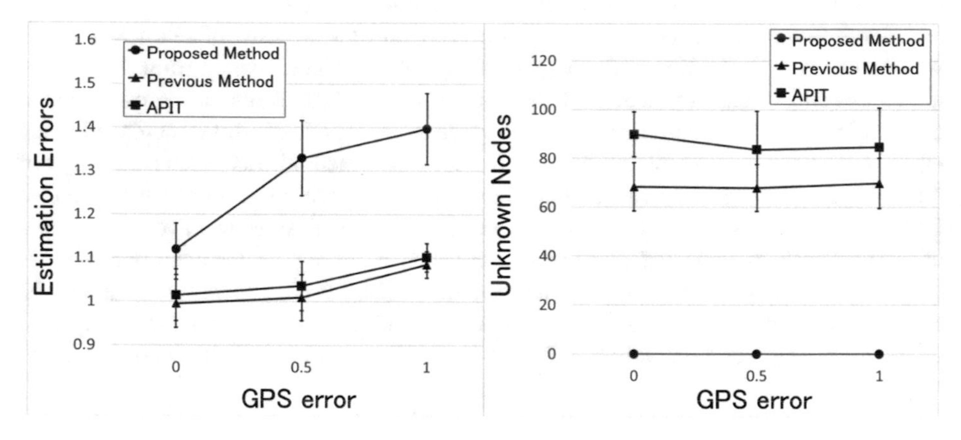

Fig. 9. Average estimation error and average number of unknown nodes when constant *e* (indicating the degree of GPS error) was varied.

whether a normal node is an unknown node depends only on the signal strengths of the beacons from anchor nodes in all the algorithms.

Figure 10 shows the average estimation error and the number of unknown nodes when the simulation area was expanded to 300 m $\times$ 300 m (compared to a 250 m $\times$ 250 m area in all of the above cases). The rest of the simulation conditions are the same those for Fig. 7. Since the node density decreased, most results became worse than those in Fig. 7 for all the algorithms. However, the difference in the average estimation error between the proposed and previous algorithms or between the proposed one and APIT was almost the same as in Fig. 7. The average number of unknown nodes when the proposed method was used was 0. These favorable results for the decreased node density can be explained by the fact that normal nodes in the proposed algorithm can obtain information about the positions of all anchor nodes that can be reached in two

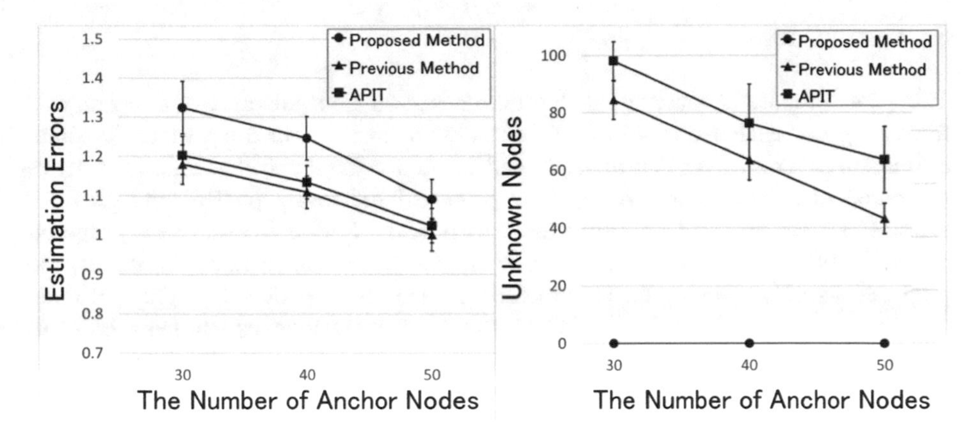

Fig. 10. Average estimation error and average number of unknown nodes when the simulation area was expanded to 300 m × 300 m.

hops rather than just one hop. Even though the simulation area was wider, the calculation time of the proposed algorithm did not increase because it depends only on the number of anchor nodes. It is important in building a sensor network that, even if the area size is changed, the ratio of the number of anchor nodes to the number of normal nodes should not be changed. As long as this ratio is maintained, the number of anchor nodes that a normal node can reach in two hops does not change.

Figure 11 shows the average estimation errors of APIT, Centroid [6], and the proposed method. The simulation conditions are the same as for Fig. 7. As mentioned in Sect. 2, the calculation load of Centroid is light, but its estimation accuracy may not be high. In fact, as shown in Fig. 11, the average estimation error was greater than that of the proposed method by about 0.2. In Centroid, each normal node can estimate its position if it can observe two or more anchor nodes. The average number of unknown nodes is not shown here because it is 0 in both Centroid and the proposed method.

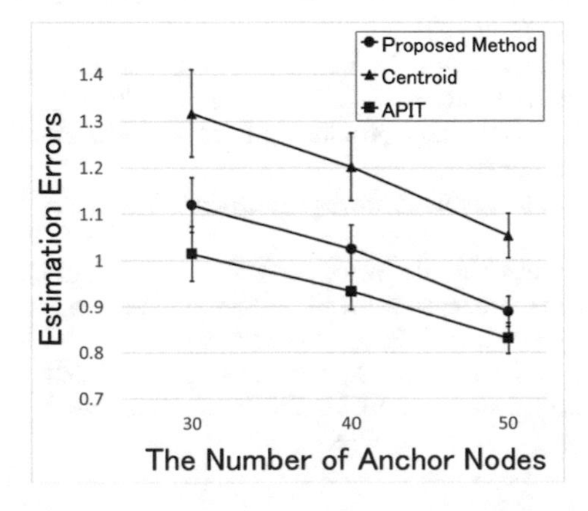

Fig. 11. Average estimation errors and average number of unknown nodes compared with those of Centroid.

5 Conclusions

This paper proposed a localization algorithm based on the availability of direct signals from neighbor anchor nodes in a sensor network in order to improve the position estimation accuracy over that of APIT and to reduce the number of unknown nodes compared to a method that the authors proposed previously (called "the previous method" in this paper). Our simulation experiments showed that the new proposed method results in a smaller estimation error and no unknown node. However, the proposed algorithm requires the GPS module in each anchor node to be highly accurate.

Going forward, the authors need to carry out simulation experiments in more network environments to confirm the effectiveness of the proposed algorithm and also need to investigate how to reduce the amount of power consumed by the proposed method.

Acknowledgment. This work was supported by JSPS KAKENHI Grant Number JP26280027.

References

1. Elkin, C., Kumarasiri, R., Rawat, D.B., Devabhaktuni, V.: Localization in wireless sensor networks: a dempster-shafer evidence theoretical approach. Ad Hoc Netw. **54**, 30–41 (2017)
2. Akcan, H., Kriakov, V., Brönnimann, H., Delis, A.: GPS-free node localization in mobile wireless sensor networks. In: Proceedings of the 5th ACM International Workshop on Data Engineering for Wireless and Mobile Access, pp. 35–42 (2006)
3. He, T., Huang, C., Blum, B.M., Stankovic, J.A., Abdelzaher, T.F.: Range-free localization and its impact on large scale sensor networks. ACM Trans. Embed. Comput. Syst. **4**(4), 877–906 (2005)
4. Yamamoto, M., Kimura, S.: Improvement of localization algorithm based on position and received signal strength indicator of surrounding nodes in sensor network. In: Proceedings of the 79th National Convention of IPSJ, vol. 2017, no. 1, pp. 329–330 (2017). (in Japanese)
5. Yamamoto, M., Kimura, S.: A localization algorithm based on availability of direct signal from neighbor anchor nodes in sensor networks. IEICE Technical report, vol. 2017, no. 1, pp. 45–50 (2018). (in Japanese)
6. Bulusu, N., Heidemann, J., Estrin, D.: GPS-less low cost outdoor localization for very small devices. IEEE Pervasive Comput. Mag. **7**(5), 28–34 (2000)
7. Niculescu, D., Nath, B.: DV based positioning in ad hoc networks. Kluwer J. Telecommun. Syst. **22**(1), 267–280 (2003)
8. Norton, K.A.: System loss in radio wave propagation. J. Res. Nat. Bur. Stand. **63D**, 53–73 (1959)
9. Ganesan, D., Krishnamachari, B., Woo, A., Culler, D., Estrin, D., Wicker, S.: Complex behavior at scale: an experimental study of low-power wire-less sensor networks. Technical report UCLA/CSD - TR02, vol. 13, pp. 1–11 (2002)
10. Zhou, G., He, T., Krishnamurthy, S., Stankovic, J.A.: Models and solutions for radio irregularity in wireless sensor networks. ACM Trans. Sens. Netw. **2**(2), 221–262 (2006)

Prediction Model of Optimal Bid Price Based on Keyword Auction Data Through Machine Learning Algorithms

Minjun Ji[1] and Hyunhee Park[2(✉)]

[1] e-Glue Communications, Seocho-gu, Seoul, Korea
inciojs@gmail.com
[2] Department of Computer Software, Korean Bible University, Nowon-gu, Seoul, Korea
parkhyunhee@gmail.com

Abstract. The RTB system is a bidding system for advertising in a specific area of on-line page. A typical RTB bidding system is a system provided by Google's search engine. In this paper, we use the data of the Naver advertisement bidding, a representative Korean search engine operated by a private bidding for the RTB system. Especially, in case of online keyword advertisement, the rank can be important factor the online page when a user enters a certain keyword into a search engine. For example, if a search keyword is ranked at the top of an online page, the probability of bid being directly connected will be increased for the link of related keyword. Therefore, the bid price of the keyword is changed according to the rank of the search keyword. In the end, it is necessary to find an appropriate bid price for registering a keyword in a private bid system. In this paper, we propose a prediction modeling mechanism to predict optimal bid price of the keyword in a specific ranking of search engine. In order to predict the optimal bid price and advertising ranking on the online page, we perform feature engineering on the related data set and define the prediction model using the machine learning algorithms for the corresponding data set.

1 Introduction

The worldwide market for sponsored search advertising and display advertising reached $61 billion and billion $57.2 billion[1], respectively, in 2015. In addition, an online advertising accounts for 49.7% of total advertising execution costs, of which 45.2% is used for the advertising of search engine. Therefore, a RTB (Real Time bidding) system is increasingly imposed as a bid solution for an advertisement execution [1,2]. If an advertiser wants to launched advertisements through the search engine, the RTB system will be used by search engines such as the Google Ad [3]. In case of South Korea, when an advertiser wants to launch

[1] https://www.zenithmedia.com/zenithoptimedia-forecasts-4-1-growth-in-global-adspend-in-2013/.

L. Barolli et al. (Eds.): IMIS 2018, AISC 773, pp. 665–674, 2019.
https://doi.org/10.1007/978-3-319-93554-6_65

a search advertisement through the Naver search engine, the advertisement bid is generally executed through the RTB system named Naver Search AD. For example, when a keyword advertisement is executed using the Naver search AD, n (e.g., 0–20) ads may be executed for one search keyword. The ranking of the search keyword can vary depending on various factors including the bid price. The search keyword of insurance, for instance, can be have various bid prices according to each rank (e.g., $6 for 1st rank of the online page, $5 for 2nd rank of the online page, and $4 for 3rd rank of the online page, etc.). However, since the advertisement bid process is conducted privately as a blind competition (i.e., bids are mutually sealed and connot be shared among advertisers), it is difficult to know how much to bid for the advertisement at the desired rank. It is also difficult to predict bid prices because bid prices can form different prices for different time zones.

In this paper, we propose a feature engineering method that predicts bid price of keyword search advertisement and a prediction model of bid price using machine learning algorithms. In particular, we present the results of prediction model for bid price using the linear regression, logistic regression, softmax, and artificial neural networks (ANNs) in the Tensor Flow. Finally, we propose the most accurate and suitable model to predict the bid price of search advertisement in real time.

The rest of this paper is organized as the followings. Section 2 reviews relevant literature from several perspectives. Section 3 illustrates the construction of the proposed prediction model based on related variables. Section 4 presents the empirical results of applying the model to predict of a search keyword and compares those results with Tensor Flow methods. The final section summarizes the findings and discusses the potential extensions of the research.

2 Related Work

The keyword auction as a form of online advertisement offers good targeting, low cost, and easy scalability [4]. The keywords selected by the advertisers to represent characteristics of their products or services are associated with an ad that appears as a sponsored link (i.e., paid listing) at the right side of the result page in a search engine in response to users query. Given a similar quality score of the landing page (used by Google or Naver) the higher the ranking, the higher the price to pay because top positions typically elicit more clicks. Now empirical models that attempt to exhibit optimization and prediction of keywords auctions have also gained attention. To our knowledge, few papers directly discuss keyword data modeling, hence greatly motivating this research. Auerbach *et al.* who state that most advertisers on Yahoo use return on investment (ROI, popular performance metric used by advertising professionals)-based strategies, but others bid with second prices may not opt for them [5].

Unlike regular media advertising, the keyword advertisement provides information tailored to the purpose of individual consumers. An advertising expense is measured as a cost per click (CPC) cost when a consumer clicks on a link in a search result. The weighted-sum method is mainly used [3]. The ranking of keyword is mainly determined by taking into account the maximum bid price, the click rate, and other related factors. The CPC-based keyword advertisement has a cost advantage that the expenditure of the advertisement cost occurs only when the user clicks [3]. The keyword advertisement that bids a relatively high amount are ranked at the top of search engine. However, the cost per click (CPC), which is the actual cost of the advertising, can be different from the bid price offered in the keyword auction. As found by Stepanchuk [6], the same bids can result in different predicted positions under different response models. In terms of prediction robustness, the semilogarithmic model is reported to be much better than others, but it is insufficient to predict ROI because of its simplicity. However, a higher position does not necessarily incur more clicks if advertisers have different reputations [7], and different impression choices result in different click-through rates (CTR) within and across different ad channels [8]. Graepel *et al.* developed a Bayesian probit regression model to predict the CTR of paid search ads and obtain a good result [9]. However, the binary feature vectors associated with ad impressions may exclude some critical category variables (e.g., position) from the model. Also, the complicated factor-graph training creates greater difficulty in the parameters estimation. The aforementioned empirical research provides a sound basis to further explore ROI prediction in keyword auctions.

3 Prediction Model of Bid Price Through Machine Learning Algorithm

3.1 Data Setup Procedure

In order to collect keyword auction data, a crawling method is used from client nodes generated keyword search results. When multiple client nodes generate the same keyword search results, they will be stored as one keyword data value. At this time, the number of occurrences of the same keyword is counted and stored. All collected keyword data is stored through the main data server.

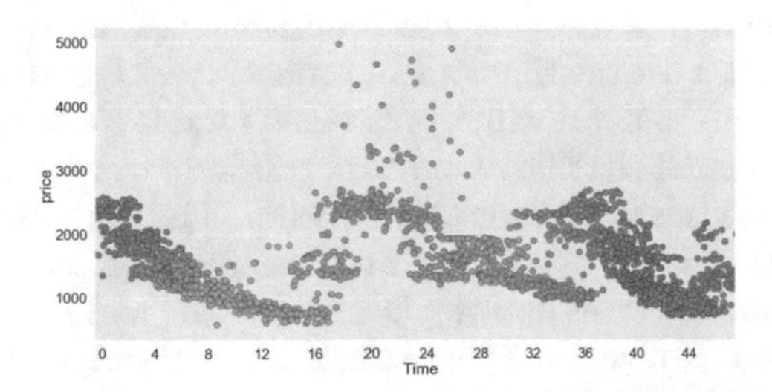

Fig. 1. Time series result of specific search keyword depends on price.

3.2 Data Feature Engineering

Even though the keyword data is the same keyword, a price differential occurs with time. In particular, keyword data varies in price depending on weekdays and weekends. In addition, it can be seen that, even if the same keyword is used for 24 h, a large price differential occurs over time. Figure 1 shows that the price of the keyword data changes over a 24-h period. As a result, it is confirmed that time is a major factor in the prediction model. In this paper, we divide 24 h into 30 min and divide 48 time units into time data sets. To do this, 48 time data sets are implemented in the form of one-hot encoders. Since the language of keyword data in Korean needs to be processed, the keyword data is converted into numerical values using the word2vec library. Through this process, the keyword data set is composed of a combination of independent variables and dependent variables. The data set of x (e.g., input), which consists of independent variables, contains the value of the price and time, and the data set of y (e.g., output), which consists of the dependent variable, contains the rank value of the keyword data.

3.3 Constructing Machine Learning Algorithms for Keyword Auction

We apply a hypothesis, loss function (or cost function) and optimizer to perform the machine learning modeling. In general, there are various optimization algorithms on which the gradient descent optimizer is based [10]. In this paper, we use the Adam algorithm for the optimization method. The learning rate is the process of finding the optimal solution that minimizes the loss function. When the learning rate is too large, overshooting can be occurred without converging to the optimal value, and when the learning rate is too small, the convergence rate will be too slow and the local minimum problem can be occurred. In this paper, we propose a process to find the appropriate learning rate with the Adam optimization algorithm.

In this paper, we analyze four machine learning algorithms to find the optimal bid price of keyword search advertisement. First, the linear regression algorithm is applied with the hypothesis and loss function. Second, we apply the logistic regression algorithm with the Sigmoid function instead of the multiple matrix. The logistic regression is, in principle, optimized for one class separation. In order to apply the multinomial classification form, third, we can apply the multinomial classification (i.e., softmax regression) algorithm. The multinomial classification algorithm is applied for the hypothesis and loss function as the same of logistic regression. However, the cross entropy is used for the loss function in the multinomial classification. Unlike these three algorithms, the ANN algorithm is based on the deep learning model. It is applied as a multilayer perceptron (MLP) with several layers in a perceptron composed of one layer. Basically, the ANN performs forward propagation by putting training data set into input to perceptron. In this case, the error value is calculated as the difference between the prediction value of the resultant Neural networks and the target value, and this can be referred to as the application of the back propagation (chain rule) algorithm. In MLP, it is expected that the accuracy will be improved by increasing the hidden layer. However, a vanishing gradient problem which may lose the previous slope value in the back propagation may occur. To do this, we apply the RELU function to solve the vanishing gradient problem. In this paper, we set the class value of input data to 151 (e.g., price (1), time set (48), weekdays or weekends (1), keyword name with word2vec (100)). When the input data comes out of the input layer and enters the first hidden layer, it comes out as 256 classes, which are then moved back into the second hidden layer. Likewise, we configure the third hidden layer to enter 256 classes, then configure the neural networks to output 15 class values as the rank of keyword search data from the output layer.

4 Extensive Simulation Results

Table 1 shows the experimental environment used to evaluate the performance of the prediction model for the keyword auction data. The keyword auction data set consists of about 9 million data, of which 70% is used for training data and 30% is used for test data. Figure 2 shows a sample for training data.

Table 1. Simulation environment

Category	Value
OS	Windows 10 Pro
CPU	Intel Core i5-8500 2.80 GHz 64bit
GPU	GeForce GTX1070 EXOC D5 8 GB
Memory	10 GB
Language	Python 3.6
Library	Tensor Flow 1.7

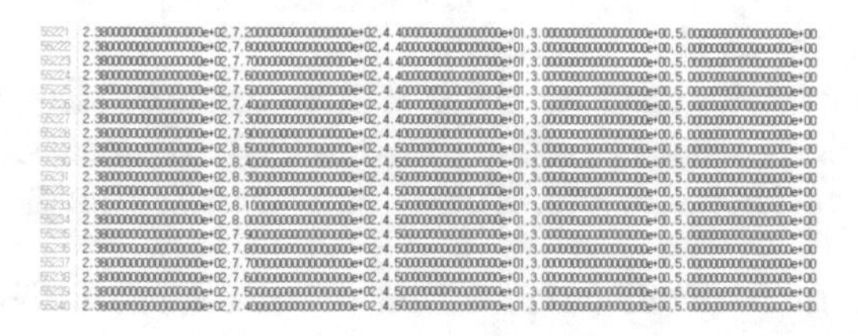

```
55221 2.380000000000000000e+02,7.200000000000000000e+02,4.400000000000000000e+01,3.000000000000000000e+00,5.000000000000000000e+00
55222 2.380000000000000000e+02,7.800000000000000000e+02,4.400000000000000000e+01,3.000000000000000000e+00,6.000000000000000000e+00
55223 2.380000000000000000e+02,7.700000000000000000e+02,4.400000000000000000e+01,3.000000000000000000e+00,5.000000000000000000e+00
55224 2.380000000000000000e+02,7.600000000000000000e+02,4.400000000000000000e+01,3.000000000000000000e+00,5.000000000000000000e+00
55225 2.380000000000000000e+02,7.500000000000000000e+02,4.400000000000000000e+01,3.000000000000000000e+00,5.000000000000000000e+00
55226 2.380000000000000000e+02,7.400000000000000000e+02,4.400000000000000000e+01,3.000000000000000000e+00,5.000000000000000000e+00
55227 2.380000000000000000e+02,7.300000000000000000e+02,4.400000000000000000e+01,3.000000000000000000e+00,5.000000000000000000e+00
55228 2.380000000000000000e+02,7.900000000000000000e+02,4.400000000000000000e+01,3.000000000000000000e+00,6.000000000000000000e+00
55229 2.380000000000000000e+02,8.500000000000000000e+02,4.500000000000000000e+01,3.000000000000000000e+00,6.000000000000000000e+00
55230 2.380000000000000000e+02,8.400000000000000000e+02,4.500000000000000000e+01,3.000000000000000000e+00,5.000000000000000000e+00
55231 2.380000000000000000e+02,8.300000000000000000e+02,4.500000000000000000e+01,3.000000000000000000e+00,5.000000000000000000e+00
55232 2.380000000000000000e+02,8.200000000000000000e+02,4.500000000000000000e+01,3.000000000000000000e+00,5.000000000000000000e+00
55233 2.380000000000000000e+02,8.100000000000000000e+02,4.500000000000000000e+01,3.000000000000000000e+00,5.000000000000000000e+00
55234 2.380000000000000000e+02,8.000000000000000000e+02,4.500000000000000000e+01,3.000000000000000000e+00,5.000000000000000000e+00
55235 2.380000000000000000e+02,7.900000000000000000e+02,4.500000000000000000e+01,3.000000000000000000e+00,5.000000000000000000e+00
55236 2.380000000000000000e+02,7.800000000000000000e+02,4.500000000000000000e+01,3.000000000000000000e+00,5.000000000000000000e+00
55237 2.380000000000000000e+02,7.700000000000000000e+02,4.500000000000000000e+01,3.000000000000000000e+00,5.000000000000000000e+00
55238 2.380000000000000000e+02,7.600000000000000000e+02,4.500000000000000000e+01,3.000000000000000000e+00,5.000000000000000000e+00
55239 2.380000000000000000e+02,7.500000000000000000e+02,4.500000000000000000e+01,3.000000000000000000e+00,5.000000000000000000e+00
55240 2.380000000000000000e+02,7.400000000000000000e+02,4.500000000000000000e+01,3.000000000000000000e+00,5.000000000000000000e+00
```

Fig. 2. An example of training dataset used experiments.

Figure 3(a) shows the accuracy of training data using the Adam optimization algorithm for the ANN. When the Adam optimization algorithm is applied, it can be seen that the accuracy increases step by step as the epoch increases.

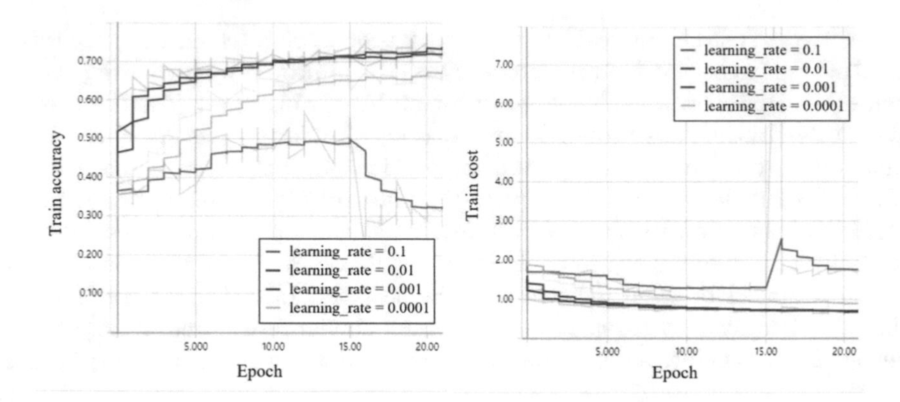

Fig. 3. (a) Simulation result of training accuracy in case of ANN with the Adam optimization algorithm, (b) Simulation result of training cost in case of ANN with the Adam optimization algorithm.

However, it can be seen that the optimum learning rate is applied to the Adam optimization algorithm to find the optimum result. For the accuracy of the training data in Fig. 3, it can be seen that the accuracy is highest when the learning rate is 0.001. In particular, when the learning rate is 0.01, the accuracy is reached quickly, but when the learning rate is 0.001, a higher accuracy is obtained.

Figure 3(b) shows the result of the cost function when the Adam optimization algorithm is applied to the ANN. It can be seen that the cost gradually reaches the minimum value as the Epoch increases. As the learning accuracy of the training accuracy is 0.001, it shows the highest accuracy and the cost result reaches the minimum value when the learning rate is 0.001. However, according to the result of Fig. 3(b), it can be seen that the learning rate converges to the minimum value faster when the learning rate is 0.01.

Figure 4(a) shows the accuracy of the test data set to determine the accuracy of the model. As shown in Fig. 3(a), we applied the Adam optimization algorithm and showed the highest accuracy when the learning rate was 0.01 and 0.001, respectively. Especially, when the learning rate is 0.01, it is found that the accuracy value is reached quickly.

Figure 4(b) shows the result of the cost function according to the test accuracy of the model. When the learning rate is 0.01 as same as the accuracy of the test data set as shown in Fig. 4(b), it can be seen that it converges to the minimum value.

Table 2 shows more detailed simulation results. As mentioned above, it can be seen that the accuracy and cost vary depending on the learning rate. In addition, when the learning rate is 0.01, the accuracy of the training data and the accuracy of the test data are the highest. Specifically, the test cost value reaches the minimum value when the learning rate is 0.01 and epochs is 10. As the epochs increase, the cost value increases again.

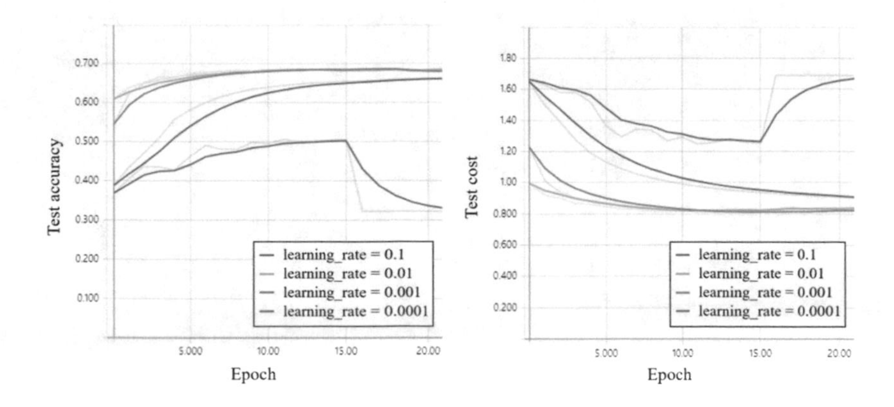

Fig. 4. (a) Simulation result of test accuracy in case of ANN with the Adam optimization algorithm, (b) Simulation result of test cost in case of ANN with the Adam optimization algorithm.

Table 2. Simulation results of ANN algorithm with the Adam optimization algorithm

Learning rate	Epoch	Training accuracy	Training Cost	Test accuracy	Test cost
0.1	5	0.461	1.523	0.463	1.365
0.1	10	0.473	1.343	0.495	1.292
0.1	15	0.520	1.259	0.505	1.255
0.01	5	0.682	0.779	0.675	0.846
0.01	10	0.719	0.706	0.686	0.819
0.01	15	0.713	0.696	0.682	0.822
0.001	5	0.679	0.837	0.671	0.863
0.001	10	0.710	0.772	0.685	0.818
0.001	15	0.705	0.756	0.685	0.812
0.0001	5	0.581	1.140	0.581	1.130
0.0001	10	0.654	1.000	0.640	0.991
0.0001	15	0.623	0.961	0.655	0.936

In this paper, the results of applying linear regression, logistic regression, and multinomial classification (softmax regression) are additionally described. In order to compare the results of the three regression algorithms, we applied the Adam optimization algorithm and applied the same learning rate of 0.01. Figure 5(a) compares the accuracy of training data according to the linear regression, logistic regression, and softmax regression. According to the results in Fig. 5(a), the accuracy of the training data shows that the multinomial classification algorithm has the highest accuracy. Figure 5(b) compares the accuracy of the test data for the three regression algorithms. According to the results

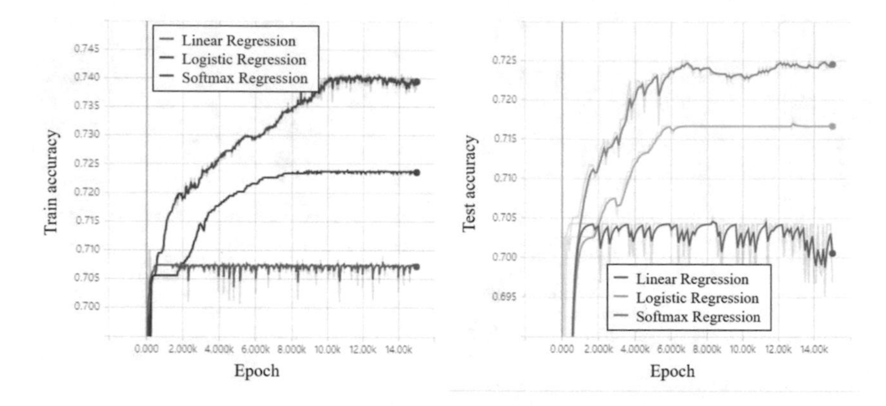

Fig. 5. (a) Simulation result of training accuracy in case of regression algorithms with the Adam optimization algorithm, (b) Simulation result of test accuracy in case of regression algorithms with the Adam optimization algorithm.

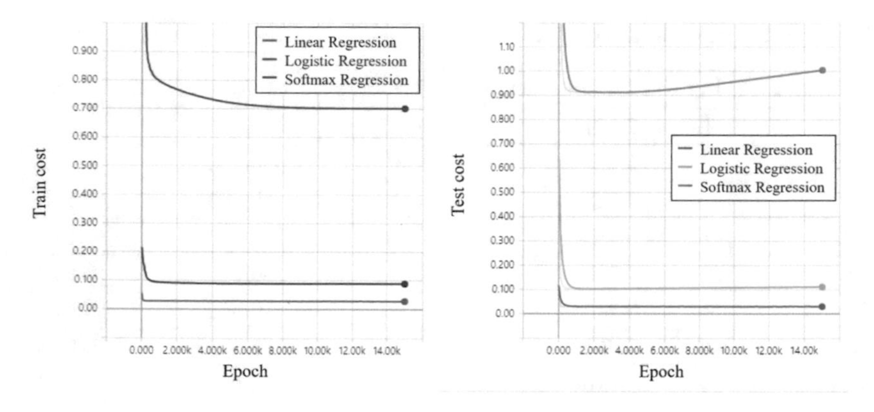

Fig. 6. (a) Simulation result of training cost in case of regression algorithms with the Adam optimization algorithm, (b) Simulation result of test cost in case of regression algorithms with the Adam optimization algorithm.

in Fig. 5(b), the accuracy of the test data and the multinomial classification algorithm show the highest accuracy.

Figure 6(a) and (b) show the cost of training data and the test data for the three regression algorithms, respectively. As a result, after finding the minimum value for each of the three regression algorithms, it can be seen that the cost increases again as the step increases. The cost values shown in Fig. 6(a) and (b) can be compared as an independent result for each algorithm. For example, the cost value for the test data of multinomial classification reaches a minimum at the step of 8000, and then the cost increases again. In addition, it can be confirmed that the cost value for the test data of the logistic regression reaches the minimum value at the step of about 670.

In this paper, we compare the price prediction results of keyword data by applying four algorithms. One of the typical algorithms of the deep learning, the ANN algorithm is applied to a total of 94,258 datasets and training data (65,980) and test data (28,278) are 70% and 30%, respectively. For the rest of the regression algorithms, we compare the results of pricing predictions using linear regression, logistic regression, and multinomial classification. The regression algorithm used 4,023 datasets in total, with training data (2,816) and test data (1,207) as 70% and 30%, respectively. As a result, in the case of the ANN algorithm, the accuracy of the training data is 77% and the accuracy of the test data is 68%. In the case of the regression algorithms, the accuracy of the training data is 74% and the accuracy of the test data is 71% on the average, depending on the algorithms. In general, the accuracy of the training data is estimated to be 75%, and the accuracy of the test data is estimated to be 70%. The accuracy of the training data is 75%, which indicates that the lack of data sets caused the underfitting problem. If we estimate the rank of the keyword data with a gap, we can derive a higher accuracy. Here, the meaning of gap means that mismatch is allowed for one rank. For example, in our prediction model that computes 15 rank outputs, it can be calculated with accuracy of zero gap to calculate exactly the third rank of the third rank prediction. If it is calculated with the accuracy of one gap, it means that the case where the third rank is predicted to the second or fourth rank also includes the accuracy. The reason why it is possible to allow a gap of one degree is because the rank of the keyword data continuously changes according to time, and the variation range shows the difference of one gap continuously. When the one gap is allowed for the accuracy, the accuracy of the ANN test data is 91.8%, the accuracy of the test data of the linear regression algorithms is 92%, the accuracy of the logistic regression algorithm and the softmax regression algorithm are 94%, respectively. As a result, it can be seen that accuracy is improved when a one gap is set for the degree of rank prediction.

Based on the results of this paper, we will extend the study to the result of solving the underfitting problem with additional data collection.

5 Conclusion

In this paper, we compare the accuracy of bid price prediction results of keyword data by applying four algorithms: the ANN algorithm, linear regression algorithm, logistic regression algorithm, and multinomial classification algorithm. As a result, in the case of the ANN algorithm, the accuracy of the training data is 77% and the accuracy of the test data is 68%. In the case of the regression algorithms, the accuracy of the training data is 74% and the accuracy of the test data is 71% on the average, depending on the algorithms. In general, the accuracy of the training data is estimated to be 75%, and the accuracy of the test data is estimated to be 70%. The accuracy of the training data is 75%, which indicates that the lack of data sets caused the underfitting problem. Despite the underfitting problem, the accuracy of the training data and the accuracy of the test data are similar in this paper. This is an indication that the prediction model of the bid price in keyword data is being trained. We will collect the data sufficiently and draw out the results of applying these prediction models. In this paper, we also applied the Adam optimization algorithm to optimize the keyword data model suitable for the prediction of bid price. Although we did not mention due to paper limitation, we concluded that the Adam optimization algorithm is the most suitable for the prediction model of bid price for keyword data.

Acknowledgements. This work was supported by the Technology development Program (C0563763) funded by the Ministry of SMEs and Startups (MSS, Korea). The keyword data set for the simulation in this paper was supported by Taeseong Kim of the *e-Glue* communications.

References

1. Hou, L.: A hierarchical bayesian network-based approach to keyword auction. IEEE Trans. Eng. Manag. **62**, 217–225 (2015)
2. Lauritzen, S.: The EM algorithm for graphical association models with missing data. Comput. Stat. Data Anal. **19**, 191–201 (1995)
3. Shuai, Y., Wang, J., Zhao, X.: Real-time bidding for online advertising: measurement and analysis. In: International Workshop on Data Mining for Online Advertising, Chicago, Illinois, USA (2013)
4. Brooks, N.: The Atlas rank report: How search engine rank impacts conversions (2004). http://www.atlasonepoint.com/pdf/AtlasRankReportPart2.pdf
5. Auerbach, J., Galenson, J., Sundararajan, M.: An empirical analysis of return on investment maximization in sponsored search auctions. In: International Workshop Data Mining and Audience Intelligence, Las Vegas, NV, USA (2008)
6. Stepanchuk, T.: An empirical examination of the relation between bids and positions of ads in sponsored search. In: 21st Bled eConference on eCollaboration: Overcoming Boundaries through Multichannel Interaction. Bled, Slovenia (2008)
7. Jerath, K., Ma, L., Park, Y., Srinivasan, K.: A position paradox in sponsored search auctions. Mark. Sci. **30**, 612–627 (2011)
8. Gopal, R., Li, X., Sankaranarayanan, R.: Online keyword based advertising impact of ad impressions on own channel and cross channel click through rates. Decis. Support Syst. **52**, 1–8 (2011)
9. Graepel, T., Candela, J., Borchert, T., Herbrich, R.: Web-scale Bayesian click through rate prediction for sponsored search advertising in Microsoft Bing search engine. In: International Conference on Machine Learning, Haifa, Israel (2010)
10. Kingma, D., Ba, J.: ADAM: a method for stochastic optimization. In: International Conference on Learning Representations, San Diego, CA, USA (2015)

Complex Activity Recognition Using Polyphonic Sound Event Detection

Jaewoong Kang, Jooyeong Kim, Kunyoung Kim, and Mye Sohn(✉)

Department of Industrial Engineering,
Sungkyunkwan University, Suwon, Korea
{kjwl727, eyesofkids, kimkun0, myesohn}@skku.edu

Abstract. In this paper, we propose a method for recognizing the complex activity using audio sensors and the machine learning techniques. To do so, we will look for the patterns of combined monophonic sounds to recognize complex activity. At this time, we use only audio sensors and the machine learning techniques like Deep Neural Network (DNN) and Support Vector Machine (SVM) to recognize complex activities. And, we develop the novel framework to support overall procedures. Through the implementation of this framework, the user can support to increase quality of life of elders'.

Keywords: Complex activity · Pattern recognition
Polyphonic sound event detection · Machine learning

1 Introduction

To recognize human activities and, moreover, provide intelligent services based on them, various sensors (for example, accelerometers, gyroscopes, or visible sensors) are used [1–4]. In case of utilizing accelerometers and gyroscopes, which are embedded in smartphones or wearable devices, it has excellent recognition performance for simple activities (e.g., standing, walking, running, etc.), which are composed of single repeated actions. However, the performance of these kinds of sensors to recognize complex activities (e.g., cooking, dishwashing, etc.) is less than expected [5]. A more serious limitation is that human must always carry a sensor or sensor embedded smart devices. The Visible sensors recognize both simple and complex activity with relatively high accuracy. However, since these sensors have a narrow recognition range, there is a problem that an enormous number of sensors must be installed to recognize human activities [1].

To overcome the limitations, we propose a novel method for recognizing the complex activity using audio sensors and the machine learning techniques. In order to recognize the complex activity of a human using the audio sensors, it is necessary to detect the sound event that occurs in the vicinity of the human being object to be recognized. Many researchers have conducted research using the audio sensors to detect the sound coming from around human. However, these researches have focused on detecting only monophonic sounds or distinguishing the various monophonic sounds contained in polyphonic sounds. Therefore, it is very difficult to recognize the complex activity that is neither the monophonic sound nor the simple combination of

L. Barolli et al. (Eds.): IMIS 2018, AISC 773, pp. 675–684, 2019.
https://doi.org/10.1007/978-3-319-93554-6_66

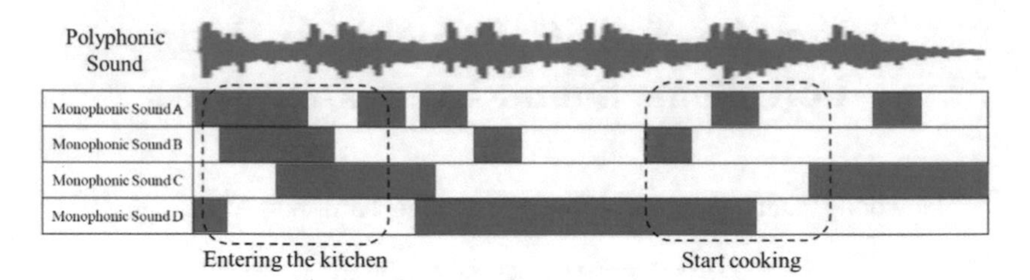

Fig. 1. Illustrative complex activity within a polyphonic sound

them. As depicted in Fig. 1, the polyphonic sound associated with the complex activity is a specific pattern of various monophonic sounds.

So, in this paper, we will look for the patterns of combined monophonic sounds to recognize complex activity. At this time, we applied Deep Neural Network (DNN) to find the patterns of combined monophonic sounds. This is the first attempt to recognize complex activities using the audio sensors and the machine learning techniques. So, in order to confirm the validity of our approach, we restricted the scope of our research to the kitchen in the home. Furthermore, it is very difficult to collect only valid sound related to the complex activity because it can generate uncontrollable and various noises in outdoor.

We performed sound events detection in order to recognize the complex activity in the kitchen. As a next step, after defining the patterns of the sound events contained in the detected sound events, we performed a first phase machine learning on them. Finally, we performed a second phase machine learning to recognize the newly occurring complex activity in the kitchen.

This paper is organized as follow. Section 2 reviews the related research papers. Section 3 offers detailed descriptions about the overall architecture and the components of the framework. In Sect. 4, we have developed an illustrative example to show the necessity of the complex activity recognition in the home. Finally, Sect. 5 presents the conclusions and further research.

2 Related Works

2.1 Sound Event Detection

Data analysis methods using audio sensors can be divided into two categories, which are classification and detection. Classification means clustering similar monophonic sounds into the same class, and classifying new monophonic sound into the appropriate class. On the other hand, detection means analyzing polyphonic sound to configure what kind of sounds existed in the certain time duration. The purpose of this study is complex activity recognition (CAR), so we can perform detection method to recognize environmental change occurred by human activity.

Unlike image data, sound data cannot be resized as a same size, or cropped near the Point of Interests (POI). Therefore, we should extract features from the sound data and analyze the features [6]. The widely used features in sound data are amplitude and frequency coefficients, spectrogram, and Mel-Frequency Cepstral Coefficients (MFCCs). Amplitude and frequency coefficients can be obtained by Fast Fourier Transform (FFT), and spectrogram and MFCCs can be obtained by Short-Time Fourier Transform (STFT) [7]. Some recent studies extracts features from sound data by deep learning, transform the data into the graph form, and apply Convolutional Neural Network (CNN) [8]. Also, Recurrent Neural Network (RNN) or Long Short-Term Memory (LSTM) network are used because of their advantage of maintaining the properties of time series data [9, 10].

To improve the efficiency of learning, methods such as Non-negative Matrix Factorization (NMF), Gaussian Mixture Model (GMM), and Random Forest (RF) are used to discriminate major factors before learning [11–13]. Also, ensemble methods which using pattern recognition methods such as Hidden Markov Model (HMM) or Support Vector Machine (SVM) with learning methods for detection is being studied [14]. However, since all of these methods use multi-label for detection, they have a drawback that they should start learning from scratch to perform polyphonic sound detection when new kind of sound is added.

2.2 Human Activity Recognition

Human Activity Recognition (HAR) is widely researched for healthcare, smart environments, and homeland security [15]. Human activity can be divided by two categories, which are simple activity and complex activity. Simple activity is repetition of monotonous actions, and complex activity is intricate composition of simple activities. Nowadays, real time recognition of simple activity become possible and research of recognizing transition activities, which occur in the very short time period while conversion between simple activities, is performed [16].

Human activity can be recognized by sensors which are installed around the human or attached to the human. Image sensors which are installed around the human are widely used to recognize human activity. The performance of perception of using image sensors with deep learning method is extremely high, but they have limitation in location and their cost is high [17]. On the other hand, using accelerometers and gyroscopes embedded in smartphones or wearable devices is cost-efficient. However, the assumption that people always equip these devices is too strong. These methods are not consistently detect human activity [1].

To overcome these location problem and consistency problem, we suggest using sound sensors which are located indoor. In this study, we does not used multi-label for sound detection. We made discriminators which learns features of each monophonic sounds and determine whether each sound exist or not in the polyphonic sound. To utilize the polyphonic sound events for the services, we proposed a method to discriminate complex activities by using SVM.

3 Complex Activity Recognition Using Polyphonic Sound Event Detection

3.1 Overall Framework

The framework for CAR implementation using audio sensor and machine learning technique is shown in Fig. 2. This framework consists of three modules: DNN-based Monophonic Sound Discriminator Generation Module (DMSM), Polyphonic Sound Event Detection Module (PSDM), and Complex Activity Reasoning Module (CARM).

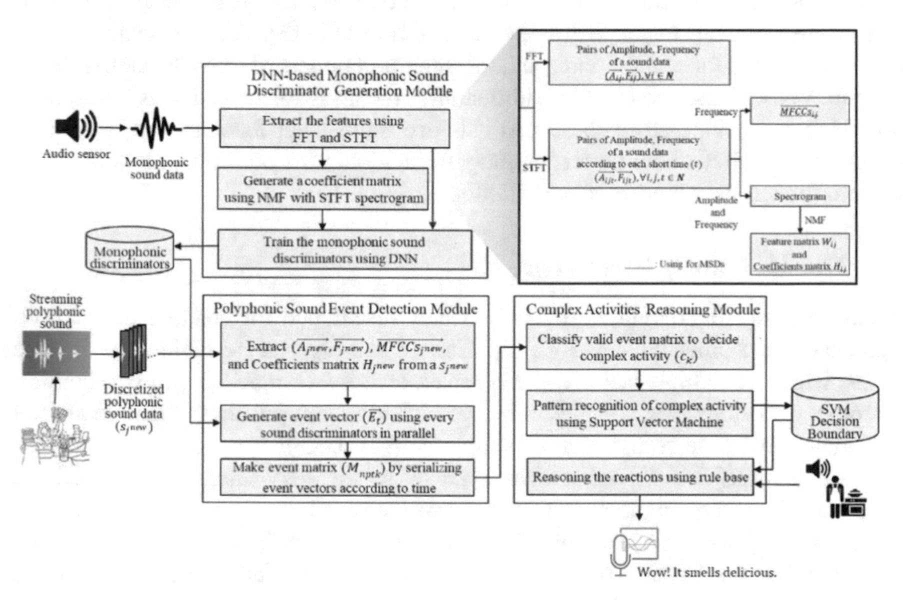

Fig. 2. Overall framework of complex activity recognition using polyphonic sound event detection

3.2 DNN-Based Monophonic Sound Discriminator Generation Module (DMSM)

To generate multiple monophonic sound discriminators (MSDs), the DMSM extracts a set of features from the monophonic sounds. As mentioned in Sect. 2, the typical features associated with monophonic sounds are a pairs of amplitude and frequency and MFCCs [18]. The process to derive the pairs of amplitude and frequency and MFCCs is depicted in the upper right rectangle of Fig. 2. The basic concepts required to describe the detail process are defined as follows.

Definition 1. i^{th} sound event (S_i) is simply represented as

$$S_i = \{s_{i1}, s_{i2}, \ldots, s_{ij}, \ldots\}$$

where s_{ij} is j^{th} occurrence of ith sound ((i, j $\geq$ 1).

Based on Definition 1, S_i is a class of ith sound data. For instance, if S_i is a class for "the tap water sounds," the sound data (elements) of the S_i are all "tap water sounds" generated in the meantime. The detailed process to extract features of the sounds is summarized as follows.

Derivation of the Pairs of Amplitude and Frequency for j^{th} Sound Data $\left(\overrightarrow{A_{ij}}, \overrightarrow{F_{ij}}\right)$. These are able to extract from each monophonic sound data using the FFT. At this time, $\left(\left(\overrightarrow{A_{ij}}, \overrightarrow{F_{ij}}\right)\right.$ vector is generated for the entire duration of jth sound data.

Derivation of MFCCs. In order to extract the MFCCs, the entire duration of jth sound data should be divided by a short time (or interval) and the frequency of the sound data at each interval measured [18]. To do so, STFT used to extract the frequency in short time interval [7]. As a result, MFCCs vector ($\overrightarrow{MFCCs_{ij}}$,) is generated for each time interval of jth sound data.

Execution of Spectrogram Analysis. In addition to the frequency of the jth sound data at each interval, we also measure their amplitude using STFT. The frequency and the amplitude of the sound data in short time interval are represented spectrogram visually. Since the spectrogram is a kind of image, and the image can be expressed a matrix, it is possible to decompose the spectrogram using NMF [19]. As a result, the spectrogram is decomposed into a basis ($\boldsymbol{W_{ij}}$) and coefficients of the basis ($\boldsymbol{H_{ij}}$) using NMF. The spectrogram of $\boldsymbol{s_{ij}}$ is simply represented as follows.

$$\text{Spectrogram of } s_{ij} = W_{ij} \times H_{ij}$$

At this time, W_{ij} is used to determine the presence of s_{ij}, and H_{ij} is used to determine the number of s_{ij}. In addition, all elements of S_i have a similar W_{ij} matrix because they are the same kind of sound. Therefore, if we can find the matrix W_i representing W_{ij} (for $\forall$j), the Spectrogram of s_{ij} can be expressed by the following approximate expression.

$$\text{Spectrogram of } s_{ij} \approx W_i \times H_{ij}$$

Using the results of spectrogram analysis, the MFCCs vector, and the pairs of amplitude and frequency for the specific sound, DNN-based training to classify the sound is performed. As results, we obtain k sound discriminators D_i (i = 1, 2, ..). D_i returns "1" if i^{th} sound event occurs in a sound, and "0" if not.

3.3 Polyphonic Sound Event Detection Module (PSDM)

Generally, audio sensors collect continuous sound data. Therefore, in order to recognize the complex activities of human beings using the continuous data, we must digitized them. We performed data digitization according to the sliding widow method proposed by Kang *et al.* (2017).

After data digitization, PSDM performs three tasks. First, it captures new sound event (s_{jnew}) and extracts features such as the pairs of $\left(\overrightarrow{A_{jnew}}, \overrightarrow{F_{jnew}}\right)$, $\overrightarrow{MFCCs_{jnew}}$, and the spectrogram of s_{jnew}. The process of extracting features is the same as described above except for the method to obtain the coefficients of the basis (H_{jnew}). We can compute H_{jnew} without decomposing the spectrogram of s_{jnew} using the representative basis (W_i) for the ith sound event derived from the learning process of ith discriminator (for $\forall i$).

$$H_{jnew} \approx \left(W_{jnew}\right)^{-1} \text{Spectrogram of } s_{jnew}$$

However, since it is not known exactly what s_{jnew} is included in the class, that is, W_{jnew} is not known, we generate n H_{jnew} for all discriminators. All features of s_{jnew} derived through the above process are used as input instances of the discriminators.

Second, the monophonic sound discriminators are simultaneously executed using the input instances. At this time, if s_{jnew} contains i^{th} monophonic sound event, i^{th} discriminator returns "1." The results of the execution of the discriminators are collected into an event vector. The event vector is defined as follows.

Definition 2. t^{th} event vector $\left(\overrightarrow{E_t}\right)$ is a result of the simultaneous execution of n monophonic sound discriminators. The ith element of $\overrightarrow{E_t}$ (e_{ti}) indicates whether i^{th} monophonic sound event is occurred in the polyphonic sound data (s_{jnew}) at time t. E_t is simply represented as

$$\overrightarrow{E_t} = \left(e_{t1}, e_{t2}, \ldots, e_{ti,\ldots}\right), e_{ti} \in \{0, 1\}, 1 \leq i \leq n$$

If i^{th} monophonic sound event occurred at time t, e_{ti} is 1, otherwise e_{ti} is 0.

Finally, to recognize the complex activities, an event matrix is generated by accumulating the event vectors along the time axis. The event matrix is defined as follows.

Definition 3. Event matrix $M_{np}^{(t)}$ is a $n \times p$ matrix that is assigned toward k^{th}complex *activity*(c_k). $M_{np}^{(t)}$ is composed of the monophonic sound vectors from $t - (p - 1)$ to arbitrary time t in s_{jnew}.

$$M_{np}^{(t)} = \left\{\overrightarrow{E_t}, \overrightarrow{E_{t-1}}, \ldots\ldots, \overrightarrow{E_{t-(p-1)}}\right\}$$

where p is arbitrary time span ($p \geq 2$). At this time, we made two assumptions to recognized the complex activities. One is that we have predefined all sorts of the complex activities (c_k, $k \geq 1$) previously, the other is that we knew exactly what combination of complex activities involved in polyphonic streaming sounds.

The relation of the event vector and the event matrix is represented in Fig. 3. The number of column p is the parameter that determines the progress time of the activity and is determined experimentally.

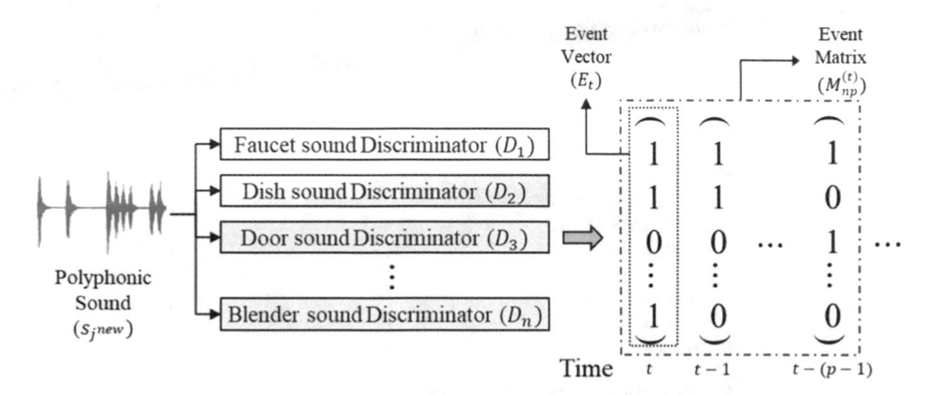

Fig. 3. Generation process of event vectors and event matrix

3.4 Complex Activity Reasoning Module (CARM)

CARM performs the task of selecting the valid event matrix that can be mapped to the complex activity among the event matrices produced PSDM. The pattern recognition boundary is decided using valid event matrices by SVM. When the complex activity of the current situation is determined by the boundary, the appropriate response is made by the predefined rules. Complex activity reasoning algorithm is in Fig. 4.

Even if the accuracy of the monophonic sound discriminator is high enough, there is no guarantee about high probability which multi discriminators will output the correct result at a single point. Therefore, it is necessary to select a good event matrix before complex activity reasoning. In this paper, we call valid event matrix which is good for reasoning. Valid event matrix is not sparse matrix, because we can't get enough information and accuracy from sparse matrix. Using valid event matrix, SVM is performed to determine decision boundary and best kernel function. When new sound sensor catch a polyphonic sound data, a proper simple reaction corresponding to the complex activity classified according to the previously defined rule is taken.

δ_k : Threshold of kth complex activity
DB: Set of decision boundaries
kf : Kernel function of SVM
EM_{new} : Event matrix of new polyphonic sound data
c_{new} : Complex activity of new polyphonic sound data

Begin Complex Activity Reasoning Process
  $V \leftarrow list()$
  For all t
    $VEM \leftarrow$ Valid Event Matrix Classification Function($M_{np}^{(t)}, \delta_k$)
    $V.append(VEM)$
  EndFor
  $DB, kf \leftarrow$ Decision Boundary Generation Function(V)
  $EM_{new} \leftarrow PSDM(s_{j^{new}})$
  $c_{new} \leftarrow$ SVM Classification(EM_{new}, DBs, kf)
  Return c_{new}
End Process

Valid Event Matrix Classification Function($M_{np}^{(t)}, \delta_k$)
  $num.of.one \leftarrow$ sum($M_{np}^{(t)}$)
  If $num.of.one > \delta_k$
    $valid \leftarrow M_{np}^{(t)}$
  Else
    $valid \leftarrow NULL$
  Return $valid$
End Function

Decision Boundary Generation Function(V)
  $DB, kf \leftarrow$ *Support Vector Machine*(V)
  Return DB, kf
End Function

Fig. 4. Complex activity reasoning algorithm

4 Illustrative Applications

According to the report 'World Population Prospects: the 2015 Revision,' United Nations insisted that the population of the elderly will dramatically increase between a few decades, and the rate of increase will become faster gradually [20]. Since a number of seniors live alone and rarely interact with people, the mental problem, such as loneliness, emerged as a social issue [21]. To cope with this problem, enterprises and researchers are doing research about artificial intelligence and developing companion or care robots to support the elderly [22]. However, the price of these robots is a bottleneck for them to be widely distributed [23]. To remove this bottleneck, we are trying to develop a cheap and small button-type device WILSON, which is composed of processor, audio sensor, and speaker. The name, WILSON the volleyball, is quoted from the movie 'Cast Away'. It is treated as a friend by a lone hero.

WILSON recognizes complex activities of the elderly, and reacts as a voice. For example, if WILSON perceives sound of water flowing on the sink, sound of lighting a gas stove, and sound of a knife on the chopping board, WILSON can figure out that the senior is cooking. By rule-based reasoning method, WILSON can find out the optimal reaction suited for the situation. In addition, WILSON can learn when and how often he or she takes meals, or goes out. By utilizing this learning data, WILSON can ask the old person whether he or she took meal, or recommend going out for a walk.

Epley [24] proved that people who lack social connection with other humans tend to rely on nonhuman agents, such as animals and gadgets, and they are very effective to overcome loneliness. Through this point, WILSON is expected to alleviate loneliness of the elderly and support them to live a regular and healthy life.

5 Conclusion and Future Works

We proposed CAR method using sound sensors instead of image sensors, accelerometers, or gyroscopes. We also proposed another event detection method by utilizing monophonic sound discriminator to recognize activity. Contributions of this research can be summarized as follows. First, we applied the audio sensors to recognize the complex activities of human being. This is the first research to apply an audio sensors. As results, this research contributed to reduce the number of sensors and computational cost and to increase to user's convenience.

However, this research has two limitations. First of all, we should use sound sensors good enough not to make any noise. Secondly, appropriate response is possible only by predefined rules. Therefore, we will future research should overcome these limitations. Moreover, we will propose the method utilizing different size of DNN according to types of event and using various boundaries of complex activity recognition to improve recognition accuracy.

Acknowledgements. This research was supported by Basic Science Research Program through the National Research Foundation of Korea (NRF) funded by the Ministry of Education, Science and Technology (NRF-2016 R1D1A1B03932110).

References

1. Attal, F., Mohammed, S., Dedabrishvili, M., Chamroukhi, F., et al.: Physical human activity recognition using wearable sensors. Sensors **15**, 31314–31338 (2015)
2. Ong, W.H., Palafox, L., Koseki, T.: Investigation of feature extraction for unsupervised learning in human activity detection. Bull. Netw. Comput. Syst. Softw. **2**, 1–30 (2013)
3. Ghosh, A., Riccardi, G.: Recognizing human activities from smartphone sensor signals. In: Proceedings of the 22nd ACM International Conference on Multimedia, pp. 865–868. ACM (2014)
4. Roshtkhari, M.J., Levine, M.D.: Human activity recognition in videos using a single example. Image Vis. Comput. **31**, 864–876 (2013)
5. Nam, Y.Y., Choi, Y.J., Cho, W.D.: Human activity recognition using an image sensor and a 3-axis accelerometer sensor. J. Internet Comput. Serv. **11**, 129–141 (2010)
6. Bregman, A.S.: Auditory scene analysis: the perceptual organization of sound. MIT Press (1994)
7. Krijnders, J., Holt, G.T.: Tone-fit and MFCC scene classification compared to human recognition. In: IEEE Workshop on Applications of Signal Processing to Audio and Acoustics (2013)

8. Parascandolo, G., Heittola, T., Huttunen, H., Virtanen, T.: Convolutional recurrent neural networks for polyphonic sound event detection. IEEE/ACM Trans. Audio, Speech, Lang. Process. **25**, 1291–1303 (2017)
9. Parascandolo, G., Huttunen, H., Virtanen, T.: Recurrent neural networks for polyphonic sound event detection in real life recordings. In: IEEE International Conference on Acoustics, Speech and Signal Processing, pp. 6440–6444 (2016)
10. Marchi, E., Vesperini, F., Eyben, F., Squartini, S., Schuller, B.: A novel approach for automatic acoustic novelty detection using a denoising autoencoder with bidirectional LSTM neural networks. In: IEEE International Conference on Acoustics, Speech and Signal Processing, pp. 1996–2000 (2015)
11. Innami, S., Kasai, H.: NMF-based environmental sound source separation using time-variant gain features. Comput. Math Appl. **64**, 1333–1342 (2012)
12. Defréville, B., Pachet, F., Rosin, C., Roy, P.: Automatic recognition of urban sound sources. In: Audio Engineering Society Convention 120 (2006)
13. Díaz-Uriarte, R., De Andres, S.A.: Gene selection and classification of microarray data using random forest. BMC Bioinform. **7**, 3 (2006)
14. Eghbal-Zadeh, H., Lehner, B., Dorfer, M., Widmer, G.: A hybrid approach with multi-channel i-vectors and convolutional neural networks for acoustic scene classification. In: Signal Processing Conference (EUSIPCO), pp. 2749–2753 (2017)
15. Aggarwal, J.K., Ryoo, M.S.: Human activity analysis: a review. ACM Comput. Surv. **43**, 16 (2011)
16. Kang J., Kim J., Lee S., Sohn M.: Recognition of transition activities of human using CNN-based on overlapped sliding window. In: 5th International Conference on Big Data Applications and Services (2017)
17. Karpathy, A., Toderici, G., Shetty, S., Leung, T., Sukthankar, R., Fei-Fei, L.: Large-scale video classification with convolutional neural networks. In: Proceedings of the IEEE Conference on Computer Vision and Pattern Recognition, pp. 1725–1732 (2014)
18. Logan, B.: Mel frequency cepstral coefficients for music modeling. In: ISMIR vol. 270, pp. 1–11 (2000)
19. Lee, D.D., Seung, H.S.: Algorithms for non-negative matrix factorization. In: Advances in Neural Information Processing Systems, vol. 13 pp. 556–562 (2001)
20. Melorose, J., Perroy, R., Careas, S.: World population prospects: the 2015 revision, key findings and advance tables. Working Paper No. ESA/P/WP. 241, pp. 1–59 (2015)
21. Abdi, J., Al-Hindawi, A., Ng, T., Vizcaychipi, M.P.: Scoping review on the use of socially assistive robot technology in elderly care. BMJ Open **8**(2), e018815 (2018)
22. Robinson, H., MacDonald, B., Broadbent, E.: The Role of Healthcare Robots for Older People at Home: A Review. Int. J. Soc. Robot. **6**(4), 575–591 (2014)
23. Kohlbacher, F., Rabe, B.: Leading the way into the future: the development of a (lead) market for care robotics in Japan. Int. J. Technol. Policy Manag. **15**(1), 21–44 (2015)
24. Epley, N., Akalis, S., Waytz, A., Cacioppo, J.T.: Creating social connection through inferential reproduction: loneliness and perceived agency in gadgets, gods, and greyhounds. Psychol. Sci. **19**, 114–120 (2008)

Dynamic Group Key Management for Efficient Fog Computing

Jiyoung Lim[1], Inshil Doh[2(✉)], and Kijoon Chae[3]

[1] Department of Computer Software, Korean Bible University, Seoul, Korea
jylim@bible.ac.kr

[2] Department of Cyber Security, Ewha Womans University, Seoul, Korea
isdohl@ewha.ac.kr

[3] Department of Computer Science and Engineering, Ewha Womans University, Seoul, Korea
kjchae@ewha.ac.kr

Abstract. Cloud system is a new computing technology for dealing with IT services at the Internet server system. In spite of its advantages in various aspects, it is not proper for IoT (Internet of Things) services that require high density and real time data processing. Fog computing is a new paradigm appropriate for decreasing data processing latency, managing the mobility, and increasing the service efficiency of IoT. However, there is a lot of security vulnerabilities for fog computing because of the lack of security related systems. In this work, we propose a group key management for the secure fog computing which can be adjusted dynamically based on the mobility of the IoT devices. Our proposed system can increase the system efficiency by controlling the IoT groups and the group keys management.

1 Introduction

IoT (Internet of Things) has been considered as the future of Internet for several years as wearable technologies, smart grid, smart home and smart connected vehicles have increased hugely. Millions or billions of smart devices in IoT have been connected to one another and have exchanged information over Internet. The limitation of computing power, battery and storage on smart devices makes the cloud computing be the future promising computing paradigm since cloud computing has elastic resources and services at low cost [1].

IoT applications such as wearable e-healthcare adapt their location and mobility while the cloud computing is not adequate to manage billions of nomadic mobile users. Emerging wearable devices such as Google glasses expect the real time sensing and data processing while mobile devices limited in resources might outsource their computing task to cloud. The high rate exchange between cloud and mobile is not acceptable for IoT applications requiring real time sensing and data processing.

IoT needs a new platform providing the mobility support, geo distribution, location awareness and low latency. Fog computing is proposed to provide elastic resource and services to users at the edge of networks and is usually cooperated to cloud computing.

L. Barolli et al. (Eds.): IMIS 2018, AISC 773, pp. 685–694, 2019.
https://doi.org/10.1007/978-3-319-93554-6_67

In fog computing, facilities and infrastructures providing services at the edge of networks are fog nodes or fog servers. They might be devices with poor resources such as set-top boxes, access points, router, switch, base station and end devices or devices with rich resources such as cloudlet [2]. Cloudlet is a computer with rich resource available by nearby mobile devices. Basic fog computing has an architecture with three level hierarchy in which each mobile device is attached to a fog device, and fog devices could be interconnected with one another and linked to the cloud [3].

Fog nodes come in different form factors and are deployed in a wide variety of environments, which requires the cooperation of different providers, the interoperation of fog nodes and the federation of services between domains. The fog collector at the edge as a role of a sink gather data generated by grid sensors and devices. The data is processed, filtered, controlled and commanded for not only human to machine interaction but also only machine to machine interaction [4].

To prevent outsiders from eavesdropping or injecting data, legitimate fog components can share group keys for encrypting or authenticating data among them. As the service domains might be different and a lot of heterogeneous fog nodes in different providers are cooperated, it is not easy to keep the communication safe. Due to high computational and communication overhead, the digital signature-based authentication is not adoptable for fog computing. If the key management process is vulnerable due to its poor security, there is a possibility that the security of the whole communication system becomes insecure. Therefore, we propose a dynamic group and group key management mechanism and an automatic group key rekeying mechanism considering the membership changes and the device mobility.

The remainder of this work is organized as follows. Section 2 describes the related work in group key related researches. System architecture for our proposal and the background technologies are described in Sect. 3. In Sect. 4, group and group key management mechanisms for IoT fog computing services are described in detail. Section 5 concludes our work.

2 Related Work

Approaches to group key management are divided into three categories [5].

Centralized Key Management Schemes: a single entity controls the generation of group key, distribution of key to authenticated group members and management of key lists.

Decentralized Key Management Schemes: a large group is divided into small groups controlled by subgroup managers, trying to minimize the problem of concentrating the work in a single place.

Distributed Key Management Schemes: members generate a key by themselves without a specific key manager, which means all members contribute some information to generate the group key, or done by one of the members.

We focus on centralized and distributed group key managements among them. There are various centralized schemes as follows. Wang et al. proposed four safe group communication methods [6]. Information for group key rekeying is unicasted to each node. This creates a heavy overload when group size grows. Broadcasting is proposed to solve the overhead problem. The broadcasting mechanism requires heavier overhead when groups are generated; however, rekeying cost is relatively low. Overlapping is also proposed to prevent flooding attack. Finally, group information pre-distribution minimizes group generation time. Karuturi et al. provide a generalized framework for centralized GKM along with a formal security model and definitions for the security properties that dynamic groups demand [7]. A lot of protocols have been researched for years. However, nomadic mobile users in fog computing environments make parent-child relationship change continuously. It is not adequate that the centralized key management is adopted in fog computing networks even if is very stable and secure.

Zhang et al. proposed a mechanism (PCGR) that pre-distributes key related information and generates group keys [8]. When group key rekeying is required, nodes cooperate and a new group key is computed. This scheme is applied in our proposal and will be more described later. Huang et al. proposed a level key infrastructure for multicast and group communication that user level keys to provide an infrastructure that lowers the cost of nodes joining and leaving [9]. This scheme has a drawback in that process delay increases even when many nodes are changed. Zhu et al. proposed a key management protocol for sensor networks designed to support in-network processing, while at the same time restricting the security impact of a compromised node [10]. This mechanism is safer, because it uses four different kinds of keys. However, key update consumes much overhead. Adusumilli et al. proposed a Distributed Group Key Distribution (DGKD) protocol which does not require existence of central trusted entities such as group controller or subgroup controllers [11]. Aparna et al. proposed a key management scheme for managing multiple groups. They use a combination of key-based and secret share-based approach for managing the keys and showed that it is possible for members belonging to two or more groups to derive the group keys with less storage [12]. Kim et al. investigated a novel group key agreement approach which blends key trees with Diffie-Hellman key exchange [13]. It yielded a secure protocol suite called Tree-based Group Diffie-Hellman (TGDH) that is both simple and fault-tolerant. Yu et al. propose a group key management mechanism [14] in which basic matrix G and secret matrices A, B are assigned to each sensor node; each matrix is used to generate group keys among nodes in the same groups and different groups, respectively. The advantage of this mechanism is that the probability of generating group keys is high. However, when the grid size is large, much energy is wasted and when the grid size is small, group keys may not be generated.

3 Proposed System Architecture and Background Technologies

3.1 IoT Communication System Architecture

In our work, for the efficient IoT group management between the cloud and IoT devices, we set hierarchical fog layers. High level fog node(FN_h) manages the group

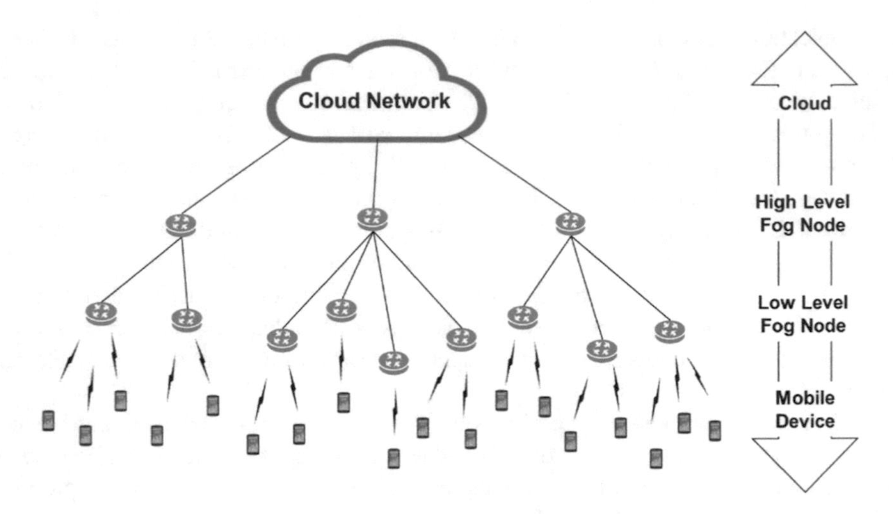

Fig. 1. Proposed IoT fog computing system architecture

keys and the low level fog nodes(FN_ls) control the IoT devices at the edge. Figure 1 shows the system architecture of our proposal.

3.2 PCGR

For the group key rekeying management in our proposal, we basically adopt a group key rekeying mechanism, PCGR [8]. The process is described in this subsection. In Fig. 2(a), a setup server assigns a unique t-degree univariate g-polynomial g(x), randomly, where g(0) is the initial group key. After deployment and neighbor discovery of a device node u, it picks a bivariate e-polynomials and generates g′-polynomial as shown in Fig. 2(b). Encryption is carried out as follows:

$$g'(x) = g(x) + e_x(x, u)\,, \tag{1}$$

where $e_u(x, y) = \sum_{0 \le i \le t, o \le j \le \mu} b_{i,j} x^i y^j$ in a parameter μ picked by u itself.

After distributing the shares of $e_u(x, y)$ to its n neighbors as shown in Fig. 2(c), a node u removes $e_u(x, y)$ and g(x). Finally a node u keeps g'(x) and a neighbor v_i receives a share $e_u(x, v_i)$ as illustrated in Fig. 2(d).

Each node maintains a rekeying timer to update its group key. When its timer expires, an innocent node u increases its c by one and returns share $e_{v_i}(c, u)$ to each trusted neighbor v_i as shown in Fig. 2(e). Having (μ + 1) shares, the node u can reconstruct a unique ρ-degree polynomial by solving the following equations:

$$\sum\nolimits_{j=0}^{\mu} (v_i)^j B_j = e_u(c, u)\,,\ i = 0, \ldots, \mu \tag{2}$$

Finally, the node u computes the new group key g(c) $= g'(c) - e_u(c,u)$ as shown in Fig. 2(f).

PCGR has an advantage that a new update group key is not revealed no matter what some nodes compromise. However, frequent group key rekeying for certain reasons causes heavy overhead since any node could initiate the rekeying process for a new group key in PCGR.

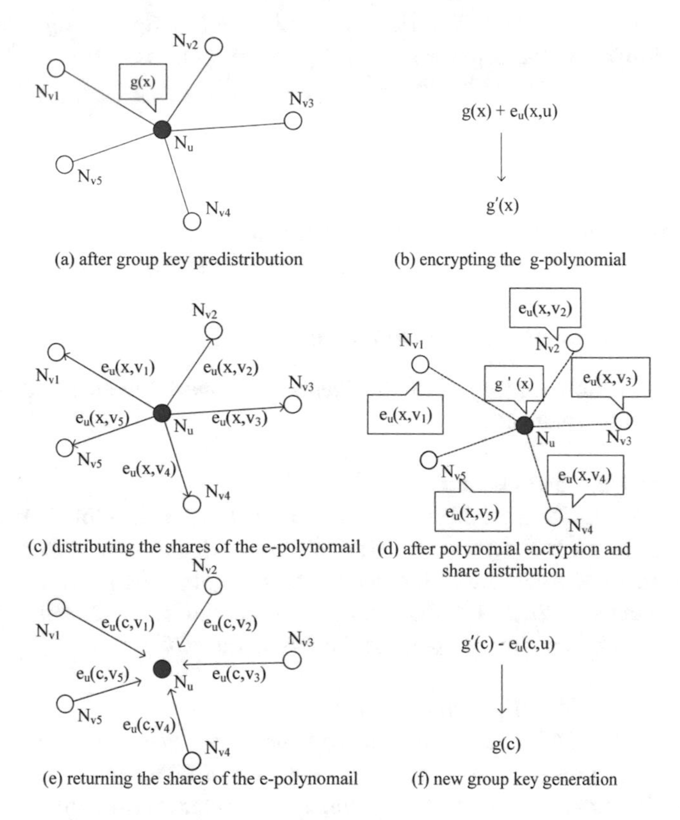

Fig. 2. PCGR: polynomial encryption, share distribution, and key updating [8]

4 Proposal of Group and Group Key Management

4.1 Characteristics of IoT Communications and Basic Assumptions

IoT devices need to communicate with one another for gathering and delivering the sensed data to the cloud. Different from other communications, they have a couple of characteristics. Some required characteristics and basic assumptions for our proposal are as follows.

- An IoT device can be controlled by one or more fog nodes.
- An IoT device needs to send data to the other IoT device or fog node for data processing.

- IoT devices need the group communication for processing a command from the fog node or from the cloud.
- Many IoT devices have mobility, and the fog node which controls the IoT device needs to be changed based on the new location.
- IoT devices receive command from the cloud through a fog node for the various environment.
- Each pair of FN_h and FN_l has pairwise keys.
- Each pair of FN_l and the IoT device has a pairwise key for sharing further security related information including the group key generation polynomials.
- Fog nodes are assumed to be stable.
- There are many different scale and functional groups and each IoT device can be included in more than one group.

Using the fog computing concept, we can expect various benefits including security especially for the group communication. In the following subsections, the dynamic group management and group key management mechanisms are described.

4.2 IoT Group Management by Fog Nodes

In IoT communication, a group of IoT devices need to work together to deliver specific data in various environment.

4.2.1 Device Registration to a Group.

When a new IoT device is installed, it needs to register for the cloud communication. As a member for an IoT service, the device may send the sensed data or receive a command from a controller, i.e., a fog node, which requires the group communication. When a new device is located in the local area, it should be pre-installed a group key information, which is group key polynomial based on PCGR.

4.2.2 IoT Device Mobility Management

In some cases, IoT devices can move around the areas. When the device moves out from the regional group, its location is sensed by the GPS, and its new location is reported to FN_l. Based on its new location, FN_h delivers the group key polynomial through the local FN_l for the newly moved IoT device to act as a member in the new group.

When many devices move very often between two areas, group key rekeying is required often which causes the high cost and system performance degradation. In this case, after checking several conditions, two groups are merged and one of the fogs becomes the FN_l and the other one becomes a supplement fog node (Fig. 3).

Conditions to be checked are as follows.

1. Same devices move between the two areas often for certain period of time.
2. Two areas need to be adjacent.
3. The number of devices in the combined group should be lower than the threshold.
4. If two groups do not show high mobility for certain period of time after group combination, they can be separated into two again for better management.

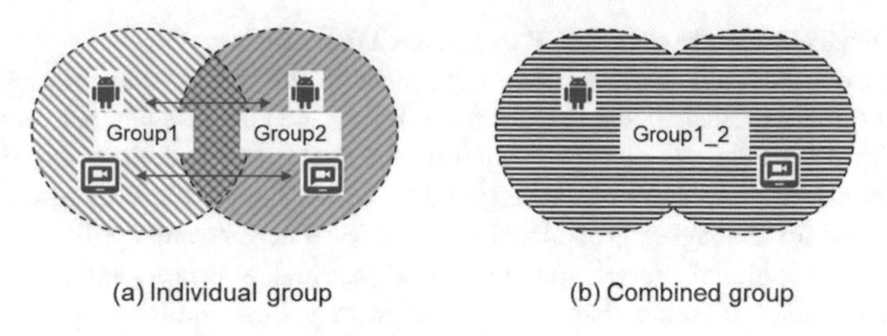

Fig. 3. IoT device group combination

4.2.3 Multiple Group Management by Fog Node

As in Fig. 4, a device can belong to more than one group, which means that more than one FN_l can control the device. If a combined group includes the new group where the device moves in, the old FN_l can distribute the device information to the new FN_l. If there is no fog node that controls the bigger group, this situation is reported to the cloud, new FN_l is adopted by the cloud, and the information of the device is delivered to the newly adopted fog node. Through this process, group keys need to be rekeyed as in the next subsection.

4.3 IoT Group Key Management

IoT devices receive the security information to generate group keys based on their location and their movement. These group keys need to be rekeyed periodically or based on the level of device movement. Information distribution is as follows.

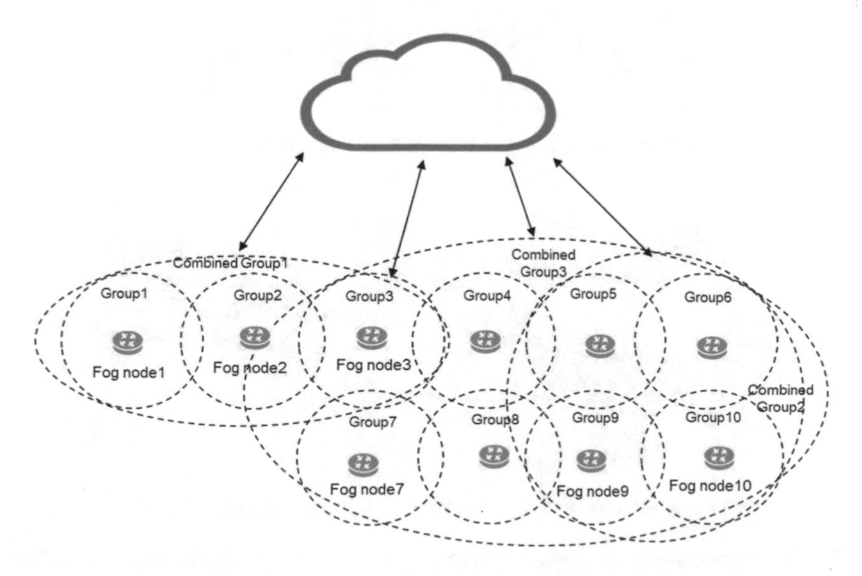

Fig. 4. Proposed multiple group architecture

4.3.1 Service Registration and Polynomial Distribution

When a device is newly located in a group, through the registration process, required pairwise keys are predistributed to the device. Using the pairwise key between FN_h and the end device, the polynomial for computing the group key is delivered to the device. One group can include one or more fog nodes, which means more than one fog node can cooperate for managing groups and group keys. There are many different kinds of groups such as regional groups, functional groups, device groups, and so on. When a device is included in more than one group, it may have multiple group keys and multiple group key polynomials that are controlled by FN_ls in each group.

4.3.2 Group Key Rekeying

Group keys are rekeyed periodically for security. When the group communication security is very import, group keys should be rekeyed on any kind of the membership change. When a device moves out from a group, a new group key should be computed for the forward security for the other devices left in the group, while the key generation polynomial needs to be delivered to a new member when there is a new device moving in a group. Rekeying decision is made by FN_h as in Fig. 5(a). In step 4, FN_l distributes a new c value, and the devices compute a new group key using the value and the polynomial.

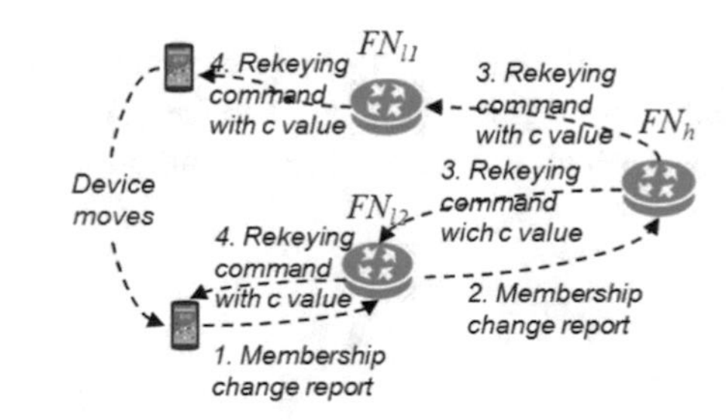

(a) Membership changes but groups are not adjusted

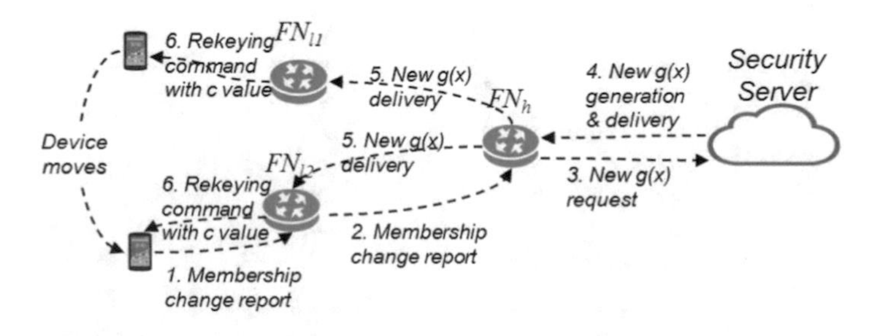

(b) Groups are adjusted based on device movement analyses by FN_h

Fig. 5. Group key rekeying on device movement (a) no group rebalancing (b) group rebalancing

4.3.3 Group Key Redistribution for Group Combination or Division

Based on the threshold values for the group rebalance, a group can be included in another group while in other cases one group can be divided into two or more. When a group is divided into two, a new group key should be delivered to members in a newly generated group. When two groups are combined, one group key information should be delivered to the member devices whose group is included into another one. When there is any change in groups, group key polynomials need to be changed by the Security Server in cloud. This is decided and reported to the Security Server by FN_h, and the new group key polynomials are generated and distributed to the end IoT device as in Fig. 5(b).

5 Conclusions

Fog computing is a new paradigm for providing the efficient cloud based IoT communication. Between the cloud and IoT devices, a fog node can play import roles for gathering and processing data. During the processes, the group communication is also required. In our work, we propose the group management and group key management processes to improve the system performance and the security. We especially manage multiple groups that IoT devices can belong to and make the group be dynamically adjusted considering IoT mobility. Some groups can be combined while others are divided based on the IoT device movement. In addition, group keys are updated periodically or dynamically when membership changes. Our proposal provides the forward and backward secrecy by making the group keys be rekeyed when required.

Acknowledgments. This work was supported by the National Research Foundation of Korea (NRF) grant funded by the Korea government (MSIP) (NRF-2016R1A2B4015899).

References

1. Aazam, M., Huh, E.N.: Fog computing and smart gateway based communication for cloud of things. In: International Conference on Future Internet of Things and Cloud (FiCloud), pp. 464–470. IEEE (2014)
2. Yi, S., Li, C., Li, Q.: A survey of fog computing: concepts, applications and issues. In: Proceedings of the 2015 Workshop on Mobile Big Data, pp. 37–42. ACM (2015)
3. Stojmenovic, I.: Fog computing: a cloud to the ground support for smart things and machine-to-machine networks. In: Telecommunication Networks and Applications Conference (ATNAC), pp. 117–122. IEEE (2014)
4. Bonomi, F., Milito, R., Zhu, J., Addepalli, S.: Fog computing and its role in the internet of things. In: Proceedings of the first edition of the MCC workshop on Mobile cloud computing, pp. 13–16. ACM (2012)
5. Rafaeli, S., Hutchison, D.: A survey of key management for secure group communication. ACM Comput. Surv. (CSUR) **35**(3), 309–329 (2003)
6. Wang, Y., Ramamurthy, B.: Group rekeying schemes for secure group communication in wireless sensor networks. In: IEEE International Conference on Communications (ICC 2007), pp. 3419–3424. IEEE (2007)

7. Karuturi, N.N., Gopalakrishnan, R., Srinivasan, R., Rangan, C.P.: Foundations of group key management-framework, security model and a generic construction. IACR Cryptology EPrint Archive, 295 (2008)
8. Zhang, W., Cao, G.: Group rekeying for filtering false data in sensor networks: a predistribution and local collaboration-based approach. In: 24th Annual Joint Conference of the IEEE Computer and Communications Societies (INFOCOM 2005), Vol. 1, pp. 503–514. IEEE (2005)
9. Huang, J.H., Buckingham, J., Han, R.: A level key infrastructure for secure and efficient group communication in wireless sensor network. In: First International Conference on Security and Privacy for Emerging Areas in Communications Networks, pp. 249–260. IEEE (2005)
10. Zhu, S., Setia, S., Jajodia, S.: LEAP+: efficient security mechanisms for large-scale distributed sensor networks. ACM Trans. Sen. Netw. (TOSN) **2**(4), 500–528 (2006)
11. Adusumilli, P., Zou, X., Ramamurthy, B.: DGKD: distributed group key distribution with authentication capability. In: Proceedings from the Sixth Annual IEEE SMC Information Assurance Workshop (IAW 2005), pp. 286–293. IEEE (2005)
12. Aparna, R., Amberker, B.B.: Key management scheme for multiple simultaneous secure group communication. In: IEEE International Conference on Internet Multimedia Services Architecture and Applications (IMSAA), pp. 1–6. IEEE (2009)
13. Kim, Y., Perrig, A., Tsudik, G.: Tree-based group key agreement. ACM Trans. Inf. Syst. Secur. (TISSEC) **7**(1), 60–96 (2004)
14. Yu, Z., Guan, Y.: A robust group-based key management scheme for wireless sensor networks. Wireless Communications and Networking Conference, vol. 4, (2005, March). pp. 1915–1920. IEEE (2005)

A Network Slice Resource Allocation Process in 5G Mobile Networks

Andrea Fendt[1(✉)], Lars Christoph Schmelz[2], Wieslawa Wajda[2], Simon Lohmüller[1], and Bernhard Bauer[1]

[1] Department of Computer Science, University of Augsburg, Augsburg, Germany
{andrea.fendt,simon.lohmueller,bauer}@informatik.uni-augsburg.de
[2] Nokia Bell Labs, Munich, Stuttgart, Germany
{christoph.schmelz,wieslawa.wajda}@nokia-bell-labs.com

Abstract. The fifth generation of mobile networks (5G) is associated with a wide spectrum of novel use cases that introduce a large number of very diverse requirements, regarding for instance throughput, latency, delay, availability and reliability. End-to-end network slicing is seen as a solution that allows to simultaneously accomplish those manifold requirements in isolated slices running on a shared network infrastructure. However, embedding those virtual end-to-end network slices into a common physical network containing wireless as well as wired network elements, while meeting all the different requirements, is still an unsolved problem. In this paper, a vision of an end-to-end network slice resource allocation process will be presented allowing to give fast feedback to a network operator or tenant on the feasibility of embedding new network slices. The associated research challenges will be discussed, especially focusing on the more complex Radio Access Network (RAN) resource allocation.

1 Introduction

The fifth generation of mobile networks (5G) covers a wide variety of novel use cases, such as the Internet of Things (IoT) and the industry of the future (Industry 4.0), requiring massive Machine Type Communication (mMTC). Furthermore, highly safety and security critical use cases, like autonomous driving and vehicular communication, require Ultra-Reliable and Low Latency Communication (URLLC). But also, traditional enhanced Mobile Broadband (eMBB) applications like HD video streaming and augmented reality must be taken into account. These diverse use cases enforce several radically different requirements on mobile networks. Network slicing is seen as the key concept of future 5G mobile networks, which aims at making the networks flexible enough to cope with those divergent requirements by dissolving the traditional concept of one monolithic mobile network serving all purposes [1]. A network slice is an isolated end-to-end virtual network containing all required resources and network functions to fulfill specific service requirements based on fixed Service Level Agreements (SLAs). Usually, several network slices share the same physical infrastructure. Network slices might be instantiated, modified or terminated dynamically or on

L. Barolli et al. (Eds.): IMIS 2018, AISC 773, pp. 695–704, 2019.
https://doi.org/10.1007/978-3-319-93554-6_68

short notice [2]. Like any network virtualization, network slicing necessitates a strong decoupling of software-based network functions from their underlying network infrastructure, as proposed by the 5G NORMA project [3]. This motivates enhancements in network programmability, flexibility and network sharing and allows to differentiate between three potentially distinct parties involved in network slice instantiation and management: The network infrastructure provider, which is the owner of the physical network elements, like antenna sites, data lines and computational hardware, the mobile service provider managing virtualized and physical resources and the tenant owning one or several network slices [4].

A customer portal provided by a mobile service provider allowing tenants to easily configure and order new network slices is envisioned. The tenant shall receive instant feedback on the feasibility of his or her request accompanied by a cost estimation for the set up and operation of the requested network slice via the portal. If the network slice request is not fully realizable, changes that would make it feasible are proposed to the tenant.

In the virtual mobile network several network slices may have been instantiated and are already running on the physical or virtualized network, also referred to as the substrate network. Therefore, a concept of how to decide whether an incoming network slice request can be accepted is necessary. To answer this question, the available resources in the network and the required resources of the running network slices as well as of the requested network slice have to be considered. Efficient Virtual Network Embedding (VNE) as well as Wireless Virtual Network Embedding (WiNE) algorithms and heuristics for embedding virtual end-to-end network slices into mixed physical networks with both wired and wireless network elements are required. Furthermore, resource and demand predictions are needed. This concerns the research areas of data analysis and Machine Learning (ML).

It is assumed that the mobile communication network is a self-optimizing mobile network (see [5]). Thus, further network optimization efforts regarding network slice deployment can be omitted. But still network slice resource allocation of the RAN and WiFi hotspot resources in an end-to-end communication system is the most demanding part of the overall VNE task [6], since the backbone of each cell usually provides enough resources for the data transport and therefore is usually not the primary bottleneck. Due to that, the main focus will be on the RAN and WiFi resource allocation, which is then extended on the core and transport network.

2 Related Work

Due to the ongoing work on network virtualization, VNE is getting increasing attention in current research. The VNE problem has been proven to be NP-hard [7], but a lot of publications can be found on heuristics for efficient solutions of the general VNE problem. However, there is a lack of concepts for embedding virtual end-to-end networks, i.e., network slices, onto a common mixed wired and wireless physical network. While the core network embedding is very well researched (see [8] for a recent survey), there are only few publications on wireless network embedding (see for instance [9] and [10]). Moreover, the survey paper [6] of Richard et al. compares several recent proposals on mobile network slice embedding. These approaches have in common that they deal with network slice resource allocation at runtime, i.e., focusing on resource partitioning and sharing as well as on network slice isolation. Network slice isolation refers to the problem of assuring that the slice specification is not violated because of changes in another slice. The authors of [6] criticize that the evaluated algorithms are described very vague and that they are based on assigning a certain number of Physical Resource Blocks (PRBs). This way no performance guarantees can be given. Varying chancel conditions make it hard to determine a suitable number of PRBs. Richard et al. [6] state that higher-level variables, such as a percentage of the total resources, should be used instead of PRBs. Considering these results, this work abstracts from PRBs and directly draws on network performance parameters, like throughput and latency.

Nevertheless, several promising algorithms have been published during the last years. For example, Esposito et al. propose a distributed approach for the NP-hard network slice resource allocation problem in form of a consensus-based auction mechanism [11]. Two possible policy configurations are analyzed. In the Single Allocation Distributed (SAD) slice embedding only one bid on one virtual node can be made per auction round. In contrast to that, the Multiple Allocation Distributed (MAD) slice embedding allows to bid on several virtual nodes simultaneously. While MAD has a lower convergence time, SAD results in a better load balancing and is faster in deciding on the feasibility of a network slice embedding. The virtual links are embedded in the next step, after the virtual nodes have been assigned completely. Yang et al. [12] use a karnaugh-map based heuristic for an efficient virtual network embedding. In a first step, the wireless resources are divided by, e.g., Time Division Multiple Access (TDMA) or Frequency Division Multiple Access (FDMA) into resource blocks. In the second step, the virtual networks are assigned to the resource blocks using a karnaugh-map based concept. The authors show that this is a feasible and efficient way of wireless virtual network embedding. However, in both approaches the actual available throughput provided to a network slice, depending on varying channel conditions, as well as important capabilities, like latency and accessibility are not considered. Moreover, they don't allow to analyze possible resource overbookings and assume steady resource provisioning and utilization. In [13] Vassilaras et al. propose an Integer Linear Programming (ILP) model for finding an optimal solution for a simplified network slice embedding problem. They state that the solution with ILP would take tens of minutes. In order to achieve faster run times, an efficient heuristic targeting at a near-optimal solution is required. However, developing such a heuristic remains an open research question. In addition to that, Vassilaras et al. mainly focus on problems related to network slicing at run time, for instance, the authors shed light on the challenges regarding end-to-end latency constraints, heterogeneous networks, multitenancy and slice fairness as well as the issue of dynamic network slicing and online optimization.

In contrast to that, this paper concentrates on challenges regarding the network slice acceptance and does not analyze questions related to the implementation, e.g., network slice isolation and fairness at runtime. Beyond that, the publications above assume steady resource demands, which is not realistic due to user mobility and fluctuations of network utilization, for instance during rush hour. In this paper a three-step process of how to make a well-founded decision on the feasibility of a network slice embedding will be given. The process estimates resource availability and demands in advance of a potential network slice deployment. Such an end-to-end view of allocating resources of a mobile network to network slices including the backbone and allowing potential overbooking of the substrate network is a novel and unique approach in literature.

3 Concept of Network Slice Resource Allocation

In this section, a process to decide whether or not a new network slice request should be allowed in a given substrate network, with already running network slices, is presented. A flowchart of this process on a high abstraction level is shown in Fig. 1. For the sake of a fast and efficient decision making on the acceptability of the network slice request, it is advisable to first of all check the technical feasibility of the network slices request (see step 1). That means to determine whether or not the SLAs of the requested network slice could be fulfilled, while in a first step only considering the network infrastructure requirements. For instance, imagine a network slice requiring a very low latency in a certain area with very low tolerance for packet loss and delay. Such an area must be fully covered with cells of 4G technology or higher. However, if this is not the case the network slice request must be considered as being technically not feasible. In such a case, it is of course important to identify the shortcomings of the substrate networks preventing the network slice acceptance. This might help to propose a mitigated version of the network slice request and its accompanying SLAs to the customer. In addition to that, the identified weak points of the current network infrastructure might also serve as an indicator for mid- and long-term network planning for the infrastructure provider. Otherwise, if the technical feasibility check is successful, the remaining resources of the substrate network have to be compared to the estimated required resources of the new network slice in the second step. Due to the large variety of partially interdependent network resources and the large number of requirements of the SLAs, usually based on different Key Performance Indicators (KPI) to be monitored during network slice operation, this is a pretty complex task. The fact that the RAN resources, e.g., throughput and latency, cannot be constantly provided, since the environment bears a lot of potential interferences that can hardly be predicted, makes this allocation task even more complex. For instance, heavy rainfall, foliage and temperature can have a significant impact on the signal quality resulting in a high effect on the radio resources available within the network. But also, the actual resource utilization of the deployed network slices usually underlies some variabilities or even unforeseen behavior. A very conservative approach to cope with these uncertainties would be assuming the guaranteed SLAs as a guideline to determine the maximum resources a network slice is allowed to occupy. However, for the physical network resource estimation such a lower benchmark does not exist. A possible approach would be to analyze the amount of resources provided over time and set up confidence intervals that could be used as a guideline. With such an approach the SLAs should be fulfilled at nearly any time with a certain confidence. However, modern RAN, especially in the Mobile Broadband (MBB) domain, are not capable of such an overprovisioning, since transmission frequencies are a limited resource and the mobile data traffic is ever increasing. Furthermore, the prices per transferred Mbit of data are falling, while the Operating Expenditures (OPEX) for the network infrastructure tend to increase with rising customer needs. To remain profitable an overbooking of the network resources seems to be unavoidable, but such overbookings must be carefully evaluated in advance (see step 3). Before being able to decide whether or not an overbooking of a certain resources can be accepted, it is essential to carefully evaluate the expected available RAN resources. Historical network performance data as well as multi-seasonal time forecasting might be suitable for this task. Also, the running network slices and their past as well as current resource utilization have to be evaluated. Moreover, predictions of the future demand of the network slices can be made often based on insider knowledge, like for example that a certain autonomous driving network slice has been set up only recently and its utilization of the booked network resources is expected to grow considerably in the next months. All those factors as well as experiences from past network slice allocations have to be taken into account to get a prediction of the remaining resources in the network that is as accurate and reliable as possible. This prediction has to be compared with the predictions for the network slice request made on the basis of the SLAs and service utilization specifications provided by the tenant. This comparison will be made with a VNE approach, which is described in more detail below.

As already explained above, the network slice resource allocation can result in overbooking of especially the RAN resources. The risk of violating SLAs can be subdivided into three main classes according to its impact on the service quality. An SLA violation with only minor impact is defined as a disruption that will hardly be noticed by the end-user of the network slice and therefore only is a Quality of Service (QoS) disturbance. Disruptions of service quality that are likely to be noticed by the end-users, i.e., affecting the Quality of Experience (QoE), are seen as major disturbances, while the third and most serious disruption would be network down times. Mobile service providers should define an upper limit of a probability for such a disruption for each of those three categories of network shortcomings. These boundaries might not only depend on the business goals of the mobile service provider, but also on the SLAs and the penalties that have been agreed on individually for each network slice. Therefore, they might also be network slice specific. To decide on an overbooking issue, the expected probability of violating the SLAs needs to be calculated for each of the tree categories and compared with the maximum tolerated probabilities of violation as defined in the business policies. Obviously, an overbooking can only be accepted when the expected probability of SLA violation in all three categories is less or equal to the maximum tolerated likelihood. When the overbooking bears an acceptable risk of SLA violation, the network slice request can be accepted, otherwise the bottlenecks should be identified in order to support future network planning. Therefore, the resource shortages should be determined, for example too little bandwidth within a certain area.

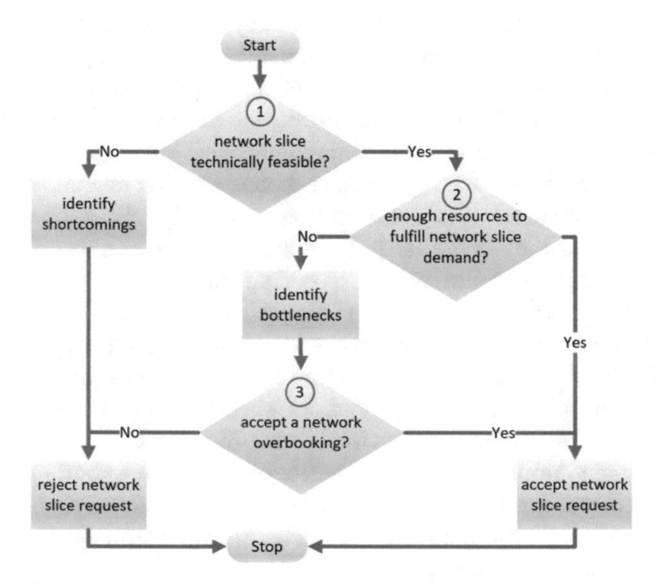

Fig. 1. Process of network slice resource allocation

4 Example of Network Slice Request Analysis

To illustrate the network slice resource allocation process, a quite academic and simplified example shall be given for the RAN part of the network. Imagine that a new MBB slice for the campus of the University of Augsburg shall be added to the existing network. Only one other network slice, also an MBB network slice covering the whole city of Augsburg, is deployed in the network. The network slice request might demand a maximum throughput per User Equipment (UE) of 200 Mbps in Uplink (UL) and an average throughput per UE of 90 Mbps in Downlink (DL) as well as a maximum throughput per UE of 50 Mbps and an average throughput per UE of 25 Mbps in UL. For the sake of simplicity, a fulfillment rate of 100% is assumed for the campus of the University of Augsburg during daytime and the evening hours (6:00–24:00 o'clock), from Mondays to Sundays with up to 1,000 concurrently connected UEs, with an assumed idle time of about 90%. The campus site is covered by three non-overlapping LTE Advanced cells providing a maximum throughput for UL of 300 Mbps and DL of 100 Mbps per UE. Each cell is able to handle up to 2,000 UEs concurrently and provides a total throughput of 3 Gbps in UL and 1.5 Gbps in DL. In order to keep this example as simple and intuitive as possible resource fluctuations are ignored here. The already running MBB city of Augsburg slice, overlapping with the requested university campus slice, is specified as follows: maximum/average throughput for the DL of 50 Mbps/45 Mbps and for the UL 25 Mbps/20 Mbps with a fulfillment rate of also 100%, 24 h a day, 7 days a week, with up to 45,000 concurrently connected UEs and the same 90% idle time as in the university campus slice. For checking the technical feasibility of the campus slice request in this simplified example we only have to compare the required and provided maximum UL and DL rates of the three cells installed at the university campus. In the second step the resources have to be evaluated. Therefore, we can focus solely on the three LTE Advanced cells on the university campus. First of all, the actual network utilization of the city of Augsburg MBB network slice at the university campus area during day and evening time have to be analyzed. It is assumed that historical data shows that in average about 2,000 UEs from the city of Augsburg network slice are connected to the campus cells and that the data traffic they produce and their idle times do not differ much from the average of the whole slice. Furthermore, it is assumed that in the future the number of users of the city of Augsburg network slice in this area will decrease by about 50% when the new campus slice is deployed, since most of the students would want to switch to the faster university campus slice. When simply multiplying the number of UEs with their average idle time and their average throughput rate, a total throughput of 4.4 Gbps in DL is expected for the city of Augsburg network slice on the campus area. The tree cells provide an overall throughput for DL of 9 Gbps. However, the new university campus slice requires a total throughput in DL of about 8.8 Gbps, which is much more than the expected available remaining resources of 4.6 Gbps.

In order to answer the question whether or not an overbooking would be acceptable, the probability of violating QoS and QoE for each network slice as well as the expected network down time need to be calculated and compared to the individual business policies of the mobile service provider. A risk estimation as shown in Table 1 can be calculated based on confidence intervals for the network slice resource provisioning and utilization. Therefore, the probability distributions of the resource availability as well as of the resource utilization for each resource have to be compared and aggregated per network slice. However, this is quite challenging, since diverse potential disruptions have to be considered and categorized. In this example an overbooking is likely to cause massive violations of QoS and significant QoE violations in the university campus network slice. According to the business policies the requested slice must be rejected.

When thinking about which changes could be made to be able to accept the university campus network slice, even in this very simplified example several alternatives are possible. The most promising options would be to drastically reduce the maximum and average throughput for the DL each UE is allowed to consume, to reduce the required fulfillment rate or to reduce the number of concurrently connected UEs that are admitted.

Table 1. Risk of SLA Violation and Business Policies

	City slice		Campus slice	
	Risk	Policy	Risk	Policy
QoS violation	5%	10%	80%	10%
QoE violation	1%	1%	60%	3%
Down time	0.09%	0.1%	0.09%	0.1%

5 Vision and Challenges of Mobile Network Slice Embedding

The second step of the network slice allocation process, presented in Sect. 3, is the most important step of the decision-making process. In this step, the predicted available network resources of the substrate network have to be allocated on the estimated resources that will be needed by the existing and the new network slice. In order to do this, the physical network resources as well as the network slice resource demands have to be modeled. Both models are represented in form of undirected graphs, consisting of nodes and edges. For the substrate networks the nodes are representing different kinds of network elements, amongst others: UEs, cells, switches, servers and cloud computers. The edges model the data connections between the network nodes. This can be mobile radio connections, transport links as well as core network links. The graphs representing the end-to-end network slices only consist of the end-nodes of the communication links required by the network slice description. In this case, an end-node can be any UE or server communicating via the network slice. For example, in a car-to-x network slices the vehicles, the roadside-infrastructure as well as services providing for example current weather information are the end-nodes of the communication network. However, modeling every single potential end-to-end connection within a network slice won't be possible for the most network slices. Therefore, it is proposed to aggregate mobile UE connections on cell level, i.e., several UEs connected to the same cell and using the same service are aggregated to one end-to-end connection. Obviously, the end-nodes and cell-nodes of the virtual networks, representing the network slices, can be directly mapped onto their according representation in the substrate network. The network slice model would usually not predefine a routing for the data between two end-nodes through intermediate nodes of the substrate network. The mapping of all end-to-end links in the virtual networks on the paths of the substrate network has to be made by a VNE algorithm regarding the resources each link and node element of the physical network provides and the expected resource utilization of the network slices. The resources involved are the throughput and latency of a communication link, but also CPU power and storage of e.g., cloud and mobile edge computing services. Furthermore, specific RAN capabilities are required, for instance, coverage of a certain area, specific services, like security services, mobility capabilities or guarantees on robustness and availability. Additionally, the duration and time window a network slice is active is an important parameter of the network slice embedding. The models will abstract from resource isolation and sharing which could be done by, e.g., using TDMA or FDMA and considers, for instance, the throughput available in a certain place as a resource to be shared among several tenants instead. Here, the varying available throughput resources depending on the channel quality and the Signal-to-Interference and Noise Ratio (SNIR) an UE experiences, which is due to the distance between the transmitter and receiver as well as obstacles like buildings, hills or vegetation and interferences by other antennas, plays an important role. The second most important resource of the physical network model is the data packet latency. The minimum time a data packet needs to be transported from the sender to the receiver is called the end-to-end latency. It depends on the used technologies and the network infrastructure hardware. The delay is the extra time a packet needs if the channels are crowded and the optimal latency cannot be achieved. However, the end-to-end latency plus delay, i.e., the actual time a data packet needs to be transported from the sender to the receiver, is also usually referred to as latency.

Network slice embedding in an end-to-end mobile network introduces specific challenges. Compared to ordinary VNE instances the nodes and links of the substrate network graph are annotated with a quite large number of different resources, capabilities and further parameters. Since the network resources can have a high variability it is helpful to annotate the graph not only with the average expected values of the resources to be allocated, but with a whole probability distribution of the resource availability. This allows to determine confidence intervals for the availability of the required amount of resources and to give the best possible indication on the probability or, in other words, the risk that resource shortages occur during network operation. This is also useful when analyzing the degree of overbooking and the associated risks in the third step of the network slice resource allocation process. Additionally, some resources are tightly coupled to each other, for instance the packet latency and delay of a mobile communication link is highly dependent on the packet size and the available throughput, since the amount of data that can be sent in parallel depends upon the available bandwidth. It is clear that a delay of zero can only be achieved when there are enough remaining throughput resources for instant data transmission using the maximum feasible bandwidth. In order to enable a correct resource allocation such interdependencies have to be considered in the network resource model. Beyond that, the vision of a customer portal allowing a tenant to configure new network slices and receive instant feedback on the feasibility of his or her request requires a very fast and efficient solution of the network slice embedding and the risk analysis. Consequently, the computational complexity of the network slice embedding heuristic has to scale for large and complex problem instances.

6 Conclusion and Outlook

Network slicing is seen as one of the key features to cope with the diverse requirements arising with the fifth generation of mobile networks. But yet, network slice resource allocation is still an unsolved issue. In this paper, a process for network slice resource allocation that allows to decide whether to accept or to reject an incoming network slice request taking into account the individual business policies and risk tolerance of the mobile service provider is presented. The approach considers network resource overbooking as a way of profitable network operation. Additionally, a vision of an end-to-end network slice resource allocation based on VNE has been presented.

The next steps will be evaluating suitable VNE and WiNE algorithms, methods of network resource estimation and network slice resource utilization as well as developing concepts for the network disruption risk estimation. Future steps will include the identification of bottlenecks and shortcomings of the network and concepts for proposing changes that transform a non-viable network slice request into a feasible one.

References

1. Nokia, Dynamic end-to-end network slicing for 5G, Espoo, Finland (2016)
2. 3GPP, Study on management and orchestration of network slicing for next generation network, TR 28.801 V15.0.0T (2017)
3. 5G NORMA. https://5gnorma.5g-ppp.eu. Accessed 22 Apr 2018
4. 5G NORMA, 5G NORMA network architecture - intermediate report, Deliverable D3.2 (2017)
5. Hämäläinen, S., et al.: LTE self-organizing networks (SON). Wiley, Great Britian (2012)
6. Richart, M., et al.: Resource slicing in virtual wireless networks: a survey. IEEE Trans. Netw. Service Manag. **13**(3) (2016)
7. Andersen, D.: Theoretical approaches to node assignment, Computer Science Department (2002)
8. Fischer, A., et al.: Virtual network embedding: a survey. IEEE Commun. Surv. Tutor. **15**(4) (2013)
9. Riggio, R.: Scheduling wireless virtual networks functions. IEEE Trans. Netw. Serv. Manag. **13**(2) (2016)
10. Tsompanidis, I., Zahran, A., Sreenan, C.: A utility-based resource and network assignment framework for heterogeneous mobile networks. In: 2015 IEEE Global Communications Conference, San Diego (2015)
11. Esposito, F., Di Paola, D., Matta, I.: A general distributed approach to slice embedding with guarantees. In: IFIP Networking Conference. Brooklyn, NY, 1–9 (2013)
12. Yang, M., Li, Y., et al.: Karnaugh-map like online embedding algorithm of wireless virtualization. In: The 15th International Symposium on Wireless Personal Multimedia Communications, Taipei, pp. 594–598 (2012)
13. Vassilaras, S., et al.: The algorithmic aspects of network slicing. IEEE Commun. Mag. **55**(8), 112–119 (2017)

The 7th International Workshop on Frontiers in Innovative Mobile and Internet Services (FIMIS-2018)

An Intelligent Opportunistic Scheduling of Home Appliances for Demand Side Management

Zunaira Nadeem[1], Nadeem Javaid[2](✉), Asad Waqar Malik[1], Abdul Basit Khan[1], Muhammad Kamran[1], and Rida Hafeez[1]

[1] School of Electrical Engineering and Computer Science (SEECS), National University of Sciences and Technology (NUST), Islamabad 44000, Pakistan
[2] COMSATS Institute of Information Technology, Islamabad 44000, Pakistan
nadeemjavaidqau@gmail.com
http://www.njavaid.com

Abstract. Demand side management plays a vital role in load shifting to off peak hours from on peak hours in response to dynamic pricing. In this paper, we propose an optimal stopping rule (OSR) and firefly algorithm (FA) for the demand response based on cost minimization. Each appliance gets the best opportunistic time to start its operation in response to dynamic electricity pricing. The threshold based cost is computed for each appliance where each appliance has its own priority and duty cycle regardless of their energy consumption profile. Numerical simulations show that our proposed scheme performed well in lowering cost, waiting time and peak to average ratio.

1 Introduction

The exponential rise in energy demand accompanied by the continuous decline in energy generation requires an ongoing up gradation in todays energy infrastructure. Academia and research industry have considered it to be the serious concern in order to address the future energy needs. The idea of a smart electricity system has moved the concept of the conventional grid to the smart grid (SG). Smart grid impersonates a vision of the future electric generation system integrated with the advanced sensing technology, two way communication at transmission and distribution level in order to efficiently supply smart electricity in a smart way. Reliability, cost saving, self-healing, self-optimization, consumer friendliness pollutant reduction are a few of the many benefits of SG. This modern grid is motivated by several social, economic and environmental factors.

Demand side management (DSM) is the modification of consumer energy demand for energy in response to variation in electricity prices. DSM programs consist of planning, implementing and monitoring activities of utility, particularly designed to encourage consumers to modify their level of using electricity. In literature, there are different DSM strategies are proposed such as load shifting,

L. Barolli et al. (Eds.): IMIS 2018, AISC 773, pp. 707–718, 2019.
https://doi.org/10.1007/978-3-319-93554-6_69

valley filling, demand response (DR) etc. Demand response is defined as changes in end-user electricity usage behaviour in response to the variation in electricity prices over time. The DR program includes price based and incentive-based DR programs.

In this study, we have proposed a mathematical technique optimal stopping rule (OSR) compared with the meta-heuristic algorithm firefly (FA) which opportunistically schedule home appliances based on priority constraint. The opportunistic scheduling refers to the best starting time of an appliance based on their priority. Each appliance has its own priority and length of operation time. The appliance with high priority has high cost and lesser waiting time and vice versa. So we can clearly see the trade off between electricity cost and waiting time.

The simulation is conducted on single home appliances. Appliances are categorized into three categories: Shiftable appliances, Non Shiftable appliances and uninterrupted appliances. The main objective of this work is the reduction in electricity cost. In addition, the yearly cost saving of appliances are demonstrated.

The remaining paper is organized in the following way. Section 2 describes the related work. Section 3 explains the system model. Section 4 presents the simulation results. In the end, the whole summary is concluded in Sect. 5.

2 Literature Review

Previously, the load scheduling problem was solved by non-integer linear programming (NILP), linear programming (LP), integer linear programming (ILP) and mixed integer linear programming (MILP). For example, in [1], a MILP is formulated to solve the scheduling problem. MILP is precise and efficient, but on the other hand, it is not capable of handling complex scenarios. The proposed technique is implemented in the realistic scenario which helps price makers in the realistic market. Today the world is more dependent on the SG so energy management is one of the serious concerns of today's world. In [2], the scheduling problem is formulated for the household task which helps the consumers to save their money. The system relies on different pricing schemes, local energy generators and flexible task based on different deadlines.

In [3] the two of the heuristic techniques, shuffled frog leaping (SFL) and teaching and learning based optimization (TLBO) algorithms are performed for modeling energy consumption scheduling problem. The results show that the proposed techniques performed well for cost reduction. The authors in [4] proposed real-time opportunistic scheduling technique based on OSR. The scheduling problem for home appliances is performed in centralized and distributed pattern. The comparison of OSR and LP is performed and the simulation results depict that OSR performed well and it is less complicated in comparison to LP. In the paper [5], the real time pricing (RTP) based environment is presented to solve opportunistic scheduling for home appliances using OSR. The purpose of this work is user comfort while considering cost reduction. The modified first come first serve scheduling algorithm is proposed which compares with first come first serve and early deadline first scheduling algorithms. The simulations validate the performance of proposed techniques of cost and user comfort. In [6], the energy management controller (EMC) is proposed based on heuristic techniques: binary particle swarm optimization (BPSO), genetic algorithm (GA) and ant colony optimization (ACO). Furthermore, multiple knapsack problem is used to formulate the problem. The price schemes used for simulation are the inclined block rate and time of use. In this paper, the users are categorized into active and passive users. The desired objectives of the work are cost reduction, peak to average ratio (PAR) minimization and user comfort maximization. The simulations conducted show that GA based management controller worked well than the ACO and BPSO based EMC. The authors in [7] consider three major divisions; commercial, residential and industrial. A large set of several types of controllable devices is handled to solve an optimization problem using heuristic techniques. Results demonstrate that reasonable cost saving can be done by considering PAR. In [8], the problem is highlighted to manage load without paying extra money, previously to manage load threshold limit is applied for each home if the consumer cross that limits additional charges are applied in their bill. To avoid this problem there is a need to propose an efficient load balancing technique which can manage the load with minimum cost while reducing the waiting time of electrical appliances. In this paper, a multi-objective optimization technique is used for load balancing. The result reveals that the proposed techniques efficiently minimize the electricity bill and reduce the waiting time of appliances.

Zunaira et al. in [9] proposed the three hybrid techniques OSR-GA, OSR-TLBO and OSR-FA. The simulations are conducted on single home as well as multiple homes. The proposed techniques are formulated via chance constrained optimization. The desired objective of this work is cost minimization while considering user comfort and PAR.

3 System Model

The proposed system model consists of home area network (HAN), advanced metering infrastructure (AMI), energy management controller (EMC), smart meter (SM), RTP signals, smart appliances (SA) and programmable logic controller (PLC). The clear picture of all these devices can be seen in Fig. 1.

3.1 HAN

A HAN is a network deployed with in the home that connects a person devices, such as computers, telephone, video games, SM, home security system and all those appliances that requires Wifi. HAN support wired and wireless technology such as Zigbee, Wifi, WiMax, etc. Typically, HAN consists of a broadband internet connection that connects multiple devices via a third party wired or wireless modem. The EMC receives the pricing information from the power grid through SM via an internet [2].

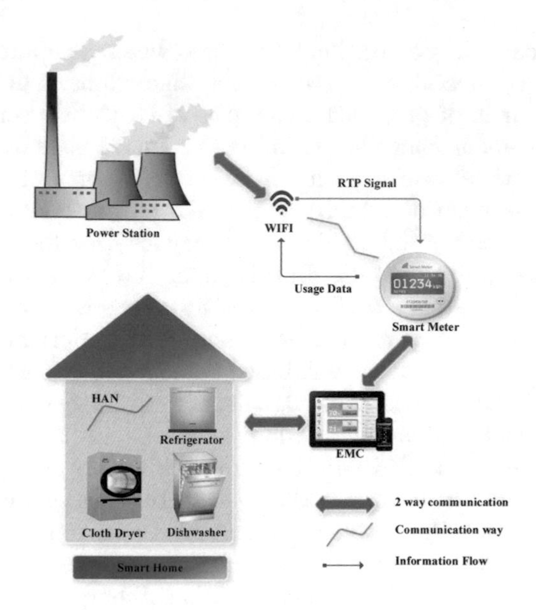

Fig. 1. Proposed system model

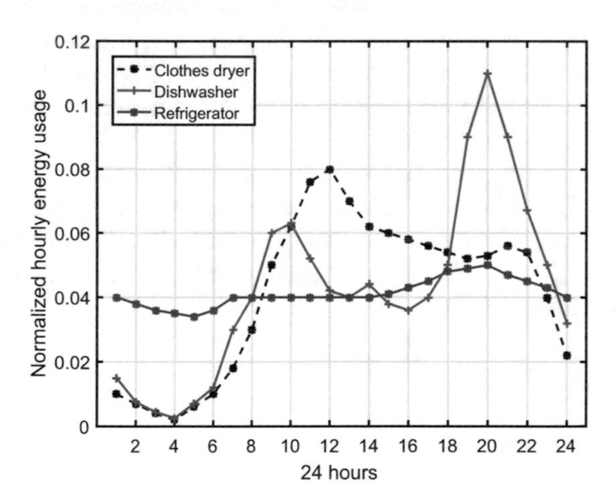

Fig. 2. Energy profile of appliances

3.2 AMI

AMI allows the utility to get energy consumption information from the smart home by a SM to utility. It is the two way communication between consumer and utility that integrates SM, communication networks, and EMC.

Electricity cost is calculated according to the price signal provided by the regulatory authority. In this work, the RTP price signal is used, which is more realistic and give accurate results for load shifting to off peak hours from on peak hours. In RTP, prices of every next hour vary while it remains the same throughout the specific time slot.

3.3 EMC

An EMC is a digital device used to monitor the appliances and their energy consumption rating. It contains all the information about appliances, power rating, priorities, threshold, length of operation and based upon these parameters and the inducted algorithm HEMC take a decision and provide low cost schedule pattern for the appliances. In this paper, OSR and FA are considered as an optimization algorithm.

3.4 SA

We divide a day into time slots, denoted as T and consider that there is a set of different appliances, denoted as A1, A2, A3, ..., An. Each appliance has its own energy profile. Energy profile of each appliance can be shown in a Fig. 1. The energy cycle of the appliance is for 24-h time slots. Each time slot is considered as 1 h which is denoted by the parameter τ. In this paper, we are concerned with Shiftable appliances. The three main appliances are cloth dryer, dishwasher, and refrigerator.

The appliances which take more time to accomplish their task can take more than one time slot. The dishwasher completes its task in three stages such as main wash, final rinse, and heated dry. The cloth dryer has less than an hour duty cycle. It accomplishes its task in only one cycle. The refrigerator has two main stages ice-making and defrost, and only defrost stage of the refrigerator can be scheduled for electricity cost reduction.

3.5 PLC

PLC helps to implement the optimization algorithm provided to it and it provides the visualization through the graphical interface between appliances, SM, and the controller.

3.6 Problem Formulation

The objective function of the propose solution is where ϵ_T stands for electricity cost:

$$minimize(\sum_{t=1}^{24}(\sum_{ap=1}^{n} \epsilon_T + WaitingTime)) \quad (1)$$

Cost. The given equation is used to calculate the cost where Pr_{hour} is power and P_r^{ap} is electricity price and $\chi(\tau)$ is on off status of appliances:

$$\epsilon_T = \sum_{hour=1}^{T}(\sum_{ap=1}^{n}(Pr_{hour} \times P_r^{ap} \times \chi(\tau))) \quad (2)$$

Load. Load can be computed by the equation below where ϵ_t stands for load and $\chi(\tau)$ indicates the on off status of appliances:

$$\epsilon_t = \sum_{\tau=1}^{T} \sum_{ap=1}^{n} P_r^{ap} \times \chi(\tau) \tag{3}$$

PAR. PAR highlights the load peaks and helps to balance the load. It can be formulated by equation below:

$$PAR = \frac{max(\varepsilon_n^{\tau})}{\frac{1}{T}\sum_{\tau=1}^{T}\sum_{n}^{An}(\varepsilon_n^{\tau})} \forall\ T = 24 \tag{4}$$

Threshold. P is the electricity price uniformly distributed between [po, pp] where ρ_o is the lower limit and ρ_p is the maximum limit. The appliances are turned on once their electricity price is lesser than the threshold. This equation is taken from [4].

$$Z^* = \sqrt{\frac{2(\rho_p - \rho_o)\mu\tau}{\varepsilon}} + \rho_o \tag{5}$$

Waiting Time. Waiting time or delay of the appliance is considered as the avg. waiting time of an appliance. It can be computed by given equation where tś is the starting time of an appliance and tr is the requested time for an appliance:

$$WaitingTime = tr - t^{'}s \tag{6}$$

4 Simulation Results and Discussion

We assume three shiftable appliances dishwasher, clothdryer and refrigerator for scheduling. The initial parameters are given in Table 1 and the energy profile of all appliances is shown in Fig. 2. All these appliances are plotted on two different priorities. To avoid randomness we have taken 10 iterations of the whole population.

Table 1. Parameters of appliances

Appliances	Average power (kW)	LOT (hours)	Priority (μ)
Clothdryer	3.0	0.75	[0.001, 0.13]
Dishwasher	0.8	1.75	[0.001, 0.015]
Refrigerator	0.089	24	[0.0033, 0.0099]

4.1 Clothdryer

The Figs. 3 and 4 show the performance of cloth dryer in terms of average cost and waiting time, respectively. The Fig. 5 shows the PAR of cloth dryer. Cloth dryer has only one main cycle. The simulations for cloth dryer is summed up in Table 2.

Table 2. Cloths dryer simulation statistics

Technique	Priority	Average cost/month	Reduction in cost (%)	Average delay in hours/day
OSR	0.001	69.5299	70	6.5706
	0.13	151.9062	35	3.5833
FA	0.001	129.7632	45	8.6357
	0.13	134.7632	43	6.6758

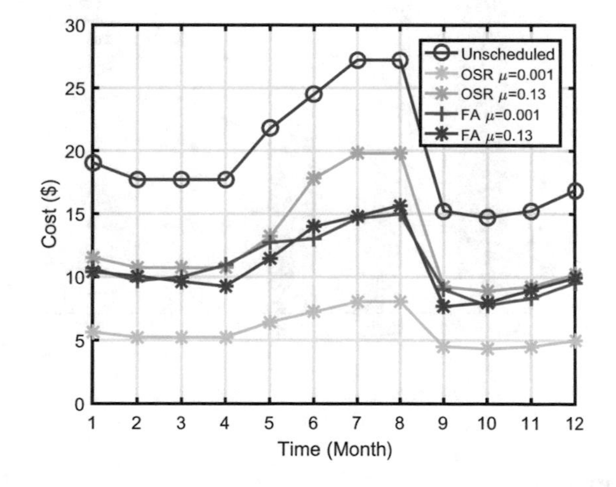

Fig. 3. Cost of cloth dryer

In Table 2, OSR and FA are compared with priority 0.001 and 0.13. In the case of OSR, the cost is 35% reduced with priority i.e., 0.13 and 70% reduced with priority i.e., 0.001. When we compare FA with unscheduled cost about 45% of the cost is saved with priority 0.001 and 43% of the cost is saved with priority 1.13. If we compare both techniques, in this case, OSR performs better than the FA. However, the delay time of OSR and FA for high priority i.e., 0.13 is comparatively less. It clearly shows the trade off between waiting time and cost.

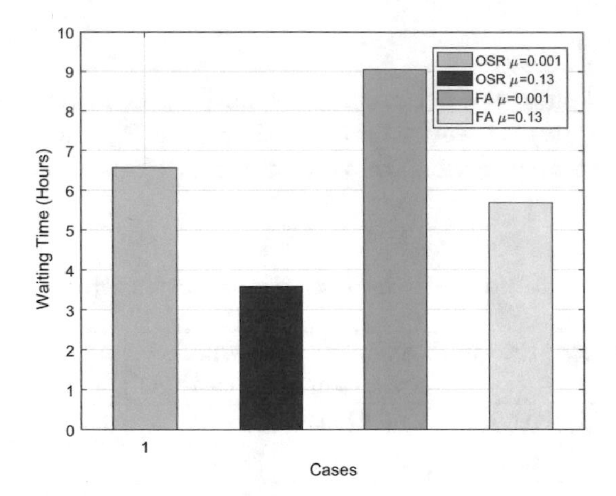

Fig. 4. Waiting time of cloth dryer

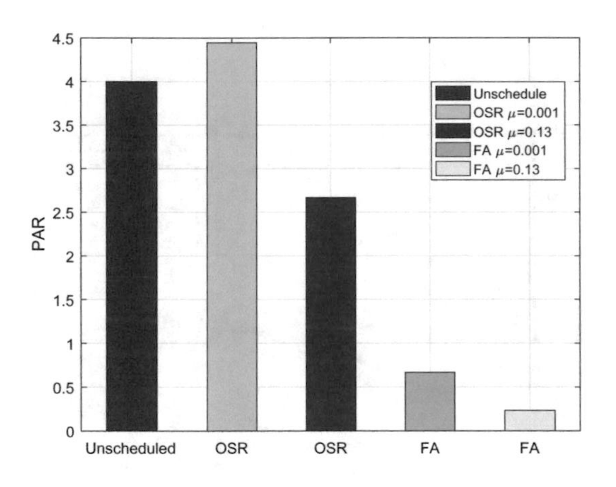

Fig. 5. PAR of cloth dryer

4.2 Dishwasher

The Figs. 6 and 7 illustrate the average cost and average waiting time of dishwasher for each month. The result of the simulation is summarized in Table 3. We compare two techniques OSR and FA for dishwasher schedule. Our target is to find the one which gives the best time schedule for a dishwasher with minimum cost and delay. Simulation results demonstrate that 15.9%, 46% of cost saving with a minimum delay of 2.02 and 4.3 respectively is calculated with priority 0.015 and 62.49%, 47.71% of cost reduction with a minimum delay of 5.0 and 6.2 respectively is calculated when priority is 0.001. By comparing their results, we noticed OSR perform optimal than FA in terms of average cost and delay. The Fig. 8 shows the PAR of the dishwasher.

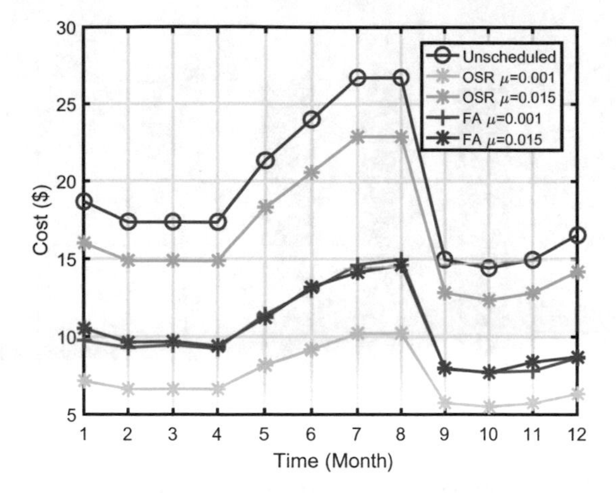

Fig. 6. Cost of dish washer

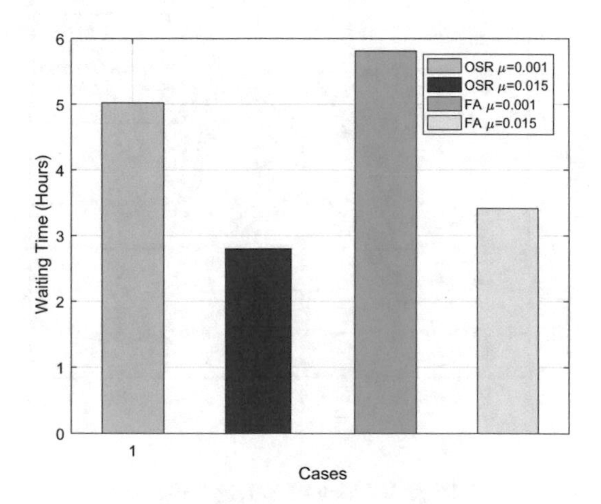

Fig. 7. Waiting time of dish washer

Table 3. Dishwasher simulation statistics

Technique	Priority	Average cost/month	Reduction in cost (%)	Average delay in hours/day
0SR	0.001	88.1046	62.49	5.0233
	0.015	197.5074	15.9	2.0244
FA	0.001	122.8069	47.71	6.2783
	0.015	124.6949	46	4.3631

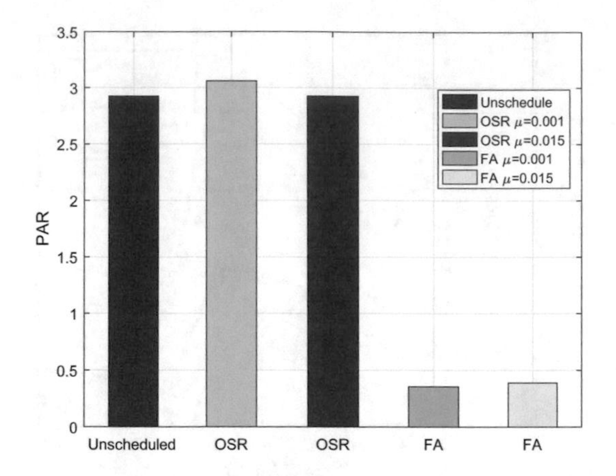

Fig. 8. PAR of dish washer

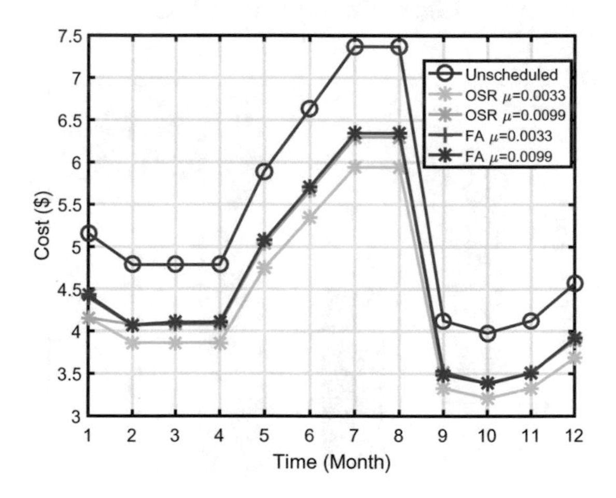

Fig. 9. Cost of refrigerator

Table 4. Refrigerator simulation statistics

Technique	Priority	Average cost/month	Reduction in cost (%)	Average delay in hours/day
OSR	0.0033	51.2714	17	6.1875
	0.0099	54.0010	15	4.1875
FA	0.0033	52	14	9.0625
	0.0099	54	12	7.0052

4.3 Refrigerator

The refrigerator is operated 24 h a day. There are two states of refrigerator: Icing state and defrost state we schedule the defrost stages of the Refrigerator.

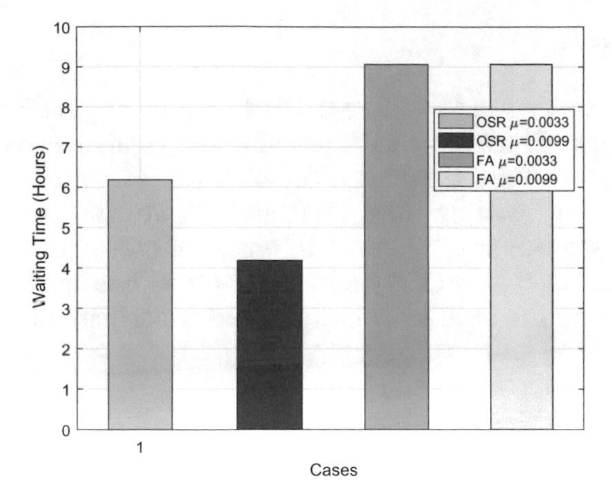

Fig. 10. Waiting time of refrigerator

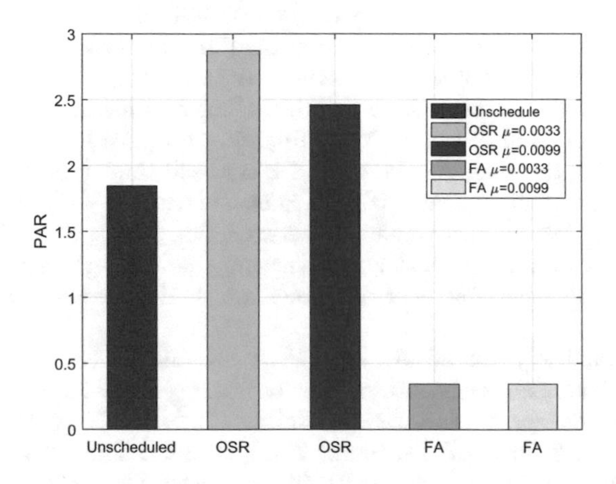

Fig. 11. PAR of refrigerator

The Figs. 9 and 10 show the average cost and average waiting time of refrigerator for each month and Fig. 11 shows the PAR. In order to obtain the required objectives, we use two techniques to see the effects on cost and average waiting time. The summary of refrigerator is summed up in Table 4. Simulation results demonstrate that TLBO with priority 0.0099 reduced 15% of cost with an average delay of 4.18 h but in the other case when the priority of refrigerator is 0.0033, 17% of the cost is reduced with an average delay of 6.1 h. Similarly, FA is compared with unscheduled cost with priority 0.0099 it saves 12% cost and when it decreases to priority 0.0033, it saves 14%. In the case of a refrigerator, OSR performs well as it saves more cost than FA.

5 Conclusion

The heuristic techniques for residential DR based on the RTP scheme are proposed in our paper. Three different types of home appliances with different priorities and duty cycle are considered to solve the cost minimization problem while considering the waiting time. OSR and FA are the proposed algorithms and their performances are compared in terms of cost and waiting time. The simulation results show that OSR performed better than the FA. Our proposed techniques facilitate the residential customer to participate in DR program.

References

1. de la Torre, S., Arroyo, J.M., Conejo, A.J., Contreras, J.: Price maker self-scheduling in a pool-based electricity market: a mixed-integer LP approach. IEEE Trans. Power Syst. **17**(4), 1037–1042 (2002)
2. Derin, O., Ferrante, A.: Scheduling energy consumption with local renewable micro-generation and dynamic electricity prices. In: First Workshop on Green and Smart Embedded System Technology: Infrastructures, Methods, and Tools, April 2010
3. Derakhshan, G., Shayanfar, H.A., Kazemi, A.: The optimization of demand response programs in smart grids. Energy Policy **94**, 295–306 (2016)
4. Yi, P., Dong, X., Iwayemi, A., Zhou, C., Li, S.: Real-time opportunistic scheduling for residential demand response. IEEE Trans. Smart Grid **4**(1), 227–234 (2013)
5. Rasheed, M.B., Javaid, N., Ahmad, A., Awais, M., Khan, Z.A., Qasim, U., Alrajeh, N.: Priority and delay constrained demand side management in realtime price environment with renewable energy source. Int. J. Energy Res. **40**(14), 2002–2021 (2016)
6. Rahim, S., Javaid, N., Ahmad, A., Khan, S.A., Khan, Z.A., Alrajeh, N., Qasim, U.: Exploiting heuristic algorithms to efficiently utilize energy management controllers with renewable energy sources. Energy Build. **129**, 452–470 (2016)
7. Logenthiran, T., Srinivasan, D., Shun, T.Z.: Demand side management in smart grid using heuristic optimization. IEEE Trans. Smart Grid **3**(3), 1244–1252 (2012)
8. Muralitharan, K., Sakthivel, R., Shi, Y.: Multiobjective optimization technique for demand side management with load balancing approach in smart grid. Neurocomputing **177**, 110–119 (2016)
9. Nadeem, Z., Javaid, N., Malik, A.W., Iqbal, S.: Scheduling appliances with GA, TLBO, FA, OSR and their hybrids using chance constrained optimization for smart homes. Energies **11**(4), 888 (2018)

Fog Computing Based Energy Management System Model for Smart Buildings

Saman Zahoor[1], Nadeem Javaid[1](✉), Adia Khalid[1], Anila Yasmeen[1], and Zunaira Nadeem[2]

[1] COMSATS Institute of Information Technology, Islamabad 44000, Pakistan
nadeemjavaidqau@gmail.com
[2] National University of Science and Technology, Islamabad 44000, Pakistan
http://www.njavaid.com

Abstract. In this article, a three layered architecture is proposed for smart buildings. A fog based infrastructure is designed and deployed on the edge of network, where fog processes the private data collected through the smart meters and stores the public data on cloud. Further, end user has facility to schedule and control the home appliances by using a centralized energy management system. Moreover, the electricity and network resources utilization charges can be calculated. We analyze the performance of cloud based centralized system, considering the fog computing as an intermittent layer between system user layer and cloud layer and without considering fog computing. Simulation results prove that fog layer enhances the efficient utilization of network resources and also reduces the bottleneck on the cloud computing.

Keywords: Smart grid · Smart building
Fog computing and Cloud computing

1 Introduction

To meet the increasing power demand, new power plants are being deployed. However, this deployment has not shown any evident change in load demand fulfillment, instead they increase the global warming and caused climate change.

To overcome the traditional grid challenges, smart grid (SG) is considered as a next generation power distribution network which consists of substations, transmission lines and transformers etc. Utilization of information communication technology (ICT) in SG offers a fully observable power distribution network, where both the utility and consumer interact with each other for information sharing. This is helpful for demand side management (DSM) i.e., load balancing and electricity cost minimization, and also effective and reliable energy transmission. However, the number of consumers increase day-by-day and the computing

L. Barolli et al. (Eds.): IMIS 2018, AISC 773, pp. 719–727, 2019.
https://doi.org/10.1007/978-3-319-93554-6_70

requirements also fluctuate. Therefore, the main issue is how to provide the ICT resources, such as data storage devices and processing units with flexibility and dynamic availability to support DSM.

In order to resolve above discussed issues, the concept of cloud computing is used which organizes the computing resources as a utility, such as power networks, to provide the ICT services to consumers. Basically, it is a distributed internet service, where the resources and shared information is provided in a cloud network (public or private) with a connection of clients, data centers and servers. The SG must consider security, reliability, privacy, scalability and flexibility when it stores its information in the cloud via smart meters. Moreover, as the number of customers connected to SG increases, the complexity and solution to these problems become difficult.

For large scale problems, the concept of fog computing is used, which provides effective solutions such as providing delay sensitive and location awareness (i.e., vehicular ad-hoc networks) [1], managing Internet of Things data [2,3], and improving website performance [4]. Fog is basically a small cloud based infrastructure established at the user end, first described by Cisco. In fog computing, decentralized devices communicate with each other to perform processing tasks and reduce the need for cloud operators. They can process information with high bandwidth and low latency. However, as compared to a cloud based system, fog system keeps private data local and allows customers to access the data in a fast and secure manner.

In this paper, we focus on energy management of SG and propose a reference model that integrates cloud computing and fog computing into SG. Fog computing provides us the facility of separate storage of public and private data in a distributed manner, which can reduce latency, provide reliability and improve security in SG. Moreover, for public data processing and storage, cloud computing is also used. We also present a scenario where different number of homes, renewable sources and energy storage systems are considered where each component information is sent to fog and fog interacts with cloud.

The remainder of the paper is organized as follows: Sect. 2 discusses the related work, which includes both the cloud-based and non-cloud based architectures. In Sect. 3 we propose system model. Simulation results are presented in Sect. 4. Finally, Sect. 5 concludes the paper.

2 Related Work

Increasing popularity of centralized systems emerged the need of energy management controller's users towards the centralized energy management system. As authors described in [5], the current centralized power systems have become an established and well-developed method that cannot be fully replaced by other new approaches. In this respect, [6] proposed the centralized energy management system for the microgrid where modeled microgrid is presented as three-phase unbalanced system. Tushar *et al.* in [7] presented the centralized energy control system for home appliance scheduling, and charging and discharging control unit for electric vehicle in order to reduce the cost and optimization of power usage. Another article [8], proposed centralized energy management system for optimal scheduling and transactive energy for direct current residential distribution systems. The targets of the proposed system are cost reduction, efficiency and stability of a residential distribution systems. Presented models in [6–8] are cost effective and efficient, however, emergent centralized systems demand for a infrastructure where in a single community multiple users can efficiently utilize an energy management system. Whereas, above mentioned systems are defined for a single user.

In order to provide a flexible power management system, a platform is required which can provide interoperability and interactivity among the devices [9] which is implementable by integration of cloud computing with an energy management system [10]. In this regard, a cost efficient cloud based information management and computational system for SG is proposed by [11]. In order to balance the electricity load and computation cost reduction of datacenter, [12] proposed a bi-level optimization problem. This is formulated as linear problem and solved by using branch and bound method. Simulation results show that proposed solution reduced up-to 46% energy cost of datacenter. Rather than using the centralized system of utilities for communication among smart energy hub and utilities, [13] proposed a cloud based communication infrastructure. The proposed system not only provides the communication medium but also a storage unit. Another cloud-based communication model is presented by [14], where objective of research is allocation of cloud resources in order to reduce the cost and improve the system efficiency.

Articles [11–13], improve the efficiency of communication network and resource allocation for energy management system, however, fast communication medium and response time is difficult to achieve. This is possible by shifting the cloud resources near the network users which is known as fog computing [15]. In this regard, [16] proposed a fog computing model for demand side management in order to reduce the communication and data transfer cost. Also, an application based fog computing model is designed to schedule the home appliances. In this system, fog computing is embedded as an intermittent layer between cloud layer and internet of energy layer. Presented system in [16] reduced the energy consumption of data transmission by using the fog computing. In the article [9], home and microgird energy management systems are implemented as a service on fog computing for efficient utilization of energy resources. Another fog computing framework is presented by [17,18]. The infrastructure of [17] is based on Intel Edison, which is efficient in terms of data storage space reduction and power consumption minimization.

3 Proposed System Model

Before the detailed explanation of emerging concept of SG with cloud computing, we discuss a short overview of SG. When we integrate information and communication technology into traditional grid, it becomes SG. For its efficient

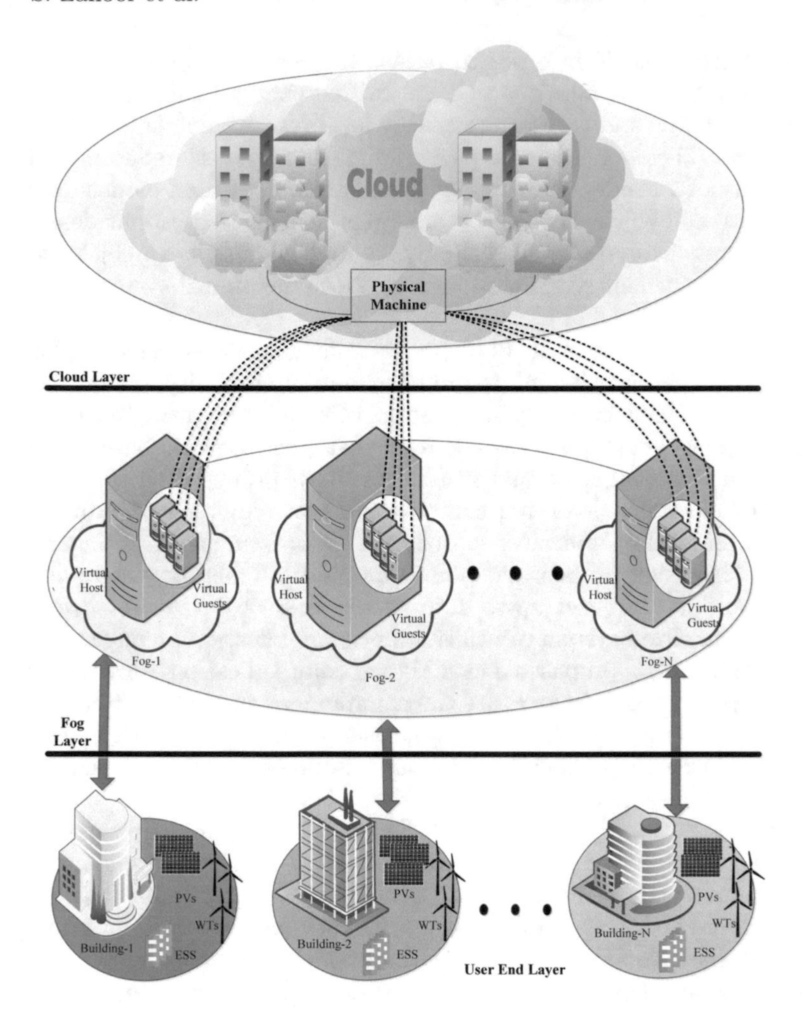

Fig. 1. SG with fog and cloud computing system model.

management, we consider a cloud computing platform which is depicted in Fig. 1. The purpose of this model is to overcome the limitations of SG such as delay, data security, optimized asset utilization, high reliability and cost minimization.

The proposed model works in three-layer architecture, where first layer is composed of various end users who want to use numerous cloud computing resources to run their applications. The second layer involves fogs where each fog has a virtual host and each host has multiple guests (virtual machine) for running first layer requests. This can be done for effective resource utilization and latency issue management. The third and last layer contains physical machines for permanent data storage. The communication topologies among layers are: device to device communication, smart devices to fog layer communication and fog layer to cloud layer communication.

At first layer or end user layer, we consider N number of buildings and each building has its on renewable power generators and energy storage systems. These type of generators have no emission and fuel cost and they are environmental friendly due to extraction of energy from natural sources. Further, the excessive generated energy is stored in energy storage systems to fulfill the load demand of building in low generation hours. All information about building consumption, generation and scheduling is sent to fog layer by smart meter. These smart meters interact via local area network, wide area nctwork or Metropolitan area network. There are numerous wireless solutions for communication links, i.e., Wi-Fi, Z-Wave or ZigBee in SG.

In second layer, we consider the M number of fogs to handle the N number of building requests. Every virtual host has multiple virtual guests that can process the different types of requests coming from end user. These requests are of multiple types: request to start an application, request for information access and electricity and resource utilization cost calculation.

The last layer is the cloud layer, where a cloud datacenter contains physical machines. These physical machines receive and send information to and from the virtual guests according to the request and save the data into datacenter. Every machine has a CPU which is usually multicore, additionally, a physical machine efficiency is measured according to its memory size, storage capacity and network I/O. A cloud datacenter often serves many simultaneous users at any given time t.

4 Simulation Results and Discussion

In order to prove the effectiveness of the proposed framework, simulations results have been discussed in this section. These results are taken while considering two example scenarios; in first scenario cloud computing infrastructures is taken for demand side management while in the second scenario fog computing. In the first scenario, we consider 20 homes which are connected with a cloud where each home requests network resources according to the demand. The cloud has two host and each host has five virtual machines. Whereas, for the second scenario we take two groups and each group contains ten homes, and each group connects with the fog.

Execution time of user queries and response time according to the packet size is shown in Figs. 2 and 3 respectively. Figure 2 illustrates execution time increases with increasing packet size. However, for packet size 0.8 (bits) execution time is 0.4 (ms) whereas, execution time for packet size 0.26 (bits) is 0.8 (ms), which is because of virtual machine size assigned for this particular task. The response time graphical representation in Fig. 3 depicts that response time of networks increases with packet size.

Execution time and response time for scenario two is shown in Figs. 4 and 5. Packet size for both scenarios are taken same except packet size 153.126 (bits) which shows very high execution as well as response time. These graphs show that execution time and network response time vary according to the packet size. However, the fluctuation found in execution time of cloud computing is not

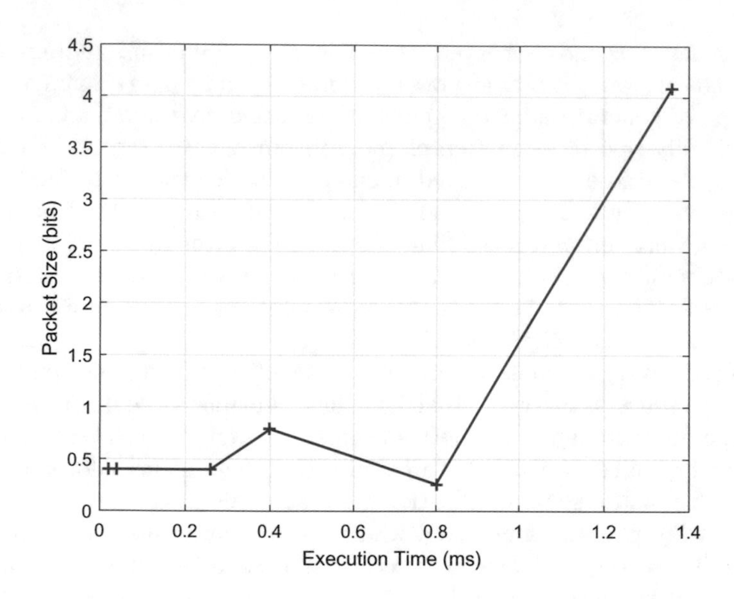

Fig. 2. Execution time according to packet size for cloud computing scenario.

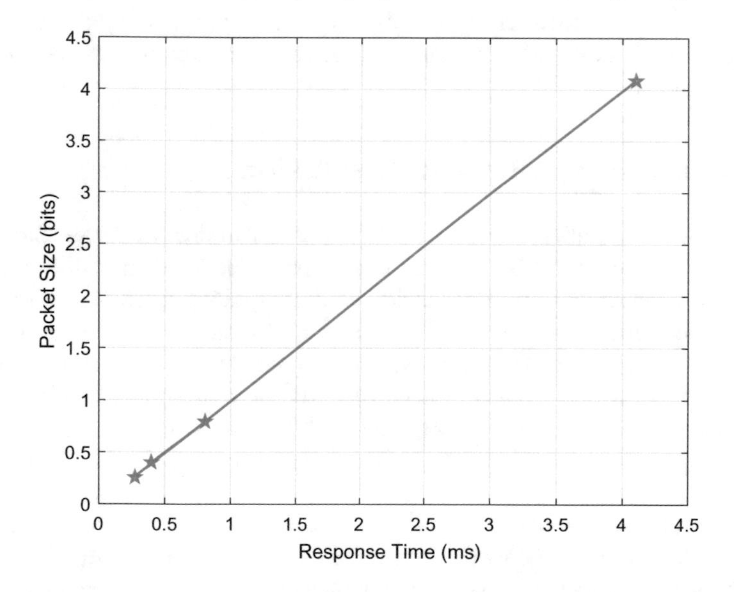

Fig. 3. Network response time according to the packet size for cloud computing.

observed in execution time of fog computing scenario. Also, average resource utilization ratio of scenario 1 and 2 is 0.222 and 0.517 respectively, which shows the effectiveness of proposed system as response time of both scenarios are almost same with respect to packet size. Whereas, resource utilization ratio of fog computing is high.

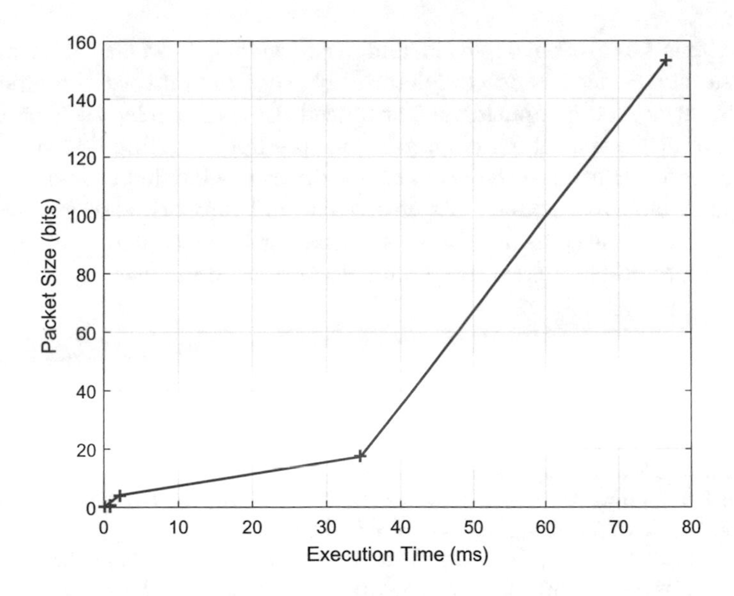

Fig. 4. Execution time according to the packet size for fog computing example scenario.

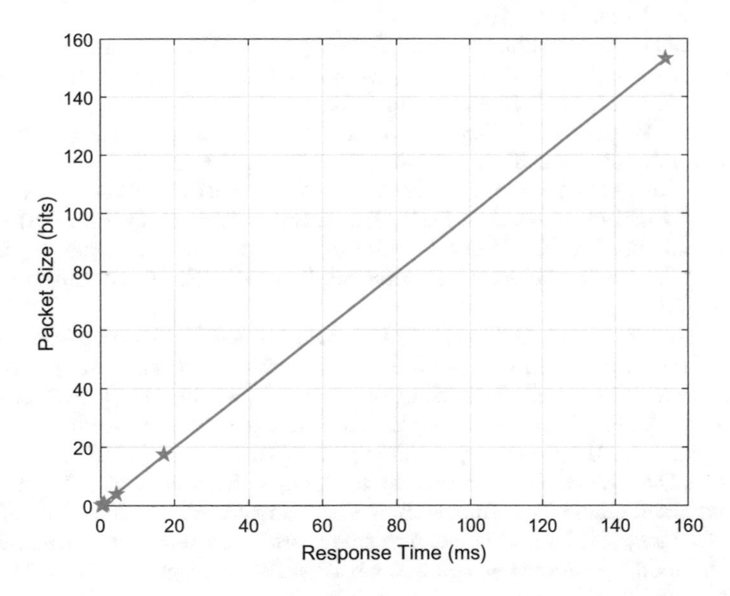

Fig. 5. Network response time along with packet size for fog computing.

5 Conclusion

In this paper, we have presented a fog and cloud based infrastructure for smart buildings. Centralized power systems are more efficient in terms of cost and resource utilization. However, growing number of users results in network bottle neck which can degrade the efficiency of network. To resolve such issues, cloud based systems have been deployed and used widely, however, it is unable to resolve the bottle neck issue completely. In order to improve the efficiency of cloud computing, fog computing is introduced. In this article, we have analyzed the affective utilization of fog computing for a smart building. We analyze how efficiently fog layer utilizes resources and reduces the burden on cloud. Further, we allow the user to calculate his electricity and network resource utilization cost and allow the user to use the energy management control unit in order to schedule and remotely control the home electricity appliances.

References

1. Rahimi, F., Ipakchi, A.: Demand response as a market resource under the smart grid paradigm. IEEE Trans. Smart Grid **1**(1), 82–88 (2010)
2. Bonomi, F., Milito, R., Zhu, J., Addepalli, S.: Fog computing and its role in the Internet of Things. In: Proceedings of the First Edition of the MCC Workshop on Mobile Cloud Computing, pp. 13–16. ACM (2012)
3. Hong, K., Lillethun, D., Ramachandran, U., Ottenwalder, B., Koldehofe, B.: Mobile fog: a programming model for large-scale applications on the Internet of Things. In: Proceedings of the Second ACM SIGCOMM Workshop on Mobile Cloud Computing, pp. 15–20. ACM (2013)
4. Zhu, J., Chan, D., Prabhu, M.M.S., Natarajan, P., Hu, H.: Improving web sites performance using edge servers in fog computing architecture. U.S. Patent Application 13/904,327, filed, 15 May 2014 (2014)
5. Pourbabak, H., Chen, T., Zhang, B., Su, W.: Control and energy management system in microgrids. arXiv preprint arXiv:1705.10196 (2017)
6. Olivares, D.E., Canizares, C.A., Kazerani, M.: A centralized energy management system for isolated microgrids. IEEE Trans. Smart Grid **5**(4), 1864–1875 (2014)
7. Tushar, M.H.K., Assi, C., Maier, M., Uddin, M.F.: Smart microgrids: optimal joint scheduling for electric vehicles and home appliances. IEEE Trans. Smart Grid **5**(1), 239–250 (2014)
8. Yue, J., Zhijian, H., Li, C., Vasquez, J.C., Guerrero, J.M.: Economic power schedule and transactive energy through an intelligent centralized energy management system for a DC residential distribution system. Energies **10**(7), 916 (2017)
9. Al Faruque, M.A., Vatanparvar, K.: Energy management-as-a-service over fog computing platform. IEEE Internet of Things J. **3**(2), 161–169 (2016)
10. Markovic, D.S., Zivkovic, D., Branovic, I., Popovic, R., Cvetkovic, D.: Smart power grid and cloud computing. Renew. Sustain. Energy Rev. **24**, 566–577 (2013)
11. Fang, X., Yang, D., Xue, G.: Evolving smart grid information management cloudward: A cloud optimization perspective. IEEE Trans. Smart Grid **4**(1), 111–119 (2013)
12. Wang, H., Huang, J., Lin, X., Mohsenian-Rad, H.: Exploring smart grid and data center interactions for electric power load balancing. ACM SIGMETRICS Perform. Eval. Rev. **41**(3), 89–94 (2014)
13. Sheikhi, A., Rayati, M., Bahrami, S., Ranjbar, A.M., Sattari, S.: A cloud computing framework on demand side management game in smart energy hubs. Int. J. Electric. Power Energy Syst. **64**, 1007–1016 (2015)
14. Cao, Z., Lin, J., Wan, C., Song, Y., Zhang, Y., Wang, X.: Optimal cloud computing resource allocation for demand side management in smart grid. IEEE Trans. Smart Grid **8**(4), 1943–1955 (2017)

15. Thakare, M.Y.A., Deshmukh, M.P.P., Meshram, M.R.A., Hole, M.K.R., Gulhane, M.R.A., Deshmukh, M.N.A.: A Review: The Internet of Things Using Fog Computing (2017)
16. Shahryari, K., Anvari-Moghaddam, A.: Demand Side Management Using the Internet of Energy based on Fog and Cloud Computing
17. Barik, R.K., Gudey, S.K., Reddy, G.G., Pant, M., Dubey, H., Mankodiya, K., Kumar, V.: FogGrid: leveraging fog computing for enhanced smart grid network arXiv preprint arXiv:1712.09645 (2017)
18. Okay, F.Y., Ozdemir, S.: A fog computing based smart grid model. In: 2016 International Symposium on Networks, Computers and Communications (ISNCC), pp. 1–6. IEEE (2016)

A Novel Indoor Navigation System Based on RFID and LBS Technology

Yizhuo Wang[1], Xuan Xu[1], Xinyu Wang[1], and He Xu[2,3](✉)

[1] Bell Honors School, Nanjing University of Posts and Telecommunications, Nanjing, China
{q16010105,q16010305,q16010101}@njupt.edu.cn
[2] School of Computer Science, Nanjing University of Posts and Telecommunications, Nanjing, China
xuhe@njupt.edu.cn
[3] Jiangsu High Technology Research Key Laboratory for Wireless Sensor Networks, Nanjing, China

Abstract. Indoor positioning technology is the key to further development of LBS system. Based on RFID technology, it is efficient and feasible to combine the indoor navigation management system with LBS technology. The LBS system provides the functions of mobile device location, communication and service, and RFID based indoor positioning provides the function of locating objects indoor environment. The combination of the two technologies can further facilitate the positioning and navigation in our lives. In this paper, a novel indoor navigation system based on RFID and LBS is presented. The implementation of this system shows that it is feasible to support service in indoor environment.

1 Introduction

With the development of mobile intelligent devices and location-based services, positioning system based on the GPS service gradually go into people's lives from the military field, now the service has become the necessary tools for people to trip, and also has a core role in all kinds of emerging industries such as bike sharing and car-hailing app. To some extent, it gives the convenience to people and meets the requirement in the outdoor travel. Location-based services have naturally become the focus of attention in increasingly complex indoor environments, especially in public places.

The national library tried to use indoor positioning and location service based on RFID as early as in 2008, but because of the limitation of the mobile intelligent device technology, the service did not achieve very good effect and not be promoted. The main reason is that the users' demand for location services is not only to know the goal's position on the map, often it also needs to be combined with its own position, or navigate, search, social networking and other additional services. This paper mainly designs an indoor positioning application based on mobile smart devices, LBS and RFID indoor location services.

L. Barolli et al. (Eds.): IMIS 2018, AISC 773, pp. 728–737, 2019.
https://doi.org/10.1007/978-3-319-93554-6_71

2 Requirements for Indoor Location Services

The core requirement of the users for indoor location service consists of two points: one is the positioning of one's position, and the other is the positioning of the location of the target [1, 2]. Since mobile smart phones are based on Wi-Fi and GPS positioning technology, there is still a lot of room for small objects such as books and commodities. For example, in the environment of a large supermarket, there are requirement of positioning the product and dividing them into regions. Shoppers can get the information of their needed goods, shopping cart's real-time location and relevant information on mobile devices, quickly finding their needed goods, making things convenient for the customers, improving the efficiency of the user and shopping experience. The characters of RFID electronic tags are cheap, passive and small size, which is suitable for a small range of objects with a wide variety of objects.

In today's Internet of things environment, indoor location services are bound to be widely used in various fields. For example, in the warehousing management of express parcels in the logistics field, electronic tags can not only carry relevant information, but also automatically locate the shelf location of the parcel. And electronic tags are easy to operate and can process a lot of information quickly. The sensitivity is high, and in an antenna range a large number of targets can be identified, which is suitable for the batch management environment and can save a large amount of human resources. In the open environment such as shopping mall and library, indoor location service also plays a role in the regulation of goods and books.

3 Overview of LBS and Indoor Positioning Technology

3.1 Concept of LBS

LBS is location-based services, and its components are usually mobile devices, location, communication networks, services and content [3]. It contains three basic contents: spatial information, social information and information inquiry. Positioning (spatial information) is the basis of LBS system. Service and content is the main body of the system, which includes bearing the weight of the function of analysis data, providing data, but also the most directly embodies the function of the system. Communication network is a bridge linking mobile device and service content, which is the key link to ensure the timeliness and feasibility of the system. Mobile devices are the direct contact with users, directly determining the value of the system in the eyes of users.

3.2 Indoor Positioning Technology

Nowadays, indoor positioning technology is developing widely, and there are several mainstream technologies, such as indoor positioning technology based on WiFi, ultrasonic and infrared. Each technology has different strengths and weaknesses and a wide range of applications. RFID technology has the advantages of low tag price, high accuracy and anti-interference [4].

Infrared has relatively mature technology. But it has relatively outstanding shortcoming, its penetration is extremely poor, and it is affected by the environment factor obvious. The complexity of the layout makes up for it, which leads to the increase its deployment costs. It is just suitable for simple trajectory identification, indoor robot positioning.

Ultrasonic technology imitates the bat positioning method, which has the cm-level accuracy. Because of ultrasonic propagation distance and penetrating, it is not suitable for larger positioning. Its layout is simple, which needs more sophisticated equipment to ensure measurement accuracy, and this greatly improves the positioning cost. At present, the system with ultrasonic localization includes Cricket location-support system and active bat system, and the application scenario has ocean survey.

The current popular WiFi location is based on the ieee802.11b standard, and the intensity of wireless signal is measured in the wireless LAN, and the method of matching signal strength is used for positioning. The location fingerprint library algorithm is generally adopted, and the typical system is the RADAR prototype system. Firstly, select a number of reference points in the system coverage area, collect its location information and signal strength, and then transmit the information to the data center to form the location fingerprint database. Then the signal intensity is measured by the nearest neighbor method. The system has low positioning accuracy and high-power consumption, which is widely applicable to indoor positioning of mobile devices.

Because the hardware cost of RFID technology is mainly in the reader, which does not need to be physically attached to the tag when reading the label, and a reader can process a large number of tags. Therefore, it has the characteristics of low cost, multi-objective and non-visual recognition. Labels can work in harsh, complex environments. Based on Landmark's algorithm location technology, the reference tag is introduced to reduce the number of readers and improve the accuracy, which is in line with the current indoor positioning technology.

4 The Principle of RFID Positioning Based on LANDMARK Algorithm

4.1 Structure of RFID System

RFID hardware system mainly consists of electronic tag, antenna and reader, which is shown in Fig. 1.

The inside of the electronic tag is the chip and the coupling element, which has the function of storing information. Tags are attached to objects, and each tag has a unique identification number, so the function of the tag is to connect the object to the system, and the positioning of the object is the positioning of the tag.

Antennas transmit signals and energy between the label and the reading device. An antenna can scan multiple tags, and a reader can connect multiple antennas. The layout of the antenna determines the range of positioning.

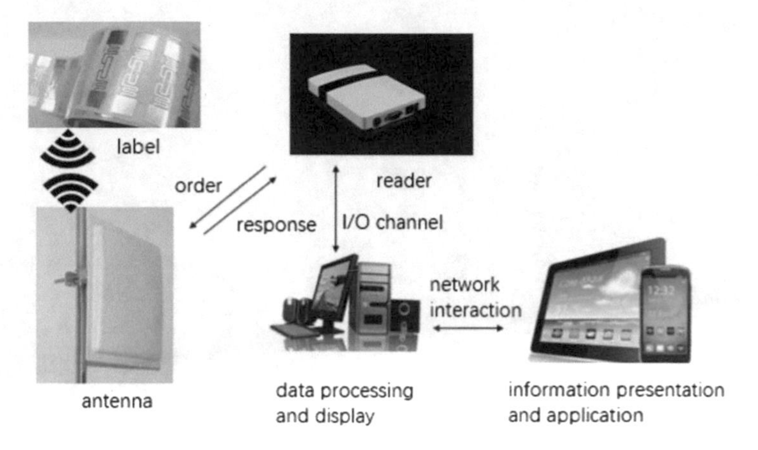

Fig. 1. The RFID system

The reader has the function of processing information, two-way communication through antennas and tags, reading and writing tag information. And it has the function of communication with computer, it is the bridge that connects RFID hardware module with computer software module.

4.2 Principle of Location Algorithm

This paper uses the LANDMARD (Location Identification Based on Dynamic Active RFID Calibration) algorithm to perform the dynamic positioning of RFID [5]. The coordinate system is set up according to the arranged antenna, and the reference tag is gridded. The system compares the detected target labels with reference labels by different antennas, and determines the reference labels which are closest to the target label signal index. The closer the reference label is to the data of the target label, the greater the weight of the reference label is. Finally, the coordinate system of the target labels is weighted according to the coordinate system of the reference labels. Figure 2 shows the LANDMARK system layout.

It is assumed that there are n antennas and m reference tags in the system. Signal intensity vector measured by reference labels on different antennas is defined as $\vec{S} = (S_1 + S_2 + S_3 + \ldots + S_n)$ and signal intensity vector measured by target labels on different antennas is defined as $\vec{Q} = (Q_1 + Q_2 + Q_3 + \ldots + Q_n)$. In this algorithm, the difference between the signal intensity of the reference label and the target label is calculated by Euclidean distance, and the Euclidean distance is defined as:

$$D_j = \sqrt{\sum_{i=1}^{n} (Q_i - S_i)^2}\,, \mathrm{j} \in (1, \mathrm{m})$$

D_j is the Euclidean distance between the target label and the jth reference label in antenna, the definition of nth reference labels' Euclidean distance vector is $\vec{D} = (D_1 + D_2 + D_3 + \ldots + D_m)$, and the k reference labels are selected with the

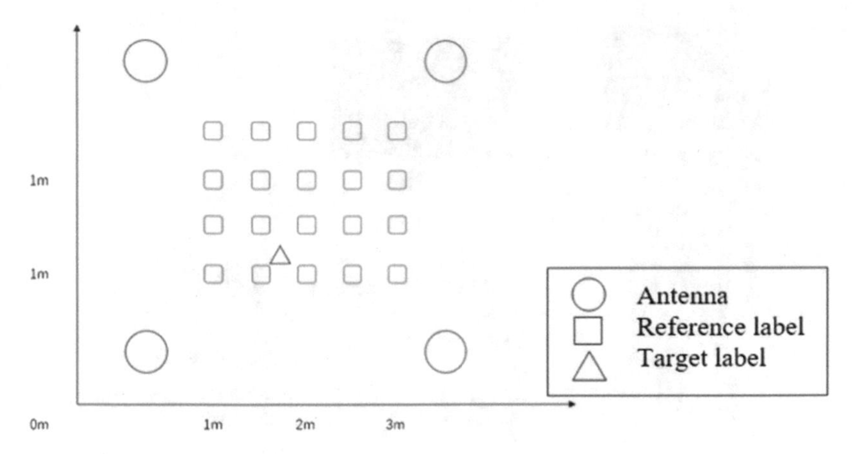

Fig. 2. Landmark system layout of a schematic diagram

smallest D value (k is usually not more than 20 integers), finally through the following formula, the label is measured coordinates:

$$(x, y) = \sum_{i=1}^{k} \frac{\frac{1}{E_i^2}}{\sum_{i=1}^{k} \frac{1}{E_i^2}} \cdot (x_i, y_i)$$

The actual coordinates of target label are set as (x_0, y_0), single positioning error is denoted as $A_{i,}$ the average measurement error of n times is denoted as $\bar{A}$. The formula of $A_{i,}$ and the average error of $\bar{A}$ is:

$$A_i = \sqrt{(x_i - x_0)^2 + (y_i - y_0)^2}$$

$$\bar{A} = \frac{\sum_{i=1}^{n} A_i}{n}$$

It is known from the above formulas that the results of the positioning are related to the value of K and the number of antennas. In general, when k = 3–6 and the number of antennas at 4 are the best.

5 The Design and Implementation of the System

5.1 An Overview of System

The system is divided into three modules: positioning hardware and layout module, information management module and application module.

The positioning module is divided into two parts, one is the location of the mobile device (the user client), and the other is the location of the target object. The location of mobile devices is integrated positioning which is usually based on the technology of

GPS, WIFI, Bluetooth, base station and so on. The technology has already been quite mature, and domestic and overseas map software also provides the corresponding API interface for developers to use [6, 7]. In addition, they are also starting to provide indoor maps of the big shopping malls. In order to locate the label, the antenna and the reference label should be arranged in advance, and the location information should be recorded.

The information management module includes the network transmission of information and the establishment of the server database, which is shown in Fig. 3. The database includes the map data, the label location data and the user location data, and the basic information of the user and the label.

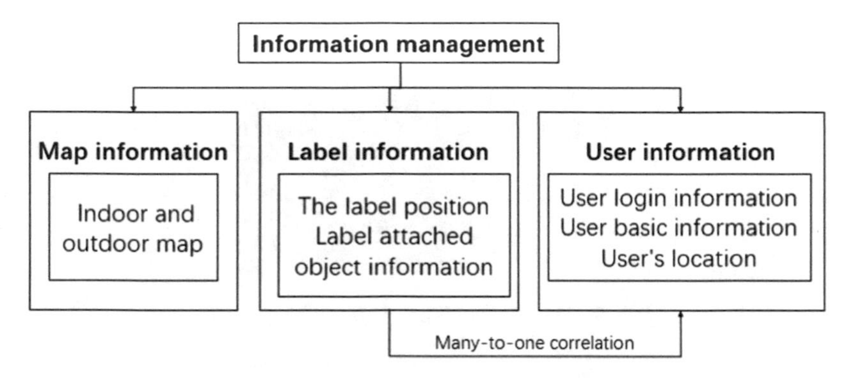

Fig. 3. Schematic diagram of information management

At present, LBS manufacturers have been collecting indoor maps of multiple floors such as shopping centers, libraries and other public places, and provide functions such as map scaling, shop search, peripheral search and so on. Using a mature LBS product for two developments is obviously more realistic and a more comfortable experience can be taken for users. Figure 4 shows the Google's indoor map.

The application module is responsible for displaying the information clearly and beautifully to the user and satisfies the operation of management, query and application of the information. The purpose of its core is to provide users with location and information of indoor objects, so that they can know the location and basic information of the goods without being at the scene, and can provide additional functions such as navigation, search, route planning and so on.

Taking the shopping mall as an example, its deployment environment has the characteristics of placing articles with fixed position, but the replacement of articles is not fixed frequently. Under the condition of satisfying the coverage area, the antenna and reference label are reasonable placed according to the space placement characteristics of the rack position. At this time, the antenna is fixed to the rack position, and the coordinate system covered by it has a fixed location information, and gets and records it. The tags need to input basic information of goods, such as the price, attributes and pictures, so that they can be reviewed by users, and at the same time, it is also convenient for merchants to manage.

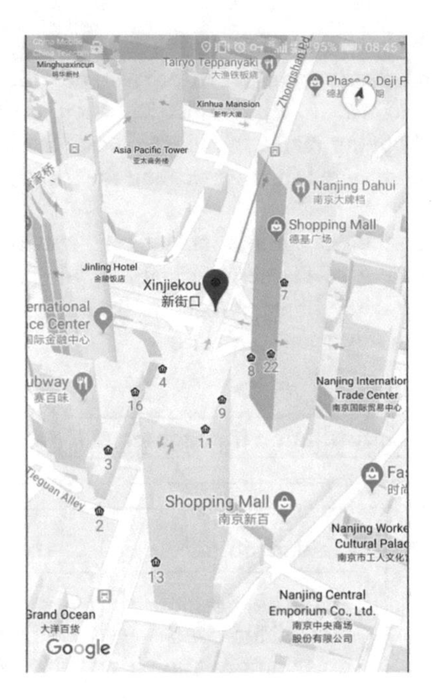

Fig. 4. Indoor navigation map in Google map

The antenna reads the RSSI value of the label, and then transfers it through the network to the cloud server. According to the location algorithm, the server calculates the relative coordinates of the measured labels relative to their relative coordinates, and the corresponding longitude and latitude values in the map, and stores them in the database. The server carries out the corresponding information processing service to the merchant and the positioning person. For business servers, the label information can be added, deleted, modified, and querying. Querying the location of the label and self-location, planning path, and providing the peripheral services can be provided for the positioning person. RFID positioning and LBS system will be combined to achieve mutual benefit.

Fig. 5. System demonstration in laboratory

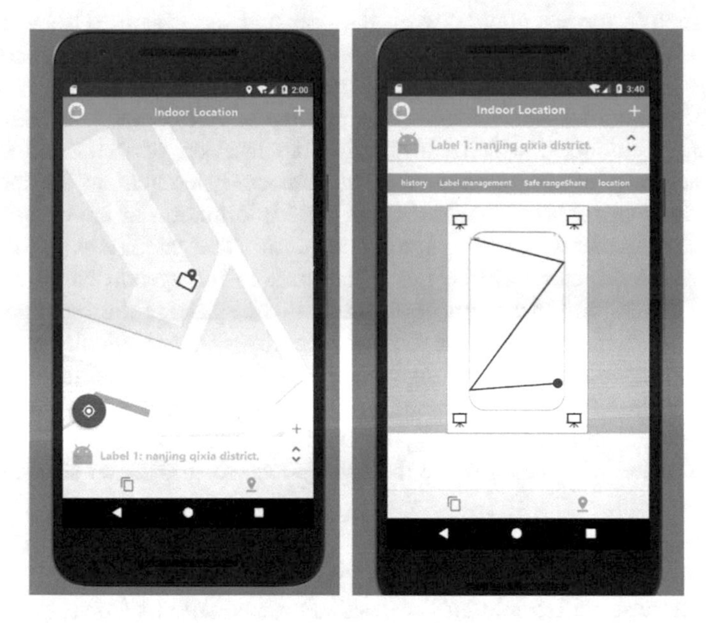

Fig. 6. Basic information, location and trajectory of the label.

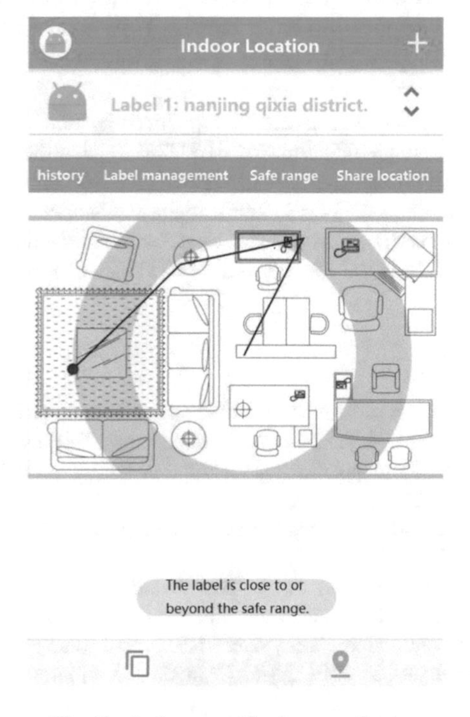

Fig. 7. Indoor positioning monitoring

Application is the window between the user and the system: User experience is critical. It should provide users with the most intuitive information query mode to attract more users and improve the applicability of the system.

The LBS system takes advantage of Baidu Map SDK to develop by using location and mapping techniques, which can provide the location of users and stores. The minimum accuracy of the general map software query content is the indoor store. Therefore, the query system should display more information of stores, such as commodity price, comments, location, introduction, etc. The positioning and navigation system based on the store will be further accurate to the system based on the commodity. For commodity management, the APP should provide the function of current position of the item, the movement track, security distance monitoring and other functions for further management. Figure 5 shows our laboratory's indoor positioning application environment. Figure 6 shows our realized application client for users. Figure 7 gives an indoor positioning monitoring application, when the monitoring target is out of the monitoring scope the application will trigger an alarm.

6 Conclusions

This paper discusses a feasible indoor positioning navigation scheme based on RFID and LBS navigation system, which can provide a great deal of convenience to both the management and the users. When it is commercially available, the information security should be further protected, and the indoor map should make reasonable adjustment according to the specific application environment. How to define a universal interface and framework for indoor position system is our future work.

Acknowledgments. This work is financially supported by the National Natural Science Foundation of P. R. China (No.: 61672296, No.: 61602261), Scientific & Technological Support Project of Jiangsu Province (No.: BE2015702, BE2016185, No.: BE2016777), China Postdoctoral Science Foundation (No.: 2014M561696), Jiangsu Planned Projects for Postdoctoral Research Funds (No.: 1401005B), Postgraduate Research and Practice Innovation Program of Jiangsu Province (No.: KYCX17_0798), and NUPT STITP(No.: XYB2017551).

References

1. Li, Y., Xu, X., Yang, Z., et al.: Location Information Service (LBS) Key Technology and Application. People's Post and Telecommunications Press, Beijing (2013)
2. Ding, Y.: Research on indoor positioning technology based on RFID. Southwest Jiaotong University, Chengdu (2014)
3. Hazas, M., Scott, J., Krumm, J.: Location-aware computing comes of age. IEEE Comput. Magaz. **37**(2), 95–97 (2004)
4. Toral, L., Gregor, D., Vargas, M., et al.: Distributed urban traffic applications-based on CORBA event services. Int. J. Space Based Situated Comput. **1**(1), 86–97 (2011)
5. Wang, J., Katabi, D.: Dude, where's my card? RFID positioning that works with multipath and non-line of sight. In: ACM SIGCOMM Computer Communication

6. Wang, E., Zhao, W., Cai, M.: Research on improving accuracy of GPS positioning -based on particle filter. In: Proceedings of 8th IEEE Conference on Industrial Electronics and Applications (ICIEA), pp. 1167–1171 (2013)
7. Sun, Y., Fan, P.: RFID technology and its application in indoor positioning. Comput. Appl. **25**(2), 1205–1208 (2005)

A Social Dimension View Model of Divergence of IoT Standardization

Toshihiko Yamakami(✉)

ACCESS, 3 Kanda-Neribei-cho, Chiyoda-ku, Tokyo 101-0022, Japan
Toshihiko.Yamakami@access-company.com

Abstract. As IoT (Internet of Things) continues to penetrate everyday life, we witness the increase in the number of IoT standardization bodies. As the coverage of standardization bodies overlap, interoperability is threatened despite of good wills of each standardization body. The author analyzes the causes of fragmentation of IoT standardization from the current landscape of IoT standardization. Then, the author presents a broken assumption model that explains the disorganized status of IoT standardization.

1 Introduction

With the progress of information communication technology, more and more things are connected to the Internet, and the trend to cooperate with each other is intensifying. It can be viewed as CPS (Cyber-Physical System).

This leverages an increasing number of IoT standardization spawning. It is an attempt to ensure interoperability Telecom industry, Internet industry, software industry, electronic industry, machinery industry, appliance industry, energy industry, healthcare industry and others attempt to bring interoperability of IoT in place. IoT standardization includes oneM2M, ITU, ISO/IEC, OMA, W3C, IEEE, IETF, ETSI, and many other standardization bodies and industrial consortia.

The random sprouts of IoT standardization may be viewed as a possible selfish act of egoistic business people. However, the author came to recognize that we need a new perspective of standardization to grasp the current landscape and cooperate to ensure better interoperability.

Without acknowledge the past assumptions and realities to have broken them, It is difficult to understand the current IoT standardization in a proper viewpoint.

The author discusses the patterns of IoT standardization. Then the author discuss the past assumptions that lead to the current situation of diverse IoT standardization spread to a wide range of different industries.

L. Barolli et al. (Eds.): IMIS 2018, AISC 773, pp. 738–747, 2019.
https://doi.org/10.1007/978-3-319-93554-6_72

2 Background

2.1 Purpose of Research

The aim of this research is to develop a framework to understand the IoT standardization landscape from the divergence viewpoint.

2.2 Related Work

Research on public transportation with IoT standardization consists of three areas: (a) IoT standardization activities, (b) landscape of IoT standardization bodies, and (c) real-world examples of standard-based IoT deployment.

First, in regard to standardization activities, Yun et al. presented interworking in oneM2M [7]. Raggett presented a W3C's plan of IoT [5]. Abou-Zahra et al. Discussed web standards for IoT [1].

Second, in regard to landscape of IoT standardization bodies, Anthopoulos et al. discussed matching smart city projects and smart city-related standardization [2]. Costa-Perez presented trends of the latest telecommunication standardization including oneM2M [3].

Third, in regard to real-world examples of standard-based IoT deployment, Kubler et al. presented a smart city example based on standards from EU Horizon 2020 IoT project [4].

The originality of this paper lies in its identification of view models to compare IoT standardization.

3 Method

The author performs the following steps:

- collect IoT standardization patterns,
- analyze the diverging and converging forces in IoT standardization,
- analyze the underlying assumptions to drive diverse IoT standardization,
- present a guideline to lead IoT standardization.

4 Observation

4.1 Landscape for Divergence

There are many standardization activities. Established standardization bodies like ITU-T and ISO/IEC have IoT standardization activities. Other academic and research fora like IEEE and W3C are also enthusiastic their takes of IoT standardization. There is an umbrella forum like OneM2M to provide a meta-framework on multiple-standardization-body activities in IoT.

Regional standardization bodies like ECMA are also active in this field. Regional industrial initiatives like Industrie 4.0 and Industrial Internet are also active. We hear major IT players' uptake in IoT standardization like Open Connectivity Foundation.

There is a continuous inflow of IoT-oriented standardization or industrial activities to push forward interoperability of IoT.

There is a certain divergence driving force to continue to push forward the more diverged IoT standardization landscape.

4.2 Driving Forces to Push Divergence

Divergence in standardization is driven by the following three dimensions:

- Social dimension
- Functional dimension
- Technical dimension

There are many functional and technical dimensions that drive divergence. Many of these dimensions are domain-specific issues. In addition, many functional and technical dimensions are shifted to convergence in a long run during the coming technical advance. In this paper, the author focuses social factors.

The author summarizes the three social factors that drive divergence in standardization:

- Business factors
- Identity factors
- Transition factors

4.3 Business Factors

One of the key components of business factors to drive divergence in IoT standardization is protection of existing business domains by the players in the market.

NTT DATA Institute of Management Consulting reported patterns of IoT standardization as depicted in Table 1 [6]. They are considered as social factors.

This illustrates different initiatives and leaderships are leading industrial IoT standardization. There are vertical forces (intra-industry) and horizontal forces (inter-industry) that drive these standardization activities.

4.4 Identity Factors

From these patterns, the author proposes a 3-dimensional view of technological components as depicted in Fig. 1.

These different factors drive the diverse standardization depicted in Table 1. Each technological component drive divergence in IoT standardization. One of the challenges in the IoT standardization is that these dimensions are interrelated in both of explicitly and implicitly manners.

Table 1. Patterns of IoT standardization

Alliance, standardization strategy	Vertical-market industrial alliance	Cross-industry Ecosystem (Companies from various industries gather together and lead the industry with competitive edge)
		Same-industry team (companies from the same industry gather together to promote standardization within the industry)
		Leader-company-centric (a leader company initiates an industrial alliance to form an ecosystem to further empower the leading company)
	Cross-industrial alliance	communication- and internet-companies(promoting specifications based on connectivity by communications)
		Home appliances- and smart device-centric (standardization driven by device companies)
		Electrical and industrial machinery-based (electrical and industrial machinery companies promote B-to-B standardization)

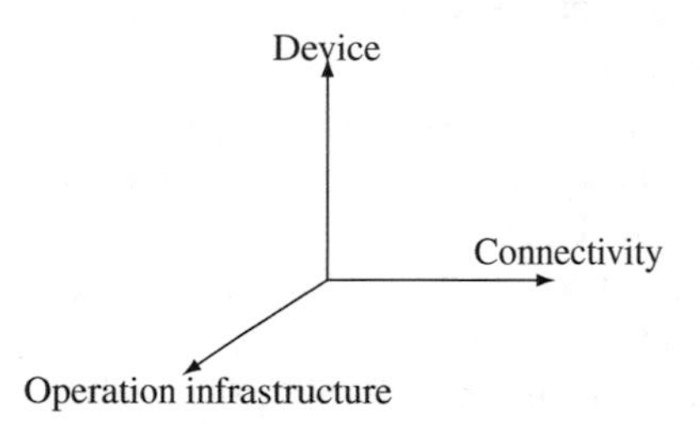

Fig. 1. Dimensional view of technological components

Stakeholders from different components drive different standardization, as depicted in Table 2.

In order to clarity, each standardization activity tends to focus each specific dimension In that process, the other dimensions are encapsulated in an abstract manner. However, in the reality of deployment and operations in cyber-physical integration, it is a challenge to completely abstract other dimensions.

Table 2. Different dimensions drive different standardization activities.

DImension	Description
Device	Device is a piece of hardware, therefore, there are hardware-specific features. When there is a need for interoperability, device-specific standardization emerges despite of other dimensions
Connectivity	Usually, connectivity needs network infrastructure investment. This investment is relatively large, and the cost and significance of the infrastructure drive standardization in spite of other dimensions
Operation infrastructure	Daily operation and fault-tolerance in a system level will be ensured in servers and databases for business execution and maintenance. Therefore, for business and service sustainability, this dimension drives standardization in spite of other dimensions

In IoT, at least two joint points, between device and connectivity, and between connectivity and operation infrastructure need interoperability. All these three dimensions can drive standardization of interoperability. This is not web-centric any more, therefore, there has provided a new standardization opportunities for existing and emerging players.

4.5 Transition Factors

IoT standardardization does not take off in a short span of time. Its tight relationship to the physical realities require significant efforts in transition factors from the existing realities. This transition factors drive existing standardizations spawn new IoT-related activities.

Transition requires deep understanding the current contexts and existing technical components. This requires involvement of a broad range of existing standard organizations and industrial fora. When some of them initiate their own IoT standardization, the coverage of wide range of physical realities of IoT easily lead to a significant level of divergence.

IoT is a process of integrating cyber- and physical spaces. This characteristic needs to address existing, legacy physical assets. And it is also applied to standardization.

5 Model of Divergence

5.1 Relationship Model

Interoperability will be maintained with healthy evolution of standardization. There are multiple patterns of relationship between standardization organizations, as depicted in Table 3.

When multiple standardization activities exist, one of the states depicted in the above table take place.

Table 3. Relationships between standardization organizations

Status	Description
Competing	Multiple organizations provide competing different standards
Conflicting	Multiple organization provide conflicting different standards
Partially shared	Multiple organizations provide partially shared standards
Technically isolated	Each organization creates a technical silo to be isolated with each other
Profiled	Multiple organizations shared the framework of standards with derivatives of different requirements
Harmonically structured	Multiple organizations co-exist under a shared framework
Complementary	Multiple organizations co-exist with complementary roles

Table 4. Reasons why multiple IoT standardization bodies emerge

Reason	Description
Scalability	Different coverage of scalability will produce different standardization activities
Industrial leadership conflict	An industry does not like an industrial standardization where their voices are not valued
Industrial protection	An industry want to decrease influence from big power of other industries. Protection against cross-industry penetration is desirable
Extension of existing standards	When a standard is viewed as extension of existing standards, the existing standardization organization want to extend it by themselves for consistency
Regulation	When there is a regulatory border, a government and an industry want a regional standardization body to build a regional ecosystem
Major certification coverage	When interoperability certification is needed to ensure interoperability, the target of certification will drive the standardization body despite of technical coverage of the standardization

5.2 Diverging Forces in Industrial Standardization

There are diverging forces in the industrial standardization.

The diverging forces are produced from multiple industrial factors. Reasons are depicted in Table 4.

These are the background factors to prevent unified standardization.

Web was a relatively coherent world compared to the current emerging IoT. In addition to the technological factors and social factors, new operation requirements also drive the standardization activity wild.

The uniqueness of the chaos of IoT standardization lies in the fact that there are so many broken assumptions and that they have difficulty to grasp this shift of landscape.

5.3 Analysis in the Social Dimension

The author summarizes the impact of three components in the social dimension as depicted in Table 5.

Table 5. Three components in the social dimension

Dimension	Description
Business	Business competition has been common in the history of standardization. This is not the dominating factor in the chaos of IoT standardization
Identity	Identity can be an important factor. It is a perceptual and subjective, therefore, it is necessary to explore more details
Transition	Cyber-physical system is a new concept, however, transition has been common in the history of standardization. Almost all successful standardization needs to early legacy things prior to standardization

The identity component can be a candidate to explore more to deal with the existing IoT standardization challenge.

5.4 Broken Assumption Awareness

Other landscape changes as well as business interests are driving the wild landscape of IoT standardization. We are witnessing many broken assumptions in Internet standardization depicted in Table 6.

These assumptions are partially or totally broken in the IoT world. These broken assumptions bring unorchestrated increases of IoT standardization activities. Each assumption kept the Internet standardization in a small circle of dense human relationship experts. This is not true in the current landscape any more.

6 Discussion

6.1 Advantages of the Proposed Method

The massive intrusion of IoT technologies provides us a vague feeling that IoT cannot be controlled by human beings. This uncontrolled-ness easily leads to a

Table 6. Broken assumptions of internet standardization in IoT

Assumptions	Past descriptions	Emerging reality
TCP/IP based	The Internet was interconnected by TCP/IP	Emerging network technology like LPWA (Low-Power, Wide Area) does not support TCP/IP
HTTP based	The web was interconnected by HTTP	Emerging technology uses CoAP or even proprietary protocols to transfer sensor data
Browser-based (Web-based)	The web was composed toward a browser user agent using HTML	Sensors do not surf the web. No browser technology is mandatory in IoT
Human-initiated	The web was coined for human interaction based on a browser	IoT will work without any human intervention
Carrier-based	Connectivity was largely supported by telecom carriers	Emerging technology like LPWA will utilize non-licensed band of wireless communication
Open source-based	Browser got fat as the browser advanced. A large chunk of sources require open community to maintain them	Light-weight vertical market does not need open source community
Cloud-based	Cloud was one of the key enabler in the Internet	AWS Greengrass, an emerging component of IoT, is still a part of cloud, but requires only once in 5 years connectivity
Instant evolution	The Internet has driven fast and instant evolution such as mobile services	This is not applicable to IoT devices which require troublesome ground installation work
Stable client and server relationship	When some communications services are designed, there are lots of assumptions about relationship between clients and servers	Fast evolution of chips is threatening the relationships and stability of relationships

casual intuition that such a chaos is difficult to manage and resulting chaos in many aspects including standardization is partially inevitable.

The author describes diversities and different patterns of IoT standardization in the social dimension. In order to provide a consistent view on IoT standardization, the author proposes a 3-dimensional view model for IoT standardization.

The dimensional model provides a framework for compare and identify overlaps and potential conflicts among IoT standardization activities.

The author discusses the broken assumption model in the social dimension. The proposed assumption provides a clear starting point to understand the increasing inconsistency in IoT standardization. It provides a social and organizational awareness that implicit assumptions of many standardization experts are broken and cannot be applied in a decent manner.

The broken assumptions will provide a difficult situation where many overlaps and conflicts are deadlocked because the broken assumptions will provide biased views from different organizations.

The proposed assumption model provides an awareness of non-technological obstacles of scattered and unfocused IoT standardization. Without the focused efforts with awareness of broken assumptions, it is difficult to organize the IoT challenges in virtual-physical integration.

When the chaos of standardization does not come from chaos of business conflicts but from chaos of standardization perspectives, it can be fixed with the increased awareness of standardization perspectives. The proposed view model of divergence provides a stepping stone for objective viewpoints of positioning standardization.

6.2 Limitations

This research is qualitative and exploratory.

Quantitative measures of each dimension are not studied in this paper. Quantitative measures such as standardization coverage, number of member companies, cognition ratios and deployment ratios are not considered in this paper.

Detailed analysis of overlapped or conflicted IoT standardization is not covered in this paper. Empirical works in specific standardization activities are future study.

7 Conclusion

Widespread chaos of IoT standardization is an early indicator that the Internet is really successful and ready to penetrate real everyday life.

In addition to technical and functional dimensions, divergence in IoT standardization takes place from the factors of the social dimensions.

The author discusses business, identity and transition factors. After identifying divergence factors in the social dimension, the author describes the relationship model.

The author proposes a broken assumption framework to understand the divergence in the social dimension. Challenges of perspective-based chaos of standardization can be fixed when the standardization players come to have a consistent view of divergence of IoT standardization.

The proposed framework can be a stepping stone for building consistent view points of divergence of IoT standardization. When there is increased awareness of convergence, it will provide a clue of convergence to fix the factors in the social dimension.

Acknowledgments. The research results have been achieved by “EUJ-02-2016: IoT/Cloud/Big Data platforms in social application contexts,” the Commissioned Research of National Institute of Information and Communications Technology (NICT), JAPAN.

References

1. Abou-Zahra, S., Brewer, J., Cooper, M.: Web standards to enable an accessible and inclusive internet of things (IoT). In: Proceedings of the 14th Web for All Conference on The Future of Accessible Work, W4A 2017, pp. 9:1–9:4. ACM, New York, NY, USA (2017)
2. Anthopoulos, L., Janssen, M., Weerakkody, V.: Smart service portfolios: do the cities follow standards? In: Proceedings of the 25th International Conference Companion on World Wide Web. WWW 2016 Companion, International World Wide Web Conferences Steering Committee, Republic and Canton of Geneva, pp. 357–362. Switzerland (2016)
3. Costa-Pérez, X., Festag, A., Kolbe, H.J., Quittek, J., Schmid, S., Stiemerling, M., Swetina, J., van der Veen, H.: Latest trends in telecommunication standards. SIGCOMM Comput. Commun. Rev. **43**(2), 64–71 (2013)
4. Kubler, S., Robert, J., Hefnawy, A., Cherifi, C., Bouras, A., Främling, K.: IoT-based smart parking system for sporting event management. In: Proceedings of the 13th International Conference on Mobile and Ubiquitous Systems: Computing, Networking and Services, MOBIQUITOUS 2016, pp. 104–114. ACM, New York, NY, USA (2016)
5. Raggett, D.: W3c plans for developing standards for open markets of services for the IoT: the internet of things (ubiquity symposium). In: Ubiquity 2015 (October), 3:1–3:8 (2015)
6. Shinichiro Sanji: Exploring IoT's supremacy conflict from standardization trends (in Japanese). http://www.keieiken.co.jp/pub/infofuture/backnumbers/52/no52_report08.html (October 2016)
7. Yun, J., Choi, S.C., Sung, N.M., Kim, J.: Demo: towards global interworking of IoT systems – oneM2M interworking proxy entities. In: Proceedings of the 13th ACM Conference on Embedded Networked Sensor Systems, SenSys 2015, pp. 473–474. ACM, New York, NY, USA (2015)

Design and Implementation of a VANET Testbed: Performance Evaluation Considering DTN Transmission over VANETs

Shogo Nakasaki[1], Yu Yoshino[1], Makoto Ikeda[2(✉)], and Leonard Barolli[2]

[1] Graduate School of Engineering, Fukuoka Institute of Technology, 3-30-1 Wajiro-higashi, Higashi-ku, Fukuoka 811-0295, Japan
tshogonakasakit@gmail.com, mgm17107@bene.fit.ac.jp

[2] Department of Information and Communication Engineering, Fukuoka Institute of Technology, 3-30-1 Wajiro-higashi, Higashi-ku, Fukuoka 811-0295, Japan
makoto.ikd@acm.org, barolli@fit.ac.jp

Abstract. In recent years, automatic driving technology and inter-vehicle communication have attracted attention because they can be applicable not only to transportation systems but also to intelligent communication systems. Delay/Disruption/Disconnection Tolerant Networking (DTN) and Vehicular Ad-hoc Networks (VANETs) are used to provide the network services as alternative network. In this work, we develop a DTN testbed for VANETs. The communication components are implemented in Raspberry Pi. We evaluate the performance of our testbed in outdoor environment within the campus. From the experimental results, we found that our implemented testbed has good performance.

Keywords: Inter-vehicle communication · DTN · VANET · Testbed

1 Introduction

A number of routing protocols for Delay/Disruption/Disconnection Tolerant Networking (DTN) have been proposed around the world [8,11,18]. Presently, IBR-DTN can run on Android devices [12]. The communication methods of vehicular DTN are important, because the conventional communication system does not work well in vehicular network due to the vehicle movement and network interference. Inter-vehicle communication has attracted attention as an alternative network in disaster situation, road safety application and so on. In these scenarios, the DTN is used as one of a key alternative option to provide the network services [10,20,22].

The DTN aims to provide seamless communications for a wide range of networks, which have not good performance characteristics [6]. DTN has the potential to interconnect vehicles in regions that current networking protocol cannot reach the destination.

L. Barolli et al. (Eds.): IMIS 2018, AISC 773, pp. 748–755, 2019.
https://doi.org/10.1007/978-3-319-93554-6_73

There are several DTN implementation for Linux [1,2], but not all of them support Vehicular Ad-hoc Networks (VANETs) [4,8,16].

In this paper, we present a design of DTN testbed for VANETs. The communication system is implemented in Raspberry Pi 3 model B. We evaluate the performance of our testbed in outdoor environment within the campus.

The structure of the paper is as follows. In Sect. 2, we present an overview of DTN. In Sect. 3 is described the design of DTN testbed. In Sect. 4, we provide the description of evaluation scenario and the experimental results. Finally, conclusions and future work are given in Sect. 5.

2 DTN

Originally, DTN was proposed for space networks [3,9,15,19]. The space networks consider the delay, disconnection and disruption. In DTN, the messages are stored and forwarded by nodes. When the nodes receive messages, they store the messages in their storage. After that, the nodes duplicate the messages to other nodes when it is transmittable. This technique is called message switching. The architecture is specified in RFC 4838 [7].

A number of DTN protocols have been proposed, such as Epidemic [14,21], Spray and Wait (SpW) [17], MaxProp [5] and so on [13].

Epidemic Routing is well-known routing protocol of DTN [14,21]. Epidemic Routing uses two control messages to duplicate messages. Nodes periodically broadcast Summary Vector (SV) message in the network. The SV contains a list of stored messages of each node. When the nodes receive the SV, they compare received SV to their SV. The nodes send REQUEST message if received SV contains unknown messages.

In Epidemic Routing, consumption of network resources and storage usage become a critical problem. Because the nodes duplicate messages to all nodes in its communication range. Moreover, unnecessary messages remain in the storage, because the messages are continuously duplicated even if destination node receives the messages. Therefore, recovery schemes such as timer or anti-packet are needed to limit the duplicate messages. In case of the timer, messages have lifetime. The messages are punctually deleted when the lifetime of the messages are expired. However, setting of a suitable lifetime is difficult. In case of anti-packet, destination node broadcasts the anti-packet, which contains the list of messages that are received by the destination node. Nodes delete the messages according to the anti-packet. Then, the nodes duplicate the anti-packet to other nodes. However, network resources are consumed by the anti-packet.

SpW achieves resource efficiency by setting a strict upper bound on the number of copies per message allowed in the network [17]. The protocol is composed of two phases: the spray phase and the wait phase. Spray phase is terminated when the number of replications reaches the upper limit. When a relay node receives the replica, it enters the wait phase, where the relay simply holds that particular message until the destination is encountered directly. In Spw, it is possible to suppress the communication cost compared with Epidemic.

MaxProp addresses the storage issue by prioritizing both the schedule of packets transmitted to other vehicles and the schedule of packets to be dropped [5]. When contacts with an adjacent vehicle, MaxProp transmits packets with high priority, which are sorted according to their hop count. When the storage is full, MaxProp dropped the packets with low priority, which are sorted according to their delivery likelihood.

3 Testbed Design

In this section, we discuss in detail the design of DTN testbed. We start from the development approach, then we introduce the implemented DTN protocol.

3.1 Development Approach

We develop a mobile device acting with on-board unit. The mobile device is composed of communication and transaction processing components.

These components are implemented in Raspberry Pi 3 model B. The devices run on Linux Raspbian. The Raspbian is an operating system based on Debian optimized for the Raspberry Pi hardware. For communication component, we use an embedded IEEE802.11g WiFi module to communicate with other nodes. We set up static IP address assignment in ad-hoc mode. We use a mobile battery to provide power supply to Raspberry Pi via USB. A snapshot of mobile device as DTN node is shown in Fig. 1. In our system, totally four mobile devices are deployed to outdoor environment, which performs exchange data with each other. For VANET, vehicle has an on-board unit for broadcasting the messages based on implemented mobile devices.

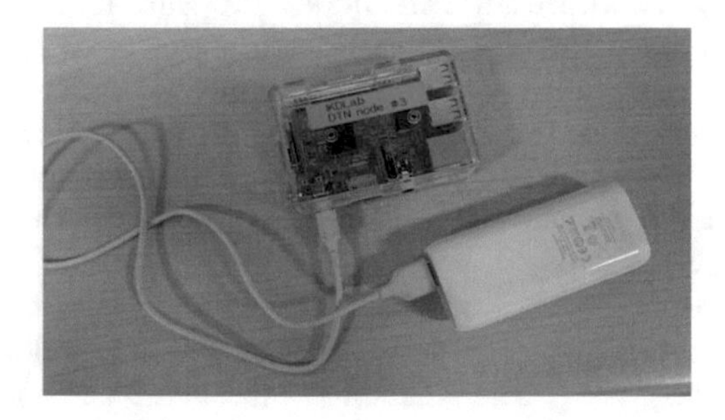

Fig. 1. Snapshot of implemented mobile device.

3.2 Epidemic Daemon

We focus on the high delivery rate characteristics of Epidemic protocol. There are several DTN implementation for Linux [1,2], but not all of them support VANET [4,8,16]. Our Epidemic daemon (epidemicd) is an implementation based on C++ code, which has been written for Linux.

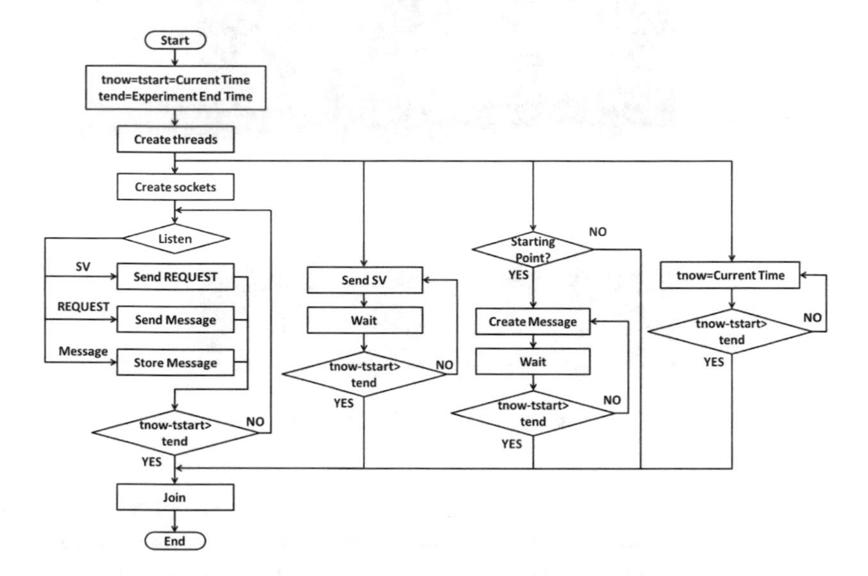

Fig. 2. Flowchart of implemented DTN protocol.

We show the flowchart of implemented Epidemic protocol in Fig. 2. First, epidemicd loads the end time of experiment, then create four threads as follows:

- Message receiving process;
- Send SV process;
- Creating message process;
- Time management process.

In thread of received messages, we consider three kind of messages: SV, REQUEST and Message. The thread of send SV periodically generates SV packets at specified interval. In thread of creating message, a message is periodically created when the message transmission condition is satisfied. Thread of time management manages elapsed time. When the elapsed time reaches the end time, the process automatically terminates.

4 Evaluation Scenario and Experimental Results

4.1 Evaluation Scenario

In this paper, we evaluate the implemented DTN testbed considering three parameters by experiments: storage usage, packet delivery ratio and delay. We use an implemented Epidemic routing protocol and send messages from starting point to end point. In this work, we consider an outdoor environment of our university campus (see Fig. 3).

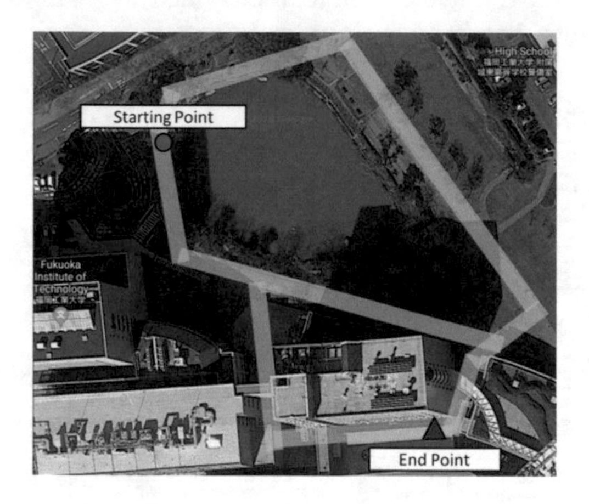

Fig. 3. Experiment scenario.

Both starting point and end point are static. The distance between starting-point to end point is 130 m. The other nodes move in the roads by considering map data and limits their mobility. In this scenario, the moving speed of mobile devices is considered as walking speed of a person. Experimental parameters are shown in Table 1.

Table 1. Experimental parameters.

Parameter	Value
Number of nodes	4
Scenario	Outdoor environment
Routing algorithm	Epidemic Routing
Number of starting point	1 (Fixed)
Number of end point	1 (Fixed)
Number of mobile nodes	2
Message starting and end time	10 - 900 sec
Message generation interval	10 sec
Message size	1000 bytes
PHY model	IEEE 802.11g
Frequency	2.4 GHz

4.2 Experimental Results

We present the experimental results of implemented testbed for different parameters. In Fig. 4 are shown the experimental results of storage usage for different intermediate nodes. The results of storage usage linearly increase with increase of the experimental time. From these results, we confirmed that intermediate nodes received the messages from starting point or other intermediate node. Then, the received messages are stored in the storage of the intermediate nodes. Our protocol does not have any recovery scheme such as timer or anti-packet, thus the storage usage is increased during experiment. Due to the positional relationship of relay node #1, we confirmed that there are 3 periods in which no message was received. The duration is less than 100 s. While for relay node #2, there are two periods. The duration is more than 100 s.

The results of packet delivery ratio is about 95.55%. End-point received totally 86 messages during experimental time. While 4 messages did not reach the end point. The results of average delay is 73.95 s. From the experimental results, there is no significant error, thus our implemented epidemic daemon can be used for different scenarios.

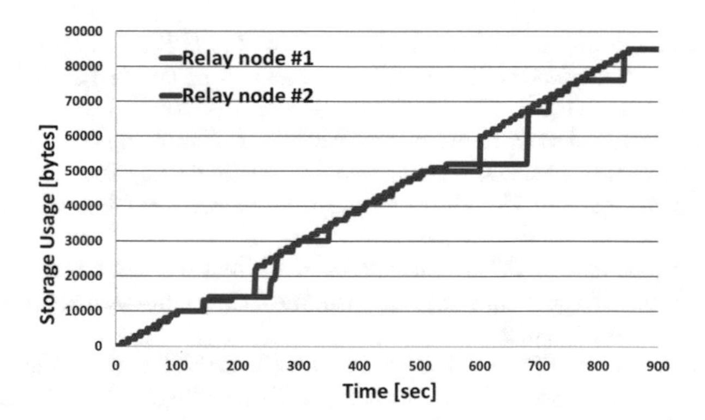

Fig. 4. Results of storage usage.

5 Conclusions

In this paper, we presented the design of DTN testbed for VANETs. We evaluated the performance of implemented Epidemic daemon in outdoor environment within the campus. The experimental results are good in this scenario, but we need to consider different environments.

In the future work, we would like to add new functions and make extensive experiments considering different parameters.

References

1. https://sourceforge.net/projects/dtn/
2. http://www.ibr.cs.tu-bs.de/projects/ibr-dtn/
3. Delay- and disruption-tolerant networks (DTNs) tutorial. NASA/JPL's Interplanetary Internet (IPN) Project (2012), http://www.warthman.com/images/DTN_Tutorial_v2.0.pdf
4. Beuran, R., Miwa, S., Shinoda, Y.: Performance evaluation of DTN implementations on a large-scale network emulation testbed. In: Proceedings of the Seventh ACM International Workshop on Challenged Networks (CHANTS-2012), pp. 39–42. August 2012
5. Burgess, J., Gallagher, B., Jensen, D., Levine, B.N.: MaxProp: Routing for vehicle-based disruption-tolerant networks. In: Proceedings of the 25th IEEE International Conference on Computer Communications (IEEE INFOCOM-2006), pp. 1–11. April 2006
6. Burleigh, S., Hooke, A., Torgerson, L., Fall, K., Cerf, V., Durst, B., Scott, K., Weiss, H.: Delay-tolerant networking: an approach to interplanetary internet. IEEE Commun. Mag. **41**(6), 128–136 (2003)
7. Cerf, V., Burleigh, S., Hooke, A., Torgerson, L., Durst, R., Scott, K., Fall, K., Weiss, H.: Delay-tolerant networking architecture, IETF RFC 4838 (Informational). April 2007
8. Doering, M., Lahde, S., Morgenroth, J., Wolf, L.: IBR-DTN: an efficient implementation for embedded systems. In: Proceedings of the 3rd ACM Workshop on Challenged Networks 2008, pp. 117–120. September 2008
9. Fall, K.: A delay-tolerant network architecture for challenged Internets. In: Proceedings of the International Conference on Applications, Technologies, Architectures, and Protocols for Computer Communications, SIGCOMM 2003, pp. 27–34 (2003)
10. Husni, E., Wibowo, A.: Delay tolerant network based e-mail system using trains. In: Proceedings of the Asian Internet Engineeering Conference (AINTEC 2012), pp. 17–22. November 2012
11. Mahmoud, A., Noureldin, A., Hassanein, H.S.: VANETs positioning in urban environments: A novel cooperative approach. In: Proceedings of the IEEE 82nd Vehicular Technology Conference (VTC-2015 Fall), pp. 1–7. September 2015
12. Morgenroth, J., Schildt, S., Wolf, L.: A bundle protocol implementation for android devices. In: Proceedings of the 18th Annual International Conference on Mobile Computing and Networking (Mobicom 2012), pp. 443–446. August 2012
13. Nelson, S.C., Bakht, M., Kravets, R.: Encounter-based routing in DTNs. In: Proceedings of the IEEE INFOCOM 2009, pp. 846–854 (2009)
14. Ramanathan, R., Hansen, R., Basu, P., Hain, R.R., Krishnan, R.: Prioritized epidemic routing for opportunistic networks. In: Proceedings of the 1st International MobiSys Workshop on Mobile Opportunistic Networking (MobiOpp 2007), pp. 62–66 (2007)
15. Schlesinger, A., Willman, B.M., Pitts, L., Davidson, S.R., Pohlchuck, W.A.: Delay/disruption tolerant networking for the international space station (ISS). In: Proceedings of the IEEE Aerospace Conference 2017, pp. 1–14 (2017)
16. Seligman, M., Walker, B.D., Clancy, T.C.: Delay-tolerant network experiments on the meshtest wireless testbed. In: Proceedings of the Third ACM Workshop on Challenged Networks (CHANTS 2008), pp. 49–56. September 2008

17. Spyropoulos, T., Psounis, K., Raghavendra, C.S.: Spray and wait: An efficient routing scheme for intermittently connected mobile networks. In: Proceedings of the ACM SIGCOMM workshop on Delay-tolerant networking 2005 (WDTN 2005), pp. 252–259 (2005)
18. Theodoropoulos, T., Damousis, Y., Amditis, A.: A load balancing control algorithm for EV static and dynamic wireless charging. In: Proceedings of the IEEE 81st Vehicular Technology Conference (VTC-2015 Spring), pp. 1–5. May 2015
19. Tsuru, M., Uchida, M., Takine, T., Nagata, A., Matsuda, T., Miwa, H., Yamamura, S.: Delay tolerant networking technology - the latest trends and prospects. IEICE Commun. Soc. Mag. **16**, 57–68 (2011)
20. Uchida, N., Ishida, T., Shibata, Y.: Delay tolerant networks-based vehicle-to-vehicle wireless networks for road surveillance systems in local areas. Int. J. Space-Based Situated Comput. **6**(1), 12–20 (2016)
21. Vahdat, A., Becker, D.: Epidemic routing for partially-connected ad hoc networks. Duke University, Technical report (2000)
22. Zaragoza, K., Thai, N., Christensen, T.: An implementation for accessing twitter across challenged networks. In: Proceedings of the 6th ACM Workshop on Challenged Networks (CHANTS 2011), pp. 71–72. September 2011

A Survey of Automated Root Cause Analysis of Software Vulnerability

JeeSoo Jurn, Taeeun Kim, and Hwankuk Kim(✉)

Korea Internet & Security Agency, 9, Jinheung-gil, Naju-si, Jeollanam-do 58324, Republic of Korea
{jjs0771, tekim31, rinyfeel}@kisa.or.kr

Abstract. In recent years, many researches on automatic exploit generation and automatic patch techniques have been published. Typically, in the CGC (Cyber Grand Challenge) competition hosted by DARPA, a hacking competition was held between machines to find vulnerabilities, automatically generate exploits and automatically patch them. In the CGC competition, they implemented themselves to work on their own platform, allowing only 7 system calls. However, in a real environment, there are much more system calls and the software works on complicated architecture. In order to effectively apply the vulnerability detection and patching process to the actual real environment, it is necessary to identify the point causing the vulnerability. In this paper, we introduce a method to analyze root cause of vulnerabilities divided into three parts, fault localization, code pattern similarity analysis, and taint analysis.

1 Introduction

As hacking techniques become advanced, vulnerabilities have been exponentially increasing. The number of vulnerabilities in which about 80,000 vulnerabilities are newly registered in CVE (Common Vulnerability Enumeration) from 2010 to 2015 is increasing [1]. While the number of vulnerabilities is increasing rapidly, response to vulnerabilities depends on manual analysis, which slows down the response speed. We need to develop techniques that can automatically detect vulnerabilities and patch them. To accurately patch the exploited results, a precise cause analysis process is needed. Figure 1 shows the techniques and methods for automated vulnerability detection, automated vulnerability analysis, and automatic patch. In order to analyze vulnerabilities, it is necessary to classify vulnerabilities preferentially as follows. After vulnerability classification, we track the location of the vulnerability and derive a scheme that can be patched for each vulnerability. Although there are various techniques for the cause analysis, Fault Localization technology is a technique of assigning weights based on test cases and analyzing points that caused errors. Code Pattern analysis is a technique of navigating the similarity analysis and outputting the part containing the affected code. To identify the cause of the vulnerability, it is classified into three methods: fault localization, code pattern analysis, and taint analysis. These three techniques will be described in detail in Sect. 3 below.

L. Barolli et al. (Eds.): IMIS 2018, AISC 773, pp. 756–761, 2019.
https://doi.org/10.1007/978-3-319-93554-6_74

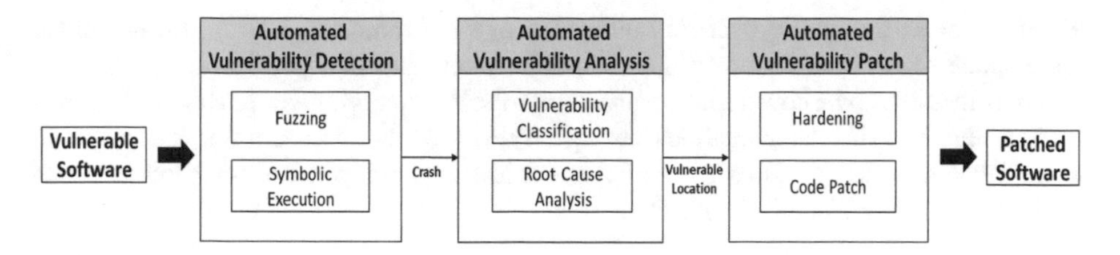

Fig. 1. Automated vulnerability remediation about vulnerable software

2 Related Work

2.1 Automated Vulnerability Detection

Automated vulnerability detection techniques can be divided into fuzzing and symbolic Execution. Fuzzing is black-box testing method that randomly change input values to find software vulnerability. Symbolic Execution is white-box testing method that find input to explore paths of the target software. Fuzzing is divided into dumb fuzzing and smart fuzzing according to input modeling. Mutation fuzzing and generation fuzzing are classified according to test case generation method. Dumb fuzzing is the simplest fuzzing technique to generate defects by randomly changing input values for the target software [2]. Because changing input values is simple, test case creation is fast, but because the scope of the code is narrow, it is difficult to find a valid crash. Smart Fuzzing is a technique for generating input values suitable for a format and generating an error through target software analysis [3–6]. Smart fuzzing has the advantage of knowing where errors can occur through software analysis. The tester can create test cases for that branch to extend the coverage of the code and generate valid conflicts. However, analyzing the target software requires expert knowledge and takes a long time to generate a template suitable for software input. Testers can create test cases for those branches to extend the coverage of the code and generate valid conflicts. However, analyzing the target software requires expert knowledge and takes a long time to generate a template suitable for software input. Mutation fuzzing is a test technique that enters the target software by modifying the data samples. Generation fuzzing is a technique for modeling the format of input values to be applied to target software and creating new test cases for that format.

Symbolic execution is largely divided into offline symbol execution and online symbol execution. Offline symbol execution is resolved by analyzing the path predicate [7] and selecting only one path to generate a new input value. The disadvantage is that the program must run from the beginning to navigate to another path, causing redo to cause overhead. Online symbolic execution is a way in which the state is replicated and the path executor is created each time the symbol executor encounters a branch statement [8, 9]. There is no overhead associated with re-execution using the online method, but significant resources are consumed because all state information must be stored and multiple states must be processed at the same time. Hybrid symbolic execution is suggested to solve this problem. Hybrid symbolic execution is performed every time a branch statement is executed, storing the status information via online symbolic execution and proceeding until memory is exhausted [10]. If you no longer have space to store, switch to offline symbolic execution and perform a path search. Fixed a memory overflow issue in online symbolic execution by saving state information and applying later methods through hybrid symbolic execution. It also eliminates the overhead of offline symbolic execution because you do not need to run it again from the start point (Table 1).

Table 1. Automated vulnerability detection techniques comparison

Techniques	Testing method	Description
Fuzzing	Mutation fuzzing	Randomly mutate-based test case generation
	Generation fuzzing	Target modeling based test case generation
	Dumb fuzzing	Test case generation without state of target
	Smart fuzzing	Test case generation with state of target
Symbolic execution	Online symbolic execution	Fork every time executor meet a branch
	Offline symbolic execution	Only resolve branch statements for one path
	Hybrid symbolic execution	Online Symbolic Execution until memory cap Offline symbolic Execution from saved point

2.2 Automated Vulnerability Patch

Binary hardening can be divided into OS level memory hardening techniques and compiler level binary enhancement techniques. First, the technique for improving the memory at the OS level is ASLR (Address Space Layout Randomization), DEP (Data Execution Prevention/Not Executable), ASCII-Armor. ASLR is a technique to randomize image-based values when programs are mapped to virtual memory. This is a security technique that prevents attackers from attacking because it is difficult for an attacker to grasp the memory structure of the target program. DEP/NX is a technique to prevent code from being executed in data areas such as stack and heap. This will restrict the execution authority on the stack or heap area to prevent attacks. ASCII - Armor is a technique for protecting the space of shared libraries by buffer overflow attack by inserting NULL (\ x00) bytes at the beginning of that address. The inserted NULL byte cannot reach the address.

The techniques that can be applied to the compiler are PIE (Position Independent Executable), SSP (Stack Smashing Protector) RELRO (relocation read only), PIE is similar to the ASLR provided by the operating system. The difference with ASLR is that ASLR applies random address to the stack area of the heap memory area and the shared library area of the heap memory area. PIE is a way to binary random randomize the logical address and to prevent overwriting attack by overwriting the SFP (saved frame pointer) by inserting a specific value (Canary) to monitor on the stack between

SFPs, especially buffers and SFPs. Buffer overflow attack overwrites SFP Attacker overwrites canary. Buffer overflow attacks are detected by monitoring the modulated canary. Finally, RELRO create a read-only ELF binary or data area for the RELRO process so that memory is not changed. There are two types of methods, depending on the state of the GOT domain, part RELRO and complete RELRO. A partial RELRO with a writable GOT domain can quickly use the entire RELRO without consuming resources, but it is vulnerable to attacks such as GOT overwriting. On the other hand, the complete RELRO of the read - only GOT domain consumes more resources and time than the partial RELRO, but you can use the GOT domain to prevent attacks (Table 2).

Table 2. Automated vulnerability patch method and protected area

Patch level	Patch method	Protected area
OS level	ASLR	Stack, heap, shared library
	DEP	Stack, heap
	ASCII-Armor	Shared library
Compiler level	PIE	Code
	SSP	Stack
	RELRO	GOT

3 Automated Vulnerability Root Cause Analysis

In order to clearly patch the vulnerability, we must correctly analyze the point that triggered this vulnerability and execute the patch. Various studies are under way to find places that cause vulnerabilities. In this section, we introduce the techniques of automated vulnerability root cause analysis classified into the following three categories of fault localization, pattern based analysis and taint analysis.

3.1 Fault Localization

Fault Localization is a technique of finding the location of this vulnerability by assigning weights using test cases. Fault localization can be divided into four types of similarity-based, statistics-based, artificial intelligence-based, program analysis-based. Similarity-based fault localization is a method that takes advantage of the frequency of execution of statements within the case of test success and failure. The Tarantula formula is representative, and means the ratio of failed cases divided by the percentage of failed cases added the percentage of normal cases [10]. Statistics based fault localization estimates the error location by measuring the probability of searching for a different path when searching for a path using conditional probability [11]. AI-based fault localization method is a method of searching a subgraph which is frequently appeared by converting a program into a behavior graph by graph mining technique. SVM is used to classify correct and incorrect execution [12]. Program-based fault localization technique measures the suspicion of a code block by calculating the edges between suspicious code blocks in the control-flow graph [13].

3.2 Signature Based Similarity Analysis

The signature-based similarity analysis is a technique to compare bug signatures with other software if there is same bug signature in target software. In the case of source code, there are many functions, such as function-based [14], instruction-based code clone and code similarity analysis, but there are many limitations in analyzing code similarity to binary. First of all, the biggest weakness is that the binary code generated for each CPU Architecture is different. In order to solve this problem, various studies have been carried out to operate on a multi-platform, analysis technique is changed according to whether a weak code pattern is analyzed based on an intermediate language, a weak code similarity analysis is compared with an object [15, 16]. Since the executable file to be generated is different, many researches have been conducted to construct an environment for pattern similarity analysis, which is possible in a multi-platform environment.

3.3 Taint Analysis

Taint analysis is a technique for analyzing how tainted input values flow from external inputs. The derivation of taint value is called 'taint propagation'. When a tainted memory or tainted register is accessed, it is judged that the area is tainted. However, in Taint analysis, there is a problem that is not tainted even though it is affected by the actual value. It is called under taint. In order to solve this problem, DTA++ suggests overcoming by offline analysis and online taint propagation technique [17]. These results cover a slightly smaller range compared to the previous study, DYTAN, which grasped all of the branch and covered a wider range than the existing taint tool [18]. In recent years, backward taint analysis technology has been introduced, which detects the location of errors by tracing contaminated values, and is now installed as a plug-in to Microsoft's Window debugger [19].

4 Conclusion

The number of vulnerabilities is increasing rapidly due to the development of new hacking techniques. However, time-consuming software analysis depending on vulnerability analyst make it difficult to respond to attacks immediately. In order to solve this, automated analytical techniques of various vulnerabilities are announced, but in order to respond positively, it is important to clearly grasp the root cause of the error and patch it. In this research, we have studied the tendency to clearly find the root cause of error. We will study techniques to classify this vulnerability by CWE via the vulnerability analysis and to find the point which caused the error fundamentally,

Acknowledgments. This work was supported by Institute for Information & communications Technology Promotion (IITP) grant funded by the Korea government (MSIT) (No. 2017-0-00184, Self-Learning Cyber Immune Technology Development).

References

1. U.S. National Vulnerability Database. http://cve.mitre.org/cve/
2. zzuf - Caca Labs. http://caca.zoy.org/wiki/zzuf
3. Peach Fuzzer. https://www.peach.tech/
4. Sulley. https://github.com/OpenRCE/sulley
5. Aitel, D.: An introduction to SPIKE, the fuzzer creation kit. In: BlackHat USA Conference (2002)
6. Bekrar, S., Bekrar, C., Groz, R., Mounier, L.: A taint based approach for smart fuzzing. In: IEEE Fifth International Conference on Software Testing, Verification and Validation, pp. 818–825 (2012)
7. Godefroid, P., Levin, M.Y., Molnar, D.A.: Automated whitebox fuzz testing. In: NDSS, vol. 8, pp. 151–166 (2008)
8. Cadar, C., Dunbar, D., Engler, D.R.: KLEE: unassisted and automatic generation of high-coverage tests for complex systems programs. In: OSDI, vol. 8, pp. 209–224 (2008)
9. Ciortea, L., Zamfir, C., Bucur, S., Chipounov, V., Candea, G.: Cloud9: a software testing service. ACM SIGOPS Operating Syst. Rev., 5–10 (2010)
10. James, A., Mary, J.: Empirical evaluation of the tarantula automatic fault-localization technique. In: ASE Proceedings of the 20th IEEE/ACM International Conference on Automated Software Engineering, pp. 273–282 (2015)
11. Liblit, B., Mayur, N., Alice X.Z., Alex, A., Micheal, I.J.: Scalable statistical bug isolation. In: Proceedings of the ACM SIGPLAN Conference on Programming Language Design and Implementation, PLDI 2005, pp. 15–26 (2015)
12. Chao, L., Long, F., Xifeng, Y., Jiawei, H., Samuel, P.M.: Statistical debugging: a hypothesis testing-based approach. IEEE Trans. Software Eng. **32**(10), 831–848 (2006)
13. Zhao, L., Lina, W., Zouting, X., Dongming, G.: Execution-aware fault localization based on the control flow analysis. In: Information Computing and Applications, ICICA, pp. 158–165 (2010)
14. Seulbae, K., Seunghoon, W., Heejo, L., Hakjoo, O.: VUDDY: a scalable approach for vulnerable code clone discovery. In: IEEE Symposium on Security and Privacy, pp. 595–614 (2017)
15. Pewny, J., Garmany, B., Gawlik, R., Rossow, C., Holz, T.: Cross-architecture bug search in binary executables. In: IEEE Symposium on Security and Privacy, pp. 709–724 (2015)
16. Feng, Q., Wang, M., Zhang, M., Zhou, R., Henderson, A., Yin, H.: Extracting conditional formulas for cross-platform bug search. In: ASIA CCS 2017 (2017)
17. MinGyung, K., Stephen, M.C., Pongsin, P., Dawn, S.: DTA++: dynamic taint analysis with targeted control-flow propagation. In: NDSS (2011)
18. James, C., Wanchun, L., Alessandro, O.: Dytan: a generic dynamic taint analysis framework. In: Proceedings of the International Symposium on Software Testing and Analysis, ISSTA 2007, pp. 196–206 (2007)
19. Weidong, C., Marcus, P., SangKil, C., Yanick, F., Vasileios, P.K.: RETracer: triaging crashes by reverse execution from partial memory dumps. In: Proceedings of the 38th International Conference on Software Engineering, ICSE 2016, pp. 820–831 (2016)

The 7th International Workshop on Sustainability Management of e-Business and Ubiquitous Commerce Engineering (SMEUCE-2018)

The Effect of Personal Moral Philosophy on Perceived Moral Intensity in Hotel Industry

Chia-Ju Lu[1], Chiung-Chi Pen[2], and Chiou-Shya Torng[1(✉)]

[1] Department of Leisure and Recreation Management, Asia University, Taichung, Taiwan, ROC
{jareelu, tcstcs}@asia.edu.tw
[2] Shin Kuang Primary School, Kaohsiung, Taiwan, ROC

Abstract. This study focuses on consumers facing controversies due to environmental protection, when they visit a leisure hotel. Specifically, how as a consumer, a teacher's moral philosophy affects the moral intensity. Two environmental ethic scenarios of leisure hotels are developed. A survey was conducted among 253 teachers from primary schools in Kaohsiung, Taiwan. This study examines and finds that moral intensity of environmental ethics can be divided into two major facets, which are "potential damage" and "validity and the level of influence". Specifically, the personal idealism in personal philosophy having positive influence on the subject-matter moral intensity's "potential damage". The findings also reveal that teachers' personal philosophy in relativism substantially affecting the "validity and the level of influence" of subject-matter moral intensity.

1 Introduction

Modern theories on business ethics generally assume that various individuals, when faced with decision making situations involving ethics, apply ethical guidelines or rules based on different moral philosophies (Ferrell et al. 1989; Muncy and Vitell 1992). These philosophies are generally classified as either idealism or relativism. As Barnett (2001) mentioned, in order to improve ethical behavior, it is critical to examine what personal, situational and issue-related factors affect individuals' decisions in situations involving ethical dilemmas. Among empirical studies, moral philosophy has been regarded as a major component in ethical decision-making since it provides individuals standards for judging the act and its consequences (Ferrell et al. 1989).

Moral intensity is partially derived from moral philosophy which is used to differentiate levels of moral responsibility (Jones 1991). Jones (1991) developed his moral intensity model includes six elements that influencing how an individual recognizes moral issues, makes moral judgment, establishes moral intent, and engages in moral behavior: magnitude of consequences, social consensus, probability of effect, temporal immediacy, proximity and concentration of effect (Singhapakdi et al. 1999). If the moral intensity of a situation is perceived strong, individuals will view the situation as having an ethical component. Singhapakdi et al. (1995) argued that dimensions of moral philosophy were found to be significantly related to the perceived importance of

L. Barolli et al. (Eds.): IMIS 2018, AISC 773, pp. 765–770, 2019.
https://doi.org/10.1007/978-3-319-93554-6_75

an ethical issue. In addition, personal moral philosophies on ethical decision making is to operate through perceptions of moral intensity (Singhapakdi et al. 1999).

This study seeks to contribute to the existing literature on ethics within the academic curriculum and focuses on ethics within leisure hotel context. Compared to the previous studies (e.g. Vitell et al. 2003; Vitell and Patwardhan 2008) emphasis on ethics among marketing/business professionals and consumers related internet issues, this study will measure the relationship between personal moral philosophy and perceived moral intensity on environmental ethic scenarios in hotel context.

2 Research Hypotheses

Difference in moral intensity was identified to correlate with the performance of individual moral philosophy. In decision-making process, greater moral intensity led to insignificance performance of idealism and relativism whereas idealism and relativism exhibit significant dominance upon situation with lesser intensity (Vitell 2003). In addition, it was proved that personal moral philosophy is a significant predictor regarding the intensity of moral issues. Namely, highly idealistic people tended to perceive situations involving problematic ethics more intense though highly relativistic people regarded the same situation less intense (Singhapakdi et al. 1999).

In a cross-cultural study, idealism exhibited significant effect on perceived moral intensity for the U.S sample whereas relativism did not. For Chinese sample, both idealism and relativism were significant determinants of moral intensity. Analyze the sample in combination, idealism in all scenarios showed stronger and greater effect on perceived moral intensity than relativism (Vitell and Patwardhan 2008).

According to empirical supports on the relationship between personal moral philosophy and perceived moral intensity, the hypothesis can be postulated in the following manner:

H1a: Idealism is positively associated with the intensity of moral issues.

H1b: Relativism is negatively associated with the intensity of moral issues.

3 Methodology

3.1 Ethics Scenarios

The scenario-based approach is used commonly and accepted widely in ethics researches because it standardizes the stimulus and approximates situations for respondents with little impression management bias (Alexander and Becker 1978). Based on the pretest results, three scenarios were dropped and phrases in the remaining scenarios were revised to make respondents get involved in scenarios easily. The scenarios exhibiting less ethicality and high intensity were selected since there may be threshold effect of ethical sensitivity (Jones 1991). Two environmental ethic scenarios of leisure hotels are developed:

Scenario 1(sand-filling issue)

One of the National Parks Tourist Hotels is famous for owning the hotel's private beach. To fill the beach lost due to the typhoon, the hotel backfilled the precious seashells of the National Park's wild sand to the outer beach to which the hotel belongs without the consent, in order to maintain the hotel's landscape.

Scenario 2(tree-cutting issue)

There is a historic tree in a hotel, which is sawn in several sections by hotel management. Hoteliers said the tree was stampeded by a large number of Chinese tourists and caused serious cracked branches. For fear of causing tourists to trip over, the trees were cut off, but by no means were the trees cut off in order to hotel activities.

3.2 Measures

Moral philosophy is identified with two underlying dimensions, idealism and relativism (Forsyth 1980). Twenty items divided into two ten-item subscales measure idealism and relativism separately.

The scale measuring moral intensity was adapted from Singhapakdi et al. (1996) and composed of six dimensions reflecting six components developed by Jones (1991). 7-point scale was utilized to assess respondents' level of agreement toward each item. Four items were reverse codes so that higher scores indicate greater intensity of moral issues respondent sense.

3.3 Sample

The exploratory nature of the study, data were accumulated via convenience sampling from several on-job-training workshops for primary schools teachers in Kaohsiung, Taiwan. A self-administration questionnaire was administrated to 268 individuals. From them, a total of 253 were valid (94.4%). The interviewers delivered the questionnaires to teachers and picked them up once they were completed.

Most respondents were female (66%). A majority (52.6%) held master's degree, others have bachelor's degree (47.4%). Eighty-three percent of respondents were under 50 years old. Among them, 31.2% were in 30s and 40.7% were in 40s. Regarding seniority, about 36% and 32.4% of the respondents have 10–19 years and 20–29 years teaching experience separately.

3.4 Analyses

Mean of moral intensity was greater than the neutral level (4.00). As a result, respondents generally recognized the existence of ethical problems and viewed depicted issues intense and morally wrong in both scenarios (see Table 1). According to different ethical issues presented to respondents, the result showed the variation of the perceived intensity components between the scenarios. Especially the scenarios 1 indicated the more clear recognition.

The $0.7 \sim 0.8$ Cronbach α in both dimensions of personal moral philosophy and moral intensity represented great homogeneity between items. The result of factor analysis suggested some of moral intensity dimensions highly loaded together. The six components can be combined into two factors. Factor one was primarily determined by

Table 1. Means and standard variations of moral intensity

	Scenario 1		Scenario 2	
Perceived moral intensity				
	Mean	SD	Mean	SD
Magnitude of consequence	6.577	0.739	5.972	1.233
Social consensus	6.387	0.797	5.842	1.287
Probability of effect	6.316	0.953	5.719	1.308
Temporal immediacy	5.123	1.749	4.767	1.857
Proximity	6.091	1.249	5.569	1.483
Concentration of effect	6.281	1.049	5.561	1.499
Mean	6.129	1.089	5.5718	1.445

magnitude of consequences, social consensus and the probability of effect and labeled as "potential damage". However, temporal immediacy, proximity and concentration of effect determined the other factor called "validity and the level of influence".

Based on the result of correlation test was performed to examine the interaction among variables, except relativism to potential damage in scenario 1, almost all variables were significantly correlated. Specifically speaking, idealism positively associated with moral intensity in both scenarios.

According the result of regression, two moral philosophy dimensions explained 9% to 17% of the variance in the dependent variable-perceived potential damage for two leisure hotels scenarios. However, only idealism showed strong effect on potential damage in both scenarios. Results indicated that both idealism and relativism are

Table 2. Regression results: effects of moral philosophy on potential damage and validity and level of influence

Dependent variable: potential damage				
Variable	Scenario 1		Scenario 2	
	β	t	β	t
Idealism	0.307	5.065***	0.361	6.231***
Relativism	0.026	0.434	−0.168	−2.902**
F		12.831***		25.669***
R^2		0.093		0.170
Dependent variable: Validity and the level of influence				
Variable	Scenario 1		Scenario 1	
	β	t	β	t
Idealism	0.189	3.051**	0.113	1.866
Relativism	−0.122	−1.967*	−0.027	−4.466***
F		7.255**		12.662***
R^2		0.055		0.092

Note: ***p < 0.001, **p < 0.01, *p < 0.05

significantly associated with variable of "perceived potential harm". However, in scenario 1 regarding sand-filling dispute, only Idealists would take perceived potential damage into consideration when encountering ethical dilemma (See Table 2). On the other hand, it was reported the 5% to 9% of the variance in perceived validity and the level of influence could be explained by personal moral philosophy. Results indicated that only relativism are significantly associated with variable of "validity and the level of influence" in both scenarios. However, in scenario 2, tree-cutting dispute, only relativist would be influenced by validity and level of influence. Therefore, hypothesis 1a: Idealism is positively associated with the intensity of moral issues is supported partially. Similarly, H1b: Relativism is negatively associated with the intensity of moral issues is supported partially.

4 Conclusion

This study aims to explore the correlation between personal moral philosophy and perceived moral intensity. The moral intensity of environmental ethics can be divided into two major facets, which are "potential damage "and "validity and the level of influence". Differing from previous studies, the scenarios adopted in present study were created according to environmental ethical issues related to leisure hotel context in Taiwan, unlike earlier research adopting scenarios related to business ethics with much verification. According to different ethical issues presented to respondents, the result showed the variation of the perceived intensity components between the scenarios. The scenarios 1 that showed the more clear recognition indicated that both idealism and relativism are significantly associated with variable of "potential damage". Both idealism and relativism have significant contribution to the "validity and level of influence" in scenario 2 that respondents perceived lower moral intensity.

The result corresponds to earlier research conducted by Singhapakdi et al. (1995). Both dimensions of moral philosophy were found to be significantly related to the construct of perceived moral intensity. Specifically, the personal idealism in personal philosophy having positive influence on the subject-matter moral intensity's "potential damage". The findings also reveal that teachers' personal philosophy in relativism substantially affecting the "validity and the level of influence" of subject-matter moral intensity. Our findings supported perceived moral intensity varying within different context (Chia and Mee 2000). However, compared with Vitell's (2003) research, there are some differences. First, the results also showed greater moral intensity in validity and the level of influence led to significance performance of idealism and relativism. Secondly, idealism and relativism exhibit significant dominance upon potential damage situation with lesser intensity.

The present study examined environmental ethical issues in leisure hotel context, scenarios contains different levels of morality might be employed to realize if different level of intensity poses influence on ethical decision-making. It can be examined in following research within the same context regarding environmental ethical issues.

This study employed primary school teachers as the sample. Teachers are considered to have higher ethical standards usually. Therefore, the further study could do the investigation randomly selected from the population at large.

References

Alexander, S.C., Becker, H.J.: The use of vignettes in survey research. Publ. Opin. Q. **42**(Spring), 93–104 (1978)
Barnett, T.: Dimensions of moral intensity and ethical decision making: an empirical study. J. Appl. Soc. Psychol. **31**(5), 1038–1057 (2001)
Chia, A., Mee, L.S.: The effects of issue characteristics on the recognition of moral issues. J. Bus. Ethics **27**(3), 255–269 (2000)
Ferrell, O.C., Gresham, L.G., Fraedrich, J.: A synthesis of ethical decision model for marketing. J. Macromarketing **11**(Fall), 55–64 (1989)
Forsyth, D.R.: A taxonomy of ethical ideologies. J. Pers. Soc. Psychol. **39**(1), 175–184 (1980)
Jones, T.M.: Ethical decision making by individuals in organizations: an issue-contingent model. Acad. Manage. **16**(2), 366–395 (1991)
Muncy, J.A., Vitell, S.J.: Consumer ethics: an investigation of the ethical beliefs of the final consumer. J. Bus. Res. **24**, 297–311 (1992)
Singhapakdi, A., Kraft, K.L., Vitell, S.J., Rallapalli, K.C.: The perceived importance of ethics and social responsibility on organizational effectiveness: a survey of marketers. J. Acad. Mark. Sci. **23**(1), 49–56 (1995)
Singhapakdi, A., Vitell, S.J., Kraft, K.L.: Moral intensity and ethical decision-making of marketing professionals. J. Bus. Res. **36**, 245–255 (1996)
Singhapakdi, A., Vitell, S., Franke, G.R.: Antecedents, consequences, and mediating effects of perceived moral intensity and personal moral philosophies. J. Acad. Mark. Sci. **27**(1), 19–36 (1999)
Singhapakdi, A., Gopinath, M., Marta, J.K., Carter, L.L.: Antecedents and consequences of perceived importance of ethics in marketing situations: a study of thai businesspeople. J. Bus. Ethics **81**(4), 887–904 (2008)
Vitell, S.J., Muncy, J.: Consumer ethics: an empirical investigation of factors influencing ethical judgments of the final consumer. J. Bus. Ethics **11**(8), 585–597 (1992)
Vitell, S.J.: Consumer ethics research: review, synthesis and suggestions for the future. J. Bus. Ethics **43**(1–2), 33–47 (2003)
Vitell, S.J., Bakir, A., Paolillo, J.P., Hidalgo, E.R., Al-Khatiband, J., Rawwas, M.Y.A.: Ethical judgments and intentions: a multinational study of marketing professionals. Bus. Ethics A Eur. Rev. **12**(2), 151–171 (2003)
Vitell, S.J., Patwardhan, A.: The role of moral intensity and moral philosophy in ethical decision making: a cross-cultural comparison of china and the European Union. Bus. Ethics A Eur. Rev. **17**(2), 196–209 (2008)

Study on Production and Marketing of Tea: A Company

Kuei-Yuan Wang[1(✉)], Xiao-Hong Lin[2], Chien-Kuo Han[3], Chi-Cheng Lin[4], Yi-Chi Liao[5], Tzu-Yun Ting[5], and Ting-Shiuan Fang[5]

[1] Department of Finance, Asia University, Taichung, Taiwan
gueei5217@gmail.com

[2] Department of International Business and Management, Fujian Hwa Nan Women's College, Fuzhou, People's Republic of China
lxh000@hotmail.com

[3] Department of Health and Nutrition Biotechnology, Asia University, Taichung, Taiwan
jackhan@asia.edu.tw

[4] Ji Cha Yuan, Nantou, Taiwan
m2750397@yahoo.com.tw

[5] National Taiwan University, Taipei, Taiwan

Abstract. Ranking first in the top three beverages in the world, tea is widely loved by the public. As the main output of Taiwan's agricultural products, tea creates many foreign exchange profits for Taiwan. With the changes of the times, changes in Taiwan's social patterns have led to the gradual decline of the tea-making industry. With the liberalization of agricultural trade, domestic tea is facing fierce competitions from a large number of imported teas. Those hot sale sceneries no longer exist. In order to enhance the advantages of Taiwan's tea industry, what we must do is to develop new industries with high added value. This study explores the evolution, development, production and sales of the tea industry. Company A is taken as the sample company, its production, sales and future development is discussed. Methods of Porter-five forces analysis and SWOT analysis are adopted in this study to develop marketing strategies for the sample company in the future. The four marketing strategies suggested are utilizing the advantages of geographical location and integrating the characteristics of local industries; developing a tourism innovation experience model; changing tea packaging creatively; increasing opportunities for tea products promotion and exposure.

Keywords: Tea industry · Industrial development · Porter-five forces analysis SWOT analysis · Marketing strategy

1 Introduction

The tea market in the world is constantly expanding and increasing. Both the total production of tea and the average amount of consumption of each person are gradually increasing. It is clear that the prospect of the tea market is still full of optimism and is expectable.

L. Barolli et al. (Eds.): IMIS 2018, AISC 773, pp. 771–782, 2019.
https://doi.org/10.1007/978-3-319-93554-6_76

In traditional Chinese culture, tea not only has its high economical, political, material and spiritual qualities, but also is an indispensable medium which gathers friends and makes good feelings among people. It transforms strangers into acquaintance, converts coldness into enthusiasm. People get closer because of tea.

Owing to the different climate and environment, and the research and development of various tea-making technologies, special and excellent tea products which have unique quality, nice fragrance and good flavor were developed gradually in different regions of Taiwan.

In addition, Taiwan has actively set up policies of the production and sales history, certification of production origin, and others for tea products in recent years. With these high standards, as well as the traits of hygiene and safety, exquisite packaging, and soft powers such as tea art culture, Taiwanese tea industry has established valuable tea gift series and tea brands, which have great potential for expansion, especially in markets such as China and Japan.

The biggest problem in the current tea-making industry is the use of traditional hand-made process. Since the process is tedious and time-consuming, most young people today are reluctant to engage in such complicated tea-making industry. The tea-making industry in Taiwan will face a crisis of human faults. The labor shortage, rising wages and other external environmental factors have contributed to the increase of the tea-making cost.

Since Taiwan's accession to the WTO (World Trade Organization) in 2002, the agricultural product market has opened up. Teas from all over the world have come to Taiwan to gain a foothold. Both the low-priced beverage tea market and the high-priced mountain tea market have been affected. The tea production and marketing market is quietly changing. Taiwan's tea has gradually lost its export competitiveness. Today's tea quality and safety and the reduction of consumer repurchase rate all affect Taiwan's tea production and sales market. The rise of low-priced chain tea beverage shops is also affecting the sales market of high-priced mountain tea.

Agricultural Council of Taiwan Executive Yuan (2015a) pointed out that the government was advising the tea industry and adopting relevant plans to develop various special tea production systems, adjusting the farming system to activate agricultural land and promoting tea growing on fallow farmland, further counseling tea producers in building "agricultural and commercial cooperation group production area", as well as actively expanding the domestic and foreign sales market, and developing more distinctive tea with different characteristics.

Data from Agricultural Council of Taiwan Executive Yuan (2012–2016) showed that, Taiwan's tea production was 59,606 metric tons in 2012, 58,871 metric tons in 2013, 60,799 metric tons in 2014, and 57,619 metric tons in 2015. In 2016, Taiwan's tea production was 52,072 metric tons. The total output value was NT$7.6 billion approximately.

Agricultural Council of Taiwan Executive Yuan (2012–2016) data also showed that in 2016, there were 11,814 he of tea gardens in Taiwan which mainly located in three areas, Nantou County, Chiayi County, and New Taipei City. Among them, Nantou County, with an area of 6,428 he, taking 54.4% of the total tea garden area, was the largest in Taiwan. Chiayi County ranked second, tea plantation area is 1,826 ha, accounting for 15.4% of Taiwan's total. New Taipei City ranked third with an area of 759 ha, accounting for about 6.4% of all Taiwan.

According to Agricultural Council of Taiwan Executive Yuan (2016), the tea-related employment population is about 36,000 in Taiwan. Department of Commerce of the Ministry of Economic Affairs (2005) summarized, viewing from the industry value chain of the tea-making industry, that the upstream industry of the entire industry is mainly engaged in the process from tea extraction to the completion of raw tea; the middle industry is mainly responsible for the procedure of fine tea production till the graded packaging; and the downstream industry is based on sales of tea products. Thus, the flourishing development of the entire tea industry is owing to the continuous value creating of the various value chains of the industries.

Therefore, the production and sales of the traditional tea industry in the Taiwan region are facing severe impacts and this study is intended to explore the motives of the tea industry in Taiwan. The research purposes of this study are as follows: (1) Analysis of the status quo of the tea industry; (2) In-depth exploration of production, sales and future development of the sample company; (3) Developing future management strategies for sample company; (4) Developing future marketing strategies for sample company.

2 Methodology

2.1 Interviewing

Bernard (1988) believes that the qualitative interview process can be divided into three types: Structured Interviewing, Unstructured Interviewing, and Semi-Structured Interviewing. Different interview methods can be used depending on the purpose and nature of the research.

Structured Interviews

Structured interviews are also referred to as standard interviews. Questionnaires or interview outlines are prepared by the researcher in advance and questions are asked one by one, and respondents are asked to respond to the design options. This method is based on questions formulated in advance. The most specific feature is that the interview process is standardized, and the data required by the research institute can be collected in a complete manner. Therefore, the respondent cannot respond to the content freely.

Unstructured Interviewing

Non-structured interviews are also referred to as non-standard interviews. Researchers do not develop questionnaires or interview outlines in advance, nor do they have a default interview theme. Instead, they are asked by the researchers and interviewees to express their opinions freely. Not limited to the answer to the question. This kind of interviewing method gets the inner thoughts and feelings expressed by interviewees during the interview process. The most obvious disadvantage is the difficulty and time-consuming of the data analysis.

Semi-Structured Interviewing

Semi-structured interviews are also called focus interviews. It has questionnaires or interview outlines for structured interviews. The researchers have some degree of control over the interview process and allow respondents to fully express their opinions and comments. The contents of the interviews prepared by the researchers in advance were adjusted at any time according to the interviewing process.

The in-depth interviewing method mainly explores the interviewees' views and ideas, and collects answers to specific questions and information. Through in-depth interviews, information required for the study can be found. These hidden contents cannot be obtained either from viewing the surface or using quantitative study method. Its content is as good as that acquired from the traditional quantitative research questionnaire.

This study integrates the advantages of Structured Interviewing and Unstructured Interviewing and uses Semi-structured Interviewing for follow-up research.

2.2 Methods of Porter-Five Forces Analysis

The MBAlib.com (2017) introduced that Porter Five forces analysis is a market structure proposed by Michael Porter in 1979, also known as the Porter Analysis Method.

Porter (1979) believes that there are five forces influencing market competition and determining the strength of exclusiveness, that is, a five force analysis framework, which belongs to the individual economics rather than the generally considered overall economics. These five kinds of strength factors will affect the trend of industrial competition. Any change in power may affect the profitability of the industry.

2.3 SWOT Analysis

The SWOT analysis method proposed by Weihrich (1982) is a method for judging the Strengths and Weaknesses of firms, as well as Opportunities and Threats in the competitive market, for the purpose of formulating corporate development strategies. Positioning is thereafter determined according to the conditions of internal resources of the company and the environment of external resources.

SWOT analysis compares the internal conditions of an enterprise with competitors and analyzes the impact of the external environmental changes on the enterprise. It finds the opportunities of the enterprise will face and the challenges that the enterprise may face, and then formulate the best strategy for the development of the enterprise.

2.4 Research Structure

This study will focus on the development of the tea industry in Lugu, Nantou County, and collect the relevant literature data to research local tea farmers and tea traders. In this study, semi-structured interviews were used to collect information provided by respondents, the sample companies. Semi-structured interviews have the advantages of structured interviews and non-structural interviews. It can avoid the deficiencies in structural interviews such as lack of flexibility and difficulty in in-depth discussion of issues, and can also avoid the disadvantage of unstructured interviews that are laborious, time-consuming, and being difficult to quantify and analyze. The outline of the interviews for this study is as follows: (1) Is the tea industry a family business, or a self-established business? (2) Which type of tea is currently planted? (3) What is the main tea making method at present? (4) What are the difficulties facing the development of the tea industry? (5) What are the current production method and sales channel? (6) What is the development of the future tea industry?

This study was based on a literature review and analysis. It started with a study of the sample company's background, motivation, purpose, scope, etc. Then, data collected from different resources and in-depth interviews was used to analyze the Strengths, Weaknesses, Opportunities, and Threats of the internal environment and the external environment for the sample companies, with the Methods of Porter-five forces analysis and SWOT analysis, for the purposes of developing marketing strategies for the sample company in the future.

3 Empirical Results

3.1 Situation of Sample Company

Background

The sample company is located in Lugu, Nantou County and has been established for 13 years. At present, the amount of capital is NT$5 million. The current number of employees is 3, and mostly family members. The sample company has obtained Taiwan GPS excellent store certification and ISO food certification. The tea sold by the sample company is qualified by the tea pesticides of Agricultural Council of Taiwan Executive Yuan and has no pesticide residue. In addition to the sales of company's self-made tea, tea equipment and refreshments sales have also been introduced.

Production

The sample company's tea production is based in the region of Alishan, Lugu, Shanlinxi and other places; teas was strictly controlled, sorted and grading packaged to strictly assure the quality so as to make consumers comfortable in tasting.

Today's tea-making industry is mostly small-scale farmers. It will be difficult to emerge in the industry and gain a stone base for sustainable development if they can neither make breakthroughs in service quality and service innovation, nor create differentiation.

Sales

All the teas sold by the sample company were qualified for the testing of tea pesticides by the Agricultural Council of Taiwan Executive Yuan and won the first prize in the Nantou Lugu Tea Competition. If they can participate in the "World Tea Expo", it is believed that the reputation of the sample company shall be highly enhanced.

Future Development

The future development of the sample companies should focus on the research and development of tea varieties, packaging design, and product development. It will be illustrated as follows: (1) Research and development of tea; (2) Packaging design; (3) Development of tea products.

3.2 Results from Porter-Five Forces Analysis

Barriers to Entry

1. Tea

Tea is a traditional beverage, which is used to quench thirst, and is also a healthy pursuit. Newcomers must also have a customer source, otherwise it is difficult to compete with the existing manufacturers; the existing manufacturers should cultivate customer loyalty to frustrate potential Newcomers' confidence.

2. Tea set

It is mainly used as tea-making tool and is complementary to tea consuming. Exquisite tea sets have their own artistic value of appreciation and collection. Newcomers need to have tea sets with unique styles to attract consumers to purchase. Otherwise, it is difficult to compete with the existing manufacturers in reputation.

3. Refreshment

Natural tea can be used as a main additive material to make refreshments by adding various kinds of food products, such as cakes, candied fruit, cereals, etc. Newcomers can strengthen this category, and develop tea food with unique flavors to attract consumers' attention.

Suppliers Bargaining Power

1. The main source of tea originates from tea farmers in all major tea regions in Taiwan. As tea plantations have been transformed into recreational farms in recent years, Taiwan's tea production has gradually declined, resulting in an increase in the proportion of imported tea raw materials. New players may try to work with the tea farmers in the producing areas to ensure source safety and the quality of raw materials.
2. The main source of tea sets is the sample company itself. The tea sets developed and designed by sample companies on their own, can be regarded as a separate market to create goods with collectible value and attract consumers' purchasing desire.
3. The main source of refreshment is the sample company also. Tea is provided as a basic material for processing the refreshment, such as tea plum, tea cake, etc.

Customer Bargaining Power

1. Tea is a high-priced luxury product that requires customers with a certain amount of purchasing power.
2. Store sales: Mostly sell tea with exquisite package, tea sets and tea food as well. Price is deductable by bargaining, giving discount, sending small gifts, etc. Therefore, buyers may pay at lower prices.

Threats from Substitutes

1. After Taiwan's accession to the WTO, a large number of foreign teas were imported, low-priced, low-cost, low-quality teas were mixed into high-priced Taiwanese tea. Taiwanese tea market was therefore partially occupied.
2. Tea beverage market is vulnerable to other alternatives, such as coffee, canned beverages, and handmade drinks.

Existing Competition

In recent years, it has been subject to competition from large-scale retail stores, supermarkets, and other emerging channels, while the tea consumption habits are relatively stable with no increase of consuming volume. Due to the advancement of Today's information technology, farmers associations and tea growers have started to work on Internet marketing, which not only drastically reduces the tasting cost for consumers, but also enables them to sell on the shelves of various countries and to expand abroad market for Taiwanese tea.

3.3 SWOT Analysis Results

The following is a detailed description of the sample company on the specific items listed in Table 1:

Strengths

1. One-stop production and sales: The sample company is running by a three-generation tea trader and has its own storefront, which facilitates consumers to purchase.
2. Tea maker's management and technical background: The head of the sample company is the third-generation tea farmer, who inherits not only the good technology, but also the good management.
3. Award, good reputation for high quality: The sample company has won the first prize in Nantou Lugu Tea Competition.
4. Continuously research and develop for new tea-making technology: The sample company leader has once spent too much time baking, resulting in the development of a new variety of tea, honey tea.
5. Stability of Quality: The sample company uses traditional dry baking methods to make tea. The quality of tea leaves is strictly controlled and graded.
6. High quality standard: All the teas sold by the sample company were all certified by the Agricultural Council of Taiwan Executive Yuan, and has passed tea pesticides test, proving that no pesticide residues were left.

Weakness

1. Far from the metropolitan area, inconvenient traffic: The sample company is located in the Nantou mountainous area and is a small-scale farmer operation type. The bird-tea tea area in China is mostly in the mountainous areas and rural areas. Inconvenient traffic became barriers for consumptions.

2. Not good at marketing: Sample company does not have good control over production costs, which resulting in deviations in pricing. Higher selling price therefore became it's disadvantage in marketing.
3. Exquisite tea, not suitable for mass production by machine automation: Sample Company is unable to use machines for mass production due to the use of traditional art of hand-making.

Opportunities

1. Tea drinking habits: Due to academic reports and news reports, people gradually accept the benefits of drinking tea. The sample company can effectively distinguish market by developing different tea products to increase consumers' purchase.
2. Recreational tourism habits: As the government actively promotes the development of leisure agriculture, people begin to travel outside in their leisure time; The sample company is located in scenic spots where more tourists could visit.
3. Tea consuming in the tea area: People gradually become accustomed to travel in the tea area, they will also buy the local tea while traveling. In addition, there are more local industries appeared, such as tourism industry, housing industry. Sample company has its own homestay that allows consumers to stay and possibly purchase tea while they travel.

Threats

1. Imported tea: The tea industry is a completely competitive market. The cost of imported tea is low, and the amount of imported tea is also increasing year by year. Inferior imported tea often serves as a local tea and imported tea is mixed with local tea in the market.
2. Competition within the industry: The geographical location of the sample company belongs to the tea area, where many other small tea farmers are existing. Therefore, competition comes from tea farmers and tea traders in the same area.
3. Hand-shaking tea drink: Most of the teas used in the market today are mostly imported from abroad. Due to the low purchase cost, hand-shaking tea drink is very competitive in price.

Table 1. SWOT analysis of sample company

(S) 1. Production and Sales 2. Tea Maker Management, Technical Origin 3. Awards and Good Reputation 4. Continuously Develop New Technology 5. Stable Tea Quality 6. Good Tea Garden Management	(W) 1. Far from the city 2. Not good in marketing 3. Dedicate Tea, not suitable for mass production
(O) 1. Tea-drinking Habit 2. Traveling Habit 3. Purchasing in Tea Area Habit	(T) 1. Imported Tea 2. Competition within the industry 3. Threats from hand-shaking tea drink

3.4 Dynamic SWOT Analysis

SO Strategy Analysis
This study uses the analysis results of "strength" (S) and "opportunity" (O) in the SWOT analysis to conduct a follow-up SO strategy analysis:

1. Tea Industry Cultural and Creative Tourism (S1, O2)

Combining the advantage of sample company's one-stop production and sales (S1), with the opportunities for leisure travel habits (O2), sample company can integrate the concept of the modern and popular cultural and creative industries to plan a highly-informed "tea industry travel itinerary". Sample company can cooperate with tour companies and collaborate on package tours to reduce the payroll cost of tour guides. It is estimated that the annual number of tourists might grow by 8-10% approximately.

2. Developing New Technology (S4, O3)

Combining R&D advantages of new tea-making technology with consumers' habits of buying tea (O3) in the tea area, sample company can develop teas that are more suitable for consumers' tastes. Sample company can upgrade the tea through technologies upgrading, such as baking, as well as the changing of the tea types. Consumers shall be willing to purchase at the price of NT$1,200 per pound, rather than NT$800 per pound. By making this decision, the sales amount might have a increase of about NT$50–100 million annually.

WO Strategy Analysis
This study uses the analysis results of the "Weakness" (W) and "Opportunities" (O) in the SWOT analysis to conduct a follow-up WO strategy analysis:

1. Diverse Alliance to Promote Leisure Life (W1, O2)

Since sample company is far away from the city (W1), it is suggested to take use of People's leisure travel habit (O2). Sample company can make alliances with local accommodation companies or provide accommodations itself and offer appropriate travel itineraries to allow consumers to stay longer in the tea area. The estimated growth of the number of tourist for each season is about 3–5%.

2. Setting Proper Price for Sales Promotion (W2, O3)

Taking full use of the consumers' habits of buying tea (O2) in the tea area, the sample company can reduce the cost of the management of the sales channel, which is estimated to be a possible deduction of 10%. Proper sales price can be set to attract consumers; sales promotion activities and other relevant publicity methods can also be adopted to promote sales at an estimated increase between the amount of NT$500,000 and NT$1,200,000.

ST Strategy Analysis
This study uses the analysis results of the "Strength" (S) and "Threats" (T) in the SWOT analysis to conduct a follow-up ST strategy analysis:

1. "Tea production and sales certification" to increase the price (S5, T1, T3)

Facing threat of imported tea (T1) and the threat of hand-shaking tea drink (T3), the case company needs to take the advantages of the company's tea quality stability (S5). Sample company can actively cooperate and support with the government's "CV production and sales history" certification system. When customers have a better idea of production history tracing, customers are definitely willing to pay higher prices for purchasing products with product history certificates based on consideration of health factors. The estimated turnover increased is about NT$600,000—NT$1,000,000.

WT Strategy Analysis
This study uses the analysis results of the "Weakness" (W) and "Threats" (T) in the SWOT analysis to conduct a follow-up SO strategy analysis:

1. "One-day Tea Farmer" Creative Life (W5, T1)

Taking sample company's disadvantage of inconvenient transportation (W1) and the threat of competition within the tea industry (T1), sample company can provide "One-day DIY Travel Itinerary". Nowadays, travel itinerary provides diversification ways. The cultural and creative industries are opening for consumers to experience and observe. Sample company can open for consumers to take on "One-day Tea Farmer" and experience the lives of tea farmers. Also, Case Company can combine with off-campus teaching courses to jointly promote the tea ceremony culture with schools. On one hand, this plan may increase the employment in the future. On the other hand, according to estimation, sightseeing plan may bring about a growth of 6–10% for adults visitors, and school plan may increase the number of people employed in the future by about 2%.

3.5 Future Marketing Strategy

This study uses case analysis, in-depth interviews, Porter's five-force analysis, and SWOT analysis to perform data analysis and comparison. The following four marketing strategies are developed and found suitable for sample company: (1) To integrate local industry characteristics, and promote cultural heritage; (2) To develop tourism innovation of experiencing model, driving the trend of experiencing travel; (3) To develop creative design to attract young consumers; (4) To develop new products, increase tea product varieties.

4 Conclusions and Suggestions

This research takes A company in Lugu Township tea industry as the research object. Suggestions on its operation and marketing strategy obtained from the study are as follows:

4.1 Research Results

1. As Taiwan's tea sales market is now nearly saturated, the threat from new entrants is fairly small.
2. As the total annual production of tea in Taiwan gradually decreases, there is no monopoly or oligopoly situation in the supplying market. Therefore, suppliers have higher negotiating power for the price of tea. So the suppliers' bargaining power is strong.
3. Due to the large number of producers in the market, such as tea merchants and tea growers, some of them have a certain degree of popularity in the market, so they will have more competition with existing producer.
4. Since Taiwan's cold drink chain shops and grocery stores provide cold hand-shaking drinks or canned teas, consumers' demand for tea can be easily satisfied. Therefore, the threat from substitutes is greater.
5. Since the market scope that can be chosen is small, the buyer's ability to negotiate prices is limited. Therefore, the buyer's bargaining power is weak.

4.2 Marketing Strategy

1. To integrate local industry characteristics, and promote cultural heritage. Depending on the good location of the sample company, which has abundant tourism resources, and the government's good policies, it is estimated that the sample company will bring about a 4–7% growth in the annual turnover.
2. To develop tourism innovation of experiencing model, driving the trend of experiencing travel. By forming an alliance with different industries such as tourism and hotel industry, sample company may have more tourists per quarter, at an estimating growth of about 2–5%.
3. To develop CREATIVE DESIGN TO ATTRACT YOUNG CONSUMERS. SAMPLE COMPANY CAN PAY MORE ATTENTION TO THE DESIGN OF THE PACKAGE. THE ESTIMATED COST INCREASE IS ABOUT NT$500 THOUSAND TO 2 MILLION, WHILE THE TURNOVER GROWTH IS ABOUT NT$1–3 MILLION.
4. TO DEVELOP NEW PRODUCTS, INCREASE TEA PRODUCT VARIETIES. COST INCREASE MIGHT BE AROUND NT$1–3 MILLION,while the turnover could grow 20–35%。

This study only carried out research on a single case company. The results of the study are less general. It is suggested that the study sample can be expanded in the future to make the research results more universal.

Acknowledgment. This study was supported by Asia University under Grant No. 105-asia-09. We acknowledge the helpful comments and suggestions from the anonymous reviewers.

References

Agricultural Council of Taiwan Executive Yuan, Promote Group Production of High-quality Tea Production Safety Taiwanese Tea (2015a) www.coa.gov.tw/ws.php?id=2503942. Accessed 20 April 2017

Agricultural Council of Taiwan Executive Yuan, Agri-food Statistics - Production of Special Crops in Year 101 (2012) agrstat.coa.gov.tw/sdweb/public/official/OfficialInformation.aspx. Accessed 20 April 2017

Agricultural Council of Taiwan Executive Yuan, Agri-food Statistics - Production of Special Crops in Year 102 (2013) agrstat.coa.gov.tw/sdweb/public/official/OfficialInformation.aspx. Accessed 20 April 2017

Agricultural Council of Taiwan Executive Yuan, Agri-food Statistics - Production of Special Crops in Year 103 (2014) www.afa.gov.tw/GrainStatistics_index.aspx?CatID=454. Accessed 20 April 2017

Agricultural Council of Taiwan Executive Yuan, Agri-food Statistics - Production of Special Crops in Year 104 (2015b) www.afa.gov.tw/GrainStatistics_index.aspx?CatID=476. Accessed 20 April 2017

Agricultural Council of Taiwan Executive Yuan, Agri-food Statistics - Production of Special Crops in Year 105 (2016) www.afa.gov.tw/GrainStatistics_index.aspx?CatID=518. Accessed 20 April 2017

Bernard, H.: Research Methods in Cultural Anthropology. Sage Publications, Newbury Park (1988)

Department of Commerce of the Ministry of Economic Affairs Research on Operation Mode and Supply & Marketing Structure of Tea Industry, Information and Communication Department of Institute of Industrial Technology (2005)

MBAlib.com, Porter Value Chain Analysis Model (2017) wiki.mbalib.com/zh-tw/%E6%B3%A2%E7%89%B9%E4%BB%B7%E5%80%BC%E9%93%BE%E5%88%86%E6%9E%90%E6%A8%A1%E5%9E%8B. Accessed 10 May 2017

Porter, M.: How competitive forces shape strategy, Harvard Business Review, pp. 137–145, March/April 1979

Weihrich, H.: The SWOT matrix - a tool for situational analysis. Long Range Plan. **15**(2), 55–66 (1982)

The Influence of U.S. FED's Interest Rate-Raising Event Announcements on the Abnormal Returns in the Taiwan Stock Market

Xiao-Hong Lin[1], Kuei-Yuan Wang[2(✉)], Chien-Kuo Han[3], Yu-Sin Huang[2], Yi-Chi Liao[4], Tzu-Yun Ting[5], and Ting-Shiuan Fang[6]

[1] Department of International Business and Management, Fujian Hwa Nan Women's College, Fuzhou, People's Republic of China
lxh000@hotmail.com
[2] Department of Finance, Asia University, Taichung, Taiwan
gueei5217@gmail.com, y9567855@kimo.com
[3] Department of Health and Nutrition Biotechnology, Asia University, Taichung, Taiwan
jackhan@asia.edu.tw
[4] Taichung Second Senior High School, Taichung, Taiwan
gigi891020@gmail.com
[5] National Chia-Yi Girls' Senior High School, Chia-Yi, Taiwan
greenstorytome@gmail.com
[6] Taichung Municipal Wen-Hua Senior High School, Taichung, Taiwan
lin2212da@gmail.com

Abstract. This study examines the influence of U.S. FED's interest rate-raising event announcements on the abnormal returns in the Taiwan stock market. The empirical results verified that, when the U.S. FED announced the interest rate decreasing, before the event announced, both the high-tech industries and traditional industries would have the opportunity to obtain significantly negative accumulated abnormal returns (CAR), after the event announced, high-tech industries still have the opportunity to obtain the significantly negative accumulated abnormal returns, while traditional industries just have the opportunity to obtain the significantly positive accumulated abnormal returns on the third day after the announcement. Whereas, when the U.S. FED announced the interest rate increasing, traditional industries have more opportunity than high-tech industries to obtain significantly positive accumulated abnormal returns. The empirical results provide references for investing decisions to the investors.

Keywords: FED · Interest rate · Event study · Abnormal return

L. Barolli et al. (Eds.): IMIS 2018, AISC 773, pp. 783–792, 2019.
https://doi.org/10.1007/978-3-319-93554-6_77

1 Introduction

Commercial Times (2017) reported that the United States Federal Reserve (Fed) announced a rate hike of 1 yard (0.25%) at 2 o'clock on the 15th of June 2017 in Taipei time and adjusted the federal funds rate from 1% to 1.25%. The United States raised interest rates twice in late 2016 and late 2015, the financial markets became more prosperous and the four major stock markets index hit record highs. Therefore, if American interest rates continue to rise in the future, funds will flow to places with high interest rates. Emerging markets might bear the brunt. Taiwan may find it hard to stay out of the situation as well.

The United States Federal Reserve FED is the central bank of U.S., mainly engaging to improve the effective operation of the U.S. economy and maintain public benefits (Fed 2017a). The FED was established under the Federal Reserve Act in 1913. While establishing the Federal Reserve System, the United States was geographically divided into 12 regions, each with its own reserve bank. The headquarters of FED is located in Aikesi Building, Washington DC, USA. The current chairman, Janet L. Yellen, took office on February 3, 2014 (FED Association 2017a).

The main function of the FED is to conduct monetary policy. The Federal Open Market Committee members formulate U.S. monetary policy according to the congressional mandate, promote maximum employment, stable prices, and long-term interest rates for the U.S. economy; promote the stability of the financial system and monitor the risks of the financial system, and participate in domestic and foreign coordination to ensure that the system provides a healthy economy for the entire American people; supervises and administers financial institutions and activities; promotes the security and soundness of individual financial institutions and monitors their impact on the entire financial system. With regard to the safety and efficiency of the settlement system, the FED is committed to promoting a safe and effective system for US dollar transactions; promoting consumer protection and community development, FED supervises, community reinvests and researches to increase consumer and community understanding of the impact of the financial service policies and practices (Federation 2017b).

Since the announcement of the FED's interest rate policy will change the investment of funds and affect the stock market, investors still need to pay close attention. In view of this, this paper explores the following two points, (1) The Fed's interest rate increase event will cause significant abnormal remuneration for Taiwan's stock market; (2) The Fed's interest rate increase event will cause significantly different abnormal returns for different industrial stocks in the Taiwan stock market, for the purpose of providing investors with an objective reference indicator.

The results of this study will enable the government to make correct actions in terms of capital investment and public debt issuance when the Fed announces its interest rate policy; will make investors more aware of the impact of the Fed's interest rate policy towards Taiwan stocks and able to make better investing decisions; will help enterprises declare best industry to invest funds into upon the announcement of FED interest rate policy. In terms of future research, this study can serve as a partial reference for the impact of the Fed's interest rate policy on Taiwan. In addition to the stock market and industry mentioned in this study, it can also extend other direct or indirect influences, so that the people may have a more comprehensive understanding of the impact of the declaration of FED interest rate policy.

2 Methodology

2.1 Empirical Period and Data Description

This study use Event Research Methods to discuss the impact of the US FED event on Taiwan stock market. The research period was from January 1, 1996 to June 30, 2017. The data on the daily remuneration of the stock price of the research sample was taken from Taiwan Economic News Database (TEJ). This study divided all the samples into three categories: traditional industries, high-tech industries, and other industries. Among the three, Yang (2007) pointed out that Taiwan's communications, information, consumer electronics, semiconductors, precision instruments and automation, aerospace, advanced Industries such as materials, special chemicals and pharmaceuticals, health care, and pollution prevention and control are the focus of high-tech industrial policy development. Therefore, this study considers and defines the following industries in the Taiwan Economic News database as a high-tech industries: motors, Electrical appliances, biotechnology, electronics, semiconductors, optoelectronics, telecommunications, telecommunications, information services, other electronics, shipping, oil and electricity, etc.; traditional industries are defined by Economic Development Committee of the Executive Yuan (2010), which includes: cement, food, plastics, textiles, rubber, automobiles, glass, paper, steel, construction, etc.; others are classified as other industries, including: tourism, finance, trade, and others.

Empirical period

The date when the FED announcement of rate increase is reported is taken as the date of the event, which is used as the base day. The estimated period of the study is 80 days, the event period is 11 days, and the total duration of the study period is 91 days.

Empirical model

(1) In this study, the expected return rate for calculating the stock price in the "market model" is as follows:

$$R_{it} = \alpha_i + \beta_i R_{mt} + \varepsilon_{it}, \mathrm{t} = -85, \ldots -6,\ \mathrm{i} = 1, 2, \ldots, \mathrm{n} \tag{1}$$

where, R_{it} indicates the return rate on investment of company i on day t; α_i indicates the interception term; β_i indicates the return rate on investment of company i changing with the market remuneration, that is, the system risk of the company; R_{mt} indicates the market remuneration on day t; ε_{it} represents the error term for day t of company i, and $\varepsilon\varepsilon_{it_{it}} \sim N(0, \sigma^2)$.

(2) Expected rate of return:

$$E(\hat{R}_{it}) = \hat{\alpha}_i + \hat{\beta}_i R_{mE}, \mathrm{E} \in \mathrm{W} \tag{2}$$

where, R_{mE} indicates the return rate of the stock market for the i day of the event; W indicates the period of the event.

(3) Abnormal return rate:

$$AR_{iE} = R_{iE} - \left(\hat{\alpha}_i + \hat{\beta}_i R_{mE}\right), E \in W \tag{3}$$

(4) Average abnormal return:

$$AAR_E = \frac{1}{N}\sum\nolimits_{i=1}^{N} AR_{iE}, E \in W \tag{4}$$

where, AAR_E represents the average abnormal return rate of all companies in the event period E; AR_{iE} represents the abnormal return rate of company i in the event period E; N represents the number of observed values.

(5) Average cumulative abnormal return rate:

$$ACAR(\tau_1, \tau_2) = \sum\nolimits_{E=\tau_1}^{\tau_2} AAR_E = \frac{1}{N}\sum\nolimits_{i=1}^{N}\sum\nolimits_{E=\tau_1}^{\tau_2} AAR_{iE}, E \in W \tag{5}$$

where, ACAR (τ_1, τ_2) represents the average cumulative abnormal return rate from the event period τ_1 to τ_2.

Statistical test

$$\mathrm{t} = \frac{AAR_E}{\sqrt{Var(AAR_E)}} = \frac{AAR_E}{\sqrt{\frac{1}{N(N-1)}\sum\nolimits_{i=1}^{N}(AR_{iE} - AAR_E)^2}} \tag{6}$$

where, AAR_E: the average abnormal return rate of a certain period E during the event period; AR_{iE}: abnormal return rate of i company in the event period E; $Var(ARR_E)$:the variance of the average abnormal rate of a certain period E during the event period

$$\mathrm{t} = \frac{ACAR_{(\tau_1,\tau_2)}}{\sqrt{Var\left(ACAR_{(\tau_1,\tau_2)}\right)}} = \frac{ACAR_{(\tau_1,\tau_2)}}{\sqrt{\frac{1}{N(N-1)}\sum\nolimits_{i=1}^{N}\left(CAR_{(\tau_1,\tau_2)} - ACAR_{(\tau_1,\tau_2)}\right)^2}} \tag{7}$$

Selection criteria of event day

The first day of the public announcement of the FED's interest rate policy being published by major domestic newspapers during the study period was consolidated as the event days. Major newspapers include China Times, Economic Daily News, United Daily News, United Evening News, and the Industrial, Commercial Times, etc.

3 Empirical Results

3.1 Impact of FED Announcement on Abnormal Return of Share Prices

Table 1 showed the results of abnormal returns on Taiwan stock market when FED announced the "decrease" in interest rates. From Table 1, we can see that market has reached a statistically significant level in the fourth to the third day before the announcement of the decline in interest rates, indicating that the information has already affected the market in advance. Base on the above reaction, it tells us that investors already have the expected mentality; at this time, investors expect to receive significantly negative abnormal returns of −24.08% and −14.94%; and the cumulative abnormal returns remained significantly negative from the fourth to the first day before the FED announcement of −26.72%, −41.65%, −31.89% and −32.56%.

On the second and the fourth to the fifth day after the announcement of the "decrease" in interest rates, the stock market reached a statistically significant level and signaled that signs were radiated; at this time, on the second day after the FED announcement, the stock price may have a positive significant abnormal remuneration of 27.82%, but from the fourth to the fifth day after the announcement, it began to reverse, and significant negative abnormal returns were obtained of −13.81% and −26.99%.

Table 1. Declaration of fed events and abnormal returns on stock prices (fall)

	AR	T	CAR	T
Day −5	−0.0263	(−0.36)	−0.0263	(−0.36)
Day −4	−0.2408	(−3.87)***	−0.2672	(−2.67)***
Day −3	−0.1494	(−2.04)**	−0.4165	(−3.46)***
Day −2	0.0976	(1.37)	−0.3189	(−2.16)**
Day −1	−0.0067	(−0.08)	−0.3256	(−1.84)*
Day 0	0.1165	(1.23)	−0.2092	(−0.99)
Day 1	0.0664	(0.79)	−0.1428	(−0.60)
Day 2	0.2782	(3.38)***	0.1354	(0.52)
Day 3	0.1286	(1.33)	0.2640	(0.94)
Day 4	−0.1381	(−1.77)*	0.1259	(0.44)
Day 5	−0.2699	(−3.02)***	−0.144	(−0.50)

Table 2 shows the results of abnormal returns on Taiwan stock market when the Fed interest rate announced "rising". From Table 2, it can be seen that on the fifth day and the second to the first day before the event day, there was a statistically significant level, indicating that the information has already reacted in advance in the market, and the investors already have expectation; at this time, the investors expected to obtain significant positive abnormal returns of 16.23%, 18.54%, and 12.23%, and receive significantly positive cumulative abnormal returns on the fifth day and the second to the first day before the event day of 16.23%, 24.45% and 36.68%. In addition, on the day when the Fed interest rate announced "rising", the accumulated abnormal return can also reach of 36.92%.

On the second and fourth day after the announcement of the "increase" in interest rates, the stock market reached a statistically significant level and signaled that signs were radiated. On the second day after the FED announcement, the stock price shows a significant positive abnormal remuneration of 9.09%, while on the fourth day after the announcement, a significant negative abnormal return of −11.05% appeared. In addition, from the first to the fourth day after the FED interest rate was declared "rising", all investors observed significant positive abnormal returns of 39.33%, 48.42%, 46.27% and 34.77%.

Table 2. Declaration of FED events and abnormal returns on stock prices (rise)

	AR	T	CAR	T
Day −5	0.1623	(3.06)***	0.1623	(3.06)***
Day −4	−0.0821	(−1.50)	0.0802	(1.05)
Day −3	−0.0211	(−0.43)	0.0592	(0.63)
Day −2	0.1854	(3.50)***	0.2445	(2.12)**
Day −1	0.1223	(2.50)**	0.3668	(2.83)***
Day 0	0.0023	(0.04)	0.3692	(2.61)***
Day 1	0.0241	(0.48)	0.3933	(2.71)***
Day 2	0.0909	(2.00)**	0.4842	(3.06)***
Day 3	−0.0215	(−0.44)	0.4627	(2.71)***
Day 4	−0.1150	(−2.32)**	0.3477	(1.91)*
Day 5	−0.0608	(−1.29)	0.2869	(1.50)

3.2 Impact of the Declaration of the Fed Event on Industrial Effects

Table 3 shows the results of the inspection of abnormal returns in high-tech industries, traditional industries, and other industries when the Fed interest rate announced "declining". From Table 3, we can see that the high-tech industry's empirical results reached a statistically significant level on the fourth day before the announcement of the event, indicating that the information has already reacted in advance in the market, and investors already have expectation. At this point, investors expect to obtain significant negative abnormal return of −27.06%. The empirical results on the second day after the announcement of the event reached a statistically significant level, which is a phenomenon of signal emission. At this point, investors can obtain significantly negative abnormal returns of −22.28%. From Table 3, it can be seen that, on the fourth day before the announcement of the event and on the second to the fifth day after the announcement of the event, all of the cumulative abnormal returns are remained significantly negative of −31.94%, −59.64%, −62.93%, −77.22% and −77.74%.

From Table 3, it can be seen that, the empirical results of the traditional industries reached a statistically significant level from the second day to the fourth day before the announcement of the event, indicating that the information has already reacted in advance in the market, and investors already have expectation. On the fourth and the fifth day before the event is announced, investors expect to receive significantly negative abnormal returns of −26.46% and −36.57%. While on the second day before the announcement of the event, the reversal begins. Investors expect to obtain significantly positive abnormal remuneration of 24.09%. The empirical results of the first, second, and the fourth and fifth day after the announcement of the event reached a statistically significant level, which is a phenomenon of signal emission. At this point, significant positive remuneration can be obtained by investors on the first and the second day after the announcement of the event while the reversal begins on the fourth and fifth days after the announcement of the event, investors expect to receive significantly negative abnormal returns of −25.68% and −40.4%. We can also see from Table 3 that, on the fourth to the second day before the announcement of the event, significant negative cumulative abnormal returns of −34.38%, −70.95% and −46.86% were maintained. And on the third day after the event was announced, a significant positive cumulative abnormal return was obtained of 92.84%.

Table 4 shows the inspected results of abnormal returns in high-tech industries, traditional industries, and other industries when the Fed's interest rate is announced "increasing". From Table 4, we can see that the high-tech industry's empirical results reached a statistically significant level on the day before the announcement of the event, indicating that the information has already reacted in advance in the market, and investors already have expectation. At this point, investors expect to obtain significantly positive abnormal remuneration of 17.41%. While on the second day after the announcement of the event, the empirical results reached a statistically significant level, which is a phenomenon of signal emission. At this point, investors can obtain significantly positive abnormal returns of 16.19%.

From Table 4, it can be seen that the empirical results of the traditional industry on the fifth and fourth day, and the second and first day prior to the event announcement reached a statistically significant level, indicating that the information had already reacted in advance in the market, and the investor already have expectation. At this point, on the fifth day before the event was announced, investors expect to receive significantly positive abnormal returns of 23.76%. While on the fourth day before the announcement, the reversal begins. Investors expect significant negative abnormal returns of −17.95%. On the second and the first day before the announcement, it reversed again. Investors expect to obtain significant positive abnormal returns of 30.3% and 12.69%. After the announcement of the event, the empirical results did not reach statistically significant levels. From Table 4, we can see that, the cumulative amount of abnormal returns on the fifth day before the announcement of the event and from the second day before the announcement to the fifth day after the announcement were maintained significantly positive of 23.76%, 37.78%, 50.47%, 52.91%, 59.67%, 67.63%, 68.6%, 59.78% and 55.44%.

Table 3. Declaration of fed event and abnormal return on stock prices (industry, decline)

	High-Tech				Traditional			
	AR	T	CAR	T	AR	T	CAR	T
Day −5	−0.0434	(−0.38)	−0.0434	(−0.38)	−0.0792	(−0.72)	−0.0792	(−0.72)
Day −4	−0.2760	(−3.08)***	−0.3194	(−2.04)**	−0.2646	(−2.68)***	−0.3438	(−2.35)**
Day −3	0.0863	(0.76)	−0.2331	(−1.29)	−0.3657	(−3.45)***	−0.7095	(−3.84)***
Day −2	−0.0802	(−0.98)	−0.3133	(−1.56)	0.2409	(1.98)**	−0.4686	(−1.95)*
Day −1	0.0067	(0.07)	−0.3066	(−1.34)	−0.0248	(−0.17)	−0.4934	(−1.64)
Day 0	0.0216	(0.18)	−0.2850	(−1.06)	0.2337	(1.53)	−0.2597	(−0.73)
Day 1	−0.0886	(−0.77)	−0.3737	(−1.16)	0.2451	(1.69)*	−0.0146	(−0.04)
Day 2	−0.2228	(−1.96)**	−0.5964	(−1.70)*	0.7034	(5.73)***	0.6888	(1.62)
Day 3	−0.0328	(−0.26)	−0.6293	(−1.73)*	0.2395	(1.46)	0.9284	(1.99)**
Day 4	−0.1429	(−1.28)	−0.7722	(−1.99)**	−0.2568	(−2.04)**	0.6716	(1.45)
Day 5	−0.0052	(−0.05)	−0.7774	(−1.99)**	−0.4040	(−2.63)***	0.2676	(0.56)

Table 4. Declaration of fed event and abnormal return on stock prices (industry, rising)

	High-Tech				Traditional			
	AR	T	CAR	T	AR	T	CAR	T
Day −5	0.0390	(0.44)	0.0390	(0.44)	0.2376	(3.14)***	0.2376	(3.14)***
Day −4	0.0341	(0.43)	0.0731	(0.62)	−0.1795	(−2.24)**	0.0581	(0.50)
Day −3	−0.104	(−1.30)	−0.0309	(−0.20)	0.0168	(0.24)	0.0749	(0.54)
Day −2	0.0468	(0.56)	0.0158	(0.08)	0.3030	(3.96)***	0.3778	(2.20)**
Day −1	0.1741	(2.35)**	0.1899	(0.93)	0.1269	(1.71)*	0.5047	(2.58)***
Day 0	0.0023	(0.03)	0.1922	(0.89)	0.0244	(0.30)	0.5291	(2.46)**
Day 1	0.0432	(0.54)	0.2355	(1.02)	0.0676	(0.92)	0.5967	(2.77)***
Day 2	0.1619	(2.44)**	0.3974	(1.55)	0.0796	(1.21)	0.6763	(2.92)***
Day 3	0.0437	(0.59)	0.4411	(1.61)	0.0097	(0.13)	0.6860	(2.70)***
Day 4	−0.1132	(−1.46)	0.3279	(1.13)	−0.0882	(−1.18)	0.5978	(2.20)**
Day 5	−0.1059	(−1.40)	0.2220	(0.74)	−0.0435	(−0.62)	0.5544	(1.94)*

4 Conclusions and Suggestions

4.1 Conclusions

This study mainly discusses the impact of the announcement of the Fed's interest rate policy on Taiwan's stock market and its industrial effect. It uses the event research method to discuss and draws the following conclusions through empirical results.

1. Impact of Fed's interest rate policy on Taiwan stock market

The declaration of interest rate policy, either rising or falling, gives investors expectation before the event day and provides signal after the event day. Abnormal negative returns can be obtained on the third and the fourth day before the interest rate reduction is declared. While, on the fifth day, the second day and the first day before the interest rate increase is declared, a significant positive remuneration can be obtained. In addition, from the fourth day to the first day before the interest rate reduction is announced, significantly negative cumulative abnormal returns are obtained. While, on the fifth day, the second day and the first day before the announcement of the interest rate increase, and on the event day and the fourth day after the announcement, significant abnormal positive remuneration can be obtained.

2. Impact of the Fed's interest rate policy on Taiwan's industries

In terms of industrial effect caused by the interest rate policy announcement, the number of days being affected for traditional industries is more than that for high-tech industries.

When interest rates are downgraded, the high-tech industry will generate significant negative abnormal returns only on the fourth day before the announcement date and the second day after the announcement date. When under the same situation, the traditional industries will generate significant negative accumulated abnormal returns on the fourth to the second day before the event is announced, and which however, will change, on the third day after the interest rate announcement, to be significantly positive. When the U.S. Fed announced that it would lower interest rates, before the announcement of the event, both the high-tech industries and the traditional industries had the opportunity to obtain significant negative accumulated abnormal returns. After the announcement of the event, the high-tech industries still maintain significant negative accumulated abnormal returns, while the traditional industries have the chance to obtain significant positive accumulated abnormal returns on the third day after the announcement.

When interest rates are increased, high-tech industries can obtain significant positive remuneration only on the first day before the announcement and on the second day after the announcement, while traditional industries can obtain significant positive abnormal remuneration on the fifth, the second to first day before the announcement of the event, though it obtained a significant negative remuneration on the fourth day before the event is declared. In terms of accumulated abnormal returns, high-tech industries cannot obtain significant accumulated abnormal returns either before or after the event is declared, while traditional industries could obtain significantly positive accumulated abnormal returns each day, except in the fourth to third day prior to the event. It shows that when the U.S. Fed announced a rising of interest rate, traditional industries are more easily to get significantly positive accumulated abnormal returns than high-tech industries.

4.2 Suggestions

Since the event research methodology is adopted in this study, this study suggests that different empirical models may be used for future research. In addition, since this study only uses Taiwan stock market as an example, it is recommended that follow-up researchers can take stock market in other countries as samples to discuss the scope or degree of impact of the announcement of the Fed's interest rate policy.

Acknowledgment. This study was supported by Asia University under Grant No. 105-asia-09. We acknowledge the helpful comments and suggestions from the anonymous reviewers.

References

Commercial Times, Fed Interest Rising Impact on Taiwan (2017). www.chinatimes.com/newspapers/20170619000028-260202. Accessed 22 July 2017

Yang, C.: High-tech Industry and Talent Innovation—Taiwan Model and Experience (2007) www.npf.org.tw/post/2/1732. Accessed 22 Dec 2017

FED: Structure of the Federal Reserve System (2017a). www.federalreserve.gov/aboutthefed/structure-federal-reserve-system.htm

FED: Purposes and Functions (2017b). www.federalreserve.gov/aboutthefed/pf.htm

The Relationship Between Dividend, Business Cycle, Institutional Investor and Stock Risk

Yung-Shun Tsai[1], Shyh-Weir Tzang[1], Chih-Hsing Hung[2], and Chun-Ping Chang[1(✉)]

[1] Department of Finance, Asia University, No. 500, Liu-Feng Road, Wu-Feng District, Taichung, Taiwan
gicha.tsai@gmail.com, swtzang@gmail.com, l2space@hotmail.com
[2] Department of Money and Banking, National Kaohsiung First University of Science and Technology, Kaohsiung, Taiwan
hunpeter65@gmail.com

Abstract. Investors usually pay more attention to stock dividend payouts and business cycle but less to investment risk. Therefore, volatility and beta, two widely used risk measures of stocks, are used to explore their relationships with dividends, business cycle and institutional ownership. We sampled 200 listed firms which have continuous records of dividend payouts and are held by institutional investors from 2008 to 2014 in Taiwan Stock market. The results show that: (1) dividend and the share ratio of institutional investors have significant positive effect on individual stock risk, (2) the relationship between business cycle and individual stock risk is negative and (3) the effect of dividend, business cycle and share ratio of institutional investor on market risk is insignificant.

Keywords: Risk · Dividend · Business cycle · Institutional investors

1 Introduction

The measure of investment risk is vital to investors' decision making and the performance of investment. Two of the most commonly used risk measures are return volatility and stock beta. A Large amount of literature has widely documented the time-varied relationship between volatility and asset returns. The clustering phenomenon has been found in financial assets such as stock prices and exchange rates ([9,12]).

The time-changing characteristics of asset risk have also drawn substantial attentions from academics and practitioners. Dividend information signal, one stream of literature that focuses on the firm-specific information, is proposed by [18] that dividend payout is not just the reflection of company's performance but also the expectation of company's profitability in the future. The increase in dividend payout implies higher expected earnings by firm managers which

L. Barolli et al. (Eds.): IMIS 2018, AISC 773, pp. 793–800, 2019.
https://doi.org/10.1007/978-3-319-93554-6_78

in turn will tend to increase the stock prices. [20] find that the dividend yield rate are significantly related with the changes of average stock returns. [17] also provides similar results based on the data of stock and corporate bonds using postwar U.S. data. He proposes that high dividends and earnings offer better predictive power for high stock returns at short horizon period.

On the other hand, macroeconomic factors also affect the stock prices. [24] apply vector error correction model (VECM) to the Japanese stock market and find that it is cointegrated with six macroeconomic factors. [22] examines the relationship between conditional stock market volatility and conditional macroeconomic volatility using monthly UK data from 1967 to 1995. He finds a significant relationship between stock market and macroeconomic volatility within the model of VAR. However, the volatility in macroeconomic variables cannot explain stock market volatility.

The relationship between institutional trading and share returns is another stream of literature. [5,29] find that the information hypothesis is supported by a positive open-to-trade market impact on purchases. However, liquidity hypothesis is supported by the observations of price reversal for sales. [21] propose that the institutional ownership may account for the rising idiosyncratic volatility among stocks which can drive abnormal returns in industry portfolios. They find that the institutional ownership can provide better predictions for the volatility of industry portfolios. [8] also document a significantly positive relationship between institutional traders and stock returns following ten days after their tradings. The growth-typed fund managers tend to follow momentum whereas value- and neutral-typed managers are contrarian traders.

In this paper we try to explore the risk of stocks based on dividend payouts, macroeconomic variables and institutional ownership. Although the Taiwan stock market is still an emerging market, however, its market capitalization grows fast in the past decades and was ranked 19th among the 20 largest stock exchanges in the world as of 2017/04 ([6]). Besides, the Taiwan stock market becomes a good candidate for study as its average dividend yield rate is around 3.26–4.60% for the past five years, and the ownership of foreign institutional investors account for 42% and 21% for listed companies in the exchange market and over-the-counter (OTC) market respectively (as of 2017/9/18 by Taiwan Financial Supervisory Committee). Therefore, we believe that the study of the relationship of stock risks with dividend payouts, business cycle and institutional ownership can provide further insights for investors in managing their portfolio risk and hence generating more flexible investment strategy in the Taiwan market. The followings are some reviews of literature associated with the correlation of stock risk and dividend payout, business cycle and institutional investors.

- ***Dividend payout***
  [27] find that the announcement of cash dividend payout increase is positively related with abnormal returns on stocks. [3] further confirm his results by showing that companies having good history of dividend announcement tend to incur positive abnormal returns whereas ones having bad history tend to incur negative abnormal returns. [14] propose theoretical support that

the variance of price change increases in the quality of disclosure. Higher quality of disclosure implies more movement in the market expectations and consequently more dramatic price shift in prices ([7]). [4] finds significant negative association between dividend yield and stock price volatility. He studied the 2,344 U.S. firms from 1967 to 1986 and the empirical results show that if dividend yield increases by 1%, the annual standard deviation of stock price changes decreases by 2.5%.

- ***Business cycle***
  [26] studies the relationship of financial volatility with macroeconomic variables. He observes higher market volatility during economic downturns by examining the period of economic recession in 1930 and during the period of WWII. [1] also propose evidence that stock index returns are negatively related with long term interest rate and budget deficit, but are positively related with inflation rate and money supply growth rate. [24] test the dynamic relationship between macroeconomic variables and the Japanese stock market. By employing a vector error correction approach, they find that there exists a long-term equilibrium relationship between the Japanese stock market and macroeconomic variables. They find that Tokyo Stock Index is positively associated with exchange rate, money supply, industrial production and average lending rate, but negatively related with long term government bond rate.
- ***Institutional investor***
  [28] find an insignificant relationship between block trading and the aggregate stock price volatility, even with a significant negative relationship between the two. Chan, Louis and Lakonishok (1993) report a 0.34% price increase that can be accounted for by institutional buyers. But [13] find no evidence supporting for the claims that foreign investors will increase the volatility of local market in Japan. [16] find that institutional investors in Japan stock market can earn higher returns from their positive feedback trading in their market timing whereas the individual investors earn lower returns from their positive feedback trading in their market timing. The assumptions of institutional investors being informed traders and individual investors being behavioral-biased traders are consistent with literature ([2,10,25]). [11] further document the positive relation between institutional ownership and future stock returns.

2 Methodology

We propose the following three hypotheses for the Taiwan stock market:

2.1 Hypothesis

- H1: Dividend payout is positively related with stock risk
- H2: Business cycle is negatively related with stock risk
- H3: Ownership of institutional investors is positively related with stock risk

2.2 Models

The study adopts panel regression ([15]) using 200 listed firms on the Taiwan stock exchange market and OTC market from 2008 to 2014.

$$Risk_t = \alpha_0 + \alpha_1\, Dividend_t + \alpha_2\, Index_t + \alpha_3\, Institution_t + \alpha_4\, Asset_t + \alpha_5\, Eps_t + \epsilon_t. \quad (1)$$

Risk is measured by the standard deviation of stock returns (*Std*) and market risk beta (β) which was originally proposed by [30], expanded by [19,23] and has been widely used by academics and practitioners to measure asset market risk. We rank βs into three groups and denote the lowest and highest group as β_L and β_H. *Dividend* denotes the rate of change in dividend payout including cash and stock dividend. A positive α_1 implies a positive relationship between stock risk and dividend payout. *Index*, the index of Taiwan stock exchange, is a proxy for business cycle. A positive α_2 implies a positive relationship between stock risk and stock index. *Institution* is the ownership of institutional investors. *Asset* is the logarithm of the value of company's total asset. *Eps* is the rate of change in earnings per share.

3 Empirical Results

Data includes 200 listed firms with continuous records of dividend payout from 2008 to 2014 in the market of Taiwan stock exchange and OTC market. Step-wise regressions are used to analyze the significance of variables.

Table 1. Descriptive statistics

	Average	Median	Maximum	Minimum	Volatility
Dividend (%)	0.2656	0.0526	14.0047	−0.9285	1.0584
Index	8296.70	8188.11	9307.26	4591.22	778.76
Institution (%)	16.9865	10.2850	79.3400	0.0008	17.4026
Asset	16.0347	15.7170	21.9082	13.1702	1.5106
Eps (%)	0.4484	0.0504	48.5000	−19.0000	3.5780
Std (%)	1.7348	1.6999	3.6450	0.4030	0.6118
β	0.8042	0.7960	1.7764	0.1005	0.3289

Table 1 shows that the index of Taiwan stock exchange (*Index*) stably stays around 7,000–8,500 and the change of dividend payout (*Dividend*) also remains stable with a mean 0.26% and standard deviation (volatility) 1.05%. The mean and median of the risk variable β are 0.80 and 0.79 respectively, indicating most of the sampled firms having betas lower than 1. The ownership of institution, however, ranges from 79.34% and 0 with a volatility of 17.40%, a highest value among the other variables excluding index.

Table 2 presents the correlation among variables. The ownership of institution is significantly correlated with asset at 51.11%. The change of Eps is also correlated with the change of dividend payout at 37.81%.

Table 2. Correlation among variables

	Std	β	Dividend	Index	Institution	Asset	Eps
Std	1						
β	0.7075	1					
Dividend	0.1944	0.0539					
Index	−0.1463	−0.3550	0.1010	1			
Institution	0.1201	0.2535	0.0065	0.0136	1		
Asset	−0.0391	0.2754	−0.0204	0.0365	0.5111**	1	
Eps	0.1116	0.1050	0.3781**	0.0541	0.0337	0.0732	1

Note: ***, ** and * denote 1, 5 and 10% significance level, respectively.

Table 3. Regression of standard deviation

	Std							
	Model-1		Model-2		Model-3		Model-4	
Variable	Coef	p-value	Coef	p-value	Coef	p-value	Coef	p-value
Constant	14.8938	0.0000***	14.6667	0.0000***	17.3719	0.0000***	16.0427	0.0000***
Asset	−0.8220	0.0000***	−0.7711	0.0000***	−0.9924	0.0000***	−0.8656	0.0000***
Eps	0.0093	0.0399**	0.0170	0.0000***	0.0162	0.0000***	0.0098	0.0252**
Dividend	0.0667	0.0000***					0.0675	0.0000***
Index			−6.95E−05	0.0001***			−8.63E05	0.0000***
Institution					0.0158	0.0000***	0.0155	0.0000***
R^2	0.5124		0.5097		0.5176		0.5364	

Note: ***, ** and * denote 1, 5 and 10% significance level, respectively.

3.1 Panel Regression on Std

By treating the asset value as a controlled variable, Model-1 in Table 3 reveals a significant positive value 0.0667 for *Dividend* which implies the existence of the strong relationship between the change of dividend and the risk of stock returns. The coefficient of *Dividend* still remains significantly stable at 0.0675 even after Model-4 accounts for the *Index* and *Institution*. This result is consistent with H1 and [27] that the announcement of dividend will affect stock price because investors will make their investment decisions based on the information of the dividend payout change.

In addition, Model-2 shows a significant negative value for *Index*. Even though this value is close to zero, the result is consistent with H2 and [26] that risk is related with business cycle. Model-3 also indicates that *Institution* is positively associated with stock risk with an estimated value of 0.0158, a result consistent H3 and [21] that institutional ownership increases in stock risk. By including all variables, Model-4 still reveals similar results for the coefficients of *Index* and *Institution*.

3.2 Panel Regression on β

Tables 4 and 5 are the regression results based on low- and high-beta firms. Table 4 shows that in Model-1, the estimated coefficient of *Dividend* is 0.0383, which is not significantly associated with the betas of firms. Even after accounting for *Index* and *Institution*, Model-4 shows a very marginal significance for *Dividend* with a value of 0.0404. This result is barely consistent with the results in Table 3. However, as all estimated coefficient of *Dividend* are positive, we still posit that the H1 is valid by the results of Model-4. In addition, similar to Table 3, the estimated coefficients of *Index* and *Institution* are found to be negatively and positively related with the beta of the firms, respectively. This is also consistent with H2 and H3.

Contrary to the results of Tables 3 and 4, the estimated coefficients of *Dividend* for two models are insignificant with values of −0.0025 and −0.0021. This is inconsistent with H1. On the other hand, similar to the results in previous two tables, the estimated coefficients of *Index* and *Institution* are also found to be negatively and positively related with the beta of the firms, respectively, but both of them are insignificant. This is inconsistent with H2 and H3.

Table 4. Regression of low market risk

	β_L							
	Model-1		Model-2		Model-3		Model-4	
Variable	Coef	*p*-value	Coef	*p*-value	Coef	*p*-value	Coef	*p*-value
Constant	13,245	0.1444	1.0509	0.2508	1.6540	0.0713*	1.3216	0.1510
Asset	−0.0345	0.5382	−0.0052	0.9275	−0.0599	0.2928	−0.0234	0.6878
Eps	0.0044	0.3718	0.0080	0.0863*	0.0067	0.1481	0.0052	0.2848
Dividend	0.0383	0.1011					0.0404	0.0966*
Index			−2.38E−05	0.0496**			−3.06E−05	0.0133**
Institution					0.0043	0.0181**	0.0039	0.0327**
R^2	0.6350		0.6359		0.6372		0.6426	

Note: ***, ** and * denote 1, 5 and 10% significance level, respectively.

Table 5. Regression of High market risk

	β_H							
	Model-1		Model-2		Model-3		Model-4	
Variable	Coef	*p*-value	Coef	*p*-value	Coef	*p*-value	Coef	*p*-value
Constant	−1.9836	0.0197**	−2.1176	0.0130**	−1.8412	0.0441**	−1.8916	0.0419**
Asset	0.1776	0.0009***	0.0192	0.0005***	0.1677	0.0041***	0.1769	0.0035***
Eps	0.0052	0.0406**	0.0050	0.0342**	0.0048	0.0388	0.0052	0.0405**
Dividend	−0.0025	0.7472					−0.0021	0.7896
Index			−1.16E−06	0.3653			−1.18E−05	0.3650
Institution					0.0011	0.6216	0.0012	0.5675
R^2	0.5572		0.5578		0.5573		0.5581	

Note: ***, ** and * denote 1, 5 and 10% significance level, respectively.

4 Conclusion

Two stock risk measures, standard deviation and stock beta, are adopted to measure their relationships between stock risk and business cycle, dividend payout and institutional ownership. We find that the change of dividend payout has positive relationship with stock price volatility and betas for firms belonging to the low beta group. But this relationship is not significant for high-beta firms. In addition, the level of index is negatively related with price volatility and betas for firms belonging to the low beta group, which, however, has only marginal significance. But this relationship is insignificant for high-beta firms. Finally, institutional ownership also positively affects the price volatility and market risk of low-beta firms. The limitation of the study is that we just select firms which have a continuous record of divided payout and are held by institution investors. For firms not being listed on the OTC market or without complete payout records, we think they may deliver more information of the investors investment behavior so as to complement our study in the future.

References

1. Abdullah, D.A., Hayworth, S.C.: Macroeconometrics of stock price fluctuations. Q. J. Bus. Econ. **32**(1), 50–67 (1993)
2. Bange, M.M.: Do the portfolios of small investors reflect positive feedback trading? J. Financ. Quantit. Anal. **35**(2), 239–255 (2000)
3. Banker, R.D., Das, S., Datar, S.M.: Complementarity of prior accounting information: the case of stock dividend announcements. Account. Rev. **68**(1), 28–47 (1993)
4. Baskin, J.: Dividend policy and the volatility of common stocks. J. Portfolio Mgmt. **15**(3), 19–25 (1989)
5. Chan, L.K., Lakonishok, J.: Institutional trades and intraday stock price behavior. J. Financ. Econ. **33**(2), 173–199 (1993)
6. Desjardins, J.: The 20 Largest Stock Exchanges in the World. http://www.visualcapitalist.com/20-largest-stock-exchanges-world/. Accessed Apr 2017
7. Fischer, P.E., Verrecchia, R.E.: Public information and heuristic trade. J. Account. Econ. **27**(1), 89–124 (1999)
8. Foster, F.D., Gallagher, D.R., Looi, A.: Institutional trading and share returns. J. Bank. Finance **35**(12), 3383–3399 (2011)
9. French, K.R., Schwert, G.W., Stambaugh, R.F.: Expected stock returns and volatility. J. Financ. Econ. **19**(1), 3–29 (1987)
10. Froot, K.A., O'Connell, P.G., Seasholes, M.S.: The portfolio flows of international investors. J. Financ. Econ. **59**(2), 151–193 (2001)
11. Gompers, P.A., Metrick, A.: Institutional investors and equity prices. Q. J. Econ. **116**(1), 229–259 (2001)
12. Hamao, Y., Masulis, R.W., Ng, V.: Correlations in price changes and volatility across international stock markets. Rev. Financ. Stud. **3**(2), 281–307 (1990)
13. Hamao, Y., Mei, J.: Living with the "enemy": an analysis of foreign investment in the Japanese equity market. J. Int. Money Finance **20**(1), 715–735 (2001)
14. Holthausen, R.W., Verrecchia, R.E.: The effect of sequential information releases on the variance of price changes in an intertemporal multi-asset market. J. Account. Res. **26**(1), 82–106 (1988)

15. Hsiao, C.: Analysis of Panel Data. Econometric Society Monographs, 2nd edn. Cambridge University Press, New York (2003)
16. Kamesaka, A., Nofsinger, J.R., Kawakita, H.: Investment patterns and performance of investor groups in Japan. Pac. Basin Finance J. **11**(1), 1–22 (2003)
17. Lamont, O.: Earnings and expected returns. J. Finance **53**(5), 1563–1587 (1998)
18. Lintner, J.: Distribution of incomes of corporations among dividends, retained earnings, and taxes. Am. Econ. Rev. **46**(2), 97–113 (1956)
19. Lintner, J.: Security prices, risk, and maximal gains from diversification. J. Finance **20**(4), 587–615 (1965)
20. Litzenberger, R.H., Ramaswamy, K.: The effect of personal taxes and dividends on capital asset prices: theory and empirical evidence. J. Financ. Econ. **7**(2), 163–195 (1979)
21. Malkiel, B.G., Xu, Y.: The structure of stock market volatility. Financ. Res. Centre Work. Paper **154**(1), 1–35 (1999)
22. Morelli, D.: The relationship between conditional stock market volatility and conditional macroeconomic volatility: empirical evidence based on UK data. Int. Rev. Financ. Anal. **11**(1), 101–110 (2002)
23. Mossin, J.: Equilibrium in a capital asset market. Econometrica **34**(4), 768–783 (1966)
24. Mukherjee, T.K., Naka, A.: Dynamic relations between macroeconomic variables and the Japanese stock market: an application of a vector error correction model. J. Financ. Res. **18**(2), 223–237 (1995)
25. Odean, T.: Are investors reluctant to realize their losses? J. Finance **53**(5), 1775–1798 (1998)
26. Officer, R.R.: The variability of the market factor of the New York Stock Exchange. J. Bus. **46**(3), 434–453 (1973)
27. Pettit, R.R.: Dividend announcements, security performance, and capital market efficiency. J. Finance **27**(5), 993–1007 (1972)
28. Reilly, F.K., Wright, D.J.: Block trading and aggregate stock price volatility. Financ. Anal. J. **40**(2), 54–60 (1984)
29. Scholes, M.S.: The market for securities: substitution versus price pressure and the effects of information on share prices. J. Bus. **45**(2), 179–211 (1972)
30. Sharpe, W.F.: Capital asset prices: a theory of market equilibrium under conditions of risk. J. Finance **19**(3), 425–442 (1964)

Commercial Real Estate Evaluation: The Real Options Approach

Shyh-Weir Tzang[1], Chih-Hsing Hung[2], Chun-Ping Chang[1], and Yung-Shun Tsai[1(✉)]

[1] Department of Finance, Asia University, No. 500, Liu-Feng Road, Wu-Feng District, Taichung, Taiwan
swtzang@gmail.com, l2space@hotmail.com, gicha.tsai@gmail.com
[2] Department of Money and Banking, National Kaohsiung First University of Science and Technology, Kaohsiung, Taiwan
hunpeter65@gmail.com

Abstract. Investors of commercial real estate tend to sell their investment property when its price rises to a level high enough to realize capital gains, and they also consider selling the investment when its price declines to a level that triggers stop-loss selling. We assume that the investors of commercial real estate have embedded call and put options in their investment when they are engaged in the transaction of commercial real estate property. By using the real options model with given risk parameters, we derive the valuation model of commercial real estate. The model can provide more insights into the evaluation of commercial real estates by considering risk factors like vacancy rate, interest and tax rate, and transaction cost. Based on this theoretical model, further analysis can also be conducted according to the different classes of commercial real estates.

Keywords: Commercial real estate · Office space · Real options

1 Introduction

The option value in evaluation investment in finance is able to account for the future value changes due to a few possible scenarios like expansion, contraction, abandonment or postponement of project. [4, 16, 18] adopted the real options approach to evaluate real estate. Furthermore, by considering the prepayment and default as options on held by borrowers, many researchers adopted real options approach to evaluate the mortgage [2, 7–9]. Additionally, [11] adopted the real options approach to evaluate the development of business centers and subcenters in the city. They proposed that urban spatial development might leapfrog assuming the irreversible development of the property by landowners. [6] collected data from transactions for land and property in Greece to estimate relevant parameters in the real options model to evaluate property. They found that the premium for the option to wait could be as high as 36.6% to 52.3%. [4] also used the real options approach to evaluate the delay of land development in the Chicago area.

Office building is usually defined in terms of its quality or so-called "class". In U.S., three classes, A, B and C, are characterized by quality of materials, design, location, age, function, tenant structure and amenities ([1, 3, 5, 15, 19, 20], among others).

L. Barolli et al. (Eds.): IMIS 2018, AISC 773, pp. 801–807, 2019.
https://doi.org/10.1007/978-3-319-93554-6_79

Similarly, in the Taiwan market, commercial real estate can also be classified into three classes: A, B and C, which are determined by their location, transportation, management, age, property rights and floor space. Contrary to the residential real estate, according to Jones Lang LaSalle in Taiwan, the supply of commercial real estate in Taipei is higher than demand, thus resulting in a continuous decline in rental income as of the first quarter of 2012. The model we derive in the paper can be applied to evaluate different classes of property in the Taiwan market.

The paper consists of four sections. Section 2 describes the model construction. Section 3 presents the model considering the optimal decision to sell and its constraint. The last section concludes.

2 Model Construction

We assume that the institutional investors are the primary investors in the commercial real estate market. To maximize the profits from their investments, investors have to consider factors such as initial purchase cost, rental yield rate (rental growth rate) and tax shields of maintenance expenses from real estate. The commercial real estate price index is assumed to be the corporate value by the model of [12]. The value of commercial real estate will be computed after considering the value of real options on investment and relevant maintenance expenses.

To calculate the rental income, we use the real options approach to obtain the optimal timing on the sale of the real estate. The model is described as follows.

- Commercial real estate price index model

The price of commercial real estate depends on the population and economics of the area being analyzed. We can therefore assume the commercial real estate price index (H) following a Geometric Brownian Motion process:

$$\frac{dH}{H} = (\mu - \delta)\,dt + \sigma dW, \tag{1}$$

where μ is the growth rate of H, δ is the deductive rate in value (considering the depreciation and maintenance expenses), σ is the standard deviation of H, and W_t is the standard Brownian Motion. Under the risk-neutral measure Q with a probability space $(\Omega, \{F_t\}_{t\geq 0}, Q)$ where $\{F_t\}_{t\geq 0}$ is the filtration, H can be expressed as follows:

$$\frac{dH}{H} = (r - \delta)\,dt + \sigma d\tilde{W}^{Q}, \tag{2}$$

where r is the risk-free rate, and $\tilde{W}_t^Q = W_t + \frac{\mu - r}{\sigma}$ is the Brownian Motion under Q measure.

- Real value of commercial real estate

The real value of commercial real estate is determined by the rental income with an assumed steady growth rate as well as by the price index (H). Therefore, under a risk-neutral measure, the real value of commercial real estate can be formulated as follows:

$$0.5H^2\sigma^2 F_{HH} + (r - \delta)\,HF_H + F_t - rF + m_t\,(1 - \tau) = 0, \tag{3}$$

where F is the real value of commercial real estate, m_t the rental income at time t of the commercial real estate with a growth rate g, r the risk-free rate, and τ the tax rate. When the price index (H) increases to a certain level (H_h), the investor will give up the rental income by selling the commercial real estate to realize his capital gains. Similarly, when the price index (H) decreases to a certain level (H_l), the investor will stop his losses by selling the property. According to [10, 12, 13, 17], the investment in commercial real estate is similar to an American option without an expiry date. Therefore, $F_t = 0$, and the generalized solution to Eq. (3) will be formulated as:

$$F = \frac{m(1-\tau)}{r-g} + X_1 H^{\lambda_1} + X_2 H^{\lambda_2}, \tag{4}$$

where

$$\lambda_1 = 0.5 - \frac{r-\delta}{\sigma^2} + \sqrt{\left(\frac{r-\delta}{\sigma^2} - 0.5\right)^2 + \frac{2r}{\sigma^2}},$$

$$\lambda_2 = 0.5 - \frac{r-\delta}{\sigma^2} - \sqrt{\left(\frac{r-\delta}{\sigma^2} - 0.5\right)^2 + \frac{2r}{\sigma^2}},$$

r : risk-free rate,

g : rental growth rate.

When $H \in (H_l, H_h)$, investors will collect rental income from commercial real estate. When $H > H_h$, however, investors will sell properties to realize their capital gains. Therefore, the real value of commercial real estate will be equal to the net present value of rental income for H within (H_l, H_h). When H is not within (H_l, H_h), the real value of commercial real estate will be equal to the present value of the market price net of trading costs. Therefore, we use formulas as follows:

$$H = H_l,\ F = (1-\alpha)H_l, \tag{5.1}$$

$$H = H_h,\ F = (1-\alpha)H_h, \tag{5.2}$$

$$H_l < H < H_h,\ F = \frac{m(1-\tau)}{r-g}, \tag{5.3}$$

where α is the trading cost, H_l the lower bound of real estate price index and H_h upper bound of real estate price index.

By the above equations, we can derive the real value of commercial real estate as follows:

$$F = \frac{m(1-\tau)}{r-g} + \frac{(1-\alpha)\left(H_l H_h^{\lambda_2} - H_h H_l^{\lambda_2}\right) - \frac{m(1-\tau)}{r-g}\left(H_l^{\lambda_2} - H_h^{\lambda_2}\right)}{H_h^{\lambda_1} H_h^{\lambda_2} - H_h^{\lambda_2} H_h^{\lambda_1}} H^{\lambda_1} + \frac{(1-\alpha)\left(H_l H_h^{\lambda_1} - H_h H_l^{\lambda_1}\right) - \frac{m(1-\tau)}{r-g}\left(H_h^{\lambda_1} - H_l^{\lambda_1}\right)}{H_l^{\lambda_2} H_h^{\lambda_1} - H_l^{\lambda_1} H_h^{\lambda_2}} H^{\lambda_2} \tag{6}$$

As the term $m(1-\tau)/(r-g)$ is the rental income of investors by holding commercial property permanently, Eq. (6) can be reconstructed as:

$$F = \frac{m(1-\tau)}{r-g} * P(H) + \text{selling value} * (1 - P(H)),$$

where $P(H)$ is the probability of holding the property. Based on the real options and income approaches in real estate valuation, Eq. (6) is the optimal value of the commercial real estate. Additionally, based on Eq. (6), we can calculate the probability of holding the commercial property by investors:

$$P(H) = 1 - \frac{\left(H_h^{\lambda_2} - H_l^{\lambda_2}\right)}{H_l^{\lambda_1} H_h^{\lambda_2} - H_l^{\lambda_2} H_h^{\lambda_1}} H^{\lambda_1} - \frac{\left(H_h^{\lambda_1} - H_l^{\lambda_1}\right)}{H_l^{\lambda_2} H_h^{\lambda_1} - H_l^{\lambda_1} H_h^{\lambda_2}} H^{\lambda_2}. \quad (7)$$

2.1 Maintenance Cost of the Property

Different from residential property, the maintenance cost of commercial property is higher due to its high utilization. The maintenance cost of the property can create tax shields to investors. The constraint on the maintenance cost can be formulated as follows:

$$\begin{aligned} &C(H) = 0 \qquad\qquad \text{if } H \geq H_h \text{ or } H \leq H_l \\ &C(H) = C(1-\tau), \text{ if } H_l < H < H_h \end{aligned} \quad (8)$$

where C is the maintenance cost for investors. The maintenance cost will be incurred only when $H_l < H < H_h$, and it will become zero when investors decide to sell the property, i.e., $H \geq H_h$ or $H \leq H_l$. The tax shields (TC) generated from maintenance costs will be as follows:

$$TC = \frac{(1-\tau)C}{r} - \frac{\frac{(1-\tau)C}{r}\left(H_h^{\lambda_2} - H_l^{\lambda_2}\right)}{H_l^{\lambda_1} H_h^{\lambda_2} - H_l^{\lambda_2} H_h^{\lambda_1}} H^{\lambda_1} - \frac{\frac{(1-\tau)C}{r}\left(H_h^{\lambda_2} - H_l^{\lambda_2}\right)}{H_l^{\lambda_2} H_h^{\lambda_1} - H_l^{\lambda_1} H_h^{\lambda_2}} H^{\lambda_2}. \quad (9)$$

2.2 The Value of Commercial Real Estate

When investors invest in commercial property for rental income, the value of the property is the market value net of maintenance costs:

$$\begin{aligned} V = &\frac{(1-\alpha)\left(H_l H_h^{\lambda_2} - H_h H_l^{\lambda_2}\right)}{H_l^{\lambda_1} H_h^{\lambda_2} - H_l^{\lambda_2} H_h^{\lambda_1}} H^{\lambda_1} + \frac{(1-\alpha)\left(H_l H_h^{\lambda_1} - H_h H_l^{\lambda_1}\right)}{H_l^{\lambda_2} H_h^{\lambda_1} - H_l^{\lambda_1} H_h^{\lambda_2}} H^{\lambda_2} \\ &+ \left(\frac{m(1-\tau)}{r-g} - \frac{(1-\tau)C}{r}\right)\left(1 - \frac{\left(H_h^{\lambda_2} - H_l^{\lambda_2}\right)}{H_l^{\lambda_1} H_h^{\lambda_2} - H_l^{\lambda_2} H_h^{\lambda_1}} H^{\lambda_1} - \frac{\left(H_h^{\lambda_1} - H_l^{\lambda_1}\right)}{H_l^{\lambda_2} H_h^{\lambda_1} - H_l^{\lambda_1} H_h^{\lambda_2}} H^{\lambda_2}\right). \end{aligned} \quad (10)$$

3 Optimal Decision to Sell

We assume there exists a relationship between H_l and H_h:

$$H_h = \kappa H_l, \tag{11}$$

where κ is a constant. Consistent with the smooth-pasting condition of [12] and [14], the lower bound of H_l^* to maximize the real estate value can be derived by setting:

$$\left.\frac{\partial V(H)}{\partial H}\right|_{H=H_l} = 0.$$

Therefore, we have

$$H_l^* = \frac{\left(\frac{m(1-\tau)}{r-g} - \frac{(1-\tau)C}{r}\right)\left[\lambda_1\left(\kappa^{\lambda_2}-1\right) - \lambda_2\left(\kappa^{\lambda_1}-1\right)\right]}{\left[\lambda_1(1-\alpha)\left(\kappa^{\lambda_2}-\kappa\right) - \lambda_2(1-\alpha)\left(\kappa^{\lambda_1}-\kappa\right)\right]}. \tag{12}$$

By Eq. (12), the investors will sell the property at H_l^* to maximize their wealth. The higher the κ is, the lower the optimal H_l^* will become. H_l^* is also inversely related to the interest rate. When the risk-free rate is low, investors will demand a higher H_l^* to sell the property. One explanation is that as the investors are paying a lower leverage cost from holding the property, they can wait for a higher price in the future. Similarly, investors will accept a lower H_l^* to sell the property due to a higher risk-free rate.

As the lower bound H_l^* is the stop-loss price borne by the investors, H_l^* will not be higher than current market price. Therefore, the following condition should be met:

$$\frac{\left[\lambda_1\left(\kappa^{\lambda_2}-1\right) - \lambda_2\left(\kappa^{\lambda_1}-1\right)\right]}{\left[\lambda_1\left(\kappa^{\lambda_2}-\kappa\right) - \lambda_2\left(\kappa^{\lambda_1}-\kappa\right)\right]} < \frac{H_0(1-\alpha)}{\left(\frac{m(1-\tau)}{r-g} - \frac{(1-\tau)C}{r}\right)}. \tag{13}$$

3.1 The Optimal Value of the Commercial Real Estate

Equation (6) is the theoretical value of the commercial real estate without accounting for the optimal decision by investors. Therefore, based on the real options approach, the optimal value of the property can be obtained when considering the current interest rate and market price. The difference between the optimal value and the market price of the commercial real estate can be regarded as the premium or discount to help investors in their risk management.

$$\begin{aligned} F = \frac{m(1-\tau)}{r-g} &+ \frac{(1-\alpha)\left(\left(H_l^*\right)^{\lambda_2+1}\kappa^{\lambda_2} - \kappa\left(H_l^*\right)^{\lambda_2+1}\right) - \frac{m}{r-g}\left(\left(H_l^*\right)^{\lambda_2}\kappa^{\lambda_2} - \left(H_l^*\right)^{\lambda_2}\right)}{\left(H_l^*\right)^{\lambda_1+\lambda_2}\kappa^{\lambda_2} - \left(H_l^*\right)^{\lambda_1+\lambda_2}\kappa^{\lambda_1}} H^{\lambda_1} \\ &+ \frac{(1-\alpha)\left(\left(H_l^*\right)^{\lambda_1+1}\kappa^{\lambda_1} - \kappa\left(H_l^*\right)^{\lambda_1+1}\right) - \frac{m}{r-g}\left(\left(H_l^*\right)^{\lambda_1}\kappa^{\lambda_1} - \left(H_l^*\right)^{\lambda_1}\right)}{\left(H_l^*\right)^{\lambda_1+\lambda_2}\kappa^{\lambda_1} - \left(H_l^*\right)^{\lambda_1+\lambda_2}\kappa^{\lambda_2}} H^{\lambda_2}. \end{aligned} \tag{14}$$

3.2 Numerical Analysis

The implications of the model can be obtained by conducting sensitivity analyses under the assumption of certain fixed parameters such as vacancy rate, interest rate, tax rate and transaction cost, which can be collected from the market. The empirical analysis can be further conducted according to the different classes of commercial real estate so as to deliver values that can be referenced by investors in managing their investment risk.

4 Conclusion

We use the real options approach to derive the value of commercial real estate. To our knowledge, this is the first paper using real options to derive commercial real estate value. We hope this model can provide more insights into the value of real estate. The model can be applied to different classes of commercial real estate by changing the setup or values of parameters in the hope of delivering different implications in the real estate risk management.

References

1. Bollinger, C.R., Ihlanfeldt, K.R., Bowes, D.R.: Spatial variation in office rents within the Atlanta region. Urban Stud. **35**(7), 1097–1118 (1998)
2. Deng, Y., Quigley, J.M., Van Order, R.: Mortgage terminations, heterogeneity and the exercise of mortgage options. Econometrica **68**(2), 275–307 (2000)
3. Gat, D.: Urban focal points and design quality influence rents: the case of the Tel Aviv office market. J. Real Estate Res. **16**(2), 229–247 (1998)
4. Grovenstein, R., Kau, J.B., Munneke, H.J.: Development value: a real options approach using empirical data. J. Real Estate Finan. Econ. **43**(3), 321–335 (2011)
5. Kahr, J., Thomsett, M.C.: Real Estate Market Evaluation and Analysis. Wiley, New York (2005)
6. Kanoutos, G., Tsekrekos, A.E.: Real options premia implied from recent transactions in the Greek real estate market. J. Real Estate Finan. Econ. **47**(1), 152–168 (2013)
7. Kau, J.B., Keenan, D.C., Muller, W.J., Epperson, J.F.: A generalized valuation model for fixed-rate residential mortgages. J. Money Credit Banking **24**(3), 279–299 (1992)
8. Kau, J.B., Keenan, D.C., Muller, W.J., Epperson, J.F.: Option theory and floating-rate securities with a comparison of adjustable- and fixed-rate mortgages. J. Bus. **66**(4), 595–618 (1993)
9. Kau, J.B., Keenan, D.C., Smurov, A.A.: Reduced form mortgage pricing as an alternative to option-pricing models. J. Real Estate Finan. Econ. **33**(3), 183–196 (2006)
10. Lambrecht, B.M., Myers, S.C.: Debt and managerial rents in a real-options model of the firm. J. Finan. Econ. **89**, 209–231 (2008)
11. Lee, T., Jou, J.B.: Urban spatial development: a real options approach. J. Real Estate Finan. Econ. **40**, 161–187 (2010)
12. Leland, H.: Corporate debt value, bond covenants, and optimal capital structure. J. Finan. **49**, 1213–1252 (1994)
13. Mauer, D.C., Sarkar, S.: Real options, agency conflicts, and optimal capital structure. J. Banking Finan. **29**, 1405–28 (2005)

14. Merton, R.: On the pricing of corporate debt: the risk structure of interest rates. J. Finan. **29**, 449–470 (1974)
15. Colwell, P.F., Cannaday, R.E.: Trade-offs in the office market. In: Clapp, J.M., Messner, S.D., (eds.) Real Estate Market Analysis, Chapter 8, pp. 172–191. Praeger, New York (1988)
16. Quigg, L.: Empirical testing of real option-pricing models. J. Finan. **48**(2), 621–640 (1993)
17. Sarkar, S.: Early and late calls of convertible bonds: theory and evidence. J. Banking Finan. **27**, 1349–1374 (2003)
18. Titman, S.: Urban land prices under uncertainty. Am. Econ. Rev. **75**(3), 505–514 (1985)
19. Tseng, I.-W., Huang, M.-Y., Chang, C.-O.: Influence of tenant structure on rental and vacancy rates of office buildings. J. City Plan. **37**(4), 481–500 (2010)
20. Vandell, K.D., Lane, J.S.: The economics of architecture and urban design: some preliminary findings. Real Estate Econ. **17**(2), 235–260 (1989)

A Dynamic Model of Optimal Share Repurchase

Chun-Ping Chang[1], Yung-Shun Tsai[1], Shyh-Weir Tzang[1(✉)], and Yong Zulina Zubairi[2]

[1] Department of Finance, Asia University, No. 500, Liu-Feng Rd., Wu-Feng District, Taichung, Taiwan
12space@hotmail.com, gicha.tsai@gmail.com, swtzang@gmail.com
[2] Mathematics Division, University of Malaya, Kuala Lumpur, Malaysia
yzulina@um.edu.my

Abstract. We propose a model of dynamic share repurchase. The model highlights the central importance of payout for corporate decisions. Our two main results are: (1) free cash flows depends on the operating cash flows changes; (2) optimal share repurchase timing is decided by the relative changes between the free cash flow and dividends.

Keywords: Dividends · Free cash flows · Share repurchase

1 Model Setup

The all-equity case has been extensively studied in the real options literature (e.g., [2]). In our model, we adopt the similar specification of [3] and [5] to measure free cash flows for the valuation. Thus, the free cash flows (FCF) correspond to the Traditional Accounting Equality. We not allow the firm to increase its free cash flow by raising outside funds in the issuance markets, not with the interest earned, and not to pay a proportional cost on the distribution and the static free cash flow should be viewed as the payouts to shareholders.

$$FCF \equiv RP + D, \tag{1}$$

where RP is the share repurchase and D is the dividends. Both sides divided by D and rearrangement would be the new calculation:

$$RP/D = FCF/D{-}1 \tag{2}$$

We recognize that variation in free cash flows and dividends implies important differences in financial relative importance dynamics. In the model, the firm determines the optimal timing of optimal share repurchase by balancing their dividends with their free cash flows. The firm is assumed to choose its distribution ratio FCF/D to maximize the value F.

L. Barolli et al. (Eds.): IMIS 2018, AISC 773, pp. 808–811, 2019.
https://doi.org/10.1007/978-3-319-93554-6_80

Assume that the free cash flows influenced by operation at date t is also uncertain and follows a geometric Brownian motion process

$$dFCF = \mu_{FCF} FCF dt + \sigma_{FCF} FCF dz^{FCF}, \tag{3}$$

where μ_{FCF} and σ_{FCF} are the expected trends and the volatility of FCF, respectively. dz^{FCF} is the increments of a Wiener process. Assume that the dividends catering to investors D_t at date t is uncertain and follows a geometric Brownian motion process:

$$dD = \mu_D D dt + \sigma_D D dz^D, \tag{4}$$

where μ_D and σ_D are the expected trends and the volatility of D_t, respectively. The derivation of F follows the standard approach (see [2,8]) using dynamic programming. The boundary conditions will be stated later.

$$F(FCF, D) = e^{-rdt} E(F(FCF, D) + dF(FCF, D)) \tag{5}$$

Applying Ito's lemma, the following partial differential equation is obtained (where r denotes the risk-free rate):

$$\begin{aligned} &\frac{1}{2}\sigma_{FCF}^2{}^2 \frac{\partial^2 F(FCF, D)}{\partial FCF^2} + \hat{\mu}_{FCF} FCF \frac{\partial F(FCF, D)}{\partial FCF} \\ &+ \frac{1}{2}\sigma_D^2 D^2 \frac{\partial^2 F(FCF, D)}{\partial D^2} + \hat{\mu}_D D \frac{\partial F(FCF, D)}{\partial D} \\ &- \frac{\partial^2 F(FCF, D)}{\partial FCF \partial D} FCF \cdot D \cdot \rho \sigma_{FCF} \sigma_D - rF(FCF, D) = 0 \end{aligned} \tag{6}$$

The necessary conditions for an optimal policy are embodied in two value-matching conditions and four smooth-pasting conditions. The key feature that keeps the model tractable is that the problem reduces to a single state variable. The key feature that keeps the model tractable is that the problem reduces to a single state variable. Share repurchase timing is subject to the usual boundary conditions. The first boundary condition is the value-matching that gives the value at which the firm should operate. The second boundary condition is the smooth-pasting that assures that the derivatives of the two functions have to be equal at the quantity point.

Similar methods can be used to obtain a closed-form solution for Eq. (8). We define $X \equiv \frac{FCF}{D}$ as the relative effect on distribution. After the appropriate substitutions, the above Eq. (6) can be written as:

$$\frac{1}{2}\sigma_F^2 X^2 \frac{\partial^2 F(X)}{\partial X^2} + \hat{\mu} X \frac{\partial F(X)}{\partial X} - rF(X) = 0, \tag{7}$$

where $\sigma_F^2 = \sigma_{FCF}^2 + \sigma_D^2 - 2\rho\sigma_{FCF}\sigma_D$ and $\hat{\mu} = \hat{\mu}_{FCF} - \hat{\mu}_D$. The solution can be written as the form $F = AX^\beta + BX^\beta$. The above Eq. (7) is an Euler's type ordinary differential equation with the following characteristic quadratic function:

$$\frac{1}{2}(\sigma_F^2 CF + \sigma_D^2 - 2\rho\sigma_F CF)\beta(\beta - 1) + (\hat{\mu}_{FCF} - \hat{\mu}_D)\,\beta - r = 0 \tag{8}$$

Equation (8) has two roots, a positive and a negative one, given by:

$$\beta^{\pm} = \frac{1}{(\sigma_{FCF}^2 + \sigma_D^2 - 2\rho\sigma_{FCF})^2} - \left(\hat{\mu}_{FCF} - \hat{\mu}_D - \frac{1}{2}(\sigma_{FCF}^2 + \sigma_D^2 - 2\rho\sigma_{FCF})\right)$$
$$\pm\sqrt{2r\left(\hat{\mu}_{FCF} - \hat{\mu}_D - \frac{1}{2}(\sigma_{FCF}^2 + \sigma_D^2 - 2\rho\sigma_{FCF})^2\right)} \tag{9}$$

To our knowledge, the ratio of free cash flows to dividends under this traditional accounting equality structure has not previously appeared in the literature. Early contributions to the financial policy literature study one barrier and similar literature typically does not consider the dynamic interaction between free cash flows and dividend jointly. We find that the optimal share repurchase policy is to retain the ratio of the free cash flows produced by operating cash flows and dividends catering to investors satisfies the second order ordinary differential equation (ODE).

2 Empirical Predictions

The model presented in the previous section has two implications and provide some ideas on related empirical analysis. The share repurchase probability decreases with operating cash flows volatility, and free cash flow volatility and the dividends As the profitability and free cash flows growth increases, and the expected dividends rate decreases, firm tends to raise the incentive to share repurchase.

3 Conclusion

Since the seminal papers by [1,4,6,7], the literature analyzing financial decisions has seen substantial achievement. In the model, firm acts in the best interest of shareholders and chooses the firm's production and payout policies to maximize the present value of future dividends to incumbent shareholders. Notably, we view distributions as options on free cash flows. We demonstrate that the ratio of financial allocation provides an incentive for decision makers to make an optimal financial policy. As shown in the paper, this incentive to realize share repurchase significantly reduces the probability of share repurchase over a given horizon and erodes the value of financial flexibility.

References

1. Dann, L.Y.: Common stock repurchases: an analysis of returns to bondholders and stockholders. J. Finan. Econ. **9**(2), 113–138 (1981)
2. Dixit, A., Pindyck, R.: Investment under uncertainty. Princeton University Press, Princeton (1994)
3. Holt, R.W.P.: Investment and dividends under irreversibility and financial constraints. J. of Econ. Dyn. Control. **27**(3), 467–502 (2003)
4. Ikenberry, D.L., Vermaelen, T.: The option to repurchase stock. Finan. Manag. **25**(4), 9–24 (1996)
5. Lambrecht, B.M., Myers, S.C.: A theory of takeovers and disinvestment. J. Finan. **62**(2), 809–845 (2007)
6. Lintner, J.: Distribution of incomes of corporations among dividends, retained earnings, and taxes. Am. Econ. Rev. **46**(2), 97–113 (1965)
7. Miller, M.H., Modigliani, F.: Dividend policy, growth and the valuation of shares. J. Finan. **34**(4), 411–433 (1961)
8. Pindyck, R.S.: Irreversibility, uncertainty, and investment. J. Econ. Lit. **29**(3), 1110–1148 (1991)

Direct and Indirect Effects of Job Complexity of Senior Managers on Their Compensation and Operating Performances

Ying-Li Lin(✉) and Yi-Jing Chen

Department of Finance, Asia University,
500, Lioufeng Rd., Wufeng, Taichung 41354, Taiwan
yllin@asia.edu.tw

Abstract. This paper explores the relationship among the job complexity of senior managers on their compensation and operating performances. The job complexity of senior managers is measured with proxy variables associated with operating activities. Simultaneous equations are then constructed to validate the endogenous relationships among these three factors. The empirical study indicates that the job complexity of senior managers is relevant to the improvement of operating performances and the level of executive compensation. This finding should serve as a reference for the information and electronics industry in the planning of executive compensation programs in order to enhance operating performances.

1 Introduction

Many studies on compensations follow the theoretic model developed by [27] and the presumption by [27] that risk-neutral or risk-averse principles hire risk-averse and hardworking agents. If all events are observable, then it is possible for the principle to mitigate agency slippage via monitoring and oversight. This allows for the differentiation of insurance effects and incentive effects as contracted. Therefore, it is not necessary to the principle to take into account the outputs or the fixed salaries paid to agents. [19] propose performance-based pay schemes by arguing that fixed compensations such as salaries and car allowances are not linked with firm performances. Therefore, fixed compensations cannot incentive managers to create value for the company, which is detrimental to the shareholders' claim for earnings. Boards usually refer to financial results in the assessment of executive performances and provide rewards accordingly. [28] suggest that in addition to financials, non-financial metrics can serve as a leading indicator for firm profitability.

The majority of the literature on executive pay focuses on the relationship between executive pay and firm performances. Studies indicate a correlation between executive pay and company performances. Therefore, managers should be held responsible for the overall performance of firms. This is why executive contracts are usually based on operating performances. Some studies argue that the better the corporate governance is, the more effective the link is between executive performances and operating results, and hence the higher the sensitivity executives pay has on firm performances. Many

L. Barolli et al. (Eds.): IMIS 2018, AISC 773, pp. 812–821, 2019.
https://doi.org/10.1007/978-3-319-93554-6_81

research papers suggest a correlation between executive pay and job descriptions. According to [26], there is a positive correlation between the level of executive pay and the amount of information to be processed, as the main responsibility of senior managers is to handle all the company information. As a result, the more operating activities there are under management (including M&A activities, investments, and share/bond issues), the greater the volume of information to be processed and communicated will be. This enhances the job complexity of senior managers and ultimately the standard of executive pays.

The literature has indicated variances in the intensity of the correlation between firm performances and different types of executive pay. [12] contend that stock compensation to external directors is positively correlated to firm performances, but cash compensation to external directors is negatively correlated with firm performances. [24] find a significant and positive correlation between options granted to the top five executives and company earnings going forward. [7, 6] suggest that firm performances are positively correlated with stock-based compensations to external directors, but not correlated with cash compensations paid to external directors. In sum, there has been no study to date on the causal relationship between executive job complexity, compensations, and company performances. Therefore, this paper sets out to examine the following issues in relation with the level of executive job complexity.

1. Whether executive pay is beneficial to the improvement of operating performances.
2. The causal relationship between executive job complexity, compensations, and operating performance by using simultaneous equations.

2 Literature Review and Research Hypotheses

2.1 Executive Job Complexity and Compensations

[33, 22] indicate that information processing is one of the main tasks for senior managers. It is also essential to the proper functioning of any organization. The types of information to be processed by executives depend on a number of factors. [11, 16, 33] note that business diversification increases the amount of information to be processed by senior managers. However, executives capable of handling different kinds of information are hard to come by [1, 16, 34]. Studies indicate that the more capable the managers are able to process the information resulting from different situations, the better off the company performances are. Therefore, companies with a large amount of information to be handled are willing to pay a high price to attract more capable managers [31]. Based on the above literature review, this paper develops the following hypotheses:

H1: With the financial performances during the past and current periods controlled, the job complexity of senior managers has a positive influence on their total compensations.

2.2 Executive Pay and Operating Performances

Studies on the correlation between firm performances and executive pay focus on the relationship between investment opportunities and executive compensations [36, 17, 18]. Most empirical results suggest that companies with a larger opportunity set for investments give higher rewards to managers. There are limited studies in the literature dealing with the relationship between executive pay and forward-looking firm performances. [9] indicate that compensation levels and firm performances are positively correlated if such compensations are aligned with shareholders' interests. In other words, higher executive pay is conducive to better operating performances. [25, 15] examine the correlation between executive compensations during the current period and operating performance in future periods. They suggest that bonuses can serve as information indicative of the performance of the agents. If operating performances going forward are related to this information, then bonuses can provide information effects regarding forward-looking firm performances. According to the theoretic model and the hypotheses developed by [3], bonuses paid to executives are inversely correlated with company performance in the past, but positively correlated with company performance in the current period. With the current and past performances controlled, there is a positive correlation between current compensations and future operating performances. However, there is no significant correlation between current bonuses and forward-looking operating performances. In sum, a good compensation system can incentive executives and help to improve company performances in the future. This paper thus develops the following hypotheses.

H2: Total executive pay has a positive effect on operating performances going forward.
H3: The job complexity of senior managers has a direct impact on operating performances going forward.

3 Research Methods

3.1 Data Sources and Sample Selection

The Regulations Governing Information to Be Published in Annual Reports of Public Companies amended in 2004 by the Financial Supervisory Commission; Executive Yuan requires the disclosure of executive remunerations during recent years. Therefore, the information on executive bonuses, salaries, and board director compensations disclosed since 2005 is available in the database maintained by the Taiwan Economic Journal. This gives a consistent basis for the comparison of remuneration structures. The Taiwan government started to encourage the independent directors system 12 years ago in 2006. In order to ensure the integrity and accessibility of research data, this paper samples companies listed on the Taiwan Stock Exchange with comprehensive financial data, equity structure, and share price information. Companies that did not disclose executive pay or have no executive pay are eliminated from the sample. Financial industries whose financial statements are different from the general industry are also removed from the sample pool. The research period covers nine years (2008–

2016). The number of observations is 6,575. All the financial data are sourced from the Taiwan Economic Journal database. Non-financial data are from annual reports and MOPS (Market Observation Post System).

3.2 Measurement of Variables

3.2.1 Dependent Variables

1. Executive pays (TOINCOMP)

The standard format for the disclosure of executive compensations pursuant to The Regulations Governing Information to Be Published in Annual Reports of Public Companies by the Financial Supervisory Commission, Executive Yuan is attached in the appendix. This paper divides total executive remunerations into incentive and non-incentive.

2. Firm performances going forward (FPF)

This paper adopts the relative performance evaluation method developed by [14]. Return on equity (ROE) is defined as the accounting performance indicator. The following period's ROE is calculated by deducting the firm's performance indicator with the industry median, in order to measure the operating performance going forward by comparing with the industry average [3].

3.2.2 Independent Variables - Job Complexity (COMPLEX)

This paper evaluates the level of executive job complexity with four factors, i.e. free cash flow; issues of long-term debt, ordinary shares, and preferred shares; the number of senior managers; and the number of divisions. It is expected that there is a positive correlation between executive job complexity and remunerations. Below is an explanation of the individual factors.

1. Free cash flows (FCF)

[23] finds that the more cash held at hand, the more proactive a company will be engaged in M&A activities. Efforts in investments enhance the job complexity of senior managers. Therefore, this paper refers to free cash flow as the proxy variable for executive job complexity.

2. Issues of long-term debt (DEBT), ordinary shares, and preferred shares (STOCK)

Executives have to make key decisions when the company issues new shares and fixed income securities. They also have to communicate and coordinate with lawyers, underwriters, and auditors. This increases the workload and task complexity of senior managers. Therefore, this paper expects a positive correlation between these two variables and executive pay.

3. Number of senior managers (TMT)

The main tasks of top management are information processing and relevant decisions. However, any manager can only handle a limited amount of information.

Efficiency can be improved by expanding the size of top management and the professional approach to information processing. However, the larger the top management team is, the more items that need to be communicated and the more complex the executive jobs. Therefore, this paper expects a positive correlation between the number of senior managers and the remunerations paid to senior managers.

4. Number of divisions (DEPARTMENT)

[21] contend that a company needs to pay a premium for diversified operations if senior managers have to take care of new business departments. Therefore, this paper measures the level of executive jobs with the number of divisions in a company.

3.2.3 Control Variables

1. Executive Pay

(1) Family enterprise (FAMILY)

If a company is family-owned, it is set at 1; otherwise, it will be set at 0 [19].

(2) Firm size (SIZE)

The larger the firm size is, the more complex its business lines and the more difficult the oversight process. Therefore, stock-based compensations should be increased as an incentive [4]. [6] indicate that the bigger the firm size is, the more supervision is required and the higher executive pay should be. This paper refers to the natural logarithm of cost of goods sold as the proxy variable for firm size.

(3) Board size (BOARD)

The larger the board is, the higher the remunerations paid to senior managers [35, 13]. Scholars have identified a negative correlation or neutrality between board size and executive pay [30, 29, 37]. This paper uses the number of board directors (and supervisors) as an indicator of board size.

(4) Debt ratio (LEV)

[7] suggest a significant and inverse correlation between debt ratios and executive pay. Debts can mitigate the agency cost associated with over-investments as debt covenants can restrain managers from over-investments in projects that eventually end up with impairments. Debt covenants also prompt managers to dispose of under-performing assets in order to avoid technical defaults. As executive remunerations undermine a company's debt service capability, the higher the leverage is, the lower executive pay should be. High leverage comes at the expense of shareholders' interest. At this juncture, more oversight is required. Therefore, there is a positive correlation between debt ratios and executive pay [6]. This paper refers to total debts over total assets as the debt ratio.

2. Operating Performances Going Forward
(1) Outsiders (OUTSIDERS)

Many studies indicate that the higher the percentage of outsiders is, the better the oversight, and the stronger the alignment between executives' interests and shareholders' interest. This improves the operating performance of the firm [8, 2, 5, 20, 10]. This paper refers to the percentage of independent directors in the board as a measurement for outsiders.

(2) Institutional ownership (INST)

The efficient monitoring hypothesis by [32] posits that institutional investors set a high bar for board functioning in order to control their own investment risks. This paper calculates institutional ownership by the number of shares held by institutional investors divided by the total number of shareholders outstanding.

3.3 Empirical Model

As the job complexity of senior managers affects the remunerations paid to executives, and such remunerations affect the company's operating performances going forward, simultaneous equations with two-stage least squares are established to test whether executive pay and financial performances are interdependent. The simultaneous equation is as follows:

$$\begin{aligned} TOINCOMP_{i,t} = {} & \pi_{11} + \pi_{21} COMPLEX_{i,t} + \pi_{31} FAMILY_{i,t} + \pi_{41} SIZE_{i,t} \\ & + \pi_{51} BOARD_{i,t} + \pi_{61} LEV_{i,t} + \varepsilon_{i1} \end{aligned} \quad (1)$$

$$\begin{aligned} FPF_{i,t} = {} & \pi_{12} + \pi_{22} TOINCOMP_{i,t} + \pi_{32} OUTSIDERS_{i,t} \\ & + \pi_{42} INST_{i,t} + \pi_{52} COMPLEX_{i,t} + \varepsilon_{i2} \end{aligned} \quad (2)$$

Where $TOINCOMP_{i,t}$ and $FPF_{i,t}$ denote the total compensations received by senior management and the financial performance of the firm. These are endogenous variables. Moreover, $COMPLEX_{i,t}$ denotes the measurement of the job complexity of senior managers, and ε_{i1} and ε_{i2} are random error terms.

4 Empirical Results

Table 1 summarizes the descriptive statistics of the sample. The minimum value of executive pays is zero, and the standard deviations of this variable are large. This suggests a large gap between the highest and the lowest remunerations paid to senior management. The maximum value of operating performances going forward is 192.39, and the minimum value is –414.77, also indicating significant variances.

The empirical results shown in Table 2 suggest that executive pay has a significant and negative impact on operating performances going forward, but the effects are close to zero. The job complexity of senior managers exhibits a significant and positive

Table 1. Descriptive statistics.

	BOARD	DEBT	DEPARTMENT	FAMILY	FCF	FPF	INST	LEV	OUTSIDERS	SIZE	STOCK	TMT	TOINCOMP
Mean	5.84	3278056.	11.51	0.11	301099.8	1.13	0.19	0.42	0.25	11126098	6715587.	6.15	18674.01
Median	6.00	103680.0	11.00	0.00	11164.00	1.81	0.14	0.40	0.29	1474777.	1850273.	5.00	12463.00
Maximum	21.00	2.20E+08	39.00	1.00	1.77E+08	192.39	8.67	1.00	0.67	1.50E+09	2.59E+08	50.00	224059.0
Minimum	2.00	0.000000	3.00	0.00	–1.63E+08	–414.77	0.00	0.01	0.00	–3081271.	0.000000	0.00	0.00
Std. Dev.	1.84	15621024	4.75	0.31	5524548.	9.82	0.27	0.17	0.17	60475585	21268736	5.49	21449.40

Table 2. Empirical results.

	FPF	TOINCOMP
TOINCOMP	−0.00*** (0.00)	
OUTSIDERS	2.99*** (0.00)	
INST	0.88* (0.09)	
FCF	3.44 (0.18)	4.97 (0.99)
DEBT	3.38** (0.02)	8.83*** (0.00)
STOCK	3.63*** (0.00)	0.00*** (0.00)
TMT	1.12*** (0.00)	2619.65*** (0.00)
DEPARTMENT	0.15*** (0.00)	249.92*** (0.00)
FAMILY		249.92*** (0.00)
SIZE		−3.11*** (0.00)
BOARD		−60.87 (0.50)
LEV		9429.54*** (0.00)
Adjusted R-squared	0.5986	

Note: The value in the parentheses of the table is the p-value. ***, **, * Denote coefficient estimates that are reliably significant at the 1%, 5%, 10% levels, respectively.

influence on executive pay and operating performances going forward. Therefore, only H1 and H3 are supported.

5 Conclusion

Scant studies have examined how the job complexity of senior managers and the level of executive pays affect a company's operating performances in the future. This paper establishes simultaneous equations in order to explore the causal relationship among the job complexity of senior managers, the level of executive pay, and a company's operating performances going forward. The empirical results suggest the correlation of the job complexity of senior managers to operating performances going forward and the level of executive pay. These findings can serve as a template for the information technology and electronics industry in the design of executive compensations in order to boost operating performances.

Acknowledgments. This research was supported by Ministry of Science and Technology of the Republic of China under contract MOST 106-2813-C-468-077-H.

References

1. Agarwal, N.C.: Determinants of executive compensation. Indus. Relat. J. Econ. Soc. **20**(1), 36–45 (1981)
2. Alexander, J.A., Fennell, M.L., Halpern, M.T.: Leadership instability in hospitals: the influence of board–CEO relations and organizational growth and decline. Adm. Sci. Q. **38**, 74–99 (1993)
3. Banker, R.D., Darrough, M.N., Huang, R., Plehn-Dujowich, J.M.: The relation between CEO compensation and past performance. Acc. Rev. **88**(1), 1–30 (2013)
4. Becher, D.A., Campbell II, T.L., Frye, M.B.: Incentive compensation for bank directors: the impact of deregulation. J. Bus. **78**(5), 1753–1778 (2005)
5. Brickley, J., Coles, J., Terry, R.: Outside directors and the adoption of poison pills. J. Financ. Econ. **35**, 371–390 (1994)
6. Brick, I.E., Palmon, O., Wald, J.K.: CEO compensation, director compensation, and firm performance: evidence of cronyism? J. Corp. Finance **12**(3), 403–423 (2006)
7. Bryan, S., Hwang, L.S., Lilien, S.: Compensation of outside directors: an empirical analysis of economic determinants. NYU Working Paper (2000)
8. Byrd, J.W., Hickman, K.A.: Do outside directors monitor managers? J. Financ. Econ. **32**, 195–221 (1992)
9. Carpenter, M.A., Sanders, W.M.: Top management team compensation: the missing link between CEO pay and firm performance? Strateg. Manag. J. **23**(4), 367–375 (2002)
10. Chatterjee, S., Harrison, J.S., Bergh, D.D.: Failed takeover attempts, corporate governance and refocusing. Strateg. Manag. J. **24**, 87–96 (2003)
11. Chandler, A.D.: Strategy and structure: Chapters in the history of the American enterprise. Massachusetts Institute of Technology Cambridge (1962)
12. Cordeiro, J., Veliyath, R., Eramus, E.: An empirical investigation of the determinants of outside director compensation. Corp. Gov. Int. Rev. **8**(3), 268–279 (2000)
13. Core, J.E., Holthausen, R.W., Larcker, D.F.: Corporate governance, chief executive officer compensation, and firm performance. J. Financ. Econ. **51**, 371–406 (1999)
14. DeFond, M.L., Park, C.W.: The effect of competition on CEO turnover. J. Acc. Econ. **27**(1), 35–56 (1999)
15. Ederhof, M.: Discretion in bonus plans. Acc. Rev. **85**(6), 1921–1949 (2010)
16. Finkelstein, S., Hambrick, D.C.: Chief executive compensation: a study of the intersection. Strateg. Manag. J. **10**(2), 121 (1989)
17. Gaver, J.J., Gaver, K.M.: Additional evidence on the association between the investment opportunity set and corporate financing, dividend, and compensation policies. J. Acc. Econ. **16**(1), 125–160 (1993)
18. Gaver, J.J., Gaver, K.M.: Compensation policy and the investment opportunity set. Financ. Manag. **24**(1), 19–32 (1995)
19. Gomez-Mejia, L., Wiseman, R.M.: Reframing executive compensation: an assessment and outlook. J. Manag. **23**(3), 291–374 (1997)
20. Goodstein, J., Gautman, K., Boeker, W.: The effects of board size and diversity on strategic change. Strateg. Manag. J. **15**, 241–250 (1994)
21. Rose, N., Shepard, A.: Firm diversification and CEO compensation: Managerial ability or executive entrenchment? RAND J. Econ. **28**(3), 489–514 (1997)

22. Haleblian, J., Finkelstein, S.: Top management team size, CEO dominance, and firm performance: the moderating roles of environmental turbulence and discretion. Acad. Manag. J. **36**(4), 844–863 (1993)
23. Harford, J.: Corporate cash reserves and acquisitions. J. Finance **54**(6), 1969–1997 (1999)
24. Hanlon, M., Rajgopal, S., Shevlin, T.: Are executive stock options associated with future earnings? J. Acc. Econ. **36**(1), 3–43 (2003)
25. Hayes, R.M., Schaefer, S.: Implicit contracts and the explanatory power of top executive compensation for future performance. RAND J. Econ. **31**(2), 273–293 (2000)
26. Henderson, A.D., Fredrickson, J.W.: Information-processing demands as a determinant of CEO compensation. Acad. Manag. J. **39**(3), 575–606 (1996)
27. Hölmstrom, B.: Moral hazard and observability. Bell J. Econ. **10**(1), 74–91 (1979)
28. Ittner, C.D., Larcker, D.F.: Are nonfinancial measures leading indicators of financial performance? an analysis of customer satisfaction. J. Acc. Res. **36**, 1–35 (1998)
29. Jensen, M.C.: The modem industrial revolution, exit, and the failure of internal control systems. J. Appl. Corp. Finance **6**, 4–23 (1993)
30. Pearce, J.A., Zahra, S.A.: Board compensation from a strategic contingency perspective. J. Manag. Stud. **29**, 411–438 (1992)
31. Pfeffer, J., Davis-Blake, A.: Understanding organizational wage structures: a resource dependence approach. Acad. Manag. J. **30**(3), 537–455 (1987)
32. Pound, J.: Proxy contests and the efficiency of shareholder oversight. Acad. Manag. J. **20** (1/2), 237–265 (1988)
33. Prahalad, C.K., Bettis, R.: The dominant logic–a new linkage between diversity and performance. Teoksessa: how organizations learn. In: Starkey, T.K. (ed.), pp. 100–122. International Thompson Business Press, London
34. Quinn, J.B.: Strategies for Change: Logical Incrementalism. Irwin Professional Publishing (1980)
35. Sanders, G., Carpenter, M.A.: Internationalization and firm performance: the roles of CEO compensation, top team composition, and board structure. Acad. Manag. J. **41**, 158–178 (1998)
36. Smith, C.W., Watts, R.L.: The investment opportunity set and corporate financing, dividend, and compensation policies. J. Financ. Econ. **32**(3), 263–292 (1992)
37. Yermack, D.: Higher market valuation of companies with a small board of directors. J. Financ. Econ. **40**, 185–211 (1996)

The Impact of Online Commentary on Young Consumer's Purchase Decision

Mei-Hua Huang, Wen-Shin Huang, Chiung-Yen Chen(✉), and An-Chi Kuan

Department of Accounting and Information Systems, Asia University, Taichung, Taiwan
{meihuang, joanchen}@asia.edu.tw

Abstract. Nowadays, young generation consumers' purchasing decision is affected tremendously by online product review, probably far more than other channels of communication. This may imply that strategies of marketing and campaign commercial products very likely need a revolutionary change. Therefore, the purpose of this study is to understand the different impacts of the traditional product endorsement and modern online commentary on young consumer trust and purchase decision. Experimental method was employed to collect data. 120 students participated in the experiment. Results show that youngsters have higher level of trust in online commentary than endorsement. Online commentary than endorsement has a greater impact on youngsters' purchase intention. As expected, trust has a significant and positive impact on purchase intention. Finally, management implications and future directions are discussed.

1 Introduction

Product endorsement or endorser's recommendation or endorsement is a traditional way of communicating and promoting products. It is popularly employed for many product types and industries. But it is generally costly to hire an endorser to speak for a product. However, this trend seems has never mitigated, because endorsement is regarded as effective and its benefits can trade off the cost of the endorsement [1]. Indeed, research indicates that endorsers possess attractiveness or credibility or both attractiveness and credibility [14]. Therefore, their endorsements can attract people's attention and persuasive people to follow their behavior. Some endorsers have the charm of impress people positively and some have specialized knowledge gain trust from people easily. In the real world, most product spokespersons are either charming people because they have the power of attractiveness or experts because they have credibility. Hence, product endorsements are considered effective by academics, business and general public.

Along with the fast growing network, personal computers, and mobile devices, especially the popularity and rapidity of social media, interactions, communication, and share information among people have tremendously changed. On one hand, people have a brand new channel to receive information. For example, people are able to get commentaries by others for a certain brand or product easily. That means audience

L. Barolli et al. (Eds.): IMIS 2018, AISC 773, pp. 822–828, 2019.
https://doi.org/10.1007/978-3-319-93554-6_82

(viewers) or consumers have more ways of obtaining product information than the common way of product endorsement in the past. This may imply the effectiveness of the traditional recommendation by endorsers are fading. On the other hand, audience and consumers can go online to express their opinions about product experiences actively, rather than receive product information passively. The transmission speed and scope of these online product information or so called word of mouth are massive than other media. This implies that internet users' commentary become reliable or even perceived with high trust for it is instantaneous and extensive and for it is coming from similar others.

Based on the above discussion, product endorsement and online commentary both possess reliability even though with different characteristics (elements). Product endorsement increases the level of trust through the spokesperson's professional knowledge or attractiveness to fans. Whereas, netizen's commentary gaining credibility for the similarity and the wisdom of crowd [15]. An interesting but not explored issue: which one will obtain a higher level of trust from consumers between product endorsement and online commentary, especially young consumers?

As such, the purpose of this study is to understand product purchase decision of the young generation in the networking age; which will have a higher impact on youngsters' decision making between traditional product endorsement and modern online commentary? The research questions are: (1) which will gain a higher trust from young consumers and (2) which will have a higher impact on young consumer's purchase intention between the two recommendation sources?

2 Literature Review

2.1 Product Recommendation

The traditional product recommendation mainly is from advertising which is one important commercial means to have spokespeople for the product recommendation or endorsement. Spokespeople include experts, scholars, celebrities, athletes, professionals and others the spokesperson is considered effective because of its attractiveness or trustworthiness. The spokespersons' attractiveness is related to a person's appearance, characteristics, status or similarity with the recipient of the message. Trustworthiness describes the evaluation of the professional knowledge or skill, objectivity and reliability of the spokesperson by the listeners or the consumers, which are related to the judgment of the endorser's credibility and whether the information is genuinely provided [14]. Research show that if a product's endorsement is considered authentic, it will gain people's trust, otherwise, it will not be trusted [5].

Online review is a new way of product recommendations recently. Consumers spontaneously tell or make comments on their product consumption experiences online. It has become one important source for young people for product information [8]. Online comments can be two-way communication. Users can exchange information with each other and discuss or ask questions. It is different from endorsement that the spokesperson does one-way communication or propaganda. Therefore the online commentary is easier to gain netizens' trust. Furthermore, the accessibility, speediness, and reach out features of the online commentary allow netizens to participate in discussions in the real time and it can gather huge amounts of people. That is the power of crowds or so called wisdom of crowds [15], a basis for making strong trust as opposed to a single or a small number of endorsers.

2.2 Product Recommendation and Trust

Trust can be learnt through social interactions. It is regarded as a key element for establishing and maintaining relations of social exchange [9]. In the purchasing decision-making process, trust is often a complementary or complementary mechanism that can replace direct search for more information; it exists in consumers' minds to overcome risks and uncertainties when environmental uncertainty is high or when there is a functional risk in selecting products [2]. Therefore, when consumers have trust in the source of product recommendations, it is likely they will generate the intention to buy or rebuy the recommended product. Based on the above discussion, the following hypothesis is proposed.

H1: Consumers have a higher level of trust in online commentary than product endorsement.

2.3 Trust and Purchase Intention

Purchase intention is the possibility of the consumer's willingness or unwillingness to buy a certain product [6]. Consumers' confidence in products are likely affected through both or either the endorser's advocacy or internet users' word-of-mouth. Many past marketing studies found a strong positive correlation between trust and purchase intention and suggested purchase intention can serve as an indicator of actual purchase [3, 4, 12]. Therefore, it can be inferred when consumers have a higher level of trust in the product recommendation message, their purchase intention for the advocated product will be greater. Hence the following is hypothesized.

H2: The higher level of consumer trust in product recommendation message, the higher the purchase intention the consumer has.

2.4 Product Recommendation and Purchase Intention

Based on the discussions for the hypothesis 1, it can be reasonably deduced that consumers' purchase intention can be directly influenced by online commentaries and product endorsements. The impact of online commentaries will be greater than that of product endorsements. Moreover, we assume that the basis of the strong positive relationship between trust and purchase intention in the hypothesis 2. It can be inferred that consumer purchase intention can also be affected indirectly (through trust) by the online comment and product endorsement. The online comment impact will be greater than the impact of product endorsement. Therefore, the hypothesis is as the follows.

H3: The impact of online commentary on consumer purchase intention is higher than that of product endorsements.

3 Method

3.1 Research Framework

The conceptual framework of this study is shown in Fig. 1. This study explores whether there is a significant difference in consumer trust and in purchase intention between the two product recommendations (product endorsement vs. netizens' commentary), and whether trust lead to affect purchase intention.

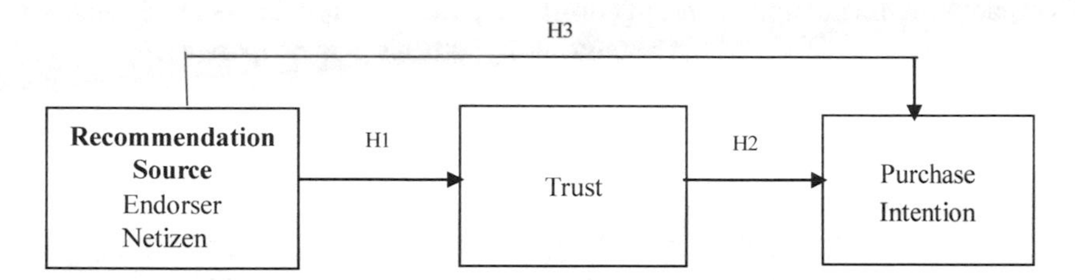

Fig. 1. The research model

3.2 Experiment and Data Collection

This study first selected 2 services and 2 tangible products that are commonly consumed by most young people. Both the endorsement film and online commentary were all found for each service and product. Therefore, a total of 8 scenarios were established. Then the measuring items for trust and purchase intention were adapted from previous studies. Trust was measured with 3 items adapted from Morgan and Hunt [10]. Purchase intention was also measured with 3 items adapted from Palmatier et al. [13]. 5 point Likert-type scales were used for measurement, where "1"is "totally disagree" and "5" is "totally agree".

The experiment was held in the classrooms and use of questionnaires to collect data. Participants are students of a university in Taichung, Taiwan. 4 sessions of the experiment was conducted. Participants were asked to answer the survey questions right each scenario was viewed. A total of 120 students were recruited to participate in the experiment. A small stationary was given as a thank you gift.

4 Data Analysis

4.1 Instrument Reliability

Reliability of the variables was first performed for the purpose of evaluation the consistency and reliability of the measuring instrument. The value of Cronbach α assures the internal reliability when it is above 0.8 [11]. The variables of this study trust and purchase intention have Cronbach α 0.827 and 0.865, respectively, which indicates the instrument is reliable.

4.2 Hypotheses Testing

Independent t was adopted to test the hypotheses 1 and 3, and simple regression to hypothesis 2. Data were separated into 2 groups, online commentary and endorsement. The two sets of data were compared to see whether a significant difference exists between them. The result in Table 1 shows that consumer trust in online commentary is much higher than that in endorsement (p = 0.000). Therefore, hypothesis 1 is supported. The same procedure was applied to test the hypothesis 3 to see whether a significant difference exists between the two groups. The result in Table 2 indicates that the impact of online commentary on consumer purchase intention is much higher than that of endorsement (p = 0.000). Therefore, hypothesis 3 is supported.

Table 1. T-test for trust

Recom. source	t	Degree of freedom	Sig. (two tails)	Average difference	95% confidence interval	
					Lower bound	Upper bound
Online commentary	64.824	119	.000	3.02639	2.9339	3.1188
Product endorsement	64.408	119	.000	2.84167	2.7543	2.9290

Table 2. T-test for purchase intention

Recom. source	t	Degree of freedom	Sig. (two tails)	Average difference	95% confidence interval	
					Lower bound	Upper bound
Online commentary	66.983	119	.000	3.16250	3.0690	3.2560
Product endorsement	55.342	119	.000	3.02986	2.9215	3.1383

Table 3. Regression results

Dependent variable	R	R^2	Adjusted R^2	Sig.	Standard error
Purchase intention	.580[a]	.337	.331	0.000	.39951

[a]Independent variable: (constant), trust

Hypothesis 2 assumes that the higher level of consumer trust in product recommendation message, the higher the purchase intention of the recommended product the consumer has. The result of regression analysis in Table 3 shows that consumer trust significantly explains the variance of purchase intention by 33.7% ($R^2 = .337$, $p = 0.000$). Therefore, as expected, hypothesis 3 is supported.

5 Conclusion

5.1 Research Findings and Discussions

The results reveal that young consumers tend to have a higher level of trust in the online comments on the consumption experience made by similar others when compared to the product recommendation by endorsers. Similarly, online commentaries have a higher impact on young consumers' purchase intention for the reviewed product than product endorsements. This result confirms the substantial influence of "wisdom of crowds" [15]. Moreover, the similar result to past research is found that trust has a significant positive correlation with purchase intention. This finding indicates that the higher the trust of young consumers have in product recommendation messages, the higher their willingness to purchase the recommended product.

5.2 Managerial Implications

This study reveals that the young generation of consumer behavior pattern is changing. Young generation tends to rely on online commentaries over celebrity or expert product endorsements; youngsters' purchase intention is also influenced by online comments over product endorsements. These findings indicate that in the era of information explosion, people paying attention to the online communication, the impact of traditional product endorsements seem to be weakening. Therefore, business practitioners probably need to reconsider the allocation of their resources. Instead of paying expensive endorsement fees, business probably want to put more resources in the enhancement of online positive word-of-mouth. For example, the use of current well-known bloggers to increase the exposure of products and form a good brand reputation. Ideally, investments in such as product improvement, service improvement, and customer relationship enhancement should guarantee the positive online word of mouth, leading to influence young generation's trust and purchase decision.

5.3 Limitations and Future Research

Although this study has some interesting and important findings, there are also some research limitations. The sample use single source. Since the participants are all from the same school students, the representative of the young generation is therefore confined. Future studies can expand the sample by including different school students and young work people. Moreover, in order to have a greater control in the experiment, endorsement advertisements and online comment messages appear only once to the subject than the actual media exposures, which may limit the research results. Finally, the experiment was conducted in classrooms. The number of students in each class was large (20–40 people) and students were ask to answer questions right after viewing the endorsement video and comment message. They were unable to proceed at their own pace. Therefore, future studies can consider to conduct experiments in small groups to enhance the validity of the study.

References

1. Agrawal, J., Kamakura, W.A.: The economic worth of celebrity endorsers: an even study analysis. J. Mark. **59**(3), 56–62 (1995)
2. Chaudhuri, A., Holbrook, M.B.: Product-class effects on brand commitment and brand outcomes: The role of brand trust and brand affect. J. Brand Manag. **10**(1), 33–58 (2002)
3. Chiu, C., Hsu, M., Lai, H., Chang, C.: Re-examining the influence of trust on online repeat purchase intention: the moderating role of habit and its antecedents. Decis. Support Syst. **53** (4), 835–845 (2012)
4. Danesh, S.N., Nasab, S.A., Ling, K.C.: The study of customer satisfaction, customer trust and switching barriers on customer retention in Malaysia hypermarkets. Int. J. Bus. Manag. **7** (7), 141–150 (2012)
5. Darke, P.R., Ritchie, R.J.B.: The defensive consumer: advertising deception, defensive processing, and distrust. J. Mark. Res. **44**(1), 114–127 (2007)
6. Guo, Y.M., Poole, M.S.: Antecedents of flow in online shopping: a test of alternative models. Inf. Syst. J. **19**(4), 369–390 (2007)
7. Huang, P., Lurie, N.H., Mitra, S.: Searching for experience on the web: an empirical examination of consumer behavior for search and experience goods. J. Mark. **73**(2), 55–69 (2009)
8. Huang, J.Y., Liu, B.Y.: Electronic WOM and online review – a literature Review. NTU Manag. Rev. Forum **26**(3), 215–256 (2016)
9. Konovsky, M.A., Pugh, S.D.: Citizenship behavior and social exchange. Acad. Manag. J. **37** (3), 656–669 (1994)
10. Morgan, R.M., Hunt, S.D.: The commitment-trust theory of relationship marketing. J. Mark. **58**(3), 20–38 (1994)
11. Nunnally, J.C., Bernstein, I.H.: Psychometric Theory. McGraw-Hill Inc., New York (1994)
12. Palmatier, R.W., Dant, R., Grewal, P.D., Evans, K.R.: Factors influencing the effectiveness of relationship marketing: a meta-analysis. J. Mark. **70**(4), 136–153 (2006)
13. Palmatier, R.W., Jarvis, C.B., Bechkoff, J.R., Kardes, F.R.: The role of consumer gratitude in relationship marketing. J. Mark. **73**(5), 1–18 (2009)
14. Solomon, M.R.: Consumer Behavior: Buying, Having, and Being. Pearson Education Inc. (2017)
15. Surowiecki, J.: The Wisdom of Crowds: Why the Many are Smarter than the Few. Anchor, New York (2004)

Penalty or Benefit? The Effect of Dividend Taxes on Stock Valuation

Wen-hsin Huang[1], Suming Lin[2], and Mei-Hua Huang[1(✉)]

[1] Department of Accounting and Information Systems, Asia University, Taichung, Taiwan
evelynkeet@asia.edu.tw, meihuang@asia.edu
[2] Department of Accounting, National Taiwan University, Taipei, Taiwan
Suming@ntu.edu.tw

Abstract. The extant literature has shown that dividends have positive valuation implications due to signaling and agency cost effects. However, under the tax systems of most countries, individual investors face a higher tax rate on dividend income than on capital gains. Therefore, individual investors pay a dividend tax penalty, which results in lower equity values. U.S Studies indicate that as the level of institutional ownership increases, the likelihood that a marginal investor is not a high-tax-rate individual increases. Consequently, the negative dividend tax penalty effect on the positive market response to dividend surprises should decrease. This study extends previous research to investigate the tax effect of dividends under Taiwan's imputation tax system. We find that dividend tax penalty partially offsets the positive effects of dividends on equity values. However, this negative tax effect of dividends can be alleviated by the presence of a marginal investor who represents a tax-exempt institution.

1 Introduction

The extant theoretical and empirical literature has shown that dividend announcements have positive valuation implications due to the signaling and agency cost effects. Consequently, the greater the level of dividends paid out by firms, the more the stock price rises. However, individuals' dividend and interest income represent ordinary taxable income that is subjected to ordinary income tax rates under the tax systems of most countries. Given that ordinary income tax rates are generally higher than the preferential tax rates applicable to capital gains, Poterba and Summers [1], Ayers et al. [2], and Li [3] have found that when firms declare dividend payments, the dividend tax penalty partially offsets their positive valuation effects. Thus, if previous studies failed to control for the impact of dividend tax, then they may have underestimated the positive effects of dividend announcements.

However, if the marginal investor who influences the stock price reaction to dividend announcements is a tax-exempt or lower-taxed institution[1] rather than an

[1] For example, for legal entities in Taiwan, the dividends received on their holdings in other companies are not included in taxable income. In the U.S, legal entities are allowed a 70%, 80%, or even 100% deduction for dividends received from other companies that they own.

L. Barolli et al. (Eds.): IMIS 2018, AISC 773, pp. 829–838, 2019.
https://doi.org/10.1007/978-3-319-93554-6_83

individual investor, the disadvantageous dividend tax effect will have little impact. Though this marginal investor cannot be identified in advance, Li [3] employs two variables in his empirical regression model to explain the cumulative abnormal returns (CARs) around dividend announcement dates. One is the highest tax penalty imposed on individuals' dividend income across different tax regimes, and the other is the stock trading situation of institutional investors. He suggests that the higher the level of institutional ownership or the frequency of institutional trading, the greater the likelihood that the marginal investor is not a fully taxable individual. Consequently, the negative dividend tax penalty effect on a positive market response to dividend should decrease. That is, while the dividend tax penalty has adverse effect on abnormal return, the presence of institutional investor holdings can mitigate this negative dividend tax effect. Li [3] empirically supports that the negative tax effect of dividends is alleviated by the presence of institutional investors.

Though Li [3] argues that his research is the first one to explore the role played jointly by dividend taxes and investors' tax attributes, he uses the data from the United States, which imposes income taxes on both corporations and individuals. Therefore, his research findings on dividend tax penalty cannot apply to countries with the imputation tax system, such as Taiwan. We extend Li's [3] research by examining Taiwan's stock market data and its unique tax system. We construct an analytical model explaining the dividend tax effect under Taiwan's imputation tax system, which simultaneously imposes income tax on stock dividends, and develop our research hypotheses based on this analytical model. As explained later, our research findings can help Taiwan's investors avoid underestimating the dividend announcement effect in Taiwan's stock market.

2 Literature

Over the past several decades, the effect of dividend announcements on stock valuations has received a lot of attention in the finance, economics, and accounting literature. Miller and Modigliani [4] state that in a perfect market without tax considerations and transaction costs, a firm's dividend policy does not affect its market value. However, Bhattacharya [5], Miller and Rock [6], John and Williams [7] show theoretically that dividend policy changes are positively correlated with stock price changes. They contend that information asymmetry between firms and outside shareholders may induce a signaling role for dividends, that is, dividends signal future cash flows. From the agency cost perspective, Jensen [8] analyzes valuation implications of dividends theoretically. He indicates that paying dividends can reduce the level of free cash flow, and mitigate potential overinvestment that managers might engage in to increase personal utility. Consequently, dividends may serve as a device to address agency problems and enhance firm value.

With regard to empirical studies, Yoon and Starks [9] show that dividend change announcements are associated with revisions in analyst forecasts of current earnings. Nissim and Ziv [10] indicate that the greater the dividend changes, the greater the future profitability, with the latter measured in terms of future earnings and future abnormal earnings. These research results support the signaling effect of dividend announcements. However, using Tobin's q as a proxy for the overinvestment problem, Lang and Litzenberger [11] conclude that their empirical results are more consistent with the free cash flow hypothesis than with the cash flow signaling hypothesis. They find a differential market reaction between firms that overinvest and firms that do not. By testing Jensen's (1986) free cash flow theory, Kallapur [12] demonstrates that earnings response coefficients depend positively on dividend payout ratios. He analyzes this is because shareholders prefer that earnings be paid out as dividends rather than be wastefully retained. These studies support the agency cost argument of dividends.

Much debate and contradictory findings exist both in the theoretical and empirical literature concerning the tax implications of dividends for equity value. Miller and Modigliani [4], while formulating their well-known dividend irrelevance propositions, observed that if dividends are taxed more heavily than capital gains, then a firm's dividend policy is relevant to its value due to the differential taxation effect. However, investors form clienteles based on their dividend preferences. That is, firms with a low dividend payout ratio may attract investors in high tax brackets, whereas investors in low tax brackets may hold high-yield stocks. Brennan [13] extends the capital asset pricing model (CAPM) to include the effects of the taxes that investors pay on dividends and capital gains. Brennan's (1970) model of these tax effects indicates that risk-adjusted pretax returns should be positively correlated to dividend yields. He suggests that investors require higher before tax, risk-adjusted returns on stocks with higher prospective dividend yields to compensate for the historically higher taxation of dividends relative to capital gains. However, Black and Scholes [14] test the Brennan's (1970) model empirically and find no difference between the pretax monthly risk-adjusted returns of high- and low-dividend-yield stocks. They suggest that if investors require higher returns for holding higher yield stocks, then corporations adjust their dividend policy to restrict the quantity of dividends paid, and thus increasing their share price. Ayers et al. [2] find that the higher a firm's dividend yield, the more negative is its stock price reaction to an increase in the individual income tax rate with regard to the enactment of the U.S. Revenue Reconciliation Act of 1993. Dhaliwal et al. [15] find that an increasing annual return premium in the dividend yield, and thus suggesting that a dividend tax penalty be incorporated into the return on a firm's stock. Using a measure of dividend tax penalty based on the relative taxation of dividends and capital gains for fully taxable individual investors, Li [3] confirms a dividend tax penalty that partially offsets the positive effects of dividend announcements.

However, taxes should not affect firm value and stock returns if investors can effectively eliminate their tax burden on stocks (Miller and Scholes 1978) [16]. This argument suggests that to detect the tax effect of dividends, it is important to consider both a firm's dividend policy and its ownership structure. Recent empirical studies in the accounting literature show that firms with a high percentage of institutional ownership are more likely to have marginal investors with low tax rates, which can mitigate the negative tax effect of dividends (Ayers et al. [2]; Dhaliwal et al. [15]; Dhaliwal et al. [17]; Dhaliwal et al. [18]; Li [3]). For two reasons, these studies use the holdings of institutional investors as a proxy for the tax status of the marginal investor who is less tax-disadvantaged. First, as indicated by Sias and Starks [19], a marginal investor is more likely to be an institutional investor for a firm with a higher level of institutional ownership. Second, Dhaliwal et al. [20] indicate that institutional investors face, on average, lower tax rates on dividend income than do individual investors. Ayers et al. [2] find that the higher a firm's dividend yield, the greater is the negative reaction of its stock price to an increase in the individual income tax, although the presence of institutional investors can mitigate this negative reaction. Li [3] uses the level of institutional ownership and the frequency of institutional trading as the proxies for the likelihood that the marginal investor is a tax-favored taxpayer. He suggests that institutional ownership is a measure of institutional holding clientele; however, the frequency of institutional trading is a dynamic measure of institutional presence around dividend announcement dates. His result also supports the view that the higher the level of institutional ownership or the frequency of institutional trading, the lower the magnitude of the negative tax effect of dividends.

3 Research Method

3.1 Hypothesis Development

We begin by clarifying the disadvantageous tax factor associated with dividends relative to capital gains under Taiwan's imputation tax system. Assume that P_{bf} denotes the closing price before ex-day, a is the percentage of stock dividends from retained earnings (Stock dividends from retained earnings are \$10a per share), b is the ratio of tax-exempt stock dividend from capital surplus (Stock dividends from capital surplus are \$10b per share), and M denotes the cash dividends per share. Thus P_{rf}, the ex-day opening reference price, is equal to $(P_{bf} - M)/(1 + a + b)$, and we rearrange the equation to get $P_{bf} = P_{rf}\ (1 + a + b) + M$. If an investor sells the stock on the date immediately preceding the ex-day, then he will receive the proceed $P_{bf} = P_{rf}(1 + a + b) + M$. In addition, assume the closing price of the ex-day is P_{af}, t_c is the effective corporate income tax rate, and t_p denotes the marginal income tax rate of individual investors. If an individual investor receives dividend payment and sells the stock (including stock dividends from retained earnings and capital surplus) at P_{af} on the ex-day, then he will receive an after-tax proceed equal to $P_{af}\ (1 + a + b) + M - [(10a + M)/(1 - t_c)] \times (t_p - t_c)$. In equilibrium, the indifferent wealth situation for an investor who is deciding whether to receive dividends or not is $P_{rf}\ (1 + a + b) + M = P_{af}\ (1 + a + b) + M - [(10a + M)/(1 - t_c)] \times (t_p - t_c)$. Consequently, the disadvantageous tax amount imposed on an investor to receive dividends is $[(10a + M)/(1 - t_c)] \times (t_p - t_c)$.

Since the independent variable in our empirical regression model is the cumulative abnormal rate of return. Thus, we divide the disadvantageous tax amount by the closing price for the three days before the dividend announcement date (P_0) to get Eq. (1) expressed in percentage term. That is, the dividend tax penalty to receive dividends, expressed in percentage term is:

$$\text{Penalty} = \text{W} = [(10a + M)/(1 - t_c)] \times [(t_p - t_c)/\text{P}_0] \quad (1)$$

Differentiating Eq. (1) with respect to the effective corporate income tax rate

$$(t_c)\text{ yields: } \partial W/\partial t_c = [(10a + M)/P_0] \times [(t_p - 1)/(1 - t_c)^2] < 0\,(t_p < 1) \quad (2)$$

After Eq. (2) is obtained, we turn to develop research hypotheses. According to the literature, dividend announcements have the signaling and agency cost effects. This leads to our first hypothesis (hereafter stated in the alternative form):

H1: Ceteris paribus, the more the change in the dividend yield (CYIELD), the higher the cumulative abnormal return (CAR) around the dividend announcement date

Hypotheses H2a and H2b are related to the dividend tax penalty.

H2a: Ceteris paribus, the dividend tax penalty (PENALTY) will reduce the positive effect of the change in the dividend yield (CYIELD) on the cumulative abnormal return (CAR)

Although there is a positive relationship between CYIELD and CAR, but based on H2a, we expect the coefficient of the interaction term CYIELD•PENALTY to be negative.

We posit H2b according to the derived result in Eq. (2). From Eq. (2), we observe that the higher the effective income tax rate of a firm (t_c), the more tax credits that shareholders receive, and the fewer taxes they pay for receiving dividends, thus resulting in a lower dividend tax penalty. Consequently, we can infer that the positive valuation implications of dividend announcements will increase. This analysis leads to the following hypothesis:

H2b: Ceteris paribus, the greater the effective corporate income tax rate (TC), the lower the dividend tax penalty (PENALTY). Therefore, the effective corporate income tax rate is positively related to the effect of the change in the dividend yield (CYIELD) on the cumulative abnormal return (CAR)

According to H2b, effective corporate income tax rate reinforces the positive relationship between CYIELD and CAR. Thus, we expect the coefficient of the interaction term CYIELD•TC to be positive. This research conducts two regressions to test H2a and H2b respectively because they are both related to the dividend tax penalty.

We consider the trading situation for institutional investors in developing hypotheses H3a and H3b. Based on the argument and empirical findings in Li (2007), we suggest that the higher the trading volume of institutional investors, the greater the likelihood that a marginal investor is not a high-tax-rate individual increases. As a result, the negative dividend tax penalty effect on the positive valuation implication of dividend announcements for a firm should decrease. Hence, H3a is developed as follows:

H3a: Ceteris paribus, the dividend tax penalty (PENALTY) reduces the positive effect of the change in the dividend yield (CYIELD) on the cumulative abnormal return (CAR), but as the trading volume of institutional investors

(INST) increases, the less the negative effect of this penalty is. That is, the coefficient of the interaction term CYIELD•INST•PENALTY is positive

Inferring from Eq. (2), there is a negative relationship between a firm's effective income tax rate and the dividend tax penalty. This means that, other things being equal, a higher effective corporate income tax rate leads to a lower dividend tax penalty. Although the coefficient of the interaction term CYIELD•TC is positive, the increasing probability that a stock's marginal investor represents institutional investors reduces the effect of the interaction term CYIELD•TC on CAR. That is, the coefficient of CYIELD•TC•INST is negative. We propose the following hypothesis:

H3b: Ceteris paribus, as the trading volume of institutional investors (INST) increases, the effect of the interaction term between the change in the dividend yield (CYIELD) and the effective corporate income tax rate (TC) on the cumulative abnormal return (CAR) will decrease. That is, the coefficient of the interaction term CYIELD•INST•TC is negative

When testing H3a and H3b, we investigate how the interaction term CYIELD•INST•PENALTY and CYIELD•INST•TC affect CAR, and the variable INST is measured as the ratio of institutional trading (explained later).

3.2 Statistical Model, Variables, and Research Period

This research tests H2a and H2b, respectively, to explore the impact of the dividend tax penalty and a firm's effective income tax rate on the CAR around dividend announcement dates. Thus, we run the following two regressions:

$$\begin{aligned} CAR_{it} = {} & \beta_0 + \beta_1 CYIELD + \beta_2 CYIELD \cdot INST_{it} + \beta_3 CYIELD \cdot PENALTY_{it} \\ & + \beta_4 CYIELD \cdot PENALTY_{it} \cdot INST_{it} + \beta_5 MV_{it} + \Sigma\beta_{\mathrm{YR}} YR + \varepsilon_{it} \end{aligned} \tag{3A}$$

$$\begin{aligned} CAR_{it} = {} & \gamma_0 + \gamma_1 CYIELD + \gamma_2 CYIELD \cdot INST_{it} + \gamma_3 CYIELD \cdot TC_{\mathrm{it}} \\ & + \gamma_4 CYIELD \cdot TC_{it} \cdot INST_{it} + \gamma_5 MV_{it} + \Sigma\gamma_{YR} YR + \varepsilon_{it} \end{aligned} \tag{3B}$$

Equation (3A) is used to test hypotheses H1, H2a, and H3a; Eq. (3B) is used to test hypotheses H1, H2b, and H3b. The definitions and measurements for variables are described below:

CAR_{it}: the five-day cumulative abnormal return for firm i; using a standard market model (Fama 1976) [21] to compute abnormal return and cumulated over the five-day period beginning two days prior to and two days subsequent to the dividend announcement date.

$CYIELD_{it}$: the change in the dividend yield for firm i; We calculate the current-year dividend yield $(P_{bf} - P_{rf}) \div P_0$ and minus its counterpart in last year to obtain CYILED. P_{bf} is the closing price before ex-day, P_{rf} is the opening reference price on ex-day, and P_0 is the closing price for the three days before the dividend announcement date (see Hypotheses Development above). According to a Taiwan Stock Exchange rule, cash dividend value = cash dividend per share, and stock dividend value = (P_{bf} - cash dividend value - P_{rf}). Therefore ($P_{bf} - P_{rf}$) represents the total value of cash dividend and stock dividend.

$\text{PENALTY}_{\text{it}} = [(10a + M)/(1 - t_c)] \times [(t_p - t_c)/\text{P}_0]$, the dividend tax penalty for investors who receive dividends, where a is the percentage of stock dividend from retained earnings, M is the cash dividend per share, and t_p denotes the marginal income tax rate of an individual investor and we assume it is the highest rate 40%[2] in this study. Also, t_c is the effective corporate income tax rate, and P_O is the closing price for three days before the dividend announcement date.

$INST_{it}$: represent the trading situation of tax-exempt institutional investors for stock i around the dividend announcement date. Following Li [3], we measure $INST_{it}$ as the ratio of trading volume bought by securities investment trust companies and dealers to the total trading volume over the 11-day period around the dividend announcement date. $INST_{it} = \Sigma_{d=[-5,5]}\ \#trades_{it,d}/\#\ all\ trades_{it,d}$

where $\#\ trades_{it,d}$ is trading volume bought by securities investment trust companies and dealers for stock i on day d ($-5 \le d \le 5$) relative to the dividend announcement date t ($d = 0$).

$\#\ all\ trades_{it,d}$ is the daily trading volume of stock i on day d relative to the dividend announcement date t.

TC_{it}: the effective income tax rate of firm i; we define TC_{it} as the income tax expense divided by the income from continuing operations before taxes.

This study covers the period for dividend distribution of fiscal year 2000 to 2014. We aim our findings to offer relevant implications to the existing tax system, and thus we do not extend our research period to the year of 1999, the first transitional year imposing a 10% surtax on undistributed profits under Taiwan's imputation tax system.

4 Empirical Results

Table 1 lists the regression result of Eq. (3A). First, we find that without control for the tax effect, the coefficient of CYIELD is significantly positive. This result is consistent with prior research findings and supports H1 that there is a positive association between the change of the dividend yield and the cumulative abnormal return. However, when we consider the dividend tax effect and the marginal investor's tax status, the positive magnitude of the coefficient of CYIELD is larger than the former. Other things being equal, if we do not control the dividend tax effect and investor's tax attributes, the signaling and agency cost effects of dividend announcements will be underestimated.

CYIELD•PENALTY has a significantly negative coefficient. The result indicates that dividend tax burden reduces the positive market reaction to dividend change announcement and thus supports H2a. Since ordinary income tax rates are generally higher than preferential tax rates for capital gains, individual investors face the disadvantageous tax factor when receiving dividends. In other words, the empirical

[2] Ayers et al. [2] and Li [3] utilize the highest individual marginal income tax rate as a starting point to conduct their theoretical analyses.

Table 1. Regression analysis—Test H1, H2a, and H3a

	With		Without	
	Tax controls		Tax controls	
Variable	Parameter estimates	*t*-statistics	Parameter estimates	*t*-statistics
Intercept	6.972***	2.85	6.1393***	2.51
CYIELD	0.2372***	4.11	0.1384***	7.08
CYIELD•INST	−1.5304**	−1.92	–	–
CYIELD•PEN	−5.4853**	−1.84	–	–
CYIELD•INST•PEN	85.7455**	1.97	–	–
MV	−0.2480***	−2.438	−0.2264**	−2.29
YR	Yes		Yes	
	Adj R^2 = 0.0301		Adj R^2 = 0.0253	

Note: a. *, **, *** indicate significance at the 1%, 5%, and 10% levels, respectively (for one-tail *t* test).
b. Year dummies (YR) are included in the regression (not tabulated).

evidence supports that negative tax effect of dividends partially offsets the positive signaling and agency cost effects of the dividend announcements. The significant and positive coefficient of the interaction term CYIELD•PENALTY•INST demonstrates that as the higher the trading volumes bought by tax-exempt institutions, the likelihood that the marginal investor is a high-tax-rate individual decreases and thus the effect of the dividend penalty on CAR is weaker, supporting H3a. We infer that control of marginal investor attributes is of critical importance when analyzing how the stock market reacts to dividend announcements.

From Eq. (2), we expect a negative relationship between effective corporate income tax rate and dividend tax penalty. Therefore, we conduct empirical tests using Eq. (3B). Table 2 shows the result. As in Table 1, H1 is supported by the significant coefficient of CYIELD. CYIELD•TC with a positive and significant coefficient renders support to

Table 2. Regression analysis—Test H1, H2b, and H3b

Variables	Parameter estimates	*t*-statistics
Intercept	6.0491***	3.12
CYIELD	0.1735***	3.65
CYIELD•INST	0.3802	0.7
CYIELD•TC	0.6689**	2.28
CYIELD•TC•INST	−9.1003**	−2.20
MV	−0.2378***	−2.35
YR	Yes	
Adj.R^2 = 0.0345		

Note: *a.**, **, *** indicate significance at the 1%, 5%, and 10% levels, respectively (for one-tail *t* test).
b. Year dummies (YR) are included in the regressions; (not tabulated).

H2b, which suggests that a firm's effective income tax rate enhances the positive relationship between CYIELD and CAR. In other words, as firms declare an increase in dividend payment, investors will react more positively to the firm with higher effective corporate income tax rate. The coefficient of CYIELD•TC•INST has a negative and significant value. This finding supports H3b. That is, the higher trading volume bought by tax-exempt institutions, the less likely that the marginal investor is a high-tax-rate individual. Hence, the presence of institutional trading reduces the positive effect of the interaction term CYIELD•TC on CAR.

5 Conclusion

The extant literature shows that dividend tax penalty has a negative effect on stock valuation. However, as the level of institutional ownership increases, the likelihood that a marginal investor is a fully taxable individual decreases. Consequently, the negative dividend tax penalty effect on the positive market response to dividend surprises should decrease. As most of the prior research examines only the distribution of cash dividends, and the tax systems of other countries are different from Taiwan's, this study investigates the tax effect of dividends on stock price at the time of dividend announcements under Taiwan's imputation tax system, which also imposes a tax on stock dividends.

We establish a theoretical model to construct our measure for the tax penalty incurred by the receipt of dividends under Taiwan's tax system and then develop our research hypotheses. We then include dividend tax penalty, the level of trading by tax-exempt institutional investors, and their interaction terms in the regression model. We find that (1) the dividend tax penalty partially offsets the positive effects of dividends on equity values; (2) however, this negative tax effect can be alleviated by the presence of a marginal investor who represents a tax-exempt institution. Both our analytic and empirical analyses provide evidence to support that the findings of the extant foreign literature could also be applied to Taiwan's imputation tax system. Most importantly, without controlling for the dividend tax effect and the tax status of investors, the signaling and agency cost effects of dividend announcements may be underestimated.

References

1. Poterba, J.M., Summers, L.H.: New evidence that taxes affect the valuation of dividends. J. Finance **39**, 1397–1415 (1984)
2. Ayers, B.C., Cloyd, C.B., Robinson, J.R.: The effect of shareholder-level dividend taxes on stock prices: evidence from the Revenue Reconciliation Act of 1993. Acc. Rev. **77**, 933–947 (2002)
3. Li, O.Z.: Taxes and valuation evidence from dividend change announcements. J. Am. Taxation Assoc. **29**, 1–23 (2007)
4. Miller, M.H., Modigliani, F.: Dividend policy, growth, and the valuation of shares. J. Bus. **34**, 411–433 (1961)

5. Bhattacharya, S.: Imperfect information, dividend Policy, and “the Bird in the Hand” fallacy. Bell J. Econ. **10**, 259–270 (1979)
6. Miller, M.H., Rock, K.: Dividend policy under asymmetric information. J. Finance **40**, 1031–1051 (1985)
7. John, K., Williams, J.: Dividends, dilution, and taxes: A signaling equilibrium. J. Finance **35**, 1053–1070 (1985)
8. Jensen, M.C.: Agency Costs of Free Cash Flows, Corporate Finance, and Takeovers. Am. Econ. Rev. **76**, 323–329 (1986)
9. Yoon, P.S., Starks, L.T.: Signaling, investment opportunities, and dividend announcements. Rev. Financial Stud. **8**, 995–1018 (1995)
10. Nissim, D., Ziv, A.: Dividend changes and future profitability. J. Finance **56**, 2111–2133 (2001)
11. Lang, L.H.P., Litzenberger, R.H.: Dividend announcements: cash flow signaling vs. free cash flow hypothesis? J. Financ. Econ. **24**, 181–191 (1989)
12. Kallapur, S.: Dividend payout ratios as determinants of earnings response coefficients. J. Account. Econ. **17**, 359–375 (1994)
13. Brennan, M.J.: Taxes, market valuation and corporate financial policy. Natl. Tax J. **23**, 417–427 (1970)
14. Black, F., Scholes, M.: The effect of dividend yield and dividend policy on common stock prices and returns. J. Financ. Econ. **1**, 1–22 (1974)
15. Dhaliwal, D., Li, O.Z., Trezevant, R.: Is a dividend tax penalty incorporated into the return on a firm’s common stock? J. Account. Econ. **35**, 155–178 (2003)
16. Miller, M.H., Scholes, M.S.: Dividends and taxes. J. Financ. Econ. **6**, 333–364 (1978)
17. Dhaliwal, D., Krull, L., Li, O.Z., Moser, W.: Dividend taxes and implied cost of equity capital. J. Acc. Res. **43**, 675–708 (2005)
18. Dhaliwal, D., Erickson, M., Li, O.Z.: Shareholder income taxes and the relation between earnings and returns. Contemp. Acc. Res. **22**, 587–616 (2005)
19. Sias, R.W., Starks, L.T.: Return autocorrelation and institutional investors. J. Financ. Econ. **46**, 103–131 (1997)
20. Dhaliwal, D., Erickson, M., Trezevant, R.: A test of the theory of tax clienteles for dividend policies. Natl. Tax J. **52**, 179–194 (1999)
21. Fama, E.F.: Foundations of Finance. Basic Books, New York (1976)

The Determinants of Admission Strategy and School Choice: A Case Study of a Private Senior Vocational School in Taiwan

Ling-Yi Chou(✉) and Yi-Yang Li

Asia University, Taichung, Taiwan
stanfordyiyi@gmail.com, sleepingleo19810818@gmail.com

Abstract. As the changing of population structure, the phenomenon of low-birth rate has been presented globally. We are also encountering a dramatically change in social structure which impacted the education development. The study aimed to discuss the relationship between the factors of business model for school and selection of students. In accordance with the business model generation that proposed by Osterwalder and Pigneur (2010), the study considers the factors as cooperation partner, operation item, resource, value position, customer relationship, channels of student recruiting strategy and the factor of student entrance. The result shows that major cooperation partner, resource, customer relation and student recruiting channels are significant positive factors to impact student's selection. The contribution of the study is to provide a reference for private educational institution and proposed correspondent strategies to reduce the impact of low-birth rate for school.

1 Introduction

From the 1980s onwards, the number of births in Taiwan has been slowly revised downwards from more than 400,000 annually, and it has been reduced by about 1 to 20,000 people each year. According to the most recent data released by the CIA World Factbook in 2017, Taiwan has the third-lowest birthrate in the world. The problem of low birthrate in Taiwan has become even more pronounced in recent 10 years. For example, the number of births in the year was 190,110 people in 2009. In 2010, the number of births in that year was only 168,866. From January to July in 2017 only 110,379 babies were born, a six percent decrease from the same period in 2016. In addition, according to the National Development Council, it is estimated that the zero growth birthrate of Taiwan's population occurred at the fastest time in 2019, and the latest in 2026. Therefore, all counties and municipalities in the world provide relevant welfare measures to stimulate fertility so as to reduce the impact of declining birthrate, especially in Taiwan (Table 1).

Y.-Y. Li—Take the master's degree in business administration at Asia University.

L. Barolli et al. (Eds.): IMIS 2018, AISC 773, pp. 839–848, 2019.
https://doi.org/10.1007/978-3-319-93554-6_84

Table 1. Birthrate in Taiwan

Year	The number of births in that year (Unit: the number of people)			The birthrate (Unit: ‰)
	Total	Man	Female	
2006	204,459	106,936	97,523	8.96
2007	204,414	106,898	97,516	8.92
2008	198,733	103,937	94,796	8.64
2009	191,310	99,492	91,818	8.29
2010	166,886	87,213	79,673	7.21
2011	196,627	101,943	94,684	8.48
2012	229,481	118,848	110,633	9.86
2013	199,113	103,120	95,993	8.53
2014	210,383	108,817	101,566	8.99
2015	213,598	111,041	102,557	9.10
2016	208,440	108,133	100,307	8.86

(Resource: Ministry of the Interior, Department of Home Affairs (2016))

It's facing a transition point for admission students with the change in population structure and fertility reduction, school need to find a good strategy to appeal students selection. Therefore, re-examine educational goals and vision for schools, school administrative effectiveness, innovate school enrollment strategies, improve teaching quality, promote teachers' professional knowledge learning, update teaching software and hardware facilities, extend admission condition, induce students' learning motivation, and improve the effectiveness of student learning, all of them have become a crucial issue that needs to be explored in all education stages. Otherwise, it will surely be dismantled under the turmoil of the new wave of educational reforms, or even shut down or close schools.

It can be seen that the seriousness of the problem of declining birth. According to the Ministry of Education (2011), a total of 503 schools in senior middle school (including higher vocational colleges) from 2014 to 2015 were enrolled. The number of enrolled students was 786,056. The enrollment of high school students in the future could not fully saturate the number of students required in high schools. Unbalanced supply and demand will become more and more obvious, and the competition for admissions between schools will become fiercer.

The purpose of this study is to analyze the current situation of the environment and to analyze the school enrollment strategies and student enrollment factors in Taiwan. We adopt a case study to understand how schools can effectively plan marketing activities and enrollment strategies. Moreover, we attempt to find a systematical plan for enhancing enrollment performance and enhancing the competitiveness after implementing them, in order to striving public recognition and affirmation.

2 Literature Review

2.1 Business Models

A business model describes the rationale of how an organization creates, delivers, and captures value (Osterwalder and Pigneur 2010). Therefore, companies can build a business model to describe new ideas and technologies to create economic value (Table 2).

Table 2. The definition of business models

Author (Year)	Definition
Viscio and Pasternack (1996)	Business models must create systemic value
Timmers (1998)	An architecture of business models for the product, service and information flows, including a description of the various business actors and their roles, and a description of the potential benefits for the various business actors; and a description of the sources of revenues
Amit and Zott (2010)	A focus on the how of doing business, and a holistic perspectives on how business is conducted; An emphasis on value creation for business model participant; A recognition that partner can help the focal firm conduct essential activities within its business model
Magretta (2002)	business model concept seems to focus more on cooperation, partnership, and joint value creation
Chesbrough and Rosenbloom (2002)	A successful business model creates a heuristic logic that connects technical potential with the realization of economic value
Morris et al. (2005)	Business models describe three categories including economic, operational, and strategic, and explore how to establish a sustainable competitive advantage in a specific market
Casadesus-Masanell and Ricart (2010)	Business Model to refer to the 'logic of the firm' how it operates and creates value for its stakeholders, and the cost structure for creating the value. Business Model is a reflection of the firm's realized strategy

According to the book "business model generation", a business model can best be described through nine basic building block, such as customer segments, value propositions, channels, customer relationship, key revenue, resource, activities, partnerships, and cost structure.

- The definition of Customer Segments: It defines the different groups of people or organizations an enterprise aims to reach and serve.
- The definition of the Value Propositions: It describes the bundle of products and services that create value for a specific Customer Segment.

- The definition of the Channels: It describes how a company communicates with and reaches its Customer Segments to deliver a Value Proposition.
- The definition of the Customer Relationships: describes the types of relationships a company establishes with specific Customer Segments.
- The definition of the Revenue Streams: It represents the cash a company generates from each Customer Segment, and costs must be subtracted from revenues to create earnings.
- The definition of the Key Resources: It describes the most important assets required to make a business model work.
- The definition of the Key Activities: It describes the most important things a company must do to make its business model work.
- The definition of the Key Partnerships: It describes the network of suppliers and partners that make the business model work.
- The definition of the Cost Structure: It describes all costs incurred to operate a business model (Fig. 1).

<table>
<tr><td rowspan="2">Key Partner</td><td>Key Activities</td><td rowspan="2" colspan="2">Value Propositions</td><td>Customer Relationships</td><td rowspan="2">Customer Segments</td></tr>
<tr><td>Key Resources</td><td>Channels</td></tr>
<tr><td colspan="3">Cost Structure:</td><td colspan="3">Revenue Streams</td></tr>
</table>

Fig. 1. The Business Model Canvas (Osterwalder and Pigneur 2010)

3 Research Method

In order to explore private vocational enrollment strategies and their effectiveness, First, we follow the business model proposed by Osterwalder and Pigneur (2010). Second, we refer the enrollment strategies from ten benchmarking schools to develop the questionnaire items of admission strategy for school. Then, we make a case study of private senior vocational school, and we conduct questionnaire item, and do in-depth interviews for increasing student willing of selection school. Our research framework as following as below (Fig. 2):

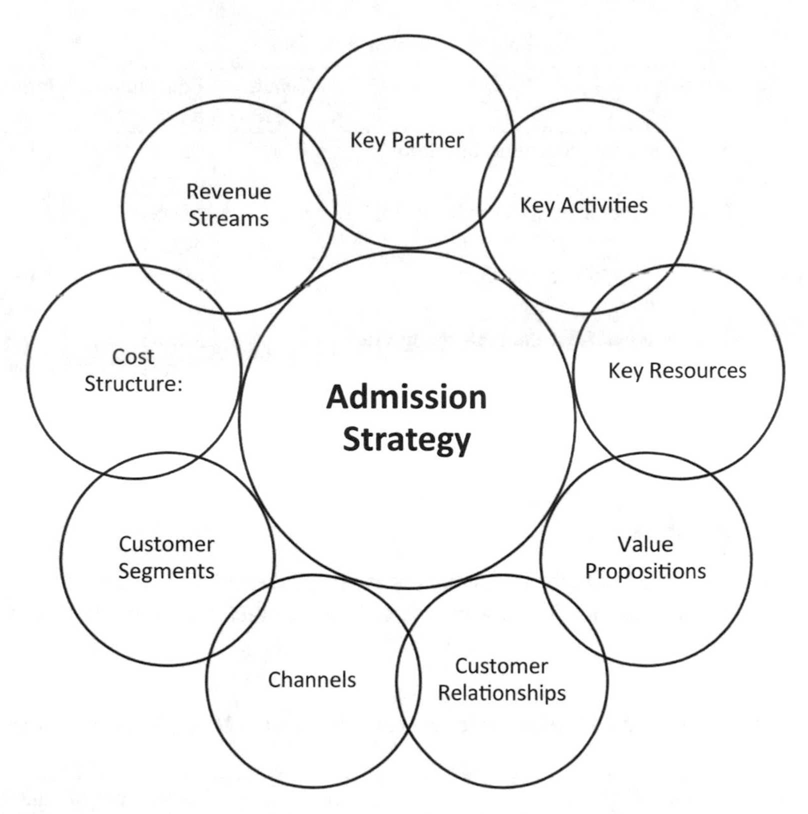

Fig. 2. Research framework of this study (It refers Osterwalder and Pigneur (2010) and this study rebuild it)

3.1 Data Collection

This study used in-depth interviews to establish interview questions through the organization of school admission documents and documents.

3.2 Sample Subjects

In the selection of interviewees, the study focused on the main executive staff who responsible for school enrollment strategies and student enrollment factors. Therefore, the case study choose the subject such as the director of education in school, the deputy director of the catering department, the head of the training in the academic affairs division, and the team leader of the performing arts division, the team leader of the drama team. We conducted an interview on November 20, 2017. The interviewee data is shown in Table 3.

Table 3. Interview subject

Name	Job title	Tenure	Education level	Interview time
OO, Wang	The director of education in school	7	Master degree	1 h
OO, Yang	Deputy director of catering in school	8	Master degree	1 h
OO, Wu	Director of training department, academic affairs office	1	Master degree	1 h
OO, Chuang	The team leader of the performing arts division	3	Master degree	1 h
OO, Chen	The team leader of the drama division	8	Bachelor degree	1 h

3.3 Interview Item

This study firstly collected the enrollment strategies of the top 10 benchmark schools in private vocational schools in Taiwan. Next, we conduct the item of quesitionare (Table 4).

Table 4. The number of students of top ten benchmark school in 2016 (Data collected from this study)

The Name of School (City)		The number of classes	The number of students		
		Total	Total	Male	Female
1	Chung Shan Industrial & Commercial School (Kaohsiung City)	182	9,736	6,061	3,675
2	Chi-Ying Senior High School (Taoyuan City)	155	7,374	3,972	3,402
3	Yu Da High School of Commerce and Home Economics (Taipei City)	146	7,203	3,067	4,136
4	Shin Shing High School (Taoyuan City)	153	6,917	3,970	2,947
5	Yu Da High School (Taoyuan City)	152	6,652	2,827	3,825
6	Chih Ping Senior High School (Taoyuan City)	147	6,642	3,650	2,992
7	Shu-Te Home economics & Commercial High School (Kaohsiung City)	154	6,539	2,195	4,344
8	Chiao Tai High School (TaiChung City)	134	6,289	3,275	3,014
9	Yung Ping Vocational High School (Taoyuan City)	130	6,259	3,438	2,821
10	Da Der Commercial and Technical Vocational School (Changhua County)	136	6,199	3,614	2,585
	The Case in the study	112	4,899	2,122	2,777

Due to cost and revenue are not public information for school, it is hard to collected the information. Thus, we delete these two constructs from questionnaire. In addition, the customer segment are similar with different school based on 12 years of national education stipulated by the Ministry of Education, therefore, we also remove it. The questionnaire are listed as below (Table 5):

Table 5. The question item for in-depth interview

Construct	The question item
Key Partner	Q1. when declining birthrate is getting serious, can you list the key partners in the case school for facing competitive environment?
Key Activities	Q2. What is the most important thing that must be done in the current main operating mode of the school?
Key Resources	Q3. Can you describe the important assets needed to support the school?
Value Propositions	Q4. What do you think that the school can meet the needs of students after graduation? What are the unique value propositions in case school?
Customer Relationships	Q5. Why the case school can appeal students select it. Can you list some ways that the case school attracts students to select?
Channels	Q6. In the age of information explosion, what are the ways to increase school exposure and popularity?

4 Analysis Results

After in-depth interviews, we attempt to record the description as following as below:

Interview content	Job title	Response
Q1. When declining birthrate is getting serious, can you list the key partners in the case school for facing competitive environment?	The director of education in school	It is import to establish a good partnership with the admission sources. We must establish a strategic alliance with the National University in Taiwan
	Deputy director of catering in school	1. College: To cooperate with strategic alliances, industry and education, etc. 2. Middle School: In conjunction with the National School of Arts and Technology Education, it cooperates with many other countries in a pull-out manner, and there are special classes in the country and cooperative teaching in associations 3. The factory: Due to the schooling system of the establishment of education classes, the Food and Beverage Branch in case school cooperated with several restaurants and restaurants to provide student recruitment options, including: Jinghua, Sheraton, Xingzi pork chops and so on

(*continued*)

(*continued*)

Interview content	Job title	Response
Q2. What is the most important thing that must be done in the current main operating mode of the school?	Deputy director of catering in school	1. Improve teachers' teaching quality and encourage students' adaptive learning: Good teaching quality can improve students' motivation for learning. There is an excellent teaching reputation in the school and good enrollment energy can be provided 2. Strengthen the teaching energy of teachers and encourage them to take on-the-job training: Teachers can cultivate the second specialty or on-the-job training, maintain the enthusiasm and motivation of teaching, and then have the energy to activate teaching
	Director of training department, academic affairs office	1. Division of professional functions of administrative divisions: Avoid overlapping of unit responsibilities, causing repeated operations or evading one another 2. Establish recruitment and improvement of recruitment assignments: (1) If a dedicated recruitment operation unit is established, after division of responsibilities is clear, each division will perform its duties without interference (2) If a project-type organization is established internally, it is necessary to carefully consider the suitability of the individual pumping personnel so as to avoid interfering with the administrative work of the original position and the difference in service characteristics of different units and inequality of work and rest
Q3. Can you describe the important assets needed to support the school?	Director of training department, academic affairs office	(1) Formal teachers (2) Informal teachers (3) Students (4) Hardware facilities
	The team leader of the performing arts division	I think the important assets for the school are: a sound board, a stable source of students, and good teachers. A sound board of directors can take care of the entire staff of the school. The board's funding and maintenance of school assets

(*continued*)

(*continued*)

Interview content	Job title	Response
		can protect the rights of staff and students. A stable source of students can create a virtuous circle for the school. When the student source is stable, the funds and the number of students in the school can also be stabilized, so that the academic tradition and quality of a school can be inherited. In the process of running a school, teachers play an important role, and teachers' professional abilities need to be reviewed and evaluated to ensure that the quality of teaching can be improved. Teachers' in-service education can also enable the development of teaching models without being eliminated by the times
Q4. What do you think that the school can meet the needs of students after graduation? What are the unique value propositions in case school?	Director of training department, academic affairs office	After completing the training of students' professional skills and verifying the training results with the certificates, it is relatively a guarantee for employment and going to college
	The team leader of the performing arts division	Two value propositions: moral education and license orientation. The school is a professional license-oriented vocational school that promotes license education to enable students to obtain national recognition after learning professional skills. Future students are more likely to receive skills recognition in the workplace
Q5. Why the case school can appeal students select it. Can you list some ways that the case school attracts students to select?	Director of training department, academic affairs office	Most graduate alumni develop into performing arts career
	The team leader of the drama division	The departments are fitting students' interesting, and the good culture between the peers, and school provide the spacious scholarships
Q6. In the age of information explosion, what are the ways to increase school exposure and popularity?	The team leader of the drama division	Online Marketing Advertising, Fb Videos, Homemade School Video and News Newspaper Media

5 Conclusion

The admission strategies of schools are not static, but are based on current circumstances. With the changes in the general environment, they continue to innovate in new enrollment modes to improve the school's institutional deficiencies. One of the factors for the survival and development of private vocational colleges is the number of students. Therefore, all private higher vocational colleges are striving to stabilize the source of students as the current primary goal, and to expand the source of students and strengthen the enrollment promotion or provide scholarships to increase students' willingness to attend school.

However, the fundamental solution is still to establish a school's competitive advantage and to develop school characteristics. Therefore, private higher vocational colleges should develop their own characteristics based on existing competitive advantages, at the same time increase revenue, reduce expenditure, create accumulation, and maintain competitive advantages. Only in the highly competitive and undergraduate education environment can we steadily move toward the school's sustainable development.

References

Amit, R., Zott, C.: Business model innovation: creating value in times of change (2010)

Casadesus-Masanell, R., Ricart, J.E.: From strategy to business models and onto tactics. Long Range Plan. **43**(2–3), 195–215 (2010)

Chesbrough, H., Rosenbloom, R.S.: The role of the business model in capturing value from innovation: evidence from Xerox Corporation's technology spin-off companies. Ind. Corp. Change **11**(3), 529–555 (2002)

Magretta, J.: Why business models matter. Harv. Bus. Rev. **80**(5), 86–92, 133 (2002)

Morris, M., Schindehutte, M., Allen, J.: The entrepreneur's business model: toward a unified perspective. J. Bus. Res. **58**(6), 726–735 (2005)

Osterwalder, A., Pigneur, Y.: Business Model Generation: A Handbook for Visionaries, Game Changers, and Challengers. Wiley, Hoboken (2010)

Timmers, P.: Business models for electronic markets. Electron. Mark. **8**(2), 3–8 (1998)

Viscio, A.J., Pasternack, B.A.: Toward a new business model. Strategy Bus. **20**(2), 125–134 (1996)

Professional Training and Operational Performance: Considering the Impact of CPA Disciplinary Incidents

Chiung-Yen Chen, Mei-Hua Huang(✉), and Zhen-Xin Xu

Department of Accounting and Information Systems,
Asia University, Taichung City, Taiwan
{joanchen,meihuang}@asia.edu.tw

Abstract. The purpose of this study has two folds: (1) to understand whether the audit failure incidents caused by the 2 major corruption cases (Procomp and Rebar) in Taiwan's history affected the strategy of employee professional training by CPA firms which provide auditing services to the public offering companies, and (2) whether those who invested more on employee professional training had resulted better operational performance. Data were collected for the period of year 1998 to 2013 from "Survey Report of Audit Firms in Taiwan," published by the Financial Supervisory Commission, Executive Yuan, from. The study used the debacles of Procomp and Rebar to separate the data sets. Results showed after the incidence of Procomp corruption, CPA firm significantly invested in employee professional training more than ever, however, they did not for the periods of before and after the Rebar incidence. This finding indicates that public accountants have different views on these two failed audit cases and come out different professional training strategies. Finally, the study found those CPA firms invested more in continuing professional development had better business performance.

1 Introduction

Successive incidences of accounting fraud by top U.S. companies such as Enron and WorldCom after 2000 have sparked debate regarding the image and quality of accounting across sectors. To address the lack of investor trust in capital markets and to reestablish public trust in CPA-issued approvals, in 2002, the American government implemented the most sweeping financial-sector reforms since the Sarbanes–Oxley Act of the 1930s. The Act strictly stipulated that the primary auditor or reviewer of firms must be a CPA and prohibits firms from employing the services of the same audit firm for 5 continuous years.

Similarly, a series of incidences of fraud, e.g., tunneling and fraudulent accounting by publicly listed companies have rocked Taiwan since 2004. Among them, the case of Procomp Informatics Ltd. tunneling NTD 7 billion in 2004 proved most shocking to investors. Before the incident, the financial statements of Procomp Informatics were audited by the Big 4 accounting firms. The scale and techniques employed in the fraud were unprecedented and led to fraudulent financial reports in the year of, and prior to,

L. Barolli et al. (Eds.): IMIS 2018, AISC 773, pp. 849–861, 2019.
https://doi.org/10.1007/978-3-319-93554-6_85

the discovery of the fraud. The four accountants who audited the financial statements were prohibited from engaging in their profession for two years, and the case also set a precedent in ruling that accountants bear the burden of responsibility for a portion of investor losses. In 2007, there was another incidence of fraud—this time in the tunneling case involving the Rebar Group—which constituted the single largest case of financial crime in Taiwan's history. During the second trial, the court ruled 132 people guilty and discovered a fraudulent balance of NTD 42.7 billion. Aside from ruling that the two CPA firms were subject to criminal and civil liability, the Financial Supervisory Commission suspended the two from auditing the financial statements of listed companies. The Procomp and Rebar cases heavily damaged public trust toward CPA firm audits. There are two differences between the cases of Procomp Informatics and the Rebar Group: the first is that the financial statements of Procomp Informatics before and after the incidents were audited by the Big 4 accounting firms, while those of the Rebar Group were not. The second difference is that the CPA firms hired by Procomp Informatics were found guilty of negligence, which is why they were subject to punishment and civil liability but not criminal liability. On the other hand, the relevant representatives of the CPA firm auditing the Rebar Group was subject to criminal liability and imprisonment in addition to punishment and civil liability.

The series of corporate scandals has led to accountants being subject to suspensions ranging from 2 months to 2 years by the Financial Supervisory Commission or being sentenced for criminal liability. The severity and extent of sentences in the cases of Procomp Informatics and the Rebar Group have set a new record in the CPA industry. Ma (2004) noted that the responsibility and punishment of accounting firms have significantly increased after the Procomp Informatics incident.

Because of the frequency of corporate fraud, the audit failure rate has increased and so have the likelihood of suspension and litigation risk for accounting firms. The relevant literature has shown that to control audit risks effectively, the strategies that accounting firms may employ include reducing the acceptable level of customer financial risk [15] or being more conservative in the possibility of issuing a standard unqualified audit opinion on audited reports [11]. However, there is a dearth of literature on the relationship of professional training, litigation risk, and operational performance from the standpoint of the internal quality control of companies. Meaning, does the risk of litigation cause firms to pay more attention to the professional training of their accountants, given the possibility of punishment or the revocation of the company's public listing, thereby increasing professionalism and decreasing the likelihood of misjudgment? The Taiwanese "Measures for Pre-employment Training and Continuing Professional Training for Accountants" stipulates that the total number of training hours of a CPA working for publicly listed companies must be more than 40 h. According to the regulations established by the competent authority, the number of study hours is the minimum requirement that practitioners must meet. In terms of corporate human resource management policies, training activities help to create and develop human capital for the company. Many scholars believe that training activities can enable companies to have more qualified and well-prepared employees on the payroll [2, 17]. At the same time, training activities can also help companies to perform more effectively at each stage of knowledge management [1, 3]. Therefore, training activities are an important part of human capital accumulation. Although there is a a wealth of literature exploring the relationship between human capital and the performance of accountants [7, 14, 18], whether professional CPA training would be adjusted relative to increases in litigation risks as a response to an environment with greater legal accountability is worth further analysis. Therefore, this study used the Procomp case of 2004 and the Rebar case of 2007 as the demarcation points to examine whether there was a significant difference in the professional training programs of publicly listed CPA firms that audited the financial statements during different periods of time. At the same time, this study examined whether CPA firms that placed greater emphasis on professional training exhibited greater operational performance to serve as a reference for practitioners and managers in formulating policies.

2 Literature Review and Research Hypotheses

In recent years, the number of lawsuits filed against accounting firms stemming from inaccurate financial reporting has increased dramatically, and some have been punished. For example, four accountants received 2-year suspensions as a result of the Procomp case in 2004. Subsequently, there have been a string of fraud cases, including Infodisc Technology, Summit Computer Technology, and ABIT Computer Corporation, resulting in the suspensions of a number of accountants by the Financial Supervisory Commission. In particular, for the first time, in 2007, the license of the firm that audited the financial statements of Rebar Group was rescinded permanently. Second, civil lawsuits followed by audit failure are an accountant's worst nightmare. Ma (2004) pointed out that there are apparent differences between the accounting firm's accountability and punishment.

In the Procomp case of 2004, the company used a number of overseas companies to arrange for improper sales, purchases, notes, accounts receivables, bank deposits, and fraudulent sales-related account receivables. The accountants did not investigate the books and simply issued a fraudulent auditor's opinion. It was evident that no professional attention was paid during the review process. Therefore, the accountants involved before and after the four cases came to light were handed 2-year suspensions by the Financial Supervisory Commission. The sentences spelled certain death for the accounting firms in a professional sense [6]. Even after the suspension runs its course, it is doubtful that any publicly listed company would entrust them with audits. The impact of punitive measures on the professional development of accountants is evident.

The similarities between the cases of Procomp Informatics in Taiwan and Enron in the United States are similar; both cases involved significant breaches of corporate governance, and the accountants who signed off on the financial statements issued their fraudulent auditor's opinions without exposing the financial fraud in those statements. It was because of this that the competent authority handed down heavy sentences to prevent similar incidents from occurring. Unexpectedly, during the 2007 expose of the Rebar Group tunneling case, the company applied for reorganization because of its inability to recover from its financial difficulties following continued losses, large borrowings, and derivative interests. The two CPA auditors entrusted with the job did not question the nature of continuing operations, considering the lack of bank agreements to extend the repayment period of loans; it was evident that the audit approval was invalid. The Financial Supervisory Commission investigated the lack of proper due diligence and investigation by the accounting firm of the fraud committed by the management of the Rebar Group, as well as its financial deterioration in the years leading up to the audit. In accordance with the law, the accountants involved were permanently suspended (this is the most substantial punishment on record) [12]. The investigators found that the two accountants working with the Rebar Group were long complicit in concealing financial reports, cooking the books, and assisting the top management of Rebar Group in tunneling assets. The four auditors were also seriously negligent in verifying the financial statements of The Chinese Bank, Lihua Bills Finance Corporation, and Union Insurance, all of which were Rebar affiliates. The various violations were not pursued in detail, and audits were not properly conducted. The Financial Supervisory Commission suspended the four accountants for 2 years. The six accountants involved in the Rebar case received, apart from administrative punishment and civil liabilities, criminal liability because of their serious violations of laws and regulations.

There have been studies on whether the decision made by a CPA to issue an audit opinion is affected by failed audit cases. Chang et al. [5] discussed whether the audit behavior of accountants was affected by the Procomp incident. Duh et al. [10] discussed whether there were significant changes in the audit decisions by using the Procomp case as a watershed incident. Liu et al. [15] used the Enron and Procomp cases as two demarcation points to discuss whether the increasing litigation risks faced by large accounting firms would reduce the acceptable level of client financial risks.

To be sustainable, companies often pursue innovative business models, production technologies, or management systems. Consequently, auditing work has become increasingly complex. It can be imagined that if accountants are required to perform this job, they must continue to pursue advanced studies. Taiwanese authorities stipulated that the number of continuous professional training hours for the accountants of a public company shall be no less than 40 hours per year and that professional assistants should also have relevant knowledge and appropriate professional training. Therefore, it can be reasonably speculated that firms will invest in more professional training to reduce the audit failure rate in the face of increasing litigation risks. Hung and Yen [13] studied whether the incidents in Taiwan where the Financial Supervisory Commission disciplined the accountants between 2000 and 2007 affected the conservative nature of the sign-off process in the aftermath. The results of the study showed that if accountants were subjected to professional suspension, then the successor CPA for the job will maintain a more conservative attitude toward the earnings management policy of the client. Although the financial statements of Procomp Informatics were audited by the Big 4 accounting firms before and during the year of the uncovering of the incident and investors generally expected the Big 4 firms' audit quality to be higher than others', the four accounting firms still signed off on accounts containing falsified auditor opinions. To trace the source of management fraud, it is inevitable that firms must strengthen the on-the-job training of their employees. Therefore, this paper speculates that the impact of the disciplinary action in the Procomp case on the continuing professional training of accountants was more significant than that in the Rebar case. Thus, hypotheses 1A and 1B are as follows:

H1A: After the Procomp case, CPA firms invested in more professional training for their employees than they did prior to the case.
H1B: The amount of professional training that CPA firms invested in for their employees was not affected by the Rebar case.

Chang et al. [4] pointed out that employees' cumulative professional experience, the average amount of professional training, education, and other such variables can enhance the professional technical standards of CPA firms and can also significantly help the firms' operating income. Lin and Chen [14] discussed whether high-quality human capital would affect firms' operating performance. They found that high-level professional firms with experienced employees are able to enhance their technical efficiency. The more firms spend on education and training the better their technical efficiency. Chen et al. [7] found that there is a clear positive correlation between the professional training of Taiwanese CPA firms and their financial performance. Even if different organizational strategies and market segments impact the firm, the relationship between the two variables remains positively correlated. The investment in professional training by firms not only meets the requirements of the competent authority but also helps to improve business performance. Therefore, hypotheses 2A and 2B are as follows:

H2A: The professional training and operating performance of accounting firms are positively correlated.
H2B: CPA firms that have invested in more professional training have better operational performance.

3 Sample Selection and Methodology

3.1 Sample Selection

This study used a sampling of accounting firms approved to audit the financial reports of public companies from 1998 to 2013. The source of the data was the "Annual Survey of the CPA Industry" of the Financial Supervisory Commission of Taiwan.

3.2 Periods Studied

This study referred to the methods of Liu et al. [14], Chen et al. [8], and other scholars and regarded the Procomp and Rebar cases as watershed incidents. The periods studied were divided into the Pre-Procomp Scandal (1998–2003), Post-Procomp Scandal (2004–2006), and Post-Rebar (2007–2013) periods. The three periods were used to explore whether the amount of professional training in CPA firms was significantly affected by the Procomp or Rebar cases. This study used dummy variables (DUM) to represent these periods.

Table 1 shows the sample size for each period. The "original number of observations" refers to the total number of sample CPA firms during the period. This study removed CPA firms that were limited to audits of private companies, those with incomplete data, or those with obvious errors in data input. Afterward, the "sampling of firms that audit financial statements of publicly listed companies" was obtained, namely, the sample used in this study. There were 866 CPA firms for the Pre-Procomp Scandal period, 360 for the Post-Procomp Scandal period, and 1029 for the Post-Rebar period, for a total of 2,255 CPA firms.

Table 1. Sample distribution

Periods	Pre-Procomp Scandal (1998–2003)	Post-Procomp Scandal (2004–2006)	Post-Rebar (2007–2013)	Total
The original number of observations	4,634	2,344	7,835	14,813
Number of CPAs firms which provide auditing services to the public offering companies	866	360	1,029	2,255

3.3 Empirical Model and Definition of Variables

This study used the difference of sample means distribution test and model (1); used regression analysis to evaluate hypotheses 1A and 1B to learn whether the differences between professional training during the Pre-Procomp, Post-Procomp, and Post-Rebar periods were statistically significant, and then used model (2) and model (3) regression analysis to evaluate 2A and 2B to understand whether the firms that invested in more professional training had better performance.

$$\text{TRN} = \alpha_0 + \alpha_1\text{DUM}_1 + \alpha_2\text{DUM}_2 + \alpha_3\text{MRK} + \alpha_4\text{FA} + \alpha_5\text{AGE} + \varepsilon \quad (1)$$

$$\text{PRM} = \beta_0 + \beta_1\text{TRN} + \beta_2\text{EXP} + \beta_3\text{EDU} + \beta_4\text{MRK} + \beta_5\text{SIZE} + \beta_6\text{AGE} + \varepsilon \quad (2)$$

$$\text{PRM} = \beta_0 + \beta_1\text{DUM}_3 * \text{TRN} + \beta_2\text{EXP} + \beta_3\text{EDU} + \beta_4\text{MRK} + \beta_5\text{SIZE} + \beta_6\text{AGE} + \varepsilon \quad (3)$$

The operational definitions of variables are depicted as follows.

Professional training (TRN) = annual expenditure on professional training ÷ year-end number of professionals.
Operational performance (PFM) = (total revenues of the audit firm – total expenses of the audit firm) ÷ year-end number of CPAs partners.
Dummy variable (DUM_1): Sample after Procomp scandal occurred = 1, otherwise = 0.
Dummy variable (DUM_2): Sample after Rebar scandal occurred = 1, otherwise = 0.
Dummy variable (DUM_3): Professional training of individual audit firm is higher than the industry average = 1, otherwise = 0.
Educational level (EDU) = (number of auditors with doctorates × 23 + number of auditors with Master's degrees × 18 + number of auditors with bachelor's degrees × 16 + number of employee with senior high school diploma × 12 + number of

employees with other education levels × 9) ÷ year-end number of professionals [Cheng *et al.* 2009].
Work experience (EXP) = final number of professionals older than 35 years) ÷ final number of professionals.
Financial audit revenue (FA) = total financial audit revenue of the audit firm ÷ year-end number of CPAs partners.
Market share (MKS) = total revenue of individual audit firm ÷ total revenue of the whole profession.
Firm size (SIZE) = nature log of total number of final number of professionals.
Firm age (AGE) = data survey year-establishment year of audit firm + 1.

4 Empirical Results and Discussion

4.1 Descriptive Statistics

Table 2 shows the descriptive statistics for each variable. It is apparent from Table 2 Panel A that the professional training for all the CPA firms in the sample (TRN) averaged NTD 4,398 per professional. Furthermore, it can be seen in Panels B ~ D that, at different times, the average amount per professional spent by the firms on professional training (TRN) per firm had gradually increased. For example, the amount during the Pre-Procomp period was NTD 4,046; NTD 5,358 during the Post-Procomp and Pre-Rebar period; and NTD 4,357 during the Post-Rebar period. The average amount spent on professional training (TRN) per person was as high as NTD 96,389 and as low as NTD 0, which clearly reflects major differences in the continuing education policies of professionals. The supervisory unit of CPA professionals has rules in place for continuing education, stating that firms can hold in-house training or participate in external training courses. Some training courses offered by outside entities are free of charge, which might explain why some firms did not have any training expenditure.

Operational performance (PRM) amounted to 1,678,100 for the entire sample. During different periods, the amount varied between NTD 1.5 million and NTD 1.7 million. The education level (EDU) of the entire sample averaged 15.50 years, near college graduate level. It increased from the 15.06 years during the Pre-Procomp period to 15.72 years during Post-Procomp period and then to 15.80 during the Post-Rebar period; this shows a gradual improvement in the education level of the employees. Employee experience (EXP) is represented by the percentage of employees over the age of 35 years in the total number of employees. In the entire sample, employees above the age of 35 years averaged 35% of firms' employee strength, meaning specifically that relatively experienced employees account for 35% of the total. With the intensification of litigation risks in various periods, employee experience has also increased gradually, from 29% to 41%. Financial income (FA) amounted to 2,711,300 for the entire sample. During the Pre-Procomp period, it was at NTD 2.47 million, which significantly declined to NTD 1.72 million during the Post-Procomp period, and which grew to NTD 2.93 during the Post-Rebar period. The market share (MRK) of the sample averaged 0.511%, with the maximum value being 27%, which is assumed to be that of the Big 4 firms. The average firm size (SIZE) was 3.49, equivalent to over 30 people when converted to headcounts, with the maximum value being over 3,400 people and the minimum value being 3 people. The average number of years of establishment (AGE) was 18 years. The maximum number of years was 94 years, and the minimum number of years was 1 year.

Table 2. Descriptive statistics

	TRN	PRM	EDU	EXP	FA	MRK	SIZE	AGE
Panel A: total number of observations (N = 2255)								
Mean	4398	1678100	15.50	0.35	2711300	0.00511	3.49	18
Max.	96389	16670939	21.25	1.00	41272172	0.27000	8.15	94
Min.	0	−1544619	11.82	0.00	9348	0.00000	1.10	1
Median	1630	1236426	15.60	0.32	1558822	0.00000	3.37	16
SD	7972	1955020	0.67	0.19	4411940	0.02617	1.06	11
Panel B: Pre-Procomp Scandal (N = 866)								
Mean	4046	1568200	15.06	0.29	2470300	0.00424	3.49	14
Max.	96389	11712462	18.19	1.00	25601986	0.24000	7.60	91
Min.	0	−897248	11.82	0.00	13881	0.00000	1.61	1
Median	1299	1211742	15.08	0.25	1436314	0.00000	3.33	12
SD	9103	1616580	0.63	0.17	3628300	0.01986	0.97	10
Panel C: Post-Procomp Scandal (N = 360)								
Mean	5358	1624700	15.72	0.33	1725670	0.00586	3.24	17
Max.	62574	11137120	17.14	1.00	26440515	0.25000	7.80	87
Min.	0	−1544619	13.91	0.00	16767	0.00000	1.10	1
Median	2146	1251287	15.79	0.31	1686251	0.00000	3.18	16
SD	8388	2658600	0.55	0.17	3986770	0.02863	1.23	10
Panel D: Post-Rebar (N = 1029)								
Mean	4357	1789300	15.80	0.41	2932600	0.00558	3.57	20
Max.	69362	16670939	21.25	1.00	41272172	0.27000	8.15	94
Min.	0	−1333607	13.64	0.00	9348	0.00000	1.10	1
Median	1692	1248634	15.88	0.38	1593592	0.00000	3.43	20
SD	6680	2262280	0.53	0.20	5094380	0.02971	1.06	11

Note: Refer to the previous section for definition of variables.

4.2 Results of the Univariate Test

Table 3 presents the t-test results of professional training in different periods (TRN). From Panel A, it can be seen that professional training (TRN) during the Post-Procomp period was significantly higher than that during the Pre-Procomp period (T = −2.432). It can be seen from Panel B that professional training (TRN) during the Post-Procomp period was significantly higher than that during the Post-Rebar period (T = 2.049). It can be seen from Panel C that professional training (TRN) during the Post-Rebar period was higher than that during the Pre-Procomp period, but the statistics had no significant differences (t = −0.835). Therefore, considering the above findings, the professional training of the firms during the Post-Procomp period was clearly significantly greater than that during the Pre-Procomp period, which supports hypothesis H1A. During the Post-Rebar period, professional training (TRN) did not increase from the two previous periods. On the contrary, it was less than that during the Post-Procomp period, which supports hypothesis H1B. The results show that CPA firms reacted differently to the threat of administrative or civil lawsuits in connection with employee professional training decisions. The Big 4 firms attached greater importance to the professional training of their employees; however, those not part of the Big 4 did not increase investment in professional training after receiving punishment.

Table 3. Univariate test results of professional training

	The number of observation	Mean	Difference	t-value
Panel A: Before and After the debacles of Procomp				
Before the debacles of Procomp	866	4,046	1312	−2.432***
After the debacles of Procomp	360	5,358		
Panel B: After the debacles of Procomp and After the debacles of Rebar				
After the debacles of Procomp	360	5,358	1001	2.049**
After the debacles of Rebar	1029	4,357		
Panel C: Before the debacles of Procomp and After the debacles of Rebar				
Before the debacles of Procomp	866	4,046	311	−0.835
After the debacles of Rebar	1029	4,357		

Note: **, *** denote significant at 10%, 5% and 1% level for two-sided hypothesis tests respectively.

4.3 Results of Regression Analysis

Table 4 shows the empirical results of testing the relationships among the Procomp case, Rebar case, and professional training (TRN). To detect whether there was serious collinearity between the variables and whether it affected the conclusion, the empirical tests in this study were all conducted using variance inflation factor (VIF), which found that the VIF value of each variable was less than 5, thus proving that no collinearity existed.

Column (A) of Table 4 shows that the Post-Procomp period (DUM1) and professional training (TRN) were significantly positively correlated ($t = 2.517$), which indicates that after the Procomp case, CPA firm investment in professional training (TRN) was significantly higher than it was before the Procomp case, thereby supporting hypothesis H1A.

Column (B) of Table 4 shows that professional training during the Post-Rebar period was less than that during the Pre-Procomp period but that it was not statistically significant. These results are consistent with expectations, meaning that the CPA industry did not increase the amount of professional training in the wake of the disciplinary action taken as part of the Rebar case, thereby supporting hypothesis H1B.

Column (C) of Table 4 shows that the Post-Procomp period (DUM1) and professional training (TRN) were significantly positively correlated (t = 2.449). The Post-Rebar period (DUM2) was also positively correlated with professional training (TRN) but not significantly (t = 0.370). These results show that professional training (TRN) during the Post-Procomp period was significantly higher than that during the Pre-Procomp period and that professional training (TRN) during the Post-Rebar period was higher than that during the Pre-Procomp period but was not statistically significant. In conclusion, the professional training (TRN) of CPA firms was the highest after the Procomp case. It was evident that the Big 4 accounting firms were severely punished for the audit failure, which had severe, widespread implications, in turn, causing these CPA firms to increase professional training. In contrast, after the Rebar case, the professional training (TRN) of CPA firms did not increase beyond the level of either the Pre-Procomp period or the Post-Procomp period, indicating that the severe punishment meted out to the accountants from non-Big-4 firms did not result in a panic over audit failure that would lead to increased professional training for employees. Therefore, hypotheses H1A and H2B were supported.

In terms of the control variables, market share (MRK) and professional training (TRN) both showed a significant positive correlation, which was consistent with the

Table 4. The relationship between professional training and CPA disciplinary incidents (Eq. 1)

	(A) Coeff.	(B) Coeff.	(C) Coeff.
Intercept	-*** (10.175)	-*** (10.865)	-*** (9.341)
DUM_1	0.052** (2.517)		0.055*** (2.449)
DUM_2		−0.014 (−0.686)	0.009 (0.370)
MRK	0.138*** (3.470)	0.141*** (3.541)	0.138*** (3.465)
FA	0.103** (2.517)	0.100** (2.442)	0.104** (2.526)
AGE	0.009 (0.419)	0.012 (0.536)	0.007 (0.312)
F-value	34.598	33.047	27.695
Adj. R^2	0.056	0.054	0.056

Note:

a. The t-value of each variable is shown in the parentheses.

b. The variable VIFs are less than 5, implying that no multi-collinearity exists among the variables.

c. Refer to the previous section for definition of variables.

expectation of this study. This indicates that the higher the market share, the greater the emphasis on employee training. Financial income (FA) and professional training (TRN) both showed a significant positive correlation, indicating that the higher the financial income of the firm the more the investment in professional continuing education, which is also consistent with the expectation of this study. The numbers of years for which the firms were in business (AGE) were positively but not significantly correlated, indicating that age was not correlated with professional training.

Table 5 shows the empirical results of the relationship between professional training and operational performance (PRM). The continuing professional education (TRN) coefficient can be seen from column (A) to be significantly positive (t = 4.496), indicating that there would be better operational performance, given more professional training, which is consistent with the expectations of this study. Therefore, hypothesis 2A was established. In terms of control variables, employee education level (EDU), market share (MRK), firm size (SIZE), and number of years in operation (AGE) are all positively correlated with operational performance, which is consistent with expectations. However, employee experience (EXP) and operational performance (PRM) are negatively related. This may be due to the higher labor costs required of the firm, given the experienced employees, thus negatively affecting performance.

Table 5. The relationship between professional training and operational performance (Eqs. 2 and 3)

	(A)	(B)
Intercept	-*** (−6.651)	-*** (−6.706)-
TRN	0.054*** (4.496)	
DUM_3*TRN		0.057*** (4.763)
EDU	0.067*** (5.582)	0.068*** (5.650)
EXP	−0.049*** (−3.886)	−0.049*** (−3.851)
MRK	0.531*** (33.549)	0.529*** (33.339)
SIZE	0.306*** (18.086)	0.307*** (18.151)
AGE2018	0.102*** (8.108)	0.102*** (8.078)
F-value	867.588	868.944
Adj. R^2	0.698	0.698

Note:

a. The t-value of each variable is shown in the parentheses.

b. The variable VIFs are less than 5, implying that no multi-collinearity exists among the variables.

c. Refer to the previous section for definition of variables.

Therefore, it was decided to test whether investments in professional training (TRN) that was greater than the industry average would result in greater operational performance; this study used model (3) for testing purposes. The results are as shown in column (B) of Table 5. The results show firms that invested in more professional training than the industry standard (DUM3) and the professional training interactive variables (DUM3*TRN) to be significantly positively correlated ($t = 4.763$), indicating that within the CPA industry, the more a firm invests in professional training over the industry standard, the better the performance of the firm. This result is consistent with the expectations of this article and also supports hypothesis 2B. With the directionality and significance of the control variables being similar to the previous ones, they are not described here.

5 Conclusions

This study was designed to investigate whether the amount of professional training received by employees in CPA firms authorized to audit the financial statements of publicly listed companies was affected by audit failure in the two major incidences of fraud—the Procomp and Rebar cases. In addition, whether greater investment in professional training translated into better operational performance was examined in the study. The data came from the "Annual Survey of the CPA Industry" of the Financial Supervisory Commission of Taiwan for the years between 1998 and 2013, and the times of occurrence of the two major fraud cases were used as the demarcation points of the sample. The empirical results showed that after the Procomp case, CPA firms naturally invested in more professional training for their employees when compared with their investment in such training in the period before the said cases. However, after the Rebar case, CPA firms did not significantly increase investment in professional training for their employees when compared with the periods before and after the Procomp case. These results indicate that the CPA industry had different views on these two failed audit cases and produced different subsequent professional training policies. The accountants punished in the Procomp case were all from notable Big 4 firms, while those involved in the Rebar case were not. The results of this study implied that the accountants were shocked by the fact that the Big 4 accounting firms were still unable to detect corporate management fraud. Therefore, after the Procomp case, they strengthened their professional training. However, after the Rebar case, they did not strengthen their professional training, which clearly shows that the industry perceived that as the audit fraud was committed by accounting firms not part of the Big 4 firms, the audit failure was unrelated to professional capabilities, and therefore, there was no need to strengthen professional training in the aftermath of the said case. Last, the study found that relatively greater investment in professional training results in relatively better operational performance. Therefore, CPA firms should put greater emphasis on professional training. On the one hand, it can meet legal and regulatory requirements and strengthen the ability of employees to serve customers, and on the other, it can also create better operational performance. The empirical results of this paper can be used as a reference for professional accounting training regulations for CPA supervisors and can also serve as a reference for firms in formulating continuing professional education policies.

References

1. Alavi, M., Leidner, D.E.: Review: knowledge management and knowledge management systems: conceptual foundations and research issues. MIS Q. **25**(March), 107–136 (2001)
2. Bartel, A.P.: Productivity gains from the implementation of employee training programs. Ind. Relat. J. Econ. Soc. **33**, 411–425 (1994)
3. Bollinger, A.S., Smith, R.D.: Managing organizational knowledge as a strategic asset. J. Knowl. Manag. **5**(1), 8–18 (2001)
4. Chang, B.G., Yang, C.C., Chen, Y.S.: The evolution of substitution in the human resources of accounting firms in Taiwan: a case of partnership pattern. J. Contemp. Account. **5**(1), 1–24 (2004)
5. Chang, R.D., Shen, W.H., Fang, C.J.: The effects of the procomp scandal on going concern audit opinions: an investigation of the moderating effects. NTU Manag. Rev. **19**(2), 75–108 (2009)
6. Chen, T.C.: The market response to the discipline on CPAs of big 4 CPA firms: evidence from the financial fraud of PROCOMP Informatics Ltd. J. Tainan Univ. Technol. **25**, 427–448 (2006)
7. Chen, Y.S., Goan, K.T., Chen, S.P.: Training and audit firms financial performance: consideration of strategy and market segment. J. Hum. Resour. Manag. **11**(4), 23–46 (2011)
8. Chen, C.Y., Huang, M.H., Chang, K.J.: The relationship between the human capital and litigation risk of audit firms. In: 2017 ATPC Conference, Chinese Culture University, Taipei (2017)
9. Cheng, Y.S., Liu, Y.P., Chien, C.Y.: The association between auditor quality and human capital. Manag. Audit. J. **6**, 523–541 (2009)
10. Duh, R.R., Lee, W.C., Lin, C.C., Chu, J.P.: An experimental study on non-audit service and auditor decisions: pre- and post-procomp scandal. Taiwan Account. Rev. **6**(2), 125–152 (2007)
11. Fu, C.J., Chang, F.H., Chen, C.L.: The impact of audit failure on auditor conservatism: is there a contagious effect of the enron case? Int. J. Account. Stud. **40**, 31–67 (2005)
12. Guan, Y.D., Chang, C.H.: An empirical study on the impact of the rebar event on the audit clients stock price. J. Account. Corp. Governance **7**(2), 19–46 (2010)
13. Hung, Y.S., Yen, S.H.: Impact of CPA discipline on auditor conservatism. J. Manag. **28**(4), 325–343 (2011)
14. Lin, C.L., Chen, Y.S.: Human capital and operating performance. Chiao Da Manag. Rev. **29**(2), 83–130 (2009)
15. Liu, C.W., Wang, T.C., Lai, S.T.: litigation risk and large audit firms' acceptable level of clients' financial risk. NTU Manag. Rev. **20**(1), 1–40 (2009)
16. Ma, H.J.: Things are unknown: procomp and certified public accountant. Account. Res. Mon. **226**, 58–84 (2004)
17. MacDuffie, J.P., Kochan, T.A.: Do U.S. firms invest less in human resources? Training in the world auto industry. Ind. Relat. J. Econ. Soc. **34**, 147–168 (1995)
18. Yang, C.C., Tsai, T.Y., Fu, C.J.: Human capital and knowledge spillover effect: evidence from Taiwan's CPA firms. Sun Yat-Sen Manag. Rev. **18**(1), 251–279 (2010)

Constructing a System for Effective Implementation of Strategic Corporate Social Responsibility

Chiung-Yao Huang(✉) and Chung-Jen Fu

Department of Accounting, National Yunlin University of Science and Technology, Douliu, Taiwan, R.O.C.
{cyhwang, fucj}@yuntech.edu.tw

Abstract. By integrating multiple management systems, this study developed a system for CSR activities, strategic management, daily operations, and sustainable management and control which allows the effective implementation of strategic CSR. The proposed strategic CSR system can help companies solve CSR reporting issues, including a lack of strategic planning, a lack of clarity of expression, and simplicity of reports. On the one hand, the proposed system may help companies comply with CSR reporting regulations introduced by the authorities. On the other hand, it may be used by managers to identify a company's key strategic areas in economic, environmental, and social aspects and incorporate CSR activities in daily operations to gain competitive advantage and create a long-term value.

1 Introduction

In recent years, there are many food safety, social, and environmental pollution incidents. Negative CSR events affect the business operations of a company, which may be even forced to stop production or terminate its business, and may cause public panic. This increases the urgency and importance of active promotion of CSR implementation. Although a number of companies voluntarily implemented CSR prior to the promulgation of CSR reporting regulations, they considered one, or slightly more, projects to be sufficient for this purpose. Integration of CSR activities and a company's strategies or daily operations can provide a more effective CSR implementation and improve a company's competitive advantage. Most companies remain unaware of how to address the issues related to constructing and implementing CSR activities, disclosing related information, and effectively integrating strategic management, daily operations, and sustainable management and control. This study proposed using the balanced scorecard (BSC) as an integrated strategic management tool to help achieve these tasks.

L. Barolli et al. (Eds.): IMIS 2018, AISC 773, pp. 862–866, 2019.
https://doi.org/10.1007/978-3-319-93554-6_86

2 Literature Review

The BSC system proposed by Kaplan and Norton (1992, 1993) provides a comprehensive implementation framework which effectively integrates strategic planning, finance budget and management control, performance evaluation, and corrective actions. The BSC has been widely approved as the most effective strategic and performance management tool which can be used to encourage organization members to join their efforts in achieving a common objective (Albertsen and Lueg 2014). First, this study applied the BSC to integrate CSR and strategic management to address the current issue of the lack of strategies in CSR implementation by most companies. As a good visualization tool, the BSC may help companies better understand the role of CSR in their development and strategies and improve both internal and external communication (Selto and Malina 2001; Banker et al. 2004). Second, the BSC was used to link the functions of operational planning and budget management in order to incorporate CSR activities into daily operations of companies. Finally, progress management and control during CSR implementation and performance after it is completed were evaluated to ensure effective CSR implementation.

In a nutshell, the BSC examines external opportunities and threats considering a company's mission and vision and evaluates its competitive advantage in terms of core abilities and resources in order to select specific strategic objectives and combine them into a strategy map based on causal relationships and financial, customer, business process, and learning and growth dimensions (Kaplan and Norton 1996, 2004). A Key Performance Index (KPI) and objective value are introduced for each objective and an appropriate action plan is developed, followed by regular examinations and revisions aimed at maintaining a good financial performance, developing and strengthening a competitive advantage, and providing a sustainable development of a company and ecological environment. Norton (2000) stated that in a modern business environment that becomes more complex and requires the ability to adapt, a system dynamics perspective can be applied when interpreting and assessing corporate strategies and can serve as a basis for strategy adjustment or modification.

3 Corporate Social Responsibility and Balanced Scorecard

In order to guarantee CSR long-term support among managers, a comprehensive management system is required that would link CSR strategic objectives to a company's vision and mission and incorporate them into daily operations. Current promotion of CSR by companies faces difficulties associated with (1) clear communication of a company's vision and mission to employees by higher-level managers; (2) establishment of CSR consensus and strategies in a company; (3) consistent inter-departmental implementation of CSR strategies and their incorporation into daily business operations; and (4) timely feedback on CSR performance and adjustment of CSR strategies in response to dynamic changes in the environment so that to establish both profitable and sustainable company.

This study developed a strategic CSR implementation system to allow companies to effectively incorporate their mission and vision into focus strategies oriented on resources and conditions, as well as the TBL approach that suggests a balance among financial, environmental, and social aspects. After implementing CSR strategies in daily business activities, companies provide stakeholders with the related information via CSR reports. This study aimed to combine multiple management systems into one practical PDCA framework that considers the needs of different stakeholders. An internal system and management procedures were built such that they developed key factors of success into the processes of value creation and risk management which were followed by information disclosure and communication.

A comprehensive sustainable management and CSR reporting framework illustrated in Fig. 1 with the BSC management and control platform at its core can, in accordance with a company's mission and strategies, systematically describe the causal relationship between strategic issues in economic, environmental, and social dimensions of a triple bottom line (TBL) and improve inter-departmental processes based on the identified KPI and objective values. The framework involves regular monitoring and control and comprehensive audit aimed at tracking improvement results and evaluating a company's performance. By integrating CSR reporting and information disclosure and interaction with stakeholders, the system provides continuous feedback and improvement using the PDCA (plan-do-check-act) model (Deming 1994), thus enhancing a company's innovation and learning development and organizational knowledge and promise mechanisms.

A special characteristic of this comprehensive framework is the integration of key environmental and social strategic issues into the process of a firm value creation which helps to resolve the conflict of interests between different stakeholders and increase their support toward the company (Huang et al. 2016). In addition, the internal business process dimension among the four BSC dimensions was renamed in this study as "business process" dimension because in the current era of division of labor and networking, research, production, and sales values are usually created in collaboration with different companies and internal processes are no longer sufficient to describe the resulting networks of shared internal and external values (Perrini and Tencati 2006). Moreover, from the strategic perspective, the BSC allows more effective allocation and utilization of resources (Karapetrovic 2002, 2003; Kerr et al. 2015).

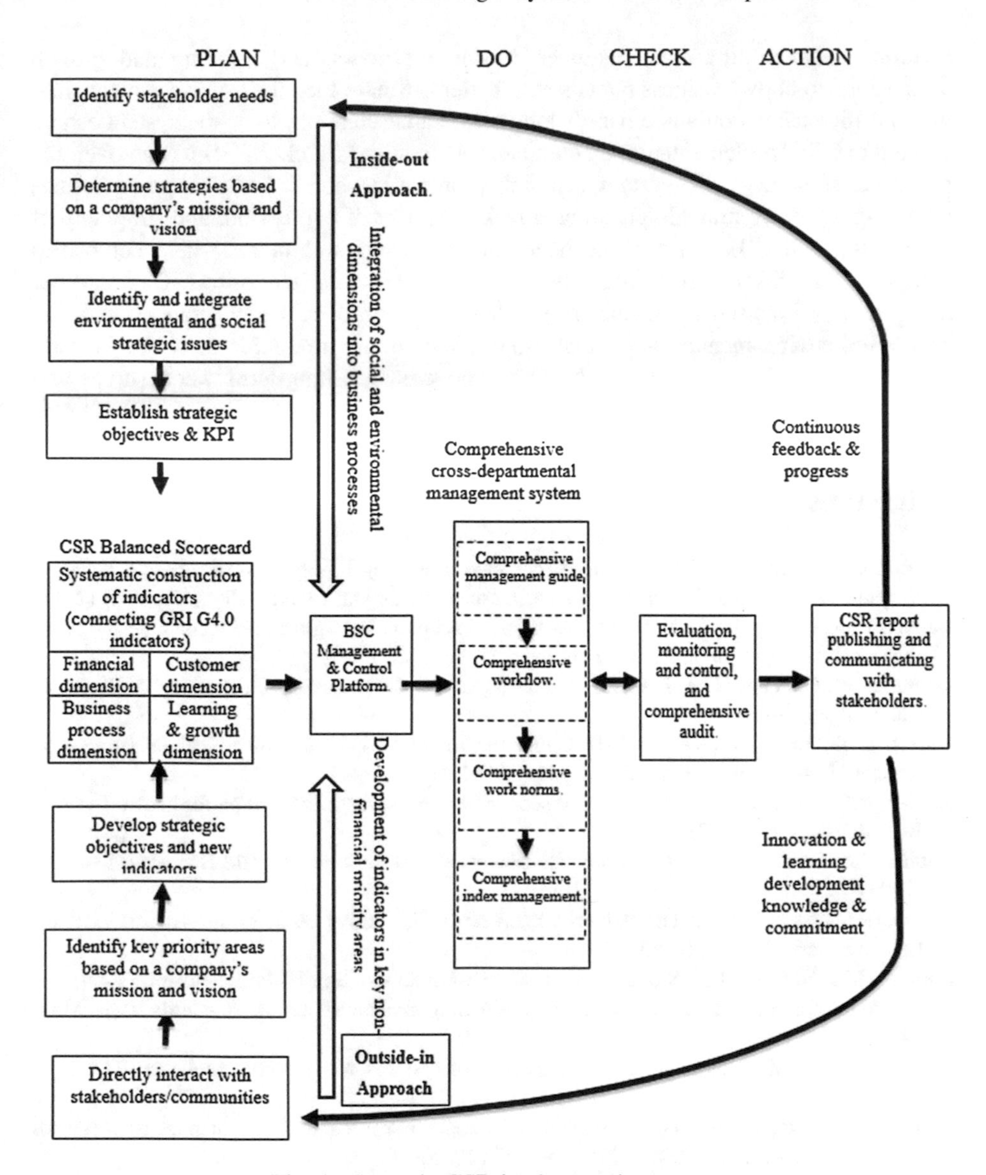

Fig. 1. Strategic CSR implementation system

4 Conclusions

this study sought to combine several systems into one practical strategic CSR system which effectively integrates disclosure of CSR actions and information and CSR strategic development, daily operations, and sustainable development and control. With regard to its practical contribution, an extended model of the traditional BSC combined with the concept of dynamic equity valuation was employed in an integrated CSR and strategic management system to provide a comprehensive understanding of dynamic relations between financial, customer, business process, and learning and growth dimensions to allow business managers to better estimate the effect of investments into the four dimensions on a company's long-term value creation; thus, the system can be applied in CSR implementation by companies (Ittner and Larcker 2001). Moreover, the proposed strategic CSR system can help companies solve CSR reporting issues, including a lack of strategic planning, a lack of clarity of expression, and simplicity of reports (Gray 2010). On the one hand, the proposed system may help companies comply with CSR reporting regulations introduced by the authorities. On the other hand, it may be used by managers to identify a company's key strategic areas in economic, environmental, and social aspects and incorporate CSR activities in daily operations to gain competitive advantage and create a long-term value (Ittner and Larcker 2005).

References

Albertsen, O.A., Lueg, R.: The balanced scorecard's missing link to compensation: a literature review and an agenda for future research. J. Account. Organ. Change **10**(4), 431–465 (2014)

Banker, R.D., Chang, H., Pizzini, M.: The balanced scorecard: judgmental effects of performance measures linked to strategy. Account. Rev. **79**(1), 1–23 (2004)

Deming, W.E.: The New Economics for Industry, Government, Education. MIT Press, Cambridge (1994)

Huang, C.Y., Lin, Y.C., Chen, C.H.: Could the impaired intention of ethical investment be recovered? J. Manag. Organ. **22**(5), 735–750 (2016)

Kaplan, R.S., Norton, D.P.: The balanced scorecard - measures that drive performance. Harvard Bus. Rev. 71–79 (1992)

Kaplan, R.S., Norton, D.P.: Putting the balanced scorecard to work. Harvard Bus. Rev. 134–142 (1993)

Kaplan, R.S., Norton, D.P.: The Balanced Scorecard: Translating Strategy into Action. Harvard Business School Press, Boston (1996)

Kaplan, R.S., Norton, D.P.: Strategy Maps. Harvard Business School Press, Boston (2004)

Karapetrovic, S.: Strategies for the integration of management systems and standards. TQM Mag. **14**(1), 61–67 (2002)

Karapetrovic, S.: Musings on integrated management systems. Meas. Bus. Excellence **7**(1), 4–13 (2003)

Kerr, J., Rouse, P., de Villiers, C.: Sustainability reporting integrated into management control systems. Pac. Account. Rev. **27**(2), 189–207 (2015)

Norton, D.P.: Is management finally ready for the "systems" approach. Balanced Scorecard Report, 3–4, 15 September 2000

Perrini, F., Tencati, A.: Sustainability and stakeholder management: The need for new corporate performance evaluation and reporting systems. Bus. Strategy Environ. **15**(5), 296–308 (2006)

Selto, F., Malina, M.: Communicating strategy: an empirical study of the effectiveness of the balanced scorecard. J. Manag. Account. Res. **13**, 47–90 (2001)

Study of the Factors that Influence Brand Loyalty Toward the Use of Tablets in Indonesia

Chien-Wen Lai[1], Candra Adi Kurnia[2], Shao-Chun Chiu[1], and Ya-Lan Chan[2](✉)

[1] Department of Accounting and Information Systems, Asia University, Taichung, Taiwan
{lcw, danielcc}@asia.edu.tw
[2] Department of Business Administration, Asia University, Taichung, Taiwan
{100183034, yalan}@asia.edu.tw

Abstract. The objective of this study is to develop and test a conceptual framework that explains and analyses the factors influencing tablet PC users to become loyal to a particular brand in Indonesia. The study utilizes a survey approach to gather data regarding factors that influence brand loyalty toward the use of tablets. The investigations of the factors that influence tablet PC users to become loyal to a particular tablet brand are anticipated to contribute toward a better understanding of consumer loyalty in tablet PC users and provide related factors that influence tablet PC users to become loyal to a particular brand. The findings supply valuable insights into which factors practitioners should focus their attention on to better tailor their approaches toward a tablet brand. This study strongly endorses the view that the loyalty intentions of tablet users are linked to the extent of their satisfaction with the brand they are using or have used.

1 Introduction

With 224 million people (2012), a third largest population in the world, Indonesia has undeniably become one of the biggest markets in the world, sought after by world-class companies including those marketing advanced technology. The growth of the Indonesian population will continue as even though the annual population growth rate is decreasing, the percentage is still high, 1.48% in 2001 to 2010. Significant economic growth and the large young population enables Indonesia to be an attractive potential market for invest and trade.

The spread of coffee shops offering free internet Wi-Fi and competition among telecommunications service and Internet service providers such as Telkomsel, ProXL, Smartfren, and Three are also the reason for the expansion of mobile computing device users (laptops, smart phones and tablet PCs) in Indonesia. Using the internet has become so simple and easy that no matter wherever we are, we may partake of these services. Many newspaper companies have also started to maintain an online presence, making reading a newspaper easier and easier.

L. Barolli et al. (Eds.): IMIS 2018, AISC 773, pp. 867–878, 2019.
https://doi.org/10.1007/978-3-319-93554-6_87

The growth of tablet computers sales in Indonesia showed a significant improvement. And, according to Growth for Knowledge (GfK), a research institution in Indonesia, tablet sales during the third quarter of 2012 reached 450,000 units with a transaction value of Rp 1.6 trillion (170 million USD). When viewed from the total volume of sales, that figure jumped 248%, while the value of transactions grew 142% over the same period of 2011. Tablet PC sales in Indonesia during January-September 2012 were 1.09 million units, a rapid increase from the previous year when only 470,000 units were sold. By the end of this year, GfK predicts sales of 1.8 million units through the tablet. Currently, there are 52 brands of tablet circulating in Indonesia. Of that amount, a total of 39 brands were local. All this rapid positive evolution has given the researcher a strong interest in seeking and investigating why a brand can survive and why consumers remain loyal to some particular brands and show the opposition to others.

The objective of this study is to develop and test a conceptual framework that explains and analyses the factors influencing tablet PC users to become loyal to a particular brand in Indonesia. This research study also aims to provide practical guidelines and suggestions regarding brand loyalty to tablet PCs for further development in the tablet PC industry.

2 Literature Review

2.1 Brand Loyalty

Building, maintaining and developing brand loyalty, a core dimension of brand equity, has been recognized as one of the key issues of marketing theory and practice if a company aspires to gain sustainable and long term competitive advantages and commercial performance. A variety of perspectives of both brand loyalty and equity concepts have been viewed during the last decades considering the dynamic marketing environment and fierce competition [15].

Oliver (1999) described brand loyalty as "a deeply held commitment to re-buy or re-patronize a preferred product/service consistently in the future [18], thereby causing repetitive same brand or same-brand-set purchasing, despite situational influences and marketing efforts having the potential to cause switching behavior".

Generally, conservative marketing literature divides the concept of brand loyalty into two different dimensions, behavioral and attitudinal. Behavioral loyalty is reflected in repeat buying behavior. While, attitudinal brand loyalty encompasses some dimensions such as cognitive, affective, and behavioral intent [8]. Mittal and Kamakura (2001) said that as the performance intention in the purchasing decision process, behavioral intent is defined as the intermediary between attitudinal and behavioral loyalty [14]. Behavioral intent may be observed as a preliminary motive in purchasing a product brand for the first time or a commitment to repurchase a current brand in the future. In strategic aims, this dimension will enable the company to maintain and enhance customer's commitment to repurchase and in the end convert this into an actual purchase [17].

Regarding the attitudinal dimension, behavioral loyalty will occur with brand loyalty. As a main prerequisite could be reached, Baldinger and Rubinson (1996) emphasize that if the attitude of the customer towards a brand is positive, then loyal customers will tend to stay loyal [3]. Meanwhile, customers who tend to switch in every purchasing period might more easily become loyal customers.

In building brand loyalty, many aspects must be considered. This is to ensure repurchasing intention stays still which also means that the consumer remains loyal. According to Knox (1996), the use of mass media communications to interpret brand image building is important [12]. By the same token, the use of promotional tools in short-term marketing activities to determine a brand's image is also necessary. It must also be followed by long-term activities such as product development and market expansion.

Davis (2002), emphasized that the main strategy to achieve brand loyalty is through a strong brand positioning [7]. With regard to this issue, a company should create and manage a unique, valued, credible and sustainable status in the customer's minds. This implementation will lead the brand to circulate around a benefit and helps the brand to survive and stand out from the competition.

Another dimension that may predominate is brand trust. Its role, which has been researched in detail for both B2C and B2B sectors [6, 9], is very important in building and maintaining both attitudinal and behavioral brand loyalty. The research indicated that brand trust improves the level of brand loyalty and also affects market share.

2.2 Customer Satisfaction

A consumer may be interested in buying a product after seeing the commercials on television that shows how great the product is. However, does the consumer immediately feel satisfied after purchasing the product? In the previous research, Oliver (1980) determined customer satisfaction as a function of expectation and expectancy disconfirmation [19]. Satisfaction, in turn, is believed to influence attitude change and purchase intention. Consumer satisfaction may be seen to represent the influence of past experience, because it is an overall evaluation of personal consumption experience. This covered the degree of realization of the product or service benefits that the customer expected. If the result exceeded the level of expectation, they were satisfied; if not, they were dissatisfied.

Prior research has suggested that customer satisfaction influences loyalty. However, Oliver (1999) argued that satisfaction does not universally translate into loyalty, but loyalty is presumably a consequence of satisfaction and brand attitudes [18]. This consequence is claims customer satisfaction as a vessel of some subsidiary dimensions that affect customer expectations. These expectations include the benefits of a product or service, brand reputation, the company's tangible and intangible commitment, and so on. To date, the study of satisfaction factors related to brand loyalty have dominated the service literature. However, research on satisfaction and brand loyalty from the context of tablet PC users in Indonesia is not well documented. Thus, the following hypothesis is arrived at:

H_1: Consumer Satisfaction is positively related to Brand Loyalty.

2.3 Service Quality

Efforts to understand and identify service quality have been undertaken in the last three decades. Service quality defines as a consumer's judgment about an entity's overall excellence or superiority [25]. The concept of service quality in marketing literature centers on perceived quality. Moreover, service quality also enables the enhancement of perceived superiority of the brand and may play a part in competitive markets by forming differentiation of the brand [1]. As time goes by and frequent studies are conducted, the nature content and number of its dimensions remain debatable. However, reviews of service quality studies explicitly indicate that European scholars have attached great influence to research on dimensions of service quality. Duo Lehtinen (1982) categorized service quality in terms of three sectors.

Physical quality refers to the tangible aspects in service provided. Interactive quality indicates how two-way interaction occurs between the customer and the service provider. For example a buyer wants to purchase a tablet PC in a store, the interaction between buyer and store employee will generate the level of satisfaction of the buyer during the interaction. This condition could lead to satisfaction or dissatisfaction and affect a buyer's willingness to purchase. Another interaction may occur in other types of representative, including both automated and animated interactions. The more convenient the information given such as touch screen information machines or attractive brochures in a museum, the more visitors' perspectives and satisfactions may increase. The last sector is corporate quality. It refers to the image attributed to a service provider by its current and potential customers, as well as other members of the publics. They also suggest that when compared with the other two quality dimensions, corporate quality tended to be more stable over time.

Another concept concerning service quality dimensions was proposed by Rust and Oliver (1994). They concluded that three components composed a model of service quality based on a customer's evaluation to service encountered.

The conclusion to both those studies, Nam et al. (2011) combined them into two direct subsidiary dimensions, physical quality plays the role as a tangible sector and staff behavior which operates as rationale in any interaction session [16].

H_2: Physical quality is positively related to Consumer Satisfaction.

H_3: Staff behavior is positively related to Consumer Satisfaction.

As a form of self-image development and due to the impact from company advertisements which constantly attempt to segment and differentiate their market share, people start to consider the symbol of the brand and the appearance of the product as a reflection of their daily life. The self-concept implements an individual's thoughts and feelings about a brand through the person as an object of thought [20]. The degree of a consumer's actual or ideal self-concept parallels with the brand image is identified as self-congruence [22, 23].

H_4: Ideal-self congruence is positively related to Consumer Satisfaction.

According to Belk (1988), "brand identification refers to consumers who identify with and associate themselves with brands that reflect and reinforce their self identities" [4]. Brand identification also strengthens the need for both social identity and self-definition [2]. For example, people buy Apple's product for a reason. Setting aside the functional intention, Apple's products satisfy buyers with the brand image that it has developed. Buyers tend to absorb the positive value of the brand as having a good reputation among the groups to which they belong or wish to belong [13]. By understanding the brand identification, we may suggest that a brand could differentiate a customer's social identity from other social identities [11]. There will be a quantifiable distance in social identity between a person who use a random brand of tablet PC made in China and a person who uses Samsung Tabs II. Hence, brand identification surely can be used as a factor that may affect brand loyalty.

H_5: Brand identification is positively related to Consumer Satisfaction.

In this modern and sophisticated era, lifestyle has been transformed as a jargon within society. Generally, lifestyle encompasses many aspects, from demographic attributes to beliefs and aspirations; it may even cover attitudes towards life [5]. Lifestyle expresses a distinctive pattern of human life through interests, opinions, activities, performances and appearances which are dissimilar among individuals [10, 24].

As lifestyle expresses an individual's character and performance, buying patterns follow toward brand image and differentiation. Consumers who are satisfied and think that brands purchased meet their needs within a particular lifestyle, then repeat the buying pattern. They will start to establish personal attachments to the brand they consume and consider it to reflect their desired lifestyle [10].

Nam et al. (2011), defined that lifestyle-congruence as the elaboration to which the brand supports the consumer's lifestyle [16]. They differentiate lifestyle-congruence from self-congruence and brand identification because in this case consumers who intend to purchase use self-conception and social group as comparison standards which are related with their goals consumption, interests, activities, and opinions. The situation may be associated with different social and personal values that are not reached by self-concept and social identity.

H_6: Lifestyle congruence is positively related to Consumer Satisfaction.

3 Research Methodology

This research aims to investigate factors that influence brand loyalty toward the use of tablets in Indonesia. Further, this research intends to identify why consumers prefer to use certain tablet brands in their daily life. In order to help the researcher properly identify and investigate the problem that he is surveying, a conceptual framework is structured from the set of dependent and independent variables. The conceptual framework is mainly based on the literature review which reflects the factors influencing brand loyalty.

In this research, the quantitative approach will be used to analyze the relationships between variables. Then, the detailed description of the research will be explained by using some measurements, numerical data, and statistic analysis. Hundreds of questionnaires will be distributed to collect information about specific required quantitative data.

The questionnaire for this research is constructed by referring to the previous research which focused on brand loyalty through customer satisfaction. The questionnaire's purpose is to identify factors that influence brand loyalty toward the use of tablets in Indonesia. As the questionnaire will be distributed in Indonesia, then the questionnaire will be administered in two languages translation, English and Indonesian. Beside the background section, all questions in the questionnaire sections are designed by using the 7 point Likert Scale (very strongly disagree to very strongly agree). The questions are generated based on study by Nam et al. (2011) and Pawar and Rant (2012).

Before distributing all questionnaires in survey areas, some sample of the questionnaires will be collected for pre-tested necessity. Furthermore, the sample will be used to measure the validity and reliability of the dependent and independent variables.

Due to the limitation of time and cost, and also considering the research needs of specific information from tablet PC consumers, the researcher used a questionnaire as the primary means of collecting data. These were distributed through social networking systems which have a connection with the purpose of the research. A total of 500 questionnaires were employed to gain the information needed from the various tablet PC brands customers.

The descriptive, reliability, factor analysis, ANOVA and regression analysis are applied for analyzing data. The analysis of the data was generated by SPSS version 20.

4 Empirical Results and Research Findings

4.1 Descriptive Statistic Analysis of Sample Demographics

Descriptive analysis is used to analyze the research results coming from characteristics and information of the sample respondents. Of the 500 questionnaires distributed, 433 were returned. Those questionnaires were completed and usable, with an overall response rate of 86.6%. Table 4.1 shows the sample demographics based on the collected data.

In order to represent the entire Indonesian demographic, six characteristics of respondents were covered: (1) Gender, (2) Age, (3) Income, (4) Occupation, (5) Education, and (6) Price Group.

Gender: Table 4.1 shows the frequency and percentage of males and females within respondents to be 62.1% male and 37.9% female.

Age: Figure 4.1 shows the frequency and percentage of age dispersion within respondents. 35.5% of the respondents were aged between 26–30 years old, followed by 20–25 years old (34.2%), 30–35 years old (18.2%), 36–40 years old (6%), over 40 years old (4.4%), and respondents under 20 years old (0.7%).

Income level: Figure 4.2 shows that 33% of respondents had an income level between Rp.1,500,001–Rp.3,000,000, followed by respondents with an income level below Rp.1,500,000 (24%), respondents with an income level of Rp.3,000,000- Rp.4,500,000 (18.2%), respondents with income level above Rp.6,000,000 (14.3%), and respondents with income level of Rp.4,500,000–Rp.6,000,000 (10.4%).

Occupation: Figure 4.3 shows that both student and employed respondents have percentages of 40.2% and self employed respondents have percentages of 19.6%.

Educational background: Figure 4.4 shows those respondents whose educational background is at Bachelor level had the biggest percentage (61.2%), followed by Master (20.8%), high school (16.9%) and PhD respondents (1.1%).

Tablet price group: Figure 4.5 shows that 37.6% of respondents buy tablet PCs within the price Rp.2,000,001–Rp.4,000,000, followed by respondents with tablets priced below Rp.2,000,000 (35.3%), respondents with the tablets priced around Rp.4,000,000–Rp.6,000,000 (15.7%), respondents with the tablets priced within Rp.6,000,000–Rp.8,000,000 (10.4%), and respondents with tablets priced above Rp.8,000,000 (0.9%).

4.2 Reliability, Validity and Correlation Analysis

The reliability of the measurement tools were assessed by the Cronbach's alpha reliability coefficient. In order to check the measurement of internal reliability, relevant analysis in SPSS was put through a run test. After the run test, the results of reliability assessments are shown in Table 1 below.

Table 1. Reliability assessment of variables

Variables	Code	No. of questions	Cronbach's α
Physical quality	PQ	4	0.789
Staff behavior	SB	3	0.892
Ideal self-congruence	ISC	3	0.880
Brand identification	BI	3	0.799
Lifestyle-congruence	LC	3	0.793

Construct validity was explored in order to confirm accuracy and adequacy within the sample, measuring the validity of each variable by the rules defined below: Eigen value must be greater than 1; KMO is suggested to be more than 0.5, but more than 0.4 is justifiable; Factor loading with cut point is less than 0.5. After running factor analysis, the results of validity assessments are shown in Table 2.

Correlation analysis was used in this study to measure relationships of independent variables and dependent variables. Correlation analysis is measured by correlation coefficients. If coefficients are positive, their relationship is positive; if coefficients are negative, relationships of variables would be the inverse. Results of correlation analysis are illustrated in Table 3.

Table 2. Validity assessment of variables

Variable	Items	Factor loading	Eigen value	Cumulative proportion	KMO	Chi-square	P value
PQ	PQ1	0.886	2.479	61.985	0.697	596.348	0.000***
	PQ2	0.78					
	PQ3	0.776					
	PQ4	0.695					
SB	SB1	0.926	2.479	82.632	0.741	781.916	0.000***
	SB2	0.892					
	SB3	0.908					
ISC	ISC1	0.864	2.419	80.642	0.718	722.898	0.000***
	ISC2	0.927					
	ISC3	0.903					
BI	BI1	0.779	2.16	72.003	0.629	490.782	0.000***
	BI2	0.917					
	BI3	0.844					
LC	LC1	0.868	2.151	71.712	0.629	502.008	0.000***
	LC2	0.748					
	LC3	0.916					
BL	BL1	0.799	3.55	59.17	0.818	1238.227	0.000***
	BL2	0.711					
	BL3	0.853					
	BL4	0.823					
	BL5	0.844					
	BL6	0.54					

***$p < 0.01$

Table 3. Correlation coefficients of research variables

Variables	PQ	SB	ISC	BI	LC	CS (DV1)	BL (DV2)
PQ	1	0.431**	0.493**	0.281**	0.494**	0.402**	0.491**
SB		1	0.422**	0.273**	0.227**	0.148**	0.329**
ISC			1	0.456**	0.612**	0.210**	0.557**
BI				1	0.401**	−0.003**	0.300**
LC					1	0.299**	0.676**
CS						1	0.356**

4.3 Multiple Regression Analysis

To test the hypotheses formulated in Chap. 3, a multiple regression is used. Multiple regression is a useful method that finds the relationship between variables. Multiple regressions are widely used to explore the relationship between variables in social science research. Hypothesis testing is based on the standardized path coefficient (r-path coefficient). To support the hypothesis, the p-value of the r-path coefficient should be significant at the 0.05 level. The two models were tested through regression analysis:

Model 1: Independent variables are Physical Quality, Staff Behavior, Ideal Self-Congruence, Brand Identification, and Lifestyle-Congruence which are correlated against the dependent variable Customer Satisfaction.

In order to examine the extent to which customers are satisfied with tablet PCs that they are using, a linear regression test was run by adding the five independent variables and the dependent variable (customer satisfaction) into the SPSS statistical area. The results are shown in Table 4.

Table 4. Result of multiple regression analysis for Model 1

Independent variables	Customer satisfaction				
	Beta (β)	t	R^2	Adjusted R^2	F ratio
Physical quality	0.432	6.579	0.202	0.193	21.678***
Staff behavior	0.001	0.024			
Ideal self-congruence	−0.006	−0.096			
Brand identification	−0.186	−3.663			
Lifestyle-congruence	0.182	3.397			

$*p < 0.10$, $**p < 0.05$, $***p < 0.01$

Table 4 shows the regression results from testing of customer satisfaction. The regression equation is significant at the 0.001 level and physical quality, staff behavior, ideal self-congruence, brand identification, and lifestyle-congruence are independent variables account for 20.2% of the variance of customer satisfaction. As shown in the table and in the order of significance, physical quality ($\beta = 0.432$), staff behavior ($\beta = 0.001$) and lifestyle-congruence ($\beta = 0.182$) are all positively related to customer satisfaction; ideal self-congruence ($\beta = -0.006$) and brand identification ($\beta = -0.186$) are negatively related to customer satisfaction. But staff behavior and ideal self-congruence are not significant (at the p-value level of 0.05).

Model 2: Figure 4.8 shows the effect of customer satisfaction on brand loyalty.

Table 5. Result of multiple regression analysis for Model 2

Independent variables	Brand loyalty				
	Beta (β)	t	R^2	Adjusted R^2	F ratio
Customer satisfaction	0.356	7.901	0.127	0.124	62.422***

$*p < 0.10$, $**p < 0.05$, $***p < 0.01$

Table 5 shows significant linear relationships between customer satisfaction and brand loyalty. 12.7% of the variance in Brand Loyalty (BL) is explained by customer satisfaction (CS).

5 Conclusion and Implication

As shown in Table 6, all the hypotheses had been accepted except for the Staff Behavior variable and the Ideal Self-Congruence variable.

Table 6. Summary of hypotheses test result

#	Hypotheses	Accepted or not
H1	Consumer Satisfaction is positively related to Brand Loyalty	Accepted
H2	Physical quality is positively related to Consumer Satisfaction	Accepted
H3	Staff behavior is positively related to Consumer Satisfaction	Not accepted
H4	Ideal self-congruence is positively related to Consumer Satisfaction	Not accepted
H5	Brand identification is positively related to Consumer Satisfaction	Accepted
H6	Lifestyle congruence is positively related to Consumer Satisfaction	Accepted

In several previous studies, the Staff Behavior variable and Ideal Self-Congruence variable have positive affects on customer satisfaction in the hotel and restaurant industry. But in the tablet PC context, the users of tablet PCs in Indonesia don't consider the Staff Behavior variable and Ideal Self Congruence variable as important factors that influence their loyalty toward the tablet brands. The most important factors affecting customer satisfaction in using tablet PCs in Indonesia is physical quality (0.432). The findings of the study also suggest that customer satisfaction positively affected the brand loyalty of tablet PC users in Indonesia.

The study's implication is that the automotive company must focus and develop on three important factors as the research result found them important for customers to be more satisfied in using tablet PCs. These factors are Physical Quality, Brand Identification, and Lifestyle-Congruence and all must be addressed in order to attract and maintain more customers.

Tablet users will make comparison in the design, features and applications that each tablet PC Company offers. Hence, tablet companies must create more attractive promotions and technology development to gain more market share. As the awareness toward the tablet brand continues to increase, brand identification is considered as an important variable in any attempt to maintain the loyalty of tablet users through the tablet brand. Tablet companies should improve the promotion which refers to the strengthening of the brand, thus all tablet users in Indonesia increasingly feel that they are fused with the purchased tablet brand. As Samsung, Apple and Asus have been acquiring good reputations the tablet PC market in Indonesia, so they need to maintain their good reputation and brand image. However, local tablet brands should also consider improving their promotional activities in view of the market share for local tablet brands such as Tabulet, Nexian and IMO evolving period to period.

References

1. Aaker, D.A.: Measuring brand equity across products and markets. Calif. Manag. Rev. **38**(3), 102–120 (1996)
2. Ahearne, M., Gruen, T., Bhattacharya, C.B.: Antecedents and consequences of customer-company identification: expanding the role of relationship marketing. J. Appl. Psychol. **90**(3), 574–585 (2005)
3. Baldinger, A., Rubinson, J.: Brand loyalty: the link between attitude and behavior. J. Adv. Res. **36**(6), 22–34 (1996)
4. Belk, R.W.: Possessions and the extended self. J. Consumer Res. **15**, 139–168 (1988)
5. Brassington, F., Pettitt, S.: Principles of Marketing. Financial Times Prentice Hall, Harlow (2003)
6. Cowles, D.: The role of trust in customer relationships: asking the right questions. Manag. Dec. **35**(3–4), 273–283 (1997)
7. Davis, M.S.: Brand Asset Management: Driving Profitable Growth through Your Brands. Jossey-Bass Publishing, San Francisco (2002)
8. Dick, A., Basu, K.: Customer loyalty: toward an integrated conceptual framework. J. Acad. Mark. Sci. **22**(2), 99–113 (1994)
9. Doney, P., Cannon, J.: An examination of the nature of trust in buyer-seller relationships. J. Mark. **61**(2), 35–51 (1997)
10. Foxall, G., Goldsmith, R., Brown, S.: Consumer Psychology for Marketing. International Thomson Business Press, London (1998)
11. Kim, C.K., Han, D.C., Park, S.B.: The effect of brand personality and brand identification on brand loyalty: applying the theory of social identification. Jpn. Psychol. Res. **43**(4), 195–206 (2001)
12. Knox, S.: The death of brand deference: can brand management stop the rot? Mark. Intel. Plann. **14**(7), 35–39 (1996)
13. Long, M.M., Shiffman, L.G.: Consumption values and relationships: segmenting the market for frequency programs. J. Consumer Mark. **17**(3), 214–232 (2000)
14. Mittal, V., Kamakura, W.: Satisfaction, repurchase intent, and repurchase behavior: investigating the moderating effect of customer characteristics. J. Mark. Res. **38**(1), 131–142 (2001)
15. Moisescu, O.I., Allen, B.: The relationship between the dimensions of brand loyalty: an empirical investigation among Romanian urban consumers. Manag. Mark. Challenges Knowl. Soc. **5**(4), 83–98 (2010)
16. Nam, J., Ekinci, Y., Whyatt, G.: Brand equity, brand loyalty and consumer satisfaction. Ann. Touris. Res. **38**(3), 1009–1030 (2011)
17. Oliva, T., Oliver, R.: A catastrophe model for developing service satisfaction strategies. J. Mark. **56**(3), 83–95 (1992)
18. Oliver, R.: Whence consumer loyalty. J. Mark. **63**, 33–44 (1999)
19. Oliver, R.L.: A cognitive model of the antecedents and consequences of satisfaction decisions. J. Mark. Res. **17**, 460–469 (1980)
20. Rosenberg, M.: Conceiving the Self. Basic Books, New York (1979)
21. Rust, R.T., Oliver, R.L.: Service quality: insights and managerial implications from the frontier. In: Rust, R.T., Oliver, R.L. (eds.) Service Quality: New Directions in Theory and Practice, pp. 1–19. Sage Publications, Thousand Oaks (1994)

22. Sirgy, M.J.: Self-concept in consumer behavior: a critical review. J. Consumer Res. **9**(3), 287–300 (1982)
23. Sirgy, M.J., Grewal, D., Mangleburg, T.: Retail environment, self congruity, and retail patronage: an integrative model and a research agenda. J. Bus. Res. **49**(2), 127–138 (2000)
24. Solomon, M.R.: Consumer Behavior: Buying, Having, and Being. Prentice Hall, New Jersey (2002)
25. Zeithaml, V.A.: Defining and Relating Prices, Perceived Quality and Perceived Value. Marketing Science Institute, Cambridge (1987)

Five Elements and Stock Market

Mei-Hua Liao[1(✉)], Shih-Han Hong[1], and Norihisa Yoshimura[2]

[1] Asia University, Taichung, Taiwan
liao_meihua@asia.edu.tw
[2] Osaka City University, Osaka, Japan
bakerstreet2003@gmail.com

Abstract. Many of Taiwan's industries are still deeply influenced by traditional customs. The characteristics of the five elements of each day should have a considerable relationship with their stock prices. This study adopts reliable quantitative indicators. Through statistical analysis of differences, correlations, regressions, etc., we observe whether the extent to which Taiwanese firms are affected by the five elements varies with different factors such as industry, company life cycle, education level of the operators, and the location of the company. So far, the relevant research on the five elements and the stock market is still rare. This topic has research value.

1 Introduction

This study which the sample period was from 2001 to 2016 explained the changes in daily returns and ten-day returns of each industry with the 24 solar terms of Ying Yong and five elements conducting a regression analysis. Through the analysis of five element solar terms and the industrial cycle, this paper observes the effect of the traditional solar terms in Taiwan stock market.

And the empirical results of this study show that on the day of cold dew, there are negative correlations with other industries excepting for 5 industries. There are negative correlations with the 10 days of cumulative return rate about cement industry, food industry, plastics industry, chemical biotechnology medical treatment industry, glass ceramics industry, rubber industry, trading merchandise industry, etc., but excluding electronics industry, computer and interface equipment industry, photoelectric industry on the day of light snow negative. The return with chemical industry, biotechnology medical industry, computer and interface equipment industry, semiconductor industry, photoelectric industry, electronics components industry, information service industry and other electronics industry are negatively related in the beginning of winter.

In addition, while the solar term becomes gold of five elements, with the daily return rate of electrical machinery industry, paper industry, electronics industry, semiconductor industry, communications network industry, trading merchandise industry, other industries, there are a significant negative association, while the financial and insurance industry are positive. We find that the cumulative return rate of the semiconductor

L. Barolli et al. (Eds.): IMIS 2018, AISC 773, pp. 879–883, 2019.
https://doi.org/10.1007/978-3-319-93554-6_88

industry, communications network industry, and other electronics industry have significant negative effects while both industries of the financial and insurance have positive on the solar term become gold. The solar term becoming fire has significant positive correlation on industries return of components and information service.

2 Background and Hypothesis

There are strong day-of-the-week effects in the Kuala Lumpur Stock Exchange (KLSE), The Stock Exchange, Bombay (SEB), the Stock Exchange of Singapore (SES) and The Stock Exchange of Thailand (SET). Strong Chinese New Year effects exist on the Stock Exchange of Singapore (SES) and the Kuala Lumpur Stock Exchange (KLSE). The empirical results show that cultural festivals evidence a stronger effect than state holidays [1].

Moreover, there are significant Lunar Year effects across Southeast Asian countries [2]. And the presence of the January effect was confirmed in the Taiwanese and South Korean stock markets [3]. Chinese New Year effects is compared in Asian stock markets [4].

Table 1. Five elements and industries.

Industry	Five element
Cement Industry	earth
Food Industry	water, fire
Plastic Industry	fire, wood
Textile Industry	water, fire, wood
Electric Machinery Industry	water, fire, metal
Electrical and Cable Industry	water, fire, metal
Chemical Biomedical Industry	water, fire
Chemical Industry	water, fire
Biotechnology and Medical Care Industry	water, fire
Glass and Ceramic Industry	fire, earth
Paper and Pulp Industry	water, wood
Iron and Steel Industry	fire, metal, earth
Rubber Industry	fire, wood
Automobile Industry	water, fire, metal
Electronic Industry	water, fire, metal
Information Service Industry	water, fire, metal
Building Material and Construction Industry	wood, earth
Shipping and Transportation Industry	water
Tourism Industry	water, fire
Financial and Insurance Industry	water, wood, metal
Trading and Consumers' Goods Industry	water, fire, wood, metal, earth
Oil, Gas and Electricity Industry	water, fire
Other Industry	water, fire, wood, metal, earth

The traditional Chinese fundamental principles described in the "Book of Changes" proposes five elements and yin-yang to explain all things. The five elements include water, fire, wood, metal and earth. Table 1 shows that each industry can be classified as those in the five elements.

We collect the daily data of 95,427 day-industry observations from 2 January 2001 to 30 December 2016. We construct our sample as follows. We begin with all publicly traded Taiwanese firms in TEJ. In order to calculate the various research variables involved in the discussion, the financial statements of the sample were included to explore the impact of 24 solar terms per year on the market.

Table 2. Descriptive statistics for daily return rate of industries.

Industry	Mean	Max	Min	SD	Median
Cement	0.04	6.87	−6.95	1.94	0.01
Food	0.05	6.72	−7.50	1.64	0.04
Plastic	0.04	6.93	−6.82	1.57	0.05
Textile	0.04	6.70	−7.55	1.69	0.05
Electric Machinery	0.04	6.18	−7.04	1.35	0.09
Electrical and Cable	0.01	6.73	−7.43	1.80	0.02
Chemical Biomedical	0.04	6.59	−6.72	1.46	0.09
Glass and Ceramic	0.02	8.16	−9.40	1.97	−0.02
Paper and Pulp	0.03	6.89	−8.03	1.74	−0.01
Iron and Steel	0.03	6.85	−8.39	1.54	0.00
Rubber	0.07	6.54	−6.90	1.74	0.04
Automobile	0.05	7.22	−7.47	1.78	−0.02
Electronic	0.02	6.75	−6.70	1.49	0.05
Information Service	0.05	6.79	−7.65	2.03	0.00
Building Material and Construction	0.02	6.83	−7.76	1.76	−0.04
Shipping and Transportation	0.03	6.96	−7.71	1.76	−0.04
Tourism	0.03	6.86	−6.84	1.64	0.01
Financial and Insurance	0.04	6.59	−6.81	1.43	0.02
Trading and Consumers' Goods	0.04	6.17	−7.24	1.26	0.07
Oil, Gas and Electricity	0.03	6.74	−6.66	1.32	0.05
Other	0.03	6.74	−6.68	1.33	0.05

Table 2 shows the average daily return rate for each industry in the sample period.

The figures in Table 2 shows that the daily returns of Information Service, Cement, Glass and Ceramic industries has changed the most. The Rubber industry has the best mean of daily returns. However, the Glass and Ceramics industry recorded the best daily earnings.

3 Results

China's traditional philosophy believes that the five elements of solar terms can explain the changes in everything in the world. Of course, the stock market is also included. Therefore, we use the twenty-four solar terms, such as the beginning of spring to the big cold, to explain the changes in the stock market.

This study sets the virtual variables for the 24 solar terms to illustrate the changes in the daily compensation of various industries. We use the following Five-Factor Asset Pricing Model [5] to observe the industrial returns.

$$\begin{aligned} \mathrm{R}_{it} - \mathrm{R}_{Ft} = & \alpha_i + \beta_i(-\mathrm{R}_{Ft}) + s_i SMB_t + h_i HML_t + r_i RMW_t \\ & + c_i CMA_t + e_{it}. \end{aligned} \quad (1)$$

In this equation R_{it} is the return on industry index i for period t, R_{Ft} is the risk free return, R_{Ft} is the return on the value-weight market portfolio, SMB_t is the return on an industry index of small stocks minus the return on an industry index of big stocks, HML_t is the difference between the returns on indicators of high and low B/M stocks, and e_{it} is a zero-mean residual. And RMW_t is the difference between the returns on indicators of stocks with robust and weak profitability, and CMA_t is the difference between the returns on indicators of the stocks of low and high investment firms, which we call conservative and aggressive. α_i is the intercept. $\beta_i, s_i, h_i, r_i, c_i$ are the respective slope coefficients from the regression.

4 Conclusion

The empirical results of this study show that the price rises in the industry with mutual producing the elements of the trading days; and that the price tends to fall in the industry with mutual conquering the elements of the trading days.

Acknowledgments. Constructive comments of editors and anonymous referees are gratefully acknowledged. This research is partly supported by the National Science Council of Taiwan (NSC 104-2815-C-468-041-H and NSC 102-2815-C-468-019-H) and by Asia University of Taiwan (ASIA-105-CMUH-10 and ASIA104-CMUH-12).

References

1. Chan, M.W.L., Khanthavit, A., Thomas, H.: Seasonality and cultural influences on four Asian stock markets. Asia Pac. J. Manag. **13**, 1–24 (1997)
2. Lee, C.F., Yen, G., Chang, C.: Business Finance in Less Developed Capital Markets. The Chinese New Year, common stock purchasing and cumulative raw returns: Is Taiwan's stock market informationally efficient? pp. 101–105. Greenwood Press, Westport (1992)
3. Tong, W.H.S.: An analysis of the January effect of the United States, Taiwan and South Korean stock returns. Asia Pac. J. Manag. **9**, 189–207 (1992)

4. Yen, C.L., Shyy, C.: Chinese New Year effect in Asian stock markets. Taiwan Natl. Univ. Manag. J. **4**, 417–436 (1993)
5. Fama, E.F., French, K.R.: A five-factor asset pricing model. J. Financ. Econ. **116**, 1–22 (2015)

Asset Structure of Long-Lived Company

Mei-Hua Liao[1(✉)], Yi-Jun Guo[1], and Hidekazu Sone[2]

[1] Asia University, Taichung, Taiwan
liao_meihua@asia.edu.tw
[2] Shizuoka University Art and Culture, Hamamatsu, Japan
h-sone@suac.ac.jp

Abstract. Sustainability is of utmost priority to most companies. The key to long-lived company success and sustainability lies in effective professional management. This paper aims to identify the characteristics of asset allocation in the long-lived company of Taiwan and observe the impact on the company's stock price, asset structure, and operating performance.

1 Introduction

This paper presents a micro-micro-econometric analysis of the relationship between the asset structure and performance for a sample of Taiwanese long-lived firms. Many firms increase input resource utilization by automation, networking and strategic alliances in delivering aspects of the service, shared facilities, franchising, etc. To maximize returns on its asset base is also important for the long-lived firm. Firms use assets to achieve business objectives for profit-making companies as increasing owner value and sustainable development.

2 Background and Hypothesis

The long-lived family firms have been successful over time, even with moderate or low levels of overall corporate entrepreneurship [1]. The reason may be found in the Australian stock market. There is a long-run equilibrium between market and accounting values of the long-lived firms listed for at least 50 years on Australian stock exchange [2].

Typically the long-lived and small firms in Scottish felt that rivalry within their market was strong and that the competitive environment had become more hostile since start-up. A highly skilled, flexible labor market may aid long-lived firms in their efforts to sustain their small-scale existences, given the observed size-performance trade-off relationship [3].

The ownership of the centuries-old firms transfer to the outside of the family often observed both in Italy and Switzerland. But this phenomenon is rarely observed in Japan [4].

A survey of companies listed on the Tehran Stock Exchange shows that there is a significant negative relationship between the asset structure and the quality of financial reports [5]. The change in debt contract enforcement costs is associated with financing and asset structure in India [6].

L. Barolli et al. (Eds.): IMIS 2018, AISC 773, pp. 884–887, 2019.
https://doi.org/10.1007/978-3-319-93554-6_89

In the United States, the management decision of the asset structure has a great impact on the company's return on assets [7]. There is a positive correlation between long-term capital structure and asset structure in Pakistani companies [8].

The purpose of this paper is to investigate the relationship between the asset structure and performance for long-lived firms listed in Taiwan stock market.

3 Results

We examine annual reports and collect the derivatives use data of 16,500 year-firm observations. We construct our sample as follows. We begin with all publicly traded Taiwanese firms in TEJ between 2008 and 2017. Our sample consists of all non-financial Taiwanese firms that have 10 years non-missing data during the sample period. There are 14,500 year-firm observations that meet this criteria.

Table 1. The sizes of industries and the numbers of long-life enterprise in the industries.

Industry	Number of samples	Number of long-life enterprise
Agricultural technology industry	1	1
Automobile industry	5	3
Biotechnology and medical care industry	8	0
Building material and construction industry	70	13
Cement industry	7	6
Chemical biomedical industry	118	22
Cultural and creative industry	16	0
Electric machinery industry	79	9
Cement industry	7	6
Chemical biomedical industry	118	22
Cultural and creative industry	16	0
Electric machinery industry	79	9
Electrical and cable industry	14	7
Electronic industry	749	6
Financial and insurance industry	42	10
Food industry	26	9
Glass and ceramic industry	5	2
Information service industry	1	0
Iron and steel industry	43	8
Oil, gas and electricity industry	12	2
Other industry	80	4
Paper and pulp industry	7	5
Plastic industry	25	11
Rubber industry	11	6
Shipping and transportation industry	26	7
Textile industry	54	12
Tourism industry	24	5
Trading and consumers' goods industry	27	4

Table 1 shows the sample sizes of industries and the numbers of long-life enterprise in the industries. In Taiwan, Electronic Industry is an important industry. There are 749 companies in the industry. But only 6 Electronic companies set up more than 50 years. In Chemical Biomedical Industries, There are 22 companies set up more than 50 years. The long-lived companies in this industry are the most. And there are the second most of long-lived companies in Building Material and Construction Industry.

Obviously, there are many long-life enterprises in the industries of chemical biomedical and construction.

We assume that longevity companies have special asset structure and corporate performance. Fixed assets by total shareholders' equity (FA/SE) shows the asset structure of company. The earnings before interest and taxes (EBIT) are assumed to measure true earnings on the project and should not be contaminated by capital charges or expenditures whose benefits accrue to future projects. So, We use EBIT by sale (EBIT/SA), return on equity (ROE) and earning per shear (EPS) to measure operating performance of company.

All of our sample consists of 1,650 companies, including 152 longevity companies. Table 2 lists and compares statistics by long-lived and other firms by the Wilcoxon test. Panel A reports the mean and median for each continuous variable for 2008. Panel B reports the figures for 2017.

Table 2. Descriptive statistics

Panel A. Financial figures for 2008					
	Long-lived firm		Other firm		Wilcoxon test
	Mean	Median	Mean	Median	Z-value
FA/SE	0.75	0.62	0.62	0.49	4.68***
EBIT/SA	−10.37	2.74	−3.96	2.78	0.45
EPS	0.59	0.63	1.18	0.87	1.96*
ROE	2.28	2.54	2.29	3.60	2.34**
Panel B. Financial figures for 2017					
	Long-lived firm		Other firm		Wilcoxon test
	Mean	Median	Mean	Median	Z-value
FA/SE	0.57	0.47	0.60	0.37	2.35**
EBIT/SA	−78.60	6.49	−15.74	4.51	3.56***
EPS	1.58	1.15	1.99	1.13	0.57
ROE	3.56	3.43	3.09	3.64	0.15

* indicates significance at 10%; ** significance at 5%; *** significance at 1%.

Panel A shows that the asset structure of long-lived firms (FA/SE = 0.62) is significantly higher than that of other firms (FA/SE = 0.49) in 2008. But the performances of other firms (EPS = 0.87; ROE = 3.6) are obviously higher than those of long-lived firms (EPS = 0.63; ROE = 2.54) in 2008. Similarly, Panel B shows that the asset structure of long-lived firms is significantly higher in 2017. And the performance of long-lived firms (EBIT/SA = 6.49) is obviously higher than that of other firms (EBIT/SA = 4.51) in 2017.

The results imply that long-lived firms have significantly higher long-term asset structure in both bear and bull markets. But the company's operating performance will vary with market conditions.

Acknowledgments. Constructive comments of editors and anonymous referees are gratefully acknowledged. This research is partly supported by the National Science Council of Taiwan (NSC 104-2815-C-468-041-H and NSC 102-2815-C-468-019-H) and by Asia University of Taiwan (ASIA-105-CMUH-10 and ASIA104-CMUH-12).

References

1. Zellweger, T., Sieger, P.: Entrepreneurial orientation in long-lived family firms. Small Bus. Econ. **38**, 67–84 (2012)
2. Clout, V.J., Willett, R.J.: Investigating the relationship between market values and accounting numbers for long-lived companies. Working Paper (2009)
3. Power, B., Reid, G.C.: Performance and strategy: simultaneous equations analysis of long-lived firms. Int. J. Econ. Bus. **22**, 345–377 (2015)
4. Goto, T.: Family business and its longevity. Kindai Manage. Rev. **2**, 78–96 (2014)
5. Hamidzadeh, S., Zeinali, M.: The asset structure and liquidity effect on financial reporting quality at listed companies in Tehran stock exchange. Arab J. Bus. Manage. Rev. **4**, 121–127 (2015)
6. Singh, M., Mukherjee, A., Gopalan, R.: Does debt contract enforcement costs affect financing and asset structure? Rev. Finan. Stud. **29**, 2774–2813 (2016)
7. Chen, G., Haque, M.: Firm's asset structure change: evidence from industrial, commercial machinery and computer equipment in the U.S. Market. China-USA Bus. Rev. **14**, 1–9 (2015)
8. Mufti, S.W., Amjad, S.: Cross industry capital structure of firm characteristics in Pakistan. Int. J. Inf. Bus. Manage. **10**, 174–188 (2018)

An Inventory Policy for Perishable Products with Permissible Delay in Payment

Ya-Lan Chan[1] and Sue-Ming Hsu[2](✉)

[1] Department of Business Administration, Asia University, Taichung, Taiwan
yalan@asia.edu.tw
[2] Department of Business Administration, Tunghai University, Taichung, Taiwan
sueming@thu.edu.tw

Abstract. For perishable products, the seller usually asks for the buyer to prepay a fraction of the purchasing cost as a good-faith deposit, to pay some cash upon the receipt of the order, and then a permissible delay is granted on the remaining of the purchasing cost. In addition, it is evident that the deterioration rate ages to 100% as time reaches the expiration date. In this paper, we incorporate the above two important and relevant facts to find the optimal cycle time and the fraction of no shortages such that the total profit is maximized. Several managerial insights are presented.

Keywords: Supply chain management · Finance · Expiration date
Perishable products · Delay in payment

1 Introduction

In the United Kingdom and the United States, about 80% of firms offer various credit terms on their products (e.g., see Seifert et al. (Seifert et al. 2013)). Hence, trade credit becomes growingly important to modern firms. For highly seasonal or perishable goods, the supplier may offer the retailer an advance-cash-credit (ACC) payment scheme. The retailer prepays a fraction of acquisition cost as a good-faith deposit when placing an order. By doing so, the retailer may obtain a price discount in return. Upon receiving the order, the retailer pays another portion of the acquisition cost in cash, and then the remaining of the acquisition cost is granted with a permissible delay. Thus, an ACC payment is a combination of the good-faith prepayment, the cash payment, and the delayed credit payment. As a result, the proposed ACC payment includes six most commonly-used previous payments as special cases. For instance, Harris (1913) used a cash payment to establish the classical economic order quantity (EOQ) model. Goyal (1985) established the EOQ model by using a credit payment. Zhang (1996) developed an optimal advanced payment scheme to pay for small-billed amount. Teng (2009) adopted a cash-credit (i.e., some in cash and the rest in credit) payment to further expand the model. Taleizadeh (2014) proposed an EOQ model with an advance-cash (i.e., some in advance and the rest in cash) payment. Concurrently, Zhang et al. (2014) built an EOQ model under an advance-credit (i.e., some in advance and the rest in

L. Barolli et al. (Eds.): IMIS 2018, AISC 773, pp. 888–902, 2019.
https://doi.org/10.1007/978-3-319-93554-6_90

credit) payment. However, none in the literature has discussed this interesting and relevant ACC payment scheme.

It is well-known that the deterioration (or degrading) rate of a perishable product increases over time and reaches 100% at its expiration date. Conversely, the demand rate for perishable goods is gradually decreasing over time and near zero when the product reaches its expiration date because the produce is not fresh and cannot be sold after its expiration date. Furthermore, the customers can easily identify the freshness of a product by checking its expiration date. As a result, the expiration date of a perishable product is an important factor in customer's purchasing decision. However, relatively little attention has been paid to the importance of the expiration date.

In this paper, we incorporate the important and relevant fact that the degrading rate of a perishable product gradually increases over time and approaches to 100% as time reaches its expiration date. Additionally, we use a generalized ACC payment scheme to explore an EOQ model, which in turn provides us the best solution among all seven potential payments. As a result, the proposed model includes many previous models as special cases. We then set up the retailer's objective to find the optimal cycle time and fraction of no shortages such that the total annual profit is maximized. Furthermore, we demonstrate that the total annual profit is strictly pseudo-convex in both the replenishment cycle time and the fraction of no shortages. This theoretical result simplifies the search for the global solution to a local maximum. In addition, we establish the optimal solution for a time-varying degrading rate related to product expiration date. Finally, several numerical examples are presented to emphasize the managerial insights. For instance, the longer the expiration date, the higher the fraction of no shortages and the total annual profit while the shorter the replenishment cycle time.

The remaining of the paper is organized as follows. In Sect. 2, the relevant literature is reviewed. In Sect. 3, the notation and assumptions used throughout the paper are introduced. The mathematical models for different cases are presented in Sect. 4. The theoretical results and the optimal solutions are derived in Sect. 5. In Sect. 6, numerical examples and sensitivity analysis are provided. Finally, the conclusions are given in Sect. 5.

2 Literature Review

In practice, there are three possible strategies for the seller to collect the buyer's acquisition cost: (i) to ask for an advance payment as a good-faith deposit or in exchange of a price discount, (ii) to get a cash-on-delivery (COD) payment in order to avoid default risks and capital costs as soon as the order quantity is delivered, (iii) to grant a short-term free-interest credit payment. Harris (1913) first established the classical EOQ model on the assumptions that the buyer must pay for the acquisition cost as soon as the order quantity is received (i.e., a COD payment). Zhang (1996) developed an EOQ model by using an advance payment. In today's markets, more than 80% of the United States companies offer their buyers a short-term interest-free credit period to stimulate sales and reduce inventory. During the credit period, the buyer earns interest on the accumulative revenue, and hence reduces the cost and increases the order quantity. However, if the buyer cannot pay off the acquisition cost after the credit period, then the buyer is charged with interest on the unpaid balance. After 2008 global financial crisis, a small business or an individual is extremely difficult to obtain a loan from banks. As a result, credit payment has become critically important since then. Grubbstrom (1980) derived an economic production quantity (EPQ) model with a credit payment by using discounted cash-flow analysis. Goyal (1985) developed an EOQ model for the buyer by taking the interest earned during credit period into consideration. Since, many perishable products (e.g., fruits, meats, baked goods, dairy products, etc.) deteriorate over time, Aggarwal and Jaggi (1995) extended the EOQ model by Goyal (1985) from non-deteriorating items to deteriorating items. Jamal et al. (1997) further generalized the model to allow for shortages. Teng (2002) analyzed the EOQ model with trade credit and concluded that it is more profitable for the buyer to order less quantity and take the benefit of trade credit more frequently. Huang (2003) expanded the model by considering both upstream and downstream trade credit in which a supplier offers its retailer an upstream credit period while the retailer in turn provides a downstream credit period to the customers. Teng (2009) established the optimal ordering policies for the retailer who asks for customers with low-credit scores to pay some cash as placing an order and offers a credit payment for the remaining cost. Teng et al. (2012) developed an EOQ model for an increasing demand instead of a constant demand. Ouyang and Chang (2013) studied an EPQ model with imperfect production process and complete backlogging. Recently, Seifert et al. (2013) provided a detailed review on the trade-credit literature, and derived a detailed agenda for future research. Several other related references are: Chen et al. (2014), Cárdenas-Barrón et al. (2014), Dye and Yang (2015), Ouyang et al. (2013), Sakar et al. (2015), and Shah and Cárdenas-Barrón (2015).

On the other hand, for highly seasonal products or flammable deteriorating items, the seller usually requests buyers to prepay a fraction of the acquisition cost as a good-faith deposit to avoid buyers' changing minds. It is also a commonly used strategy for the seller to offer credit-risk buyers an advance payment in exchange of a price discount to avoid default risks. Furthermore, for small payments such as electricity, water, or telephone bills, the customer can opt to send a check for an amount larger than billed to save the cost of time, writing a check, postage, etc. Zhang (Zhang 1996) developed the optimal advanced payment scheme to save time and money by paying a larger payment in advance than the billed amount. Zhang et al. (2014) studied an advance-credit payment scheme in which the buyer prepays a portion of purchasing cost to get a price discount and then receives a permissible delay in payment for the remaining amount. At the same time, Taleizadeh (2014) constructed a slightly different advance-cash payment for an evaporating product with partial backordering. However, most researchers in the area of advance payments assumed the product can be sold indefinitely. They did not incorporate the fact that many products (e.g., bakeries, fruits, meat, milk, vegetables, etc.) cannot be sold after their expiration dates into models.

Most fresh or fashion goods are perishable and deteriorating continuously due to several reasons such as evaporation, spoilage, and obsolescence. For detailed reviews in the area of deteriorating items, we refer to the reader to consult the works of Raafat (1991), Goyal and Giri (2001) and Bakker et al. (2012). Also, for a complete survey with regard to deterioration and lifetime constraints in production and supply chain planning, we refer to the reader to see the paper of Pahl and Voß (2014). In a pioneer work, Ghare and Schrader (1963) proposed an EOQ model with a constant deterioration rate. Covert and Philip (1973) then extended the constant deterioration rate to a two-parameter Weibull distribution. Dave and Patel (1981) generalized the model for deteriorating items from a constant demand to a linearly increasing demand pattern. Teng et al. (1999) further extended the model with complete backlogging from an increasing demand to a continuously fluctuated demand. Teng (2002) then explored the model to allow for partial backlogging and lost sales. Dye (2013) studied the effect of technology investment (e.g., refrigeration) on deteriorating items. Wu et al. (2014b) derived optimal replenishment policies for non-instantaneous deteriorating items (i.e., the items start deteriorating after a period of no-deterioration due to refrigeration). Wu et al. (2014a) then utilized the fact that the deterioration rate increases over time and reaches to 100% as time is near to its expiration date, and then derived the optimal credit period and cycle time. Trade credit is measured based on time value of money on the acquisition cost. Hence, for a sound and rigorous analysis, Chen and Teng (2015) adopted a discounted cash-flow analysis to calculate the revenue and all relevant costs, and derive the optimal inventory and credit policies for time-varying deteriorating items.

In this paper, we develop the inventory model by incorporating the fact that degrading rate is increasing with time and reaching 100% at the expiration date. In addition, we construct the model in a general framework, which permits or prohibits shortages. Finally, we propose a brand-new ACC payment, which contains all six commonly-used and previously-proposed payment schemes as special cases.

3 Notation and Assumptions

The following notation and assumptions are introduced to model the EOQ model for perishable products with an advance-cash-credit payment scheme.

3.1 Notation

The following parameters and variables are used in developing the problem.

α The fraction of the acquisition cost to be prepaid before the time of delivery $0 \leq \alpha \leq 1$.

β The fraction of the acquisition cost to be paid at the time of delivery, $0 \leq \beta \leq 1$.

χ The fraction of the acquisition cost to be granted a permissible delay in payment after delivery, $0 \leq \chi \leq 1$ and $\alpha + \beta + \chi = 1$.

δ The downstream credit period by the retailer to its customers, $\delta \geq 0$.

λ The fraction of shortages to be backordered, $0 \leq \lambda \leq 1$.

A The acquisition cost per cycle time in dollars, A > 0.

b The backordering cost per unit per year, b > 0.

B The backordering cost per cycle time in dollars.

c The purchasing cost per unit in dollars, c > 0.

C_a The capital cost for the advance payment per cycle time in dollars.

C_d The capital cost for the delay or credit payment per cycle time in dollars.

D The market annual demand rate in units.

h	The holding cost excluding interest charge per unit per year in dollars.
H	The holding cost excluding interest charge per cycle time in dollars after receiving order quantity.
I_c	The interest charged by the supplier per dollar per year.
I_e	The interest earned per dollar per year.
l	The length of time in years during which the prepayments are paid, l > 0.
m	The expiration date or the maximum lifetime in years, m > 0.
n	The number of equal prepayments before receiving the order quantity.
O	The ordering cost in dollars per order, O > 0.
p	The price per unit in dollars, p > c > 0.
R	The revenue received from sales.
s	The cost of lost sales in dollars per unit, s > c.
S	The cost of lost sales per cycle time in dollars.
t	The time in years, $t \geq 0$.
$\theta(t)$	The degrading rate at time t, $0 \leq \theta(t) \leq 1$.
E	The partially backordered quantity.
$I(t)$	The inventory level in units at time t.
K	The fraction of no shortages, $0 \leq K \leq 1$ (a decision variable).
Q	The order quantity in units.
T	The length of cycle time in years, $KT \leq m$ (a decision variable).
TP	The total annual profit in dollars.

Note that KT is K times T, and stands for the length of inventory period (a.k.a., the ending selling time). Likewise, (1- K)T represents the length of shortage period. A good example for this problem is a distributor of a chemical material as shown in Taleizadeh (2014). For convenience, the asterisk symbol on a variable is denoted the optimal solution of the variable. For instance, T^* is the optimal solution of T. Next, the necessary assumptions to build up the mathematical model are given below.

3.2 Assumptions

Next, the following assumptions are made to establish the mathematical inventory model.

All perishable products continuously degrade over time, and cannot be sold when time is approaching to the expiration date m. For simplicity, we assume the same as in Wang et al. (2014) and Chen and Teng (2014) that the degrading (or decaying, deterioration) rate is

$$\theta(t) = \frac{1}{1+m-t}, 0 \leq t \leq KT \leq m. \tag{1}$$

Note that it is clear from Fig. 1 that the ending selling time KT is less than or equal to the product maximum lifetime m (i.e., $KT \leq m$).

There is no replacement, repair, financing, or salvage value of perished items during the replenishment cycle [0, T].

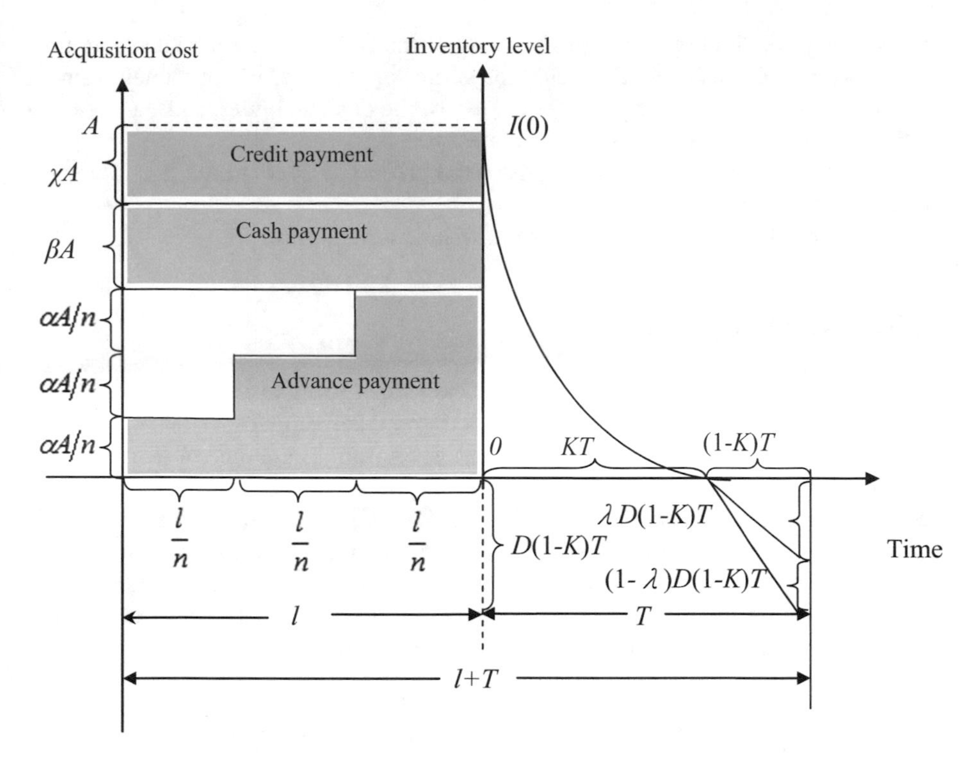

Fig. 1. Graphical representation for the inventory system with shortages

For highly seasonal products or perishable products, the supplier usually demands the retailer (i) to prepay α fractions of the acquisition cost A (i.e., αA) by n equal installments in l years prior to the time of delivery, (ii) to pay another β percentage of the acquisition cost (i.e., βA) at the time of delivery (i.e., t = 0), and (iii) to offer an upstream credit period of μ years on the remaining χ portions of the acquisition cost (i.e., χA, with $0 \leq \alpha, \beta, \chi \leq 1$, and $\alpha + \beta + \chi = 1$). The graphical representation of the system is depicted in Fig. 1. Notice that if the supplier does not request an advance payment, then $\alpha = 0$. Similarly, if the supplier does not ask for an instant cash payment upon the receipt of the order quantity, then $\beta = 0$. Finally, if the supplier does not offer a credit (or delay) payment, then $\chi = 0$. Consequently, the proposed model has 7 possible combination models.

The retailer receives an upstream credit period of μ years on the χ portions of the acquisition cost (i.e., χA) from the supplier (i.e., the retailer receives all items at time 0, and must pay the supplier χA at time μ without interest charges), and in turn provides a downstream credit period of δ years to his/her customers (i.e., the customer orders and receives items at time t, and must pay the credit payment at time $t + \delta$).

If $\mu \geq \delta$, then the retailer deposits the sales revenue into an interest bearing account during the time interval $[\delta, \mu]$. The retailer sells the last unit at time KT as shown in Fig. 1. If $\mu \geq KT + \delta$ (i.e., the upstream credit period is longer than the time at which the retailer receives the last payment from the customer), then the retailer receives all revenue and pays off the entire acquisition cost at the end of the permissible delay time μ. Otherwise (if $\mu \leq KT + \delta$), the retailer pays the supplier the sum of money from all units sold by $\mu - \delta$, keeps the profit for the use of the other activities, and starts paying for the interest charges on the items sold after $\mu - \delta$.

If $\mu \leq \delta$, the retailer must finance χA for the credit payment at time μ, and pay off the loan at time $KT + \delta$.

The replenishment rate is infinite, and the lead time is zero.

4 Mathematical Model

Given the above notation and assumptions, the proposed EOQ model for perishable products is described as follows. The retailer pays the supplier α fractions of the acquisition cost A by n equal installments in l years prior to the time of delivery. The order quantity (i.e., Q units) arrives at time 0. The quantity received is used partly to meet the accumulative backorders (i.e., $E = \lambda D(1 - K)T$ units) in the previous cycle. The remaining quantity is gradually depleted to zero at time KT due to the combination of demand and deterioration. Then shortages are partially backordered at the rate of λ during the time interval [KT, T]. For illustration, please see Fig. 1. Consequently, the inventory level at time t is governed by the following differential equation:

$$\frac{dI(t)}{dt} = -D - \theta(t)I(t) = -D - \frac{1}{1 + m - t}I(t), 0 \leq t \leq KT, \tag{2}$$

with boundary condition $I(KT) = 0$. The solution of the above differential equation is:

$$I(t) = D(1 + m - t)\int_t^{KT}\left(\frac{dt}{1 + m - t}\right) = D(1 + m - t)\ln\left(\frac{1 + m - t}{1 + m - KT}\right), 0 \leq t \leq KT. \tag{3}$$

It is clear that the partially backordered quantity is $E = \lambda D(1 - K)T$, and the lost sales quantity is $(1 - \lambda)D(1 - K)T$. Hence, the sales volume is $DKT + \lambda D(1 - K)T$, and thus the revenue received from sales is

$$R = p[DKT + \lambda D(1 - K)T] = pDT[K + \lambda(1 - K)]. \tag{4}$$

The order quantity per replenishment cycle time in this case is:

$$Q = I(0) + E = D(1 + m)\ln\left(\frac{1 + m}{1 + m - KT}\right) + \lambda D(1 - K)T. \tag{5}$$

Then the acquisition cost per replenishment cycle is:

$$A = cQ = cD\left[(1 + m)\ln\left(\frac{1 + m}{1 + m - KT}\right) + \lambda(1 - K)T\right]. \tag{6}$$

The holding cost per cycle from time 0 to time KT is given by:

$$H = h\int_{0}^{KT} I(t)dt = hD\left[\frac{(1+m)^2}{2}\ln\left(\frac{1+m}{1+m-KT}\right) + \frac{(KT)^2}{4} - \frac{(1+m)KT}{2}\right]. \quad (7)$$

It is clear from Fig. 1 that the partially backordered cost per cycle is

$$B = \frac{1}{2}b\lambda D(1-K)^2T^2, \quad (8)$$

and the cost of lost sales per cycle is

$$S = s(1-\lambda)D(1-K)T. \quad (9)$$

We know from Fig. 1 that the capital cost per cycle before the delivery for the advance payment αA is:

$$\begin{aligned} C_a &= I_c\left[\frac{\alpha A}{n}\left(\frac{l}{n}\right)(1+2+\ldots+n)\right] \\ &= \alpha lcDI_c\left(\frac{n+1}{2n}\right)\left[(1+m)\ln\left(\frac{1+m}{1+m-KT}\right) + \lambda(1-K)T\right]. \end{aligned} \quad (10)$$

Next, we derive the capital cost per cycle after the delivery, which can be divided by two parts: the advance and cash payment (i.e., $(\alpha+\beta)A$), and the credit payment (i.e., χA). The retailer offers the customers a downstream credit period of δ, and hence the interest charged on the inventory level on-hand for the advance and cash payment (i.e., $(\alpha+\beta)cDKT$) during $[0, KT+\delta]$ is shown in Fig. 2 as

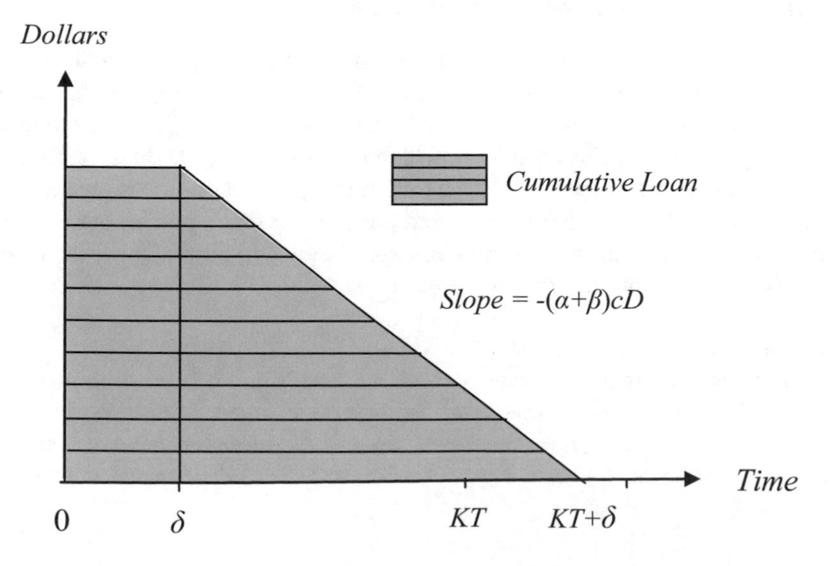

Fig. 2. The interest charged for the advance and cash payment

$$(\alpha+\beta)\delta\, cI_cDKT + \frac{(\alpha+\beta)cI_cD}{2}(KT)^2 = (\alpha+\beta)cI_cDKT\left(\delta+\frac{KT}{2}\right). \tag{11}$$

As to the capital cost per cycle for the credit payment χA, from the values of the upstream and the downstream credit periods μ and δ, we have two potential cases: (1) $\delta \le \mu$, and (2) $\delta \ge \mu$. Let us discuss them separately.

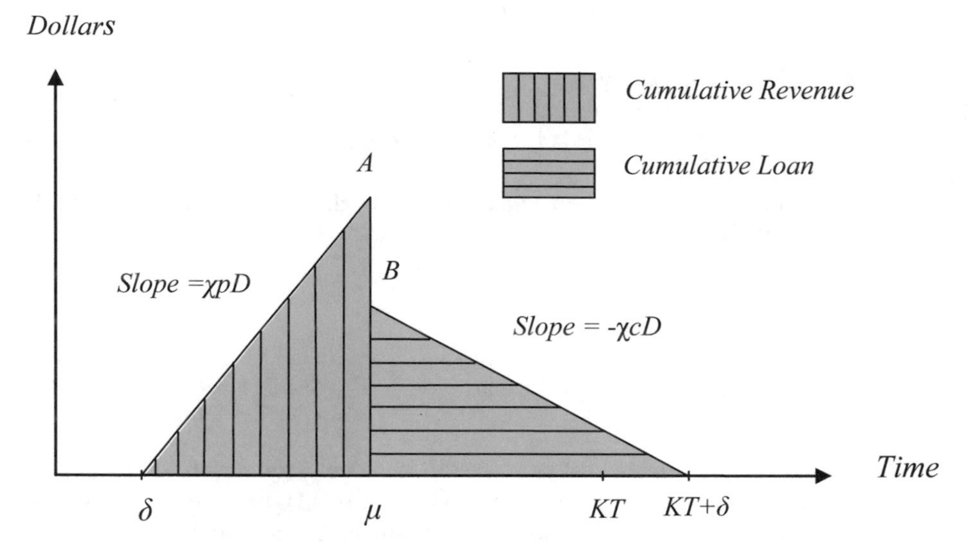

Fig. 3. The interest earned and charged for the case of $\delta \le \mu$ and $\mu \le KT+\delta$

4.1 Downstream Credit $\delta \le$ Upstream Credit μ

The retailer sells the last unit at time KT, and receives the money at time $KT+\delta$. Hence, we have two possible sub-cases. If $\mu \le KT+\delta$ (i.e., the time at which the retailer receives the last-sales payment), then the retailer pays $cD(\mu-\delta)$ at time μ, and starts paying the interest charges after μ, which is shown in Fig. 3. On the other hand, if $\mu \ge KT+\delta$, then the retailer receives the total revenue $pDKT$(which is greater than the purchasing cost $cDKT$) at time $KT+\delta$, and pays off the entire purchasing cost at time μ. The graphical representation of this sub-case is shown in Fig. 4. Now, let us discuss the detailed formulation in each sub-case.

4.1.1 Sub-case 1-1: $\mu \le KT+\delta$

In this sub-case, the retailer cannot pay off the acquisition cost at time μ, and must finance on-hand items after time μ at an interest charged I_c per dollar per year. Therefore, the interest charged per cycle time is I_c times the area of the triangle $\mu B(KT+\delta)$ as shown in Fig. 3. Hence, the interest charged per cycle is given by

$$\frac{c\chi I_c D}{2}(KT+\delta-\mu)^2. \tag{12}$$

On the other hand, the retailer sells products from time 0, but receives the money from time δ. Consequently, the retailer accumulates revenue in an account that earns Ie per dollar per year from δ through μ. Therefore, the interest earned per cycle is Ie multiplied by the area of the triangle $\delta A\mu$ as shown in Fig. 3. Hence, the interest earned per cycle is

$$\frac{p\chi I_e D(\mu-\delta)^2}{2}. \tag{13}$$

From (11), (12) (13), the capital cost after the delivery per replenishment cycle is:

$$C_d = (\alpha+\beta)cI_c DKT\left(\delta+\frac{KT}{2}\right)+\frac{\chi}{2}cI_c D(KT+\delta-\mu)^2-\frac{\chi}{2}pI_e D(\mu-\delta)^2. \tag{14}$$

Combining the above results, and simplifying terms, we know that the retailer's total profit per cycle for this case is as follows:

$$\begin{aligned} & R-O-A-H-B-S-C_a-C_d \\ & = pDT[K+\lambda(1-K)]-O-cD\left[(1+m)\ln\left(\frac{1+m}{1+m-KT}\right)+\lambda(1-K)T\right]\left(1+\alpha lI_c\frac{n+1}{2n}\right) \\ & -hD\left[\frac{(1+m)^2}{2}\ln\left(\frac{1+m}{1+m-KT}\right)+\frac{(KT)^2}{4}-\frac{(1+m)KT}{2}\right] \\ & -\frac{1}{2}b\lambda D(1-K)^2T^2-s(1-\lambda)D(1-K)T-(\alpha+\beta)cI_c DKT\left(\delta+\frac{KT}{2}\right) \\ & -\frac{\chi}{2}cI_c D(KT+\delta-\mu)^2+\frac{\chi}{2}pI_e D(\mu-\delta)^2. \end{aligned} \tag{15}$$

Consequently, if $\mu-\delta \le KT$, then the total annual profit is a function of the replenishment cycle time T and the fraction of no shortages K as follow:

$$\begin{aligned} & TP_1(K,T)=\frac{1}{T}\{pDT[K+\lambda(1-K)]-O \\ & -D(1+m)\ln\left(\frac{1+m}{1+m-KT}\right)\left[c\left(1+\alpha lI_c\frac{n+1}{2n}\right)+\frac{h}{2}(1+m)\right] \\ & -(1-K)DT\left[c\lambda\left(1+\alpha lI_c\frac{n+1}{2n}\right)+\frac{b\lambda}{2}(1-K)T+s(1-\lambda)\right]-hDT\left[\frac{K^2T}{4}-\frac{(1+m)K}{2}\right] \\ & -(\alpha+\beta)cI_c DKT\left(\delta+\frac{KT}{2}\right).-\frac{\chi}{2}cI_c D(KT+\delta-\mu)^2+\frac{\chi}{2}pI_e D(\mu-\delta)^2\Big\}. \end{aligned} \tag{16}$$

Next, we derive the retailer's total annual profit for the other sub-case in which $\mu \geq KT + \delta$.

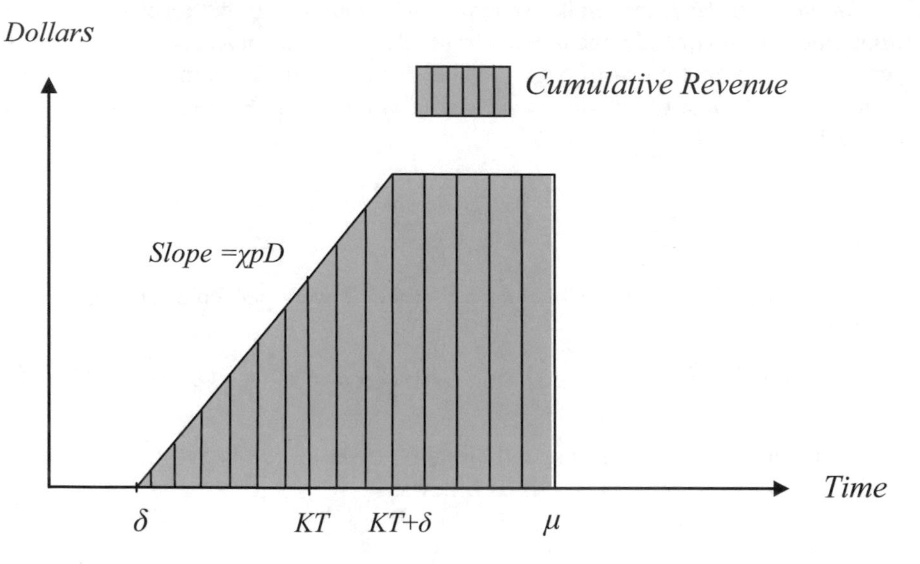

Fig. 4. The interest earned for the case of $\delta \leq \mu$ and $KT + \delta \leq \mu$

4.1.2 Sub-case 1-2: $\mu \geq KT + \delta$

In this sub-case, the retailer receives the total revenue at time $KT + \delta$, and is able to pay off the total purchasing cost at time μ. Hence, there is no interest charge while the interest earned per cycle is Ie multiplied by the area of the trapezoid on the interval $[\delta, \mu]$ as shown in Fig. 4. Hence, the interest earned per cycle is given by

$$\frac{\chi pI_e D(KT)^2}{2} + \chi pI_e DKT(\mu - KT - \delta) = \chi pI_e DKT\left(\mu - \delta - \frac{KT}{2}\right).$$

Consequently, the retailer's capital cost per cycle time after the delivery is

$$C_d = (\alpha + \beta)cI_c DKT\left(\delta + \frac{KT}{2}\right) - \chi pI_e DKT\left(\mu - \delta - \frac{KT}{2}\right). \tag{17}$$

Similar to (16), if $KT \leq \mu - \delta$, then we know that the retailer's total annual profit is

$$\begin{aligned} TP_2(K,T) = \frac{1}{T}\{ & pDT[K+\lambda(1-K)] - O \\ & - D(1+m)\ln\left(\frac{1+m}{1+m-KT}\right)\left[c\left(1+\alpha lI_c\frac{n+1}{2n}\right)+\frac{h}{2}(1+m)\right] \\ & - (1-K)DT\left[c\lambda\left(1+\alpha lI_c\frac{n+1}{2n}\right)+\frac{b\lambda}{2}(1-K)T+s(1-\lambda)\right] - hDT\left[\frac{K^2T}{4}-\frac{(1+m)K}{2}\right] \\ & -(\alpha+\beta)cI_cDKT\left(\delta+\frac{KT}{2}\right)+\chi pI_eDKT\left(\mu-\delta-\frac{KT}{2}\right)\Big\}. \end{aligned} \tag{18}$$

It is clear from (16) and (18) that

$$TP_1\left(K,\frac{\mu-\delta}{K}\right) = TP_2\left(K,\frac{\mu-\delta}{K}\right). \tag{19}$$

Then we formulate the retailer's total annual profit for the case of $\delta \geq \mu$ below.

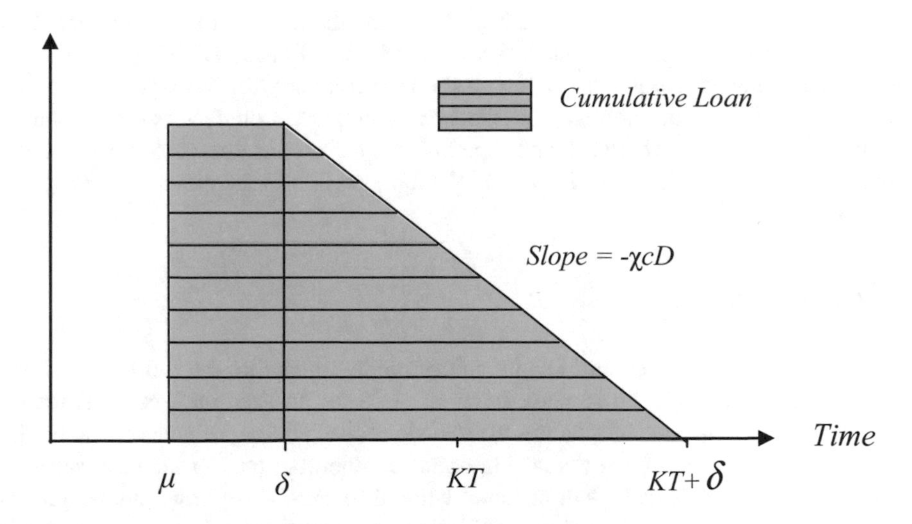

Fig. 5. The interest charged for the case of $\delta \leq \mu$

4.2 Downstream Credit $\delta \geq$ Upstream Credit μ

There is no interest earned for the retailer in this case. In addition, the retailer must finance the portions of the acquisition cost χA at time μ, and pay off the loan from time δ to time KT+δ. As a result, the interest charged per cycle is I_c multiplied by the area of the trapezoid on the interval $[\mu, \text{KT}+\delta]$, as shown in Fig. 5. Thus, the retailer's interest charged per cycler after the delivery for the credit payment is

$$\frac{\chi}{2} c I_c D K T [2(\delta - \mu) + KT].$$

Therefore, the retailer's capital cost per cycle time after the delivery in this case is

$$C_d = (\alpha + \beta) c I_c D\, KT \left(\delta + \frac{KT}{2} \right) + \frac{\chi}{2} c I_c D K T [2(\delta - \mu) + KT]. \tag{20}$$

Hence, the retailer's total annual profit is

$$\begin{aligned} TP_3(K,T) = & \frac{1}{T} \{ pDT[K + \lambda(1-K)] - O \\ & - D(1+m) \ln\left(\frac{1+m}{1+m-KT} \right) \left[c\left(1 + \alpha I I_c \frac{n+1}{2n} \right) + \frac{h}{2}(1+m) \right] \\ & - (1-K)DT \left[c\lambda \left(1 + \alpha I I_c \frac{n+1}{2n} \right) + \frac{b\lambda}{2}(1-K)T + s(1-\lambda) \right] - hDT \left[\frac{K^2 T}{4} - \frac{(1+m)K}{2} \right] \\ & - (\alpha + \beta) c I_c D K T \left(\delta + \frac{KT}{2} \right) - \frac{\chi}{2} c I_c D K T [2(\delta - \mu) + KT] \Big\}. \end{aligned} \tag{21}$$

Therefore, the retailer's objective is to determine the fraction of no shortages K and the replenishment cycle time T such that the total annual profit $TP_i(K,T)$ for i = 1, 2, and 3 is maximized. Notice that if K = 1, then the case of with shortages is simplified to the case of without shortages. In the next section, we characterize the retailer's optimal fraction of no shortages and replenishment cycle time in each case, and then obtain the certain conditions in which the optimal T^* is in either $\mu \le KT + \delta$ or $\mu \ge KT + \delta$.

5 Conclusions

This paper have developed an EOQ inventory model to capture the following relevant and important facts: (1) An increase in shelf space for an item induces consumers to buy more, (2) Consumers' purchasing decisions are vitally influenced by product freshness, (3) Pricing strategy is an important competitive tool to increase sales and profits, and (4) The demand is diminished when time is reaching toward its expiration date. For generality, it is relaxed the traditional assumption of zero ending inventories to non-zero ending inventory in order to boost the total profit. Then, this paper have demonstrated that the total profit is strictly pseudo-concave in all three decision variables (i.e., unit price, replenishment time, and ending-stock level), which reduces the search for the global solution to a local optimal. Finally, some numerical examples to illustrate the proposed model and its managerial approaches are provided.

An applied mathematical model is always a simplification of the complicated real-life problem. To strengthen the applicability, this paper can be extended in several forms. For instance, it may be added advertising strategy, quantity discount, and time-varying demand into the model. Furthermore, there is usually a significant cost to account for the additional effort of refilling the shelf space from the backroom storage. It should take the extra cost into consideration in the future study. Finally, it could be expanded the single player local optimal solution to an integrated cooperative solution for multiple players in a supply chain.

Acknowledgement. This is part of an Accepted Manuscript of an article published by Taylor & Francis in International Journal of Systems Science : Operations & Logistics on May, 2017, available online: https://doi.org/10.1080/23302674.2017.1308038.

References

Aggarwal, S.P., Jaggi, C.K.: Ordering policies of deteriorating items under permissible delay in payments. J. Oper. Res. Soc. **46**(5), 658–662 (1995)

Bakker, M., Riezebos, J., Teunter, R.H.: Review of inventory systems with deterioration since 2001. Eur. J. Oper. Res. **221**(2), 275–284 (2012)

Cambini, A., Martein, L.: Generalized convexity and optimization: Theory and Application. Lecture Notes in Economics and Mathematical Systems, vol. 616. Springer-Verlag, Heidelberg (2009). https://doi.org/10.1007/978-3-540-70876-6

Cárdenas-Barrón, L.E., Chung, K.J., Treviño-Garza, G.: Celebrating a century of the economic order quantity model in honor of Ford Whitman Harris. Int. J. Prod. Econ. **155**, 1–7 (2014)

Chen, S.-C., Cárdenas-Barrón, L.E., Teng, J.T.: Retailer's economic order quantity when the supplier offers conditionally permissible delay in payments link to order quantity. Int. J. Prod. Econ. **155**, 284–291 (2014)

Chen, S.C., Teng, J.T.: Retailer's optimal ordering policy for deteriorating items with maximum lifetime under supplier's trade credit financing. Appl. Math. Model. **38**(15–16), 4049–4061 (2014)

Chen, S.C., Teng, J.T.: Inventory and credit decisions for time-varying deteriorating items with up-stream and down-stream trade credit financing by discounted cash-flow analysis. Eur. J. Oper. Res. **243**(2), 566–575 (2015)

Covert, R.B., Philip, G.S.: An EOQ model for items with Weibull distribution deterioration. AIIE Trans. **5**(4), 323–326 (1973)

Dave, U., Patel, L.K.: (T, Si) policy inventory model for deteriorating items with time proportional demand. J. Oper. Res. Soc. **32**(2), 137–142 (1981)

Dye, C.Y.: The effect of preservation technology investment on a non-instantaneous deteriorating inventory model. Omega **41**(5), 872–880 (2013)

Dye, C.Y., Yang, C.T.: Sustainable trade credit and replenishment decisions with credit-linked demand under carbon emission constraints. Eur. J. Oper. Res. **244**(1), 187–200 (2015)

Ghare, P.M., Schrader, G.P.: A model for an exponentially decaying inventory. J. Ind. Eng. **14**(5), 238–243 (1963)

Goyal, S.K.: Economic order quantity under conditions of permissible delay in payments. J. Oper. Res. Soc. **36**(4), 335–338 (1985)

Goyal, S.K., Giri, B.C.: Recent trends in modeling of deteriorating inventory. Eur. J. Oper. Res. **134**(1), 1–16 (2001)

Grubbstrom, R.W.: A principle for determining the correct capital costs of work-in-progress and inventory. Int. J. Prod. Res. **18**(2), 259–271 (1980)

Harris, F.W.: How many parts to make at once. Factory Magaz. Manage. **10**(2), 135–136, 152 (1913)

Huang, Y.F.: Optimal retailer's ordering policies in the EOQ model under trade credit financing. J. Oper. Res. Soc. **54**(9), 1011–1015 (2003)

Jamal, A.M., Sarker, B.R., Wang, S.: An ordering policy for deteriorating items with allowable shortage and permissible delay in payment. J. Oper. Res. Soc. **48**(8), 826–833 (1997)

Ouyang, L.Y., Chang, C.T.: Optimal production lot with imperfect production process under permissible delay in payments and complete backlogging. Int. J. Prod. Econ. **144**(2), 610–617 (2013)

Ouyang, L.-Y., Yang, C.-T., Chan, Y.L., Cárdenas-Barrón, L.E.: A comprehensive extension of the optimal replenishment decisions under two levels of trade credit policy depending on the order quantity. Appl. Math. Comput. **224**, 268–277 (2013)

Pahl, J., Voß, S.: Integrating deterioration and lifetime constraints in production and supply chain planning: a survey. Eur. J. Oper. Res. **238**(3), 654–674 (2014)

Raafat, F.: Survey of literature on continuously deteriorating inventory model. J. Oper. Res. Soc. **42**(1), 27–37 (1991)

Sarkar, B., Saren, S., Cárdenas-Barrón, L.E.: An inventory model with trade-credit policy and variable deterioration for fixed lifetime products. Ann. Oper. Res. **229**(1), 677–702 (2015)

Seifert, D., Seifert, R.W., Protopappa-Sieke, M.: A review of trade credit literature: opportunity for research in operations. Eur. J. Oper. Res. **231**(2), 245–256 (2013)

Shah, N.H., Cárdenas-Barrón, L.E.: Retailer's decision for ordering and credit policies for deteriorating items when a supplier offers order-linked credit period or cash discount. Appl. Math. Comput. **259**, 569–578 (2015)

Taleizadeh, A.A.: An EOQ model with partial backordering and advance payments for an evaporating item. Int. J. Prod. Econ. **155**, 185–193 (2014)

Teng, J.T.: On the economic order quantity under conditions of permissible delay in payments. J. Oper. Res. Soc. **53**(8), 915–918 (2002)

Teng, J.T.: Optimal ordering policies for a retailer who offers distinct trade credits to its good and bad credit customers. Int. J. Prod. Econ. **119**(2), 415–423 (2009)

Teng, J.T., Chang, H.J., Dye, C.Y., Hung, C.H.: An optimal replenishment policy for deteriorating items with time-varying demand and partial backlogging. Oper. Res. Lett. **30**(6), 387–393 (2002)

Teng, J.T., Chern, M.S., Yang, H.L., Wang, Y.J.: Deterministic lot-size inventory models with shortages and deterioration for fluctuating demand. Oper. Res. Lett. **24**(1–2), 65–72 (1999)

Teng, J.T., Min, J., Pan, Q.: Economic order quantity model with trade credit financing for non-decreasing demand. Omega **40**(3), 328–335 (2012)

Wang, W.C., Teng, J.T., Lou, K.R.: Seller's optimal credit period and cycle time in a supply chain for deteriorating items with maximum lifetime. Eur. J. Oper. Res. **232**(2), 315–321 (2014)

Wu, J., Ouyang, L.Y., Cárdenas-Barrón, L.E., Goyal, S.K.: Optimal credit period and lot size for deteriorating items with expiration dates under two-level trade credit financing. Eur. J. Oper. Res. **237**(3), 898–908 (2014a)

Wu, J., Skouri, K., Teng, J.T., Ouyang, L.Y.: A note on "optimal replenishment policies for non-instantaneous deteriorating items with price and stock sensitive demand under permissible delay in payment". Int. J. Prod. Econ. **155**, 324–329 (2014b)

Zhang, A.X.: Optimal advance payment scheme involving fixed pre-payment costs. Omega **24**(5), 577–582 (1996)

Zhang, Q., Tsao, Y.C., Chen, T.H.: Economic order quantity under advance payment. Appl. Math. Model. **38**(24), 5910–5921 (2014)

The Research on Lifestyle, Physical and Mental Health, and Potential Consumption for Elderly

Sue-Ming Hsu[1] and Ya-Lan Chan[2(✉)]

[1] Department of Business Administration, Tunghai University, Taichung, Taiwan
sueming@thu.edu.tw
[2] Department of Business Administration, Asia University, Taichung, Taiwan
yalan@asia.edu.tw

Abstract. Sustainable development contains three aspects of development including social, economic, and environmental objectives. For the sustainable social issues, the aging society has considered as the major issue and challenging problem facing by the countries of the world. In addition to the diseases and functional decline, the interpersonal interaction, social participation, and mental health would affect the lifestyle of elderly. Given the increasing numbers of aging population, the mature market is developed globally, which changes the structure of elderly consumption and commercial opportunity. With the help of Taichung City Government, a total of 600 effective samples of the survey of social and living status of the 65–75 years old elderly in Taichung City from the project of Global Research & Education on Environment and Society (GREEnS) of Tunghai University were used to study the lifestyle of elderly and their values in life. The current study used cluster analysis to determine the homogenous characteristics of elderly. Furthermore, influencing factors of daily life conditions, diet, and health status of the elderly were investigated to discuss the differences of consumption patterns among elderly. Measurement of the AIO (activities, interests and opinions) was used as a basis to access the status of lifestyle of elderly in Taiwan, and further to demonstrate the economic activities and market segmentation within the different characteristics of elderly.

Keywords: Mature market · Lifestyle of elderly · Cluster analysis AIO measurement

1 Introduction

In the "MANAGING IN THE NEXT SOCIETY: BEYOND THE INFORMATION REVOLUTION", Peter Drucker mentioned big three trends,which would change future for next 50 years. In Another book "INNOVATION AND ENTREPRENEURSHIP" he indicated that there was seven source of innovation. Among of these seven source, the changing of demographic structure had highest opportunity to culture an enormous market. Now, in Taiwan, we are facing the double threaten, high aging of population and low birth rate. They will lead to many social problems. Government need to face the challenge these problems actively, and also need to insight the opportunity of the processing of social changing.

L. Barolli et al. (Eds.): IMIS 2018, AISC 773, pp. 903–909, 2019.
https://doi.org/10.1007/978-3-319-93554-6_91

These days the increasing of aging population not only affect society nut also influence industry. Industry need to face the changing, and find the niche market. Find a way to build a successful model, in other hand, find a way to undertake this social responsibility issue. This paper will focus on the behavior of aging population, and try to categorize different type of behavior.

2 Literature Review

2.1 Senior-Oriented Industry Challenges and Opportunities

Currently, there are two major trends in the demographic change in Taiwan: the aging and the declining birth rate, which will make Taiwan face considerable impact in the future, such as labor supply, market consumption, government finance, social support and health care, etc. However, They believe that the structural changes caused by these impacts will also become the opportunities for future industrial innovation.

When it comes to the impact of population aging on the national society, past researches have mostly focused on the restructuring of existing social care and security mechanisms. Including how to implement the management of medical and nursing institutions, how to integrate limited resources, and how to maintain the most basic living needs of the elderly through retirement systems, annuities and social insurance schemes. Most of the above-mentioned views focus on the challenges posed by aging.

This paper hopes to explore the innovative business opportunities of the Senior-oriented Industry to under the change of population structure based on the views of the integrate market mechanism and relevant policies.

The current government needs to guide more manufacturers willing to invest in the Senior-oriented Industry and to develop a more profitable and forward-looking operation mode to help solve the problems in the current aging society. Based on the foregoing discussion, this article defines the Senior-oriented Industry for the Senior-oriented derived health care, trust management, leisure and entertainment, mobile technology, accessibility and other related industries. Next, we analyze the possible challenges posed by the demographic transition by comparing the four aspects of labor market, market consumption, government finance, social support and health care respectively, and we will focus on the health care and market consumption.

2.2 Social Support and Health Care

In the aging society, how to construct a complete social support and health care system? How enough people and businesses are willing to invest? How to maintain the most basic health, economic security and social security for the elderly with limited resources? It will be a major challenge for the future government to cope with the aging society. In our long-term care promotion program, the Welfare Law for the Elderly and its implementation were formally approved in 1980, and the Executive Yuan approved the “Ten-Year Long-Term Care Plans in China” (Executive Yuan 2007) in April 2007 and in May 2015 Third Reading of the Legislative Yuan, with a view to building a complete long-term care service system, strengthening the hardware and software infrastructure required for long-term care services, ensuring adequate services for the elderly and persons with mental and physical disabilities, enhancing their independent living ability and quality of life In order to maintain dignity and autonomy. In addition, the government is planning long-term care insurance law, another important act that is crucial for long-term care. However, due to the scope of services, the level of benefits and the financial resources of the government are concerned, and the scope involved is very wide. At present, the parties can not reach a consensus.

According to estimates by Wu (1998), the number of elderly people aged 65 and above in Taiwan is about 170,000 with long-term care needs due to obstacles in their daily activities (bathing, eating, etc.). In addition, due to the instrumental daily routine (laundry, Shopping, going out, preparing food, etc.) and the number of people requiring long-term care to as high as 240,000. If the number of people with cognitive impairment increases, the number of people in need of care will be as high as 270,000 (Shuqiong 2002). However, on the supply side, according to the survey conducted by the Ministry of the Interior in 2011, the capacity of long-term care institutions is only about 46,000 and the actual number of people staying is about 35,000. The reason for the relatively low number of long-term care providers is that they are related to the traditional family and social values on one hand, these care providers on the other. At present, most elderly care institutions rely more on government subsidies and most of them can not afford to be self-sufficient.The resources of government subsidies are rather limited, and there is lack of lucrative operating modes and the low willingness of the private sector to invest, so the supply of elderly Long-term care capacity is far from adequate. In addition, some institutions, due to their poorly allocated geography, poor standard of care, poor service procedure or other problems, only led to a utilization rate of about 75% for these institutions, descripting the inadequate and uneven investment in care resources for the elderly.

With reference to the information published publicly by the domestic and foreign literature and the Ministry of the Interior in the past, Lin and Li (2015) study calculates the operating efficiency of these institutions through Data Envelopment Analysis (DEA). The input variables include direct service staff, indirect Service staff and Jian Ping as input variables, the actual number of recipients as output variables. Through examining the relationship between operational efficiency and quality of care, the research team mainly found the following findings: (1) There is a significant difference between the input-output structure and operational efficiency of the integrated and conservation types. (2) Apart from Taipei City Socialist Department elderly self-care centers, the remaining 12 are private-owned ones, which shows that the overall incentive mechanism for private-care institutions is superior to that of the public sector, making them There is a bigger incentive to improve its operational efficiency. (3) On average, there is still room for improvement in the allocation of resources in Taiwan's elderly welfare institutions by 36% to 45%. (4) there is a significant relationship between the quality of care and operational efficiency, and the two are not easy to compromise. However, we also found that this intractable phenomenon mainly occurs in comprehensive and appraisal-first-class care institutions and is not significant in other types of care institutions. Therefore, the results of this study only partially support the inefficiency of quality competition Hypothesis (Cherong, Kaiyi, Shuming, Huiwen, 2014).

In addition to institutional care, community and home care may be more generally accepted by the silver-haired and also in line with the concept of "aging on the ground." However, the strength that social support systems and related supporting mechanisms have not yet been able to exert is also affected limit. For example, the current shortage of local workforce in nursing care, lack of training, and foreign workers. Those are the challenges that can not be avoided in promoting social support policies in the future.

There is same condition with medical care. The current medical system will flood more elders who need medical resources and special-need care. This will inevitably have a considerable impact on the current health insurance system. For some medical institutions where the current labor of doctors and nurses are already quite tight, it will be worse when the time pass. To put it simply, in a future society called "young, old and poor", the government is bound to face a serious shortage of resources in promoting social support and health care policies. It must also have more systematic planning forces. Combining the power of the government, the private sector, the individual

In the aging society, except basic social safety and health, we must also take into consideration the different levels of mental health, interpersonal interaction and social participation. Coupled with the current structural core of family structure, the employment of women has increased and the family care function has become increasingly weaker and other factors will make the problems facing the aging society more complicated. If they are not properly handled, they will have an unmanageable impact on the entire national financial and social security system.

2.3 Market Consumption

Since low birth rate and prolong lifespan, the current aging group of aging consumers is significantly higher than the previous generation in terms of assets and consuming power because it coincides with the rapid economic take-off of Taiwan. Take Japan as an example. At present, the output value of the Senior-oriented family accounts for more than 20% of Japan's overall consumption, which is a very large market. However, with the advent of aging, the consumption behavior of most consumers will also become more and more conservative. It may also give rise to something different from the past in terms of "what to buy?," "where are you going to buy?" Whether elderly people have lifestyle preferences and consumer preferences different from those of ordinary consumers needs to be further explored. To allow such relatively "more affluent" elderly people to spend their pensions out is a considerable challenge. The future depends on a more in-depth study on the life style and consumption patterns of elderly people in order to further grasp the business opportunities of the increasingly important Senior-oriented industry.

In Taiwan, the forthcoming age group to join the ranks of the silver-hawk community is the so-called "post-war baby boomers", middle-aged ethnic groups aged 45 to 64. The ethnic groups of this generation have been experiencing a period of rapid economic growth in Taiwan. In the past, they focused on their career development and accumulated considerable wealth. The potential business opportunities for the future include at least the following: (1) Senior-oriented catering service: including catering provided by ordinary restaurants in response to the special dietary preferences and nutritional needs of elder, Lock the elderly for the positioning of the restaurant, and even provide food delivery services or health care consulting catering services. (2) Elderly-friendly buildings: most of the existing apartments or homes are not fully in line with the needs of the elderly. Elderly friendly residential not only to plan the home and the surrounding environment, but also need to ensure that the various nursing, medical, home services can be provided nearby. (3) A combination of information and mobile technologies in aging products/services, including products such as home care, medical care, nutrition and health care, leisure travel, etc. In the future, if a user-oriented design can be achieved and elderly people are willing to purchase, they will become Fast growing market. (4) Various types of elderly medical care institutions, home care services for the elderly and transport products and services for the elderly.

3 Research Methods

It is imperative for the elder industry to fully understand this consumer group of elderly people. For this reason, the life style of the elderly has become an important research topic. This study adopts the research framework, which is commonly used in the field of marketing management and designs different problems based on the AIO scale, the activities of the actors, the interests and the Opinions, Out of the commonality of different actors, and can analyze the different life style of actors, whether there are different propensity to consume, health status and life satisfaction. This study aims to develop appropriate scales for elderly people in terms of their possible activities, interests, and opinions, and conduct a study on the life style of the elderly in China. Through the questionnaire survey results collected from the “Survey of Social Living Conditions of the 65–75-year-old Population in Taichung City” (GREEnS total 4, 2014) collected by the GREEnS Project, the lifestyle of elderly people in Taichung City is analyzed.

From the answers of elderly people in Taichung to the various issues they participated in, activities such as “leisure activities” and “work inputs” can be further identified through factor analysis, and among the issues of interest, they can be identified There are two types of “home-oriented” and “leisure-oriented” variables that include three different opinions on “education in culture”, “social policy” and “economic consumption”

4 Conclusion

Here are three categories: “cultural education”, “social policy” and “economic consumption” could be identified from the questions about personal opinions. Acceptance of different opinions.

(1) Group of Negative silence
This group who account 41.7% of the total elderly population. Their participation in leisure activities and work inputs are not high, and their interest in family and

leisure is also the lowest among the three groups. At the same time, their values are conservative and traditional, and they agree the pro-consumption and investment concepts and less trust in the government's policy on senior citizens. Because of their low participation in various family, leisure and work activities, they are not accustomed to expressing their opinions voluntarily. We can call this a "Group of Negative Silence." In the past, consumers in the study, the more negative among the overall consumer, not stand, and less consumption of the population is usually about 3 percent, compared with the findings of this survey can be seen that the proportion of senior citizens belonging to the negative silent family, May be higher in other age groups.

(2) Group of Active defensive
This group of elderly people, accounting for about 27.2% of the total elderly, in terms of activity and interest, this group is relatively more investment in family and work, the highest score of the three clusters. Although they are getting older, but still pay a lot of attention to their family and work. Specifically, in terms of opinions on culture, education, social policies and economic consumption, the scores of this group are also significantly higher than those of other ethnic groups, which indicates that these groups of people have more innovative ideas and are brave in expressing opinions as compared with other ethnic groups Independent and active. Therefore, we can name it "Group of Active Defensive".

(3) Leisure Group
The elderly population of this cluster, accounting for about 31.1% of the total elderly population, their participation in leisure activities and interests are the highest among the three groups. The participation in family and work activities is slightly lower than that of "Group of Active defensive", which we can name as "Leisure Group". On the whole, the leisurely people who participated in the activities were both very interested and interested in the activities, but they were less active in expressing their opinions moderately than in the active ones. As they grow older, elders of this group gradually shift their focus from work and family to leisure activities.

This rearch further compared elderly people of different life styles with five aspects of life quality satisfaction, physical activity function, daily life function, religious participation and retirement preparation, in order to outline the elderly with different lifestyle. We are not yet able to judge whether the lifestyle affects these variables or whether the life style is affected by functions such as physical health. The future of this part remains to be further studied. However, the results recovered by this questionnaire, there are quite significant differences in the appearance of elderly people of different lifestyle. For example: (1) The "Leisure Group" are more prepared for various activities in order to engage in more leisure activities. Therefore, this group are more prepared. (2) "Group of Negative silence" have the highest level of religious participation, while "Group of Active Defensive" is slightly higher than "Leisure Group". (3) "Group of Negative silence" unexpectedly outperformed the other two groups in their scores on the functions of the body and the functions of daily life. (4) Compared with the other two ethnic groups, "Leisure Group" are more satisfied with their own quality of life.

We found that the different lifestyles of elderly people, there are quite significant differences in the quality of life satisfaction, physical activity, function, activities of daily living, religion investment, aged preparation. It shows that "life style" is a direction worthy of further research for advanced research, which is worth further study by future researchers.

References

行政院. 。我國長期照顧十年計畫-大溫暖社會福利套案之旗艦計畫. *台北:* 行政院 (2007)

吳淑瓊 。全國長期照護需要評估. *行政院衛生署九十年度委託研究計劃成果報告* (DOH90-MA-5L01) (2002)

吳淑瓊.。老人長期照護政策評估 (1998)

李穎彥 。台灣老人福利機構財務收支與照護品質之研究(碩士論文)。取自 (2015). http://handle.ncl.edu.tw/11296/ku68t5

林灼榮、黃開義、許書銘、陳惠雯台灣老人福利機構營運類型, 照護品質與生產效率: EBM-Metafrontier DEA 模型. In: *東海大學主辦*. (2014)

東海大學 。「全球環境暨永續社會發展」(Global Research & Education on Environment and Society,簡稱GREEnS) (2014)

Drucker, P.: Innovation and Entrepreneurship. Routledge (2014)

Drucker, P.H.: Managing in the Next Society: Beyond the Information Revolution. Martin Press, New York (2003)

The 4th International Workshop on Big Data and IoT Security (BDITS-2018)

A Cooperative Evaluation Approach Based on Blockchain Technology for IoT Application

Hsing-Chung Chen[1,2,3](✉), Bambang Irawan[1,2,3], and Zon-Yin Shae[1,2,3](✉)

[1] Department of Computer Science and Information Engineering, Asia University, Taichung, Taiwan
shin8409@ms6.hinet.net, {cdma2000, zshae1}@asia.edu.tw
[2] Department of Medical Research, China Medical University Hospital, China Medical University, Taichung, Taiwan
[3] Department of Computer Science, Esa Unggul University, West Jakarta, Indonesia
bambang.irawan@esaunggul.ac.id, parikesit.irawan@gmail.com

Abstract. The Blockchain is the world's leading software platform for digital assets. The development of Blockchain technology is now growing very rapidly. Blockchain technology could be also deployed in the Internet of Things (IoT) Networks during their transaction processes. However, safe methods for different types of transactions still have major problems. A good trust management system (TMS) is essential for success between IoT devices and Blockchain node during transaction processes. This paper illustrates how IoT devices could be evaluated by the sink nodes acted as a blockchain nodes in order to give the contribution for cooperative evaluation in the blockchain for the integration IoT application. The cooperative evaluation method is required while executing transaction process in Blockchain network, which could validate IoT devices by these collaboration blockchain agent nodes. Finally, the scheme we proposed cooperative evaluation for private blockchain IoT application, which could give trust evaluation for IoT devices by the blockchain nodes during the blockcahin transaction processes.

Keywords: IoT · Blockchain · Cooperative evaluation · Trust management

1 Introduction

Since the blockchain technology was introduced by Swan [1], blockchain technology is now growing very rapidly in various fields, among industry, social, health, agriculture and others [2–5]. The blockchain is a technology that became the precursor of distributed ledger technology. It was originally developed to support the cryptocurrency, such as Bitcoin, Ripple, Litecoin, and others [2–5]. Peer-to-peer transactions could occur in the absence of a third party to ensure validity and security by applying this technology. There are still many problems appeared in the integrated Internet of Things (IoT) application based on blockchain technology, *e.g.* one of them is how to secure

L. Barolli et al. (Eds.): IMIS 2018, AISC 773, pp. 913–921, 2019.
https://doi.org/10.1007/978-3-319-93554-6_92

each IoT device status [1]. Blockchain technology is now widely integrated with a set of IoT device, but there are still many weaknesses in implementing security in IoT devices, blockchain technology seeks to address this growing security problem in a better way [5]. In this paper, we describe a blockchain agent as a node that manages blockchain software and as an interface of a set of IoT sensor devices. The blockchain agent node will also monitor the condition of the IoT sensor device, whether the condition of the IoT sensor device is "normal" or "abnormal". The condition of the IoT sensor device in this scenario consists of three conditions and can be developed again according to the requirement as described below. Blockchain agent nodes will ensure that all IoT sensor devices will be in "normal" condition, by collecting information provided by IoT sensor devices and performing a calculation operation before transacting and placing in the Blockchain network. The next step is each blockchain agent node to evaluate each other based on trust value obtained from the previous calculation stage. In this paper, we propose a cooperative evaluation method in order to support cooperative evaluation function which co-works by a set of registered IoT devices can work together in a Blockchain network.

The remainder of this paper is organized as follows: in Sect. 2, we introduce the related works. In Sect. 3, we first formalize a cooperative evaluation approach based on Blockchain technology for IoT application. Then, we present discussions in Sect. 4. Finally, we draw our conclusions and examine future work in Sect. 5.

2 Related Works

There are two subsections illustrated in this section. First, the basic Blockchain technology is described in Subsect. 2.1. The IoT wireless sensor network is addressed in Subsect. 2.2.

2.1 Basic Blockchain Technology

Blockchain or distributed ledger technology (DLT) is a protocol technology that data will be directly exchanged between two distinct users in a network without the need for intermediaries [1]. Network participants interact with an encrypted identity (unnamed); each transaction is then added to the unchanged transaction chain and distributed to all network nodes [3–5]. The blockchain is a distributed data structure that is replicated and shared among its network members. Blockchain was originally introduced as a solution for double spending on coin [6]. There are three types of blockchain technology: public blockchain, private blockchain, and consortium Blockchain [3]. Blockchain protocol is a transaction procedure stored in a distributed database that is used to maintain a growing list of records, called blocks. Each block contains a timestamp and a link to the previous block *e.g.* Bitcoin, Ethereum, Hyperledger, *etc.* [2]. In general, blockchain is managed by peer-to-peer networks that collectively adhere to certain protocols to validate new blocks [3–5, 7]. Interaction with blockchain using pairs of public key and private key. The private key is used to mark (sign) the transaction.

2.2 IoT Wireless Sensor Network

With the IoT technology, various devices can be connected to a network and controlled remotely or automatically to get the information needed [8–10]. One that is developing today is the WSN (Wireless Sensor Network) [13]. A WSN could generally be described as a network of nodes that cooperatively sense and control the environment, allowing interaction between people or computers and the surrounding environment [11, 12]. WSNs current usually includes sensor nodes, actuator nodes, gateways, and clients.

3 Our Approach

The cooperative evaluation approach based on Blockchain technology is proposed in this section for IoT application in WSNs. The system architecture is assumed and presented in Subsect. 3.1. Moreover, the system processes and procedures are presented in Subsect. 3.2.

3.1 System Architecture

The access control handling and authentication issues are the most challenging problems in IoT currently [4]. We want to propose cooperative evaluation approach to measure the trust level of blockchain agent nodes that communicate each other. In general, we give an overview of the system architecture shown in Fig. 1, which there are 4 components that will be used in the system architecture are as follows:

1. Blockchain agent node;
2. Wireless sensor network;
3. Blockchain database ("distributed ledger");
4. Private blockchain network.

The blockchain agent node is a node that can be a sink. It is an interface to connect IoT wireless sensor devices that receive information encoded by a protocol, *e.g.* IAPF (Constrained Application Protocol) IETF RFC [5]. The blockchain agent node will connect to other blockchain agent nodes to authenticate the trust level before forwarding the transaction to the Blockchain network. The wireless sensor network is a set of IoT sensors that can be controlled remotely or automatically connected to the network. Each device will be uniquely identified to interact with the Blockchain network through the blockchain agent node. Then, it obtains a public and private key pair from the public key generator. The task of IoT sensor device provides information to blockchain agent node about the monitored area condition. It consists of three symbolized conditions as mentioned below. One binary bit represents a condition where bit "1" denotes an abnormal condition and the bit "0" indicates a normal state. Transactions performed by an IoT sensor device will be stored on the ledger distribution at the Blockchain network. A Blockchain database is the Blockchain distributed ledger stored all transactions performed by IoT sensor devices in a private Blockchain network. This Blockchain database could be also used as a corresponding blockchain agent node to

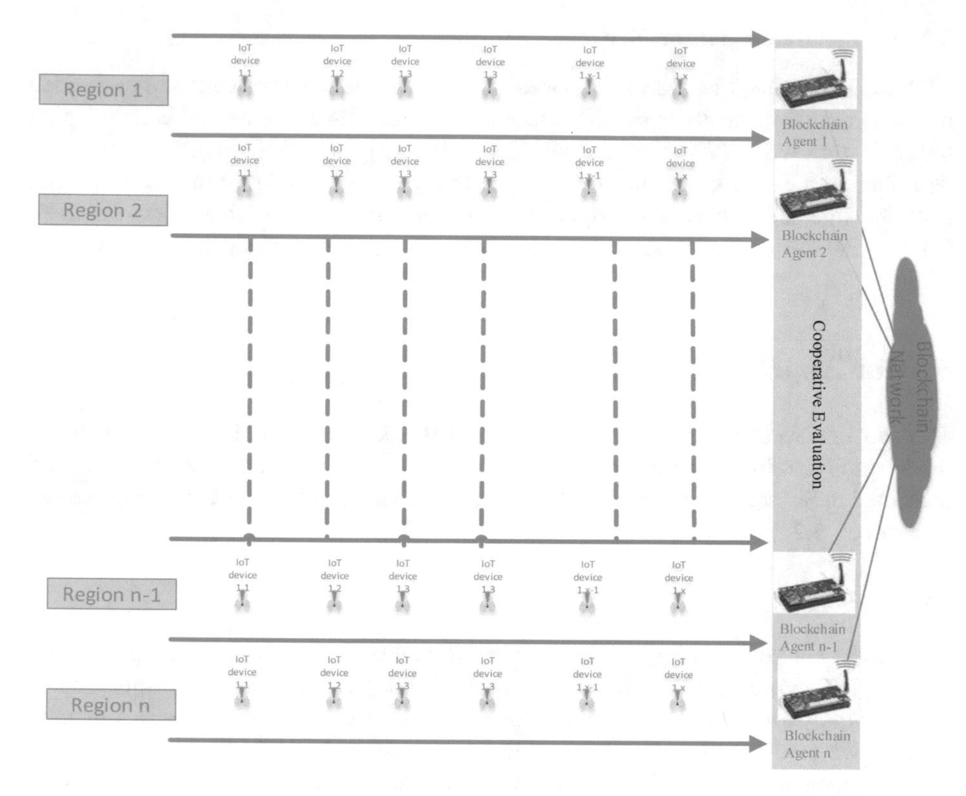

Fig. 1. The overview of the system architecture.

determine trust level in the Blockchain network. The private Blockchain network is one type of blockchain technology that exists. Thus, we propose a system based on private blockchain and it can be extended to a public blockchain that has its own cryptocurrency. As explained previously, in private blockchain, write-permissions are stored centrally in one node.

3.2 The System Processes and Procedures

The system processes and procedures are presented as followings. Each region has several IoT sensor devices that will provide information or events about the environment. Previously, the IoT sensor device performs registration phase on the blockchain agent node in each region. The blockchain agent node will collect the information and events provided by each IoT sensor device. After collecting information stored in the blockchain agent node ledger database, among the blockchain agent nodes will do a cooperative evaluation. It sees the behavior of the condition of each IoT sensor device. Whether the IoT sensor device is in abnormal condition before any further transactions between blockchain agent nodes. Blockchain agent nodes will perform authentication based on trust level before forwarding the transaction to Blockchain network. Each IoT sensor device sends a subregion monitored information signal to the blockchain agent node in each region. The blockchain agent node will collect and calculate the information obtained by using the formulas listed in definition 1 to monitor the condition of each IoT sensor device as in **Stage 1**. Next, the blockchain agent node will evaluate each blockchain which are the transaction records stored in Blockchain database by using some trust evaluation algorithms [14–16] in order to detect out the untrusted the Blockchain agent node and his IoT devices. It will be described in **Stage 2**.

Stage 1

In this case, the normal conditions can be explained as follows. Each IoT sensor device provides information to the blockchain agent node that is the use of the subregion sent at a certain time. Then the blockchain agent node will collect and write the data into the Blockchain database within a specified time period *e.g.* in every 5 min. The IoT sensor device will send an information signal consisting of 3 bits where the first bit (1st) represents *Condition* 1, the second bit (2nd) represents *Condition* 2, and the third bit (3rd) represents *Condition* 3. The bit "0" represents satisfied and "1" for an unsatisfied condition respectively. Finally, the three IoT sensor device information conditions are shown Fig. 2 below. The trust operation by adopting 'Θ' trust operation according to the trust evaluation algorithms in Ref. [14–16] is used to define *Definition* 1 and Eq. (1) shown below. From the result of sending information transmitted by IoT sensor device to blockchain agent node in each region. Then, an example is given that the trust value for the IoT sensor devices calculated by the blockchain agent node shown in Table 1.

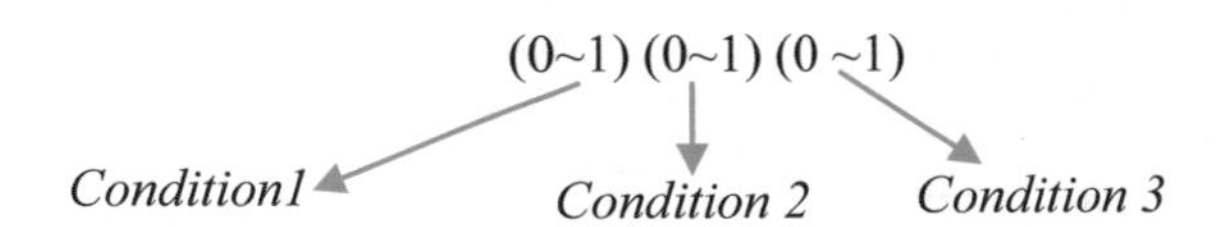

Fig. 2. Three condition information for IoT device.

Definition 1. *Assume that the trust operation for the IoT devices calulated by blickchain agent noode is represented as* $t_{S_i}^{BA_j} = t_{i,1}^{j}||t_{i,2}^{j}||t_{i,3}^{j}$, *where* $t_{i,1}^{j}$ *represents as the condition* 1, $t_{i,2}^{j}$ *represents as the condition 2, and* $t_{i,3}^{j}$ *represents as the condition 3. For example, first,* $t_{S_2}^{BA_3}$ *represents that the trust values are evaluated by a blockchain agent node* BA_3 *which evaluates an IoT device* S_2*. Second,* $t_{2,1}^{3}||t_{2,2}^{3}||t_{2,3}^{3} = 0||1||0$ *means that the trust value* $t_{2,1}^{3}$ *is the condition* 1 *for IoT device* S_2 *which is belonging to the blockchain agent node* BA_3*, the trust value* $t_{2,2}^{3}$ *is the condition* 2 *for IoT device* S_2 *which is belonging to the blockchain again* BA_3 *and trust value* $t_{2,3}^{3}$ *is the condition 3 for IoT device* S_2 *which is belonging to the blockchain again* BA_3*. Therefore, the final trust operation for all blockchain agent node during a blockchain transaction is defined as Eq.* (1) *below.*

Table 1. The trust value of bit combination of the IoT sensor device to the blockchain agent node.

IoT device S_i Blockchain agent node BA_j	IoT devices						Trust values
	S_1	S_2	S_3	…	S_{m-1}	S_m	$\Theta\left(t_{i,1}^{j} \| t_{i,2}^{j} \| t_{i,3}^{j}\right), i = 1,2,\ldots,m; j = 1,2,\ldots n$
	$c_1c_2c_3$	$c_1c_2c_3$	$c_1c_2c_3$	$c_1c_2c_3$	$c_1c_2c_3$	$c_1c_2c_3$	
Blockchain agent node BA_1	000	000	000	…	000	000	$\Theta\left(t_{i,1}^{1} \| t_{i,2}^{1} \| t_{i,3}^{1}\right), {}_{i=1,2,\ldots,m}$
	001	001	001	…	001	001	
	010	010	010	…	010	010	
	011	011	011	…	011	011	
	100	100	100	…	100	100	
	101	101	101	…	101	101	
	110	110	110	…	110	110	
	111	111	111	…	111	111	
Blockchain agent node BA_2	000	000	000	…	000	000	$\Theta\left(t_{i,1}^{2} \| t_{i,2}^{2} \| t_{i,3}^{2}\right), {}_{i=1,2,\ldots,m}$
	001	001	001	…	001	001	
	010	010	010	…	010	010	
	011	011	011	…	011	011	
	100	100	100	…	100	100	
	101	101	101	…	101	101	
	110	110	110	…	110	110	
	111	111	111	…	111	111	
⋮	⋮	⋮	⋮	…	⋮	⋮	⋮
Blockchain agent node BA_n	000	000	000	…	000	000	$\Theta\left(t_{i,1}^{n} \| t_{i,2}^{n} \| t_{i,3}^{n}\right), {}_{i=1,2,\ldots,m}$
	001	001	001	…	001	001	
	010	010	010	…	010	010	
	011	011	011	…	011	011	
	100	100	100	…	100	100	
	101	101	101	…	101	101	
	110	110	110	…	110	110	
	111	111	111	…	111	111	

$$\underset{i=1,2,\ldots,m;j=1,2,\ldots n}{\Theta\left(t_{S_i}^{BA_j}\right)} = \Theta\left(t_{i,1}^{j} \| t_{i,2}^{j} \| t_{i,3}^{j}\right). \tag{1}$$

Normal information can be explained as follows, information provided by an IoT sensor device on transactions occurring within a specified time frame corresponding to the number of costs obtained and stored to be a blockchain block stored in the Blockchain database. The transaction could be known from the public key which is generated into an address key of any IoT sensor device. The transaction is valid after it is signed by the sender's private account key. Suppose, the 6 installed IoT device sensors will have 42 states, the sensors will send data every 1 min, and those deliver packet every 60: 6 = 10 s.

Abnormal conditions of IoT sensor device are illustrated below.

```
Condition 1:
    begin
        If IoT sensor device is always used or occupied and it is compared to the
        payment status of receiving cryptocurrency value stored in the
        blockchain agent database ledger, it is not in accordance with lease time.
    End;
Condition 2 :
    begin
        If IoT sensor device always sends plentiful of abnormal information
    End;
Condition 3 :
    Begin
        IoT sensor device never sends information.
    End.
```

Stage 2

In this stage, the Blockchain database will be deal with the data mining together with trust evaluation processes. Each blockchain agent node in the private Blockchain network will evaluate each transaction record not only the record from his own managed IoT devices, but also the records from the others' blockchain agent nodes blockchain which are stored in Blockchain database by using some trust evaluation algorithms [14–16] in order to detect out the untrusted the Blockchain agent node and his IoT devices. The detail is presented below.

Each blockchain agent node receives information transmitted by IoT sensor devices as follows $\Theta\left(t_{i,1}^{j} \,||\, t_{i,2}^{j} \,||\, t_{i,3}^{j}\right)_{,\, i=1,2,\ldots,m;\, j=1,2,\ldots n}$. Then, it will launch a data mining process and trust evaluation algorithm to mine according to the three condition mentioned above. Then, it will get the result whether the IoT sensor device is under one of the three abnormal conditions or not. Finally, each blockchain agent node could perform cooperative evaluation process by correspondence with other blockchain agent nodes using the trust values collected from the IoT sensor device. Moreover, the transaction information in the distributed ledger could be also accessed by all blockchain agent nodes connected to the private Blockchain network in order to evaluate each others' trust values.

4 Discussions

In this section, the current IoT and blockchain by exploring recent research and up to date trends are disscussed below. Dorri et al. [9], they conducted systematic research, claiming that the IoT-based blockchain architecture handles most security and privacy threats, while considering the resource constraints of many IoT devices. Ammar et al. [6] conducted a systematic literature review on the IoT, their surveies had covered a subset of the commercially available framework and platform for developing industrial and consumer based IoT applications. Bahga et al. [12] they presented a Blockchain Platform for Industrial Internet of Things (BPIIoT). The BPIIoT platform could enable a marketplace of manufacturing services where the machines have their own Blockchain accounts and the users who are able to provision and transact with the machines directly to avail manufacturing services. In [4], they had identified the key security and trust related challenges and shown how blockchains could be used to overcome them. Also presented the design of a blockchain assisted information distribution system for the IoT and analyzed how the key security mechanisms could be built by leveraging blockchain technology. Thereefore, IoT application development has been done with various technologies to improve service and security. In this paper, the proposed approach compared with the related works mentioned above, which could be the highlighted common themes are integration IoT application based on private Blockchain network, and given trust evaluation between IoT devices and Blockchain node during the blockcahin transaction processes.

5 Conclusion

The IoT application based on blockchain technology is highly credible and developed. Due to cooperative evaluation method is getting more and more important requirement in developing the system with an integration IoT application based on Blockchain technology. Therefore, the cooperative evaluation approach proposed in this paper, it will improve the value of trustworthiness among the blockchain agent nodes to increase the degree of a successful transaction in the Blockchain network. In addition, the blockchain is typically managed by a peer-to-peer network collectively adhering to a protocol for validating new blocks. Once recorded, the data in any given block cannot be altered retroactively without the alteration of all subsequent blocks, which requires collusion of the network majority. Thus, the sink node in this paper acts as a blockchain agent node which could evaluate the behaviours of the managed and monitored IoT devices. It could also evaluate another blockchain agent node based on the transaction history or events logged in private blockchain network. Finally, the cooperative evaluation method proposed in this paper has an impact on enhancing security in IoT application based on blockchain technology.

Acknowledgments. This work was supported by the Ministry of Science and Technology (MOST), Taiwan, Republic of China, under Grant MOST 106-2632-E-468-003.

References

1. Swan, M.: Blockchain: Blueprint for a New Economy. 1st Edition. O'Reilly Media, Inc., 1005 Gravenstein Highway North, Sebastopol (2015)
2. Cag, D. https://richtopia.com/emerging-technologies/review-6-major-blockchainprotocols. Accessed 20 Apr 2018
3. Voshmgir, S., Kalinov V. https://blockchainhub.net/blockchains-and-distributed-ledger-technologies-ingeneral/. Accessed 20 Apr 2018

4. Polyzos, G.C., Fotiou, N.: Blockchain-assisted information distribution for the internet of things. In: IEEE International Conference on Information Reuse and Integration, San Diego, pp. 75–78 (2017)
5. Nakamoto, S.: Bitcoin : A Peer-to-Peer Electronic Cash System, pp. 1–9 (2008)
6. Ammar, M., Russello, G., Crispo, B.: Internet of Things: a survey on the security of IoT framework. J. Inf. Secur. Appl. **38**, 8–27 (2018)
7. de Kruijff, J., Weigand, H.: Understanding the blockchain using enterprise ontology. In: Dubois, E., Pohl, K. (eds.) CAiSE 2017. LNCS, vol. 10253, pp. 29–43. Springer, Cham (2017). https://doi.org/10.1007/978-3-319-59536-8_3
8. Sun, Y., Song, H., Jara, A.J., Bie, R.F.: Internet of things and big data analytics for smart and connected communities. IEEE Access **4**, 766–773 (2016)
9. Dorri, A., Kanhere, S.S., Jurdak, R.: Blockchain in Internet of Things: Challenges and Solutions. eprint arXiv:1608.05187 (2016)
10. Lin, Y.P., Petway, J.R., Anthony, J., Mukhtar, H., Liao, S.W., Chou, C.F., Ho, Y.F.: Blockchain: the evolutionary next step for ICT E-agriculture. Environments **4**, 1–13 (2017)
11. Wang, Y., Varadharajan, V.: Interaction trust evaluation in decentralized environments. In: Bauknecht, K., Bichler, M., Pröll, B. (eds.) EC-Web 2004. LNCS, vol. 3182, pp. 144–153. Springer, Heidelberg (2004). https://doi.org/10.1007/978-3-540-30077-9_15
12. Bahga, A., Madisetti, V.K.: Blockchain platform for industrial internet of things. J. Softw. Eng. Appl. **9**, 533–546 (2016)
13. Yinbiao, S., et al.: Internet of Things: Wireless Sensor Network. International Electrotechnical Commission, Switzerland Geneva (2014)
14. Chen, H.C.: TCABRP: a trust-based cooperation authentication bit-map routing protocol against insider security threats in wireless ad hoc networks. IEEE Syst. J. **11**(02), 449–459 (2017)
15. Chen, H.C.: A negotiation-based cooperative RBAC scheme. Int. J. Web Grid Serv. **13**(1), 94–111 (2017)
16. Chen, H.C.: A cooperative RBAC-based IoTs server with hierarchical trust evaluation mechanism. In: The 3rd EAI International Conference on IoT as a Service (IoTaaS 2017), Taiwan, Taichung City (2017)

The Study and Realization of Vulnerability-Oriented Fuzzing Technology for ActiveX Plug-Ins

Baojiang Cui[1,2] and Pin Mao[1,2(✉)]

[1] School of Cyberspace Security, Beijing University of Posts and Telecommunications, Beijing, China
cuibj@bupt.edu.cn, pinkomeo@gmail.com
[2] National Engineering Laboratory for Mobile Network, Beijing, China

Abstract. With the development of internet technology, more and more browsers have introduced third-party plug-ins to add additional features to attract users, which bring more potential risks to browsers. This paper presents a vulnerability-oriented security detection methods for IE browser ActiveX plug-ins. By using technologies such as dynamic binary instrumentation and vulnerability-oriented reverse analysis, the framework can assign different risk factors to each function and parameter inside the ActiveX plug-ins. In this way, this paper build an automated fuzz framework which can quickly generate fuzzing samples focusing on the fragile functions and fragile parameters. The experimental results show that our framework makes the efficiency of ActiveX plug-ins vulnerability detection significantly improved.

1 Introduction

The browser is our window to access the Internet. With the continuous advancement of technology, browsers play a pivotal role in our daily life and work. Due to the use of more and more users, a large number of hackers begin to pay attention to browser vulnerabilities. And IE browser, as Windows's own browser, has become the main target for most attackers and therefore the security of ActiveX [1] plug-ins has been receiving great attention.

ActiveX technology is based on the Component Object Model (COM) [2] and is an independent and complete code unit. It provides external services through a set of well-defined interfaces. The external application can call functions to achieve code reuse and function. Extensions. While ActiveX technology has been widely used, it has become a large attack surface for IE browsers due to the support of IE browsers for plug-ins and scripting languages.

In the case where the vulnerability of the ActiveX plug-in gradually becomes serious, many fuzzing tools for ActiveX plug-ins such as COMRaider [3], Dranzer [4], Axman [5], have begun to appear at home and abroad. Although these tools can be targeted to test the code in the ActiveX plug-in, but when there are too many functions in the ActiveX plug-in, the tool can only sort the function names in alphabetical order and then test the functions in turn. In addition, when one function contains multiple

L. Barolli et al. (Eds.): IMIS 2018, AISC 773, pp. 922–931, 2019.
https://doi.org/10.1007/978-3-319-93554-6_93

parameters, these tools will perform many "blind" tests because it cannot determine the priority of different parameters, which will undoubtedly reduce the efficiency of the fuzzing framework.

For the reasons mentioned above, this paper proposes a vulnerability-oriented security detection methods for IE browser ActiveX plug-ins which mainly focusing on solving two problems: 1. In the process of fuzzing plug-ins, since there are many interface functions exist in the plug-in, how to find potentially vulnerable interfaces efficiently and prioritize them. 2. In the process of fuzzing functions, since there are many parameters, how to assign different risk factors to each parameter and fuzz certain parameter specifically to improve the efficiency of the framework.

2 Architecture

By studying the related principles and techniques of fuzzing [7, 8], this paper proposes a more efficient fuzzing framework for ActiveX plug-ins. This framework first preprocesses the entire fuzzing object, then put the priority information obtained during the preprocessing into the fuzzing loop. The flow chart of the overall fuzzing framework is shown in Fig. 1.

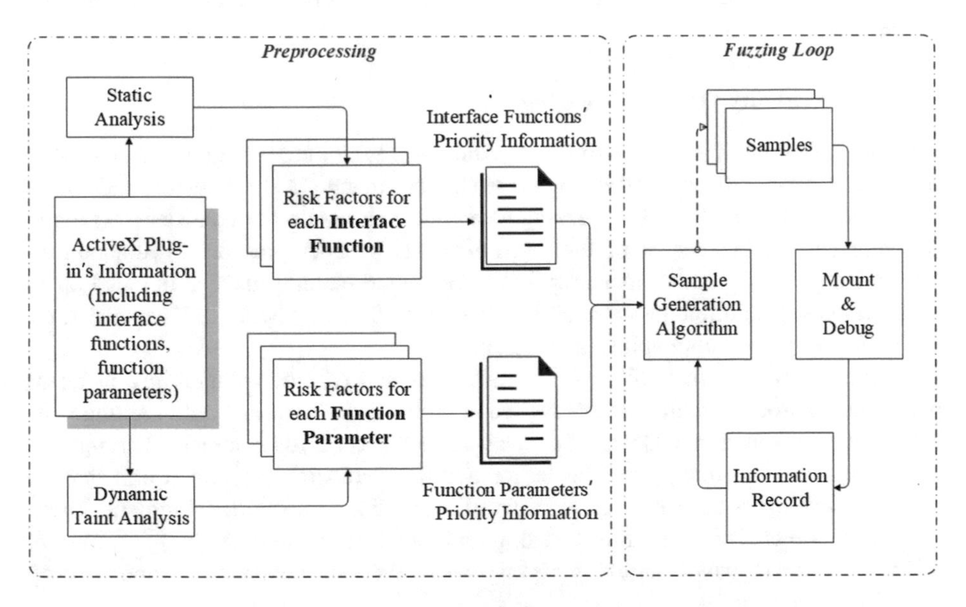

Fig. 1. The flow chart of the whole fuzzing framework.

Through static analysis, this framework can identify fragile instructions and functions from assembly instruction level and set different risk factors for each interface function inside the ActiveX plug-in. Then through dynamic taint analysis, this framework can identify which parameter of the function directly reached the path where the fragile instruction is located, and set different risk factors for each parameter.

After preprocessing, the framework will set different fuzzing priorities for each interface function and parameter by the result of the risk factor obtained in the previous step.

Due to the priority-driven fuzzing algorithm, the framework can more effectively and quickly discover the security vulnerability of ActiveX plug-ins and therefore reduce security events and accidents caused by the vulnerability of ActiveX plug-ins. This paper will then detail the framework of the construction ideas, algorithm model and application technology.

3 Technical Details

This paper uses static analysis and dynamic taint analysis to construct a vulnerability-oriented ActiveX plug-in fuzzing framework. And this framework is divided into two parts: the preprocessing and the fuzzing loop. In this section, we will introduce the specific technical details of each part according to the framework structure. Besides, we will explain some keywords proposed in the framework, such as vulnerability-oriented fuzzing and risk factors. At the same time, we will carry out a detailed description according to the implementation steps of the framework, and use some mathematical expressions to better illustrate the accuracy and effectiveness of the framework.

3.1 Vulnerability-Oriented Fuzzing

Generally speaking, there are two types of fuzzing: White-box fuzzing and Black-box fuzzing. White-box fuzzing means that researchers can get the entire source code of the test object, while the Black-box fuzzing means researchers do not consider the internal structure and characteristics of the test object. This paper presents a vulnerability-oriented framework model, meaning that a framework should make different fuzzing strategies based on different vulnerability characteristics. Following this paper will use the buffer overflow vulnerability as an example.

Buffer overflow vulnerability is the most common type of vulnerability in binary programs. Through buffer overflow vulnerability, an attacker can arrange his well-written instruction code on the stack to achieve his own attack. Through the exploration of the principle of the buffer overflow vulnerability, we found that the program developers used non-standard **fragile functions** during development. Therefore, when we analyze the buffer overflow vulnerability in ActiveX plug-ins, the first thing we need to pay attention to is a set of functions which are prone to cross boundaries such as:

```
char *strcpy(char* dest, const char *src);
char *strcat(char *dest, const char *src);
void *memcpy(void *dest, const void *src, size_t n);
```

In the actual analysis of reverse analysis, we found that in most of the plug-ins, because of the optimization options during development, IDA cannot identify this type of fragile function calls without a symbol table. So the framework also needs to be able to analyze from the instruction level, focusing on the core instructions of the fragile function, such as String Instruction, etc.

This paper will take buffer overflow vulnerability in ActiveX plug-ins as an example, so as a vulnerability-oriented fuzzing framework, it will focus more on functions or instructions that are vulnerable to buffer overflow vulnerability.

3.2 Risk Factor

The risk factor is a quantified value. Every time a fragile function is called in an interface function, the framework will increase the risk factor of this interface function. The more **fragile points** (fragile functions or instructions) found during the preprocessing, the greater the risk factor of this interface function will have. Assuming that there are x functions in the ActiveX plug-in, and the risk factor for each function is r_x.

Through the value of the risk factor, the framework will assign different fuzzing priorities to each interface function according to the algorithm. The higher priority interface functions will be tested by the framework first and give more fuzzing time.

3.3 Assign Risk Factors to Interface Functions in Preprocessing

The ActiveX plug-in is a set of independent code units which can provide external services through internal interface functions. Inside an ActiveX plug-in, there are usually many different interface functions. When the total number of the interface functions is small, the fuzzing sequence of the interface functions hardly affects the efficiency of the fuzzing test. But when the total number of interface functions reaches a certain level, if these interface functions have no priority characteristics, blind fuzzing tests will bring not only unnecessary computational overhead but also unnecessary time overhead.

This paper mainly uses dynamic binary instrumentation technology and IDA Python script to analyze the interface functions of the target plug-in. First, find all the interface functions' entry points, and then give the IDA Python script the address range of the function. Then, recursively collect all the assembly instruction of the interface function, and record all fragile points as the basis for generating the risk factor of the interface function.

Without loss of generality, assuming that the probability that a vulnerable function is distributed in x positions constitute a vector $P = (P_1, P_2, \ldots, P_x)^T$, and the number of functions processed before each position constitute a vector $K = (K_1, K_2, \ldots, K_x)^T$. So the average number of function processed by the framework to find the vulnerability is:

$$K^T P \tag{1}$$

And because the framework in this paper maintains a priority sequence according to the risk factor of the interface functions, all probabilities in P satisfy a descending order relationship, so the result of $K^T P$ is a reverse order sum.

However, the traditional fuzzing algorithm does not perform the test in descending order, which is equivalent to making an out-of-order processing of K . So the average number of function processing of the traditional fuzzing strategy is $K'^T P$, which is an out-of-order sum and is greater than or equal to the reverse sum all the time.

In order to implement the vulnerability-oriented strategy, this paper decided to prioritize all the interface functions inside an ActiveX plug-in before fuzzing. The priority is determined by the risk factor introduced in the previous section. This paper defines n different fragile points for the same vulnerability, and assign different weights to each fragile point. This results in an n-column weight vector $W = (w_1, w_2, \ldots, w_n)^T$. For each function F_x in the plug-in, a count vector $C_x = (c_1, c_2, \ldots, c_n)^T$ is set, where c_n represents the number of the nth fragile point in function F_x.

Let $Y = (C_1, C_2, \ldots, C_x)$, then the risk vector R for all interface functions can be expressed as:

$$R = W^T Y = (r_1, r_2, \ldots, r_x)^T \tag{2}$$

The framework arranges the priority information of all the interface functions according to the descending order of the risk factor of each function.

3.4 Dynamic Taint Analysis

This paper chooses to use Intel's Pin binary platform [9] to analyze the entire ActiveX plug-in. The framework use Dynamic Taint Analysis mainly to solve these problems: When there are a large number of parameters in an interface function, the framework can connect the input parameters and the fragile point in the function. When the input parameter and the fragile point are in one execution path, the framework will give this parameter a greater Risk Factor.

In the preprocessing, the framework collects all the interface functions and function parameters' information about the plug-in, and then generates the initial test sample and a corresponding file for each parameter. Then the initial test sample will read the content of the file as an input to the interface function. In this way, the dynamic taint analysis module can mark the content of the file as a taint by monitoring the file read system call.

There are three kinds of data flows related to the transmission of taint data: the first type is direct data flow, that is, the taint data itself and its copy move between memory and registers; the second type is taint data pointer spread; the third type is control flow, that is, conditional jumps and calls affected by the taint data. By recording the direct taint data, it can be judged whether the internal access address is taint data when the pointer differences the access, so the second type of data stream does not need to be tracked. The tracking of the control flow will introduce wrong taint data, so it will not be tracked either. Therefore, all the taint data flow in this paper is the direct taint data transmission. During the taint analysis process, all taint data and its taint attribute labels need to be recorded and updated. Taint data is divided into two parts, register taint and memory taint.

Taint Register. The framework maintains a collection of taint registers to indicate the status of registers. In the instruction sequences with register participation, the framework tracks the spread of taint data by looking for registers' states in the collection.

Taint Memory. Because of the large memory space and the characteristics of data continuity, the framework maintains a structure that represents a collection of tainted memory areas, and performs addition, and elimination of taint memory according to the operations of search, add, and delete of collections.

3.5 Assign Risk Factors to Parameters in Preprocessing

After the priorities of the ActiveX plug-in interface functions are determined, the framework need to preprocess all the parameters of the interface function. In the previous step, we obtained fragile instructions in the interface function through static analysis. In order to get to the potentially vulnerable path faster, this paper uses the taint tracking technology of binary instrumentation to monitor the path of parameter spread and bind parameters to vulnerable locations. With this binding relationship, we can set a priority for each parameter too. In this way, in the final fuzzing, these parameters will be given different weight according to the priority to achieve better fuzzing results. The priority is also determined by the risk factor. The risk factor depends on whether the parameters have passed our custom fragile functions or instructions.

Assuming that the interface function F_x in the plug-in has a set of parameters called $O_x = \{o_1, o_2, \ldots, o_y\}$, and the risk factor of the parameter is defined as r_y. If the function parameter passes the fragile point, r_y will be set to a positive integer 1. So the framework will use the indicative variable I to represent the risk factor of the parameter:

$$r_y = I_{o_y}\{o_y \in O_x\} \tag{3}$$

By ranking the risk factors of each parameter in descending order, priority information of function parameters is obtained.

3.6 Sample Generation Algorithm in Fuzzing Loop

After getting all the priority information, this article uses the priority information to **sort** the interface functions and function parameters to implement the following test sample generation algorithm. In the process of sorting, the framework will maintain a priority queue $Q = (F_1, F_2, \ldots, F_x), \{r_1 \geq r_2 \geq \ldots \geq r_x\}$ for interface functions. Each interface function in the queue also corresponds to a parameter priority queue $Q'_x = (o_1, o_2, \ldots, o_y), \{r_1 \geq r_2 \geq \ldots \geq r_y\}$.

(1) For each fuzzing test, take the front interface function from the priority queue to generate the test sample.
(2) During the generation process, follow the priority queue of the parameter and set different test weight for different parameters.

(3) After the sample is generated, the framework mounts and debugs the entire running process. The debugger module and variant character set used by the framework are stripped from the Peach Fuzzer [10]. Once the crash information is found, the crash and test samples are immediately recorded. The time of how long this interface function has been fuzzed will also be recorded as T_x.
(4) The framework will send this time to the test sample generation algorithm as a feedback. The time threshold is proportional to the priority of the function. If the proportion is S, the time threshold of the interface function F_x with priority R_x will be:

$$T_{threshold} = S \cdot \frac{R_x}{\overline{R}} \tag{4}$$

S is an observation that can be adjusted based on resources and experimental results, so then the time T_x of fuzzing the interface function F_x should satisfy:

$$T_x \leq T_{threshold} \tag{5}$$

The algorithm will decide whether the interface function should be queued to the priority queue according to the priority of the interface function and the tested time, and assign the next interface function as the new fuzzing object.

4 Implementation and Evaluation

In order to verify the efficiency of the vulnerability-oriented fuzzing framework for ActiveX plug-ins and the efficiency of the fuzzing algorithm in which risk factor determines different priorities of different testing objects, research selected three main ActiveX plug-in fuzzing tools named COMRaider, Dranzer, and Axman. Experiments compare the tools in terms of function and operating efficiency. Here are the results of comparing the tools with VoFuzzer (Vulnerability-oriented Fuzzer, the framework in this paper):

(1) **Fuzzing tools comparison**

This study focuses on comparing several major features of the ActiveX Fuzz tool. Its main features are:

(a) Fuzzing Order
Use what kind of order to fuzz all the functions in the plug-in.
(b) Output
Ability to support crash output and fuzz record output.
(c) User Interaction

Ability to close the window during the test automatically and if a large amount of manual clicks are required during use

(d) Mutation
Mutation strategy used during fuzzing (Table 1).

Table 1. Fuzzing tools comparison.

Tool name	Fuzzing order	Output	User interaction	Mutation
Axman	Order read from the interface	None	High	Random
Dranzer	Order read from the interface	Text	Medium	Fixed
COMRaider	Alphabetical order	Database	Medium	Fixed
VoFuzzer	Vulnerability-oriented priority order	Text	Low	Boundary value

(2) **Effectiveness comparison**
In order to evaluate the effectiveness of the framework, this paper selected a common dynamic link library SkinCrafter.dll as a test object for testing. The experiment focuses on the following two aspects: the time from the beginning of the obfuscation to the discovery of the vulnerability and the total number of samples already running (Table 2).

Table 2. Average crash time and total samples run for a published vulnerability.

Tool name	Average crash time (100 fuzzing tests)	Total samples
Axman	85.6 s	9688
Dranzer	none	none
COMRaider	799.4 s	167
VoFuzzer	4.3 s	4

From the test results, we can see that our framework found a crash in the plug-in with fewer test cases and a shorter time. After later verification, this crash is an exploitable N-day buffer overflow vulnerability that has been disclosed on Exploit-Database [11]. As for the experimental results, the reason why Dranzer didn't participate in the comparison is because, first of all, Dranzer doesn't support loop fuzzing, and all functions only test once. Second, the Dranzer uses only a fixed string as mutated parameters. The published vulnerability in the evaluation section cannot crash with only a very long string input. This also explains the importance of the mutation strategy and the importance of the structure of the fuzzing framework on the other hand.

5 Related Work

The framework of this article was developed due to the limitations of other publicly available ActiveX fuzzing tools. The following is a brief description of other ActiveX tools.

COMRaider. COMRaider is a plug-in fuzzing tool written in VB language. COMRaider can display a good list of the interface functions in the plug-in with a graphical interface, but in the actual fuzzing process, COMRaider just test all the functions in alphabetical order, and requires a high level of user interaction. In addition, COMRaider does not well recognize the presence of multiple classes in one plug-in. When the ActiveX plug-in contains multiple classes, each class should use different 'CLSID' [6] to distinguish when called.

Dranzer. Dranzer is a plug-in fuzzing tool written in C++ language. This tool can collect crashes caused by functions' parameters and crashes when the plug-in loads into Internet Explorer for the first time. However, this tool does not make much breakthrough in parameter mutation algorithms. Dranzer uses only a simple 10 k string of lowercase 'x' characters or a-1 integer for different parameter types when fuzzing ActiveX plug-ins.

Axman. Axman is a fuzzing tool developed by JS. Because Axman is a web-based testing tool, it also requires a high level of user interaction. For example, a debugger must be attached manually to Internet Explorer to retrieve the result of a crash, and when a crash occurs, the test process must be restarted manually too.

6 Conclusion

This paper presents a vulnerability-oriented Fuzzing technology for ActiveX plug-ins, and completes an automated ActiveX plug-in security detection framework based on this vulnerability-oriented technology. Vulnerability-oriented fuzzing technology solves two efficient problems when fuzzing ActiveX plug-ins, and gives different priorities to interface functions and function parameters in plug-ins base on different risk factors. The experimental results show that the framework can well detect security holes in some ActiveX plug-ins and is more efficient and effective than the mainstream vulnerability detection tools. This paper will keep focusing on how to detect plug-ins' vulnerabilities in the future and try to explore the relationship between the crash and the integrity of the exception handling in the plug-in [13]. We will also use new ideas to explore the permissions rules and access control policies that the plug-ins give to different users and roles [14, 15] to make our framework an intelligent and evolutionary [12] fuzzing tool for ActiveX plug-ins.

Acknowledgments. This work is supported by National Natural Science Foundation of China (No. U1536122, No. 61502536).

References

1. Wikipedia. ActiveX-Wikipedia, the Free Encyclopedia (2010-0815). http://en.wikipedia.org/wiki/ActiveX
2. Mircosoft Corporation. Micrsoft Component Object Model (COM); A Technical Overview of COM. http://www.cs.umd.edu/pugh/com
3. iDefense Labs. COMRaider: A tool designed to fuzz COM object Interfaces. http://labs.idefense.com/software/fuzzing.php
4. Dormann, W., Plakosh, D.: Vulnerability Detection in ActiveX Controls Through Automated Fuzz Testing (2008). http://www.cert.org/vuls/discovery/dranzer.html
5. Moore, H.: AxMan ActiveX Fuzzer (2006). http://digitaloffense.net/tools/axman
6. CLSID Key. http://msdn2.microsoft.com/en-us/library/aa908849.aspx
7. Rawat, S., Jain, V., Kumar, A., Cojocar, L., Giuffrida, C., Bos, H.: VUzzer: application-aware Evolutionary Fuzzing. In: NDSS (2017)
8. Li, Y., Chen, B., Chandramohan, M., Lin, S.W., Liu, Y., Tiu, A.: Steelix: program-state based binary fuzzing. In: Joint Meeting on Foundations of Software Engineering, pp. 627–637. ACM (2017)
9. Pin - A Dynamic Binary Instrumentation Tool. https://software.intel.com/en-us/articles/pin-a-dynamic-binary-instrumentation-tool
10. Peach Fuzzer. http://www.peachfuzzer.com/
11. The Exploit Database – ultimate archive of Exploits, Shellcode, and Security Papers. https://www.exploit-db.com/
12. American fuzzy lop. A security-oriented fuzzer. http://lcamtuf.coredump.cx/afl/
13. Zhai, X., Hu, X., Jia, X., et al.: Verifying integrity of exception handling in service-oriented software. Int. J. Grid Util. Comput. **8**(1), 7 (2017)
14. Nakamura, S., Duolikun, D., Enokido, T., et al.: A read-write abortion protocol to prevent illegal information flow in role-based access control systems. J. Intell. Inf. Syst. **22**(1), 89–109 (2004)
15. Xu, L., Liu, Z., Luo, J.: A fine-grained attribute-based authentication for sensitive data stored in cloud computing. Int. J. Grid Util. Comput. **7**(4), 237 (2016)

An Open Source Software Defect Detection Technique Based on Homology Detection and Pre-identification Vulnerabilitys

Jun Yang(✉), Xuyan Song, Yu Xiong, and Yu Meng

Beijing University of Posts and Telecommunications, Beijing, China
{junyang,bearsmall}@bupt.edu.cn, song_xuyan@163.com,
mengyu7183@sina.com

Abstract. Homology detection technology plays a very important role in the copyright protection of computer software. Homology detection technology mainly includes text based technology token, based technology and abstract syntax tree based technology. This paper introduces a method of defect detection based on homology detection technology for open source software. This detection method will collect the code fragments with vulnerabilities and the source code in open source software to compare, through three levels of comparison, to find because of plagiarism code introduced by the vulnerability fragment. After that, the vulnerability fragment is compared with the trigger condition of the vulnerability, and the judgment result is obtained. Finally, the superiority of this technique is verified by experiments.

1 Introduction

Open source software (OSS) is software that runs under open source tags. Its source code should be available and can be modified. In an OOS, a software suite must contain freely accessible source code that allows users to modify and redistribute. Some OOS may retain permission to republish, but in other cases it may be free. Distributors or developers may charge for services, including special training, installation, programming, and technical support. In general, the term open source software refers to software that is freely available, widely accessible and reusable [1].

With the development of the Internet, there are more and more resources that can be shared, and more and more resources are available to people. However, as the threshold for software development becomes lower and lower, a large number of junior developers often release code to the open source community without checking the security of the code, which is often a security concern [2].

The code in open source software becomes a resource for developers to plagiarize. For reasons of intellectual property rights, developers usually do not specify the part of the code they use, and make changes to the source code that do not affect the function of the code, such as changing the function name or variable name, disrupting the order of statements, Type redefinition, etc. [3, 4]. If there are some security vulnerabilities in the source code of open source software, when developers use the code, the vulnerabilities spread to the new software along with the code. Hackers' sense of smell is

L. Barolli et al. (Eds.): IMIS 2018, AISC 773, pp. 932–940, 2019.
https://doi.org/10.1007/978-3-319-93554-6_94

extremely sensitive, and the response is extremely fast. When a vulnerability is discovered, there will be an attack on that day, there will be tools developed against the vulnerability on that day, and a large-scale attack will soon reach a peak, leaving the security community with a very short response time [5].

The innovation of this paper is to apply the idea of plagiarism detection to the defect detection of source code. In the course of the experiment, we collected vulnerabilities with code fragments, and used homology detection technology to match them with the source code. To find a vulnerability caused by plagiarism code.

2 Materials and Methods

In this section, we will show how to obtain preidentity vulnerabilities, how to construct lexical parsers and syntax parsers through JavaCC, and how to use parsers to construct homology matching tools. And how to use the homology ratio tool for vulnerability detection.

2.1 Vulnerability Pre-identification

We use Common Vulnerabilities & Exposures (CVE) to obtain a sample of the vulnerability. CVE is like a dictionary table, giving a common name for widely accepted information security vulnerabilities or vulnerabilities that have been exposed. If a vulnerability is specified in a vulnerability report, if there is a CVE name, You can quickly find patched information in any other CVE compatible database to solve security problems [6].

Based on the patch information, we can find the source code of the vulnerability, and pre-identify the trigger condition of the code. After analyzing the data in CVE, we find that, the trigger conditions of the vulnerability are divided into three categories. In the first case, a single fragment of vulnerability code can be triggered, such as CVE-2016-0705 and CVE-2016-1901. The second case is where there are multiple pieces of vulnerability code that can be triggered at the same time, such as CVE-2015-2692. The third scenario is where there are multiple pieces of vulnerability code, but one of them triggers a vulnerability, such as CVE-2016-2175. We finally found 62 qualifying vulnerabilities after 2015.

2.2 Application of JavaCC

Java Compiler Compiler (JavaCC) is the most popular parser generator for Java applications. A parser generator is a tool that reads the syntax specification and converts it into an Java program that recognizes syntax matching. In addition to the parser generator itself, JavaCC provides other standard functions related to parser generation, such as tree building (via the JJTree tool that ships with JavaCC), operations, debugging, etc. [7, 8].

JavaCC provides three tools to complete the unload of programs. The complete JavaCC tool consists of three parts: javacc, JJTree, and JJDoc. Javacc is one of the

Table 1. JavaCC tool composition

Tool name	Tool function
JJTree	Process the jjt file, generate the JJ file, and generate the code for the tree node
javacc	Process JJ files generated by JJTree, generate syntax parser files
JJDoc	Generating BNF normal form files based on JJ files

main JJTree tools used to help generate abstract syntax trees and JJDoc tools to generate BNF paradigms for source programs [9] (Table 1).

To complete the lexical analysis and syntax analysis of the input target language, we need to write a jjt file according to the structural characteristics of the target language. In this file, you define the name of the parser class, lexical rules, and jj files with lexical and syntax rules. Next, the JJ file is processed with javacc, generating seven Java source files, which contain the main parsing file [10].

This paper uses JavaCC to generate lexical parsers and syntax parsers for C/C++ and Java.

2.3 Create Homology Matching Tools

The homology detection tool used in this paper consists of three parts: text based homology detection, Token based homology detection and abstract syntax tree based detection. Each of the three technologies has different priorities. At the same time, the accuracy of detection can be improved by using three detection techniques. Three techniques are described below.

Text Based Detection Technology

Text based homology alignment is a rapid detection and detection technology, which can efficiently realize the source code homology alignment under massive data. Compared with the comparison technology based on file hash, which is used by some testing tools at this stage, this technology has been put forward in comparison accuracy.

In this paper, Simhash algorithm is used for text alignment. The approximate detection algorithm based on Simhash is widely used by Google to detect the re-checking of web files. Text detection and image detection. The algorithm can convert a piece of document into n-bit signature. We can regard this process as the operation of dimensionality for high-dimensional data, that is, the high-dimensional vector is transformed into a signature with fewer digits [11] (Fig. 1).

The implementation process of the algorithm is as follows:

(a) The input text is processed with corresponding features. This is mainly to extract the keywords of the text, then calculate the weight of the feature words according to the word frequency measured by the feature, and finally get n binary groups, which are recorded as *feature_weight = (fi, wi), i = 1, 2, 4 ... n.*
(b) By calculating the hash value of the feature words, n *feature_weight* were transformed into *(hash, weiight)*, which was recorded as *hash_weight = (hash(fi), wi), i = 1, 2, 3 ... n.*

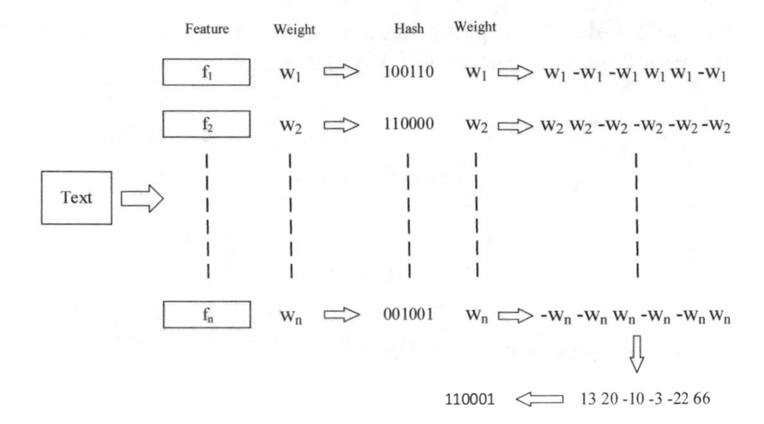

Fig. 1. The implementation process of the algorithm

(c) The hash value corresponding to the feature word is represented by the weight value. If the corresponding bit in hash is 1, the *weight* of that bit is weighted. If 0, the weight of the bit is *-weight*.
(d) Add the weights of each column to the final weights according to the calculation rules of the vertical addition.
(e) Convert the result to a 0–1 digital signature. The conversion rule is: if the corresponding weight value is >0, then the bit is recorded as 1, or 0.

Token Based Detection Technology
The homology detection technology based on Token preprocesses the identifiers, keywords, operators, delimiters, constants and annotators in code, and transforms the source code into Token sequences by lexical analyzer for homology detection. This method can effectively identify the program code replacement problem and improve the accuracy of source code homology detection [12–15].

In this paper, JavaCC is used to generate Token sequences, and the longest common substring (LCS) algorithm is used to match. If we want to complete the homology detection scenario in this paper, the traditional LCS algorithm needs to be modified in business logic.

LCS has the following problems.

- The purpose of homology detection is to detect all Token sequence values common substring, not necessarily the oldest substring.
- The traditional LCS method is used to match in the matching matrix, so it is difficult to extract the subsequences other than the longest common subsequences.
- The LCS algorithm allows for certain discontinuities and cannot define the correlation boundaries within the code logic structure, so it can not fully adapt to the business scenario of code-level homology detection.
- LCS lacks the condition of stopping matching and does not specify the discontinuity of common subsequences.

To solve these problems, this paper made some modifications to the LCS algorithm so that it can meet the objectives proposed in this paper.

$$L(i,j) = \begin{cases} \max\left\{L(h,k)|(h,k)_{\in w(i,j)}\right\} + 1 \; if \; A_i = B_j \wedge max\{L(h,k)\} > 0 \wedge \\ \quad (h,k) \in \Omega \wedge diagonal\left(S_{w(i,j)}\right) \leq th_{win} \\ 0 \quad if \; A_i \neq B_j \\ 0 \quad if \; i = 0 \vee j = 0 \\ 1 \quad if \; A_i = B_j \wedge max\{L(h,k)\} = 0 \wedge \\ \{L(h,k)\} \notin \Omega \wedge diagonal\left(S_{w(i,j)}\right) > th_{win} \end{cases} \quad (1)$$

$$S_w(\mathrm{i,j}) = \begin{cases} S_w(\mathrm{i,j}) + 1 \; if \; max\{L(h,k)\} = 0 \vee (h,k)_{L(h,k) \neq 0} \notin \Omega \vee \\ \quad diagonal\left(S_{w(i,j)}\right) < th_{win} \\ 1 \quad else \end{cases} \quad (2)$$

$$\Omega = \begin{cases} \Omega \backslash (h,k) \quad if \; A_i = B_j \wedge max\{L(h,k)\} > 0 \wedge \\ \quad (h,k) \in \Omega \\ \{(h,k)|1 \leq h \leq m, |1 \leq k \leq n\} \; if \; i = 0 \wedge j = 0 \end{cases} \quad (3)$$

The similarity matrix is generated according to formula 1. First, the first row and the first column of the initialization matrix are 0. In a row-first manner, the matching of the column summary and the row summary is checked row by line.

In order to achieve a specified range of code matching, the elements that match successfully need to be calculated. The method is to search for the maximum matching count of values in the valid range of the square window, as shown in formula 2.

To avoid repeated use of the cumulative match count. An unmarked set is used to eliminate the cumulative count that has been matched to reduce repeated matches, as shown in formula 3.

Abstract Syntax Tree Based Detection Technology

The detection algorithm based on abstract abstract syntax tree (AST) is a detection method based on syntax structure [16, 17]. It depends on the AST node information of source code to judge the plagiarism of code.

```
void funA(int a)
{
    a = 0;
    int b = 0;
    fun2(a, b);
}
```

First of all, the AST structure of the source code file is obtained by calling the lexical and syntax parser. Above is a program source code in the function, Fig. 2 is the corresponding syntax tree structure.

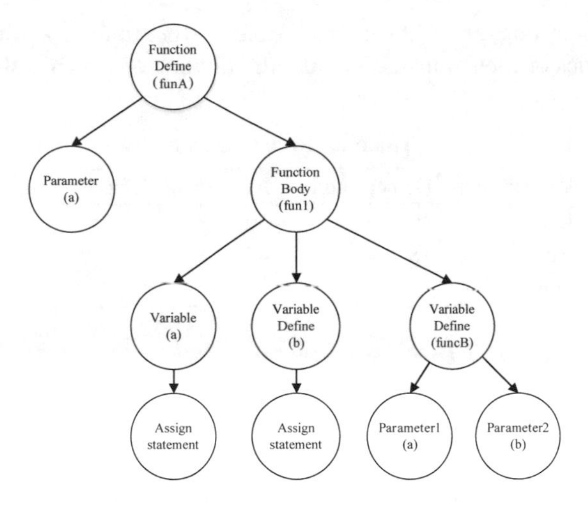

Fig. 2. Abstract syntax tree structure

Secondly, the Hash generation algorithm of the abstract syntax tree is called to get the Hash value of each node of the file syntax tree. The Hash value and the number of child nodes are calculated for all nodes of the syntax tree during the traversal. The Hash matching algorithm is called to compare the Hash values of each node in the file syntax tree, and the similarity is calculated.

Finally, what we need to do is compare the syntax tree. We calculate the hash value based on each node of the syntax tree and the type of its subtree, and then compare it. The main comparison steps are as follows:

(a) Call the hash value generation algorithm of the abstract syntax tree to get the hash value and other relevant information of the syntax tree of the vulnerability code and the original code.
(b) Each subtree is classified by its child node number. Store the information in an array of linked lists and then call the hash comparison algorithm for comparison.
(c) Record a subtree with the same hash value and its corresponding source code location.

Because subtrees with different number of child nodes represent different structures, only pairs of subtrees with the same number of child nodes need to be compared. This not only reduces the number of comparisons, but also reduces the complexity of the algorithm.

3 Result

In order to test the defect detection effect of this method to open source project. This paper mainly selected Apache, Linux, Spring and other open source projects as the test object. By counting the 1045 CVE vulnerabilities in these projects since 2014, there are fewer than 100 officially available patches. In the experiment, 62 typical vulnerabilities were selected for testing, and 142 suspected defects were found by homology detection of the defect codes of each vulnerability, the distribution of which is shown in Table 2.

Table 2. Result statistic

Project name	Defect number	Number of defects found
Linux	29	59
Apache	18	44
Spring	15	39

Table 3 shows some of the test results, including the CVE number of the known defect, the location of the defect, and so on. For security reasons, details and versions of these suspected defects are blocked.

Table 3. Localization defect

Project name	CVE number	Defect position
Linux kernel	CVE-2016-4470	b/security/keys/key.c
	CVE-2016-44569	sound/core/timer.c
	CVE-2017-1000112	kernel/exit.c
Apache server	CVE-2017-9798	server/core.c
	CVE-2016-5387	server/util_script.c
Spring	CVE-2017-4971	org/springframework/webflow/mvc/view/AbstractMvcView.java
	CVE-2016-4977	org/springframework/security/oauth2/provider/endpoint/SpelView.java

We took a lot of code samples from the open source community and tested them from different sizes of code. As shown in Fig. 3, the horizontal coordinates are the capacity of the code, and the vertical coordinates are the time consumed. Four broken lines represent the results of the detection of different numbers of vulnerabilities. As you can see from Fig. 3, when the number of vulnerabilities is the same, the cost of implementation increases linearly as the amount of code increases. With the same amount of code, the time spent grows faster as the number of vulnerabilities increases.

In the process of identifying defects, we still face the problems of high false alarm rate in static analysis and limitation of defect type. For example, some types of defects are difficult to describe in code snippets. Even so, the method proposed in this paper has greatly reduced the scope of defect detection, reduced the amount of manual participation, and effectively reduced the appearance of multiplexing defects.

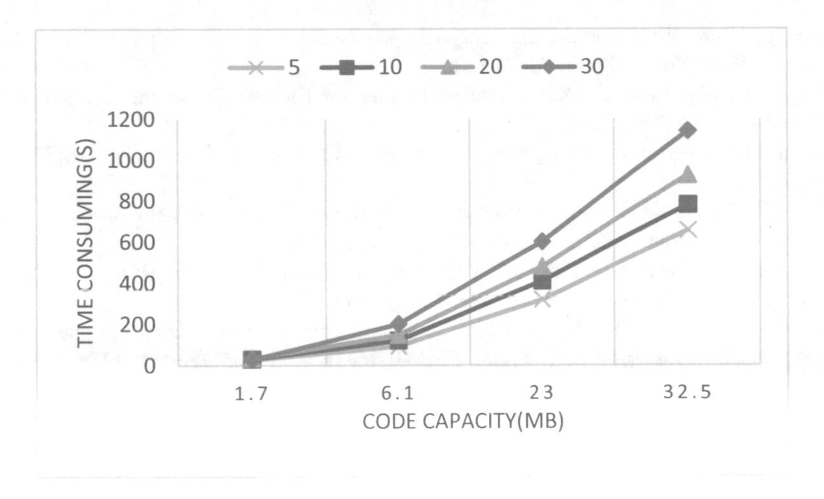

Fig. 3. Time efficiency contrast diagram

4 Conclusion

Aiming at the defects caused by code reuse, this paper presents a new method of defect detection, which is independent of defect type. Taking the code fragment as the analysis unit, the feature extraction and quantization of the code are carried out from many levels. By calculating the similarity between codes, you can get the areas of code that may be flawed. Through the testing of several open source projects, the ideal experimental results are obtained, and the comparison experiment further verifies that the method proposed in this paper has certain advantages in the detection of reuse defects.

Acknowledgment. This work is supported by National Natural Science Foundation of China (No. U1536122, No. 61502536).

References

1. Ettrich, M.: Open Source Software: A Guide for Small and Medium Enterprises. Thormann & Goetsch GmbH, Berlin (2001)
2. Tan, L., Liu, C., Li, Z., Wang, X., Zhou, Y., Zhai, C.: Bug characteristics in open source software. Empir. Softw. Eng. **19**(6), 1665–1705 (2014)
3. Mirza, O.M., Joy, M., Cosma, G.: Style analysis for source code plagiarism detection - an analysis of a dataset of student coursework. In: IEEE International Conference on Advanced Learning Technologies, pp. 296–297. IEEE (2017)
4. Mirza, O.M., Joy, M.: Style analysis for source code plagiarism detection (2015)
5. Ryoo, J., et al.: The use of security tactics in open source software projects. IEEE Trans. Reliab. **65**(3), 1195–1204 (2016)
6. Surhone, L.M., et al.: Common Vulnerabilities and Exposures. Betascript Publishing (2010)

7. Kodaganallur, V.: Incorporating language processing into Java applications: a JavaCC tutorial. IEEE Softw. **21**(4), 70–77 (2004)
8. Gupta, K., Nandivada, V.K.: Lexical state analyzer for JavaCC grammars. Softw. Pract. Exp. **46**(6), 751–765 (2016)
9. Ruan, M.: Research of testing tool parasoft Jtest. Comput. Knowl. Technol. **8**(32), 7724–7727 (2012)
10. Chen, T.: There Search and Application of Automatic Scoring System Based on Abstract Syntax Tree. Dalian Maritime University, Liaoning (2011)
11. Johnson, J.H.: Identifying redundancy in source code using fingerpringts. In: IBM Centre for Advanced studies Conference, pp. 171–183 (1993)
12. Jiang, L., et al.: DECKARD: scalable and accurate tree-based detection of code clones. In: International Conference on Software Engineering, pp. 96–105. IEEE (2007)
13. Cui, B., et al.: Code syntax-comparison algorithm based on type-redefinition-preprocessing and rehash classification. J. Multimed. **6**(4), 320–328 (2011)
14. Toomey, W.: Ctcompare: code clone detection using hashed token sequences. In: International Workshop on Software Clones, pp. 92–93. IEEE (2012)
15. Xu, J.H.: Research on anti-cheating technology of massive documents based on Simhash algorithm. Comput. Technol. Dev. **9**, 103–107 (2014)
16. Wibowo, A.T., Sudarmadi, K.W., Barmawi, A.M.: Comparison between fingerprint and winnowing algorithm to detect plagiarism fraud on Bahasa Indonesia documents. In: Information and Communication Technology, pp. 128–133. IEEE (2013)
17. Chan, P.P.F., Hui, L.C.K., Yiu, S.M.: Heap graph based software theft detection. IEEE Trans. Inf. Forensics Secur. **8**(1), 101–110 (2013)

Analysis on Mobile Payment Security and Its Defense Strategy

Simin Yin[1], Jingye Sheng[1], Tong Wang[1], and He Xu[1,2](✉)

[1] School of Computer Science, Nanjing University of Posts and Telecommunications, Nanjing 210023, China
{B15040208, B15040209, B15040301, xuhe}@njupt.edu.cn

[2] Jiangsu High Technology Research Key Laboratory for Wireless Sensor Networks, Nanjing 210003, China

Abstract. In recent years, with the rapid development of mobile Internet technology, and the increasing popularity of mobile payment, the security of mobile payment terminal is becoming more and more significant. Even if mobile payment has gained popularity worldwide due to its convenience, it is also facing many threats and security challenges. This paper is targeted at mobile payment security on Android platforms. In detail, we first introduce the current development of mobile payment application, and then we analyze two main threats of mobile payment security, namely near-field payment security and remote payment security. We also focus on phishing attacks on mobile platforms, by analyzing the behavior, characteristics, common techniques and attack methods of these attacks. Finally, we develop a defense strategy based on monitoring running applications, aimed at alerting users when malicious applications are leaking payment information, and we test the feasibility of this strategy on a dataset of Alipay application.

1 Introduction

With the rapid development of mobile Internet, mobile terminals have become an important tool for people's daily life and work. According to the 40th "statistical report on China Internet development status" released by CNNIC (China Internet network information center), as of 2017, the number of netizen in China has broken through 750 million. The Internet penetration was 54.3%, while the percentage of mobile netizen is as high as 96.3%. The dominant position of mobile Internet was further strengthened.

From communication to transportation and sports, even financial management and shopping, there is no shortage of applications on mobile terminals. In particular, mobile payments applications are increasingly more popular. In fact, with these applications, users can perform mobile payment through mobile phones, pads and other mobile terminals. Mobile payment is a complex integration of terminal equipment, mobile Internet, and financial institutions [1].

On August 9, 2017, the world's leading new economic data mining and analysis institutions, iiMedia Research released the 'Research report on China mobile payment market in the first half of 2017'. According to the report, online payment transaction volume reached 2085.0 trillion yuan in 2016, a 3.3 percent increase from 2015. Mobile

L. Barolli et al. (Eds.): IMIS 2018, AISC 773, pp. 941–946, 2019.
https://doi.org/10.1007/978-3-319-93554-6_95

payment in 2016 reached 157.6 trillion yuan, up 45.6% from 2015. In addition, according to relevant data, mobile payment users reached 460 million in 2016, up 30% from 360 million in 2015. With the emergence of third-party payment software, mobile payment has been closely related to People's Daily life. With the help of the remote payment, users can use mobile phone online shopping, booking tickets to the scenic spot, pay the electricity and water, and through the near field payment (i.e., scan to pay) users can deal with others anytime and anywhere. It can be said that third-party payment applications make smart mobile terminals a "second wallet".

While mobile payments are making life more convenient, numerous security issues have emerged. In the initial stage of mobile payment development, developers focused more on platform construction and user experience, without paying too much attention to its security. Mobile payment security technology is still immature [2], and this has resulted in cases of leakage of payment information. In addition, the weak security awareness of the users also makes some criminals take advantage of it.

2 Security Threats of Mobile Payments

The security of mobile payment can be divided into two aspects: near-field payment security and remote payment security. The security threat of near-field payment is mainly focused on QR code payment [3], which has been popular in recent years. The security of remote payment is more heterogeneous, and includes threats of the wireless payment environment, the threat posed by software viruses, the threat of application tier and so on [4]. In the following, we mainly analyze the behavior of phishing to obtain user's payment information, which is one of the threats at application-level.

Phishing is not only one of the most common forms of network fraud on personal computers, it also is extended with the mobile networks development, and the total number of Chinese phishing websites exceeds the PC. According to APAC (anti-phishing alliance of China) fishing website processing bulletin, the number of fishing websites identified and handled in China is more than 410 000 in 2017, including the main counterfeiters: Taobao, ICBC and JD. Among them, the payment transaction category of fishing websites accounted for the highest amount of processing this month, 78.9%.

Phishing is very deceptive and diverse. Traditional phishing is often used to send text messages and emails, using false information to trap Internet users. In the early days of Internet development, it is easy for Internet users to be fooled by these messages asking users to provide their card number. Nowadays, as it is not always easy to be fooled by similar cases, fraudsters have also increased the sophistication of phishing attacks. For example, in the iOS systems, users often see pop-up windows asking for an Apple ID, and some of the third-party malicious fishing apps mimic a window that is identical to the system's window to entice the user to enter the password. In this case, a malicious application can obtain a password whenever the user presses "confirm" or "cancel". Since these pop-ups often appear when the system has just been updated, such as when user logins iCould, and installs the app, they enter their Apple ID password without thinking. If there's a malicious app designed to do a "fishing" with a pop-up window that's identical to the payment interface, the consequences can be dire.

The main steps of a phishing attack are the following: (1) the first step is to form a fishing interface; (2) the user inputs information on the fishing interface; (3) the attacker obtains user's information and uses the database technology to save information.

To prevent phishing, programmers can apply the following defense strategy [5]:

(i) create an anti-phishing URL database: for phishing websites, this requires comparing the address of the phishing site with the address of the existing anti-phishing database. If the correct address of the official website in the database is not the same, it should be stopped immediately or issued a warning to the user to avoid being deceived.
(ii) real-time protection: intelligently and dynamically evaluate the objects accessed by users. If the application UI jumps out of a non-system pop page or interface the application will be terminated.
(iii) cloud computing strategy: integrate the fishing web information into the cloud security database and share it with the network users.

3 A Defense Strategy for Phishing

Current popular phishing attacks on Android to obtain the user's payment password perform these steps: when a user's Android phone is installed with an app that carries a phishing attack, the attack will firstly silently start a system service. This service will monitor which applications the phone is running, and once the fraud target application is matched, Phishing will use the switching time of the Activity to automatically switch

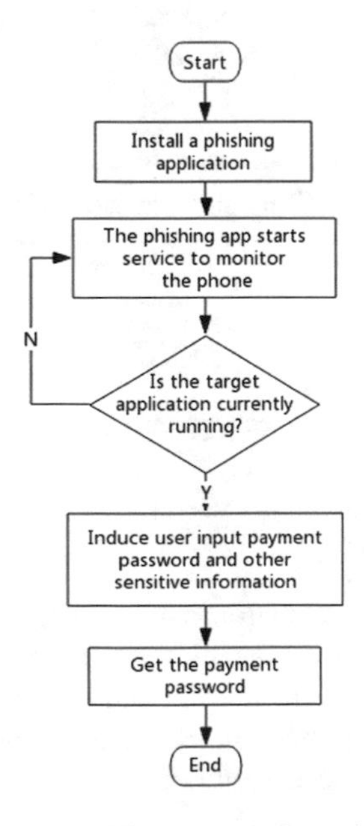

Fig. 1. Phishing attack flow chart

out the same spoofing interface as the target application interface [6]. Users can be enticed to enter information such as their account password if their unable to distinguish between the real or fake interface. Therefore, user's payment information may be leaked: this process is described in Fig. 1.

To deal with these phishing attacks, we consider adopting the method of application layer monitoring method, so as to timely alert users of leakage of payment information. To do this, we firstly set the target application (such as Alipay) as protected. Then, we monitor the application that the phone is currently running, and when it is overlapped in the protected application run, start the notification and open the phone's vibration to attract the attention of the user. Figure 2 shows the diagram process of the proposed defensive solution.

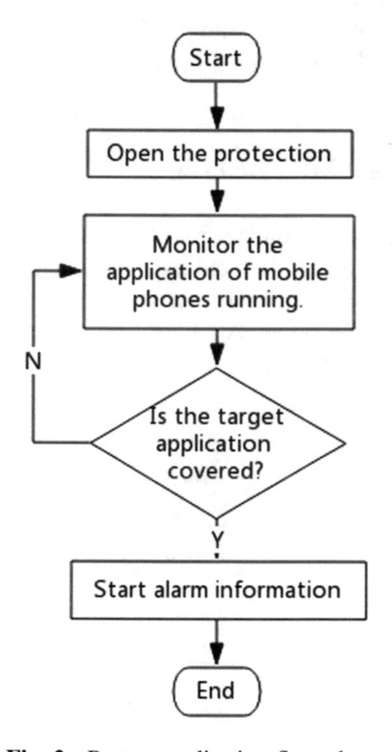

Fig. 2. Protect application flow charts

4 Experiment

Figure 3 shows that our implementation a fraud page for deceiving someone application password page. When it uses our protection strategy, the protection mechanism can detect the Phishing and ensure the application security, the result is shown in Fig. 4. If the user is not to use protection mechanism, the user may input the password and we can get the password successfully, those results are shown in Figs. 5 and 6.

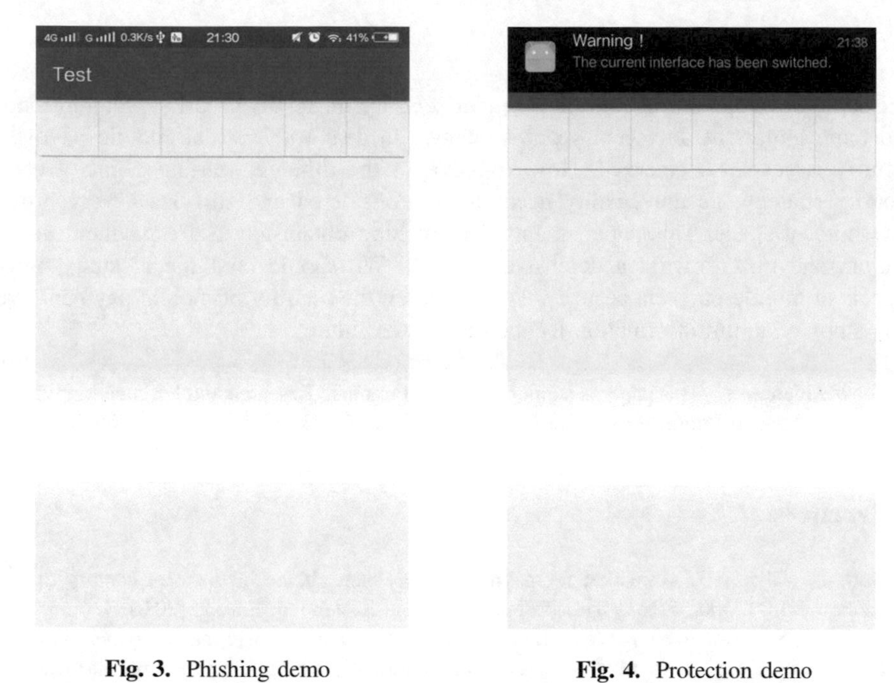

Fig. 3. Phishing demo

Fig. 4. Protection demo

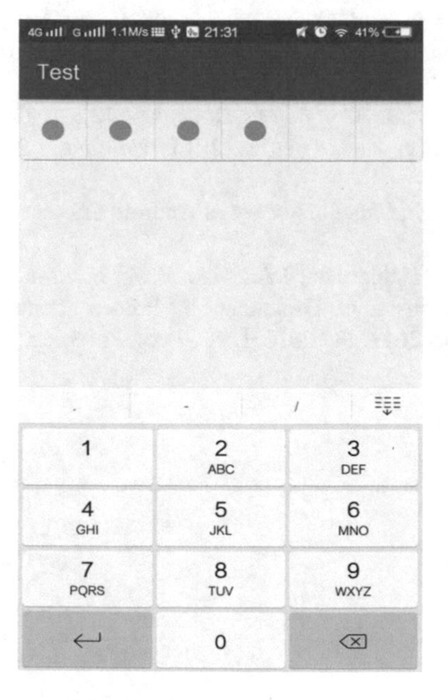

Fig. 5. Without protection

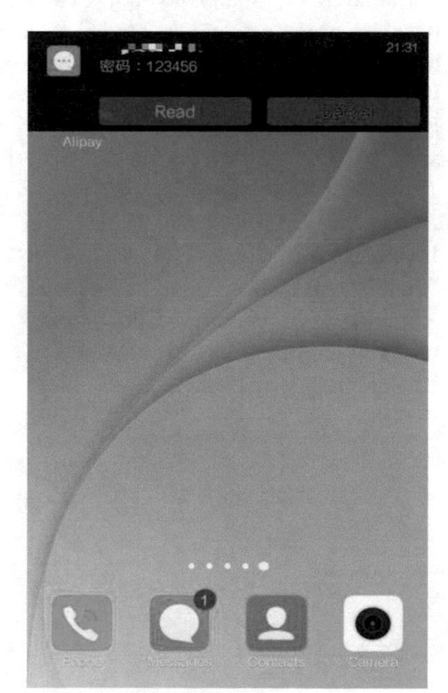

Fig. 6. Get password

5 Conclusion

The rapid development of mobile payment security makes its security problem more and more important. In recent years, strategies to deal with various mobile payment security issues have emerged. However, due to the different starting point of each security strategy, its universality needs to be considered and still has a very broad development space. This paper mainly for phishing obtain the user's payment information, and puts forward a defensive strategy. We should continue to study other aspects of mobile payment security. To strengthen the security of mobile payment, we may study quantum information technology in the future.

Acknowledgements. This paper is supported by Postgraduate Research and Practice Innovation Program of Jiangsu Province (KYCX17_0798), and NUPT STITP (No. XZD2017057).

References

1. Bott, J., Milkau, U.: A market for payments—payment choice in the 21st century digital economy. In: Górka, J. (ed.) Transforming Payment Systems in Europe (2016)
2. Murdoch, S.J., Anderson, R.: Security protocols and evidence: where many payment systems fail. In: Christin, N., Safavi-Naini, R. (eds.) International Financial Cryptography Association. FC 2014. LNCS, vol. 8437, pp. 21–32 (2014)
3. Purnomo, A.T., Gondokaryono, Y.S., Kim, C.-S.: Mutual authentication in securing mobile payment system using encrypted QR code based on public key infrastructure. In: 2016 IEEE 6th International Conference on System Engineering and Technology (ICSET), Bandung, Indonesia, 3–4 October 2016 (2016)
4. Faruki, P., Bharmal, A., Ganmoor, V., Gaur, M.S., Conyi, M., Rajarajan, M.: Android security: a survey of issues, malware penetration, and defenses. IEEE Commun. Surv. Tutorials, **17**(2), 998–1022 (2015)
5. Zhang, Y., Wang, K., Yang, H., Fang, Z., Wang, Z., Cao, C.: Survey of Android OS security. J. Comput. Res. Dev. **51**(7), 1385–1396 (2014)
6. Fernandes, E., Chen, Q.A., Paupore, J., Essl, G., Halderman, J.A., Mao, Z.M., Prakash, A.: Android: UI deception revisited: attacks and defenses. In: Grossklags, J., Preneel, B. (eds.) International Financial Cryptography Association 2017. FC 2016, LNCS, vol. 9603, pp. 41–59 (2017)

Author Index

L. Barolli et al. (Eds.): IMIS 2018, AISC 773, pp. 947–950, 2019.
https://doi.org/10.1007/978-3-319-93554-6

Printed in the United States
By Bookmasters